ENCYCLOPEDIA OF
ENVIRONMENTAL
CHANGE

ENCYCLOPEDIA OF
ENVIRONMENTAL CHANGE

EDITOR-IN-CHIEF

JOHN A. MATTHEWS

$SAGE reference

Los Angeles | London | New Delhi
Singapore | Washington DC

Los Angeles | London | New Delhi
Singapore | Washington DC

FOR INFORMATION:

SAGE Publications Ltd.
1 Oliver's Yard
55 City Road
London, EC1Y 1SP
United Kingdom

SAGE Publications, Inc.
2455 Teller Road
Thousand Oaks, California 91320
United States of America

SAGE Publications India Pvt. Ltd.
B 1/I 1 Mohan Cooperative Industrial Area
Mathura Road, New Delhi 110 044
India

SAGE Publications Asia-Pacific Pte. Ltd.
3 Church Street
#10-04 Samsung Hub
Singapore 049483

Publisher: Delia Martinez Alfonso
Acquisitions Editor: Robert Rojek
Editorial Assistant: Colette Wilson
Developmental Editor: Judi Burger
Production Editor: Tracy Buyan
Copy Editor: QuADS Prepress (P) Ltd.
Typesetter: Hurix Systems (P) Ltd.
Proofreaders: Lawrence W. Baker, Kate
              Peterson, Kristin Bergstad
Indexer: Julie Grayson
Cover Designer: Wendy Scott
Marketing Manager: Teri Williams

Printed in the United Kingdom.

Library of Congress Cataloging-in-Publication Data

Encyclopedia of environmental change / John A. Matthews, Swansea University, general editor.

pages cm
Includes bibliographical references and index.

ISBN 978-1-4462-4711-2 (cloth)

1. Global environmental change—Encyclopedias. I. Matthews, John A. (John Anthony), 1947–

GE149.E442 2014
363.703—dc23   2013017218

13 14 15 16 17 10 9 8 7 6 5 4 3 2 1

Contents

How to Use This Encyclopedia

1. First consult the **headwords** for over 4,000 terms and topics that are covered in alphabetical order. All such headwords appear in **bold** in the text.
2. If a term is not found as an entry in the body of the encyclopedia, consult the *index*, which contains terms that are explained in other entries even though they do not have their own headwords. These terms appear in the text in *italics*.
3. Words in SMALL CAPITALS are used throughout the work for cross-referencing. They indicate links to headwords under which further related information can be found and are used both within and at the end of entries.
4. References are listed at the end of most entries. These provide the reader with exemplars of the use of each term, research papers on the subject and/or further general reading. They include the sources of information cited in the text, figures and tables.

List of Entries

Ga A *giga-annum*: an abbreviation for thousands of millions (billions) of years before the present. For example, 4.6 Ga is 4,600 million years ago (Ma). *GO*

gabion A mesh basket (usually of wire) filled with rocks to prevent erosion or to serve as a foundation. It is usually an inexpensive, temporary structure, much used in COASTAL (SHORE) PROTECTION. *HJW*

Jackson CW, Bush DM and Neal WT (2006) Gabions, a poor design for shore hardening: The Puerto Rico experience. *Journal of Coastal Research* 39: 852–857.

Gaia hypothesis The concept, developed by James Lovelock, that suggests the Earth is a complex entity in a state of equilibrium, HOMEOSTASIS or HOMEORHESIS, maintained by FEEDBACK MECHANISMS. The physical and chemical 'health' of the planet, especially ATMOSPHERIC COMPOSITION, is seen as the result of SELF-REGULATION throughout geological time. Contrary to more conventional ideas that see life as adapting to the ABIOTIC environment, the Gaia concept stresses control by the BIOSPHERE. *Weak Gaia* asserts a substantial influence of life over the abiotic environment whereas *strong Gaia* emphasises that the EARTH SYSTEM itself is analogous to an organism (the *organismic analogue*; see COMMUNITY CONCEPTS). This should not be confused with the word 'Gaia' being derived from the Greek goddess of the Earth! In his latest book, Lovelock (2009) sees the self-regulating properties of the Earth system being threatened by HUMAN IMPACT ON CLIMATE, especially GLOBAL WARMING. *JAM*

[*See also* DYNAMICAL SYSTEMS, EARTH-SYSTEM ANALYSIS (ESA), GLOBAL ENVIRONMENTAL CHANGE, HOLISTIC APPROACH]

Kleidon A (2004) Beyond Gaia: Thermodynamics of life and Earth system functioning. *Climatic Change* 66: 271–319.

Lovelock JE (1995) *Gaia: A new look at life on Earth, 2nd edition*. Oxford: Oxford University Press.

Lovelock JE (2009) *The vanishing face of Gaia: A final warning*. New York: Basic Books.

Lovelock JE and Margulis L (1974) Atmospheric homeostasis by and for the biosphere. *Tellus* 26: 1–10.

Margulis L (1998) *Symbiotic planet: A new look at evolution*. London: Weidenfeld and Nicolson.

Moody DE (2012) Seven misconceptions regarding the Gaia hypothesis. *Climatic Change* 113: 277–284.

Schneider SH (ed.) (2004) *Scientists debate Gaia: The next century*. Cambridge: MIT Press.

gallery forest A narrow strip of EVERGREEN or *semi-deciduous* closed-canopy forest along a water course draining an area of DECIDUOUS forest or SAVANNA vegetation. In the tropics, gallery forests are often similar to TROPICAL RAIN FOREST in structure and physiognomy and some have been regarded as the RELICTS of more extensive evergreen forests in the past. However, continuously high SOIL MOISTURE permits an evergreen habit in regions that are seasonally dry. Gallery forests are often species-rich ecosystems and fire is believed to promote this BIODIVERSITY. The boundaries of gallery forests are fire-prone, but they contain core zones into which fire very rarely intrudes. *Fire-tolerant trees* in the outer zone may protect a forest interior of low *flammability*. Although of high CONSERVATION value, many gallery forests are threatened by *irrigated agriculture* and *flooding* for HYDROPOWER generation. *NDB/JLI*

[*See also* IRRIGATION, WILDLIFE CORRIDOR]

Miguel A, Marimon BS, de Oliveira EA et al. (2011) Woody community dynamics of a gallery forest in the transition Cerrado-Amazon Forest in Eastern Mato Grosso, over a seven year period (1999 to 2006). *Biota Neotropica* 11: 53–61.

Parron LM, Bustamente MMC and Markewitz D (2011) Fluxes of nitrogen and phosphorus in a gallery forest in the Cerrado of central Brazil. *Biogeochemistry* 105: 89–104.

Seaman BS and Schulze CH (2010) The importance of gallery forests in the tropical lowlands of Costa Rica for understorey forest birds. *Biological Conservation* 143: 391–308.

VanderWeide BL and Hartnell DC (2011) Fire resistance of tree species explains historical gallery forest community composition. *Forest Ecology and Management* 261: 1530–1538.

game management The management of essentially wild animals primarily for economic (HARVESTING) purposes. It includes, for example, the management of MOORLAND for grouse or deer in Scotland, and it contrasts with the raising of domesticated animals on farms, although the distinction is blurred in *game ranching*, which includes, for example, Australian crocodile farming and South African ostrich farming. *JAM*

[*See also* RANGE MANAGEMENT, WILDLIFE CONSERVATION]

Hudson D (2006) *Gamekeeping*. Shrewsbury: Swan Hill Press.
Leopold A (1933) *Game management*. Madison: University of Wisconsin Press.
Suich H and Child B (eds) (2008) *Evolution and innovation in wildlife conservation: Parks and game ranches to transform conservation areas*. London: Earthscan.

gamma correction In IMAGE ENHANCEMENT, a non-linear point-wise transformation used to control the overall *brightness* in images. It is based on the relationship between the actual values in the pixels, the voltage into which these values will be translated in the display device and the brightness that will be shown to the user for every voltage. *ACF*

Jain AJ (1989) *Fundamentals of digital image processing*. Englewood Cliffs, NJ: Prentice Hall.

gamma ray attenuation porosity evaluator (GRAPE) A rapid, non-destructive method for assessing the *water content* of SEDIMENTS by measuring the attenuation of a beam of gamma rays projected through the sediment. It is widely used for CORE logging, as GRAPE records from some MARINE SEDIMENT CORES display cyclicities characteristic of MILANKOVITCH THEORY. *MRT*

Mayer LA, Jansen E, Backman J and Takayama T (1993) Climatic cyclicity at Site 806: The GRAPE record. *Proceedings of the Ocean Drilling Program. Scientific Results* 130: 623–639.

gap analysis A technique used to identify *biota* and BIOTIC communities that are not adequately represented in an existing series or network of PROTECTED AREAS. The technique can help locate areas for CONSERVATION and can also be used to prioritise planning with respect to HABITAT protection and management. Elements of the analysis typically include vegetation classification and the mapping of the geographical distribution of vegetation types, plant and animal species and areas set aside for protection of nature and of the habitats of greatest importance to THREATENED SPECIES. GEOGRAPHICAL INFORMATION SYSTEMS are often employed in gap analysis. The term is also used for the analysis of gaps in VEGETATION CANOPY and their effects. *IFS*

[*See also* CONSERVATION BIOLOGY, PROTECTED AREA APPROACH, SINGLE LARGE OR SEVERAL SMALL RESERVES (SLOSS DEBATE)]

Langhammer PF, Bakarr MI and Bennus L (2007) *Identification and gap analysis of key biodiversity areas: Targets for comprehensive protected area systems*. Gland: International Union for the Conservation of Nature.
Maxted N, Dulloo E, Ford-Lloyd BV et al. (2008) Gap analysis: A tool for complementary genetic conservation assessment. *Diversity and Distributions* 141: 1018–1030.
Spellerberg IF and Sawyer JWD (1999) *An introduction to applied biogeography*. Cambridge: Cambridge University Press.

garden An area (normally outdoors) set aside for the CULTIVATION and/or enjoyment of *plants* or other forms of NATURE. Gardens may be located near homes or in *parks*, but there are many different types. *JAM*

[*See also* BOTANICAL GARDENS, HORTICULTURE, SHIFTING CULTIVATION, ZOOLOGICAL GARDENS (ZOOS)]

Turner T (2011) *European gardens: History, philosophy and design*. Abingdon: Routledge.

gas hydrates CRYSTALLINE mixtures of methane and water, stable at high ambient pressure (>50 bar) and low temperature (<7°C), which are widespread in CONTINENTAL SLOPE and CONTINENTAL RISE sediments and in PERMAFROST areas. They are also known as *methane hydrates* or *clathrates*. The gas is thought to have been released from organic-rich sediments at shallow burial depths and is trapped in sediment pore spaces within about 1 km of the sea floor.

The sudden release of large volumes of methane gas into the oceans and atmosphere, where it is a potent GREENHOUSE GAS, is a form of DEGASSING and is due to the DISTURBANCE of *sea-floor sediments* or a lowering of HYDROSTATIC PRESSURE, triggered by EARTHQUAKES, SEA-LEVEL CHANGE or MASS MOVEMENT PROCESSES of sediment, such as TURBIDITY CURRENTS. Sudden degassing produces large pockmarks on the sea floor and is thought by some to be responsible for the sinking of ocean-going vessels in the 'Bermuda Triangle', an area above the Blake Ridge and Carolina Rise where gas hydrate production in marine sediments is known to be very high. The release of gas hydrates may have contributed to the global firestorm at the K-T BOUNDARY. Abrupt BOTTOM WATER warming at the end of the PALAEOCENE may have caused widespread methane release, resulting in a worldwide negative *carbon isotopic excursion* (see COMPOUND-SPECIFIC CARBON ISOTOPE ANALYSIS). The change from solid to gas phase also raises sediment *pore pressure* which may destabilise *slope deposits*, causing large-scale mass movement. This may be particularly prevalent at times of falling sea level and there is some evidence of increased frequency of slumping in hydrate-prone areas during sea-level lowstands. De-gassing may cause a fall in sea level, through a reduction in the volume of sea-bed sediments.

Gas hydrates may be underlain by sediment containing large volumes of gaseous methane, which could be exploited as an ENERGY RESOURCE. It has been estimated that gas hydrates are associated with at least twice as much combustible carbon as in all other FOSSIL FUELS.

BTC/MRT/JP

Bratton JF (1999) Clathrate eustasy: Methane hydrate melting as a mechanism for geologically rapid sea-level fall. *Geology* 27: 915–918.

Carozza DA, Mysak LA and Schmidt GA (2011) Methane and environmental change during the Paleocene-Eocene thermal maximum (PETM): Modeling the PETM onset as a two-stage event. *Geophysical Research Letters* 38: L05702.

Dickens GR, O'Neil JR, Rea DK and Owen RM (1995) Dissociation of oceanic methane hydrate as a cause of the carbon isotope excursion at the end of the Paleocene. *Paleoceanography* 10: 965–971.

Henriet J-P and Mienert J (eds) (1998) *Gas hydrates: Relevance to world margin stability and climatic change.* Bath: Geological Society.

Lorenson TD, Collett TS and Hunter RB (2011) Gas geochemistry of the Mount Elbert gas hydrate stratigraphic test well, Alaska North Slope: Implications for gas hydrate exploration in the Arctic. *Marine and Petroleum Geology* 28: 343–360.

Max MD, Dillon WP, Nishimura C and Hurdle BG (1999) Sea-floor methane blow-out and global firestorm at the K-T boundary. *Geo-Marine Letters* 18: 285–291.

Paull CK, Buelow W, Ussler W and Borowski WS (1996) Increased continental-margin slumping frequency during sea-level lowstands above gas hydrate-bearing sediments. *Geology* 24: 143–146.

Thakur NK and Rajput S (2010) *Exploration of gas hydrates: Geophysical techniques.* Dordrecht: Springer.

Gauss-Gilbert geomagnetic boundary

The geomagnetic polarity boundary between the Gilbert (reverse polarity) and Gauss (NORMAL POLARITY) POLARITY CHRONS, dated by POTASIUM-ARGON (K-Ar) DATING to ca 3.58 Ma BP.

DH

[*See also* GEOMAGNETIC POLARITY REVERSAL, GEOMAGNETIC POLARITY TIMESCALE (GPTS)]

Løvlie R (1989) Palaeomagnetic stratigraphy: A correlation method. *Quaternary International* 1: 129–149.

Gaussian filter

In IMAGE PROCESSING, an implementation in the FILTERING process, where the values in the neighbouring pixels are multiplied by constants proportional to the exponential of the square of the distance to the central pixel (i.e. the *Gaussian function*). It is a *noise reduction* technique that induces a blur in the original image.

ACF

Gaussian model

The most widely used PROBABILITY DISTRIBUTION function for real-world data, including image data from OPTICAL REMOTE-SENSING INSTRUMENTS. Conformity to this model, which is manifested as a bell-shaped curve, is a prerequisite for many techniques of PARAMETRIC STATISTICS, such as the (Gaussian) maximum likelihood classification.

ACF

[*See also* NORMAL DISTRIBUTION]

Simon MK (2006) *Probability distributions using Gaussian random variables: A handbook for engineers and scientists.* Berlin: Springer.

Gauss-Matuyama geomagnetic boundary

The geomagnetic boundary between the Gauss (NORMAL POLARITY) and Matuyama (*reverse polarity*) POLARITY CHRONS or epochs. It has been variously dated by POTASSIUM-ARGON (K-Ar) DATING to 2.48 million years (Ma) BP, 2.589 Ma BP and, most recently, *astronomically tuned* to 2.61 Ma BP (see ORBITAL TUNING). This boundary has been suggested as a MARKER HORIZON for determining the PLIOCENE-PLEISTOCENE TRANSITION.

MHD/CJC

[*See also* GEOMAGNETIC POLARITY REVERSAL, GEOMAGNETIC POLARITY TIMESCALE (GPTs)]

Deino AL, Kingston JD, Glen JM et al. (2006) Precessional forcing of lacustrine sedimentation in Late Cenozoic Chemeron Basin, Central Kenya Rift and calibration of the Gauss/Matuyama boundary. *Earth and Planetary Science Letters* 247: 41–60.

Valet J-P and Meynadier L (1993) Geomagnetic field intensity and reversals during the past four million years. *Nature* 366: 234–238.

gelifluction

In PERIGLACIAL ENVIRONMENTS, the slow downslope flow of soil or surface sediments through the release of MELTWATER from thawing ice lenses. It can occur in SEASONALLY FROZEN GROUND as well as in ground underlain by PERMAFROST. Together with *frost creep*, it constitutes SOLIFLUCTION. Where ice lenses develop both near the surface and at the base of the ACTIVE LAYER, solifluction also involves a *plug-like flow*.

RAS

Benedict JB (1976) Frost creep and gelifluction: A review. *Quaternary Research* 6: 55–76.

Harris C (2000) Gelifluction: Observations from large-scale laboratory simulations. *Arctic, Antarctic and Alpine Research* 32: 202–207.

Matthews JA, Harris C and Ballantyne CK (1986) Studies on a gelifluction lobe, Jotunheimen, Norway: ^{14}C chronology, stratigraphy, sedimentology and palaeoenvironment. *Geografiska Annaler* 68A: 345–360.

gelifraction/gelivation

The mechanical breakdown of rock through FROST WEATHERING acting on pores, cracks or bedding planes. *Macrogelifraction* (also known as *macrogelivation*, *frost shattering*, *frost splitting*, *frost bursting* or *frost wedging*) may be distinguished from *microgelifraction* (or *microgelivation*) which involves GRANULAR DISINTEGRATION resulting from frost action. There is controversy over the extent to which these constitute distinct processes and the involvement of processes other than freezing and thawing (e.g. *hydration shattering* and SALT WEATHERING) in PERIGLACIAL ENVIRONMENTS.   *JAM/RAS*

[*See also* FREEZE-THAW CYCLES, PHYSICAL WEATHERING]

Hall K (2006) Perceptions of rock weathering in cold regions: A discussion on space and time attributes of scale. *Géomorphologie: Relief, Processus, Environnement* 3: 187–196.
Matsuoka N (2001) Microgelivation versus macrogelivation: Towards bridging the gap between laboratory and field frost weathering. *Permafrost and Periglacial Processes* 12: 299–313.

gene The molecular unit of HEREDITY, consisting of sequences of DNA (or RNA in some VIRUSES). Genes carry the genetic code of organisms from generation to generation. There are many genes, each of which may influence a specific biological trait. It is the genes rather than the traits that are inherited.   *JAM*

[*See also* GENOME]

Dawkins R (1990) *The selfish gene*. Oxford: Oxford University Press.
Pearson H (2006) Genetics: What is a gene? *Nature* 441: 398–401.

gene pool A collective term for all the GENES in the organisms of a genetic POPULATION.   *KDB*

genecology The study of genetic differences between individuals in POPULATIONS, in relation to their ENVIRONMENT.   *KDB*

[*See also* GENETIC FITNESS, GENOTYPE]

St Clair JB, Mandel NL and Vance-Borland KW (2005) Genecology of Douglas fir in western Oregon and Washington. *Annals of Botany* 96: 1199–1214.
Turesson G (1922) The genotypic response of the plant species to the habitat. *Hereditas* 3: 211–350.

general circulation models (GCMs) Mathematical models of the GENERAL CIRCULATION OF THE ATMOSPHERE based on fundamental principles and developed using the largest and fastest digital computers. They were first developed in the 1960s. The central element of a GCM is a three-dimensional, time-evolving model of the atmosphere, usually represented as an array of 'grid boxes'. Beyond this, there is a set of BOUNDARY CONDITIONS at the top and bottom of the model atmosphere, which include a specification of the physical character of the Earth's surface and the ocean. The mathematical formulation of the model involves a set of time-dependent governing equations, which describe

the dynamics of the atmosphere. The better the spatial resolution of the model, the greater the requirement for high computer power. Despite their complexity, different GCMs continue to yield widely different predictions from the same questions, such as the character of past climates or the likely scale of enhanced GLOBAL WARMING in the near future, indicating the need for further improvements.   *AHP*

[*See also* CLIMATIC MODELS, COUPLED OCEAN-ATMOSPHERE MODELS, FUTURE CLIMATE, REGIONAL CLIMATIC MODELS (RCMs)]

Knutti R (2008) Should we believe model predictions of future climate change? *Philosophical Transactions of the Royal Society A* 366: 4647–4664.
Randall DA (2000) *General circulation model development: Past, present and future*. San Diego, CA: Academic Press.
Randall DA, Wood RA, Bony S et al. (2007) Climate models and their evaluation. In Solomon S, Qin D, Manning M et al. (eds) *Climate change 2007: The physical science basis* [Contribution of Working Group 1 to the Fourth Assessment Report of the Intergovernmental Panel on Climate Change]. Cambridge: Cambridge University Press, 590–662.
Reichert BK, Bengtsson L and Oerlemanns J (2001) Midlatitude forcing mechanisms for glacier mass budget investigated using general circulation models. *Journal of Climate* 14: 3767–3784.

general circulation of the atmosphere The average long-term state of the planetary *wind systems*, which are driven by the distribution of heat received from the Sun, and the Earth's rotation. The general circulation is three-dimensional and includes global-scale phenomena such as the INTERTROPICAL CONVERGENCE ZONE (ITCZ), HADLEY CELLS, LONG WAVES, MONSOONS, POLAR FRONTS, TRADE WINDS and the WESTERLIES. It is summarised in the Figure, the upper part of which shows the generalised *meridional circulation* (i.e. the average pattern of winds in relation to latitude and altitude). Because of the variability of the CLIMATIC SYSTEM, however, this averaged pattern is strongly expressed in reality only in the case of the Hadley cell. In the OCEAN, the equivalent meridional component is known as the MERIDIONAL OVERTURNING CIRCULATION (MOC).   *JAM/AHP*

Lorenz EN (1967) *The nature and theory of the general circulation of the atmosphere*. Geneva: World Meteorological Organization.
Palmén E and Newton CW (1969) *Atmospheric circulation systems*. New York: Academic Press.
Schneider T and Sobel AH (eds) (2007) *The global circulation of the atmosphere*. Princeton, NJ: Princeton University Press.
Smagorinsky J (1972) The general circulation of the atmosphere. In McIntyre DP (ed.) *Meteorological challenges: A history*. Ottawa: Information Canada.

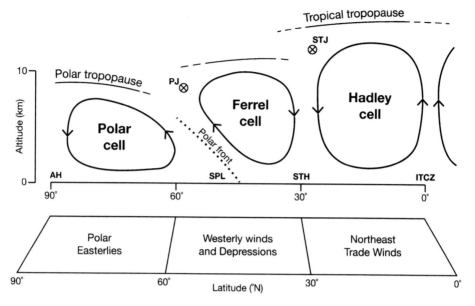

General circulation of the atmosphere *The meridional component is shown for the Northern Hemisphere in relation to latitude and altitude in the upper part of the figure, where PJ = polar jet; STJ = subtropical jet; AH = Arctic high; SPL = subpolar low; STH = subtropical high; ITCZ = intertropical convergence zone. The lower part of the figure shows the average directions of the major wind systems at the Earth's surface (based on various sources).*

general system(s) theory (GST) a concept developed by Ludwig von Bertalanffy and introduced to the English-speaking world in 1951. von Bertalanffy had two primary objectives: (1) to resolve the problems biologists were having accommodating the OPEN SYSTEMS of physicists (necessitating the invocation of 'supposedly vitalistic characteristics') and (2) to minimise the inefficiencies associated with repetitive discoveries stemming from the isolated nature of individual scientific disciplines. GST was promoted as a discipline akin to logic and mathematics, and some general principles (sometimes called LAWS by von Bertalanffy) such as the exponential and logistic curves were emphasised. von Bertalanffy dismissed *analogy* as an appropriate transfer mechanism from one discipline to another, while emphasising *homologues* and *isomorphs* which he conflated. SYSTEMS concepts predate GST and are pervasive in science; consequently, the benefits of invoking GST (as opposed to systems in general) has been questioned.

CET

von Bertalanffy L (1951) An outline of general system theory. *The British Journal for the Philosophy of Science* 1: 134–165.

generalised additive models (GAM) A class of statistical models that are *non-parametric* extensions of GENERALISED LINEAR MODELS (GLM). They use *smoothing techniques* such as *locally weighted regression* or SPLINE FUNCTIONS to identify and represent possible non-linear relationships between the *response (y)* and the *explanatory variables* as an alternative to the parametric models implicit in GLM. In GAM, the *link function* of the expected value of the response function is modelled as the sum of a number of smooth functions of the explanatory variables rather than in terms of the explanatory variables themselves as in GLM. The basic GAM equation is thus as follows:

$$\text{link function } (y) = (\text{explanatory variables})$$
$$(\text{smooth functions}) + \text{error function}$$

GAM allow the data to determine the shape of the response function. As a result, *bimodality* and pronounced SKEWNESS can be easily detected. GAM often provide a better tool for EXPLORATORY DATA ANALYSIS than GLM and are increasingly used in ECOLOGY.

HJBB

Faraway JJ (2006) *Extending the linear model with R: Generalized linear, mixed effects and nonparametric regression.* Boca Raton, FL: Chapman and Hall.

Hastie TJ and Tibshirani RJ (1990) *Generalized additive models.* London: Chapman and Hall.

Wood SN (2006) *Generalized additive models.* Boca Raton, FL: Chapman and Hall.

Yee TW and Mitchell ND (1991) Generalized additive models in plant ecology. *Journal of Vegetation Science* 2: 587–602.

generalised linear models (GLM)   A framework for all REGRESSION ANALYSIS and associated statistical MODELS that involve *linear models*, namely, equations that contain mathematical variables, parameters and random variables, that are linear in the parameters and the random variables. GLM consist of an *error function*, a *linear predictor* and a *link function*.

LEAST SQUARES REGRESSION analysis assumes a NORMAL DISTRIBUTION error function. Many kinds of environmental data have non-normal errors. In GLM, the error function can be expressed as one of the members of the exponential function of PROBABILITY DISTRIBUTIONS (e.g. *Gamma, exponential* and *normal* for *continuous probability distributions*; *binomial, multinomial* and *Poisson* for *discrete probability distributions*). The structure of the GLM relates each observed *response value* (*y*) to a *predicted value*. The predicted value is obtained by transforming the value derived from the linear predictor, which is a linear combination of one or more *explanatory variables*. There are as many terms in the linear predictor as there are parameters to be estimated from the data. To determine the fit of a given model, the linear predictor is evaluated for each *y* value and the predicted value is then compared with a transformed value of the response variable. The transformation applied is specified by the *link function*. This relates the mean value of the response variable to its linear predictor. The use of a *non-linear link function* allows the model to use response and explanatory variables measured in different scales as it maps the linear predictor onto the scale of the response variable. The basic or core equation of a GLM is thus

link function (*y*) = linear predictor + error function

With an appropriate choice of link function (e.g. LOG, *logit, identity, reciprocal, exponential* and *probit*) and error function, a wide range of statistical models can be developed that are appropriate for different data types (e.g. counts, percentages or proportions and binary data). Techniques such as least squares regression analysis, ANALYSIS OF VARIANCE (ANOVA), ANALYSIS OF COVARIANCE (ANCOVA), *logit regression, probit regression, multiple linear regression* and *contingency table analysis* are all types of GLM.

Parameter estimation is by *maximum likelihood estimation*. *Goodness-of-fit* is assessed by *deviance*, a measure of the extent to which a particular model differs from the full or the *saturated model* for a data set where there is one parameter for every data point. Deviance is analogous to residual sum-of-squares in least squares regression and has a *chi-squared distribution*. Care is required to find the minimal adequate model, namely the model that produces the minimal residual deviance, subject to the constraint that it has the smallest number of statistically significant parameters.

Advantages of GLM are the following: (1) they have great versatility; (2) they can cope with different data types and associated error distributions; (3) they provide a common and powerful framework for REGRESSION ANALYSIS, ANOVA and, if covariates are included, ANCOVA; and (4) it is not necessary to transform the data as the regression is transformed through the link function. GLM can be fitted by software packages such as GLIM, GenStat, SAS and S-PLUS.

HJBB

[*See also* GENERALISED ADDITIVE MODELS (GAM)]

Crawley MJ (1993) *GLIM for ecologists*. Oxford: Blackwell.
Crawley MJ (2007) *The R book*. Chichester: Wiley.
Dobson AJ (1990) *An introduction to generalized linear models*. London: Chapman and Hall.
Faraway JJ (2005) *Linear models with R*. Boca Raton, FL: CRC Press.
Faraway JJ (2006) *Extending the linear model with R: Generalized linear, mixed effects and nonparametric regression*. Boca Raton, FL: Chapman and Hall.
Fox J (2008) *Applied regression analysis and generalized linear models*. Thousand Oaks, CA: Sage.
McCullagh P and Nelder JA (1989) *Generalized linear models*. London: Chapman and Hall.
O'Brian L (1992) *Introducing quantitative geography: Measurement, methods and generalised linear models*. London: Routledge.

genetic algorithms   ALGORITHMS that carry out simulated EVOLUTION on a population of numbers, which are called *chromosomes*. They were first postulated by Holland in 1975. *Simulated evolution* takes place by modifying the chromosomes which consist of sequences of genes. During evolution, NATURAL SELECTION favours 'successful' chromosomes, which reproduce more frequently. Evolution is mainly based on the process of *reproduction*, whereas MUTATIONS occur relatively rarely.

A genetic algorithm has at least five components: chromosome/gene representation, initialisation of the population, evaluation function determining the fitness for survival, genetic operators altering chromosomes (reproduction, crossover and mutation), parameters for population size and PROBABILITIES of genetic operators. Genetic algorithms provide robust estimates in complex spaces. They have been applied in REMOTE SENSING (e.g. for automatic identification of spectral signatures).

HB

Davis L (1987) *Genetic algorithms and simulated annealing*. London: Pitman.
Dwivedi S and Pandey AC (2011) Forecasting the Indian summer monsoon intraseasonal oscillations using genetic algorithm and neural network. *Geophysical Research Letters* 38: L15801.
Karimi A, Haddad OB and Shadkam S (2010) Rainfall network optimization using transinformation entropy and genetic

algorithm. *Proceedings of the Conference on 21st Century Watershed Technology: Improving Water Quality and Environment, Costa Rica, 2010.* St. Joseph, MI: American Society of Agricultural and Biological Engineers..

genetic drift   Changes in the GENE POOL of a POPU-LATION by RANDOM processes. These include the spread of an ALLELE because it is neutral (and hence neither eliminated nor favoured), or changes in gene frequency as populations diminish. Such processes are only likely to be significant in small populations, being swamped by non-random processes in larger populations.   *KDB*

[*See also* FOUNDER EFFECT]

Sutuyma DJ (1998) *Evolutionary biology.* Sunderland, MA: Sinauer.
Young A, Boyle T and Brown T (1996) The population genetic consequences of habitat fragmentation for plants. *Trends in Ecology and Evolution* 11: 413–418.

genetic engineering   The artificial manipulation of the genetic make-up or genome of an organism; the molecular aspect of BIOTECHNOLOGY. Genetic engineering, also known as *recombinant DNA technology, gene cloning* and in vivo (in cell) *genetic manipulation,* has already been widely used to introduce desirable traits in crops and domesticated animals and to produce antibiotics and hormones. Some of the most important applications of genetic engineering relate to GENETICALLY MODIFIED ORGANISMS (GMOs), including crops and livestock, which improve productivity either directly or indirectly through engineered resistance to, for example, PESTS or frost. Genetic engineering modifies organisms to suit the environment rather than vice versa: the latter being the traditional approach to AGRICULTURAL INTENSIFICATION. The potential benefits are clear but the environmental risks are largely unknown. There are lessons to be learnt from the release of *introduced species* in the past, but conclusions from these may not be readily transferred, as species behave individualistically; and many genetically modified organisms are TRANSGENIC ORGANISMS (derived from more than one species), which adds to the uncertainty in their behaviour when released into the environment.   *JAM*

[*See also* DNA, GENETIC POLLUTION]

Fincham JRS and Ravetz JR (1991) *Genetically-engineered organisms: Benefits and risks.* New York: Wiley.
Lindow SE (1990) Use of genetically altered bacteria to achieve plant frost control. In Nikas J and Hagerdorn C (eds) *Biotechnology of plant-microbe interaction.* New York: McGraw-Hill, 85–110.
Lycett G and Grierson D (eds) (1990) *Genetic engineering of crop plants.* London: Butterworth.
Mannion AM (1992) Biotechnology and genetic engineering: New environmental issues. In Mannion AM and Bowlby SR (eds) *Environmental issues in the 1990s.* Chichester: Wiley, 147–160.

McHughen A (2000) *Pandora's picnic basket: The potential and hazards of genetically modified foods.* Oxford: Oxford University Press.

genetic fitness   Typically, the *genetic diversity* of a population which, in turn, is assumed to be a measure of its ability to adapt to a variable environment. Conservation biologists often worry about the loss of genetic fitness in captive or otherwise small populations of ENDANGERED SPECIES.   *MVL*

[*See also* ADAPTATION, CONSERVATION BIOLOGY, INBREEDING DEPRESSION, POPULATION VIABILITY ANALYSIS (PVA)]

Black S, Yamaguchi N, Harland A and Groombridge J (2010) Maintaining the genetic health of putative Barbary lions in captivity: An analysis of Moroccan Royal Lions. *European Journal of Wildlife Research* 56: 21–31.
Fraser DJ (2008) How well can captive breeding programs conserve biodiversity? A review of salmonids. *Evolutionary Applications* 1: 535–586.

genetic pollution   The process by which genes from domesticated organisms, or INVASIVE SPECIES, or GENETICALLY MODIFIED ORGANISMS (GMOs) become incorporated into wild or NATIVE species. Genes conferring resistance to PESTICIDES or DISEASE in genetically modified crops may, for example, be transferred in pollen from the crop plants to related WEED species.   *JAM*

Porteous A (2000) *Dictionary of environmental science and technology, 3rd edition.* Chichester: Wiley.

genetically modified (GM) foods   Food made in whole or in part from GENETICALLY MODIFIED ORGANISMS. Developments in BIOTECHNOLOGY since the 1990s allow the transfer of GENES across species boundaries, which has huge potential and possibly risks in the context of food. The potential ranges from greater ability to feed the world from more productive and/or nutritious crops (e.g. *golden rice*) and foods to faster development of new varieties of crops and livestock, the development of crops that are resistant to environmental hazards (e.g. DROUGHT) and foods without allergic reactions, PEST CONTROL without PESTICIDES and NITROGEN FIXATION without FERTILISERS. Possible risks include the loss of traditional varieties leading to BIODIVERSITY LOSS, unforeseen long-term effects on HUMAN HEALTH HAZARDS (short-term effects seem negligible) and the increasing dependency of growers on the large companies that bio-engineer these foods, all of which raise questions of ENVIRONMENTAL ETHICS. Currently, the main *genetically modified crops* involved in producing GM foods are soya, maize, salad vegetables, rapeseed and cottonseed. Currently, around 50 per cent of the World's GM crops are grown in the United States: the Grocery Manufacturers of America estimate that 75 percent of all processed foods in the United States contain GM ingredients.   *CJB/JAM*

Chrispeels MJ and Sadava DE (2003) *Plants, genes, and crop biotechnology*. Boston: Jones and Bartlett.
De la Perriere RAB and Seuret F (2000) *Brave new seeds: The threat of GM crops to farmers*. London: Zed Books.
Pinstrup-Andersen P and Schiøler E (2001) *Seeds of contention: World hunger and the global controversy on GM crops*. Baltimore, MD: Johns Hopkins University Press.
Toke D (2004) *The potential of GM foods: A comparative study of the UK, USA and EU*. London: Routledge.

genetically modified organism (GMO) An organism, the DNA of which has been altered by GENETIC ENGINEERING. The first GMOs were BACTERIA, produced in the 1970s. Small-scale planting of genetically modified plants began in the 1980s, when the first guidelines on their use were produced by the United Nations. Commercial CULTIVATION of *genetically modified crops* began in the 1990s and has continued to expand rapidly, especially in North and South America. However, there is considerable resistance to their use in Europe. The main advantage of GMOs over traditional methods of genetic improvement, such as SELECTIVE BREEDING, is the ability to precisely control desired characteristics with great potential benefits for *food production*. The main disadvantage lies in the possibility of undesirable effects with the release of the products of this largely untested *technology* into the environment. There are additional objections, particularly to the genetic modification of animals, on religious, ethical and animal welfare grounds.   *JAM*

[*See also* BIOTECHNOLOGY, GENETIC POLLUTION, GENETICALLY MODIFIED (GM) FOODS, TRANSGENIC ORGANISM]

Bodiquel L and Cardwell M (eds) (2010) *The regulation of genetically modified organisms: Comparative approaches*. Oxford: Oxford University Press.
European Food Safety Authority (EFSA) (2011) *Genetically modified organisms: The risk assessment of genetically modified plants and derived food and feed*. Saarbrücken: Dictus Publishing.
Halford N (2011) *Genetically modified crops, 2nd edition*. London: Imperial College Press.
Nelson GL (ed.) (2001) *Genetically modified organisms in agriculture: Economics and politics*. San Diego, CA: Academic Press.
Parekh SR (ed.) (2004) *The GMO handbook: Genetically modified animals, microbes and plants in biotechnology*. Totowa, NJ: Human Press.
Thomson JA (2002) *Genes for Africa: Genetically modified crops in the developing world*. Cape Town: University of Cape Town Press.

genetics The part of the BIOLOGICAL SCIENCES that deals with the study of HEREDITY and variation within and between individuals and POPULATIONS.   *KDB*

Klug WS, Cummings MR, Spencer CA and Palladino MA (2009) *Essentials of genetics, 7th edition*. Abingdon: Pearson.

genome The complete set of GENES in an organism. It defines the genetic characteristics of the species, including the variation between individuals. The genome is stored on *chromosomes* (long strands of DNA). *Genomics* is the study of the genome. Applications are developing rapidly, especially in relation to MEDICAL SCIENCE, DISEASE, ENVIRONMENTAL ARCHAEOLOGY and the study of EVOLUTION.   *JAM*

Benfry PN and Protopapas AD (2004) *Essentials of genomics*. San Francisco: Benjamin Cummins.
Lander ES (2011) Initial impact of the sequencing of the human genome. *Nature* 470: 187–197.
Rasmussen M, Guo X, Wang Y et al. (2011) An aboriginal Australian genome reveals separate human dispersals into Asia. *Science* 334: 94–98.
Van Straalen NM and Roelofs D (2012) *An introduction to ecological genomics, 2nd edition*. Oxford: Oxford University Press.

genotype The total *genetic constitution* of an individual organism. Thus, the genotype is the totality of those characteristics of the organism that are genetically determined. The GENOME (the physical MOLECULES of DNA inherited from its parents) is the material basis of the genotype.   *JAM/KDB*

[*See also* GENE POOL, PHENOTYPE, SPECIES CONCEPT]

gentrification The replacement of traditional inner-city, working-class, residential areas with improved housing, more affluent inhabitants and up-market services.   *JAM*

Lees L, Slater T and Wyly E (2007) *Gentrification*. New York: Routledge.
Lees L, Slater T and Wyly E (eds) (2010) *The gentrification reader*. New York: Routledge.

geo-archaeology The application of principles and techniques from the EARTH SCIENCES to the study of the human past.   *SPD*

[*See also* ARCHAEOLOGICAL GEOLOGY, ENVIRONMENTAL ARCHAEOLOGY, LANDSCAPE ARCHAEOLOGY]

Brown AG (1997) *Alluvial geoarchaeology: Floodplain archaeology and environmental change*. Cambridge: Cambridge University Press.
French CAI (2003) *Geoarchaeology in action: Studies in soil micromorphology and landscape evolution*. London: Routledge.
Goldberg P and Macphail RI (2006) *Practical and theoretical geoarchaeology*. Oxford: Blackwell.
Rapp Jr G and Hill CL (2006) *Geoarchaeology, 2nd edition*. New Haven, CT: Yale University Press.
Waters MR (1992) *Principles of geoarchaeology: A North American perspective*. Tuscon: University of Arizona Press.

Wilson L (ed.) (2011) *Human interactions with the geosphere: The geoarchaeological perspective.* Bath: Geological Society.

geobotany   Variously used for (1) the historical study of the distribution of plants (PLANT GEOGRAPHY), (2) the study of the linkages between plants and soils (GEO-EDAPHICS) and (3) the largely applied field of using plant indicators (e.g. so-called copper mosses) as indicators of RESOURCES (e.g. HEAVY METALS).   *JAM*

[*See also* BIO-INDICATORS]

Rübel E (1927) Ecology, plant geography, and geobotany: Their history and aim. *Botanical Gazette* 84: 428–439.

geochemical proxies   Indirect measures of environmental changes and processes derived from the chemical analysis of geological samples. Sediment sequences do not directly record environmental parameters. Palaeoenvironmental and palaeoclimatic information can, however, be derived from both inorganic and organic sources in terrestrial and marine environments (NATURAL ARCHIVES). The geochemical proxies can provide information about, for example, PALAEOTEMPERATURE, VEGETATION HISTORY, cycling of NUTRIENTS and OCEAN CIRCULATION and SALINITY. Although the proxies can be used independently, a combination of several proxies in a MULTIPROXY APPROACH helps compensate for the effects of alteration by DIAGENESIS, and it improves the PALAEOENVIRONMENTAL RECONSTRUCTION.

Inorganic geochemical proxies are derived from bulk components such as *biogenic opal*, CARBONATES and QUARTZ or are expressed as ratios of *metals*. ELEMENTS that are sensitive to changes in PRODUCTIVITY, such as CADMIUM (Cd), *barium* and SILICON (Si), can be normalised with respect to the *detrital content*, which is derived from ALUMINIUM (Al) or POTASSIUM concentrations. Inorganic proxies are often measured on the remains of calcitic PLANKTON, such as FORAMINIFERA, as the rate of incorporation of metal ions compared with the principal constituent, CALCIUM, varies according to temperature, salinity and nutrient availability. Therefore, ratios of Mg/Ca, Sr/Ca and Cd/Ca are indicators of past temperatures, salinity and nutrient availability. K/Al ratios are used to indicate chemical weathering as K is highly soluble and readily leached by chemical weathering while Al is not. Thus, low K/Al values indicate a strong chemical weathering signal. The ratio of the radiogenic isotopes, $^{87}Sr/^{86}Sr$ is a proxy for *silicate weathering* and allows the PROVENANCE of the source material to be determined.

Organic geochemical proxies can identify the general source of organic matter by analyzing bulk *elemental composition* and STABLE ISOTOPES or its detailed origins using BIOMARKERS and COMPOUND-SPECIFIC CARBON ISOTOPE ANALYSIS. Palaeoenvironmental studies routinely apply bulk geochemical properties such as *total organic carbon, carbon:nitrogen ratios,* CARBON ISOTOPES and NITROGEN ISOTOPES. Examination of organic matter at a molecular level enables the separation of terrestrial, aquatic and sedimentary components and simultaneous examination of environmental conditions in the water column and surrounding DRAINAGE BASIN. Combined with compound-specific isotope analysis, geochemical proxies can be used to determine PRIMARY PRODUCTIVITY (identification of biomarkers, e.g. *sterols*), SEA-SURFACE TEMPERATURE (SST) (U_{37}^{k} index), salinity (%C$_{37:4}$ alkenone), *hydrological variability* (compound specific δD), *redox conditions, vegetation type* (P$_{aq}$, δ1^{3c} of *plant pigments*), *soil organic matter input* to aquatic environments (BIT index), *soil temperature* and *soil* pH (methylation index of branched tetraeters and cyclisation ratio of branched tetraethers, TEX$_{86}$), BIOGEOCHEMICAL CYCLING and *carbon storage* and preservation of organic matter.   *KJF*

[*See also* ALKENONES, CARBON SEQUESTRATION, LAKE-CHEMISTRY RECONSTRUCTIONS, MAGNESIUM:CALCIUM RATIO (Mg:Ca), PALAEOCEANOGRAPHY: PHYSICAL AND CHEMICAL PROXIES, PALAEOTHERMOMETRY, PIGMENTS: FOSSIL, STRONTIUM/CALCIUM PALAEOTHERMOMETRY]

Babek O, Famera M, Hilscherova K et al. (2011) Geochemical traces of flood layers in the fluvial sedimentary archive: Implications for contamination history analyses. *Catena* 87: 281–290.

Buggle B, Glaser B, Hambach U et al. (2011) An evaluation of geochemical weathering indices in loess-paleosol studies. *Quaternary International* 240: 12–21.

Castañeda IS and Schouten S (2011) A review of molecular organic proxies for examining modern and ancient lacustrine sediments. *Quaternary Science Reviews* 30: 2851–2891.

Eglinton TI and Eglinton G (2008) Molecular proxies for palaeoclimatology. *Earth and Planetary Science Letters* 25: 1–16.

Gerbersdorf SU, Jancke T and Westrich B (2007) Sediment properties for assessing the erosion risk of contaminated riverine sites. *Journal of Soils and Sediments* 7: 25–35.

Killiops S and Killops V (2005) *Introduction to organic geochemistry.* Malden, MA: Blackwell.

Peters KE, Walters CC and Moldowan JM (2005) *The biomarker guide: Biomarkers and isotopes in petroleum exploration and earth history.* New York: Cambridge University Press.

geochemistry   The study of the chemistry of Earth materials and processes. *Environmental geochemistry* focuses on the geochemistry at the Earth's surface, including interactions between the Earth, OCEAN and ATMOSPHERE. Within SEDIMENTOLOGY, much geochemical study is focused on DIAGENESIS. Approaches include the study of MINERAL and aqueous-phase chemistry, STABLE ISOTOPE ANALYSIS (e.g. for *palaeotemperature* determination) and *radioisotope* investigation (e.g. for RADIOMETRIC DATING). Analytical techniques include ATOMIC ABSORPTION SPECTROPHOTOMETRY (AAS), ENERGY DISPERSIVE SPECTROMETRY (EDS), INDUCTIVELY COUPLED PLASMA ATOMIC EMISSION SPECTROMETRY (ICP-AES), INDUCTIVELY COUPLED PLASMA MASS SPECTROMETRY (ICP-MS), OPTICAL EMISSION SPECTROSCOPY (OES) and X-RAY FLUORESCENCE ANALYSIS (XRF). A key technique for mineralogical determination is X-RAY DIFFRACTION ANALYSIS (XRD), and microanalytical techniques include use of the ELECTRON MICROPROBE ANALYSIS (EMPA) and SCANNING ELECTRON MICROSCOPY (SEM). TY

[*See also* BIOGEOCHEMICAL CYCLES, CHEMICAL ANALYSIS OF SOILS AND SEDIMENTS, ENVIRONMENTAL CHEMISTRY, ENVIRONMENTAL GEOLOGY]

Appelo CAJ and Postma D (2005) *Geochemistry, groundwater and pollution, 2nd edition.* Leiden: Taylor and Francis.

Chester R and Jickells T (2012) *Marine geochemistry, 3rd edition.* Oxford: Blackwell.

De Lacerda LD, Santelli RE, Duursma EK and Abrao JJ (eds) (2012) *Environmental geochemistry in tropical and subtropical environments.* Berlin: Springer.

Drever JI (2005) *Surface and ground water, weathering, and soils, 2nd edition.* Amsterdam: Elsevier.

Eby GE (2003) *Principles of environmental geochemistry.* Belmont, CA: Brooks-Cole.

Elderfield H (2005) *The oceans and marine geochemistry, 2nd edition.* Amsterdam: Elsevier.

Keeling RK (2006) *The atmosphere, 2nd edition.* Amsterdam: Elsevier.

Killops SD and Killops VJ (1993) *An introduction to organic geochemistry.* Harlow: Longman.

Mackenzie FT (ed.) (2005) *Sediments, diagenesis, and sedimentary rocks, 2nd edition.* Amsterdam: Elsevier.

Rudnick RL (2006) *The crust, 2nd edition.* Amsterdam: Elsevier.

Schlesinger WH (2005) *Biogeochemistry, 2nd edition.* Amsterdam: Elsevier.

Sherwood LB (2005) *Environmental geochemistry, 2nd edition.* Amsterdam: Elsevier.

geochronology The study of GEOLOGICAL TIME. Successions of SEDIMENTS and SEDIMENTARY ROCKS clearly record a history of events comprising the GEOLOGICAL RECORD, and during the nineteenth century, a timescale of RELATIVE AGES was built up without any idea of the actual time involved (see STRATIGRAPHICAL COLUMN). Similar successions can be recognised through sequences of volcanic rocks, which may consist of LAVA FLOWS, VOLCANIC ASH or other PYROCLASTIC material. CHRONOSTRATIGRAPHY is concerned with the application of time to rock successions, usually through the use of FOSSILS in SEDIMENTARY ROCKS that provide divisions between *chronostratigraphical units*. Research continues on the establishment of a series of international *standard reference sections* that will ultimately provide a boundary marker point at a *Global Stratotype Section and Point* for worldwide geochronological CORRELATION (STRATIGRAPHICAL) for each chronostratigraphical unit.

Geochronometry, the direct measurement of geological time (see CHRONOMETRY), is an important application of geochronology. Early attempts, for example, comparing the thickness of a sedimentary succession with an 'average' rate of sedimentation, failed because of the incomplete nature of rock successions and widely varying sediment ACCUMULATION RATES in different environments. Calculations by the physicist William Thomson (Lord Kelvin) in the AD 1890s, assuming that the Earth's internal heat was a relic from a once-molten state, gave a maximum age for the Earth of 60 million years. Kelvin's stature ensured respect for his calculations and provided comfort for anti-evolutionists as Darwin had stated that EVOLUTION of life would require at least 120 million years. The discovery of RADIOACTIVITY in the early twentieth century showed that the Earth had its own internal heat source, thus negating Kelvin's calculations. The discovery that the rate of radioactive decay was constant provided the potential for accurate geochronometry, and the results of the first *absolute-age dating* using radioactive minerals were published in 1907. Ages obtained in this way are referred to as *isotopic dates* or *radiometric dates* (see RADIOMETRIC DATING) and are quoted in millions of years (Ma) or thousands of millions of years (Ga) before the present.

ISOTOPES suitable for dating PRECAMBRIAN and most PHANEROZOIC rocks include *rubidium* (^{87}Rb), which decays to *strontium* (^{87}Sr) with a HALF-LIFE of 48,800 million years; POTASSIUM (^{40}K), which decays to *argon* (^{40}Ar) with a half-life of 11,930 million years; and *uranium* (^{238}U), which decays to LEAD (^{206}Pb) with a half-life of 4,469 million years. Different dating methods may yield different ages, and the UNCERTAINTY associated with radiometric dates is commonly millions or tens of millions of years. The minerals that contain these elements are mostly found in IGNEOUS ROCKS, in which case the age obtained represents the cooling of the rock through some critical temperature, but the mineral *glauconite* (a hydrous silicate of iron and potassium) forms in shallow water marine sediments and provides an excellent method of dating sedimentary rocks directly by the POTASSIUM-ARGON (K-Ar) DATING method. Where rocks have subsequently experienced *metamorphism* (see METAMORPHIC ROCKS), the radiometric 'clock' may be reset and care must be taken over the meaning of any radiometric age. Rock successions in which igneous rocks such as LAVA FLOWS are interbedded with fossiliferous sedimentary rocks have been used to calibrate the stratigraphical column, producing the GEOLOGICAL TIMESCALE. Geochronology can also be used to determine the age of the Earth, currently estimated at between 4,500 and 4,600 Ma.

More recently, developed methods of geochronology, some of which are directly applicable to sediments and sedimentary rocks, include FISSION-TRACK DATING, COSMOGENIC-NUCLIDE DATING and ELECTRON SPIN RESONANCE (ESR) DATING. Dates of Late PLEISTOCENE and HOLOCENE organic samples may be determined using RADIOCARBON DATING. Their age is usually quoted as years before the present, using 1950 as the baseline.   *JCWC/GO*

[*See also* DATING TECHNIQUES]

Berggren WA, Kent DV, Aubry M-P and Hardenbol J (eds) (1995) *Geochronology, time scales and global stratigraphic correlation.* Tulsa, OK: Society of Economic Paleontologists and Mineralogists [Special Publication 54].
Dunay RE and Hailwood EA (eds) (1995) *Non-biostratigraphical methods of dating and correlation.* Bath: Geological Society.
Geyh MA and Schleicher H (1990) *Absolute age determination: Physical and chemical dating methods and their application.* Berlin: Springer.
Gradstein FM, Ogg JG and Smith AG (eds) (2005) *A geologic time scale 2004.* Cambridge: Cambridge University Press.
Wagner GA (1998) *Age determination in young rocks and artifacts: Physical and chemical clocks in Quaternary geology and archaeology.* Berlin: Springer.
Wells JW (1963) Coral growth and geochronometry. *Nature* 197: 948–950.

geocryology   This field is now regarded as the study of perennially frozen ground, or *permafrost science*. However, the term has been used more broadly in the past, ranging from the study of the whole CRYOSPHERE (including GLACIERS) to the study of PERIGLACIAL ENVIRONMENTS (SEASONALLY FROZEN GROUND as well as PERMAFROST).   *JAM/HMF*

[*See also* CRYOSTRATIGRAPHY]

Ershov E (1998) *Geocryology.* Cambridge: Cambridge University Press.
Washburn AL (1979) *Geocryology: A survey of periglacial processes and environments, 2nd edition.* London: Arnold.
Williams PJ and Smith MJ (1989) *The frozen earth: Fundamentals of geocryology.* Cambridge: Cambridge University Press.

geodesy   The study of the Earth's *gravitational field* and related topics including the shape of the Earth (see GEOID), rotation of the Earth, TIDES and the precise measurement and mapping of points on the Earth's surface (*geodetic surveying*).   *DNT*

[*See also* GEOMATICS, SURVEYING]

Horwath M, Legrésy B, Rémy F et al. (2012) Consistent patterns of Antarctic ice sheet interannual variations from ENVISAT radar altimetry and GRACE satellite gravimetry. *Geophysical Journal International* 189: 863–876.

Ivins ER and James TS (1999) Simple models for late Holocene and present-day Patagonian glacier fluctuations and predictions of a geodetically detectable isostatic response. *Geophysical Journal International* 138: 601–624.
Keay J (2000) *The great arc: The dramatic tale of how India was mapped and Everest was named.* London: HarperCollins.
Stacey FD and Davis PM (2008) *Physics of the Earth.* Cambridge: Cambridge University Press.

geodiversity   The variety within ABIOTIC nature: the abiotic counterpart of BIODIVERSITY. Geologists and geomorphologists started to use the term in the 1990s to correct the imbalance given, within the fields of the CONSERVATION of nature and WILDLIFE CONSERVATION and MANAGEMENT, to biodiversity. Geodiversity (the quality being conserved) may be distinguished from *geoconservation* (the endeavour of conservation) and GEOHERITAGE (the significant examples worthy of conservation). Geodiversity is a *georesource* that is being threatened by human activities and should be valued, whether for cultural, aesthetic, functional, economic, research and/or educational reasons. Geodiversity therefore needs to be conserved by, for example, conferring PROTECTED AREA status (*geoprotection*) or RESTORATION where damaged (*georestoration*). Geodiversity also needs to be managed sustainably, which includes *geomanagement* beyond protected areas, where principles of *geodesign* and sustainable use of *geomaterials* should be universally applicable.   *JAM*

[*See also* GEOLOGICAL CONSERVATION (GEOCONSERVATION), GEOTOURISM, HERITAGE, RESOURCE]

Gray JM (2004) *Geodiversity: Valuing and conserving abiotic nature.* Chichester: Wiley.
Gray JM (2008) Geodiversity: Developing the paradigm. *Proceedings of the Geologists Association* 119: 287–298.
Ruban DA (2010) Quantification of geodiversity and its loss. *Proceedings of the Geologists Association* 121: 326–333.
Thomas MF (2012) A geomorphological approach to geodiversity: Its applications to geoconservation and geotourism. *Questiones Geographicae* 31: 81–89.

geodynamics   (1) In GEOLOGY, the study of dynamic processes affecting the Earth, particularly those relating to TECTONICS, including PLATE TECTONICS. (2) In ECOLOGY, *geodynamic factors* are physical ENVIRONMENTAL FACTORS involving soil DISTURBANCE, such as FROST HEAVE, SOLIFLUCTION and CRYOTURBATION.   *GO/JAM*

geo-ecology   The interdisciplinary area of GEOGRAPHY and ECOLOGY that focuses on the interrelations and interactions of BIOTIC and ABIOTIC components of NATURAL and CULTURAL LANDSCAPES.   *JAM*

[*See also* GEO-ECOSYSTEM, LANDSCAPE ECOLOGY, LANDSCAPE SCIENCE]

Huggett RJ (1995) *Geoecology: An evolutionary approach.* London: Routledge.

Matthews JA (1992) *The ecology of recently-deglaciated terrain: A geoecological approach to glacier forelands and primary succession.* Cambridge: Cambridge University Press.

Troll C (1971) Landscape ecology (geoecology) and biogeocenology: A terminological study. *Geoforum* 8: 43–46.

geo-ecosphere   The global geoecosystem: the totality of those landscapes of the Earth in which organisms are present. As a concept of LANDSCAPE ECOLOGY, it is the *landscape sphere*, not only the terrestrial BIOSPHERE but also the interacting parts of the upper LITHOSPHERE and lower ATMOSPHERE (see Figure).   *JAM*

[*See also* ANTHROPOSPHERE, EARTH SPHERES, EARTH SYSTEM, GEO-ECOLOGY, NOÖSPHERE]

Matthews JA, Bartlein PJ, Briffa KR et al. (2012) Background to the science of environmental change. In Matthews JA, Bartlein PJ, Briffa KR et al. (eds) *The SAGE handbook of environmental change, volume 1.* London: Sage, 1–33.

geo-ecosystem   A *landscape system* including the *ecosystem* in its LANDSCAPE context. Geo-ecosystems are dynamic spatial entities, the components of which interact and are continually responding to environmental change, but the behaviour of which is holistic and therefore not necessarily predictable from the behaviour of their component parts.   *JAM*

[*See also* BIOGEOCENOSIS/BIOGEOCOENOSIS, GEO-ECOLOGY, GEO-ECOSPHERE, HOLISM, LANDSCAPE ECOLOGY, LANDSCAPE GEOCHEMISTRY]

Demek J (1978) The landscape as a geosystem. *Geoforum* 9: 29–34.

Morgan LA (ed.) (2007) Integrated geoscience studies in the Greater Yellowstone area: Volcanic, tectonic and hydrothermal studies in the Yellowstone geoecosystem. *United States Geological Survey Professional Paper 1717.*

geo-edaphics   The study of the interactions between plants (and VEGETATION) and soils, including their broader interrelationships with topography and lithology.   *JAM*

[*See also* GEOBOTANY, GEO-ECOLOGY]

Kruckeberg AR (2002) *Geology and plant life: The effects of landforms and rock types on plants.* Seattle: University of Washington Press.

geo-engineering   (1) The application of geological knowledge and techniques to engineering solutions to ENVIRONMENTAL PROBLEMS (e.g. the safe disposal of RADIOACTIVE WASTE, the use of COASTAL ENGINEERING STRUCTURES in reducing coastal EROSION and the implementation of FLOOD CONTROL MEASURES). (2) The deliberate manipulation at the global scale of the Earth's environment. The term has become associated particularly with GLOBAL WARMING, how the FORCING FACTORS might be abated, how people and economic systems might be adapted to reduce the impact and how the climate system might be manipulated to counteract anthropogenic impact (see CLIMATIC ENGINEERING). Examples include spreading DUST or SULFUR DIOXIDE in the STRATOSPHERE to mitigate expected global temperature rise and increasing the REFLECTANCE of marine stratocumulus clouds.   *RAS*

[*See also* ENVIRONMENTAL ENGINEERING]

Aoki K and Shiogama Y (1993) Geoengineering techniques used in the construction of underground openings in jointed rocks. *Engineering Geology* 35: 167–173.

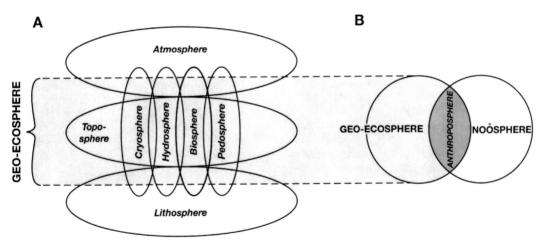

Geo-ecosphere *(A) The interacting components of the landscape sphere at the Earth's surface. (B) The anthroposphere, the human-modified geo-ecosphere, is shown as the area of interaction between the geo-ecosphere and the noösphere, the sphere of the human mind (Matthews et al., 2012).*

Hale B and Dilling L (2011) Geoengineering, ocean
    fertilisation, and the problem of permissible pollution.
    Science Technology and Human Values 36: 190–212.
Schneider SH (1996) Geoengineering: Could we or should
    we do it? *Climatic Change* 33: 291–302.
Watts RG (1998) *Engineering response to global climate
    change.* London: CRC Press.

geofact   A pseudo-artefact. Geofacts are objects created by natural processes (e.g. frost-shattered pebbles),
which may be mistaken for archaeological ARTEFACTS
(e.g. primitive axe heads).   *JAM*

[*See also* ECOFACT, EOLITHS]

Haynes CV (1973) The Calico site: Artefacts or geofacts?
    Science 181: 305–310.

geoglyph   Ground figures carved from rock, such as
those on Easter Island (although they need not be of
such an immense size). Such archaeological features,
also known as *intaglios* or *earthen art*, should be distinguished from PETROGLYPHS, which are carvings into
rock surfaces.   *JAM*

Flenley J and Bahn P (2003) *The enigmas of Easter
    Island: Island on the edge, 2nd edition.* Oxford: Oxford
    University Press.

geographical information system (GIS)   A
suite of computer ALGORITHMS designed to store, query,
manipulate and display data that are intrinsically geographical. The GIS was developed in the 1960s by
government agencies as part of the prevailing paradigm shift towards more systematic and automated
collection and collation of *environmental inventories*.
Its main features are shown in the Figure. The Canadian GIS led by Roger Tomlinson is widely regarded as
the first operational GIS, which utilised computers to
create a central DATABASE for storing digital representations of NATURAL RESOURCES. Such representations are
known as *spatial data* and include geographical features such as spot heights, river lengths, forest expansion and lake capacity, which are stored in a GIS in one
of four ENTITIES: POINTS, *lines* (see ARC), *areas* (sometimes known as *polygons*) or three-dimensional *volumes*, respectively. Each entity is identified by LABELS,
which can be either text (spot height of Ben Nevis,
the River Severn, the New Forest and Lake Windermere) or values (spot height of 1,335 m, river velocity
of 5 m/s, forest area of 350 km^2 and lake volume of
251,321 acre-feet). Names and values of entities are
often termed ATTRIBUTES. Collections of related entities and their associated attributes are stored in a GIS
as layers or COVERAGES. For example, elevation samples
would be combined to produce a relief coverage (see
DIGITAL ELEVATION MODEL (DEM)), roads would contribute to a transport network coverage, forests part of an
ecological habitat coverage and lakes to a hydrology
coverage. A coverage can best be viewed as a *digital
map* and, like all MAPS, is bound by conventional cartographic rules that dictate that geographical information is symbolised, generalised to scale and referenced
to the Earth by CO-ORDINATES and MAP PROJECTION systems. Coverages can then be stored in one of three
main formats, VECTOR, RASTER and QUAD TREE. Coverages based on the VECTOR DATA MODEL use geometric
co-ordinates to locate individual points, which can be
interconnected to represent lines, start and end at the
same location to represent areas, and include height/
depth values to represent volume. Raster-based coverages, on the other hand, are composed of a matrix of
GRID CELLS, where points are represented by a single
occupied cell, and where lines are a series of adjacent
occupied cells, areas are a mass of neighbouring cells
and volumes are tagged with height/depth values.
Lastly, quad trees are less popular and use a hierarchical structure based on quadrants to recursively pinpoint
occupied cells. Associated coverages representing the
same geographical area or thematic features (e.g. agricultural LANDUSE CHANGE) are combined and stored in
a digital database (see RELATIONAL DATABASE) and are
controlled by a *database management system*. One of
the strengths of a GIS is its ability to interrogate coverages held in a database simultaneously, using SPATIAL
ANALYSIS routines, which include OVERLAY ANALYSIS
(for a diagram), POLYGON ANALYSIS and the generation
of BUFFER ZONES. The efficient handling and manipulation of digital geographical information can answer
many questions dealing with *environmental change*.
For instance, HABITAT LOSS assessments may involve
questions such as what impact will the location of a
new power station have on an environmentally sensitive area. Questions such as the impact of effluents on
WATER RESOURCES, SOIL and CONTAMINATED LAND can be
answered by a GIS within a DECISION-MAKING process.
Coverages representing information on waterbodies,
soil types, woodland land cover, agricultural landuse,
relief and prevailing winds would be captured, spatially
registered to common geometric co-ordinates and held
within a database. A QUERY LANGUAGE would then be
used to interrogate these coverages and reveal answers
to the likely effects of a power station on the environment. The process would start with the generation of
buffer zones to determine the proportion of land within
some distance of the power plant deemed to be at risk.
Polygon analysis would then calculate affected areal
units of land, and overlay analysis would finally compare all at-risk land at common locations, along with
some indications of the degree of risk to each parcel
of land and water. Over time, further data would reveal
changes in water and soil acidity, tree damage and any
fall in crop yield.

For most people, a GIS is proprietary computer software. There are many GIS packages available on the market, including ArcGIS (from the Environmental Systems Research Institute, USA), Idrisi (from Clark Laboratories, USA) and Geographic Resources Analysis Support System (an open source software). A broader definition of a GIS would include data and people. Data and DATA ACQUISITION are becoming increasingly more important to a GIS as many environmental change projects are now global in nature and data can be transferred around the world via the *Internet*. As computer software and hardware costs have decreased, *data quality* is now seen as a critical consideration in GIS environmental measurement and monitoring. Data can be obtained by *digitising paper maps*, transferable *secondary digital data*, surveys, GROUND MEASUREMENTS, pinpoint locations from a GLOBAL POSITIONING SYSTEM and data from AIRBORNE and SATELLITE REMOTE SENSING. People too are fast becoming valuable components of a successful GIS. Skilled analysts, efficient managers, and focussed users are essential for a structured and co-ordinated GIS. A GIS is also known as a *land information system*, particularly when applied solely to the monitoring of ENVIRONMENTAL INDICATORS. Recently, theory in geographic analysis has led to the adoption of geographical information science (GIScience), which links EPISTEMOLOGY and ONTOLOGY to traditional GIS computer analysis. Popularity has also led to distributed and web-based GIS, *participatory* GIS and *mobile* and CLOUD COMPUTING.   VM

[*See also* BOOLEAN LOGIC, CARTOGRAPHY]

Brimicombe A (2010) *GIS, environmental modeling and engineering.* Boca Raton, FL: CRC Press.

Heywood I, Cornelius S and Carver S (2011) *An introduction to geographical information systems, 4th edition.* Harlow: Pearson.

Horner MW, Zhao T and Chapin TS (2011) Towards an integrated GIScience and energy research agenda. *Annals of the Association of American Geographers* 101: 764–774.

Kozak KH, Graham CH and Wiens JJ (2008) Integrating GIS-based environmental data into evolutionary biology. *Trends in Ecology and Evolution* 23: 141–148.

Longley PA and Barnsley MJ (2004) The potential of geographical information systems and Earth observation. In Matthews JA and Herbert DT (eds) *Unifying geography: Common heritage, shared future.* Abingdon: Routledge, 62–80.

Longley PA, Goodchild MF, Maguire DJ and Rhind DW (eds) (2010) *Geographic information systems and science, 3rd edition.* Chichester: Wiley.

Scally R (2006) *GIS for environmental management.* Redlands: ESRI Press.

Skidmore AK, Franklin J, Dawson TP and Pilesjo P (2011) Geospatial tools address emerging issues in spatial ecology: A review and commentary on the Special Issue. *International Journal of Geographical Information Science* 25: 337–365.

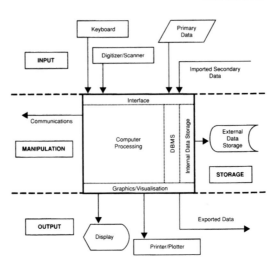

Geographical information system *The main stages and components of a geographical information system.*

geographical sciences   The *transdisciplinary* use by scientists of geographical concepts and techniques in their work, including the reciprocal links between people and NATURE, and spatial analysis (e.g. GEOGRAPHICAL INFORMATION SYSTEMS (GIS) and REMOTE SENSING) of the Earth's biophysical and human environment. Geographical sciences are increasingly concerned with the ENVIRONMENTAL ISSUES and ENVIRONMENTAL PROBLEMS associated with *environmental change*. Hence, this may be the *era of the geographical sciences*.   JAM

National Research Council (2010) *Understanding the changing planet: Strategic directions for the geographical sciences.* Washington, DC: National Academies Press.

geography   The study of the surface of the Earth. It involves the phenomena and processes of the Earth's natural and human environments and LANDSCAPES at local to global scales. Its basic division is between PHYSICAL GEOGRAPHY, which is unambiguously a science and analyses the physical make-up of the Earth's surface (including the BIOSPHERE and lower ATMOSPHERE), and HUMAN GEOGRAPHY where the focus is upon the human occupants of this area. There are key unifying themes across the physical and human aspects of geography. One is the link between the NATURAL ENVIRONMENT, with its LANDFORMS, VEGETATION, SOILS and CLIMATES, and the patterns and processes of human settlement and activities. ENVIRONMENTALISM focuses on the nature of this link and its study takes many forms, including investigating the consequences of the human use and transformation of the Earth. Another connecting theme is the cartographic tradition and the geographer as mapmaker; these draw together the many elements of the Earth's surface into a visible and unified form, and exemplify the central importance of location, spatial variation and spatial relationships to the discipline. It is this nexus of space, place and environment with all their interconnections and integrations that give geography both its identity and its distinctiveness (see Figure). With the advent of satellite imagery, EARTH OBSERVATION (EO) has added new dimensions to the cartographic tradition and, allied with the new technologies of GEOGRAPHICAL INFORMATION SYSTEMS (GIS), lays claim to be regarded as a subdiscipline of geography in its own right.

Geography carries forward its historic identifiers—EXPLORATION, DISCOVERY and FIELD RESEARCH—together with its central interests in concepts such as landscape, region and place. Its *interdisciplinary* character is seen in its interactions with allied disciplines in the physical, biological and EARTH SCIENCES; HISTORY; the HUMANITIES and SOCIAL SCIENCES. There is a strong applied dimension that ranges from ENVIRONMENTAL IMPACT ASSESSMENT (EIA) and ENVIRONMENTAL MANAGEMENT to URBAN AND RURAL PLANNING and the definition of political space. Geography addresses many of the major issues of our time that are associated with *environmental change*, such as POLLUTION, GLOBAL WARMING, CONSERVATION, SUSTAINABILITY and the manifestations of GLOBALISATION. The surface of the Earth is complex and ever-changing; geography provides the methodologies needed to understand, portray and predict these complexities.   *DTH/JAM*

[*See also* CARTOGRAPHY, ENVIRONMENTAL GEOGRAPHY, INTERDISCIPLINARY RESEARCH, REMOTE SENSING]

Bonnett A (2008) *What is geography?* London: Sage.

Castree N, Rogers A and Sherman D (eds) (2005) *Questioning geography: Fundamental debates*. Oxford: Blackwell.

Clifford NJ, Holloway SL, Rice SP and Valentine G (eds) (2009) *Key concepts in geography, 2nd edition*. London: Sage.

Douglas I, Huggett R and Perkins C (eds) (2007) *Companion encyclopedia of geography: From local to global*. London: Routledge.

Gaile GL and Willmott CJ (eds) (2003) *Geography in America at the dawn of the 21st century*. Oxford: Oxford University Press.

Gomez B and Jones III JP (eds) (2010) *Research methods in geography*. Chichester: Wiley-Blackwell.

Hanson S (ed.) (2001) *Ten geographic ideas that changed the world*. New Brunswick, NJ: Rutgers University Press.

Johnston R and Williams M (eds) (2003) *A century of British geography*. Oxford: Oxford University Press.

Matthews JA and Herbert DT (eds) (2004) *Unifying geography: Common heritage, shared future*. London: Routledge.

Matthews JA and Herbert DT (2008) *Geography: A very short introduction*. Oxford: Oxford University Press.

National Research Council (1997) *Rediscovering geography: New relevance for science and society*. Washington, DC: National Academies Press.

Pacione M (ed.) (1999) *Applied geography: Principles and practice. An introduction to useful research in physical, environmental and human geography*. London: Routledge.

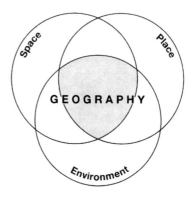

Geography *Geography, defined as the nexus between three core interacting concepts (Matthews and Herbert, 2008).*

geohazard   A category of NATURAL HAZARDS relating to internal or surface Earth processes. Geohazards, or *geological hazards* (including *geomorphological hazards*) include EARTHQUAKES, VOLCANIC IMPACTS ON PEOPLE, TSUNAMIS, *magnetic storms*, LANDSLIDES, hazardous geological materials, ground SUBSIDENCE, river and coastal FLOODS, SEA-LEVEL CHANGE and METEORITE IMPACT. *Meteorological hazards* (e.g. TROPICAL CYCLONES, TORNADOES, STORMS and DROUGHT) and *biological hazards* (e.g. PESTS, EPIDEMICS) are normally excluded.   *GO*

[*See also* RAPID-ONSET HAZARDS]

Alcantara-Ayala I and Goudie A (eds) (2010) *Geomorphological hazards and disaster prevention*. Cambridge: Cambridge University Press.

Cloetingh S, Tibaldi A and Burov E (2012) Coupled deep Earth and surface processes and their impact on geohazards. *Global and Planetary Change* 90–91: 1–19.

Cui P, Chen X-Q, Zhu Y-Y et al. (2011) The Wenchuan earthquake (May 12, 2008), Sichuan Province, China, and resulting geohazards. *Natural Hazards* 56: 19–36.

Maslin M, Owen M, Betts R et al. (2010) Gas hydrates: Past and future geohazard? *Philosophical Transactions of the Royal Society A* 368: 2369–2393.

Maund JG and Eddleston M (eds) (1998) *Geohazards in engineering geology*. Bath: Geological Society [Geological Society Engineering Geology Special Publication 15].

McGuire B, Betts R, Kilburn C et al. (eds) (2010) *Climate forcing of geological and geomorphological hazards*. London: Royal Society Publishing.

Smith K and Petley DN (2009) *Environmental hazards: Assessing risk and reducing disaster.* London: Routledge.

Terry JP and Goff J (eds) (2012) *Natural hazards in the Asia-Pacific region: Recent advances and emerging concepts.* Bath: Geological Society.

geoheritage Geological and geomorphological features of the physical LANDSCAPE that are intrinsically important in terms of beauty or rarity, or are scientifically important for the insights they provide into the evolution and/or dynamics of the Earth. Geoheritage provides an incentive for *geoconservation* (see GEOLOGICAL CONSERVATION (GEOCONSERVATION)) and has re-emerged relatively recently as a basis for GEOTOURISM, especially as a component of SUSTAINABLE DEVELOPMENT in DEVELOPING COUNTRIES where it is providing a much needed source of revenue.   *JAM*

[*See also* CONSERVATION, ECOTOURISM, GEODIVERSITY, HERITAGE, WORLD HERITAGE SITES]

Asrat A, Demissie M and Mogessie A (2012) Geoheritage conservation in Ethiopia: The case of the Simien Mountains. *Questiones Geographicae* 31: 7–23.

Badman T (2010) World heritage and geomorphology. In Migoń P (ed.) *Geomorphological landscapes of the world.* Dordrecht: Springer, 357–368.

Brocx M (2008) *Geoheritage: From global perspective to local principles for conservation and planning.* Perth: Western Australian Museum.

Brocx M and Semeniuk V (2007) Geoheritage and geoconservation: History, definition, scope and scale. *Journal of the Royal Society of Western Australia* 90: 53–87.

geoid The shape of the Earth's sea surface without tides, currents, water-density variations and atmospheric effects. It is an *equipotential surface* that is uneven due to regional variations in the Earth's *gravity field*.   *RAS*

[*See also* GEOIDAL EUSTASY]

Mörner NA (1976) Eustasy and geoid changes. *Journal of Geology* 84: 123–152.

geoidal eustasy Variations in *sea level* caused by the unevenness of the sea surface which is caused by regional variations in the Earth's *gravitational field*. Ocean surfaces represent an *equipotential surface* of the Earth's gravitational field, known as the GEOID. There are large regional differences in *geoidal sea-surface altitudes*. For example, there is a 180 m difference in sea level between the low level off the Maldives in the Indian Ocean and the high level near New Guinea. In the Gulf of Corinth, Greece, the geoidal sea surface varies by as much as 12 m over a distance of only ~150 km. During ICE AGES, the distribution of ICE SHEETS caused considerable changes in the Earth's gravity field and

therefore also large changes in the TOPOGRAPHY of the geoidal sea surface.   *AGD*

[*See also* EUSTASY, GLACIO-ISOSTASY, ISOSTASY]

Cavallotto JL, Violante RA and Parker G (2004) Sea-level fluctuations during the last 8600 years in the de la Plata river (Argentina). *Quaternary International* 114: 155–65.

Devoy RJN (ed.) (1987) *Sea surface studies.* London: Croom Helm.

Mörner N-A (1980) Eustasy and geoid changes as a function of core/mantle changes. In Mörner N-A (ed.) *Earth rheology, isostasy and eustasy.* Chichester: Wiley, 535–553.

Woodroffe S and Horton BP (2005) Holocene sea-level changes in the Indo-Pacific. *Journal of Asian Earth Sciences* 25: 29–43.

geo-indicator A high-resolution measure of short-term change in the geological environment, which is important for ENVIRONMENTAL MONITORING and ENVIRONMENTAL IMPACT ASSESSMENT. Geo-indicators (or *geo-environmental indicators*) are the magnitudes, rates and trends of the near-surface geological processes and phenomena that vary appreciably over timescales of <100 years. Most involve local (0.1–10.0 km scale) to mesoscale (10–100 km) landscapes though some, such as RELATIVE SEA LEVEL and volcanic activity, have regional to global dimensions. Among the characteristics of a good geo-indicator are the following: scientific validity, geographic scope, responsiveness to change, relevance and utility to users and capability of forward projection. These attributes mean that they should be sensitive to HUMAN IMPACT ON ENVIRONMENT on the environment and also relevant to PALAEOENVIRONMENTAL RECONSTRUCTION.   *JAM*

[*See also* BIO-INDICATORS, ENVIRONMENTAL INDICATOR, INTERACTION INDICATOR]

Berger AR and Iams WJ (eds) (1996) *Geoindicators: Assessing rapid environmental changes in Earth systems.* Rotterdam: Balkema.

Berger AR and Satkunas J (eds) (2002) Special Issue on geoindicators. *Environmental Geology* 42(2): 709–772.

geological conservation (geoconservation) The CONSERVATION of sites and features of significance to GEOLOGY and GEOMORPHOLOGY. MINERALS, FOSSILS, rock types, GEOLOGICAL STRUCTURES, significant sections of the STRATIGRAPHICAL RECORD, SOILS, LANDFORMS, LANDSCAPES and sites of educational value or historical significance need to be conserved against damage, overcollecting, EROSION or development. In the United Kingdom, geological conservation comes under the remit of the national conservation bodies and is incorporated into the system of SITES OF SPECIAL SCIENTIFIC INTEREST (SSSIS). A further, non-statutory status of conservation is awarded to sites formerly (and still commonly) known as REGIONALLY

IMPORTANT GEOLOGICAL AND GEOMORPHOLOGICAL SITES (RIGS); these are now known in England as *Local Geological Sites*, in Wales as *Regionally Important Geodiversity Sites* and in Scotland as *Local Geodiversity Sites*. GO

[*See also* GEODIVERSITY]

Burek CV and Prosser CD (eds) (2008) *The history of geoconservation.* Bath: Geological Society [Special Publication 300].

Ellis N, Bowen DQ, Campbell S et al. (1996) *An introduction to the geological conservation review.* Peterborough: Joint Nature Conservation Committee.

Nature Conservancy Council (1990) *Earth science conservation in Great Britain: A strategy.* Peterborough: Nature Conservancy Council.

geological controls on environmental change

Geological processes exert a major influence on the environment at all scales. As controls on environmental change, they may be divided into two categories: *external controls* that bring about ALLOCYCLIC CHANGE, and *internal controls* that bring about AUTOCYCLIC CHANGE. The latter are typically local in scope and include events such as DELTA lobe switching and river-channel AVULSION.

Many large-scale, allocyclic controls are related to PLATE TECTONICS. At the global scale, the migration of continental areas across climatic zones (CONTINENTAL DRIFT) leads to a gradual change in climate, as happened in the case of Spitsbergen as it moved from equatorial latitudes in the DEVONIAN period to its present Arctic position. The formation and break-up of SUPERCONTINENTS, such as PANGAEA, also has major environmental consequences. The interiors of these enormous land areas develop extreme continental climates (see CONTINENTALITY), while break-up involves the formation of RIFT VALLEYS and ultimately *seaways*, opening up new routes for moisture supply. The collision of continental fragments leads to mountain building (OROGENESIS). Physical UPLIFT results in environmental change, but in addition, the creation of RELIEF can influence the GENERAL CIRCULATION OF THE ATMOSPHERE. Uplift of the Tibetan Plateau, for example, may have affected the course of the subtropical JET STREAM, changing the extent and intensity of the Asian MONSOON. Global tectonics also influences *sea level* by changing the volume of the ocean basins (*tectono-eustasy*). Formation and decay of the MID-OCEAN RIDGE (MOR) systems and *continental collision* are especially significant.

Regional tectonics (see TECTONICS) affect the environment in a variety of ways. *Tectonic closure* of the Strait of Gibraltar in the Late Miocene isolated the Mediterranean, leading to the *Messinian salinity crisis* (see MESSINIAN) and ultimately to complete DESICCATION. Formation of the Isthmus of Panama separated the Atlantic and Pacific Oceans, changing the equatorial

OCEAN CURRENT system, with dramatic consequences for the North Atlantic. *Rift-shoulder uplift* resulting from continental extension has dammed and diverted major drainage systems in East Africa.

CATASTROPHIC EVENTS can bring about profound local change. LANDSLIDES, DEBRIS FLOWS, LAVA or PYROCLASTIC MATERIAL deposits may dam valleys or fill them in. Subaquatic SLUMPS can cause loss of coastline. VOLCANIC ERUPTIONS cause physical and chemical changes to the ATMOSPHERE, which may lead to CLIMATIC CHANGE. Longer-term variations in the composition of the atmosphere are also subject to geological control, particularly the abundance of two of the most important GREENHOUSE GASES, CARBON DIOXIDE and *methane*. The former is added to the atmosphere by volcanic activity and probably reaches its highest concentrations when SEA-FLOOR SPREADING is at a maximum. CHEMICAL WEATHERING and LIMESTONE formation remove carbon dioxide from the atmosphere. Large volumes of methane are trapped in deep-sea sediments as GAS HYDRATE. Episodic gas release from this repository could increase the greenhouse effect and may be a cause of ABRUPT CLIMATIC CHANGE.   *MRT*

[*See also* TECTONIC UPLIFT–CLIMATIC CHANGE INTERACTION]

Berner RA and Kothaval Z (2001) GEOCARB III: A revised model of atmospheric CO_2 over Phanerozoic time. *American Journal of Science* 301: 182–204.

Bishop P (2012) Plate tectonics, continental drift, volcanism and mountain building. In Matthews JA, Bartlein PJ, Briffa KR et al. (eds) *The SAGE handbook of environmental change, volume 1*. London: Sage, 363–383.

Fort M (1996) Late Cenozoic environmental changes and uplift on the northern side of the central Himalayas: A reappraisal from field data. *Palaeogeography, Palaeoclimatology, Palaeoecology* 120: 123–145.

Haq BU (1998) Natural gas hydrates: Searching the long-term climatic and slope-stability records. In Henriet JP and Mienert J (eds) *Gas hydrates: Relevance to world margin stability and climate change*. Bath: Geological Society, 303–318.

Keefer DK (1999) Earthquake-induced landslides and their effects on alluvial fans. *Journal of Sedimentary Research* 69: 84–104.

Klein GD (ed.) Pangea: Paleoclimate, tectonics and sedimentation during accretion, zenith and breakup of a supercontinent. *Geological Society of America Special Paper* 288: 41–55.

Plint AG, Eyles N, Eyles CH and Walker RG (1992) Controls of sea level change. In Walker RG and James NP (eds) *Facies models: Response to sea level change*. St. John's: Geological Association of Canada, 15–25.

Reading HG and Levell BK (1996) Controls on the sedimentary rock record. In Reading HG (ed.) *Sedimentary environments: Processes, facies and stratigraphy, 3rd edition*. Oxford: Blackwell, 5–36.

geological evidence of environmental change

The GEOLOGICAL RECORD is the primary source of information on prehistoric environmental change. Information about PALAEOENVIRONMENTS is recorded in many different ways. The most widespread is *sedimentological evidence* which reflects the response of DEPOSITIONAL ENVIRONMENTS to changing conditions of sediment formation and accumulation. Changes in LITHOLOGY, ACCUMULATION RATE, FACIES or PALAEOCURRENT direction may be a result of local or global environmental change. FOSSILS can be excellent indicators of environmental change, as many organisms are sensitive to variations in their HABITAT. Fossil evidence may be preserved as unlithified SUBFOSSIL remains (shell, bone, wood, pollen, etc.); as lithified BODY FOSSILS of the original animal or plant, including *molds* or CASTS of their organic remains; or as TRACE FOSSILS.

The mineralogical and chemical composition of SEDIMENTS and fossils also provide important palaeoenvironmental information. DETRITAL *sediment mineralogy* may reflect the relative effectiveness of physical and chemical WEATHERING processes in the source area, which are a function of CLIMATE and RELIEF. *Syndepositional minerals* typically reflect the composition of the water from which they precipitate. Variations in the mineralogy of, for example, CARBONATE or EVAPORITE minerals in a sequence of lagoonal or lacustrine deposits are commonly the result of changes in water chemistry and SALINITY which may in turn be related to changes in HYDROLOGICAL BALANCE. *Fluid inclusions* trapped within minerals can be used to determine the temperature of precipitation as well as water composition. GEOCHEMISTRY has been central to many modern studies of environmental change. Cyclic variations in the OXYGEN ISOTOPE composition of marine fossils, for example, have been used to confirm the MILANKOVITCH THEORY of climatic change. TRACE ELEMENTS may reveal variations in environmentally significant parameters such as organic compounds, particularly BIOMARKERS, or the temperature or oxygen content of waterbodies.

Geomorphological features such as former glaciated valleys (see U-SHAPED VALLEY), MELTWATER CHANNELS and ROCK PLATFORMS are significant indicators of environmental change and have considerable historical importance as some of the earliest recognised evidence of PLEISTOCENE glaciation and sea-level change (see GLACIAL THEORY). Ancient examples of GLACIAL LANDFORMS provide vital evidence in support of postulated ICE AGES of PRECAMBRIAN and PALAEOZOIC age.   *MRT*

Broecker WS (1995) *The glacial world according to Wally.* Palisades: Eldigio Press.

Bromley RG (1996) *Trace fossils: Biology, taphonomy and applications.* London: Chapman and Hall.

Francis J, Haywood AM, Hill D et al. (2012) Environmental change in the geological record. In Matthews JA, Bartlein PJ, Briffa KR et al. (eds) *The SAGE handbook of environmental change, volume 1.* London: Sage, 165–180.

Goodwin ID and Howard WR (2012) Evidence of environmental change from the marine realm. In Matthews JA, Bartlein PJ, Briffa KR et al. (eds) *The SAGE handbook of environmental change, volume 1.* London: Sage, 181–210.

Hinnov LA and Park J (1998) Detection of astronomical cycles in the stratigraphic record by frequency modulation (FM) analysis. *Journal of Sedimentary Research* 68: 524–539.

Martini IP (ed.) (1997) *Late Glacial and Postglacial environmental changes: Quaternary, Carboniferous-Permian, and Proterozoic.* Oxford: Oxford University Press.

Maynard JR and Leeder MR (1992) On the periodicity and magnitude of Late Carboniferous glacio-eustatic sea-level changes. *Journal of the Geological Society* 149: 303–311.

Pettijohn FJ, Potter PE and Siever R (1987) *Sand and sandstone.* Berlin: Springer.

Tyson RV (1995) *Sedimentary organic matter: Organic facies and palynofacies.* London: Chapman and Hall.

geological record

The total of all materials that preserve information about the Earth's past, including particularly SEDIMENTARY DEPOSITS of all ages (see SEDIMENTOLOGICAL EVIDENCE OF ENVIRONMENTAL CHANGE), but also IGNEOUS ROCKS, METAMORPHIC ROCKS, FOSSILS and GEOLOGICAL STRUCTURES. The geological record, particularly the STRATIGRAPHICAL RECORD of materials that accumulated on the Earth's surface, is the main source of information about pre-QUATERNARY environmental change. *GO*

[*See also* FOSSIL RECORD, GEOLOGICAL EVIDENCE OF ENVIRONMENTAL CHANGE, GEOLOGICAL RECORD OF ENVIRONMENTAL CHANGE]

Ager DV (1993) *The nature of the stratigraphical record, 3rd edition.* Chichester: Wiley.

Francis J, Haywood A, Hill D et al (2012) Environmental change in the geological record. In Matthews JA, Bartlein PJ, Briffa KR et al. (eds) *The SAGE handbook of environmental change, volume 1.* London: Sage, 165–180.

Stanley SM (2009) *Earth system history.* New York: Freeman.

geological record of environmental change

Geological evidence is vital in understanding environmental change because of the time limitations of other data sources. HISTORICAL EVIDENCE and DOCUMENTARY EVIDENCE provide data for only the very recent historical past; ENVIRONMENTAL ARCHAEOLOGY extends the record to not much further than 20,000 years. In contrast, GEOLOGICAL EVIDENCE OF ENVIRONMENTAL CHANGE extends as far back as the oldest rocks identified on Earth, about 3,800 million years ago (Ma). Given the 4.5 billion year age of the Earth, this means that only the first 700–800 million years of Earth history are without evidence (see HADEAN (*Priscoan*)). Additionally, the rapid climatic fluctuations of the Late

PLEISTOCENE and HOLOCENE suggest that the time period covered by non-geological evidence is not typical of Earth history generally (see ACTUALISM).

SEDIMENTOLOGICAL EVIDENCE OF ENVIRONMENTAL CHANGE is considerable, because many SEDIMENTARY DEPOSITS record evidence of their DEPOSITIONAL ENVIRONMENT and, in the case of TERRIGENOUS clastic sediments and SEDIMENTARY ROCKS, that of the hinterland from where the sediments were eroded.

FOSSILS represent some of the most unequivocal and potent evidence of former environments and environmental change. The FOSSIL RECORD, however, poses many problems for PALAEOENVIRONMENTAL RECONSTRUCTION. Some of these are palaeobiological, such as the loss of soft-tissued organs and organisms (see TAPHONOMY). Other problems are the mixing together of fossils from environments that were separated in space (see DEATH ASSEMBLAGE) or time (see DERIVED FOSSIL). Some groups of fossils are good PALAEOENVIRONMENTAL INDICATORS. For example, colonial *scleractinian corals* are confined to warm, shallow marine waters. TRACE FOSSILS provide direct evidence of the activity of organisms. *Plants* are excellent environmental indicators for non-marine settings and POLLEN ANALYSIS in particular is a sensitive tool for defining palaeoclimates. In general, the confidence and the precision with which a fossil type can be used for palaeoenvironmental analysis decreases with age; the environmental significance of a CENOZOIC (CAINOZOIC) mollusc is easier to interpret than a CAMBRIAN one. Equally, the PALAEOECOLOGY of fossils with close relations to living taxa can be more confidently interpreted than those without.

A correct palaeoenvironmental interpretation is critical if rocks and fossils are to be used as data for environmental change. This is sometimes difficult. Sand bodies formed in the deep OCEAN and in *fluvial systems* may be superficially similar. Furthermore, DIAGENESIS can remove, modify or disguise environmental information in both sediments and fossils. Ideally, sedimentological, palaeontological and FACIES information obtained at the *thin section, hand specimen* and *outcrop* scales should be integrated in any palaeoenvironmental analysis. A necessary component is the unravelling of lateral, vertical and temporal relationships: this is the province of STRATIGRAPHY. Increasingly, such data are tied in, at a still higher level, with the general depositional framework of the SEDIMENTARY BASIN in the context of BASIN ANALYSIS.

On a regional scale, any geological interpretation of environmental change must take into account the PLATE TECTONICS context. Replacement of warm-water sediments over time by a glacial TILLITE, for example, could indicate either global cooling or a regional poleward shift in PALAEOLATITUDE as a result of CONTINENTAL DRIFT.

A major problem with interpreting environmental change in the geological record is that the *resolution* of DATING TECHNIQUES is such that the timescale of change is often imprecisely known. Despite progress in rocks older than about 30 Ma, the greatest precision using BIOSTRATIGRAPHY and RADIOMETRIC DATING is around 0.5 million years and frequently it is no better than 2–3 million years. One implication is that it is hard to determine the exact duration of most geological events unless they are many hundred times longer than all recorded human history. Another is that it is difficult to be certain that particular short-lived EVENTS were truly synchronous across wide areas.

The GEOLOGICAL RECORD sets limits on the range of environmental fluctuations that have occurred on the Earth. Thus, at the most general level, the continuous fossil record of life from about 3,800 Ma onwards indicates that at no time did all surface water become either ice or steam as appears to have happened on Mars and Venus, respectively (see PLANETARY ENVIRONMENTAL CHANGE). Given the variability of SOLAR RADIATION as the Sun evolved, this is remarkable and is one line of evidence for some sort of planetary HOMEOSTASIS as proposed by the GAIA HYPOTHESIS. Within this gross stability, the geological record does, however, indicate considerable variation. For instance, during the JURASSIC and CRETACEOUS periods, there is very limited evidence for polar *ice caps*, suggesting that global temperatures were such that GREENHOUSE CONDITIONS prevailed. In contrast, during the Late PROTEROZOIC, ice sheets appear to have been widespread, extending to low latitudes, and the Earth experienced ICEHOUSE CONDITIONS (see SNOWBALL EARTH).

The geological record also shows evidence of environmental conditions and processes very different from those prevailing at present. One example is OCEAN CIRCULATION patterns that allowed the development of widespread OCEANIC ANOXIC EVENTS. Geology also suggests that short-term biotic and environmental crises have occurred periodically in MASS EXTINCTIONS (see K-T BOUNDARY).

Finally, geological evidence is vital in both the creation and VALIDATION of GENERAL CIRCULATION MODELS (GCMs) for the ancient atmosphere and reconstructions of PALAEOCEANOGRAPHY. These models are developed from regional geological studies and based on PLATE TECTONICS, and their outputs can be tested by examining sedimentological and palaeontological data in contemporaneous rocks. Geology similarly provides data for testing models of the past and future evolution of the ATMOSPHERE. Thus, the idea that the Earth's atmosphere became oxygen-rich during the PRECAMBRIAN is supported by the change from BANDED IRON FORMATIONS (BIFs) to RED BEDS from 2,000 to 1,500 Ma.   CDW

[*See also* EARTH LAWS, EARTH REVOLUTIONS]

Bender ML (2011) *Paleoclimate*. Princeton, NJ: Princeton University Press.

Cronin TM (2009) *Paleoclimates: Understanding climate change past and present*. New York: Columbia University Press.

Fischer AG (1981) Climatic oscillations in the biosphere. In Nitecki M (ed.) *Biotic crises in ecological and evolutionary time*. New York: Academic Press, 103–131.

Francis J, Haywood A, Hill D et al. (2012) Environmental change in the geological record. In Matthews JA, Bartlein PJ, Briffa KR et al. (eds) *The SAGE handbook of environmental change, volume 1*. London: Sage, 165–180.

Miall AD (2010) *The geology of stratigraphic sequences*. New York: Springer.

Nichols G (2009) *Sedimentology and stratigraphy*. Oxford: Blackwell.

Pirrie D (1998) Interpreting the record: Facies analysis. In Doyle P and Bennett MR (eds) *Unlocking the stratigraphical record: Advances in modern stratigraphy*. Chichester: Wiley, 395–420.

Stanley SM (2009) *Earth system history*. New York: Freeman.

geological structures   Features produced by the permanent DEFORMATION of sediments or rocks, mainly in response to TECTONIC stresses while rocks are buried in the Earth's CRUST. FAULTS and JOINTS are *brittle* structures produced by the *elastic deformation* and *fracturing* of rocks in *tension* or COMPRESSION. FOLDS are *ductile* structures formed by the bending of rock *strata*, usually in compression. Geological structures range in scale from microscopic features to regional structures tens or hundreds of kilometres across. They are responsible for the irregularities of outcrop patterns on a geological map. The orientation of geological structures is described by the STRIKE and DIP of planar features (e.g. tilted BEDS, fault surfaces, fold axial surfaces) and by the *plunge* of linear features (e.g. fold *hinge lines* and STRIATIONS).

The analysis of geological structures can provide information about the magnitude and orientation of stresses in the geological past, allowing the deformation history of an area to be reconstructed. This is important in reconstructing past PLATE TECTONICS, unravelling the STRATIGRAPHY of an area and understanding the RHEOLOGY of the CRUST and LITHOSPHERE. *Structural geology* is the branch of GEOLOGY concerned with the interpretation of geological structures: it overlaps with tectonics, GEODYNAMICS, rheology and *materials science*.   *GO*

[*See also* SEDIMENTARY STRUCTURES, SOFT-SEDIMENT DEFORMATION, UNCONFORMITY]

Bennison GM, Oliver PA and Moseley KA (2011) *An introduction to geological structures and maps, 8th edition*. London: Hodder Education.

Coe AL (ed.) (2010) *Geological field techniques*. Milton Keynes: Wiley-Blackwell and Open University.

Fossen H (2010) *Structural geology*. Cambridge: Cambridge University Press.

Twiss RJ and Moores EM (2006) *Structural geology*. New York: Freeman.

geological time   Geological time is the same as time in the conventional sense, but there are difficulties in measuring the immense spans of time represented in the GEOLOGICAL RECORD, which exceeds 4,000 million years (the age of the Earth is estimated at 4,540 million years). The ages of geological materials and events are expressed in two distinct ways: as RELATIVE AGES, in which they are placed in sequence; and as *absolute ages* measured in years. *Absolute-age dating* (see CALIBRATED-AGE DATING) only became possible on geological timescales with the discovery of radioactivity in the early twentieth century. RELATIVE-AGE DATING, primarily using FOSSILS (see STRATIGRAPHICAL COLUMN), is the only method directly applicable in fieldwork and is capable of identifying more precise time 'slices' than most absolute dating methods, although those slices may be of uncertain absolute age.   *GO*

[*See also* GEOCHRONOLOGY, GEOLOGICAL TIMESCALE, STRATIGRAPHY]

Gould SJ (1987) *Time's arrow, time's cycle: Myth and metaphor in the discovery of geological time*. London: Penguin Books.

Ogg JG, Ogg G and Gradstein FM (2008) *The concise geologic time scale*. Cambridge: Cambridge University Press.

geological timescale   An absolute timescale, or *geochronologic scale*, for the GEOLOGICAL RECORD, expressed in years (*chronometric units*). The scale has been developed by calibrating the RELATIVE AGES of the STRATIGRAPHICAL COLUMN (*chronostratic scale*) with *absolute-age dating* from appropriate sediments and rocks, using RADIOMETRIC DATING or other methods of GEOCHRONOLOGY. The calibration of these two distinct methods of measuring GEOLOGICAL TIME is subject to revision and refinement, so dates applied to specific boundaries vary in different versions of the geological timescale. For example, the beginning of the CAMBRIAN period was placed at 600 million years ago (Ma) in the AD 1960s, 570 Ma in the 1980s, 544 Ma in 1998 and 542 Ma in 2009. The position of the boundary in terms of BIOSTRATIGRAPHY, or its physical position in a rock succession has not, however, changed.

A *relative timescale* is more appropriate than an absolute scale with regard to field observations, where a reference to rocks as 'Early Jurassic' is more appropriate than a statement that they are 'approximately 200 million years old', since absolute-age dating can only be carried out through complex laboratory analyses of certain types of rock. This is analogous to the usage of terms such as BRONZE AGE on the ARCHAEOLOGICAL TIMESCALE. Moreover, radiometric dates are generally not as precise as some CORRELATION (STRATIGRAPHICAL) using BIOSTRATIGRAPHY. For example, the correlation of some JURASSIC ammonite faunas allows the discrimination of rock units representing intervals of the order of 50,000

Eon	Era	Period			Epoch	age (Ma)
PHANEROZOIC	Cenozoic	Quaternary			Holocene	0.01
					Pleistocene	1.64
		Tertiary	Neogene		Pliocene	5.2
					Miocene	23.3
			Palaeogene		Oligocene	35.4
					Eocene	56.5
					Palaeocene	65.0
	Mesozoic	Cretaceous			Senonian	
					Gallic	
					Neocomian	144
		Jurassic			Malm	
					Dogger	
					Lias	205
		Triassic			Late	
					Middle	
					Scythian	248
	Palaeozoic	Late	Permian		Zechstein	
					Rotliegendes	295
			Carboniferous	Pennsylvanian	Stephanian	
					Westphalian	
					Namurian	
				Mississipian	Visean	
					Tournaisian	354
			Devonian		Late	
					Middle	
					Early	416
		Early	Silurian		Pridoli	
					Ludlow	
					Wenlock	
					Llandovery	442
			Ordovician		Ashgill	
					Caradoc	
					Llandeilo	
					Llanvirn	
					Arenig	
					Tremadoc	495
			Cambrian		Merioneth	
					St. David's	
					Caerfai	544
PRECAMBRIAN		PROTEROZOIC				2500
		ARCHAEAN				3800
		PRISCOAN				4560

Geological timescale *The geological timescale: ages in millions of years before present (Ma) (Haq and van Eysinga, 1998).*

years, which is much more precise than radiometric dates can provide for the same period.

A parallel geological timescale is provided by the GEOMAGNETIC POLARITY TIMESCALE (*magnetostratigraphic timescale*), which uses the pattern of GEOMAGNETIC POLARITY REVERSALS recorded in rock successions. Such changes are synchronous and can be correlated worldwide, although some overall biostratigraphical or radiometric control is also usually necessary.

The principal divisions of geological time and dates for the boundaries are shown in the Table. No rocks are preserved from HADEAN (*PRISCOAN*) time. The ARCHAEAN and PROTEROZOIC eonothems were originally classed together as PRECAMBRIAN, and are still often referred to as such. This major division separates rocks in which FOSSIL content is obvious (PHANEROZOIC, Greek for 'visible life') from those in which it is much less obvious and often absent. The boundary between the Archaean and the Proterozoic is taken arbitrarily at 2,500 Ma. The PALAEOZOIC (Greek for 'ancient life') begins rapidly with the CAMBRIAN EXPLOSION—a sudden appearance of invertebrate taxa, with hard parts, although soft-bodied faunas are now widely known from the latest Neoproterozoic and constitute the Ediacaran fauna that ranges from 547 to 543 Ma in age. Palaeozoic faunas are dominated by such fossil groups as trilobites, brachiopods and graptolites, together with crinoids and rugose and tabulate corals. Following the end-Permian MASS EXTINCTION, in which some 90 per cent of species became extinct, the MESOZOIC (Greek for 'middle life') began. Mesozoic marine faunas are dominated by molluscs, especially ammonoids and belemnoids, as well as large marine reptiles such as ichthyosaurs, plesiosaurs and pliosaurs. On land, dinosaurs were the dominant vertebrates, but the first mammals and birds appeared in the TRIASSIC and JURASSIC, respectively. The first flowering plants (angiosperms) appeared in the CRETACEOUS. At the end-Cretaceous mass extinction (K-T BOUNDARY), the ammonites, belemnites, large marine reptiles and dinosaurs disappeared. CENOZOIC (*CAINOZOIC*) (Greek for 'recent life') marine invertebrate faunas are dominated by molluscs, especially bivalves and gastropods. Mammals evolved rapidly on land, the evolution of the GRAZING habit following on from the appearance of grasses in the PALAEOGENE. HOMINIDS appeared in the PLIOCENE.

JCWC/GO

Gradstein FM, Ogg JG and Smith AG (eds) (2005) *A geologic time scale 2004*. Cambridge: Cambridge University Press.

Hailwood EA (1989) *Magnetostratigraphy*. Bath: Geological Society.

Hailwood EA and Kidd RB (1993) *High resolution stratigraphy*. Bath: Geological Society.

Haq BU and van Eysinga WB (1998) *Geological time table, 5th edition*. Amsterdam: Elsevier.

International Commission on Stratigraphy (2010) *International stratigraphic chart* [Available at http://www.stratigraphy.org/ICSchart/ChronostratChart2012.pdf]

Vickers-Rich P and Komarower P (eds) (2007) *The rise and fall of the Ediacaran biota*. Bath: Geological Society.

Walker JD and Geissman JW (eds) (2009) *Geologic time scale*. Boulder, CO: Geological Society of America. doi:10.1130/2009.CTS004R2C

geology The investigation of the solid Earth, its composition, structure, processes and history, through the study of materials such as *rocks*, MINERALS and FOSSILS—the GEOLOGICAL RECORD. The discipline can be usefully split into *physical geology* (the study of present-day processes and products) and *historical geology* (study of the Earth's past). The principal subdivisions of geology are *crystallography, mineralogy*, PETROLOGY (including IGNEOUS, SEDIMENTARY and METAMORPHIC ROCKS), PALAEONTOLOGY, STRATIGRAPHY, SEDIMENTOLOGY, *structural geology*, TECTONICS, GEOPHYSICS, GEOCHEMISTRY, HYDROGEOLOGY, MARINE GEOLOGY, *applied geology* and ENVIRONMENTAL GEOLOGY. Unlike EARTH SCIENCE, geology does not normally cover OCEANOGRAPHY or CLIMATOLOGY. In recent years, the methods of geology have been applied to other bodies in the solar system (see PLANETARY GEOLOGY).

CDW

[*See also* GEOLOGICAL EVIDENCE OF ENVIRONMENTAL CHANGE, GEOLOGICAL RECORD OF ENVIRONMENTAL CHANGE]

Lutgens FK, Tarbuck EJ and Tasa D (2011) *Essentials of geology, 11th edition*. Upper Saddle River, NJ: Pearson.

Lyell C (1830–1833) *Principles of geology, being an attempt to explain the former changes of the Earth's surface by reference to causes now in operation, 3 volumes*. London: John Murray.

Macdougall D (2011) *Why geology matters: Decoding the past, anticipating the future*. Berkeley: University of California Press.

Park G (2010) *Introducing geology: A guide to the world of rocks, 2nd edition*. Edinburgh: Dunedin Press.

Rothery D (2008) *Teach yourself geology*. London: Hodder Education.

Tarbuck EJ, Lutgens FK and Tasa D (2011) *Earth science, 13th edition*. Upper Saddle River, NJ: Pearson.

geomagnetic polarity reversal A change in the orientation of the Earth's magnetic field (see GEOMAGNETISM) such that north and south magnetic poles swap positions. Geomagnetic polarity reversals (also known as *field reversals, geomagnetic reversals, magnetic reversals, magnetic polarity reversals* or *polarity reversals*) have occurred roughly every few hundred thousand years over the past 100 million years. Polarities like that existing today are known as NORMAL POLARITY, in contrast to episodes of *reversed polarity* (see POLARITY CHRON). During a reversal, the intensity (strength) of the Earth's magnetic field gradually declines, the polarity suddenly flips and the intensity gradually increases, the entire event apparently lasting a few thousand years. Reversals are attributed to fluctuations in flow in the Earth's outer CORE. Reversals are globally synchronous, providing a means for CORRELATION (STRATIGRAPHICAL) in MAGNETOSTRATIGRAPHY, and their dating provides the basis for the GEOMAGNETIC POLARITY TIMESCALE.

GO

[*See also* MAGNETIC ANOMALY, PALAEOMAGNETISM]

Backus G, Parker R and Constable C (1996) *Foundations of geomagnetism.* Cambridge: Cambridge University Press.

Campbell WH (2003) *Introduction to geomagnetic fields.* Cambridge: Cambridge University Press.

Courtillot V and Olson P (2007) Mantle plumes link magnetic superchrons to Phanerozoic mass depletion events. *Earth and Planetary Science Letters* 260: 495–504.

Kageyama A, Ochi MM and Sato T (1999) Flip-flop transitions of the magnetic intensity and polarity reversals in the magnetohydrodynamic dynamo. *Physical Review Letters* 82: 5409–5412.

Singer BS, Hoffman KA, Coe RS et al. (2005) Structural and temporal requirements for geomagnetic field reversal deduced from lava flows. *Nature* 434: 633–636.

Willis DM, Holder AC and Davis CJ (2000) Possible configurations of the magnetic field in the outer magnetosphere during geomagnetic polarity reversals. *Annales Geophysicae: Atmospheres, Hydrospheres and Space Sciences* 18: 11–27.

geomagnetic polarity timescale (GPTS) A

GEOLOGICAL TIMESCALE based on GEOMAGNETIC POLARITY REVERSALS, calibrated by *absolute-age dating* of reversals (see CALIBRATED-AGE DATING) (see Figure). The geomagnetic polarity timescale, or *magnetostratigraphic timescale*, is principally used where a distinctive set of reversals in a rock sequence can be reliably matched to a section of the timescale. It provides a detailed record of geomagnetic polarity reversals from ~250 million years ago to the present through the dating of MAGNETIC ANOMALIES in ocean-floor sediments (see MARINE SEDIMENT CORES). The numbers given to ocean-floor anomalies and the names given to the most recent POLARITY CHRONS are also applied to the timescale. The reversal pattern is calibrated with BIOSTRATIGRAPHY and RADIOMETRIC DATING.   *DNT*

[*See also* DATING TECHNIQUES, MAGNETOSTRATIGRAPHY]

Muttoni G, Ravazzi C, Breda M et al. (2007) Magnetostratigraphic dating of an intensification of glacial activity in the southern Italian Alps during Marine Isotope Stage 22. *Quaternary Research* 67: 161–173.

Nichols G (1999) Sedimentology and stratigraphy. Oxford: Blackwell.

Ogg JG, Ogg G and Gradstein FM (2008) *The concise geologic time scale.* Cambridge: Cambridge University Press.

Parés JM, Pérez-González A, Rosas A et al. (2006) Matuyama-age lithic tools from the Sima del Elefante site, Atapuerca (northern Spain). *Journal of Human Evolution* 50: 163–169.

Szurlies M (2007) Latest Permian to Middle Triassic cyclo-magnetostratigraphy from the Central European Basin, Germany: Implications for the geomagnetic polarity timescale. *Earth and Planetary Science Letters* 261: 602–619.

geomagnetism The study of the Earth's MAGNETIC

FIELD (*geomagnetic field*). This is best approximated by

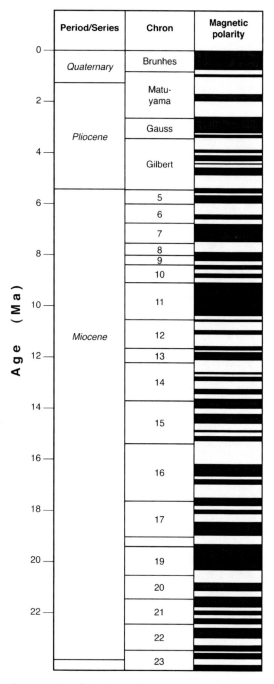

Geomagnetic polarity timescale *The geomagnetic polarity timescale for the past 24 million years (Ma). Black bands are episodes of normal polarity; white bands are episodes of reversed polarity (Nichols, 1999).*

a bar magnet at the centre of the Earth, its axis aligned with the Earth's axis of rotation. This *Geocentric Axial Dipole* model represents the time-averaged geomagnetic field throughout geological time. The geomagnetic field originates in the iron-rich fluid outer CORE of the Earth (see EARTH STRUCTURE) and is maintained by convective motions in this region. Changes in the MAGNETIC INTENSITY (strength), DECLINATION (D) (AZIMUTH; orientation relative to true north) and INCLINATION (I) (DIP; orientation in a vertical plane relative to the Earth's surface) define the SECULAR VARIATION in time of the geomagnetic field, and they are attributed to physical and chemical changes in the core and the core-MANTLE boundary region. Over longer timescales, the field experiences GEOMAGNETIC POLARITY REVERSALS when north and south magnetic poles reverse.   *DNT*

[*See also* PALAEOMAGNETISM]

Backus G, Parker R and Constable C (1996) *Foundations of geomagnetism.* Cambridge: Cambridge University Press.

Campbell WH (2003) *Introduction to geomagnetic fields.* Cambridge: Cambridge University Press.

Courtillot V, Gallet Y, Le Mouël J-L et al. (2007) Are there connections between the Earth's magnetic field and climate? *Earth and Planetary Science Letters* 253: 328–339.

Juarez MT and Tauxe L (2000) The intensity of the time-averaged geomagnetic field: The last 5 Myr. *Earth and Planetary Science Letters* 175: 169–180.

Lee YS and Kodama K (2009) A possible link between the geomagnetic field and catastrophic climate at the Paleocene-Eocene thermal maximum. *Geology* 37: 1047–1050.

Merrill RT (2010) *Our magnetic Earth: The science of geomagnetism.* Chicago: University of Chicago Press.

geomatics An overarching term for the field of *geospatial technology*, which derives from the combination of GEODESY (emphasising survey) with *geoinformatics* (emphasising information). It includes CARTOGRAPHY (especially *mathematical cartography*), GEOGRAPHIC INFORMATION SYSTEMS (GIS) (including *geographical information science*), GLOBAL POSITIONING SYSTEMS (GPSs), GLOBAL NAVIGATION SATELLITE SYSTEMS (GNSSs), MAP PROJECTION, PHOTOGRAMMETRY, REMOTE SENSING and SURVEYING. *Hydrogeomatics* involves similar technologies applied in or over the OCEAN and other waterbodies.   *JAM/TF*

Ghilani CD and Wolf PR (2012) *Elementary surveying: An introduction to geomatics, 13th edition.* Harlow: Pearson.

Meyer HT (2009) *Introduction to geometrical and physical geodesy: Fundamentals of geomatics, 3rd edition.* Charleston, SC: ESRI Press.

Sarasua W and McCormac JC (2012) *Geomatics.* Chichester: Wiley.

geomedicine The science of the influence of ENVIRONMENTAL FACTORS on the distribution of HUMAN HEALTH HAZARDS and on human and animal DISEASE. Examples include the effects of TRACE ELEMENT deficiencies and excesses, such as *Keshan disease* in China (caused by a selenium deficiency) and arsenic poisoning in the Indian subcontinent (caused by excess ARSENIC (As)).   *JAM*

Lag J (ed.) (1990) *Geomedicine.* Boca Raton, FL: CRC Press.

Steinnes E (2009) Soils and geomedicine. *Environmental Geochemistry and Health* 31: 523–535.

geometric transformation A mathematical adjustment applied in the rectification of errors and restoration of images to geographical correctness or specific MAP PROJECTIONS. Each process of *data collection*, whether a manually digitised map, sensor or scanner, generates singular characteristics in the raw data. Errors and distortion which occur may be constant, systematic or random. These are generally checked and adjusted prior to the imagery being released for wider use and analysis. The corrected image is then supplied in a recognised CO-ORDINATE SYSTEM format. Further computation and adjustment may be required to match other images generated on different coordinate systems or lacking well-defined geometry.   *TF*

Kraak M-J and Ormeling F (2010) *Cartography: Visualisation of spatial data, 3rd edition.* Harlow: Pearson.

geomorphology The study of LANDFORMS from microscale to global scale, including their formative ENDOGENETIC and, especially, EXOGENETIC processes, and their evolution in the face of environmental change. This includes the interrelationships between landforms, sediments and processes in LANDSYSTEMS, and *applied geomorphology* (geomorphology in the service of society).   *JAM*

Anderson RS and Anderson SP (2010) Geomorphology: The mechanics and chemistry of landscapes. Cambridge: Cambridge University Press.

Goudie AG (2010) *Landscapes and geomorphology: A very short introduction.* Oxford: Oxford University Press.

Gregory KJ (2010) *The Earth's land surface: Landforms and processes in geomorphology.* London: Sage.

Gregory KJ and Goudie AS (2011) *The SAGE handbook of geomorphology.* London: Sage.

Harvey A (2012) *Introducing geomorphology: A guide to landforms and processes.* Edinburgh: Dunedin Academic Press.

Oldroyd DR and Grapes RH (2011) Contributions to the history of geomorphology and Quaternary geology: An introduction. In Grapes RH, Oldroyd D and Grigelis A (eds) *History of geomorphology and Quaternary geology.* London: Geological Society, 1–12 [Special Publication 301].

Shroder JF (ed.) (2013) *Treatise on geomorphology*, 14 vols. San Diego, CA: Academic Press.

Summerfield MA (1991) *Global geomorphology.* Harlow: Longman.

geomorphology and environmental change Since many geomorphic processes are conditioned by changes in CLIMATE and in the nature of the ground surface (both natural and human-induced), GEOMORPHOLOGY is intimately linked with *environmental change*. The Earth's land surface comprises POLYGENETIC landscapes reflecting the influence of processes that either no longer operate or operate at spatial or temporal scales different from earlier ones. Some geomorphic systems and processes are more sensitive to change than others making it difficult to generalise about LANDFORM development in response, for example, to a particular magnitude of change in temperature or precipitation.

The importance of environmental change in geomorphological research is now recognised more widely than formerly, but much remains to be understood about the nature and speed of geomorphic response to natural and/ or human-induced change. Attempts to assess the impact of environmental change on geomorphic systems and processes have frequently adopted one or more of the following three distinct approaches. First, in the *palaeogeomorphological approach*, PROXY EVIDENCE and historical records of past changes ranging from long to short timescales have been investigated. The resulting *palaeogeomorphological data* may enable RESPONSE TIME and *recurrence interval* to be assessed, but the *temporal resolution* is often comparatively coarse (i.e. recording CENTURY- TO MILLENNIAL-SCALE VARIABILITY). The resolution required for future PREDICTIONS (i.e. DECADAL-SCALE VARIABILITY and ANNUAL VARIABILITY) is seldom achieved. Second, the *monitoring approach* to current *geomorphic processes* has been carried out (e.g. using EROSION PLOTS, EROSION PINS and MICRO-EROSION METER), providing valuable information at a higher temporal resolution about the physical basis of processes and site-specific quantifiable data. Such data inevitably suffer, however, from the brevity of the monitoring periods and limited scope for spatial and temporal EXTRAPOLATION (but see the ERGODIC HYPOTHESIS), and they are likely to omit EXTREME EVENTS with a long recurrence interval. Third, the *modelling approach*, which involves *simulation of geomorphic systems* and processes in order to assess *impact-process-response relationships*, has been applied. Data input can be drawn from the first two approaches, though substitute parameters and data may have to be used if field data are unavailable. Criticisms of MODELS include oversimplification because of the limited number of parameters and the strong dependence of the output on the quality of the input. Ideally, more than one approach may be appropriate in evaluating geomorphic response to environmental change, but its understanding also requires recognition and comprehension of the inter-relationships existing between various processes and individual system components. This is more likely to be achieved through the adoption of a HOLISTIC APPROACH than a strategy based on REDUCTIONISM.

SENSITIVITY of geomorphic systems and processes to change is complex. It can be defined in different ways: as the recognisable response of a system to a change in system controls (FORCING FACTORS), or the susceptibility of a system to DISTURBANCE. Whether geomorphic systems and processes change following an environmental PERTURBATION depends on the spatial and temporal balance of the resisting and disturbing forces. *Landscape stability* is maintained if the resisting forces prevent environmental perturbations from having any persistent effect.

Landscape response to environmental change varies through time and space and operates at different scales. Geomorphic systems and processes may respond slowly, adjusting to change over several millennia. Alternatively, response may be sudden, even catastrophic (see CATASTROPHE), with recognisable change occurring in minutes only. In reality, both levels of response condition all systems. For much of the time, geomorphic systems evolve slowly through processes of moderate intensity and FREQUENCY. Major modifications are often associated with low frequency–high intensity events to the extent that it has often been assumed that most change is effected during these events. This is not necessarily so because *geomorphic sensitivity* is not simply a function of the magnitude of the change but depends on whether, large or small, it crosses important geomorphic THRESHOLDS.

An important focus for geomorphology in the past decade or so, as with other environmental sciences, has been the prediction of sensitivity of systems and processes to GLOBAL WARMING–induced CLIMATIC CHANGE. Projected magnitudes and likely impacts of temperature and precipitation changes from GENERAL CIRCULATION MODELS (GSMs) are critical, as are any predictions of increased frequency of high magnitude extreme climatic events. Many palaeo-studies documenting geomorphic response to climate shifts since the Late PLEISTOCENE provide a useful basis for developing and exploring future scenarios, although they usually relate to times and/or situations in which *human agency* in environmental change was either lacking or inconsequential. For most areas of the world, the history of HUMAN IMPACT ON LANDFORMS AND GEOMORPHIC PROCESSES is well documented, but the response of geomorphic systems and processes to future climatic change in landscapes currently strongly affected by human action is by no means clearly understood.   *PW*

[*See also* ENVIRONMENTAL MODELLING, ENVIRONMENTAL MONITORING, EQUILIBRIUM CONCEPTS, FEEDBACK MECHANISMS, LAG TIME, LANDSCAPE EVOLUTION, MAGNITUDE-FREQUENCY CONCEPTS, PALAEOENVIRONMENTAL RECONSTRUCTION, PHYSIOGRAPHY, RATES OF ENVIRONMENTAL CHANGE, RELAXATION TIME, SCALE CONCEPTS IN ENVIRONMENTAL CHANGE]

Baker VR and CR Twidale (1991) The reenchantment of geomorphology. *Geomorphology* 4: 73–100.
Brunsden D (1990) Tablets of stone: Towards the ten commandments of geomorphology. *Zeitschrift für Geomorphologie, Supplement Band* 79: 1–37.

Brunsden D (2001) A critical assessment of the sensitivity concept in geomorphology. *Catena* 42: 99–123.

Bull WB (1991) *Geomorphic responses to climatic change.* New York: Oxford University Press.

Higgitt DL and Lee EM (eds) (2001) *Geomorphological processes and landscape change.* Oxford: Blackwell.

Kennedy BA (2005) *Inventing the Earth: Ideas on landscape development since 1740.* Oxford: Blackwell.

Lewin J and Woodward J (2009) Karst geomorphology and environmental change. In Woodward J (ed.) *The physical geography of the Mediterranean.* Oxford: Oxford University Press, 287–317.

Phillips JD (2011) Disturbance and responses in geomorphic systems. In Gregory KJ and Goudie AS (eds) *The SAGE handbook of geomorphology.* London: Sage, 555–566.

Scheidegger AE (1987) The fundamental principles of landscape evolution. *Catena Supplement* 10: 199–210.

Slaymaker O (ed.) (2000) *Geomorphology, human activity and global environmental change.* Chichester: Wiley.

Slaymaker O, Spencer T and Embleton-Hamann C (eds) (2009) *Geomorphology and global environmental change.* Cambridge: Cambridge University Press.

Thomas DSG and Allison RJ (eds) (1993) *Landscape sensitivity.* Chichester: Wiley.

Thomas MF (2001) Landscape sensitivity in time and space: An introduction. *Catena* 42: 83–98.

Williams M (2011) Environmental change. In Gregory KJ and Goudie AS (eds) *The SAGE handbook of geomorphology.* London: Sage, 535–554.

geomorphometry   The science of quantitative *land-surface analysis.* Straddling the disciplines of EARTH SCIENCE, mathematics and computer science, geomorphometry employs a blend of mathematical, statistical and image-processing techniques to quantify the form of the Earth's surface at various spatial scales.   *JPT*

[*See also* DIGITAL ELEVATION MODEL/MATRIX (DEM), GEOMORPHOLOGY, HYPSOMETRY, LANDFORM, MORPHOMETRY]

Evans IS (2012) Geomorphometry and landform mapping: What is a landform? *Geomorphology* 137: 94–106.

Hengl T and Reuter H (eds) (2008) *Geomorphometry: Concepts, software, applications.* Amsterdam: Elsevier.

geophysical surveying   The practical application of GEOPHYSICS to investigate the Earth's interior, particularly shallow levels in the CRUST. Geophysical surveying (*geophysical exploration* or *exploration geophysics*) is comparable with REMOTE SENSING, in that measurements made at the Earth's surface are used to investigate the subsurface. Techniques of geophysical surveying include *electrical surveying, electromagnetic surveying, gravity surveying* (see GRAVITY ANOMALY), GROUND-PENETRATING RADAR (GPR), *magnetic surveying* (see MAGNETIC ANOMALY) and SEISMIC SURVEYING (including SEISMIC REFLECTION SURVEYING and SEISMIC REFRACTION SURVEYING). Geophysical surveys can detect objects or GEOLOGICAL STRUCTURES in the subsurface that have contrasting physical properties to their surroundings, and together with data from surface exposures and BOREHOLES are important in the study of EARTH STRUCTURE, in ARCHAEOLOGICAL PROSPECTION, in the exploration for geological NATURAL RESOURCES (e.g. PETROLEUM) and in surveys to assess ground conditions for *engineering.*   *GO*

Gaffney C (2008) Detecting trends in the prediction of the buried past: A review of geophysical techniques in archaeology. *Archaeometry* 50: 313–336.

Gowda BMR, Ghosh N, Wadhwa RS et al. (1998) Seismic refraction and electrical resistivity methods in landslide investigations in the Himalayan foothills. *Environmental and Engineering Geoscience* 4: 130–135.

Kearey P, Brooks M and Hill I (2002) *An introduction to geophysical exploration.* Oxford: Blackwell.

geophysics   The application of principles of physics to the study of the Earth. Within EARTH SCIENCE, geophysics involves studies of the Earth's gravity, magnetism (GEOMAGNETISM), behaviour with respect to elastic waves (SEISMOLOGY), *heat flow* and RADIOACTIVITY to investigate the structure, physical and environmental condition and geological evolution of the Earth. *Pure geophysics* (*global geophysics*) is the study of processes occurring within the whole planet (see EARTH STRUCTURE). Some methods of pure geophysics can also be applied to the study of PLANETARY GEOLOGY. *Applied geophysics* (*exploration geophysics* or GEOPHYSICAL SURVEYING) is the study of the Earth's CRUST, particularly at shallow depths, for practical purposes such as *mineral exploration, engineering* and ARCHAEOLOGICAL PROSPECTION. *Environmental geophysics* is the study of near-surface physical and chemical interactions and the evaluation of their implications for ENVIRONMENTAL MANAGEMENT.   *DNT*

Kearey P, Brooks M and Hill I (2002) *An introduction to geophysical exploration.* Oxford: Blackwell.

Kearey P, Klepeis KA and Vine FJ (2009) *Global tectonics.* Chichester: Wiley.

Reynolds JM (1997) *An introduction to applied and environmental geophysics.* Chichester: Wiley.

Stacey FD and Davis PM (2008) *Physics of the Earth.* Cambridge: Cambridge University Press.

geophyte   A terrestrial plant that can survive ENVIRONMENTAL STRESS (e.g. *summer drought* or *winter frosts*) through underground food-storage organs, such as *bulbs, corms, tubers* and *rhizomes.*   *JLI*

[*See also* LIFE FORM]

Hoffmann AJ, Liberona F and Hoffmann AE (2010) Distribution and ecology of geophytes in Chile: Conservation threats to geophytes in Mediterranean-type regions. In Rundel PW, Montenegro G and Jaksic FM (eds) *Landscape disturbance and biodiversity in Mediterranean-type ecosystems.* Berlin: Springer, 231–256.

georeferencing   The process of linking selected data with their location element to make the data mappable. The location definition may be a *grid reference* position or may employ another method such as *postcode* or *address*.   *TF*

Guo Q, Liu Y and Wieczorek J (2008) Georeferencing locality descriptions and computing associated uncertainty using a probabilistic approach. *International Journal of Geographical Information Science* 22: 1067–1090.
Kraak M-J and Ormeling F (2010) *Cartography: Visualisation of spatial data, 3rd edition*. Harlow: Pearson.

geoscience   A synonym for EARTH SCIENCE sometimes criticised on etymological, if not other grounds.   *CDW*

Edwards D and King C (1999) *Geoscience: Understanding geological processes*. London: Hodder and Stoughton.

geosol   In part synonymous with the term PALAEOSOL, geosols include all soil stratigraphical units (see SOIL STRATIGRAPHY) formed in past environments. Recent geological research has shown that soils are preserved in the FOSSIL RECORD with examples from most geological formations. *Seat earths* have been long recognised in the CARBONIFEROUS succession in association with COAL. Many soil characteristics survive LITHIFICATION and hence may enable an explanation and interpretation of otherwise inexplicable details in SEDIMENTARY ROCKS. CALCRETES and other DURICRUSTS, *mottle patterns*, *vertic structures* and various CASTS and PSEUDOMORPHS are recognised in beds from the Carboniferous to the TERTIARY. BURIED SOILS are particularly well known from the QUATERNARY, where they can often be linked to the large-scale CLIMATIC VARIATIONS during GLACIAL-INTERGLACIAL CYCLES.   *EMB*

[*See also* FOSSIL SOIL, LOESS]

Jacobs PM, Konen ME and Curry BB (2009) Pedogenesis of a catena of the Farmdale-Sangamon Geosol complex in the north central United States. *Palaeogeography, Palaeoclimatology, Palaeoecology* 282: 119–132.
Morrison RB (1978) Quaternary soil stratigraphy: Concepts, methods and problems. In Mahaney WC (ed.) *Quaternary soils*. Norwich: Geo Abstracts, 77–108.
Retallack GJ (2001) *Soils of the past: An introduction, 2nd edition*. Oxford: Blackwell.
Wright VP (ed.) (1986) *Palaeosols: Their recognition and interpretation*. Oxford: Blackwell.

geosphere   It has been used in many different ways, including (1) the totality of the EARTH SPHERES from Earth's CORE to the outer layers of the atmosphere; (2) any one of these spheres, such as the ATMOSPHERE, BIOSPHERE, CRYOSPHERE, HYDROSPHERE, LITHOSPHERE or PEDOSPHERE; (3) The spheres comprising the solid Earth only (or various combination of those from the lithosphere to the core) and (4) The geophysical spheres only (i.e. including the atmosphere but excluding the biosphere).   *JAM*

geostrophic current   An OCEAN CURRENT driven by pressure gradients related to variations in water-surface level, balanced by the CORIOLIS FORCE so that the direction of flow is at right angles to the horizontal pressure gradient.   *GO*

[*See also* EKMAN MOTION, GEOSTROPHIC WIND]

Denny M (2008) *How the ocean works: An introduction to oceanography*. Princeton, NJ: Princeton University Press.

geostrophic wind   A horizontal wind blowing parallel to the isobars, which is determined by the balance between the CORIOLIS FORCE and the pressure gradient force. The speed of the geostrophic wind is inversely related to the spacing of the isobars. It is an approximation to the actual wind, except at the Equator (where the Coriolis force is zero) and near the ground (where frictional effects dominate producing a *gradient wind* that crosses the isobars). The geostrophic wind may be considered the top (non-frictional) level of the *Ekman spiral* (see EKMAN MOTION). A similar concept applies to ocean currents (see GEOSTROPHIC CURRENT).   *JAM*

Hess SL (1957) *Introduction to theoretical meteorology*. New York: Holt.

geosyncline   A long-lived, regional- to continental-scale depression in the Earth's CRUST in which thick piles of SEDIMENTARY ROCKS accumulated before DEFORMATION, UPLIFT and the formation of an OROGENIC BELT (OROGEN). Geosyncline theory has been abandoned since the 1960s as a result of developments in the understanding of PLATE TECTONICS, CONTINENTAL MARGINS and SEDIMENTARY BASINS.   *GO*

Dott Jr RH and Shaver RH (eds) (1974) *Modern and ancient geosynclinal sedimentation*. Tulsa, OK: Society of Economic Paleontologists and Mineralogists.
Mitchell AHG and Reading HG (1986) Sedimentation and tectonics. In Reading HG (ed.) *Sedimentary environments and facies, 2nd edition*. Oxford: Blackwell, 471–519.

geothermal   Associated with heat from the Earth's interior. The *geothermal gradient* is the increase in temperature with depth in the Earth's CRUST. Typical values are 20–30°C/km, with a range from less than 10°C/km (e.g. on CRATONS) to over 300°C/km (e.g. at MID-OCEAN RIDGES (MORs)). *Geothermal energy* is obtained by transferring underground heat to the surface using heated GROUNDWATER or by pumping water down from the surface. It is considered a relatively clean source of energy but there are technical problems in its extraction, and its use is restricted to areas of high *heat flow*, particularly areas of active or recently active VOLCANISM, such as Iceland and North Island, New Zealand.   *GO*

[*See also* ALTERNATIVE ENERGY, EARTH STRUCTURE, HYDROTHERMAL, HYDROTHERMAL VENT]

Huenges E (ed.) (2010) *Geothermal energy systems: Exploration, development, and utilization.* Weinheim: Wiley-VCH.

geotourism The physical LANDSCAPE as a tourist venue. It is the physical equivalent of biologically based ECOTOURISM. Arguably, it has had a long history as an integral component of NATIONAL PARKS and other PROTECTED AREAS, but it has re-emerged recently as the specific focus of singular features (e.g. Iguazu Falls, Argentina; Uluru/Ayer's Rock, central Australia; and Mount Kilimanjaro, Tanzania) and *geoparks*.   *JAM*

[*See also* GEODIVERSITY, GEOHERITAGE, GEOLOGICAL CONSERVATION (GEOCONSERVATION)]

Dowling R and Newsome D (eds) (2010) *Geotourism.* Amsterdam: Elsevier.
Hose TA (2008) Towards a history of geotourism: Definition, antecedents and the future. In Burek CV and Prosser CD (eds) *History of geoconservation.* Bath: Geological Society, 37–60.

Gerlach trough A type of container dug into the surface on a hillslope to catch OVERLAND FLOW and trap most of the SEDIMENT being transported. There are various modified versions of this type of SEDIMENT TRAP since the original device was first described by T. Gerlach. Most versions comprise some form of metal or plastic box or trough (typically up to 50 cm wide) with, importantly, a lip (separate or integral) on the upslope side flush with the ground surface to ensure that overland flow and transported sediment are guided into the container. A lid prevents splashed material from entering the box, if only overland flow-transported sediment is required. As the contributing area of overland flow is not usually known (cf. EROSION PLOTS), Gerlach troughs tend to be used to indicate relative amounts of EROSION during a given period. They have the advantage over erosion plots of not interfering with upslope overland flow processes.   *RAS*

[*See also* EROSION PIN, SOIL EROSION, SOIL MICROPROFILING DEVICE]

Gerlach T (1967) Hillslope troughs for measuring sediment movement. *Revue de Géomorphologie Dynamique* 17: 132–140.
McDonald MA, Lawrence A and Shrestha PK (2003) Soil erosion. In Schroth G and Sinclair FL (eds) *Trees, crops and soil fertility: Concepts and research methods.* Cambridge, MA: CABI Publishing, 325–343.
Shakesby RA, Walsh RPD and Coelho COA (1991) New developments in techniques for measuring soil erosion in burned and unburned forested catchments, Portugal. *Zeitschrift für Geomorphologie, Supplementband* 83: 161–174.

Ghijben-Herzberg principle In a coastal AQUIFER, where the difference in DENSITY between *freshwater* and SALINE WATER ensures that a freshwater GROUNDWATER lens overlies saltwater at depth, this principle defines the quantitative relationship between the height of the fresh WATER TABLE above sea level and the depth of the interface between freshwater and saltwater. Because the difference in density between freshwater and saltwater is so small, ABSTRACTION of fresh GROUNDWATER leading to a DRAWDOWN of the water table will be compensated by a relatively large upward SALTWATER INTRUSION. Thus, for every 1 m drawdown in a WELL, for example, there is likely to be a corresponding rise of around 40 m in the saltwater table beneath the well.   *JAM/RPDW*

Thomas DSG and Goudie A (eds) (2000) *The dictionary of physical geography, 3rd edition.* Oxford: Blackwell.
Ward RC and Robinson M (2000) *Principles of hydrology, 4th edition.* London: McGraw-Hill.

gibber An aboriginal Australian term for a DESERT or *semi-desert* plain covered with a layer of pebbles or boulders. It is a type of DESERT PAVEMENT or *stone pavement* thought to be formed from the break-up of a siliceous surface crust (SILCRETE). Sturt's Stony Desert is an example.   *MAC*

[*See also* DURICRUST]

gibbsite A crystalline *aluminium hydroxide* found in soils, particularly in highly weathered soil material in *tropical environments* (e.g. FERRALSOLS, ACRISOLS and PLINTHOSOLS). It is mined as an ore of ALUMINIUM (Al).   *EMB*

[*See also* CLAY MINERALS, FERRALITISATION]

gigantism Also known as *giantism*, the apparent tendency for some isolated populations to undergo significant increases in body size in comparison to their *conspecifics* on the mainland. Gigantism has been particularly observed amongst murid *rodents* on islands, whereas larger mammals, especially carnivores, heteromyid rodents and artiodactyls, show evidence of insular *dwarfism*. Various hypotheses have been proposed to explain insular gigantism, including reduced PREDATION pressure, increased availability of RESOURCES, reduced interspecific COMPETITION and *immigrant selection*. Gigantism and NANISM form the two components of the so-called *island rule*. However, this may simply be an artefact associated with comparisons of distantly related groups that have responded in clade-specific ways to *insularity*.   *JLI*

[*See also* EVOLUTION, ISLAND BIOGEOGRAPHY, ISOLATION]

Li YM, Xu F, Guo ZW et al. (2011) Reduced predator species richness drives the body gigantism of a frog species on the Zhoushan Archipelago in China. *Journal of Animal Ecology* 80: 171–180.

Lyras GA, van der Geer AAE and Rook L (2010) Body size of insular carnivores: Evidence from the fossil record. *Journal of Biogeography* 37: 1007–1021.

Meik JM, Lawing AM and Pires-daSilva A (2010) Body size evolution in insular Speckled Rattlesnakes (Viperidae: *Crotalus mitchellii*). *PLOS ONE* 5: Article # e9524.

Meiri S, Raia P and Phillimore AB (2011) Slaying dragons: Limited evidence for unusual body size evolution on islands. *Journal of Biogeography* 38: 89–100.

Gilbert-type delta   A rivermouth DELTA characterised by coarse-grained SEDIMENT forming a steep (up to 35°) delta-front slope (*foreset beds*) overlain and underlain, respectively, by more gently sloping *topset beds* and *bottomset beds*. Gilbert-type deltas develop at steep basin margins and are typical of relatively small deltas, such as those at the heads of FJORDS.   GO

Breda A, Mellere D and Massari F (2007) Facies and processes in a Gilbert-delta-filled incised valley (Pliocene of Ventimiglia, NW Italy). *Sedimentary Geology* 200: 31–55.

Dorsey RJ, Umhoefer PJ and Falk PD (1997) Earthquake clustering inferred from Pliocene Gilbert-type fan deltas in the Loreto basin, Baja California Sur, Mexico. *Geology* 25: 679–682.

Edmonds DA, Shaw JB and Mohrig D (2011) Topset-dominated deltas: A new model for river delta stratigraphy. *Geology* 39: 1175–1178.

Gilbert GK (1885) The topographic features of lake shores. *Annual Reports of the United States Geological Survey* 5: 75–123.

gilgai   An Australian aboriginal word used to describe the undulating microrelief found on clay-rich VERTISOLS. It is a type of PATTERNED GROUND commonly taking the form of near-circular depressions a few metres in diameter but they may reach many tens of metres in diameter. The undulations are produced by the *shrinking and swelling* activity of SMECTITE clays during *wetting and drying cycles*. In the dry season, the soil cracks as the clays lose moisture and crumbs of soil fall down the cracks. When the soil is rewetted, the clays expand but the additional material at depth causes heaving and the development of trapezoidal structures, often separated by curved, slickensided, thrust planes. Linear forms also exist, which may also involve RUNOFF and/or AEOLIAN processes.   EMB/JAM

Goudie AS, Sands MJS and Livingstone I (1992) Aligned linear gilgai in the west Kimberley District, Western Australia. *Journal of Arid Environments* 23: 157–167.

Hallsworth EG, Robertson GK and Gibbons FR (1955) Studies in pedogenesis in New South Wales, Part VII: The 'gilgai' soils. *Journal of Soil Science* 6: 1–34.

Verger F (1964) Mottureux et gilgais. *Annales de Géographie* 73: 413–430.

glacial   (1) An adjective meaning pertaining to GLACIERS (e.g. GLACIAL EROSION, GLACIAL LAKE). (2) A noun meaning a GLACIAL EPISODE and contrasting with INTERGLACIAL.   JAM

glacial deposition   The release of SEDIMENTS by GLACIER ice, or in close proximity to a glacier. A wide range of processes is involved in glacial deposition, reflecting not only the interaction of glaciers with rivers, lakes and the sea but also processes of *reworking* (e.g. by MASS MOVEMENT PROCESSES) following deposition. Consequently, the deposits resulting from glacial activity are among the most complex to be found in the natural environment. The processes and products of glacial deposition in different environments are summarised in the Figure (next page).   MJH

[*See also* GLACIAL LANDFORMS, GLACIAL SEDIMENTS, REDEPOSITION]

Benn DI and Evans DJA (2010) *Glaciers and glaciation, 2nd edition*. London: Hodder Education.

Bennett MR and Glasser NF (2009) *Glacial geology: Ice sheets and landforms, 2nd edition*. Chichester: Wiley.

Evans DJA (eds) (2003) *Glacial landsystems*. London: Arnold.

Hambrey MJ (1999) The record of the Earth's glacial climates over the last 3000 Ma. In Barrett PJ and Orombelli J (eds) *Geological records of global change*. Siena: Terra Antarctica, 73–108.

glacial episode   A protracted cold phase or interval of CONTINENTAL GLACIATION marked by the expansion of ICE SHEETS and glaciers. Glacial episodes, *glacial periods*, GLACIATIONS or GLACIALS are subdivided into STADIALS and INTERSTADIALS and separated from each other by INTERGLACIALS. The latest glacial or LAST GLACIATION terminated about 11,700 years ago and was known in the British Isles as the DEVENSIAN, in northern Europe as the *Weichselian* and in America as the *Wisconsinan*.

JAM/DH

[*See also* ICE AGES, MARINE ISOTOPIC STAGE (MIS), TERMINATION]

glacial erosion   Glaciers erode their beds directly by ABRASION and QUARRYING, and indirectly through the action of glacial MELTWATER. Abrasion produces STRIATIONS, FRICTION CRACKS and *polished surfaces*. Quarrying or *plucking* involves a twofold process in which cracks, joints and other discontinuities within a rock mass are first propagated, before loosened blocks are removed via glacial ENTRAINMENT. *Fracture propagation* is assisted by repetitive LOADING and unloading, and by fluctuations in subglacial *hydraulic pressure*. *Glacial entrainment* may be achieved via *ice shear* and the *freezing-on* of debris to the sole of a glacier. Freezing-on may be achieved via variations in basal ice temperature and pressure (see REGELATION). *Glacial meltwater* may cause both SUBGLACIAL and *proglacial* erosion. The effects of glacial erosion are widespread in glaciated LANDSCAPES, ranging

(a) Terrestrial temperate/polythermal glacier

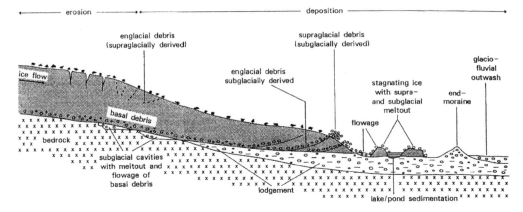

(b) Temperate tidewater glacier in fjord

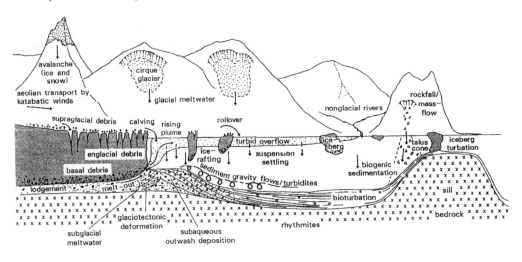

(c) Ice shelf and continental shelf (Antarctica)

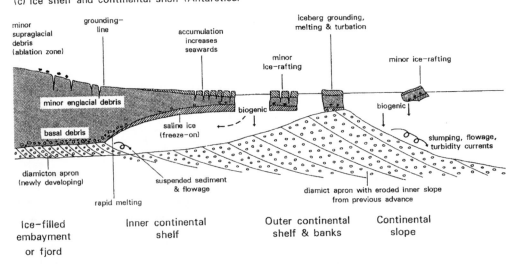

Glacial deposition *Examples of processes and facies in three contrasting glacial environments: (A) a temperate glacier terminating on land; (B) a temperate glacier terminating in an Alaskan fjord and (C) a cold-based glacier/ice shelf terminating on the continental shelf of Antarctica (Hambrey, 1999).*

from striations to large-scale *areal scouring* (see KNOCK AND LOCHAN TOPOGRAPHY), U-SHAPED VALLEYS and FJORDS.

Rates of glacial erosion have rarely been measured directly but indirect methods include estimates based on the SUSPENDED LOAD, DISSOLVED LOAD and BEDLOAD of glacial MELTWATERS; the volume of sediment in GLACIO-LACUSTRINE DEPOSITS and MORAINES; the reconstruction of *preglacial surfaces* and the COSMOGENIC-NUCLIDE DATING of glaciated terrain. However, the EROSION RATES of glaciers are highly variable, depending on environmental factors such as the GLACIER THERMAL REGIME, LITHOLOGY, TOPOGRAPHY and glacier size. Under optimum conditions, rates of glacial erosion exceed those of subaerial processes but COLD-BASED GLACIERS generally protect the LANDSCAPES to which they are frozen. For TEMPERATE GLACIERS, the zone of maximum erosion tends to be associated with maximum *ice flow* beneath the EQUILIBRIUM LINE, which may change position in response to GLACIER VARIATIONS, with areas of *convergent flow*, and over relatively steep slopes (see ICE STREAM). OVERDEEPENING of *glacial valleys* and the existence of HANGING VALLEYS provide additional evidence of the efficacy of glacial erosion over the long term.   *MRB/JAM/JS*

[*See also* GLACIAL PROTECTION CONCEPT, ROCK FLOUR]

Benn DI and Evans DJA (2010) *Glaciers and glaciation, 2nd edition*. London: Hodder Education.

Braun DD (1989) Glacial and periglacial erosion of the Appalachians. *Geomorphology* 2: 233–256.

Colgan PM, Bierman PR, Mickelson DM and Caffee M (2002) Variation in glacial erosion near the southern margin of the Laurentide Ice Sheet, south-central Wisconsin, USA: Implications for cosmogenic dating of glacial terrains. *Geological Society of America Bulletin* 114: 1581–1591.

Delmas M, Calvet M and Gunnell Y (2009) Variability of Quaternary glacial erosion rates: A global perspective with special reference to the Eastern Pyrences. *Quaternary Science Reviews* 28: 484–498.

Duhnforth M, Anderson RS, Ward D and Stock G M (2010) Bedrock fracture control of glacial erosion processes and rates. *Geology* 38: 423–42.

Hallet B, Hunter L and Bogen J (1996) Rates of erosion and sediment evacuation by glaciers: Review of field data and their implications. *Global and Planetary Change* 12: 213–235.

Krabbendam M and Glasser NF (2011) Glacial erosion and bedrock properties in NW Scotland: Abrasion and plucking, hardness and joint spacing. *Geomorphology* 130: 374–383.

Owen G, Matthews JA and Albert PG (2007) Rates of Holocene chemical weathering, 'Little Ice Age' glacial erosion, and implications for Schmidt-hammer dating at a glacier-foreland boundary, Fåbergstølsbreen, southern Norway. *The Holocene* 17: 829–834.

glacial lake   A body of water formed in a SUPRA-GLACIAL, ENGLACIAL, SUBGLACIAL or PROGLACIAL position and dammed by GLACIER ice (ICE-DAMMED LAKE) or by a MORAINE (*moraine-dammed lake*). It may undergo progressive or catastrophic drainage through breaching of the dam, the latter type of drainage, known as a *glacier outburst flood* or JÖKULHLAUP, potentially leading to downstream damage to property and risk to life. Within the Himalayas, there are some 20,000 glacial (or glacier) lakes, many forming since the 1950s as a result of widespread glacial retreat. There are concerns that a major EARTHQUAKE could trigger catastrophic release of water from moraine-dammed lakes and lead to consequent danger to people and property. Probably the most famous example of a glacial lake worldwide is Lake Missoula (a glacial MEGALAKE), which existed in western Montana, USA, between about 15,000 and 13,000 years ago. It is the largest known ICE-DAMMED LAKE and measured about 7,700 km^2 in size and contained 2,100 m^3 of water at its maximum. Numerous Jökulhlaups caused by breaching of the ice dam led to huge floods resulting in spectacular erosional and depositional landforms in the CHANNELED SCABLANDS.

                                                                            RAS

[*See also* SUBGLACIAL LAKE]

Bajracharya SR and Mool P (2009) Glaciers, glacial lakes and glacial lake outburst floods in the Mount Everest region, Nepal. *Annals of Glaciology* 50: 81–86.

Mangerud J, Jakobsson M, Alexanderson et al. (2004) Ice-dammed lakes and rerouting of the drainage of northern Eurasia during the Last Glaciation. *Quaternary Science Reviews* 23: 1313–1332.

glacial landforms   Morphological features produced by GLACIAL EROSION or GLACIAL DEPOSITION at a range of spatial scales by the direct action of GLACIER ice. Microscale erosional forms include STRIATIONS, CHATTERMARKS and FRICTION CRACKS. Mesoscale features include ROCHES MOUTONNÉES, *whalebacks*, *rock grooves*, *rock basins*, CRAG AND TAIL and KNOCK-AND-LOCHAN TOPOGRAPHY, while macroscale features include CIRQUES, U-SHAPED VALLEYS (*glacial troughs*) and *flyggbergs* (mega roches moutonnées). Depositional landforms can be classified in a range of different ways. Landforms that are orientated transverse to ice-flow direction include end and recessional MORAINES, ROGEN MORAINES and those parallel to it include *fluted moraine* and DRUMLINS, while some landforms have no particular orientation with respect to ice flow (e.g. HUMMOCKY MORAINE).

Glacial LANDFORMS have always been a major focus of *glacial geomorphology*. At various points in the history of the study of landforms, it can be argued that their precise role has varied. Glacial landforms were instrumental in establishing the GLACIAL THEORY in the nineteenth century. For much of the twentieth century, however, the *morphology* of landforms seems to have been an end in itself. By the 1970s, a process-based approach can be said to have dominated, not least so that the landforms could be better understood. Most recently, glacial geomorphology appears to be reorienting towards the *interdisciplinary* study of GLOBAL ENVIRONMENTAL CHANGE, with glacial landforms being used to reconstruct the past as a guide to the future. In particular, RELICT glacial landforms provide the clues in the LANDSCAPE from which the lateral and the vertical extent of former GLACIERS and ICE SHEETS can be reconstructed—a process known as *inversion modelling* or PALAEOGLACIOLOGY.

MRB/JAM

[*See also* GLACIAL SEDIMENTS, GLACIOFLUVIAL LANDFORMS, ICE-MARGIN INDICATORS]

Benn DI and Evans DJA (2010) *Glaciers and glaciation, 2nd edition.* London: Hodder Education.
Bennett MR and Glasser NF (2009) *Glacial geology: Ice sheets and landforms, 2nd edition.* Chichester: Wiley-Blackwell.
Greenwood SL and Clark CD (2009) Reconstructing the last Irish Ice Sheet 1: Changing flow geometries and ice flow dynamics deciphered from the glacial landform record. *Quaternary Science Reviews* 28: 3085–3100.
Menzies J (2011) Glacial geomorphology. In Gregory KJ and Goudie AS (eds) *The SAGE handbook of geomorphology.* London: Sage, 378–392.
Sugden DE and John BS (1976) *Glaciers and landscape.* London: Arnold.

glacial maximum   The maximum extent of ICE-SHEET coverage during a GLACIAL EPISODE, or GLACIATION, often defined by MORAINE limits. The LAST GLACIAL MAXIMUM (LGM), which occurred some 20,000 years ago, was so extensive in many regions that earlier limits were overridden. Hence, defining the maximum extent of ice reached during some earlier glaciations is proving difficult. However, the timings of glacial maxima are now well established from the continuous STRATIGRAPHICAL RECORD preserved in MARINE SEDIMENT CORES and ICE CORES.   *DH*

Barrows TT, Stone JO, Fifield LK and Cresswell RG (2002) The timing of the Last Glacial Maximum in Australia. *Quaternary Science Reviews* 21: 159–173.
Dyke AS, Andrews JT, Clark PU et al. (2002) The Laurentide and Inuitian Ice Sheets during the Last Glacial Maximum. *Quaternary Science Reviews* 21: 9–31.

glacial protection concept   An early view that GLACIERS generally protect the LANDSCAPE from ERO-SION is no longer seen as applicable. A modern view is that COLD-BASED GLACIERS, and ICE SHEETS that are cold-based in parts, may be protective. This is suggested by the preservation of such delicate landforms as TORS (e.g. Cairngorms, UK) and sorted PATTERNED GROUND in formerly glaciated areas as well as the preservation of GLACIAL LANDFORMS from earlier glacial episodes. The thermal conditions at the ice bed have important

implications for ICE-SHEET DYNAMICS and PALAEOENVI-RONMENTAL RECONSTRUCTION which, in turn, are essential elements of CLIMATIC MODELS for the period of GLACIATION.   *MRB/JS*

Kleman J and Glasser NF (2009) The subglacial thermal organisation (STO) of ice sheets. *Quaternary Science Reviews* 26: 585–597.
Kleman J and Stroeven A P (1997) Preglacial surface remnants and Quaternary glacial regimes in northwestern Sweden. *Geomorphology* 19: 35–54.

glacial sediments   Materials that are being or have been transported by GLACIER ice usually excluding debris that is subsequently reworked (see REDEPOSI-TION). *Glacigenic sediment* is usually used in a broader sense to include sediments with a greater or lesser component derived from glacier ice. A comprehensive classification of terrestrially deposited glacigenic sediments was developed for the International Union for Quaternary Research (INQUA), based on our understanding of the processes of transport and deposition but is constantly being refined. Sediment carried by the glacier is located (1) in the basal ice layer, where it is referred to as *basal glacial debris*; (2) on the surface, where it is derived primarily from ROCKFALL, referred to as SUPRAGLACIAL debris and (3) in the interior if ingested from the surface via *crevasses* or *meltwater conduits*, or from the bed via FOLDS and THRUST FAULTS, where it is referred to as ENGLACIAL debris. Commonly, ice also moves over a soft *deformable bed* of SUBGLA-CIAL debris.

The principal genetic types of subglacially deposited sediment are as follows. Sediment deposited by uniquely *glacial processes* without subsequent *disaggregation* and reworking is termed TILL. This is further subdivided into LODGEMENT TILL, which results from active plastering onto the bed of sediment by the glacier, and MELTOUT TILL, which is released by melting from relatively inactive debris-rich ice. This debris is highly mobile and the term *glacigenic sediment flow* (formerly referred to as FLOW TILL) is applied. *Deformation till* comprises weak rock and sediment that has been detached by the glacier from its source and its original structural integrity destroyed, although it retains many of its primary characteristics. Together, all these are grouped under the name *subglacial till* (or *basal till*). In recent years, there has been a tendency to avoid the terms lodgement and meltout tills and to refer to them together as *subglacial traction tills*. This is because the glacier bed experiences a wide range of processes and the resulting sediments are a complex mosaic of different depositional products.

Sediment released directly by glaciers or from a concentrated cluster of ICEBERGS into a waterbody is referred to as *ice-proximal* GLACIOMARINE DEPOSITS (or

GLACIOLACUSTRINE DEPOSITS), formerly termed *water-lain till*. Farther from the ice in the waterbody can be found *ice-distal glaciomarine (or glaciolacustrine) sediment*.

All these sediments are generally poorly sorted, with material potentially ranging in size from clay or silt to boulders or cobbles. Except for ice-distal glaciomarine sediment, all these types of glacigenic sediment usually lack BEDDING, and can only be distinguished on the basis of detailed analysis of sediment TEXTURE (e.g. *grain-size analysis*), particle SHAPE and ROUNDNESS, clast surface features (e.g. STRIATIONS and FRICTION CRACKS) and FABRIC ANALYSIS. Ice-distal deposits include material derived directly from icebergs, mixed with sediment formerly in suspension, and other types of MARINE SEDIMENTS or LACUSTRINE SEDIMENTS. The well-defined practice of FACIES ANALYSIS has been established to interpret sedimentary sequences and, in particular, glacigenic sediments. Different sediment types (LITHOFACIES) are first described objectively and rigorously, and only then interpreted (e.g. a DIAMICTON may be interpreted as a till). Lithofacies are grouped into FACIES ASSOCIATIONS to aid interpretation of the overall environment. By combining facies analysis with descriptions of LANDFORM morphology, *sediment/landform associations* can be formally defined. From these, in specific geographical contexts, and based on modern examples, a variety of *glacial landsystems* have been defined.   *MJH*

[*See also* GLACIAL DEPOSITION, GLACIATION, GLACIOFLUVIAL SEDIMENTS]

Benn DI and Evans DJA (2010) *Glaciers and glaciation, 2nd edition*. London: Hodder Education.
Bennett MR and Glasser NF (2009) *Glacial geology: Ice sheets and landforms, 2nd edition*. Chichester: Wiley.
Dowdeswell JA and Cofaigh CO (eds) (2002) *Glacier-influenced sedimentation on high-latitude continental margins*. Bath: Geological Society.
Evans DJA (ed.) (2003) *Glacial landsystems*. London: Arnold.
Hambrey MJ, Christoffersen P, Glasser NF and Hubbard B (eds) (2007) *Glacial sedimentary processes and products*. Malden, MA: Blackwell.
Maltman AJ, Hubbard B and Hambrey MJ (eds) (2000) *Deformation of glacial materials*. Bath: Geological Society.

Glacial Theory

A theory developed in the early to mid-nineteenth century largely by members of the Helvetic Society, including Perraudin, Venetz, de Charpentier and Louis Agassiz. Their theory proposed the existence of a former ICE AGE when greatly expanded glaciers and ice sheets produced many of the erosional landforms and especially the sedimentary deposits in and around the European Alps. Critical evidence in establishing the Glacial Theory by the application of the principles of UNIFORMITARIANISM included the recognition of (1) ERRATICS; (2) STRIATIONS, grooves and glacially polished rock surfaces; (3) MORAINES; (4) unsorted drift (TILL) and (5) oversteepened U-SHAPED VALLEYS and, in Norway, FJORDS. Soon afterwards, multiple GLACIATIONS and their global implications were recognised (see Table, next page). The Glacial Theory provided a new PARADIGM, which replaced the DILUVIAL THEORY, and can now be seen as the beginning of modern ideas on *environmental change* and QUATERNARY SCIENCE.   *JAM*

[*See also* ENVIRONMENTAL CHANGE: HISTORY OF THE FIELD, ICEBERG DRIFT THEORY, MILANKOVITCH THEORY]

Bowen DQ (1978) *Quaternary geology: A stratigraphic framework for multidisciplinary work*. Oxford: Pergamon Press.
Flint RF (1957) *Glacial and Pleistocene geology, 1st edition*. New York: Wiley.
Imbrie J and Imbrie KP (1979) *Ice ages: Solving the mystery*. London: Macmillan.
North FJ (1943) Centenary of the Glacial Theory. *Proceedings of the Geologists Association* 54: 1–28.

glacial-interglacial cycle

The repeated alternation of cold (GLACIAL) and warm (INTERGLACIAL) stages within an ICE AGE. These cycles appear to have a PERIODICITY of ca 100,000 years between the major interglacials, and also exhibit periodicities at ca 43 ka and 23–19 ka. The MILANKOVITCH THEORY of variations in global INSOLATION receipt due to the changing astronomical position of the Earth relative to the Sun is generally acknowledged as the dominant driving force behind the cycles.   *DH*

[*See also* INTERGLACIAL CYCLE, MARINE ISOTOPIC STAGE (MIS), ORBITAL FORCING, ORBITAL TUNING]

Porter SC (2001) Chinese loess record of monsoonal climate during the last glacial-interglacial cycle. *Earth-Science Reviews* 54: 115–128.
Tudhope AW, Chilcott CP, McCulloch MT et al. (2001) Variability in the El Niño-Southern Oscillation through a glacial-interglacial cycle. *Science* 291: 1511–1517.

glaciation

The term 'glaciation' is used in several senses. At its most general, it refers to occupancy of part or all of a landscape by GLACIERS and/or ICE SHEETS. Glaciation is also used to refer to a glacier advance-retreat cycle, sometimes synonymously with *glacial period*, GLACIAL, or GLACIAL EPISODE, although this is not the correct formal usage. The term, however, may also refer to periods of ice occupancy on both longer and shorter timescales, in these cases synonymous with ICE AGE and STADIAL, respectively. Periods of glaciation normally occurred during even-numbered MARINE ISOTOPIC STAGES in the OXYGEN

Date (AD)	Personality	Contribution
1779	De Saussure	Coined the term ERRATIC; attributed them to floods
1787	Kuhn	Interpreted erratics as evidence of ancient GLACIATION
1795	Hutton	Recognised erratics are glacially transported
1802	Playfair	Supported Hutton
1815	Perraudin	Recognised an alpine glaciation
1821	Venetz	First scientific account of an alpine glaciation
1823	Buckland	Publication of *Reliquae Diluvianae* (see DILUVIAL THEORY)
1824	Esmark	Recognised Norwegian *mountain glaciation*
1829	Venetz	Recognised extensive European glaciation beyond the Alps
1832	Bernhardi	Recognised separate CONTINENTAL GLACIATION in Germany
1834	De Charpentier	Recognised glacial STRIATIONS
1837	Agassiz	Discourse of Neuchâtel; strongly advocates the Glacial Theory
1838	Buckland	'Converted' to the Glacial Theory
1839	Conrad	First acceptance of the theory in America
1840	Agassiz	Publication of *Études sur les glaciers*; elaborates the theory
1840	Agassiz	Visits Britain; lectures with Buckland and Lyell
1840	Schimper	Coined the term 'Eiszeit' (ICE AGE)
1841	De Charpentier	Publication of *Essai sur les glaciers*
1841	Maclaren	Recognised GLACIO-EUSTASY (sea-level fall)
1847	Collomb	Recognized two glaciations in the Vosges Mountains
1851	Godwin-Austen	Recognised the PERIGLACIAL origin of much COLLUVIUM
1852	Ramsay	Recognised two glaciations in Wales
1853	Chambers	Recognised two glaciations in Scotland
1856	Morlot	Recognized two glaciations in Switzerland
1858	Heer	Recognised INTERGLACIAL deposits
1863	Jamieson	Recognised high lake levels in ARIDLANDS during glacial episodes
1865	Jamieson	Recognised GLACIO-ISOSTASY (depression of the Earth's crust)
1868	Tylor	Coined the term PLUVIAL
1868	Whittlesey	Calculated approximate value of glacio-eustatic sea-level fall
1872	Von Richthofen	Recognised the AEOLIAN origin of LOESS
1882	Penck	Recognised three glaciations in the Alps
1905	Howorth	Last major scientific opposition to the Glacial Theory in Britain
1909	Penck and Bruckner	Recognised four glaciations in the Alps

Glacial Theory *Some personalities and their landmark contributions in the development of the Glacial Theory, before and after AD 1840 (Bowen, 1978; Flint, 1957).*

ISOTOPE timescale, although these should more appropriately be seen simply as *cold stages*. In METEOROLOGY, the term 'glaciation' refers to the formation of ice crystals in clouds.

Glaciations (Ice Ages) have occurred on Earth during many geological periods, notably the Early and the Late PROTEROZOIC, ORDOVICIAN, *Carboniferous-Permian* and CENOZOIC (*CAINOZOIC*), persisting for many millions of years on each occasion. The reasons why major Ice Ages occurred at these times are still unclear, although the distribution of land masses and mountain belts, and concentrations of greenhouse gases are likely to have been major factors. The most recent Ice Age is the Cenozoic, which is still ongoing. Significant ice cover began to form in the Late MIOCENE (about 10 million years ago) and has been a particular characteristic of the QUATERNARY.

Numerous glaciations (glacial episodes or glacials) during the Quaternary have been interspersed with INTERGLACIALS forming GLACIAL-INTERGLACIAL CYCLES.

Each glacial was probably composed of a number of relatively short-lived stadials or *stades* (not always involving extensive glaciation), with intervening periods of comparatively mild climate as INTERSTADIALS or *interstades*. During the most recent glacial episode, large ice sheets formed over North America (the LAURENTIAN/LAURENTIDE ICE SHEET), Northern Europe (the FENNOSCANDIAN ICE SHEET and British ice sheets) and the islands and shelves surrounding the ARCTIC (e.g. the *Barents Sea ice sheet*). Considerable ice cover also developed on all of the world's great mountain ranges, including the North American Cordillera, Andes, European and New Zealand Alps and the Himalayas.

ICE-SHEET GROWTH and decay during glaciation is associated with major environmental changes on global and regional scales, including major reorganisations of the ATMOSPHERIC and OCEANIC CIRCULATIONS.

DIB

[*See also* GLACIERISATION, GLACIERS AND ENVIRONMENTAL CHANGE, ICEHOUSE CONDITION, QUATERNARY ENVIRONMENTAL CHANGE, TERMINATION]

Benn DI and Evans DJA (2010) *Glaciers and glaciation, 2nd edition*. London: Hodder Education.
Bennett MR and Glasser NF (2009) *Glacial geology: Ice sheets and landforms, 2nd edition*. Chichester: Wiley-Blackwell.
Bose M, Luthgens C, Lee JR and Rose J (2012) Quaternary glaciation history of Northern Europe. *Quaternary Science Reviews* 44: 1–240 [Special Issue].
Ehlers J, Gibbard PL and Hughes PD (eds) (2011) *Quaternary glaciations—extent and chronology: A closer look*. Amsterdam: Elsevier.
Evans D (2012) *Atlas of glaciation*. London: Hodder Education.
Haug GH, Ganopolski A, Sigman DM et al. (2005) North Pacific seasonality and the glaciation of North America 2.7 million years ago. *Nature* 433: 821–825.
Lowe JJ and Walker MJC (1997) *Reconstructing Quaternary environments, 2nd edition*. Harlow: Longman.
Nesje A and Dahl SO (2000) *Glaciers and environmental change*. London: Arnold.
Pagani M, Huber M, Liu Z et al. (2011) The role of carbon dioxide during the onset of Antarctic glaciation. *Science* 334: 1261–1264.
Thomson SN, Brandn MT, Tomkin JH et al. (2010) Glaciation as a destructive and constructive control on mountain building. *Nature* 467: 313–317.

glaciation threshold   The glaciation threshold, *glaciation limit* or *glaciation level* is the critical elevation above which GLACIERS can exist. The concept is operationalised as the mean elevation between the highest topographically suited summit without a glacier and the lowest summit hosting a glacier in the same geographical region.   *AN*

Østrem G (1966) The height of the glaciation limit in southern British Columbia and Alberta. *Geografiska Annaler* 48A: 126–138.

glaciel   A term of French-Canadian origin covering PERIGLACIAL phenomena formed by floating ice in fluvial, lacustrine and marine environments. For example, glaciel STRIATIONS (*stries glacielles*) formed by drift ice contrast with glacial striations formed by GLACIER ice. Other glaciel phenomena include BOULDER PAVEMENTS, *ice-shove ridges*, ICEBERG PLOUGH MARKS and ICE-RAFTED DEBRIS (IRD).   *JAM*

[*See also* FROST WEATHERING, PERIGLACIOFLUVIAL SYSTEM, POLAR SHORE EROSION]

Dionne J-C (1973) Distinction entre stries glacielles et stries glaciaires. *Revue de Géographie de Montréal* 27: 185–190.
Hamelin L-E (1976) La famille du mot 'glaciel'. *Revue de Géographie de Montréal* 30: 233–236.

glacier   A mass of ice, irrespective of size, derived largely from snow, and continuously moving from higher to lower ground, or spreading over the sea. Glaciers may be classified according to their size and morphology. The largest are ICE SHEETS, defined as more than 50,000 km^2 in size, and so today they are limited to the ANTARCTIC and Greenland. Morphologically similar, but covering <50,000 km^2 are *ice caps*. Ice sheets and ice caps commonly have multiple elevated areas referred to as ICE DOMES, the *East Antarctic Ice Sheet* being a good example of this. Within ice sheets are zones of fast-flowing ice called ICE STREAMS, which sometimes extend into the sea as so-called *glacier tongues* or *ice tongues*. Ice sheets also feed large floating slabs of ice called ICE SHELVES, although these are mainly restricted to Antarctica today. Extensive areas of undulating ice which broadly mirror the underlying topography, and through which mountains project as NUNATAKS, are referred to as *highland ice fields* or *plateau glaciers*. Glaciers flowing between rock walls, generally within a U-SHAPED VALLEY, are known as *valley glaciers*, whether or not they have well-defined local accumulation areas, or emanate as OUTLET GLACIERS from an ice cap or ice sheet. Where valley glaciers spread out as wide lobes onto flat land beyond the confines of valleys, *piedmont glaciers* are formed. Small glaciers occupying hollows carved out high in the mountains or uplands (CIRQUES) are referred to as *cirque glaciers*. Steep areas of ice clinging to precipitous mountainsides are known as *ice aprons*, *niche glaciers* or *glacierets*. Glaciers that form from *ice avalanche* (including SNOW AVALANCHE material) at the foot of a slope are known as *rejuvenated* or *regenerated glaciers*. *Hanging glaciers* are small unstable ice masses, prone to ice avalanching, that adhere to precipitous mountain sides. *Transection glaciers* flow across topographic divides.   *MJH*

[*See also* ANTARCTIC ENVIRONMENTAL CHANGE, CRYOSPHERE, MARINE-BASED ICE SHEET, TIDEWATER GLACIER]

Benn DI and Evans JA (2010) *Glaciers and glaciation, 2nd edition*. London: Hodder Education.
Cuffey KM and Paterson WSB (2010) *The physics of glaciers, 4th edition*. Burlington, MA: Butterworth-Heinemann.
Hambrey MJ and Alean JC (2002) *Glaciers, 2nd edition*. Cambridge: Cambridge University Press.
Knight PG (1999) *Glaciers*. Cheltenham: Stanley Thornes.
Knight PG (ed.) (2006) *Glacier science and environmental change*. Malden, MA: Blackwell.
Singh VJ, Singh P and Haritashya UK (eds) (2011) *Encyclopedia of snow, ice and glaciers*. Dordrecht: Springer.

glacier foreland   The recently deglacierised zone in front of a retreating glacier (see Figure). Derived from the German *Gletschervorfeld*, and introduced into the English language by R.E. Beschel, the term 'glacier foreland' is usually restricted to the area deglacierised in historical time since the maximum glacier extent of the LITTLE ICE AGE. Thus, it is a visually distinct zone of relatively bare terrain, and a generally immature LANDSCAPE and GEO-ECOSYSTEM. Glacier forelands are increasingly used as *field laboratories* or natural MICRO-COSMS for investigating ecological and other aspects of *landscape change* over timescales of tens and hundreds of years. In this respect, they are particularly useful because the increasing age of the landscape with distance from the glacier can be used to infer temporal change using the CHRONOSEQUENCE approach. This has been used as a conceptual framework in the development of DATING TECHNIQUES (e.g. LICHENOMETRIC DATING) and in studies of PRIMARY SUCCESSION in plant and animal communities.   *JAM*

Beschel RE (1961) Dating rock surfaces by lichen growth and its application to glaciology and physiography (lichenometry). In Raasch GO (ed.) *Geology of the Arctic, volume 2*. Toronto: University of Toronto Press, 1044–1062.
Engstrom DR, Fritz SC, Almendinger JE and Juggins S (2000) Chemical and biological trends during lake evolution in recently deglaciated terrain. *Nature* 408: 161–166.
Matthews JA (1992) *The ecology of recently-deglaciated terrain: A geoecological approach to glacier forelands and primary succession*. Cambridge: Cambridge University Press.
Owen G, Matthews JA and Albert PG (2007) Rates of Holocene chemical weathering, 'Little Ice Age' glacial erosion, and implications for Schmidt-hammer dating at a glacier: Foreland boundary, Fåberstølsbreen, southern Norway. *The Holocene* 17: 829–834.

Glacier foreland *The Storbreen glacier foreland, Jotunheimen, Norway: isochrones (lines of equal age) indicate the pattern of glacier retreat since the middle of the eighteenth century.*

glacier milk   The sediment-laden MELTWATER emanating from GLACIERS. The cloudy appearance of the water is due to the SUSPENDED LOAD, especially silt. In distal GLACIAL LAKES that receive meltwater, the finest of the particles still in suspension result in the emerald-green colour of the water.   *MJH*

[*See also* GLACIOFLUVIAL SEDIMENTS, ROCK FLOUR]

Keller WD and Reesman AL (1963) Glacier milks and their laboratory-simulated counterparts. *Geological Society of America Bulletin* 74: 61–76.
Singh P (2006) Estimates and analysis of suspended sediment from a glacierized basin in the Himalayas. In Rowan JS, Duck RW and Werrity A (eds) *Sediment dynamics and hydromorphology of fluvial systems*. Wallingford: IASH Press, 21–27.

glacier modelling   The numerical simulation of GLACIERS or ICE SHEETS, or some aspect of their behaviour. Because glaciers and ice sheets evolve over long timescales, cycles of growth and decay cannot be observed directly. Understanding of such cycles, therefore, largely depends upon numerical modelling, in which glaciers are simulated (see SIMULATION MODEL) by sets of equations. Furthermore, models are widely used in developing theories of SUBGLACIAL processes, which generally operate in inaccessible environments. Several types of glacier model have been developed, with varying degrees of detail and sophistication.

A widely used modelling procedure attempts to reconstruct the climatic inputs required to 'grow' a glacier or ice sheet of a given size. Inputs for such models include a representation of topography, annual temperature and PRECIPITATION values (and their variation with altitude), and equations describing ice flow, and yield outputs such as glacier thickness, extent and velocity. Successive model runs with varying climatic inputs can be compared to derive the most likely climatic conditions associated with observed glacier limits. Alternatively, such models can be used to explore the future behaviour of ice masses under given climatic inputs to determine, for example, glacier response to GLOBAL WARMING. At a more detailed level, glacier ABLATION rates can be modelled by simulating the ENERGY BALANCE at the ice surface, using inputs such as SOLAR RADIATION, air temperature, HUMIDITY and windspeed. Other types of glacier models explore the evolution of ice temperatures and GLACIER THERMAL REGIME, HYDROLOGY, surging behaviour, SEDIMENT TRANSPORT and EROSION RATES.   *DIB*

Bougamont M, Price S, Christoffersen P and Payne AJ (2011) Dynamic patterns of ice stream flow in a 3-D higher-order ice sheet model with plastic bed and simplified hydrology. *Journal of Geophysical Research* 116: F04018.

Cuffey KM and Paterson WSB (2010) *The physics of glaciers, 4th edition.* London: Butterworth-Heinemann.

Greve R and Blatter H (2009) *Dynamics of ice sheets and glaciers.* Berlin: Springer.

Marshall SJ, Pollard D, Hostetler S and Clark PU (2004) Coupling ice-sheet and climate models for simulation of former ice sheets. In Gillespie AR, Porter SC and Atwater BF (eds) *The Quaternary period in the United States.* Amsterdam: Elsevier, 105–126.

Ridley KJ, Huybrechts P, Gregory JM and Lowe JA (2005) Elimination of the Greenland Ice Sheet in a high CO_2 climate. *Journal of Climate* 18: 3409–3427.

Siegert MJ (2007) A brief review of modelling sediment erosion, transport and deposition by former large ice sheets. In Hambrey MJ, Christoffersen P, Glasser NF et al. (eds) *Glacial sedimentary processes.* Malden, MA: Blackwell, 53–64.

glacier surge   A short-lived phase of accelerated GLACIER flow (up to 10–100 times faster than previously) during which the surface becomes broken into a maze of *crevasses* and the terminus advances rapidly. Only about 4 per cent of glaciers surge, and they tend to be concentrated geographically (notably Svalbard, Alaska, Yukon and Karakorum). For TEMPERATE GLACIERS flowing over a hard bed, the *trigger mechanism* is thought to be a change in the SUBGLACIAL *drainage system* leading to increased basal HYDROSTATIC

PRESSURE. For *polythermal glaciers* (see COLD-BASED GLACIERS) resting on a soft, *deformable bed*, the surge may be induced by a change in the thermal state of subglacial sediment from frozen to thawed. Surges tend to occur cyclically over varying timescales. The most frequent occur at 10- to 20-year intervals and last a few months (Alaska), whilst the slowest surges (e.g. in Svalbard) have rarely had more than one documented event, but last for several years.   *MJH*

[See also GLACIER THERMAL REGIME*]*

Cuffey KM and Paterson WSB (2010) *The physics of glaciers, 4th edition.* Burlington, MA: Butterworth-Heinemann.

Frappé-Sénéclauze T-P and Clarke GKC (2007) Slow surge of Trapridge Glacier, Yukon Territory, Canada. *Journal of Geophysical Research* 112: F03S32.

Kamb B, Raymond CF, Harmon WD et al. (1985) Glacier surge mechanism: 1982-1983 surge of Variegated Glacier, Alaska. *Science* 227: 469–479.

Meier MF and Post A (1969) What are glacier surges? *Canadian Journal of Earth Sciences* 6: 807–817.

Murray T, Stuart GW, Miller PJ et al. (2000) Glacier surge propagation by thermal evolution at the bed. *Journal of Geophysical Research* 105: 13491–13507.

Sharp M (1988) Surging glaciers: Behaviour and mechanisms. *Progress in Physical Geography* 12: 349–370.

glacier thermal regime   The state of a GLACIER as determined by its temperature distribution. TEMPERATURE is one of the most important parameters controlling glacier behaviour. Temperature affects glacier morphology, *ice flow*, *water flow*, debris ENTRAINMENT, SEDIMENTATION and the development of GLACIAL LANDFORMS. Based on thermal regime, there are three types of glacier: (1) *warm*, *wet-based* or TEMPERATE GLACIERS in which the ice is mainly at the PRESSURE MELTING POINT (PMP) throughout, except for a cold surface layer (about 10–15 m thick) which develops in winter; (2) *cold, polar* or COLD-BASED GLACIERS in which the bulk of the ice is below the PMP; and (3) *polythermal glaciers* (formerly termed *subpolar glaciers*), which are a transitional form, in which the ice in the upper and marginal parts of the glacier is below the PMP, whereas ice at depth is at the PMP.

The fundamental difference between these glacier types is that temperate glaciers slide on their beds and therefore erode them effectively, whereas cold-based glaciers are frozen to their beds and only erode them in exceptional circumstances. Polythermal glaciers are *wet-based glaciers* in parts and dry in others, and thus have a more complex interaction with the substrate. Since the thermal regime is a balance between temperature at the surface and the input of GEOTHERMAL heat at the base, it is important to understand the sediments and landforms produced by the different types of glacier in order to make sound judgements about past glacial climates. As more becomes known about the Antarctic and Greenland ice sheets, it is evident that there are extensive areas, notably ICE STREAMS, that are polythermal in character.   MJH

[*See also* COLD-BASED GLACIERS, GLACIAL EROSION, GLACIAL PROTECTION CONCEPT, GLACIER SURGE, ICE-SHEET GROWTH]

Benn DI and Evans DJA (2010) *Glaciers and glaciation, 2nd edition*. London: Hodder Education.

Bennett MR and Glasser NF (2009) *Glacial geology: Ice sheets and landforms, 2nd edition*. Chichester: Wiley.

Cuffey KM and Paterson WSB (2010) *The physics of glaciers, 4th edition*. Burlington, MA: Butterworth-Heinemann.

Evans DJA (ed.) (2003) *Glacial landsystems*. London: Arnold.

glacier variations   Changes in the size of glaciers, commonly their *frontal variations*. Glacier variations provide information about CLIMATIC CHANGE and rates of change with respect to short- and long-term *energy fluxes* at the GLACIER surface. Historical and longer-term glacier variations reconstructed from *direct measurements*, *paintings*, DOCUMENTARY EVIDENCE, MORAINES, GLACIOFLUVIAL SEDIMENTS and GLACIOLACUSTRINE DEPOSITS, indicate that the glaciers in many mountain regions have fluctuated considerably in extent (see Figure). During the HOLOCENE, for example, the range of variability is defined by the HOLOCENE THERMAL OPTIMUM and the maximum extent of glaciers during the LITTLE ICE AGE.

Glacier margins advance or retreat, with variable time lags, in response to variations in glacier MASS BALANCE. ABLATION removes ice from the glacier and the horizontal velocity component carries ice forward. A glacier margin remains in the same position when the horizontal velocity component is equal to the horizontal component of ablation. Although the frontal position is stationary, the ice is in motion, but is removed from the glacier at a rate equal to the velocity. Frontal *glacier retreat* takes place when the horizontal velocity component is less than the horizontal ablation component, whereas *glacier advance* occurs when the horizontal velocity component is larger than the horizontal ablation component. During the winter season, glacier-sliding velocities of temperate glaciers tend to be low due to little meltwater at the glacier base. Commonly, *winter advances* start late in the ablation season when melting at the margin does not exceed the forward flow of glacier ice. Commonly, the horizontal ablation component is low in late winter, causing the small winter flow velocities to produce small glacier advances. Despite higher summer flow velocities than in winter, high summer ablation rates cause net retreat of the glacier (see ANNUAL MORAINES).

Advance and retreat of the glacier front normally lags behind the CLIMATIC FLUCTUATIONS because the signal must be transferred from the *accumulation area* to the glacier snout. This is referred to as the LAG TIME or RESPONSE TIME, which is longest for long, low-gradient and slowly moving glaciers, and shortest for short, steep and fast-flowing glaciers. *Kinematic wave theory* has been applied to calculating response times. However, physically based three-dimensional *flow models* may help determine the response times more precisely.   JAM

[*See also* CENTURY- TO MILLENNIAL-SCALE VARIABILITY, GLACIERS AND ENVIRONMENTAL CHANGE]

Cuffy KM and Paterson WSB (2010) *The physics of glaciers, 4th edition*. Kiddlington: Butterworth-Heinemann.

Frenzel B, Boulton GS, Gläser B and Huckreide U (eds) (1997) *Glacier fluctuations during the Holocene*. Stuttgart: Gustav Fischer Verlag.

Johannesson T, Raymond C and Waddington E (1989) Timescale for adjustment of glaciers to changes in mass balance. *Journal of Glaciology* 35: 355–369.

Kaiser G (1999) A review of the modern fluctuations of tropical glaciers. *Global and Planetary Change* 22: 93–103.

Nesje A (2009) Latest Pleistocene and Holocene alpine glacier fluctuations in Scandinavia. *Quaternary Science Reviews* 28: 2119–2136.

Nye JF (1960) The response of glaciers and ice sheets to seasonal and climatic changes. *Proceedings of the Royal Society A* 256: 559–584.

Roe GH (2011) What do glaciers tell us about climate variability and climate change? *Journal of Glaciology* 57: 567–578.

Rohrhofer F (1954) Untersuchungen am Ötztaler Gletschern über den Ruckgang 1850-1950 [Investigations on the Oetztal glaciers jerk response 1850-1950]. *Geographischer Jahresbericht aus Österreich* 25 (1953/1954): 57–84.

Winkler S and Matthews JA (2010) Holocene glacier chronologies: Are high-resolution global and inter-hemispheric comparisons possible? *The Holocene* 20: 1137–1147.

glacierisation   The progressive covering of a landscape by glacier ice. Glacierisation and its antonym, DEGLACIERISATION, should be distinguished from GLACIATION and DEGLACIATION, which refer to these processes in the context of GLACIAL-INTERGLACIAL CYCLES.   JAM

glaciers and environmental change   GLACIERS and ICE SHEETS will grow if the *accumulation* of snow and ice exceeds ABLATION by melting and/or CALVING. Conversely, shrinkage and retreat occur when ablation exceeds accumulation. Thus, glaciers and ice sheets will respond, sometimes dramatically, to any climatic or other environmental changes that alter the local MASS BALANCE. This, in turn, may precipitate further environmental changes, including CLIMATIC CHANGE due to the action of FEEDBACK MECHANISMS. Consequently, it is important to understand glacier-environmental relationships to be able to predict the role of glaciers and

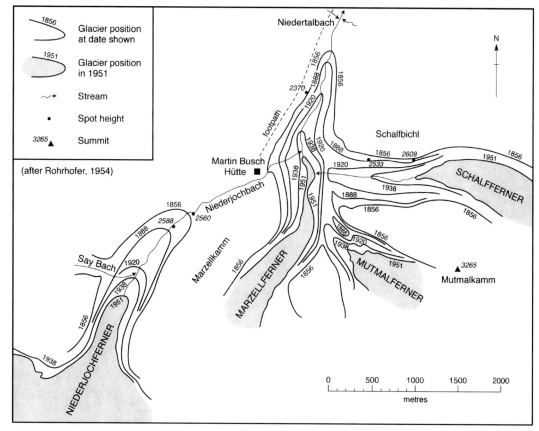

Glacier variations *The pattern of glacier retreat AD 1856–1951 in the vicinity of Martin Busch Hütte, Niedertal, Ötztal Alps, Austria: further retreat has occurred in the past 50 years (Rohrhofer, 1954).*

ice sheets in current and future environmental change. Furthermore, the past fluctuations of glaciers and ice sheets can be used to reconstruct past climates, providing an important source of evidence for long-term environmental change.

The rate at which a glacier or ice sheet advances or retreats following climatic change depends on its RESPONSE TIME, which generally increases with the dimensions of the ice mass. Although in many cases glacier response will be predictable, some glaciers may exhibit NON-LINEAR BEHAVIOUR, with delayed or disproportionate responses to climatic inputs. Non-linear behaviour is particularly characteristic of GLACIER SURGES, CALVING GLACIERS and DEBRIS-MANTLED GLACIERS. Complex responses are also thought to be characteristic of ICE SHEETS due to feedbacks involving global *sea level* and changes in surface ALBEDO.

Glacier advance can result in the destruction of farmland and settlements, as happened in Norway, Iceland and the European Alps during the LITTLE ICE AGE. *Glacier retreat* is currently occurring throughout the world (see Figure), with a few local exceptions due to increased snowfall. Glacier retreat may trigger wide-ranging environmental changes. First, the frequency of slope failures tends to increase due to the destabilisation of rock slopes by the removal of ice support and the exposure of unconsolidated sediments on steep hillsides. An interval of accelerated sediment yield, known as a PARAGLACIAL phase, thus tends to be associated with glacier retreat. Second, river DISCHARGE will increase as snow and ice are removed from storage. With the disappearance of glacier ice from a catchment, river discharges may dramatically decrease, especially in summer, with potentially damaging consequences for AGRICULTURE. Third, unstable moraine-dammed or ICE-DAMMED LAKES may form, which are prone to glacier lake outburst floods (JÖKULHLAUPS). Dangerous lakes associated with retreating glaciers pose a particularly severe environmental hazard on debris-mantled glaciers in high mountain regions. Fourth, the return of MELTWATER to the oceans will contribute to SEA-LEVEL RISE. It is estimated that the retreat of valley glaciers and ice caps may cause a rise in global sea level of up to 12 cm during the present century.

Changes in the magnitude of ice sheets have a potentially large environmental impact. ICE-SHEET GROWTH has, on several occasions during the QUATERNARY, engulfed large areas of the mid-latitudes, removing huge volumes of water from the oceans and lowering global sea level by up to about 120 m (see EUSTASY). The formation of successive LAURENTIAN and FENNOSCANDIAN ICE SHEETS was particularly important in this respect. Large ice sheets also affect regional sea levels through the effects of ISOSTASY, causing the underlying LITHOSPHERE to sag into the MANTLE due to the additional imposed weight of the ice. Subsequent retreat of the ice results in GLACIO-ISOSTATIC REBOUND which, in much of the area within the ice limits, is large enough to counteract eustatic sea-level rise, resulting in local sea-level fall. Some areas of northern Canada and Scandinavia are still experiencing such a fall in RELATIVE SEA LEVEL as a result of residual isostatic uplift. The retreat and disappearance of the mid-latitude ice sheets was associated with a complex series of environmental changes with both regional and global consequences.

In recent years, much attention has focused on the possible response of the Antarctic and Greenland ice sheets to recent global warming, particularly the possibility of significant sea-level rise. The Greenland Ice Sheet is undergoing net mass loss by a combination of ice melt and rapid calving, and is currently contributing ~0.5 mm/year to global sea-level rise. Parts of the West Antarctic Ice Sheet are losing mass by calving and bottom melting of ice shelves, whereas the East Antarctic Ice Sheet is gaining mass due to increased snowfall. On balance, however, Antarctica is contributing ~0.2 mm/year to global sea level. Dynamic changes may cause these figures to increase in the near future, although such changes cannot be predicted accurately at present.

Evidence for former glacier limits, such as MORAINES, GLACIOMARINE DEPOSITS and GLACIOLACUSTRINE DEPOSITS, can give information on long-term climatic fluctuations, provided that the glacier limits can be dated and glacier-climate relationships are known. The climatic significance of former glaciers is commonly estimated in EQUILIBRIUM-LINE ALTITUDE (ELA) RECONSTRUCTION, or the elevation at which glacier ablation and accumulation are in balance. Such reconstructions are most accurate in mountain regions, where the three-dimensional form of glaciers can be determined with greatest ACCURACY, and where some form of independent data are available from PALAEOCLIMATOLOGY (e.g. from faunal or pollen evidence).   *DIB*

[*See also* GLACIER VARIATIONS, GLACIOSEISMOTECTONICS]

Beniston M (2012) Environmental change in mountain regions. In Matthews JA, Bartlein PJ, Briffa KR et al. (eds) *The SAGE handbook of environmental change, volume 2.* Sage: London, 262–281.

Benn DI and Evans DJA (2010) *Glaciers and glaciation, 2nd edition.* London: Hodder Education.

Bindoff NL, Willebrand J, Artale V et al. (2007) Observations: Ocean climate change and sea level. In Solomon S, Qin D, Manning M et al. (eds) *Climate change 2007: The physical science basis* [Contribution of Working Group I to the Fourth Assessment of the Intergovernmental Panel on Climate Change]. Cambridge: Cambridge University Press, 385–432.

Cazenave A and Llovel W (2010) Contemporary sea level rise. *Annual Review of Marine Science* 2: 145–173.

Kaser G, Großhauser M and Marzeion B (2010) Contribution potential of glaciers to water availability in different climate regimes. *Proceedings of the National Academy of Sciences* 107: 20223–20227.

Nesje A and Dahl SO (2000) *Glaciers and environmental change.* London: Arnold.

Radić V and Hock R (2011) Regionally differentiated contribution of mountain glaciers and ice caps to future sea-level rise. *Nature Geoscience* 4: 91–94.

Vaughan DG (2008) West Antarctic Ice Sheet collapse: The fall and rise of a paradigm. *Climate Change* 91: 65–79.

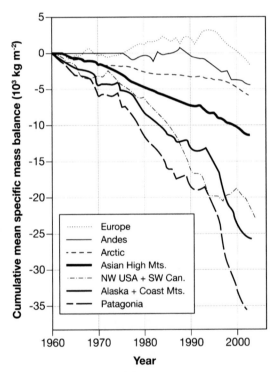

Glaciers and environmental change *Changes in the average mass balance of mountain glaciers in different regions (Beniston, 2012).*

glacio-aeolian processes   EROSION and DEPOSITION by *wind action* on unconsolidated SEDIMENTS during GLACIATION and DEGLACIATION, and on GLACIER FORELANDS. Open LANDSCAPES, sediment availability and *glacier winds* are important factors favouring glacio-aeolian processes.   *JAM/RAS*

[*See also* COVERSAND, LOESS, PARAGLACIAL, SAND DUNES, WIND EROSION]

Derbyshire E and Owen LA (1996) Glacioaeolian processes, sediments and landforms. In Menzies J (ed.) *Past glacial environments: Sediments, forms and techniques*. Oxford: Butterwort-Heinemann, 213–237.

glaciochemistry   The study of the chemistry of ice in GLACIERS, MELTWATER and, especially, ICE CORES. The latter focuses on the investigation of past atmospheric conditions from the trapped air bubbles, which contain information on GREENHOUSE GASES (e.g. CARBON DIOXIDE and METHANE), and from both the soluble AEROSOLS and insoluble PARTICULATES trapped in PRECIPITATION. Glaciochemistry is also important in understanding subglacial CHEMICAL WEATHERING and GLACIAL EROSION.   *JAM*

Kreutz KJ (2007) Glaciochemistry. In Elias SA (ed.) *Encyclopedia of Quaternary science*. Amsterdam: Elsevier, 1193–1199.

glacio-eustasy   The addition and removal of water to the total ocean volume as a result of climate-induced changes in the dimensions of GLACIERS and ICE SHEETS. A reasonably accurate indirect record of glacio-eustatic ocean volume changes has been derived from records of OXYGEN ISOTOPES in MARINE SEDIMENT CORES using FORAMINIFERA: ANALYSIS, since the measured oxygen isotope changes mostly record long-term changes in the isotopic composition of seawater. These changes, in turn, are dependent on temporal changes in the volume of ice locked up in the world's ice sheets and glaciers. The difference in *sea level* between GLACIALS and INTERGLACIALS caused by glacio-eustasy was up to ~120 m. Currently, worldwide *glacier retreat* makes the major contribution to SEA-LEVEL RISE.   *AGD*

[*See also* EUSTASY, GLACIER VARIATIONS, GLACIO-ISOSTASY, OXYGEN ISOTOPES, SEA-LEVEL CHANGE]

Bard E, Hamelin B and Fairbanks RG (1990) U-Th ages obtained by mass spectrometry in corals from Barbados: Sea level during the past 130,000 years. *Nature* 346: 456–458.
Meier M, Dyurgerov MB, Rick UK et al. (2007) Glaciers dominate eustatic sea-level rise in the 21st century. *Science* 317: 1064-1067.
Mörner N-A (1995) Sea level and climate: The decadal to century signals. *Journal of Coastal Research* 11: 261–268.

glaciofluvial   Pertaining to water flowing on, in, under, against or away from GLACIERS or ICE SHEETS. The adjective is more widely used than the alternatives (*glacifluvial, fluvioglacial* or *fluviglacial*) for the HYDROLOGY, LANDFORMS, PROCESSES and ENVIRONMENTS associated with streams and rivers fed by melting glacier ice. This produces large volumes of MELTWATER, a distinctive RIVER REGIME with high SEASONALITY, and large SUSPENDED LOADS and BEDLOADS. These characteristics affect the otherwise FLUVIAL PROCESSES of EROSION and DEPOSITION for long distances downstream from glaciers. The glaciofluvial processes operating in SUPRAGLACIAL, ENGLACIAL, SUBGLACIAL or ice-marginal positions are further modified by having to operate between channel walls, and sometimes roofs or floors, composed of ICE.   *JAM*

[*See also* PERIGLACIOFLUVIAL SYSTEM]

Brennand TA (2004) Glacifluvial (glaciofluvial). In Goudie AS (ed.) *Encyclopedia of geomorphology*. London: Routledge, 459–465.
Gurnell AM and Clark MJ (eds) (1987) *Glacio-fluvial sediment transfer: An alpine perspective*. Chichester: Wiley.
Lundqvist J (1985) What should be called glaciofluvium? *Striae* 22: 5–8.

glaciofluvial landforms   LANDFORMS that result from the action of glacial MELTWATER in SUPRAGLACIAL, SUBGLACIAL, ice-marginal and proglacial positions. Landforms produced by glaciofluvial EROSION include MELTWATER CHANNELS (*ice marginal, submarginal*, SUBGLACIAL, *lake-overflow or overspill and proglacial*), TUNNEL VALLEYS and P-FORMS (various sculpted bedrock features). Depositional forms include valley trains and outwash plains (see OUTWASH DEPOSIT and SANDUR), KAMES, KAME TERRACES and ESKERS.   *MRB*

[*See also* GLACIAL LANDFORMS, GLACIOFLUVIAL SEDIMENTS, ICE-MARGIN INDICATORS]

Benn DI and Evans DJA (2010) *Glaciers and glaciation, 2nd edition*. London: Hodder Education.
Bennett MR and Glasser NF (2009) *Glacial geology: Ice sheets and landforms, 2nd edition*. Chichester: Wiley-Blackwell.
Price RJ (1973) *Glacial and fluvioglacial landforms*. Edinburgh: Oliver and Boyd.

glaciofluvial sediments   MELTWATER discharging from a GLACIER carries a large volume of sediment as SUSPENDED LOAD, including sand, silt and clay (see ROCK FLOUR), as well as rock fragments up to BOULDER size, that roll or bounce (see SALTATION) on the bed as BEDLOAD. This sediment is deposited as OUTWASH DEPOSITS in *proglacial* areas, where it forms an extensive *braidplain*. Glacial meltwater streams generally rework and sort *glacigenic sediment* as it is transported downstream, with the finer fraction being carried farthest. The seasonal and diurnal variations in discharge from glaciers control the nature of sedimentation on, and the morphology of, braidplains. Modifications of the *fluvial system* occur as a result of EROSION of previously deposited glacial materials, and of the transport and burial of DEAD ICE. Although glaciofluvial sediments are best preserved in the proglacial area, they are also deposited in contact with the glacier, including supraglacially, englacially, subglacially and ice marginally.

Braidplains resulting from glaciofluvial sedimentation are known as *outwash plains*, *sandar* (singular: SANDUR) or, where constrained by steep mountain-sides, *valley trains*. Sandar are typical of the southern coastal fringe of Iceland (whence the name is derived), and valley trains are common in Alaska and New Zealand. The channel system is constantly changing in braidplains. Characteristic forms are BARS, which are ridges of sediment that form between channels (see CHANNEL BAR). There are *longitudinal, point* and *linguoid bars*. CROSS-BEDDING may be developed in bars and channel-fill sequences, while smaller-scale SEDIMENTARY STRUCTURES such as RIPPLES, CROSS-LAMINATION, *desiccation cracks* (see DESICCATION CRACKING), *mud films* and SCOUR MARKS may also be found. Organic matter may be trapped in the sediment, which presents an opportunity to apply RADIOCARBON DATING. Glaciofluvial sediments can also be dated directly using OPTICALLY STIMULATED LUMINESCENCE (OSL) DATING.

The principal sedimentary types or FACIES are well-sorted sands and gravels, often arranged in *fining-upward cycles*. In *backwater* areas, finer sediments may also be deposited. In many outwash plains, ice blocks carried by the river from collapsed ice tunnels or remnant dead ice, may become buried. Later, as the blocks slowly melt, depressions called KETTLE HOLES form in the braidplain, which, if numerous, give rise to *pitted outwash plains*.

Beyond some GLACIERS, major catastrophic flood events can heavily modify the braidplain. Such floods may be generated by glacial lake outburst floods (GLOFs) or JÖKULHLAUPS from the drainage of ICE-DAMMED LAKES or *moraine-dammed lakes*.   *MJH*

[*See also* GLACIAL SEDIMENTS, GLACIOFLUVIAL LANDFORMS, ICE-CONTACT FAN/RAMP]

Benn DI and Evans DJA (2010) *Glaciers and glaciation, 2nd edition.* London: Hodder Education.
Bennett MR and Glasser NF (2009) *Glacial geology: Ice sheets and landforms, 2nd edition.* Chichester: Wiley.
Gurnell AM and Clark MJ (eds) (1987) *Glacio-fluvial sediment transfer: An alpine perspective.* Chichester: Wiley.
Luethgens C, Boese M and Preusser F (2011) Age of the Pomeranian ice-marginal position in northeastern Germany determined by optically stimulated luminescaence (OSL) dating of glaciofluvial sediments. *Boreas* 40: 598–615.
Russell AJ, Fay H, Marren PM et al. (2005) Icelandic jökulhlaup impacts. *Developments in Quaternary Sciences* 5: 153–203.
Thrasher IM, Mauz B, Chiverrell RC et al. (2009) Testing an approach to OSL dating of Late Devensian glaciofluvial sediments of the British Isles. *Journal of Quaternary Science* 24: 785–801.

glacio-isostasy Deformation of the LITHOSPHERE due to the LOADING and UNLOADING of ICE SHEETS. This was a prominent process at the regional to continental scale during the QUATERNARY. Glacio-isostatic changes in RELATIVE SEA LEVEL were particularly large in association with the LAURENTIAN and FENNOSCANDIAN ICE SHEETS, which were the thickest and heaviest.   *AGD*

[*See also* FOREBULGE, GLACIO-ISOSTATIC REBOUND, ISOSTASY, SEA-LEVEL CHANGE]

Andrews JT (ed.) (1974) *Glacio isostasy.* Stroudsburg, PA: Dowden.
Klemann V, Martinec Z and Ivins ER (2008) Glacial isostasy and plate motion. *Journal of Geodynamics* 46: 95–103.
Lambeck K (1995) Late Devensian and Holocene shorelines of the British Isles and North Sea from models of glacio-hydro-isostatic rebound. *Journal of the Geological Society of London* 152: 437–448.
Peltier WR (2004) Global glacial isostasy and the surface of the ice-age Earth: The ICE-5G (VM2) model and GRACE. *Annual Review of Earth and Planetary Science* 32: 111–149.
Sabadini R, Lambeck K and Boschi E (eds) (1991) *Glacial isostasy, sea level, and mantle rheology.* Dordrecht: Kluwer.

glacio-isostatic rebound Upward deformation of the LITHOSPHERE due to the unloading of ICE SHEETS as a result of DEGLACIATION. Investigations of patterns of change in RELATIVE SEA LEVEL in areas affected by glacio-isostatic rebound have been used to reconstruct former patterns of *isostatic deformation* and regional patterns of *ice thinning* and *retreat* (see Figure). In areas previously covered by ice sheets of the LAST GLACIATION, glacio-isostatic rebound is still taking place. For example, in Angermanland in northern Scandinavia, the present rate of glacio-isostatic rebound is about 9 mm/year.   *AGD*

[*See also* GLACIO-ISOSTASY, SEA-LEVEL CHANGE]

Goudie AS (1992) *Environmental change, 3rd edition.* Oxford: Blackwell.
Gray JM (1995) *Influence of Southern Uplands ice on glacio-isostatic rebound in Scotland: The Main Rock Platform in the Firth of Clyde.* Boreas 24: 30–36.
Ivins ER and James TS (2005) Antarctic glacial isostatic adjustment: A new assessment. *Antarctic Science* 17: 541–553.
Shennan I, Bradley S, Milne G et al. (2006) Relative sea-level changes, glacio-isostatic modeling and ice-sheet reconstructions from the British Isles since the Last Glacial Maximum. *Journal of Quaternary Science* 21: 585–599.

glaciolacustrine deposits Sediments deposited in lake water, with a significant input of glacially derived material. Glaciolacustrine deposits can be (1) *glacier-fed*, in which glacial debris is transported to the lake via subaerial meltstreams, or (2) *glacier-contact*, in which sediment is delivered directly to

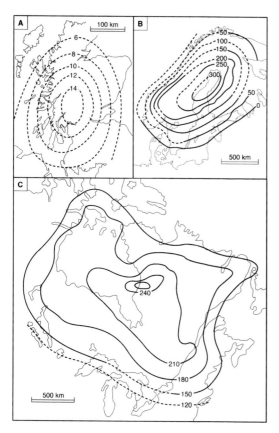

Glacio-isostatic rebound *Effects of glacio-isostatic rebound on different parts of the Earth's land surface: (A) generalised isobases in metres for the main postglacial raised shoreline in Scotland; (B) the amount of isostatic recovery in metres of Scandinavia during the Holocene; and (C) the maximum postglacial rebound in metres in northeastern North America (based on Goudie, 1992).*

the lake from glacier ice. Deposits laid down in *ice-proximal environments* (i.e. closest to the glacier) typically have complex geometries, and may take the form of DELTAS, MORAINES and subaqueous *outwash fans*. In *distal environments*, glaciolacustrine deposits tend to have more uniform geometry, forming extensive blankets of fine-grained sediments with thickness depending on basin topography and proximity to sediment influx points. They may contain a record of GLACIER VARIATIONS. Annual cycles of sedimentation in distal environments may be recorded in VARVES, rhythmically bedded sands, silts and clays that can be used to establish VARVE CHRONOLOGIES, allowing correlation of successions with and between large basins. Studies of MICROFOSSILS in distal glaciolacustrine deposits provide important data on climatic and other environmental changes.   *DIB*

Ashley GM (1995) Glaciolacustrine environments. In
   Menzies J (ed.) *Glacial environments, volume 1: Modern*

glacial environments: Processes, dynamics and sediments.
   Oxford: Butterworth-Heinemann, 417–444.
Bakke J, Dahl SO, Paasche Ø et al. (2010) A complete record
   of Holocene glacier variability at Austre Okstindbreen,
   northern Norway: An integrated approach. *Quaternary
   Science Reviews* 29: 1246–1262.
Benn DI and Evans DJA (2010). *Glaciers and glaciation,
   2nd edition*. London: Hodder Education.
Matthews JA, Dahl SO, Nesje A et al. (2000) Holocene
   glacier variations in central Jotunheimen, southern
   Norway based on distal glaciolacustrine sediment cores.
   Quaternary Science Reviews 19: 1624–1647.
Murton DK and Murton JB (2012) Middle and Late
   Pleistocene glacial lakes of lowland Britain and the
   southern North Sea Basin. *Quaternary International*, 260:
   115–142.

glaciological volcanic index (GVI)   A measure of volcanic activity derived from samples of acidity or sulfate concentrations contained in ICE CORES. Unlike the DUST VEIL INDEX (DVI) or the VOLCANIC EXPLOSIVITY INDEX (VEI), a GVI has the potential to provide a continuous record of past volcanic activity and to provide a direct indication of the concentrations of SULFUR DIOXIDE in the past ATMOSPHERE. The index is most frequently used in the field of PALAEOCLIMATOLOGY for assessing VOLCANIC FORCING.   *RC/JAM*

[**See also** ACIDITY RECORD OF VOLCANIC ERUPTION, VOLCANIC AEROSOLS, VOLCANIC ERUPTIONS]

Gao C, Robock A and Ammann C (2008) Volcanic forcing of
   climate over the past 1500 years: An improved ice core-
   based index for climate models. *Journal of Geophysical
   Research* 113: D23111.
Legrand M and Delmas RJ (1987) A 220-year continuous
   record of volcanic H_2SO_4 in the Antarctic Ice Sheet.
   Nature, 327: 671–676.
Robock A and Free MP (1995) Ice cores as an index of
   global volcanism from 1850 to the present. *Journal of
   Geophysical Research* 100: 11549–11567.

glaciology   The study of ICE in all its manifestations within the CRYOSPHERE. Thus, glaciology can be envisaged as equivalent to *cryospheric science*, including the study of ice in the ATMOSPHERE, SNOW AND SNOW COVER, PERMAFROST and seasonal types of ice in the ground, and SEA ICE and other types of ice associated with waterbodies. A narrower definition might focus on GLACIERS and ICE SHEETS (*glacier science*) excluding, for example, *permafrost science* from its remit. Major areas of investigation include the behaviour of ice, glacier flow, MASS BALANCE, processes of GLACIAL EROSION and DEPOSITION, and the reconstruction and modelling of past glaciers in relation to climate. One of the major developments of the past 50 years, which has been defined as a PARADIGM shift, has been the recognition of the existence of deformable beds at the ice/sediment interface at the base of glaciers, which has provided linkage between the fields of glaciology and QUATER-NARY SCIENCE. *JAM*

[*See also* GEOCRYOLOGY, PALAEOGLACIOLOGY]

Boulton GS (1986) A paradigm shift in glaciology. *Nature* 322: 18.

Knight PG (ed.) (2006) *Glacier science and environmental change*. Oxford: Blackwell.

Knight PG (2011) Glaciology. In Singh VP, Singh P and Haritashya UK (eds) *Encyclopedia of snow, ice and glaciers*. Dordrecht: Springer, 440–443.

Murray T (1997) Assessing the paradigm shift: Deformable glacier beds. *Quaternary Science Reviews* 16: 995–1016.

glaciomarine deposits Sediments deposited in the sea, containing a significant amount of glacially derived material. Glaciomarine or *glacimarine sediments* laid down in close proximity to glacier ice may be similar in overall geometry to GLACIOLACUSTRINE DEPOSITS, although sediment tends to be more widely distributed due to the buoyancy of glacial meltwater in seawater, and the action of TIDES. Distal glaciomarine deposits commonly take the form of extensive MUD DRAPES, which may contain a rich archive of MICROFOSSILS. ICEBERGS can carry glacial debris to considerable distances in high-latitude oceans. ICE-RAFTED DEBRIS may be preserved in distinctive layers in MARINE SEDIMENT CORES, allowing periods of heavy iceberg passage to be identified. Several extensive layers of ice-rafted debris below the North Atlantic record a succession of intense ice-rafting events (HEINRICH EVENTS) that occurred during the LAST GLACIATION. *DIB*

[*See also* GLACIOMARINE HYPOTHESIS, ICEBERG DRIFT THEORY]

Dowdeswell JA and Scourse JD (eds) (1990) *Glacimarine environments: Processes and products*. Bath: Geological Society.

Hambrey MJ (1994) *Glacial environments*. Boca Raton, FL: CRC Press.

Kehrl LM, Hawley RL, Powell RD and Brigham-Grette J (2011) Glaciomarine sedimentation processes at Kronebreen and Longsvegen, Svalbard. *Journal of Glaciology* 57: 841–847.

Polyak L, Murdmaa I and Ivanova E (2004) A high-resolution, 800-year glaciomarine record from Russkaya Gavan', a Novaya Zemlya fjord, eastern Barents Sea. *The Holocene* 14: 628–634.

Powell RD and Domack E (1995) Modern glaciomarine environments. In Menzies J (ed.) *Glacial environments, volume 1: Modern glacial environments: Processes, dynamics and sediments*. Oxford: Butterworth-Heinemann, 445–486.

glaciomarine hypothesis The disputed hypothesis that many sediments previously interpreted as glacial TILL are glaciomarine MUDS. The best-known examples are in the Irish Sea Basin, where DIAMICTONS containing marine shells and FORAMINIFERA occur on both coasts. Though first interpreted as evidence of the Biblical Flood, these have long been regarded as tills, with the mud and shells being reworked from deposits that accumulated in the Irish Sea before the LAST GLACIATION. They were reinterpreted as GLACIOMARINE DEPOSITS: sediments deposited both proximal and distal to the ice margin as it retreated northwards. During the Last Glaciation, world sea levels were eustatically depressed (see GLACIO-EUSTASY), so the glaciomarine hypothesis requires very substantial *isostatic depression* to raise local relative sea levels to far above the present height (see GLACIO-ISOSTASY). Similar sediments in East Anglia and elsewhere have been reinterpreted in the same way.

The main evidence used in support of the glaciomarine hypothesis is sedimentological. The marine shells and foraminifera show mixed assemblages, but the pre-Quaternary forms and those with warm affinities are regarded as the product of REDEPOSITION whilst the cold-water forms are considered to be in situ. Sand and gravel deposits around the Irish Sea basin have been interpreted as deltas marking RELATIVE SEA LEVEL during deglaciation. The varying altitudes, which do not increase from south to north, are explained using the concept of *piano-key tectonics*, whereby different fault-bounded parts of the crust were depressed by varying amounts and rebounded at varying speed.

All of the evidence on which the glaciomarine hypothesis is based has been disputed. At many sites, alternative sedimentological models have been used to interpret the DEPOSITIONAL ENVIRONMENTS of the sediments. It has been argued that all of the shells and foraminifera are derived and reworked from marine sediments that predate the Last Glaciation. The delta deposits may represent local ponding of water at the ice margin or where glaciers from different source areas uncoupled. There is no evidence for substantial movements on reactivated faults during the Quaternary. No consensus has been reached. *DMcC*

[*See also* DILUVIAL THEORY, QUATERNARY TIMESCALE]

Eyles N and McCabe AM (1989) The Late Devensian (<22,000 BP) Irish Sea Basin: The sedimentary record of a collapsed ice sheet margin. *Quaternary Science Reviews* 8: 307–351.

McCabe AM (1997) Geological constraints on geophysical models of relative sea-level change during deglaciation of the western Irish Sea Basin. *Journal of the Geological Society of London* 154: 601–604.

McCarroll D (2001) Deglaciation of the Irish Sea basin: A critique of the glacimarine hypothesis. *Journal of Quaternary Science* 16: 393–404.

Scourse JD and Furze MFA (2001) A critical review of the glaciomarine model for Irish Sea deglaciation: Evidence from southern Britain, the Celtic shelf and adjacent continental slope. *Journal of Quaternary Science* 16: 419–434.

glacioseismotectonics   The interplay between glacial dynamics, structural deformation and SEISMICITY. Also termed *deglaciation seismotectonics*, it operates at the scale of the Earth's CRUST as distinct from the largely surficial direct effects of glacier movement and loading at the base or margin of a glacier. Important aspects include (1) the role of GLACIO-ISOSTATIC REBOUND in modulating crustal DEFORMATION, FAULTS and EARTHQUAKE generation; (2) the possible triggering of HEINRICH EVENTS; (3) the broader interactions with climate, TECTONICS and TOPOGRAPHY and (4) the implications for the disposal of RADIOACTIVE WASTE in relatively 'stable' continental regions.   *JAM*

[*See also* FOREBULGE, GLACIO-ISOSTASY, GLACIOTECTONICS, NEOTECTONICS]

Hunt AG and Malin PE (1998) Possible triggering of Heinrich events by ice-load-induced earthquakes. *Nature* 393: 155–158.
Stewart IS, Sauber J and Rose J (2000) Glacioseismotectonics: Ice sheets, crustal deformation and seismicity. *Quaternary Science Reviews* 19, 1367–1389.

glaciotectonics   The dislocation of sediment or rock masses under glacially applied stresses. This may occur beneath a glacier (*subglacial tectonics*) or at the ice margin (*proglacial tectonics*), and can involve extensional or compressional DEFORMATION. The term is normally restricted to the surficial deformation as a direct result of glacial movement or loading in the SUBGLACIAL and proximal *proglacial* domains rather than the deeper and more far-reaching crustal effects (see GLACIOSEISMOTECTONICS). Subglacial tectonics may result in the detachment and transport of intact *megablocks* or rafts, the excavation of *ice-scooped basins*, the deformation of bedrock or SOFT-SEDIMENT DEFORMATION. Proglacial tectonic deformation typically occurs in association with high subsurface *pore water pressures*, steep *stress gradients*, and rapid LOADING associated with glacier advances or SURGING GLACIERS. Submarginal excavation combined with proglacial upthrusting produces *thrust* or *push moraines*, which may be several tens of metres in height.   *DIB*

[*See also* GLACIER VARIATIONS, MORAINES]

Aber JS, Croot DG and Fenton MM (1989) *Glaciotectonic landforms and structures*. Dordrecht: Kluwer.
Andersen LY, Hansen DL and Huuse M (2005) Numerical modelling of thrust structures in unconsolidated sediments: Implications for glaciotectonic deformation. *Journal of Structural Geology* 27: 587–596.
Benn DI and Evans DJA (2010) *Glaciers and glaciation, 2nd edition*. London: Hodder Education.
Hart JK and Boulton GS (1991) The inter-relation of glaciotectonic and glaciodepositional processes within the glacial environment. *Quaternary Science Reviews* 10: 335–350.
Thorson RM (2000) Glacial tectonics: A deeper perspective. *Quaternary Science Reviews* 19, 1391–1398.

Van der Wateren DFM (1995) Processes of glaciotectonism. In Menzies J (ed.) *Glacial environments, volume 1: Modern glacial environments: Processes, dynamics and sediments*. Oxford: Butterworth-Heinemann, 309–335.

glacis   A French term for a low-angle slope segment of transport and EROSION located at the foot of a *hillslope*. The term *glacis d'erosion* is used especially to describe PIEDMONT slopes in ARIDLANDS and SEMI-ARID regions. It may be composed of bedrock, COLLUVIUM and/or ALLUVIUM.   *MAC*

[*See also* PEDIMENT]

White K (2004) Glacis d'erosion. In Goudie AS (ed.) *Encyclopedia of geomorphology*. London: Routledge, 469–470.

gleying   A process of REDUCTION of IRON (Fe), and its *segregation*, which occurs in poorly drained soils as a result of ANAEROBIC conditions. The result is that the soil has a grey appearance when saturation is continuous and a *mottled* character when saturation alternates with AEROBIC conditions. Iron segregation into CONCRETIONS may also occur. These features of gleying persist after drainage has removed the cause of their development. Features of reduction are described as *stagnic* if they occur within 50 cm of the surface and *gleyic* if they relate to saturation associated with a high *groundwater table*.   *EMB*

[*See also* GLEYSOLS, GROUNDWATER, HYDRIC SOILS]

Buol SW, Hole FD and McCracken RJ (1988) *Soil genesis and classification*. Ames: Iowa State University Press.
Crompton E (1952) Some morphological features associated with poor soil drainage. *Journal of Soil Science* 3: 277–289.
International Union of Social Sciences (IUSS) Working Group World Reference Base (WRB) (2006) *World reference base for soil resources, 2nd edition*. Rome: Food and Agriculture Organization of the United Nations.

Gleysols   Soils present in all continents with temporary or permanent waterlogging within the soil profile. As a result of the GLEYING process, iron compounds may be segregated into mottles or concretions of ferrihydrite, lepidocrocite or goethite, most likely where there is temporary wetness and hence OXIDATION for some of the time (*oxymorphic* conditions), or in extreme cases leached from the profile leaving a (reduced) grey soil matrix. (SOIL TAXONOMY: Aquic great groups of *Entisols*, *Afisols*, *Inceptisols* and *Mollisols*). Movement of ferrous iron and oxidation to the ferric state may lead to the segregation of irregular masses of so-called *bog iron ore*. Climate change with increased wetness could lead to extension of gleying as could higher sea levels in coastal areas. Gleysols are little used unless drained.   *EMB*

[*See also* REDUCTION, WETLAND, WORLD REFERENCE BASE FOR SOIL RESOURCES (WRB)]

Bedard-Haughn A (2011) Gleysolic soils of Canada: Genesis, distribution and classification. *Canadian Journal of Soil Science* 91: 763–779.
Duntze D, Watermann F and Giani L (2005) Reconstruction of the palaeoenvironment and geopedogenesis of non-calcareous marshland soils (Eutric Gleysols) of Lower Saxony. *Journal of Plant Nutrition and Soil Science* 168: 53–59.
Fominykh LA and Zolotareva BN (2004) Ecological peculiarities of gleysols in the Russian Arctic. *Eurasian Soil Science* 37: 122–130.

global capitalism The notion that the economic system of the world is predominantly a unitary, global one in which national controls play a decreasing role. Only half of the 100 largest economic entities in the world are nation states: the rest are *international corporations*, which are increasingly influential in the political, social, cultural, scientific and environmental fields. Notable adverse environmental impacts include those of emissions from the FOSSIL FUEL industry on AIR POLLUTION and CLIMATIC CHANGE, and the inequitable use of tropical genetic resources in agriculture. International corporations are able to play off nation states against one another in seeking lowest costs, thus putting pressure on countries to reduce the attention they give to ENVIRONMENTAL PROTECTION. There are, however, growing constraints on the inherent lack of social and environmental responsibility of global capitalism, which are exemplified by the parallel growth of NON-GOVERNMENTAL ORGANISATIONS (NGOs). *JAM*

[*See also* CAPITALISM, CIVIL SOCIETY, GLOBALISATION, NATURAL CAPITALISM]

Hutton W and Giddens A (eds) (2000) *On the edge: Living with global capitalism*. London: Jonathan Cape.
Norberg J (2003) *In defense of global capitalism*. Washington, DC: Cato Institute.

global change 'Transformation processes that operate at a truly planetary scale plus processes that operate at smaller spatial scales (local, regional and continental) but that are so ubiquitous and pervasive as to assume global proportions' (Grübler, 1998: 3). Global change research began as a field of the NATURAL ENVIRONMENTAL SCIENCES but it is now recognised that the 'global environment' includes not only (1) NATURAL processes of the environment, their impacts on humans and the ANTHROPOGENIC causes of change in these processes but also (2) a wide range of human processes, such as those directly and indirectly affecting TECHNOLOGICAL CHANGE, and the many economic, social and political factors controlling production and CONSUMPTION. Global change is synonymous with GLOBAL ENVIRONMENTAL CHANGE, although the latter is sometimes still confined to global aspects of NATURAL ENVIRONMENTAL CHANGE (a position that is no longer tenable). *JAM*

[*See also* EARTH SYSTEM, GLOBALISATION, INTERNATIONALISATION OF RESEARCH, WORLD SYSTEM]

Barange M, Field JG, Harris RP et al. (eds) (2010) *Marine ecosystems and global change*. Oxford: Oxford University Press.
Bresmann F and Sietenhüber B (eds) (2009) *Managers of global change: The influence of international bereaucracies*. Cambridge: MIT Press.
Cuff DJ and Goudie AS (eds) (2009) *The Oxford companion to global change*. Oxford: Oxford University Press.
Garrido A and Dinar A (eds) (2010) *Managing water resources in a time of global change*. London: Routledge.
Goudie AS and Cuff DJ (eds) (2002) *Encyclopedia of global change: Environmental change and human society*, 2 volumes. New York: Oxford University Press.
Grübler A (1998) *Technology and global change*. Cambridge: Cambridge University Press.
National Research Council (1999) *Global environmental change: Research pathways for the next decade*. Washington, DC: National Academies Press.
National Research Council (2007) *Analysis of global change assessments: Lessons learned*. Washington, DC: National Academies Press.
Organisation for Economic Co-operation and Development (OECD) (1994) *Global change of planet Earth*. Paris: OECD.
Proctor JD (1998) The meaning of global environmental change: Retheorising culture in human dimensions research. *Global Environmental Change* 8: 227–248.
Taylor PJ, Watts MJ and Johnston RJ (eds) (2002) *Geographies of global change: Remapping the world, 2nd edition*. Oxford: Blackwell.

global cooling Any interval in Earth history when GLOBAL MEAN SURFACE AIR TEMPERATURE, which combines the surface air temperature over land with SEA-SURFACE TEMPERATURE (SST) over the OCEANS, exhibits a declining TREND. The last episode of global cooling in terms of annual mean temperatures was ca AD 1940–1975. It interrupted the GLOBAL WARMING trend of the twentieth century (see the Figure associated with the entry on GLOBAL WARMING, which shows this cooling to have been about −0.1°C relative to the 1961–1990 average). The LITTLE ICE AGE of ca AD 1570–1900 was the latest of several *neoglacial events* (see CENTURY- TO MILLENNIAL-SCALE VARIABILITY), some of which may have been global and involved annual mean temperatures of 1–2°C colder than average temperatures during the twentieth century. Major GLACIATIONS (see also GLACIAL EPISODE) were certainly caused by global coolings of 10°C or more. Longer-term coolings are well known from PALAEOCLIMATOLOGY. Shorter-term, interannual coolings also occur, as followed the 1991 VOLCANIC ERUPTION of Mount Pinatubo. The most severe periods of global cooling in the GEOLOGICAL RECORD were those associated with SNOWBALL EARTH.   *JAM*

[*See also* ANTARCTIC COOLING AND WARMING]

Jones PD, Briffa KR, Osborn TJ et al. (2009) High-resolution palaeoclimatology of the last millennium: A review of current status and future prospects. *The Holocene* 19: 3–49.

Matthews JA and Briffa KR (2005) 'The Little Ice Age': Re-evaluation of an evolving concept. *Geografiska Annaler* 87A: 17–36.

Soden BJ, Weatherald RJ, Stenchikov GL and Robock A (2002) Global cooling after the eruption of Mount Pinatubo: A test of climate feedback by water vapour. *Science* 296: 727–730.

global dimming

The reduced levels of solar IRRADIANCE received in recent years at the Earth's surface. This results from PARTICULATES or AEROSOLS in the atmosphere. Particular importance might be attached to sulfate AEROSOLS that are one of the by-products of industrial processes, although similar aerosols are derived from oceanic plankton in the form of DIMETHYL SULFIDE (DMS). *Black carbon*, from the burning of wood in particular and *contrails* from high-flying aircraft, and DUST from explosive VOLCANIC ERUPTIONS, add to the ATMOSPHERIC LOADING of aerosols. These aerosols both absorb *radiation*—thereby warming the atmosphere at that point—and reflect it. They may also act as CONDENSATION NUCLEI, the greater concentrations of which will increase the density of cloud droplets making those clouds more reflective and adjusting yet further the global ENERGY BUDGET. Estimates suggest that since the middle of the twentieth century, the radiation received at the surface has fallen by 4 per cent. This has had a cooling effect but the degree to which it has offset the GLOBAL WARMING of the same period is difficult to quantify reliably. Nevertheless, this uncertainty has not prevented suggestions that artificial STRATOSPHERIC AEROSOL INJECTION (SAI) could provide an approach to CLIMATIC ENGINEERING. *DAW*

[*See also* SULFUR CYCLE, VOLCANIC IMPACTS ON CLIMATE]

Alpert P, Kishcha R, Kaufman YJ and Schwarzbard R (2005). Global dimming or local dimming? Effect of urbanization on sunlight availability. *Geophysical Research Letters* 32: 17802.

Kerr RA (2007) Climate change: Is a thinning haze unveiling the real global warming? *Science* 315: 1480.

Stjern CW, Kristjansson JE and Hansen AW (2009) Global dimming and global brightening: An analysis of surface radiation and cloud cover data in northern Europe. *International Journal of Climatology* 29: 643–653.

Wild M (2009) Global dimming and brightening: A review. *Journal of Geophysical Research* 114: D00D16.

Global Earth Observation System of Systems (GEOSS)

Designed to provide decision-support tools to a wide variety of users through a flexible global network of EARTH OBSERVATION (EO) content providers. It links together existing and planned SATELLITE REMOTE SENSING observing systems and programmes around the world and supports the development of new systems where gaps exist. It promotes common technical standards so that data from diverse set of sensors can be combined into coherent data sets. The *GEO portal* offers access to data, imagery and software tools of global relevance.

The *Group on Earth Observation (GEO)* is co-ordinating the building of the GEOSS. It was launched after the 2002 *World Summit on Sustainable Development* by the G8 leading industrialised countries as an international collaboration to exploit the growing potential of EO data to support decision-making in an increasingly complex and environmentally stressed world. It is a voluntary partnership of governments and NON-GOVERNMENTAL ORGANISATIONS (NGOs), and provides a framework within which new projects, co-ordinated strategies and investments can be developed. GEO is constructing GEOSS on the basis of a 10-year Implementation Plan (2005–2015), which outlines the vision, purpose, scope and expected benefits of system. GEOSS will address the nine SOCIETAL BENEFIT AREAS (SBAs) of disasters, health, energy, climate, water, weather, ecosystems, agriculture and biodiversity.   *GMS*

[*See also* ENVIRONMENTAL MONITORING]

Group on Earth Observation (2009) *GEO 2009-2011 work plan. Revision 2.* [Available at http://www.earthobservations.org/documents/work%20plan/geo_wp0911_rev2_091210.pdf]

global environmental change

Directional *environmental changes* that are experienced in most regions of the world at approximately the same time and so might be considered 'global' in extent or scale, such as GLACIAL–INTERGLACIAL CYCLES in climate. Possible examples since the LAST GLACIATION include the YOUNGER DRYAS STADIAL and the LITTLE ICE AGE, for which there is evidence of global TELECONNECTIONS. The term is *not* usually applied to regional, continental or hemispherical phenomena, such as EL NIÑO-SOUTHERN OSCILLATION (ENSO) events or to the NORTH ATLANTIC OSCILLATION (NAO), nor to short-lived climatic fluctuations associated with single explosive volcanic eruptions. Causes, *trigger factors* or FORCING FACTORS of global NATURAL ENVIRONMENTAL CHANGE include variations in receipt of SOLAR RADIATION (see SOLAR FORCING and MILANKOVITCH THEORY), changes in the THERMOHALINE CIRCULATION of the OCEAN, reorganisation of atmospheric circulation (AUTOVARIATION), VOLCANIC ERUPTIONS (see VOLCANIC FORCING) and solar variability on SUB-MILANKOVITCH timescales.

It is well established that natural environmental change has affected the Earth since its formation and that the effects of environmental change are ubiquitous. Many disciplines, especially the NATURAL ENVIRONMENTAL SCIENCES, investigate such changes at a range of temporal and spatial scales. The concept of global environmental change is newer and has become increasingly important for two main reasons. First, there is the realisation that the complex, interactive, physical, chemical and biological processes of the total EARTH SYSTEM are poorly understood. This means there are fundamental weaknesses in explanations of not only the long-term environmental changes that are prominent on GEOLOGICAL and QUATERNARY TIMESCALES but also HOLOCENE ENVIRONMENTAL CHANGE and the related short-term changes that are likely to affect human society in the immediate future. Second, there is the realisation that HUMAN IMPACTS ON ENVIRONMENT, which have increased during the Holocene, have the potential to affect the Earth system and are already visible at the global scale. Indeed, the human impact has arguably become dominant during the past few centuries (see ANTHROPOCENE).

Some recognise a distinction between *systemic* global change and *cumulative* global change, at least in the context of human impacts—systemic referring to operation at the global scale (e.g. enhanced GLOBAL WARMING); cumulative implying accumulation of local effects until the extent is global (e.g. DEFORESTATION and URBANISATION). *FMC/JAM*

[*See also* CLIMATIC CHANGE, EARTH-SYSTEM ANALYSIS (ESA), GAIA HYPOTHESIS, GLOBAL CHANGE, HUMAN IMPACT ON CLIMATE]

Alverson KD, Bradley RS and Pedersen TF (eds) (2003) *Paleoclimate, global change and the future.* Berlin: Springer.

Bradley RS (2000) Past global changes and their significance for the future. *Quaternary Science Reviews* 19: 391–402.

Human Dimensions of Global Environmental Change Programme (HDP) (1996) *A framework for research on the human dimensions of global environmental change.* Barcelona: HDP Secretariat.

Mackay A, Battarbee R, Birks J and Oldfield F (eds) (2003) *Global change in the Holocene.* London: Arnold.

Munn T (ed.) (2002) *Encyclopedia of global environmental change, 5 volumes.* Chichester: Wiley.

Slaymaker O and Kelly R (2007) *The cryosphere and global environmental change.* Chichester: Wiley.

Steffen W, Sanderson A, Tyson PD et al. (2004) *Global change and the Earth system.* Berlin: Springer.

Turner BL, Kasperson RE, Meyer WB et al. (1991) Two types of global environmental change: Definitional and spatial-scale issues in their human dimensions. *Global Environmental Change* 1, 14–22.

global mean surface air temperature   The commonly used measure of the Earth's surface temperature for CLIMATE CHANGE studies. It is based on an area-weighted global average of the SEA-SURFACE TEMPERATURE (SST) and the land-surface air temperature, and is usually expressed as annual ANOMALIES (see the Figure in the GLOBAL WARMING entry). *JAM*

Thompson DWJ, Kennedy JJ, Wallace JM and Jones PD (2008) A large discontinuity in the mid-twentieth century in observed global-mean surface temperature. *Nature* 453: 646–649.

global navigation satellite system (GNSS)   A generic term applied to the several constellations of satellites used to fix geographical location anywhere on the Earth's surface. GNS systems are now provided by the United States (GPS), Russia (GLONASS), China (COMPASS), Europe (GALILEO) and other contributors such as India and Japan. Instruments to gather the data range from inexpensive handheld types to more sophisticated and expensive *differential* systems. *TF*

[*See also* GLOBAL POSITIONING SYSTEM (GPS)]

UN Office for Outer Space Affairs (2010) *Current and planned global and regional navigation satellite systems and satellite-based augmentation systems.* New York: UN International Committee on Global Navigation Satellite Systems Providers Forum.

global positioning system (GPS)   A free-to-use *location determination system* using a radio receiver to translate signals transmitted from orbiting satellites. The result of the signal measurement is that the geographic CO-ORDINATES of any location on the Earth can be rapidly computed. Results may be provided in any of a selection of CO-ORDINATE SYSTEMS, for example, latitude and longitude, UK *Ordnance Survey National Grid* or *Universal Transverse Mercator (UTM)* co-ordinates. Developed by the US Department of Defense, the satellites orbit at about 20,000 kilometres above the Earth and numbered 30 in 2008. Initially intended for military and other selected users, the accuracy of GPS has improved since the withdrawal of so-called selective availability in 2001. Additional features including differential systems are now operational and hence location fixes can be improved to practical survey levels by drawing on differential GPS methods. The technique uses fixed control station(s) in conjunction with a mobile unit. Readings and timings are compared and adjusted with respect to the fixed receiver. Height accuracy is less reliable than for planimetric information and may be no closer than 50 m compared with better than 10 m (or 1 m differential) achievable in the horizontal mode. *TF*

UN Office for Outer Space Affairs (2010) *Current and planned global and regional navigation satellite systems and satellite-based augmentation systems.* New York: UN International Committee on Global Navigation Satellite Systems Providers Forum.

global warming The recent rise in global temperature caused, at least in part, by an anthropogenically enhanced GREENHOUSE EFFECT is frequently referred to as global warming. Global warming has become one of the most important environmental issues of our time. Over the past century or so, the Earth's GLOBAL MEAN SURFACE AIR TEMPERATURE has increased in the order of 0.8°C. Most of this rise occurred in two steps: first, between about AD 1910 and 1940, and second, between about AD 1975 and 2010. The 1980s, 1990s and 2000s were the warmest decades and each warmer than the previous (see Figure). Global CLIMATIC MODELS suggest that, with a BUSINESS-AS-USUAL SCENARIO, global temperatures will continue to rise by between 1.5°C and 3.5°C over the next century. This rise of temperature exceeds the fastest global temperature trend since the INSTRUMENTAL RECORD began and is probably greater than anything NATURAL that has occurred over the past 10,000 years. A consensus now exists amongst scientists that global warming is taking place, although a small but vocal (often non-scientific) minority continue to question its reality and/or its ATTRIBUTION to human activities. However, following improved understanding of ANTHROPOGENIC warming and cooling influences on climate, the INTERGOVERNMENTAL PANEL ON CLIMATE CHANGE (IPCC) has over the past few decades strengthened its conclusions, 'leading to very high confidence [with at least 90 per cent certainty] that the global average net effect of human activities has been one of warming'(IPCC, 2007: 3).

The continuing media and public scepticism over the reality of global warming is scientifically unjustified. Similar global temperature trends have been revealed by three relatively long-established *climate science* groups: the United States National Oceanic and Atmospheric Administration (NOAA), the NASA Goddard Institute for Space Studies and the UK Meteorological Office Hadley Centre in collaboration with the Climatic Research Unit of the University of East Anglia. Criticisms of their results, including, for example, the supposed subjective selection of the *meteorological stations*, spurious effects of URBAN CLIMATE and faulty analytical procedures, have been thoroughly investigated by several independent enquiries. Furthermore, a fourth group—the University of California, Berkeley Earth Surface Temperature (BEST) project—has recently reassessed the situation using the data from >39,000 stations on land worldwide (over five times the number of stations used in the other analyses). The main result of the BEST analysis is a temperature rise of 0.9 ± 0.04°C since the mid-1950s, which confirms the work of the other three groups and, most importantly, showed that there was no BIAS introduced from urban sites.

Further analyses indicate that the world will not warm uniformly: warming is likely to be greatest in the ARCTIC regions and over land masses. The predicted warming is likely to contribute to a SEA-LEVEL RISE in the order of 50 cm. In the next century, global warming could trigger large and unpredictable changes to global environmental phenomena (see TIPPING POINT). Confidence in the climatic PROJECTIONS is less at the regional level than at the global scale, and less in relation to associated changes in PRECIPITATION. The impacts of global warming are complex and far from uniform over the globe.   *JAM/PDJ/AHP*

[***See also*** ARCTIC WARMING, CLIMATE OF THE LAST MILLENNIUM, CLIMATIC CHANGE, FUTURE CLIMATE, GLOBAL DIMMING, GREENHOUSE CONDITION, HUMAN IMPACT ON CLIMATE, POLAR AMPLIFICATION]

Burroughs W (ed.) (2003) *Climate into the 21*st *century.* Cambridge: World Meteorological Organisation and Cambridge University Press.

Forster P, Ramaswamy V, Artaxo P et al. (2007) Changes in atmospheric constituents and in radiative forcing. In Solomon S, Qin D, Manning M et al. (eds) *Climate change 2007: The physical science basis* [Contribution of Working Group 1 to the Fourth Assessment Report of the Intergovernmental Panel on Climate Change]. Cambridge: Cambridge University Press, 129–234.

Harvey D (1999) *Global warming: The hard science.* London: Prentice Hall.

Houghton J (2009) *Global warming: The complete briefing, 4th edition.* Cambridge: Cambridge University Press.

Huber M and Knutti R (2012) Anthropogenic and natural warming inferred from changes in Earth's energy balance. *Nature Geoscience* 5: 31–36.

Intergovernmental Panel on Climate Change (IPPC) (2007) Summary for policymakers. In Solomon S, Qin D, Manning M et al. (eds) *Climate change 2007: The physical basis* [Contribution of Working Group 1 to the Fourth Assessment Report of the Intergovernmental Panel on Climate Change]. Cambridge: Cambridge University Press, 1–18.

Philander SG (ed.) (2008) *Encyclopedia of global warming and climate change,* 2 volumes. London: Sage.

Schwartz SE, Charlson RJ, Kahn RA et al. (2010) Why hasn't Earth warmed as much as expected? *Journal of Climate* 23: 2453–2464.

Trenberth KE (2010) The ocean is warming, isn't it? *Nature* 465: 304.

Wentz FJ, Ricciardulli L, Jilburn K and Mears C (2007) How much more rain will global warming bring? *Science* 317: 233–235.

Witze A (2012) Climate change confirmed... again. *Nature Geoscience* 5: 4.

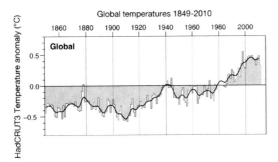

Global temperatures 1849-2010

Global warming *Global warming over land and sea (AD 1849–2010) according to the HadCRUt3 data set. Shaded areas show years above or below the 1961–1990 average; the solid line is a 10-year running mean (from http://www.cru.uea.ac.uk/cru/data/temperature).*

global warming impacts The complex and wide-ranging influence of GLOBAL WARMING not only on the CLIMATE but also on all related physical, social and economic SYSTEMS and component BIOGEOCHEMICAL CYCLES, amongst the most important of which is the CARBON CYCLE and its constituent SINKS and sources. Global warming impacts are all the more challenging as they can be direct and indirect in character and can vary depending upon the local or regional setting in which they take place. Thus, for example, some features of global warming can be more adequately confronted in advanced than in DEVELOPING COUNTRIES. All components of the environment are, however and importantly, sensitive to GLOBAL WARMING to a greater or lesser degree. Amongst these, the HYDROSPHERE is one of the most important and in many ways typical of the interconnectedness of all the EARTH SPHERES. The HYDROLOGICAL CYCLE determines PRECIPITATION and, in combination with temperatures, SOIL MOISTURE. In this way soil biota are influenced leading to changes in plant communities and also agricultural YIELDS. Global warming is also placing water supply under stress with social and political as well as physical systems experiencing similar stresses. The key question is, however, as much one of the ability of people to adapt as to the question of the intrinsic nature of the impact, and it is here that the question of the political, social and economic milieu in which the impacts are manifest assumes importance.   *DAW*

[*See also* ADAPTATION OF PEOPLE, ATTRIBUTION, CLIMATE CHANGE, CLIMATIC CHANGE: PAST IMPACT ON HUMANS, CLIMATIC CHANGE: POTENTIAL FUTURE ECONOMIC IMPACTS, CLIMATIC IMPACT ASSESSMENT, HUMAN IMPACT ON CLIMATE, HUMAN IMPACT ON ENVIRONMENT]

Arctic Climate Impact Assessment (2004) *Impacts of a warming Arctic*. Cambridge: Cambridge University Press.
Finney BO, Gregory-Eaves I, Sweetman et al. (2000) Impacts of climate change and fishing on Pacific salmon abundance over the past 300 years. *Science* 290: 795–799.
Hergerl GC, Zwiers FW, Braconnot P et al. (2007) Understanding and attributing climate change. In

Intergovernmental Panel on Climate Change (ed.) *Climate change 2007: The physical science basis*. Cambridge: Cambridge University Press, 663–745.
Letcher TM (ed.) (2009) *Climate change: Observed impacts on planet Earth*. Amsterdam: Elsevier.
Parry M and Carter T (1998) *Climate impact and adaptation assessment*. London: Earthscan.
Parry ML, Canziani OF, Palutikof JP et al. (eds) (2007) *Impacts, adaptation and vulnerability* [Contribution of Working Group II to the Fourth Assessment Report of the Intergovernmental Panel on Climate Change]. Cambridge: Cambridge University Press.

globalisation The term has moved from a simple descriptor of a process that becomes worldwide or 'makes global' to a more complex concept. This transformation arises in large part from the rapid development of systems of communications and the impact of trends, policies and role models on a global scale. Robertson (1992) used the phrase 'the scope and depth of consciousness of the world as single place', and this usefully captures the potential universality of many global processes. There are key actors in globalisation. Pragmatically, the multinational companies have major influences on economies at all scales, marketing organisations export brand images to many parts of the world, the media invade even the remoter parts of the Earth's surface and global forums (e.g. the United Nations) regard the world as their parish. Conceptually, the globalisation theme raises key questions. Giddens (1990) theorised ways in which space and time were compressed by what he termed 'distanciation' and 'disembedding'. The significance of locality is, however, strongly contested. *Cultural diversity* is a strong and resilient quality, while *community* has its roots in tradition and history as well as in the practice of everyday life. All the conditions for greater globalisation are present; the evidence to measure its impacts remains to be realised. As Western financial systems experienced severe problems of debt and insolvency in the twenty-first century, the reality of globalisation and the interdependence of global financial and economic systems have become starkly apparent. The collapse of the American Lehmann's bank had massive impacts; the rising problems of the Eurozone extend far beyond its boundaries and the roles of nations such as China attain new significance on the world stage.   *DTH*

[*See also* GLOBAL CHANGE]

Bagwati J (2007) *In defense of globalisation: With an afterword*. New York: Oxford University Press.
Bordo MD, Taylor AM and Williamson JG (2004) *Globalization in historical perspective*. Chicago: University of Chicago Press.
Eade J (ed.) (1997) *Living the global city: Globalisation as a local process*. London: Routledge.
Giddens A (1990) *The consequences of modernity*. Cambridge: Polity.

Hoekstra AY and Chapagain AK (2008) *Globalization of water: Sharing the planet's freshwater resources.* Oxford: Blackwell.

Kiely R and Marfleet P (eds) (1999) *Globalisation and the Third World.* London: Routledge.

Kilpatrick AM (2011) Globalization, land use, and the invasion of West Nile virus. *Science* 334: 323–327.

Osterhammel J and Peterson NP (2005) *Globalization: A short history.* Princeton, NJ: Princeton University Press.

Robertson R (1992) *Globalisation.* London: Sage.

Sparke M (2006) *Introduction to globalization: The ties that bind us.* Oxford: Blackwell.

Steger M (2009) *Globalisation: A very short introduction.* Oxford: Oxford University Press.

glow curve   A curve showing the light emission or *thermoluminescence* given off by SEDIMENTS or CERAMICS as they are heated rapidly to 500°C and measured by a *photomultiplier*. It is used in THERMOLUMINESCENCE (TL) DATING, as on first heating the curve represents both the red-hot glow due to the heating and the inherent thermoluminescence derived from RADIATION from surrounding radioactive impurities in minerals over the period since deposition or firing, whereas on second heating only the glow due to heating is recorded. The amount of thermoluminescence represented in the glow curve is a function of the radiation flux, the *dose rate* and the susceptibility of the minerals to acquire thermoluminescence.   *CJC*

Aitken MJ (1985) *Thermoluminescence dating.* London: Academic Press.

Gondwana   A former SUPERCONTINENT comprising Western Europe, South America, Africa, Antarctica, Australia, parts of China and peninsular India. In ORDOVICIAN times, it included equatorial (Antarctica and Australia) and southern Polar regions (north Africa) where a short-lived Late-Ordovician ICE AGE was centred that was responsible for major faunal extinctions. The migration of Gondwana across the *South Pole* in CARBONIFEROUS and PERMIAN times gave rise to a more extensive ice age. During the Permian period, Gondwana formed the southern part of the supercontinent of PANGAEA. It is named after a region of India.   *JCWC*

[*See also* CONTINENTAL DRIFT, PLATE TECTONICS]

Audley-Charles MG and Hallam A (eds) (1988) *Gondwana and Tethys.* Bath: Geological Society.

Caputo MV and Crowell JC (1985) Migration of glacial centers across Gondwana during Paleozoic Era. *Geological Society of America Bulletin* 96: 1020–1036.

Crowley TJ, Mengel JG and Short DA (1987) Gondwanaland's seasonal cycle. *Nature* 329: 803–807.

Pankhurst RJ, Trouw RAJ, de Brito Neves BB and de Wit MJ (eds) (2008) *West Gondwana: Pre-Cenozoic correlations across the South Atlantic Region.* Bath: Geological Society.

Smith AG (1999) Gondwana: Its shape, size and position from Cambrian to Triassic times. *Journal of African Earth Sciences* 28: 71–97.

Sutcliffe OE, Dowdeswell JA, Whittington RJ et al. (2000) Calibrating the Late Ordovician glaciation and mass extinction by the eccentricity cycles of Earth's orbit. *Geology* 28: 967–970.

Yoshida M, Windley BF and Dasgupta S (eds) (2002) *Proterozoic East Gondwana: Supercontinent assembly and breakup.* Bath: Geological Society.

gorge   An *incised valley* or small CANYON characterised by a relatively high depth:width ratio, usually with precipitous rocky walls and steep channel gradient. Gorges are generally considered to originate where fluvial *downcutting* greatly exceeds *valley widening*. In some environments, however, gorge excavation by purely FLUVIAL PROCESSES may be augmented by other processes, such as DISSOLUTION and collapse in KARST landscapes, and FROST WEATHERING in alpine PERIGLACIOFLUVIAL SYSTEMS. The deepest gorges have been excavated by rivers in regions of rapid tectonic UPLIFT. In some areas, gorge sections of rivers have been attributed to the HEADWARD EROSION of KNICKPOINTS marking REJUVENATION of rivers following uplift. Some gorges have been attributed to glaciofluvial erosion in the SUBGLACIAL environment where MELTWATER, highly charged with coarse sediment and sometimes under HYDROSTATIC PRESSURE, may be capable of enhanced incision. In recent years, COSMOGENIC-NUCLIDE DATING has led to renewed interest in the age and development of gorges, as it has allowed the age of rock surfaces within gorges to be determined. Conclusions remain inconclusive, however, because of the possibility of phases of gorge infilling and re-excavation during GLACIAL-INTERGLACIAL CYCLES, as well as higher rates of fluvial incision during glacial-interglacial transitions than today.   *JAM/RPDW*

[*See also* SUBMARINE CANYON, V-SHAPED VALLEY]

Hantke R and Scheidegger A (1993) On the genesis of the Aare gorge, Berner Oberland, Switzerland. *Geographica Helvetica* 48: 120124.

Hayakawa YS and Matsukura Y (2009) Factors influencing the recession rate of Niagara Falls since the 19th century. *Geomorphology* 110: 212–216.

McEwen LJ, Matthews JA, Shakesby RA and Berrisford MS (2002) Holocene gorge excavation and boulder-fan formation related to rates of frost weathering in a Norwegian alpine periglacio-fluvial system. *Arctic, Antarctic and Alpine Research* 34: 345–357.

Montgomery DR and Korup O (2010) Preservation of inner gorges through repeated alpine glaciations. *Nature Geoscience* 4: 62–67.

Valla PG, van der Beek and Carcaillet J (2010) Dating bedrock gorge incision in the French Western Alps (Ecrins-Pelvoux massif) using cosmogenic ^{10}Be. *Terra Nova* 22: 18–25.

graben A GEOLOGICAL STRUCTURE comprising a *downthrown block* bounded on each side by FAULTS. At a given level, the rock between the faults is younger than that on each side. A RIFT VALLEY is the corresponding landform: the topographic expression of a graben. Grabens usually result from extensional TECTONICS. A *half graben* is a tilted downthrown block adjacent to a single extensional fault.   *GO*

[*See also* HORST]

Chorowicz J (2005) The East African rift system. *Journal of African Earth Sciences* 43: 379–410.
Peakall J (1998) Axial river evolution in response to half-graben faulting: Carson River, Nevada, USA. *Journal of Sedimentary Research* 68: 788–799.
Seidel M, Seidel E and Stoeckhert B (2007) Tectono-sedimentary evolution of Lower to Middle Miocene half-graben basins related to an extensional detachment fault (western Crete, Greece). *Terra Nova* 19: 39–47.

grade (of particles) A general term for GRAIN SIZE in a SEDIMENT or SEDIMENTARY ROCK, such as 'a sediment of medium sand grade'.   *GO*

grade concept An old term in *fluvial geomorphology* used to indicate that a river channel is in equilibrium or balance with its environment. A *graded stream* or stream reach was defined by Mackin (1948: 471) as one 'in which, over a period of years, slope is delicately adjusted to provide, with available discharge and prevailing channel characteristics, just the velocity required for the transportation of the load supplied from a DRAINAGE BASIN'. The concept was closely associated with the idea that a smooth concave longitudinal profile (*thalweg*) indicated a LANDSCAPE in which its *drainage system* was adjusted to the stage of LANDSCAPE EVOLUTION; and that, conversely, irregularities in a longitudinal profile (e.g. KNICKPOINTS) indicated interruptions to an erosion cycle (e.g. UPLIFT and SEA-LEVEL CHANGE) and a river that was unadjusted. Such irregularities were widely used in DENUDATION CHRONOLOGY in reconstructions of the long-term history of landscapes. Chorley (2000) explains why Mackin's qualitative explication is essential to understanding subsequent quantitative developments.   *RPDW*

[*See also* BASE LEVEL, EQUILIBRIUM CONCEPTS IN GEOMORPHOLOGICAL AND LANDSCAPE CONTEXTS, LANDSCAPE EVOLUTION]

Chorley RJ (2000) Classics in physical geography revisited: 'Mackin, J. H. 1948: Concept of the graded river. *Bulletin of the Geological Society of America* 59: 463–512'. *Progress in Physical Geography* 24: 563–578.
Knox JC (1975) Concept of the graded stream. In Mulhorn WN and Flemel RC (eds) *Theories of landform development.* London: Allen and Unwin, 169–198.
Mackin JH (1948) Concept of the graded river. *Bulletin of the Geological Society of America* 59: 463–512.

graded bedding A SEDIMENTARY STRUCTURE characterised by a consistent variation in GRAIN SIZE from bottom to top of a single BED. *Normal grading* passes from coarser at the base to finer at the top, and is commonly the product of DEPOSITION from a waning flow such as a TURBIDITY CURRENT or other depositional EVENT, although thin normally *graded laminae* (see LAMINATION) can be produced by sediment settling from *suspension* (see SUSPENDED LOAD). *Reverse grading* or *inverse grading* passes from fine up to coarse within a single bed, and is commonly characteristic of the deposits of DEBRIS FLOWS.   *BTC/JP*

Kuenen PhH (1953) Significant features of graded bedding. *Bulletin of the American Association of Petroleum Geologists* 37: 1044–1066.
Monteith H and Pender G (2005) Flume investigations into the influence of shear stress history on a graded sediment bed. *Water Resources Research* 41: W12401.
Nichols G (2009) *Sedimentology and stratigraphy, 2nd edition.* Oxford: Wiley-Blackwell.

gradient analysis The study and analysis of species occurrences and abundances along gradients. The gradients may be unknown a priori and derived from the data by ORDINATION using, for example, CORRESPONDENCE ANALYSIS (CA) or PRINCIPAL COMPONENTS ANALYSIS (PCA) (*indirect gradient analysis*); or known and measured by independent analysis (*direct gradient analysis*). The latter involves REGRESSION ANALYSIS (one species) or canonical ordination techniques, such as CANONICAL CORRESPONDENCE ANALYSIS (CCA) or REDUNDANCY ANALYSIS (many species).   *HJBB*

ter Braak CJF and Prentice IC (1988) A theory of gradient analysis. *Advances in Ecological Research* 18: 271–317.
Whittaker RH (1967) Gradient analysis of vegetation. *Biological Reviews* 42: 207–264.

gradualism The concept that the GEOLOGICAL RECORD is dominated by the products of slow, steady and gradual processes, as opposed to the cataclysmic EVENTS preferred by CATASTROPHISM. A component of Lyell's UNIFORMITARIANISM, gradualism alone is now considered an inadequate explanation for the geological record. The modern consensus on the nature of Earth history could be described by analogy with biology's PUNCTUATED EQUILBRIUM as *punctuated gradualism* with a background of slow, long-term processes periodically affected by dramatic, short-lived, major interruptions.   *CDW*

[*See also* NEOCATASTROPHISM, PHYLETIC GRADUALISM]

Ager DV (1993) *The nature of the stratigraphical record, 3rd edition.* Chichester: Wiley.
Gould SJ 91984) Towards the vindication of punctuational change. In Berggren WA and Couvering JA (eds) *Catastrophes and Earth history: The new*

uniformitarianism. Princeton, NJ: Princeton University Press, 9–34.

Saint-Ange F, Savoye B, Michon L et al. (2011) A volcaniclastic deep-sea fan off La Réunion Island (Indian Ocean): Gradualism versus catastrophism. *Geology* 39: 271–274.

grain   (1) A general term for a particle of rock or soil. In SEDIMENTOLOGY, it may refer to particles, CLASTS or mineral grains, and is sometimes restricted to particles smaller than a few millimetres. (2) More specifically, a grain is a particle in a sediment or SEDIMENTARY ROCK, which is of the nominal or larger GRAIN SIZE (i.e. GRAVEL, SAND or MUD) in contrast to MATRIX, which is of finer grain size. (3) Grain is also sometimes used as a synonym for TEXTURE in the broader context of LANDFORMS and LANDSCAPES, referring to the scale of the components of the landscape (e.g. landscapes with much detail may be termed 'fine grained'). It also tends to be used in this sense when referring to images in REMOTE SENSING.   *TY*

grain size   The size (diameter) of particles (particularly CLASTS) in a SOIL, SEDIMENT or SEDIMENTARY ROCK; one of the most important attributes of sediment TEXTURE and a key property in the definition of sedimentary FACIES. Grain size tends to diminish with distance of sediment transport. Variations in grain size of SEDIMENTARY DEPOSITS at one place reflect the COMPETENCE of agents of sediment transport: it can therefore be used as a proxy indicator of the energy of a DEPOSITIONAL ENVIRONMENT (see ENERGY (OF ENVIRONMENT)). Patterns of grain-size frequency distributions—mean sizes and other moment measures of the distribution such as SORTING, SKEWNESS and KURTOSIS—have been used to characterise different processes of SEDIMENT TRANSPORT and DEPOSITION and attempts have been made to relate grain-size characteristics to depositional environments in sedimentary deposits.

The most commonly used scale of grain size among Earth scientists is the *Udden-Wentworth scale* (see Figure 1) which recognises GRAVEL, SAND, MUD and subdivisions. Other scales are in use, notably for PYROCLASTIC MATERIAL and in disciplines such as *engineering*. A modification of the Udden-Wentworth scale has been proposed to encompass particles with diameters up to 1,075 km (= very coarse megalith). Figure 2 shows the names given to mixtures of gravel, sand and mud. The large range in naturally occurring grain sizes can be dealt with more conveniently for graphical purposes by converting grain diameter in millimetres into PHI (φ) units.

The size of GRAVEL particles can be directly measured. It is common to measure the long, intermediate and short axes of a number of clasts, giving both a size frequency distribution and a measure of particle SHAPE. Grain size of unconsolidated SAND is commonly determined by

sieve analysis—passing the sediment through a stack of sieves with mesh diameter decreasing downwards. The mass retained on each sieve after shaking gives a *mass frequency distribution*. Grain size in lithified SANDSTONES can be estimated visually against a reference scale to give a representative *grade* on the Udden-Wentworth scale, or else clasts are measured in THIN-SECTION ANALYSIS to give a *number frequency distribution*. *Mud* particles are difficult to resolve by eye. In unconsolidated or poorly consolidated samples, the ratio of SILT to CLAY can be estimated by grinding a small wet sample between the fingers: silt is silky whereas clay is sticky. In the laboratory, a *grain-size distribution* can be obtained by SEDIMENTATION methods that use the relationship between *settling velocity* and *particle mass*. Grain size of unconsolidated sand and mud can also be determined using a *Coulter counter*, which uses the electrical properties of particles carried in a fluid, or a *Sedigraph*, which uses X-rays, or by *laser-diffraction* techniques. There are difficulties in comparing frequency distributions obtained by different methods, particularly where particles are non-spherical.

particle diameter			SEDIMENT	SEDIMENTARY ROCK
mm	ø units			
— 256	— -8	BOULDER		
		COBBLE	GRAVEL	CONGLOMERATE (rounded CLASTS) BRECCIA (angular CLASTS)
— 64	— -6			
		PEBBLE		RUDACEOUS
— 4	— -2			
— 2	— -1	GRANULE		
— 1.0	— 0	very coarse	SAND	
— 0.5	— 1	coarse		SANDSTONE
— 0.25	— 2	medium		ARENACEOUS
— 0.125	— 3	fine		
— 0.063	— 4	very fine		
		SILT	MUD	MUDROCK (MUDSTONE, SHALE)
— 0.004	— 8			ARGILLACEOUS
		CLAY		

Grain size (1) *The Udden-Wentworth scale for grain size.*

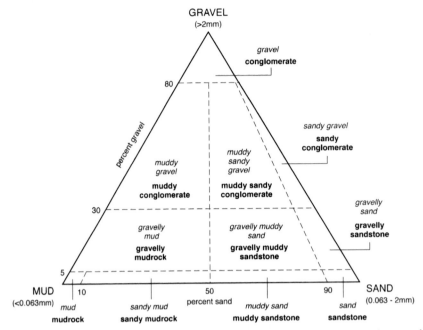

Grain size (2) *A scheme for naming sediments (names in italics) and sedimentary rocks (names in bold) that are combinations of gravel, sand and mud.*

Grain size can be expressed as a typical value on the Udden-Wentworth scale, as a HISTOGRAM, a frequency distribution curve or a *cumulative frequency curve*. When plotted on probability graph paper, cumulative percentage curves of grain-size distributions commonly approximate to a straight line or a line made up of straight segments. Attempts have been made to relate these straight line segments to depositional processes.

GO/TY

[*See also* SOIL TEXTURE]

Anthony EJ and Hequette A (2007) The grain-size characterisation of coastal sand from the Somme estuary to Belgium: Sediment sorting processes and mixing in a tide- and storm-dominated setting. *Sedimentary Geology* 202: 369–382.

Bianchi GG, Hall IR, McCave IN and Joseph L (1999) Measurement of the sortable silt current speed proxy using the Sedigraph 5100 and Coulter Multisizer IIe: Precision and accuracy. *Sedimentology* 46: 1001–1014.

Blair TC and McPherson JG (1999) Grain-size and textural classification of coarse sedimentary particles. *Journal of Sedimentary Research* 69: 6–19.

Folk RL (1980) *Petrology of the sedimentary rocks.* Austin, TX: Hemphill.

Harrell J and Eriksson KA (1979) Empirical conversion equations for thin-section and sieve derived size distribution parameters. *Journal of Sedimentary Petrology* 49: 273–280.

Konert M and Vandenberghe J (1997) Comparison of laser grain size analysis with pipette and sieve analysis: A

solution for the underestimation of the clay fraction. *Sedimentology* 44: 523–535.

Macquaker JHS and Adams AE (2003) Maximizing information from fine-grained sedimentary rocks: An inclusive nomenclature for mudstones. *Journal of Sedimentary Research* 73: 735–744.

McManus J (1988) Grain size determination and interpretation. In Tucker ME (ed.) *Techniques in sedimentology.* Oxford: Blackwell, 63–85.

Morton RA, Goff JR and Nichol SL (2008) Hydrodynamic implications of textural trends in sand deposits of the 2004 tsunami in Sri Lanka. *Sedimentary Geology* 207: 56–64.

Syvitski JPM (ed.) (1991) *Principles, methods and applications of particle size analysis.* Cambridge: Cambridge University Press.

Wentworth CK (1922) A scale of grade and class terms for clastic sediments. *Journal of Geology* 30: 377–392.

graminoid   A *grass* or grass-like HERB including, for example, sedges and rushes.   JAM

granular disintegration   WEATHERING of rock at the scale of individual *mineral grains* and *rock crystals*, which leaves a rough, pitted surface. It is caused by a range of processes, possibly in combination in some environments, including FROST WEATHERING, INSOLATION WEATHERING and SALT WEATHERING.   RAS/JAM

[*See also* GELIFRACTION, GRUSS/GRUSSIFICATION]

Eppes MC and Griffing D (2010) Granular disintegration of marble in nature: A thermal-mechanical origin for grus and corestone landscape. *Geomorphology* 117: 170–180.

Hall K, Guglielmin M and Stini A (2008) Weathering of granite in Antarctica: II. Thermal stress at the grain scale. *Earth Surface Processes and Landforms* 33: 475–493.

granule A sedimentary particle of GRAIN SIZE 2–4 mm. *TY*

[*See also* GRAVEL, SEDIMENT]

graphic log A diagrammatic representation of a vertical succession of SEDIMENTS, SEDIMENTARY ROCKS or PYROCLASTIC MATERIAL in which the vertical scale represents thickness, the horizontal scale represents GRAIN SIZE and the resulting column of variable width is ornamented to illustrate features such as SEDIMENTARY STRUCTURES, contacts between BEDS, FOSSILS or sampling points. Grain size reflects the ENERGY of the DEPOSITIONAL ENVIRONMENT, and the log gives a clear visual impression of *environmental change* through time. There has been some criticism of over-reliance on graphic logs as a tool in FACIES ANALYSIS as their one-dimensional form detracts from a careful three-dimensional reconstruction of PALAEOENVIRONMENT (see FACIES ARCHITECTURE and ALLUVIAL ARCHITECTURE). *GO*

[*See also* ALLUVIAL ARCHITECTURE, FACIES ARCHITECTURE]

Bouma AH (1962) *Sedimentology of some flysch deposits: A graphic approach to facies interpretation.* Amsterdam: Elsevier.

Bridge JS (1985) Paleochannel patterns inferred from alluvial deposits: A critical evaluation. *Journal of Sedimentary Petrology* 55: 579–589.

Coe AL (ed.) (2010) *Geological field techniques.* Milton Keynes: Wiley-Blackwell and Open University.

graphical user interface (GUI) A set of programs that act as mediators between the user and other computer programs (usually referred to as '*commands*'). They usually rely on windows (different information can be displayed simultaneously), iconic representation of entities (e.g. files, processes, etc.), pull-down or pop-up menus (commands or other options are selected from lists rather than from their actual names) and pointing devices (the mouse can be used for selection). These programs aim at helping the user in the pursuit of tasks, and they should take into account cognitive factors, such as the size of short-term memory, colour perception, etc. The interface should be based on user-oriented terms and concepts rather than on computer concepts, and typical users should be involved in the design of GUIs. *ACF*

Fisher P (2005) *An introduction to graphical user interfaces with Java swing.* London: Pearson.

grass cuticle analysis The use of SUBFOSSIL grass cuticles, extracted from SEDIMENTS, for PALAEOENVIRONMENTAL RECONSTRUCTION. Subfossil grass cuticles are the resistant surface remains of grass leaves, retaining the micromorphological features of the cells they once covered, and consist principally of complex *polymers* (*cutin, cutan* and *suberin*), cuticular *leaf waxes* and PHYTOLITHS embedded in the cuticle complex. Grass cuticles are relatively inert and well preserved in sediments, due to the chemical stability of their constituent hydroxy monocarboxylic acids—*phloinic acid* ($C_{18}H_{34}O_{16}$), *cutinic acid* ($C_{13}H_{22}O_3$) and *cutic acid* ($C_{26}H_{50}O_6$)—and retain micromorphological features that can aid identification to taxonomic levels below family, thus providing valuable BIO-INDICATORS. Grass cuticle analysis therefore helps overcome the limited taxonomic resolution associated with *grass pollen* (often only identifiable to family level). Grass cuticle analysis is particularly suited to *tropical environments* where the BIODIVERSITY of grasses is high but can also be applied at high-latitude sites. Grass cuticles are also used for monitoring shifts in woodland-grassland ECOTONES and for determining the *photosynthetic pathways* used by past grasslands. *MJW*

[*See also* CHARRED-PARTICLE ANALYSIS, PALAEOECOLOGY]

Gaglioti B, Severin K and Wooller MJ (2010) Developing graminoid cuticle analysis for application to Beringian paleoecology. *Review of Palaeobotany and Palynology* 162: 95–110.

Mworia-Maitima J (1997) Prehistoric fires and land-cover change in western Kenya: Evidence from pollen, charcoal, grass phytoliths, and grass cuticle analyses. *The Holocene* 7: 409–417.

Palmer PG (1976) Grass cuticles: A new paleoecological tool for East African lake sediments. *Canadian Journal of Botany* 54: 1725–1734.

Wooller MJ, Zazula G, Blinnikov M et al. (2011) The detailed paleoecology of a mid-Wisconsinan interstadial (ca. 32,000 ^{14}C BP) vegetation surface from interior Alaska. *Journal of Quaternary Science* 26: 746–756.

grasses and grasslands Fossil evidence suggests that grasses (*Poaceae,* formerly *Gramineae*) are one of the most recent ANGIOSPERM families to evolve. The first unequivocal MACROFOSSIL evidence for grasses has been found in deposits from North America dated approximately 60 million years ago (Ma). *Grasslands* as a BIOME are more recent with palaeoecological evidence suggesting that open grasslands did not come into existence until approximately 15 Ma. There are a number of suggestions to account for their relatively late EVOLUTION and ADAPTIVE RADIATION. These include increased global ARIDITY, increased FIRE FREQUENCY and an increase in FAUNA (e.g. hoofed mammals) physiologically adapted to a diet rich in *cellulose* and SILICA. Despite their late appearance, the radiation of grasslands has been rapid. Presently, there are more than 10,000 species of grasses on Earth with estimates indicating that modern grasslands cover more than 30 per cent of the land surface providing up to 52 per cent of the *protein* in human diets worldwide. *KJW*

[*See also* GRAMINOID, SAVANNA, STEPPE, TEMPERATE GRASSLANDS, TROPICAL GRASSLAND]

Axelrod DI (1985) Rise of the grassland biome, central North America. *Botanical Review* 51: 163–201.
Blinnikov MS, Gaglioti BV, Walker DA et al. (2011) Pleistocene graminoid-dominated ecosystems in the Arctic. *Quaternary Science Reviews* 30: 2906–2929.
Crepet WL and Feldman GD (1991) The earliest remains of grasses in the fossil record. *American Journal of Botany* 78: 1010–1014.
Hubbard CE (1992) *Grasses: A guide to their structure, identification, uses and distribution, 3rd edition.* London: Penguin Books.
Stebbins GL (1981) Coevolution of grasses and herbivores. *Annals of the Missouri Botanical Gardens* 68: 75–86.

graticule   The pattern generated by intersecting lines of *meridians* (*longitude*) and *parallels* (*latitude*) on a globe or map. The pattern will vary according to the MAP PROJECTION employed and the map area covered. *TF*

Heywood I, Cornelius S and Carver S (2011) *An introduction to geographical information systems, 4th edition.* Harlow: Pearson.

gravel   A SEDIMENT with a dominant GRAIN SIZE >2 mm. The term is occasionally used as a synonym for GRANULE. *TY*

gravel-bed river   A river with a sediment load dominated by GRAVEL moving as BEDLOAD, although such rivers also transport significant quantities of sediment of finer GRAIN SIZE. Gravel-bed rivers tend to dominate under conditions of high CHANNEL GRADIENT and high, often variable DISCHARGE. They have unstable, mobile channels due to a lack of cohesive bank material. BEDFORMS are dominated by longitudinal CHANNEL BARS and most gravel-bed rivers have a low-sinuosity, braided CHANNEL PATTERN, although some meandering rivers have gravel beds. Many streams in *glacial outwash* settings (see SANDUR) and in SEMI-ARID settings are gravel-bed rivers. Before the evolution of land plants in the Early PALAEOZOIC, which led to the stabilisation of river banks, most rivers may have shared many of the attributes of modern gravel-bed rivers, regardless of the prevailing climate. *GO*

[*See also* BEDROCK CHANNEL, BRAIDING, SAND-BED RIVER, SUSPENDED-LOAD RIVER]

Ashmore P (1991) How do gravel-bed rivers braid? *Canadian Journal of Earth Sciences* 28: 326–341.
Billi P, Hey RD, Thorne CR and Tacconi P (eds) (1992) *Dynamics of gravel-bed rivers.* Chichester: Wiley.
Brasington J, Rumsby BT and McVey RA (2000) Monitoring and modelling morphological change in a braided gravel-bed river using high resolution GPS-based survey. *Earth Surface Processes and Landforms* 25: 973–990.
Bridge JS and Demicco RV (2008) *Earth surface processes, landforms and sediment deposits.* New York: Cambridge University Press.
Eaton BC (2006) Bank stability analysis for regime models of vegetated gravel bed rivers. *Earth Surface Processes and Landforms* 31: 1438–1444.
Gibling MR and Davies NS (2012) Palaeozoic landscapes shaped by plant evolution. *Nature Geoscience* 5: 99–105.
Karaus U, Alder L and Tockner K (2005) 'Concave islands': Habitat heterogeneity of parafluvial ponds in a gravel-bed river. *Wetlands* 25: 26–37.
Wooldridge CL and Hickin EJ (2005) Radar architecture and evolution of channel bars in wandering gravel-bed rivers: Fraser and Squamish rivers, British Columbia, Canada. *Journal of Sedimentary Research* 75: 844–860.

gravity anomaly   A local variation in the strength of the Earth's gravity field that can be attributed to variations in the density of buried materials. In GEOPHYSICAL SURVEYING, a *Bouguer anomaly* is a gravity reading that has been corrected for the predictable effects of altitude and latitude: residual variations in Bouguer anomalies can be interpreted in terms of buried geological or archaeological features. *GO*

Milsom J (2000) Gravity measurement and interpretation of anomalies. In Hancock PL and Skinner BJ (eds) *The Oxford companion to the Earth.* Oxford: Oxford University Press, 470–475.

grazing   The CONSUMPTION of green plants by VERTEBRATES, *invertebrates* and some *micro-organisms*. Sometimes the distinction is made between the grazing of *grasses* and the *browsing* of *trees* and SHRUBS. *IFS*

Gordon IJ and Prins HHT (eds) (2010) *The ecology of browsing and grazing.* Berlin: Springer.

grazing history   Although grazing animals are often believed to be one of the primary mechanisms of FOREST CLEARANCE, detecting their presence from the palaeoecological record can be problematic. Some ANTHROPOGENIC INDICATOR species are closely associated with grazed HABITATS. During the MESOLITHIC, palaeoecological evidence for woodland DISTURBANCE and from FIRE HISTORY are often interpreted in terms of human activity designed to open up the VEGETATION CANOPY and encourage plentiful new growth of ground-level vegetation, in order to make hunting easier by concentrating populations of native grazing animals (e.g. red deer in Britain) in one location, or by increasing the

CARRYING CAPACITY of a given tract of woodland. Franz Vera offered an alternative interpretation of early Holocene woodlands in northern Europe which argues for a greater role of indigenous grazing animals and a relatively open park-like landscape before the arrival of agriculturalists; this model has led to extensive debate and re-examination of the evidence.

The NEOLITHIC TRANSITION includes two major changes with regard to grazing animals, the DOMESTICATION of native species (e.g. *cattle*) and the INTRODUCTION of new species (e.g. *sheep*, *goats*). Once the location and nature of grazing herds was controlled to some extent by human activity, landscape impacts often become significant.

The nature and severity of grazing-related *environmental change* varies in space and time and depends on the HERBIVORE, since different grazing animals have different tastes and requirements, and thus different effects on the landscape. Intensification of grazing activity is often believed to have been an important factor in the initiation of BLANKET MIRE growth and the development of MOORLAND and HEATHLAND, and therefore in profound changes in the use and value of LANDSCAPES for prehistoric communities. In upland Britain, *heather moorland* is an important resource for grazing sheep, *red deer* and *grouse*, and this use seems to have continued sustainably for several thousand years. However, intensification of *burning* (designed to improve the grazing) and higher stocking levels of sheep in the past 200 years is leading to a reduction in *heather moorland*, and shows that *grazing pressure* is an important factor in environmental change. *MJB*

[*See also* OVERGRAZING, PASTORALISM, TRANSHUMANCE]

Moore PD (1993) The origin of blanket mire, revisited. In Chambers FM (ed.) *Climate change and human impact on the landscape*. London: Chapman and Hall, 133–145.
Stephenson AC and Thompson DBA (1993) Long term changes in the extent of heather moorland in upland Britain and Ireland: Palaeoecological evidence for the importance of grazing. *The Holocene* 3: 70–76.
Svenning JC (2002) A review of natural vegetation openness in north-western Europe. *Biological Conservation* 104: 133–148.
Vera FWM (2000) *Grazing ecology and forest history*. Abingdon: CABI Publishing.

grazing: impacts on ecosystems Throughout EVOLUTION, grazing has led to interactions between grazing animals and plants. Interactions between herbivorous insects and plants include toxic chemicals being produced by the plants and *selective grazing* by the insects. Plants grazed by vertebrates range from the palatable to those that are avoided. Selective grazing affects the floristic composition of PASTURES, with some

species declining and some becoming abundant. For example, CONSERVATION of *floristic diversity* of *chalk grasslands* in the United Kingdom is dependent on the selective grazing of domestic HERBIVORES. Grazing by wild herbivores (ALIEN SPECIES) has halted ECOLOGICAL SUCCESSION in some communities and elsewhere has had major implications for conservation of NATIVE or indigenous flora. OVERGRAZING is a major process of ecosystem and SOIL DEGRADATION. *IFS*

Danell K, Bergström R, Duncan P and Pastor J (eds) (2006) *Large herbivore ecology, ecosystem dynamics and conservation*. Cambridge: Cambridge University Press.
Hodgson J and Illius AW (eds) (1996) *The ecology and management of grazing systems*. Wallingford: CABI Publishing.
Vera FWM (2000) *Grazing ecology and forest history*. Wallingford: CABI Publishing.

(the) Great Acceleration The sharp increase in the rate of growth in the human impact on the EARTH SYSTEM, which occurred from AD 1950 onwards. It constitutes the second phase of the ANTHROPOCENE, the first phase having seen much lower, relatively steady growth in many indicators (e.g. population growth, carbon emissions, water use and fertiliser consumption) since the beginning of the INDUSTRIAL REVOLUTION. Initially, the Great Acceleration was almost entirely driven by developed countries but since the start of the twenty-first century, the largest DEVELOPING COUNTRIES (especially China but also India, Brazil, South Africa and Indonesia) account for almost all of the growth. *JAM*

Hibbard KA, Crutzen PJ, Lambin EF et al. (2006) Decadal interactions of humans and the environment. In Constanza R, Graumlich L and Steffen W (eds) *Sustainability or collapse? An integrated history and future of people on Earth*. Cambridge, MA: MIT Press, 341–375.
Steffen W, Grinevald J, Crutzen P and McNeill J (2011) The Anthropocene: Conceptual and historical perspectives. *Philosophical Transactions of the Royal Society A* 369: 842–867.

(the) Great Migration The mass international HUMAN MIGRATION from European countries to Canada and the United States and, to a lesser extent, Latin America, Australia, New Zealand and South Africa, that occurred during the nineteenth and early twentieth centuries. Around 50 million people were involved, driven mainly by *poverty*, unemployment, greater religious or political freedom and/or the desire to start a new life. *JAM*

Baines D (1991) *Emigration from Europe 1815–1930*. London: Macmillan.

green belts Areas of land, usually around cities, which are deliberately left undeveloped except from AGRICULTURE and *forestry*. They have been developed as planning devices with the prime purpose of controlling *urban growth*. Their use around cities has much older historical precedents than the twentieth century, but in 1935 a *Metropolitan Green Belt* was proposed for London and the principle of green belts as part of development control was incorporated into the *UK Town and Country Planning Act, 1947*. By 2008, green belts had emerged around 22 cities in the United Kingdom. There were green belts elsewhere such as Canada (Ottawa and Toronto) and European countries such as Sweden, though the names for these differed (e.g. *green spaces, green wedges* and *green parks*).

Green belts had a number of specified purposes. They were intended to place limits on urban sprawl and to avoid cities merging together, they were meant to provide accessible open space for *leisure* and RECREATION to city dwellers, they provided sanctuaries for the CONSERVATION of natural environments with their flora and fauna and they would provide relief from HUMAN HEALTH HAZARDS in areas close to cities. Green belts have been criticised on several counts. If employed literally as a belt, they are indiscriminate and may include land unsuitable for recreation, leisure or health promotion. They can be restrictive on genuine growth needs and sometimes are used for political purposes. They may increase commuting distances and create satellite towns outside the designated area. These aspects need to be managed but the principle of providing some balance in urban growth and ensuring access to open spaces is well founded.   *DTH*

[*See also* URBAN AND RURAL PLANNING, URBAN ENVIRONMENTAL CHANGE, URBAN SUSTAINABILITY, URBANISATION]

Amati M (ed.) (2008) *Urban green belts in the twenty-first century*. Aldershot: Ashgate Publishing.
Gallent N (2010) Greenbelts. In Warf B (ed.) *Encyclopedia of geography*. Thousand Oaks, CA: Sage, 1363–1367.
Maathai W (2004) *The green belt movement: Sharing the approach and the experience, 2nd edition*. Herndon, VA: Lantern Books.
Munton R (2006) *London's green belt: Containment in practice, 2nd edition*. London: Routledge.
Thompson CW and Travlou P (eds) (2007) *Open space, open people*. London: Taylor and Francis.

green economy   An economy where growth and development is accompanied by reduced carbon EMISSIONS and POLLUTION, providing *societal benefits*, improved human well-being and reduced negative environment impacts. The aim of a green economy is low emissions of GREENHOUSE GASES, efficient use of RESOURCES, low WASTE generation and sustainable economic growth reliant on ECOLOGICAL SERVICES provided by NATURAL CAPITAL. The transition to a green economy requires developing *green infrastructure* and markets for resource-efficient and low-carbon technologies. In the United Kingdom, the establishment or development of *Nature Improvement*

Areas (NIAs) stimulates growth and promotes large-scale or cross-boundary support for NATURAL NETWORKS. NIAs require development, restoration or reconnecting to nature on a significant scale. Partnership working is recognised as the best way to achieve effective action. The government's Natural Environment White Paper promotes the establishment of *Local Nature Partnerships (LNPs)*, which are seen as an effective method for bringing society together to promote local action towards protecting and improving the environment and stimulating growth of a green economy. It is expected that LNPs will bring together a diverse range of individuals, businesses and organisations at a local level and enable society to work to influence local decisions as well as benefit from promoting an *ecosystem* approach to CONSERVATION and ENVIRONMENTAL MANAGEMENT. Society will in turn benefit from improved social and economic conditions stimulated by local enterprise and an improved environment.
                                                                AEV

[*See also* CITIZEN SCIENCE, SOCIETAL BENEFIT AREAS (SBAs)]

Secretary for the Environment, Food and Rural Affairs (2011) *The natural choice: Securing the value of nature*. London: Her Majesty's Government.

green lists   Those species which are known to be secure and not in need of CONSERVATION are on green lists. However, if a species is not in a RED DATA BOOK, it does not necessarily mean that it is not in need of conservation. The use of green lists has been suggested as an alternative to *red lists*.   *IFS*

Keith M and Van Jaarsveld AS (2001) Revisiting Green Data lists. *Biodiversity and Conservation* 11: 1313–1316.

green manure   Plant material incorporated with the soil while green, or soon after maturity, to improve the soil. Typically, *legumes* are grown for their *nitrogen-fixing* properties, while non-legumes are used for the control of WEEDS and the addition of *organic matter* to the soil; both are grown for a period and then ploughed in. Green manure has a particularly important role in ORGANIC FARMING where it replaces chemical FERTILISERS.   *RAS*

[*See also* SOIL CONSERVATION]

Eilittä M, Mureithi J and Derpsch R (ed.) (2004) *Green manure/cover crop systems of smallholder farmers: Experiences from tropical and subtropical regions*. Berlin: Springer.
Nyberg G, Ekblad A, Buresh RJ and Hogberg P (2000) Respiration from C-3 plant green manure added to a C-4 plant carbon dominated soil. *Plant and Soil* 218: 83–89.

green politics   The injection of increased environmental awareness into political thinking and policymaking since the late 1980s. Although environmental concerns have a long history, the realisation that humanity must learn to conserve and protect the NATURAL ENVIRONMENT rather than exploit it in a non-sustainable way, only became widespread through the publicity of ENVIRONMENTAL MOVEMENTS and radical PRESSURE GROUPS. Recognition of the very real threat to humanity posed by increased GREENHOUSE GAS emissions, for example, led to shifts in the political agenda with ENVIRONMENTAL ISSUES taking a high profile.

In the United Kingdom, the government environmental white paper of 1990 prompted a restructuring of central government to include a Minister for the Environment and two 'green' cabinet committees responsible for issuing guidance documents for civil servants. It also encouraged corporate efforts to integrate environmental concerns and sustainable practices into the day-to-day organisation of business, industrial and domestic life. Apart from government policy changes, green politics incorporates political parties like the *Green Party*, whose principal objective is to make a 'green society' and promote environmental awareness and sustainable business practices.   *JGS*

[*See also* CONSERVATION, ENVIRONMENTAL POLICY, ENVIRONMENTAL PROTECTION, GREENING OF SOCIETY, POLITICAL ECOLOGY, SUSTAINABILITY, SUSTAINABLE DEVELOPMENT]

Carter N (2007) *The politics of the environment: Ideas, activism, policy, 2nd edition.* Cambridge: Cambridge University Press.
Dobson A (2007) *Green political thought, 4th edition.* New York: Routledge.
Jordan A (1998) The construction of a multi-level environmental governance system. *Environment and Planning* 17: 227–235.
Weale A (1998) Environmental policy. In Budge I, Crewe I, McKay D and Newton K (eds) *The new British politics.* Harlow: Longman, 171–129.

Green Revolution   The rapid increase in crop yields of the AD 1950s brought about by the introduction of high-yielding varieties of grain, particularly rice and wheat, combined with heavy inputs of artificial FERTILISERS. Especially in the developing world, where the 'revolution' has supported rapid POPULATION GROWTH, the SUSTAINABILITY of such systems is open to question for both ecological and economic reasons. For example, a large number of resistant varieties of grain have been replaced by a small number of varieties that are vulnerable to PESTS and DISEASE; and the price of fertilisers and PESTICIDES is both high and variable. Until the 1970s, it focused on wheat, rice and maize and aimed mainly for yield increase; since the 1970s a broader range of foods, including livestock, have been given attention and efforts are being made to reduce environmental and social impacts and reliance on oil. The goal today is generally held to be a *Doubly Green Revolution*, one that emphasises CONSERVATION as well as PRODUCTIVITY.   *JAM/CJB*

[*See also* AGRICULTURAL REVOLUTION]

Brown LR (1970) *Seeds of change.* New York: Praeger.
Conway GR and Barbier ER (1990) *After the Green Revolution: Sustainable agriculture for development.* London: Earthscan.
Das RJ (2001) The green revolution and poverty: A theoretical and empirical examination of the relation between technology and society. *Geoforum* 33: 55–72.
Evenson RE and Gollin D (2003) Assessing the impact of the Green Revolution. *Science* 300: 758–762.
Glaesser B (ed.) (1987) *The Green Revolution revisited: Critique and alternatives.* London: Allen and Unwin.
Ruttan V, Serageldin I and Conway G (1999) *The Doubly Green Revolution: Food for all in the twenty-first century.* Ithaca, NY: Cornell University Press.

greenhouse condition   A term used to describe a long interval of global warmth in the Earth's past, associated with high *sea level*, an absence of ICE SHEETS or ICE CAPS other than at polar latitudes and a prominence of sedimentary FACIES that indicate a warm palaeoclimate (see PALAEOCLIMATOLOGY). At least five oscillations between greenhouse and cooler ICEHOUSE CONDITIONS can be recognised in the GEOLOGICAL RECORD of the Late PROTEROZOIC and PHANEROZOIC. Greenhouse conditions characterised the Late Cambrian to Late Devonian and Mid-Triassic to Early Cretaceous.   *GO*

[*See also* GEOLOGICAL RECORD OF ENVIRONMENTAL CHANGE, GREENHOUSE EFFECT, HYPERTHERMALS, SEQUENCE STRATIGRAPHY]

Katz ME, Miller KG, Wright JD et al. (2008) Stepwise transition from the Eocene greenhouse to the Oligocene icehouse. *Nature Geoscience* 1: 329–334.
Kidder DL and Worsley TR (2012) A human-induced hothouse climate? *GSA Today* 22: 4–11.
Lenz OK, Wilde V, Riegel W and Harms F-J (2010) A 600 ky record of El Niño-Southern Oscillation (ENSO): Evidence for persisting teleconnections during the Middle Eocene greenhouse climate of Central Europe. *Geology* 38: 627–630.
Retallack GJ (2009) Greenhouse crises of the past 300 million years. *Geological Society of America Bulletin* 121: 1441–1455.

greenhouse effect   The imperfect analogy between the Earth's ATMOSPHERE and a greenhouse. Incoming short-wave SOLAR RADIATION penetrates the atmosphere, whereas outgoing long-wave TERRESTRIAL RADIATION heats the lower atmosphere as it is *absorbed* by certain TRACE GASES (GREENHOUSE GASES). This NATURAL greenhouse effect results in the air temperature close to the Earth's surface being some 30°C higher than it would otherwise be and should therefore be differentiated from any *enhanced* greenhouse effect being produced by anthropogenic EMISSIONS of greenhouse gases.   *JAM/EZ*

[*See also* GLOBAL WARMING]

Arrhenius S (1896) On the influence of carbonic acid in the air upon the temperature on the ground. *Philosophical Magazine* 41: 237–276.

Fleming JR (2007) *The Callendar effect: The life and work of Guy Stewart Callendar (1898–1964), the scientist who established the carbon dioxide theory of climate change.* Boston: American Meteorological Society.

Ramanathan V (1999) Trace-gas greenhouse effect and global warming. *Ambio* 27: 187–197.

greenhouse gases   Atmospheric TRACE GASES that allow short-wave SOLAR RADIATION to pass through unaffected but which absorb long-wave TERRESTRIAL RADIATION. The net effect is to trap part of the *infrared radiation* that would otherwise escape to space and to warm the Earth's surface and lower ATMOSPHERE. Greenhouse gases naturally present in the ATMOSPHERE raise the GLOBAL MEAN SURFACE AIR TEMPERATURE by about 21°C more than it would be if the NATURAL greenhouse gases were not present. This effect is known as the GREENHOUSE EFFECT. Human activities are causing concentrations of some greenhouse gases to increase and this is causing an additional anthropogenic or *enhanced greenhouse effect* estimated by GENERAL CIRCULATION MODELS to be currently around 0.2°C per decade.

Greenhouse, *radiatively active gases* or *climate-active gases* include principally WATER VAPOUR (H_2O) but also CARBON DIOXIDE (CO_2), *methane* (CH_4), *nitrous oxide* (N_2O), OZONE (O_3), HALOGENATED HYDROCARBONS (HALOCARBONS) and halocarbon substitutes. Halogenated hydrocarbons include CHLOROFLUOROCARBONS (CFCs), *bromofluorocarbons* (BFCs), *perfluorocarbons* (PFCs), *sulfur hexaflouride* (SF_6), *methyl chloroform* (CH_3CCl_3) and *carbon tetrachloride* (CCl_4). Halocarbon substitutes include HYDROCHLOROFLUORO-CARBONS (HCFCs) and *hydrofluorocarbons* (HFCs). Carbon dioxide is produced by the combustion of FOSSIL FUELS and LANDUSE change (especially tropical DEFORESTATION). Methane is released by AGRICULTURE (rice PADDY SOILS, animal husbandry), waste disposal (LANDFILL), BIOMASS BURNING and FOSSIL FUEL production and use. Nitrous oxide sources are mainly agriculture (development of PASTURE in tropical regions), biomass burning and some industrial processes (*nitric acid production*). Tropospheric ozone is a SECONDARY POLLUTANT formed by the action of sunlight on NITROGEN OXIDES and volatile organic compounds. The increase in carbon dioxide concentrations is held responsible for about two-thirds of the enhanced radiative forcing by greenhouse gases.   *DME/PU/CW*

[*See also* CARBON BALANCE, CARBON SEQUESTRATION]

Boag S, White DH and Howden SM (1994) Monitoring and reducing greenhouse gas emissions from agricultural, forestry and other human activities. *Climatic Change* 27: 5–11.

Galford GL, Melillo J, Mustard JF et al. (2010) The Amazon frontier of land-use change: Croplands and consequences for greenhouse gas emissions. *Earth Interactions* 14: 1–24.

Hannachi A and Turner AG (2008) Preferred structures in large-scale circulation and the effect of doubling greenhouse gas concentration in HadCM3. *Quarterly Journal of the Royal Meteorological Society* 134: 469–480.

Houghton JT (2009) *Global warming: The complete briefing,* 4th edition. Cambridge: Cambridge University Press.

Jouzel J and Masson-Delmotte V (2010) Deep ice cores: The need for going back in time. *Quaternary Science Reviews* 29: 3683–3689.

Montzka SA, Dlugokencky EJ and Butler JH (2011) Non-CO_2 greenhouse gases and climate change. *Nature* 476: 43–50.

Ramanathan V and Feng Y (2009) Air pollution, greenhouse gases and climate change: Global and regional perspectives. *Atmospheric Environment* 43: 37–50.

Raynaud D, Barnola J-M, Chappellaz J et al. (2000) The ice record of greenhouse gases: A view in the context of future changes. *Quaternary Science Reviews* 19: 9–17.

Solomon SD, Qin D, Manning M et al. (eds) (2007) *Climate change 2007: The physical basis* [Contribution of Working Group I to the Fourth Assessment Report of the Intergovernmental Panel on Climate Change]. Cambridge: Cambridge University Press.

greenhouse-gas forcing   The role played by GREENHOUSE GASES in controlling the emission of TERRESTRIAL RADIATION (i.e. *long-wave radiation*) from the lower atmosphere into space, and hence CLIMATE. Greenhouse-gas forcing has played a major role throughout much of Earth's history and has done much to provide an environment within which life can flourish. Indeed, without the so-called GREENHOUSE EFFECT, it has been estimated that the average global surface temperature would be >30°C lower than currently, making life all but impossible. However, recent concerns have arisen over the input of anthropogenically derived greenhouse gases, mostly derived from the burning of fossil fuels. The consequent increasing concentrations of greenhouse gases are widely believed to be raising global temperatures (and bringing about GLOBAL WARMING) at a rate not witnessed before. This is threatening livelihoods, the *ecosystems* of the world, WATER RESOURCES and the productivity of AGRICULTURE, to mention only a few of the anxieties.   *DAW*

[*See also* CLIMATE OF THE LAST MILLENNIUM, FORCING FACTOR, FOSSIL FUELS: IN CLIMATE CHANGE, FUTURE CLIMATE, GREENHOUSE EFFECT]

Forster P, Ramaswamy V, Artaxo P et al. (2007) Changes in atmospheric constituents and in radiative forcing. In Intergovernmental Panel on Climate Change (ed.) *Climate change 2007: The physical science basis.* Cambridge: Cambridge University Press, 129–234.

Harries JE (1996) The greenhouse earth: A view from space. *Quarterly Journal of the Royal Meteorological Society* 122: 799–818.

Houghton J (2004) *Global warming: The complete briefing,*
  2nd edition. Cambridge: Cambridge University Press.

greening of society The realignment of societal
values since the 1970s in response to the growing threat
of ENVIRONMENTAL DEGRADATION on a global scale. The
greening of society has been exemplified at many levels, but mainly through changing CONSUMPTION habits in
response partly to government legislation and more so
to the influence of ENVIRONMENTAL MOVEMENTS and PRES-
SURE GROUPS. There is a trend towards low consumption and sustainable practices such as paper and glass
RECYCLING and a growing sense of responsibility felt by
society for limiting further damage to the environment.
In addition, environmental education has grown in status at all levels with the next generation of adults being
particular targets for media coverage of ENVIRONMEN-
TAL ISSUES such as DEFORESTATION and OZONE DEPLETION.
Also, the expansion of environmental awareness into
the political arena has seen the development of GREEN
POLITICS and ENVIRONMENTAL LAW to monitor compliance
to regulations for potentially damaging practices.   *JGS*

[*See also* ENVIRONMENTAL MOVEMENT,
ENVIRONMENTALISM, GREEN POLITICS, POLITICAL
ECOLOGY, SUSTAINABILITY]

Carley M and Spapens M (1998) *Sharing the world:*
  Sustainable living and global equality in the 21st century.
  London: Earthscan.
Driessen PT and Glasbergen P (eds) (2002) *Greening*
  society: The paradigm shift in Dutch environmental
  politics. Dordrecht: Kluwer.
Sachs W, Laske R and Linz M (1998) *Greening the north: A*
  post-industrial blueprint for ecology and equity. London:
  Zed Books.

Greenland Ice-Core Chronology The timescale
developed for the Greenland ICE-CORE records and
referred to as Greenland Ice-Core Chronology 2005 or
GICC05. The chronology is based on a multiparameter
approach of counting annually resolved variations (see
ANNUALLY RESOLVED RECORD) in high-resolution data
sets (see HIGH-RESOLUTION RECONSTRUCTIONS) such as
isotopic measurements, continuous flow analysis data
(e.g. chemical impurities, DUST and *electrolytical conductivity*) and ELECTRICAL CONDUCTIVITY MEASUREMENTS
(ECM) of solid ice as well as visual stratigraphical data
in three ice cores (Dye 3, GRIP and NGRIP). All three
cores provide a record to ~8,000 year ago, GRIP and
NGRIP to ~15,000 years ago and only NGRIP beyond
that to ~60,000 years ago. An ongoing challenge is the
SYNCHRONISATION of the ice-core chronology with other
records of PALAEOCLIMATOLOGY such as records from
MARINE SEDIMENT CORES, terrestrial long-core sequences,
such as those from LACUSTRINE SEDIMENTS, and SPELEO-
THEM records.   *SMD/GO*

Andersen KK, Svensson A, Johnsen SJ et al. (2006) The
  Greenland Ice Core Chronology 2005, 15-42 ka. Part 1:
  Constructing the time scale. *Quaternary Science Reviews*
  25: 3246–3257.
Rasmussen SO, Andersen KK, Svensson AM et al. (2006)
  A new Greenland ice core chronology for the last glacial
  termination. *Journal of Geophysical Research* 111:
  D06102.
Svensson A, Bigler M, Clausen HB et al. (2008) A 60000
  year Greenland stratigraphic ice core chronology. *Climate
  Past* 4: 47–57.
Vinther BM, Clausen HB, Johnsen SJ et al. (2006) A
  synchronized dating of three Greenland ice cores
  throughout the Holocene. *Journal of Geophysical
  Research* 111: D13102.

greenstone belt An area of deformed and metamorphosed SEDIMENTARY ROCKS and IGNEOUS ROCKS
(including KOMATIITE lavas) of ARCHAEAN age. Many
greenstone belts have been intruded by granite, forming *granite-greenstone belts*, which form the cores of
many CRATONS. It is generally agreed that greenstone
belts formed through SEA-FLOOR SPREADING and provide
evidence for the initiation and nature of PLATE TECTONICS
on the early Earth, which may have operated in rather
different ways than in the PROTEROZOIC and PHANEROZOIC
(see UNIFORMITARIANISM). Greenstone belts demonstrate
the existence in the Archaean of many of the aqueous
processes and environments that operate today and preserve the earliest traces of life on Earth in the form of
STROMATOLITES and filamentous MICROFOSSILS in CHERT,
both with a record to 3,500 million years ago.   *GO*

Banerjee NR, Furnes H, Muehlenbachs K et al. (2006)
  Preservation of ~3.4-3.5 Ga microbial biomarkers in
  pillow lavas and hyaloclastites from the Barberton
  Greenstone Belt, South Africa. *Earth and Planetary
  Science Letters* 241: 707–722.
De Wit MJ (1998) On Archean granites, greenstones, cratons
  and tectonics: Does the evidence demand a verdict?
  Precambrian Research 91: 181–226.
Hofmann A (2005) The geochemistry of sedimentary rocks
  from the Fig Tree Group, Barberton Greenstone Belt:
  Implications for tectonic, hydrothermal and surface
  processes during mid-Archaean times. *Precambrian
  Research* 143: 23–49.
O'Neil J, Francis D and Carlson RW (2011) Implications of
  the Nuvvuagittuq Greenstone Belt for the formation of
  Earth's early crust. *Journal of Petrology* 52: 985–1009.
Schopf JW (2006) Fossil evidence of Archaean life.
  Philosophical Transactions of the Royal Society B 361:
  869–885.
Stanley SM (2009) *Earth system history.* New York: Freeman.

greigite Ferrimagnetic *iron sulfide* (Fe_3S_4) occurring in various forms in freshwater and BRACKISH WATER
or MARINE SEDIMENTS, as well as in GLEYSOLS and PEAT
AND PEATLANDS.   *UBW*

[*See also* GEOMAGNETISM, LAMINATED LAKE
SEDIMENTS, MAGNETIC INTENSITY, MAGNETIC
SUSCEPTIBILITY, MAGNETITE, PALAEOMAGNETISM]

Babinszki E, Marton E, Marton P and Kiss LF (2007)
Widespread occurrence of greigite in the sediments
of Lake Pannon: Implications for environment
and magnetostratigraphy. *Palaeogeography,
Palaeoclimatology, Palaeoecology* 252: 626–636.
Oldfield F (2007) Sources of fine-grained magnetic minerals
in sediments: A problem revisited. *The Holocene* 17:
1265–1271.
Walden J, Oldfield F and Smith J (eds) (1999) *Environmental
magnetism: A practical guide*. London: Quaternary
Research Association.

Grentzhorizont   An abrupt reduction in SPHAG-
NUM decomposition around 2,500 radiocarbon years
BP reported from northwestern and Central European
MIRES. It was originally used to link PEAT AND PEATLANDS
from different geographical areas to the same fluctuations in climate, but evidence from PEAT STRATIGRAPHY
and RADIOCARBON DATING shows that the surfaces are
neither as regular nor as synchronous as was originally
believed, and the term can be misleading.   *MJB*

[*See also* PEAT HUMIFICATION, RECURRENCE SURFACE]

Weber CA (1900) Über die Moore, mit besonderer
Berucksichtigung der awischen Underweser und
Underelbe liegenden [Lying on the moors, with special
regard to awischen Under Weser and Elbe Under].
Jahresbericht der Manner von Morgenstern 3: 3–23.

grey-scale analysis   A measure of variations in the
relative *reflectivity* of SEDIMENTS, normally obtained by
scanning the freshly cut surface of a CORE. The digital
records can be treated statistically and are a powerful
tool in the analysis and correlation of VARVES and other
laminated sediments.   *MRT*

[*See also* CORE SCANNING, IMAGE INTERPRETATION,
X-RADIOGRAPHY]

Hughen KA, Overpeck JT, Peterson LC and Tumbore S
(1996) Rapid changes in the tropical Atlantic region
during the last deglaciation. *Nature* 380: 51–54.

greywacke   A largely outdated term for a WACKE
(i.e. a MUD-rich SANDSTONE, with >15 per cent MATRIX).
The term, sometimes spelled 'graywacke', became
closely associated with sandstones interpreted as having been deposited by TURBIDITY CURRENTS and hence
was too tainted with process implications for continued
employment as a formal descriptive term.   *TY*

Dott Jr RH (1964) Wacke, graywacke and matrix: What
approach to immature sandstone classification? *Journal of
Sedimentary Petrology* 34: 625–632.
Dzulynski S and Walton EK (1965) *Sedimentary features of
flysch and greywackes*. Amsterdam: Elsevier.

grèzes litées   A regional name given to the type of
rhythmically stratified slope-waste deposit, of Pleistocene age, that occur beneath degraded LIMESTONE
cliffs in the region of Charente, France. Characteristics include relatively small angular clasts and crude
parallel BEDDING with alternating *clast-supported* and
matrix-supported beds of thickness 10–25 cm. FROST
WEATHERING, SOLIFLUCTION, NEEDLE ICE, DEBRIS FLOW
and SLOPEWASH processes have all been implicated in
their formation. The more general term for this type
of stratified sediment is *stratified slope deposit*. Such
deposits occur mostly in mid-latitudes but are also
known to be forming today in cold humid regions. In
general, they are interpreted as the result of intense
frost weathering acting upon frost-susceptible bedrock assisted by MASS MOVEMENT PROCESSES. There has
been some confusion with the term ÉBOULIS ORDONNÉS,
which some view as synonymous with grèzes litées.
    HMF

[*See also* COLLUVIUM, PERIGLACIAL SEDIMENTS]

García-Ruiz JM, Valero B, González-Sampériz P et al.
(2001) Stratified scree in the Central Spanish Pyrenees:
Palaeoenvironmental implications. *Permafrost and
Periglacial Processes* 12: 233–242.
Guillien Y (1951) Les grèzes litées de Charente [The layered
grèzes Charente]. *Revue de Géographie des Pyrénées et
du Sud-Ouest* 22: 154–162
Ozouf JC, Coutard JP and Lautridou JP (1995) Grèzes,
grèzes litées: Historique des définitions [Grèzes,
grèzes layered: Historical definitions]. *Permafrost and
Periglacial Processes* 6: 85–87.
Van Steijn H (2011) Stratified slope deposits: Periglacial
and other processes involved. In Martini IP, French HM
and Perez-Alberti A (eds) *Ice-marginal and periglacial
processes and sediments*. Bath: Geological Society,
213–226.

grid   A square network formed by two sets of intersecting equidistant parallel straight lines. Grids are
used as the basis of reference location system such as
the *UK Ordnance Survey National Grid* or *US State
Plane System*.   *TF*

[*See also* CO-ORDINATE, GRID CELL]

grid cell   A single enclosed element in a GRID
structure. Grid cells are associated with RASTER-type
images where a grid cell may equate to a PIXEL (picture element).   *TF*

[*See also* GRID]

grit   Formerly used in an ill-defined sense to describe
a 'gritty' SANDSTONE; in most cases a tough, coarse-
grained sandstone with angular CLASTS. The term is
obsolete in a descriptive sense because of a lack of clar-
ity and consistency in its use. However, it is retained

in many formal names in STRATIGRAPHY (e.g. Ystrad Meurig Grits FORMATION (STRATIGRAPHICAL)).   *GO*

gross primary productivity (GPP)   The total fixation rate of energy by PHOTOSYNTHESIS. GPP is the production rate before heat losses from RESPIRATION are accounted for; it is the *photosynthetic production rate* for plants and *metabolisable production rate* for animals.   *RJH*

[*See also* NET PRIMARY PRODUCTIVITY (NPP), PRODUCTIVITY]

Williams M, Rastetter EB, Fernandes DN et al. (1997) Predicting gross primary productivity in terrestrial ecosystems. *Ecological Applications* 7: 882–894.

ground ice   A general term used to refer to the many types of ice formed, or preserved, in freezing and frozen ground. It is one of the most important components of PERMAFROST and is an essential component of THERMO-KARST. The main types of ground ice are (1) PORE ICE, (2) SEGREGATION ICE, (3) *vein ice* (in cracks or fissures), (4) *wedge ice* (repeated vein-ice formation producing ICE WEDGES), (5) *intrusive ice* (which is injected under pressure, e.g. into PINGOS) and (6) *thermokarst-cave ice* (formed from the winter freezing of pooled water in poorly drained TUNDRA). Buried GLACIER ice is not generally regarded as a type of ground ice sensu stricto.
                                                          HMF

[*See also* CRYOSTRATIGRAPHY, GEOCRYOLOGY]

French HM (2007) *The periglacial environment, 3rd edition.* Chichester: Wiley.
Harris SA, French HM and Heginbottom JA (eds) (1988) *Glossary of permafrost and related ground-ice terms.* Ottawa: National Research Council of Canada.

ground measurement   Measurement of variables on the Earth's surface in order to calibrate or validate data derived by REMOTE SENSING. It should be carried out at the same time as the data acquisition from the sensor, at accurate sampling locations.   *HB*

ground temperature   The temperature (1) at the ground surface or (2) near the ground surface at specified shallow depths. At *meteorological stations,* minimum temperatures are measured at the ground surface: *grass minimum temperature* is recorded a few millimetres above a grass surface. At such a height, temperatures below freezing are known as *ground frosts.* In addition, *soil temperatures* are recorded at depths of 50, 100 and 200 mm.

Ground temperature may differ considerably from the AIR TEMPERATURE above, which may be a poor predictor of it, even though smoothing of data from daily to monthly scales may improve the correlation between the two. Ground temperatures tend to be more extreme and variable than the corresponding air temperature. *Mean annual ground temperature (MAGT)* estimated from measurements based on a thermistor cable to 10 m depth in an area of MOUNTAIN PERMAFROST in southern Norway was about 2.5°C lower than the MEAN ANNUAL AIR TEMPERATURE (MAAT), but this may vary greatly between sites as a result of factors such as vegetation cover, topography and snow depth.   *JGT/JAM*

[*See also* MICROCLIMATE, SUBSURFACE TEMPERATURE]

Ødegård RS, Sollid JL and Liestøl O (1992) Ground temperature measurements in mountain permafrost, Jotunheimen, southern Norway. *Permafrost and Periglacial Processes* 3: 231–234.
Thorn CE, Darmody RG and Allen CE (2008) Ground temperature variability on a glacier foreland, Storbreen, Jotunheimen, Norway. *Norsk Geografisk Tidsskrift* 62: 290–302.

ground-based remote sensing   REMOTE SENSING of environmentally relevant variables carried out from the ground. Ground-based or *surface-based remote sensing* of the ATMOSPHERE with LIDARS, RADARS, RADIOMETERS and other instruments is used to retrieve the distribution and profile of different atmospheric quantities such as TEMPERATURE, PRECIPITATION, WATER VAPOUR, OZONE and AEROSOLS. Ground-based remote sensing is also relevant for the establishment of useful relationships between specific terrestrial parameters and the energy received by THERMAL, MICROWAVE or OPTICAL REMOTE-SENSING INSTRUMENTS and for the proper choice of the instrument's parameters (e.g. wavelength, incidence angle and polarisation) to be used from space. Ground-based remote sensing is also used to capture the vertical aspects of structures and environmental features. For instance, oblique photography is used to capture building facades and ground-based LiDARs are used to capture the three-dimensional structure of urban areas and forests.   *TS*

Cimini D, Marzano FS and Visconti G (eds) (2010) *Integrated ground-based observing systems: Applications for climate, meteorology and civil protection.* Heidelberg: Springer.
Emeis S (2010) *Surface-based remote sensing of the atmospheric boundary layer.* Dordrecht: Springer.

grounding line   The boundary between the part of an ICE SHEET that is floating and the part that is in contact with bedrock. It lies below sea level and is an important control on the dynamics and stability of the terminal zone of MARINE-BASED ICE SHEETS.   *JAM*

[*See also* GLACIOMARINE DEPOSITS]

Katz RF and Worster MG (2010) Stability of ice-sheet grounding lines. *Proceedings of the Royal Society A* 466: 1597–1620.
MacGregor JA, Anandakrishnan S, Catania GA and Winebrenner DP (2011) The grounding zone of the Ross

Ice Shelf, West Antarctica, from ice-penetrating radar. *Journal of Glaciology* 57: 917–928.

grounding-zone wedge   A ridge or fan of primarily SUBGLACIAL sediment deposited at the front of many ICE STREAMS and TIDEWATER GLACIERS normally at approximately right angles to ice flow direction when the GROUNDING LINE is stationary. The volume of sediment in the wedge is linked to the rate of *sediment delivery* and the time that the grounding line remains stationary. Wedges can range from 10 to 100 m in height and in some situations are thought to stabilise the grounding line for many decades, buffering the ICE SHEET from FORCING FACTORS such as SEA-LEVEL RISE.   *BTIR*

[*See also* MARINE-BASED ICE SHEET]

Alley RB, Anandakrishnan S, Dupont TK et al. (2007) Effect of sedimentation on ice-sheet grounding-line stability. *Science* 315: 1838–1841.
Dowdeswell JA, Ottesen D, Evans J et al. (2008) Submarine glacial landforms and rates of ice-stream collapse. *Geology* 36: 819–822.

ground-penetrating radar (GPR)   A method of GEOPHYSICAL SURVEYING that uses *electromagnetic radiation* to provide detailed images of the subsurface structure of SOILS and SEDIMENTS to depths of several metres.   *GO*

[*See also* ELECTROMAGNETIC SPECTRUM, RADAR REMOTE-SENSING INSTRUMENTS]

Baker GS and Jol HM (eds) (2007) *Stratigraphic analyses using GPR.* Boulder, CO: Geological Society of America.
Bristow CS, Lancaster N and Duller GAT (2005) Combining ground penetrating radar surveys and optical dating to determine dune migration in Namibia. *Journal of the Geological Society of London* 162: 315–321.
Kramer N, Wohl EE and Harry DL (2012) Using ground penetrating radar to 'unearth' buried beaver dams. *Geology* 40: 43–46.
Sass O and Krautblatter M (2007) Debris flow-dominated and rockfall-dominated talus slopes: Genetic models derived from GPR measurements. *Geomorphology* 86: 176–192.

groundwater   The part of the HYDROSPHERE beneath the ground surface. Groundwater, under half of which is *fresh groundwater*, comprises about 1.6 per cent of all the Earth's water. Fresh groundwater comprises about 30 per cent of the Earth's freshwater (excluding frozen groundwater in PERMAFROST environments). Most is contained within PERMEABLE rocks (AQUIFERS) in the saturated or PHREATIC ZONE, where it forms contiguous bodies of water that move. Groundwater is a geological and geomorphological agent (particularly apparent in KARST landscapes), is an important supplier of BASEFLOW to streams and rivers and is an important *reservoir* or *store* within the global HYDROLOGICAL CYCLE. Groundwater may have a considerable age: commonly hundreds of years and in some cases thousands of years or more in deep aquifers,

but often a much shorter residence time in shallow aquifers (especially in relatively wet environments).

Groundwater is a major source for urban and rural *water supplies* and hence is a major NATURAL RESOURCE. In the developed world, URBANISATION is a major cause of GROUNDWATER DEPLETION through excessive ABSTRACTION, which often leads to a lowering of the WATER TABLE. The impact and use of water in cities in ARIDLANDS may, somewhat perversely, lead to a localised rise in the water table as surplus water descends through the VADOSE ZONE to the phreatic zone by PERCOLATION. Elsewhere in aridlands, however, abstraction of groundwater for IRRIGATION generally leads to a lowering of the water table. Recent modelling has indicated that 42 per cent of the ~8 cm SEA-LEVEL RISE observed between AD 1961 and 2003 could be accounted for by changes in terrestrial water storage, particularly groundwater use. Natural GROUNDWATER RECHARGE of the depleted aquifers may take a very long time to the extent that, in some areas at least, groundwater should be regarded as a NON-RENEWABLE RESOURCE. In coastal areas, abstraction may also lead to SALTWATER INTRUSION and hence *groundwater salinisation*. In areas affected by AGRICULTURAL INTENSIFICATION, groundwater POLLUTION from FERTILISERS is an additional concern.

Groundwater is emerging as a NATURAL ARCHIVE for PALAEOHYDROLOGY and PALAEOCLIMATOLOGY. PROXY DATA available from large groundwater bodies tend to be of low *temporal resolution* (typically ±1,000 years). Nevertheless, climatic and vegetational history may be reconstructed with the aid of various chemical and isotopic signals; especially in *confined aquifers*, where sequential changes may be recorded along flow lines or in the stratification of the phreatic zone. Under favourable circumstances, moisture in the unsaturated vadose zone also contains records of DECADAL-SCALE VARIABILITY or CENTURY- TO MILLENNIAL-SCALE VARIABILITY.   *JAM/RPDW*

[*See also* ARTESIAN, DRY VALLEY, GROUNDWATER MINING, HYDROGEOLOGY]

Alley WM, Healy RW, LaBaugh JW and Reilly TE (2002) Flow and storage in groundwater systems. *Science* 296: 1985–1990.
Brown AG (ed.) (1995) *Geomorphology and groundwater.* Chichester: Wiley.
Dragoni W and Sukhija BS (eds) (2008) *Climate change and groundwater.* Bath: Geological Society.
Edmunds WM and Tyler SW (2002) Unsaturated zones as archives of past climates: Towards a new proxy for continental regions. *Hydrogeology Journal* 10: 216–228.
Edmunds WM, Dodo A, Djoret et al. (2004) Groundwater as an archive of climatic and environmental change: Europe to Africa. In Battarbee RW, Gasse F and Stickley CE (eds) *Past climate variability through Europe and Africa.* Dordrecht: Springer, 279–306.
Fitts CR (2002) *Groundwater science.* San Diego, CA: Academic Press.

Kresic N (2006) *Quantitative solutions in hydrogeology and groundwater modelling, 2nd edition.* Boca Raton, FL: CRC Press.

Pokhrel YN, Hanasaki N, Yeh PJ-F et al. (2012) Model estimates of sea-level change due to anthropogenic impacts on terrestrial water storage. *Nature Geoscience* 5: 389–392.

Price M (1996) *Introducing groundwater, 2nd edition.* London: Chapman and Hall.

Younger PL (2007) *Groundwater in the environment: An introduction.* Malden, MA: Blackwell.

groundwater depletion   Reduction in the volume of water stored in an AQUIFER and consequent lowering of the WATER TABLE through water ABSTRACTION. Groundwater depletion may be the direct effect of *groundwater abstraction* or a more complex, indirect effect of the abstraction of river water, or the result of LANDUSE CHANGE or a drier climate, leading to reduced GROUNDWATER RECHARGE. It results, therefore, when abstraction and natural losses to rivers, the sea or adjacent basins exceed groundwater recharge from PERCOLATION of soil water or INFILTRATION of river water into beds and banks. It has been estimated that around 12.6 mm (or 6 per cent) of global SEA-LEVEL RISE over the period AD 1900–2008 has been due to groundwater depletion.

JAM/RPDW

[*See also* GROUNDWATER MINING]

Konikow LF (2011) Contribution of groundwater depletion since 1900 to sea-level rise. *Geophysical Research Letters* 38: L17401.

Nelson RL (2012) Assessing local planning to control groundwater depletion: California as a microcosm of global issues. *Water Resources Research* 48: W01502.

Sarkar A (2012) Sustaining livelihoods in face of groundwater depletion: A case study of Punjab, India. *Environmental Development and Sustainability* 14: 183–195.

groundwater mining   Excessive groundwater ABSTRACTION to the extent that SUSTAINABILITY of supply is threatened. The term is used in particular in situations where (1) GROUNDWATER RECHARGE fails to keep pace with current abstraction and hence GROUNDWATER DEPLETION occurs and/or (2) the groundwater is ancient and hence can be considered as a NON-RENEWABLE RESOURCE. The Ogallala Aquifer, 1,800 m beneath the Great Plains, USA, provides a good example of groundwater mining because the abstraction rate appears to be 100 times the current recharge rate and much of the water in the aquifer dates from the LAST GLACIATION. The term is increasingly used in ARIDLANDS and SEMI-ARID regions, such as the Middle East, where groundwater abstraction is accelerating and recharge is slow.   *JAM*

Brikowski TH (2008) Doomed reservoirs in Kansas, USA? Climate change and groundwater mining on the Great Plains lead to unsustainable surface water storage. *Journal of Hydrology* 354: 90–101.

Moore JE, Raynolds RG and Barkmann PE (2004) Groundwater mining of bedrock aquifers in the Denver Basin: Past, present and future. *Environmental Geology* 47: 63–68.

Sloggett G and Dickason C (1986) *Groundwater mining in the United States.* Washington, DC: Government Printing Office.

groundwater pollution   The POLLUTION of water stored beneath the ground surface in soil pore spaces and in pores, cracks and fissures in rocks. GROUNDWATER is in direct contact with the *subsoil* or BEDROCK, and hence *groundwater quality* is affected by the interaction of water, minerals, gases, pollutants and microbes. Physical and chemical theories can be applied to explain observed water qualities in groundwater and variations over space and time. POLLUTANTS and CONTAMINANTS can be introduced into groundwater in different ways. Pollution can occur when HAZARDOUS SUBSTANCES, including HAZARDOUS WASTE come into contact with soils. Rainfall or surface water seeps through soils, dissolving the contaminants which can then enter groundwater. Activities with potential to cause contaminated soils include URBANISATION, *mining* and AGRICULTURE including FERTILISERS and PESTICIDES. In addition, mining activities at depth, such as FRACKING and the extraction of SHALE GAS, run the risk of groundwater pollution. Some hazardous liquids do not mix with the groundwater but remain pooled within the soil or bedrock. These then act as long-term sources of groundwater pollution as water flows through the soil or rock and comes into contact with them. There are debates about the speed with which *pollution fronts* (e.g. NITRATES from agriculture) can move down to the groundwater resource. Polluted groundwater can resurface through artificial or natural processes (e.g. SPRINGS). Groundwater pollution can be a major HUMAN HEALTH HAZARD through drinking water supplies (see POTABLE WATER) and impacts on the FOOD CHAIN (e.g. via animals drinking polluted water). *Groundwater vulnerability* describes the susceptibility of groundwater to pollution from human activities. This is affected by the thickness and permeability of overlying strata and the depth of the *saturated zone* (see PHREATIC ZONE), which influence the movement of contaminants from surface sources to underlying AQUIFERS.   *LJMcE*

[*See also* ACID MINE DRAINAGE, CONTAMINATED LAND, METAL POLLUTION, SOIL POLLUTION]

Appelo CAJ and Postma D (2005) *Geochemistry, groundwater and pollution, 2nd edition.* London: Taylor and Francis.

Lerner DN (ed.) (2007) *Urban groundwater pollution.* Lisse: Swets and Zeitlinger.

Ramachandra TV (2009) *Soil and groundwater pollution from agricultural activity.* New Delhi: The Energy and Resources Institute (TERI).

US Environmental Protection Agency (2012) *Groundwater contamination* [Available at http://www.epa.gov/superfund/students/wastsite/grndwatr.htm].

Xu Y and Usher B (eds) (2006) *Groundwater pollution in Africa.* Leyden: Taylor and Francis.

groundwater recharge The replenishment of GROUNDWATER that has been depleted by natural means or human ABSTRACTION. It is an important part of the HYDROLOGICAL CYCLE, by which SURFACE WATER moves down into the subsurface. Surface water for groundwater recharge can be supplied naturally by rain, snowmelt, rivers, streams and lakes. The recharge rate depends on environmental and artificial controls. It is important to understand how much water is entering the groundwater supply as this influences how much water can safely be abstracted from groundwater supplies for human uses, such as *drinking water* (see POTABLE WATER) or IRRIGATION. Studies of groundwater recharge tend to focus on SEMI-ARID regions because here groundwater is often the main source of water, is vulnerable to CONTAMINANTS and is prone to *depletion.* Key research areas include current understanding of recharge processes, identification of recurring recharge-evaluation problems and groundwater recharge estimation techniques. Groundwater recharge can be managed artificially as a groundwater storage technique to restore or stabilise groundwater levels. *LJMcE*

[*See also* GROUNDWATER DEPLETION, GROUNDWATER MINING, GROUNDWATER POLLUTION]

De Vries J and Simmers I (2002) Groundwater recharge: An overview of processes and challenges. *Hydrogeology Journal* 10: 5–17.
Healy RW (2010) *Estimating groundwater recharge.* Cambridge: Cambridge University Press.
McMahon PB, Plummer LN, Boehlke JK et al (2011) A comparison of recharge rates in aquifers of the United States based on groundwater-age data. *Hydrogeology Journal* 19: 779–800.

group A unit in LITHOSTRATIGRAPHY comprising two or more FORMATIONS (STRATIGRAPHICAL) linked by significant characteristics of LITHOLOGY or origin (e.g. *Torridon Group, Lias Group*). The boundaries of a group should mark changes in lithology, depositional environment or process or a stratigraphical break. Group is used in SEQUENCE STRATIGRAPHY to link genetically related successions of formations between major bounding unconformities. *LC*

Barnes RP, Branney MJ, Stone P and Woodcock NH (2006) The Lakesman terrane: The Lower Palaeozoic record of the deep marine Lakesman Basin, a volcanic arc and a foreland basin. In Brenchley PJ and Rawson PF (eds) *The geology of England and Wales, 2nd edition.* Bath: Geological Society, 103–129.

growing season The period of a year when plants will grow, commonly operationalised as the period when mean daily temperatures exceed the temperature at which plants will grow. Where temperature is the LIMITING FACTOR in TEMPERATE CLIMATES and at high latitudes, this can be approximated by such indices as the period between the first spring and last autumn frost (the *frost-free season*) or, alternatively, a system of ACCUMULATED TEMPERATURE based on DEGREE DAYS. The growing season determines the mix of natural vegetation and types of crop that will grow. In DRYLANDS, moisture may be the limiting factor. *RJH/JAM*

[*See also* PHENOLOGY, PHOTOSYNTHESIS]

Thompson R and Clark RM (2008) Is spring starting earlier? *The Holocene* 18: 95–104.

growth form A group of organisms (usually plants) classified according to their morphological characteristics. Examples include *broad-leaved evergreen trees, needle-leaved evergreen trees, palms, lianas, succulents, dwarf shrubs* and *cushions.* Such *vegetation classification* based on growth form or LIFE FORM is complementary to *taxonomic classification.* *JLI*

Dorrepaal E (2007) Are plant growth-form-based classifications useful in predicting northern ecosystem carbon cycling feedbacks to climate change? *Journal of Ecology* 95: 1167–1180.
Rowe N and Speck T (2005) Plant growth forms: An ecological and evolutionary perspective. *New Phytologist* 166: 61–72.

groyne A structure, commonly made of wood or concrete, built perpendicular or subperpendicular to a SHORELINE to intercept sediment. Groynes (American spelling: *groins*) are built to reduce EROSION and/or to increase SEDIMENTATION. On *beaches,* groynes intercept sediment moving laterally along the beach by LONGSHORE DRIFT and thus there is often enhanced erosion (sediment starvation) beyond the *downdrift* end of a groyne or series of groynes. On shallow and more sheltered coasts, groynes may be used to capture fine sediment (SILT and CLAY) in the process of LAND RECLAMATION. *HJW/IM*

[*See also* COASTAL ENGINEERING STRUCTURES, JETTY]

Dornbusch U, Robinson DA, Moses CA and Williams RBG (2008) Variation in beach behaviour in relation to groyne spacing and groyne type for mixed sand and gravel beaches, Saltdean, UK. *Zeitschrift für Geomorphologie* 52: 125–143.

Grundhoecker A German term coined by Julius Büdel to describe a joint-controlled convexity or *basal knob* of the *basal surface of weathering.* Such knobs form shield INSELBERGS with knobbly small-scale relief when cropping out at the ground surface. *MAC*

[*See also* DOUBLE PLANATION SURFACES]

Büdel J (1982) *Climatic geomorphology.* Princeton, NJ: Princeton University Press [Translated from the German].

gruss/grussification   Grussification is the process of PHYSICAL WEATHERING by GRANULAR DISINTEGRATION of coarse-grained crystalline rock (e.g. granite), whereby gruss is produced. Gruss or *grus* usually consists almost entirely of SAND grains and coarser clasts with little finer material because there is little chemical change to the rock minerals. *Growan* is a related weathering product associated with Dartmoor granite TORS, with evidence of chemical decomposition including up to 20 per cent SILT and a small proportion of CLAY.   *PW/JAM/RAS*

[*See also* GELIFRACTION, SAPROLITE]

Ballantyne CK and Harris C (1994) *The periglaciation of Great Britain.* Cambridge: Cambridge University Press.
Migoń P and Thomas MF (2002) Grus weathering mantles: Problems of interpretation. *Catena* 49: 5–24.

Gulf Stream   A warm OCEAN CURRENT that originates in the eastern Gulf of Mexico, travels north along the east coast of the United States to about 40° N, then crosses the Atlantic Ocean to reach the British Isles at 50° N as the *North Atlantic Drift (NAD)*, which flows on to the northern reaches of the Atlantic. The Gulf Stream and its continuation as the NAD has a considerable influence on the climate of Western Europe.   *GS*

Duplessy J-C (1999) Climate and the Gulf Stream. *Nature* 403: 594–595.
Lynch-Stieglitz J, Curry WB and Slowey N (1999) Weaker Gulf Stream in the Florida Straits during the Last Glacial Maximum. *Nature* 402: 644–648.
Minobe S, Kuwano A, Komori N et al. (2008) Influence of the Gulf Stream on the troposphere. *Nature* 452: 206–209.
Sieger R, Battisti DS, Yin J et al. (2002) Is the Gulf Stream responsible for Europe's mild winters? *Quarterly Journal of the Royal Meteorological Society* 128: 2563–2586.

gull   A deep, expanded bedrock joint (on the *dip slope* of a CUESTA and parallel to the *escarpment*), which results from CAMBERING of valley-side CAPROCKS probably during PERMAFROST thaw (see Figure in the entry on VALLEY BULGING).   *PW*

gully   A steep-sided EPHEMERAL channel, often formed in poorly consolidated sediments or soils by concentrated OVERLAND FLOW or ephemeral streams. On agricultural land the term is used for features that are too large to be removed by PLOUGHING. Gullies are often initiated by the reduction or complete removal of protective VEGETATION and may enlarge rapidly and extend to form *gully networks.*   *SHD*

[*See also* ARROYO, BADLANDS, DONGA, EROSION, HEADCUT EROSION, RILL, WADI]

Bocco G (1991) Gully erosion: Processes and models. *Progress in Physical Geography* 15: 392–406.
Evans M and Lindsay J (2010) Impact of gully erosion on carbon sequestration in blanket peatlands. *Climate Research* 45: 31–41.
Herzig A, Dymond JR and Marsden M (2011) A gully-complex model for assessing gully stabilisation strategies. *Geomorphology* 133: 23–33.
Wells NA and Andriamihaja B (1993) The initiation and growth of gullies in Madagascar: Are humans to blame? *Geomorphology* 8: 1–46.

gully control dam   A small structure, usually up to about 2 m in height, constructed from wood, rocks, GABIONS or other materials and aligned across a GULLY to trap sediment and control EROSION.   *SHD*

Heede BH (1976) *Gully development and control: The status of our knowledge* [USDA Forest Service Research Paper: RM-196]. Fort Collins, CO: US Department of Agriculture.

guyot   A flat-topped SEAMOUNT. The flat top is considered to have been eroded by *waves* when the guyot was an island close to the MID-OCEAN RIDGE, before it subsided and moved away from the ridge due to SEA-FLOOR SPREADING.   *CDW*

[*See also* ATOLL]

Robinson SA (2011) Shallow-water carbonate record of the Paleocene-Eocene Thermal Maximum from a Pacific Ocean guyot. *Geology* 39: 51–54.

gymnosperms   A subdivision of the *Spermatophyta (seed plants)*, which literally means 'naked' (gymno) 'seed' (sperm), in which the seeds are borne in an exposed position on the cone scale or equivalent structure. Originating over 350 million years ago, they dominated on Earth before the ANGIOSPERMS *(flowering plants)* and are a major component of COAL deposits. The most successful today in terms of ecological abundance are the *coniferous trees* of the BOREAL FOREST.   *JCMCE*

[*See also* TREE OF LIFE]

Beck C (ed.) (1988) *Origin and evolution of gymnosperms.* New York: Columbia University Press.

gypcrete   A DURICRUST cemented with GYPSUM. Gypcretes are largely confined to regions of extreme ARIDITY, where the mean annual precipitation is <250 mm. The primary sources of gypsum are PANS and SABKHAS, where EVAPORITES accumulate. Gypcrete may be powdery, nodular or massive with the crystals of gypsum varying from microcrystalline up to 30 cm in size in *desert rose* forms. Gypcrete is more soluble in water than other duricrusts.   *MAC/JAM*

[*See also* CALCRETE, FERRICRETE, SILCRETE]

Aref MAM (2003) Classification and depositional
environments of Quaternary pedogenic gypsum crusts
(gypcretes) from east of the Fayum Depression, Egypt.
Sedimentary Geology 155: 87–108.
Eckardt FD, Drake NA, Goudie AS et al. (2001) The role of
playas in the formation of pedogenic gypsum crusts of
the Central Namib Desert. *Earth Surface Processes and
Landforms* 26: 1177–1193.
Watson A (1983) Gypsum crusts. In Goudie AS and Pye K
(eds) *Chemical sediments and geomorphology:
Precipitates and residua in the near-surface environment.*
London: Academic Press, 92–123.

Gypsisols   Soils with a secondary accumulation of
GYPSUM in a *gypsic* or *petrogypsic horizon*. They are
characteristic of ARIDLAND in North Africa, the Middle
East and Central Asia with a net annual water deficit
of atmospheric PRECIPITATION in relation to EVAPOTRAN-
SPIRATION (SOIL TAXONOMY: *Gypsiorthids*). There are two
potential sources for gypsum in soils, the most common
is for it to be precipitated from gypsum-rich GROUNDWA-
TER; the alternative source is as DUST blown out of sedimentary basins as, for example, the Great Chott Lake in
Tunisia. IRRIGATION of gypsisols can lead to solution of
gypsum and beneficial *de-gypsification* of the surface
layer but collapse of irrigation structures.   *EMB*

[*See also* GYPCRETE, WORLD REFERENCE BASE FOR
SOIL RESOURCES (WRB)]

Alphen JG and de los Rios F (1991) *Gypsiferous soils.*
Wageningen: International Institute for Land Reclamation
and Improvement (ILRI).
Herrero J and Poch RM (eds) (1998) Soils with gypsum.
Geoderma 87: 1–135 [Special Issue].

gypsum   A widely distributed soft mineral consisting of hydrous *calcium sulfate* ($CaSO_4 \cdot 2H_2O$). It is
the most common sulfate mineral and is frequently
associated with *halite* and *anhydrite* in EVAPORITES,
forming thick, extensive beds commonly interstratified
with limestone, shale and clay. Where evaporites of
gypsum harden, they are termed GYPCRETE.   *MAC*

gyre   A large-scale ocean circulatory system generated by the atmospheric circulation. Gyres circulate
clockwise in the Northern Hemisphere and counter-
clockwise in the Southern Hemisphere. The most
important are the five mid-ocean *subtropical gyres*,
each bounded by an *equatorial current*, a *western
boundary current* and an *eastern boundary current*
and located between latitudes 20° and 40° N or S. The
position of the oceanic boundary between the North

Atlantic subtropical gyre and subpolar North Atlantic
water (the OCEANIC POLAR FRONT (OPF)) is particularly
sensitive to GLACIAL-INTERGLACIAL CYCLES and higher
frequency CLIMATIC CHANGES. *Subpolar gyres* exist only
in the North Atlantic and North Pacific Oceans.   *JAM*

[*See also* OCEAN CURRENTS]

Hátún H, Sandø AB, Drange H et al. (2005) Influence of the
Atlantic subpolar gyre on the thermohaline circulation.
Science 309: 1841–1844.
Palter JB, Lozier MS and Barber RT (2005) The effect of
advection on the nutrient reservoir in the North Atlantic
subtropical gyre. *Nature* 437: 687–692.
Pediosky J (1990) The dynamics of the ocean subtropical
gyres. *Science* 248: 316–322.
Reverdin G (2010) North Atlantic subpolar gyre surface
variability (1895–2009). *Journal of Climate* 23:
4571–4584.

gyttja   A term introduced by H. von Post to describe
a brown LACUSTRINE SEDIMENT formed mainly under
ANAEROBIC and EUTROPHIC conditions and with an
ORGANIC CONTENT of >30 per cent. It consists of a mixture of ALLOCHTHONOUS and decomposed plant and animal fragments, minerogenic particles (see MINEROGENIC
SEDIMENT) and AUTOCHTHONOUS material. Coarse detrital gyttja is rich in MACROFOSSILS and MICROFOSSILS and
occurs in shallow water, whereas homogeneous fine
detrital gyttja is deposited in deeper water. Drift gyttja
with larger, rounded plant fragments and sand is found
close to the shore. Algal gyttja is mainly composed of
AUTOCHTHONOUS algal detritus and to a minor extent of
minerogenic particles and forms in shallow, high-pro-
ductivity lakes. Calcareous gyttja, which is rich in shell
fragments, has a calcium carbonate content of 20–80
per cent and deposits in shallow water. Shell gyttja
is mainly composed of shell fragments but may have
a matrix of algae or fine detritus. Gyttja clay has an
organic *carbon content* of 3–6 per cent and clay gyttja
6–30 per cent.   *UBW*

[*See also* DETRITUS, DIATOMITE, DY, MARL, SEDIMENT
TYPES]

Bohncke SJP and Hoek WZ (2007) Multiple oscillations
during the Preboreal as recorded in a calcareous gyttja,
Kingbeekdal, the Netherlands. *Quaternary Science
Reviews* 26: 1965–1974.
Von Post H (1862) *Kungliga Svenska Vetenskapsakademiens
Handlingar* [The Royal Swedish Academy of Sciences
Documents], *Volume 1: Studier öfver Nutidens koprogena
Jordbildningar, Gyttja, Dy, Torf och Mylla* [Studies of
mud formations of coprogenic origin, gyttja, dy, turf/sod
and soil/loam]. Stockholm: Norsteds.

habitat The physical space, place or ECOTOPE, occupied by organisms during one or more parts of their life cycle. Habitats are characterised by distinct suites of ENVIRONMENTAL FACTORS. *Habitat structure* refers to the physical arrangement or 'architecture' of the environment of organisms. Various measures of habitat are often used as ECOLOGICAL INDICATORS because they are much easier to measure and monitor than populations of individual species. *JLI*

[*See also* HABITAT DEGRADATION, HABITAT LOSS, NICHE]

Bell SS, McCoy CD and Mushinsky HR (1991) *Habitat structure: The physical arrangement of objects in space.* London: Chapman and Hall.
Bunnell FL and Dunsworth GB (2009) *Forestry and biodiversity: Learning how to sustain biodiversity in managed forests.* Vancouver: University of British Columbia Press.
Rondinini C and Chiozza F (2010) Quantitative methods for defining percentage area targets for habitat types in conservation planning. *Biological Conservation* 143: 1646–1653.

habitat degradation Decline in the quality and distribution (see FRAGMENTATION) of BIOTIC environments. It is measured as a quality reduction in ecosystem structural (e.g. BIODIVERSITY, VEGETATION, SOILS) and functional (e.g. ENERGY FLOW, MINERAL CYCLING) components. *Forest degradation* has been recognised as a critical source of carbon EMISSIONS from TROPICAL AND SUBTROPICAL FORESTS, and is a major component of REDD+ (reducing emissions from deforestation and forest degradation). *GOH/JLI*

[*See also* FOREST DECLINE, HABITAT LOSS]

Brook BW, Sodhi NS and Bradshaw CJA (2008) Synergies among extinction drivers under global change. *Trends in Ecology and Evolution* 23: 453–460.
Fischer J and Lindenmayer DB (2007) Landscape modification and habitat fragmentation: A synthesis. *Global Ecology and Biogeography* 16: 265–280.

habitat island A patch or fragment of a formerly more extensive and relatively continuous *ecosystem* or LANDSCAPE. Habitat islands may be formed by natural processes, such as ENVIRONMENTAL CHANGE during the PLEISTOCENE, or by ANTHROPOGENIC activity in more recent times. They include, for example, isolated woodlands in a 'sea' of grassland, and mountains surrounded by lowland. The term may also be used in relation to *habitat refuges* on islands. *MVL*

[*See also* CONSERVATION BIOLOGY, FRAGMENTATION, ISLAND BIOGEOGRAPHY, REFUGE THEORY IN TEMPERATE REGIONS, REFUGE THEORY IN THE TROPICS, REFUGIUM, SINGLE LARGE OR SEVERAL SMALL RESERVES (SLOSS DEBATE)]

Estrada A and Coates-Estrada R (2011) Dung beetles in continuous forest, forest fragments and in an agricultural mosaic habitat island at Los Tuxtlas, Mexico. *Biodiversity and Conservation* 11: 1903–1918.
Pryde PR and Cocklin C (1998) Habitat islands and the preservation of New Zealand's avifauna. *The Geographical Review* 88: 86–113.

habitat loss Reduction in the area of NATURAL and SEMI-NATURAL VEGETATION and ecosystems as a result of human activities. It is distinct from FRAGMENTATION, which does not necessarily involve loss of HABITAT, although loss of habitat often involves fragmentation of the remaining habitat. This reduces the number and diversity of habitats available for NATIVE species, is a major factor in BIODIVERSITY LOSS and may also precipitate BIOLOGICAL INVASIONS by ALIEN SPECIES. An example is shown in the Figure, which highlights the rapid loss of hedges, woodland patches and individual trees from an area of the Netherlands over a 20-year interval. Forest losses are of increasing concern because of the role that forests play in CARBON SEQUESTRATION. *JAM/JLI*

[*See also* CONVERSION, DEFORESTATION, HEDGEROW REMOVAL, LAND DRAINAGE, LANDUSE CHANGE, WETLAND CONSERVATION AND MANAGEMENT, WETLAND LOSS]

Arnold R and Villain C (1990) *New directions for European agricultural policy.* Brussels: Centre for European Policy Study.

Didham RK, Kapos V and Ewers RM (2012) Rethinking the conceptual foundations of habitat fragmentation research. *Oikos* 121: 161–170.

Gavish Y, Ziv Y and Rosenzweig M (2012) Decoupling fragmentation from habitat loss for spiders in patchy agricultural landscapes. *Conservation Biology* 26: 150–159.

Mantyka-Pringle C, Martin TG and Rhodes JR (2012) Interactions between climate and habitat loss effects on biodiversity: A systematic review and meta-analysis. *Global Change Biology* 18: 1239–1252.

Pimm SL (2007) Biodiversity: Climate change or habitat loss: Which will kill more species? *Current Biology* 18: 117–119.

Swift T and Hannon SJ (2010) Critical thresholds associated with habitat loss: A review of the concepts, evidence, and applications. *Biological Reviews* 85: 35–53.

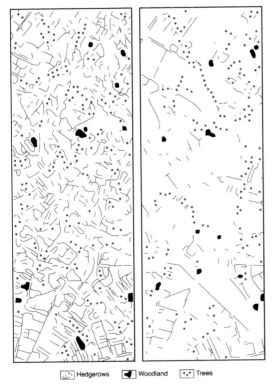

Hedgerows   Woodland   Trees

Habitat loss *Loss of three habitat types between AD 1950 and 1970 near Groenlo, the Netherlands (Arnold and Villain, 1990).*

hadal Depths and environments in the OCEAN deeper than ABYSSAL (i.e. >6,000 m), corresponding principally to OCEANIC TRENCHES. The term is derived from the French, *Hadés* (hell).   *BTC*

[*See also* OCEAN BASIN]

Vinogradova NG (1997) Zoogeography of the abyssal and hadal zones. *Advances in Marine Biology* 32: 325–387.

Hadean (*Priscoan*) That part of the early history of the Earth from which no rock assemblages are definitely known. The Earth is estimated to have formed ~4,540 million years ago (Ma), and the oldest known rocks are just over 4,000 Ma, although there is debate about rocks from northern Canada interpreted to have formed 4,280 Ma (see ARCHAEAN). Much evidence for conditions during the Hadean is derived from studies of other PLANETS and *satellites* (see PLANETARY GEOLOGY), the surfaces of which have not been modified by the water-related processes, PLATE TECTONICS, *life* and WEATHERING that occur on Earth. The Hadean is thought to have been characterised by frequent, high-energy METEORITE IMPACT events, rapid MANTLE convection, abundant VOLCANISM and an unstable, sometimes molten, surface. An early, reducing ATMOSPHERE, possibly rich in CARBON DIOXIDE, probably developed from the release of VOLATILES during volcanism.   *GO*

[*See also* GEOLOGICAL TIMESCALE, PLANETARY ENVIRONMENTAL CHANGE]

Bowring SA and Williams IS (1999) Priscoan (4.00-4.03 Ga) orthogneisses from northwestern Canada. *Contributions to Mineralogy and Petrology* 134: 3–16.

Iizuka T, Komiya J, Johnson SP et al. (2009) Reworking of Hadean crust in the Acasta gneisses, northwestern Canada: Evidence from in-situ Lu-Hf isotope analysis of zircon. *Chemical Geology* 259: 230–239.

Kemp AIS, Wilde SA, Hawkesworth CJ et al. (2010) Hadean crustal evolution revisited: New constraints from Pb-Hf isotope systematics of the Jack Hills zircons. *Earth and Planetary Science Letters* 296: 45–56.

Morse JW and Mackenzie FT (1998) Hadean ocean carbonate geochemistry. *Aquatic Geochemistry* 4: 301–319.

Rasmussen B, Fletcher IR, Muhling JR et al. (2011) Metamorphic replacement of mineral inclusions in detrital zircon from Jack Hills, Australia: Implications for the Hadean Earth. *Geology* 39: 1143–1146.

Hadley cell The direct thermally driven, mean *meridional circulation* of the tropical ATMOSPHERE which consists of rising air in low latitudes, at the INTERTROPICAL CONVERGENCE ZONE (ITCZ), and sinking air in the subtropics. Poleward transport aloft carries latent and sensible heat from the tropics to mid-latitudes. Part of the GENERAL CIRCULATION OF THE ATMOSPHERE (see Figure under that entry) and of MACROCLIMATE, the existence of Hadley cells was first proposed by George Hadley in AD 1735. There is one Hadley cell in each hemisphere and their intensity and position can vary seasonally, with the one in the winter hemisphere being the strongest. Again based on zonally averaged observations, a thermally indirect FERREL CELL (named after William Ferrel) is recognised polewards of each Hadley cell with ascending air in the mid-latitude region of the POLAR FRONT.   *AHP/JAM*

[*See also* ANTITRADES, MERIDIONAL OVERTURNING
CIRCULATION (MOC), TRADE WINDS]

Barry RG and Chorley RJ (1998) *Atmosphere, weather and
climate, 7th edition*. London: Methuen.
Diaz HF and Bradley RS (eds) (2004) *The Hadley
circulation: Past present and future*. Dordrecht: Kluwer.
Hadley G (1735) Concerning the cause of the general trade
wind. *Philosophical Transactions of the Royal Society of
London* 39: 58–73.
Hastenrath SL (1968) On mean meridional circulations in the
tropics. *Journal of Atmospheric Science* 25: 979–983.

haematite   An *iron oxide* (αFe_2O_3) with a hexago-
nal crystal structure. It is found in igneous rocks and
sediments formed under oxidising conditions, generally giving a blood-red colour and often responsible for
the magnetisation observed in RED BEDS. Adjacent iron
layers in haematite are coupled antiferromagnetically,
resulting in imperfect antiferromagnetic behaviour and
low values of MAGNETIC SUSCEPTIBILITY between approximately 0.3 and 2.0×10^{-6} m^3/kg.   *AJP/WZ*

Evans ME and Heller F (2003) *Environmental magnetism:
Principles and applications of enviromagnetics*. San
Diego, CA: Academic Press.
Walden J, Oldfield F and Smith JP (eds) (1999)
Environmental magnetism: A practical guide. London:
Quaternary Research Association.

Hale cycle   The periodic switch in polarity of *sunspot pairs* connected by *magnetic loops* that extend
through the *solar corona*. For each pair, one has a
polarity in one direction, the other in the opposite
direction. The period of change is approximately
22 years and matches the 11-year SUNSPOT CYCLE. It has
been suggested, but never wholly proven, to correlate
with a number of climatic changes, most importantly,
perhaps being the DROUGHTS that periodically strike the
American Midwest.   *DAW*

Burroughs WJ (1992) *Weather cycles: Real or imaginary?*
Cambridge: Cambridge University Press.
Raspopov OM, Dergachev VA and Kolström T (2004) Hale
cyclicity of solar activity and its relationship to climatic
variability. *Solar Physics* 224: 455–463.

half-life   A measure of the rate of disintegration of a
RADIONUCLIDE or radioactive ISOTOPE. It is the time taken
for a quantity of a radioactive isotope to reduce by a
half by *radioactive decay*. Radioactive decay is exponential; after two half-lives a quarter of the original
remains, and after 10 half-lives 0.1 per cent remains.
Half-lives of individual radionuclides vary from small
fractions of a second to 10^{10} years.   *PQD*

[*See also* BIOLOGICAL HALF-LIFE, LIBBY HALF-LIFE]

halocline   A steep vertical gradient in SALINITY in the
OCEAN. For example, there is a halocline in the upper

part of the Arctic Ocean where relatively fresh water
overlies saltier water. As salinity affects DENSITY, the
halocline plays a role in the stratification and stability
of the water column.   *JAM*

[*See also* CHEMOCLINE]

halogenated   hydrocarbons   (halocarbons)
Hydrocarbons into which one or more of the *halogen*
elements—chlorine (Cl), bromine (Br) or fluorine
(Fl)—have been introduced. In chlorinated hydrocarbons, for example, the halogens are *added* to hydrocarbon molecules. In CHLOROFLUOROCARBONS (CFCs or
freons), *bromofluorocarbons* (*halons*), *methyl chloride*
(CH_3Cl) and *methyl bromide* (CH_3Br), the halogens
replace hydrogen atoms.

Both categories are important POLLUTANTS: chlorinated hydrocarbons have been extensively used as synthetic PESTICIDES; CFCs, widely used in refrigerants and
air-conditioning systems, are important GREENHOUSE
GASES and probably the major cause of OZONE DEPLETION;
halons, used mainly in fire extinguishers, are actually
more effective than freons in the breakdown of ozone;
and methyl bromide, used as a fumigant in the fruit and
vegetable industry since the 1960s, has a comparable
ozone depletion potential (ODP) to many CFCs.

Measurements of TRACE GASES trapped in FIRN
(consolidated snow) from sites in the Antarctic and
Greenland have demonstrated that natural sources of
chlorofluorocarbons, bromofluorocarbons and most
other halogenated hydrocarbons are minimal or non-
existent. Reconstructions from these firn samples of
atmospheric concentrations back to the late nineteenth
century are consistent with anthropogenic emission
rates and their known PERSISTENCE in the atmosphere.
The firn samples also show that methyl chloride and
methyl bromide were the only species of halogenated
hydrocarbons to exhibit unequivocal natural BACK-
GROUND LEVELS prior to the twentieth century. It appears
that most methyl chloride in the atmosphere is of natural origin, whereas only half of the methyl bromide can
be attributed to natural sources.   *JAM*

[*See also* CONVENTION, HYDROCHLOROFLUOROCARBONS
(HCFCs), ORGANOCHLORIDES, PERSISTENT ORGANIC
COMPOUNDS (POCs), POLYCHLORINATED BIPHENYLS
(PCBs)]

Butler JH, Battle M, Bender ML et al. (1999) A record of
atmospheric halocarbons during the twentieth century
from polar firn air. *Nature* 399: 749–755.
Lovelock JE, Maggs RJ and Wade RJ (1973) Halogenated
hydrocarbons in and over the Atlantic. *Nature* 241:
194–196.

halophyte   A *salt-tolerant* plant species defined as
having an ability to grow and complete its life cycle
at salt concentrations in excess of 100–200 Mm NaCl.

Halophytes are commonly found in SALTMARSH, DRYLANDS and SALINE-LAKE environments.   MLC

[*See also* MANGROVE]

Flowers TJ, Hajibaghri MA and Clipson NJW (1986) Halophytes. *Quarterly Review of Biology* 61: 313–317.
Lieth H, Sucre MG and Herzog B (eds) (2008) *Mangroves and halophytes: Restoration and utilisation*. Dordrecht: Springer.
Reimold RJ and Queen WH (eds) (1974) *Ecology of halophytes*. New York: Academic Press.

haloturbation   The heaving (*salt heave*) and churning of SOILS or SEDIMENTS due to the alternating PRECIPITATION and DISSOLUTION of SALTS (e.g. *halite* or GYPSUM). The process occurs in ARIDLANDS and may lead to the development of SOFT-SEDIMENT DEFORMATION structures, crude PATTERNED GROUND and damage to engineering structures.   JAM/SHD

Ferrarese F, Macaluso T, Madonia G et al. (2003) Solution and recrystallisation processes and associated landforms in gypsum outcrops of Sicily. *Geomorphology* 49: 25–43.
Horta JC de OS (1985) Salt heaving in the Sahara. *Géotechnique* 35: 329–337.

hanging valley   In glaciated terrain, a *tributary valley* the mouth of which ends abruptly part way up the side of a *trunk valley*, formed as a result of the greater amount of GLACIAL EROSION of the latter. There are several possible explanations for the lesser amount of erosion of the hanging valley: it may be attributed, for example, to a smaller glacier that is less erosive, the absence of a glacier when the trunk valley contained ice, or the presence of a COLD-BASED GLACIER when the trunk valley contained a TEMPERATE GLACIER.   MJH/JAM

Haralick texture measures   A means of quantifying the spatial relationships among grey levels (or IMAGE TEXTURE) in a digital image such as those formed by REMOTE SENSING. These measures, which are one of many methods used to quantify image texture, are derived from the grey-level co-occurrence matrix of the input image and can be successfully used as new features for IMAGE CLASSIFICATION.   ACF

Haralick RM, Shanmugam K and Dinstein I (1973) Textural features for image classification. *IEEE Transactions on Systems, Man and Cybernetics* 3: 610–621.

hard engineering   The emplacement of ARTIFICIAL structures (e.g. SEA WALLS, GROYNES) along a SHORELINE with the intention of stabilising the coastline. The term can also be applied in non-coastal environments where engineering structures are used to disrupt NATURAL processes. Hard engineering tends to be more expensive, less ecologically based, aesthetically intrusive and non-sustainable compared with SOFT ENGINEERING solutions.   HJW/IM

[*See also* COASTAL ENGINEERING STRUCTURES, COASTAL (SHORE) PROTECTION]

French PW (2001) *Coastal defences: Processes, problems, and solutions*. London: Routledge.
Walker HJ (ed.) (1988) *Artificial structures and shorelines*. Dordrecht: Kluwer.

hardpan   A SOIL HORIZON or layer that has become hardened by CEMENTATION of the soil particles by *organic matter*, SILICA, *iron sesquioxides* or *calcium carbonate*. The hardness does not change with varying moisture content and fragments do not disintegrate when placed in water. Development of hardpans may follow FOREST CLEARANCE on sandy soils which are subsequently podzolised with an *iron-humus pan*. In Alaska, development of a hardpan has led to vegetation change as woodland turns to MUSKEG.   EMB

[*See also* FRAGIPAN, INDURATION, PADDY SOILS, PODZOLISATION]

Thompson CH, Bridges EM and Jenkins DA (1993) An exploratory examination of relict hardpans in the coastal lowlands of southern Queensland. In Ringrose-Voase AJ and Humphreys GS (eds) *Soil micromorphology: Studies in management and genesis*. Amsterdam: Elsevier, 233–245.
Ugolini FC and Mann DH (1979) Biopedological origin of peatland in south east Alaska. *Nature* 281: 366–368.

hard-water effect/error   The effect or ERROR in RADIOCARBON DATING of plant material from AQUATIC ENVIRONMENTS resulting from plants that assimilated inert (dead) carbon. This carbon, dissolved in the lake water, originated from ancient carbonate-bearing rocks (e.g. LIMESTONE). The contemporary ^{14}C is diluted and the resulting age is too great. Lake mud (GYTTJA) contains decayed plant material from *macrophytes*, *mosses* and PHYTOPLANKTON that may have utilised 'dead' carbon to a greater or lesser degree. The effect can be estimated by measuring ^{14}C activity in lake water and aquatic plants. *Submerged plants* are more likely to have a *reservoir age* than *emergent aquatics* that photosynthesise atmospheric CARBON DIOXIDE. A correction can be made for dates from a lake whose modern reservoir age has been estimated. The effect may be avoided by ACCELERATOR MASS SPECTROMETRY (AMS) DATING terrestrial MACROFOSSILS that have been incorporated in the gyttja, ensuring that no aquatic material is included. The hard-water effect should be distinguished from the marine RESERVOIR EFFECT.   HHB

Andrée M, Oeschger H, Siegenthaler U et al. (1986) ^{14}C dating of macrofossils in lake sediments. *Radiocarbon* 28: 411–416.
Deevey ES, Gross MS, Hutchinson GE and Henry L (1954) The natural ^{14}C content of materials from hard-water lakes. *Proceedings of the National Academy of Sciences* 40: 285–288.

harrowing   The mechanical process of breaking up and pulverising the soil after PLOUGHING (by dragging a harrow) to remove WEEDS and prepare a seed bed.   *JAM*

harvest records   Quantitative or qualitative records of harvest yields and/or dates for crops, vineyards or fruit that are of potential use as PROXY DATA for reconstructions of past climate. Harvest yields fluctuate from year to year and these fluctuations can be interpreted as an expression of the yield-forming factors which mould them. Yield can be seen as a function of work, soil, technology, capital and climate. Although the weighting of these factors can vary greatly both in space and through time, climate plays an important role. It becomes noticeable in the ANNUAL VARIABILITY, while economic innovations tend to become effective in the longer term, and changes in field boundaries must be taken into consideration. Also, if the yields in MARGINAL AREAS decrease then fields may be abandoned, as happened, for example, at northerly latitudes and high altitudes in the Alps and other mountain areas in the LITTLE ICE AGE. The decoding of fluctuations is a difficult operation which must be solved by a variety of statistical methods. Results both between and within studies do not always agree. Inconsistencies in the flora-climate relations can, however, often be explained by regional or even local models.   *RG*

[*See also* PHENOLOGY, VINE HARVESTS]

Mozny M, Brazdil R, Dobrovdny P and Trnka M (2012) Cereal harvest data in the Czech Republic between 1501 and 2008 as a proxy for March-June temperature reconstruction. *Climatic Change* 110: 801–821.
Neumann J and Sigrist RM (1978) Harvest dates in ancient Mesopotamia as possible indicators of climatic variations. *Climatic Change* 1: 239–252.
Parry ML (1981) Climatic change and the agricultural frontier: A research strategy. In Wigley T, Ingram M and Farmer G (eds) *Climate and history: Studies in past climates and their impact on man*. Cambridge: Cambridge University Press, 319–376.
Rowntree LB (1985) A crop-based rainfall chronology for preinstrumental record southern California. *Climatic Change* 7: 327–341.

harvesting   Exploitation of biological RENEWABLE RESOURCES as practised, for example, in AGRICULTURE, forestry (SILVICULTURE) and fish farming (AQUACULTURE, WHALING AND SEALING and GAME MANAGEMENT). A central idea is that of maximum SUSTAINABLE YIELD, which, if exceeded, will lead to population decline and possible harvest failure. The term is also used in the sense of energy and WATER HARVESTING and, in MEDICAL SCIENCE, for *organ harvesting* for transplants.   *JAM*

Gunnell Y, Anupama K and Sultan B (2007) Response of South Indian runoff-harvesting civilization to northeast monsoon rainfall variability during the last 2000 years:

Instrumental records and indirect evidence. *The Holocene* 17: 207–215.
Holtsmark B (2012) Harvesting in boreal forests and the biofuel carbon debt. *Climatic Change* 112: 415–428.

hazard   A phenomenon that presents a RISK of harm or damage to humans, other organisms or their environment. For example, the hazard may be drowning—but the risk would vary between a rowing boat and a liner in a rough sea.   *JAM*

[*See also* ANTHROPOGENIC HAZARDS, DISASTER, NATURAL HAZARDS, RAPID-ONSET HAZARDS]

hazardous substances   Substances with properties capable of causing harm to humans or other organisms, or damage to ecosystems. Such substances may be harmful because they are toxic, flammable, explosive, corrosive or have high chemical reactivity. They are often WASTE products. In the Unites States, the most important sources of HAZARDOUS WASTE, in decreasing order by volume, are industrial organics, chemical manufacturing, petroleum refining, explosives manufacture, plastics and resins, refuse, agricultural chemicals, inorganic pigments and alkaline substances. Hazardous WASTE MANAGEMENT may involve recovery and reuse, incineration, detoxification, BIOREMEDIATION, CHEMICAL REMEDIATION or long-term storage.   *JAM*

[*See also* RADIOACTIVE WASTE]

Nemerow NL and Dasgupta A (1991) *Industrial and hazardous waste*. New York: Van Nostrand.
Zirm KK and Mayer J (eds) (1990) *The management of hazardous substances in the environment*. London: Elsevier and International Society for Environmental Protection.

hazardous waste   Any WASTE substance with the potential to have an adverse effect upon people, plant and animal life, or the environment. As new chemical compounds are being continuously produced, it is impossible to produce a definitive list of hazardous wastes. Hazardous waste was viewed as a localised issue throughout the nineteenth century before the rapid growth of the *chemical industry*. It became a *second-generation environmental concern* in all sectors of industrialised societies by the late 1970s. This can be attributed partly to the number and diversity of potentially hazardous substances in common use: approximately half of the ~80,000 synthetic chemicals in common use has some associated characteristic (e.g. *toxicity, flamability or corrosivity*) which renders them hazardous. Other factors include their association with some of the most feared HUMAN HEALTH HAZARDS (e.g. CANCER and *birth defects*), and the CHEMICAL TIME BOMB effect stemming from the lack of WASTE MANAGEMENT in the past.

Categories of hazardous waste include *inorganic acids, organic acids,* ALKALIS, *toxic metal compounds, metals, metal oxides, inorganic compounds*, other inorganic materials (e.g. ASBESTOS), *organic compounds, polymers, fuels, oils and greases, pharmaceutical chemicals, biocides, filter materials, treatment sludge, tars, paints, dyes, pigments, tannery waste, timber preservatives, soap, detergents* and *animal-processing wastes.*   *EMB/JAM*

[*See also* HAZARDOUS SUBSTANCES, HEAVY METALS, INDUSTRIAL WASTE, PERSISTENT ORGANIC COMPOUNDS (POCS), RADIOACTIVE WASTE]

Budd WW (1999) Hazardous waste. In Alexander DE and Fairbridge RW (eds) *Encyclopedia of environmental science.* Dordrecht: Kluwer, 311–312.
Clapp J (2001) *Toxic exports: The transfer of hazardous wastes and technology from rich to poor countries.* Ithaca, NY: Cornell University Press.
Cope CB, Fuller WH and Willetts SL (1983) *The scientific management of hazardous wastes.* Cambridge: Cambridge University Press.
Nemerow NL and Dasgupta A (1991) *Industrial and hazardous waste.* New York: Van Nostrand Reinhold.
Vallero D (2003) *Engineering the risks of hazardous wastes.* Amsterdam: Elsevier.
Wang LK, Hung YT, Lo H and Yapijakis C (eds) (2004) *Handbook of industrial and hazardous wastes treatments.* Boca Raton, FL: CRC Press.

head   A regional term originally used to describe RELICT periglacial MASS MOVEMENT deposits (typically DIAMICTONS) capping many coastal cliff sections in parts of southwest England. Head is usually regarded as deposits formed by SOLIFLUCTION, but usage is sometimes broadened to include a wider range of COLLUVIAL PROCESSES.   *HMF*

[*See also* PERIGLACIAL SEDIMENTS]

Dines HG, Hollingworth SE, Edwards W et al. (1940) The mapping of head deposits. *Geological Magazine* 77: 198–226.
Harris C (1987) Solifluction and related periglacial deposits in England and Wales. In Boardman J (ed.) *Periglacial processes and landforms in Britain and Ireland.* Cambridge: Cambridge University Press, 209–224.
Harris C (1998) The micromorphology of paraglacial and periglacial slope deposits: A case study from Morfa Bychan, west Wales, UK. *Journal of Quaternary Science* 13: 78–84.

headcut erosion   Water EROSION at the upslope limit of a GULLY system, resulting in a steep wall, which gradually migrates upslope. Headcut erosion may be aided by *piping.*   *SHD*

[*See also* KNICKPOINT]

Archibold OW, DeBoer DH and Delanoy L (1996) A device for measuring gully headward morphology. *Earth Surface Processes and Landforms* 21: 1001–1005.
Kumar DA, Tsujimoto T and Tadanori K (2007) Experimental investigations on different modes of headcut migration. *Journal of Hydraulic Research* 45: 333–346.

headward erosion   The extension of the channel network or an individual first-order channel by EROSION of the *channel head.* In studies of long-term LANDSCAPE EVOLUTION and DENUDATION CHRONOLOGY, headward erosion is seen as an important process in accomplishing RIVER CAPTURE. The processes involved in headward erosion may include erosion by OVERLAND FLOW, GULLY erosion, SAPPING and *piping.* Rapid rates (1.9 m/year) of headward erosion of tidal creeks have also been recorded in recent decades at Cape Romain in South Carolina in response to sea-level rise.   *RPDW*

[*See also* DRAINAGE NETWORK, HEADCUT EROSION]

Bishop P (1995) Drainage rearrangement by river capture, beheading and diversion. *Progress in Physical Geography* 19: 449–473.
Dohrenwend JC, Abrahams AD and Turrin BD (1987) Drainage development in basaltic lava flows, Cima Volcanic Field, southeast California, and Lunar Crater Volcanic Field, south-central Nevada. *Bulletin of the Geological Society of America* 99: 405–413.
Hughes ZJ and Fitzgerald DM (2009) Rapid headward erosion of marsh creeks in response to relative sea-level rise. *Geophysical Research Letters* 36: L03602.

heat balance/budget   The balance of gains and losses of heat for a SYSTEM over a specified time period. For example, it forms part of the ENERGY BALANCE of the ATMOSPHERE.   *AHP*

Rapp D (2008) *Assessing climate change: Temperatures, solar radiation and heat balance, 2nd edition.* Berlin: Springer.
Sellers WD (1965) *Physical climatology.* Chicago: University of Chicago Press.

heat island   The area of localised higher temperatures that occurs, especially under clear, calm conditions and at night, over urban areas. The effect is caused by ANTHROPOGENIC heat and the heat absorbed by structures and urban surfaces. There is a relationship between city size and heat island intensity. The difference in MEAN ANNUAL TEMPERATURE from the surrounding rural areas is typically 1.0°C for small cities up to about 4.0°C for large cities. The effects of the development of heat islands on meteorological stations have been one of the factors raised by climate-change sceptics in relation to GLOBAL WARMING.   *AHP/JAM*

[*See also* URBAN CLIMATE]

Gartland L (2008) *Heat islands: Understanding and mitigating heat in urban areas.* London: Earthscan.
Goward SN (1981) Thermal behaviour of urban landscapes and the urban heat island. *Physical Geography* 2: 19–33.
Oke TR (1982) The energetic basis of the urban heat island. *Quarterly Journal of the Royal Meteorological Society* 108: 1–24.
Wong NH and Chen Y (2009) *Tropical urban heat islands: Climate, buildings and greenery.* Abingdon: Taylor and Francis.

Zeng Y, Qui XF, Gu LH et al. (2009) The urban heat island
in Nanjing. *Quaternary International* 208: 38–43.

heathland   A vegetation community and/or landscape developed on relatively dry, acidic, sandy and/
or podzolic soils, dominated by low SHRUBS such as
heather (*Calluna vulgaris* and *Erica* spp.) and *gorse*
(*Ilex europaeus*). It usually occurs <200 m above sea
level in Britain where, in previous INTERGLACIAL CYCLES,
heathland development occurred towards the end of the
INTERGLACIAL (in the *telocratic phase*). SOIL EXHAUSTION
and climatic cooling contribute to the development of
acidic *pozols*, which favour colonisation by coniferous
trees (e.g. pine, *Pinus*) and by heathland vegetation.
In the HOLOCENE, human activities, especially GRAZING
of livestock, have had a marked effect on the timing
and extent of heathland and MOORLAND development in
northwest Europe.   *MJB*

[*See also* CULTURAL LANDSCAPE, FIRE HISTORY,
GRAZING HISTORY, HUMAN IMPACT ON VEGETATION
HISTORY, SEMI-NATURAL VEGETATION]

Jeffers ES, Bonsall MB, Watson JE and Willis KJ (2012)
Climate change impacts on ecosystem functioning:
Evidence from an *Empetrum* heathland. *New Phytologist*
193: 150–164.
Newton AC, Stewart GB, Myers G et al. (2009) Impacts
of grazing on lowland heathland in north-west Europe.
Biological Conservation 142: 935–947.
Odgaard BV (1988) Heathland history in western Jutland,
Denmark. In Birks HH, Birks HJB, Kaland PE and Moe
D (eds) *The cultural landscape: Past, present and future*.
Cambridge: Cambridge University Press, 311–319.
Specht RL (ed.) (1979) *Heathlands and related shrublands:
Descriptive studies*. Amsterdam: Elsevier.
Specht RL (ed.) (1981) *Heathlands and related shrublands:
Analytical studies*. Amsterdam: Elsevier.
Thompson DBA (ed.) (1995) *Heaths and moorland:
Cultural landscapes*. Edinburgh: Her Majesty's
Stationery Office.

heavy metals   A disputed term still widely used for
metallic TRACE ELEMENTS in SOILS and elsewhere in the
environment. Some elements such as ARSENIC (As), CAD-
MIUM (Cd), chromium, LEAD (Pb), MERCURY (Hg), nickel
and uranium may be described as *toxic metals*; but others such as cobalt, COPPER (Cu), manganese, selenium
and ZINC (Zn) are biologically essential at low concentrations (MICRONUTRIENTS). However, in all cases, high
concentrations are toxic and reduce the activity of soil
organisms. The main ANTHROPOGENIC source of heavy
metals is INDUSTRIAL WASTE.   *EMB*

[*See also* METAL POLLUTION, SOIL POLLUTION, TOXIN]

Alloway BJ (ed.) (2012) *Heavy metals in soils, 3rd edition*.
Berlin: Springer.
Bradi HB (2005) *Heavy metals in the environment: Origin,
interaction and remediation*. Amsterdam: Elsevier.

Huebner R, Astin KB and Herbert RJH (2010) 'Heavy
metal': Time to move from semantics to pragmatics?
Journal of Environmental Monitoring 12: 1511–1514.
Nieboer E and Richardson DHS (1980) The replacement
of the nondescript term 'heavy metals' by a biologically
and chemically significant classification of metal ions.
Environmental Pollution 1: 3–26.
Wang LK, Chen JP, Hung Y-T and Shammas NZ (eds) (2009)
Heavy metals in the environment. Boca Raton, FL: CRC Press.

heavy mineral analysis   The investigation of
those MINERALS in a sample of SEDIMENT, SEDIMENTARY
ROCK or SOIL, which are denser than, and can be separated using, bromoform (relative density = 2.9). Heavy
minerals usually comprise <1 per cent of the sample
and include *apatite, epidote, garnet, rutile, stauro-
lite, tourmaline* and *zircon*. Their study is particularly
useful in determining the PROVENANCE of SANDSTONES.
Several minerals of economic importance are heavy
minerals, such as *diamond, gold* and *casserite* (an ORE
of TIN), and can be exploited from PLACER DEPOSITS.   *TY*

Blatt H, Middleton G and Murray RC (1980) *Origin of
sedimentary rocks, 2nd edition*. Englewood Cliffs, NJ:
Prentice Hall.
Dill HG (1998) A review of heavy minerals in clastic
sediments with case studies from the alluvial-fan through
the nearshore-marine environments. *Earth-Science
Reviews* 45: 103–132.
Lubke RA and Avis AM (1998) A review of the concepts and
application of rehabilitation following heavy mineral dune
mining. *Marine Pollution Bulletin* 37: 546–557.
Mange MA and Wright DT (eds) (2007) *Heavy minerals in
use*. Amsterdam: Elsevier.
Pinto L, Herail G, Fontan F et al. (2007) Neogene erosion
and uplift of the western edge of the Andean Plateau
as determined by detrital heavy mineral analysis.
Sedimentary Geology 195: 217–237.
Rasmussen B and Buick R (1999) Redox state of the
Archean atmosphere: Evidence from detrital heavy
minerals in ca. 3250-2750 Ma sandstones from the Pilbara
Craton, Australia. *Geology* 27: 115–118.

heavy oil   A form of PETROLEUM that is too viscous to
flow, causing difficulties in extraction. Many heavy oil
deposits have formed from a reaction between CRUDE
OIL and GROUNDWATER. The Orinoco Oil Belt in Venezuela is a major source, producing an oil-water emulsion known as *orimulsion*. Toxic, HAZARDOUS WASTE
products from the burning of orimulsion have led to
environmental objections to its use.   *GO*

Gluyas J and Swarbrick R (2004) *Petroleum geoscience*.
Oxford: Blackwell.
Ji G, Sun T and Ni J. (2007) Impact of heavy oil-polluted soils
on reed wetlands. *Ecological Engineering* 29: 272–279.
Wang Y-F, Chao H-R, Wang L-C et al. (2010) Characteristics
of heavy metals emitted from a heavy oil-fueled power
plant in northern Taiwan. *Aerosol and Air Quality
Research* 10: 111–118.

hedgerow removal Hedges are valuable for conserving BIODIVERSITY, providing shelter and reducing SOIL DEGRADATION. In many countries, they are being removed to facilitate mechanised farming. Government incentives in the United Kingdom and Europe have encouraged the loss in recent decades. Eastern England has been especially badly affected, and in the United Kingdom as a whole roughly one-quarter of hedgerows were lost between AD 1945 and 1985 (the loss has continued since).   *CJB*

[*See also* CONSERVATION]

Dowdeswell WH (1987) *Hedgerows and verges*. London: Allen and Unwin.
Hooper MD (1979) Hedges and small woodlands. In Davidson J and Lloyd R (eds) *Conservation and agriculture*. Chichester: Wiley.

Heinrich events Short-lived, widely spaced events characterised by periods of ice rafting related to DISCHARGE predominantly from the LAURENTIDE ICE SHEET. Each Heinrich event lasts for around 750 years while sharp basal contacts indicate rapid onset and other characteristics in MARINE SEDIMENT CORES including low FORAMINIFERA abundance amongst ICE-RAFTED DEBRIS (IRD). Six events have been defined between 70,000 and 14,000 years ago, labelled H6 to H1, and the IRD peak in the YOUNGER DRYAS STADIAL is sometimes defined as H0. Ice rafting can be linked to global cold palaeoclimatic events via the THERMOHALINE CIRCULATION.
                                                          LJW/WENA/JP

[*See also* BINGE-PURGE MODEL, BOND CYCLE, DANSGAARD-OESCHGER (D-O) EVENTS]

Alvarez-Solas J (2011) Heinrich event 1: An example of dynamical ice-sheet reaction to oceanic changes. *Climate of the Past* 7: 1297–1306.
Alvarez-Solas J, Charbit S, Ritz C et al. (2010) Links between ocean temperature and iceberg discharge during Heinrich events. *Nature Geoscience* 3: 122–126.
Alvarez-Solas J and Ramstein G (2011) On the triggering mechanism of Heinrich events. *Proceedings of the National Academy of Sciences* 108: E1359–E1360.
Heinrich H (1988) Origin and consequences of cyclic ice-rafting in the Northeast Atlantic Ocean during the past 130,000 years. *Quaternary Research* 29: 142–152.
Stanford JD, Rohling EJ, Bacon S et al. (2011) A new concept for the paleoceanographic evolution of Heinrich event 1 in the North Atlantic. *Quaternary Science Reviews* 30: 1047–1066.

hemlock decline A marked decline in the abundance of the tree *Tsuga canadensis* (hemlock) throughout eastern North America in the mid-HOLOCENE. SUBFOSSIL pollen data from numerous sites indicate that this tree species declined sharply in abundance about 5,000 years ago. As the hemlock decline appears to have been rapid, more or less synchronous, species specific, and over a large area, it is thought to have been the result of a *pathogenic outbreak*. Recent evidence from Québec suggests that *insect activity* may have been responsible for the decline. Several other forest taxa appear to have increased in abundance following the hemlock decline, suggesting that the demise of this long-lived, highly shade-tolerant conifer may have had a major impact on regional *forest dynamics*. This dramatic decline of a highly competitive tree species indicates the importance of pathogenic outbreaks as a rare but potentially catastrophic DISTURBANCE mechanism in forested ecosystems.   *JLF*

[*See also* ECOSYSTEM COLLAPSE, ELM DECLINE, PATHOGENS, PLANT DISEASES, POLLEN ANALYSIS]

Bhiry N and Filion L (1996) Mid-Holocene hemlock decline in eastern North America linked with phytophagous insect activity. *Quaternary Research* 45: 312–320.
Evans DM, Aust WM, Dolloff CA et al. (2011) Eastern hemlock decline in riparian areas from Maine to Alabama. *Northern Journal of Applied Forestry* 28: 97–104.
Fuller JL (1998) Ecological impact of the mid-Holocene hemlock decline in southern Ontario, Canada. *Ecology* 79: 2337–2351.
Zhao Y, Yu ZC and Zhao C (2010) Hemlock (*Tsuga canadensis*) declines at 9800 and 5300 cal yr BP caused by Holocene climatic shifts in northeastern North America. *The Holocene* 20: 877–886.

hepatitis Several types of infectious liver DISEASES are involved, of which three are now well known. *Hepatitis A* is an acute disease, rarely fatal, that affects tens of millions of people each year in regions with poor hygiene. The *HAV virus* is spread by ingestion of contaminated food or water by contact with those infected. *Hepatitis B* has affected about one-quarter of the world's population, mostly in Asia and Africa, and may become chronic, especially in children. It is spread by the *HBV virus* from infected blood or other body fluids but not by casual contact. Although the majority of infected people recover from the acute form, there are around 350 million chronic carriers of the HBV virus, which result in at least 500,000 deaths each year.

Hepatitis C was discovered in 1989 and is spread by blood-to-blood contact (mainly through transfusions, shared needles and reused medical supplies). Up to 200 million people are now estimated to be infected with the *HCV virus*, and some 350,000 people die from HCV-related liver disease each year. It is spreading rapidly in the developing world, where it threatens to rival AIDS in its virulence, but transmission rates are declining in the developed world thanks to screening of blood supplies, disposable instruments and use of expensive treatments. Egypt has the highest rate of transmission (14 per cent). In the United States, an estimated 1.6 per cent of the general population and 90 per cent of long-term injection drug users are infected.   *JAM*

Gravitz L (2011) A smouldering public-health crisis. *Nature* 474: S2–S4.

Worman HJ (2006) *The liver disorders and hepatitis handbook*. New York: McGraw-Hill.

herb   Any non-woody terrestrial plant, either annual, biennial or perennial. Herbs are commonly FORBS or GRAMINOIDS. In POLLEN ANALYSIS, pollen from herbs is often summarised as NON-ARBOREAL POLLEN (NAP), in contrast to ARBOREAL POLLEN (AP) from trees and shrubs.   *BA/RMF*

herbivore   Any animal that is adapted to eating plant material (i.e. exhibits *herbivory*). *Insect herbivores* evolved shortly after the EVOLUTION of land plants in the DEVONIAN; later herbivore evolutions were dependent on adapting to the nature of the evolving food plants. *Vertebrate herbivores* evolved much later, around 300 million years ago, in the CARBONIFEROUS. Specialised herbivore NICHES include grazers, browsers, grain and seed eaters (*granivores*), leaf eaters (*folivores*), fruit and berry eaters (*frugivores*), nut eaters (*nucivores*), nectar eaters (*nectarivores*) and root eaters.   *JAM/RJH*

[*See also* CARNIVORE, FOOD CHAIN/WEB, HETEROTROPHIC ORGANISM, TROPHIC LEVEL]

Danell K, Bergström R, Duncan P and Pastor J (2006) *Large herbivore ecology, ecosystem dynamics and conservation*. Cambridge: Cambridge University Press.

Labandeira CC (2007) The origin of herbivory on land: Initial patterns of plant tissue consumption by arthropods. *Insect Science* 14: 259–275.

Owen-Smith N (ed.) (2010) *Dynamics of large herbivore populations in changing environments*. Chichester: Wiley-Blackwell.

Sues H-D (ed.) (2000) *Evolution of herbivory in terrestrial vertebrates: Perspectives from the fossil record*. Cambridge: Cambridge University Press.

Wielgolaski FE, Karlsson S, Neuvonen S and Thannheiser D (eds) (2005) *Plant ecology, herbivory and human impact in northern mountain birch forests*. Berlin: Springer.

heredity   The process by which *genetic variation* is passed from one generation to another. It can eventually lead to SPECIATION. GENETICS is the study of heredity.   *JAM*

[*See also* GENE, GENOME]

heritage   Something that is inherited from the past or which a past generation has preserved and handed on to the present and which a significant part of the population wishes to hand on to the future. Heritage takes many forms but the broad division is between natural and cultural. *Natural heritage* refers to valued environments that can be found in LANDSCAPE, scenery and areas or SITES OF SPECIAL SCIENTIFIC INTEREST (SSSIs). Various types of PROTECTED AREAS, such as NATIONAL PARKS, are all designated to preserve or conserve environments that a society classifies as valued. *Cultural heritage* centres on the imprint of people on landscape and includes physical remains such as archaeological sites or historic monuments, and significant places such as battlefields or the residences of major figures in history. Such sites or monuments often have great symbolic significance that reach deep into a society's past and give meaning to its identity. These forms of heritage are found in *heritage places* and have a geographical expression but other forms of heritage, such as art, cultural events and rituals, are more aspatial.

All forms of heritage raise issues of preservation and CONSERVATION and many societies have laws enshrined in their constitutions to support these functions. Concepts such as STEWARDSHIP and *custodianship* are often linked with the responsibilities of government and special agencies are created to preserve, conserve and manage heritage. There are large numbers of WORLD HERITAGE SITES, designated by UNESCO, which cover both cultural and natural features. Among the criteria for a recognised cultural site is 'masterpiece of human creative genius' and for a natural site 'superlative natural phenomena or exceptional natural beauty and aesthetic importance'.   *DTH*

Herbert DT (ed.) (1995) *Heritage, tourism and society*. London: Cassells.

Lowenthal D (1996) *Possessed by the past: The heritage crusade and the spoils of history*. New York: Free Press.

Sabbioni C, Brimblecombe P and Cassar M (eds) (2010) *The atlas of climate change impacts on European cultural heritage*. London: Anthem Press.

Skeates R (2000) *Debating the archaeological heritage*. London: Duckworth.

UNESCO (2010) *World heritage series 26: Cultural landscapes*. Brussels: UNESCO.

Whelan Y and Moore N (eds) (2007) *Heritage, memory and the politics of identity: New perspectives on the cultural landscape*. Aldershot: Ashgate Publishing.

herpetology   The study of amphibians and reptiles; from the Greek for 'crawling thing'.   *JAM*

Beebee TJC and Griffiths RA (2000) *Amphibians and reptiles: A natural history of the British herpetofauna*. London: HarperCollins.

Vitt LJ and Caldwell JP (2009) *Herpetology: An introductory biology of amphibians and reptiles, 3rd edition*. Burlington, MA: Academic Press.

herringbone cross-bedding   A style of CROSS-BEDDING in which closely associated *sets* indicate opposed (bipolar) PALAEOCURRENT directions, commonly taken as a good indicator of sediment movement by TIDES. Associated MUD DRAPES and TIDAL BUNDLES can provide important information about past tidal conditions.   *GO*

[*See also* FLASER BEDDING, TIDAL RHYTHMITE]

Le Bot S and Trentesaux A (2004) Types of internal structure and external morphology of submarine dunes under the influence of tide- and wind-driven processes (Dover Strait, northern France). *Marine Geology* 211: 143–168.

heterotrophic organism   An organism that cannot make its own organic food. Heterotrophic organisms or *heterotrophs* therefore obtain their requirements from other living (e.g. HERBIVORES, CARNIVORES and PARASITES) or dead (SAPROPHYTES) organisms.   *JAM*

[*See also* AUTOTROPHIC ORGANISM]

Anderson JW (1980) *Bioenergetics of autotrophs and heterotrophs*. London: Arnold.
Hodkinson ID, Webb NR and Coulson SJ (2002) Primary community assembly on land—the missing stages: Why are the heterotrophic organisms always there first? *Journal of Ecology* 90: 569–577.

heterozygosity   The proportion of genetic loci, or GENES, that are heterozygous (possessing different ALLELES on paired *chromosomes* of the same individual) in the average individual.   *MVL*

[*See also* GENOTYPE, INBREEDING DEPRESSION, POLYMORPHISM]

Grueber CE, Wallis GP and Jamieson IG (2008) Heterozygosity-fitness correlations and their relevance to studies of inbreeding depression in threatened species. *Molecular Ecology* 17: 3978–3984.
Harrison XA, Bearhop S, Inger R et al. (2011) Heterozygosity-fitness correlations in a migratory bird: An analysis of inbreeding and single-locus effects. *Molecular Ecology* 20: 4786–4795.

heuristic algorithm   A computational procedure that uses *trial and error*, or methods based on PROBABILITY, to approximate a solution for statistically difficult problems. It is sometimes used in a GEOGRAPHICAL INFORMATION SYSTEM (GIS).   *VM*

hexapod   A standard concrete armour unit with six 'legs' designed to be used in numbers to protect a SHORELINE from EROSION.   *HJW*

[*See also* COASTAL ENGINEERING STRUCTURES, RIPRAP, TETRAPOD]

hiatus   A cessation of deposition resulting in a gap in the STRATIGRAPHICAL RECORD which is not represented by any SEDIMENTS. For example, in the QUATERNARY of the British Isles, there is a major hiatus in deposition between ca 1.6 Ma BP and ca 0.65 Ma BP.   *DH*

[*See also* CROMERIAN INTERGLACIAL, UNCONFORMITY]

hibernation   The reduction of metabolic rates that enables some mammals to survive cold conditions, generally over an entire winter season, without (true hibernation), or with, very limited food supplies.   *JLI*

[*See also* AESTIVATION]

Roots C (2006) *Hibernation*. Westport, CT: Greenwood Press.

hidden surface removal   A technique in the projection of a three-dimensional object onto a two-dimensional computer screen that entails displaying only those parts of surfaces that would be naturally visible to the user. It forms part of the study of *computer graphics* for the modelling and visualisation of objects such as DIGITAL ELEVATION MODELS.   *ACF*

Foley JD, van Dam A, Feiner SK et al. (1994) *Introduction to computer graphics*. Reading, MA: Addison-Wesley.

hierarchy theory   A theoretical framework proposed for SYSTEMS ANALYSIS in general and *ecosystem analysis* in particular, based on the idea that organisation results from differences in *process rates*. SYSTEMS can be decomposed into different units based on differences in process rates. Each stratum that is defined can be divided into a number of units, termed *holons*, which are separated by gradients in process rates. The method is highly applicable to studies of environmental change because of the emphasis that it places on *scale concepts*. It views ecosystems as hierarchical structures, with each stratum responding differently to changes in the environment. The overall response of the system can then be determined by examining the cumulative effects of changes in individual strata.   *JLI*

[*See also* ECOSYSTEM CONCEPT, FRACTAL ANALYSIS]

Allen TFH and Starr TB (1982) *Hierarchy: Perspectives for ecological complexity*. Chicago: University of Chicago Press.
Hewitt JE and Thrush SF (2009) Reconciling the influence of global climate phenomena on macrofaunal temporal dynamics at a variety of spatial scales. *Global Change Biology* 15: 1911–1929.
May RM (1989) Levels of organization in ecology. In Cherrett JM (ed.) *Ecological concepts*. Oxford: Blackwell, 339–363.
O'Neill RV, DeAngelis DL, Waide JB and Allen TFH (1986) *A hierarchical concept of ecosystems*. Princeton, NJ: Princeton University Press.
Turnbull L, Wainwright J and Brazier RE (2008) A conceptual framework for understanding semi-arid land degradation: Ecohydrological interactions across multiple-space and time scales. *Ecohydrology* 1: 23–34.
Wagner C and Adrian R (2009) Exploring lake ecosystems: Hierarchy responses to long-term change? *Global Change Biology* 15: 1104–1115.

hieroglyph   A character or 'letter', often pictorial, in early Egyptian (ca 3200 BC to AD 400) and Mesoamerican writing. Also called *glyphs* in the context of Mesoamerican archaeology, hieroglyphs are mostly known from the decoration carved or painted on architecture and ARTEFACTS.   *JAM*

Rice P (1999) Hieroglyphs. In Shaw I and Jameson R (eds)
   A dictionary of archaeology. Oxford: Blackwell, 275–277.

High Arctic   A general term used especially by ecolo-
gists to differentiate the various islands within the *Arc-
tic Basin*, such as the Canadian Arctic islands, Svalbard,
Franz Josef Land, northern Novaya Zemblya and northern
Greenland from the LOW ARCTIC. The transition between
these two divisions of the ARCTIC is not sharply delineated.
The High Arctic is characterised by an extreme, desert-
like environment with a sparse vegetation cover, either
POLAR DESERT or *polar semi-desert*, and a very short GROW-
ING SEASON (typically <2 months).   *HMF*

[*See also* SUBARCTIC, TUNDRA]

Lévesque E and Svoboda J (2005) High Arctic. In
   Nuttall M (ed.) *Encyclopedia of the Arctic.* New York:
   Routledge, 863–867.
Meltofte H, Christensen TR, Elberling B et al. (eds) (2008)
   High-Arctic ecosystem dynamics in a changing climate.
   London: Academic Press.

high nature value farmland   A category of agricultural LANDUSE defined in the European Union, which
recognises ECOLOGICAL VALUE as grounds for CONSERVA-
TION. Some 15–25 per cent of the European countryside
qualifies, especially in eastern and southern Europe.
It includes semi-natural *grassland*, DEHESAS, STEPPE,
alpine PASTURE and MEADOWS. These generally exhibit
extensive, low-intensity agriculture and high BIODIVER-
SITY, and are vulnerable to change. High nature value
farmland is part of the so-called second pillar of the
Common Agricultural Policy, namely, support for less-
favoured areas and agri-environment schemes.   *JAM*

European Environmental Agency (EEA) (2004) *High nature
   value farmland: Characteristics, trends and potential
   challenges.* Luxembourg: Office of Official Publications
   of the European Communities [EEA Report 1/2004].

high-activity clays   Clays with a high CATION
EXCHANGE CAPACITY (CEC) (>24 cmol$_c$/kg clay), usually developed over base-rich rocks. Such clays are of
considerable importance in soils of the *humid tropics*
as they have relatively great potential for SUSTAINABLE
AGRICULTURE. The high-activity clays are, however,
less common than the LOW-ACTIVITY CLAYS. A study of
the clay fractions from *topsoils* in seven West African
countries showed that over 68 per cent of the samples
consisted of low-activity clays (e.g. KAOLINITE) whereas
only 27 per cent consisted of high-activity clays (e.g.
SMECTITE and VERMICULITE).   *JAM/EMB*

[*See also* LUVISOLS]

Abe SS, Masunaga T, Yamamoto S et al. (2006)
   Comprehensive assessment of the clay mineralogy
   composition of lowland soils in West Africa. *Soil Science
   and Plant Nutrition* 52: 479–488.

high-precision dating   A small number of RADIO-
CARBON DATING laboratories worldwide employ techniques producing results with small precisions (± error
terms). Conventional RADIOMETRIC DATING TECHNIQUES
measure *residual radioactivity* and the precision is
dependent on the number of *decay events* recorded.
To obtain smaller precisions, more events must be
recorded, either by using larger samples or by measuring
the sample for a longer time. For routine dating laboratories, sample size is often limited and greater counting
times would reduce sample throughput rates, increasing
expense. Most specialist high-precision dating laboratories use techniques where larger sample sizes are used,
and these laboratories have been involved in the establishment of the CALIBRATION data set.   *PQD*

[*See also* HIGH-RESOLUTION RECONSTRUCTIONS,
WIGGLE MATCHING]

Pearson GW (1980) High precision radiocarbon dating
   by liquid scintillation counting applied to radiocarbon
   timescale calibration. *Radiocarbon* 22: 337–345.
Stuiver M, Robinson SW and Yang IC (1979) ^{14}C dating
   to 60,000 years BP with proportional counters.
   In Berger R and Suess HE (eds) *Proceedings of the
   9th International ^{14}C Conference on radiocarbon
   dating.* Berkeley: University of California Press, 202–215.
Tans PP and Mook WG (1978) Design, construction and
   calibration of a high accuracy carbon-14 counting set up.
   Radiocarbon 21: 22–40.

high-resolution reconstructions   The phrase
'high-resolution reconstruction' normally means the
reconstruction of different environmental parameters
with high temporal *resolution* (seasonal, annual, decadal, or centennial). As the general concern with the
issues surrounding GLOBAL ENVIRONMENTAL CHANGE
increases, the pressure on the scientific community to produce MODELS and PREDICTIONS of CLIMATIC
CHANGE increases. While recent meteorological and
oceanographic observations have paid attention to the
processes and mechanisms of atmospheric and oceanographic circulation, this has produced only short-term
perspectives (in general less than 200–300 years) of
global change, limited by the range of INSTRUMENTAL
DATA and HISTORICAL EVIDENCE. Studies in PALAEOCLI-
MATOLOGY and PALAEOCEANOGRAPHY have, on the other
hand, been mainly on coarser (in general greater than
century to millennial) timescales. The palaeorecords
that have the yielded temporal (annual to decadal)
resolution are, in addition to historical evidence, TREE
RINGS, ICE CORES, CORALS and laminated MARINE and
LACUSTRINE SEDIMENTS.

   High-resolution palaeoclimatic records from high
latitudinal regions are important for understanding the behaviour of the GENERAL CIRCULATION OF THE
ATMOSPHERE. The patterns and causes of climatic

change at these timescales are poorly understood, even though the magnitude and change recorded are significant. Annual to decadal records are necessary because changes in FUTURE CLIMATE will take place on these timescales. We must therefore know the range of natural CLIMATIC VARIABILITY against which to measure future change.

Several initiatives, including IGBP-PAGES (*International Geosphere-Biosphere Programme–Past Global Changes*) and NSF-PALE (*National Science Foundation–Paleoclimate of Arctic Lakes and Estuaries*), focus on obtaining high-resolution records of climate change for the past 1,000–2,000 years. The instrumental climate record of the past 50–150 years underestimates the range of natural climate variability across all temporal scales. It is becoming apparent that significant regional, spatial and temporal variability characterised the CLIMATE OF THE LAST MILLENNIUM. Networks of annually dated time series from trees, sediments, corals, ice cores and historical documents reveal that globally synchronous cold periods longer than a decade or two did not occur within the past 500 years. Emerging data also suggest that the largest and most extensive temperature shift of the past 1,000 years occurred between AD 1850 and today. Greenland ice-core data and instrumental records have revealed large decadal-scale climatic variations over the North Atlantic that can be related to a major source of high-frequency variability, the NORTH ATLANTIC OSCILLATION (NAO).
JAM

[*See also* ANNUAL VARIABILITY, ANNUALLY RESOLVED RECORD, DECADAL-SCALE VARIABILITY, NATURAL ARCHIVES, SUB-MILANKOVITCH]

Alverson KD, Bradley RS and Pedersen TF (eds) (2003) *Paleoclimate, global change and the future*. Berlin: Springer.

Battarbee RW (2008) Holocene climate variability and global warming. In Battarbee RW and Binney HA (eds) *Natural climate variability and global warming: A Holocene perspective*. Chichester: Wiley-Blackwell, 1–6.

Moberg A, Sonechkin DM, Holmgren K et al. (2005) Highly variable Northern Hemisphere temperatures reconstructed from low- and high-resolution proxy data. *Nature* 433: 613–617.

National Research Council (1995) *Natural climate variability on decade-to-century time scales*. Washington, DC: National Academies Press.

Neukom R and Gergis J (2012) Southern Hemisphere high-resolution palaeoclimate records of the last 2000 years. *The Holocene* 22: 501–524.

Thompson LG, Moseley-Thompson E, Davis ME et al. (2013) Annually resolved ice core records of tropical climate variability over the past ~1800 years. *Science* 340: 945–950.

Wahl ER and Frank D (2012) Evidence of environmental change from annually resolved proxies with particular reference to dendrochronology and the last millennium. In Matthews JA, Bartlein PJ, Briffa KR et al. (eds)

The SAGE handbook of environmental change, volume 1. London: Sage, 320–344.

Winkler S and Matthews JA (2010) Holocene glacier chronologies: Are high-resolution global and inter-hemispheric comparisons possible? *The Holocene* 20: 1137–1147.

highstand The maximum of an environmental cycle, fluctuation or oscillation. It may leave visible evidence in the LANDSCAPE, such as a RAISED SHORELINE at a *sea-level highstand* or a LAKE TERRACE following a lake-level highstand (see LAKE-LEVEL VARIATIONS). There may be more than one highstand close in time to the maximum, such as the three LITTLE ICE AGE highstands in glacier extent recognised in the Alps. The complementary term, LOWSTAND, is used for the minimum point in the cycle: for example, the sea-level lowstand at the LAST GLACIAL MAXIMUM (LGM) was some 120 m below present sea level.
JAM

[*See also* CLIMATIC FLUCTUATION, SEQUENCE STRATIGRAPHY, TRANSGRESSION, MARINE]

Bacon SN, Burke RM, Pezzopane SK and Jayko AS (2006) Last Glacial Maximum and Holocene lake levels of Owens Lake, eastern California, USA. *Quaternary Science Reviews* 25: 1264–1282.

Dorale JA, Onac DP, Fornós JR et al. (2010) Sea-level highstand 81,000 years ago in Mallorca. *Science* 327: 860–863.

Himalayan uplift The Himalayan mountain range is the result of the ongoing collision between the Indo-Australian and Asian tectonic plates. The resulting rapid uplift of 1,000–2,500 m in a few million years had created a mountain barrier by the Late MIOCENE (11–7.5 Ma). The elevation of the Himalayas by 5 km resulted in complex atmosphere-ocean-lithosphere interlinkages that have been a FORCING FACTOR in GLOBAL COOLING and the onset of the Late Cenozoic ICE AGE. GENERAL CIRCULATION MODELS (GCMs) have been used to model the complex FEEDBACK MECHANISMS initiated by Himalayan uplift.

It is suggested that the creation of the Himalayas and Tibetan plateau caused perturbations of the JET STREAM, which resulted in hemispheric changes in the GENERAL CIRCULATION OF THE ATMOSPHERE and CLIMATE. Himalayan uplift may have induced strong convective atmospheric circulation on the Tibetan Plateau, and increased temperature differences between summer and winter, leading to a regionally intense eastern Asia MONSOON circulation. LAND UPLIFT and a wet humid climate on the southern and eastern flanks of the Himalayas would have induced accelerated WEATHERING and EROSION RATES, leading to a reduction in atmospheric CARBON DIOXIDE concentrations, forcing global cooling.
MHD/CJC

[*See also* TECTONIC UPLIFT–CLIMATIC CHANGE
INTERACTION]

Aitchison JC, Ali JR and Davis AM (2007) When and
   where did India and Asia collide? *Journal of Geophysical
   Research: Solid Earth* 112: B5. doi:10.1029/2006JB004706.
Clift PD and Blusztajn J (2007) Reorganization of the
   western Himalayan river system after five million years
   ago. *Nature* 438: 1001–1003.
Raymo ME (1994) The initiation of Northern Hemisphere
   glaciation. *Annual Review of Earth and Planetary Science*
   22: 353–383.
Royden LH, Burchfiel BC and van der Hilst RD (2008) The
   geological evolution of the Tibetan Plateau. *Science* 321:
   1054–1058.
Ruddiman WF and Kutzbach JE (1991) Plateau uplift and
   climatic change. *Scientific American* 264: 42–50.
Vance D, Bickle M, Ivy-Ochs S and Kubik PW (2003)
   Erosion and exhumation in the Himalayas from
   cosmogenic isotope inventories of river sediments.
   Earth and Planetary Science Letters 206: 273–288.

histogram   A graph designed to quantify the FRE-
QUENCY of occurrence of data values within specific
CLASS INTERVALS, which is used to visualise the PROB-
ABILITY DISTRIBUTION of large data sets (e.g. to determine if they follow the GAUSSIAN MODEL). *Frequency
histograms* are often used in the preliminary, *graphical
analysis* of data, and in IMAGE ENHANCEMENT of images
acquired through REMOTE SENSING.   *ACF*

historical archaeology   The archaeological study
of people in historical times, that is, since the appearance of the earliest written records and extending to
the present day. It is distinguished by focusing on the
post-PREHISTORIC past, utilising a greater diversity of
sources than 'prehistoric' archaeology and impinging
on the modern, globalised world. The subdiscipline
is well developed in America, where it is sometimes
known as *historic sites archaeology* and the combined
archaeological/anthropological/historical study of the
effects of the spread of European culture is a strong
theme.   *JAM*

Hall M and Silliman SW (eds) (2006) *Historical
   archaeology*. Oxford: Blackwell.
Hicks D and Beaudry MC (eds) (2008) *The Cambridge
   companion to historical archaeology*. Cambridge:
   Cambridge University Press.
Little J (ed.) (1992) *Text-aided archaeology*. Boca Raton,
   FL: CRC Press.
Majewski T and Gaimster D (eds) (2011) *International
   handbook of historical archaeology*. Berlin: Springer.
Orser Jr CE (2002) *Encyclopedia of historical archaeology*.
   London: Routledge.
Schuyler RL (1970) Historical archaeology and historic
   sites archaeology as anthropology: Basic definitions and
   relationships. *Historical Archaeology* 4: 83–89.

historical climatology   The time period of relevance to 'historical' climatology has not been precisely
defined. A conservative interpretation regards it as
covering the period of written HISTORY of the past 2,000
years or so. Scientists such as H.H. Lamb have enjoyed
success in using the sparse historical records of both
the *written* and *oral traditions* to deduce broad features
of climates in 'pre-instrumental' times. Many descriptions and accounts of weather provide DOCUMENTARY
EVIDENCE of environmental change.

   Weather-dependent natural phenomena, both
physical (PARAMETEOROLOGY) and biological (PHENOL-
OGY), are also used in historical climatology. VINE
HARVEST dates, for example, provide good phenological indicators of summer heat and rainfall. Weather
information for oceanic areas can be found in SHIP LOG
BOOK RECORDS which date back to the earliest years of
European voyages of discovery. Looking back yet further, Mediaeval *estate accounts, diaries, grain-price
records*—all shed further light on climate through its
influence on AGRICULTURE. Such sources need, however, to be used with care as other, non-climatic,
elements can have influence on agriculture. By these
means, information has been gathered to provide a
more detailed picture of the two most climatically
distinctive intervals of historical times, the MEDIAEVAL
WARM PERIOD and the LITTLE ICE AGE.

   PROXY CLIMATIC INDICATORS provide helpful indications of climatic change for those millennia for which
written evidence is sparse or non-existent. DENDRO-
CHRONOLOGY, VARVE, ICE and MARINE SEDIMENT CORE data
provide information for much of the HOLOCENE and help
corroborate documentary evidence from more recent
centuries.

   Only since the middle of the nineteenth century
have reliable instrumental data been available from
organised networks. Some INSTRUMENTAL DATA exist for
earlier times but are of more limited scientific value
because of the lack of standardisation of instruments
and of their exposure. The British Isles have some of
the oldest INSTRUMENTAL RECORDS and these have been
used to reconstruct the CENTRAL ENGLAND TEMPERATURE
RECORD (CET). Rainfall records present more problems
because of variations of catch over short distances,
but both Holland and England have continuous series,
which date from the start of the eighteenth century.
Wind data, which require no complex instruments,
exist in various forms in England, most notably, from
the MIDDLE AGES.   *DAW*

Bradley RS and Jones PD (eds) (1992) *Climate since AD
   1500*. London: Routledge.
Brázdil R, Dubrovolný P, Luterbacher J et al. (2010)
   European climate of the past 500 years: New challenges
   for historical climatology. *Climatic Change* 101: 7–40.

Brázdil R, Pfister C, Wanner H et al. (2005) Historical climatology in Europe: The state of the art. *Climatic Change* 70: 363–430.

Brown N (2001) *History and climate change: A Eurocentric perspective.* London: Routledge.

Dawson AG (2009) *So foul and fair a day: A history of Scotland's weather and climate.* Edinburgh: Berlinn Limited.

Fleming JR (1998) *Historical perspectives on climate change.* New York: Oxford University Press.

Jones PD, Ogilvie AEJ, Davies TD and Briffa KR (eds) (2001) *History and climate: Memories of the future?* New York: Kluwer.

Lamb HH (1995) *Climate, history and the modern world.* London: Routledge.

Landsberg, HE (1985) Historic weather data and early meteorological observations. In Hecht AD (ed.) *Paleoclimate analysis and modelling.* New York: Wiley, 27–70.

Ogilvie AE (2010) Historical climatology, climatic change, and implications for climate science in the twenty-first century. *Climatic Change* 100: 33–47.

Peterson TC and Vose RS (1997) An overview of the Global Historical Climatology Network temperature data base. *Bulletin of the American Meteorological Society* 78: 2837–2849.

Pfister C (2010) The vulnerability of past societies to climatic variation: A new focus for historical climatology in the twenty-first century. *Climatic Change* 100: 25–31.

historical ecology   The study of the changes to *ecosystems* and LANDSCAPES that have occurred over the historical period. It is exemplified by Oliver Rackham's ecological studies of the status and history of British woodlands based on DOCUMENTARY EVIDENCE and FIELD RESEARCH, including investigation of the structure and composition of remaining fragments of ANCIENT WOOD-LAND. Such woodlands may be located on early maps and remain in areas unsuitable for agriculture, possess sinuous boundaries and trees that are irregularly spaced with straight trunks; they may also be recognised by the mixed-age stands, a rich epiphytic lichen flora and an absence of light-demanding species, though none of these characteristics are conclusive. Historical ecology necessitates an approach involving INTERDISCIPLINARY RESEARCH, and often a more geographical or anthropological interpretation is taken of its meaning. Indeed, William Balée (2000: 13) defines it as the investigation of 'interrelationships between human beings and the biosphere'.   *JAM*

[*See also* CULTURAL LANDSCAPE, ETHNOBIOLOGY, HUMAN ECOLOGY, LANDSCAPE ECOLOGY]

Balee W (ed.) (2000) *Advances in historical ecology.* New York: Columbia University Press.

Balee WL and Erickson CL (eds) (2006) *Time and complexity in historical ecology: Studies in the neotropical lowlands.* New York: Columbia University Press.

Rackham O (1986) *The history of the countryside.* London: Dent.

Rackham O (2010) *Woodlands.* London: Collins.

historical evidence   The term 'historical evidence' is used in at least three ways in the context of investigating past environmental phenomena: (1) a synonym for DOCUMENTARY EVIDENCE, written, graphical and numerical; (2) for evidence from the *historical period*, irrespective of its type, and especially including evidence from HISTORICAL ARCHAEOLOGY and (3) for any evidence from the past, irrespective of age or context. Here, the second use of the term is preferred in which documentary evidence is an important component, and it is proposed that historical evidence sensu stricto occupies the niche between the direct measurements of the INSTRUMENTAL RECORD and the proxy data from NATURAL ARCHIVES, the latter being the only source relating to PREHISTORIC times. Historians, archaeologists, scientists and others have developed rigorous methodologies to check for the veracity and reliability of historical evidence from a variety of sources both human and natural.

The reconstruction, analysis and interpretation of climate in historical times prior to standardised instrumental meteorological records (HISTORICAL CLIMATOLOGY) is the field in which the use of historical evidence is most highly developed. Historical written records or documentary evidence are the main data source. Widely used sources include EARLY INSTRUMENTAL METEOROLOGICAL RECORDS, WEATHER DIARIES and PROXY DATA yielding climatic information, such as HARVEST RECORDS, TREE RINGS, *cherry blossom dates* and other *phenological indicators*, data on FLOOD HISTORY, SEA ICE information and dates of FAMINE.

An important aim of historical climatology has been to extend the modern INSTRUMENTAL DATA time series backwards through the construction of quantitative, continuous and HOMOGENEOUS SERIES of climatic elements that ideally overlap with and have been correlated and calibrated with modern data. Because of the nature, comprehensiveness, time span and degree of continuity of the historical information available, this is often not possible. Considerable attention has been given to detailed reconstruction of the LITTLE ICE AGE and the preceding MEDIAEVAL WARM PERIOD in Europe and North America, but much longer reconstructions have been undertaken using Chinese, Japanese and Arabian historical sources. Although most attention has focussed on the development of time series of monthly or seasonal indices of temperature and precipitation, some studies have focussed on EXTREME CLIMATIC EVENTS, such as FLOODS and TROPICAL CYCLONES. The METHODOLOGY of historical climatology draws upon the disciplines of HISTORY and language in the translation and evaluation of historical source material, on HISTORICAL GEOGRAPHY in considerations of the impact upon and interrelationships with society, as well as statistics and climatology in their climatic evaluation and analysis. There are also considerable efforts being made to link such historical evidence to HIGH-RESOLUTION RECON-STRUCTIONS of climate from natural archives (see CLIMATE OF THE LAST MILLENNIUM).   *RG/JAM*

[*See also* CALIBRATION, FIELD RESEARCH, PALAEOCLIMATOLOGY]

Bradley RS and Jones PD (eds) (1992) *Climate since AD 1500*. London: Routledge.

Brazdil R, Wheeler D and Pfister C (2010) European climate of the past 500 years based on documentary and instrumental data. *Climate Change* 101: 1–6.

Diaz HF and Stahle DW (2007) Climate and cultural history in the Americas: An overview. *Climatic Change* 83: 1–8.

Glaser R (1997) Data and methods of climatological evaluation in historical climatology. *Historical Social Research* 22: 59–87.

Glaser R and Walsh RPD (eds) (1991) Historische Klimatologie in verschiedenen Klimazonen [Historical climatology in different climatic zones]. *Würzburger Geographische Arbeiten* 80.

Jones PD, Ogilvie AEJ, Davies TD and Briffa KR (eds) (2001) *History and climate: Memories of the future?* New York: Kluwer.

Lamb HH (1977) *Climate: Present, past and future, volume 2*. London: Methuen.

Ogilvie AEJ (2010) Historical climatology, climatic change and implications for climate science in the twenty-first century. *Climatic Change* 100: 33–47.

Wahl ER and Frank D (2012) Evidence of environmental change from annually resolved proxies with particular reference to dendrochronology and the last millennium. In Matthews JA, Bartlein PJ, Briffa KR et al. (eds) *The SAGE handbook of environmental change, volume 1*. London: Sage, 320–344.

historical geography   Traditionally defined as the study of the geographies of past periods, historical geography focuses on the interplay between time and place. Key geographical concepts of LANDSCAPE and ENVIRONMENT can only be understood as the products of ongoing processes of change and interaction. Historical geography is largely practised by human geographers in relation to the human events of HISTORY.   *DTH*

Baker AHR (2003) *Geography and history: Bridging the divide*. Cambridge: Cambridge University Press.

Butlin RA (1993) *Historical geography: Through the gates of space and time*. London: Arnold.

history   The academic discipline of understanding and interpreting past events. Events in the past prior to the written record are referred to as PREHISTORY. Maintaining the historical record is a key role for historians. The events normally relate to people, nations and GLOBAL CHANGE but in one sense everything on the Earth's surface has a history that is essential to its understanding. Intended as a true narrative of the past, historical accounts vary with the interpretations put upon them by different writers. These interpretations may reflect known biases or the amount of information available to historians at particular points in time.   *DTH*

[*See also* ENVIRONMENTAL HISTORY, ORAL HISTORY, PROTOHISTORY]

Carr EJ (2001) *What is history?* (with a new introduction by R.J. Evans). Basingstoke: Palgrave Macmillan.

Lukacs J (2012) *The future of history*. New Haven, CT: Yale University Press.

histosols   Formed in organic materials such as upland or lowland peat MIRES, histosols are characteristically found in wet climates or in wet declivities in the landscape. Variable amounts of accessory mineral material may be present. Upland peats and RAISED BOGS are normally extremely acid and are rarely cultivated, but the lowland peats of FENS are neutral or even calcareous in reaction. Histosols, soils with more than 40 cm depth of organic matter, are most extensive in boreal, ARCTIC and especially SUBARCTIC regions but there is a scatter of these soils throughout the world in association with other soil groups. The diagnostic feature of histosols is the presence of a *histic horizon* of organic material in various stages of DECOMPOSITION, which is water saturated for at least some of the year. Drainage and aeration of these soils result in loss of peat through OXIDATION and shrinkage. When cultivated, histosols are also subject to WIND EROSION after drying out. As a result, the area of valuable lowland histosols is decreasing through human activities (SOIL TAXONOMY: histosols).   *EMB*

[*See also* PEAT AND PEATLANDS, WORLD REFERENCE BASE FOR SOIL RESOURCES (WRB)]

Andriesse JP (1988) *Nature and management of tropical peat soils*. Rome: FAO [FAO Soils Bulletin 59].

Bridgham SD, Ping C-L, Richardson JL and Updegraf K (2000) Soils of northern peatlands: Histosols and Gelisols. In Richardson JL and Vepraskas MJ (eds) *Wetland soils: Genesis, hydrology, landscapes and classification*. Boca Raton, FL: Lewis Publishers, 343–370.

Lucas RE (1983) Organic soils (Histosols): Formation, distribution, physical and chemical properties and management for crop production. *Michigan Agricultural Experiment Station Research Report 1982*.

holism   The view that natural SYSTEMS constitute *wholes*, which are more than the sum of their parts and exhibit *emergent properties*. Holism contrasts with REDUCTIONISM, which denies *emergence* in this sense. The debate between holism and reductionism is very relevant to the field of environmental change as it is highly questionable whether the systems of interest, such as *ecosystems*, GEO-ECOSYSTEMS and the EARTH SYSTEM, can be understood without a HOLISTIC APPROACH. The term was coined by the South African statesman Jan Smuts in the early twentieth century.   *JAM*

[*See also* COMPLEXITY, DYNAMICAL SYSTEMS, EMERGENCE AND EMERGENT PROPERTIES]

Esfield M (2001) *Holism in philosophy of mind and philosophy of physics*. Berlin: Springer.

Smuts J (1927) *Holism and evolution*. London: Macmillan.

holistic approach   An approach to SCIENCE that treats phenomena (e.g. *ecosystems*) as entire entities, without trying to identify the individual processes operating within the SYSTEM. It may be justified in terms of *emergent properties* that cannot be predicted from the behaviour of the component parts of the system. EARTH-SYSTEM SCIENCE and SUSTAINABILITY SCIENCE are explicit holistic approaches to *environmental change*. Many INDIGENOUS PEOPLES adopt a holistic approach to the environment, in marked contrast to the *reductionist approach* generally adopted by Western science.

JLl/JAM

[*See also* EARTH SYSTEM, ECOSYSTEM CONCEPT, EMERGENCE AND EMERGENT PROPERTIES, GAIA HYPOTHESIS, GENERAL SYSTEM(S) THEORY (GST), HOLISM, REDUCTIONISM, WORLD SYSTEM]

Atleo ER (2004) *Tsawalk: A Nuu-chah-nulth worldview*. Vancouver: University of British Columbia Press.

Cajete G (2000) *Native science: Natural laws of interdependence*. Santa Fe, NM: Clear Light Publishers.

Castaños H and Lomnitz C (2012) *Earthquake disasters in Latin America: A holistic approach*. Dordrecht: Springer.

Sumi A, Mimura N and Masui T (eds) (2010) *Climate change and global sustainability: A holistic approach*. Tokyo: United Nations University Press.

Holocene   Also known as the *postglacial*, *recent*, FLANDRIAN, or *present interglacial*, the Holocene is the latest EPOCH of the GEOLOGICAL TIMESCALE, which has so far lasted an estimated 11,700 years (~10,000 radiocarbon years) with a maximum counting error of 99 years. Taken literally, the term means 'wholly recent', which originally referred to Holocene fossil assemblages containing only MODERN species. Together with the preceding PLEISTOCENE epoch, it completes the QUATERNARY.

JAM

[*See also* ANTHROPOCENE, HOLOCENE ENVIRONMENTAL CHANGE, HOLOCENE TIMESCALE]

Walker M, Johnsen S, Rasmussen SO et al. (2009) Formal definition and dating of the GSSP (Global Stratotype Section and Point) for the base of the Holocene. *Journal of Quaternary Science* 24: 3–17.

Holocene climatic mode   The concept of coherent reorganisations of the CLIMATIC SYSTEM during the HOLOCENE. Wanner and Brönnimann (2012) have suggested that such a mode is evident in multidecadal- to millennial-scale CLIMATIC VARIABILITY. They postulate that two submodes can be linked to large-scale climatic patterns during the HOLOCENE THERMAL OPTIMUM (before ~4.5 ka) and the following *Neoglacial* (after ~4.5 ka), respectively. Similar submodes are recognised in PROXY

RECORDS at multidecadal to century timescales. The HTM submode is characterised, for example, by a northerly position of the INTERTROPICAL CONVERGENCE ZONE (ITCZ), relatively high activity of the Afro-Asian MONSOON (transporting moisture to continental areas), low EL NIÑO frequency and positive NORTH ATLANTIC OSCILLATION indices. An almost opposite pattern existed during the Neoglacial submode. They also question whether one of these submodes could dominate in the ANTHROPOCENE under the influence of GLOBAL WARMING.   *JAM*

[*See also* CENTURY- TO MILLENNIAL-SCALE VARIABILITY, CLIMATIC MODES, DECADAL-SCALE VARIABILITY, NEOGLACIATION]

Wanner H, Beer J, Bütikofer J et al. (2008) Mid- to late-Holocene climate change: A review. *Quaternary Science Reviews* 27: 1791–1828.

Wanner H and Brönnimann S (2012) Is there a global Holocene climatic mode? *PAGES News* 20: 44–45.

Wanner H, Solomina O, Grosjean M et al. (2011) Structure and origin of Holocene cold events. *Quaternary Science Reviews* 30: 3109–3123.

Holocene domain   The environmental envelope characteristic of the HOLOCENE epoch, within which contemporary CIVILISATION has developed and thrived. It includes the variability inherent in HOLOCENE ENVIRONMENTAL CHANGE. Currently, there is concern that HUMAN IMPACTS ON ENVIRONMENT are driving some parts of the EARTH SYSTEM out of the Holocene domain.   *JAM*

[*See also* ANTHROPOCENE, PLANETARY BOUNDARIES]

Rockström J, Steffen W, Noone K et al. (2009) Planetary boundaries: Exploring the safe operating space for humanity. *Ecology and Society* 14: Article No. 32.

Holocene environmental change   Major global environmental changes occurred at the start of the Present Interglacial. ABRUPT CLIMATIC CHANGE saw withdrawal of ICE SHEETS from the mid-latitude continental land masses and triggered important changes in the oceans and on land, such as an increase in ocean volume, SEA-LEVEL CHANGE, marine TRANSGRESSION, PARAGLACIAL geomorphological activity, plant and animal MIGRATION, ECOLOGICAL SUCCESSION and SOIL DEVELOPMENT. Beyond the limits of the ice sheets, tropical and subtropical regions were also affected. The Sahara DESERT, for example, which had been more extensive than at present in the Late PLEISTOCENE, was effectively replaced by SAVANNA in the early Holocene and TROPICAL RAIN FORESTS from their Late-Pleistocene REFUGIA as conditions became warmer and wetter.

For the remainder of the Holocene, the Earth's present landscapes evolved under climatic conditions comparable to those of today with relatively small-scale natural climatic changes, such as temperature variations with an amplitude of a few degrees Celsius and frequencies of decades to millennia. In addition to this NATURAL background, however, the Holocene has witnessed unique and marked ANTHROPOGENIC environmental changes with increasing magnitude, frequency and complexity towards the present day. Indeed, it is the presence of the human agency that provides an important justification for differentiating the Holocene from the other INTERGLACIALS that characterised the PLEISTOCENE. Identifying and separating the natural from the anthropogenic effects are important research foci in the study of Holocene environmental change (see DETECTION and ATTRIBUTION).

After the attainment of an early- to mid-Holocene *Climatic Optimum* or HOLOCENE THERMAL OPTIMUM, when, for at least some of the time, temperate latitudes were 2–3°C warmer than today, the major trend in Holocene climate has been a LATE-HOLOCENE CLIMATIC DETERIORATION when temperatures have tended to decline and precipitation to increase in temperate latitudes. Irregular and/or quasi-cyclic PERIODICITIES superimposed on these relatively long-term trends include the prominent, cold event (termed the *Finse Event* in southern Norway) at about 8,200 years ago, which has been recognised in ICE CORES, MARINE SEDIMENT CORES and other PROXY RECORDS, the MEDIAEVAL WARM PERIOD and the LITTLE ICE AGE. Both long- and short-term patterns can be seen in the CLIMATIC RECONSTRUCTION of summer temperatures and winter precipitation for southern Norway, based on evidence from GLACIER VARIATIONS and TREE-LINE VARIATIONS (see Figure). Orbital forcing, as explained by the MILANKOVITCH THEORY, accounts for the long-term trend but the shorter-term SUB-MILANKOVITCH variations are poorly understood and probably involved several causes, namely, SOLAR FORCING, VOLCANIC FORCING, AUTO-VARIATION (including fluctuations in the THERMOHALINE CIRCULATION of the oceans) and, in recent times, forcing by GREENHOUSE GASES.

Prior to the Holocene, *Homo sapiens* had little more effect on the landscape than any other animal: environment affected humans far more than humans affected their environment. With the onset of the Holocene, and particularly after the NEOLITHIC REVOLUTION, this balance began to change. Major anthropogenic effects on vegetation and soils were evident first in the Old World over 7,000 years ago. Technological innovation has since continually improved the ability of an expanding human population to occupy, use and transform more of the Earth's surface, reversing the balance. However, some areas remained relatively immune from human impacts, as illustrated by the dramatic discovery that dwarf mammoths survived on Wrangel Island in the Siberian Arctic until about 3,000 years ago. HUMAN IMPACTS ON ENVIRONMENT reached their apotheosis after the advent of European IMPERIALISM and especially since the INDUSTRIAL REVOLUTION. Only recently has the extent of these human impacts been acknowledged in the wide recognition of a need for CONSERVATION and ENVIRONMENTAL PROTECTION to be taken seriously on a global scale, together with the exploitation of resources associated with economic development.

An increasing recognition of the importance of the study of Holocene environmental change can be attributed to what may be termed the *social imperative* and the *scientific imperative*. First, there are reciprocal relations between human societies and the natural environment, which affect wealth creation and quality-of-life considerations. This social imperative requires an understanding of both natural background levels and likely future human impacts on environment, which can be illuminated by knowledge of what has happened in the recent past. Second, there are fundamental aspects of the science of the natural environment that can only be fully understood by investigating environmental change over the Holocene timescale. This scientific imperative includes the unique opportunities afforded for developing theory about events that are rare spatially, exhibit a low frequency of occurrence through time or have a duration of decades to millennia. The variety and quality of NATURAL ARCHIVES available for the Holocene is higher than for earlier parts of the Quaternary or longer geological timescales. Many sources of PROXY DATA for the Holocene, such as ICE CORES, TREE RINGS, LACUSTRINE SEDIMENTS, PEAT STRATIGRAPHY and CORALS are capable of yielding continuous records with high (often annual or seasonal) temporal *resolution*. By adopting a MULTIPROXY APPROACH to PALAEOENVIRONMENTAL RECONSTRUCTION in the Holocene, the combined information abstracted from the evidence covers an impressive range of environmental variables in unprecedented detail. Spatial coverage on local, regional and global scales for the Holocene may approach those available at present, and Holocene records overlap with observations, the INSTRUMENTAL RECORD and DOCUMENTARY EVIDENCE, which are invaluable for CALIBRATION and CONFIRMATION (VALIDATION) of environmental reconstructions. These attributes also confer on the Holocene considerable potential as a source of MODERN ANALOGUES for interpreting the pre-Holocene. Last, but not least, they also render Holocene data invaluable for testing the GENERAL CIRCULATION MODELS (GCMs) that are currently the most effective basis for predicting FUTURE CLIMATIC CHANGE. Thus, the scientific importance of the environmental changes of the Holocene is immense as, in several respects, Holocene environmental change holds the key to a fundamental understanding of the past, present and future environments of the Earth.

JAM

[*See also* CENTURY- TO MILLENNIAL-SCALE VARIABILITY, ENVIRONMENTAL ARCHAEOLOGY, NEOGLACIATION, VEGETATION HISTORY]

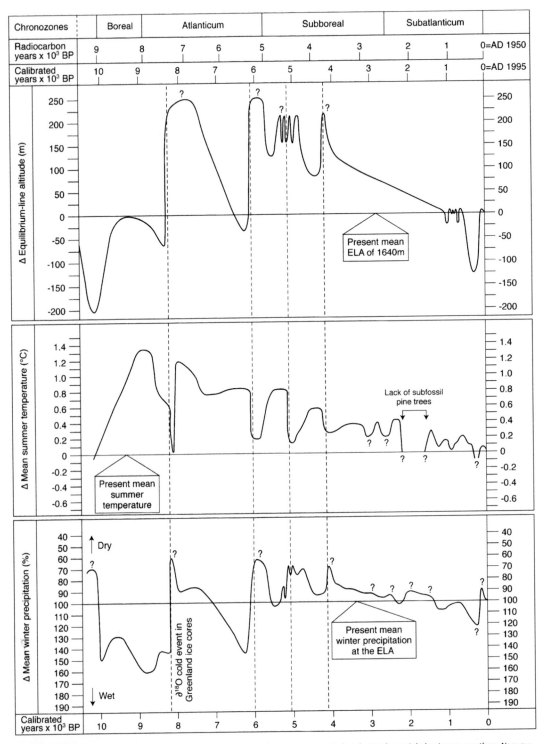

Holocene environmental change *Holocene glacier and climatic changes reconstructed at the Hardangerjøkulen ice cap, southern Norway:
variations in equilibrium-line altitude (ΔELA) on Hardangerjøkulen (upper graph) were combined with variations in mean summer temperature
derived from tree-line variations (middle) to derive the variations in mean winter precipitation (lower graph) (Dahl and Nesje, 1996).*

Battarbee RW and Binney HA (eds) (2008) *Natural climate variability and global warming: A Holocene perspective.* Chichester: Wiley-Blackwell.

Bentley MJ, Hodgson DA, Smith JA et al. (2009) Mechanisms of Holocene palaeoenvironmental change in the Antarctic Peninsula region. *The Holocene* 19: 51–70.

Bond G, Showers W, Cheseby M et al. (1997) A pervasive millennial-scale cycle in North Atlantic Holocene and Glacial climates. *Science* 278: 1257–1266.

Dahl S-O and Nesje A (1996) A new approach to calculating Holocene winter precipitation by combining glacier equilibrium-line altitudes with pine-tree limits: A case study from Hardangerjøkulen, central southern Norway. *The Holocene* 6: 381–398.

Issar AS (2003) *Climatic changes during the Holocene and their impact on hydrological systems.* Cambridge: Cambridge University Press.

Mackay A, Battarbee R, Birks J and Oldfield F (eds) (2003) *Global change in the Holocene.* London: Arnold.

Matthews JA (1998) The scientific and geographical importance of Holocene environmental change. *Swansea Geographer* 33: 1–6.

Matthews JA and Dresser PQ (2008) Holocene glacier variation chronology of the Smørstabbtinden massif, Jotunheimen, southern Norway, and the recognition of century- to millennial-scale European Neoglacial Events. *The Holocene* 18: 181–201.

Mayewski PA, Rohling EE, Stager JC et al. (2004) Holocene climate variability. *Quaternary Research* 62: 243–255.

Naussbaumer SU, Steinhilber F, Trachsel M et al. (2011) Alpine climate during the Holocene: A comparison between records of glaciers, lake sediments and solar activity. *Journal of Quaternary Science* 26: 703–713.

Roberts N (1998) *The Holocene: An environmental history, 2nd edition.* Oxford: Blackwell.

Turvey ST (ed.) (2009) *Holocene extinctions.* Oxford: Oxford University Press.

Vartanyan SL, Garutt VE and Sher AV (1993) Holocene dwarf mammoths from Wrangel Island in the Siberian Arctic. *Nature* 362: 337–340.

Wanner H, Beer J, Bütikofer J et al. (2008) Mid- to late-Holocene climate change: A review. *Quaternary Science Reviews* 27: 1791–1828.

Wanner H, Solomina O, Grosjean M et al. (2011) Structure and origin of Holocene cold events. *Quaternary Science Reviews* 30: 3109–3123.

Holocene thermal optimum

Various terms have been used to describe the interval of relatively mild climate that characterised the early to mid-HOLOCENE following the LAST GLACIATION, including *climatic optimum, thermal maximum, megathermal, xerothermic, altithermal* and *hypsithermal*. Some include a relatively long time interval; others restrict these terms to a shorter time interval when thermal conditions were indeed 'optimal', especially for vegetation development. Average summer temperatures may have reached 2–3°C warmer than today, depending on location. *JAM*

[*See also* HOLOCENE ENVIRONMENTAL CHANGE, LATE-HOLOCENE CLIMATIC DETERIORATION]

Berke MA, Johnson TC, Werne JP et al. (2012) A mid-Holocene thermal maximum at the end of the African Humid Period. *Earth and Planetary Science Letters* 351–352: 95–104.

Deevey ES and Flint RF (1957) Postglacial Hypsithermal interval. *Science* 125: 182–184.

Jansen E, Andersson C, Moros M et al. (2008) The early- to mid-Holocene thermal optimum in the North Atlantic. In Battarbee RW and Binney HA (eds) *Natural climate variability and global warming: A Holocene perspective.* Chichester: Blackwell, 123–137.

Kaufmann DS, Ager TA and Anderson NJ et al. (2004) Holocene thermal maximum in the western Arctic (0-180 degrees W). *Quaternary Science Reviews* 23: 529–560.

Renssen H, Seppä H, Crosta X et al. (2012) Global characterization of the Holocene thermal maximum. *Quaternary Science Reviews* 48: 7–19.

Holocene timescale

The Holocene is the second series or EPOCH of the QUATERNARY System/Period following the PLEISTOCENE. It is more commonly viewed as merely the most recent INTERGLACIAL or MARINE ISOTOPIC STAGE (MIS) 1. The opening of the HOLOCENE was conventionally drawn at 10,000 radiocarbon years BP, approximating to 11,500–11,700 years ago (calendar years BP) as defined in ICE CORES and VARVES, but has recently been formally defined as starting at 11,700 calendar years b2k based on a *Global Stratotype Section and Point* (*GSSP*) in the Greenland ice core from North GRIP (NGRIP).

The traditional subdivision of the Holocene was based on the BLYTT-SERNANDER TIMESCALE, originating in the latter part of the nineteenth century. Using evidence from MACROFOSSIL remains in *peat bogs*, they defined an alternating series of zones reflecting climatic changes through the period: PREBOREAL (increasing warmth but no trees), BOREAL (warm and dry), ATLANTIC (warm and wet), SUBBOREAL (warm and dry) and SUBATLANTIC (cool and wet). This pattern was later equated with the divisions of Iversen's INTERGLACIAL CYCLE, a pattern assumed to have been characteristic of all recent interglacials. Furthermore, early studies of vegetation history in the British Isles by Godwin and Iversen also utilised these periods as a basis for a scheme of pollen zonation applicable across the whole area as shown in the Figure.

In the absence of RADIOCARBON DATING, the ages of the boundaries between zones were approximate, based in the later Holocene on archaeological correlations. As radiocarbon dating developed, ages were derived from key sites and adopted over wide areas, hence the timescale was based on a very limited number of age determinations. As the number of horizons dated increased, it was realised that the vegetational changes on which the boundaries were founded were highly DIACHRONOUS. As the environmental characteristics and changes described by the zones were, however, still considered largely valid the terms were retained, and are still used

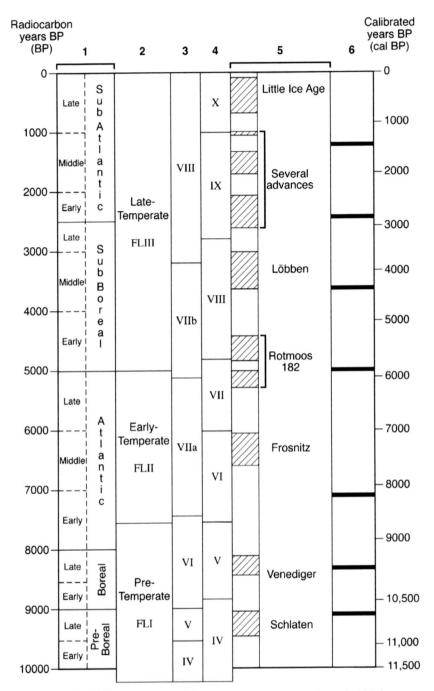

1. North-west European chronozones (after Mangerud, *et al.*, 1974)
2. British subdivisions of the Flandrian (after Sparks and West, 1972)
3. Pollen zones in the British Isles (after Godwin, 1975)
4. Alpine pollen zones (after Burga, 1988)
5. Alpine glacier advances (after Patzelt, 1974)
6. N. Atlantic ice rafting events (after Bond *et al.*, 1997)

Holocene timescale *The Holocene timescale in radiocarbon years (left, BP) and in calendrical or calendar years (right, cal. BP) as defined by a variety of approaches: (1) Northwest European chronozones (Mangerud et al., 1974); (2) British subdivisions of the Flandrian (Sparks and West, 1972); (3) Pollen zones in the British Isles (Godwin, 1975); (4) alpine pollen zones (Burga, 1988); (5) alpine glacier advances (Patzelt, 1974) and (6) North Atlantic ice-rafting episodes (Bond et al., 1997).*

to describe the broad subdivisions of the Holocene as CHRONOZONES. A widely used scheme in northwestern Europe is that of Mangerud et al. (1974). Defining ages within the Holocene is still very largely based on radiocarbon, although DENDROCHRONOLOGY and ICE-CORE DATING now provide much more precise dating techniques, and THERMOLUMINESCENCE (TL) DATING, OPTICALLY STIMULATED LUMINESCENCE (OSL DATING), URANIUM-SERIES DATING, TEPHROCHRONOLOGY and even some forms of COSMOGENIC-NUCLIDE DATING have been used, albeit usually with far less PRECISION. In the later Holocene, dating associated with archaeological evidence has also been used, and eventually human observation and documentary records can be used to improve or provide a timescale. In the absence of such evidence, and as radiocarbon dating is influenced by the SUESS EFFECT and BOMB EFFECT, dating events over the past 200 years has often proved relatively imprecise relying on techniques such as LEAD-210 and LICHENOMETRIC DATING. Over the latter half of the past century, nuclear weapons testing and NUCLEAR ACCIDENTS such as Chernobyl have allowed use of CAESIUM-137 for dating.

Definition of regional terms for periods is not such a problem in the Holocene as it is over the QUATERNARY TIMESCALE, although a number of ill-defined terms are widely used. The period of maximum warmth experienced in the Holocene is defined as the *climatic optimum* or HOLOCENE THERMAL OPTIMUM, although the timing and length of such a period varies considerably. The *Neoglacial* (see NEOGLACIATION), a term originally coined in North America, is sometimes used for the period after the thermal optimum when glaciers redeveloped in mountain areas, culminating in the LITTLE ICE AGE when glaciers in many areas reached their maximum Holocene limits. The Little Ice Age is difficult to define and is very variable in timing and length globally, although tends to centre of the middle and later part of the last millennium. On the continent of Europe, the term *Neuzeitlich* is also used in a chronological sense to identify renewed glaciation.

Because the Holocene has been a time of relative climatic stability, at least within a Quaternary perspective, changes have not been rapid, making boundaries difficult to specific dates. 'Events' have been identified and suggested to represent relatively RAPID ENVIRONMENTAL CHANGE, notably at 8,200 cal. BP and 2,700 cal. BP, but the extent to which these provide widely identifiable, precise time markers is uncertain. The recognition and *correlation* of such SUB-MILANKOVITCH (decadal to millenial) events represents one of the main tasks of those researching environmental changes on the Holocene timescale.

With expanding human populations, especially in the later Holocene, and the increasing HUMAN IMPACT ON ENVIRONMENT, it has become difficult to separate such effects

from natural, largely, CLIMATIC FORCING. The development of an ARCHAEOLOGICAL TIMESCALE has meant that division of time according to cultural parameters is of importance for most populated areas. Thus, the timescale originally developed on CLIMATOSTRATIGRAPHIC grounds utilising biological evidence exists in parallel with a timescale based entirely on TECHNOLOGICAL CHANGE or CULTURAL CHANGE criteria for recent millennia.   *CJC*

Bond G, Showers W, Cheseby M et al. (1997) A pervasive millenial-scale cycle in North Atlantic Holocene and glacial climates. *Science* 278: 1257–1266.

Burga CA (1988) Swiss vegetation history during the last 18,000 years. *New Phytologist* 110: 581–602.

Godwin H (1975) *History of the British flora: A factual basis for phytogeography, 2nd edition.* Cambridge: Cambridge University Press.

Iversen J (1958) The bearing of glacial and interglacial epochs on the formation and extinction of plant taxa. *Uppsala Universitet Årsskrift* 6: 210–215.

Mangerud J, Andersen ST, Berglund BE and Donner JJ (1974) Quaternary stratigraphy of Norden: A proposal for terminology and classification. *Boreas* 3: 109–127.

Matthews JA (2013) Neoglaciation in Europe. In Elias SA (ed.) *Encyclopedia of Quaternary science, 2nd edition, volume 2.* Amsterdam: Elsevier, 257–268.

Patzelt B (1974) Holocene variations of glaciers in the Alps. *Colloques Internationaux du CNRS* 219: 51–59.

Roberts N (1998) *The Holocene: An environmental history, 2nd edition.* Oxford: Blackwell.

Sparks BW and West RG (1972) *The Ice Age in Britain.* London: Methuen.

Walker M, Johnsen S, Rasmussen SO et al. (2009) Formal definition and dating of the GSSP (Global Stratotype Section and Point) for the base of the Holocene. *Journal of Quaternary Science* 24: 3–17.

Wanner H, Solomina O, Grosjean M et al. (2011) Structure and origin of Holocene cold events. *Quaternary Science Reviews* 30, 3109–3123.

holocoenotic environment The principle that ENVIRONMENTAL FACTORS are interdependent and mutually interacting. This was depicted in a classic diagram by W.D. Billings in 1952 (see Figure) showing the large number of factors that can potentially influence a single plant. If all the factors were equally important, this would create an immense problem for understanding *communities* and *ecosystems*, and environmental SYSTEMS more generally. However, very few factors may be influential at a particular place or time and a single factor (the limiting factor) tends to be critical. This is the principle of LIMITING FACTORS, which is a general formulation of Liebig's famous *law of the minimum* in relation to crop PRODUCTIVITY (which recognises that removal of a deficiency in a single nutrient may be all that is necessary to improve YIELDS). The two principles are linked by a third, the principle of *trigger factors*, which acknowledges that when environmental changes (natural or anthropogenic) remove (or overcome) the limiting factor this triggers

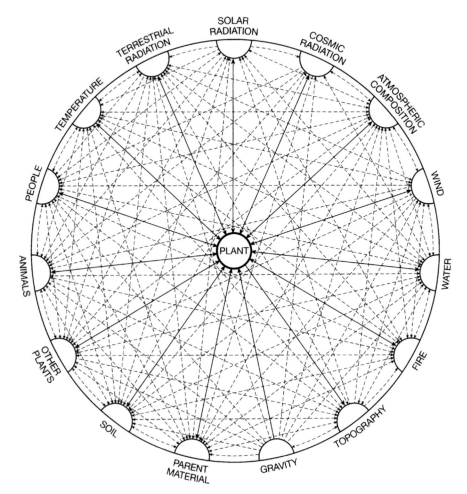

Holocoenotic environment *Schematic representation of potential direct interactions (solid lines) and indirect interactions (dashed lines) between environmental factors and a single plant; arrows indicate the direction of each effect (based on Billings, 1952, 1965).*

far-reaching effects throughout the holocoenotic environment until a new factor becomes limiting. The new limiting factor may be the trigger factor itself or something not apparent in the original system.   *JAM*

Billings WD (1952) The environmental complex in relation to plant growth and distribution. *Quarterly Review of Biology* 27: 251–265.
Billings WD (1965) *Plants and the ecosystem.* London: Macmillan.

holokarst   An area of KARST landscape that contains the full range of karst landforms (including POLJES) formed in thick, pure LIMESTONE beds, as exemplified by the Dinaric area of the former Yugoslavia. It contrasts with the relatively poorly developed MÉROKARST.   *RAS*

Cvijić J (1925) *Types morphologiques des terrains calcaires: Le holokarst* [Morphological types of limestone terrain: The holokarst]. *Comptes Rendus de l'Académie des Sciences* 180: 592–594.

holomictic lake   A lake which has unrestricted circulation through the whole water column after autumn OVERTURNING.   *HHB*

[*See also* MEROMICTIC LAKE]

homeorhesis   A concept of SELF-REGULATION whereby a DYNAMICAL SYSTEM returns to a trajectory. It should be distinguished from HOMEOSTASIS, whereby the SYSTEM returns to a particular state.   *JAM*

[*See also* FEEDBACK MECHANISMS, GAIA HYPOTHESIS]

homeostasis   The maintenance of a *steady-state equilibrium* by SYSTEM SELF-REGULATION. Homeostasis is brought about by negative FEEDBACK MECHANISMS ensuring that inputs and outputs of a system counterbalance. It confers RESILIENCE in the face of PERTURBATIONS from external FORCING FACTORS. In organisms, homeostasis is a *physiological equilibrium* produced by a balance of functions and of chemical compositions. Homeostatic mechanisms also occur in *ecosystems* and at the level of the BIOSPHERE (see GAIA HYPOTHESIS).   *JAM/RJH*

[*See also* COMPLEXITY, DISTURBANCE, EQUILIBRIUM CONCEPTS, HOLOCOENOTIC ENVIRONMENT, STABILITY CONCEPTS: ECOLOGICAL CONTEXTS]

Bradshaw SD (1997) *Homeostasis in desert reptiles*. Berlin: Springer.
Ernest SKM and Brown JH (2001) Homeostasis and compensation: The role of species and resources in ecosystem stability. *Ecology* 82: 2118–2132.
Reynolds CS (2002) Resilience in aquatic ecosystems: Hysteresis, homeostasis, and health. *Aquatic Ecosystem Health and Management* 5: 3–17.
Trojan P (1984) *Ecosystem homeostasis*. The Hague: Junk.

hominid   Of the family Hominidae. The term *Hominidae* was previously used to refer only to humans and their immediate FOSSIL or SUBFOSSIL ancestors and although scientifically questionable, the term 'hominid' is still frequently applied in this narrower sense with the term *hominoid* reserved for the broader grouping. The Hominidae in fact comprise two subfamilies, the *Ponginae* and the *Homininae*. The Ponginae migrated into Asia and are now represented solely by the *orangutan*, whereas the Homininae are represented by *chimpanzees* and *gorillas*, as well as humans. This latter group remained in Africa, dividing again into African apes and the human LINEAGE. Humans and their fossil ancestors are distinguished from other apes by larger brain sizes permitting complex communication such as speech, an upright bipedal gait and by a slow rate of postnatal growth and development, favouring the development of a more elaborate social organisation with concomitant complex technological and cultural behaviours.   *DCS*

[*See also* ANATOMICALLY MODERN HUMANS (AMH), ARCHAIC HUMANS, HOMINISATION, HUMAN EVOLUTION, HUMAN MIGRATION: 'OUT OF AFRICA']

Bonnefille R (2010) Cenozoic vegetation, climate changes and hominid evolution in tropical Africa. *Global and Planetary Change* 72: 390–411.
Cameron DW (2004) *Hominid adaptations and extinctions*. Sydney: University of New South Wales Press.
Scally A, Dutheil JY, Hillier LW et al. (2012) Insights into hominid evolution from the gorilla genome sequence. *Nature* 483: 169–175.
Spassov N, Geraads D, Hristova L et al. (2012) A hominid tooth from Bulgaria: The last pre-human hominid of continental Europe. *Journal of Human Evolution* 62: 138–145.

hominin   Any species of the *Homininae*: the subfamily of the HOMINIDS that contains ANATOMICALLY MODERN HUMANS, ancestral humans, *gorillas* and *chimpanzees*.   *JAM*

Carbonell E, Bermúdez de Castro JM, Parés JM et al. (2008) The first hominin of Europe. *Nature* 452: 465–469.

Cerling TE, Wynn JG, Andanje SA et al. (2011) Woody cover and hominin environments in the past 6 million years. *Nature* 476: 51–56.
Wood B and Harrison T (2011) The evolutionary context of the first hominins. *Nature* 470: 347–352.

hominisation   The progression from apes to ANATOMICALLY MODERN HUMANS whereby more 'human-like' characteristics, both biological and cultural, obtain. The process has arguably taken at least 5 million years. Pagel (2012a, 2012b) has argued that the development of a unique capacity for CULTURE around 200,000 years ago was the defining event in the evolution of modern humans. Culture provided strategies for survival based, in particular, on the ability to learn from others, to negotiate using *language* and to transmit KNOWLEDGE and TECHNOLOGICAL CHANGE. Unlike other species, this enabled humans to become adapted to, and dominate, almost every environment on Earth. Their co-operative as well as competitive systems were based on a flexibility of *mind* that enabled a large range of abilities and skills to be developed by different individuals. Ultimately, the process of hominisation resulted in *art*, RELIGIOUS BELIEFS and SCIENCE, all dependent on the EVOLUTION of the brain and the development of the mind. Allenby and Sarewitz (2011) argue, however, that terms such as *human*, technology and NATURE are outmoded, and should be rethought within a new concept of an integrated *techno-human condition*.   *JAM*

[*See also* ADAPTATION OF PEOPLE, CAVE ART, HOMINID, HUMAN EVOLUTION, TRANSHUMANITY]

Allenby BR and Sarewitz D (2011) *The techno-human condition*. Cambridge: MIT Press.
Klein RG (1999) *The human career: Human biological and cultural origins*. Chicago: Chicago University Press.
Mithen S (1996) *The prehistory of the mind*. London: Thames and Hudson.
Pagel M (2012a) Adapted to culture. *Nature* 482: 297–299.
Pagel M (2012b) *Wired for culture: Origins of the human social mind*. New York: Norton.
Tattersall I (1998) *Becoming human: Evolution and human uniqueness*. New York: Harcourt Brace.
Tobias PV (1979) Men, minds and hands: Cultural awakenings over two million years of humanity. *South African Archaeological Bulletin* 34: 92–95.
Williams MAJ (1985) On becoming human: Geographical background to cultural evolution. *Australian Geographer* 16: 175–184.

homogeneous series   A climatic *time series* containing data drawn from a single population. It may consist of a sequence of values recorded at one meteorological station, or a sequence constructed from observations made under closely similar conditions that have been corrected statistically (*homogenised*) for any differences that may be attributable to changes of station location or of instrument.   *AHP/JAM*

[*See also* CENTRAL ENGLAND TEMPERATURE RECORD (CET), TIME-SERIES ANALYSIS]

Hall MJ (2003) The interpretation of non-homogeneous hydrometeorological time series: A case study. *Meteorological Applications* 10: 61–67.
Toreti A, Kuglitsch FG, Xoplaki E et al. (2010) A novel method for the homogenization of daily temperature series and its relevance for climate change analysis. *Journal of Climate* 23: 5325–5331.

homoiothermic organism A 'warm-blooded' organism (*mammals* or *birds*) that, unlike *poikilothermic organisms* (e.g. *invertebrates, fishes, amphibians* and *reptiles*) has the ability to vary, and to some extent maintain its body temperature independent of the temperature of its environment.   *JAM*

horizon scanning A *futures technique*, which attempts to identify gaps in KNOWLEDGE and/or policy; anticipate future needs, opportunities and threats; reduce RISK and inform *strategic planning*. It may involve *brainstorming*, widespread *consultation, expert groups* and *agenda setting* for policy, practice and RESEARCH.   *JAM*

[*See also* FUTURES STUDIES]

Sutherland WJ, Bailey MJ, Bainbridge P et al. (2008) Future novel threats and opportunities facing UK biodiversity identified by horizon scanning. *Journal of Applied Ecology* 45: 821–833.
UK Environment Agency (2004) *Horizon scanning: Air pollution abatement*. London: Department for Environment, Food and Rural Affairs (DEFRA).

horst A GEOLOGICAL STRUCTURE comprising an uplifted block bounded on either side by FAULTS. At a given level, the rock between the faults is older than that on either side. Examples of topographically expressed horst blocks include the upland areas of the Vosges and the Black Forest, separated by the GRABEN of the Rhine Valley.   *GO*

[*See also* RIFT VALLEY]

Graversen, O. (2009) Structural analysis of superimposed fault systems of the Bornholm horst block, Tornquist Zone, Denmark. *Bulletin of the Geological Society of Denmark* 57: 25–49.

horticulture The HARVESTING or CULTIVATION of *garden plants*, including *fruit trees, vegetables, flowers* and *ornamental plants*. It was important in AGRICULTURAL ORIGINS. Today it takes many forms ranging from the harvesting of *fruit* and *nuts*, and SHIFTING CULTIVATION in the absence of *cultivated fields*, to specialised and industrialised form of modern AGRICULTURE involving large acreages of *polytunnels*.   *JAM*

[*See also* ARBORICULTURE, SILVICULTURE, VERTICAL FARMING]

Levinson D (1995) *Human environments: A cross-cultural encyclopedia*. Santa Barbara, CA: ABC-CLIO.
Piperno DR, Ranere AJ, Holst I and Hansell P (2000) Starch grains reveal early root crop horticulture in the Panamanian tropical forest. *Nature* 407: 894–897.
Preece JE and Read PE (2005) *The biology of horticulture, 2nd edition*. Hoboken, NJ: Wiley.

Hortonian overland flow Water that flows over the ground surface (but not in defined channels) when *rainfall intensity* exceeds the INFILTRATION CAPACITY of the surface soil, sediment or rock. First proposed by R.E. Horton, it is sometimes termed *infiltration-excess overland flow*. Hortonian overland flow tends to be important in ARIDLANDS and SEMI-ARID environments, urban areas and areas subjected to OVERGRAZING and/or SOIL EROSION. It can also occur on soils that exhibit high HYDROPHOBICITY, possess a high silt-clay content, are covered by a matted leaf LITTER or are characterised by a fine-root mat.   *ADT/RPDW*

[*See also* RUNOFF PROCESSES, SATURATION OVERLAND FLOW, WATER REPELLENCY]

Horton RE (1933) The role of infiltration in the hydrologic cycle. *Transactions of the American Geophysical Union* 14: 446–460.
Sidle RC, Hirono T, Gomi T and Terajima T (2007) Hortonian overland flow from Japanese forest plantations: An aberration, the real thing, or something in between? *Hydrological Processes* 21: 3237–3247.

hotspot, in biodiversity Areas with exceptional concentrations of ENDEMIC species and where HABITAT LOSS is a threat to their survival. Thus, it may be argued that hotspots are where there is the greatest need for CONSERVATION and where CONSERVATION efforts might be appropriately concentrated in proportion to the share of the world's species at risk. The term 'biodiversity hotspot' is sometimes used to refer exclusively to a specific set of 34 *conservation priority sites* identified by *Conservation International*, but is also used to refer to other sites with very high levels of biodiversity. An estimated 50 per cent (150,000 species) of all *vascular plant species* and 42 per cent (11,980 species) of all *terrestrial vertebrate species* are endemic to 34 hotspots identified by Conservation International, occupying only 2.3 per cent of the Earth's land surface. These hotspots have been defined on the basis of two criteria: first, they contain at least 0.5 per cent of the world's plant species as ENDEMIC (i.e. at least 1,500 endemic species); second, they have lost at least 70 per cent or more of their *primary vegetation* (indeed, 11 have lost at least 90 per cent).

TROPICAL AND SUBTROPICAL FORESTS and MEDITERRANEAN REGIONS account for most hotspots. Leading hotspots according to one scheme is shown in the Figure. The five leading hotspots—the tropical Andes, Sundaland (centred on the Sunda Straits of Indonesia),

Madagascar, the Atlantic forests of Brazil and the Caribbean—each contain endemic plants and vertebrates amounting to at least 2.0 per cent and together account for about 20 per cent and 16 per cent, respectively, of all plant and vertebrate species. Some 86 per cent by area of the hotspots' habitat has been destroyed, reducing their area from 15.7 per cent of the Earth's land surface to 2.3 per cent. The average protected area coverage of hotspots is about 10 per cent of their original extent. These protected areas will not guarantee the survival of species since many of the protected areas are *paper parks*, with no effective protection. As biodiversity hotspots are the principal *Holocene refugia* for the Earth's biota, they deserve better protection.   *JAM/JLI*

[*See also* BIODIVERSITY, REFUGIUM]

Cincotta RP, Wisnewski J and Engelman R (2000) Human population in the biodiversity hotspots. *Nature* 404: 990–992.
Conservation International (2012) *Biodiversity hotspots.* [Available at http://www.biodiversityhotspots.org]
Malcolm JR, Liu C, Neilson RP et al. (2006) Global warming and extinctions of endemic species from biodiversity hotspots. *Conservation Biology* 20: 538–548.
Mittermeier RA, Gil PR, Hoffmann M and Pilgrim J (2004) *Hotspots revisited.* Chicago: University of Chicago Press.
Mittermeier RA, Myers N and Mittermeier CG (2000) *Hotspots: Earth's biologically richest and most endangered terrestrial ecoregions.* Chicago: University of Chicago Press.
Myers N, Mittermeier RA, Mittermeier CG et al. (2000) Biodiversity hotspots for conservation priorities. *Nature* 403: 853–858.

Orme CDL, Davies RG, Burgess M et al. (2005) Global hotspots of species richness are not congruent with endemism or threat. *Nature* 436: 1016–1019.
Sander J and Wardell-Johnson G (2011) Fine-scale patterns of species and phylogenetic turnover in a global biodiversity hotspot: Implications for climate change vulnerability. *Journal of Vegetation Science* 22: 766–780.
Zachos FE and Habel JC (eds) (2011) *Biodiversity hotspots: Distribution and protection of conservation priority areas.* Berlin: Springer.

hotspot, in geology   A localised, long-lived area of VOLCANISM, commonly characterised by FLOOD BASALTS or, on CONTINENTAL CRUST, very high-volume explosive eruptions of *rhyolite* (VOLCANIC EXPLOSIVITY INDEX (VEI) = 8). The locations of hotspots are independent of the PLATE TECTONICS system; most are not situated at PLATE MARGINS and produce *intraplate volcanism* in the form of *oceanic islands* (e.g. Hawaii and Galapagos) or localised continental activity (e.g. Yellowstone). Where a hotspot does coincide with a plate margin, it gives rise to greater melting of the MANTLE, and more voluminous volcanic products than would otherwise occur (e.g. the Icelandic hotspot at a CONSTRUCTIVE PLATE MARGIN). As plates move, VOLCANOES formed at a hotspot move away from the site of high heat flow, so that an active hotspot such as Hawaii lies at one end of a chain of extinct VOLCANOES, SEAMOUNTS and GUYOTS. Dating these *hotspot tracks* provides a means of determining rates and directions of *plate movements*, and their distribution provides a fixed frame of reference

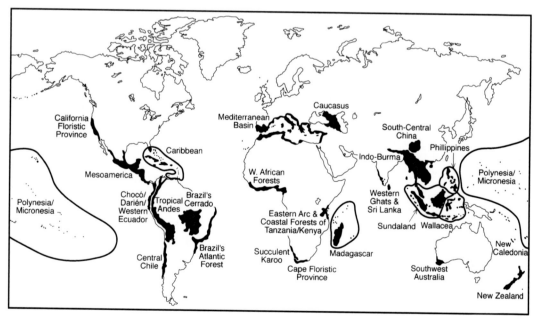

Hotspot, in biodiversity *The 25 leading biodiversity hotspots occupying 1.4 per cent of the Earth's surface and containing 44 per cent of all vascular plant species and 35 per cent of mammals, birds, reptiles and amphibians (Myers et al., 2000).*

for determining the absolute motion of the Earth's tectonic plates. Hotspots on thicker CONTINENTAL CRUST cause regional UPLIFT, influence patterns of EROSION, and may lead to the break-up of continents and the development of SEA-FLOOR SPREADING. Hotspot activity is understood by many Earth scientists to represent the surface expression of MANTLE PLUMES, although this PARADIGM has recently been hotly debated.   *GO*

[*See also* RIFT VALLEY, WILSON CYCLE]

DiVenere V and Kent DV (1999) Are the Pacific and Indo-Atlantic hotspots fixed? Testing the plate circuit through Antarctica. *Earth and Planetary Science Letters* 170: 105–117.

Farnetani CG and Samuel H (2005) Beyond the thermal plume paradigm. *Geophysical Research Letters* 32: L07311.

Foulger GR and Anderson DL (2005) A cool model for the Iceland hotspot. *Journal of Volcanology and Geothermal Research* 141: 1–22.

Mason BG, Pyle DM and Oppenheimer C (2004) The size and frequency of the largest explosive eruptions on Earth. *Bulletin of Volcanology* 66: 735–748.

hotspot, in remote sensing   The direction in which light SCATTERING occurs most strongly from a surface relative to the source of illumination. For a VEGETATION CANOPY, this is typically back in the direction of the illumination since this is the viewing angle at which shadows cast by canopy elements are obscured from view by the elements themselves.   *TLQ*

[*See also* BIDIRECTIONAL REFLECTANCE DISTRIBUTION FUNCTION (BRDF)]

Hoxnian Interglacial   Named after the Hoxne type site in Suffolk, England, the Hoxnian Interglacial followed the *Anglian glaciations* and preceded the *Wolstonian glaciations*. Occurring somewhere between ca 430 and 300 ka BP, it broadly corresponds to the Holsteinian Interglacial of Northern Europe and almost certainly to MARINE ISOTOPIC STAGE (MIS) 11 (425–395 ka BP), although in the British Isles because of the difficulties in dating interglacials, biostratigraphical records believed to be Hoxnian may date from more than one interglacial, interstadial or MIS.   *MHD/CJC*

[*See also* QUATERNARY TIMESCALE]

Ashton N, Lewis SG, Penkman KEH and Coope GR (2008) New evidence for complex climate change in MIS 11 from Hoxne, Suffolk, UK. *Quaternary Science Reviews* 27: 652–668.

Roe HM, Coope GR, Devoy RJN et al. (2009) Differentiation of MIS9 and MIS11 in the continental record: Vegetational, faunal, aminostratigraphic and sea-level evidence from coastal sites in Essex, UK. *Quaternary Science Reviews* 28: 2342–2373.

hum   A residual hill or positive landform in a KARST landscape produced by the DISSOLUTION of surrounding rock material.   *SHD*

human dimensions of environmental change
The interactions between environmental change and people. They include both the impacts of people on the environment and the human response to ENVIRONMENTAL CHANGE. Studies of both aspects have received renewed interest since the 1970s with a focus on the social causes and consequences of large-scale environmental transformations, such as OZONE DEPLETION, GLOBAL WARMING and LANDUSE CHANGE.

The HUMAN IMPACT ON ENVIRONMENT became significant early in the HOLOCENE and dominant over NATURAL environmental change during the ANTHROPOCENE. But it was not until the late twentieth century that it became a topic of importance to individuals, businesses, governments and NON-GOVERNMENTAL ORGANISATIONS (NGOs). The human (ANTHROPOGENIC) drivers of the impacts include POPULATION GROWTH, CONSUMPTION, GLOBALISATION, URBANISATION and TECHNOLOGICAL CHANGE. Attitudes and ENVIRONMENTAL ETHICS underlie these causes and hence shape both human impacts and the human response. As people have become more aware of their impact, the fields of ENVIRONMENTAL POLICY, ENVIRONMENTAL ECONOMICS, ENVIRONMENTAL LAW and ENVIRONMENTAL PROTECTION have become more important.

Understanding the human dimensions of environmental change requires an INTERDISCIPLINARY approach. This is necessary for both the SCIENCE of environmental change and its application to the real world. However, managing environmental change demands education, communication and political skill as well as APPLIED SCIENCE. Ideas of altruism, equity and social justice, for example, vital for ensuring future generations do not suffer a degraded environment (see STEWARDSHIP CONCEPT), does not spontaneously develop. Furthermore, reducing negative impacts and supporting positive impacts is not costless to planners, managers and citizens who must accept trade-offs. There is now realisation that such approaches, values and behaviours are needed and this has been reflected in the rapid development of the concept of SUSTAINABLE DEVELOPMENT and the growing number of international CONVENTIONS relating to environmental change and the human impact.

MITIGATION of unwanted human-induced environmental change and ADAPTATION to environmental change are other important human dimensions of environmental change. Both sets of adjustments involve DETECTION of change, recognition of critical THRESHOLDS, ENVIRONMENTAL MONITORING, ENVIRONMENTAL MODELLING and FORECASTING to provide early warning and implement effective ENVIRONMENTAL MANAGEMENT. Some decisions have to be made about future environmental change (e.g. FUTURE CLIMATE) based on inadequate research and poorly tested, uncertain theory. Short-term gains almost certainly have to be deferred in order to achieve long-term goals. Efforts to adapt to or mitigate GLOBAL

ENVIRONMENTAL CHANGE may require new types of governance structures and co-operative agreements.

The response of people to environmental change has been debated for at least as long as human impacts on environment. Between the 1860s and 1940s, researchers placed too much emphasis on simplistic environmental DETERMINISM, which has since been discredited. A succession of more complex views then developed (see ENVIRONMENTAL IMPACT ON PEOPLE). Since the 1980s, there has been interest in learning from past human-environment interactions in order to establish realistic theories, apply them to future environmental changes and guide future human responses. Caution is needed to ensure correlations are not read as causation. A climate change may seem to coincide with decline of a culture but more than one line of evidence may be needed to be sure, and it must not be forgotten that past human (and perhaps natural) responses may not be repeated in the future. For example, the development aid sector has explored the relationship between human poverty and environmental change, concluding that poverty causes degradation and that poor people are vulnerable. However, modern humans are possibly more vulnerable and less adaptable to environmental change than peoples in the past.
CJB

[*See also* COLLAPSE OF CIVILISATIONS, EARLY-WARNING SYSTEMS, ENVIRONMENTALISM, ENVIRONMENT-HUMAN INTERACTIONS, VULNERABILITY]

Barrow CJ (2003) *Environmental change and human development: Controlling nature?* London: Arnold.

Committee on the Human Dimensions of Global Change and Committee on Global Change Research (1999) *Human dimensions of global environmental change: Research priorities for the next decade.* Washington, DC: National Academies Press.

Cooperrider DL and Dutton JE (eds) (1999) *Organisational dimensions of global change: No limits to cooperation.* London: Sage.

International Human Dimensions Programme (IHDP) (2011) *International human dimensions programme on global environmental change.* Tokyo: United Nations University.

Leichenko RM and O'Brien KL (2010) Human dimensions of global environmental change. In Warf B (ed.) *Encyclopedia of geography.* Thousand Oaks, CA: Sage, 1450–1455.

Stern PC (1992) Psychological dimensions of global environmental change. *Annual Review of Psychology* 43: 269–302.

Stern PC, Young OR and Drukman D (eds) (1992) *Global environmental change: Understanding the human dimensions.* Washington, DC: National Academy Press.

human ecology   The investigation of how humans relate to, interact with and impact upon their surroundings. It has also been defined in terms of the complex interactions between ecological systems and human social systems. Intellectually, human ecology emerged following the application of ECOLOGY to human populations in the late nineteenth century but it has dimensions focusing on CULTURE, human attitudes, behaviour and belief systems, which place it as much in the SOCIAL SCIENCES as in the natural or ENVIRONMENTAL SCIENCES. Indeed, multidisciplinary 'roots' to human ecology can be recognised in *biology, sociology,* ANTHROPOLOGY, GEOGRAPHY, *psychology* and other disciplines. *Cultural ecology* is sometimes used as a synonym.
MLW/JAM

[*See also* DETERMINISM, ENVIRONMENTALISM, ENVIRONMENT-HUMAN INTERACTIONS, HUMAN GEOGRAPHY, HUMAN IMPACT ON ENVIRONMENT, POLITICAL ECOLOGY]

Butzer KW (1990) The realm of cultural-human ecology: Adaptation and change in historical perspective. In Turner BC, Clark WC, Kates RW et al. (eds) *The Earth as transformed by human action.* New York: Cambridge University Press, 685–701.

Hawley AH (1986) *Human ecology: A theoretical essay.* Chicago: University of Chicago Press.

Miller JR, Lerner RM, Schiamberg LD and Anderson PM (eds) (2003) *Encyclopedia of human ecology.* Santa Barbara, CA: ABC-CLIO.

Young GL (1974) Human ecology as an interdisciplinary concept: A critical enquiry. *Advances in Ecological Research* 8: 1–105.

human environment   Two different common meanings of the term are (1) all the influences on people from their surroundings, whether those influences are of biophysical or social/cultural origin; and (2) those influences of human social/cultural origin that affect the NATURAL ENVIRONMENT. In both senses, the human environment can be interpreted as interacting with but distinct from the natural environment. Furthermore, distinct impacts of the natural environment upon people and of people upon the Earth's surface can be identified. There has been considerable debate about the significance of the ENVIRONMENTAL IMPACT ON PEOPLE, and the nature and extent of ADAPTATION OF PEOPLE to the environments, landforms, climates, soils and vegetation that nature has provided. The ENVIRONMENTALISM debate with its positions of DETERMINISM and POSSIBILISM centred around these questions and remain largely unresolved.

HUMAN IMPACT ON ENVIRONMENT has clearly been profound. The world's great cities stand as icons of human endeavour, sometimes evolving over long periods of time; major transport systems from canals and railroads to highways and air routes have transformed both places and spaces. There are many positive impacts such as the systems of IRRIGATION that allow marginal lands to be brought into CULTIVATION and sustain large populations or engineering projects that provide heat and light in hostile environments. There are also, however, negative impacts such as the many forms of POLLUTION from OIL SPILLS to AIR POLLUTION, and human-induced ENVIRONMENTAL DISASTERS, such as the NUCLEAR ACCIDENT at Chernobyl in Russia and many disease PANDEMICS. The clearance of large areas of TROPICAL RAIN FOREST continues as does the depletion of *fisheries* in the seas. These negative impacts have prompted political movements with green agendas and conservation targets brought increasingly to the forefronts of governments' agendas if not actions. Human environments are integral parts of the world we inhabit and as population continues to rise and multiply, the need to manage, control and conserve becomes increasingly apparent.   *DTH/JAM*

[*See also* COLLAPSE OF CIVILISATIONS, CONSERVATION, ENVIRONMENT, ENVIRONMENTAL HISTORY, ENVIRONMENTAL MANAGEMENT, ENVIRONMENT-HUMAN INTERACTIONS, GLOBAL ENVIRONMENTAL CHANGE, HUMAN DIMENSIONS OF ENVIRONMENTAL CHANGE, HUMAN ECOLOGY, TECHNOLOGICAL CHANGE, WORLD SYSTEM]

Ausubel JH and Langford HD (eds) (1997) *Technological trajectories and the human environment.* Washington, DC: National Academies Press.
Constanza R, Graumlich LJ and Steffen W (eds) (2007) *Sustainability or collapse? An integrated history and future of people on Earth.* Cambridge: MIT Press.
Hornborg A, McNeill J and Martinez-Alier J (2007) *Rethinking environmental history: World system history and global environmental change.* Lanham, MD: AltaMira Press.
Levinson D (1995) *Human environments: A cross-cultural encyclopedia.* Santa Barbara, CA: ABC-CLIO.
Rayner S and Malone EL (2002) Social science and global environmental change. In Timmerman P (ed.) *Encyclopedia of global environmental change, volume 5: Social and economic dimensions of global environmental change.* Chichester: Wiley, 109–123.
Young GL (1989) A conceptual framework for an integrated human ecology. *Acta Oecologiae Hominis* 1: 1–135.

human evolution   A long sequence of events leading to the emergence of ANATOMICALLY MODERN HUMANS. Divergence of the *Hominidae* from ancestral gibbons is placed at ~17 million years ago (Ma), with the earliest fossil evidence suggesting an African origin (see Figure). The genetic proximity of humans to other great apes suggests a recent separation (5–6 Ma), which corresponded to a time of major environmental change in Africa as TROPICAL GRASSLANDS replaced TROPICAL AND SUBTROPICAL FORESTS due to increasing ARIDITY. The opening of the East African RIFT VALLEY separated forest in the west with the tree-living *apes* (*gorillas* and *chimpanzees*) and *grassland* in the east with the remaining apes, which eventually evolved to *Homo sapiens*, and *Ardpithecus ramidus,* thought to lie close to the point of divergence of the humans and other great apes.

Footprints found in ash layers dating to ~4 Ma heralded the development of bipedal locomotion, and the

appearance of *Australithicus afarensis*, the oldest well-preserved skeleton being dated at 3.2 Ma. The diversification of the HOMINID lineage occurred as *A. afarensis*, which had many ape-like characteristics evolved into two new genera *Paranthropus* and *Homo*, both of which were fully adapted to life on the ground. *Paranthropus* were heavily built *vegetarians* with small brains whereas *Homo* were more lightly built *omnivores* that were the first to evolve a large brain and use *tools*. The genus *Homo* evolved through the species *Homo habilis*, *Homo ergaster* and *Homo erectus* into *H. sapiens*.

H. habilis and *H. ergaster* are only known from Africa whereas *H. erectus* is known in Africa and Asia, and is thought to be the first species to have migrated out of Africa ~1 Ma. Descendant populations of *H. erectus* in the Old World (ARCHAIC HUMANS) evolved into Neanderthals in Europe and western Asia before worldwide replacement of ancestral populations by anatomically modern *H. sapiens*. *Homo heidelbergensis* originated from *H. erectus* in an unknown location and dispersed widely across Africa, southern Europe and southern Asia. After early modern humans migrated out of Africa around 60,000 years ago, there was some interbreeding with other descendants of *H. heidelbergensis*, including *Homo neanderthalis*. DNA studies of modern humans show that Eurasian groups are more similar to those found in southern India than to populations in Africa suggesting that HUMAN MIGRATION: 'OUT OF AFRICA' followed a 'southern route' through Arabia rather than via a more northerly route through Egypt.   *KJF/DCS*

[*See also* HOMINISATION, HUMAN MIGRATION: 'OUT OF AFRICA', NEANDERTHAL DEMISE]

Chamberlain AT (1999) Human evolution. In Barker G (ed.) *Companion encyclopaedia of archaeology*. London: Routledge, 757–796.
Melé M, Javed A, Pybus M et al. (2012) Recombination gives a new insight in the effective population size and the history of the Old World human populations. *Molecular Biology and Evolution* 29: 25–30.

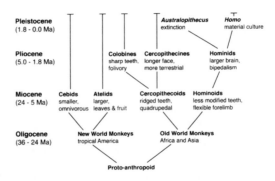

Human evolution *Human evolution in the context of the divergence of anthropoid primates since the Oligocene and an indication of the major morphological adaptations in each lineage (Chamberlain, 1999).*

Palmer D (2010) *Origins: Human evolution revealed.* London: Mitchell Bleazley.

Roberts A (2011) *Evolution: The human story.* London: Dorling Kindersley.

Stringer C (2012) What makes a modern human? *Nature* 485: 33–35.

Stringer C and Andrews P (2011) *The complete world of human evolution.* London: Thames and Hudson.

Stringer C and McKie R (1996) *African exodus: The origins of modern humanity.* London: J. Cape.

Wood B (2005) *Human evolution: A very short introduction.* Oxford: Oxford University Press.

human evolution: climatic influences

Environmental change, particularly climate, has underpinned many of the models of human evolution, involving MIGRATION, ADAPTATION, SPECIATION and EXTINCTION of early HOMINIDS. Some early theories relating to specific adaptations, such as the first appearance of *bipedalism* being driven by SAVANNISATION, do not stand up to scrutiny in the light of modern knowledge. A plausible case can be made, however, for the rapid turnover of early hominids (*Australopithicines* to *Homo*) and their subsequent sporadic spread out of Africa being associated with climatic shifts to increasing ARIDITY, and associated increases in the drying and extension of TROPICAL GRASSLANDS, between about 4.0 and 1.0 Ma. Nevertheless, there are significant limitations imposed on all such inferences by the nature of the evidence. Potts (1999) suggested, in the *variability selection hypothesis,* that rapid environmental changes associated with GLACIAL-INTERGLACIAL CYCLES, rather than general trends in climate, were the driving force of human evolution; and the *turnover pulse hypothesis* implied that dramatic and frequent climate changes drove adaptations that enabled individuals to cope with a wide range of environmental conditions. Subsequently, the *pulsed climate variability hypothesis* has been proposed as a mechanism for focused periods of innovation, whereby ABRUPT CLIMATIC CHANGES caused by ORBITAL FORCING are separated by periods of relatively reduced amplitude change. Climatic change has also been implicated in NEANDERTHAL DEMISE and the dispersal of ANATOMICALLY MODERN HUMANS across the globe.   *JAM/CJC*

[*See also* ARCHAIC HUMANS, FLORES MAN, HUMAN MIGRATION: 'OUT OF AFRICA']

Blockley SPE, Candy I and Blockley SM (2012) Testing the role of climate change in human evolution. In Matthews JA, Bartlein PJ, Briffa KR et al. (eds) *The SAGE handbook of environmental change, volume 2.* London: Sage, 301–327.

Bobe R and Behrensmeyer K (2004) The expansion of grassland ecosystems in Africa in relation to mammalian evolution and the origin of the genus Homo. *Palaeogeography, Palaeoclimatology, Palaeoecology* 207: 399–420.

De Menocal PB (2004) African climate change and faunal evolution during the Pliocene-Pleistocene. *Earth and Planetary Science Letters* 220: 3–24.

Foley R (2002) Adaptive radiations and dispersals in hominin evolutionary ecology. *Evolutionary Anthropology* 1: 32–37.

Maslin MA and Christensen B (2007) Tectonics, orbital forcing, global climate change and human evolution. *Journal of Human Evolution* 53: 443–464.

Potts R (1999) Variability selection in hominid evolution. *Evolutionary Anthropology* 7: 81–96.

Vrba ES, Denton GH and Prentice ML (1989) Climatic influences on early hominid behaviour. *Ossa* 14: 127–156.

human geography

That part of GEOGRAPHY that studies the ways in which people occupy and interact with the surface of the Earth. Drawing upon the arts, HUMANITIES and SOCIAL SCIENCES, human geography is concerned with the patterns and processes that typify people on Earth. Its older PARADIGMS include EXPLORATION, ENVIRONMENTALISM and REGIONALISM, all permeated by the cartographic tradition. Key modern concepts include the understanding of relationships between people, environments and LANDSCAPES, the meanings of place and the geometry of space and spatial interactions. The many subdivisions include HISTORICAL GEOGRAPHY, urban geography, economic geography, cultural geography and political geography, although these 'adjectival' subdivisions have become overwritten by broad unifying themes such as GLOBALISATION, locality effects and the complexities of human relationships. Human geography has always applied qualities that range from its early formative links with URBAN AND RURAL PLANNING to its methodologies of SPATIAL ANALYSIS and GEOGRAPHICAL INFORMATION SYSTEMS (GIS), and its wider inputs to policy formulation and the understanding of modern patterns of CONSUMPTION. *DTH*

[*See also* ENVIRONMENTAL GEOGRAPHY, PHYSICAL GEOGRAPHY]

Agnew J, Livingstone DN and Rogers A (1996) *Human geography: An essential anthology.* Oxford: Blackwell.

Aitken S and Valentine G (2006) *Approaches to human geography.* London: Sage.

Benko G and Strohmayer U (eds) (2004) *Human geography: A history for the 21st century.* London: Arnold.

Cloke P, Crang P and Goodwin M (2005) *Introducing human geographies, 2nd edition.* London: Arnold.

Gregory D, Johnston R, Pratt G et al. (eds) (2009) *The dictionary of human geography, 5th edition.* Chichester: Wiley-Blackwell.

Johnston RJ and Sidaway JD (2004) *Geography and geographers: Anglo-American geography since 1945, 6th edition.* London: Arnold.

human health hazards

Environmental changes, such as GLOBAL WARMING, OZONE DEPLETION and a greater frequency of EXTREME CLIMATIC EVENTS, have the potential to pose both direct and indirect hazards to human health. Direct stress may be imposed by EXTREME WEATHER EVENTS, whereby abnormally hot or cold spells can increase the risk of heat stroke and hypothermia, respectively. Health hazards may also be an indirect effect of, for example, STORMS, TROPICAL CYCLONES and SEA-LEVEL CHANGE, with consequent FLOODS, LANDSLIDES, population displacement (HUMAN MIGRATION) or disease OUTBREAKS, which can lead to injury, illness, stress or fatality. An increase in ambient temperature and moisture often facilitates the reproduction and spread of PATHOGENS and the vectors involved in DISEASE transmission. CHOLERA and MALARIA distributions are linked to temperature and wetness and both diseases may be becoming more widespread in some tropical and subtropical zones. Pathogens that cause diseases in crops and livestock may also proliferate, reducing the volume and quality of food production. However, uncertainties in understanding both CLIMATE CHANGE and specific diseases make prediction of future disease and human health effects complex and difficult. A framework for understanding the effects of climate change on disease transmission is shown in the Figure.   MLW

[*See also* CARCINOGEN, EPIDEMIC, HUMAN MIGRATION, INDOOR ENVIRONMENTAL CHANGE]

Baxter PJ (1990) The medical consequences of volcanic eruptions: I. Main causes of death and injury. *Bulletin of Volcanology* 52: 532–544.

Baylis M and Morse AP (2012) Disease, human health, and environmental change. In Matthews JA, Bartlein PJ, Briffa KR et al. (eds) *The SAGE handbook of environmental change*. London: Sage, 387–405.

Gosling SN, Lowe JA, McGregor GR et al. (2009) Associations between elevated atmospheric temperature and human mortality: A critical review of the literature. *Climatic Change* 92: 299–341.

Jha P, Mills A, Hanson K et al. (2002) Improving the health of the global poor. *Science* 295: 2036–2039.

Kalkstein LS (2005) Human health and climate. In Oliver J (ed.) *Encyclopedia of world climatology*. Dordrecht: Springer, 407–411.

Langford IH and Bentham G (1995) The potential effects of climate change on winter mortality in England and Wales. *International Journal of Biometeorology* 38: 141–147.

Lines J (1995) The effects of climate change and land-use changes on insect vectors of human disease. In Harrington R and Stork NE (eds) *Insects in a changing environment*. London: Academic Press, 158–175.

McMichael AJ, Haines A, Sloof R and Kovats S (eds) (1996) *Climate change and human health*. Geneva: World Health Organisation.

Patz J (1996) Global climate change and emerging infectious diseases. *Journal of the American Medical Association* 275: 217–223.

Patz JA, Campbell-Lendrum D, Holloway T and Foley JA (2005) Impact of regional climate change on human health. *Nature* 438: 310–317.

Takken W, Martens P and Bogers RJ (eds) (2005) *Environmental change and malaria risk: Global and local implications*. Dordrecht: Springer.

Thomson MC, Garcia-Herrera R and Beniston M (eds) (2008) *Seasonal forecasting, climatic change and human health*. Berlin: Springer.

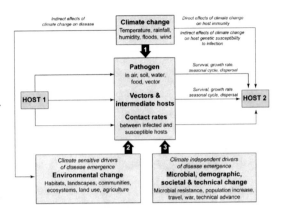

Human health hazards *Schematic framework of the effects of climate change on disease transmission in people and animals. Climate change may affect pathogens, their hosts or their environments directly or indirectly. Climate-independent effects include environmental, social, demographic and technological change. The significance of climatic change as a driver of disease depends on the relative scales of arrows 1–3 (Baylis and Morse, 2012).*

human impact on climate   The impacts of humans on climate can be intentional or inadvertent. The former are usually small scale and include the construction of walled gardens in temperate regions to create an artificial MICROCLIMATE to grow plants, such as fruit trees, that would not normally flourish. Unintentional impacts include the effects of URBANISATION on URBAN CLIMATES and the effects of large reservoirs on land and water breezes. Such CLIMATIC MODIFICATION has occurred since the beginning of human occupancy of the Earth, particularly since the earliest settled communities and the NEOLITHIC TRANSITION. A very substantial body of evidence now exists from CLIMATIC MODELS, which suggests that relatively small PERTURBATIONS at the surface can impact on regional-scale CLIMATOLOGY.

Changes in LAND COVER and LANDUSE, from natural SAVANNA to CROPLAND, for example, can alter SOIL MOISTURE and RUNOFF and affect precipitation amounts. Changes in surface characteristics may also have important climatic consequences in ARIDLANDS: removal of vegetation decreases soil-water storage and increases ALBEDO, affecting surface temperatures. DEFORESTATION, for example, in the Amazon Basin, may influence the HYDROLOGICAL BALANCE and hence climate on a regional scale. Furthermore, model results suggest that lower rates of EVAPOTRANSPIRATION feed less moisture into the tropical WALKER CIRCULATION, which can weaken and produce extraregional effects.

On a global scale, the modification of the composition of the atmosphere by the addition of GREENHOUSE GASES to create the enhanced GREENHOUSE EFFECT and GLOBAL WARMING is probably the single most important contemporary human impact on climate. Even with global agreement to limit the EMISSION of such gases, there is every sign that the human impact on climate will become more marked in the future.   *AHP*

[*See also* AGRICULTURAL IMPACT ON CLIMATE, CLIMATIC CHANGE, DEFORESTATION: CLIMATIC IMPACTS, FUTURE CLIMATE, GLOBAL DIMMING, GLOBAL ENVIRONMENTAL CHANGE, HUMAN IMPACT ON ENVIRONMENT]

Bonan GB (1997) Effects of land use on the climate of the United States. *Climatic Change* 37: 449–486.

Cotton WR and Pielke Sr RA (2007) *Human impact on weather and climate, 2nd edition.* Cambridge: Cambridge University Press.

Gornall JL, Wiltshire AJ and Betts RA (2012) Anthropogenic drivers of environmental change. In Matthews JA, Bartlein PJ, Briffa KR et al. (eds) *The SAGE handbook of environmental change, volume 1.* London: Sage, 517–535.

Harvey D (2000) *Climate and global environmental change.* Harlow: Addison Wesley Longman.

Noone KJ (2012) Human impacts on the atmosphere. In Matthews JA, Bartlein PJ, Briffa KR et al. (eds) *The SAGE handbook of environmental change, volume 2.* London: Sage, 95–110.

Solomon S, Qin D, Manning M et al. (eds) (2007) *Climate change 2007: The physical science basis.* [Contribution of Working Group 1 to the Fourth Assessment Report of the Intergovernmental Panel on Climate Change.] Cambridge: Cambridge University Press.

human impact on coasts   With approximately 60 per cent of the world's population living in the coastal zone, and coastal population rise exceeding the global average, exploitation of *coastal resources* has been extensive and is likely to increase. Many competing uses such as MINING, LAND RECLAMATION, AGRICULTURE, INDUSTRY, TOURISM, *residential* and *transport infrastructure*, and *waste disposal*, combined with COASTAL (SHORE) PROTECTION have drastically changed *coastal morphology*. On many shorelines of the developed world (most notably, the Netherlands and Japan), erosion protection structures have converted natural SHORELINES into artificial ones. Modern developments are the culmination of a long history of similar direct and indirect impacts that have occurred since PREHISTORY. Until recently, the various commercial and residential pressures have led to uncontrolled development of coastal areas but, fortunately, in developed countries at least, national and local ENVIRONMENTAL LAWS and regulations have been enacted to control future planning, and COASTAL ZONE MANAGEMENT is in place. However, many DEVELOPING COUNTRIES are not learning from the well-documented mistakes of the past.   *HJW/IM/JAM*

[*See also* ARTIFICIAL SHORELINE, HARD ENGINEERING, SEA-LEVEL RISE, SOFT ENGINEERING]

Davis Jr. RA and Barnard PL (2000) How anthropogenic factors in the back-barrier area influence tidal inlet stability: Examples from the Gulf Coast of Florida, USA. In Pye K and Allen JRL (eds) *Coastal and estuarine environments: Sedimentology, geomorphology and geoarchaeology.* Bath: Geological Society, 293–303.

French PW (2001) *Coastal defences: Processes, problems, and solutions.* London: Routledge.

Kelletat D (1989) Biosphere and man as agents in coastal geomorphology and ecology. *Geoökodynamik* 10: 215–252.

Stauble DM (ed.) *Barrier Islands: Process and management.* New York: American Society of Civil Engineers.

Viles HA and Spencer T (1995) *Coastal Problems: Geomorphology, ecology and society at the coast.* London: Arnold.

human impact on environment   The actions of people have dramatically modified the NATURAL ENVIRONMENT and have affected all parts of the Earth on a variety of temporal and spatial scales. ENVIRONMENTAL CHANGE due to human activity has diversified and intensified in parallel with DEMOGRAPHIC CHANGE and has become a major focus of public concern in the late twentieth century. CULTURAL CHANGE has been accompanied by TECHNOLOGICAL CHANGE which has increased human potential to modify the environment and has demanded more energy to sustain. Human-induced environmental change may be conceptualised on a spatial-hierarchical basis as in Figure 1 whereby local impacts collectively determine regional impacts, which, in turn, contribute to the planetary FEEDBACK MECHANISMS that drive GLOBAL ENVIRONMENTAL CHANGE. The three major subdivisions of human activity (agricultural, industrial and recreational) are convenient units of study for considering a potentially vast subject area.

The development of agriculture since its inception some 10,000 years ago (see AGRICULTURAL HISTORY) has resulted in a huge increase in the land area occupied by AGRO-ECOSYSTEMS, which are typically low diversity—artificial systems designed to keep up with food demands from the growing world population. In the developed world, agricultural impacts have stemmed mainly from the gradual mechanisation and labour intensification of farming over the past two centuries. The most obvious expression of environmental change in the temperate mid-latitudes, for example, has been DEFORESTATION and the FRAGMENTATION of natural and semi-natural vegetation cover with the result that British vegetation, in particular, can be viewed as completely cultural in origin. Injudicious practices in Europe, parts of Australia and the United States have caused an alarming decline in the organic component of soils, resulting in LAND DEGRADATION and major SOIL EROSION problems. The addition of synthetic FERTILISERS and PESTICIDES has led to significant WATER POLLUTION, with the runoff of NITRATES from agricultural land into rivers and GROUNDWATER aquifers being particularly problematic.

Agricultural impacts have been even more pronounced in the DEVELOPING COUNTRIES, where concern for the environment remains secondary to the necessity to produce food for an expanding population. Money for ENVIRONMENTAL PROTECTION and CONSERVATION measures is severely limited and land management strategies are non-sustainable due to lack of education. DEFORESTATION in the tropics and humid subtropics is perhaps the greatest cause for concern, not least because of the effects on the CARBON CYCLE and possible contributions to GLOBAL WARMING. Locally, increased frequency of downstream flooding, accelerated soil erosion through unsuitable felling practices and declining productivity through removal of the BIOMASS nutrient store are the dominant impacts. It should be noted that agriculture is not the only source of deforestation with the timber industry being a major secondary source. AFFORESTATION and REFORESTATION, although conservational and sustainable in nature, not only cause changes in flora and fauna but also modify water, sediment and nutrient transfer mechanisms. Ecologically speaking, the creation of managed AGROFORESTRY drastically modifies the functioning of the existing ecosystem.

In many arid and semi-arid areas, ENVIRONMENTAL DEGRADATION, notably DESERTIFICATION, has emerged as a major problem in *land management*, as has SALINISATION and waterlogging of irrigated agricultural land. The advent of BIOTECHNOLOGY and GENETIC ENGINEERING is an additional recent agent of environmental change, which permanently affects species composition and diversity. The overall impact of agriculture has been immense and has affected all aspects of the environment.

The exponential growth of industry since the INDUSTRIAL REVOLUTION has been accompanied by diverse and damaging environmental impacts on a global scale. The provision of accommodation and services for the growing number of industrial employees during the Industrial Revolution was, in itself, an agent of environmental change. The emplacement of artificial land surfaces and buildings altered albedo, wind patterns, vegetation composition and runoff. The principal effects of industry, however, have been AIR POLLUTION and ACIDIFICATION. The emission of particulate matter from FOSSIL FUEL consumption has resulted in high DUST concentrations in the troposphere which contribute to the formation of PHOTOCHEMICAL SMOG and HEAT ISLAND effects. Similarly, CARBON DIOXIDE and other GREENHOUSE GASES released into the atmosphere is a significant factor in global warming. The formation of dilute nitric and sulfuric acids in the atmosphere has been cited as the dominant cause of worldwide lake and stream acidification. ACID RAIN and runoff have caused a significant decrease in PRODUCTIVITY in AQUATIC ENVIRONMENTS.

METAL POLLUTION is another significant human impact in the industrialised world. The dumping of waste metal products has caused widespread industrial LAND DEGRADATION and serious water pollution through HEAVY METAL contamination. Similarly, contaminated runoff from agricultural land, which transports artificial fertilisers, pesticides and plant nutrients into lake basins, has been the dominant factor in cultural EUTROPHICATION. Initially, eutrophication may stimulate productivity, but it eventually leads to serious oxygen depletion and declining productivity. In many areas, the discharge of sewage and industrial waste into the sea has been responsible for MARINE POLLUTION and OIL SPILLS, which have caused long-term damage to marine, shoreline and ESTUARINE ENVIRONMENTS. The nuclear energy industry has also been highly publicised as a source of environmental pollution, particularly in relation to NUCLEAR ACCIDENTS.

Although pollution is normally judged to be the main industrial impact, there are others. Ground SUBSIDENCE is common in MINING areas and often results in the formation of large surficial chasms. The provision of transport routes is another example: the growing demand for roads and railways since the industrial revolution has meant that more and more land has been given over to this type of landuse. Roads in particular often require the destruction of natural habitat and modifications to the landscape in the form of hillside cuttings and tunnels. COASTAL (SHORE) PROTECTION works are a further example with the emplacement of groynes, jetties, breakwaters, sea walls and riprap often causing disruption to sediment transport patterns and erosion through wave reflection. Similarly, FLOOD CONTROL MEASURES to modify river channels by artificial straightening, concrete channelling, dams, barrages, weirs, diversions and dredging may significantly alter flow regime, discharge characteristics and upset the balance between erosion and deposition.

As the standard of living in the developed world has increased, it has been paralleled by higher disposable income and more leisure time. As a result, the diversity of leisure activities has increased and RECREATION has become an additional agent of environmental change. Vegetation TRAMPLING is the most obvious direct impact of recreation. Skiing, walking, boating, camping, fishing, riding and picnicking all cause DISTURBANCE, which may result in local extinction of species and patch generation, as well as accelerated soil erosion. Motorcross, rallying and mountain biking may also result in denuded vegetation cover, soil compaction, intensified runoff, waterlogging or deflation. MILITARY IMPACTS ON ENVIRONMENT produce similar effects.

The diversification and intensification of human impact on the environment lead to speculation over the SUSTAINABILITY of human activities in all walks of life. In the future, further population growth and the increasing sophistication and pervasiveness of technological advances will surely further intensify and magnify human impacts which, arguably already outweigh NATURAL ENVIRONMENTAL CHANGE (see Figure 2).   JGS

[*See also* AGRICULTURAL IMPACT ON SOILS, ANTHROPOCENE, CLIMATE CHANGE, CONTAMINATED LAND, DEFORESTATION, DESERTIFICATION, ENVIRONMENTAL DEGRADATION, GLOBAL WARMING IMPACTS, HUMAN IMPACT ON CLIMATE, HUMAN IMPACT ON VEGETATION HISTORY, LAND DEGRADATION, POLLUTION].

Arctic Climate Impact Assessment (ACIA) (2005) *Arctic Climate Impact Assessment: Scientific report*. Cambridge: Cambridge University Press.

Bell M and Walker JC (2005) *Late Quaternary environmental change: Physical and human perspectives, 2nd edition*. Harlow: Pearson Education.

Goudie AG (2006) *The human impact on the natural environment: Past, present and future, 6th edition*. Oxford: Blackwell.

Head L (2008) Is the concept of human impacts past its use-by date? *The Holocene* 18: 373–377.

Intergovernmental Panel on Climate Change (2007) *Climate change 2007: Impacts, adaptations and vulnerability*. Cambridge: Cambridge University Press.

Jackson ARW and Jackson JM (2000) *Environmental science: The natural environment and human impact, 2nd edition*. Harlow: Pearson Education.

Letcher TM (ed.) (2009) *Climate change: Observed impacts on planet Earth*. Amsterdam: Elsevier.

Matthews JA, Bartlein PJ, Briffa KR et al. (2012) *The SAGE handbook of environmental change, volume 2: Human impacts and responses*. London: Sage.

Messerli B, Grosjean M, Hofer T et al. (2000) From nature-dominated to human-dominated environmental changes. *Quaternary Science Reviews* 19, 459–579.

Meyer WB (1999) *Human impact on the Earth*. Cambridge: Cambridge University Press.

Oldfield F, Wake R, Boyle J et al. (2003) The late-Holocene history of Gormire Lake (NE England) and its catchment: A multiproxy reconstruction of past human impact. *The Holocene* 13: 677–690.

Redman CL (1999) *Human impacts on ancient environments*. Tucson: University of Arizona Press.

Roberts N (1998) *The Holocene: An environmental history, 2nd edition*. Oxford: Blackwell.

Simmons IG (1996) The modification of the Earth by humans in pre-industrial times. In Douglas I, Huggett R and Robinson M (eds) *Companion encyclopedia of geography: The environment and humankind*. London: Routledge, 137–156.

Turner II BL, Clark WC, Kates RW et al. (1990) *The Earth as transformed by human action*. Cambridge: Cambridge University Press.

Wright L (1993) *Environmental systems and human impact*. Cambridge: Cambridge University Press.

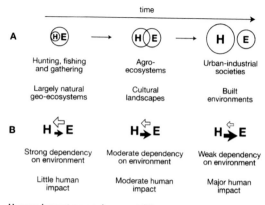

Human impact on environment (2) *Changing relationships between humans (H) and the natural environment (E) over the course of the Holocene including (A) the nature of interactions and (B) relative impacts (based on Roberts, 1998).*

human impact on hydrology The direct and indirect effects of human activities on hydrology are diverse. All components of the HYDROLOGICAL CYCLE are potentially subject to human modification with the speeds, magnitude and quality of water following different hydrological pathways all affected.

LANDUSE CHANGE has had radical effects on EVAPOTRANSPIRATION and its main component processes INTERCEPTION and *transpiration*, on INFILTRATION and RUNOFF PROCESSES and on river DISCHARGE and the STORM HYDROGRAPH. DEFORESTATION and URBANISATION tend to lead to reduced interception and transpiration, increased STREAMFLOW and larger and quicker streamflow responses to rainstorms and increased flood frequency. Urbanisation, industrial activity and MINING all tend to lead to WATER POLLUTION. *Forest fires* (often the result of human action) can also lead to enhanced HORTONIAN OVERLAND FLOW and *storm peaks* as well as a flush of *nutrient losses*.

Intensive AGRICULTURE can greatly affect the HYDROLOGICAL CYCLE by ABSTRACTION from GROUNDWATER and reduction of river flow, especially in summer. OVERLAND FLOW and THROUGHFLOW from agricultural land can also contain high levels of NUTRIENTS and PESTICIDES leading to potential WATER POLLUTION and EUTROPHICATION problems.

Structural changes within catchments can also induce important hydrological changes. Perhaps the most visible example of this is DAMS. Dam construction increased markedly across the world during the second half of the twentieth century in order to provide water for industry, HYDROPOWER generation, domestic CONSUMPTION and IRRIGATION. Dams and reservoirs also result in (sometimes by design) the regulation of river flows, particularly in the form of reduced peak flows and the maintenance of stable, and often increased, low flows. The modification of flow regime may also therefore lead to changes in the size, shape and sinuosity of downstream channels, in part due to the reduction

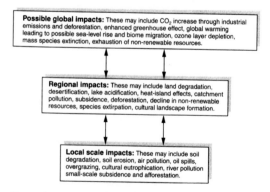

Human impact on environment (1) *The spatial-hierarchical basis of human impacts on environment.*

in SEDIMENT LOAD (see RESERVOIRS: ENVIRONMENTAL EFFECTS). Many channels have been subject to FLOOD CONTROL MEASURES with the construction of LEVÉES or through river CHANNELISATION, though frequently such measures merely transfer flooding problems further downstream. *ADT*

[*See also* AFFORESTATION: IMPACTS ON HYDROLOGY AND GEOMORPHOLOGY, DEFORESTATION: HYDROLOGICAL AND GEOMORPHOLOGICAL IMPACTS, RIVER DIVERSION, TROPICAL FOREST FIRES]

Burns DA, McHale MR, Driscoll CT and Roy KM (2006) Response of surface water chemistry to reduced levels of acid precipitation: Comparison of trends in two regions of New York, USA. *Hydrological Processes* 20: 1611–1627.

Cerdà A, Imeson AC and Calvo A (1995) Fire and aspect induced differences on the erodibility and hydrology of soils at La Costera, Valencia, southeast Spain. *Catena* 24: 289–304.

Church M, Burt TP, Galay VJ and Kondolf GM (2009) Rivers. In Slaymaker O, Spencer T and Embleton-Hamann C (eds) *Geomorphology and global environmental change*. Cambridge: Cambridge University Press, 98–129.

Goudie AS (2005) *Human impact on the natural environment, 6th edition*. Chichester: Wiley-Blackwell, 143–158.

Goudie AS (2006) Global warming and fluvial geomorphology. *Geomorphology* 79: 384–394.

Gregory KJ (2006) The human role in changing river channels. *Geomorphology* 79: 172–191.

Jones JAA (2010) *Water sustainability: A global perspective*. London: Hodder.

May L, Defew LH, Bennion H and Kirika A (2012) Historical changes (1905-2005) in external phosphorus loads to Loch Leven, Scotland, UK. *Hydrobiologia* 681: 11–21.

Milliman JD and Farnsworth K (2011) *River discharge to the coastal ocean: A global synthesis*. Cambridge: Cambridge University Press.

Nilsson C, Ekblad A, Gardfjell M and Carlberg B (1991) Long-term effects of river regulation on river margin vegetation. *Journal of Applied Ecology* 28: 963–987.

Peiry J-L, Givel J and Panton G (1999) Hydroelectric developments, environmental impact. In Alexander DE and Fairbridge RW (eds) *Encyclopaedia of environmental science*. Dordrecht: Kluwer, 332–336.

Petts GE and Gurnell AM (2005) Dams and geomorphology: Research progress and future directions. *Geomorphology* 71: 27–47.

Van Dijk AIJM and Keenan RJ (2007) Planted forests and water in perspective. *Forest Ecology and Management* 251: 1–9.

Williams GE and Wolman MG (1984) Downstream effects of dams on alluvial rivers. *United States Geological Survey Professional Paper* 1286: 1–83.

human impact on landforms and geomorphic processes

Geomorphological impact by our ancestors for most of the 5 million years or so of human development was negligible. From about 1 Ma, and particularly from the beginning of the HOLOCENE, the scope and magnitude of impacts have multiplied as a result of *population growth* and TECHNOLOGICAL CHANGE.

The earliest significant impacts by HUNTING, FISHING AND GATHERING societies resulted from deliberate BIOMASS BURNING chiefly to encourage *regrowth* of new shoots and hence attract GRAZING animals for easier killing. This burning led to accelerated SOIL EROSION. The development of AGRICULTURE in different subtropical centres (e.g. FERTILE CRESCENT) in the early Holocene coincided broadly with a substantial increase in world population. In NEOLITHIC times, hafted polished *stone axes* allowed efficient FOREST CLEARANCE for CULTIVATION. Development of the *plough* in the Old World meant that soils hitherto too heavy for agriculture could now be cultivated. VALLEY-FILL DEPOSITS with sediment dating from BRONZE and IRON AGES in Europe indicate the accelerated erosion brought about by AGRICULTURAL INTENSIFICATION. Requirements of large amounts of wood for *metal smelting* doubtless contributed significantly to the loss of woodland and increased soil erosion. Cultivation requires sedentary lifestyle (SEDENTISM), which leads to *settlements* and URBANISATION. The emergence of urban CIVILISATIONS in Egypt, Mesopotamia, India and China in mid- to late-Holocene times was conditioned by specific physical conditions—major *alluvial river valleys* which had low and/or unreliable rainfall. Here, successful agriculture on these fertile lands could only be achieved with major manipulation of WATER RESOURCES, involving modification of stream courses.

From the fifteenth century onwards, European COLONIALISM caused the rapid transformation of large tracts of the world (e.g. North America, Australia) from often a relatively unmodified state to one highly altered through intensive agriculture by technologically advanced societies. The most notable geomorphological consequence was a rapid and marked increase in soil erosion. The mechanical technology available over the past two centuries has enabled the Earth's land surface to be transformed on a scale and at a rate not possible under pre-industrial societies. Such changes (e.g. MINING, LAND DRAINAGE, FLOOD CONTROL MEASURES and COASTAL ENGINEERING STRUCTURES), impressive though they may be, are often localised or at most regional in their extent. Anthropogenic atmospheric GREENHOUSE GAS modifications, on the other hand, are anticipated to make future indirect human impact on geomorphology potentially global in extent, as the precipitation and temperature changes resulting from CLIMATIC CHANGE cross geomorphological thresholds in certain environmental conditions. *RAS*

[*See also* AGRICULTURAL IMPACT ON GEOMORPHOLOGY, ANTHROPOGEOMORPHOLOGY, ARTIFICIAL GROUND, CLIMATIC CHANGE: PAST IMPACT ON LANDFORMS AND GEOMORPHOLOGICAL PROCESSES, CLIMATIC CHANGE: POTENTIAL FUTURE GEOMORPHOLOGICAL IMPACTS, DISTURBANCE, PERMAFROST DEGRADATION]

Brown EH (1970) Man shapes the earth. *Geographical Journal* 136: 74–85.

Goudie A (1993) Human influence in geomorphology. *Geomorphology* 7: 37–59.

Jennings JN (1966) Man as a geological agent. *Australian Journal of Science*, 28: 150–156.

Marsh GP (1864) *Man and nature*. New York: Scribner.

Nir D (1983) *Man, a geomorphological agent: An introduction to anthropic geomorphology*. Jerusalem: Keter Publishing.

Price SJ, Ford JR, Cooper AH and Neal C (2011) Humans as major geological and geomorphological agents in the Anthropocene: The significance of artificial ground in Great Britain. *Philosophical Transactions of the Royal Society A* 369: 1056–1084.

Roberts N (1996) The human transformation of the Earth's surface. *International Social Science Journal* 150: 493–510.

Rozsa P and Novak T (2011) Mapping anthropic geomorphological sensitivity on a global scale. *Zeitschrift für Geomorphologie* 55, Supplement 1: 109–117.

Slaymaker O (ed.) (2000) *Geomorphology, human activity and global environmental change*. Chichester: Wiley.

human impact on soil   Soils are the basis of all terrestrial ecosystems and are the main medium for human food production. They also provide fodder for animal husbandry, as well as timber for shelter and fuel. As a consequence, human impact on soils ranges from minimal changes to profound alteration. In most areas of the inhabited world, there has been a history of human soil modification over thousands of years. In terms of site characteristics, the land surface may be altered directly, for example, by *levelling* or TERRAC-ING. The land surface may have been raised or lowered indirectly by ANTHROPOGENIC changes in SEDIMENTATION or human-induced EROSION. Changes to SOIL QUALITY have been even more widespread. In some places, soils have been created where none existed previously by importation of soil material; in others, pre-existing soils have been rendered unrecognisable or have been buried producing PALAEOS.   *EMB*

[*See also* AGRICULTURAL IMPACT ON SOILS, ANTHROPOGENIC SOIL HORIZONS, ANTHROSOLS, CONTAMINATED LAND, METAPEDOGENESIS, SUSTAINABLE AGRICULTURE]

Bidwell OW and Hole FD (1965) Man as a factor in soil formation. *Soil Science* 99: 65–72.

Bridges EM (1978) Interaction of soil and mankind. *Journal of Soil Science* 29: 125–139.

Bridges EM and de Bakker H (1998) Soil as an artifact: Human impact on the soil resource. *Land* 1: 197–215.

Lahmar R (1998) *Des sols et des hommes: récits authentiques de gestion de la ressource sol* [Soils and men: True stories of soil resource management]. Paris: Éditions Charles Léopold Mayer.

McNeill JR and Winiwarter V (eds) (2006) *Soils and societies: Perspectives from environmental history*. Strond, Isle of Harris: White Horse Press.

Montgomery DR (2007) *Dirt: The erosion of civilizations*. Berkeley: University of California Press.

Yaalon DH (2007) Human-induced ecosystem and landscape processes always involve soil change. *BioScience* 57: 918–919.

human impact on terrestrial vegetation   Con-temporary impacts vary in scale and intensity. The most intense impacts are found in highly managed AGRO-ECOSYSTEMS where the original vegetation has been replaced by CULTIVARS, followed by areas with substantially modified vegetation, such as the TROPICAL GRASSLANDS that have replaced TROPICAL AND SUBTROPICAL FORESTS. Finally, there are environments where the vege-tation has not yet been substantially altered, as in certain remote TROPICAL RAIN FOREST, BOREAL FOREST and TUNDRA regions in northern latitudes. Impact can be direct (e.g. LOGGING) or indirect (e.g. CONTROLLED FIRE, GRAZING and POLLUTION). One of the main characteristics of human-disturbed vegetation is structural and functional sim-plification. When PRIMARY WOODLAND is removed (see FOREST CLEARANCE), it is replaced, in different HABITATS, by SECONDARY WOODLAND, wet and dry *grasslands* and HEATHLANDS, and semi-arid SCRUB. Vegetation replace-ment in this way typically involves a reduction in *plant diversity, plant height* and *layering*, and simplified *age profiles*. The same applies to the replacement of veg-etation by plant cultivars and the human degradation of grass and scrub vegetation in SEMI-ARID areas to DESERT landscapes in the process of DESERTIFICATION.

Many other aspects of the environment are transformed by the interference with the vegetation cover, including associated FAUNA, SOIL, CLIMATE, HYDROLOGICAL CYCLE and LANDSCAPE. Degraded vegetation with reduced BIODIVER-SITY and more open, simplified habitats has encouraged BIOLOGICAL INVASIONS. Soil organic content is lost if grass-land is degraded to desert scrub in semi-arid environments (e.g. in the SAHEL). When the moisture-recycling capacity of forest is lost during FOREST CLEARANCE (as in parts of the Tropical Rain Forest in Brazil), rainfall amounts can be reduced by 50 per cent. The removal of the protective vegetation cover can accelerate SOIL LOSS by WIND EROSION and WATER EROSION within and outside the deforested area. In the humid tropics, vegetation removal can lead to other sterile landscapes (e.g. DURICRUSTS).

In terms of ECOLOGICAL SUCCESSION, humans have altered the so-called CLIMAX VEGETATION communities and maintained many of them, by fire, GRAZING and direct cutting, at earlier PLAGIOCLIMAX stages. The mainte-nance by fire and/or grazing of well-adapted vegetation types including tropical and TEMPERATE GRASSLANDS, MEDITERRANEAN-TYPE VEGETATION (e.g. maquis and gar-rigue scrub) and cool temperate MOORLAND and HEATH-LAND provide examples of this mechanism.   *GOH*

[*See also* ANTHROBLEME, ANTHROME, DEFORESTATION, HUMAN IMPACT ON VEGETATION HISTORY, LANDUSE IMPACTS ON SOILS]

Goudie AS (2006) *The human impact on the natural environment, 6th edition.* Oxford: Blackwell.

Holzner W, Werger MJA and Ikusima I (eds) (1983) *Man's impact on vegetation.* The Hague: Junk.

Kharin N (2002) *Vegetation degradation in Central Asia under the impact of human activity.* Dordrecht: Kluwer.

Kirkpatrick J (2000) *A continent transformed: Human impact on the natural vegetation of Australia, 2nd edition.* Oxford: Oxford University Press.

Moeyersons J, Nyssen J, Poesen J et al. (2004) Human impact on the environment in the Ethiopian and Eritrean highlands: A state of the art review. *Earth-Science Reviews* 64: 273–320.

Provost S, Jones MLM and Edmondson SE (2011) Changes in landscape and vegetation of coastal dunes in northwest Europe: A review. *Journal of Coastal Conservation* 15: 207–226.

Shakesby RA (2011) Post-wildfire soil erosion in the Mediterranean: Review and future research directions. *Earth-Science Reviews* 105: 71–100.

human impact on vegetation history Human activity has been a driving force in vegetation change in many parts of the world in the latter half of the HOLO-CENE. Human impacts have included BIOMASS BURNING, FOREST CLEARANCE, FOREST MANAGEMENT, FIELD DRAIN-AGE, IRRIGATION and the INTRODUCTION of plants to areas beyond their original geographical range. The main sources of evidence for human impacts on long-term vegetation change are POLLEN ANALYSIS, PLANT MACRO-FOSSIL ANALYSIS and CHARRED-PARTICLE ANALYSIS.

HUNTING, FISHING AND GATHERING peoples have had relatively minor effects, involving, for example, peri-odic burning of vegetation or creation of small tem-porary forest clearings. Agricultural peoples produced more pronounced changes, connected with the need to create fields for crop cultivation. In northwest Europe, the original postglacial woodland began to be cleared in the NEOLITHIC period, although much of the landscape remained densely wooded until later prehistory. The spread of agriculture was accompanied not only by the expansion in the geographical range of domesticated plants and animals but also by WEEDS, creating a vari-ety of new plant communities of periodically disturbed soils. Shade-intolerant plants such as *grasses*, docks (*Rumex*) and ribwort plantain (*Plantago lanceolata*) were favoured by the creation of openings in the wood-land, and those with distinctive pollen have been used in pollen analysis as ANTHROPOGENIC INDICATORS to assist in identifying human impacts in the pollen record.

DEFORESTATION allowed an expansion of communi-ties of open ground, such as *grassland*, HEATHLAND and MOORLAND, often maintained by GRAZING and/or burn-ing. Sometimes it is difficult to disentangle the roles of human activity and natural factors, such as CLIMATIC

Period	Impact	Category
Early Mesolithic	Apparently transient, but hints of locally severe impacts on woodland and soils	1
Later Mesolithic	Widespread management of woodlands and their edges; consequent paludification	2
Agricultural prehistory	Loss of woodland, introduction and expansion of field systems temporary and permanent, domestic cattle, etc; expanded population's need for wood; paludification	5
Roman	Demands for corn, road building	3
Mediaeval	Permanent parcelling of landscape: common lands and grazing management systems emplaced; monasteries add to grazing, ironworking; also remove some settlements. Some evidence for continued paludification	6
Early modern	Steady state with cumulative effects of grazing and metal extraction; sheep progressively replace cattle	4
Nineteenth century	Industrialisation, especially of extractive processes; moor management for sport in east; much vegetation acidified from rainout	10
Twentieth century	Collapse of some industries; forestry and recreation gain in importance; sheep grazing progressively on economic knife-edge; expansion of bracken very rapid	7

Column 3 is a comparative scale from 0 to 10, where 0 = no detectable alteration at all and 10 = the heaviest impacts that can be envisaged.

Human impact on vegetation history *Relative human impact on vegetation and other aspects of the environment in the uplands of England and Wales since Early Mesolithic times (Simmons, 1996b).*

CHANGE and SOIL DEVELOPMENT, in the creation of such communities. For example, BLANKET MIRES in north-west Europe began to form at various dates, from the MESOLITHIC period onward. While in the highest rain-fall areas PEAT formation probably began naturally, at some sites woodland clearance may have caused the GROUNDWATER level to rise, triggering the onset of peat accumulation. The long history of human impacts on the vegetation of the uplands in England and Wales is summarised in the Table.

It is clear that many of today's vegetation communi-ties result from an interaction of human activity and NATURAL ENVIRONMENTAL CHANGE operating on a variety of timescales. Understanding the long-term history of these communities is of particular importance for CON-SERVATION, and in appreciating the potential effects of future changes in climate and human activity.   *SPD*

[*See also* AGRICULTURAL HISTORY, BOREAL FOREST HISTORY, BRONZE AGE, COPPER AGE, FIRE IMPACTS: ECOLOGICAL, GRASSES AND GRASSLANDS, GRAZING HISTORY, HOLOCENE ENVIRONMENTAL CHANGE, IRON AGE, LANDNÁM, LANDUSE CHANGE, MIDDLE AGES, PALAEOLITHIC: HUMAN-ENVIRONMENT RELATIONS, ROMAN PERIOD, TUNDRA VEGETATION: HUMAN IMPACT, VEGETATION HISTORY]

Birks HH, Birks HJB, Kaland PE and Moe D (eds) (1988) *The cultural landscape: Past, present and future.* Cambridge: Cambridge University Press.

Chambers FM (ed.) (1993) *Climate change and human impact on the landscape.* London: Chapman and Hall.

Goudie A (2006) *The human impact on the natural environment, 6th edition.* Oxford: Blackwell.

Huntley B and Webb III T (eds) (1988) *Vegetation history.* Dordrecht: Kluwer.

Roberts N (1998) *The Holocene: An environmental history, 2nd edition.* Oxford: Blackwell Science.

Simmons IG (1996a) *Changing the face of the Earth: Culture, environment, history, 2nd edition.* Oxford: Blackwell.

Simmons IG (1996b) *The environmental impact of later Mesolithic cultures: The creation of moorland landscape in England and Wales.* Edinburgh: Edinburgh University Press.

human migration   All forms of human spatial mobility that involve long-term relocation of place of residence. The statistical threshold for 'long-term' is generally taken to be one year but this excludes *seasonal migrants*, such as those associated with employment in AGRICULTURE and TOURISM. Three broad types of migration can be recognised, which are not mutually exclusive categories: (1) *population redistri-bution* (internal movements within national boundaries, especially rural to urban migration), (2) *international migration* (*emigration* and *immigration* of economic migrants) and (3) *forced movements* (including politi-cal refugees or ENVIRONMENTAL REFUGEES). These are amongst the most powerful forces of GLOBAL CHANGE.

Some important international MIGRATIONS of the past include prehistoric HUMAN MIGRATION: 'OUT OF AFRICA'; Greek and Phoenician *colonial migration* within the Mediterranean basin (1050–550 BC); migration during the spread of the Roman Empire (510 BC to AD 117); migrations of Europeans throughout the world associ-ated with COLONIALISM; slave migration from Africa, especially to the Americas, during the *slave trade* (fifteenth–nineteenth centuries); migration of *inden-tured labour* from China and India to European colo-nies after the abolition of slavery in the nineteenth century and the GREAT MIGRATION of Europeans to the Americas during the nineteenth and early twentieth centuries. Historically, international migration was dominated by movements out of Europe but as FERTIL-ITY declined and the economy boomed in the late twen-tieth century, it became a continent of net immigration.

Currently, the international migration system is Pacific-centred, rather than Atlantic-centred with many other important migration foci. In 2010, there were 214 million international migrants (around 3 per cent of the world's population). Paradoxically, international migration is rising, despite the increasing political and legal restrictions placed upon it. Major factors that have influenced the pattern and increasing rate of global migration are the expansion of free-market economies, the spread of democracy, the export of labour-intensive industries to DEVELOPING COUNTRIES, the import of migrant workers to more advanced countries, the move-ment of skilled workers, the movement of family mem-bers to join immigrants (the so-called *quiet migration*), all facilitated by GLOBALISATION, the development of air transport and improvements to transport infrastructure. Issues and consequences of these changes include, for example, the ensuing DEMOGRAPHIC CHANGES, the spread of DISEASES, problems of cultural integration, the so-called *brain-drain*, and the problems of LAND DEGRADA-TION, POLLUTION and WASTE MANAGEMENT associated with transportation and new concentrations of people.

Population redistribution is, however, quantita-tively more important in the contemporary world than international migration. In 2009, the United Nations Population Division estimated the total number of internal migrants to be 740 million (over 10 per cent of the world's population). The largest single internal migration of recent years is the >100 million people who have relocated from rural provinces to the cities of China. Accelerated human migration of all types over the past 20 or so years has led to the so-called *Age of Migration* in which new social, educational and demographic characteristics are evident amongst, for example, women (more so than men) migrating for *marriage*, men (more so than women) migrating for *skilled work*, *child migration*, *student migration*, *return migration* and *retirement migration*.   *JAM*

[*See also* COLUMBIAN EXCHANGE, COUNTER-URBANISATION, DIASPORA, (THE) GREAT MIGRATION, IMPERIALISM, TRANSPORT POLICY, URBANISATION]

Bates DC (2002) Environmental refugees? Classifying human migrations caused by environmental change. *Population and Environment* 23: 465–477.

Black R, Bennett SR, Thomas SM and Beddington JR (2011) Migration as adaptation. *Nature* 478: 447–449.

Boyle P, Halfacree K and Robinson V (1998) *Exploring contemporary migration.* Harlow: Pearson Educational.

Castles S and Miller MJ (2009) *The age of migration: International population movements in the modern world.* Basingstoke: Palgrave Macmillan.

Curtin PD (1969) *The Atlantic slave trade: A census.* Madison: University of Wisconsin Press.

International Organization for Migration (IOM) (2005) *World migration report 2005: Costs and benefits of international migration.* Geneva: IOM.

King R (2007) *The history of human migration.* London: New Holland.

King R, Black R, Collyer M et al. (eds) (2011) *Atlas of human migration: Global patterns of people on the move.* London: Earthscan.

Ness I and Bellwood P (eds) (2013) *The encyclopedia of global human migration, 5 volumes.* Chichester: Wiley.

Skelton R (2002) Migration. In Goudie AS and Cuff DJ (eds) *Encyclopedia of global change, volume 2.* Oxford: Oxford University Press, 92–95.

Stalker P (2000) *Workers without frontiers: The impact of globalization on international migration.* Boulder, CO: Lynne Reinner.

human migration: 'out of Africa'   According to the generally favoured theory, ANATOMICALLY MODERN HUMANS evolved as the southern descendants of *Homo heidelbergensis* in East Africa around 200,000 years ago. The subsequent migration of *Homo sapiens* 'out of Africa' no later than 50,000 years ago is depicted in the Figure. Their spread was rapid, especially in tropical and temperate environments: by 30,000 years ago they had reached the ARCTIC but they arrived relatively late in the Americas (ca 15,000 years ago). During their spread, they appear to have interbred with ARCHAIC HUMANS, as evidenced by the 2.0–7.5 per cent of today's human GENOME that contains remnants of archaic DNA (including some 2.5 per cent attributed to interbreeding of native Eurasian, American and Australian people with Neanderthals). The migration of earlier HOMINIDS out of Africa had been much slower and less extensive. The unprecedented ability of modern humans to thrive in a very wide range of environments and habitats can be attributed to their intelligence and creativity signalled, not least, by CAVE ART. The alternative, now-discredited *multiregional theory*, suggested that modern humans did not evolve in a single African centre but evolved gradually in several isolated regions from an earlier ancestor.   *JAM*

[*See also* DNA: ANCIENT, HUMAN EVOLUTION]

Blockley SPE, Candy I and Blockley S (2012) Testing the role of climate change in human evolution. In Matthews JA, Bartlein PJ, Briffa KR et al. (eds) *The SAGE handbook of environmental change, volume 2.* London: Sage, 301–327.

Goebel T, Waters MR and O'Rourke DH (2008) The Late Pleistocene dispersal of modern humans in the Americas. *Science* 319: 1497–1502.

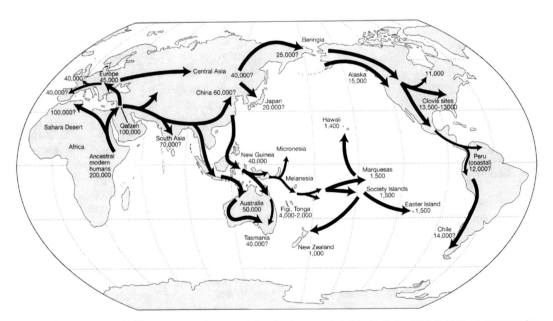

Human migration: 'out of Africa' *The timing of the spread of modern humans out of Africa (Hoffecker, 2009). Numbers are years; question marks indicate uncertainty.*

Hoffecker J (2009) The human story. In Fagan B (ed.) *The complete Ice Age: How climate change shaped the world.* London: Thames and Hudson, 93–141.

Müller UC, Pross J, Tzedakis PC et al. (2011) The role of climate in the spread of modern humans into Europe. *Quaternary Science Reviews* 30: 273–279.

Oppenheimer S (2011) The great arc of dispersal of modern humans: Africa to Australia. *Quaternary International* 202: 2–13.

Stewart JR and Stringer CB (2012) Human evolution out of Africa: The role of refugia and climate change. *Science* 335: 1317–1321.

Stringer C (2012) What makes a modern human? *Nature* 485: 33–35.

humanism   A philosophical approach which emphasises the distinctively human value, quality and subjectivity in people's lives. This is an approach distinguished by the central and active role given to human awareness, consciousness, creativity and human agency. The rise of humanism in the 1970s resulted especially from a dissatisfaction with mechanistic models of social science. Humanism offers another way of understanding people and their environments in contrast to more structural and positivist approaches. Principal philosophies in humanism include IDEALISM, PHENOMENOLOGY and REALISM.   *ART*

Law S (2011) *Humanism: A very short introduction.* Oxford: Oxford University Press.

Mann N (1996) *The origins of humanism.* Bloomington: Indiana University Press.

Norman R (2004) *On humanism.* London: Routledge.

humanities   The set of academic disciplines belonging to the 'arts' and typically including HISTORY, languages and philosophy. They use a wide range of methodologies that include analysis, criticism and speculation. The term is most commonly used in relation to the organisation of subjects or disciplines within university faculties, where they tend to be in the front line of financial cutbacks. The humanities remain resilient and true to their traditional areas of research (often helped by the popular appeal of historical narratives and historical novels) but have adapted to modern trends with publications such as the *International Journal of Humanities and Arts* published by the Edinburgh University Press.   *DTH*

Aldama FL (2008) *Why the humanities matter: A commonsense approach.* Austin: University of Texas Press.

Humboldtian science   A type of science practised in the nineteenth century epitomised by the researches of Alexander von Humboldt. It was characterised by accurate observation and measurement collected during FIELD RESEARCH, analysis involving MAP GENERALISATION (see CARTOGRAPHY) especially using ISOPLETHS, and an overall HOLISTIC APPROACH to the LANDSCAPE that emphasised the interactions between the biophysical and human environment.   *JAM*

[*See also* GEO-ECOLOGY, HOLISM, PHYSICAL GEOGRAPHY]

Bravo M (1998) Humboldtian science. In Good GA (ed.) *Sciences of the Earth: An encyclopedia of events, people and phenomena.* New York: Garland Publishing, 430–433.

Nicolson M (1987) Alexander von Humboldt, Humboldtian science, and the origins of the study of vegetation. *History of Science* 25: 167–194.

humic acids   A complex mixture of dark-coloured organic substances precipitated by acidification of a dilute alkali extract from soil. Humic acids have a higher *molecular weight* than FULVIC ACIDS. *Grey humic acids* and *brown humic acids* are sometimes recognised.
                                                          EMB

[*See also* SOIL DATING]

Biagorri R, Fuentes M, González-Gaitano G et al. (2009) Complementary multianalytical approach to study the distinctive structural features of the main humic fractions in solution: Gray humic acid, brown humic acid, and fulvic acid. *Journal of Agricultural and Food Chemistry* 57: 3266–3272.

Chefetz B, Tarchhitzky J, Desmukh AP et al. (2002) Structural characterization of soil organic matter and humic acids in particle-size fractions of an agricultural soil. *Soil Science Society of America Journal* 66: 129–141.

humic substances   Major components of the *organic matter* in soil and water as well as in geological organic deposits such as LACUSTRINE SEDIMENTS, PEAT and SHALE. In soils, humic substances are derived from HUMUS following the decay and transformation of plant remains (HUMIFICATION) and are a significant part of the CARBON CONTENT OF SOILS. Conventionally, they are defined as yellow to black coloured, complex, high molecular weight and heterogeneous, refractory naturally occurring organic compounds which persist in soils but are eventually degradable. They are operationally defined as three primary fractions in terms of the methods used to isolate them from the soil: HUMIC ACIDS, FULVIC ACIDS and *humin*. Humic substances are highly chemically reactive yet recalcitrant with respect to BIODEGRADATION.   *SN*

Clapp CE, Hayes MHB and Senesi N (eds) (2001) *Humic substances and chemical contaminants.* Madison, WI: Soil Science Society of America.

Ghabbour EA and Davies G (eds) (2001) *Humic substances: Structures, models and functions.* London: Royal Society of Chemistry.

Hatcher P (2001) Modern analytical studies of humic substances. *Soil Science* 166: 770–794.

Hessen DO and Tranvik LJ (eds) (2010) *Aquatic humic substances: Ecology and biogeochemistry.* Berlin: Springer.

McCarthy P (2001) The principles of humic substances. *Soil Science* 166: 738–751.

Tipping E (2002) *Cation binding by humic substances.* Cambridge: Cambridge University Press.

humid tropics: ecosystem responses to environmental change

Many coastal ecosystems in humid *tropical environments* are vulnerable to SEA-LEVEL RISE and an increase in WAVE ENERGY that may result from GLOBAL WARMING. Rapid rates of rise may reduce the extent of MANGROVE SWAMPS. Many mangrove species are extremely sensitive to variation in their *hydrological regimes* or *tidal regimes*. CORAL REEFS have been shown to be extremely sensitive to changes in water temperature. Increased water temperatures during the AD 1997–1998 EL NIÑO-SOUTHERN OSCILLATION (ENSO) event resulted in widespread CORAL BLEACHING.

Evidence of significant changes in the distribution and composition of TROPICAL AND SUBTROPICAL FORESTS during periods of CLIMATIC CHANGE during the QUATERNARY suggests that present-day forests are unlikely to be resistant to future climatic change. Evidence from long-term ENVIRONMENTAL MONITORING of forest plots throughout the humid tropics indicates that forests are dynamic, and are responding to environmental change. Humid tropical ecosystems are likely to be more sensitive to changes in *soil-water availability* than to temperature. Large areas of southern and eastern Amazonia have a climate that is marginal for TROPICAL RAIN FOREST and survive by having very deep *root systems* that can tap GROUNDWATER sources. Forest FRAGMENTATION and DEGRADATION exacerbate the vulnerability to DROUGHT and fire. The high resistance of humid tropical ecosystems to BIOLOGICAL INVASION is now believed to be attributable primarily to rapid rates of recovery after disturbance. Significant ENVIRONMENTAL CHANGE observed in forest fragments has made them much more vulnerable to invasions.

Tropical rain forests may respond to rising CARBON DIOXIDE levels by exhibiting faster turnover. Tropical forest may be an important CARBON SINK/SOURCE, rather than simply a store, and there is some evidence that sink strength may be increasing as a result of the fertilisation effect of elevated levels of atmospheric carbon dioxide (see CARBON DIOXIDE FERTILISATION). As humid tropical PLANTATIONS are among the most productive and tropical rain forests contain the largest total stock of BIOMASS carbon, there is growing interest in the potential for managing humid tropical forests for CARBON SEQUESTRATION.

Changes inflicted on humid tropical ecosystems may also have a profound effect on the global environment. Annual global EMISSIONS of carbon from tropical DEFORESTATION are estimated to be about 2.9 ± 0.5 Pg C. In most humid tropical countries, which are DEVELOPING COUNTRIES, low incomes, high national debt and a dependency on AGRICULTURE make large *human populations* vulnerable to climatic variation.   *NDB/JLI*

[*See also* BIOMASS BURNING, CARBON CYCLE, REFUGIUM, TROPICAL CYCLONE, TROPICAL PEATLANDS]

Asner GP, Rudel TK, Aide TM et al. (2009) A contemporary assessment of change in humid tropical forests. *Conservation Biology* 23: 1386–1395.

Bush MB and Gosling WD (2012) Environmental change in the humid tropics and monsoonal regions. In Matthews JA, Bartlein PJ, Briffa KR et al. (eds) *The SAGE handbook of environmental change, volume 2.* London: Sage, 113–140.

Cleveland CC, Wieder WR, Reed SC and Townsend AR (2010) Experimental drought in a tropical rain forest increases soil carbon dioxide losses to the atmosphere. *Ecology* 91: 2313–2323.

Gardner TA, Barlow J, Sodhi NS and Peres CA (2010) A multi-region assessment of tropical forest biodiversity in a human-modified world. *Biological Conservation* 143: 2293–2300.

Heubes J, Kuhn I, König K et al. (2011) Modelling biome shifts and tree cover change for 2050 in West Africa. *Journal of Biogeography* 38: 2248–2258.

Morton RA (2002) Coastal geoindicators of environmental change in the humid tropics. *Environmental Geology* 42: 711–724.

Pan YD, Birdsey RA, Fang JY et al. (2011) A large and persistent carbon sink in the world's forests. *Science* 333: 988–993.

Phillips OL and Gentry AH (1994) Increasing turnover through time in tropical forests. *Science* 263: 954–958.

Raneesh KY and Santosh GT (2011) A study on the impact of climate change on streamflow at the watershed scale in the humid tropics. *Hydrological Sciences Journal* 56: 946–965.

humidity

The amount of WATER VAPOUR in the air, which is controlled by EVAPORATION and CONDENSATION. The *absolute humidity* is the mass of the water vapour in a given volume of air. The *specific humidity* is the ratio of the mass of the water vapour in the air to the combined mass of the water vapour and the air; it is approximated by the *mixing ratio* (the ratio of the mass of water vapour to the mass of dry air). The *relative humidity* (usually expressed as a percentage) is the ratio of the water vapour content of the air to the maximum amount the air could hold at that temperature and volume. This is the common usage of the term 'humidity': 0 per cent is 'dry' air; 100 per cent is totally saturated with water vapour. Relative humidity may also be expressed as the ratio of the mixing ratio of the air to the saturation mixing ratio at the same temperature and pressure. Humidity is measured using a *hygrometer* and is a fundamental ENVIRONMENTAL FACTOR of both the outdoor and INDOOR ENVIRONMENTAL CHANGE.   *JAM*

Fairbridge RW and Oliver JE (1987) Humidity. In Oliver JE and Fairbridge RW (eds) *The encyclopedia of climatology.* New York: Van Nostrand Reinhold, 479–483.

humification   The processes of breakdown and synthesis of complex organic RESIDUES, through biological activity, microbial synthesis and chemical reactions, to HUMUS. In SOILS, as plant material is utilised by successive groups of *soil animals* for their nutrition, it is broken into smaller fragments and finally invaded by FUNGI and BACTERIA which complete the breakdown. In the process, the organic material is changed from green to brown and eventually to black *amorphous humus*. Once formed, the humus breaks down relatively slowly, offering the possibility of CARBON SEQUESTRATION in soils, which helps mitigate GLOBAL WARMING by abstracting CARBON DIOXIDE from the ATMOSPHERE and improves overall SOIL QUALITY.   *EMB*

[*See also* DECOMPOSITION, HUMIC SUBSTANCES, LITTER, PEAT HUMIFICATION]

Balser TC (2005) Humification. In Hillel D (ed.) *Encyclopedia of soils in the environment.* Amsterdam: Elsevier, 195–207.
Tan KH (2003) *Humic matter in soil and the environment: Principles and controversies.* New York: Marcel Dekker.
Wild A (ed.) (1988) *Russell's soil conditions and plant growth, 11th edition.* Harlow: Longman.

humin   One of the three broad fractions identified as HUMIC SUBSTANCES. Humin is not soluble in water at any pH value and is alkaline. Humins are black in colour. The humin fraction, the residual fraction after PRETREATMENT, tends to be the preferred fraction for RADIOCARBON DATING of soils.   *SN*

[*See also* SOIL DATING]

Rice JA (2001) Humin. *Soil Science* 166: 848–857.

hummocky cross-stratification (HCS)   A SEDIMENTARY STRUCTURE comprising gently undulating LAMINATION in fine- to medium-grained sand, characterised by convex-upward laminae with no preferred direction of DIP, that define hummocks and troughs (swales) with an amplitude of a few centimetres and spacing of a metre or so. It is generally agreed that HCS forms in response to combined flows. Most is reported from deposits that accumulated below fairweather WAVE BASE in response to STORMS, in which a storm-induced *unidirectional current* is combined with a storm wave-induced *oscillatory current*, although there has been no unequivocal recognition of appropriate BEDFORMS in modern sediments. A related structure in which the swales are more prominent than the hummocks is recognised as *swaley cross-stratification (SCS)*.   *GO*

Quin JG (2011) Is most hummocky cross-stratification formed by large-scale ripples? *Sedimentology* 58: 1414–1433.
Swift DJP, Figueiredo AG, Freeland GL and Oertel GF (1983) Hummocky cross-stratification and megaripples: A geological double standard? *Journal of Sedimentary Petrology* 53: 1295–1317.

Yang BC, Dalrymple RW and Chun S (2006) The significance of hummocky cross-stratification (HCS) wavelengths: Evidence from an open-coast tidal flat, South Korea. *Journal of Sedimentary Research* 76: 2–8.

hummocky moraine   An apparently chaotic assemblage of MORAINE mounds and ridges. The term has been used in a wide range of senses to refer to glacial landforms of different origins, although most commonly it is used to refer to moraines deposited by the ABLATION of DEBRIS-MANTLED GLACIERS.   *DIB*

Benn DI and Evans DJA (2010) *Glaciers and glaciation, 2nd edition.* London: Hodder-Arnold.

humus   The relatively resistant, usually dark brown to black fraction of SOIL *organic matter* that results from the biological breakdown and synthesis of organic residues (HUMIFICATION). It is composed mainly of carbon and nitrogen in complex organic molecules, based on a chemical structure similar to the benzene ring; other elements are present in low amounts. Humus is capable of forming an intimate association with CLAY MINERALS to form the CLAY-HUMUS COMPLEX, the most reactive part of the soil. In acid conditions, beneath needle-leaved, coniferous BOREAL FOREST or HEATHLAND, *collembola* and *mites* are the most important animals involved, together with FUNGI, in the process, which takes place above the mineral soil. The resulting layers of LITTER, FERMENTATION and HUMUS are called MOR. In base-rich conditions, beneath broad-leaved deciduous TEMPERATE FOREST and TEMPERATE GRASSLAND, the organic matter is broken down and the humus is incorporated into the surface soil as MULL. BACTERIA are more important than fungi, and EARTHWORMS actively draw organic matter down into the soil so there is no surface *accumulation*. MODER is an intermediate form of humus, often occurring on acid and poorly drained sites, having litter and fermentation layers of approximately equal thickness.
                          EMB

[*See also* FULVIC ACIDS, HUMIC ACIDS, HUMIC SUBSTANCES, HUMIN]

Kononova MM (1975) Humus of virgin and cultivated soils. In Gieseking JE (ed.) *Soil components, volume 1: Organic components.* Berlin: Springer Verlag.
Krosshavn M, Bjorgum JO, Krane J and Steinnes E (1990) Chemical structure of terrestrial humus materials formed from different vegetation characterized by solid-state ^{13}C NMR with CP-MAS techniques. *Journal of Soil Science* 41: 371–377.
Wild A (ed.) (1988) *Russell's soil conditions and plant growth, 11th edition.* Harlow: Longman.

hunger   Persistent consumption of a diet that, in terms of quantity (*undernutrition* or *undernourishment*) or quality (*malnutrition*), is inadequate to maintain good health, normal activity and growth. The symptoms of hunger are physical discomfort, pain or deficiency DIS-EASES. In the nineteenth century, hunger was viewed as an unavoidable natural phenomenon but during the twentieth century, the hungry were increasingly viewed as the product of human political and economic forces that required government intervention in developed countries (e.g. introduction of the welfare state in the UK) and NONGOVERNMENTAL ORGANISATIONS (NGOs) in developing countries. Despite increases in food production, and medical and social improvements, the number of undernourished people in the world continued to rise. The undernourished now exceeds 1,000 million, and is rising more rapidly than ever (see Figure A). The greatest numbers of undernourished are in Asia and sub-Saharan Africa. Although the percentage of undernourished in DEVELOPING COUNTRIES fell significantly during the past three decades of the twentieth century, there was a worrying upturn after about 2005 (see Figure B). GLOBALISATION, URBANISATION and the shift of responsibility for alleviating hunger to NGOs (the *depoliticisation* of hunger) are reshaping the availability of food and changing patterns of CONSUMPTION.

JAM

[*See also* FAMINE, FOOD SECURITY]

Bassett TJ and Winter-Nelson A (2010) *The atlas of world hunger*. Chicago: University of Chicago Press.

Deegan LA, Johnson DS, Warren RS et al. (2012) Coastal eutrophication as a driver of salt marsh loss. *Nature* 490: 388–392.

European Environment Agency (EEA) (2010) *The European environment: State and outlook 2010, synthesis*. Copenhagen: EEA.

Food and Agriculture Organization of the United Nations (FAO) (2009) *The state of food security in the world 2009*. Rome: FAO.

Lang T, Barling D and Caraher M (2009) *Food policy: Integrating health, environment and society*. Oxford: Oxford University Press.

Patel R (2008) *Stuffed and starved: The hidden battle for the world food system*. Brooklyn, NY: Melville House.

Vernon J (2007) *Hunger: A modern history*. Cambridge, MA: Harvard University Press.

hunting, fishing and gathering A subsistence economy and lifestyle based on hunting wild animals (e.g. deer) for meat, hide and bone; fishing in coastal and fresh waters, and collecting wild plants (e.g. nuts, seeds and fungi) for food. The balance between each type of activity in the economy can vary (e.g. seasonally or geographically). Unwooded areas were more likely to have been used for hunting (e.g. Late Pleistocene STEPPE of Europe) whilst gathering probably dominated in wooded landscapes (e.g. MESOLITHIC lowland Britain) where mobility would have been more restricted. The success of such an economy depends on an intimate knowledge of the environment and the resources available. Such communities could have had a significant impact on the environment and Late Pleistocene extinctions of MEGAFAUNA have been attributed to OVERHUNTING. Extensive burning of woodland was often undertaken to drive game, to ease movement by increasing the extent of open land, to increase the quality of browse or to encourage certain food plants (e.g. hazel nuts).

LD-P/RMF

[*See also* MEGAFAUNAL EXTINCTION, NOMADISM, PASTORALISM, SEDENTISM]

Bettinger RL (1991) *Hunter-gatherers: Archaeology and evolutionary theory*. New York: Plenum Press.

Domínguez-Rodrigo M (2002) Hunting and scavenging by early humans: The state of the debate. *Journal of World Prehistory* 16: 1–54.

Enright NJ and Thomas I (2008) Pre-European fire regimes in Australian ecosystems. *Geography Compass* 2: 979–1011.

Ingold T (1980) *Hunters, pastoralists and ranchers*. Cambridge: Cambridge University Press.

A

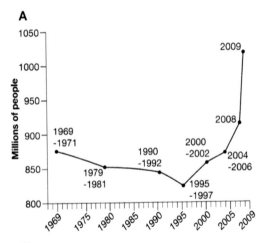

B

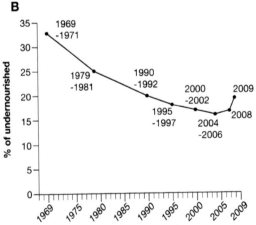

Hunger *Recent trends in (A) the number of undernourished people in the world and (B) the percentage of the people in developing countries who are undernourished (data from EEA, 2010; FAO, 2009).*

West D (1997) *Hunting strategies in Central Europe during the last glacial maximum*. Oxford: British Archaeological Reports, International Series 672.

hybridisation The process of crossing between individuals of genetically distinct POPULATIONS (e.g. *species* or *subspecies*), resulting in individuals (hybrids) that may be fertile or sterile. The offspring are more likely to be sterile with greater genetic distance between the parents. Hybridisation is important as a process in natural EVOLUTION and in *breeding* new varieties artificially.

KDB

Arnold ML (1997) *Natural hybridization and evolution*. New York: Oxford University Press.

hydration The ADSORPTION of water by certain minerals causing a change in volume and setting up stresses within a rock. Examples include the hydration of *calcium sulfate* to GYPSUM, and the conversion of *iron oxides* into *iron hydroxides*. It is a type of CHEMICAL WEATHERING process.

RAS

[*See also* DISSOLUTION, SALT WEATHERING]

Sperling CHB and Cook RU (1985) Laboratory simulation of rock weathering by salt crystallization and hydration processes in hot, arid environments. *Earth Surface Processes and Environments* 10: 541–555.
White SE (1976) Is frost weathering really only hydration shattering? *Arctic and Alpine Research* 8: 1–6.

hydraulic conductivity A measure of the ability of rock or soil to transmit water at a specified state of wetness. It is affected by *porosity* (see POROUS), the number and size of structural fractures and *fluid viscosity*. The term is normally applied to flow in saturated soil (*saturated hydraulic conductivity*); if used for flows in unsaturated soil then values of hydraulic conductivity will vary with the soil *water content*.

ADT

[*See also* INFILTRATION, PERMEABLE, THROUGHFLOW]

hydraulic geometry The analysis of the hydraulic characteristics of stream channels in relation to channel geometry, a concept that was first developed by L.B. Leopold and T. Maddock. *At-a-station hydraulic geometry* defines how width (w), mean depth (d) and mean velocity (v) in particular, but also other hydraulic variables, at a specific cross section change over time as discharge (Q) rises or falls. *Downstream hydraulic geometry* describes how, along a given river and for a given flow frequency (e.g. *mean annual discharge*), these properties adjust in a downstream direction to accommodate (in most, but not all cases) the increasing DISCHARGE. The original fitted relations were *power functions*:

$$w = aQ^b; \; d = cQ^f; \; v = kq^m$$

where the coefficients a, c and k and exponents (slopes) b, f and m are derived from least squares fits to the empirical data. Given that $Q = w \cdot d \cdot v$, then $b + f + m = 1$ and $a \cdot c \cdot k = 1$. In downstream hydraulic geometry, *catchment area* is often used as a surrogate for discharge.

A certain consistency in the values implies that channels attain an equilibrium size and shape which minimises total work. In at-a-station hydraulic geometry, the exponent m is generally >0.20 (indicating that velocity is much higher at high flows), but the exponents b and f tend to vary with CHANNEL SHAPE, with values of b being small for box-shaped channels with vertical banks, but higher for wide-shallow channels associated with less cohesive banks. For the downstream case, $b \sim 0.55$, $f \sim 0.35$ and $m \sim 0.1$. The WIDTH-DEPTH RATIO therefore increases downstream, and width adjusts most effectively to increasing flow (and is the most responsive to environmental change). A novel finding ($m \sim +0.1$) of the approach was that mean velocity increases downstream, a declining channel slope being more than offset by decreases in channel roughness and increases in channel depth and efficiency. The approach has also been applied to numerous other fluvial variables such as channel roughness, SUSPENDED LOAD, *stream power, shear stress, boundary sediment*, BANK EROSION and, as the *river continuum concept*, to freshwater ecology. Relationships are not always log-linear and shallow polynomials sometimes provide better fits. In environmental situations where discharge is lost along river channels (as in the Macquarie River in southeastern Australia) and/or there are changes in channel bank stability, standard relationships will not apply. Climatic changes that involve changes in high flow magnitude-frequency and/or affect RIPARIAN vegetation will tend to result in changes in hydraulic geometry. The downstream hydraulic geometry approach has been widely and very successfully used to detect, assess and predict downstream impacts of URBANISATION, and other catchment changes on river channels.

DML/RPDW

[*See also* RESERVOIRS: ENVIRONMENTAL EFFECTS, URBANISATION IMPACTS ON HYDROLOGY]

Dury GH (1969) Hydraulic geometry. In RJ Chorley (ed.) *Water, earth and man*. London: Methuen.
Ferguson RI (1986) Hydraulics and hydraulic geometry. *Progress in Physical Geography* 10: 1–31.
Leopold LB and Maddock T (1953) The hydraulic geometry of stream channels and some physiographic implications. *United States Geological Survey Professional Paper* 252: 1–57.
Ralph TJ and Hesse PP (2010) Downstream hydrogeomorphic changes along the Macquarie River, southeastern Australia, leading to channel breakdown and floodplain wetlands. *Geomorphology* 118: 48–64.
Richards KS (1977) Channel and flow geometry. *Progress in Physical Geography* 1: 65–102.

Wehl E and Dust D (2012) Geomorphic response of a headwater channel to augmented flow. *Geomorphology* 138: 329–338.

hydric soils Reduced soils under ANOXIC conditions produced by water saturation and characteristic of WETLANDS. Typically, they include GLEYSOLS (where the WATER TABLE is close to the surface in topographic lows), FLUVISOLS (on poorly drained FLOODPLAINS and tidal coastal areas) and SULFIDIC SOILS (where sulfate REDUCTION occurs). *JAM*

[*See also* ACID SOILS, HYDROMORPHIC SOILS]

Mausbach MJ and Richardson JL (1994) Biogeochemical processes in hydric soils. *Current Topics in Wetland Biogeochemistry* 1: 68–127.
Richardson JL and Vepraskas MJ (2001) *Wetland soils: Genesis, hydrology, landscapes and classification.* Boca Raton, FL: Lewis Publishers.

hydrocarbons Complex organic compounds of carbon and hydrogen that make up PETROLEUM. The burning of hydrocarbons as FOSSIL FUELS releases into the atmosphere GREENHOUSE GASES that were formerly locked up in SEDIMENTARY ROCKS, contributing to GLOBAL WARMING. Incomplete COMBUSTION and the presence of impurities, such as *sulfur*, contribute to other ENVIRONMENTAL PROBLEMS such as PHOTOCHEMICAL SMOG and ACID RAIN. Hydrocarbons are also the basis for the *petrochemical industries* producing, for example, FERTILISERS, PESTICIDES, *pharmaceuticals* and PLASTICS, with associated POLLUTION and WASTE MANAGEMENT problems. *GO/JAM*

[*See also* CHLOROFLUOROCARBONS (CFCs), ORGANOCHLORIDES, PERSISTENT ORGANIC COMPOUNDS (POCs)]

Gluyas J and Swarbrick R (2004) *Petroleum geoscience.* Oxford: Blackwell.
Schobert HH (1991) *The chemistry of hydrocarbon fuels.* London: Newnes.

hydrochlorofluorocarbons   (HCFCs) Compounds of carbon in which some of the hydrogen atoms have been replaced by *chlorine* and *fluorine*. They are less stable than CHLOROFLUOROCARBONS (CFCs) with limited atmospheric lifetimes of 1–20 years and tend to break down in the TROPOSPHERE before diffusing into the STRATOSPHERE. They are therefore around 95 per cent less damaging to the OZONE layer and are widely used as substitutes for CFCs but are nevertheless to be phased out following on from the *Montreal Protocol* (see CONVENTION). *Hydrofluorocarbons (HFCs)* are viewed as more appropriate substitutes because they contain no chlorine and hence do not attack ozone. *JAM*

[*See also* HALOGENATED HYDROCARBONS (HALOCARBONS)]

Kanakidou M, Dentener FJ and Crutzen PJ (1995) A global three-dimensional study of the fate of HCFCs and HFC-134a in the troposphere. *Journal of Geophysical Research* 100: 18781–18801.
Monzka SA, Hall BD and Elkins JW (2009) Accelerated increases observed for hydrochlorofluorocarbons since 2004 in the global atmosphere. *Geophysical Research Letters* 36: L03804.

hydroclimatology The study of the interaction between climatic processes and the components of the HYDROLOGICAL CYCLE. *JBE/NP*

Shelton ML (2008) *Hydroclimatology: Perspectives and applications.* Cambridge: Cambridge University Press.

hydrocompaction COMPACTION resulting from the infiltration of water into moisture-deficient sediments. The term is also sometimes used with the opposite sense, to refer to compaction resulting from the loss of water from sediments and rocks in the subsurface due to *drainage*, GROUNDWATER extraction or natural processes, which may result in SUBSIDENCE of the ground surface. *GO*

Psimoulis P, Ghilardi M, Fouache E and Stiros S (2007) Subsidence and evolution of the Thessaloniki plain, Greece, based on historical leveling and GPS data. *Engineering Geology* 90: 55–70.

hydrogen isotopes Hydrogen has two STABLE ISOTOPES (natural abundance, ^1H = 99.985 per cent and ^2H or *deuterium*, D = 0.015 per cent) and one *radioisotope* (^3H or *tritium*, T; half-life = 12.5 years). The stable ISOTOPE RATIO is expressed as $^2H/^1H$ relative to the *Vienna Standard Mean Ocean Water (VSMOW)* standard (see REFERENCE STANDARD). As hydrogen stable isotopes have the largest *relative mass difference*, hydrogen exhibits the largest variation in its stable isotope ratio. A high-resolution hydrogen isotope record has been determined for the 800 ka European Project for Ice Coring in Antarctica (EPICA) Dome C ice core from Antarctica. The influence of ORBITAL FORCING on global temperature variations is evident in the eight GLACIAL-INTERGLACIAL CYCLES covered by the core. Millennial scale changes in δD in the core are influenced by *North Atlantic Deep Water (NADW)* formation supporting the thermal *bipolar seesaw* hypothesis. The hydrogen and OXYGEN ISOTOPE values of *meteoric water* are strongly associated. PHOTOSYNTHESIS causes a large ISOTOPIC DEPLETION in ^2H of non-exchangeable hydrogen relative to metabolic water. Post-photosynthetic processes cause a large ISOTOPIC ENRICHMENT. Small changes in metabolic activity can therefore strongly influence plant δD values. TREE-RING δD values of non-exchangeable hydrogen have been used to reconstruct both spatial and temporal variations in climate. *IR*

[*See also* ISOTOPES AS INDICATORS OF ENVIRONMENTAL CHANGE]

Filot MS, Leuenberger M, Pazdur A and Boettger T (2006) Rapid online equilibration method to determine the D/H ratios of non-exchangeable hydrogen in cellulose. *Rapid Communications in Mass Spectrometry* 20: 3337–3344.
Jouzel J, Masson-Delmotte V, Cattani O et al. (2007) Orbital and millennial Antarctic climate variability over the past 800,000 years. *Science* 317: 793–797.
Loader NJ, Santillo PM, Woodman-Ralph JP et al. (2008) Multiple stable isotopes from oak trees in southwestern Scotland and the potential for stable isotope dendroclimatology in maritime climatic regions. *Chemical Geology* 252: 62–71.
Petit JR, Jouzel J, Raynaud D et al. (1999) Climate and atmospheric history of the past 420,000 years from the Vostok ice core, Antarctica. *Nature* 399: 429–436.

hydrogeology

hydrogeology   The branch of GEOLOGY dealing with GROUNDWATER. It differs from *geohydrology*, which covers the geological aspects of HYDROLOGY.   *CDW*

[*See also* AQUIFER, CONTAMINANT, SALINISATION, SALTWATER INTRUSION]

Domenico PA and Schwartz FW (1998) *Physical and chemical hydrogeology, 2nd edition.* New York: Wiley.
Fetter Jr CW (2003) *Applied hydrogeology, 4th edition.* Upper Saddle River, NJ: Pearson Education.
Poehls DJ and Smith GJ (2009) *Encyclopedic dictionary of hydrogeology.* San Diego, CA: Academic Press.

hydrograph   A graphical representation of channel DISCHARGE from a DRAINAGE BASIN over time. Analysis of the low-flow and storm-flow characteristics of hydrographs is useful in assessing impacts of LANDUSE CHANGE, CLIMATIC CHANGE, river CHANNELISATION or DAMS and reservoirs on the catchment WATER BALANCE and responses to PRECIPITATION events.   *ADT/RPDW*

[*See also* AFFORESTATION: IMPACTS ON HYDROLOGY AND GEOMORPHOLOGY, STORM HYDROGRAPH]

hydro-isostasy   Isostatic LOADING and UNLOADING by water upon ocean floors and CONTINENTAL SHELVES due to changes in RELATIVE SEA LEVEL. A similar process may occur in response to LAKE-LEVEL VARIATIONS, producing warped shorelines.   *AGD*

[*See also* ISOSTASY, SEA-LEVEL CHANGE]

Dickinson WR (2004) Impacts of eustasy and hydro-isostasy on the evolution and landforms of Pacific atolls. *Palaeogeography, Palaeoclimatology, Palaeoecology* 213: 251–269.
Smith DE and Dawson AG (1983) *Shorelines and isostasy.* London: Academic Press.

hydrological balance/budget   An account-book approach to gains and losses (or inputs, storages and outputs) in any hydrological SYSTEM. WATER BALANCE/BUDGET and *moisture balance/budget* are widely used synonyms. It may be applied at any spatial scale, from water exchange at a small area of the Earth's surface to the inputs and outputs of a LAKE or DRAINAGE BASIN, and the global water balance (see Figure). Seasonal fluctuations in rainfall, temperature and EVAPOTRANSPIRATION result in parallel changes in storage of soil water and GROUNDWATER *and lags* between rainfall inputs and streamflow outputs. Hence, at the DRAINAGE BASIN scale, meaningful hydrological budgets tend to be calculated on annual timescales, and time series extending over several years or more are needed for accuracy.   *JAM/RPDW*

[*See also* HYDROLOGICAL CYCLE, LAKE-LEVEL VARIATIONS]

Baumgartner A and Reichel E (1975) *The world water balance: Mean annual global, continental and maritime precipitation and runoff.* Amsterdam: Elsevier.
Henshaw PC, Charlson RJ and Burges SJ (2000) Water and the hydrosphere. In Jacobson MC, Charlson RJ, Rodhe H and Orians GH (eds) *Earth system science: From biogeochemical cycles to global change.* San Diego, CA: Academic Press, 109–131.
Shiklomanov IA, Street-Perrott AF, Beran MA and Ratcliffe RAS (eds) (1983) *Variations in the global water budget.* Dordrecht: Reidel.
Widen-Nilsson E, Halldin S and Xu CY (2007) Global water-balance modelling with WASMOD-M: Parameter estimation and regionalisation. *Journal of Hydrology* 340: 105–118.

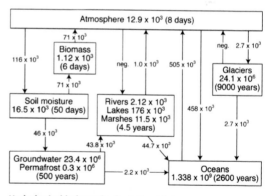

Hydrological balance/budget *The global water balance: fluxes are shown as arrows (km^3/year); storages (km^3) and turnover times (years) are indicated in the boxes (Henshaw et al., 2000).*

hydrological cycle   The complex of processes by which water circulates through the Earth-atmosphere-ocean SYSTEM. It can be applied at different spatial scales, the two most important being the *global hydrological cycle*, which is a CLOSED SYSTEM, and the *basin hydrological cycle*, which is an OPEN SYSTEM where there is no or only indirect causal links between outputs (river DISCHARGE, GROUNDWATER outputs and EVAPOTRANSPIRATION) and inputs (PRECIPITATION). Also known as the *water cycle*, the hydrological cycle not only is a central concept of HYDROLOGY but also constitutes a BIOGEOCHEMICAL CYCLE. In general, precipitation from the atmosphere may generate STREAMFLOW; may be stored for varying times in soil, lakes and groundwater; or may be returned to the ATMOSPHERE by EVAPOTRANSPI-RATION, but there are many complexities to any simple model. Because most of the Earth's surface is water, a high proportion of the global hydrological cycle occurs over the OCEANS involving only EVAPORATION and PRECIP-ITATION. The relative proportions of the world's water that lie in the oceans (97 per cent) and in ice sheets and glaciers (most of the remainder) have changed sub-stantially with Quaternary CLIMATIC CHANGE with dra-matic consequences for SEA-LEVEL CHANGE and ocean SALINITY change. The proportion of the Earth's water in the oceans is currently rising and is predicted to rise farther with continued GLOBAL WARMING. It is also predicted that in a 2–3°C warmer world, the pattern of salinity changes in the ocean will lead to a 16–24 per cent intensification of the global water cycle, driven by enhanced *surface water fluxes* involving evaporation and precipitation. *JAM/RPDW*

[*See also* HYDROLOGICAL BALANCE]

Berner EK and Berner RA (1987) *The global water cycle: Geochemistry and environment.* Englewood Cliffs, NJ: Prentice Hall.
Chahine MT (1992) The hydrologic cycle and its influence on climate. *Nature* 359: 373–380.
Durack PJ, Wijffels SE and Matear RJ (2012) Ocean salinities reveal strong global water cycle intensification during 1950 to 2000. *Science* 336: 455–458.
Elahire EAB and Bras RL (1996) Precipitation recycling. *Reviews of Geophysics* 34: 367–378.
Oki T and Kanae S (2006) Global hydrological cycles and world water resources. *Science* 313: 1068–1072.
Ramanathan V, Crutzen PJ, Kiehl JT and Rosenfeld D (2001) Aerosols, climate, and the hydrological cycle. *Science* 294: 2119–2124.

hydrology The study of water in all its phases within the *Earth-atmosphere-ocean system*, espe-cially the nature of *hydrological processes* (*physical hydrology*) and their application (*applied hydrology*). It includes not only RUNOFF on the Earth's surface but also GROUNDWATER and water in the ATMOSPHERE and CRYOSPHERE. *JAM*

[*See also* HYDROGEOLOGY, HYDROLOGICAL BALANCE, HYDROLOGICAL CYCLE, HYDROSPHERE, PALAEOHYDROLOGY]

Jones JAA (1997) *Global hydrology: Processes, resources and environmental management.* Harlow: Longman.
Ward AC and Elliot WJ (eds) (1995) *Environmental hydrology.* Boca Raton, FL: CRC Press.
Ward RC and Robinson M (2000) *Principles of hydrology, 4th edition.* London: McGraw-Hill.

hydrolysis (1) A type of CHEMICAL WEATHERING involving the formation of both an ACID and a BASE from a SALT when it dissociates with water. Hydrolysis affects particularly FELDSPAR and *mica* minerals and is thus an especially important process affecting IGNEOUS ROCKS. In *humid tropical environments*, hydrolysis-induced weathering of igneous rocks can proceed to depths of several tens of metres. (2) The disintegra-tion of *organic compounds* through their reaction with water. *SHD*

White AF and Brantley SL (eds) (1995) *Chemical weathering rates of silicate minerals.* Washington, DC: Mineralogical Society of America.

hydrometeor Any product of CONDENSATION or DEP-OSITION of WATER VAPOUR from the ATMOSPHERE as liquid *water droplets* or *ice particles*. It may fall to the Earth's surface as PRECIPITATION (*rain, snow or hail*), remain suspended (fog or CLOUD) or form at the surface (*dew, frost or rime*). *JAM*

hydromorphic soils Soils with features of SOIL TEXTURE and SOIL STRUCTURE that are characteristic of poor drainage and waterlogging, such as CONCRETIONS of iron or manganese. *JAM*

[*See also* GLEYING, GLEYSOLS, HYDRIC SOILS]

Hurt GW, Whited PM and Pringle RF (eds) (2002) *Field indicators of hydric soils in the United States.* Fort Worth, TX: USDA-NRCS.

hydrophobicity Literally, 'water-fearing' behav-iour of substances. It leads to non-mixing of hydrophobic liquids with water, and the non-spreading (or non-occurrence of INFILTRATION) of water when in contact with solid (or POROUS) surfaces. It occurs when the attraction of water MOLECULES to each other is stronger than the attraction of water molecules to another sub-stance. There remains some attraction of hydrophobic particles or powders to water so that these will nor-mally adhere to the surface of water droplets. In the natural environmental context, soil hydrophobicity (or WATER REPELLENCY) has many consequences. *SHD*

Doerr SH and Shakesby RA (2011) Soil water repellency. In Huang P, Levi G and Sumner ME (eds) *Handbook of soil science, 2nd edition.* Boca Raton, FL: CRC Press.

hydrophyte A waterplant, either submerged (e.g. *Chara, Najas* and *Potamogeton* spp.) or with float-ing leaves (e.g. *Potamogeton natans, Nymphaea* spp. and *Nuphar* spp.). According to water depth, different species may form belts or zones, the MACROFOSSILS of which can be used for the reconstruction of LAKE-LEVEL VARIATIONS. *BA*

[*See also* AQUATIC ENVIRONMENT, HYGROPHYTE, PALAEOLIMNOLOGY, ZONATION]

Tiner RW (1991) The concept of a hydrophyte for wetland
identification. *BioScience* 41: 236–247.

hydropower   Energy generated from flowing water.
The use of *hydromechanical power* has a long history: it
was used for IRRIGATION in ancient Egypt and the FERTILE
CRESCENT, while *water mills* and associated small-scale
damming of rivers date back to pre-Roman times. *Hydroelectric power*, which uses flowing water to generate
electricity using turbines, is an important modern form
of ALTERNATIVE ENERGY or RENEWABLE ENERGY. Hydro-
electricity was first produced in the 1880s but large-
scale DAMS for hydropower generation are a post-1930
phenomenon, which spread to DEVELOPING COUNTRIES
post-1960 with some unexpected impacts. Although it
provides only a minority of human needs at the global
scale (about 20 per cent of total electricity consumed),
it meets almost all the requirements of certain countries,
such as Norway and Switzerland, where most of the
available capacity has been developed. Future expansion of hydropower may depend on sources of *marine
energy* (e.g. TIDAL ENERGY and WAVE ENERGY).   *JAM*

McCully P (2001) *Silenced rivers: The ecology and politics
of large dams*. London: Zed Books.
Reaney SM (2010) Hydroelectric power. In Warf B (ed.)
Encyclopedia of geography. Thousand Oaks, CA: Sage,
1506–1512.
Reynolds TS (2003) *Stronger than a hundred men: A history
of the vertical water wheel*. Baltimore, MD: Johns
Hopkins University Press.

hydrosphere   The sum total of all the Earth's water,
including the OCEANS, liquid water on the continents
(*rivers*, LAKES and GROUNDWATER), SEA ICE and land ice
(ICE SHEETS, GLACIERS and SNOWBEDS) and *atmospheric
moisture*. The oceans comprise most of the hydrosphere (some 97 per cent by volume) and around 2 per
cent of the remainder is frozen as glacier ICE. Some
would exclude water in the ATMOSPHERE and CRYOSPHERE
in a narrower definition of the term.   *JAM*

[*See also* HYDROLOGICAL CYCLE]

Henshaw PC, Charlson RJ and Burges SJ (2000) Water and
the hydrosphere. In Jacobson MC, Charlson RJ, Rodhe
H and Orians GH (eds) *Earth system science: From
biogeochemical cycles to global change*. San Diego, CA:
Academic Press, 109–131.

hydrostatic pressure   The pressure exerted in a
fluid that depends on the depth or vertical head of the
fluid and its density. It is important in understanding
the flow of water from (1) ARTESIAN springs and wells,
and other aspects of GROUNDWATER movement; (2) soil
piping systems and (3) some ENGLACIAL and SUBGLA-
CIAL meltwater streams. In the latter two cases, it can
enhance the erosional and sediment transport capabili-
ties of the waters involved.   *RPDW/JAM*

Scheffers A (2008) Rock sculpturing by extreme
meltwater flow in western Greenland. *Zeitschrift für
Geomorphologie* 52: 145–167.

hydrothermal   Activity and processes associated
with the movement of GROUNDWATER heated by GEO-
THERMAL activity, usually due to proximity to MAGMA
in areas of active or dormant VOLCANISM. Hydrothermal
activity causes the chemical alteration of rocks, PRECIPI-
TATION of minerals (including ORE minerals) and surface
features such as *hot springs*, *geysers* and submarine
HYDROTHERMAL VENTS.   *JBH*

hydrothermal vent   An opening in the Earth's
CRUST that emits a jet of hot water containing dissolved
compounds. The term is most commonly used in relation to hydrothermal vents on the OCEAN floor that were
discovered in the late 1970s along the axial regions of
MID-OCEAN RIDGES. Hydrothermal vents include BLACK
SMOKERS in which the jet, at a temperature greater
than 350°C, is blackened by the precipitation of *metal
sulfides* that can build chimneys or columns up to several metres high, and *white smokers* at temperatures
between 100°C and 350°C that precipitate minerals
such as SILICA (SiO_2) and *barytes* ($BaSO_4$). The areas
of deep sea floor around many hydrothermal vents are
inhabited by communities of *chemosynthetic organisms* (see CHEMOSYNTHESIS) that obtain energy ultimately from a GEOTHERMAL source. Hydrothermal vent
processes have important implications for EVOLUTION,
the origin of life on Earth and ORE genesis.   *GO/JBH*

Campbell KA (2006) Hydrocarbon seep and hydrothermal
vent paleoenvironments and paleontology: Past
developments and future research directions.
Palaeogeography, Palaeoclimatology, Palaeoecology
232: 362–407.
Kormas KA, Tivey MK, Von Damm K and Teske A (2006)
Bacterial and archaeal phylotypes associated with distinct
mineralogical layers of a white smoker spire from a
deep-sea hydrothermal vent site (9 degrees N, East Pacific
Rise). *Environmental Microbiology* 8: 909–920.
Kump LR and Seyfried WE (2005) Hydrothermal Fe
fluxes during the Precambrian: Effect of low oceanic
sulfate concentrations and low hydrostatic pressure on
the composition of black smokers. *Earth and Planetary
Science Letters* 235: 654–662.
Parson LM, Walker CL and Dixon DR (eds) (1995)
Hydrothermal vents and processes. Bath: Geological
Society.
Prieur D, Erauso G and Jeanthon C (1995)
Hyperthermophilic life at deep-sea hydrothermal vents.
Planetary and Space Science 43: 115–122.
Russell MJ (1996) The generation at hot springs of
sedimentary ore deposits, microbialites and life.
Ore Geology Reviews 10: 199–214.
Svensen H, Planke S, Malthe-Sørensson A et al. (2004)
Release of methane from a volcanic basin as a mechanism
for initial Eocene global warming. *Nature* 429: 542–545.

hydrovolcanic eruption   An explosive VOLCANIC ERUPTION driven by the interaction between hot MAGMA and water, including GROUNDWATER or ice, which is converted explosively to steam. *Phreatic eruptions* produce steam and may eject fragmented country rock (VOLCANICLASTIC debris). *Phreatomagmatic eruptions* also eject magmatic products (PYROCLASTIC debris). GO

[*See also* MAAR]

Carmona J, Romero C, Dóniz J and García A (2011) Characterization and facies analysis of the hydrovolcanic deposits of Montana Pelada tuff ring: Tenerife, Canary Islands. *Journal of African Earth Sciences* 59: 41–50.

Francis P and Oppenheimer C (2003) *Volcanoes*. Oxford: Oxford University Press.

Keszthelyi LP, Jaeger WL, Dundas CM et al. (2010) Hydrovolcanic features on Mars: Preliminary observations from the first Mars year of HiRISE imaging. *Icarus* 205: 211–229.

hygrophyte   A WETLAND plant or a plant of humid HABITATS. Hygrophytes often grow at transitions between water (inhabited by HYDROPHYTES) and land plants on drier ground. Such wetland plants can survive ANOXIA of the roots: some form PEAT in *fens* or *bogs*. BA

[*See also* AQUATIC ENVIRONMENT, ECOTONE, MANGROVE, MIRES, TERRESTRIAL ENVIRONMENT]

hygroscopicity   The ability of small particles to attract moisture from the atmosphere caused by SORPTION and CAPILLARY ACTION. *Hygroscopic moisture* is determined by the weight loss of air-dried soil after oven drying at 105–110°C. JAM

hyperconcentrated flow   A flow of sediment and fluid that is intermediate in terms of sediment concentration between DEBRIS FLOW and STREAMFLOW with suspended sediment particles. Also known as *hyperconcentrated flood-flow* and *hyperconcentrated streamflow*, such flow carry very high sediment loads of up to 80 per cent by weight and 60 per cent by volume, can occur during the downstream evolution of debris flows and produce MASSIVE to weakly stratified deposits. GO

Beverage JP and Culbertson JK (1964) Hyperconcentrations of suspended sediment. *ASCE Journal of the Hydraulics Division* 90 (HY6): 117–126.

Russell HAJ and Arnott RWC (2003) Hydraulic-jump and hyperconcentrated-flow deposits of a glacigenic subaqueous fan: Oak Ridges Moraine, southern Ontario, Canada. *Journal of Sedimentary Research* 73: 887–905.

Sohn YK, Rhee CW and Kim BC (1999) Debris flow and hyperconcentrated flood-flow deposits in an alluvial fan, northwestern part of the Cretaceous Yongdong Basin, central Korea. *Journal of Geology* 107: 111–132.

hyperthermals   Sudden and extreme GLOBAL WARMING events superimposed on a gradual long-term warming trend in the GEOLOGICAL RECORD. Such *hothouse conditions*, which were more extreme than likely to be experienced under current GLOBAL WARMING scenarios, have been approached or achieved more than a dozen times in PHANEROZOIC history, each lasting for <1 million years. A series of hyperthermals occurred, for example, between 55.5 and 52 million years ago (Ma). The first and largest of these events (the *Palaeocene-Eocene Thermal Maximum*) was characterised by an increase in global temperature of about 5°C within a few thousand years, a massive input of carbon into the ATMOSPHERE and OCEAN ACIDIFICATION. This and several successively smaller hyperthermals may have been caused by the release of soil organic carbon from PERMAFROST areas (each successive hyperthermal being smaller because of the reducing area of permafrost in response to long-term warming). The hothouse condition of hyperthermals should be distinguished from the much longer intervals of GEOLOGICAL TIME when the Earth existed under a milder GREENHOUSE CONDITION. However, current GLOBAL WARMING is taking place more rapidly than the atmospheric warming associated with hyperthermals. JAM

DeConto RM, Galeotti S, Pagani M et al. (2012) Past extreme warming events linked to massive carbon release from thawing permafrost. *Nature* 484: 87–91.

Kidder DL and Worsley TR (2010) Phanerozoic large igneous provinces (LIPs), HEATT (Haline Euxinic Acidic Thermal Transgression) episodes, and mass extinctions. *Palaeogeography, Palaeoclimatology, Palaeoecology* 295: 162–191.

Kidder DL and Worsley TR (2012) A human-induced hothouse climate? *GSA Today* 22: 4–11.

Nicolo MJ, Dickens GR, Hollis CJ and Zachos J (2007) Multiple Early Eocene hyperthermals: Their sedimentary expression on the New Zealand continental margin and in the deep sea. *Geology* 35: 699–702.

hypothesis   A conjecture that possesses generality and is testable or has the potential to be tested. In SCIENCE, progress is made by testing hypotheses or attempting to refute hypotheses (FALSIFICATION). The more general the hypothesis and the more severe the test, the greater is the likelihood of scientific advance (i.e. the development of reliable THEORY and KNOWLEDGE). Hypotheses are an essential part of SCIENTIFIC METHOD, which may be conceived as a continuous cycle of comparing hypotheses (ideas) against OBSERVATIONS (data), often involving the progressive elimination of MULTIPLE WORKING HYPOTHESES. JAM

[*See also* CONFIRMATION, VERIFICATION]

Baker VR (1996) Hypotheses and geomorphological reasoning. In Rhoads BL and Thorn CE (eds) *The scientific nature of geomorphology*. Chichester: Wiley, 57–85.

Battarbee RW, Flower RJ, Stevenson J and Rippy B (1985) Lake acidification in Galloway: A palaeoecological test of competing hypotheses. *Nature* 314: 350–352.

Carpinter SR, Cole JT, Essington TE et al. (1998) Evaluating alternative explanations in ecosystem experiments. *Ecosystems* 1: 335–344.

Chamberlin TC (1965) The method of multiple working hypotheses. *Science* 148: 754–759.

Turner RE (1997) Wetland loss in the northern Gulf of Mexico: Multiple working hypotheses. *Estuaries* 20: 1–13.

hypsometry   The measurement of land elevation relative to *sea level*. The underwater equivalent is BATHYMETRY. A plot of the cumulative distribution of elevations in a defined area, such as a DRAINAGE BASIN or the whole Earth (see Figure), is termed a *hypsometric curve* and can be used as a basis for further geomorphological or geological comparison and analysis.

JPT

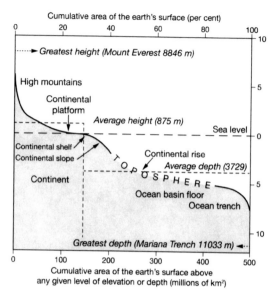

Hypsometry *A cumulative hypsometric curve of the Earth's land surface area above and below sea level: major areas are occupied by the continental platform and continental shelf (close to sea level) and the ocean basin floor, separated by the continental rise and continental slope. The average height of the land (875 m) and the average ocean depth (3,729 m) are also shown (Huggett and Cheesman, 2002).*

Brocklehurst SH and Whipple KX (2004) Hypsometry of glaciated landscapes. *Earth Surface Processes and Landforms* 29: 907–926.

Huggett R and Cheesman J (2002) *Topography and the environment*. Harlow: Pearson Education.

Singh O (2009) Hypsometry and erosion proneness: A case study in the lesser Himalayan Watersheds. *Journal of Soil and Water Conservation* 8: 53–59.

hysteresis   Partial or incomplete reversibility produced by LEAD AND LAG relationships between variables. Hysteresis describes a phenomenon occurring in many fields of science whereby a physical quality is dependant not only on another variable but also on the prior history of that relationship (e.g. at a given moisture content, it matters whether a soil is in a drying or wetting phase). *Hysteresis loops* are commonly displayed in graphs relating sediment to discharge through time.

CET/JAM

Smith HG and Dragovich D (2009) Interpreting sediment delivery processes using suspended sediment-discharge hysteresis patterns from nested upland catchments, south-eastern Australia. *Hydrological Processes* 23: 2415–2426.

Sternberg LDC (2001) Savanna-forest hysteresis in the tropics. *Global Ecology and Biogeography* 10: 369–378.

ice The solid phase of water (H_2O), and a *mono-mineralic rock* with an unusually low melting point. Several other important properties include (1) its *polycrystalline* structure with the ability to behave as a brittle solid yet also deform plastically under pressure; (2) its lower DENSITY than water, so that ice floats on water; (3) its volume change on formation from water, which results in expansion on freezing; (4) its low *thermal conductivity*, which makes ice and snow excellent *insulators*; (5) the bonding and shearing properties of snow, which are important in understanding the behaviour of SNOWPACK and SNOW AVALANCHES; and (6) the high REFLECTANCE of snow, hence its high ALBEDO and involvement in FEEDBACK MECHANISMS affecting CLIMATE. The presence of ice in various forms is the defining material characteristic of the CRYOSPHERE.   *JAM*

[*See also* FROST WEATHERING, GLACIER, GROUND ICE, ICEBERGS, SEA ICE, SNOW AND SNOW COVER]

Furukawa Y (2011). Ice. In Singh VP, Singh P and Haritashya UK (eds) *Encyclopedia of snow, ice and glaciers*. Dordrecht: Springer, 557–560.
Gosnell M (2005) *Ice: The nature, the history, and the uses of an astonishing substance*. New York: Knopf.
Marshall SJ (2012) *The cryosphere*. Princeton, NJ: Princeton University Press.

ice ages Phases in Earth's history when ICE SHEETS expanded to cover large areas of the globe in the form of large, continental glaciers. The term was first used by K. Schimper in 1837 and is commonly used to describe time intervals on two different scales. First and foremost, it describes long, generally cool, intervals of Earth history (tens to hundreds of millions of years) during which GLACIERS repeatedly expanded and contracted. Second, the term is also used to describe shorter time intervals (tens of thousands of years) during which glaciers were at times at, or near, their maximum extent. These shorter intervals are more appropriately known as GLACIATIONS, GLACIAL EPISODES or, more correctly, STADIALS. Ice ages in the first sense

have affected the Earth on numerous occasions during its history. These glaciations have resulted in significant lowering of temperatures and are not randomly distributed in time. There are records of Early and Late PROTEROZOIC (650–700 million years ago [Ma]), ORDOVICIAN (ca 450 Ma BP), *Carboniferous-Permian* (250–300 Ma BP) and CENOZOIC (*CAINOZOIC*) (past 15 Ma) ice ages.

The QUATERNARY has been considered to be synonymous with the 'Ice Age'. Sir Edward Forbes wrote in 1846 that the PLEISTOCENE equated to the *Glacial Epoch*. If ice age is used to refer to long, generally cool, intervals during which the CRYOSPHERE waxes and wanes, we are still in one today. Our modern climate represents a short, warm interval between glacial advances. An ice age comprises several glacials interspersed with INTERGLACIAL phases. The GLACIAL-INTERGLACIAL CYCLE appears to have a PERIODICITY of approximately 100,000 years. It is now accepted that variations in SOLAR RADIATION receipt according to MILANKOVITCH THEORY are responsible for global cooling as the Earth enters each glacial episode, and also for the fluctuations in ice extent during ice ages, but not for the existence of ice ages. CONTINENTAL DRIFT, PLATE TECTONICS, LAND UPLIFT and the reduction of CARBON DIOXIDE in the atmosphere may be cited as causal factors in the onset of ice ages. The evidence for ancient ice ages is clear in many localities and can be seen from striations and grooves in rocks and the presence of TILLITES. Basal temperatures and water conditions can be inferred from the types of TILL identified in QUATERNARY stratigraphic sequences, and SEA-LEVEL CHANGES from rhythmic MARINE SEDIMENTS.

Penck and Bruckner in their major work published in 1909 in which they defined four glaciations used the term *Eiszeitalter* to describe the glacial episodes, and the term 'ice age' remains in common usage to define a single period within which cold and/or cool conditions prevailed. Here the recommendation is to restrict the term 'ice age' to the first sense defined above (i.e. an ice age includes many glacial-interglacial cycles).

DH/CJC/JAM

[*See also* ICEHOUSE CONDITION, SNOWBALL EARTH]

Crowell JC (1999) *Pre-Mesozoic ice ages: Their bearing on understanding the climate system.* Boulder, CO: Geological Society of America [Memoir No. 192].

Imbrie J and Imbrie KP (1986) *Ice ages: Solving the mystery, 2nd edition.* Cambridge, MA: Harvard University Press.

John BS (ed.) (1979) *The winters of the world: Earth under the ice ages.* Newton Abbot: David and Charles.

Murdoch TQ, Weaver AJ and Fanning AF (1997) Palaeoclimatic response of the closing of the Isthmus of Panama in a coupled ocean-atmosphere model. *Geophysical Research Letters* 24: 253–256.

Penck A and Bruckner E (1909) *Die Alpen in Eiszeitalter* [The Alps in the Ice Age]. Leipzig: Tachnitz.

Wilson RCL, Drury SA and Chapman JL (1999) *The great Ice Age: Climate change and life.* London: Routledge.

ice cores   Cylindrical cores of ice drilled from ICE SHEETS and GLACIERS, which preserve records of annual snow and ice *accumulation*, providing a wealth of palaeoenvironmental data with up to annual and even subannual resolution. Dating of ice layers is straightforward in the upper parts of cores, where annual layers are clearly distinguishable from dirt or ICE-MELT LAYERS. In the deeper parts of cores, ice layers may be hard to detect, and dating must be achieved by indirect means, such as models of ice-column thinning by compaction and flow. Long cores have been obtained from Antarctica (e.g. Vostok, Dome C and Byrd), Greenland (e.g. the *Greenland Ice-core Project* [*GRIP*] and *Greenland Ice Sheet Project* [*GISP*] cores), which span the last GLACIAL-INTERGLACIAL CYCLE, including the LAST INTERGLACIAL (*Eemian*). Shorter cores have been drilled from Arctic *ice caps* such as the *Agassiz Ice Cap*, Ellesmere Island, and low-latitude, high-altitude ice masses such as the *Dunde Ice Cap*, Tibet, and the *Quelccaya Ice Cap*, Peru.

The EPICA (European Project for Ice Coring in *Antarctica*) ice core from Dome C in East Antarctica has allowed the extension of the ice-core record of atmospheric composition and climate over the past seven glacial cycles, indicating sequences of CLIMATIC CHANGE, in which the effects of ORBITAL FORCING are amplified by GREENHOUSE GASES and *ice-albedo feedbacks* leading to full INTERGLACIAL conditions.

Many types of information can be obtained from ice cores. First, the ice layers are composed of a range of ISOTOPES of oxygen and hydrogen, the relative abundances of which reflect global reservoirs and local air temperatures at the time of their deposition as snow. Variations in the isotopic composition of ice within a core thus provide detailed records of climatic change, at a much higher resolution than almost all other types of evidence. The ice-core oxygen isotope record has revealed patterns of climatic change during the last glacial cycle, including the quasi-periodic DANSGAARD-OESCHGER (D-O) EVENTS. The high resolution of the record has revealed extremely rapid rates of climatic warming at the onset of the HOLOCENE, of the order of 5°C within a few decades.

Second, glacier ice contains air bubbles, sealed off from the atmosphere during compaction of snow layers, providing samples of the atmosphere at the time of sealing (within a few decades for high latitude cores, but possibly over 4,000 years during glacial maxima in Antarctica). Bubbles in Antarctic and Greenland ice cores have provided long records of trace greenhouse gases such as CARBON DIOXIDE and *methane*, giving a unique record of changing atmospheric composition over the last glacial-interglacial cycle. This evidence has shown that greenhouse gas concentrations and the oxygen isotope record are closely correlated, with high carbon dioxide and methane levels occurring at times of high atmospheric temperatures (see Figure). This correlation has suggested that greenhouse gases may participate in natural FEEDBACK MECHANISMS serving to amplify climatic cycles, and has prompted the search for causal mechanisms.

A third type of evidence contained in ice cores is the concentrations of wind-blown PARTICULATES, such as DUST, SODIUM CHLORIDE and TEPHRA, contained within the ice. Such impurities can reveal changing patterns of atmospheric circulation and provide evidence of PALAEOWIND erosion, STORMS and VOLCANIC ERUPTIONS. In the upper parts of cores, industrial POLLUTANTS and RADIONUCLIDES can be used to reconstruct POLLUTION HISTORY and dispersal and fallout patterns. Additionally, ACIDITY levels can be detected using ELECTRICAL CONDUCTIVITY MEASUREMENTS. High concentrations of acids such as HNO_3 and H_2SO_4 in ice cores are generally indicative of volcanic AEROSOLS, and therefore provide a record of past eruptions.

The presence of many forms of evidence within a single, high-resolution record means that ice cores are among the most valuable of all palaeoclimatic records, allowing many aspects of environmental changes to be interrelated. Furthermore, the relative ease of ICE-CORE DATING, at least for the upper parts of cores, places these environmental changes within an absolute timescale, independent of RADIOMETRIC DATING methods. As a result of these advantages, many now regard ice cores as standard yardsticks of climatic change, with which other forms of evidence—such as MARINE SEDIMENT CORES and terrestrial POLLEN ANALYSES—can be correlated, compared and calibrated.   *DIB*

[*See also* HIGH-RESOLUTION RECONSTRUCTIONS]

Alley RB (2000) *The two-mile time machine: Ice cores, abrupt climate change and our future.* Princeton, NJ: Princeton University Press.

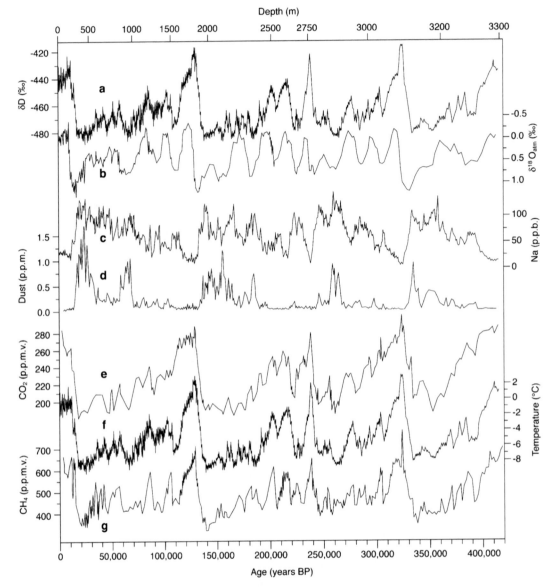

Ice cores Records of environmental change reconstructed from the Vostoc ice core, Antarctica. The ice core is over 3 km long and covers the past 4,400,000 years. The past four glacial-interglacial cycles can be clearly seen in these data relating to (a) deuterium, (b) oxygen-18, (c) sodium, (d) dust, (e) carbon dioxide, (f) isotopic temperature and (g) methane (Petit et al., 1999).

Alley RB (2011) Reliability of ice-core science: Historical insights. *Journal of Glaciology* 56: 1095–1103.

Dansgaard W, Johnsen SJ, Clausen HB et al. (1993) Evidence for general instability of past climate from a 250-kyr ice-core record. *Nature* 364: 218–220.

Dansgaard W and Oeschger H (1989) Past environmental long-term records from the Arctic. In Oeschger H and Langway CC (eds) *The environmental record in glaciers and ice sheets*. Chichester: Wiley, 287–317.

EPICA community members (2004) Eight glacial cycles from an Antarctic ice core. *Nature* 429: 623–628.

Johnsen SJ, Dahl-Jensen D, Gundestrup N et al. (2001) Oxygen isotope and palaeotemperature records from six Greenland ice-core stations: Camp Century, Dye-3, GRIP, GISP2, Renland and NorthGRIP. *Journal of Quaternary Science* 16: 299–307.

Mayewski PA and White F (2002) *The ice chronicles: The quest to understand global climatic change*. Lebanon, NH: University Press of New England.

Orombelli G, Maggi V and Delmonte B (2010) Quaternary stratigraphy and ice cores. *Quaternary International* 219: 55–65.

Petit JR, Jozel J, Raynaud D et al. (1999) Climate and atmospheric history of the past 420,000 years from the Vostok ice core, Antarctica. *Nature* 399: 429–436.

Raynaud D, Barnola JM, Chappellaz J et al. (2000) The ice record of greenhouse gases: A view in the context of future changes. *Quaternary Science Reviews* 19: 9–17.

Thompson LG (2000) Ice core evidence for climate change in the Tropics: Implications for our future. *Quaternary Science Reviews* 19: 19–35.

ice dome A symmetrical dome-shaped mass of ice, forming parts of both ICE SHEETS (>50,000 km² and up to several kilometres thick) and *ice caps* (<50,000 km²). The term is also used to describe elevated areas within the *East Antarctic Ice Sheet* (e.g. *Dome C*, also known as *Dome Charlie* and *Dome Circe*). Such ice domes are favoured for ICE CORE drilling because *ice-flow* disturbance of the *ice stratigraphy* is minimised at the *ice shed*.   *MJH/JAM*

[*See also* GLACIER, OUTLET GLACIER]

Aciego S, Bourdon B, Schwander J et al. (2011) Towards a radiometric ice clock: Uranium ages of the Dome C ice core. *Quaternary Science Reviews* 30: 2389–2397.

Vallelonga P, Gabrielli P, Balliani E et al. (2010) Lead isotopic composition in the EPICA Dome C ice core. *Quaternary Science Reviews* 29: 247–255.

Zirizzotti A, Cafarelli L and Urbini S (2012) Ice and bedrock characteristics underneath Dome C (Antarctica) from radio echo sounding data analysis. *IEEE Transactions on Geoscience and Remote Sensing* 50: 37–43.

ice foot A seasonal feature of cold-climate SHORE-LINES that forms during winter as wave spray and swash freezes. Its thickness and width depend on *tidal range* and *storm wave activity*. It protects the SHORELINE from wave action, has hydrological and ecological implications for the *intertidal zone* and the *cliff base* and may be a locus for POLAR SHORE EROSION.   *JAM/HJW*

[*See also* FROST WEATHERING, STRANDFLAT]

Allard M, Michaud Y, Ruz MH and Hequet A (1998) Ice foot, freeze-thaw of sediments, and platform erosion in a subarctic microtidal environment, Manitounuk Strait, northern Quebec, Canada. *Canadian Journal of Earth Science* 35: 965–979.

Scrosali R and Eckersberg LK (2007) Thermal insulation of the intertidal zone by the ice foot. *Journal of Sea Research* 58: 331–334.

ice scars Evidence of physical damage seen on the trunks of trees growing on the margins of lakes or rivers, caused by ice killing part of the CAMBIUM of the tree. These can often be dated to the precise year by dendrochronological study of the year or years of cambial injury.   *KRB/APP*

[*See also* DENDROCHRONOLOGY, FIRE SCAR, ICE-FLOOD HISTORY, TREE RING]

Mickael L and Yves B (2008) Hydroclimatic analysis of an ice-scar tree-ring chronology of a high-boreal lake in Northern Quebec, Canada. *Hydrology Research* 29: 451–464.

Tardif JC, Kames S and Bergeron Y (2010) Spring water levels reconstructed from ice-scarred trees and cross-sectional area of the earlywood vessels in tree rings from eastern boreal Canada. In Stoffel M, Bollschweiler M, Butler DR and Luckman BH (eds) *Tree rings and natural hazards: A state-of-the-art.* Dordrecht: Springer, 257–262.

ice sheet An extensive dome-shaped mass of ice and snow up to several kilometres thick and occupying an area of at least 50,000 km². The two contemporary ice sheets are the composite one in the ANTARCTIC and that in Greenland, but during the PLEISTOCENE several other ice sheets existed, the largest of which were in the Northern Hemisphere (the LAURENTIAN ICE SHEET and the FENNOSCANDIAN ICE SHEET). Very little was known about the world's extant ice sheets until the 1957–1958 International Geophysical Year, when large-scale in-situ scientific investigations began.

Recent improvements in scientific understanding have depended upon TECHNOLOGICAL CHANGE, which have improved methods of measuring present and past ice sheets and ice-sheet change. These methods include those carried out from the surface of today's ice sheets (e.g. ICE-CORE drilling and *over-snow studies*), SATELLITE REMOTE SENSING, AIRBORNE REMOTE SENSING and *shipborne studies* (both REMOTE SENSING using TOWED VEHICLES and the drilling of MARINE SEDIMENT CORES). Deciphering past natural transitions during GLACIAL-INTERGLACIAL CYCLES, including the most recent DEGLACIATION, which began prior to the beginning of the HOLOCENE, may help clarify likely future changes by throwing light on the FORCING FACTORS and *forcing mechanisms*. Although rapid changes, especially in relation to the MARINE-BASED ICE SHEETS, are being recorded now, it remains to be seen whether the short-term behaviour (e.g. ANNUAL VARIABILITY and DECADAL-SCALE VARIABILITY) are precursors of major ice-sheet loss (and SEA-LEVEL RISE) over the medium-term future.   *JAM/MJH*

[*See also* ANTARCTIC ENVIRONMENTAL CHANGE, GLACIATION, GLACIER, GLACIER THERMAL REGIME, ICE DOME, ICE SHELF, ICE-SHEET GROWTH]

Anderson JB, Shipp SS, Lowe AL et al. (2002) The Antarctic ice sheet during the Last Glacial Maximum and its subsequent retreat history: A review. *Quaternary Science Reviews* 21: 49–70.

Bamber JL and Payne AJ (2004) Mass balance and the cryosphere. Cambridge: Cambridge University Press.

Benn DI and Evans DJA (2010) *Glaciers and glaciation, 2nd edition.* London: Hodder Education.

Cuffey KM and Paterson WSB (2010) *The physics of glaciers, 4th edition.* Burlington, MA: Butterworth-Heinemann.

Graham AGC (2011) Ice sheet. In Singh VJ, Singh P and Haritashya UK (eds) *Encyclopedia of snow, ice and glaciers.* Dordrecht: Springer, 592–608.

Siegert MJ, Barrett P, DeConto R et al. (2008) Recent advances in understanding Antarctic climate evolution. *Antarctic Science* 4: 313–325.

Vaughan DG (2008) West Antarctic ice sheet collapse: The fall and rise of a paradigm. *Climatic Change* 91: 65–79.

ice shelf   A large slab of *floating ice* (floating on the sea), up to several hundred metres thick, but remaining attached to, and partly fed by, *land-based ice*. Ice shelves commonly restrain *inland ice* (grounded on land) from discharging rapidly into the ocean. Many ice shelves, notably those in the *Antarctic Peninsula* region and the Canadian ARCTIC in recent decades have undergone collapse in response to REGIONAL CLIMATIC CHANGE (warming) and oceanographic changes. Removal of the buttressing effect of ice shelves has allowed interior ice to discharge more rapidly into the ocean. *MJH*

[*See also* ANTARCTIC ENVIRONMENTAL CHANGE, GLOBAL WARMING, GROUNDING LINE, MARINE-BASED ICE SHEET]

Pritchard HD, Ligtenberg SRM, Fricker HA et al. (2012) Antarctic ice-sheet loss driven by basal melting of ice shelves. *Nature* 484: 502–505.

Scambos TA, Bohlander JA, Shuman CA and Skvarca P (2004) Glacier acceleration and thinning after ice shelf collapse in the Larsen B embayment, Antarctica *Geophysical Research Letters* 31: L18402.

Vincent WF, Gibson JAE and Jeffries MO (2001) Ice-shelf collapse, climate change, and habitat loss in the Canadian high Arctic. *Polar Record* 37: 133–142.

ice stream   Part of an ICE SHEET, typically >20 km in width and >150 km in length, where ice flows more rapidly than, though not necessarily in the same direction as, the surrounding ice. Ice streams discharge much of the ice and SEDIMENT within an ICE SHEET and their location is normally at least partly topographically controlled. Enhanced flow velocity (typically between 100 and 800 m/year) can be caused by changes in the substrate, *bed roughness*, ice RHEOLOGY, changes in *shear stress* and changes in the *hydrological conditions* at the bed. During the LAST GLACIAL MAXIMUM (LGM), *palaeo-ice streams* extended via bathymetric troughs to the edge of the CONTINENTAL SHELF adjacent to the *Antarctic Peninsula* and the *West Antarctic Ice Sheet*, and typically to the mid-shelf position adjacent to the *East Antarctic Ice Sheet*. Considerable *asynchrony* is indicated in the timing of GROUNDING-LINE retreat since the LGM, which can be attributed to internal factors such as bed characteristics and DRAINAGE BASIN size.   *BTIR*

[*See also* ANTARCTIC, ICE-SHEET DYNAMICS, MARINE-BASED ICE SHEET, SUBGLACIAL BASIN]

Alley RB, Anandakrishnan S, Dupont TK and Parizek BR (2004) Ice streams: Fast, and faster? *Comptes Rendus Physique* 5: 723–734.

Bennett M (2003) Ice streams as the arteries of an ice sheet: Their mechanics, stability and significance. *Earth-Science Reviews* 61: 309–339.

Livingstone SJ, Ó'Cofaigh C, Stokes CP et al. (2012) Antarctic palaeo-ice streams. *Earth-Science Reviews* 111: 90–128.

Stokes CR and Clark CD (2001) Paleo-ice streams. *Quaternary Science Reviews* 20: 1437–1457.

Tulaczyk S (2006) Fast glacier flow and ice streaming. In Knight P (ed.) *Glacier science and environmental change.* Oxford: Blackwell, 353–359.

ice wedge   A wedge-shaped body of ice present in near-surface PERMAFROST caused by THERMAL CONTRACTION CRACKING of the ground in winter, and the subsequent infilling of these FROST CRACKS in early summer by water which then freezes. If this process is repeated for a number of years, a wedge-shaped body of foliated ice develops. The size of ice wedges depends largely upon the availability of water and the age of the ice wedge. In parts of central Siberia, and the western North American Arctic, wedges in excess of 1–3 m in width and 5–10 m in depth can be observed. Most wedges are EPIGENETIC but some large ones may be either SYNGENETIC or *antisyngenetic* (form in surfaces that are lowering). Where AEOLIAN transport is locally dominant, or in extremely arid regions such as Antarctica, the thermal-contraction cracks are filled largely with mineral soil particles rather than with ice, forming *sand wedges*. In plan, ice wedges join predominantly at right angles to form polygonal, chiefly tetragonal, nets of PATTERNED GROUND that cover large areas of the ARCTIC and SUBARCTIC. The average dimensions of the *ice-wedge polygons* are 15–40 m compared with 10–30 m for less common sand wedges.

When permafrost thaws, ice wedges melt and the void is filled with mineral soil and/or organic material which slumps down from the sides and from above. Such structures are termed ice-wedge CASTS or PSEUDOMORPHS. Thermal-contraction cracking has climatic and palaeoclimatic significance. In general, it requires a mean annual air temperature of −6°C or colder. Provided they are correctly identified, pseudomorphs provide incontrovertible evidence of the former existence of permafrost.   *HMF*

Ghysels G and Heyse I (2006) Composite-wedge pseudomorphs in Flanders, Belgium. *Permafrost and Periglacial Processes* 17: 145–162.

Mackay JR (2000) Thermally-induced movements in ice-wedge polygons, western Arctic Coast: A long term study. *Géographie Physique et Quaternaire* 54: 41–68.

Murton JB (2007) Ice wedges and ice wedge casts. In Elias SA (ed.) *Encyclopedia of Quaternary science.* Amsterdam: Elsevier, 2153–2170.

Murton JB and Bateman M (2007) Syngenetic sand veins and anti-syngenetic sand wedges, Tuktoyaktuk

Coastlands, western Arctic Canada. *Permafrost and Periglacial Processes* 18: 33–48.

Wertowski M (2009) Ice-wedge pseudomorphs and frost cracking structures in Weichselian sediments, Central-West Poland. *Permafrost and Periglacial Processes* 20: 316–330.

Iceberg Drift Theory

A 'hybrid' between the DILUVIAL THEORY and the GLACIAL THEORY, which merged aspects of the catastrophic origin of DILUVIUM with the existence of a relatively cold climate. The theory, championed by Charles Lyell and others for a time in the mid- to late nineteenth century, attributed DRIFT to the deposition of debris transported by ICEBERGS during marine submergence. It delayed acceptance of the Glacial Theory, partly because of greater compatibility with religious beliefs. It also appeared to be supported by geological evidence of SEA-LEVEL CHANGE and reports of icebergs in high-latitude oceans from whalers, explorers and Charles Darwin's voyage in the Beagle.   *JAM*

Walsh SL (2008) The Neogene: Origin, adoption, evolution, and controversy. *Earth-Science Reviews* 89: 42–72.

iceberg plough marks

Elongated furrows formed by the *keels* of grounded ICEBERGS dragged across the soft SEDIMENTS of a lake- or sea-floor by TIDES and/or *wind action*. Such marks are up to 20 m deep and 250 m wide, and may extend for several kilometres. They are also sometimes termed *iceberg scour marks* but are examples of TOOL MARKS. Their form depends on the nature of the sediment, the shape of the iceberg keel(s) and the nature of the motion of the iceberg. In soft sediments, plough marks may be regular and continuous, curved or straight, flat-bottomed troughs or furrows. The microtopography of the plough marks tends to vary with the characteristics of movement of the iceberg. Linear *scour berms* (i.e. ridges) of displaced blocky material at the sides of the plough marks can attain heights of 6 m. The term *iceberg grounding structures* is used to refer to the *deformed sediments* associated with the plough marks as well as the marks themselves. Positive identification of RELICT or FOSSIL iceberg plough marks may provide evidence of PALAEO-CURRENT or PALAEOWIND directions and, where applicable, the GLACIOMARINE HYPOTHESIS.   *RAS*

[*See also* SOFT-SEDIMENT DEFORMATION]

Bennett MR and Bullard JE (1991) Iceberg tool marks: An example from Heinabersjökull, South East Iceland. *Journal of Glaciology* 37: 181–183.

Delage M and Gangloff P (1993) Relict iceberg marks near Montreal, Quebec. *Géographie Physique et Quaternaire* 47: 69–80.

Dowdeswell JA, Villinger H, Whittington RJ and Marienfeld P (1993) Iceberg scouring in Scoresby Sund, east Greenland. *Sedimentology* 41: 21–35.

Gebhardt AC, Jokat W, Niessen F et al. (2011) Ice sheet grounding and iceberg plow marks on the northern and central Yermak Plateau revealed by geophysical data. *Quaternary Science Reviews* 30: 1726–1738.

Thomas GSP and Connell RJ (1985) Iceberg drop, dump and grounding structures from Pleistocene glaciolacustrine sediments, Scotland. *Journal of Sedimentary Petrology* 55: 243–249.

icebergs

Large floating masses of ice that have broken off from the seaward front of a GLACIER or ICE SHELF during the CALVING process. They vary in shape and size and the larger examples can be a hazard to shipping. In the Southern Ocean, there are an estimated 300,000 icebergs greater than 10 m in width. The production of huge icebergs from the recent disintegration of ice shelves (e.g. *Larsen B ice shelf* in 2002) has been suggested as a MODERN ANALOGUE for oceanic HEINRICH EVENTS during the QUATERNARY. Icebergs have also been suggested as a source of freshwater, if they could be dragged to lower latitudes before melting.   *AHP/JAM*

[*See also* ICEBERG DRIFT THEORY, ICE-RAFTED DEBRIS (IRD), SEA ICE]

Broecker W (1994) Massive iceberg discharges as triggers for global climate change. *Nature* 372: 421–424.

Hulbe CL, MacAyeal DR, Denton GH et al. (2004) Catastrophic ice shelf breakup as the source of Heinrich event icebergs. *Paleoceanography* 19: PA1004.

ice-contact fan/ramp

A latero-frontal depositional LANDFORM distal to the margin of a GLACIER or ICE SHEET formed by GLACIOFLUVIAL and/or MASS MOVEMENT PROCESSES. The outer slope has a relatively shallow gradient, whereas the proximal slope, an ICE-CONTACT SLOPE, is relatively steep. They commonly include *reworked sediments* from MORAINES, may act as a barrier to glacier advance and may contribute to moraines if overridden. Consequently, they may contain complex sedimentary sequences and exhibit GLACIOTECTONIC structures. Where deposition is predominantly glaciofluvial, the outer slope may grade imperceptibly into OUTWASH DEPOSITS. Krzyszkowski and Zielinski (2002) recognised three types: (1) Type A fans contain exclusively DIAMICTONS, (2) Type B fans contain interfingering of diamictons and stratified sands and gravels and (3) Type C fans contain exclusively stratified sands and gravels.   *JAM*

[*See also* ICE-MARGIN INDICATORS]

Krzyszkowski D (2002) Sedimentary successions in ice-marginal fans of the Late Saalian glaciation, southwestern Poland. *Sedimentary Geology* 149: 93–109.

Krzyszkowski D and Zielinski T (2002) The Pleistocene end moraine fans: Controls on their sedimentation and location. *Sedimentary Geology* 149: 73–92.

Pisarska-Jamrozy M (2006) Transitional deposits between the end moraine and outwash plain in the Pomeranian glaciomarginal zone of NW Poland: A missing component of ice-contact sedimentary models. *Boreas* 35: 126–141.

ice-contact slope Retreat of a glacier or the melting of DEAD ICE commonly involves collapse of sediment shored up by the ice to create steep slopes often characterised by KETTLE HOLES and with extensional faults in associated sediments. Extensive ICE-CONTACT FANS/RAMPS may lie distal to the ice-contact slope. Where *ice-marginal deposition* is dominated by GLACI-OFLUVIAL processes, ice-contact slopes may be the main evidence for former *glacial limits*.   *MRB*

[*See also* GLACIER VARIATIONS, KAME TERRACE, ICE-MARGIN INDICATORS, MORAINES]

Zielinski T and van Loon AJ (2000) Subaerial terminoglacial fans. III: Overview of sedimentary characteristics and depositional model. *Geologie en Mijnbouw* 79: 93–107.

ice-core dating Counting of annual layers, which can be discerned due to seasonal variations in the properties of snow, is the most accurate method for establishing CHRONOLOGY in ICE CORES. Because of uncertainties due to thinning of the layers, the dating of deeper ice cores (beyond the LAST INTERGLACIAL) is mainly based on *ice-flow modelling* and SYNCHRONISATION with MARINE ISOTOPIC STAGES using, for example, the *isotopic composition* of gas bubbles. Ages are now defined according to the GREENLAND ICE-CORE CHRONOLOGY (GICC05) and expressed as years b2k, the ice-core zero age of AD 2000.   *JAM/CJC/SMD*

Rasmussen SO, Andersen KK, Svensson AM et al. (2006) A new Greenland ice core chronology for the Last Glacial termination. *Journal of Geophysical Research* 111: D06102.
Schwander J (2007) Ice-core methods: Chronologies. In Elias SA (ed.) *Encyclopedia of Quaternary science*. Amsterdam: Elsevier, 1173–1181.

ice-cored moraine Although the term has been used in a variety of contexts, it is best retained as a distinctive type of ice-marginal MORAINE characteristically formed by COLD-BASED GLACIERS in areas of PERMAFROST. An ice-cored moraine contains a discrete body of (glacier) ice buried beneath TILL. However, the origin of the buried ice is problematic: theories vary from overridden proglacial snowbanks to an isolated part of glacier terminus protected by supraglacial sediment of various possible origins. Ice-cored moraines are common on modern GLACIER FORELANDS and may survive for thousands of years before the ice core melts out. They may be regarded on a continuum of landforms between DEBRIS-MANTLED GLACIERS and ROCK GLACIERS.   *JAM*

Lukas S (2011) Ice-cored moraines. In Singh VP, Singh P and Haritashya UK (eds) *Encyclopedia of snow, ice and glaciers*. Dordrecht: Springer, 616–619.

Østrem G (1964) Ice-cored moraines in Scandinavia. *Geografiska Annaler* 46A: 282–337.
Shakesby RA, Matthews JA and Winkler S (2004) Glacier variations in Breheimen, southern Norway: Relative-age dating of Holocene moraine complexes at six high-altitude glaciers. *The Holocene* 14: 899–910.

ice-dammed lake A lake in glaciated terrain created where a GLACIER or ICE SHEET has moved against the regional slope or across a valley entrance, forming a barrier to drainage. Ice-dammed lakes are noted for their sudden, often catastrophic drainage via JÖKULHLAUPS. Evidence for former ice-dammed lakes includes SHORELINES, lake-floor RHYTHMITES or VARVES, DELTAS and OVERFLOW CHANNELS.   *MRB*

[*See also* CHANNELED SCABLANDS, GLACIAL LAKE, GLACIOLACUSTRINE DEPOSITS, MEGALAKE]

Carrivick JL (2011) Jökulhlaups: Geological importance, deglacial association and hazard management. *Geology Today* 27: 133–140.
Margold M, Jansson KN, Stroeven AP and Jansen JD (2011) Glacial Lake Vitim, a 3000-km³ outburst flood from Siberia to the Arctic Ocean, *Quaternary Research* 76: 393–396.
Shakesby RA (1985) Geomorphological effects of jökulhlaups and ice-dammed lakes, southern Norway. *Norsk Geografisk Tidsskrift* 39: 1–16.
Sissons JB (1977) Former ice-dammed lakes in Glen Moriston, Inverness-shire, and their significance in upland Britain. *Transactions of the Institute of British Geographers NS* 2: 224–242.

ice-directional indicators Forms of GLACIAL EROSION and GLACIAL DEPOSITION which show former directions of ice flow, including STRIATIONS, FLUTES and DRUMLINS. Such features occur at widely differing spatial scales and have varying regional significance. Ice-directional indicators are commonly TIME TRANSGRESSIVE, relating to different periods of ice build-up and decay. At the very largest scales, several generations of ice-directional indicators may be distinguished by SATELLITE REMOTE SENSING, and can be used to establish aspects of ICE-SHEET evolution.   *DIB*

Benn DI and Evans DJA (2010) *Glaciers and glaciation, 2nd edition*. London: Hodder Education.
Greenwood SL and Clark CD (2009) Reconstructing the last Irish Ice Sheet 1: Changing flow geometries and ice flow dynamics deciphered from the glacial landform record. *Quaternary Science Reviews* 28: 3085–3100.
Kleman A, Hättestrand H, Borgström I and Stroeven A (1997) Fennoscandian palaeoglaciology reconstructed using a glacial geological inversion model. *Journal of Glaciology* 43: 283–299.

ice-flood history The banks of lakes and rivers at high latitudes are commonly flooded annually in the spring when damage is caused by floating ice and debris.

Analysis of ice-scarred trees using DENDROCHRONOLOGY enables the reconstruction of ice-flood history and, in certain regions of Arctic Canada, has demonstrated an increase in magnitude and frequency of ice floods since the end of the LITTLE ICE AGE and, especially during the twentieth century.   *JAM*

[*See also* GLACIEL, ICING, PERIGLACIOFLUVIAL SYSTEM]

Tardif J and Bergeron Y (1997) Ice-flood history reconstructed with tree-rings from the southern boreal forest limit, western Québec. *The Holocene* 7: 291–300.

icehouse condition   A term used to describe long intervals of global cooling in the Earth's past, characterised by relatively low *sea level*, a scarcity of sedimentary FACIES characteristic of hot palaeoclimates (see PALAEOCLIMATOLOGY), and evidence of ICE-AGE conditions for some of the time. At least five oscillations between icehouse and warmer GREENHOUSE CONDITIONS can be recognised in the GEOLOGICAL RECORD of the Late PROTEROZOIC and PHANEROZOIC. Icehouse conditions characterised the Late Proterozoic to Late CAMBRIAN, Late DEVONIAN to Mid-TRIASSIC and Early CENOZOIC (*Cainozoic*) to the present day.   *GO*

[*See also* GEOLOGICAL RECORD OF ENVIRONMENTAL CHANGE]

Fairchild IJ and Kennedy MJ (2007) Neoproterozoic glaciation in the Earth system. *Journal of the Geological Society of London* 164: 895–921.
Price GD, Valdes PJ and Sellwood BW (1998) A comparison of GCM simulated Cretaceous 'greenhouse' and 'icehouse' climates: Implications for the sedimentary record. *Palaeogeography, Palaeoclimatology, Palaeoecology* 142: 123–138.
Saltzman MR and Young SA (2005) Long-lived glaciation in the Late Ordovician? Isotopic and sequence-stratigraphic evidence from western Laurentia. *Geology* 33: 109–112.

Iceman   The Alpine Iceman is the well-preserved body of a prehistoric man that was discovered in 1991 melting out of glacier ice on the Tisa Pass (3,280 m) close to the border between Austria and Italy. RADIOCARBON DATING revealed a NEOLITHIC age (4,500 radiocarbon years BP, about 5,200 years ago), which was older than expected given the accompanying clothing and equipment, including a copper axe, bow, arrows and backpack. Nicknamed 'Ötzi', a conflation of 'Ötztal' and 'yeti', the find is of considerable environmental as well as archaeological importance. The degree of preservation means that the body must have been buried very rapidly and remained buried in an environment conducive to its preservation until discovery: this is consistent with a rapid CLIMATIC CHANGE leading to NEOGLACIATION and with the glacier remaining no smaller than today throughout the period of burial. Soils at the periphery of the site have yielded dates as old as 5,600 radiocarbon years and are indicative of the

more favourable climate prior to burial of the Iceman. The lack of distortion of the corpse is explained by an absence of glacier flow or creep, which in turn indicates a thin ice mass frozen to the ground (COLD-BASED GLACIER) in a PERMAFROST environment.   *JAM*

[*See also* COPPER AGE, GLACIER VARIATIONS, HOLOCENE CLIMATIC OPTIMUM, HOLOCENE ENVIRONMENTAL CHANGE, LATE-HOLOCENE CLIMATIC DETERIORATION, MUMMY]

Baroni C and Orombelli G (1996) The Alpine 'Iceman' and Holocene climatic change. *Quaternary Research* 46: 78–83.
Bortenschlager S and Oeggl K (eds) (2000) *The iceman in his natural environment: Palaeobotanical results.* New York: Springer.
Rom W, Golser R, Kutschera W et al. (1999) AMS ^{14}C dating of equipment from the iceman and of spruce logs from the prehistoric salt mines of Hallstatt. *Radiocarbon* 41: 183–197.
Spindler K (1994) *The man in the ice.* New York: Harmony Books.

ice-margin indicators   Geomorphological evidence that indicates the former position of the margin of a GLACIER or ICE SHEET including, for example, many types of MORAINES, KAME TERRACES, ICE-CONTACT SLOPES and associated ICE-CONTACT FANS/RAMPS, SANDUR and TRIMLINES. Ice-margin indicators are used in the reconstruction of the maximum extent of former glaciers and also subsequent recessional stages.   *JAM*

[*See also* LAST GLACIAL MAXIMUM (LGM)]

Evans DJA (ed.) (2003) *Glacial landsystems.* London: Hodder-Arnold.
Lowe JJ and Walker MJC (1997) *Reconstructing Quaternary environments, 2nd edition.* Harlow: Addison Wesley Longman.
Pisarska-Jamrozy M (2006) Transitional deposits between the end moraine and outwash plain in the Pomeranian glaciomarginal zone of NW Poland: A missing component of ice-contact sedimentary models. *Boreas* 35: 126–141.

ice-melt layer   Layer of clear, bubble-free glacier ice, marking the refreezing of water in the near-surface snowpack. Ice-melt layers or *melt layers* are also referred to as SUPERIMPOSED ICE, and are considered a form of glacier *accumulation.* Ice-melt layers can allow annual increments of accumulation to be distinguished in ICE CORES, although it is possible for multiple layers to form in a single year. They are also used in PALAEOCLIMATOLOGY as indicators of summer temperature.   *DIB*

Fisher DA, Koerner RM and Reeh N (1995) Holocene climatic records from Agassiz Ice Cap, Ellesmere Island, NWT, Canada. *The Holocene* 5: 119–124.
Wadham J, Kohler J, Hubbard A et al. (2006) Superimposed ice regime of a high Arctic glacier inferred using ground-penetrating radar, flow modeling, and ice cores, *Journal of Geophysical Research* 111: F01007.

ice-rafted debris (IRD)   Particles or CLASTS in GLACIOLACUSTRINE DEPOSITS or GLACIOMARINE DEPOSITS, either from ICEBERGS or, passively or actively, by lake ice or SEA ICE. The size of the ice-rafted debris will depend on the agent by which it has been deposited. In LACUSTRINE SEDIMENTS, ice-rafted debris is normally relatively coarse grained, angular COLLUVIUM transported onto lake ice by DEBRIS FLOWS or SNOW AVALANCHES. In the glaciomarine environment, icebergs enable the transportation of particles from CLAY (<0.002 mm) to BOULDER size; particles transported by active sea ice do not normally include boulders, whereas passive sea ice rarely includes clay or SILT. SEDIMENT may be deposited as a single particle (DROPSTONE), agglomerations of more than one particle (*dump*), frozen aggregates or sediment-laden ice.

Deep-ocean sediments are rich in TERRIGENOUS sediment with estimates of 40 per cent for the amount of sediment deposited in the oceans during the QUATERNARY consisting of IRD (see Figure). High-resolution MARINE SEDIMENT CORES show numerous episodes of North Atlantic ice-rafted debris deposition within the last GLACIAL EPISODE. These are seen as layers of coarse-grained LITHIC particles and clasts within the marine sediment which were transported from Quaternary ICE SHEETS to the North Atlantic by CALVING mechanisms. These clasts therefore allow studies of PALAEOCEANOGRAPHY and PALAEOCLIMATOLOGY, and source region identification of IRD on the basis of mineral composition. IRD in palaeoceanographic studies is often defined as lithic material >125 μm.

Marine records of the last glaciation exhibit two notable PERIODICITIES of increased IRD content, at intervals of 2,000–3,000 years and 7,000–10,000 years. These frequencies correspond, respectively, with DANSGAARD-OESCHGER (D-O) EVENTS and HEINRICH EVENTS. IRD defining Heinrich Events in the mid-latitude North Atlantic contain lithic material found to originate predominantly from the LAURENTIAN ICE SHEET and to a lesser extent from the FENNOSCANDIAN, Icelandic and British ice sheets.

This multiple-source origin of the Heinrich Events suggests that the discharge from the ice sheets may have been triggered by a common mechanism, although asynchrony in their timing in different parts of the ocean is a complication. Debate continues over the triggering mechanism for widespread IRD on these millennial timescales; these include internal ice-sheet dynamics (the BINGE-PURGE MODEL), eustatic SEA-LEVEL CHANGE and a linked MELTWATER-THERMOHALINE CIRCULATION system. External ORBITAL FORCING may provide a driving mechanism, but the *power spectra* are weak for the relevant PERIODICITIES.

LJW/WENA/JP

[*See also* GLACIOMARINE HYPOTHESIS]

Andrews JT (2009) Seeking a Holocene drift ice proxy: Non-clay mineral variations from SW to N-central Iceland shelf: trends, regime shifts and periodicities. *Journal of Quaternary Science* 24: 664–676.

Bond G, Showers W, Cheseby M et al. (1997) A pervasive millennial-scale cycle in North Atlantic Holocene and glacial climates. *Science* 278: 1257–1266.

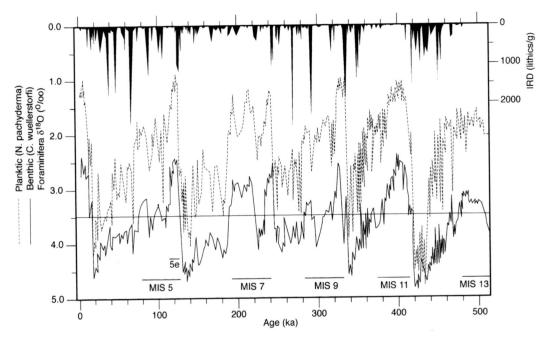

Ice-rafted debris Variations in an index of ice-rafted debris over the past 500,000 years from the North Atlantic (number of lithic grains larger than 150 mm per bulk sample weight, solid black) in relation to oxygen isotope variations in planktic (dashed) and benthic (lower solid line) foraminifera. Interglacial marine isotope stages (MIS) are numbered (McManus et al., 1999).

Dowdeswell JA, Elverhoi A, Andrews JT and Hebbeln D (1999) Asynchronous deposition of ice-rafted layers in the Nordic seas and North Atlantic Ocean. *Nature* 400: 348–351.

Haapaniemi AI, Scourse JD, Peck VL et al. (2010) Source, timing, frequency and flux of ice-rafted detritus to the Northeast Atlantic margin, 30-12 ka: Testing the Heinrich precursor hypothesis. *Boreas* 39: 576–591.

McManus JF, Oppo DW and Cullen JL (1999) A 0.5 million year record of millennial-scale climatic variability in the North Atlantic. *Science* 283: 971–975.

Passchier S (2011) Linkages between East Antarctic Ice Sheet extent and Southern Ocean temperatures based on a Pliocene high-resolution record of ice-rafted debris off Prydz Bay, East Antarctica. *Paleoceanography* 26: PA4204.

Stickley CE, St John K, Koç N et al. (2009) Evidence for middle Eocene Arctic sea ice from diatoms and ice-rafted debris. *Nature* 460: 376–379.

Zimmerman SRH, Pearl C, Hemming SR et al. (2011) Freshwater control of ice-rafted debris in the last glacial period at Mono Lake, California, USA. *Quaternary Research* 76: 264–271.

ice-sheet dynamics   The behaviour of ICE SHEETS, especially the ice-velocity characteristics, which govern the rate at which ice and sediment are transported from interior regions towards the margins and OCEANS, the location of preferred channels of ice transport and ice-mass evolution. Ice discharges or perimeter fluxes are now known around almost the entire peripheries of the *Greenland* and *Antarctic Ice Sheets* from SATELLITE REMOTE SENSING of ice velocity and airborne measurements of ice thickness. In the ANTARCTIC, ice is lost primarily by *basal melting* and the CALVING of ICEBERGS from ICE SHELVES. In contrast, the loss of ice from the Greenland Ice Sheet is primarily by *surface melting* in addition to calving. The MASS BALANCES of the Antarctic and Greenland Ice Sheets have been largely controlled by the evolution of their *outlet glaciers* and ICE STREAMS. In Greenland, drainage of SUPRAGLACIAL lakes and surface melt to the bed of the ice sheet is thought to increase *lubrication* by weakening the contact between the ice and its bed leading to increased ice-flow velocity. However, the period of accelerated flow is reduced during years of rapid melting because, as the amount of MELTWATER increases, SUBGLACIAL drainage takes less time to become more efficient and ice is once more frozen to its bed. In Antarctica, increased heat from the ocean in contact with the glacier GROUNDING LINES melts ice from underneath, disengages the glaciers from their bed, reduces buttressing of the inland ice and allows faster rates of ice discharge to the sea. Melt and increased *ice discharge* from the major outlet glaciers of both the Antarctic and Greenland ice sheets contribute directly to current SEA-LEVEL RISE. Large variations in sea level during the QUATERNARY have been controlled by ice-sheet mass balance, with rates of sea-level rise at least one order of magnitude larger than at present during times of rapid DEGLACIATION.   *BTIR*

[*See also* MARINE-BASED ICE SHEET]

Alley RB, Spencer MK and Anandakrishnan S (2007) Ice-sheet mass balance: Assessment, attribution and prognosis. *Annals of Glaciology* 46: 1–7.

Rignot E, Mouginot J and Scheuchl B (2011) Ice flow of the Antarctic Ice Sheet. *Science* 333: 1427–1430.

Rignot E and Thomas R (2002) Mass balance of the polar ice sheets. *Science* 297: 1502–1506.

Sundal AV, Shepherd A, Nienow P et al. (2011) Melt-induced speed-up of Greenland ice sheet offset by efficient subglacial drainage. *Nature* 469: 521–524.

ice-sheet growth   The build-up of an ICE SHEET due to persistent positive MASS BALANCE. Due to erosion and deposition during the later stages of an ice sheet, little geological evidence is preserved of periods of ice-sheet growth, and current understanding is based largely on theoretical GLACIER MODELLING and indirect evidence, such as evidence from MARINE SEDIMENTS for changing global ice volume. Rapid ice sheet growth is encouraged by low summer insolation and high precipitation rates. The modern climate of the Earth (characterised by polar ice sheets and large temperature gradients between the equator and the poles) began with the onset of GLACIATION in the ANTARCTIC at ~33.7 million years ago.   *DIB/JAM*

Alonso-Garcia M, Sierro FJ, Kucera M et al. (2011) Ocean circulation, ice sheet growth and interhemispheric coupling of millennial climate variability during the mid-Pleistocene (ca 800–400 ka). *Quaternary Science Reviews* 30: 3234–3247.

Cutler KB, Edwards RL, Taylor FW et al. (2003) Rapid sea-level fall and deep-ocean temperature change since the last interglacial period. *Earth and Planetary Science Letters* 206: 253–271.

Pagani M, Huber M, Liu Z et al. (2011) The role of carbon dioxide during the onset of Antarctic glaciation. *Science* 334: 1261–1264.

Pollard D and DeConto RM (2009) Modeling West Antarctic ice sheet growth and collapse through the last 5 million years. *Nature* 458: 329–332.

ichnofacies   A body of sediment or rock characterised by its TRACE FOSSIL assemblage. Ichnofacies analysis can be combined with more conventional sedimentary FACIES ANALYSIS to enhance PALAEOENVIRONMENTAL RECONSTRUCTION.   *GO*

[*See also* BIOFACIES, FACIES, LITHOFACIES]

McIlroy, D. (2008) Ichnological analysis: The common ground between ichnofacies workers and ichnofabric analysts. *Palaeogeography, Palaeoclimatology, Palaeoecology* 270: 332–338.

icing   A sheet-like, tabular mass of ice, also known as a *naled* and *Aufeis* that forms at the surface in winter wherever water issues from the ground. Icings occur in the ARCTIC and SUBARCTIC. Most are small but some in central Alaska and Siberia associated with perennial SPRINGS and sub-PERMAFROST assume considerable dimensions, exceeding 10 km in length and 10 m in thickness. *River icings* form at localities where the river freezes to its bottom, thereby forcing water out of the river bed. Evidence of possible RELICT icings has been identified in former periglacial environments.

HMF

[*See also* FROST MOUNDS]

Bennett MR, Huddart D, Hambrey MJ and Ghienne JF (1998) Modification of braided outwash surfaces by Aufeis: An example of Pedersenbreen, Svalbard. *Zeitschrift für Geomorphologie NF* 42: 1–20.
Hu XG and Pollard WH (1997) The hydrologic analysis and modelling of river icing growth, North Fork Pass, Yukon Territory, Canada. *Periglacial and Permafrost Processes* 8: 279–294.
Van Everdingen RO (1990) Ground-water hydrology. In Prowse TD and Ommanney CSL (eds) *Northern hydrology: Canadian perspectives*. Saskatoon: National Hydrology Research Institute, 77–101.

idealism   A philosophical position which views the 'real world', as created by the human mind. The existence of objects and a reality outside of a personal consciousness is questioned and considered inherently untrustworthy. Reality is seen as contingent on a mind and thought processes for its existence; there is no such thing as a detached observer. The term should be associated primarily with 'ideas', rather than with 'ideals'. The primary contrast in the philosophy of science is with REALISM. Immanuel Kant proposed although it is impossible to experience the external world (sensory experience is not the world itself), it is nevertheless possible to know, through reason, that such a world exists (*transcendental idealism*).   *CET/MAU/JAM*

[*See also* KNOWLEDGE, OBJECTIVE KNOWLEDGE, SOCIAL CONSTRUCTION, TRUTH]

Harrison S and Dunham P (1998) Decoherence, quantum theory and their implications for the philosophy of geomorphology. *Transactions of the Institute of British Geographers NS* 23: 501–514.
Pritchard D (2006) *What is this thing called knowledge?* Abingdon: Routledge.

identification (ID)   A *tag*, LABEL or ATTRIBUTE identifying an ENTITY or COVERAGE in a GEOGRAPHICAL INFORMATION SYSTEM (GIS). It is also the primary FIELD for a *tuple* in a RELATIONAL DATABASE.   *VM*

[*See also* OVERLAY ANALYSIS]

idiographic science   Science that is concerned with developing comprehensive explanations of individual cases (*case studies*), as opposed to seeking *generalisations* or LAWS (NOMOTHETIC SCIENCE). It is a rather dated concept but has relevance particularly in the SOCIAL SCIENCES and HUMANITIES, where approaches based on the perceptions and behaviour of particular individuals are common. For example, *regional geography* was considered to be idiographic as it emphasised the unique characteristics of particular areas of the Earth's surface, whereas *systematic geography* focused on general principles applicable anywhere.

JAM/CET

Cone JD (1986) Idiographic, nomothetic and related perspectives in behavioural assessment. In Nelson RO and Hayes SC (eds) *Conceptual foundations of behavioural assessment*. New York: Guilford Press, 111–128.
Salvatore S and Valsiner J (2009) Idiographic science on its way: Towards making science of psychology. *Yearbook of Idiographic Science* 1: 9–19.

igneous rocks   Rocks formed by the cooling and solidification of molten MAGMA, either below the Earth's surface as INTRUSIONS, or at the surface through VOLCANIC ERUPTIONS of LAVA and PYROCLASTIC MATERIAL. Their characteristic feature is a crystalline *texture*. Igneous rocks are subdivided into four categories according to their chemistry: (1) *ultrabasic* (silica content, SiO_2 <45 per cent), (2) *basic* (SiO_2 45–53 per cent), (3) *intermediate* (SiO_2 53–66 per cent) and (4) *acidic* (or silicic, SiO_2 >66 per cent). Mineral content or percentage of *dark minerals* are proxies for chemistry. Crystal size ranges from *fine* (<1 mm-*aphanitic*; fast cooling as lava or high-level intrusion) through *medium* to *coarse* (>3 mm-*phaneritic*; slow-cooling as large, deep plutonic intrusion). VOLCANIC GLASS is a product of very rapid cooling. Common igneous rocks are defined in the Table. Pyroclastic rocks are named according to different criteria.

Minerals in igneous rocks are important for RADIOMETRIC DATING. Studies of igneous rocks contribute to understanding the behaviour and evolution of VOLCANOES, which constitute a major NATURAL HAZARD. Sites of igneous activity are closely related to PLATE TECTONIC activity (see HOTSPOT, IN GEOLOGY), and igneous rocks in the GEOLOGICAL RECORD can help reconstruct past plate positions and movements. Pyroclastic BEDS form important, synchronous marker horizons of wide extent (see TEPHROCHRONOLOGY) and igneous rocks provide evidence of past episodes of VOLCANISM, which may have contributed to environmental change on a variety of timescales.   *JBH*

[*See also* METAMORPHIC ROCKS, SEDIMENTARY ROCKS, VOLCANIC IMPACTS ON CLIMATE]

Crystal size	Igneous setting	Igneous form	Rock type				
			Silica content				
			<45%	45–53%	53–60%	60–66%	>66%
			Ultrabasic	Basic	Intermediate		Acidic
Fine (*glassy*)	Volcanic	Lava		Basalt	Andesite	Dacite	Rhyolite (*obsidian*)
Medium	Hypabyssal	Dykes, sills		Dolerite			Microgranite
Coarse	Plutonic	Deep intrusions, batholiths	Peridotite	Gabbro	Diorite	Granodiorite	Granite
Dominant mineral phases			Olivine, pyroxene	Pyroxene, plagioclase	Plagioclase, hornblende	Plagioclase, K-feldspar, quartz	K-feldspar, quartz

Igneous rocks Characteristics of the principal igneous rock types.

Best MG (2003) *Igneous and metamorphic petrology.* Oxford: Blackwell.

Cas RAF and Wright JV (1987) *Volcanic successions: Modern and ancient.* London: Allen and Unwin.

Francis P and Oppenheimer C (2003) *Volcanoes.* Oxford: Oxford University Press.

Gill R (2010) *Igneous rocks and processes: A practical handbook.* Chichester: Wiley-Blackwell.

Le Maitre RW (ed.) (1989) *A classification of igneous rocks and glossary of terms: Recommendations of the International Union of Geological Sciences Subcommission on the Systematics of Igneous Rocks.* Oxford: Blackwell.

Middlemost EAK (1997) *Magmas, rocks and planetary development: A survey of magma/igneous rock systems.* Harlow: Longman.

ignimbrite A deposit of a PYROCLASTIC FLOW. Because of confusion in past use, the term is best restricted to deposits rich in PUMICE. Some of the largest single eruptive units known (>1,000 km^3) are ignimbrites, associated with major explosive VOLCANIC ERUPTIONS such as Santorini (1470 BC), Krakatau (AD 1883) and Mount St. Helens (AD 1980). Many ignimbrite deposits are preserved as extensive sheets of WELDED TUFF. *JBH*

Fedele FG, Giaccio B and Hajdas I (2008) Timescales and cultural process at 40,000 BP in the light of the Campanian Ignimbrite eruption, western Eurasia. *Journal of Human Evolution* 55: 834–857.

Gravley DM, Wilson CJN, Leonard GS and Cole JW (2007) Double trouble: Paired ignimbrite eruptions and collateral subsidence in the Taupo Volcanic Zone, New Zealand. *Geological Society of America Bulletin* 119: 18–30.

Legros F and Druitt TH (2000) On the emplacement of ignimbrite in shallow-marine environments. *Journal of Volcanology and Geothermal Research* 95: 9–22.

illite A 2:1 CLAY MINERAL of the *mica* family having only slight expansion characteristics, as a result of *potassium ions* occupying sites in the interlayer space. It is sometimes called *fine-grained mica*. ISOMORPHIC SUBSTITUTION in the clay lattice results in a CATION EXCHANGE CAPACITY (CEC) between that of KAOLINITE and SMECTITE. *EMB*

Brady NC and Weil RR (2007): *The nature and properties of soils, 14th edition.* New York: Prentice Hall.

illuviation Following ELUVIATION from the upper A and E horizons of a soil, the complementary process by which colloidal material (CLAY and HUMUS) is redeposited in a lower, B horizon. Illuviation is involved in the formation of the *argic horizon* present in LUVISOLS, ALBELUVISOLS, ACRISOLS, ALISOLS, LIXISOLS and NITISOLS, and the *spodic horizon* of PODZOLS. *EMB*
[*See also* PODZOLISATION]

Buol SW, Hole FD and McCracken RJ (1988) *Soil genesis and classification.* Ames: Iowa State University Press.

image classification The procedure carried out to generate a *thematic map* from remotely sensed data. An image consists of a number of PIXELS, each characterised by a vector of spectral reflectance values (OPTICAL REMOTE-SENSING INSTRUMENTS) and/or microwave BACKSCATTERING values (RADAR REMOTE-SENSING INSTRUMENTS). *Pixel-based classification* assigns a class to each individual pixel, while *object-oriented* *classification* assigns a class to a polygon (or landscape object), containing a group of neighbouring pixels with similar properties. Linear features in the landscape can be used to create parcels from a pixel-based image. A *supervised classification* uses user-defined TRAIN-ING AREAS to estimate spectral signatures of individual classes. The classification assigns a class to each pixel/polygon by comparing the spectral information of that pixel/polygon with the estimated signature. An *unsupervised classification* uses the spectral content of the whole image to estimate spectral signatures automatically. The two most commonly used iterative *unsupervised classification* algorithms are the *K-means* and the *isodata algorithm*. K-means requires a predefined number of classes, while isodata is able to split and merge clusters of pixels in an iterative procedure and arrive at an optimal number of classes.

A widely used classification algorithm is the *maximum likelihood classification*. A pixel is assigned the class with the highest likelihood of causing the observed REFLECTANCE or backscatter given this class. BAYESIAN STATISTICS have provided the *maximum a posteriori classifier*. According to Bayes' theorem, the a posteriori PROBABILITY is calculated from the likelihood and the a priori probability. The likelihood contains information about spectral similarity, while the a priori probability includes previous knowledge about the region. A pixel is assigned the class with the highest a posteriori probability of belonging to this class given the observed reflectance or backscatter.

Different methods of contextual classification have been developed in order to correct pixels that have been misclassified due to RANDOM variability in the data, by using information about spatially neighbouring pixels. FUZZY LOGIC has also been applied to image classification, as it enables quantification of the pixel-wise UNCERTAINTY of the thematic map. Because of the variability in the data, a classification always has errors associated with it. An ACCURACY assessment of the thematic map, in which the map is compared with GROUND MEASUREMENTS, is the final step of image classification.   *HB*

[*See also* ELECTROMAGNETIC SPECTRUM, IMAGE ENHANCEMENT, IMAGE PROCESSING, IMAGE TEXTURE]

Binaghi E, Madella P, Montesano MG and Rampini A (1997) Fuzzy contextual classification of multisource remote sensing images. *IEEE Transactions on Geoscience and Remote Sensing* 35: 326–340.
Gibbes C, Adhikari S, Rostant L et al. (2010) Application of object based classification and high resolution satellite imagery for savanna ecosystem analysis. *Remote Sensing* 2: 2748–2772.
Jensen JR (2004) *Introductory digital image processing: A remote sensing perspective, 3rd edition.* Upper Saddle River, NJ: Prentice Hall.
Jensen JR, Im J, Hardin P and Jensen RR (2009) Image classification. In Warner TA, Nellis MD and Foody GM (eds) *The SAGE handbook of remote sensing.* London: Sage, 269–281.
Kwarteng AY, Dobson MC, Kellndorfer J and Williams R (2008) SAR-based land cover classification of Kuwait. *International Journal of Remote Sensing* 29: 6739–6778.
Perko R, Raggam H, Deutscher J et al. (2011) Forest assessment using high resolution SAR data in X-band. *Remote Sensing* 3: 792–815.

image enhancement   Operations that aim at augmenting the visibility of desired features within an image (e.g. in REMOTE SENSING). These features may be the contrast, some or all of the colours, the borders, large areas etc, and the particular techniques to be used depend upon the input data and on the desired effect. Image enhancement is usually applied during IMAGE INTERPRETATION, and some of the techniques it relies on are contrast stretching, HISTOGRAM equalisation and FILTERING.   *ACF*

Velho L, Frery AC and Miranda J (2008) *Image processing for computer graphics and vision, 2nd edition.* Berlin: Springer.

image interpretation   The transformation of data, presented as images, into information. This transformation can be performed by means of two (complementary rather than disjoint) approaches: NUMERICAL ANALYSIS and *photo-interpretation*. In the former, the digital essence of the data is exploited through the use of computer-based techniques, while the latter presents the data to the human interpreter in the form of images, so a visual inspection can be carried out. Quantitative *image analysis* is optimum at PIXEL level, it is accurate for *area estimation*, can perform true *multispectral analysis* and can make use of all available *brightness* levels. Photo-interpretation is best suited for shape and contextual (spatial) determination and assessment.   *ACF*

Drury SA (2001) *Image interpretation in geology, 3rd edition.* Cheltenham: Nelson Thornes.
Francus P (ed.) (2004) *Image analysis of sediments and paleoclimates.* Dordrecht: Springer.
Lillesand TM, Kiefer RW and Chipman JW (2008) *Remote sensing and image interpretation, 6th edition.* Hoboken, NJ: Wiley.
Paine DP and Kiser JD (2012) *Aerial photography and image interpretation, 3rd edition.* Hoboken, NJ: Wiley.
Richards JA (2005) *Remote sensing digital image analysis: An introduction, 4th edition.* Berlin: Springer.

image processing   The manipulation of pictures by means of digital computers. In order to be able to use digital machines, pictures have to be in a proper format: a finite array of real, complex or multidimensional data, represented by a finite number of bits. In this manner, a *digital image* is defined in a discrete domain and as having discrete values in each position. Amongst the possible applications of digital image processing three very important ones are (1) FILTERING, (2) ENHANCEMENT and (3) RESTORATION. The techniques involved in image processing can be classified according to the kind of function: linear and non-linear and, depending on the number of input data, point-wise (single-valued functions), local (functions of a few variables) and global (functions that use the whole image as input). Image processing is a successful melting pot of disciplines: it uses concepts, techniques and ideas coming from diverse areas such as statistics, theoretical computing, artificial intelligence, psychometry and information theory, to name a few. Its success is greatly owed to the advent of fast computers with massive storage capacity. A current trend in image processing is the development of systems that ease the user's burden, providing sharp tools that require little or no specialised training to obtain the desired results.   *ACF*

Gonzalez RC and Woods RE (2008) *Digital image processing, 3rd edition*. London: Pearson.
Petrou M and Petrou C (2010) *Image processing: The fundamentals, 2nd edition*. Chichester: Wiley.
Velho L, Frery AC and Miranda J (2008) *Image processing for computer graphics and vision, 2nd edition*. Berlin: Springer.

image texture   Syncretism between vision and touch; that is, the tactile response analogy to a visual stimulus. Areas with similar mean colour properties can be distinguished by their *roughness* properties (e.g. forest and non-forest in a REMOTE-SENSING image) since texture is related to the *spatial organisation* of colours. Many techniques have been devised aiming at the quantification of this perceptual feature, HARALICK TEXTURE MEASURES being among the most well-known and useful ones. Two main categories of textures are usually considered: *structural textures* (where repetition of exact patterns is observed) and *statistical textures* (where the similarity is in the distributional sense).   *ACF*

Luckman AJ, Frery AC, Yanasse CCF and Groom GB (1997) Texture in airborne SAR imagery of tropical forest and its relationship to forest regeneration stage. *International Journal of Remote Sensing* 18: 1333–1349.

imbrication   (1) A FABRIC in SEDIMENTS and SEDIMENTARY ROCKS in which particles overlap like tiles: their long axes have a consistent DIP. Due to shear stresses exerted by a CURRENT during DEPOSITION, the *maximum projection planes* (*a*–*b* planes), or apparent long axes in cross section, of non-spherical particles are inclined gently (usually 10°–30°) in the up-current direction relative to the BEDDING surface. It is a useful indicator of PALAEOCURRENT direction, particularly for

GRAVEL, which may lack other SEDIMENTARY STRUCTURES. (2) In tectonically deformed rocks or sediments, *imbricate structure* describes a series of high-angle reverse FAULTS between two THRUST FAULTS.   *TY/GO*

Collinson JD, Mountney NP and Thompson DB (2006) *Sedimentary structures*. Harpenden: Terra Publishing.
Fossen H (2010) *Structural geology*. Cambridge: Cambridge University Press.
Millane RP, Weir MI and Smart GM (2006) Automated analysis of imbrications and flow direction in alluvial sediments using laser-scan data. *Journal of Sedimentary Research* 76: 1049–1055.
Qin J, Zhong D, Wang G and Ng SL (2012) On the characterization of the imbrication of armoured gravel surfaces. *Geomorphology* 159–160: 116–124.

impactite   A rock formed as a result of a METEORITE IMPACT, with properties indicative of *shock metamorphism* or *impact metamorphism* (see METAMORPHIC ROCKS) such as *shocked quartz grains*, TEKTITES or included *diamonds*. An example of an impactite is *suevite*, a BRECCIA formerly interpreted as a TUFF, originally described from the Ries impact crater at Nördlingen, southern Germany. Suevite is similar in characteristics to lunar REGOLITH.   *GO*

Ding Y and Veblen DR (2004) Impactite from Henbury, Australia. *American Mineralogist* 89: 961–968.
Kalleson E, Dypvik H and Nilsen O (2010) Melt-bearing impactites (suevite and impact melt rock) within the Gardnos structure, Norway. *Meteoritics and Planetary Science* 45: 798–827.

impacts of environmental change   The effects of environmental change on natural systems, human-social systems and coupled human-ecological systems. However, the term 'impacts' is sometimes incorrectly confined to HUMAN IMPACT ON ENVIRONMENT and/or human impacts in the context of GLOBAL ENVIRONMENTAL CHANGE. Past, present and future impacts result from NATURAL or ANTHROPOGENIC causes, they may be direct or indirect, single or multiple, and they vary in their speed of onset and their magnitude. Commonly, there are complex chains of CAUSAL RELATIONSHIPS and FEEDBACK MECHANISMS. These result in indirect cumulative or synergistic effects, which are difficult to predict. Although techniques such as ENVIRONMENTAL MODELLING and ENVIRONMENTAL IMPACT ASSESSMENT (EIA) can help predict impacts, there is UNCERTAINTY in PREDICTION and unexpected ENVIRONMENTAL SURPRISES occur. Furthermore, human impacts are less easy to predict than biophysical environmental changes because human behaviour changes with experience.

In the context of predicting the impact of CLIMATIC CHANGE, a distinction can be made between potential and residual impacts, and market and non-market impacts: *potential impacts* include 'all impacts that may occur given a projected change in climate, without considering ADAPTATION'; *residual impacts* are 'the impacts of climate change that would occur after adaptation'; *market impacts* are 'impacts that can be quantified in monetary terms and directly affect Gross Domestic Product'; *non-market impacts* are 'impacts that affect ecosystems and human welfare, but that are not easily expressed in monetary terms' (Parry et al., 2007: 876–878).   *CJB/JAM*

[*See also* CLIMATIC CHANGE: PAST IMPACT ON ANIMALS, CLIMATIC CHANGE: POTENTIAL FUTURE IMPACTS ON SOILS, HUMAN IMPACT ON CLIMATE, HUMAN IMPACT ON ENVIRONMENT]

Fry C (2008) *The impact of climate change: The world's greatest challenge in the twenty-first century.* London: New Holland.
Hall CM, Gossling S and Scott D (2011) *Tourism and climate change: Impacts, adaptation and mitigation.* London: Routledge.
Matthews JA, Bartlein PJ, Briffa PJ et al. (eds) (2012) *The SAGE handbook of environmental change, volume 2: Human impacts and responses.* London: Sage.
Parry ML, Canziani JP, Palutikof JP et al. (eds) (2007) *Impacts, adaptation and vulnerability* [Contribution of Working Group II to the Fourth Assessment Report of the Intergovernmental Panel on Climate Change]. Cambridge: Cambridge University Press.
Whyte I (2008) *World without end? Environmental disaster and the collapse of empires.* London: IB Taurus.

imperialism   The control and exploitation of one state by another based on ideas of superiority and usually in pursuit of empire. Examples include the Roman Empire, the Ottoman Empire, the British Empire and the communist Russian Empire. The term has tended to be replaced, at least in modern times, by the closely related term, COLONIALISM which, taken literally, involves establishing permanent colonies. The term *new imperialism* or *Age of Imperialism* is used to distinguish the phase of territorial conquest beyond Europe in the latter half of the nineteenth century from the earlier phase of European expansion beginning in the fifteenth century. In a remarkably short time, Britain, France and Russia (and to a lesser extent Belgium, Germany and Italy) took control of much of the globe, which is exemplified by the *scramble for Africa* (see Figure). The success of European imperialism has a biological component, which has been termed *ecological imperialism*, reflecting the introduced plants, animals and DISEASES that suppressed, if not devastated, the native inhabitants and their *ecosystems*.
                                                        JAM

Crosby AW (1986) *Ecological imperialism: The biological expansion of Europe, 900–1900.* Cambridge: Cambridge University Press.
Johnson R (2003) *British imperialism.* Basingstoke: Palgrave Macmillan.
Wolfe P (2004) Twentieth-century theories of imperialism. In Duara P (ed.) *Decolonization: Perspectives from now and then.* London: Routledge, 101–117.

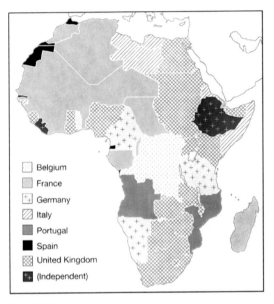

Imperialism The division of Africa amongst European states ca AD 1914 after the so-called *scramble for Africa* (based on an original map by Declan Graham).

impermeable   Having a structure and/or texture that does not allow the transmission of liquids or gases. The term can be applied to rock strata, sediments or layers within a soil profile that either do not allow or slow considerably the passage of water through pores or fissures. There is, however, a more restricted usage of the term to materials, including rocks and soils, that do not allow the DIFFUSION of fluids through the pores.   *JAM/ADT*

[*See also* HYDRAULIC CONDUCTIVITY, IMPERVIOUS, INFILTRATION, PERMEABLE]

impervious   (1) A synonym for IMPERMEABLE. (2) It is sometimes used in a more restricted sense of materials (e.g. rocks, sediments or soils) that do not permit the flow of fluids through fissures, joints or cracks. (3) Urban areas within DRAINAGE BASINS where there is little or no INFILTRATION of rainfall are also said to be impervious.   *ADT/JAM*

[*See also* PERMEABLE, PERVIOUS]

impoundment   The act of impounding, the state of being impounded, or the physical structure (e.g. a DAM or *weir*) that blocks the passage of surface water creating an artificial body of water (e.g. a *reservoir*). The term is also used for structures built to contain INDUSTRIAL WASTE slurries. A structure raising the level of an existing natural LAKE is also considered an impoundment. Any waterbody created by excavation below the pre-existing ground level, and ponds related to artificial treatment systems (e.g. *sustainable urban drainage systems*) are excluded. Impoundments can present an artificial HAZARD.   *LJMcE*

Ferris JA and Lehman JT (2008) Nutrient budgets and river impoundments: Interannual variation and implications for detecting future changes. *Lake and Reservoir Management* 24: 273–281.

National Research Council (2002) *Coal waste impoundments: Risks, responses, and alternatives.* Washington, DC: National Academies Press.

Vorosmarty CJ, Meybeck M, Fekete B et al. (2003) Anthropogenic sediment retention: Major global impact from registered river impoundments. *Global and Planetary Change* 39: 169–190.

impulse response   The changes in a SYSTEM following from a sudden input to the system. An example is the variation through time in the amount of CARBON DIOXIDE that remains in the ATMOSPHERE following a sudden injection of CO_2 into the atmosphere.   *JAM*

inbreeding depression   The loss in *genetic diversity* and GENETIC FITNESS that may result from breeding among close relatives. Inbreeding depression is especially problematic for POPULATIONS with relatively few individuals, which unavoidably must comprise close relatives. Evidence for inbreeding depression is often based on a correlation between HETEROZYGOSITY and fitness, although heterozygosity may be a poor proxy measure of inbreeding. *Genetic rescue* occurs when an unrelated individual is introduced to the inbred population, resulting in an improvement in genetic fitness. In some cases, this individual may be so much fitter that a *genomic sweep* of the entire inbred population occurs. Small populations may be as susceptible to *outbreeding depression* (reduced fitness from cross-breeding) as to inbreeding depression.   *JLI/MVI.*

[*See also* BOTTLENECK, ENDANGERED SPECIES, POPULATION VIABILITY ANALYSIS (PVA)]

Adams JR, Vucetich LM, Hedrick PW et al. (2011) Genomic sweep and potential genetic rescue during limiting environmental conditions in an isolated wolf population. *Proceedings of the Royal Society B* 278: 3336–3344.

Finger A, Kettle CJ, Kaiser-Bunbury CN et al. (2011) Back from the brink: Potential for genetic rescue in a critically endangered tree. *Molecular Ecology* 20: 3773–3784.

Houde ALS, Fraser DJ, O'Reilly P and Hutchings JA (2011) Relative risks of inbreeding and outbreeding depression in the wild in endangered salmon. *Evolutionary Applications* 4: 634–647.

Johnson HE, Mills LS, Wehausen JD et al. (2011) Translating effects of inbreeding depression on component vital rates to overall population growth in endangered Bighorn Sheep. *Conservation Biology* 25: 1240–1249.

Walling CA, Nussey DH, Morris A et al. (2011) Inbreeding depression in red deer calves. *BMC Evolutionary Biology* 11: Article No. 318.

incised meanders   MEANDERS produced by *river downcutting* into BEDROCK; also known as *valley meanders*. There are two types of incised meanders: (1) *entrenched meanders* and (2) *ingrown meanders*. Entrenched meanders develop where meandering alluvial rivers are superimposed onto bedrock during tectonic UPLIFT without significant tilting. Entrenched meanders possess a symmetrical cross section and occur if valley sides are resistant to EROSION, or when there has been rapid *incision* by the river (e.g. entrenched meanders on the San Juan River, Utah). They occur when a river is confined to a CANYON or GORGE, with little to no FLOODPLAIN development. In contrast, ingrown meanders are associated with less rapid uplift of land, which allows the river to move laterally and erode laterally into bedrock, thus modifying meander form as it does so.   *LJMcE*

[*See also* ANTECEDENT DRAINAGE, MISFIT MEANDER, SUPERIMPOSED DRAINAGE]

Rogers RD, Kárason H and van der Hilst RD (2002) Epeirogenic uplift above a detached slab in northern Central America. *Geology* 30: 1031–1034.

Rosgen DL (1994) A classification of natural rivers. *Catena* 22: 169–248.

Stark CP, Barbour JR, Hayakawa YS et al. (2010) The climatic signature of incised river meanders. *Science* 327: 1497–1501.

Tinkler KJ (1971) Active valley meanders in south-central Texas and their wider implications. *Geological Society of America Bulletin* 81: 1873–1899.

inclination (I)   The angle of dip between the lines of force of the Earth's MAGNETIC FIELD and the horizontal. It is a parameter of the geomagnetic field used in PALAEOMAGNETIC DATING, which varies between 90° at the magnetic equator and 0° at the magnetic poles.   *MHD*

[*See also* DECLINATION (D)]

Thompson R (1991) Palaeomagnetic dating. In Smart PL and Frances PD (eds) *Quaternary dating methods: A user's guide.* Cambridge: Quaternary Research Association, 177–198.

incremental dating methods   DATING TECHNIQUES based on incremental growth of inorganic and organic deposits. This type of dating may use ANNUALLY RESOLVED RECORDS (see, e.g., CORAL AND CORAL REEFS: ENVIRONMENTAL RECONSTRUCTION ANALYSIS, DENDRO-CHRONOLOGY, ICE-CORE DATING, SPELEOTHEMS and VARVE CHRONOLOGY) or growth rates calibrated by independent dating methods (e.g. LICHENOMETRIC DATING, WEATHER-ING RINDS). Annually resolved records can provide age estimates of greater PRECISION than non-incremental, CALIBRATED-AGE DATING techniques (e.g. ±9 years for dendrochronology and ±1 per cent or less for ice-core years over the HOLOCENE), whereas incremental tech-niques requiring some form of AGE CALIBRATION, as in the case of lichenometry, are usually of relatively low precision, especially with increasing age. *DAR/CJC*

Baillie MGL (1995) *A slice through time: Dendrochronology and precision dating.* London: Batsford.

Andersen KK, Svensson A, Johnsen SJ et al. (2006) The Greenland ice-core chronology 2005, 15-42 ka. Part 1: Constructing the time scale. *Quaternary Science Reviews* 25: 3246–3257.

indeterminacy One element of UNCERTAINTY found in scientific inquiry derived from Heisenberg's famous observation that the position and momentum of a par-ticle cannot be measured both simultaneously and exactly (*Heisenberg's uncertainty principle*). In *com-plex systems*, it can be interpreted as the impossibility of specifying all exact causal chains relating variables to one another (*indeterminacy of state*) or the impos-sibility of measuring all relationships exactly (*indeter-minacy of measurement*). *MAU/CET*

Cassidy DC (1998) Answer to the question: When did the indeterminacy principle become the uncertainty principle? *American Journal of Physics* 66: 278–279.

index fossil A FOSSIL or SUBFOSSIL species that char-acterises a biozone in BIOSTRATIGRAPHY. *GO*

[*See also* ENVIRONMENTAL INDICATOR, INDICATOR SPECIES]

Rawson PF, Allen PM, Brenchley PJ et al. (eds) (2002) *Stratigraphical procedure.* Bath: Geological Society.

Indian Ocean Dipole (IOD) A CLIMATIC MODE of coupled atmosphere-ocean variability in the tropical Indian Ocean producing *climatic anomalies* in precipi-tation and low-level winds. Events associated with the IOD, also termed the *Indian Ocean Zonal Mode* (*IOZM*), often coincide with EL NIÑO-SOUTHERN OSCILLATION (ENSO) phenomena, are strongly linked to the annual cycle of the MONSOON and affect precipitation in many regions around the Indian Ocean and beyond. *JAM*

[*See also* ATMOSPHERE-OCEAN INTERACTION, CLIMATIC OSCILLATION]

Abram NJ, Gagan MK, Liu Z et al. (2007) Seasonal characteristics of the Indian Ocean Dipole during the Holocene epoch. *Nature* 445: 299–302.

Hastenrath S (2007) Circulation mechanisms of climate anomalies in East Africa and the equatorial Indian Ocean. *Dynamics of Atmospheres and Oceans* 43: 25–35.

indicator species Species that are indicative of a particular set of environmental conditions. They include UMBRELLA SPECIES, KEYSTONE SPECIES, FLAGSHIP SPECIES and a range of other surrogates. Certain plants are widely used as indicator species for determining land suitability for FOREST MANAGEMENT and other uses. Certain microorganisms, such as coliform BACTERIA, are useful indicators of the potential for WATERBORNE DISEASE outbreaks. Subfossil DIATOMS, *pollen* and other organic remains are used as indicators for PALAEOENVI-RONMENTAL RECONSTRUCTION. *JLI*

[*See also* BIO-INDICATORS, CLIMATIC RECONSTRUCTION, ECOLOGICAL INDICATOR, NICHE, POLLEN ANALYSIS, TRACE FOSSIL]

Caro T (2010) *Conservation by proxy: Indicator, umbrella, keystone, flagship, and other surrogate species.* Washington, DC: Island Press.

Goodsell PJ, Underwood AJ and Chapman MG (2009) Evidence necessary for taxa to be reliable indicators of environmental conditions or impacts. *Marine Pollution Bulletin* 58: 323–331.

Klinka K, Krajina VJ, Ceska A and Scagel AM (1989) *Indicator plants of coastal British Columbia.* Vancouver: University of British Columbia Press.

indigenous peoples Ethnic groups comprising the original inhabitants of a territory or state, as opposed to the later colonisers. *Native Americans* (USA), *First Nations* (Canada) and *aborigines* (Australia) are spe-cific examples, the last being derived from the Latin *ab origine* (from the beginning). Aspects relevant to environmental change include preservation of *cultural identity* and *indigenous knowledge*, *land rights*, owner-ship of and access to NATURAL RESOURCES, ENVIRONMENTAL DEGRADATION, *poverty*, health and discrimination. *JAM*

Clark JS and Royall PD (1995) Transformation of a northern hardwood forest by aboriginal (Iroquois) fire: Charcoal evidence from Crawford Lake, Ontario, Canada. *The Holocene* 5: 1–9.

Head L (1999) *Second nature: The history and implications of Australia as aboriginal landscape* New York: Syracuse University Press.

Huntington H and Fox S (2005) The changing Arctic: Indigenous perspectives. In Arctic Climate Impact Assessment (eds) *ACIA scientific report.* Cambridge: Cambridge University Press, 61–98.

Johansen BE (2003) *Indigenous peoples and environmental issues: An encyclopedia.* Westport, CT: Greenwood Press.

Sanders D (1999) Indigenous peoples: Issues of definition. *International Journal of Cultural Property* 8: 4–13.

individualistic concept   The idea that individual species within a community behave differently in relation to ENVIRONMENTAL FACTORS, such that any community is essentially a chance combination of species and individuals. The idea was put forward by H.A. Gleason, and contrasts with the CLIMAX VEGETATION concept of F.E. Clements, which saw communities as eventually developing a stable and repeatable species composition determined principally by climate. The difference in approach is very important when modelling the possible responses of plants and plant communities to environmental change.   *JLI*

[*See also* COMMUNITY CONCEPTS, ECOLOGICAL SUCCESSION]

Clements FE (1916) *Plant succession: An analysis of the development of vegetation*. Washington, DC: Carnegie Institution.

Gleason HA (1917) The structure and development of the plant association. *Bulletin of the Torrey Botanical Club* 44: 463–481.

Gleason HA (1926) The individualistic concept of the plant association. *Bulletin of the Torrey Botanical Club* 53: 7–26.

Gleason HA (1927) Further views on the succession concept. *Ecology* 8: 299–326.

Matthews JA (1996) Classics in physical geography revisited 'Gleason, H.A. 1939: The individualistic concept of the plant association'. *Progress in Physical Geography* 20: 193–203.

McIntosh RP (1995) H.A. Gleason's 'individualistic concept' and theory of animal communities: A continuing controversy. *Biological Reviews* 70: 317–357.

indoor environmental change   Environmental change inside buildings, which impacts on HUMAN HEALTH HAZARDS, ENVIRONMENTAL QUALITY, thermal comfort and AESTHETICS. The indoor environment essentially insulates people from the WEATHER and CLIMATE, including SEASONALITY, of the *outdoor environment*, and reducing the environmental differences normally experienced from place to place. The indoor environment at home and in the workplace has evolved in response to changes in building design and TECHNOLOGICAL CHANGE. Modern building at its best is improving temperature control, noise suppression, ventilation and air quality. Globally, many buildings are not up to standard, and there is much that can be done to reduce energy leakage through better insulation, air conditioning, flexible shading, through-flow ventilation and heat exchangers. AIR QUALITY is a key facet of the indoor environment. Since the 1950s, LEAD (Pb) pipes, lead paint additives, some PESTICIDES and ASBESTOS have been banned but there is some way to go in reducing EMISSIONS from solvents in glues, paints and furnishings. Some current ventilation systems pose a risk from the bacterial *Legionella disease*. *Dust mites*, *asthma* and *allergies* are increasing in developed countries, due to indoor fumes and/or too 'clean' an environment, which weakens the immune system. In regions with high RADON EMANATION, the risk of cancer has prompted passive and active ventilation of basements and use of gas-proof membranes. Increased use of *central heating* and also GLOBAL WARMING may pose new problems from TERMITES, *bedbugs*, *cockroaches* and *rodents*. Future developments in lighting or transmission of sunlight by fibre optic cables may have beneficial effects. Indoor AIR POLLUTION from primitive household *cooking fires* in DEVELOPING COUNTRIES has been listed by the World Health Organisation as the leading cause of death in the world (it contributes to about 2 million deaths per year). Women and children are most at risk from *pulmonary diseases* caused by smoke-filled kitchens.   *CJB/JAM*

[*See also* DISEASE, ENVIRONMENTAL QUALITY]

Bluyssen, PM (2009) *The indoor environment handbook: How to make buildings healthy and comfortable*. London: Earthscan.

Committee on the Effect of Climate Change on Indoor Air Quality and Public Health (2011) *Climate change, the indoor environment and health*. Washington, DC: National Academies Press.

Flannigan B (ed.) (2001) *Microorganisms in home and indoor work environments*. London: Taylor and Francis.

Hitchings R (2010) Seasonal climate change and the indoor environment. *Transactions of the Institute of British Geographers NS* 35: 282–298.

Humphries C (2012) Indoor ecosystems. *Science* 335: 648–650.

Indoor Air (2011) Commemorating 20 years of Indoor Air. *Indoor Air: International Journal of Indoor Environment and Health* 21: 177–230 [Special Section].

Martin II WJ, Glass RI, Balbus JM and Collins FS (2011) A major environmental cause of death. *Science* 334: 180–181.

induction   To reason from particular instances to a general TRUTH, or from *individual cases* to *universal* generalisations. Induction develops a conclusion that is not a logical necessity from the premises. Inductive reasoning brings into doubt many empirical claims of SCIENCE, which is why C.D. Broad described the problem of induction as the glory of science but the scandal of *philosophy*. One modern view is that, in practice, induction is used successfully in science by taking account of the UNCERTAINTY involved and by making inferences based on a combination of induction with DEDUCTION.   *JAM/CET*

[*See also* CRITICAL RATIONALISM, LOGICAL POSITIVISM, REALISM, SCIENTIFIC METHOD]

Feeney A and Heit (eds) (2007) *Inductive reasoning: Experimental, developmental and computational approaches.* New York: Cambridge University Press.

Lipton P (2004) *Inference to the best explanation, 2nd edition.* Abingdon: Routledge.

inductively coupled plasma atomic emission spectrometry (ICP-AES) A technique for elemental analysis used in GEOCHEMISTRY and related fields involving OPTICAL EMISSION SPECTROSCOPY (OES) of the sample as a *plasma* (a gas produced from the sample, held at around 10,000K by a radiofrequency generator). Instruments can typically analyse for over 50 *spectral lines* simultaneously (cf. ATOMIC ABSORPTION SPECTROPHOTOMETRY, AAS) with a very small sample size (1–10 mg of material in solution) and the accuracy is about ±5 per cent. Detection limits for most ELEMENTS are below single parts per billion (ppb) level, much lower than for AAS but higher than for INDUCTIVELY COUPLED PLASMA MASS SPECTROMETRY (ICP-MS). *TY*

Butler OT and Howe AM (1999) Development of an international standard for the determination of metals and metalloids in workplace air using ICP-AES: Evaluation of sample dissolution procedures through an interlaboratory trial. *Journal of Environmental Monitoring* 1: 23–32.

Fairchild IJ, Hendry G, Quest M and Tucker ME (1988) Chemical analysis of sedimentary rocks. In Tucker ME (ed.) *Techniques in sedimentology.* Oxford: Blackwell, 274–354.

Linderholm J and Lundberg E (1994) Chemical characterization of various archaeological soil samples using main and trace-elements determined by inductively-coupled plasma-atomic emission-spectrometry. *Journal of Archaeological Science* 21: 303–314.

Szaloki I, Somogyi A, Braun M and Toth A (1999) Investigation of geochemical composition of lake sediments using ED-XRF and ICP-AES techniques. *X-Ray Spectrometry* 28: 399–405.

inductively coupled plasma mass spectrometry (ICP-MS) A technique for elemental analysis used in GEOCHEMISTRY and related fields involving *mass spectrometry* of a *plasma.* Instruments can typically analyse for over 50 ELEMENTS simultaneously, with detection limits at or below the single part per trillion (ppt) level for much of the *periodic table,* except for several of the important *major elements,* most notably SILICON (Si), with a small sample size (routinely 100–200 mg of material in solution is used to allow repeat measurements). With multicollector techniques, ICP-MS may be used for the determination of isotopic abundances for geochronologic, radiogenic isotopic and stable isotopic studies. As an alternative to the introduction of samples in solution, ICP-MS systems are often run with a *laser ablation sampling system,* allowing precisely localised chemical and isotopic analyses from solid samples. *TY*

[*See also* GEOCHRONOLOGY, INDUCTIVELY COUPLED PLASMA ATOMIC EMISSION SPECTROMETRY, MASS SPECTROMETER, STABLE ISOTOPE ANALYSIS, TEPHRA ANALYSIS]

Dai XX, Chai ZF, Mao XY et al. (2000) An alpha-amino pyridine resin preconcentration method for iridium in environmental and geological samples. *Analytica Chimica Acta* 403: 243–247.

Durrant SF and Ward NI (2005) Recent biological and environmental applications of laser ablation inductively coupled plasma mass spectrometry (LA-ICP-MS). *Journal of Analytical Atomic Spectrometry* 20: 821–829.

Golub MS, Keen CL, Commisso JF et al. (1999) Arsenic tissue concentration of immature mice one hour after oral exposure to gold mine tailings. *Environmental Geochemistry and Health* 21: 199–209.

Jarvis KE, Gray AL and Houk RS (1991) *Handbook of inductively coupled plasma mass spectrometry.* London: Blackie.

induration A SOIL HORIZON, or part of a soil horizon, that is cemented with *calcium carbonate;* the oxides of SILICON (Si), IRON (Fe) or ALUMINIUM (Al); or HUMUS. It is difficult to dig or for plant roots to penetrate. Induration also refers to the results of the hardening processes of CEMENTATION and COMPACTION, which apply to both SOILS and unconsolidated SEDIMENTS. Induration may occur in various types of soils, including PODZOLS, GLEYSOLS, but particularly in DURISOLS. *EMB*

[*See also* DURICRUST, HARDPAN]

Kubotera H and Yamada I (2000) Characteristics of the induration of tephra-derived soils in Kyushu, Japan: (3) properties and typology of five indurated soils. *Soil Science and Plant Nutrition* 46: 365–379.

Thompson CH, Bridges EM and Jenkins DA (1993) An exploratory examination of relict hardpans in the coastal lowlands of southern Queensland. In Ringrose-Voase AJ and Humphreys GS (eds) *Soil micromorphology: Studies in management and genesis.* Amsterdam: Elsevier, 233–246.

industrial archaeology The study of the legacy of the INDUSTRIAL REVOLUTION from an archaeological perspective. It includes investigation of abandoned industrial workings, their landscape settings, environmental impacts and broader social significance. The subdiscipline developed in post-war Britain following the recognition of a need to preserve some relics of the industrial HERITAGE. *JAM*

[*See also* HISTORICAL ARCHAEOLOGY]

Casella EC and Symonds J (eds) (2005) *Industrial archaeology: Future directions.* New York: Springer.

Jones WR (2006) *Dictinary of industrial archaeology, 2nd edition.* Stroud: Sutton.

Palmer M (1999) The archaeology of industrialization. In Barker G (ed.) *Companion encyclopedia of archaeology.* London: Routledge, 1160–1197.

Trinder B (ed.) (1992) *The Blackwell encyclopedia of industrial archaeology*. Oxford: Blackwell.

industrial ecology   An integrated approach to the organisation of human industrial activities that mimics natural *ecosystems* in the sense that there is little or no WASTE, and the outputs from each process is used as inputs to another process. This complementarity can be sought at various scales with the objective of approaching and maintaining SUSTAINABILITY in production and CONSUMPTION. Such industrial systems tend to optimise energy and material MANAGEMENT throughout the entire life cycle of a product, use RESOURCES efficiently, reduce POLLUTION, extend the life of products and exist in concert with surrounding systems. The full implications of industrial ecology are exemplified by *eco-industrial parks*, such as Kalundborg, Denmark.   *AJDF*

[*See also* ECODESIGN, LIFE-CYCLE ASSESSMENT (LCA), SUSTAINABLE CONSUMPTION, SUSTAINABLE PRODUCTION, WASTE MANAGEMENT, WASTE MINING, WASTE RECYCLING]

Allenby BR and Richards DJ (eds) (1994) *The greening of industrial ecosystems*. Washington, DC: National Academies Press.
Graedel TE and Allenby BR (2009) *Industrial ecology and sustainable engineering*. Upper Saddle River, NJ: Prentice Hall.

Industrial Revolution   The period of rapid transition from an agricultural to an industrial society characterised by the use of inanimate energy, technological innovation, the factory system and increased productivity. It is generally considered to have begun in Great Britain around the mid-eighteenth century. It continued into the late nineteenth century and was characterised especially by the use of coal as an energy source powering heavy engineering industries dominated by steam, railways and steel. The Industrial Revolution had profound effects on the global economy and the natural environment and created the modern working class. Other 'industrial revolutions' have been proposed, such as the *new industrial revolution*, defined by Marsh (2012) as the transformation of MANUFACTURING INDUSTRY by GLOBALISATION, the characteristics of which include a broader participation in manufacturing across the world, a greater focus on *specialisation* and the growing importance of SUSTAINABLE PRODUCTION. Markillie (2012) recognised a *second industrial revolution*, which began in the early twentieth century with *mass production* made possible by the *assembly line* and developed into modern manufacturing dominated by highly automated *machine tools*; and a *third industrial revolution*, currently underway, brought about by the widespread use of digital electronic technology, which permits *additive manufacturing* (three-dimensional printing of materials layer by layer).   *JAM*

[*See also* INDUSTRIALISATION, MANUFACTURING INDUSTRY, TECHNOLOGICAL CHANGE]

Allen RC (2009) *The British Industrial Revolution in global perspective*. Cambridge: Cambridge University Press.
Ashton TS (1967) *The Industrial Revolution: 1760-1830*. Oxford: Oxford University Press.
Crump T (2010) *A brief history of how the Industrial Revolution changed the world*. London: Constable and Robinson.
Markillie P (2012) A third industrial revolution. *The Economist* [Special Report].
Marsh P (2012) *The new industrial revolution*. New Haven, CT: Yale University Press.

industrial waste   Industrial activities produce a wide variety of *natural wastes* from *mining* and QUARRYING, and *by-products* from the processes of INDUSTRY. Some of these wastes are reusable, others are classified as HAZARDOUS WASTE and require special treatment to isolate them from the environment or render them innocuous. Reject material from COAL, *china clay*, *slate* and *gravel extraction* amount to many millions of tonnes and *tipping* is mainly on land adjacent to the site worked. Material from these industries can be used for *aggregate*, *brick making*, *concrete blocks* and also used for *road-building* material or as a filter medium for *sewage works*. Wastes from power stations include *pulverised fuel ash* and *clinker*, some of the former is used in cement as a filler and with the latter can be made into concrete blocks and aggregate. Similarly, *slag* from blast furnaces and steel making can be used for aggregate and *basic slag* is used as a FERTILISER, especially for PASTURES. *Mining spoil* and slag from *non-ferrous metal industries* usually contains *toxic metallic elements* that limit their usefulness, but TIN (Sn) and COPPER (Cu) slags have been used for *sandblasting*. *Red mud* from ALUMINIUM (Al) manufacture has been used as *pigment* in paints and plastics. SLUDGES from the PETROLEUM industry are usually consigned to special LANDFILLS or incinerators. *Carbonisation plants* produce *gas*, *coke*, *coal tar* and *ammoniacal liquor*, all of which are useful materials, but in the process of *gas purification*, *ferrous iron oxides* were used which became highly acidic and contaminated with *cyanide*. NATURAL GAS has made most of these plants redundant. *Chemical production* and *pharmaceutical production* also results in sludges that require special treatment to make them innocuous. Dumping in the nineteenth century from *alkali works* and *chromate works* still remains as visual intrusions in the British landscape. Production of *munitions* and *explosives* for two World Wars and *radioactive leakage* from NUCLEAR ENERGY stations have caused ENVIRONMENTAL PROBLEMS and HUMAN HEALTH HAZARDS. CONTAMINATED LAND is a legacy of poor WASTE MANAGEMENT during the past 100 years.   *EMB*

[*See also* AGRICULTURAL WASTE, DOMESTIC
WASTE, RADIOACTIVE WASTE, WASTE MINING, WASTE
RECYCLING]

Brunner CR (1991) *Handbook of incineration systems.* New
York: McGraw-Hill.
Nigam Singh P and Pandey A (eds) (2009) *Biotechnology for
agro-industrial residues utilisation.* Berlin: Springer.
Samuelson JP (2009) *Industrial waste: Environmental
impact, disposal and treatment.* New York: Nova Science
Publishers.
Wang LK, Hung YT, Lo H and Yapijakis C (eds) (2004)
Handbook of industrial and hazardous wastes treatments.
Boca Raton, FL: CRC Press.
Water Environment Federation (WEF) (2008) *Industrial
wastewater management, treatment and disposal.*
Alexandria, VA: WEF/McGraw-Hill.

industrialisation   The process of change from a
predominantly agricultural society to one dominated by
industrial activity, particularly EXTRACTIVE and MANU-
FACTURING INDUSTRY. It began with the INDUSTRIAL REVO-
LUTION in Great Britain, which is dated, conventionally,
from around the mid-eighteenth century. This was the
culmination of many innovations initiated in the late
seventeenth and early eighteenth centuries. Rapid
advances in engineering that occurred in the water-
based *textile industry* and the charcoal-based *iron
smelting* industry resulted in experiments in the use of
coal for smelting, while coal had also been used in the
early development of the non-ferrous metal industry.
Together, such advances combined to set the scene for
the period of rapid change, which followed from the
middle of the eighteenth century onwards. The second
phase of industrialisation occurred around the mid-
nineteenth century and had a much greater environmental impact, saw the application of the *steam engine*
powered by coal to pumps, machines, ships and then
railways, which spread rapidly throughout Europe.
Industrialisation continued in the late nineteenth century with the rise of *steel* and *chemical industries* and
the application of *electrical power*: Germany and the
United States surpassed Great Britain in industrial
output at this time, while Russia and Japan began to
industrialise. These countries enhanced their position
at the expense of Europe in the fourth phase, which
saw a further revolution in transport, the invention of
the *internal combustion engine*, its application to shipping and *railways* and the ascendancy of *road* and
air transport. GLOBALISATION continues in the present
phase, which began after World War II, with major
centres of industrial activity developing around the
Pacific rim, in the Middle East and in Asia based less
on 'heavy engineering' and more on electronics and
SERVICE INDUSTRIES.

Differences remain between the older centres of
industrialisation and those that are 'catching up' but

overall industrial expansion has been enormous. This
has been possible only through the successive replacement of technologies during TECHNOLOGICAL CHANGE.
These changes have yielded major productivity gains
in energy use, materials and labour which have sustained the increasing levels of output and incomes,
eased demands on NATURAL RESOURCES, reduced traditional environmental impacts such as AIR POLLU-
TION and created more leisure time and a RECREATION
industry. However, new environmental concerns have
arisen, such as enhanced GLOBAL WARMING from burning FOSSIL FUELS, EUTROPHICATION and use of PESTICIDES,
from the industrialisation of agriculture, synthetic
chemicals leading to *ozone holes* and the uncertain
environmental impact of BIOTECHNOLOGY and GENETIC
ENGINEERING.

Since the eighteenth century, *labour productivity*
has risen by a factor of 200 in industry and at least a
factor of 20 in agriculture, while productivity in the
use of natural resources and in energy use per unit
of economic output has risen by a factor of 10. Historically, such economic or *technological productivity*
gains have been outpaced by output and consumption
leading to heavier ENVIRONMENTAL BURDENS. However,
in the past two decades, demand for bulk materials and
energy use per unit of economic output has stabilised
in the most advanced industrialised countries. *Environmental productivity* gains can therefore be seen to be
an unplanned side effect of recent technological productivity gains and it has been suggested that there is
a large potential for further environmental productivity
gains in the future.   *JAM*

[*See also* DE-INDUSTRIALISATION, INFORMATION AGE
(IA), NEWLY INDUSTRIALISING COUNTRIES (NICS),
POSTINDUSTRIALISATION]

Bairoch P (1982) International industrialization levels from
1750 to 1980. *Journal of European Economic History* 11:
269–333.
Headrick DR (1990) Technological change. In Turner II BL,
Clark WC, Kates RW et al. (eds), *The Earth transformed
by human action.* Cambridge: Cambridge University
Press, 55–86.
Ilbery BW and Bowler IR (1996) Industrialization and world
agriculture. In Douglas I, Huggett R and Robinson M
(eds) *Companion encyclopedia of geography.* London:
Routledge, 228–248.
Pomeranz K (2000) *The great divergence: China, Europe
and the making of the modern world economy.* Princeton,
NJ: Princeton University Press.
Weiss J (2002) *Industrialization and globalization: Theory
and evidence from developing countries.* London:
Routledge.

industry   (1) The production of goods and services
including, for example, MANUFACTURING INDUSTRIES and
SERVICE INDUSTRIES, which are characteristic of developed

societies. (2) In the context of archaeology, a set of ARTE-FACTS drawn from different ASSEMBLAGES but related in terms of technology, style or other context: LITHIC industries and CERAMIC industries are examples.   *JAM*

[*See also* CULTURE, INDUSTRIALISATION, TRADITION]

infant mortality rate   The average number of deaths of children under one year of age per 1,000 live births. Currently and historically, high infant mortality rates tend to reflect limited medical services.   *JAM*

[*See also* DEATH RATE]

infauna   Bottom-dwelling organisms (BENTHOS) that live in rather than on the SEDIMENT or rock. Infauna include *burrowers* (soft sediment) and *borers* (hard substrates). Trophic habits (see TROPHIC LEVEL) vary from *suspension-feeding* (exploiting food resources immediately above the sediment-water interface) or *deposit- or detritus-feeding* (mining the sediment for organic particles), to PREDATION and *scavenging*. Soft-bodied *worms* are important among the infauna as agents of reworking of surface sediment through BIOTURBATION. In modern marine settings, soft-bodied organisms comprise the majority of species in *benthic faunas* and are increasingly dominant in FAUNAS offshore shelf settings. In the FOSSIL RECORD, such organisms are mostly represented only by TRACE FOSSILS or BIOTURBATION fabrics in the sediment, except in FOSSIL LAGERSTÄTTEN. Organic material is concentrated in near-surface sediment (<5 cm) and decreases rapidly downwards. The PHANEROZOIC history of infaunal TIERING, or *vertical partitioning* of food resources, starts with only very shallow burrowing in the Early PALAEOZOIC, expanding to a depth of about 1 m from Late Palaeozoic onwards. Evolutionary ADAPTIVE RADIATION of infauna is a feature of the MESOZOIC *Marine Revolution*, shown notably by NICHE diversification of *bivalves*, *echinoids*, *crustaceans* and *soft-bodied organisms* and well-developed infaunal tiering.   *LC*

[*See also* EPIFAUNA]

Aberhan M, Kiessling W and Fürsich FT (2006) Testing the role of biological interactions in the evolution of mid-Mesozoic marine benthic ecosystems. *Paleobiology* 32: 259–277.

Ausich WI and Bottjer DJ (2001) Sessile invertebrates. In Briggs DEG and Crowther PA (eds) *Palaeobiology II.* Malden, MA: Blackwell, 384–386.

Bottjer DJ and Ausich WI (1986) Phanerozoic development of tiering in soft substrata suspension-feeding communities. *Paleobiology* 12: 400–420.

inferential statistic   A statistic that involves the concept of PROBABILITY to draw conclusions from samples about statistical POPULATIONS. Inferential statistics or *probabilistic statistics* provide a means of measuring the UNCERTAINTY associated with SAMPLING and a precise measure of the statistical confidence that can be placed in results based on representative samples.   *JAM*

[*See also* DESCRIPTIVE STATISTIC, NUMERICAL ANALYSIS, STATISTICAL ANALYSIS]

Casella G (2008) *Statistical inference, 2nd edition.* Pacific Grove, CA: Brooks Cole.

infiltration   (1) The process by which rainfall enters the soil surface. (2) The amount of water that enters the soil over a period of time.   *ADT/RPDW*

[*See also* HORTONIAN OVERLAND FLOW, HYDROPHOBICITY, INFILTRATION CAPACITY, RUNOFF PROCESSES, WATER REPELLENCY]

Knapp BJ (1978) Infiltration and storage of soil water. In Kirkby MJ (ed.) *Hillslope hydrology.* Chichester: Wiley, 43–72.

infiltration capacity   The maximum rate at which rainfall can enter the surface soil under given conditions. If it is exceeded, excess water ponds on the surface and HORTONIAN OVERLAND FLOW is generated. The concept formed the cornerstone of R.E. Horton's theory of RUNOFF in 1933, which dominated HYDROLOGY for 40 years. Infiltration capacity under natural conditions only tends to be lower than characteristic *rainfall intensities* in some ARIDLANDS, where the lack of vegetation allows CRUSTING OF SOIL, and in areas with soils rich in silt and clay. In most vegetated areas, infiltration capacities are relatively high. Infiltration capacities may be radically reduced by many forms of human interference, notably DEFORESTATION, OVERGRAZING, ANTHROPOGENIC fires, SOIL EROSION and URBANISATION. Infiltration capacities are also low in areas where soils exhibit HYDROPHOBICITY (or WATER REPELLENCY).   *RPDW*

[*See also* RUNOFF PROCESSES]

Horton RE (1933) The role of infiltration in the hydrologic cycle. *Transactions of the American Geophysical Union* 14: 446–460.

influenza   An infectious DISEASE caused by the influenza VIRUS, which is present in the guts of domestic and wild birds (the natural reservoir for the disease). *Flu* has probably affected humans since the DOMESTICATION of water fowl and pigs in China around 9,500 years ago. OUTBREAKS commonly occur each winter, killing up to half a million people globally, EPIDEMICS are frequent, and there were three PANDEMICS in the twentieth century. The 1918–1919 *Spanish flu* pandemic (also known as *La Grippe*), which killed an estimated 50–100 million people, may have been the deadliest NATURAL DISASTER of human history. *Bird flu* strains commonly swap GENES with human strains in domestic pigs before they can 'jump' the species barrier and infect humans efficiently. New strains of the virus can spread rapidly through human populations but lose virulence as people gain immunity. Such strains circulate in the community slowly accumulating mutations by GENETIC DRIFT until they have changed sufficiently to be unrecognisable to the human immune system and may then infect again.   *JAM*

[*See also* HUMAN HEALTH HAZARDS, ZOONOSES]

Crosby AW (2003) *America's forgotten pandemic: The influenza of 1918, 2nd edition.* Cambridge: Cambridge University Press.

Olsen B, Munster VJ, Wallenstein A et al. (2006) Global patterns of influenza: A virus in wild birds. *Science* 312: 384–388.

Phillips H and Killingray D (eds) (2003) *The Spanish influenza pandemic of 1918-19: New perspectives.* New York: Routledge.

Quinn T (2008) *Flu: A social history of influenza.* London: New Holland.

Tambyah P and Leung P-C (eds) (2006) *Bird flu: A rising pandemic in Asia and beyond?* Singapore: World Scientific Publishing.

infochemical   A chemical that, in the natural context, conveys information between organisms and evokes a behavioural or physiological response. A *pheromone* evokes the response in an individual of the same species, whereas an *allelochemical* affects a different species (see ALLELOPATHY). Interactions between organisms are also mediated by other classes of chemicals, such as NUTRIENTS and TOXINS.   *JAM*

[*See also* ECOPHYSIOLOGY]

Dicke M and Takken W (eds) (2006) *Chemical ecology: From gene to ecosystem.* Dordrecht: Springer.

Information Age (IA)   Also known as the *Computer Age*, the Information Age is seen by some as the modern successor to the STONE, BRONZE and IRON AGES. It is a period characterised by INFORMATION TECHNOLOGY (IT), which has enabled individuals and organisations to access and transfer large amounts of information almost instantaneously. The Information Age began with the development of telecommunications and computers. Arguably, it began with the invention of the telegraph in the nineteenth century but it has been facilitated by further developments, including radio and television, computers, the Internet, search engines and mobile phones.   *CJB*

Castells M (1996-2010) *Information Age: Economy, society and culture, volumes 1–3.* Chichester: Wiley-Blackwell.

information technology (IT)   The gathering, processing and distribution of all forms of qualitative and quantitative information by use of computers and telecommunications technology. Adoption of IT marks a profound cultural, social and economic shift: information is power and a means to earn a living. Instead of MANUFACTURING INDUSTRY, richer nations are increasingly making a living from information processing: in the INFORMATION AGE (IA), the *information economy* is replacing the industrial economy. Employment patterns are shifting, some countries operate call centres or process data for others, there is increasing potential for many to work from home and reduce their CARBON FOOTPRINT and computing and robotics have already replaced large numbers of manufacturing and construction workers. Similarly, we are seeing the rise of the *information society* as IT influences almost all areas of life and, in Western cultures, there is increasing legislation relating to access to data (*freedom of information* rules). Even warfare and power relations are shifting as *cyberwarfare* develops, and events in North Africa and the Middle East in 2011 show how rapidly unrest can spread through the Internet and use of *mobile phones*. In addition, ENVIRONMENTAL MODELLING relies on IT and rapid data transmission supports REMOTE SENSING and ENVIRONMENTAL MONITORING activities. General purpose computing capacity grew at an annual rate of 58 per cent between 1986 and 2007, while bidirectional telecommunication grew by 28 per cent and stored information by 23 per cent. A recent IT advance is the practice of linking large networks of computers to model or analyse (see CLOUD COMPUTING, GLOBALISATION).   *CJB/JAM*

Hilbert M and López P (2011) The world's technological capacity to store, communicate, and compute information. *Science* 332: 60–65.

Turban E, Rainer Jr RK and Potter RF (2003) *Introduction to information technology, 2nd edition.* Chichester: Wiley.

Webster F (2006) *Theories of the information society, 3rd edition.* London: Routledge.

Zook M (2005) *The geography of the Internet industry: Venture capital, dot-coms, and local knowledge.* Boston: Blackwell.

infrared-stimulated luminescence (IRSL) dating   A form of LUMINESCENCE DATING based on stimulation by infrared wavelengths in FELDSPARS. Its advantages over OPTICALLY STIMULATED LUMINESCENCE (OSL) DATING are that (1) a wider wavelength region is available for detection and (2) IRSL in minerals is more effectively bleached during deposition.   *DAR/CJC*

Berger GW, Perz-Gonzalez A, Carbonell E et al. (2008) Luminescence chronology of cave sediments at the Atapuerca paleoanthropological site, Spain. *Journal of Human Evolution* 55: 300–311.

Lang A and Wagner GA (1996) Infrared stimulated luminescence dating of archaeosediments. *Archaeometry* 38: 129–141.

Tsukamoto S, Duller GAT, Wintle AG and Muhs D (2011)
Assessing the potential for luminescence dating of basalts.
Quaternary Geochronology 6: 61–70.
Wiggenhorn H, Lang A and Wagner GA (1994) Infrared
stimulated luminescence: Dating tool for archaeosediments.
Naturwissenschaften 81: 556–558.

inheritance, in evolutionary context

Biological inheritance, or HEREDITY, is the genetically controlled process of passing on of *heritable characteristics* (*genetic traits*) from one generation of an organism to the next generation. Through inheritance, variations exhibited by individuals (which are determined by the GENES of the parents) can accumulate in POPULATIONS, leading to EVOLUTION. *Heritability* refers to the proportion of the observable variations between individuals (the PHENOTYPE) that is due to genetic differences (the GENOTYPE), and allows a comparison of the relative importance of genes and the environment to the phenotypic variation.   *JAM*

[*See also* DNA, GENETICS]

Roberts WJC (1994) *Biological inheritance: An introductory genetics text*. Brighton: Book Guild Publishing.
Visscher PM, Hill WG and Wray NR (2008) Heritability in the genomics era: Concepts and misconceptions. *Nature Reviews Genetics* 9: 255–266.

inheritance, in landscape context

In GEOMORPHOLOGY, *landscape inheritance* is the concept that LANDSCAPES can retain LANDFORMS produced under climatic conditions different to the prevailing ones. The PERSISTENCE and prominence of the inherited landscape elements vary according to LANDSCAPE SENSITIVITY to *geomorphological change*. Similar concepts apply to other aspects of the landscape, such as *soil inheritance*. In PEDOLOGY, the question also arises as to whether particular features of a soil are inherited from the PARENT MATERIAL or produced by PEDOGENESIS.   *RAS/JAM*

[*See also* EQUILIBRIUM CONCEPTS IN GEOMORPHOLOGICAL AND LANDSCAPE CONTEXTS, GEOMORPHOLOGY AND ENVIRONMENTAL CHANGE, PALIMPSEST, RELICT]

Brunsden D (2001) A critical assessment of the sensitivity concept in geomorphology. *Catena* 42: 99–123.
Palumbo B, Angelone M and Bellanca A (2000) Influence of inheritance and pedogenesis on heavy metal distribution in soils of Sicily, Italy. *Geoderma* 95: 247–266.

inhibition model

A MODEL suggesting that the net effect of early colonising species on later arrivals is negative, delaying establishment and growth of the next stage in ECOLOGICAL SUCCESSION.   *LRW*

[*See also* COLONISATION, FACILITATION, TOLERANCE MODEL]

Connell JH and Slatyer RO (1977) Mechanisms of succession in natural communities and their role in
community stability and organization. *American Naturalist* 111: 1119–1144.

inlier

An area of older rock completely surrounded by younger, contrasting with the normal situation where a rock outcrop is bounded by older rocks on one side and younger on the other. Inliers can form as a result of EROSION or DEFORMATION.   *GO*

[*See also* OUTLIER]

inorganic matter

Substances which are devoid of plant material, animal material and *organic compounds*.   *UBW*

[*See also* MINEROGENIC SEDIMENT, SEDIMENT TYPES]

insect analysis

Insect remains are relatively rare in the SUBFOSSIL and especially the FOSSIL RECORD. The majority of fossil insect remains (except those in AMBER) consist of compressed and sometimes carbonised fragments or, more rarely, whole insects. Material may be found in a wide range of situations where preservation conditions are favourable, such as LACUSTRINE SEDIMENTS, CAVES, ARIDLANDS, PERMAFROST environments and also archaeological sites, such as WELLS and refuse MIDDENS. Aspects of insect analysis may be related to ENVIRONMENTAL ARCHAEOLOGY, particularly in the recognition of ANTHROPOGENIC modification of the natural environment. Assemblages associated with early occupation sites may provide information concerning *insect infestations* in ancient *granaries* or crops under CULTIVATION, or even suggest directions of early *trading systems* or HUMAN MIGRATION. A major use of fossil insects, particularly beetles, is also in PALAEOENVIRONMENTAL and *palaeoclimatic reconstruction* and in ZOOGEOGRAPHY, on account of their COLONISATION abilities, apparent *evolutionary stability* and specific HABITAT and TEMPERATURE preferences.   *DCS*

[*See also* BEETLE ANALYSIS, MUTUAL CLIMATIC RANGE (MCR) METHOD]

Brooks SJ (2006) Fossil midges (Diptera: Chironomidae) as palaeoclimatic indicators for the Eurasian region. *Quaternary Science Reviews* 25: 1894–1910.
Coope GR (1994) The response of insect faunas to glacial-interglacial climatic fluctuations. *Proceedings of the Royal Society B* 344: 19–26.
Elias SA (1994) *Quaternary insects and their environments*. Washington, DC: Smithsonian Institution Press.
Elias SA (2001) Coleoptera and Trichoptera. In Smol JP, Birks HJB and Last WM (eds) *Tracking environmental change using lake sediments, volume 4*. Dordrecht: Kluwer, 67–80.
Elias SA (2010) *Advances in Quaternary entomology*. Amsterdam: Elsevier.
Velle G, Brodersen KP, Birks HJB and Willassen E (2010) Midges as quantitative temperature indicator species: Lessons for palaeoecology. *The Holocene* 20: 989–1002.

inselberg A prominent, isolated, residual hill, usually smoothed and rounded. Inselbergs rise abruptly from and are surrounded by extensive lowland ERO-SION SURFACES and are characteristic of ARIDLANDS or SEMI-ARID landscapes in a late stage of evolution. The term, from the German meaning 'island mountain', was proposed by Bornhardt in 1900 to describe the abrupt, rocky hills that commonly interrupt tropical plains, but use of the term was soon extended to include a variety of isolated hill forms and its exact meaning became confused.

Three main mechanisms have been proposed for the origin of inselbergs: (1) the exposure of CORESTONES by STRIPPING PHASES brought about by ENVIRONMENTAL CHANGE over GEOLOGICAL TIME (see Figure), (2) PARAL-LEL SLOPE RETREAT across bedrock or (3) *scarp retreat* across deeply weathered rocks. The longevity and size of these landforms mean that they can occur in climatic environments different from those of their formation and also that they may have developed over a long sequence of climatic changes. Their development has often been considered to be favoured by a change of climate from prolonged warm and wet conditions towards a drier, more seasonal climate. Theories concerning their origin can be divided into four groups: (1) non-climatic theories that emphasise LITHOLOGY or GEOLOGICAL STRUCTURES or *scarp retreat* under moist climates; (2) single-climate theories that link inselberg development to a particular climate (see CLIMATIC GEO-MORPHOLOGY); (3) POLYGENETIC theories that invoke CLIMATIC CHANGE or a sequence of climates, in which a long period of deep weathering and ineffective erosion under a hot-wet climate is followed by a stripping phase under a less humid climate; and (4) polygenetic theories that invoke a combination of climates and geological factors of UPLIFT or tilting.

Single-climate theories have suggested that contrasting climates followed rather than caused inselberg formation. J. Büdel's DOUBLE PLANATION SURFACES theory favoured the *seasonally wet tropics*, whereas H. Bremer's DIVERGENT WEATHERING theory considered the *humid tropics* as more favourable for inselberg formation. Similarly, authors proposing climatic change theories have also disagreed about the sequence of climates that may be involved, some suggesting dissection and stripping during more humid phases, whereas others

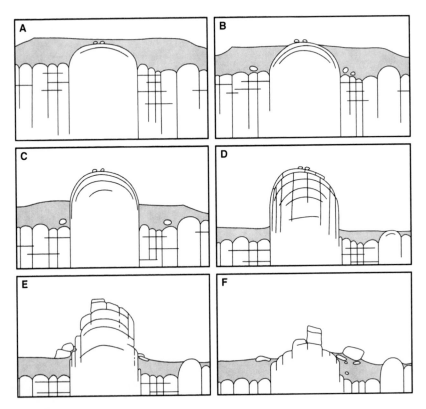

Inselberg The evolution of an inselberg: each stripping phase, during which the products of deep weathering (*shaded*) are removed from around the domed inselberg or bornhardt, is accompanied by lowering of the basal weathering surface, culminating in the formation of a residual koppie or tor (Thomas, 1994).

suggest that semi-arid phases favour stripping. With really large inselbergs (BORNHARDTS), a polygenetic origin over geological time would seem the only logical explanation.   *MAC/RPDW*

[*See also* BORNHARDT, ETCHPLAIN, LANDSCAPE EVOLUTION, PEDIMENT, TOR]

Bremer H and Jennings J (eds) (1978) Inselbergs. *Zeitschrift für Geomorphologie Supplementband* 31 [Special Issue].
Bremer H and Sander H (2002) Inselbergs: Geomorphology and geoecology. In Porembski S and Barthlott W (eds) *Inselbergs: Biodiversity of isolated rock outcrops in tropical and temperate regions.* Berlin: Springer, 7–35.
Ollier CD (1960) The inselbergs of Uganda. *Zeitschrift für Geomorphologie NF* 4: 43–52.
Piotr M (2009) Are any granite landscapes distinctive of the humid tropics? Reconsidering multiconvex topographies. *Singapore Journal of Tropical Geography* 30: 327–342.
Thomas MF (1994) *Geomorphology in the tropics: A study of weathering and denudation in low latitudes.* Chichester: Wiley.
Twidale CR (1990) The origin and implications of some erosional landforms. *Journal of Geology* 98: 343–364.

in-situ conservation 'On-site' CONSERVATION of animals and plants in their natural HABITAT. It depends on the availability of sufficient natural habitat to support the threatened POPULATION, which needs to have a viable *population size* and sufficient *genetic diversity*, and it may require measures such as *habitat protection* or *predator control*. In-situ conservation and its counterpart, EX-SITU CONSERVATION, are becoming increasingly interdependent approaches to the conservation of ENDANGERED SPECIES.   *JAM*

[*See also* BIODIVERSITY, HABITAT LOSS, WILDLIFE CONSERVATION]

Clayton LM, Milner-Gulland EJ, Sinaga DW and Mustari AH (2000) Effects of a proposed ex-situ conservation program on in-situ conservation of Babirosa, an endangered suid. *Conservation Biology* 14: 382–385.
Gusset M and Dick G (2010) 'Building a future for wildlife'? Evaluating the contribution of the world zoo and aquarium community to in-situ conservation. *International Zoo Yearbook* 44: 183–191.
Nabham GP and Tuxill J (2001) *People, plants and protected areas: A guide to in-situ management.* Abingdon: Routledge.
Volis S and Blecher M (2010) Quasi in situ: A bridge between ex-situ and in-situ conservation of plants. *Biodiversity and Conservation* 19: 2441–2454.

insolation The flow of direct SOLAR RADIATION intercepted by a unit area of a horizontal surface either at or above the Earth's surface. It varies with season, latitude, slope, ASPECT and atmospheric TURBIDITY and also on longer timescales due to SOLAR CYCLES, ORBITAL FORCING and other factors.   *JCS*

[*See also* MILANKOVITCH THEORY, SEASONALITY, SOLAR ENERGY]

Cruz FW, Burns SJ, Karmann I et al. (2005) Insolation-driven changes in atmospheric circulation over the past 116,000 years in subtropical Brazil. *Nature* 434: 63–66.

insolation weathering Rock breakdown resulting from direct solar heating of rock surfaces with consequent large diurnal temperature changes and differential expansion and contraction of different parts of the rock. Also known as *thermoclasty* or *thermal stress fatigue*, insolation weathering is a feature of ARIDLANDS. The diurnal range in AIR TEMPERATURE may exceed 50°C in hot DESERTS and the *daily maximum rock temperature* may reach 80°C. The related process of *dirt cracking* involves the cracking of rock as a result of the thermal expansion of material lodged within rock fissures. However, as with many other forms of PHYSICAL WEATHERING, CHEMICAL WEATHERING processes associated with moisture (e.g. HYDRATION and SALT WEATHERING) may also be involved. In particular, laboratory experiments involving heating and cooling of rock fragments with and without the presence of moisture have led to considerable doubt about heating and cooling alone being sufficient to account for rock breakdown.   *JAM*

Blackwelder F (1933) The insolation hypothesis in rock weathering *American Journal of Science* 26: 97–113.
Dorn RI (2011) Revisiting dirt cracking as a physical weathering process in warm deserts. *Geomorphology* 135: 129–142.
Goudie AS (2004) Insolation weathering. In Goudie AS (ed.) *Encyclopedia of geomorphology.* London: Routledge, 566–567.
Weiss T, Siegesmund S, Kirchner D and Sippel J (2004) Insolation weathering and hygric dilatation: Two competing factors in stone degradation. *Environmental Geology* 46: 402–413.

instantaneous field of view (IFOV) In REMOTE SENSING, the cone-shaped region of space from which radiation is collected for each discrete measurement, or PIXEL. The IFOV is defined by the solid angle subtended by a combination of the detector dimensions and the instrument focussing system. The sensitivity of an instrument is not usually uniform over the maximum extent of the IFOV, resulting in a two-dimensional *point spread function*. The nominal IFOV is usually reported for some fraction of the maximum value of the point spread function. The field of view of an instrument is the sum of the IFOVs recorded for a single line of the RASTER image. The size of the *IFOV footprint* on the surface is proportional to the distance between the point on the surface where the radiation is reflected and the instrument: the size of the IFOV footprint will therefore increase as the viewing direction moves away from the NADIR POINT. *GMS*

Schowengerdt RA (2006) *Remote sensing: Models and methods for image processing, 3rd edition*. New York: Academic Press.

instrument calibration   The act of associating an instrument's measurement to a sensor-independent geophysical quantity. For the comparison of measurements of the same instrument gathered at different times or of values from different parts of the same environment or image, *relative calibration* may be sufficient. *TS*

[*See also* CALIBRATION, EARLY INSTRUMENTAL METEOROLOGICAL RECORDS]

instrumental data   Derived from appliances with a standardised construction, exposure and observation times that respond to environmental conditions in a constant manner in order to acquire data that are precise and comparable from place to place and over periods of time. They normally have a *scale of measurement* that is calibrated against a known standard and often internationally agreed, as in the *Système Internationale d'Unités* (SI UNITS) agreed in 1948. Where instruments have a source of power (mechanical or electrical), they may record the environment in remote areas such as mountains, or continuously over long periods of time (e.g. AUTOMATIC WEATHER STATIONS (AWS) and DATA LOGGERS). They contrast with visual OBSERVATIONS, data from PARAMETEOROLOGY and PHENOLOGY, and PROXY DATA that depend on the responses of humans, other organisms or NATURAL ARCHIVES to environmental conditions that are more difficult to standardise. The data from early instruments lack the PRECISION and comparability of modern data because designs were unique and experimental. Instrumental data are now available for almost every aspect of the environment and environmental change. *JGT*

[*See also* EARLY INSTRUMENTAL METEOROLOGICAL RECORDS, INSTRUMENTAL RECORD, MEASUREMENT, REMOTE SENSING]

instrumental record   A record of an environmental variable involving MEASUREMENT on a quantitative scale using an instrument. In the context of ENVIRONMENTAL CHANGE, instrumental records are important for understanding *variability*, for detecting relatively short-term ANNUAL VARIABILITY and decadal PERIODICITIES and TRENDS, and for the CALIBRATION of PROXY DATA in PALAEOENVIRONMENTAL RECONSTRUCTION. An exceptionally early example of an instrumental record exists for the height of NILE FLOODS from about 5,000 years ago. Currently, the instrumental record of environmental change extends to data obtained by ENVIRONMENTAL MONITORING

and REMOTE SENSING, some of which is global in extent and covers the land, sea and air.

The instrumental record of CLIMATE is particularly important and illustrative. The network of regular meteorological observations, which reached its peak with the massive expansion of *meteorological stations* on land in the last half of the nineteenth century, dwindles as one goes back in time towards the invention of the thermometer by Galileo in AD 1597. The British Isles has the longest systematic record of air temperature in the world and precipitation was first measured at Kew in AD 1697. Atmospheric pressure was first recorded at Trondheim, Norway, in 1762 and London in 1787. By 1750, several meteorological records are available from northwestern Europe and the first early records have survived from 1850 in the eastern USA as well as from Eastern Europe and parts of Asia. From most other parts of the world, records stretch back for about a century and a half (e.g. Australasia, Auckland 1853; South America, Rio de Janeiro 1832). EARLY INSTRUMENTAL METEOROLOGICAL RECORDS were often taken in unrepresentative locations: for example, temperature readings in Britain were recorded for a time in the eighteenth century in north-facing rooms. Factors such as changes in instrumentation, exposure, location and methodological practises can all cause *inhomogeneities* in records. *AHP/JAM*

[*See also* CENTRAL ENGLAND TEMPERATURE RECORD (CET), CLIMATIC VARIABILITY, CLIMATIC VARIATION, HISTORICAL CLIMATOLOGY, HOMOGENEOUS SERIES]

Alverson K (2012) Direct observation and monitoring of climate and related environmental change. In Matthews JA, Bartlein PJ, Briffa KR et al. (eds) *The SAGE handbook of environmental change, volume 1*. London: Sage, 53–66.
Jones PD and Bradley RS (1992) Climatic variations in the longest instrumental records. In Bradley RS and Jones PD (eds) *Climate since A.D.1500*. London: Routledge, 246–268.
Middleton WEK (1969) *The history of the thermometer and its use in meteorology*. Baltimore: Johns Hopkins University Press.

interaction indicator   An ENVIRONMENTAL INDICATOR, such as a *sustainability indicator*, that signals the nature of the interaction between people and ecosystems, and hence enables an assessment of how human activities should be modified to avoid undesirable outcomes. *JAM*

[*See also* ECOSYSTEM INDICATOR, GEO-INDICATOR, SYNTHESIS INDICATOR]

Hák T, Moldan B and Dahl AL (eds) (2007) *Sustainability indicators in scientific assessment*. Washington, DC: Island Press.

Hodge RA (1996) Indicators and their role in assessing progress towards sustainability. In Berger AR and Iams WJ (eds) *Geoindicators: Assessing rapid environmental changes in earth systems*. Rotterdam: Balkema, 19–24.

Ward TJ (2000) Indicators for assessing the sustainability of Australia's marine ecosystems. *Marine and Freshwater Research* 51: 435–446.

Interactive Graphics Retrieval System (INGRES)

A DATABASE MANAGEMENT SYSTEM for the efficient handling of RELATIONAL DATABASES. It is used especially in education and business, and capable of supporting structured QUERY LANGUAGES.   *VM*

interception   The temporary storage of a proportion of PRECIPITATION, usually by VEGETATION, prior to it either evaporating back to the atmosphere or falling to the ground as *throughfall* or STEMFLOW. *Interception loss* is the amount (mm) or proportion (%) of precipitation that fails to reach the ground and is evaporated back to the atmosphere. Interception tends to be of greatest importance in densely vegetated areas with high annual rainfall mainly falling in small rainstorms of low intensity, but of less importance in climates where a high proportion of the annual rainfall falls in large, intense rainstorms. In the UK, interception is greatest in *conifer forest*, intermediate in *broad-leafed forest* and least in MOORLAND and *grassland*. Interception is greatly affected by LANDUSE CHANGE, with consequences for river DISCHARGE, WATER RESOURCES and *water supply*.   *ADT/RPDW*

Durocher MG (1990) Monitoring spatial variability of forest interception. *Hydrological Processes* 4: 215–219.

Vernimmen RRE, Bruijnzeel LA, Romdoni A and Proctor J (2007) Rainfall interception in three contrasting lowland rain forest types in Central Kalimantan, Indonesia. *Journal of Hydrology* 340: 217–232.

interdisciplinary research   Research that exists between or that transcends the boundaries of conventional *disciplines* but that involves the merging of KNOWLEDGE, the development of common concepts and the devising of unified METHODOLOGIES to investigate and solve problems. *Interdisciplinarity* has been variously defined as a methodology, a concept, a process, a way of thinking, a philosophy and a reflexive ideology. True interdisciplinary research may be distinguished from the usual practice for issues that transcend conventional discipline boundaries, which would involve a range of specialists bringing their expertise and their own methodologies to bear: this is MULTIDISCIPLINARY research. O'Riordan (1999: 16) argued, 'True interdisciplinarity has probably never existed, because the phenomenon involves the unification of concepts that are designed to be conceived as separate entities'. Nevertheless, he identified four promising concepts

that 'embrace both the social and natural sciences': (1) CHAOS THEORY; (2) *social learning*; (3) *dynamic equilibrium*; and (4) CARRYING CAPACITY, which he saw as akin to SUSTAINABILITY.   *FMC*

Frodeman R, Mitcham C and Klein JT (eds) (2010) *The Oxford handbook of interdisciplinarity*. New York: Oxford University Press.

Jacobs JA and Frickel S (2009) Interdisciplinarity: A critical assessment. *Annual Review of Sociology* 35: 43–65.

Klein JT (1990) *Interdisciplinarity: History, theory and practice*. Detroit, MI: Wayne State University Press.

Moran J (2010) *Interdisciplinarity, 2nd edition*. Abingdon: Routledge.

O'Riordan T (ed.) (1999) *Environmental science for environmental management, 2nd edition*. Harlow: Prentice Hall.

Repko AF (2008) *Interdisciplinary research: Process and theory*. Thousand Oaks, CA: Sage.

Weart S (2013) Rise of interdisciplinary research on climate. *Proceedings of the National Academy of Sciences USA* 110, Supplement 1: 3657–3664.

interferometry   The process of combining two coherent measurements of the same surface to detect differences in *phase*, from which *differential range* and *range change* may be inferred. A coherent measurement is one that retains information on both the phase and the amplitude (brightness) of scattered or reflected RADIATION from the ELECTROMAGNETIC SPECTRUM. The phase information is often discarded but under certain conditions the relative difference in phase between two measurements of the same surface can provide information on surface change, precise to a fraction of the wavelength of the radiation. In REMOTE SENSING, interferometry is often undertaken using spaceborne *synthetic aperture radar* (SAR) instruments (in a technique known as *SAR interferometry* or INSAR). In this technique, two images of the same scene are acquired from slightly different positions (by a platform carrying two antennas in either across- or along-track configuration), by a single-antenna SAR imaging at two different times (*repeat-pass interferometry*) or by identical sensors flown in tandem, such as DLR (Deutsches Zentrum für Luft- und Raumfahrt)-Astriums's *TerraSAR-X* (launched in 2007) and TerraSAR-X add-on for Digital Elevation Measurement *TanDEM* (launched in 2010). This mission comprises two identical satellites' sensors that are placed in the same orbit configuration but between 250 and 500 m apart. An *interferogram* is produced by constructing a map of the phase differences but requires that both images have been co-registered to *sub-pixel accuracy* (see PIXEL).

Provided that the baseline (distance between two detectors) and orientation between the two antenna positions is known precisely, the phase difference between two measurements can be directly related to the range (distance) to scatterers in the scene and hence the topography of a surface mapped. Applications have thus far centred around topographic mapping over land for DIGITAL ELEVATION MODELS/MATRIX (DEMs), monitoring ice margins and the dynamics of GLACIERS. Relative changes in topography can be mapped using *differential interferometry* (*Differential* Synthetic Aperture Radar *Interferometry;* DifSAR) where commonly a pair of interferograms of the same area are differenced to reveal any changes that have occurred in the Earth's surface. Successful interferometry often has to compete against *decoherence* which degrades the quality of the phase information. Decoherence is caused by *spatial decorrelation* (where the terrain is viewed from different directions and therefore has slightly different scattering properties) and also by *temporal decorrelation* (where changes in the scattering surface result from different weather conditions or movement of the surface). In some cases, propagation effects in the ATMOSPHERE may also randomise phases and lead to patches of low coherence (*artefacts*) in the image.   *MEJC/PJS*

[*See also* MICROWAVE REMOTE SENSING (MRS), RADAR REMOTE-SENSING INSTRUMENTS]

Prati C, Ferretti A and Perissin D (2010) Recent advances on surface ground deformation measurement by means of repeated space-borne SAR observations. *Journal of Geodynamics* 49: 161–170.
Quincey DJ and Luckman A (2009) Progress in satellite remote sensing of ice sheets. *Progress in Physical Geography* 33: 547–567.
Zhou X, Chang N-B and Li S (2009) Applications of SAR interferometry in earth and environmental science research. *Sensors* 9: 1876–1912.

intergenerational equity   A concept of ENVIRONMENTAL ETHICS that current generations should not prejudice the rights of future generations by degrading environmental systems. Instead, by application of the PRECAUTIONARY PRINCIPLE in relation to HUMAN IMPACTS, the environment should be held in trust for future generations. It is important, for example, in relation to the depletion of RESOURCES, environmental SUSTAINABILITY and the maintenance of *quality-of-life*.   *JAM*

[*See also* QUALITY-OF-LIFE INDICATORS, STEWARDSHIP CONCEPT]

Parfit D (1982) Future generations: Further problems. *Philosophy and Public Affairs* 11: 113–172.
Roemer J and Suzumura K (eds) (2007) *Intergenerational equity and sustainability*. Basingstoke: Palgrave Macmillan.

interglacial   A period of thermal improvement separating cold GLACIAL EPISODES, when climatic conditions were similar to, or warmer than, those experienced today. The current interglacial, known as the HOLOCENE,

has lasted approximately 11,700 years (10,000 radiocarbon years). A large number of interglacials are recognised for the QUATERNARY in MARINE SEDIMENT CORES and given odd numbers in terms of oxygen MARINE ISOTOPIC STAGES (MIS). However, fewer interglacial periods are precisely identified and dated in the terrestrial record, and some longer periods encompassing several interglacials are defined as *interglacial complexes* as for the CROMERIAN INTERGLACIAL in the British Isles. Interglacials are usually considered to have been periods of relatively stable climate showing a consistent pattern of climatic warming and subsequent cooling, as defined in the INTERGLACIAL CYCLE. Data from ICE CORES from Greenland have been taken to represent brief cooling episodes. Although these data have been discounted, there is still some debate over interglacial climate stability. Interglacials MIS 11 and 19 have been studied in some detail as possible analogues for the current interglacial, allowing some estimation of the degree of anthropogenic climate influence and duration of 'warm' conditions (i.e. estimating the length of time before the onset of conditions leading to the next GLACIATION).   *CJC*

[*See also* QUATERNARY TIMESCALE]

Lowe JJ and Walker MJC (1997) *Reconstructing Quaternary environments, 2nd edition*. Harlow: Longman.
Penkman KEH, Preece RH, Bridgland DR et al. (2011) A chronological framework for the British Quaternary based on *Bithynia opercula*. *Nature* 476: 446–449.
Schreve D and Candy I (2010) Intergalcial climates: Advances in our understanding of warm climate episodes. *Progress in Physical Geography* 34: 845–856.
Sirocko F, Claussen M, Litt T and Sánchez-Goñi MF (eds) (2007) *The climate of past interglacials*. Amsterdam: Elsevier.
Tzedakis PC (2011) The MIS11-MIS1 analogy, southern European vegetation, atmospheric methane and the 'Early Anthropogenic hypothesis'. *Climate of the Past* 6: 131–144.

interglacial cycle   The cycle of responses of FLORA, VEGETATION and SOILS to changes in seasonal insolation according to MILANKOVITCH THEORY (see Figure). Originally proposed by Iversen (1958), and modified by Andersen (1966), the idea has been reviewed by Birks (1986). Iversen proposed four stages to the cycle: (1) *cryocratic*, (2) *protocratic*, (3) *mesocratic* and (4) *telocratic*. There are three key concepts: (1) a rise and subsequent fall in mean annual temperature, (2) progressive vegetation development (MIGRATION and ECOLOGICAL SUCCESSION) in the early stages and (3) RETROGRESSIVE SUCCESSION in the later stages (see Figure). This concept has rather been sidelined as higher-resolution studies have allowed more detailed determination of patterns within individual INTERGLACIALS, although very similar vegetation dynamics appear to be recorded during recent interglacials despite differences in ORBITAL FORCING.

BA/CJC

Andersen S Th (1966) Interglacial succession and lake development in Denmark. *Palaeobotanist* 15: 117–127.

Birks HJB (1986) Late-Quaternary biotic changes in terrestrial and aquatic environments, with particular reference to north-west Europe, In Berglund BE (ed.) *Handbook of Holocene palaeoecology and palaeohydrology*. Chichester: Wiley, 3–65.

Cheddadi R, de Beaulieu J-L, Jouzel J et al. (2005) Similarity of vegetation dynamics during interglacial periods. *Proceedings of the National Academy of Sciences* 102: 13939–13943.

Iversen J (1958) The bearing of glacial and interglacial epochs on the formation and extinction of plant taxa. *Uppsala Universiteit Årsskrift* 6: 210–215.

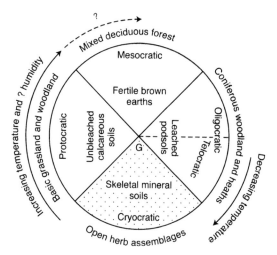

Interglacial cycle An interglacial cycle showing interrelated changes in climate, vegetation and soils (Birks, 1986).

Intergovernmental Panel on Climate Change (IPCC)

The IPCC was established jointly by the World Meteorological Organisation (WMO) and the United Nations Environment Programme (UNEP) in 1988. It was chaired by B. Bolin (Sweden); other members of the panel were A. Al Gain (Saudi Arabia), who was vice chairman; J.A. Adejokun (Nigeria), who was rapporteur; and N. Sundararaman (WMO), who was secretary. The panel had two objectives: (1) to assess the scientific information related to the issue of CLIMATE CHANGE and to evaluate its environmental and socio-economic consequences and (2) to formulate realistic response strategies for the MANAGEMENT of climate change.

To carry out these objectives, the panel set up three Working Groups to assess, in relation to climate change, the available scientific information (Working Group I, chaired by J.T. Houghton, UK) and its impacts (Working Group II, Y. Israel, USSR), and to formulate policy responses (Working Group III, F. Bernthal, USA). In addition, a special committee (chaired by M.J. Ripert, France) was established to promote the participation of DEVELOPING COUNTRIES in these activities.

In 1990, the initial reports of Working Groups I and II were published in time for the Second World Climate Conference held in Geneva. Working Group I (scientific assessment) concluded that there is a natural GREENHOUSE EFFECT, which keeps the Earth warmer than it would otherwise be and that human activities result in EMISSIONS that have enhanced, and will increasingly enhance, this effect. It confidently calculated that immediate reductions by 60 per cent in the emissions of long-lived gases (CARBON DIOXIDE, *nitrous oxide* and CHLOROFLUOROCARBONS (CFCs)) were necessary to keep concentrations at current levels and predicted that under the BUSINESS-AS-USUAL SCENARIO (i.e. no steps will be taken to limit emissions) global mean temperatures would rise by 0.3°C per decade over the next 100 years. Working Group II (impacts and adaptations) concluded that changes in climate would have important effects, amongst others, on AGRICULTURE, natural *ecosystems* and WATER RESOURCES. Working Group III (mitigation, economic and social dimensions of climate change) found that climate change is a global issue and will require a global effort, but the onus will be on the *developed countries*. Updates are ongoing and the various assessment reports for the Working Groups and some synthesis reports have all been published by either Cambridge University Press or other publishers. (All the reports can be accessed at http://www.ipcc.ch/ publications_and_data/publications_and_data_reports .shtml.) The four reports so far have been controversial with some atmospheric scientists (*climate change skeptics* or *deniers*) casting doubt on either the data or the results or both. But they are a vociferous minority, and generally speaking, ANTHROPOGENIC climate change is now accepted. The *Fifth Assessment Report (AR5)* is in progress and will again consist of three Working Group reports and a synthesis report. It should be completed in 2013/2014.

BDG/DAW

Houghton JT, Ding Y, Griggs DJ et al. (2001) *Climate change 2001: The scientific basis.* Cambridge: Cambridge University Press.

Houghton JT, Jenkins GJ and Ephraums JJ (eds) (1990) *Climate change. The IPCC scientific assessment.* Cambridge: Cambridge University Press.

Houghton JT, Meira Filho LG, Callander N et al. (1996) *Climate change 1995: The science of climate change.* Cambridge: Cambridge University Press.

Solomon S, Qin D, Manning M et al. (2007) *Climate change 2007: The physical science basis* [Contribution of

Working Group I to the Fourth Assessment Report of the Intergovernmental Panel on Climate Change]. Cambridge: Cambridge University Press.

interhemispheric see-saw

The concept of *antiphase* CLIMATIC VARIABILITY and/or CLIMATIC CHANGE between the Northern and Southern Hemispheres (also known as the *bipolar see-saw*). Although the timing of GLACIAL-INTERGLACIAL CYCLES between the two hemispheres are generally in-phase, shorter-term CENTURY-TO MILLENNIAL-SCALE VARIABILITY may exhibit antiphase temporal variations, at least for some of the time. During the last DEGLACIATION, for example, there appears to have been an antiphased millennial-scale temperature response of the two hemispheres superimposed on a globally coherent warming. Differences between the respective temperature changes in the two hemispheres between the LAST GLACIAL MAXIMUM (LGM) and the early HOLOCENE correlate with the strength of the Atlantic MERIDIONAL OVERTURNING CIRCULATION (MOC), which suggests that the antiphased temperature response reflects changes in the OCEAN CIRCULATION.   *JAM*

[*See also* DANSGAARD-OESCHGER (D-O) EVENTS, THERMOHALINE CIRCULATION, YOUNGER DRYAS STADIAL]

Barker S, Diz P, Vautravers MJ et al. (2009) Interhemispheric Atlantic seesaw response during the last deglaciation. *Nature* 457: 1097–1102.

Newnham RM, Vandergoes MJ, Sikes E et al. (2012) Does the bipolar seesaw extend to the terrestrial southern mid-latitudes? *Quaternary Science Reviews* 36: 214–222.

Shakun JD, Clark PU, He F et al. (2012) Global warming preceded by increasing carbon dioxide concentrations during the last deglaciation. *Nature* 484: 49–54.

Weldeab S (2012) Bipolar modulation of millennial-scale West African monsoon variability during the last glacial (75,000-25,000 years ago). *Quaternary Science Reviews* 40: 21–29.

Wolff EW (2012) A tale of two hemispheres. *Nature* 484: 41–42.

intermediate disturbance hypothesis

Originally proposed by J.H. Connell and M. Huston, this hypothesis states that species richness in a habitat is highest when disturbances occur with intermediate frequency or intensity. As DISTURBANCE frequency or intensity increases beyond this level, richness declines because only species that colonise and grow very rapidly are able to establish or maintain themselves successfully. At low levels of disturbance, *competitive exclusion* (see COMPETITION) reduces local species richness. This hypothesis may apply at a local scale, but recent evidence suggests that it is infrequently important. There is no evidence to suggest that it can explain large-scale patterns such as the *latitudinal gradient* BIODIVERSITY. MANAGEMENT of *disturbance frequency* is important in CONSERVATION.

When humans alter natural *disturbance regimes*, ecosystems may undergo major structural changes, such as when *fire suppression* causes *prairie* habitats to change into forests through ECOLOGICAL SUCCESSION. Such HABITAT LOSS is a cause of EXTINCTION. CLIMATIC CHANGE may change disturbance regimes at a very large scale.   *JTK*

[*See also* DIVERSITY CONCEPTS, DIVERSITY-STABILITY HYPOTHESIS]

Connell JH (1978) Diversity in tropical rain forests and coral reefs. *Science* 199: 1302–1310.

Dos Santos FS, Johst K, Huth A and Grimm V (2010) Interacting effects of habitat destruction and changing disturbance rates on biodiversity. Who is going to survive? *Ecological Modelling* 221: 2776–2783.

Feyer LJ and Duffus DA (2011) Predatory disturbance and prey species diversity: The case of gray whale (*Eschrichtius robustus*) foraging on a multi-species mysid (family Mysidae) community. *Hydrobiologia* 678: 37–47.

Fox JF (1981) Intermediate levels of soil disturbance maximize alpine plant diversity. *Nature* 293: 564–565.

Huston M (1979) A general hypothesis of species diversity. *American Naturalist* 113: 81–101.

Mackey RL and Currie DJ (2000) A re-examination of the expected effects of disturbance on diversity. *Oikos* 88: 483–492.

Mackey RL and Currie DJ (2001) The diversity-disturbance relationship. Is it generally strong and peaked? *Ecology* 82: 3479–3492.

Miller AD, Roxburgh SH and Shea K (2011) How frequency and intensity shape diversity-disturbance relationships. *Proceedings of the National Academy of Sciences* 108: 5643–5648.

Rixen C, Haag S, Kulakowski D and Bebi P (2007) Natural avalanche disturbance shapes plant diversity and species composition in subalpine forest belt. *Journal of Vegetation Science* 18: 735–742.

Svensson JR, Lindegarth M and Pavia H (2010) Physical and biological disturbances interact differently with productivity: Effects on floral and faunal richness. *Ecology* 91: 3069–3080.

Wang SL and Chen HYH (2010) Diversity of northern plantations peaks at intermediate management intensity. *Forest Ecology and Management* 259: 360–366.

intermediate technology

Technology that is more productive than inefficient *traditional technologies* but less costly and environmental damaging than *industrialised technologies*. Originating from the ideas of E.F. Schumacher, use of intermediate technologies implies ECOLOGICAL MODERNISATION without additional unnecessary complexity. The term is a predecessor and subset of APPROPRIATE TECHNOLOGY.   *CJB/JAM*

[*See also* ALTERNATIVE TECHNOLOGY]

Schumacher EF (1973) *Small is beautiful: A study of economics as if people mattered*. London: Blond and Briggs.

intermediate water   WATER MASSES that are found at intermediate depths in the OCEANS, below SURFACE WATERS and above DEEP WATERS. The intermediate waters of the world ocean are usually traced by their SALINITY values. Intermediate waters with salinity maxima, *Mediterranean Sea Water* and *Red Sea Water*, are formed by high surface water evaporation processes. Intermediate water masses with salinity minima are formed at high subtropical or subpolar latitudes from surface water that has relatively low salinity. *North Pacific Intermediate Water* is formed in the Okhotsk Sea, *Labrador Sea Water* forms in the western Labrador Sea and *Antarctic Intermediate Water* forms in the region of *Antarctic Circumpolar Current* system. Intermediate water is sometimes incorrectly used interchangeably with the term *mode water*. *Mode waters* are waters of extremely uniform properties that occur over an extensive depth range, usually formed by convection, and often contributing to more than one water mass.   *JP*

Glickman TS (2000) *Glossary of meteorology, 2nd edition.* Chicago: University of Chicago Press.

Graham JA, Stevens DP, Heywood KJ and Wang Z (2011) North Atlantic climate responses to perturbations in Antarctic Intermediate Water. *Climate Dynamics* 37: 297–311.

Hartin CA, Fine RA, Sloyan BM et al. (2011) Formation rates of subantarctic mode water and Antarctic intermediate water within the South Pacific. *Deep-Sea Research I* 58: 524–534.

Louarn E and Morin P (2011) Antarctic intermediate water influence on Mediterranean Sea water outflow. *Deep-Sea Research I* 58: 932–942.

intermediate-complexity modelling   An approach to modelling any complex SYSTEM that aims to bridge the gap between conceptual, more inductive and simplistic models and the more complex, three-dimensional comprehensive models. In EARTH-SYSTEM ANALYSIS (ESA), these models of intermediate complexity (known as EMICs or most recently as NEMICs (*natural Earth systems models of intermediate complexity*) are part of a spectrum of models that aim to simulate the FEEDBACK MECHANISMS between the components of the CLIMATIC SYSTEM. The purpose of EMICs is to avoid the problems of both oversimplification and oversophistication, providing a pragmatic solution to modelling the complexities of planet Earth as a whole.

EMICs are designed for a broad range of purposes and can include most of the processes described in comprehensive models in a more parameterised form with an aim of simulating the interactions among several components of the natural Earth system, including BIOGEOCHEMICAL CYCLES. Other EMICs have a more defined purpose having a lesser degree of integration and thus tend to be used for long-term ensemble simulations. There are over 10 EMICs models in existence with some

general in nature and others focussed on specific components of the Earth system, including the ANTHROPOSPHERE and GEOGRAPHY.   *DSB/JAM*

Claussen M, Mysak LA, Weaver AJ et al. (2002) Earth system models of intermediate complexity: Closing the gap in the spectrum of climate system models. *Climate Dynamics* 18: 579–586.

Ganopolski A, Rahmstorf S, Petoukhov V and Claussen M (1998) Simulation of modern and glacial climates with a coupled global model of intermediate complexity. *Nature* 391: 351–356.

Huybrechts P, Goelzer H, Janssens I et al. (2011) Response of the Greenland and Antarctic ice sheets to multi-millennial greenhouse warming in the Earth system model of intermediate complexity LOVECLIM. *Surveys in Geophysics* 42: 397–416.

International Tree-Ring Database (ITRDB)   Part of the World Data Center—for Paleoclimatology at the National Geophysical Data Center (NGDC) housed at Boulder, Colorado, USA. It comprises many different types of palaeoclimate data besides those derived from TREE RINGS. The ITRDB grew out of a databank originally established at the Laboratory of Tree-Ring Research in Tucson, Arizona, but today provides a central storage and distribution centre for tree-ring data from thousands of locations around the world. These data are freely searchable and downloadable to anyone via www.ncdc.noaa.gov/paleo/treering .html. A useful guide can also be found at http://web.utk .edu/~grissino/itrdb.htm.   *KRB/APP*

Grissino-Mayer HD and Fritts HC (1997) The International Tree-Ring Data Bank: An enhanced global database serving the global scientific community. *The Holocene* 7: 235–238.

Mann ME, Bradley RS and Hughes MK (1999) Northern Hemisphere temperatures during the past millennium: Inferences, uncertainties, and limitations. *Geophysical Research Letters* 26: 759–762.

internationalisation of research   As the nature of *environmental change* research has itself changed from a largely local or regional pursuit to encompass GLOBAL CHANGE, it has become increasingly internationalised. Internationalisation of environmental change research is affecting disciplines, institutions, resources and methodologies. For example, the International Council of Scientific Unions, known, since 1998 as the International Council for Science with the same acronym (ICSU), recognises four GLOBAL ENVIRONMENTAL CHANGE research programmes: (1) the *World Climate Research Programme* (WCRP), which investigates the physical CLIMATIC SYSTEM; (2) the *International Geosphere-Biosphere Programme* (IGBP), which is concerned with interactions between living and non-living systems; (3) the *International Human Dimensions Programme on*

Global Environmental Change (IHDP), which considers interactions between human society and the environment at a global scale; and (4) the *Diversitas Programme*, focused on the structure and function of BIODIVERSITY. Such *supranational science programmes* are a relatively new dimension of environmental change research, and raise important questions about how research problems are identified, pursued and utilised. While co-ordination is increasing at the international level, most funding and operations are still carried out at the national level.

JAM

[*See also* GLOBALISATION, INFORMATION AGE (IA), INTERGOVERNMENTAL PANEL ON CLIMATE CHANGE (IPCC)]

DeSombre E (2006) *Global environmental institutions*. New York: Routledge.
Ehlers E (1999) Environment and geography: International programmes on global environmental change. *International Geographical Union Bulletin* 49: 5–18.
Jasanoff S and Wynne B (1998) Science and decision making. In Rayner S and Malone EL (eds) *Human choice and climate change, volume 1. The societal framework*. Columbus, OH: Battelle Press, 1–88.

interpluvial A relatively dry time interval between two PLUVIAL phases.

JAM

interpolation Methods that estimate intermediate values of a *dependant variable* as a function of *independent variables*. These techniques are used, for instance, to ZOOM images and, more generally, to derive *missing data values*.

ACF

Massopust P (2010) *Interpolation and approximation with splines and fractals*. New York: Oxford University Press.

interrill erosion Removal of SOIL or SEDIMENT in areas between RILLS (i.e. over most of the hillslope). It is generally caused by the combined action of OVER-LAND FLOW and *rainsplash*.

SHD

Cassol EA and de Lima VS (2003) Interrill soil erosion under different tillage and management systems. *Pesquisa Agropecuaria Brasiliera* 38: 117–124.
Morgan RPC (2004) *Soil erosion and conservation, 3rd edition*. Oxford: Blackwell.

interstadial A relatively short-lived period of thermal improvement during a GLACIAL EPISODE. Although considered to reflect amelioration to conditions not as warm as at present, some interstadials probably experienced rapid changes to temperatures as high as during the current INTERGLACIAL, albeit for very short periods. Whilst terrestrial records had suggested about eight interstadials within the last glacial phase (MARINE ISOTOPIC STAGES 5d–2), 24 interstadials are now recognised from the Greenland ICE-CORE record for the same period.

CJD/DH

[*See also* ALLERØD INTERSTADIAL, CENTURY- TO MILLENNIAL-SCALE VARIABILITY, DANSGAARD-OESCHGER (D-O) EVENTS, LATE GLACIAL ENVIRONMENTAL CHANGE, QUATERNARY TIMESCALE, WINDERMERE INTERSTADIAL]

Lowe JJ, Rasmussen SO, Björck S et al. (2008) Synchronisation of palaeoenvironmental events in the North Atlantic region during the Last Termination: A revised protocol recommended by the INTIMATE group. *Quaternary Science Reviews* 27: 6–17.
Wolf EW, Chappellaz J, Blunier T et al. (2010) Millennial-scale variability during the Last Glacial: The ice core record. *Quaternary Science Reviews* 29: 2928–2938.

intertropical convergence zone (ITCZ) The belt of low pressure found near the Equator and formerly known as the *intertropical front* (see FRONT) or *equatorial low* and, by mariners, as the *doldrums*. The ITCZ is a region of ascending air, visible on cloud images as a band of discontinuous convective cloud between the convergent TRADE WIND systems. Rainfall typically occurs from cloud clusters about 100 km across. Over the Western Hemisphere, the ITCZ moves only a few degrees of latitude during the year but in the Eastern Hemisphere, variable seasonal shifts up to some 40° of latitude contribute to the problem of DROUGHT, especially in the African SAHEL and associated with the Asian monsoon (see the Figure in the MONSOON entry).

Moisture and ATMOSPHERE-OCEAN INTERACTIONS are important in the formation of the ITCZ: hence, there is no exact correspondence between its location and the warmest land and SEA-SURFACE TEMPERATURES (SSTs) of the tropics. In the eastern Pacific, UPWELLING causes the formation of a northern and southern branch of the ITCZ and there are complex relationships during EL NIÑO-SOUTHERN OSCILLATION (ENSO) events. *Easterly waves* (also known as *tropical waves* or *African waves*) propagate along the ITCZ, travel towards the west and occasionally intensify into TROPICAL CYCLONES (hurricanes, typhoons). Longer-term variations in the average position of the ITCZ are very important in understanding *environmental change* in the tropics. Major southward shifts in the ITCZ in southeast Africa since the LAST GLACIAL MAXIMUM (LGM), which were important controls on HYDROCLIMATOLOGY, phases of high PRECIPITATION and high river DISCHARGE, appear to have been driven primarily by high-latitude climatic changes, especially cold phases (e.g. the YOUNGER DRYAS STADIAL), rather than by sea-surface temperatures in the Indian Ocean. However, local summer INSOLATION variations have been important during the past 4,000 years of the late HOLOCENE.

JAM/AHP

[*See also* HADLEY CELL]

Hamilton RA and Archbold JW (1945) Meteorology over Nigeria and adjacent territory. *Quarterly Journal of the Royal Meteorological Society* 71: 231–262.

Nicholson SE and Flohn H (1980) African environmental and climatic changes and the general atmospheric circulation during the Late Pleistocene and Holocene. *Climatic Change* 2: 313–348.

Riehl H (1979) *Climate and weather in the tropics.* New York: Academic Press.

Schefuß E, Kuhlmann H, Mollenhauer G et al. (2011) Forcing of wet phases in southeast Africa over the past 17,000 years. *Nature* 480: 509–512.

Waliser DE (2003) Inter Tropical Convergence Zones. In Holton JR, Curry JA and Pyle JA (eds) *Encyclopedia of atmospheric sciences.* San Diego, CA: Academic Press, 2325–2334.

Yancheva G, Nowaczyk NR, Mingram J et al. (2007) Influence of the intertropical convergence zone on the East Asian monsoon. *Nature* 445: 74–77.

intraplate earthquake   An EARTHQUAKE originating along a FAULT that is distant from, and unrelated to, a PLATE MARGIN. Because stress builds up more slowly in the interior of a plate than at its margin (which is where most earthquakes are generated) the RETURN PERIOD of intraplate earthquakes (estimated using PALAEOSEISMICITY) may be of the order of several hundred years. The EARTHQUAKE MAGNITUDE may be very high, and as a result intraplate earthquakes represent a significant NATURAL HAZARD because protection measures may not be in place. The Wenchuan earthquake in Sichuan province, China, in AD 2008, which killed over 70,000, was an intraplate earthquake of magnitude 8.0. The largest historical earthquake in the United States (with the exception of Alaska) was the New Madrid earthquake swarm in Missouri in AD 1811–1812, the largest of which is estimated to have been of magnitude 8. Earthquakes in the British Isles, which reach a maximum magnitude in the order of 5.5, are intraplate earthquakes.   *GO*

Calais E, Freed AM, van Arsdale R and Stein S (2010) Triggering of New Madrid seismicity by Late-Pleistocene erosion. *Nature* 466: 608–611.

Westaway R (2006) Investigation of coupling between surface processes and induced flow in the lower continental crust as a cause of intraplate seismicity. *Earth Surface Processes and Landforms* 31: 1480–1509.

intrazonal soil   Soils with more-or-less well-developed characteristics that reflect the dominating influence of a local factor of RELIEF, PARENT MATERIAL or SOIL AGE over the 'normal' zonal effect of climate and vegetation. It was also a *soil order* in the system of classification used in the USA before 1965 but is not used in SOIL TAXONOMY or the WORLD REFERENCE BASE FOR SOIL RESOURCES (WRB).   *EMB*

[*See also* ZONAL SOIL]

introduction   The human-assisted movement of individual species into a site that lies outside their natural, historical RANGE. It may be purposeful or accidental.

Examples include the introduction of *rabbits* and *Opuntia* into Australia. *Reintroduction* involves the similar movement of species into sites inside their natural range but from which they have been previously excluded by direct or indirect human impacts. Introductions, in the form of *assisted migration* or *assisted colonisation*, may help alleviate some of the impacts of CLIMATE CHANGE on VULNERABLE SPECIES. There may be significant risks associated with INBREEDING DEPRESSION while the populations of introduced species remain small, or a risk that the population explodes, becoming *invasive*, as with the *cane toad (Bufo marinus)* in Australia.   *MVL/JLI*

[*See also* BIOLOGICAL CONTROL, BIOLOGICAL INVASION, BIOSECURITY, ENDANGERED SPECIES, EXOTIC SPECIES, SPECIES TRANSLOCATION]

Alho CJR, Mamede S, Bifencourt K and Benites M (2011) Introduced species in the Pantanal: Implications for conservation. *Brazilian Journal of Biology* 71: 321–325.

Kreyling J, Bittner T, Jaeschke A et al. (2011) Assisted colonization: A question of focal units and recipient localities. *Restoration Ecology* 19: 433–440.

Rout TM, Moore JL, Possingham JP and McCarthy MA (2011) Allocating biosecurity resources between preventing, detecting, and eradicating island invasions. *Ecological Economics* 71: 54–62.

Shine R (2010) The ecological impact of invasive cane toads (*Bufo marinus*) in Australia. *Quarterly Review of Biology* 85: 253–291.

Weeks AR, Sgro CM, Young AG et al. (2011) Assessing the benefits and risks of translocations in changing environments: A genetic perspective. *Evolutionary Applications* 4: 709–725.

intrusion   A structural feature produced by the movement of MAGMA, ICE or SEDIMENT beneath the Earth's surface, driven by contrasts in density or temperature between the intruding mass and its host (often termed *country rock*). *Igneous intrusions* are classified according to their geometry. Near-surface, small intrusions (*hypabyssal*) solidify as medium-grained IGNEOUS ROCKS. *Dykes* cut across BEDDING and are commonly vertical or nearly so. *Sills* are intruded parallel to bedding. *Plugs* are the central remnants of eroded VOLCANOES, and *ring dykes* are concentric intrusions around volcanic centres which may lead to faulting and CALDERA formation. Deep-seated, large intrusions (*plutonic, plutons*) are preserved as coarse-grained igneous rocks. They may take thousands of years to cool completely. *Magma chambers* form reservoirs for VOLCANIC ERUPTIONS. *Batholiths* are very large granitic bodies, often several hundred kilometres long, intruded along destructive PLATE MARGINS. *Non-igneous intrusions* include SALT, mud DIAPIRS or *domes*, and the injection of ice to form PERMAFROST MOUNDS.   *JBH*

Dziak RP, Bohnenstiehl DR, Cowen JP et al. (2007) Rapid dike emplacement leads to eruptions and hydrothermal plume release during seafloor spreading events. *Geology* 35: 579–582.

Francis P and Oppenheimer C (2003) *Volcanoes*. Oxford: Oxford University Press.

Johnson AM (1970) *Physical processes in geology*. San Francisco: Freeman and Cooper.

Ramberg H (1981) *Gravity, deformation and the Earth's crust, 2nd edition*. London: Academic Press.

invasive species   Plants or animals that are able to rapidly expand their RANGE. They may be ALIEN SPECIES that are introduced beyond their natural range (see INTRODUCTION), which sometimes cause CONSERVATION problems, especially if they displace the NATIVE species. Invasive species tend to be adapted to rapid DISPERSAL and COLONISATION, and are favoured by ecosystem DISTURBANCE.   *JAM*

[*See also* ALIEN SPECIES, BIOLOGICAL INVASION, COLONISATION, EXOTIC SPECIES, PEST, WEED]

Clout MN and Williams PA (eds) (2009) *Invasive species management: A handbook of principles and techniques*. Oxford: Oxford University Press.

Francis RA (ed.) (2012) *A handbook of global freshwater invasive species*. Abingdon: Earthscan.

Mooney HA and Hobbs RJ (eds) (2000) *Invasive species in a changing world*. Washington, DC: Island Press.

Woodward SL and Quinn JA (2011) *Encyclopedia of invasive species: From Africanized honey bees to zebra mussels*. Westport, CT: Greenwood Press.

inversion   A layer of the ATMOSPHERE in which the temperature of the air rises with height; an inversion of the normal positive LAPSE RATE. Inversions may form at the surface by nocturnal *radiative cooling*, above the surface due to *air subsidence* and *adiabatic heating* in ANTICYCLONES, or associated with the passage of a FRONT. Persistent inversions are usually associated with stable anticyclonic conditions, leading to the restriction of atmospheric dispersion mechanisms and thus AIR POLLUTION episodes.   *JCS*

[*See also* ADIABATIC PROCESSES]

Bailey A, Chase TN, Cassano JJ and Noone D. (2011) Changing temperature inversion characteristics in the U.S. Southwest and relationships to large-scale atmospheric circulation. *Journal of Applied Meteorology and Climatology* 50: 1307–1323.

inverted relief   Upstanding LANDFORMS (positive topographic features) developed from former depressions (negative topographic features) in the Earth's surface. For example, a LAVA FLOW on a valley floor might eventually form an upstanding ridge if it is more resistant to WEATHERING and EROSION than the surrounding valley-side bedrock. Similarly, DURICRUSTS formed on a former lake bed may resist erosion and weathering better than the surrounding topography and thus become positive landscape features (e.g. MESAS). Large-scale *relief inversion* can occur in terrain dominated by FOLD structures if there is differential erosion of *anticlines* relative to *synclines*.   *RAS*

[*See also* RELIEF, TOPOGRAPHY]

Pain CF and Ollier CD (1995) Inversion of relief: A component of landscape evolution. *Geomorphology* 12: 151–165.

invisible drought   A type of DROUGHT characterised by suboptimal crop yields in areas where moisture is normally sufficient for crop growth; it can be eliminated easily by IRRIGATION.   *JAM/JET*

[*See also* CONTINGENT DROUGHT, PERMANENT DROUGHT, SEASONAL DROUGHT]

Nagarajan R (2009) *Drought assessment*. London: Springer.

involutions   Disturbed, distorted and deformed structures occurring in unconsolidated sediments, usually of PLEISTOCENE age. There are two types. The first is frequently associated with PERIGLACIAL ENVIRONMENTS where involutions form through frost action within SEASONALLY FROZEN GROUND, or the ACTIVE LAYER if PERMAFROST is present. The second type is a THERMOKARST involution produced by *loading* and density differences that develop during the thaw of ice-rich permafrost.   *HMF*

[*See also* CRYOTURBATION, SOFT-SEDIMENT DEFORMATION]

Harris C, Murton JB and Davies MCR (2000) Soft-sediment deformation during thawing of ice-rich frozen soils: Results of scaled centrifuge modeling experiments. *Sedimentology* 47: 687–700.

Vandenberghe J (2007) Cryoturbation structures. In Elias SA (Ed.) *Encyclopedia of Quaternary science*. Amsterdam: Elsevier, 2147–2153.

ion   An electrically charged particle, which may be atomic or polyatomic (formed from an ATOM or a group of atoms). Ions are formed by *ionisation* processes whereby electrons are added or removed from particles. In solution, many compounds become dissociated into their component ions, which may be positively (*anions*) or negatively (*cations*) charged, depending on whether an electron is removed or added, respectively. The process of *ion exchange* used, for example, in water purification, water softeners, some DESALINATION plant and sewage-treatment works, involves the removal or replacement of anions or cations as a solution passes through a medium or filter. Gases may become ionised when an electrical charge passes through them or by *ionising radiation*, such as ultraviolet RADIATION; this commonly occurs in the *ionosphere* (above about 80 km in the upper ATMOSPHERE). Ions formed from metals are generally cations.   *JAM/AHP*

[*See also* CATION EXCHANGE CAPACITY (CEC), CATION-RATIO DATING]

Boekker E and van Grondelle R (1999) *Environmental physics, 2nd edition.* Chichester: Wiley-Blackwell.
Kelley MC (2009) *The Earth's ionosphere.* San Diego, CA: Elsevier.
Ratcliffe JA (1972) *An introduction to the ionosphere and magnetosphere.* Cambridge: Cambridge University Press.
Wachinski AM and Etzel JE (1997) *Environmental ion exchange: Principles and design.* Boca Raton, FL: CRC Press.

iron (Fe)   The most important element of the Earth's CORE, and one of the major constituents of the Earth's CRUST (estimated at about 5 per cent). It constitutes about 3.5 per cent of SOILS. The behaviour of Fe in terrestrial environments is complex and is largely determined by the relative ease with which it changes its oxidation state in response to environmental conditions. Fe is considered a very reactive metal.

The two most common states in soils and near-surface rocks are Fe^{2+} and Fe^{3+} although it may occur in various complexes as Fe^{4+} and Fe^{6+}. The behaviour and fate of Fe in WEATHERING and SOIL-FORMING PROCESS is dependent largely on the Eh-pH system and the stage of oxidation of the Fe compounds involved. In broad terms, Fe is precipitated in alkaline and oxidising conditions, whereas acid and reducing conditions increase the mobility of Fe compounds. The Fe released readily precipitates as oxides and hydroxides, but it often substitutes for *magnesium* (Mg) and ALUMINIUM (Al) in other minerals and is often involved in COMPLEXING with *organic ligands.* The most common Fe minerals found in soil environments are HAEMATITE (α-Fe_2O_3); MAGNETITE (Fe_3O_4); *ferrihydrite* ($Fe_2O_3 \cdot nH_2O$); *goethite* (*a*FeOOH); *lepidocrocite* (γ-FeOOH); *pyrite* (FeS_2) and *siderite* ($FeCO_3$). The red-orange-yellow *soil colours* are largely associated with amounts and forms of the Fe compounds present. The distribution of Fe minerals and compounds in soils is variable and reflects the many different SOIL-FORMING PROCESSES.

Iron is an essential TRACE ELEMENT for almost all living things but is toxic in excess. It has many functions within the human body where iron deficiency leads to *anaemia.* In PREHISTORY, invention of iron smelting led to the IRON AGE and its many environmental impacts. Large-scale iron and *steel making* was one of the foundations of the INDUSTRIAL REVOLUTION and its aftermath. The first iron bridge, for example, was built in Ironbridge, Shropshire, England, in AD 1778, while the largest boost to the *iron industry* was the development of *rail transport* from 1830 onwards. World production of iron is now >500 million tonnes/year, with RECYCLING adding another 300 million tonnes.   *SN*

[*See also* FERRALSOLS, FERRICRETE, IRON CYCLE, LATERITE, PLINTHOSOLS]

Alloway BJ (ed.) (2012) *Heavy metals in soils, 3rd edition.* Berlin: Springer.
Cornell RM and Schwertmann U (1996) *The iron oxides.* Weinheim: VCH.
Kabata-Pendias A (2011) *Trace elements in soils and plants.* Boca Raton, FL: CRC Press.

Iron Age   The Iron Age was the archaeological period following the BRONZE AGE in which tools and weapons were primarily produced from IRON (Fe). In Eurasia, the Iron Age began ca 1000 BC and persisted until the ROMAN PERIOD, although iron continued to be the primary tool-making material up to recent times. Dating of events in the earlier part of the Iron Age can be problematical due to the presence of a radiocarbon plateau (sometimes called the *Hallstatt Plateau*) at ca 800–400 BC. FOREST CLEARANCE was widespread in the Iron Age due to the requirement of wood for fuel and increasing agricultural activity in the larger, more permanent settlements. In some areas of Central Europe, these activities favoured faster-growing forest taxa such as *Quercus* (oak) and *Carpinus betulus* (hornbeam). In Scandinavia, the species composition of present-day grassland carries a signal of Iron Age LANDUSE CHANGE. In addition, the construction of *hillforts* and *earthworks* during the Iron Age required large quantities of timber and left scars on the landscape that remain visible in the modern age. Outside of Eurasia, the Iron Age is generally thought to have occurred much later. Iron appeared in the Americas only after European settlement. However, recent geochemical analyses of MARINE SEDIMENTS suggest that climate was not the sole driver of the so-called *African rain forest crisis* around 3,000 years ago, when mature rain forest was replaced in West Central Africa by a lighter type of forest with more pioneer trees. This vegetation disturbance appears to have been associated with the MIGRATION of Bantu-speaking farmers together with their agricultural and iron-smelting technologies.   *ARG/JAM*

[*See also* THREE-AGE SYSTEM]

Bayon G, Dennielou B, Etoubleau et al. (2012) Intensifying weathering and land use in Iron Age Central Africa. *Science* 335: 1219–1222.
Bruun HH, Fritzbøger B, Rindel PO and Hansen UL (2001) Plant species richness in grasslands: The relative importance of contemporary environment and land-use history since the Iron Age. *Ecography* 24: 569–578.
Cunliffe BW (2005) *Iron Age communities in Britain, 4th edition.* London: Routledge.
Guttmann EB, Simpson IA, Nielsen N and Dockrill SJ (2008) Anthrosols in Iron Age Shetland: Implications for arable and economic activity. *Geoarchaeology* 23: 799–823.

Küster H (1997) The role of farming in the postglacial expansion of beech and hornbeam in the oak woodlands of Central Europe. *The Holocene* 7: 239–242.

Ralston IBM (1999) The Iron Age: Aspects of human communities and their environments. *Quaternary Proceedings* 7: 501–512.

Turner J (1981) The Iron Age. In Simmons I and Tooley M (eds) *The environment in British prehistory.* London. Duckworth, 250–281.

Wells PS (2002) The Iron Age. In Milisauskas S (ed.) *European prehistory: A survey.* New York: Kluwer, 335–383.

iron cycle The natural BIOGEOCHEMICAL CYCLE of iron within the Earth-atmosphere-ocean system. Iron is delivered to the open ocean as SUSPENDED LOAD via rivers, ATMOSPHERIC DUST and HYDROTHERMAL inputs (see Table). Particulate SEDIMENTS from rivers are usually efficiently deposited on the CONTINENTAL SHELF and hydrothermal iron is rapidly precipitated at depth in the ocean, so the dominant source of *bio-available dissolved iron* in ocean surface waters is AEOLIAN dust. DUST production depends on the supply of wind-erodible sediments. *Fluvial erosion* from high-altitude ground transports iron-bearing sediment to lower altitudes where, if it dries out and vegetation cover is sparse or absent, it is lifted into the ATMOSPHERE by the wind and transported to the OCEANS where it is predominantly deposited by WET DEPOSITION. Single-celled organisms in the surface of the ocean, such as BACTERIA and DIATOMS, require dissolved iron for cellular functions. PHYTOPLANKTONS, such as diatoms, have a physiological requirement for iron in order to carry out PHOTOSYNTHESIS. A lack of dissolved iron in the surface waters will restrict phytoplankton growth, even if other essential nutrients are available in abundance. This leads to large regions of the world oceans, such as the Southern Ocean, being described as *high nutrient– low chlorophyll (HNLC) regions* that exhibit low PRIMARY PRODUCTIVITY. Iron-containing particles (*biogenic and abiogenic*) sink to the ocean depths and gradually undergo *remineralisation*, releasing the iron back into soluble forms. In regions of ocean UPWELLING, this dissolved iron-rich deeper water returns to the surface to fertilise primary productivity, in the same way as the new dissolved iron derived from dust. Reproducing this natural process of OCEAN FERTILISATION, by adding dissolved iron to the surface waters in HNLC regions, has been suggested as a way of reducing atmospheric CARBON DIOXIDE concentrations. This human manipulation of the planetary environment is known as GEO-ENGINEERING.   *JP*

[*See also* EARTH SYSTEM, GEO-ECOSPHERE]

Boyd PW and Elwood MJ (2010) The biogeochemical cycle of iron in the ocean. *Nature Geoscience* 3: 675–682.

Jickells TD, An ZS, Andersen KK et al. (2005) Global iron connections between desert dust, ocean biogeochemistry, and climate. *Science* 308: 67–71.

Zeebe RE and Archer D (2005) Feasibility of ocean fertilization and its impact on future atmospheric CO_2 levels. *Geophysical Research Letters* 32: L09703.

Source	Delivered to open ocean	Flux (Tg C/year)
Fluvial particulate total iron	No[a]	625–962
Fluvial dissolved iron	Yes	1.5
Glacial sediments	No[a]	34–211
Atmospheric	Yes	16
Coastal erosion	No[a]	8
Hydrothermal	Yes	14
Authigenic[b]	Yes	5

iron cycle Global iron fluxes to the oceans (Jickells et al., 2005).

a. Sedimented in coastal zone.

b. Released from deep-sea sediments during diagenesis.

irradiance The amount of *radiant flux* per unit of surface that flows across or into a surface. The usual unit is watt per square metre (W/m^2).   *TS*

[*See also* RADIANCE]

irrigation The addition of water to enhance *crop production* where water shortage is a LIMITING FACTOR. Thus, irrigation is most common as a CULTIVATION technique in DRYLANDS. By AD 1989, there was an estimated 233 million hectares of irrigated land worldwide, of which 73 per cent was located in DEVELOPING COUNTRIES. Although some irrigation systems (e.g. using AQUEDUCTS and QANATS) are ancient, global expansion of irrigation in the period 1950–1980 was implemented mainly by the construction of large DAMS.

The environmental costs of irrigation may be high and depend on the efficiency and management of the irrigation system. NUTRIENT loss, SALINISATION, ALKALISATION, *waterlogging* and pollution of GROUNDWATER are some of the main detrimental environmental impacts. In the worst cases, LAND DEGRADATION may be so far developed that land is abandoned. Some of these problems are reduced or eliminated by modern systems of *drip irrigation* or *spray irrigation*. Dam construction often involves the displacement of people. For example, the impoundment of Lake Nasser behind the Aswan Dam completed in 1964 required the displacement of about 100,000 people. In addition, if control measures are not adhered to in TROPICAL CLIMATES, *canals* and *ditches* may help spread various DISEASES (e.g. MALARIA, YELLOW FEVER, *filariasis, dengue fever, river blindness, sleeping sickness, schistosomiasis or bilharzia, liver fluke* and *guinea worm*).   *RAS*

[*See also* AQUIFER, DESERTIFICATION, DRY FARMING, PIOSPHERE, WATER RESOURCES]

Agnew C and Anderson E (1992) *Water resources in the arid realm*. London: Routledge.
Barrow CJ (1987) *Water resources and agricultural development in the tropics*. London: Longman.
Ciracono S (1998) *Land drainage and irrigation*. Farnham: Ashgate.
Hansen VE, Israelsen OW and Stringham GE (1980) *Irrigation principles and practices, 4th edition*. New York: Wiley.
Hillel D (1987) *The efficient use of water in irrigation: Principles and practice for improving irrigation in arid and semiarid regions*. Washington, DC: World Bank.
Postel S (1993) Water and agriculture. In Gleick PH (ed.) *Water in crisis: A guide to the world's fresh water resources*. Oxford: Oxford University Press, 56–59.
Sacks WJ, Cook BI, Buenning N et al. (2009) Effects of global irrigation on the near-surface climate. *Climate Dynamics* 33: 159–175.
Samad M, Merrey D, Vermillion D et al. (1992) Irrigation management strategies for improving the performance of irrigated agriculture. *Outlook on Agriculture* 21: 279–286.

island arc   An arcuate chain of islands in the OCEAN, usually of volcanic origin and associated with an OCEANIC TRENCH, formed above a *subduction zone* at a DESTRUCTIVE PLATE MARGIN (e.g. the Aleutian Islands).   *CDW*

Burbank DW and Anderson RS (2011) *Tectonic geomorphology, 2nd edition*. Oxford: Wiley-Blackwell.
Hamilton WB (1988) Plate tectonics and island arcs. *Geological Society of America Bulletin* 100: 1503–1527.
Kearey P, Klepeis KA and Vine FJ (2009) *Global tectonics, 3rd edition*. Chichester: Wiley.

island biogeography   The study of the ecological and evolutionary characteristics of isolated biotas. Most science historians agree that island studies played a central role in the development of Darwin and Wallace's theory of NATURAL SELECTION. Their observations, along with those of others studying *insular ecosystems*, also became the foundations of the fields of BIOGEOGRAPHY and ECOLOGY. The *equilibrium theory of island biogeography* is an important theory put forward in 1963 by R.H. MacArthur and E.O. Wilson, which posits that island species number represents a *dynamic equilibrium* (see EQUILIBRIUM CONCEPTS IN ECOLOGICAL AND EVOLUTIONARY CONTEXTS) between *immigration* to an island and EXTINCTION from it. Island biogeographers study an impressive diversity of patterns, but most of them deal with the relationships between physical characteristics of islands (e.g. size and degree of isolation) and the

characteristics of *biotic communities* (e.g. number and types of species, their distributions among islands and variation in their morphological, behavioural or genetic characteristics). Island biogeography continues to play an important role in providing insights for evolutionary biologists, ecologists and biogeographers, in general. Perhaps just as important, because NATURE RESERVES and fragments of native ecosystems share many island-like characteristics (see HABITAT ISLAND), island biogeography theory has become an especially important tool for CONSERVATION of BIODIVERSITY.   *MVL*

[*See also* CONTINENTAL ISLAND, EQUILIBRIUM CONCEPTS, METAPOPULATION MODEL, RELAXATION, SINGLE LARGE OR SEVERAL SMALL RESERVES (SLOSS DEBATE), TAXON CYCLE]

Bramwell D and Caujapé-Castells J (eds) (2011) *The biology of island floras*. Cambridge: Cambridge University Press.
Lomolino MV and Brown JH (2009) The reticulating phylogeny of island biogeography theory. *Quarterly Review of Biology* 84: 357–390.
Losos JB and Ricklefs RE (2009) *The theory of island biogeography revisited*. Princeton, NJ: Princeton University Press.
MacArthur RH and Wilson EO (1967) *The theory of island biogeography*. Princeton, NJ: Princeton University Press.
Whittaker RJ and Fernández-Palacios JM (2007) *Island biogeography: Ecology, evolution and conservation*. Oxford: Oxford University Press.

isochron technique   A method used in RADIOMETRIC DATING to correct for the effect of initial DAUGHTER ISOTOPES (e.g. inherited *argon* in POTASSIUM-ARGON (K-Ar) DATING and ARGON-ARGON (Ar-40/Ar-39) DATING or detrital *thorium* in URANIUM-SERIES DATING). Coeval samples from a deposit are analysed to produce a straight-line relationship between ISOTOPE RATIOS such that the gradient can be used to calculate the age.   *DAR*

Faure G (1986) *Principles of isotope geology*. New York: Wiley.

isochrone   A line joining points of equal age on a map. This particular kind of ISOPLETH provides a means of representing an *areal chronology* depicting, for example, areal patterns of DEGLACIATION (see the Figure in the GLACIER FORELAND entry), land emergence associated with EMERGENT COASTS and TREE MIGRATION.   *JAM*

isolated system   A SYSTEM in which the boundaries are closed to the import and export of both mass and energy. It is a concept most applicable to LABORATORY SCIENCE or theoretical ENVIRONMENTAL MODELLING.   *JAM*

[*See also* CLOSED SYSTEM, OPEN SYSTEM]

Chorley RJ and Kennedy BA (1971) *Physical geography: A systems approach*. London: Prentice Hall.

isolation The process of separation of POPULATIONS of species to the extent that *interbreeding* is no longer possible. Isolation may arise in a number of ways, and may have a geographic or ecological basis. It is probably the process that leads to most SPECIATION events. 'Rapidly evolving peripherally isolated populations may be the place of origin of many evolutionary novelties. Their isolation and comparatively small size may explain phenomena of rapid evolution and lack of documentation in the FOSSIL record, hitherto puzzling to the paleontologist' (Mayr, 1954: 179).   *KDB*

[*See also* FOUNDER EFFECT, ISLAND BIOGEOGRAPHY]

Mayr E (1954) Change of genetic environment and evolution. In Huxley J, Hardy AC and Ford EB (eds) Evolution as a process. London: Allen and Unwin, 157–180.

isolation basin A terrestrial basin that has previously been submerged below sea level. MARINE SEDIMENTS accumulate in isolation basins while they are submerged below sea level. Subsequently, a fall in RELATIVE SEA LEVEL results in the isolation of the basin from marine influences, and a shift to accumulation of freshwater LACUSTRINE SEDIMENTS. Isolation basins may later be re-submerged and re-emerge. Dating the changes from freshwater to marine or BRACKISH WATER sedimentation in isolation basins may provide *sea-level index points* for reconstructing regional patterns of SEA-LEVEL CHANGE.   *MHD/CJC*

[*See also* SEAWATER COMPOSITION]

Balascio NL, Zhang ZH, Bradley RS et al. (2011) A multi-proxy approach to assessing isolation basin stratigraphies from the Lofoten Island, Norway. *Quaternary Research* 75: 288–300.
Long AJ, Woodroffe DH, Roberts DH and Dawson S (2011) Isolation basins, sea-level changes and the Holocene history of the Greenland Ice Sheet. *Quaternary Science Reviews* 30: 3748–3768.

isomer Two or more identical molecular counterparts of a COMPOUND, each with different structural or constitutional arrangements of the atoms. Different isomers of the same compound are represented by the same *chemical formula* but have different *structural formulae*. AMINO ACID DATING is based on the time-dependent transformation of amino acid molecules from one isomeric form to another.   *MHD*

isomorphic substitution The substitution of one atom for another of similar size but lower valence in a crystal without disrupting or changing the structure of the mineral. In soils, Al^{3+} is frequently substituted for Si^{4+} and Mg^{2+} for Al^{3+} which leaves a deficit of positive charge on the crystal and allows the development of the CATION EXCHANGE CAPACITY (CEC).   *EMB*

isopach A line on a map joining points of equal thickness, usually of a distinctive BED or other stratigraphical unit (see LITHOSTRATIGRAPHY). The analysis of isopachs (also known as *isopachytes*) is important in the reconstruction of PALAEOGEOGRAPHY.   *GO*

isopleth (1) A line on a map connecting points of equal value in relation to a particular VARIABLE; also known as an *isoline* or *isarithm*. Some examples are given below:

isanomaly	line of equal departure from the norm
isobar	line of equal pressure
isobase	line of equal land uplift
isobath	line of equal water depth
isochrone	line of equal time lapse
isohaline	line of equal salinity
isohel	line of equal sunshine
isohyet	line of equal precipitation
isohypse	line of equal altitude (a contour line)
isallobar	line of equal pressure change (a measure of atmospheric disturbance)
isoneph	line of equal cloudiness
isonif	line of equal snowfall
isonomaly	line of equal anomalies
isopach	line of equal (geological) bed thickness
isophene	line of equal seasonal (biological) phenomena
isophyte	line of equal vegetation height
isopoll	line of equal percentage of a pollen taxon
isoterp	line of equal (human) comfort
isotherm	line of equal temperature

(2) A graph illustrating the frequency or intensity of a particular phenomenon as a function of two variables.   *JCM*

isostasy The condition of *equilibrium* caused by the Earth's LITHOSPHERE which essentially 'floats' on the less dense underlying ASTHENOSPHERE. DEFORMATION of the lithosphere due to the *loading* and UNLOADING of ice sheets during the QUATERNARY is termed GLACIO-ISOSTASY. Owing to spatial variations in former ICE-SHEET thickness, patterns of crustal rebound rate varied for different areas, the greatest rates occurring in areas where the ice had been the thickest. During ice-sheet thinning and melting, however, most GLACIO-ISOSTATIC REBOUND takes place while rapidly thinning ice still covers the landscape.

Isostatic loading and *unloading* by water on ocean floors and continental shelves due to changes in RELATIVE SEA LEVEL is termed HYDRO-ISOSTASY. For example, a long-term rise in sea level during the last DEGLACIATION was associated with a compensatory depression of the crust beneath major waterbodies. This process is greatly affected also by the tectonic setting of individual regions and by the flexural rigidity of the lithosphere. That the magnitude of changes can be considerable is demonstrated by the fact that during the past 7,000 years there has been an overall depression of 8 m in the ocean floors and an average uplift of adjacent continents by 16 m. It has also been argued, however, that whereas *hydro-isostatic loading* during eustatic SEA-LEVEL RISE may induce ocean-floor SUBSIDENCE, it does not follow that *hydro-isostatic unloading* due to regional sea-level fall would cause uplift of the ocean floor.   *AGD*

[*See also* EUSTASY, SEA-LEVEL CHANGE]

Chappell J (1974) Late Quaternary glacio- and hydro-isostasy on a layered Earth. *Quaternary Research* 4: 405–428.
Dawson AG (1992) *Ice Age Earth: Late Quaternary geology and climate*. London: Routledge.
Lyustikh EN (1960) *Isostasy and isostatic hypotheses*. New York: American Geophysical Union.
Nielsen SB, Gallagher K, Leighton C et al. (2009) The evolution of western Scandinavian topography: A review of Neogene uplift versus the ICE (isostasy-climate-erosion) hypothesis. *Journal of Geodynamics* 47: 72–95.
Smith DE and Dawson AG, edis (1983) *Shorelines and isostasy*. London: Academic Press.
Walcott RJ (1972) Past sea levels, eustasy, and deformation of the Earth. *Quaternary Research* 2: 1–14.
Watts AD (2001) *Isostasy and flexure of the lithosphere*. Cambridge: Cambridge University Press.

isostatic decantation   Displacement (or decant-ing) of seawater into the OCEANS from shallow SEAS or CONTINENTAL SHELVES as a result of GLACIO-ISOSTATIC REBOUND following melting of an ice mass.   *MJH*

[*See also* GLACIO-EUSTASY, HYDRO-ISOSTASY, ISOSTASY, SEA-LEVEL CHANGE]

Stoddart DR, McLean RF, Scoffin TP et al. (1978) Evolution of reefs and islands, northern Great Barrier Reef: Synthesis and interpretation. *Philosophical Transactions of the Royal Society B* 284: 149–159.

isothermal remanent magnetisation (IRM)   The remanent magnetisation induced by exposure to an applied MAGNETIC FIELD at a given temperature, measured in Am^2/kg. IRM varies according to the strength of the applied field but the maximum remanence which can be produced is called the SATURATION ISOTHERMAL REMANENT MAGNETISATION (SIRM), which is normally considered to be that remaining after exposure to a magnetic field of the order of 1 T. SIRM is influenced mainly by the con-centrations of remanence-carrying magnetic minerals, especially ferro(i) magnetic minerals.   *WZ/AJP*

[*See also* MAGNETIC REMANENCE, MINERAL MAGNETISM]

Walden J, Oldfield F and Smith JP (eds) (1999) *Environmental magnetism: A practical guide*. London: Quaternary Research Association.

isothermal remanent magnetism (IRM)   One of the magnetic components comprising the NATURAL REMANENT MAGNETISM (NRM) of rocks and sediments, it is magnetisation which remains in the absence of an applied magnetic field; that is, magnetisation remain-ing after the application and subsequent removal of a magnetic field.   *CJC*

Thompson R (1991) Palaeomagnetic dating. In Smart PL and Frances PD (eds) *Quaternary dating methods: A users guide*. Cambridge: Quaternary Research Association, 177–198.

isotope   The isotopes of an element are ATOMS hav-ing the same number of *protons* in the nucleus but a different number of *neutrons* and hence a different atomic mass. Isotopes are two or more *nuclides* that belong to the same ELEMENT. Different isotopes of the same element have similar chemical properties as they contain the same number of *electrons*. Isotopes can be divided into two forms, STABLE ISOTOPES and unstable *radioisotopes* (see RADIONUCLIDE).   *IR*

[*See also* COSMOGENIC-NUCLIDE DATING, ISOTOPES AS INDICATORS OF ENVIRONMENTAL CHANGE]

Dicken AP (2005) *Radiogenic isotope geology, 2nd edition*. Cambridge: Cambridge University Press.
Faure G and Mensing TM (2004) *Isotopes: Principles and applications*. Hoboken, NJ: Wiley.
Hoefs J (2009) *Stable isotope geochemistry, 6th edition*. Berlin: Springer.

isotope dendrochronology   A branch of TREE-RING studies concerned with the variations either of STABLE or radioactive ISOTOPES contained in the rings of trees. The absolute timescale that underpins dendrochrono-logical studies means that measurements of the iso-topic composition of wood may be assigned accurate dates. Perhaps the best-known example of the value of such studies is the *radiocarbon anomaly curve*—a record of short (decadal) and longer (millennial) vari-ations in the content of the ATMOSPHERE, assembled from measurements of the $^{14}C/^{12}C$ ratios in thousands of wood samples spanning the past 10,000 years. The variations must be accounted for when interpreting ^{14}C dates on organic matter. STABLE ISOTOPE ANALYSIS (e.g. studies of the ISOTOPE RATIOS $^{18}O/^{16}O$, $^1H/^2H$ and $^{13}C/^{12}C$) also provide evidence of changing physiological activ-ity in trees through time and are often interpreted as evidence of CLIMATIC CHANGE.   *KRB/APP*

[*See also* DENDROCHRONOLOGY, MICRODENDROCLIMATOLOGY, RADIOCARBON DATING]

Gagen M, McCarroll D, Loader NJ and Robertson I (2011) Stable isotopes in dendroclimatology: Moving beyond 'potential'. In Hughes MK, Swetnam TW and Diaz HF (eds) *Dendroclimatology: Progress and prospects*. Dordrecht: Springer, 147–172.
Leavitt SW (1993) Environmental information from $^{13}C/^{12}C$ ratios of wood. *Geophysical Monographs* 78: 325–331.

McCarroll D and Loader NJ (2008) Stable isotopes in tree rings. *Quaternary Science Reviews* 23: 771–801.

Stuiver M and Braziunas TF (1987) Tree cellulose $^{13}C/^{12}C$ isotope ratios and climatic change. *Nature* 328: 58–60.

isotope ratio Usually the relationship between the two main STABLE ISOTOPES of an ELEMENT expressed as the proportion of the 'heavier' to the 'lighter' isotope relative to a REFERENCE STANDARD. The differences between samples and reference standards are usually small and expressed as parts per thousand or per mille (‰). *IR*

[*See also* ISOTOPE, ISOTOPIC FRACTIONATION, STABLE ISOTOPE ANALYSIS]

isotopes as indicators of environmental change

Changes in the natural abundance of CARBON ISOTOPES, HYDROGEN ISOTOPES, NITROGEN ISOTOPES, OXYGEN ISOTOPES and SULFUR ISOTOPES are those most frequently used in studies of environmental change. ISOTOPIC FRACTIONATION is caused by different *kinetic rate constants* during physical processes and chemical reactions. Rather than measure absolute values, the difference between a sample and a REFERENCE STANDARD enables the determination of ISOTOPE RATIOS to a high degree of PRECISION.

The theoretical basis for the use of STABLE ISOTOPES as indicators of environmental change was established in the late AD 1940s. Carbon isotope values can be used to elucidate plant photosynthetic pathways as the $\delta^{13}C$ values of C-3 *plants* exhibit greater ISOTOPIC DEPLETION than C-4 *plants*. Reconstructions of PALAEOECOLOGY may be inferred from $\delta^{13}C$ values from LACUSTRINE SEDIMENTS, PALAEOSOLS, RODENT MIDDENS and MIRES. Concerns over anthropogenic CARBON DIOXIDE emissions have led to the establishment of a worldwide network to monitor the concentration and $\delta^{13}C$ values of atmospheric carbon dioxide. Weekly air samples are analysed from remote sites to represent the large, well-mixed air masses of the TROPOSPHERE. These values have revealed that there is a 'missing' CARBON SINK/SOURCE, which has been allocated to increased CARBON SEQUESTRATION in the terrestrial BIOSPHERE. The response of plants to increasing atmospheric carbon dioxide concentration has also been estimated using *growth chamber* experiments and theoretical MODELS. The problems of UPSCALING from growth chamber results to represent an *ecosystem* may be overcome using intrinsic *water-use efficiency* values derived from $\delta^{13}C$ values in TREE RINGS to provide a time-integrated measure. These results confirm that the water-use efficiency of C-3 plants has increased together with the atmospheric CARBON DIOXIDE concentration.

The oxygen isotope value of FORAMINIFERA in MARINE SEDIMENT CORES represents one of the most frequently used indirect measures of long-term climatic change. PERIODICITIES in $\delta^{18}O$ values can be related to the changes in ORBITAL FORCING with frequencies at about 23, 41 and 100 ka (see MILANKOVITCH THEORY). ORBITAL TUNING may be used to date cores indirectly by matching properties with the predictable changes in the Earth's orbit and axis. Global changes in $\delta^{18}O$ values are well replicated and a composite 780 ka record forms the basis of the widely adopted SPECTRAL MAPPING PROJECT TIMESCALE (SPECMAP). Terrestrial records of CLIMATIC CHANGE derived, for example, from POLLEN ANALYSIS, LOESS STRATIGRAPHY, LAKE-LEVEL VARIATIONS and ICE CORES are compared with this standard. Stable isotope records have been obtained from ice cores taken from Antarctica, the Greenland ICE SHEET and high-altitude ice caps. The high-resolution HYDROGEN ISOTOPE record from the 800 Kyr EPICA Dome C ice core was found to be similar to other East Antarctic ice cores over their common period and broadly reflects Antarctic temperature. Although, the influence of orbital forcing was clearly evident in the eight GLACIAL-INTERGLACIAL CYCLES covered by the core, CENTURY- TO MILLENNIAL-SCALE VARIABILITY was influenced by *North Atlantic Deep Water* (NADW) formation supporting the thermal *bipolar seesaw* hypothesis.

The hydrogen isotope and oxygen isotope values of *meteoric water* are strongly associated and represented by the *Meteoric Water Line*. EVAPORATION causes an ISOTOPIC DEPLETION in the 'heavier' isotopes in *cloud water* (^{1}H and ^{16}O) with CONDENSATION causing an ISOTOPIC ENRICHMENT of the 'heavier' isotopes of the resulting precipitation (^{2}H and ^{18}O). Initial PRECIPITATION will have an isotopic composition similar to the source, but this will become progressively depleted in the 'heavier' isotopes and therefore precipitation is usually isotopically depleted compared with ocean water. As there is no isotopic fractionation during water uptake through the roots of terrestrial plants, the interpretation of non-exchangeable δD and $\delta^{18}O$ values from organic matter requires an understanding of isotopic fractionation during PHOTOSYNTHESIS and post-photosynthetic exchanges. Despite these concerns, stable isotopes from absolutely dated tree rings have been used successfully to reconstruct past climate over the past millennia.

Stable isotopes determined on *animal tissues* may be used to study *diet* and mobility between geologically distinct regions. Carbon isotope values from HERBIVORES may be used to determine the contribution of C-3, C-4 AND CAM PLANTS to the diet. As carbon and nitrogen isotope values are isotopically enriched at each TROPHIC LEVEL in the FOOD CHAIN, carbon and nitrogen isotope values from CARNIVORES will be enriched (with the level of enrichment depending upon the position in the food chain). Marine inputs to the diet can be estimated from a combination of nitrogen, sulfur and strontium isotope values. If the SEASONALITY of precipitation is taken into consideration, oxygen isotopes have the potential to provide information about the region where an individual lived.

The isotope ratio of atmospheric SULFUR DIOXIDE (SO_2) originating from ANTHROPOGENIC activities is related to the source from which it was originally derived and can be used to monitor AIR POLLUTION. Similarly, LEAD ISOTOPE values have been used to evaluate changing sources of human exposure. Although the $\delta^{15}N$ values of nitrate have been used to demonstrate POLLUTION originating from the application of nitrogen-based FERTILISERS, the use of $\delta^{15}N$ of nitrate in pollution studies is limited as the largest isotopic fractionation of nitrogen isotopes is from metabolic reactions.   *IR*

[***See also*** COMPOUND-SPECIFIC CARBON ISOTOPE ANALYSIS]

Cerling TE, Wittemyer G, Rasmussen HB et al. (2006) Stable isotopes in elephant hair document migration patterns and diet changes. *Proceedings of the National Academy of Sciences* 103: 371–373.

Heaton, THE (1986) Isotopic studies of nitrogen pollution in the hydrosphere and atmosphere: A review. *Chemical Geology* 59: 87–102.

Imbrie J, Hays JD, Martinson DG et al. (1984) The orbital theory of Pleistocene climate: Support from a revised chronology of the marine $\delta^{18}O$ record. In Berger A, Imbrie J, Hays J et al. (eds) *Milankovitch and climate: Understanding the response to astronomical forcing. Part I.* Dordrecht: Reidel, 269–305.

Jouzel J, Masson-Delmotte V, Cattani O et al. (2007) Orbital and millennial Antarctic climate variability over the past 800,000 years. *Science* 317: 793–797.

Leng MJ and Marshall JD (2004) Palaeoclimate interpretation of stable isotope data from lake sediment archives. *Quaternary Science Reviews* 23: 811–831.

Leng MJ, Street-Perrott FA and Barker PA (eds) (2008) Isotopes in biogenic silica. *Journal of Quaternary Science* 23: 311–400 [Special Issue].

McCarroll D and Loader NJ (2004) Stable isotopes in tree rings. *Quaternary Science Reviews* 23: 771–801.

McDermott F (2004) Palaeoclimate reconstruction from stable isotope variations in speleothems: A review. *Quaternary Science Reviews* 23: 901–918.

Müller W, Fricke H, Halliday AN et al. (2003) Origin and migration of the Alpine Iceman. *Science* 302: 862–866.

Savard MM (2010) Tree-ring stable isotopes and historical perspectives on pollution: An overview. *Environmental Pollution* 158: 2007–2013.

Shackleton NJ and Opdyke ND (1973) Oxygen isotope and palaeomagnetic stratigraphy of equatorial Pacific core V28-238: Oxygen isotope temperatures and ice volumes on a 10^5 year and 10^6 year scale. *Quaternary Research* 3: 39–55.

Treydte K, Schleser GH, Helle G et al. (2006) Millennium-long precipitation record from tree ring oxygen isotopes in northern Pakistan. *Nature* 440: 1179–1182.

Tzedakis PC, Andrieu V, de Beaulieu J-L et al. (1997) Comparison of terrestrial and marine records of changing climate of the last 500,000 years. *Earth and Planetary Science Letters* 150: 171–176.

Urey HC (1947) The thermodynamic properties of isotopic substances. *Journal of the Chemical Society of London* 85: 562–581.

isotopic depletion   The reduction in the abundance of the 'heavier' ISOTOPE during both *thermodynamic reactions* and *kinetic reactions* as a result of ISOTOPIC FRACTIONATION.   *IR*

[***See also*** ISOTOPIC ENRICHMENT]

isotopic enrichment   The enhancement in the abundance of the 'heavier' ISOTOPE during both *thermodynamic reactions* and *kinetic reactions* as a result of ISOTOPIC FRACTIONATION.   *IR*

[***See also*** ISOTOPIC DEPLETION]

isotopic fractionation   The separation of ISOTOPES of an ELEMENT during physico-chemical processes primarily due to differences in their relative masses. In chemical systems, isotopic fractionation may be due to *isotopic exchange reactions* operating under equilibrium conditions or to *unidirectional processes*. In isotopic exchange reactions there is no net reaction, but a redistribution of isotopes amongst different chemicals, between different phases or between individual MOLECULES. Unidirectional processes are caused by unequal *kinetic rate constants* for the different isotopes of the reactants and usually result in ISOTOPIC DEPLETION of the 'heavier' isotope in the product. In a CLOSED SYSTEM, the reactant concentration falls progressively causing the isotopic concentration of reactant and product to change with time. If the kinetic rate constant remains constant and the product is removed immediately under equilibrium conditions, the isotopic separation process may be modelled by the *Rayleigh equation*. The main physical processes that influence isotopic fractionation are DIFFUSION, EVAPORATION and CONDENSATION.   *IR*

[***See also*** ISOTOPES AS INDICATORS OF ENVIRONMENTAL CHANGE]

Gat JR (1981) Isotopic fractionation. In Gat JR and Gonfiantini R (eds) *Stable isotope hydrology: Deuterium and oxygen-18 in the water cycle*. Vienna: International Atomic Energy Authority, 21–33.

Hoefs J (2009) *Stable isotope geochemistry, 6th edition*. Berlin: Springer.

Urey HC (1947) The thermodynamic properties of isotopic substances. *Journal of the Chemical Society of London* 85, 562–581.

isotopologues   MOLECULES that differ only in their *isotopic composition*. For example, $^{13}C^{18}O^{16}O$ and $^{12}C^{17}O^{16}O$ are different isotopologues of CARBON DIOXIDE (CO_2).   *HKC*

[***See also*** STRONTIUM/CALCIUM PALAEOTHERMOMETRY]

Elsner M (2010) Stable isotope fractionation to investigate natural transformation mechanisms of organic contaminants: Principles, prospects and limitations. *Journal of Environmental Monitoring* 12: 2005–2031.

Jaramillo polarity event A NORMAL POLARITY event during the *Matuyama Chron* of the palaeomagnetic timescale, originally dated by POTASSIUM-ARGON (K-Ar) DATING to between 0.90 and 0.97 million years ago (Ma), and, following ORBITAL TUNING to between 0.99 and 1.07 Ma BP, with Hawaiian data suggesting a duration of 67 k years between 1.053 and 0.986 Ma BP. *MHD/CJC*

[*See also* GEOMAGNETIC POLARITY REVERSAL, MARKER HORIZON]

Singer BS, Hoffmann KA, Chauvin A et al. (1999) Dating transitionally magnetized lavas of the late Matuyama Chron: Toward a new Ar-40/Ar-39 timescale of reversals and events. *Journal of Geophysical Research: Solid Earth* 104: B1. doi:10.1029/1998JB900016.

jet streams Strong, narrow air currents in the TROPOSPHERE near the *tropopause*. They extend longitudinally for thousands of kilometres (length), latitudinally for hundreds of kilometres (width) and vertically for several kilometres (depth). Vertical wind shear ranges from 5 to 10 m/s/km and horizontal wind shear is 5 m/s/100 km. Arbitrary minimum speeds of 30 m/s are used to define the three-dimensional extent of jet streams, whilst core speeds (*jet maxima*) are often in excess of 70 m/s. Jet streams are analogous to a series of concentric hosepipes that exhibit LONG WAVES, they may bifurcate and they occur in both hemispheres. On the cold-air side of each jet are areas of clear air turbulence, which are usually avoided by aircraft. The *polar jet* (or *polar front jet*) flows from west to east in the mid-latitudes and its position varies widely from day to day. It is found near the 300 HPa level (about 10 km). The *subtropical jet* flows from west to east at about 30° latitude and its position is more constant in a given season. Found near 200 HPa (about 12 km), the winds tend to be less strong than those of the polar jet. There is evidence that the jet streams have moved polewards since 1970. They are caused by a combination of the rotation of the Earth and heating of the ATMOSPHERE.

There are also jet streams near the *stratopause* and *mesopause* and sometimes *low-level jets* occur at INVERSIONS above the planetary BOUNDARY LAYER. They also occur on other planets. *BDG/JAM*

Archer CL and Caldeira LC (2008) Historical trends in the jet streams. *Geophysical Research Letters* 35: LO8803.
Muñoz E and Enfield D (2011) The boreal spring variability of the Intra-Americas low-level jet and its relation with precipitation and tornadoes in the eastern United States. *Climate Dynamics* 36: 247–259.
Nakamura H and Shapiro A (2004) Seasonal variations in the Southern Hemisphere storm tracks and jet streams as revealed by a reanalysis data set. *Journal of Climate* 17: 1828–1844.
Reiter ER (1963) *Jet-stream meteorology*. Chicago: University of Chicago Press.

jetty An artificial structure built out from the shore adjacent to the mouth of a river or tidal inlet to prevent *shoaling* (i.e. reduction in water depth) by LONGSHORE DRIFT and enhance *hydraulic flushing*. Jetties are primarily constructed at navigational entrances. *HJW*

[*See also* BREAKWATER, COASTAL ENGINEERING STRUCTURES, GROYNE, HARD ENGINEERING]

Seabergh WC and Kraus NC (2003) Progress in management of sediment bypassing of coastal inlets: Natural bypassing, weir jetties, jetty spurs, and engineering aids in design. *Coastal Engineering Journal* 45:533–563.

join operation A procedure for connecting two or more relations (or tables) in a DATABASE using common FIELDS or ATTRIBUTES. *VM*

[*See also* RELATIONAL DATABASE]

joint A FRACTURE in rocks with no relative displacement between the two sides. Joints are ubiquitous in rocks exposed at the Earth's surface. They can develop because of stresses during DEFORMATION, contractional stresses associated with the cooling of a LAVA (*columnar jointing*) or the release of stress due to EROSION of *overburden* or UNLOADING following the melting of valley-filling

GLACIERS (DILATION or *pressure-release jointing*). Joints influence the WEATHERING of rock masses, and their distribution and orientation can affect SLOPE STABILITY. MINERAL material may grow in the cavity where the walls of a joint have separated, forming a *vein*.   GO

Fossen H (2010) *Structural geology.* Cambridge: Cambridge University Press.
Jain A, Guzina BB and Voller VR (2007) Effects of overburden on joint spacing in layered rocks. *Journal of Structural Geology* 29: 288–297.

jökulhlaup   A catastrophic FLOOD resulting from the sudden release of stored water associated with a GLACIER or ICE SHEET. Jökulhlaups can be caused by (a) the sudden drainage of an ICE-DAMMED LAKE below or through the *ice dam*; (b) *water overflow* accompanied by rapid fluvial *downcutting* of ice, bedrock or sediment barriers; or (c) the build-up and release of SUBGLACIAL reservoirs. Jökulhlaup discharges can exceed ABLATION-related flows by several orders of magnitude and, in so doing, effect considerable *geomorphological change*, damage to buildings and other structures as well as deaths. The retreat of glaciers in the European Alps following their LITTLE ICE AGE maximum extent caused a number of them to reach critical positions in tributary valleys such that they could provide effective barriers to streams in the main valleys. The timing of the *catastrophic drainage* and refilling of the resulting ice-dammed lakes are well documented for a number of glaciers in the Alps because of the concern generated by jökulhlaup destruction. Jökulhlaup size tended to attenuate through time until the retreating glacier could no longer form an effective barrier, reflecting the decreasing effectiveness of the ice dam. The term 'jökulhlaup' is Icelandic and means *glacier flood.* Alternative terms are *aluvión* (a moraine-dammed lake outburst) in South America and *débâcle* in Europe. RAS

[*See also* CHANNELED SCABLANDS]

Benn DI and Evans DJA (2010) *Glaciers and glaciation, 2nd edition.* London: Arnold.
Bjørnsson H (2011) Understanding jökulhlaups: From tale to theory. *Journal of Glaciology* 56: 1002–1010.
Fowler AC (1999) Breaking the seal at Grimsvötn. *Journal of Glaciology* 45: 506–516.
Grove JM (2004) *Little Ice Ages: Ancient and modern.* London: Routledge.
Marren PM (2005) Magnitude and frequency in proglacial rivers: A geomorphological and sedimentological perspective. *Earth-Science Reviews* 70: 203–251.
Shakesby RA (1985) Geomorphological effects of jökulhlaups and ice-dammed lakes, southern Norway. *Norsk Geografisk Tidsskrift* 39: 1–16.
Tufnell L (1984) *Glacier hazards.* London: Longman.
Tweed FS and Russell AJ (1999) Controls on the formation and sudden drainage of glacier-impounded lakes:

Implications for jökulhlaup characteristics. *Progress in Physical Geography* 23: 79–110.

jump dispersal   A form of *long-distance dispersal* in which a species colonises a patch of suitable habitat that is geographically separated from the area from which it dispersed. Such dispersal is frequently aided, knowingly or unwittingly, by humans.   JTK/JLI

[*See also* BIOLOGICAL INVASION, DISPERSAL, RANGE ADJUSTMENT, VICARIANCE]

Henne DC, Johnson SJ and Cronin JT (2007) Population spread of the introduced red imported fire ant parasitoid, *Pseudacteon tricuspis* Borgmeier (Diptera: Phoridae), in Louisiana. *Biological Control* 42: 97–104.
Kullman L (2008) Thermophilic tree species reinvade subalpine Sweden: Early responses to anomalous late Holocene climate warming. *Arctic, Antarctic and Alpine Research* 40: 104–110.
Mineur F, Davies AJ, Maggs CA et al. (2010) Fronts, jumps and secondary introductions suggested as different invasion patterns in marine species, with an increase in spread rates over time. *Proceedings of the Royal Society B* 277: 2693–2701.

Junge layer   A layer in the STRATOSPHERE at around 20–25 km above the Earth's surface characterised by a permanent high concentration of sulfate AEROSOLS, mainly sulfuric acid droplets. The sulfates are VOLCANIC AEROSOLS, the concentration of which increases markedly after large, explosive, VOLCANIC ERUPTIONS, influences the temperature of the STRATOSPHERE and affects the Earth's RADIATION BALANCE.   JAM

Junge CE and Manson JF (1961) Stratospheric aerosol studies. *Journal of Geophysical Research* 66: 2163–2182.
Oppenheimer C (2011) *Eruptions that shook the world.* Cambridge: Cambridge University Press.

Jurassic   A SYSTEM of rocks, and a PERIOD of geological time from 201.6 to 145.5 million years ago. Terrestrial environments during the Jurassic were dominated by *dinosaurs* while marine *reptiles*, such as *ichthyosaurs* and *plesiosaurs*, were an important part of the *marine fauna*. SEA-FLOOR SPREADING began in the Atlantic Ocean.   GO

[*See also* GEOLOGICAL TIMESCALE]

Crichton M (1993) *Jurassic Park.* London: Arrow.
Gutierrez K and Sheldon ND (2012) Paleoenvironmental reconstruction of Jurassic dinosaur habitats of the Vega Formation, Asturias, Spain. *Geological Society of America Bulletin* 124: 596–610.
Hesselbo SP, Jenkyns HC, Duarte LV and Oliveira LCV (2007) Carbon-isotope record of the Early Jurassic (Toarcian) Oceanic Anoxic Event from fossil wood and marine carbonate (Lusitanian Basin, Portugal). *Earth and Planetary Science Letters* 253: 455–470.

Loope DB, Steiner MB, Rowe CM and Lancaster M (2004) Tropical westerlies over Pangaean sand seas. *Sedimentology* 51: 315–322.

juvenile water   Water originating from the Earth's interior that has not previously participated in the HYDROLOGICAL CYCLE. The term was proposed by Meinzer (1923), who contrasted juvenile water with both surface-derived *meteoric water* and *connate water*, which was trapped in the interstices of SEDIMENTARY ROCKS during their formation. *JAM*

Greeley R (1987) Release of juvenile water on Mars: Estimated amounts and timing associated with volcanism. *Science* 236: 1653–1654.

Meinzer OE (1923) Outline of ground-water hydrology. *United States Geological Survey Water-Supply Paper* No. 494.

ka A *kilo-annum*: An abbreviation for thousands of years before the present.

GO

Kalman filter A powerful statistical method for FILTERING time series and used widely in detecting and characterising time-dependent patterns of behaviour in TREE RINGS. In the form developed for DENDROCLI- MATOLOGY, Kalman filtering is essentially a *multiple regression modelling* procedure in which the regression coefficients of the predictor variables are allowed to vary with time. This is done in a recursive fashion by casting the regression problem into a *state-space model*, which explicitly allows for timewise changes in the coefficients to be modelled. In essence, a series of *one-step-ahead predictions* are made using the available predictors, and their coefficients are allowed to change to improve the overall predictive ability of the model results. This is accomplished in an objective fashion using *maximum likelihood estimation*. The principal application of the Kalman filter in DEN- DROCHRONOLOGY has been in the detection of FOREST DECLINE through modelled changes in the response of tree growth to climate. The major difficulty in using the Kalman filter method lies in the interpretation of any time dependence detected by the technique. Unless there are strong a priori reasons to expect a certain pattern or direction of time dependence between tree growth and some predictor variable(s), the mere presence of time dependence may be difficult to interpret in a causal sense. However, when applied to a well-constructed EXPERIMENTAL DESIGN, the Kalman filter can be extremely useful for detecting changes in tree growth that may be induced by climatic and environmental change.

ERC/APP

[*See also* REGRESSION ANALYSIS, TIME-SERIES ANALYSIS]

Cook ER, D'Arrigo RD and Mann ME (2002) A well-verified, multiproxy reconstruction of the winter North Atlantic Oscillation Index since AD 1400. *Journal of Climate* 15: 1754–1764.

Harvey AC (1984) A unified view of statistical forecasting procedures. *Journal of Forecasting* 3: 245–275.

Rozas V (2005) Dendrochronology of pedunculate oak (*Quercus robur* L.) in an old-growth pollarded woodland in northern Spain: Tree-ring growth responses to climate. *Annals of Forest Science* 62: 209–218.

Van Deusen PC (1990) Evaluating time-dependent tree ring and climate relationships. *Journal of Environmental Quality* 19: 481–488.

Visser H and Molenaar J (1988) Kalman filter analysis in dendroclimatology. *Biometrics* 44: 929–940.

Visser H, Büntgen U, D'Arrigo RD and Petersen AC (2010) Detecting instabilities in tree-ring proxy calibration. *Climate of the Past* 6: 367–377.

kame A steep-sided mound or short irregular ridge comprising mainly sand and gravel, and formed by SUPRAGLACIAL or *ice-contact* glaciofluvial deposition. It is, therefore, a GLACIOFLUVIAL LANDFORM. The name derives from the Scottish words *cam* or *kaim*, meaning crooked and winding or steep-sided mound. The deposits are typically bedded with faulting and folding structures, which can be found in examples that are RELICT, particularly in the margins, reflecting removal of supporting ice. In association with KETTLE HOLES, kames may form *kame and kettle* TOPOGRAPHY, and they can easily be confused with HUMMOCKY MORAINE.

RAS

[*See also* ESKER, GLACIOFLUVIAL SEDIMENTS, KAME TERRACE]

Benn DI and Evans DJA (2010) *Glaciers and glaciation, 2nd edition.* London: Arnold.

kame terrace A GLACIOFLUVIAL LANDFORM composed of sediments deposited by a stream or river flowing between a glacier and the valley side (or between the glacier and a MORAINE). The terrace form results when a steep and irregular ICE-CONTACT SLOPE is produced as the glacier melts.

JAM

Gray M (1995) The kame terraces of lower Loch Etive. *Scottish Geographical Magazine* 111: 113–118.

Kampfzone   Literally, the 'battlezone'. The term is widely used in the European Alps for the *altitudinal zone* between the TREE LINE and the *tree-species limit*. In the Kampfzone, tree species 'struggle' to attain tree growth form, and are usually stunted.   *HHB*

[*See also* ALTITUDINAL ZONATION, KRUMMHOLZ]

kaolinite   A 1:1 CLAY MINERAL having a basic structure of one sheet of *silicon* atoms and one sheet of *aluminium atoms* held together by shared oxygen atoms in a fixed lattice structure that does not expand when wetted. There is little ISOMORPHIC SUBSTITUTION and the CATION EXCHANGE CAPACITY (CEC) is low. Kaolinite is a widely distributed clay mineral but is dominant in moderately to strongly weathered soil environments where LIXISOLS, ACRISOLS and FERRALSOLS have developed.   *EMB*

[*See also* LOW-ACTIVITY CLAYS]

Brady NC and Weil RR (2007) *The nature and properties of soils, 14th edition*. New York: Prentice Hall.

karoo   The BIOME of the SEMI-ARID region of southern Africa bordering the Kalahari Desert and containing wide environmental and biotic diversity. The *succulent karoo*, characterised by relatively short-lived shrubs, occupies a broad coastal belt with a less variable annual rainfall but sparse winter rains and summer drought ameliorated by coastal fog. The *Nama-karoo*, characterised by more grasses and by longer-lived shrubs, lies further inland where the interannual rainfall variability is more extreme and low summer rainfall is received from the east and north.   *JAM*

Dean WRJ and Milton SJ (1999) *The Karoo: Ecological patterns and processes*. Cambridge: Cambridge University Press.
Thomas DSG and Shaw PA (1991) *The Kalahari environment*. Cambridge: Cambridge University Press.

karren   A term of German origin relating to a variety of small-scale KARST landforms (including surface and subsurface pits, steps, grooves and channel forms), typically produced on LIMESTONE rock surfaces by CHEMICAL WEATHERING (DISSOLUTION). Although they are most commonly developed on carbonate and sulfate rocks, they have also been found on sandstone, quartzite and granite. The equivalent French term is *lapiés*. There are a series of specialised terms to differentiate the various forms, some of the larger and more complex of which may reflect past environmental conditions and may therefore be RELICT features:

- *Rillenkarren:* small (finger-like) parallel runnels, 1–3 mm wide and <30 cm long, separated by sharp ribs best developed near rock edges and oriented down slope
- *Rinnenkarren:* larger, vertical, linear grooves separated by sharp dividing ridges, commonly about 30 cm wide and often tens of metres long

- *Rundkarren:* rounded grooves formed under a soil cover
- *Spitzkarren:* sharp residual pinnacles up to tens of metres in height
- *Trittkarren:* step-like features, sometimes associated with horizontal lithological differences reflected in the tread of the step (10–40 cm in diameter)
- *Trichterkarren:* funnel-like forms, possibly associated with snow accumulation
- *Kluftkarren:* fissures or *grikes* where vertical joints have been widened, mostly <0.5 m wide but may be several metres deep and often partly infilled
- *Flachkarren:* upstanding blocks (*clints*), typically about 1 m wide and several metres long, separated by grikes (forming *limestone pavement*)
- *Kamenitza:* shallow (typical depth, 3 cm) *weathering pits* or *solution pans* tending towards a circular shape (typical diameter, 10–100 cm)
- *Mäanderkarren:* meandering channels reflecting surficial water flow.   *JAM/SHD*

Ford DC and Lundberg J (1987) A review of dissolutional rills in limestone and other soluble rocks. *Catena Supplement* 8: 119–140.
Ginés À (2004) Karren. In Gunn J (ed.) *Encyclopedia of caves and karst science*. New York: Fitzroy Dearborn, 470–473.
Goldie HS and Cox NJ (2000) Comparative morphometry of limestone pavements in Switzerland, Britain and Ireland. *Zeitschrift für Geomorphologie NF Supplement band* 122: 85–112.
Mottershead D and Lucas G (2001) Field testing of Glew and Ford's model of solution flute evolution. *Earth Surface Processes and Landforms* 26: 839–846.
Veress M and Toth G (2004) Types of meandering karren. *Zeitschrift für Geomorphologie NF* 48: 53–77.

karst   The LANDFORMS and LANDSCAPES characterised by solutional WEATHERING along surface and subsurface pathways, typically leading to the progressive replacement of surface with underground drainage (*karstification*). The term derives from a LIMESTONE region in western Slovenia, in the northern Dinaric Mountains near the Italian border. Karst systems are generally associated with rocks of high *solubility* and well-developed secondary *porosity* (e.g. limestone and GYPSUM). An estimated 7–10 per cent of the Earth's ice-free land surface comprises karst terrain which, when karstification is well advanced, exhibits a scarcity of flat agricultural land. Key elements of karst landscapes are solutional features, such as CAVE PASSAGES, DOLINES, POLJES, KARREN and *sinkholes*, and depositional forms, such as SPELEOTHEMS and TUFA. These and other datable CAVE SEDIMENTS can be used in various ways to reconstruct changes in GROUNDWATER conditions, CLIMATIC CHANGE and SEA-LEVEL CHANGE. Speleothems in particular are important for HIGH-RESOLUTION RECONSTRUCTION in PAL-AEOCLIMATOLOGY. The term PSEUDOKARST is sometimes used for similar landforms produced in non-carbonate rocks.   *SHD*

[*See also* AGGRESSIVE WATER, BIOKARST, CHEMICAL WEATHERING, COCKPIT KARST, DISSOLUTION, DROWNED KARST, RESURGENCE, THERMOKARST, TOWER KARST, TROPICAL KARST]

Cvijić J (1893) Das Karstphänomen [The karst phenomenon]. *Geographisches Abhandlung* 5: 218–329.

De Waele J, Gutierrez F, Parise M and Plan L (eds) (2011) Geomorphology and natural hazards in karst areas. *Geomorphology* 134: 1–170 [Special Issue].

Drew D and Hötzl H (eds) (1999) *Karst hydrology and human activities: Impacts, consequences and implications.* Rotterdam: Balkema.

Ford DC and Williams PW (2007) *Karst hydrogeology and geomorphology, 2nd edition.* Chichester: Wiley.

Goldschneider N and Drew D (eds) (2007) *Methods in karst hydrology.* London: Taylor and Francis.

Gunn J (ed.) (2004) *Encyclopedia of caves and karst science.* New York: Fitzroy Dearborn.

Jennings JN (1985) *Karst geomorphology.* Oxford: Basil Blackwell.

Kresic N (2012) *Water in Karst: Management, vulnerability and restoration.* New York: McGraw-Hill.

Onac BP and Constantin S (eds) (2008) Archives of climate and environmental change in karst. *Quaternary International* 187: 1–116 [Special Issue].

Vereis M (2010) *Karst environments: Karren formation in high mountains.* Dordrecht: Springer.

Kastanozems

Soils with a thick, dark brown topsoil, a *mollic horizon*, rich in organic matter and having a calcareous or gypsiferous subsoil. They are common in central areas of North and South America, and Eurasia (SOIL TAXONOMY: Ustic and Boric great groups of *Mollisols*). Typically, Kastanozems develop in a warmer climate on the drier side of CHERNOZEMS under short-grass *prairie* or STEPPE. The soil is denser packed with smaller pore space, less capacity for water storage and lower *permeability* than Chernozems. They dry out to considerable depth in the dry season so that, below the WETTING FRONT, there may be a dry layer that fails to receive rain-fed moisture from above or capillary-fed moisture from below (even after heavy rain in the wet season). DROUGHT is the main agricultural HAZARD; there is also a risk of SALINISATION following IRRIGATION and WIND EROSION following OVERGRAZING. In earlier literature, these soils are referred to as *Chestnut Soils*.   *EMB*

[*See also* WORLD REFERENCE BASE FOR SOIL RESOURCES (WRB)]

Alekseeva T, Alekseeva A and Maher B (2003) Late-Holocene climate reconstruction for the Russian steppe, based on mineralogical and magnetic properties of buried palaeosols. *Palaeogeography, Palaeoclimatology, Palaeoecology* 201: 321–341.

Maki A, Kenji T, Kiyokazu K and Teruo H (2007) Morphological and physico-chemical characteristics of soils of the steppe region of the Kherla river basin, Mongolia. *Journal of Hydrology* 333: 100–108.

K-cycles

A concept relating to SOILS and LANDSCAPE, in which there is an alternation of stable and unstable phases of SOIL DEVELOPMENT. The concept was developed following observations on hillslopes in Australia where soils are formed in stable phases and eroded or covered by SEDIMENTATION in unstable phases. Each couplet of instability (Ku) followed by stability (Ks) constitutes a single K-cycle, and they are numbered consecutively. Evidence may be present in the landscape as various types of *soil layering*, which, in favourable circumstances, may permit the reconstruction of K-cycles. Soils of differing age may coexist in the same landscape as a result of the EROSION and REDEPOSITION of slope deposits.   *JAM*

[*See also* BIOSTASY, DEPOSITION, EROSION, LANDSCAPE EVOLUTION, RHEXISTASY, SOIL AGE, SOIL STRATIGRAPHY]

Butler BE (1959) *Periodic phenomena in landscapes as a basis for soil studies.* Melbourne: CSIRO Publishing.

Phillips JD and Lorz C (2008) Origins and implications of soil layering. *Earth-Science Reviews* 89: 144–155.

Vreeken WJ (1996) A chronogram for postglacial soil-landscape change from Palliser Triangle, Canada. *The Holocene* 6: 433–438.

kettle hole

A self-contained bowl-shaped depression, formed as a result of melting of a buried ice block. Kettle holes, also known as *kettles*, are especially common in glacier OUTWASH DEPOSITS (*kettled outwash*) and in MORAINE complexes. The ice blocks may be derived as remnants of a GLACIER snout or as ICEBERGS transported onto an *outwash plain* (see SANDUR) by FLOODS and JÖKULHLAUPS. Kettle holes are commonly occupied by lakes (*kettle lakes*) and the sediments they contain are often used in PALAEOENVIRONMENTAL RECONSTRUCTION. These organic and minerogenic sediments are younger than the surrounding deposits and tend to survive because of the enclosed nature of the kettle hole. In some cases, the resulting stratigraphical NATURAL ARCHIVES have survived subsequent GLACIATION.   *MJH*

[*See also* GLACIOFLUVIAL SEDIMENTS, ICE-CONTACT FANS/RAMPS, KAME, KAME TERRACE]

Benn DI and Evans DJA (2010) *Glaciers and glaciation, 2nd edition.* London: Hodder Education.

Maizels JK (1992) Boulder ring structures produced during Jökulhlaup flows: Origin and hydraulic significance. *Geografiska Annaler* 74A: 21–33.

Matthews JA, Seppälä M and Dresser PQ (2005) Holocene solifluction, climatic variation and fire in a subarctic landscape at Pippokangas, Finnish Lapland, based on radiocarbon-dated buried charcoal. *Journal of Quaternary Science* 20: 533–548.

keystone species   Many ecologists believe that a few species have a disproportionately strong influence on others in their community such that, if removed, the structure of that community would change dramatically (see ECOSYSTEM COLLAPSE), much like the collapse of the stones forming a Roman arch once the central stone ('keystone') has been removed. While there is much debate over this concept, examples of likely keystone species include nitrogen-fixing BACTERIA, DECOMPOSERS, army ANTS, some top CARNIVORES, *beavers, prairie dogs* and, at least in recent times, humans. *Keystone structures* have also been identified, representing HABITAT structures such as *tree cavities* that are important for multiple species. *Reverse keystone species* are those whose presence may have an adverse effect on other species, with the *noisy miner* (*Manorina melanocephala*) an Australian *honey eater*, being a good example.

MVL/JLI

[*See also* COMMUNITY CONCEPTS, NITROGEN FIXATION]

Berg S, Christianou M, Jonsson T and Ebenman B (2011) Using sensitivity analysis to identify keystone species and keystone links in size-based food webs. *Oikos* 120: 510–519.

Green K (2011) The transport of nutrients and energy into the Australian Snowy Mountains by migrating bogong moths *Agrostis infusa*. *Austral Ecology* 36: 25–34.

MacNally R and Timewell CAR (2005) Resource availability controls bird-assemblage composition through interspecific aggression. *Auk* 122: 1097–1111.

Montague-Drake RM, Lindenmayer DB, Cunningham RB and Stein JA (2011) A reverse keystone species affects the landscape distribution of woodland avifauna: A case study using the Noisy Miner (*Manorina melanocephala*) and other Australian birds. *Landscape Ecology* 26: 1383–1394.

Remm J and Lohmus A (2011) Tree cavities in forests: The broad distribution pattern of a keystone structure for biodiversity. *Forest Ecology and Management* 262: 579–585.

kinetic temperature   The temperature that indicates the average *kinetic energy* of the MOLECULES or ATOMS of a substance (i.e. that which is measurable by an instrument in direct contact with an object). It is also known as the *physical temperature*.   TS

[*See also* RADIANT TEMPERATURE]

Kirchoff radiation law   Under conditions of local *thermodynamic equilibrium*, thermal emission has to be equal to absorption. A system is in thermodynamic equilibrium if mechanical, chemical and thermal equilibria are held.   TS

Ulaby FT, Moore RK and Fung AK (1981) *Microwave remote sensing: Active and passive*. Norwood, MA: Artech House.

***k*-means clustering**   A method of non-hierarchical CLUSTER ANALYSIS or *partitioning* in which from an initial random or other partition of the objects into *k* clusters, each object is examined in turn and reassigned, if appropriate, to a different cluster in an attempt to optimise some predefined numerical criterion that quantifies in some way the 'quality' of the cluster solution. Many clustering criteria have been proposed, but the most commonly used in the environmental sciences consider features of the within-groups ($\mathbf{W}$), between-groups ($\mathbf{B}$), and total ($\mathbf{T}$) matrices of sum of squares and cross-products that can be defined for every partition of the objects into a particular number of clusters or groups. The two most commonly used clustering criteria involve minimisation of the trace of matrix $\mathbf{W}$ or minimisation of the determinant of matrix $\mathbf{W}$. The first criterion tends to produce 'spherical' clusters, whereas the second tends to produce clusters of all the same shape. The most useful criterion for deciding on the final optimal number of groups to adopt is the so-called *Calinski-Harabasz criterion* which is simply the F statistic of multivariate ANALYSIS OF VARIANCE (ANOVA) and *canonical variates analysis* (see DISCRIMINANT ANALYSIS).   HJBB

Borcard D, Gillet F and Legendre P (2011) *Numerical ecology with R*. New York: Springer.

Hand DJ and Krzanowski WJ (2005) Optimising *k*-means clustering results with standard software packages. *Computational Statistic and Data Analysis* 49: 969–973.

Legendre P and Birks HJB (2012) Clustering and partitioning. In Birks HJB, Lotter AF, Juggins S and Smol JP (eds) *Tracking environmental change using lake sediments, volume 5: Data handling and numerical techniques*. Dordrecht: Springer, 167–200.

knickpoint   A sharp break of slope in a river long profile characterised by a steepening of channel gradient. It may be marked by rapids or a WATERFALL and a GORGE. Also known as a *nickpoint* (American) or a *rejuvenation head*, a knickpoint tends to form and move upstream by HEADWARD EROSION following a negative change in BASE LEVEL. Some, however, represent the occurrence of resistant geological structures. The term has usually been used in the context of long-term LANDSCAPE EVOLUTION and the evolution of DRAINAGE NETWORKS, but more recently more subtle knickpoints have been recognised in response to human interference in river channels. River CHANNELISATION involving straightening of MEANDERS introduces an increased channel gradient, leading to DEGRADATION, which progresses upstream as a knickpoint. In recent years, there has been a resurgence of interest with the ability to date knickpoint retreat using COSMOGENIC- NUCLIDE DATING.   RPDW/JAM

[*See also* DENUDATION CHRONOLOGY, REJUVENATION]

Brookes A (1985) River channelization: Traditional
   engineering methods, physical consequences and
   alternative practices. *Progress in Physical Geography* 9:
   44–73.
Jansen JD, Fabel D, Bishop P et al. (2011) Does decreasing
   paraglacial supply slow knickpoint retreat? *Geology* 39:
   543–546.
Lanue J-P (2011) Longitudinal profiles and knickzones: The
   examples of the rivers of the Cher basin in the northern
   Massif Central. *Proceedings of the Geological Association*
   122: 125–142.
Poiraud A (2013) Knickpoints from watershed scale
   to hillslope scale: A key to landslide control
   and geomorphological resilience. *Zeitschrift für
   Geomorphologie Supplementband* 56: 19–35.
Ye F-Y, Barriot J-P and Carretier S (2013) Initiation and
   recession of the fluvial knickpoints of the Island of Tahiti
   (French Polynesia). *Geomorphology* 186: 162–173.

knock and lochan topography   A Scottish term
for an extensive area of glacially scoured bedrock terrain comprising eroded *rock knobs* and *rock basins*,
many of which are occupied by LAKES. It has also been
suggested that the form of some such terrain may owe
much to preglacial WEATHERING in certain locations. *RAS*

[*See also* GLACIAL EROSION]

Johansson M, Olvmo M and Lidmar-Bergstrom K (2001)
   Inherited landforms and glacial impact of different
   palaeosurfaces in southwest Sweden. *Geografiska Annaler*
   83A: 67–89.
Rea BR and Evans DJA (1996) Landscapes of areal scouring
   in NW Scotland. *Scottish Geographical Magazine* 112:
   47–50.

knowledge   The sum of all that is known, including
theoretical and practical knowledge. EPISTEMOLOGY is
the THEORY of knowledge. The two basic requirements
of knowledge are the *truth* of something and *belief* that
it is true. However, it is difficult to establish what is
true. This requires independent evidence and a rational
approach to assessing the evidence which, at least
from the perspective of SCIENCE, is best provided by
SCIENTIFIC METHOD. Knowledge may increase through
time, but what is regarded as true, reliable or OBJECTIVE
KNOWLEDGE may change in the light of new evidence or
new insights related to existing evidence. *JAM*

Pritchard D (2006) *What is this thing called knowledge?*
   London: Routledge.
Ziman J (1978) *Reliable knowledge: An exploration of the
   grounds for belief in science.* Cambridge: Cambridge
   University Press.

knowledge-based systems   Computer programmes
capable of advising and problem solving in a particular
field of expertise on a level comparable to a human
expert. During the past 20 years, two paradigms have
evolved—namely, *knowledge transfer* and *knowledge
modelling.* The *knowledge-transfer paradigm* states
that human knowledge is transferred rapidly into
implemented computer systems and made available in
the DECISION-MAKING process. These knowledge-based
systems could be developed quickly but failed to be
reliable and maintainable. The *knowledge-modelling
paradigm* regards knowledge acquisition not as the
elicitation and collection of existing knowledge but
the process of creating a *knowledge model*, which did
not exist beforehand. The design of such a knowledge-
based system requires several steps: elicitation, interpretation, formalisation and operationalisation, design
and implementation.   *HB*

Akerkar R (2009) *Knowledge-based systems*. Boston: Jones
   and Bartlett.
Davis R (1986) Knowledge-based systems. *Science* 231:
   957–963.
Humphries HC, Bourgeron PS and Reynolds KM
   (2010) Sensitivity analysis of land unit suitability
   for conservation using a knowledge-based system.
   Environmental Management 46: 225–236.

komatiite   An IGNEOUS ROCK with unusually high
magnesia (MgO) content, representing very high-
temperature and low-viscosity LAVA FLOWS, named
after the Komati River in Swaziland, southern Africa.
Komatiites are restricted to the GEOLOGICAL RECORD of
the ARCHAEAN, forming an important constituent of
many GREENSTONE BELTS. They pose problems for the
strict application of UNIFORMITARIANISM and date from
a period in the Earth's history when more internal heat
was generated, with important implications for the
nature of early PLATE TECTONICS.   *GO*

Kamenetsky VS, Gurenko AA and Kerr AC (2010)
   Composition and temperature of komatiite melts from
   Gorgona Island, Colombia, constrained from olivine-
   hosted melt inclusions. *Geology* 38: 1003–1006.
Parman SW, Dann JC, Grove TL and de Wit MJ (1997)
   Emplacement conditions of komatiite magmas from the 3.49
   Ga Komati Formation, Barberton Greenstone Belt, South
   Africa. *Earth and Planetary Science Letters* 150: 303–323.
Williams DA, Wilson AH and Greeley R (2000) A komatiite
   analog to potential ultramafic materials on Io. *Journal of
   Geophysical Research: Planets* 105: 1671–1684.

koniology   The scientific study of ATMOSPHERIC DUST
and other suspended PARTICULATES in the ATMOSPHERE,
such as soot, pollen and spores. The *konisphere* or
staubosphere is the part of the atmosphere where such
particles occur.   *AHP*

[*See also* AEROBIOLOGY]

Fairbridge RW (1987) Koniology. In Oliver JE and
   Fairbridge RW (eds) *The encyclopedia of climatology.*
   New York: Van Nostrand Reinhold, 530–531.

kriging A family of statistical techniques for optimal spatial INTERPOLATION between discrete measurement points, minimising the error VARIANCE. It was developed by the mining engineer D.G. Krige in South Africa. The interpolation of a random variable $Z(X)$ is based on the empirical VARIOGRAM $2\gamma(h)$. A *theoretical variogram model* is then fitted to the empirical variogram. From this model, local weights are estimated. These are used to predict $Z(x_0)$ at the unsampled location x_0 by calculating a local weighted mean. Kriging techniques include ordinary kriging, simple kriging, block kriging, universal kriging, disjunctive kriging, indicator kriging and co-kriging. *HB*

Cressie NAC (1993) *Statistics for spatial data*. New York: Wiley.
Montes F and Ledo A (2010) Incorporating environmental and geographical information in forest data analysis: A new fitting approach for universal kriging. *Canadian Journal of Forest Research* 40: 1852–1862.
Zhu Q and Lin HS (2010) Comparing ordinary kriging and regression kriging for soil properties in contrasting landscapes. *Pedosphere* 20: 594–606.

Krummholz The stunted growth form of tree species beyond the TREE LIMIT. Buds projecting beyond the climatically warmer BOUNDARY LAYER or the SNOWPACK are killed. Trees may develop if conditions ameliorate (e.g. *flag-form trees* of *Picea* spp.). Some recognise a difference between *Krüppelholz* ('climatically determined' crippled form) and Krummholz ('genetically determined', dwarfed form as exhibited by *Pinus mugo* in the Alps). *HHB*

[*See also* GENOTYPE, KAMPFZONE, PHENOTYPE]

Holttmeier F-K (2003) *Mountain timberlines: Ecology, patchiness and dynamics*. Dordrecht: Kluwer.

***K*-selection** A conceptual, *evolutionary strategy* that emphasises species' competitive abilities to maintain populations near CARRYING CAPACITY (K) in *stable habitats*. Theory suggests that *K-strategists* are slow growing, characterise mature patches of VEGETATION and the later stages of ECOLOGICAL SUCCESSION, and utilise RESOURCES for sustained existence at a *steady-state equilibrium*. *LRW/JAM*

[*See also* COMPETITION, EQUILIBRIUM CONCEPTS, R-SELECTION]

Pianka ER (1970) On r- and *K*-selection. *American Naturalist* 104: 592–597.

K-T boundary The boundary between the CRETACEOUS and TERTIARY periods (~65 million years ago), which is marked by a major MASS EXTINCTION event. Groups of organisms that became extinct include, most famously, the *dinosaurs*, but also other reptile groups including the flying *pterosaurs* and the marine plesiosaurs, marine *molluscs* including the *ammonites*, *belemnites* and the *rudist bivalves*, and many groups of land plants and FORAMINIFERA.

The cause of the extinctions has long been a matter of controversy. Explanations that have been put forward include SEA-LEVEL CHANGE and the effects of VOLCANIC ERUPTIONS, notably those of the Deccan traps in northwest India (see LAVA PLATEAU). Explanations specific to the extinction of the dinosaurs include disease, poisoning, suicide, constipation and sterility. Detailed studies of rock sections with continuous sedimentation from Cretaceous to Tertiary times, notably at Gubbio in central Italy, have revealed the presence of a distinct *K-T boundary layer*. Alvarez et al. (1980) demonstrated that CLAY in this layer was enriched in the *rare earth* element *iridium*, and attributed this *iridium anomaly* to an extraterrestrial source, specifically a major METEORITE IMPACT event that would also have caused the extinctions. This *impact hypothesis* was greeted with much controversy by the geological community. Further evidence to support the hypothesis has since been collected, including TSUNAMI deposits in the boundary layer, shocked QUARTZ grains and the identification of an impact site at Chicxulub on the Yucatán peninsula in southern Mexico. The theory of a major impact event at the K-T boundary is now widely accepted and the iridium-enriched layer had been widely recognised around the globe as an isochronous event. It now provides the basis for the GSSP (*Global Stratotype Section and Point*) for the base of the PALAEOGENE at El Kef, Tunisia.

The problem remains, however, of whether or not the impact caused the extinctions. A major review of biotas across the boundary by MacLeod et al. (1997) revealed a surprising lack of detailed information for many groups. Some were in decline through the latest Cretaceous, before their final extinction (which in some cases occurred before, rather than at the boundary), while others crossed the boundary with little apparent change. The nature of events at the K-T boundary remains a topic of lively debate. *JCWC/GO*

Alvarez LW, Alvarez W, Asaro F and Michel HV (1980) Extraterrestrial cause for the Cretaceous-Tertiary extinction. *Science* 208: 1095–1108.
Alvarez W (1997) *T. rex and the crater of doom*. London: Penguin Books.
Frankel C (1999) *The end of the dinosaurs: Chicxulub crater and mass extinctions*. Cambridge: Cambridge University Press.
Hildebrand AR, Penfield GT, Kring DA et al. (1991) Chicxulub crater: A possible Cretaceous/Tertiary boundary impact crater on the Yucatán Peninsula, Mexico. *Geology* 19: 867–871.
Hudson JD and MacLeod N (1998) Discussion on the Cretaceous-Tertiary biotic transition. *Journal of the Geological Society* 155: 413–419.

MacLeod N, Rawson PF, Forey PL et al. (1997) The Cretaceous-Tertiary biotic transition. *Journal of the Geological Society, London* 154: 265–292.

Molina E, Alegret L, Arenillas I et al. (2006) The Global Boundary Stratotype Section and Point for the base of the Danian Stage (Paleocene, Paleogene, "Tertiary", Cenozoic) at El Kif, Tunisia: Original definition and revision. *Episodes* 29: 263–273.

Nichols DJ and Johnson KR (2008) *Plants and the K-T boundary*. Cambridge: Cambridge University Press.

Norris RD, Huber BT and Self-Trail J (1999) Synchroneity of the K-T oceanic mass extinction and meteorite impact: Blake Nose, western North Atlantic. *Geology* 27: 419–422.

Ryder G, Fastovsky D and Gartner S (eds) (1996) *The Cretaceous-Tertiary event and other catastrophes in Earth history.* Geological Society of America Special Paper No. 307.

Ward PD (1995) After the fall: Lessons and directions from the K/T debate. *Palaios* 10: 530–538.

kurtosis   The extent to which the peak of a unimodal frequency or PROBABILITY DISTRIBUTION departs from the shape of a NORMAL DISTRIBUTION by either being more pointed (*leptokurtic*) or flatter (*platykurtic*). The *coefficient of kurtosis* is the fourth moment about the mean divided by the variance squared (often with three subtracted so that a NORMAL DISTRIBUTION has zero kurtosis).   *HJBB*

[*See also* SKEWNESS]

Sokal RR and Rohlf FJ (1995) *Biometry.* New York: Freeman.

La Niña A period of strong TRADE WINDS and unusually low SEA-SURFACE TEMPERATURE (SST) in the central and eastern tropical Pacific. It is the negative, opposite or 'cold' phase of EL NIÑO-SOUTHERN OSCILLATION (ENSO) to EL NIÑO. *AHP/JAM*

Philander SGH (1990) *El Niño, La Niña and the Southern Oscillation.* San Diego, CA: Academic Press.

label (1) A textual description of a geographic feature or object on a paper MAP. (2) The name and description (ATTRIBUTE) of an ENTITY in a GEOGRAPHICAL INFORMATION SYSTEM (GIS). *VM*
[*See also* IDENTIFICATION (ID), OVERLAY ANALYSIS]

labile pool Materials (organic and organic) which are readily decomposed. It is not the same as the *plant-available nutrient pool*; rather it is the pool that has the potential to be made rapidly available to plants. In terms of *soil organic matter*, the labile pool is that which has a rapid turnover rate and includes *sugars* and *starches* which are easily decomposed by *soil microbes* and other SOIL BIOTA. The recalcitrant substrate including *lignins, tannins* and compounds rich in *polyphenols* can only be used by specialised organisms. For *inorganic materials*, the labile pool includes those materials that can be solubilised rapidly and exchange with ions in solution to replace those taken up by plants or lost from the system through LEACHING. *SN*
[*See also* DECOMPOSITION, NUTRIENT]

Ebelhar SA (2008) Labile pool. In Chesworth W (ed.) *Encyclopedia of soil science.* Dordrecht: Springer, 425–426.

laboratory science The use of laboratories for analysis and EXPERIMENT. Predominantly, but not entirely, the domain of the physical and biological sciences, the distinctive quality of the use of laboratories is the ability to restrict 'outside' *environmental effects*, to reduce *contamination* and to conduct tightly *controlled experiments*. Thus, for example, in conducting laboratory experiments, CAUSAL RELATIONSHIPS may be firmly established by varying one causal factor at a time, an approach that is difficult, if not impossible, to achieve by FIELD RESEARCH. A broader definition of 'laboratory' is sometimes interpreted to include computer experiments, especially in the SOCIAL SCIENCES. *JAM*

Janssen MA, Holahan R, Lee A and Ostrom E (2010) Lab experiments for the study of social-ecological systems. *Science* 328: 613–617.

lacustrine sediments Sediments deposited in lakes, which are important NATURAL ARCHIVES of environmental change. They can be sampled by a variety of CORERS used from open water, winter ice, or the surface of an overgrown or drained lake. SEDIMENTARY STRUCTURES include RHYTHMITES (LAMINATIONS resulting from rhythmic deposition cycles). Annual laminations are VARVES. *Minerogenic varves* produced by glacial MELTWATER are widespread around the Baltic Sea, and have been used to construct the *Swedish Varve Timescale* (see VARVE CHRONOLOGY). Organic and chemical (e.g. carbonate) laminations are produced in certain lake types, and are best preserved in MEROMICTIC LAKES where BIOTURBATION is minimal.

ALLOCHTHONOUS lake sediments originate from the *catchment.* They may be organic (e.g. terrestrial MACROFOSSILS, SOIL and *peat*) or minerogenic, and are derived by catchment EROSION. High minerogenic content indicates catchment disturbance that can result from GLACIAL EROSION, PERIGLACIAL activity, ARIDITY (e.g. LOESS), *human impact* and CATASTROPHIC EVENTS (e.g. SNOW AVALANCHES, LANDSLIDES, FLOODS and *fires*). In special circumstances, a *trash layer* of terrestrial material is deposited during lake formation at the melting of a buried ice block that had supported vegetation and soil.

AUTOCHTHONOUS sediments originate within the lake and are the net deposition of lacustrine *biogenic material* (less decay and outflow loss). LOSS ON IGNITION (LOI) at 550°C reflects the balance between minerogenic and organic material. It can be a sensitive synthetic environmental indicator. LOI at 950°C estimates CARBONATES in *calcareous sediments*, including biogenic MARL, *shells* and CLASTIC carbonates.

Sediment chemistry can be used to infer environmental change. X-RAY FLUORESCENCE ANALYSIS (XRF) is a newly developed non-destructive method that measures amounts of elements. For example, Ca, Mg and Ti are indicators of mineral erosion in the catchment; Pb, Zn, Cu and Hg are indicators of POLLUTION by HEAVY METALS; and N, P, C and organic compounds (pigments derived from chlorophyll, lipids etc.) are indicators of algal or bacterial types and of EUTROPHICATION.

STABLE ISOTOPES (δ^{13}C, δ^{18}O, δ^2H) can indicate TEMPERATURE, PRECIPITATION and/or HUMIDITY. *Radioisotopes* are used for dating. For example, RADIOCARBON DATING covers ~50,000 years; ^{210}Pb covers the past ~200 years; and peaks of ^{137}Cs are time markers for nuclear testing, which culminated in 1963, and the 1986 Chernobyl NUCLEAR ACCIDENT. TEPHRA horizons are also chronological markers. All these can be used for CORE CORRELATION. The analysis of the plant and animal remains in lacustrine sediments results in a variety of environmental reconstructions through time. Ideally, a study of environmental change should use evidence from both sediments and biota in a MULTIPROXY APPROACH.   *HHB*

[*See also* LIMNOLOGY, PALAEOLIMNOLOGY]

Appleby PG (1993) Foreward to the ^{210}Pb dating anniversary series. *Journal of Paleolimnology* 9: 155–160.
Battarbee RW, Bennion H, Gell P and Rose N (2012) Human impacts on lacustrine ecosystems. In Matthews JA, Bartlein PJ, Briffa KR et al. (eds) *The SAGE handbook of environmental change, volume 2*. London: Sage, 47–70.
Birks HH, Battarbee RW and Birks HJB (2000) The development of the aquatic ecosystem at Kråkenes Lake, western Norway, during the Late Glacial and early Holocene: A synthesis. *Journal of Paleolimnology* 23: 91–114.
Boyle JF (2000) Rapid elemental analysis of sediment samples by isotope source XRF. *Journal of Paleolimnology* 23: 213–221.
Last WM and Smol JP (eds) (2001) *Tracking environmental change using lake sediments, volumes 1 and 2*. Dordrecht: Kluwer.
Litt T, Brauer A, Goslar T et al. (2001) Correlation and synchronisation of Lateglacial continental sequences in northern Central Europe based on annually laminated lacustrine sediments. *Quaternary Science Reviews* 20: 1233–1249.
Smol JP, Birks HJB and Last WM (eds) (2001) *Tracking environmental change using lake sediments, volumes 3 and 4*. Dordrecht: Kluwer.
Verrecchia EP (2007) Lacustrine and palustrine geochemical sediments. In Nash DJ and McLaren SJ (eds) *Geochemical sediments and landscapes*. Malden, MA: Blackwell, 298–329.

White JDL and Riggs NR (eds) *Volcaniclastic sedimentation in lacustrine settings*. Oxford: Blackwell.
Wohlfarth B, Björck S and Possnert G (1995) The Swedish Time Scale: A potential calibration tool for the radiocarbon time scale during the Late Weichselian. *Radiocarbon* 37: 347–359.

lag deposit   A sedimentary deposit of relatively coarse CLASTS produced in situ by the removal of finer interstitial particles. Examples include DESERT PAVEMENTS formed from the winnowing away of fines by wind, and periglacial BOULDER PAVEMENTS produced in pronival locations by snowmelt. Once formed, such pavements may be relatively stable and resistant to erosion.   *JAM*

lag time   The time lapse between the occurrence of an event (e.g. a CLIMATIC CHANGE, LANDUSE CHANGE, a human intervention or an EXTREME CLIMATIC EVENT) and the resulting effect. The concept is important, for example, in understanding the CAUSAL RELATIONSHIPS involved in environmental change and in the prediction and mitigation of NATURAL HAZARDS. In the context of GLACIER VARIATIONS in response to a climatic change, it is the time interval between the mass balance change and the maximum (or minimum) of the resulting glacier advance (or retreat).   *MAB/JAM*

[*See also* LEAD AND LAG, REACTION TIME, RELAXATION TIME, RESPONSE TIME]

Allen JRL (1974) Reaction, relaxation and lag in natural sedimentary systems: General principles, examples and lessons. *Earth-Science Reviews* 10: 263–342.

lagoon   A shallow body of water connected permanently or intermittently with a larger body of water. Many lagoons lie parallel to the COAST and are separated from the sea by a BARRIER ISLAND, BARRIER BEACH or SPIT. Others are associated with CORAL REEF ISLANDS. Lagoons differ from LAKES, which are completely enclosed, and *bays*, which are substantially open to the sea.   *HJW*

Oertel GF (2005) Coastal lakes and lagoons. In Schwartz ML (ed.) *Encyclopedia of coastal science*. Dordrecht: Springer, 263–266.

lahar   A DEBRIS FLOW composed of volcanic debris. Such flows are common around VOLCANOES because of the combination of steep slopes and abundant, unconsolidated PYROCLASTIC MATERIAL debris. Lahars may occur during or immediately after a VOLCANIC ERUPTION, or by secondary mobilisation weeks to years later. Lahars are a potentially devastating NATURAL HAZARD: 23,000 people died in a lahar following the AD 1985 eruption of Nevado del Ruiz (Colombia). Some authors advocate abandoning the term because the flow processes and characteristics of the deposits are no different from debris flows in non-volcanic settings. *JBH*

Barclay J, Alexander J and Susnik J (2007) Rainfall-induced lahars in the Belham Valley, Montserrat, West Indies. *Journal of the Geological Society of London* 164: 815–827.

Bollschweiler M, Stoffel M, Vazquez-Selem L and Palacios D (2010) Tree-ring reconstruction of past lahar activity at Popocatepetl volcano, Mexico. *The Holocene* 20: 265–274.

Lowe DR, Williams SN, Leigh H et al. (1986) Lahars initiated by the 13 November 1985 eruption of Nevado del Ruiz, Colombia. *Nature* 324: 51–53.

Procter J, Cronin DJ, Fuller IC et al. (2010) Quantifying the geomorphic impacts of a lake-breakout lahar, Mount Ruapehu, New Zealand. *Geology* 38: 67–70.

Voight B (1990) The 1985 Nevado del Ruiz Volcano catastrophe: Anatomy and retrospection. *Journal of Volcanology and Geothermal Research* 42: 151–188.

lake A body of (normally fresh) water occupying a depression in the Earth's continental surface. *JAM*

lake stratification and zonation *Water-column stratification* occurs typically in temperate regions where water temperatures range either side of 4°C. Surface water warms in spring, and when wind action becomes insufficient to mix the water column, *summer stratification* develops. The warm, oxygenated *epilimnion* overlies the denser, cool, de-oxygenated *hypolimnion* (see Figure). Autumn cooling induces OVERTURNING. *Winter stratification*, with ice and water at 0°C overlying water at 4°C, overturns in spring.

Macrophytes and PHYTOPLANKTON inhabit the *photic zone*. Macrophytes occupy the LITTORAL ZONE, their depth zonation being determined mainly by light penetration, and their distribution by wave exposure, sediment type and deposition rate. *Emergent plants* grow nearest the shore. Turf-like *isoetids* prefer

wave-exposed shores but can extend (e.g. *Isoetes lacustris*) to ~6 m depth. Floating-leaved *nyphaeids* are restricted to ~2 m depth, but submerged *elodeids* can reach 11 m or deeper in clear water. *Free-floating plants* (e.g. *Lemna, Azolla*) can cover a lake. Animal diversity is greatest in the epilimnion, but several invertebrate types thrive in the anoxic hypolimnion. *HHB*

[**See also** LIMNOLOGY, PALAEOLIMNOLOGY]

Hutchinson GE (1957) *Treatise on limnology, volume 1: Geography, physics, and chemistry.* New York: Wiley.

Lewis WM (1983) A revised classification of lakes based on mixing. *Canadian Journal of Fisheries and Aquatic Science* 40: 1779–1787.

Spence DHN (1964) The macrophytic vegetation of lochs, swamps and associated fens. In Burnett JH (ed.) *The vegetation of Scotland*. Edinburgh: Oliver and Boyd, 306–425.

lake terrace A *terrace* at a lake margin formed during a *regressive phase* when there is a reduced volume of water in the lake. They mark former SHORELINES, and if mapped and dated can provide information on changing precipitation levels, particularly in SEMI-ARID regions and ARIDLANDS. In PERIGLACIAL environments, they can be formed or enhanced by FROST WEATHERING of underlying bedrock, as in the case of the *parallel roads* of Glen Roy. *DH/CJC*

[**See also** LAKE-LEVEL VARIATIONS, POLAR SHORE EROSION]

Benson L and Thompson RS (1987) The physical record of lakes in the Great Basin. In Ruddiman WF and Wright Jr HE (eds) *North America and adjacent oceans during the last deglaciation*. New York: Geological Society of America, 241–260.

Fabel D, Small D, Miguens-Rodriguez M and Freeman SPHT (2010) Cosmogenic nuclide exposure ages from the 'Parallel Roads' of Glen Roy. *Journal of Quaternary Science* 25: 597–603.

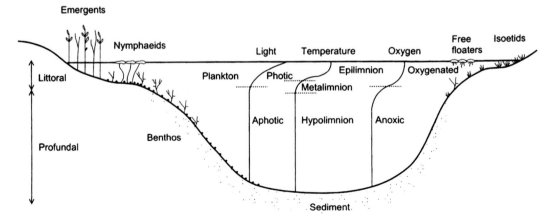

Lake stratification and zonation *Zones of stratification of light, temperature, and oxygen in a lake, and vegetation zonation.*

lake-chemistry reconstructions   The chemical constituents in lake water are closely related to the geology and biology of the aquatic HABITATS, and the analysis of these chemical components form an important part of LIMNOLOGY. The relative abundance of the major IONS in lake waters varies from region to region and is a reflection of the local climate, geology, topography, biotic activity and time. WEATHERING of soil and rocks are the main source of ions in lake waters.

PALAEOENVIRONMENTAL RECONSTRUCTION within PALAEOLIMNOLOGY uses multiple approaches to determine the timing, magnitude, causes and sources of changes in past lake chemistry. Lake water chemistry has a wide-ranging effect on both *aquatic organisms* and *inorganic processes*, so analysis of these can help in the reconstruction of past conditions. No HISTORICAL EVIDENCE is available for many components of lake chemistry, so signatures left in *sediment geochemistry* and in *macrofossil communities* are crucial to enable inferences to be made of past lake chemical conditions. Characteristics such as ACIDITY, *trophic status*, SALINITY, *dissolved organic carbon*, PERSISTENT ORGANIC COMPOUNDS (POCs) and *metal concentrations* can be reconstructed from the sediment record. Many lake-chemistry reconstructions have used inorganic components but organic components such as *plant pigments* can imply much about past PRODUCTIVITY of ALGAE, and certain algal compounds, LOSS ON IGNITION (LOI) and chitinised *head capsules* of *midge larvae* can give an indication of EUTROPHIC or *oligotrophic* conditions. High quantities of sodium, potassium, calcium, magnesium and low dissolved organic carbon suggest in-wash of unweathered, clastic material into lakes, thus reflecting local climate conditions.   *KJF*

[*See also* EUTROPHICATION, GEOCHEMICAL PROXIES, LACUSTRINE SEDIMENTS, OLIGOTROPHICATION, PIGMENTS: FOSSIL]

Chou L, Davison W, Eisenreich SJ et al. (1995) *Physics and chemistry of lakes*. Berlin: Springer.

Fott J (ed.) (2011) *Limnology of mountain lakes*. Berlin: Springer.

Hausmann S and Pienitz R (2009) Seasonal water chemistry and diatom changes in six boreal lakes of the Laurentian Mountains (Quebec, Canada): Impacts of climate and timber harvesting. *Hydrobiologia* 635: 1–14.

Lerman A (1978) *Lakes: Chemistry, geology, physics*. Berlin: Springer.

Rosen P, Bindler R, Korsman T et al. (2011) The complementary power of pH and lake-water organic carbon reconstructions for discerning the influences on surface waters across decadal to millennial time scales. *Biogeosciences* 8: 2717–2727.

Smol JP, Birks HJB and Last WM (eds) (2001) *Tracking environmental change using lake sediments, volume 3: Terrestrial, algal and siliceous indicators*. Dordrecht: Kluwer.

Vallet-Coulomb C, Gasse F, Robison L et al. (2006) Hydrological modeling of tropical closed Lake Ihotry (SW Madagascar): Sensitivity analysis and implications for paleohydrological reconstructions over the past 4000 years. *Journal of Hydrology* 331: 257–271.

Zhu BQ, Yu JJ, Qin XG et al. (2012) Climatic and geological factors contributing to the natural water chemistry in an arid environment from watersheds in northern Xinjiang, China. *Geomorphology* 153: 102–114.

lake-level variations   Past lake levels are an important source of PROXY DATA for PALAEOCLIMATIC RECONSTRUCTION. Lake-level variations reflect the HYDROLOGICAL BALANCE of the lake and its catchment, and may be a particularly valuable indicator of changes in *effective moisture*. Regional- or continental-scale synchroneity in lake-level variations points to a climatic cause. Water-level changes can be reliably reconstructed from deep, *closed-basin lakes*, where water level may be strongly correlated with water SALINITY and the lake does not dry out. Reconstructions from *open-basin lakes*, where water is lost through surface outflows or GROUNDWATER seepage, may be more problematic. Nevertheless, water-level changes can be reconstructed from open basins by analysing (1) *plant macrofossils* to identify changes in the distribution of lake-shore vegetation, (2) coarse minerogenic sedimentary composition to identify changes in the distribution of near-shore sediments affected by waves and (3) changes in the position of the organic 'sediment limit' (the highest level for permanent deposition of predominantly organic sediments or GYTTJA. At least 15 century- to millennial-scale intervals of relatively high lake levels have been recognised in the Jura Mountains, the northern French Pre-Alps and the Swiss Plateau over the Holocene (see Figure). The shorelines of former PLUVIAL lakes in the southwest USA and Lake Chad in North Africa indicate that the surface area of these lakes were 10–20 times larger during the Last Glacial than at present.   *JAM*

[*See also* LACUSTRINE SEDIMENTS, PALAEOLIMNOLOGY, PLANT MACROFOSSIL ANALYSIS]

Almendinger JE (1993) A groundwater model to explain past lake levels at Parkers Prarie, Minnesota, U.S.A. *The Holocene* 3: 105–115.

Benson LV, Lund SP, Smoot JP et al. (2011) The rise and fall of Lake Bonneville between 45 and 105 ka. *Quaternary International* 235: 57–69.

Digerfeldt G (1986) Studies on past lake-level fluctuations. In Berglund BE (ed.) *Handbook of Holocene palaeoecology and palaeohydrology*. Chichester: Wiley, 127–143.

Enzel Y, Bookman R, Sharon D et al. (2003) Late Holocene climates of the Near East deduced from Dead Sea level variations and modern regional winter rainfall. *Quaternary Research* 60: 263–273.

Gasse F (2000) Hydrological changes in the African tropics since the Last Glacial Maximum. *Quaternary Science Reviews* 19: 189–211.

Harrison SP, Frenzel B, Huckriede U and Weiss MM (eds) (1998) *Palaeohydrology as reflected in lake-level changes as climatic evidence for Holocene times.* Stuttgart: Gustav Fischer [*Paläoklimaforschung* Band 25].

Li Y and Morrill C (2013) Lake levels in Asia at the Last Glacial Maximum as indicators of hydrological sensitivity to greenhouse gas concentrations. *Quaternary Science Reviews* 60: 1–12.

Magny M (2007) Lake-level studies: West-Central Europe. In Elias SA (ed.) *Encyclopedia of Quaternary science.* Elsevier: Amsterdam, 1389–1399.

Street-Perrott FA and Harrison SP (1985) Lake levels and climatic reconstruction. In Hecht AD (ed.) *Paleoclimatic analysis and modeling.* New York: Wiley, 291–340.

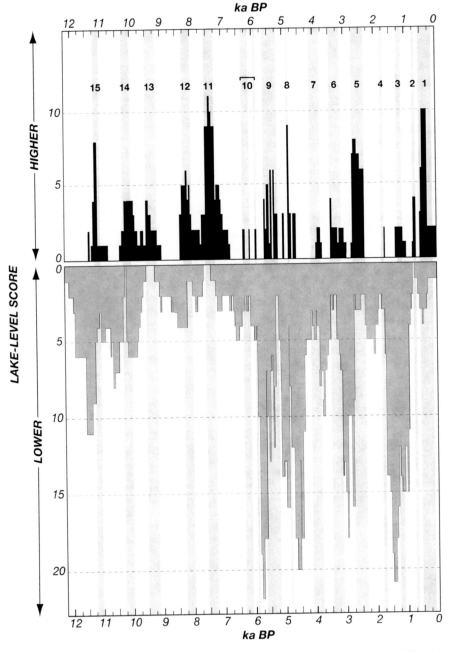

Lake-level variations *Holocene episodes of high* (upper histogram) *and low* (lower histogram) *lake levels in west-central Europe based on the frequency of archaeological, dendrochronological and radiocarbon-dated evidence. Length of each column is proportional to the number of dates in 50-year age classes (Magny, 2007).*

lakes as indicators of environmental change

Lakes respond to environmental changes over time. These can be observed directly or reconstructed in the past from the sediment NATURAL ARCHIVE using PROXY DATA. *Closed-basin lakes* indicate *precipitation:evaporation balance* by changes in water level, detectable by former SHORELINES, sediment transects from shallow to deep water associated with changes in *aquatic macrophyte* distribution, and by SALINITY changes inferred from, for example, diatoms and sediment chemistry.

Lake biota respond to changes in water chemistry. DIATOM ANALYSIS reveals changes in pH and dissolved organic carbon resulting from ACID RAIN. EUTROPHICATION is indicated by changes in sediment composition, chemistry (e.g. C, N, P, pigments, lipids) and organism fossils (e.g. *chironomids, cladocera,* DIATOMS, other ALGAE, CYANOBACTERIA, aquatic macrophytes). HEAVY-METAL concentrations in sediment indicate METAL POLLUTION levels. Lakes also respond to temperature changes. INDICATOR SPECIES and STABLE ISOTOPES in lake sediments provide a record of *water* and AIR TEMPERATURE, *ice cover, light penetration*, PRODUCTIVITY, etc. (e.g. chironomids, cladocera, *coleoptera*, diatoms).

Lake sediments also contain evidence of *catchment* changes. MINEROGENIC SEDIMENT input reflects EROSION. This may be climatic (e.g. YOUNGER DRYAS STADIAL, GLACIER VARIATIONS), catastrophic (e.g. AVALANCHES, DEBRIS FLOWS, FLOODS, FIRE FREQUENCY) or human-induced SOIL EROSION (from DEFORESTATION, AGRICULTURE, modification of the LANDSCAPE) exacerbated by ARIDITY or IRRIGATION. Lakes indicate terrestrial ecosystem changes through sedimentary records of terrestrial biota. *HHB*

[*See also* BEETLE ANALYSIS, CHIRONOMID ANALYSIS, CLADOCERA ANALYSIS, DIATOM ANALYSIS, LACUSTRINE SEDIMENTS, LAKE-LEVEL VARIATIONS, LIMNOLOGY, MULTIPROXY APPROACH, PALAEOLIMNOLOGY, PLANT MACROFOSSIL ANALYSIS, POLLEN ANALYSIS, SEDIMENT, SEDIMENTOLOGICAL EVIDENCE OF ENVIRONMENTAL CHANGE]

Battarbee RW (2000) Palaeolimnological approaches to climate change, with special regard to the biological record. *Quaternary Science Reviews* 19: 107–124.

Fritz SC (1989) Lake development and limnological response to prehistoric and historic land-use in Diss, Norfolk, UK. *Journal of Ecology* 77: 182–202.

Fritz SC (1996) Paleolimnological records of climatic change in North America. *Limnology and Oceanography* 41: 882–889.

Gaillard M-J and Birks HH (2007) The contribution of plant macrofossil analysis to paleolimnology. In Elias SA (ed.) *Encyclopedia of Quaternary science.* Amsterdam: Elsevier, 2337–2355.

Laird KR, Fritz SC, Grimm EC and Mueller PG (1996) Century-scale paleoclimate reconstruction from Moon Lake, a close-basin lake in the northern Great Plains. *Limnology and Oceanography* 41: 890–902.

Psenner R and Schmidt R (1992) Climate driven pH control of remote alpine lakes and effects of acid deposition. *Nature* 356: 781–783.

Smith AJ (2012) Evidence of environmental change from terrestrial and freshwater palaeoecology. In Matthews JA, Bartlein PJ, Briffa KR et al. (eds) *The SAGE handbook of environmental change, volume 1.* London: Sage, 254–283.

Lamarckism   The view of Jean Baptiste Pierre Antione de Monet, Chevalier de Lamarck (1744–1829) ascribing EVOLUTION to inheritable modification in the individual by habit, behaviour and the ENVIRONMENT. Thus, Lamarckism suggests that characteristics acquired during an individual's lifetime can be passed on to future generations. The classic example is the proposal that the giraffe acquired its long neck from continually stretching to reach leaves in the tree canopy. Also known as *soft inheritance*, Lamarckism has been disproved. Modern evolutionary theory only allows *genotypic variation* to be inherited. *Neo-Lamarckism* describes any later theory that has supported Lamarck's ideas.   *JAM/KDB*

[*See also* ADAPTATION, CREATIONISM, DARWINISM, GENOTYPE, INHERITANCE IN EVOLUTIONARY CONTEXT, PHENOTYPE]

Knapp S (2007) Jean-Baptiste Lamarck: The inheritance of acquired characteristics. In Huxley R (ed.) *The great naturalists.* London: Thames and Hudson, 190–196.

laminar flow   Flow dominated by fluid viscosity and characterised by individual fluid elements (e.g. MOLECULES) moving parallel to the flow direction. At a value of *Reynolds number* (ratio of inertial to viscous forces) between 500 and 2,000, laminar flow (also known as *viscous flow*) transforms into TURBULENT FLOW. Examples of laminar flows include GLACIERS, DEBRIS FLOWS and LAVA FLOWS. *Water flows* are laminar only at very low velocities or very shallow depths. *GO*

Bridge JS and Demicco RV (2008) *Earth surface processes, landforms and sediment deposits.* New York: Cambridge University Press.

Leeder MR and Pérez-Arlucea M. (2005) *Physical processes in Earth and environmental sciences.* Oxford: Wiley-Blackwell.

laminated lake sediments   Laminated sediments often occur in deep ANOXIC lakes, as there is no BIOTURBATION of the sediment by bottom-dwelling fauna. There are four main types: (1) *ferrogenic laminations*, (2) *calcareous laminations*, (3) *biogenic laminations* and (4) *clastic laminations*. Often *laminae* are arranged in *couplets*, with relatively coarse-grained layers alternating regularly with finer-grained bands (RHYTHMITES); if the LAMINATIONS arise because of annual variations in the supply of sediments, they are termed VARVES. The formation of annual laminations in lakes is governed by a number of factors, including lake MORPHOMETRY and SEA-SONALITY of sediment supply, which may be controlled by physical, biological or cultural processes in the lake, or in its DRAINAGE BASIN. Annually laminated lake sediments combine an internal, independent chronology with preservation of discrete, visible increments of seasonal and annual sedimentation. As they are deposited annually, varves can be used as a DATING TECHNIQUE.

Clastic varves are the classic example of laminated sediments and are commonly found in QUATERNARY deposits in glaciated areas but have also been identified in sediments of older GLACIATIONS. *Glacial varves* form from the input of *glacial meltwater* which deposits coarse-grained silt and sand during the summer months and the settling out of the finer clay particles during the winter months when the lakes are frozen. The coarse summer band is graded and there is a sharp contact between the fine winter band and the following summer band. Varve thickness can vary from year to year due to variations in climate and local factors such as the material supply, intensity of melting and the velocity of meltwater flow. Colour, grain size, chemical composition and the thickness of the varves form the basis for CORRELATION (STRATIGRAPHICAL) of sediment sequences. Changes in *biogenic production* or *water chemistry* can be recorded as varves and represent seasonal changes such as DIATOM blooms and *carbonate content* in the lakes.

Chemical varves may be formed by PRECIPITATION and EVAPORATION, such as SPELEOTHEMS and GYPSUM deposits. *Iron-rich varves* show contrasting *laminae* that denote the mixing of lakes. When oxygen is abundant during spring and autumn due to the mixing of the water column, *ferrous iron* is oxidised to *ferric iron* and is precipitated as a pale-coloured layer whereas during the winter and summer, ANAEROBIC conditions prevail due to stagnation and stratification and ferrous *iron sulfides* are deposited resulting in a dark-coloured layer.

Varved sediment sequences are often used in palaeoenvironmental studies to reconstruct the age of the sediment and to determine annual sediment ACCUMULATION RATE. Continuous varve sequences typically record 10,000 years or less of deposition but the varve sequence from Lago Grande di Monticchio in southern Italy extends back more than 100 ka and is the longest-known sequence. Recently, the correlation of varve thickness with summer temperatures has enabled the effects of CLIMATIC FORCING during the HOLOCENE to be studied and past climate conditions to be reconstructed.   *KJF*

[*See also* LAMINATED MARINE SEDIMENTS, PALAEOCLIMATOLOGY]

Alliksaar T and Veski S (2003) Comparison of different dating methods in a lake with annually laminated sediments. *Geochimica et Cosmochimica Acta* 67: A13–A13.

Brauer A, Mingram J, Frank U et al. (2000) Abrupt environmental oscillations during the Early Weichselian recorded at Lago Grande di Monticchio, Southern Italy. *Quaternary International* 73/74: 79–90.

Brauer A and Negendank JFW (eds) (2002) The value of annually laminated lake sediments in palaeoenvironment reconstruction. *Quaternary International* 88: 1–80.

Cohen AS (2003) *Paleolimnology: The history and evolution of lake systems*. New York: Oxford University Press.

Lowe JJ and Walker MJC (1997) *Reconstructing Quaternary environments, 2nd edition*. Harlow: Longman.

O'Sullivan PE (1983) Annually-laminated lake sediments and the study of Quaternary environmental changes: A review. *Quaternary Science Reviews* 1: 245–313.

Tian JA, Nelson DM and Hu FS (2011) How well do sediment indicators record past climate? An evaluation using annually laminated sediments. *Journal of Paleolimnology* 45: 73–84.

laminated marine sediments   Marine sediments comprising thin layers, or LAMINATIONS, often millimetre-scale in thickness, typically with contrasting sediment colours and textures. These sediments often comprise seasonally separated, alternating layers of biological origin (i.e. plankton MICROFOSSILS) and TERRIGENOUS origin and are called *marine varves* (see VARVE). *Seasonally laminated marine sediments* can accumulate on the sea floor where there is a seasonal variation in the nature of the SEDIMENT supplied, and sea-floor dissolved oxygen concentration is very low, inhibiting BIOTURBATION of the sediments. Laminated marine sediments comprise tangled *planktonic* diatom colonies (or mats) which can accumulate in open-ocean, well-oxygenated sea-floor environments, beneath convergent (DOWNWELLING) *oceanic frontal systems* where the flux of sediment at the sea floor overwhelms the BENTHIC burrowers.   *JP*

[*See also* LAMINATED LAKE SEDIMENTS, RHYTHMITE, SEASONALITY]

Kemp AES and Baldauf JG (1993) Vast Neogene laminated diatom mat deposits from the eastern equatorial Pacific Ocean. *Nature* 362: 141–144.

Maddison EJ, Pike J and Dunbar R (2012) Seasonally-laminated diatom-rich sediments from Dumont d'Urville Trough, East Antarctic Margin: Late Holocene Neoglacial sea-ice conditions. *The Holocene* 22: 857–875.

Pike J and Stickley CE (2007) Diatom records: Marine laminated sequences. In Elias SA (ed.) *Encyclopedia of Quaternary science*. Amsterdam: Elsevier, 557–567.

Pilskaln CH and Pike J (2001) Formation of Holocene sedimentary laminae in the Black Sea and the role of the benthic flocculent layer. *Paleoceanography* 16: 1–19.

lamination   The smallest scale of STRATIFICATION commonly visible in sediments and sedimentary rocks. A *lamina* (plural: *laminae*) defines a sedimentation unit thinner than 1 cm. *Horizontal lamination* is formed by the vertical accumulation of sediment (see PARALLEL LAMINATION); *inclined lamination* is produced by deposition on laterally accreting surfaces such as the downstream faces of migrating RIPPLES (see CROSS-LAMINATION). The term *microlamination* is sometimes used for submillimetre-scale lamination present in some very fine-grained sediments.   *MRT*

[*See also* HIGH-RESOLUTION RECONSTRUCTIONS, LACUSTRINE SEDIMENTS, LAMINATED MARINE SEDIMENTS, ROCK-VARNISH MICROLAMINATION (VML) DATING]

Collinson JD, Mountney N and Thompson DB (2006) *Sedimentary structures, 3rd edition.* Harpenden: Terra Publishing.
Pike J and Stickley CE (2007) Diatom records: Marine laminated sequences. In Elias SA (ed.) *Encyclopedia of Quaternary science.* Amsterdam: Elsevier, 557–567.
Segall MP and Kuehl SA (1994) Sedimentary structures on the Bengal shelf: A multiscale approach to sedimentary fabric interpretation. *Sedimentary Geology* 93: 165–180.

land 'A delineable area of the Earth's terrestrial surface, encompassing all attributes of the biosphere immediately above or below this surface, including those of the near-surface climate, the soil and terrain forms, the surface hydrology (including shallow lakes rivers and swamps), the near-surface sedimentary layers and associated groundwater reserve, the plant and animal populations, the human settlement pattern and physical results of past and present human activity (terracing, water storage or drainage structures, roads, buildings etc.' (FAO, 1995: 6). This definition refers to a natural unit (not an administrative area). It conforms to LANDSCAPE UNITS, *land system units* or *landscape-ecological units* as building blocks of an approach to the Earth's surface based on DRAINAGE BASINS (*catchments* or *watersheds*) and GEO-ECOSYSTEM. The concept of a land is useful for LANDUSE planning purposes.   *EMB*

[*See also* LANDSCAPE ECOLOGY]

Food and Agriculture Organization of the United Nations (FAO) (1995) *Planning for sustainable use of land resources: Towards a new approach.* Rome: FAO.
National Committee for Soil and Terrain (2009) *Australian soil and land survey field handbook, 3rd edition.* Canberra: CSIRO Publishing.

land bridge A land connection between CONTI-NENTS, between parts of continents or between *islands*. The fluctuating presence of land bridges has played a crucial biogeographical role in the history of Earth's FAUNA (particularly *mammals*) and FLORA. The availability of such 'highways' provides WILDLIFE CORRIDORS for MIGRATION and DISPERSAL and directly affects terrestrial BIODIVERSITY in the various FAUNAL PROVINCES through faunal interchange and COMPETITION.

Conversely, the absence of land bridges has a profound influence on the EVOLUTION of geographically isolated populations (see ISOLATION). Prior to the MESO-ZOIC, *biotic homogeneity* was largely maintained due to presence of land connections throughout PANGAEA (PANGEA). The subsequent disruption of these terrestrial links led to a veritable explosion of *faunal diversity* in the TERTIARY through SPECIATION of isolated groups on continental landmasses and the evolution of orders such as *marsupials* and *edentates*. During the PLIOCENE, the formation of the central American (*Panamanian*) land connection permitted *faunal interchange* between North America and South America but concomitantly led to the EXTINCTION of many South American groups through the dominance of *placental mammals* from the north. Thus, the presence of land bridges may facilitate dispersal and accordingly increase overall species diversity in any given area but may also ultimately precipitate the extinction of other species in the face of newly arrived competitors.

The effects of ice build-up and associated lowered *sea level* during the various GLACIATIONS of the PLEISTO-CENE led to the periodic reconnection of North America and Eurasia across the land bridge of *Beringia*. This permitted the migration of many plants and animals, including humans, into the Americas from the Old World (see HUMAN MIGRATION: 'OUT OF AFRICA'). The disappearance of Pleistocene land bridges due to SEA-LEVEL RISE following DEGLACIATION also affected populations on small islands, leading to either *dwarfism* (NANISM) of some large mammal species or GIGANTISM of *micromammals*. In the Mediterranean, 16 islands (or former islands) have been found to contain fossils of dwarf ENDEMIC mammals, including *elephants, hippopotami* and antelope-like *bovids*. Further afield, fossil *dwarf elephants* are known from many islands in Southeast Asia, while miniature MAMMOTHS have been found on Wrangel Island off Siberia and on the Californian channel islands. Cases of gigantism, such as that of *dormice* on Malta, have also been noted.   *DCS*

[*See also* BIOGEOGRAPHY, ISLAND BIOGEOGRAPHY]

Darlington PJ (1957) *Zoogeography: The geographical distribution of animals.* New York: Wiley.
Denk T, Grimsson F and Zetter R (2010) Episodic migration of oaks to Iceland: Evidence for a North Atlantic 'land bridge' in the latest Miocene. *American Journal of Botany* 97: 276–287.
Goebel T and Buvit I eds (2011) *From Yenisei to the Yukon.* College Station: Texas A & M University Press.
Hopkins DM, Matthews Jr JV, Schweger CE and Young SB (eds) (1982) *Paleoecology of Beringia.* New York: Academic Press.
Lozhkin AV, Anderson P, Eisner WR and Solomatkina TB (2011) Late Glacial and Holocene landscapes of central Beringia. *Quaternary Research* 76: 383–392.

O'Neill D (2004) *The last giant of Beringia: The mystery of the Bering land bridge.* Boulder, CO: Westview Press.

Pearson GA and Cooke RG (2002) The role of the Panamanian land bridge during the initial colonization of the Americas. *Antiquity* 76: 931–932.

Szalay FS, Novacek MJ and McKenna MC (eds) (1993) *Mammal phylogeny, volume 1: Mesozoic differentiation, multituberculates, monotremes, early therians, and marsupials.* Berlin: Springer.

Szalay FS, Novacek MJ and McKenna MC (eds) (1993) *Mammal phylogeny, volume 2: Placentals.* Berlin: Springer.

land cover A classification of feature characteristics of the surface layer of the land. For example, woodland, cropland or buildings are construed as land cover. Some define vegetation and artificial constructions as constituting land cover while an alternative view limits land cover to 'the physical state of the land' of particular interest to natural environmental scientists. *Land-cover change* is an important element in ENVIRONMENTAL MONITORING and is often a primary application of REMOTE SENSING. It is important, for example, in providing information on BIODIVERSITY and *climatic impacts*, and for understanding the HYDROLOGICAL CYCLE, BIOGEOCHEMICAL CYCLES, ENERGY BUDGETS, ENVIRONMENTAL DEGRADATION and FUTURE CLIMATE.

TF

[*See also* CONVERSION, LANDUSE]

Di Gregorio A and Jansen LJM (2000) *Land cover classification system (LCCS): Classification concepts and user manual.* Rome: Environment and Natural Resources Service, Food and Agriculture Organization of the United Nations.

Faddema JJ, Oleson KW, Bonan GB et al. (2005) The importance of land-cover change in simulating future climates. *Science* 310: 1674–1678.

Giri CP (ed.) (2012) *Remote sensing of land use and land cover: Principles and applications.* Boca Raton, FL: CRC Press.

Herzschuh U, Birks HJB, Ni J et al. (2010) Holocene land-cover change on the Tibetan Plateau. *The Holocene* 20: 91–104.

Lambin EF and Geist HJ (eds) (2006) *Land-use and land-cover change: Local processes and global impacts.* Berlin: Springer.

Los SO and Williams J (2012) Monitoring global land cover. In Matthews JA, Bartlein PJ, Briffa KR et al. (eds) *The SAGE handbook of environmental change, volume 2.* London: Sage, 3–24.

land degradation A temporary or permanent decline in the *productive capacity* or *resource potential* of the land. Some types of degradation are irreversible, such as extensive GULLY FORMATION or extreme SALINISATION; other types are reversible and can be changed by improved *farming practices*. Land degradation includes SOIL DEGRADATION (through SOIL EROSION or loss of FERTILITY), the removal of natural VEGETATION (accompanied by HABITAT LOSS and reduced BIODIVERSITY) combined with a general deterioration of the LANDSCAPE. It can be caused by both natural and anthropogenic factors, although the latter are normally viewed as the main cause. Land degradation may result from the effects of OVERGRAZING, excessive TILLAGE, FOREST CLEARANCE, soil erosion, disposal of INDUSTRIAL WASTE and other wastes causing CONTAMINATED LAND. Soil contamination also affects the plants and animals that are capable of exploiting the degraded land situation.

EMB

[*See also* DESERTIFICATION]

Abel JOJ and Blakie PM (1989) Land degradation, stocking rates and conservation policies in the communal rangelands of Botswana and Zimbabwe. *Land Degradation and Rehabilitation* 1: 101–123.

Advances in Geoecology (2009) Land degradation and rehabilitation: Dryland ecosystems. *Advances in Geoecology* 40: 1–432 [Special Issue].

Chisholm A and Dumsday R (1987) *Land degradation: Problems and policy.* Cambridge: Cambridge University Press.

Genske DD (2003) *Urban land: Degradation, investigation, remediation.* Berlin: Springer.

Johnson DL and Lewis LA (1995) *Land degradation: Creation and destruction.* Oxford: Blackwell.

Maconnachie R (2007) *Urban growth and land degradation in developing countries: Change and challenges in Kano Nigeria.* Aldershot: Ashgate Publishing.

Mannava MVK and Ndiang'ui N (eds) (2007) *Climate and land degradation.* Berlin: Springer.

Meshesha D (2009) *Spatial analysis of land use change and land degradation.* Saarbrücken: VDM Verlag.

Scherr SJ and Yadev S (1996) *Land degradation in the developing world: Implications for food, agriculture, and the environment to 2020.* Washington, DC: International Food Policy Research Institute.

land drainage The removal of water from the land by *artificial drainage systems* for a variety of reasons, including (1) CONVERSION of WETLANDS to agricultural use, (2) improvement of existing agricultural land (FIELD DRAINAGE), (3) preservation of irrigated land from waterlogging, GLEYING and SALINISATION, (4) dewatering and DESALINISATION of empoldered land reclaimed from the sea, (5) preparation of land for AFFORESTATION, (6) disposal of sewage WASTE and surface water from urban areas, (7) FLOOD control and (8) protection from GROUNDWATER contamination.

Historically, land drainage has been seen as part of a process of progressive land improvement, but detrimental environmental effects are increasingly recognised. Particularly in relation to WETLAND LOSS, conflicts often arise between drainage interests and CONSERVATION interests. It has been estimated that the United

States has lost 54 per cent of its original wetlands since European settlement, whereas in Italy about 94 per cent has been lost since Roman times. Much of the remaining wetland has suffered alteration, DEGRADATION and loss of functional integrity. The AGRICULTURAL INTENSIFICATION that often follows wetland drainage can produce *secondary ecological impacts* on the remaining wetland area, such as reduced WATER QUALITY and changes in species composition. This has occurred, for example, in the Everglades National Park, Florida, following elevated phosphorus levels draining from the Everglades Agricultural Area.   *JAM*

[*See also* ACID MINE DRAINAGE, ACID SULFATE SOILS]

Armstrong AC and Garwood EA (1991) Hydrological consequences of artificial drainage of grassland. *Hydrological Processes* 5: 157–174.
Baldock D (1984) *Wetland drainage in Europe.* London: International Institute for Environment and Development.
Framji KK and Mahajan IK (1969) *Irrigation and drainage in the world: A global view.* New Delhi: International Commission on Irrigation and Drainage.
Hill AR (1976) The environmental impact of agricultural land drainage. *Journal of Environmental Management* 4: 251–274.
Holden J, Evans MG, Burt TP and Horton M (2006) Impact of land drainage on peatland hydrology. *Journal of Environmental Quality* 35: 1764–1778.

land evaluation   A semi-quantitative process in which site and soil characteristics are assessed for specific purposes. In a land evaluation, land is considered to be more than SOIL. It includes all the reasonably stable attributes of the LANDSCAPE above and below an area, including geology, hydrology, plant and animal populations, and the results of past and present human activity. *Soil maps* are commonly interpreted in the light of these other environmental features to indicate the relative *suitability class* (highly, moderately, marginally) or unsuitability (currently not suitable, permanently not suitable) for a particular use, or crop under well-defined conditions of MANAGEMENT. As decisions about LANDUSE are a policy-driven activity, it is essential that all the factors involved are discussed with the stakeholders before landuse is changed.   *EMB*

[*See also* ENVIRONMENTAL MANAGEMENT SYSTEM (EMS), LAND, LANDUSE CAPABILITY CLASSIFICATION]

Costantini EAC (2009) *Manual of soil and land evaluation.* Enfield, NH: Science Publishers.
Food and Agriculture Organization of the United Nations (FAO) (1993) *FELSM, an international framework for evaluating sustainable land management.* Rome: FAO.
McRae SG and Burnham CP (1981) *Land evaluation.* Oxford: Clarendon Press.

land reclamation   In historical times, the process of bringing land under CULTIVATION from a NATURAL VEGETATION or SEMI-NATURAL VEGETATION. The draining of MARSHES, clearing of HEATHLAND and FOREST CLEARANCE were described as reclaiming land for agriculture. Currently, the term is used for the REHABILITATION of derelict, contaminated or otherwise despoiled land mainly resulting from industrial activity. However, the term is also used for the *reclamation* of land from the sea, for example, *empoldering* in the Netherlands, whereby new land with new uses results.

Remedial measures should always attempt to solve the problem once and for all as temporary solutions inevitably result in further work and greater expense at a later date. The process of RESTORATION will, in many cases, be determined by the future use of the land. Reuse of land for industrial purposes, where a solid concrete floor and tarmac parking places around the factory seal the ground surface, does not require an expensive reclamation as for housing or for a return to productive AGRICULTURE. A frequently employed alternative use for former *derelict land* is to provide *amenity open space* in urban areas. Since the mid-twentieth century, considerable experience has been gained reclaiming land despoiled by opencast coal and ironstone *mining* or gravel QUARRYING. Such schemes require re-creating a SOIL from the available geological materials to enable plants to make satisfactory growth.

Reclamation of IRRIGATION land that has been subject to SALINISATION in areas of DESERTIFICATION is a growing problem in many DEVELOPING COUNTRIES with a SEMI-ARID environment. CALCIUM salts can usually be leached from PERMEABLE soils, but the presence of *sodium* salts causes greater problems and both must be dealt with in association with improved SOIL DRAINAGE. A rise of sea level through global warming may lead to increased problems of coastal FLOODS in low-lying areas in which case reclamation of the salt-affected land must take place before crops can be grown.   *EMB*

[*See also* CONTAMINATED LAND, LAND DEGRADATION, LAND DRAINAGE, LAND RESTORATION, SOIL DEGRADATION, SOIL RECLAMATION]

Bradshaw AD (1998) Land reclamation. In Calow R (ed.) *The encyclopedia of ecology and environmental management.* Oxford: Blackwell Science, 394–396.
Bridges EM (1987) *Surveying derelict land: The ecology and reclamation of derelict and degraded land.* Oxford: Clarendon Press.
Hebbink AJ (1999) Reclamation, polders. In Alexander DE and Fairbridge RW (eds) *Encyclopedia of environmental science.* Dordrecht: Kluwer, 367–369.

land restoration   Land restoration sensu stricto may be described as the process of bringing back *disused land* or CONTAMINATED LAND to a pre-existing LANDUSE. Whereas LAND RECLAMATION is the general term for bringing back to use, land restoration implies full reconditioning and land REHABILITATION implies only partial success. RESTORATION may be for either 'hard' or 'soft' uses. *Hard use* includes uses where people are in close daily contact with a site, such as children playing, or where there is a high dependence on garden produce grown on the site; *soft use* is where land is restored to public open spaces and playing fields where contact is less intensive. Plans for site restoration would place considerable emphasis upon the after use of the site. Thus, a restoration for agricultural use would imply a complete cleansing of the site to provide healthy conditions for plants and animals. The restored land would maximise gently sloping surfaces and a *topsoil* would be carefully replaced that was freely draining and composed of loam, sandy loam, sandy clay loam or silt loam. Special care should be taken to avoid loss of soil by EROSION and *graded waterways* should be provided for the safe disposal of excess water. Similar qualifications are imposed where land is restored to *forestry*.

Restoration of *derelict land* and contaminated land for housing raises many problems as people are in close contact with the ground around their houses. It is essential that all toxic materials are removed and that EMISSIONS of potentially dangerous gases are eliminated. Where houses have been built upon CONTAMINATED LAND (e.g. Lekerkerke, the Netherlands) or even near industrial *toxic waste* dumps (e.g. Love Canal, USA) human health suffers, especially that of children, and expensive *remedial measures* become necessary. If an industrial after use is planned it may be that *hazardous materials* can be surrounded by barriers below car parks or roadways and sealed by *tarmacadam*. Where land is restored to *amenity use*, less rigorous standards can be applied as human contact is not so intense, however, the well-being of the NATURAL ENVIRONMENT must still be considered. *EMB*

Bradshaw AD and Chadwick MJ (1980) *The restoration of land: The ecology and reclamation of derelict and degraded land.* Oxford: Blackwell.
Cramer VA, Hobb RJ and Falk DA (eds) (2007) *Old fields: Dynamics and restoration of abandoned farmland.* Washington, DC: Island Press.
Harris JA, Palmer J and Birch P (1996) *Land restoration and reclamation: Principles and practice.* Harlow: Longman.
Jordan III WR, Gilpin ME and Aber JE (eds) (1987) *Restoration ecology.* Cambridge: Cambridge University Press.
Jordan III WR and Lubick GM (2011) *Making nature whole: A history of ecological restoration.* Washington, DC: Island Press.
Paignen B, Goldman LR, Highland JH et al. (1985) Prevalence of health problems in children living near Love Canal. *Hazardous Waste and Hazardous Materials* 2: 23–43.
Schuuring C (1981) Dutch dumps. *Nature* 289: 340.
Smith MA (ed.) (1985) *Contaminated land: Treatment and reclamation.* New York: Plenum Press.
Wong MH and Bradshaw AD (eds) (2003) *The restoration and management of derelict land: Modern approaches.* Singapore: World Scientific Publishing.

land subsidence The sinking or foundering of an area of the Earth's surface. Tectonic effects, such as *downwarping*, are normally excluded (see SUBSIDENCE). Land subsidence, or *ground subsidence*, can be the result of natural processes (e.g. the thawing of PERMAFROST, DISSOLUTION in KARST landscapes and the DESICCATION of PEAT AND PEATLANDS), but the term is more widely applied to the results of human activities (e.g. the extraction of GROUNDWATER, oil or natural gas, the *mining* of coal, ores or salt, and the IRRIGATION of certain soils). Hence, subsidence can be either an ANTHROPOGENIC or NATURAL HAZARD. *JAM*

[*See also* AUTOCOMPACTION]

Hotzer TL (1984) *Man-induced land subsidence.* Boulder, CO: Geological Society of America.
Hu RL, Yue ZQ, Wang LC and Wang SJ (2004) Review on current status and challenging issues of land subsidence in China. *Engineering Geology* 76: 65–77.
Johnson AL (ed.) (1991) Land subsidence. *International Association of Scientific Hydrology (IAHS) Publication* 200: 1–690.

(the) land system The coupled socio-environmental terrestrial system that includes LANDUSE, LAND COVER and *ecosystems* (Global Land Project, 2005). It is a subset of the EARTH SYSTEM and the focus of LAND-CHANGE SCIENCE (see Figure). *The Global Land Project (GLP)* of the *International Geosphere-Biosphere Programme (IGBP)* and the *International Human Dimensions Programme on Global Environmental Change (IHDP)* have three objectives in relation to the land system: firstly, to identify the agents, structures and nature of change in coupled human-environmental systems; secondly, to assess how the provision of ecosystem services is affected; and thirdly, to understand the dynamics of vulnerable and sustainable human-environmental systems to PERTURBATIONS, including CLIMATIC CHANGE. The central research challenges are (1) UPSCALING local and regional understanding to the global level and (2) integrating the societal and environmental dimensions of the problem. *JAM*

[*See also* EARTH-SYSTEM SCIENCE, ECOLOGICAL GOODS, ECOLOGICAL SERVICES, ENVIRONMENT-HUMAN INTERACTIONS, SOCIO-ENVIRONMENTAL DYNAMICS, SUSTAINABILITY]

Dearing JA, Braimoh AK, Reenberg A et al. (2010) Complex land systems: The need for long time perspectives in order to assess their future. *Ecology and Society* 15: Article 21 [online].

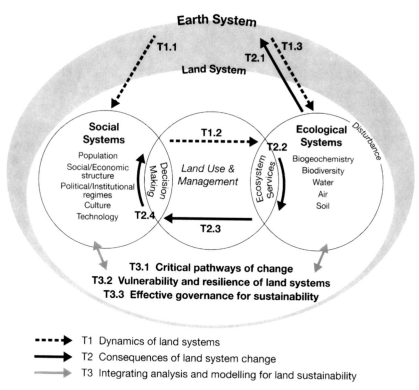

The land system *Schematic representation of the global land system involving interactions between landuse, land cover and ecosystems (Global Land Project, 2005).*

Global Land Project (GLP) (2005) *Science plan and implementation strategy* (IGBP Report 53/IHDP Report 19). Stockholm: IGBP Secretariat.

Ostrom E (2009) A general framework for analyzing sustainability of socio-ecological systems. *Science* 325: 419–422.

land transformation   Changes in LAND COVER and LANDUSE CHANGE: the former involves the physical state of the land, changes of which may be caused by NATURAL and/or human agency; the latter involves changes in human use of the land.   *JAM*

[*See also* CONVERSION]

Richards JF (1991) Land transformation. In Turner II BL, Clark WC, Kates RW et al. (eds) *The Earth as transformed by human action.* Cambridge: Cambridge University Press, 163–173.

Wolman MG and Fournier FGA (eds) (1987) *Land transformation in agriculture.* Chichester: Wiley.

land uplift   Land elevation occurs on a variety of spatial and temporal scales. Uplift may arise from tectonic OROGENESIS (e.g. collision and rifting of tectonic plates) and *tectono-isostasy* (e.g. unloading of crust by erosion or melting ice). *Environmental changes that may be triggered are interlinked and largely determined by the nature of uplift.* RAISED SHORELINES

may result from lowering RELATIVE SEA LEVEL. Adjustment of DRAINAGE BASIN to falling BASE LEVEL may lead to enhanced EROSION, *incision* and RIVER TERRACE formation.

It is argued that land uplift may lead to local, regional and global CLIMATIC CHANGE, due to OROGRAPHIC effects (e.g. RAIN SHADOW or WIND SHADOW), major perturbations of the ocean and GENERAL CIRCULATION OF THE ATMOSPHERE and reduced atmospheric CARBON DIOXIDE levels due to enhanced WEATHERING and EROSION RATES. Tectonic uplift around the North Atlantic and in the Himalayas during the CENOZOIC (*CAINOZOIC*) may be an important driving force in GLOBAL COOLING at the onset of the Late Cenozoic ICE AGE and the development of MONSOON climates in the Northern Hemisphere. Land uplift and increasing dryness in Eastern Africa is considered to have influenced HOMININ evolution.

MHD/CJC

[*See also* HIMALAYAN UPLIFT, ISOSTASY, TECTONIC UPLIFT–CLIMATIC CHANGE INTERACTION, TECTONICS]

An SZ, Wang SM, Wu XH et al. (1999) Eolian evidence from the Chinese Loess Plateau: The onset of the Late Cenozoic Great Glaciation in the Northern Hemisphere and Qinghai-Xizang Plateau uplift forcing. *Science in China Series D* 42: 258–271.

Bailey GN, Reynolds SC and King GCP (2011) Landscapes of human evolution: Models and methods of tectonic geomorphology and the reconstruction of hominin landscapes. *Journal of Human Evolution* 60: 257–280.

Raymo ME and Ruddiman WF (1992) Tectonic forcing of Late Cenozoic climate. *Nature* 359, 117–122.

land-change science The science that seeks a theoretical and practical understanding of the dynamics of LAND COVER and LANDUSE CHANGE from the perspective of ENVIRONMENT-HUMAN INTERACTION. It is a fundamental aspect of GLOBAL ENVIRONMENTAL CHANGE and part of SUSTAINABILITY SCIENCE.   *JAM*

[*See also* EARTH SYSTEM, EARTH-SURFACE SYSTEM, (THE) LAND SYSTEM, LANDSCAPE ECOLOGY, LANDSCAPE SCIENCE]

Gutman G, Janetus AC and Justine CO (eds) (2004) *Land change science: Observing, monitoring and understanding trajectories of change on the Earth's surface.* Dordrecht: Kluwer.

Gutman G and Reissell A (eds) (2010) *Eurasian Arctic land cover and land use in a changing climate.* Dordrecht: Springer.

Millington A and Jepson W (eds) (2008) *Land change science in the tropics: Changing agricultural landscapes.* New York: Springer.

Rindfuss RR, Walsh SJ, Turner II BL et al. (2004) Developing a science of land change: Challenges and methodological issues. *Proceedings of the National Academy of Sciences* 101: 13976–13981.

Turner II BL, Lambin EF and Reenberg A (2007) The emergence of land change science for global environmental change and sustainability. *Proceedings of the National Academy of Science USA* 104: 20666–20671.

Turner II BL and Robbins P (2008) Land change science and political ecology: Similarities, differences, and implications for sustainability. *Annual Review of Environment and Resources* 33: 295–316.

landfill A method of waste disposal, known as *sanitary landfill* in North America, that involves *dumping* above or below ground. Disposal of DOMESTIC WASTE and INDUSTRIAL WASTE on land usually takes place at designated sites, the purpose of which is to contain the wastes without contaminating the surrounding environment. Older landfill sites were uncontrolled and so may contain HAZARDOUS SUBSTANCES, but environmental legislation during the past 30 years or so in most European and other industrialised countries has segregated HAZARDOUS WASTES from relatively harmless materials for disposal at designated facilities where they can be effectively made harmless.

Accepted procedure is that ordinary DOMESTIC WASTE should be dumped in compartments, preferably on a *puddled clay floor*, in layers that are level and no more than 2.5 m deep. Each layer should be compacted and covered as soon as possible with inert material or subsoil. The capping of the landfill should comprise a metre of soil material. Co-disposal of industrial wastes and domestic waste is not considered to be a satisfactory means of dispersion of hazardous materials. Industrial waste more often than not includes toxic materials (see TOXICANT, TOXIN) and these have to be treated or consigned to lined landfill sites, the aim of which is to contain the toxicity within the site. Lining of these landfill sites is by thick *polypropylene* material, and a suitable capping should be in place to limit ingress of rainwater or GROUNDWATER. LEACHATE from both domestic and special landfill sites should be collected and treated before release into the environment. Emission of *methane* and other gases also takes place to the ATMOSPHERE unless these are collected and burnt-off, or used for local heating schemes. Where landfill sites are badly sited, such as on very PERMEABLE rocks or below the *groundwater table*, there will be a rapid transfer of POLLUTION into the rivers or the groundwater. Where a slowly draining unsaturated zone lies beneath a landfill, there is attenuation of the plume of pollution with distance from the site.

In the light of these ENVIRONMENTAL PROBLEMS, the permanence of existing landfills has been questioned and their removal by *landfill mining* has been advocated. Landfill mining may be defined as the excavation, processing, treatment and RECYCLING of the deposited materials.   *EMB*

[*See also* WASTE MANAGEMENT, WATER POLLUTION]

Krook J, Svensson N and Eklund M (2011) Landfill mining: A critical review of two decades of research. *Waste Management* 32: 513–520.

Laner D, Crest M, Scharff H et al. (2011) A review of approaches for the long-term management of municipal solid waste landfills. *Waste Management* 32: 498–512.

Lisk DJ (1991) Environmental effects of landfills. *Science of the Total Environment* 100: 415–468.

Wong MH (1999) Landfill, leachates, landfill gases. In Alexander DE and Fairbridge RW (eds) *Encyclopedia of environmental science.* Dordrecht: Kluwer, 356–361.

Wong MH and Leung CK (1989) Landfill leachates as irrigation water for tree and vegetable crops. *Waste Management Research* 7: 311–324.

landform The form and nature of a particular topographic feature on the Earth's surface (or the surface of another planet). Landforms range from microscale (e.g. STRIATIONS) to macroscale (*mountain ranges*).   *RAS/JAM*

[*See also* GEOMORPHOLOGY, LANDSCAPE, LANDSYSTEM]

landnám Danish for taking possession of the land, landnám refers to the first FOREST CLEARANCE in PREHISTORY. The first DEFORESTATIONS coincided with the first findings of CEREAL POLLEN, implying a change from

MESOLITHIC hunting and gathering to NEOLITHIC farming. The term has been widely applied beyond its area of first use in northwest Europe and especially for the Norse colonisation of Iceland and Greenland.   *BA/CJC*

Caseldine CJ and Fyfe RM (2006) A modeling approach to locating and characterizing elm decline/landnam landscapes. *Quaternary Science Reviews* 25: 632–644.

Edwards KJ, Schofield EJ and Mauquoy D (2008) High resolution paleoenvironmental and chronological investigations of Norse landnam at Tasiusaq, Eastern Settlement, Greenland. *Quaternary Research* 69: 1–15.

Iversen J (1941) Landnám i Danmarks Stenalder [Landnám in Denmark's Stone Age]. *Danmarks Geologiske Undersøgelse II Raekke* Nr 66: 1–68.

Lawson IT, Gathorne-Hardy FJ, Church MJ et al. (2007) Environmental impact of the Norse settlement: Palaeoenvironmental data from Myvatnssveit, northern Iceland. *Boreas* 36: 1–19.

Rasmussen P (2005) Mid- to late-Holocene land-use change and landscape development at Dallund So, Denmark: Vegetation and land use history inferred from pollen data. *The Holocene* 15: 116–129.

landscape   In the scientific and environmental sense, the landscape is a spatial concept relating to the interacting complex of systems on and close to the Earth's surface, including parts of the lower ATMOSPHERE, the upper LITHOSPHERE, HYDROSPHERE, CRYOSPHERE, BIOSPHERE and PEDOSPHERE. The *natural landscape* may be differentiated from the CULTURAL LANDSCAPE, the latter encompassing the modifications and creations of human activities. *Farmscape, townscape* and *wildscape* may also be recognised as components of the landscape, depending on whether the LANDUSE is predominantly rural, urban or 'unproductive', respectively. Although many disciplines investigate the landscape, Matthews and Herbert (2004) considered it to be the unifying object of study in GEOGRAPHY. There are many different interpretations of the meaning of landscape ranging from a physical PALIMPSEST, which supplies evidence of how the landscape evolved to a social process or a SOCIAL CONSTRUCTION.   *JAM*

[*See also* LANDSCAPE ECOLOGY, LANDSCAPE SCIENCE, LANDSCAPE UNITS]

Atkins P, Simmons I and Roberts B (1998) *People, land and time: An historical introduction to the relations between landscape, culture and environment*. London: Arnold.

Chambers FM (ed.) (1993) *Climatic change and human impact on the landscape*. London: Chapman and Hall.

Farina A (2007) *Principles and methods of landscape ecology: Towards a science of landscape*. Dordrecht: Kluwer.

Head L (2000) *Cultural landscapes and environmental change*. London: Arnold.

Isachenko AG (1977) L.S. Berg's landscape: Geographical ideas, their origins and their present significance. *Soviet Geography* 18: 13–18.

Matthews JA and Herbert DT (2004) Landscape: The face of geography. Introduction. In Matthews JA and Herbert DT (eds) *Unifying geography: Common heritage, shared future*. London: Routledge, 217–223.

Muir R (1999) *Approaches to landscape*. London: Macmillan Press.

Muir R (2004) *Landscape encyclopaedia: A reference guide to the historical landscape*. Macclesfield: Windgather Press.

Whyte ID (2002) *Landscape and history since 1500*. London: Reaktion Books.

Wilkinson TJ (2004) The archaeology of landscape. In Bintliff J (ed.) *A companion to archaeology*. Oxford: Oxford University Press, 334–356.

landscape archaeology   An approach to archaeology that emphasises the topographic setting and environmental characteristics of the archaeological site, including the ways in which such sites were perceived by people in the past.   *JAM*

[*See also* GEO-ARCHAEOLOGY, SITE-CATCHMENT ANALYSIS]

Aston M (1985) *Interpreting the landscape: Landscape archaeology and local history*. London: Routledge.

Eveson P and Williamson T (eds) (1998) *The archaeology of landscape*. Manchester: Manchester University Press.

Higham NJ and Ryan MJ (eds) (2010) *Landscape archaeology of Anglo-Saxon England*. Woodbridge: Boydell Press.

Yamin R and Metheny KB (eds) (1996) *Landscape archaeology: Reading and interpreting the American historic landscape*. Knoxville: University of Tennessee Press.

landscape architecture   The modification of landscapes to make them more aesthetically pleasing, enjoyable or useful. Early examples are seen in the layout of extensive gardens and parklands around stately homes in England. Modern landscape architecture is an integral part of designing and managing the BUILT ENVIRONMENT: the broader environmental context and ecological functionality in the design of buildings, highways, golf courses, monuments, etc.   *JAM*

[*See also* AESTHETICS, URBAN AND RURAL PLANNING, URBAN ENVIRONMENTAL CHANGE]

Waterman T (2009) *The fundamentals of landscape architecture*. Lausanne: AVA Publishing.

landscape ecology   The ECOLOGY and MANAGEMENT of distinct areas of the Earth's surface up to regional scale. There are different schools of LANDSCAPE ecology, including (1) the spatial arrangements of landscape elements and the ecological and cultural mechanisms that result in ecological change at a landscape scale; (2) the study of the form, structure, function and evolution of the visual aspects of landscapes; (3) the attributes and spatial arrangements of attributes in landscapes; and

(4) the landscape as a GEO-ECOSYSTEM. Landscape ecology has both a strong theoretical basis and an applied aspect used in areas such as planning and NATURAL RESOURCES MANAGEMENT. European studies of *regional geography* and *vegetation science* led to the use of the term by Troll in AD 1939.   *IFS*

[*See also* AUTOECOLOGY, COMMUNITY ECOLOGY, ECOSYSTEM CONCEPT, GEO-ECOLOGY, LANDSCAPE GEOCHEMISTRY, LANDSCAPE SCIENCE, NATURAL AREAS CONCEPT]

Burel F and Baudry J (2003) *Landscape ecology: Concepts, methods and applications.* Enfield, NH: Science Publishers.
Dramstad WE, Olson JD and Forman TT (1996) *Landscape ecology principles in landscape architecture and landscape planning.* Washington, DC: Island Press.
Turner MG and Gardner RH (eds) (1991) *Quantitative methods in landscape ecology: The analysis and interpretation of landscape heterogeneity.* New York: Springer.
Wiens JA and Moss MR (eds) (2005) *Issues and perspectives in landscape ecology.* Cambridge: Cambridge University Press.
Wiens JA, Moss RR, Turner MG and Mladenoff DJ (eds) (2007) *Foundation papers in landscape ecology.* New York: Columbia University Press.
Wu J and Hobbs RJ (eds) (2007) *Key topics in landscape ecology.* Cambridge: Cambridge University Press.

landscape evaluation Quantitative or semi-quantitative evaluation of the 'qualities' of landscape for planned development or CONSERVATION purposes. Landscape evaluation is primarily concerned with the visual, aesthetic, cultural and HERITAGE values of landscape rather than the ecological aspects, such as biodiversity, rarity and complexity.   *JAM*

[*See also* AESTHETICS, ENVIRONMENTAL IMPACT ASSESSMENT (EIA), GEODIVERSITY, LAND EVALUATION]

Brabyn L (1996) Landscape classification using GIS and national digital databases. *Landscape Research* 21: 277–287.
Burton R (1999) Landscape evaluation. In Pacione M (ed.) *Applied geography: Principles and practice.* London: Routledge, 236–245.
Wilson S (2002) *Guidelines for landscape and visual impact assessment.* London: Spon Press.

landscape evolution The nature and speed by which a LANDSCAPE changes through time. The knowledge and understanding of how landscapes evolve over a GEOLOGICAL TIMESCALE under different climatic and geological BOUNDARY CONDITIONS formed the main objective of GEOMORPHOLOGY until the 1960s, and it remains an important aim particularly in continental Europe. In the English-speaking world, timescales of study then shortened considerably with the increased emphasis on process measurement and attempts to relate form to process and, in longer-term studies, an increased emphasis on Late QUATERNARY and HOLOCENE landscape development.

Studies up to the 1960s sought to reconstruct the history of landscape development (or DENUDATION CHRONOLOGY) of parts of the Earth's surface using morphological (and increasingly later sedimentological) evidence in the current landscape. The timescale encompassed by such evidence (and ultimately the age of the landscape) largely determined the timescale covered by such studies. Thus, in Europe the timescales involved generally went back to the TERTIARY, but studies in parts of Africa and Australia encompassed even longer timescales. Morphological evidence used in denudation chronological studies included details of *drainage patterns* (e.g. orientation of river valleys, elbows of RIVER CAPTURE, longitudinal profiles, KNICK-POINTS, GORGES and *misfit streams*), the heights and extents of EROSION SURFACES and dry cols in CUESTAS. In Britain, one of the reasons for the abandonment of such studies was the absence of means of dating and hence proving or disproving the chronological schemes that were proposed. In continental Europe, in contrast, the availability of TEPHRA (distinctive marker horizons associated with particular major volcanic eruptions) provided a means of RELATIVE-AGE DATING during the Tertiary and this partly explains the continued dominance of long-term geomorphology in Germany into the 1980s.

The first 60 years of the twentieth century were dominated by models of landscape evolution seeking to explain how initially high relief terrain was progressively reduced during periods of crustal stability to produce plains (e.g. ETCHPLAINS, PENEPLAINS or PEDIPLAINS), before renewed OROGENY produced new high-relief terrain. The most influential scheme was that of W.M. Davis, who described a 'normal' cycle (see DAVISIAN CYCLE OF EROSION) operating in climatic environments dominated by fluvial activity, but with GLACIAL and *arid cycles* applying in terrain dominated by ice and wind action, respectively. A *periglacial cycle* was added by Peltier (1950), who also envisaged that the relative importance of different processes of WEATHERING and EROSION would vary with annual temperature and precipitation within the 'normal' cycle area. The alternative schemes of Walther Penck and Lester King differed from that of Davis in terms of the SLOPE EVOLUTION MODELS involved. The main problem with the cyclical concept, however, was the assumption that short periods of orogeny or crustal instability (i.e. *landscape construction*) alternated with long periods of crustal stability, in which landscapes could evolve under the influence of climate and lithology and inherited structure.

Schumm (1963) demonstrated this assumption to be false and that a more realistic scenario is that significant mountain building (often exceeding EROSION) is not confined to short periods, but is characteristic of landscapes most of the time. Likewise, assumptions of a single climate or narrow range of climates operating unchanged on a land surface throughout a cycle (or even for long periods during a cycle) have had to be rejected even for the inner tropics with the emergence of evidence in the latter part of the twentieth century of relatively frequent and large-scale climatic change over most of the Earth's surface.

Although landscape evolution is influenced by the interplay of a range of factors, notably climate and climatic history, lithology, structure and the history of earth movements (uplift, subsidence, tilt, folding, etc.), geomorphologists have tended to approach the issue by placing one of the factors in a primary position and treating the other factors as subsidiary, thus leading to CLIMATIC GEOMORPHOLOGY, *lithological geomorphology* and *structural geomorphology*, respectively. Büdel (1963) made an early conceptual attempt to broaden such a view in proposing CLIMATOGENETIC GEOMORPHOLOGY, in which climate was seen to be operating within a long-term wider framework incorporating the other factors to produce generations of relief development.

Within climatic geomorphology, many attempts have been made to define morphogenetic regions or MORPHOCLIMATIC ZONES on the basis of climatic and ecological controls over geomorphic processes and landforms and some of these have associated particular types of landscape evolution with particular climates or groups of climates. Peltier (1950) developed a scheme proposing how the relative importance of different weathering and erosional processes would vary with annual temperature and precipitation within the 'normal' cycle area. These early attempts, however, were largely based upon a mixture of deduction about process based on climatic parameters and simple (and often unwarranted) linkages between landforms and landscape and current process. This is demonstrated most strikingly by the various attempts to link INSELBERG landscapes to climate, with different workers linking its development at different times to SEMI-ARID, *seasonal tropical* and *humid tropical* environments. More recent studies of inselbergs have acknowledged that the timescales required for development of such LANDFORMS are much longer than the period of operation of any single climate or any period of crustal stability. Theories incorporating alternating deep weathering and STRIPPING PHASES linked to either a combination of sequences of different climates or tectonic changes have thus become more prevalent. More recent climatic geomorphology has thus tended to focus on the influence of climatic factors and climate-linked processes on current and past landforms, landscape development and EROSION RATES in the context of a frequently changing climate. It has also focussed more on the influence of particular elements of climate, the magnitude-frequency of climatic events, and the relative roles of EXTREME EVENTS and more frequent events, rather than on crude climatic mean values.

EQUILIBRIUM CONCEPTS within long-term landscape evolution, which became important in the late 1960s and 1970s, have also had to be revised with the realisation that landscapes are subject to more frequent climatic change and geological disruption than previously thought. Landscapes are probably in the process of adjusting to changed conditions (a state of disequilibrium) far more of the time than they are in a state of adjustment to a particular climate. Also, it is arguable that far more of a landscape and many of its individual landforms are the product of changes in climate than of the individual climates themselves. The shortening of timescales of interest—and in particular the growth in interest and focus on *human impacts*—has reinforced the importance of concepts of landscapes in transition (or adjustment) and therefore often in *disequilibrium* or *non-equilibrium* rather than *equilibrium*. Also important is the concept that there is a hierarchy of adjustment times of different components of a landscape. In rivers, bed configuration and channel width and depth may respond almost immediately to changed DISCHARGE or sediment transport regime, MEANDER wavelengths within a century, but *longitudinal profile* gradient and concavity may take 1,000–10,000 years to respond. Similarly, evidence from stream networks developed on volcanic centres of contrasting age in the eastern Caribbean demonstrates how the speed of adjustment (RELAXATION TIME) of DRAINAGE-BASIN shape, *bifurcation ratios* and *stream-length ratios* from the initial radial volcanic values to more typical values of dendritic networks falls with basin order, with higher-order basins and ratios retaining atypical values for much longer than first- or second-order basins.

Thus, attention is increasingly focussed on how, by how much and how quickly landscapes and processes respond to changes in climate, geological movements and human disturbance. Brunsden (1980) considers that geomorphic time can be divided into (a) REACTION TIMES (times for a landscape or individual landscape components to react to a change in conditions), (b) *relaxation times* (times for a landscape or component landforms to attain a new characteristic equilibrium state) and (c) *characteristic-form times* (times over which those new states may be expected to persist).

In the past decade, there has been a remarkable renaissance in interest and progress in long-term landscape evolution. This has been facilitated by the advent of COSMOGENIC-NUCLIDE DATING of land surfaces, increased knowledge of the history of PLATE TECTONICS and orogeny (including the importance of flexure and denudational ISOSTASY) and the development of *numerical modelling* techniques applied to landscape development under varying tectonic and denudational controls. Bishop (2007) considers that both the Davisian *cycle of erosion* and Hack's *dynamic equilibrium* concepts appear to have considerable applicability in these studies, depending on the balance between erosion, orogeny and denudational isostasy.   *RPDW*

[*See also* MAGNITUDE-FREQUENCY CONCEPTS]

Bishop P (2007) Long-term landscape evolution: Linking tectonics and surface processes. *Earth Surface Processes and Landforms* 32: 329–365.

Bracken LJ and Wainwright J (2006) Geomorphological equilibrium: Myth and metaphor. *Transactions of the Institute of British Geographers NS* 31: 167–178.

Brunsden D (1980) Applicable models of long term landform evolution. *Zeitschrift für Geomorphologie Supplementband* 36: 16–26.

Büdel J (1948) Das System der klimatischen Geomorphologie [The system of climatic geomorphology]. *Verhandlungen Deutscher Geographie* 27: 65–100.

Büdel J (1963) Klimagenetische geomorphologie [Climatogenetic geomorphology]. *Geographische Rundschau* 15: 269–285.

Chorley RJ, Schumm SA and Sugden DE (1984) *Geomorphology*. London: Methuen.

Davis WM (1899) The geographical cycle. *Geographical Journal* 14: 481–504.

Hack JT (1960) Interpretation of erosional topography in humid temperate regions. *American Journal of Science* 258A: 80–97.

Heimsath AM and Korup O (2012) Quantifying rates and processes of landscape evolution. *Earth Surface Processes and Landforms* 37: 249–251.

King LC (1953) Canons of landscape evolution. *Bulletin of the Geological Society of America* 64: 721–752.

Knighton DA (1998) *Fluvial forms and process*. London: Arnold.

Peltier LC (1950) The geographic cycle in periglacial regions as it is related to climatic geomorphology. *Annals of the Association of American Geographers* 40: 214–236.

Penck W (1924) *Die morphologische Analyse* [Morphological analysis]. Stuttgart: Geographische Abhandlungen.

Schumm SA (1963) The disparity between present rates of denudation and orogeny. *United States Geological Survey Professional Paper* 454-H: 1–13.

Tricart J and Cailleux J (1965) *Introduction à la géomorphologie climatique* [Introduction to climatic geomorphology]. Paris: SEDES.

Tucker GE and Hancock GR (2010) Modelling landscape evolution. *Earth Surface Processes and Landforms* 35: 28–50.

Walsh RPD (1996) Drainage density and network evolution in the humid tropics: Evidence from the Seychelles and the Windward Islands. *Zeitschrift für Geomorphologie Supplementband* 103: 1–23.

landscape geochemistry An approach to LANDSCAPE description and evolution developed from within the GEOSCIENCES in the former Soviet Union, notably by B.B. Polynov and A.I. Perel'man. Central ideas include the importance of geochemical and biological WEATHERING during LANDSCAPE EVOLUTION and the controlling position of the WATER TABLE.   *JAM*

[*See also* BIOGEOCHEMISTRY, LANDSCAPE SCIENCE]

Fortescue JAC (1980) *Environmental geochemistry: A holistic approach*. Berlin: Springer Verlag.

Nash DJ and McLaren SJ (eds) (2007) *Geochemical sediments and landscape*. Oxford: Blackwell.

Perel'man AI (1966) *Landscape geochemistry*. Moscow: Vysshaya Shkola [Geological Survey of Canada, Translation No. 676, 1972].

Snytko VA, Semenov YM and Davydova ND (1981) A landscape-geochemical evaluation of geosystems for purposes of rational nature management. *Soviet Geography* 22: 569–578.

landscape management Applied LANDSCAPE ECOLOGY, or the integrated management of whole LANDSCAPE UNITS.   *JAM*

Burel F and Baudry J (2003) *Landscape ecology: Concepts, methods and applications*. Enfield, NH: Science Publishers.

Vink APA (1983) *Landscape ecology and land use*. London: Longman.

landscape mosaic The mosaic of *patches* (woods, fields, ponds, rock outcrops, houses), *corridors* (roads, hedgerows, rivers) and *matrices* (background ecosystems or landuse types) that form landscapes.   *RJH*

[*See also* LANDSCAPE ECOLOGY, PATCH DYNAMICS, WILDLIFE CORRIDOR]

Collinge SK (2009) *Ecology of fragmented landscapes*. Baltimore: Johns Hopkins University Press.

Forman RTT (1996) *Land mosaics: The ecology of landscapes and regions*. Cambridge: Cambridge University Press.

landscape science The study of LANDSCAPE in the scientific sense. Whether considered as a part of ECOLOGY or PHYSICAL GEOGRAPHY (LANDSCAPE ECOLOGY or GEO-ECOLOGY) or of the GEOSCIENCES (LANDSCAPE GEOCHEMISTRY), it is characterised by a HOLISTIC APPROACH to the interacting environmental processes at the surface of the Earth.   *JAM*

[*See also* GEO-INDICATOR, LAND-CHANGE SCIENCE]

Fortescue J (1996) Guidelines for a 'systematic landscape geoscience'. In Berger AR and Iams WJ (eds) *Geoindicators: Assessing rapid environmental changes in earth systems*. Rotterdam: Balkema, 351–364.

Isachenko AG (1973) *Principles of landscape science and physical-geographical regionalisation*. Melbourne: Melbourne University Press.

Kupfer JA (1995) Landscape ecology and biogeography. *Progress in Physical Geography* 19: 18–34.

Shaw DJB and Oldfield J (2007) Landscape science: A Russian geographical tradition. *Annals of the Association of American Geographers* 97: 111–126.

Zonneveld IS (1979) *Land evaluation and landscape science.* Enschede: Enschede International Training Centre.

landscape sensitivity The magnitude of the response of a landscape to change in an external ENVIRONMENTAL FACTOR or DISTURBANCE. High sensitivity implies a large response to relatively small NATURAL disturbances and/or *human impacts*. JAM

[*See also* FORCING FACTOR, PERTURBATION, RESILIENCE, RESPONSE TIME]

Brunsden D (2001) A critical assessment of the sensitivity concept in geomorphology. *Catena* 42: 99–123.

Brunsden D and Thornes J (1979) Landscape sensitivity and change. *Transactions of the Institute of British Geographers, NS* 4: 463–484.

Thomas MF (2001) Landscape sensitivity in time and space. *Catena* 42: 83–99.

Verleysdonk S, Krautblatter M and Dikau R (2011) Sensitivity and path dependence of mountain permafrost systems. *Geografiska Annaler* 93A: 113–135.

landscape units Various systems of units have been proposed, mostly with an underlying hierarchical structure, for use in the description and investigation of the LANDSCAPE MOSAIC. Several schemes based on morphology or PHYSIOGRAPHY are summarised in the Table. JAM

[*See also* LANDSCAPE, LANDSCAPE GEOCHEMISTRY, LANDSCAPE SCIENCE]

Fenneman NM (1916) Physiographic divisions of the United States. *Annals of the Association of American Geographers* 6: 19–98.

Huggett RJ (1995) *Geoecology: An evolutionary approach.* London: Routledge.

Linton DL (1949) The delimitation of morphological regions. *Transactions of the Institute of British Geographers* 14: 86–87.

Whittlesey D (1954) The regional concept and the regional method. In James PE and Jones C (eds) *American geography, inventory and prospect.* Syracuse, NY: Syracuse University Press, 19–68.

landslide Although this term is in popular usage, a variety of definitions and classifications exists. Because few of these are clear and unambiguous, there are conflicting applications of terms. Strictly speaking, a landslide is a type of MASS MOVEMENT PROCESS in which FAILURE occurs on a distinct *zone of sliding* (a *shear plane*) and the displaced material moves with uniform velocity throughout its mass. This excludes fall, flow, *topple* and *creep*. However, with the exception of creep, these latter mechanisms are frequently included in schemes of landslide classification resulting in more broad-based definitions in which process is not inferred. Thus, a landslide is often defined as a perceptible downslope displacement of rock or REGOLITH under the influence of gravity. As such, the term encompasses most forms of mass movement. This type of overarching definition can be justified on the grounds that processes initiating movement are often complex and difficult to identify, and that debris often undergo transformation during movement, for example, from slide to flow or AVALANCHE, depending on water content and degree of debris break-up.

Falls are free-fall movements of material from steep slopes or *cliffs*; topples involve a pivoting action at the base of the FAILURE and *flows* occur when the displaced units of material move as viscous substances in which air or water are significant components. *Slides* are subdivided into *rotational slides* and *translational slides* depending on the form of the shear plane; the former involves a curved (concave upwards) *shear surface*, the latter has a planar slip face roughly parallel to the ground surface. *Complex landslides* are a combination of two or more of these movements acting simultaneously in different parts of the feature or

Scale	Approximate area (km²)	Fenneman (1916)	Linton (1949)	Whittlesey (1954)
Micro (small)	$<10^0$	—	Site	—
Meso (medium)	$10^0–10^1$	—	—	—
	$10^1–10^2$	—	Stow	Locality
	$10^2–10^3$	District	Tract	District
	$10^3–10^4$	Section	Section	—
Macro (large)	$10^4–10^5$	Province	Province	Province
	$10^5–10^6$	Major division	Major division	Realm
Mega (very large)	$>10^6$	—	Continent	—

Landscape units *Terminology of landscape units in relation to spatial scale (Huggett, 1995).*

sequentially downslope. Many landslides begin as slides but become flows in their terminal zones.

Several factors influence landslide activity. A *trigger factor* may be recognised but seldom can a landslide be attributed to a single cause. Important factors may include VEGETATION, SEISMICITY, *water content*, WEATHERING, CLIMATE and *human impact*. The probability of LANDSLIDES occurring in response to CLIMATIC CHANGE and increasing levels of human activity is high, and it is important to identify potential instability.

Some types of RELICT or fossil landslides can be difficult to distinguish from other accumulations of deposits found at the bases of steep slopes (e.g. PRONIVAL RAMPARTS, ROCK GLACIERS and MORAINES). It has even been suggested that they may produce suitably shaped hollows for CIRQUE glaciers to form. Incorrect diagnosis of a landslide origin can therefore lead to incorrect palaeoenvironmental interpretation.   *PW*

[*See also* DEBRIS FLOW, LANDSLIDE FREQUENCY, LANDSLIDE HAZARD, RUNOUT DISTANCE, SLOPE STABILITY, STURZSTRÖM]

Antinao JL and Gosse J (2009) Large rockslides in the southern central Andes of Chile (32–34.5°S): Tectonic control and significance for Quaternary landscape evolution. *Geomorphology* 104: 117–133.

Bromhead E, Dixon N and Ibsen M-L (eds) (2000) *Landslides in research, theory and practice, 3 volumes.* London: Thomas Telford.

Clague J and Stead D (2012) *Landslides: Types, mechanisms and modelling.* Cambridge: Cambridge University Press.

Crozier MJ (2004) Landslide. In Goudie AS (ed.) *Encyclopedia of geomorphology.* London: Routledge, 605–608.

Crozier MJ (2010) Deciphering the effect of climate change on landslide activity: A review. *Geomorphology* 124: 260–267.

Dikau R, Brunsden D, Schrott L and Ibsen M-L (eds) (1996) *Landslide recognition: Identification, movement and causes.* Chichester: Wiley.

Hewitt K, Clague JJ and Orwin JF (2008) Legacies of catastrophic rock slope failures in mountain landscapes. *Earth-Science Reviews* 87: 1–38.

Jarman D (2006) Large rock slope failures in the Highlands of Scotland: Characterisation, causes and spatial distribution. *Engineering Geology* 83: 161–182.

Jarman D, Agliardi F and Crosta GB (2011) Megafans and outsize fans from catastrophic slope failures in alpine glacial troughs: The Malser Haide and the Val Venosta cluster, Italy. In Jaboyedoff M (ed.) *Slope tectonics.* Bath: Geological Society, 253–277.

Mitchell WA, McSaveney MJ, Zondervan A et al. (2007) The Keylong Serai rock avalanche, NW Indian Himalaya: Geomorphology and palaeoseismic implications. *Landslides* 4: 245–254.

Oppikofer T, Jaboyedoff M and Keusen HR (2008) Collapse at the eastern Eiger flank in the Swiss Alps. *Nature Geoscience* 1: 531–535.

Soldati M, Corsini A and Pasuto A (2004) Landslides and climate change in the Italian Dolomites since the Late Glacial. *Catena* 55: 141–161.

Winter MG, Dixon N, Wasowski J and Dijkstra TA (eds) (2010) Land-use and climate change impacts on landslides. *Quarterly Journal of Engineering Geology and Hydrogeology* 43: 367–496 [Special Issue].

landslide dam   A *natural dam* to a river caused by any form of MASS WASTING. The resulting LAKE may be short- or long-lived. Because of the loose nature of the dam sediment, such dams can fail suddenly causing downstream FLOODS, which constitute a NATURAL HAZARD. Flooding upstream may also occur in the form of a wave immediately following the landslide event itself. Landslide dams are generally found in valleys flanked by steep terrain from which the landslides originate. There are many examples worldwide but the forms and processes involved are not well understood.   *RAS*

Costa JE and Schuster RL (1988) The formation and failure of natural dams. *Geological Society of America Bulletin* 100: 1054–1068.

Korup O (2002) Recent research on landslide dams: A literature review with special attention to New Zealand. *Progress in Physical Geography* 26: 206–235.

landslide frequency   The total geomorphic effect of an erosional process or event is determined partly by how often it occurs. LANDSLIDE frequency is therefore an important consideration in estimating the degree of risk involved and is fundamental to understanding stability conditions. Frequency is usually stated in terms of the PROBABILITY of recurrence, expressed as a probability that an event will occur in a stated number of years. Unfortunately, the detailed history of landslide activity in a given area is rarely well enough known for assessments of *recurrence interval* to be made by direct means. The factors that influence landslide frequency in the short term are those that vary substantially such as MICROCLIMATE, SOIL MOISTURE conditions and LANDUSE CHANGES. Over longer time intervals, WEATHERING and CLIMATIC CHANGE become significant determinants of frequency.   *PW*

[*See also* LANDSLIDE, LANDSLIDE HAZARD, SLOPE STABILITY]

Hermanns RL, Blikra LH, Naumann M et al. (2006) Examples of multiple rock-slope collapses from Köfels (Ötz valley, Austria) and western Norway. *Engineering Geology* 83: 94–108.

Huggel C, Clague JJ and Korup O (2012) Is climate change responsible for changing landslide activity in high mountains? *Earth Surface Processes and Landforms* 37: 77–91.

Lee J, Davies T and Bell D (2009) Successive Holocene rock avalanches at Lake Coleridge, Canterbury, New Zealand. *Landslides* 6: 287–297.

Matthews JA, Dahl SO, Dresser PQ et al. (2009) Radiocarbon chronology of Holocene colluvial (debris-flow) events at Sletthamn, Jotunheimen, southern Norway: A window on the changing frequency of extreme climatic events and their landscape impact. *The Holocene* 19: 1107–1129.

Van Asch TWJ (1998) The temporal activity of landslides and its climatological signals. In Matthews JA, Brunsden D, Frenzel B et al. (eds) *Rapid mass movement as a source of climatic evidence for the Holocene.* Stuttgart: Gustav Fischer, 7–16.

landslide hazard   A LANDSLIDE is only considered a HAZARD if people, property and/or utilities are at RISK of injury, damage or loss. Recognition of a landslide hazard is not confined to the more rapid and violent events; relatively slow movements of SOIL, SEDIMENT and ROCK may cause significant disruption to the societies affected. LANDSLIDE FREQUENCY is also of considerable importance. The landslide scale, the terrain in which it occurs, and the antecedent and prevailing climatic conditions are important factors influencing the level of impact. Increasingly, landslide hazard is seen as being as much a human-induced as a natural one. In some locations, modification of the land surface has decreased SLOPE STABILITY and allowed NATURAL events (e.g. rainfall) to trigger FAILURE. It is now a requirement in many areas for urban or economic development to be preceded by *landslide hazard assessment* and *landslide hazard mapping*. This entails investigation and evaluation of slope stability, a statement concerning the severity of the hazard and recommendations to reduce the hazard. *Landslide management* may involve, for example, GEO-ENGINEERING and landuse ZONING. A LAHAR (a volcanic mudflow) is a particularly hazardous landslide: a *cold lahar* results from heavy rains during an eruption whereas a *hot lahar* results from the emptying of a crater lake.   *PW*

[*See also* ENVIRONMENTAL MANAGEMENT, RISK ASSESSMENT, RISK MANAGEMENT]

Evans SG, Mugnozza GS, Strom A and Hermanns RL (eds) (2006) *Landslides from massive rock slope failures.* Dordrecht: Springer.
Lee EM and Jones DKC (2004) *Landslide risk assessment.* London: ICE Publications.
Sassa K, Rouhban B, Briceno S et al. (eds) (2012) *Landslides: Global risk preparedness.* Berlin: Springer.

landsystem   A large-scale association of SEDIMENTS and LANDFORMS which may be linked genetically to a specific depositional setting in the LANDSCAPE. At the lowest level are *land elements* or individual landforms; at the intermediate level are *land facets* or groups of land elements; and landsystems are composites of linked land facets. Although a *landsystem approach* can be applied to a range of *geomorphological systems*, it has been particularly well developed in relation to *glacial environments*. For example, a landscape comprising areas of ice-scoured bedrock, subglacial sediment, DRUMLINS, FLUTES and ESKERS belongs to the *subglacial landsystem*.   *MRB*

[*See also* GLACIAL LANDFORMS, GLACIAL SEDIMENTS]

Evans DJA (2005) *Glacial landsystems.* London: Arnold.
Evans DJA and Twigg DR (2002) The active temperate glacial landsystem: A model based on Breiðamerkurjökull and Fjallsjökull, Iceland. *Quaternary Science Reviews* 21: 2143–2177.

landuse   An expression applied to the employment of land in terms of its human-related utilisation. While it has, in the past, been applied as an overarching term which included LAND COVER, today it applies only to the human aspects. While it is entirely possible for the two to be associated on a one-to-one category basis, it is also feasible for landuse to extend over a variety of different LAND COVERS. For example, a new development site may incorporate former built-up land and extend onto *green-field sites*.   *TF*

[*See also* BROWN FIELD SITES, CONVERSION]

Foley JA, DeFries R, Asner GP et al. (2005) Global consequences of landuse. *Science* 309: 570–574.
Geist HJ (ed.) (2005) *Our Earth's changing land: An encyclopedia of land-use and land-cover change, 2 volumes.* Westport, CT: Greenwood Press.

landuse capability classification   An approach to LAND EVALUATION for general use, also called land capability. In this system, land is placed in classes depending upon the severity of the limitations on its use. Limitations include EROSION hazard, excess water, *soil limitations* in the *rooting zone* and climatic limitations. Thus, in the United Kingdom, class I land has no limitations and can be employed for any form of landuse, whereas class VIII land has several limitations and may have only the capability for supporting *wildlife* and *watershed protection*.   *EMB*

[*See also* LAND]

Bibby JS and Mackney D (1977) *Land-use capability classification.* Harpenden: Soil Survey of Great Britain.
Klingebiel AA and Montgomery PH (1958) *Land-capability classification.* Washington, DC: US Department of Agriculture.

landuse change   Changes in human use of the land are caused largely by a variety of social, economic and political factors. For most of the time, the long history of landuse change involved impacts on vegetation and soil as, first, humans occupied most of the Earth's surface and, second, use of biotic resources was intensified. The degree and rate of change of this transformation have varied greatly in different regions. Agricultural DEFORESTATION, for example, began first in the Near East and East Asia, affected North Africa and Central Europe at 7,000–8,000 radiocarbon years ago, but affected the Americas much later (see Figure).

The net loss of forested area since pre-agricultural times is about 8 million km², of which more than 75 per cent has been cleared since the end of the seventeenth century. Over the past three centuries, 19 per cent of the forests remaining in AD 1700 were removed: 8 per cent of the world's grasslands and pastures were removed over the same period while there was a >400 per cent increase in the area of cropland. This agricultural expansion and related processes, such as LAND DRAINAGE and IRRIGATION of grassland, has for the most part accelerated under European political and economic control. In Europe itself, a common pattern of intensifying landuse developed around the major cities: intensive market gardening (HORTICULTURE) pushed outwards extensive cereal growing and livestock rearing. Shorter-term landuse changes have been many and varied. In England and Wales since the World War II, for example, there has been loss of ANCIENT WOODLAND and SEMI-NATURAL VEGETATION (heathland, wetlands, dunes and bracken), while the total area of woodland has increased as a result of AFFORESTATION (mainly coniferous plantations). Most recently, the area of agricultural land has decreased as a result of SET-ASIDE SCHEMES.

Although replacement of forests by agricultural systems and subsequent landuse changes throughout history have been viewed as 'land improvement', they have led in many cases to SOIL DEGRADATION. Where land has remained productive for a long period with increasing yields this has only been possible with large inputs of human energy and materials. Landuse change accelerated and the direct and indirect environmental impacts increased in scale and complexity with AGRICULTURAL INTENSIFICATION and INDUSTRIALISATION. Almost all the world's lands are now used to some extent and landuse change is a major aspect of GLOBAL ENVIRONMENTAL CHANGE. Cumulative effects, such as reduced BIODIVERSITY and modification of the CARBON SEQUESTRATION capacity of the BIOSPHERE, have become global in extent. The historical CONVERSION of forest to agricultural and urban landuse has resulted in release of CARBON DIOXIDE comparable in amount to that released by FOSSIL FUELS. Only since the mid-twentieth century has the rate of carbon release from fossil fuels exceeded that from land transformation.   *JAM*

[*See also* AGRICULTURAL HISTORY, AGRICULTURAL IMPACT ON SOILS, AGRICULTURAL REVOLUTION, FOREST CLEARANCE, GREEN REVOLUTION, HUMAN IMPACT ON ENVIRONMENT, LAND COVER, LAND DRAINAGE, LAND TRANSFORMATION]

Bell M (1983) Valley sediments as evidence of prehistoric land-use on the South Downs. *Proceedings of the Prehistoric Society* 49: 119–150.

Berglund BE (1994) Methods for quantifying prehistoric deforestation. In Frenzel B, Andersen ST, Berglund BE and Gläser B (eds) *Evaluation of land surfaces cleared from forests in the Roman Iron Age and the time of migrating German tribes based on regional pollen diagrams* [Paläoklimaforschung Band 12]. Stuttgart: Gustav Fischer, 5–11.

Bradshaw EG, Rasmussen P and Odgaard BV (2005) Mid- to late-Holocene land-use change and lake development at Dallund Sø, Denmark: Synthesis of multiproxy data, linking land and lake. *The Holocene* 15: 1152–1162.

Foley JA, DeFries R, Asner GP et al. (2005) Global consequences of land use. *Science* 309: 570–574.

Lambin EF and Geist HJ (eds) (2006) *Land-use and land-cover change: Local processes and global impacts.* Berlin: Springer.

Mannion AM (2002) *Dynamic world: Land-cover and land-use change.* London: Hodder Arnold.

Meyer WB and Turner II BL (eds) (1994) *Changes in land use and land cover: A global perspective.* Cambridge: Cambridge University Press.

Rasmussen P (2005) Mid- to late-Holocene land-use change and lake development at Dallund Sø, Denmark: Vegetation and land-use history inferred from pollen data. *The Holocene* 15: 1116–1129.

Walker B, Steffen W, Canadell J and Ingram J (eds) (1999) *The terrestrial biosphere and global change: Implications for natural and managed ecosystems.* Cambridge: Cambridge University Press.

Walker D and Singh G (1994) Earliest palynological records of human impact on the world's vegetation. In Chambers FM (ed.) *Climate change and human impact on the landscape.* London: Chapman and Hall, 101–108.

Watson RT, Noble IR, Bolin B et al. (eds) (2000) *Land use, land-use change, and forestry: Special report of the Intergovernmental Panel on Climate Change.* Cambridge: Cambridge University Press.

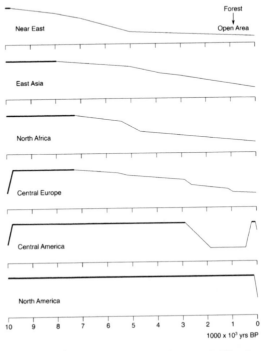

Landuse change *Trends in agricultural deforestation in different parts of the world over the past 10,000 radiocarbon years. The thin line indicates human-induced deforestation (Berglund, 1994).*

landuse impacts on hydrology Changes in water, solute and sediment movement through catchments produced deliberately or inadvertently by shifts in LANDUSE. Impacts have been evaluated at many timescales, using before-and-after, paired catchment, palaeoenvironmental reconstruction or modelling approaches. Disentangling the effects of landuse shift from CLIMATIC CHANGE impacts is challenging, however, especially given cyclical landuse changes and complex hydrological RESPONSE TIMES. Typical foci include URBANISATION IMPACTS ON HYDROLOGY, the effects of artificial LAND DRAINAGE, AFFORESTATION, DEFORESTATION and shifts to intensive arable CULTIVATION, and palaeohydrological effects of long-term changes in AGRICULTURE. Hydrological effects of landuse change partly depend on its location with respect to DRAINAGE NETWORKS and hydrologically sensitive areas. Impacts of FIELD DRAINAGE include faster runoff response, higher flood peaks and frequencies, increased EROSION RATES and SEDIMENT YIELDS, and reduced *solute fluxes*. AFFORESTATION can reduce DRAINAGE BASIN water and sediment yields, and modulate flows, though pre-afforestation *ditching* may enhance stream sediment loads and flood responses initially. DEFORESTATION leads to enhanced erosion rates and downstream FLOODS. Shifts to intensive arable cultivation methods, if associated with bare soils stripped of protective surface vegetation, regular PLOUGHING and application of FERTILISERS and PESTICIDES, may result in higher flood risk, soil erosion rates and *pollutant loads* (e.g. NITRATES (NO_3^-)). *DML*

[*See also* URBANISATION IMPACTS ON HYDROLOGY]

Brown AG and Quine TA (1999) *Fluvial processes and environmental change*. Chichester: Wiley.
Church M, Burt TP, Galay VJ and Kondolf GM (2009) Rivers. In Slaymaker O, Spencer T and Embleton-Hamann C (eds) *Geomorphology and global environmental change*. Cambridge: Cambridge University Press, 98–129.
Hollis GE (ed.) (1979) *Man's impact on the hydrological cycle*. Norwich: GeoBooks.
Latubresse EM, Amsler ML, de Morais RP and Aquino S (2009) The geomorphological response of a large pristine alluvial river to tremendous deforestation in the South American tropics: The case of the Araguaia River. *Geomorphology* 113: 239–252.
Newson MD (1997) *Land, water and development, 2nd edition*. London: Routledge, 70–80.

landuse impacts on soils Interruption of natural MINERAL CYCLING of NUTRIENTS. CULTIVATION reduces *organic matter* content and FERTILITY and promotes SOIL EROSION; IRRIGATION may raise the WATER TABLE and cause SALINISATION in DRYLANDS; HARVESTING reduces the plant nutrient content of the soil and roads and buildings mean biological use of soil is ended. Positive impacts include practices such as increasing fertility and organic matter content by manuring and FERTILISER application, reduction of ACIDITY by LIMING and SOIL DRAINAGE by lowering the WATER TABLE, which helps with DESALINISATION of soils in ARIDLANDS and SEMI-ARID regions. However, increasing population in many DEVELOPING COUNTRIES forces farmers onto MARGINAL AREAS for subsistence. With greater production risks in these areas, SOIL DEGRADATION is encouraged. There are numerous examples of landuse impacts on soils in HISTORY and PREHISTORY. *EMB/JAM*

[*See also* AGRICULTURAL IMPACT ON SOILS, COLLAPSE OF CIVILISATIONS, DESERTIFICATION, HUMAN IMPACT ON SOIL]

Braimoh AK and Vlek PLG (eds) (2008) *Land use and soil resources*. Berlin: Springer.
Brye KR and West CP (2001) Grassland management effects on soil surface properties in the Ozark Highlands. *Soil Science* 170: 63–73.
Lucke B (2008) *Demise of the Decapolis: Past and present desertification in the context of soil development, land use, and climate*. Saarbrücken: VDM Verlag.
Noel H, Garbolino E, Brauer A et al. (2001) Human impact and soil erosion during the last 5000 years as recorded in lacustrine sedimentary organic matter, Lac d'Annecy, the French Alps. *Journal of Paleolimnology* 25: 229–244.

landuse planning and management The set of public policies that seek to establish order and regulate the efficient and ethical use and protection of land parcels. For environmental landuse planning and MANAGEMENT, this aims for a state of SUSTAINABILITY that attempts to balance economic progress with the exploitation of NATURAL RESOURCES. Agencies at the national and local levels focus on ENVIRONMENTAL PROTECTION and WILDLIFE CONSERVATION, including NATIONAL PARKS and other PROTECTED AREAS. They sometimes work with planning agencies to monitor the ZONING of landuses. They are also concerned with the management of soil toxicology, stability of topography, flooding, runoff and stormwater discharge and groundwater hydrology; maintaining BIODIVERSITY, HABITATS, GEODIVERSITY, LAND RESTORATION, AIR QUALITY and WATER QUALITY; access to public land and the MITIGATION of NATURAL HAZARDS and CLIMATE CHANGE. *VM*

Beinat E and Nijkamp P (eds) (2010) *Multicriteria analysis for land-use management*. Dordrecht: Kluwer.
Holcombe RG and Staley SR (eds) (2001) *Smarter growth: Market-based strategies for land-use planning in the 21st century*. Westport, CT: Greenwood Press.
Randolph J (2012) *Environmental land use planning and management, 2nd edition*. Washington, DC: Island Press.
Silberstein J and Maser C (2000) *Land-use planning for sustainable development*. Boca Raton, FL: CRC Press.

lapilli PYROCLASTIC MATERIAL in the form of fragments between 2 and 64 mm in GRAIN SIZE (i.e. PEBBLES). The name is from the Italian 'lapillus', meaning little stone. *GO*

[*See also* ASH FALL, GRAIN SIZE, TEPHRA]

Brown RJ, Branney MJ, Maher C and Davila-Harris P (2010)
Origin of accretionary lapilli within ground-hugging
density currents: Evidence from pyroclastic couplets on
Tenerife. *Geological Society of America Bulletin* 122:
305–320.

lapse rate The variation of an atmospheric variable
with height in the ATMOSPHERE; most commonly, the
temperature lapse rate. In the TROPOSPHERE, temperature
decreases with height (average lapse rate of 6.5°C/km):
this is a positive lapse rate with cooler air above. A negative lapse rate (air warmer above) is an INVERSION, whilst
an *isothermal layer* occurs when there is no change of
temperature with height. A RADIOSONDE measures the
environmental lapse rate (the actual lapse rate). This is
compared with two theoretical lapse rates to assess the
stability of an atmospheric layer: these are the *dry adiabatic lapse rate* (9.8°C/km) and the *saturated adiabatic lapse rate*. The latter depends on humidity but varies
from 5°C/km in the lower troposphere to close to the
dry adiabatic lapse rate in the dry upper troposphere. An
adiabatic process is one in which heat does not enter or
leave a system. The STRATOSPHERE is either isothermal or
has a negative lapse rate.   *BDG*

Linacre E and Geerts B (1997) *Climates and weather
explained*. London: Routledge.
McIlveen R (2010) *Fundamentals of weather and climate,
2nd edition*. Kettering: Oxford University Press.

large igneous province (LIP) A large volume of
LAVA, commonly dominated by FLOOD BASALTS, erupted
over a relatively short interval of geological time
(probably 1–2 million years), forming a LAVA PLATEAU
on land (a *Continental Flood Basalt Province, CFBP*)
or in the ocean. Examples include the flood basalts of
the *Deccan Traps* of India and the Kerguelen Plateau
in the southern Indian Ocean, both formed during the
CRETACEOUS period. A large igneous province may form
when the head of a MANTLE PLUME first reaches the base
of the LITHOSPHERE. The plume tail is then responsible
for HOTSPOT IN GEOLOGY volcanoes, which remain active
over a long period. Volcanic activity related to the
formation of large igneous provinces may have implications for climatic change (see VOLCANIC IMPACTS ON
CLIMATE).   *GO*

Arndt N (2000) Hot heads and cold tails. *Nature* 407: 458–461.
Bryan SE and Ernst RE (2008) Revised definition of Large
Igneous Provinces (LIPs). *Earth-Science Reviews* 86:
175–202.
Rao NVC and Lehmann B (2011) Kimberlites, flood basalts
and mantle plumes: New insights from the Deccan Large
Igneous Province. *Earth-Science Reviews* 107: 315–324.

laser A device (*light amplification by stimulated
emission of radiation*) that emits a high-intensity beam
by exciting electronic, ionic or molecular transitions to
higher energy levels and then allowing these to fall to
lower energy levels. Generally, the beam is coherent
(in phase), has narrow spectral width (or is monochromatic) and is highly directional.   *PJS*

Heritage GL and Large ARG (eds) (2009) *Laser scanning for
the environmental sciences*. Chichester: Wiley-Blackwell.
Silfvas WT (2008) *Laser fundamentals, 2nd edition*.
Cambridge: Cambridge University Press.

laser-induced fluorescence (LIF) The process
of exciting molecules within an object or substance to
higher energy levels using a LASER and detecting the
emitted RADIATION (*fluorescence*) which follows when
energy is released and the molecules fall back to their
ground (lower energy) state. In REMOTE SENSING, this
phenomenon has been used in applications such as
detecting OIL SPILLS in water and estimating the CHLO-
ROPHYLL concentration of leaves with LiDAR.   *MEJC*

Thornton JA, Wooldridge PJ and Cohen RC (2000)
Atmospheric NO_2: In situ laser-induced fluorescence
detection at parts per trillion mixing ratios. *Atmospheric
Chemistry* 72: 528–539.

Last Glacial Maximum (LGM) The oxygen
isotope signal in MARINE SEDIMENT CORES and ICE
CORES indicates that the maximum ICE-SHEET volume
occurred during oxygen MARINE ISOTOPIC STAGE MIS 2,
towards the end of the LAST GLACIATION. The LGM
coincided with an estimated global RELATIVE SEA LEVEL
of ca −120 m OD (ordnance datum) and atmospheric
CARBON DIOXIDE levels below 190 ppmv (parts per million by volume). It is approximately bracketed by
HEINRICH EVENTS 2 and 3 in the North Atlantic, dated to
21–19 ka (radiocarbon years BP) and 28–27 ka (years
ago), respectively. The low sea levels lasted from 26.5
to 19 ka and terminated with a rapid decrease in ice
volume by about 10 per cent within a few hundred
years reflected as a meltwater pulse. These events
apparently coincide with low SEA-SURFACE TEMPERA-
TURES (SSTs) and a reorganisation in the THERMOHA-
LINE CIRCULATION, and reflect a period of 7,500 years
when there was a near equilibrium between global ice
sheets and CLIMATE.

At this time, the spread of ice sheets across large
areas of the Earth's surface in the high latitudes was
accompanied by significant modification of geomorphological and ecological systems as a result of
changes in the GENERAL CIRCULATION OF THE ATMOSPHERE
and the OCEAN CIRCULATION. For example, the LGM is
associated with increased LOESS mobilisation and DEPO-
SITION on the Loess Plateau of China and PERIGLACIATION
and AEOLIAN activity in northwest Europe (see Figure).
Low levels of PLUVIAL lakes in tropical Africa and Australasia suggest regional ARIDITY in low latitudes during
the LGM, although the question of whether equatorial

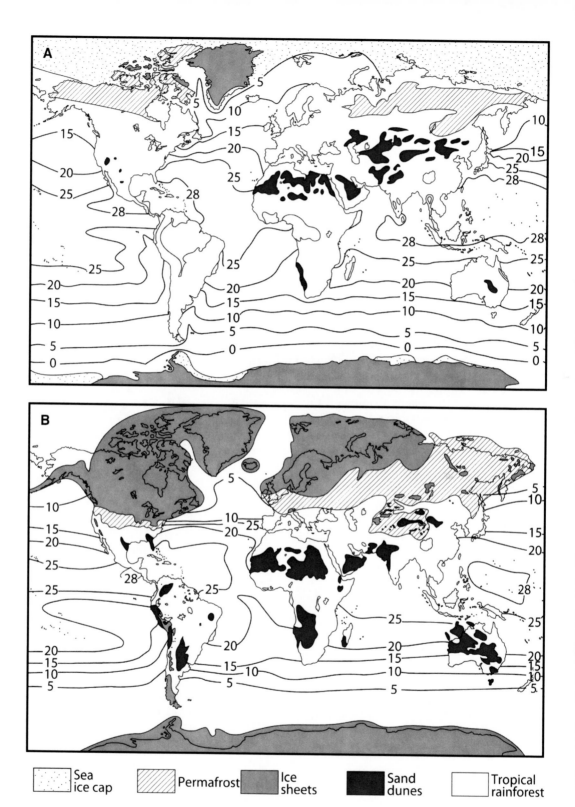

Sea ice cap ⋯⋯ **Permafrost** ▨ **Ice sheets** ▨ **Sand dunes** ■ **Tropical rainforest** ☐

—10— August sea-surface temperature (°C)

Last Glacial Maximum (LGM) *The Earth (A) at present and (B) at the Last Glacial Maximum, ca 18–21 ka: on land the distribution of ice sheets, perennially frozen ground (permafrost), active desert sand dunes and tropical rain forest are shown; sea-surface temperatures are shown in the oceans. Note that the reduction in area of tropical rain forest in (B) is probably overestimated (Wilson et al., 2000).*

regions of the Amazon Basin and Africa experienced aridity is still uncertain with estimates of much greater aridity in Africa than in South America.

In North America, during the LGM, the confluent LAURENTIAN and *Cordilleran ice sheet* covered ~16 million km², resulting in the developments of ICE-DAMMED LAKES (e.g. Lake Agassiz) and MEGAFLOODS beyond the ice margins (see CHANNELED SCABLANDS). The Laurentian Ice Sheet split the JET STREAM, enhancing the cooling of the continent and extensive PERIGLACIATION. LAKE-LEVEL VARIATIONS indicate that displacement of storm tracks (see STORMS) by the jet stream caused wetter and stormier conditions in the southwestern USA.

The CLIMAP (Climate: Long range Investigation, Mapping, and Prediction) and COHMAP (Cooperative Holocene Mapping Project) projects originally placed the LGM at 18 ka BP, implying a global synchroneity in maximum ice-sheet extent. It is now clear that there was considerable regional variability when ice sheets reached their initial LGM maxima, possibly between 33 and 29 ka BP in the case of *mountain glaciation* and the *West Antarctic Ice Sheet* (*WAIS*), although by 26.5 ka BP, all ice masses appeared to have been at their maximum extent. Despite the large degree of synchroneity of retreat in the Northern Hemisphere, the WAIS only began to retreat between 15.2 and 13.9 ka BP leading to *Meltwater Pulse* 1a.   *MHD/CJC*

Anhuf D, Ledru M-P, Behling H et al. (2006) Paleo-
    environmental change in Amazonian and African
    rain forest during the LGM. *Palaeogeography,
    Palaeoclimatology and Palaeoecology* 239: 510–527.
Bond G, Broecker W, Johnsen S et al. (1993) Correlations
    between climate records from North Atlantic sediments
    and Greenland ice. *Nature* 365: 143–147.
Clark PU, Dyke AS, Shakun JD et al. (2009) The Last
    Glacial Maximum. *Science* 325: 710–714.
COHMAP Members (1988) Climatic changes of the last
    18,000 years: Observation and model simulations.
    Science 421: 1043–1052.
Colinvaux PA, De Oliveira PE and Bush MB (2000)
    Amazonian and neotropical plant communities on
    glacial time-scales: The failure of the aridity and refuge
    hypotheses. *Quaternary Science Reviews* 19: 141–169.
McIntyre A, Kipp NG, Bé A et al. (1976) Glacial North
    Atlantic 18,000 years ago: A CLIMAP reconstruction.
    Geological Society of America Memoir 145: 43–47.
Wilson RCL, Drury SA and Chapman JL (2000) *The great
    Ice Age: Climate change and life*. London: Routledge.
Wright Jr HE, Kutzbach JE, Webb III T et al. (eds) (1993)
    Global climates since the Last Glacial Maximum.
    Minneapolis: University of Minnesota Press.

Last Glaciation   The Last Glaciation was the DEVENSIAN in Britain, otherwise known as the *Weichselian* in northern Europe, the *Würm* in the Alps, the *Valdai* in Russia and the *Wisconsinan* in North America. It comprised MARINE ISOTOPIC STAGES 4–2 (ca 74–11.7 ka BP).   *DH*

[See also LAST GLACIAL MAXIMUM (LGM), PLENIGLACIAL, QUATERNARY TIMESCALE]

Last Interglacial   The penultimate INTERGLACIAL, that is, before the present interglacial (the HOLOCENE), also known as the *Ipswichian* (Britain), *Eemian* (northern Europe) and *Sangamon* (North America), and correlated with MARINE ISOTOPIC STAGE (MIS) 5e. The transition from the SAALIAN glaciation to the Last Interglacial is dated to ca 128 ka BP and the end to ca 115 ka BP on the SPECTRAL MAPPING PROJECT TIMESCALE (SPECMAP) (see QUATERNARY TIMESCALE). This chronology is, however, debated and present dates for MIS 5e are 130 ± 2 ka BP to 119 ± 2 ka BP. Both terrestrial and marine PROXY DATA indicate that during the Last Interglacial, mid- and high latitudes were generally at least 2°C warmer than at present, enabling species such as *hippopotamus* and *lion* to colonise Britain, with *global sea level* higher by 4–6 m. Rates of SEA-LEVEL RISE were as high as 1.6 m per 100 years during the interglacial. Evidence from ICE CORES was once thought to indicate that the interglacial climate was more unstable than previously thought, but this proved to be an artefact of the lowest part of the ice core. However, some European SPELEOTHEM evidence is interpreted as representing up to three CLIMATIC OSCILLATIONS within the interglacial.

                                                            MHD/CJC

[See also DANSGAARD-OESCHGER (D-O) EVENTS, RAPID ENVIRONMENTAL CHANGE]

Aalbersberg G and Litt T (1998) Multiproxy climate
    reconstructions for the Eemian and Early Weichselian.
    Journal of Quaternary Science 13: 367–390.
Couchoud I, Genty D, Hoffmann D et al. (2009) Millennial-
    scale climate variability during the Last Interglacial
    recorded in a speleothem from south-west France.
    Quaternary Science Reviews 28: 3263–3274.
GRIP (Greenland Ice-core Project) Members (1993) Climate
    instability during the Last Interglacial period recorded in
    the GRIP ice core. *Nature* 364: 203–207.
McManus JF, Bond GC, Broecker WS et al. (1999) Climate
    and atmospheric history of the past 420,000 years from
    the Vostok ice core, Antarctica. *Nature* 399: 429–436.
Rohling E, Grant K, Hemleben Ch et al. (2008) High rates of
    sea-level rise during the Last Interglacial period. *Nature
    Geoscience* 1: 38–42.

Late Glacial environmental change   The Last Glacial-Interglacial transition, the *Late Weichselian* or *Late Devensian Lateglacial*, the boundary between MARINE ISOTOPIC STAGES (MIS) 1 and 2 (TERMINATION I), often just referred to as the *Late Glacial*, marks the boundary between the PLEISTOCENE and HOLOCENE epochs. Although traditionally dated to about 13–10 ka (radiocarbon years BP), Late Glacial environmental changes span the time interval of 15–11.7 ka BP based on CALENDRICAL DATES. The period following Termination I was characterised by a series of rapid environmental changes between STADIAL and INTERSTADIAL climatic regimes, accompanied by a major reorganisation of both the GENERAL CIRCULATION OF THE ATMOSPHERE and the OCEAN CIRCULATION. Proxy evidence for this is best found in the HIGH-RESOLUTION RECONSTRUCTIONS based on, for example, ICE CORES, MARINE SEDIMENT CORES and LACUSTRINE SEDIMENTS. A relatively robust temporal framework is provided by, for example, RADIOCARBON DATING, VARVE CHRONOLOGIES, DENDROCHRONOLOGY and ICE-CORE DATING.

In northern Europe, the Late Glacial comprises four main *biozones* (see BIOSTRATIGRAPHY) that have subsequently been defined as CHRONOZONES and climatostratigraphic subdivisions (in radiocarbon years): (1) BØLLING INTERSTADIAL (13–12 ka BP), (2) OLDER DRYAS STADIAL (12–11.8 ka BP), (3) ALLERØD INTERSTADIAL (11.8–11 ka BP) and (4) YOUNGER DRYAS STADIAL (11–10 ka BP). The Younger Dryas ended with rapid warming into the Holocene interglacial (10 ka BP to present). The TIME TRANSGRESSIVE nature of environmental changes means that this system is rather unsatisfactory and a more informal tripartite division is often adopted: initial post–LAST GLACIAL MAXIMUM (LGM) warming (prior to 13 ka BP), *Late Glacial Interstadial* (LGI: 13–11 ka BP) and Younger Dryas (11–10 ka BP). With the latest ice-core chronology, based on NGRIP (GICC05), a more detailed interstadial/stadial pattern is defined between 14,692 b2k (the earliest warming) and 11,703 b2k (the end of the Younger Dryas) and opening of the Holocene (zones GS-1e–a and GS-1).

Terrestrial evidence for Late Glacial environmental change includes biological remains (e.g. BEETLE ANALYSIS, POLLEN ANALYSIS and PLANT MACROFOSSIL ANALYSIS), AEOLIAN SEDIMENTS (e.g. COVERSAND and LOESS), LACUSTRINE SEDIMENTS (e.g. DIATOMITE, VARVES and RHYTHMITES), *peat* and various LANDFORMS (especially ICE-DIRECTIONAL INDICATORS, ICE-MARGIN INDICATORS and PERIGLACIAL TRIMLINES) and PALAEOSOLS.

The timing of DEGLACIATION was strongly dependent on latitude and altitude, with the earliest terrestrial evidence for climatic amelioration in the Iberian Peninsula. Deglaciation was generally more widespread and marked by 13 ka BP. Although ICE SHEETS are thought to have been largely absent from Britain, Denmark and the southern Baltic Basin by 13 ka BP, there were periods of DIACHRONOUS ice advances and retreats during the whole period of the Late Glacial, and in some parts of Scotland, ice may have persisted as it did in Scandinavia. For the period preceding 13–13.5 ka BP, PROXY DATA generally suggest POLAR DESERT and steppe TUNDRA environments, with a cold and continental climatic

regime for northern and Central Europe, including widespread PERIGLACIAL activity facilitated by winter temperatures of −20°C to −25°C.

SUBFOSSIL beetle evidence (see BEETLE ANALYSIS), now supported by temperatures inferred from CHIRONOMID ANALYSIS, which indicate very rapid warming at ca 13 ka BP (estimated at 2.6–7.2°C per century) with the thermal maximum of the LGI being reached within the first 500 years. During the early LGI, summer temperatures in much of Europe were at least as warm as today, if not warmer, but with more continental winters (see CONTINENTALITY). This led to the replacement of open-ground communities by scrub vegetation, and ultimately probably open woodland, soil formation, restricted SOLIFLUCTION and a shift towards organic sediment accumulation in lakes. Increased river discharge, resulting from enhanced MELTWATER, and slope stabilisation led to a transition from BRAIDING to MEANDER stream patterns, *channel downcutting* and RIVER TERRACE formation.

From ca 12.5 ka BP there was a stepwise climatic downturn (*revertence episodes*), and an abrupt fall at ca 12 ka BP (equivalent to the Older Dryas) to maximum summer temperature in northwest Europe of less than 11–15°C, leading to increased aeolian sand activity, minerogenic inwash into lakes, vegetation disturbance and glacial readvances. In the Greenland ice-core record, these are represented by GI-1d and 1b; stadials lasting 100–200 years, with onsets at 14,075 and 13,311 b2k, respectively.

The onset of the Younger Dryas saw marked climatic cooling in Europe from ca 11 ka BP, with temperatures 4–8°C lower than at the LGI maximum, mean July temperatures below 10°C and extremely continental winters (i.e. at least ca −20°C). This was accompanied by LANDSCAPE instability and evidence for SEA ICE off the West coast of Ireland. The Younger Dryas was characterised by a pronounced periglacial regime throughout northwest Europe, with extensive discontinuous PERMAFROST, ice-sheet expansion, LOCAL GLACIATION (development of mountain glaciers and ice-caps) and ICE-DAMMED LAKES (e.g. the *Baltic Ice Lake*). The brevity of the Younger Dryas, together with a less severe climate than the LGM, restricted the build-up of ice. Trees and HEATHLAND plant communities of the LGI were replaced by TUNDRA or alpine scrub, while THERMOPHILOUS beetles were replaced by Arctic *stenothermic* species. Renewed aeolian activity in the lowlands (e.g. coversand deposition) and LAKE-LEVEL VARIATIONS suggest the later part of the Younger Dryas (10.5–10 ka BP, 12,896–11,703 b2k) was perhaps markedly colder and more arid.

The timing of the Late Glacial/Younger Dryas-Holocene transition is now taken as the formal boundary between the PLEISTOCENE and Holocene, and fixed on the basis of the ice-core record at 11,700 calendar years b2k. Biological evidence indicates a 6–8°C increase in mean summer temperatures during the first few hundred years of the Holocene (estimated rate of change: 1.7–2.8°C per century). All proxy records suggest that a thermal regime equitable with that of today was reached by 9.5 ka BP throughout northwest Europe. These changes led to a reduction in mineral inwash into lake basins, melting of glaciers and permafrost, restricted periglaciation, increased river discharge and soil formation. Cold-adapted, open-ground and heath communities of the Younger Dryas were replaced by various woodland communities, themselves also changing as species migrated from warmer REFUGIA during the early Holocene.

Late Glacial environmental changes in Europe have potential correlatives with climatic oscillations reconstructed throughout the North Atlantic region and beyond (e.g. Older Dryas, *Killarney, Gerzensee, Amphi-Atlantic* Oscillation). An EVENT STRATIGRAPHY in the isotope record from the GRIP ice core has been suggested as the *global stratotype section* for the *Last Termination* with correlation over long distances being aided by an expanding TEPHROCHRONOLOGY based on up to 21 tephras represented in the North Atlantic region between 18 and 10,000 b2k. This approach has also been used for New Zealand. There is some indication that the climate cooling during the Late Glacial (i.e. Younger Dryas) may have been a global climatic event, although variations between hemispheres seem more likely.

Both ice-core and fossil-beetle evidence suggest that the shift from interstadial to stadial conditions at the onset of the Younger Dryas occurred in under a century, while climate warming at the Younger Dryas/Holocene boundary occurred within a few decades. The rapid and almost synchronous changes in ocean circulation, sea-surface temperatures, snow accumulation rates, atmospheric temperatures over ice sheets and terrestrial climatic regimes during the Late Glacial are best explained by the 'open and shut door' modes of the OCEANIC POLAR FRONT (OPF). Marine sediment evidence indicates during the LGM the southerly limit of the Oceanic Polar Front and winter pack ice was on the same latitude as Lisbon, Portugal (20–13 ka BP). The front migrated rapidly (estimated rate ca 2 km/yr) northwards to a position off Iceland during the Late Glacial Interstadial (13–11 ka BP), resulting in a 7–9°C warming in sea surface temperatures. An equally rapid southerly shift in position to the same latitude as southwest Ireland for the duration of the Younger Dryas (11–10 ka BP) resulted in a 5–8°C decline in SEA-SURFACE TEMPERATURES (SSTs) during the Younger Dryas. By 10 ka BP, the Oceanic Polar Front had retreated northwards (estimated rate ca 5 km/year) to its current position between Greenland and Iceland.

Variation in the strength of THERMOHALINE CIRCULATION has been invoked to explain the shift in the Oceanic Polar Front and resulting changes in Late Glacial climate. The onset of the Younger Dryas apparently coincides with THERMOHALINE CIRCULATION and there is evidence for a series of cooling episodes during the LGI corresponding with ice melting pulses. It is suggested that enormous meltwater and iceberg discharges from collapsing ice sheets may have stalled the thermohaline circulation and triggered the climatic downturn of the Younger Dryas. Whether or not Heinrich Events are ultimately climatically forced or result from internal ICE-SHEET DYNAMICS is uncertain.   *MHD/CJC*

Alloway BV, Lowe DJ, Barrell DJA et al. (2007) Towards a climate event stratigraphy for New Zealand over the past 30,000 years (NZ-INTIMATE Project). *Journal of Quaternary Science* 22: 9–35.

Ammann B (ed.) (2000) Biotic responses to rapid climatic changes around the Younger Dryas. *Palaeogeography, Palaeoclimatology, Palaeoecology* 159: 191–361 [Special Issue].

Andersen E (1997) Younger Dryas research and its implications for understanding of abrupt climatic change. *Progress in Physical Geography* 21: 230–249.

Birks HH and Ammann B (2000) Two terrestrial records of rapid climatic change during the Glacial-Holocene transition (14,000-9,000 calendar years B.P.) from Europe. *Proceedings of the National Academy of Sciences* 97: 1390–1394.

Bjørk S, Kromer B, Johnsen S et al. (1996) Synchronised terrestrial-atmospheric deglacial records around the North Atlantic. *Science* 274: 1155–1160.

Lowe JJ, Ammann B, Birks HH et al. (1994) Climatic changes in areas adjacent to the North Atlantic during the Last Glacial-Interglacial transition (14-9 ka BP). *Journal of Quaternary Science* 9: 185–198.

Lowe JJ, Rasmussen SO, Bjorck S et al. (2008) Synchronisation of palaeoenvironmental events in the North Atlantic region during the Last Termination: A revised protocol recommended by the INTIMATE group. *Quaternary Science Reviews* 27: 6–17.

Markgraf V, Baumgartner TR, Bradley JP et al. (2000) Paleoclimate reconstruction along the Pole-Equator-Pole transect of the Americas (PEP 1). *Quaternary Science Reviews* 19: 125–140.

Marshall JD, Jones RT, Crowley SF et al. (2002) A high resolution Late-Glacial isotopic record from Hawes Water, Northwest England: Climatic oscillations, calibration and comparison of palaeotemperature proxies. *Palaeogeography, Palaeoclimatology, Palaeoecology* 185: 25–40.

late-Holocene climatic deterioration The concept of a climatic cooling trend and/or increasing wetness following the HOLOCENE THERMAL OPTIMUM or *Hypsithermal* of the early to mid-HOLOCENE. Terms for the time interval encompassed include *hypothermal, katathermal, medithermal* and *neoglacial*. It may be accounted for by reduced insolation receipt on Earth as predicted by MILANKOVITCH THEORY.   *JAM*

[*See also* HOLOCENE ENVIRONMENTAL CHANGE, NEOGLACIATION]

latent heat   The energy absorbed or released when a substance changes its phase: *latent heat of vaporisation* is involved during changes between gaseous and liquid state; *latent heat of fusion* is involved during liquid to solid transformations. Latent heat is released during the change from gas to liquid and from liquid to solid states: latent heat is absorbed in overcoming intermolecular bonds during the change from solid to liquid or liquid to gas. During SUBLIMATION, the change from a solid to a gas, the latent heat released is the sum of the latent heat of fusion and the latent heat of vaporisation. It is called 'latent heat' because there is no change in the temperature of the substances involved. In the environmental context, latent heat is particularly important in understanding EVAPORATION, CONDENSATION, *freezing* and *thawing processes*, LAPSE RATES, ENERGY BUDGETS and heat transfer in the atmosphere and at the Earth's surface.   *AHP/JAM*

Lock GSH (1994) *Latent heat transfer: An introduction to fundamentals.* New York: Oxford University Press.

lateral accretion   The accumulation of sediment on an inclined surface, resulting in the lateral migration of that surface, in contrast to the vertical AGGRADATION of a near-horizontal surface. The term is commonly applied to the inner bank (*point bar*) of a MEANDER in a river channel, which migrates laterally across the FLOODPLAIN. In the geological record, EPSILON CROSS-BEDDING has been interpreted as evidence of lateral accretion.   *GO*

Davies NS and Gibling MR (2010) Paleozoic vegetation and the Siluro-Devonian rise of fluvial lateral accretion sets. *Geology* 38: 51–54.

laterite   A formerly widely used term for a highly weathered red and often mottled subsoil, rich in *secondary oxides* of IRON (Fe), ALUMINIUM (Al), or both, nearly devoid of BASES and *primary silicates*, and commonly contained quartz and KAOLINITE. The term was first used by F. Buchanan at the beginning of the nineteenth century for the material cut into bricks, allowed to dry and harden, and used for building in India.

The term has now largely been replaced by the term *plinthite* or FERRICRETE, though the latter is only the indurated form of laterite.

Laterite develops in *seasonal tropical* or *subtropical* climates, and is a RESIDUAL product of large-scale, deep CHEMICAL WEATHERING where SILICA and bases are leached from the PARENT MATERIAL, creating a concentration of iron and aluminium *sesquioxides*. It is associated mainly with mature surfaces of low relief, and occurs in three forms: (1) a soft lateritic clay at depth, which hardens on exposure to air; (2) a tough indurated layer forming a HARDPAN or DURICRUST at the surface, where the superficial material has been stripped off; or (3) a horizon of *nodules* or *lenses*. *Laterisation* is a general term for the process that converts a rock to laterite. SECONDARY LATERITES tend to form in valley-bottom areas through the *reprecipitation* of iron that has been transported downslope by THROUGHFLOW. Laterite crusts may play a major role in long-term LANDSCAPE EVOLUTION.   *MAC*

[*See also* CEMENTATION, INDURATION, PLINTHIC HORIZON, PLINTHOSOLS, SAPROLITE, WEATHERING PROFILE]

Bourman RP (1993) Perennial problems in the study of laterite: A review. *Australian Journal of Earth Science* 40: 387–401.

Bourman RP and Ollier CD (2002) A critique of the Scellmann definition and classification of "laterite". *Catena* 47: 117–131.

McFarlane MJ (1976) *Laterite and landscape.* London: Academic Press.

McFarlane MJ (1991) Some sedimentary aspects of lateritic weathering profile development in the major bioclimatic zones of tropical Africa. *Journal of African Earth Sciences* 12: 267–282.

Ollier CD and Sheth HC (2008) The High Deccan duricrusts of India and their significance for the 'laterite' issue. *Journal of Earth System Science* 117: 537–551.

Tardy Y (1992) Diversity and terminology of laterite profiles. In Martini IP and Chesworth W (eds) *Weathering, soils and paleosols.* Amsterdam: Elsevier, 379–405.

Tardy Y (1997) *Petrology of laterites and tropical soils.* Rotterdam: Balkema.

Widdowson M (2007) Laterite and ferricrete. In Nash DJ and McLaren SJ (eds) *Geochemical sediments and landscapes.* Oxford: Blackwell, 46–94.

Widdowson M (2009) Laterite. In Gornitz V (ed.) *Encyclopedia of paleoclimatology and ancient environments.* Dordrecht: Springer, 514–517.

Laurentian/Laurentide Ice Sheet   The ICE SHEET that formed over the Canadian SHIELD on many occasions during the QUATERNARY. It is named after the St. Lawrence River. The *Laurentide Ice Sheet* had two major *accumulation* areas, located over Keewatin and Labrador, from which it advanced. Lobes of the last (*Wisconsinan*) Laurentide Ice Sheet extended over the Great Lakes, and into the Northern American Plains.   *DIB*

[*See also* GLACIATION, LAST GLACIAL MAXIMUM (LGM), LAST GLACIATION]

Andrews JT (2006) The Laurentide Ice Sheet: A review of history and processes. In Knight PG (ed.) *Glacier science and environmental change.* Oxford: Blackwell, 201–207.

Mickelson DM and Colgan PM (2004) The southern Laurentide Ice Sheet. In Gillespie AR, Porter SC and Atwater BF (eds) *The Quaternary period in the United States.* Amsterdam: Elsevier, 1–16.

Stokes CR and Tarasov L (2010) Ice streaming in the Laurentide Ice Sheet: A first comparison between data-calibrated numerical model output and geological evidence. *Geophysical Research Letters* 37: L01501.

lava   The liquid component of MAGMA emitted at the surface of the Earth (or other PLANET) in a VOLCANIC ERUPTION. Lava properties and behaviour depend on the magma composition: *low-silica magma* has a low viscosity, forming fast, mobile LAVA FLOWS of BASALT, which cools as a fine-grained, dark IGNEOUS ROCK. Eruptions of *high-silica magma*, which is highly viscous and inhibits the release of VOLATILES, commonly represent a significant NATURAL HAZARD. Flows are very slow and *lava domes* or *plugs* develop. These can collapse because of volatile pressure or SEISMICITY, leading to *Plinian* eruptions or PYROCLASTIC FLOWS.   *JBH*

[*See also* IGNEOUS ROCKS]

Chester D (1993) *Volcanoes and society.* London: Arnold.
Druitt TH and Kokelaar BP (eds) (2002) *The eruption of Soufriere Hills volcano, Montserrat, from 1995 to 1999.* Bath: Geological Society [Memoir 21].
Francis P and Oppenheimer C (2003) *Volcanoes.* Oxford: Oxford University Press.
Sparks RSJ, Young SR, Barclay J et al. (1998) Magma production and growth of the lava dome of the Soufriere Hills volcano, Montserrat: November 1995 to December 1997. *Geophysical Research Letters* 25: 3421–3424.
Wright R, Glaze L and Baloga SM (2011) Constraints on determining the eruption style and composition of terrestrial lavas from space. *Geology* 39: 1127–1130.

lava dome   A build-up of very viscous, high-silica LAVA around the vent of a VOLCANO during a VOLCANIC ERUPTION. The collapse of a lava dome can cause a PYROCLASTIC FLOW.   *GO*

Druitt TH and Kokelaar BP (eds) (2002) *The eruption of Soufrière Hills volcano, Montserrat, from 1995 to 1999.* Bath: Geological Society [Memoir 21].
Sparks RSJ, Young SR, Barclay J et al. (1998) Magma production and growth of the lava dome of the Soufrière Hills volcano, Montserrat: November 1995 to December 1997. *Geophysical Research Letters* 25: 3421–3424.

lava flow   A flow of liquid LAVA from a VOLCANIC ERUPTION at the surface of the Earth or another PLANET. Lava flows vary from metre-scale effusions to volumes measured in tens of cubic kilometres (see FLOOD BASALT). Flows of basalt are rarely hazardous to life, although a basaltic flow from Nyiragongo (Democratic Republic of Congo) in AD 1977 travelling at 11 m/s overran 400 houses in the city of Goma, killing 72 people, and at least 45 people died in a similar event in AD 2002. *Flow diversion* is often attempted to prevent structural damage (see MITIGATION). Calculations suggest that a 3-m wall thickness is needed to deflect each 1 m depth of lava, as attempted successfully in the AD 1983 eruption of Etna (Sicily). Blocking of the harbour on Heimay (Iceland) in AD 1973 was prevented by *spraying* 1,200 L/s of *seawater* onto the 1055°C *basaltic flow*. *Bombing* of lava flows has also been attempted. Surfaces of basaltic lava flows are commonly described by the Hawaiian terms AA (blocky or rubbly) or PAHOEHOE (smooth or rope-like).   *JBH*

[*See also* IGNEOUS ROCKS, PLANETARY ENVIRONMENTAL CHANGE]

Chirico GD, Favalli M, Papale P et al. (2009) Lava flow hazard at Nyiragongo Volcano, D.R.C. 2. Hazard reduction in urban areas. *Bulletin of Volcanology* 71: 375–387.
Favalli M, Chirico GD, Papale P et al. (2009) Lava flow hazard at Nyiragongo Volcano, D.R.C. 1. Model calibration and hazard mapping. *Bulletin of Volcanology* 71: 363–374.
Favalli M, Tarquini S, Fornaciai A and Boschi E (2009) A new approach to risk assessment of lava flow at Mount Etna. *Geology* 37: 1111–1114.
Leverington DW (2011) A volcanic origin for the outflow channels of Mars: Key evidence and major implications. *Geomorphology* 132: 51–75.
Lockwood JP and Torgerson FA (1980) Diversion of lava flows by aerial bombardment: Lessons from Mauna Loa, Hawaii. *Bulletin Volcanologique* 43: 727–741.
Williams RS and Moore JG (1973) Iceland chills a lava flow. *Geotimes* 18: 14–17.

lava plateau   An elevated tableland mainly constructed by FLOOD BASALTS, with intercalated TUFFS and PALAEOSOLS. The CRETACEOUS *Deccan Traps* in northwest India cover an area greater than 500,000 km^2 and have a volume over 1 million km^3. Lava plateaux are constructed over a geologically short time interval, and may affect ATMOSPHERIC COMPOSITION and CLIMATE through the release of VOLATILES. This has been suggested as a mechanism for MASS EXTINCTIONS such as at the K-T BOUNDARY.   *JBH*

[*See also* LARGE IGNEOUS PROVINCE (LIP)]

Caldwell WGE and Young GM (2011) The Early Carboniferous volcanic outliers of Little Cumbrae and south Bute: Implications for westward attenuation of the Clyde Lava Plateau. *Transactions of the Royal Society of Edinburgh: Earth and Environmental Sciences* 102: 59–76.
Courtillot V, Jaupart C, Manighetti I et al. (1999) On causal links between flood basalts and continental breakup. *Earth and Planetary Science Letters* 166: 177–195.
Khadkikar AS, Sant DA, Gogte V and Karanth RV (1999) The influence of Deccan volcanism on climate: Insights from lacustrine intertrappean deposits, Anjar, western India. *Palaeogeography, Palaeoclimatology, Palaeoecology* 147: 141–149.
Rao NVC and Lehmann B (2011) Kimberlites, flood basalts and mantle plumes: New insights from the Deccan Large Igneous Province. *Earth-Science Reviews* 107: 315–324.

lava tube   An elongated CAVE in a solidified LAVA FLOW that formed when the surface of the flow solidified but the LAVA beneath was still molten and drained away.   *GO*

Francis P and Oppenheimer C (2003) *Volcanoes.* Oxford: Oxford University Press.
Sarkar PK, Friedman GM and Karmalkar N (1998) Speleothem deposits developed in caves and tunnels of Deccan-Trap basalts, Maharashtra, India. *Carbonates and Evaporites* 13: 132–135.

law   Universally accepted scientific statements that are believed to be true. If most scientists are convinced that a HYPOTHESIS or THEORY will pass every conceivable test, whether or not such tests are applied, and is unlikely ever to be falsified under real or known conditions, then it can be considered a *scientific law* (or *natural law*). At their indefinite lower limit, such laws degenerate into HYPOTHESES, and at their upper indefinite limit they grow into THEORY (a web of laws) for many, but not all, scientists.   *MAU/CET*

[*See also* EARTH LAWS, ENVIRONMENTAL LAW, FALSIFICATION, KNOWLEDGE, SCIENTIFIC METHOD, TRUTH]

Lee JA (2000) *The scientific endeavor: A primer on scientific principles and practice.* San Francisco: Addison Wesley.
Murray Jr BG (2000) Universal laws and predictive theory in ecology and evolution. *Oikos* 89: 403–408.

leachate   A *solution* containing ELEMENTS and/or COMPOUNDS in solution (or in *suspension*) taken up on its passage through a substance. For example, rainwater leaches soluble substances from soils and, in the laboratory, various solutions are used to extract substances from soil or sediment samples.   *EMB*

[*See also* SOLUTE]

leaching   The process by which soluble materials are *dissolved*, carried in aqueous *solution* down the SOIL PROFILE and ultimately lost from the soil. After soluble SALTS, GYPSUM and CARBONATES have been removed, leaching continues as base *cations* are removed from the CLAY-HUMUS COMPLEX and replaced by *hydrogen ions.* IRRIGATION is sometimes employed to prevent the build-up or leach-out undesirable accumulations of soluble salts at or near the soil surface. Leaching can become an ENVIRONMENTAL PROBLEM if harmful substances are leached into GROUNDWATER. The process is also employed widely in INDUSTRY for removing soluble CONTAMINANTS from substances using a wide variety of *solvents.*   *EMB/JAM*

[*See also* ACIDIFICATION, CATION EXCHANGE CAPACITY (CEC), ELUVIATION, ILLUVIATION, SOIL-FORMING PROCESSES]

Buol SW, Hole FD and McCracken RJ (1988) *Soil genesis and classification.* Ames: Iowa State University Press.

lead (Pb)   A TRACE ELEMENT, HEAVY METAL and POLLUTANT, lead has been locally important in the environment and as a factor in human health since the onset of lead mining in the ROMAN PERIOD. There has been major release of lead into the environment and extensive pollution even of remote regions since the INDUSTRIAL REVOLUTION. Soils have been polluted by lead from industrial activity and sewage SLUDGE, and there is long-term release from lead mine TAILINGS into GROUNDWATER. A further escalation of AIR POLLUTION by lead followed the widespread use of lead *additives* in *petrol*, but atmospheric concentrations fell slightly by the AD 1990s with the introduction of *unleaded petrol*. HUMAN HEALTH HAZARDS have been generated directly and indirectly by release of lead into the environment through industrial activity.

The normal range of lead in soil is 30–100 mg/kg. Uncontaminated soils in remote areas may contain less than 30 mg/kg. In areas rich in metalliferous deposits, soils may contain up to 300 mg/kg. Lead links firmly with organic substances and mineral silicates in soils and so has a long RESIDENCE TIME compared with other metal POLLUTANTS. It accumulates mainly in the organic-rich surface horizons and there is little evidence of LEACHING by natural means. Alongside major roads, soils are contaminated with lead emitted from vehicle engines, and soils in the gardens of old houses usually contain lead from *paints*. Despite the value of plant NUTRIENTS in sewage sludge, the presence of lead and other heavy metals limits its disposal to farmland and forests. Many uses of lead declined in the twentieth century following recognition of its effects on human health and the environment, leaving today's main uses in the manufacture of automobile batteries and glass screens for computers and televisions (the lead creating a radiation shield in the latter).   *EMB*

[*See also* CONTAMINATED LAND, LEAD ISOTOPES, LEAD-210 DATING, METAL POLLUTION, POLLUTION HISTORY]

Boutron CF, Gorlach U, Candelone JP et al. (1991) Decrease in anthropogenic lead, cadmium and zinc in Greenland snows since the late 1960s. *Nature* 353: 153–156.
Cook CA, Abbott MB and Wolfe AP (2008) Late-Holocene atmospheric lead deposition in the Peruvian and Bolivian Andes. *The Holocene* 18: 353–359.
Flora SJS, Flora G and Saxene G (2006) Environmental occurrence, health effects and management of lead poisoning. In Casas JS and Sordo J (eds) *Lead: Chemistry, analytical aspects, environmental impact and health effects.* Amsterdam: Elsevier, 158–228.
Hutchinson TC and Meerna KM (eds) (1987) *Lead, mercury, cadmium and arsenic in the environment.* Oxford: Blackwell.

Kramers JD, Reese S and Van Der Knaap WO (1998)
History of atmospheric lead deposition since 12,370 ^{14}C yr
BP from a peat bog, Jura Mountains, Switzerland. *Science*
281: 1635–1640.
Nriagu JO (ed.) (1978) *The biogeochemistry of lead in the
environment.* Amsterdam: Elsevier.

lead and lag   Leads and lags are central to understanding *synchroneity* and *phase relationships* between EARTH SYSTEM components during environmental change and hence the causal mechanisms of change. If it can be demonstrated that a particular phenomenon consistently leads another, then it is a potential PROXIMATE CAUSE or may be causally related. Similarly, if a phenomenon lags behind another, it can be ruled out as a potential causal factor. For example, based on cross-dated oxygen isotope records from the Last Glacial episode in Greenland and Antarctic ice cores, it has been suggested that rapid warming events in the Southern Hemisphere lead those in the Northern Hemisphere by a little more than a millennium. This, however, has been disputed, which emphasises the need for accurate CHRONOLOGY and HIGH-RESOLUTION RECONSTRUCTIONS.

JAM

Blunier T, Chapellaz J, Schwander J et al. (1998)
Asynchrony of Antarctica and Greenland climate change
during the last glacial period. *Nature* 394: 739–743.
Brauer A, Allen JRM, Mingram J et al. (2007) Evidence for
last interglacial chronology and environmental change
from southern Europe. *Proceedings of the National
Academy of Science USA* 104: 450–455.
White JWC and Steig EJ (1998) Timing is everything in a
game of two hemispheres. *Nature* 394: 717.

lead isotopes   Three RADIONUCLIDES of lead, [206]Pb, [207]Pb and [208]Pb, occur naturally in the environment as DAUGHTER ISOTOPES of [238]U, [235]U and [232]Th, respectively. The partitioning of U/Th and Pb between different MINERALS and variations in the age of their formation give rise to systematic differences in the ratio of radiogenic isotopes and non-radiogenic [204]Pb between ORE bodies and rock units throughout the GEOSPHERE. [210]Pb is a member of the naturally occurring [238]U decay series with a half-life of 22.22 years. [210]Pb can be used as a DATING TECHNIQUE in the context of NATURAL ARCHIVES, such as LACUSTRINE SEDIMENTS and MIRES, for a period covering approximately the past 150 years. Ambiguous dates, often resulting from the different models used to explain the natural sources of ^{210}Pb, can be resolved by determining *artificial isotopes* (e.g. CAESIUM-137) in the record. Lead ISOTOPE RATIO measurements have been used to record the sources of lead fluxes into the ATMOSPHERE, the OCEANS, SOILS and organisms. Particular attention has been given to the identification of ANTHROPOGENIC (ore body) lead in PROXY EVIDENCE for past atmospheric composition from the analysis of polar ICE CORES, lacustrine sediments and mires. Lead isotope measurements of contemporary and archaeologically preserved tissue have been used to determine migration patterns and to assess changing sources of human exposure.

PB/IR

[*See also* ISOTOPES AS INDICATORS OF ENVIRONMENTAL CHANGE, LEAD-210 DATING]

Appleby PG (2008) Three decades of dating recent
sediments by fallout of radionuclides: A review.
The Holocene 18: 83–93.
Faure G and Mensing TM (2005) *Isotopes: Principles and
applications, 3rd edition.* Hoboken, NJ: Wiley.
Müller W, Fricke H, Halliday AN et al. (2003) Origin and
migration of the Alpine Iceman. *Science* 302: 862–866.
Rosman KJR, Chrisholm W, Boutron CF et al. (1993)
Isotopic evidence for the source of lead in Greenland
snows since the 1960s. *Nature* 362: 333–335.

lead-210 dating   A RADIOMETRIC DATING technique used to establish chronologies for sediments aged less than about 150 years and therefore too young for RADIOCARBON DATING. Lead-210 (^{210}Pb) is present in the ATMOSPHERE as a DAUGHTER ISOTOPE of sedimentary *radon*. This ^{210}Pb is deposited to accumulating sediments in excess of any in-situ ^{210}Pb, and decays according to radiometric law. Reliable estimates of initial sedimentary ^{210}Pb activity plus measurements of current isotope activity are used to date the sediments.

There are two approaches to ^{210}Pb dating. The *constant initial concentration (CIC) model* is suited to homogeneous sediments with little variation in sediment ACCUMULATION RATE and provides linear ^{210}Pb profiles which are fairly simple to interpret. Linear ^{210}Pb profiles are, however, uncommon due to changes in sediment accumulation rate, BULK DENSITY or other factors. The *constant rate of supply (CRS) model* is therefore normally preferred, in which initial ^{210}Pb activity does not necessarily decline monotonically with depth.

The interpretation of ^{210}Pb profiles may be complicated by their distortion or remobilisation through *hydrological change* within both peat and LACUSTRINE SEDIMENTS. The ^{210}Pb record in *peat* is particularly susceptible to degradation and the accuracy of *peatland* [210]Pb inventories is often significantly lower than those of lakes. The CRS ^{210}Pb dating model can often account for such discrepancies, although the ^{210}Pb profile in lake sediments may be further affected by SEDIMENT FOCUSING or LAKE-LEVEL VARIATIONS. Ambiguities in ^{210}Pb dating can nonetheless often be resolved by independent radiometric dating techniques (e.g. CAESIUM-137 or AMERICIUM-241 DATING).

DZR

Appleby PG (2001) Chronostratigraphic techniques in recent
sediments. In Last WM and Smol JP (eds) *Tracking
environmental change using lake sediments, volume 1.*
Dordrecht: Kluwer, 171–203.

Appleby PG (2008) Three decades of dating recent sediments by fallout radionuclides: A review. *The Holocene* 18: 83–94.

Appleby PG and Oldfield F (1992) Application of lead-210 to sedimentation studies. In Ivanovich M and Harmon RS (eds) *Uranium series disequilibrium*. Oxford: Oxford University Press, 731–778.

Oldfield F, Richardson N and Appleby PG (1995) Radiometric dating (^{210}Pb, ^{137}Cs, ^{241}Am) of recent ombrotrophic peat accumulation and evidence for changes in mass balance. *The Holocene* 5: 141–148.

Varvas M and Punning JM (1993) Use of the ^{210}Pb method in studies of the development and human impact history of some Estonian lakes. *The Holocene* 3: 34–44.

leaf area index (LAI)   One-half of the total green leaf area per unit ground surface area or, for coniferous species, the total needle surface area per unit ground area. LAI ranges from 0 (bare ground) to over 10 (dense conifer forests). It can be measured in the field or by REMOTE SENSING and is a principal determinant of SOLAR RADIATION interaction with the BIOSPHERE, from which estimates of carbon and water exchange with the ATMOSPHERE via PHOTOSYNTHESIS and RESPIRATION can be produced.   *DSB*

[*See also* VEGETATION INDICES]

Jockheere I, Fleck S, Nackaerts K et al. (2004) Review of methods for insitu leaf area index determination. Part 1: Theory, sensors and hemispherical photographs. *Agricultural and Forest Meteorology* 121: 19–35.

leaf physiognomy   Leaf or *foliar physiognomy* describes the character of leaves from their general shape and form. Various classification schemes have been proposed, most notably by Raunkiaer (1934), later modified by Webb (1959). Leaf physiognomy is strongly controlled by environment and plays an important role in whole-plant ADAPTATION and survival, as the size, thickness, toughness and shape of leaves can affect the rate at which plants take up carbon, exchange heat and lose water. Broad patterns of leaf physiognomy have been observed in vegetation demonstrating that generally large leaves occur in wet, hot environments and small leaves in hot or cold dry environments. Hence, analysis of FOSSIL leaf physiognomy has emerged as an important technique in PALAEO-BOTANY and in PALAEOENVIRONMENTAL RECONSTRUCTION, based on strong correlations which have been defined between extant ANGIOSPERM leaf characteristics and present-day climatic variables.

As early as 1916, Bailey and Sinott demonstrated that as one moves from warmer to colder climate regimes, there is a marked decrease in the proportion of tree and shrub species with entire-margined leaves. This characteristic, along with *leaf size*, CUTICLE thickness and the presence or absence of a *drip tip*, were the

first leaf physiognomy characteristics used to reconstruct *palaeoclimates* of the CRETACEOUS and TERTIARY from angiosperm *fossil floras*. Since the 1940s, the method has been much refined and MULTIVARIATE ANALYSIS has been applied to leaf physiognomy and modern climatic data sets to define models based on REGRESSION ANALYSIS with which to reconstruct climatic variables such as MEAN ANNUAL AIR TEMPERATURE (MAAT), mean *temperature range* and *mean annual precipitation* of Tertiary and Cretaceous times.   *JCMCE*

[*See also* EVAPOTRANSPIRATION, PHOTOSYNTHESIS, RESPIRATION]

Bailey IW and Sinnott EW (1916) The climate distribution of certain types of angiosperm leaves. *American Journal of Botany* 3: 24–39.

Chaloner WG and Creber GT (1990) Do fossil plants give a climate signal? *Journal of the Geological Society* 147: 343–350.

Givinish TJ (1979) On the adaptive significance of leaf form. In Solbrig OT, Jain S and Raven PH (eds) *Topics in plant population biology*. New York: Columbia University Press, 375–407.

Jacobs BF (1999) Estimation of rainfall variables from leaf characters in tropical Africa. *Palaeogeography, Palaeoclimatology, Palaeoecology* 145: 231–251.

Raunkiaer C (1934) *The life-forms of plants and statistical plant geography*. Oxford: Oxford University Press.

Webb LJ (1959) A physiognomic classification of Australian rain forests. *Journal of Ecology* 47: 551–570.

Wolfe JA (1993) A method of obtaining climatic parameters from leaf assemblages. *United States Geological Survey Bulletin* 2040: 71.

least squares regression   A method of fitting a statistical MODEL, in the simplest case a line, to data. Parameters of the model equation are estimated so that the sum of squared differences between the model and the observations of the *dependent variable* is minimised. It may be used as the basis for the prediction of the dependent variable, or merely to describe the relationship between dependent and *independent variables*. Multiple *least-squares regression* involves more than one independent variable. *Stepwise multiple regression* adds the independent variables one at a time (maximising the explanatory power of the *regression equation* at each step).   *HB*

[*See also* NUMERICAL ANALYSIS, REGRESSION ANALYSIS]

Leptosols   Shallow soils of all regions over hard rock or in unconsolidated, very gravelly material mainly occurring in mountainous areas and deserts (SOIL TAXONOMY: Lithic subgroups of *Entisols, Rendolls*). In other systems of SOIL CLASSIFICATION, they are commonly termed *lithosols*, and they include RANKERS and RENDZINAS. Leptosols are the most extensive of the major soils of the world, diverse in character and subject to erosion, particularly in mountainous areas. Historically, limitations to use, including shallowness, stoniness and DROUGHT, have encouraged extensive TERRACING.   *EMB*

[*See also* AZONAL SOIL, WORLD REFERENCE BASE FOR SOIL RESOURCES (WRB)]

Bridges EM and Creutzberg D (1994) Leptosols and Fluvisols. *Transactions of the 15th International Congress of Soil Science* 6A: 868–872.
Pereverzev VN (2010) Genetic features of soils in altitudinal zones of the Khibiny Mountains. *Eurasian Soil Science* 43: 509–518.

lessivage   A French term used to describe the SOIL-FORMING PROCESS whereby particulate CLAY is mechanically transported by percolating water from the upper part of the SOIL PROFILE to its lower part. Evidence of this process are the *clay skins* or *argillans* in *argic horizons*. The importance of the process in the formation of LUVISOLS and ALBELUVISOLS is disputed. Horizons of clay accumulation may, in extreme cases, restrict SOIL DRAINAGE sufficiently to cause GLEYING.   *SN/JAM*

[*See also* ELUVIATION, LEACHING, PERCOLATION, PERVECTION]

Quenard L, Samouelian A, Laroche B and Corn S (2011) Lessivage as a major process of soil formation: A revisitation of existing data. *Geoderma* 167–168: 135–147.
Zaidel'man FR (2007) Lessivage and its relation to the hydrological regime of soils. *Eurasian Soil Science* 40: 115–125.

levée   A ridge of fine-to-coarse SEDIMENT deposited on a FLOODPLAIN, on an ocean ABYSSAL plain, or alongside a DEBRIS-FLOW track. It may be produced naturally as an OVERBANK DEPOSIT or artificially as a FLOOD CONTROL MEASURE.   *PW*

Cazanach D and Smith ND (1998) A study of the morphology and texture of natural levees: Cumberland Marshes, Saskatchewan, Canada. *Geomorphology* 25: 43–55.
Kane IA, McCaffrey WD and Peakall J (2010) On the origin of paleocurrent complexity within deep marine channel levees. *Journal of Sedimentary Research* 80: 54–66.

Libby half-life   The HALF-LIFE of *radiocarbon* (^{14}C) determined by W.F. Libby to be 5,568 years and used in the calculation of CONVENTIONAL RADIOCARBON DATES, even though a revised figure of 5,730 years has since been determined.   *PQD*

[*See also* RADIOCARBON DATING]

Stuiver M and Polach HA (1977) Discussion: Reporting of ^{14}C data. *Radiocarbon* 19: 355–363.

lichenometric dating   The use of lichen size or related indices of lichen growth for dating rock surfaces. The technique, also known as *lichenometry*, was first developed and applied by Roland Beschel in the 1950s to dating MORAINES in the Austrian Alps. It is still most widely used in alpine and polar environments where crustose lichens, especially those of the relatively slow-growing, yellow-green *Rhizocarpon* subgenus (widely known as the *Rhizocarpon geographicum* group), commonly dominate on rock outcrops and boulders. In principle, any abundant species that colonises rock surfaces rapidly after exposure, and grows steadily in an approximately circular fashion, is potentially useful for lichenometric dating. Applications have included dating a wide range of glacial, periglacial, lacustrine and coastal landforms and the technique has also been used on archaeological structures.

The technique has traditionally been based on the maximum diameter of the largest thallus growing on each surface on the grounds that the largest is the oldest specimen growing under optimal environmental conditions. In order to avoid the pitfall of anomalous single thalli, the average size of several 'largest lichens' (commonly five) is often employed instead of the single largest. Other, more time-consuming approaches have, however, been developed, such as use of lichen size-frequency distributions. As older surfaces tend to be characterised by larger lichens, lichen size can be used as a method of RELATIVE-AGE DATING. Given an adequate number of surfaces of known age to establish a *lichenometric-dating curve* (a numerical relationship between lichen size and surface age), it becomes a CALIBRATED-AGE DATING technique. This constitutes so-called *indirect lichenometry*, which has been most effective in dating surfaces up to about 500 years old to an accuracy of up to about 10 per cent where accurate control points are available and lichen growth rates approach 1.0 mm/year. Justification of indirect lichenometry by direct measurement of lichen growth rates in real time (and the construction of lichenometric dating curves by *direct lichenometry*) has met with only limited success, largely due to high variability and slow rates of growth of the relevant taxa.   *JAM*

Armstrong TE and Bradwell T (2010) Growth of crustose lichens: A review. *Geografiska Annaler* 92A: 3–17.
Benedict JB (2009) A review of lichenometric dating and its application to archaeology. *American Antiquity* 74: 143–172.
Beschel RE (1961) Dating rock surfaces by lichen growth and its application to glaciology and physiography (lichenometry). In Raasch GO (ed.) *Geology of the Arctic, volume 1*. Toronto: University of Toronto Press, 1044–1062.

Bradwell T and Armstrong RA (2007) Growth rates of
Rhizocarpon geographicum lichens: A review with new
data from Iceland. *Journal of Quaternary Science* 22:
311–320.

Innes JI (1985) Lichenometry. *Progress in Physical
Geography* 9: 187–254.

Jomelli V, Grancher D, Naveau P and Colley D (2007)
Assessment study of lichenometric methods for dating
surfaces. *Geomorphology* 93: 2001–2012.

Matthews JA (2005) Little Ice Age glacier variations in
Jotunheimen, southern Norway: A study in regional
lichenometric dating with implications for climate and
lichen growth rates. *The Holocene* 15: 1–19.

Matthews JA and Trenbirth HE (2011) Growth rate of a
very large crustose lichen (*Rhizocaron* subgenus) and its
implications for lichenometry. *Geografiska Annaler* 93A:
27–39.

Trenbirth HE and Matthews JA (2010) Lichen growth rates
on glacier forelands in southern Norway: Preliminary
results from a 25-year monitoring programme.
Geografiska Annaler 92A: 19–39.

LiDAR An active REMOTE-SENSING system (*Light
Detection and Ranging*) based on LASERS, which send
out a series of very short pulses of a very narrow beam
of coherent light in a precise waveband. The time difference between pulse emission and the BACKSCATTER-
ING of that pulse and the strength of that backscatter
recorded by the sensor can be used to build a three-
dimensional (3-D) structure of the terrain. The LiDAR
can operate in profiling, scanning and full waveform
modes and employs *satellite*, *airborne* and *terrestrial
platforms*. Depending on their sophistication, LiDAR
instruments may record just the first pulse (return),
first and last pulse or the detailed series of pulses (full
waveform) between the first and last pulses.

The principal applications of LiDAR remote-sensing
systems are for the profiling of water depths, the
heights of tree stands and 3-D structure of urban environments. The materials in urban environment provide a strong first return from which a *digital surface
model* (*DSM*) can be computed. In the case of water,
the strong first return is from the water surface, which
is then followed by a weaker return from the bottom
of the waterbody. This allows the water depth to be
determined from the two-way travel time that the pulse
is in the water. Determining the heights of tree stands
is afforded because the laser pulse is predominantly
scattered at the top of the VEGETATION CANOPY, but some
may also reach the ground surface and, over *clearings*,
the pulse is scattered from the ground surface only.
Full waveform instruments allow the accuracy description of the density of the canopy and any *understory*
vegetation. Estimates of forest BIOMASS are possible
given *tree height*, *crown density* and species data.
Further applications include their use in atmospheric

studies and for the measurement of the LASER-INDUCED
FLUORESCENCE (LIF) properties of the Earth's surface.

GMS/DSB

Fujii T and Fukuchi T (2005) *Laser remote sensing*.
Boca Raton, FL: CRC Press.

Hofle H and Rutzinger M (2011) Topographic airborne
LiDAR in geomorphology: A technological perspective.
Zeitschrift für geomorphologie Supplementband 55: 1–29.

Hyyppä J, Wagner W, Hollaus M and Hyyppä H (2009)
Airborne laser scanning. In Warner TA, Nellis MD and
Foody GM (eds) *The SAGE handbook of remote sensing*.
London: Sage, 199–211.

life assemblage FOSSILS or SUBFOSSILS preserved
essentially where they lived, that is, AUTOCHTHONOUS.
Some may retain life orientations, and be in original clustered groups, with hinged or plated skeletons
intact (see DEATH ASSEMBLAGE). The fossil assemblage
is modified from the original *community* through the
effects of TAPHONOMY including the loss of *soft-bodied
organisms*, but the *shelled assemblage* remains similar
(i.e. exhibits high *fidelity*). Life assemblages are represented in many *conservation Lagerstätten* (see FOSSIL
LAGERSTÄTTEN). They are also commonly preserved by
rapid burial in EVENT DEPOSITS generated by STORMS or
SLUMPS, which prevents later DISTURBANCE of shells by
burrowing (BIOTURBATION) and *disarticulation*. STORM
shell beds occur throughout the PHANEROZOIC, with a
notable increase in thickness from the Mid-CENOZOIC
(CAINOZOIC). *Cemented organisms* (see CEMENTA-
TION) such as *oysters* or modern reef CORALS may be
preserved in situ. Life assemblages are important for
PALAEOECOLOGY and PALAEOBIOLOGY. *LC*

Cherns L, Wheeley JR and Wright VP (2008) Taphonomic
windows and molluscan preservation. *Palaeogeography,
Palaeoclimatology, Palaeoecology* 270: 220–229.

Hendy AW, Kamp PJJ and Vonk AJ (2006) Cool-water shell
bed taphofacies from Miocene-Pliocene shelf sequences
in New Zealand: Utility of taphofacies in sequence
stratigraphic analysis. In Pedley M and Carannante G
(eds) *Cool-water carbonates*. Bath: Geological Society,
283–305.

Kidwell SM and Brenchley PJ (1994) Patterns in bioclastic
accumulation through the Phanerozoic: Changes in input
or in destruction. *Geology* 22: 1139–1143.

life expectancy The average age to which a person
is expected to live. It varies according to age of individual, sex and aspects of the human environment,
and is calculated using *life tables*. In AD 1800, global
life expectancy was around 30 years; by 2000, it was
67 years; and it may be 76 years by 2050. This improvement (the *health transition*) has been described as the
crowning achievement of the modern era. Life expectancy is still generally lower in DEVELOPING COUNTRIES,

due to factors such as war, poverty and DISEASE. In 2008, life expectancy varied from 82 years in Japan to 33 years in Swaziland. The term is also used in the context of other organisms and manufactured goods.

JAM

[*See also* BIRTH RATE, DEATH RATE, DEMOGRAPHIC CHANGE, LIFE-CYCLE ASSESSMENT (LCA)]

Riley JC (2001) *Rising life expectancy: A global history.* Cambridge: Cambridge University Press.

life form The characteristic overall form, or morphology or physiognomy of an organism, most widely used in relation to mature plants. A particularly well-known scheme, *Raunkiaer's life-form classification*, which is based on the position of *perennating buds* (*growth points*) in relation to the ground surface. The five main categories are (1) *phanerophytes* (trees), (2) *chamaephytes* (dwarf shrubs), (3) *hemicryptophytes* (buds at ground level), (4) *cryptophytes* (buds below ground or in water) and (5) *therophytes* (ANNUALS and EPHEMERALS). *JLI*

[*See also* FUNCTIONAL TYPE, GEOPHYTE, GROWTH FORM]

Harrison SP, Prentice IC, Barboni D et al. (2010) Ecophysiological and bioclimatic foundations for a global plant functional classification. *Journal of Vegetation Science* 21: 300–317.
Kloeke AEEV, Douma JC, Ordonez JC et al. (2012) Global quantification of contrasting leaf life span strategies for deciduous and evergreen species in response to environmental conditions. *Global Ecology and Biogeography* 21: 224–235.
Raunkaier O (1934) *The life forms of plants and statistical plant geography.* Oxford: Clarendon Press.

life zone A concept first developed by C.H. Merriam describing the latitudinal and altitudinal similarities in the ZONATION of plant and animal communities. Life zones are one of many ways of subdividing the Earth's surface based on the major climatic controls on life. The best known is the *Holdridge system*, in which life zones are defined in terms of MEAN ANNUAL AIR TEMPERATURE (MAAT), PRECIPITATION and POTENTIAL EVAPOTRANSPIRATION (PET). Life zones correspond broadly with ECOZONES, BIOMES and VEGETATION FORMATION TYPES, but include the whole ecosystem and are more precisely defined in climatic terms. *JAM*

Holdridge LR (1967) *Life zone ecology.* San José, Costa Rica: Tropical Science Center.
Kendeigh SC (1954) History and evaluation of various concepts of plant and animal communities in North America. *Ecology* 35: 152–171.
Merriam CH (1898) *Life zones and crop zones of the United States.* Washington, DC: Government Printing Office [US Department of Agriculture Bulletin 10].

life-cycle assessment (LCA) A technique that addresses the potential environmental impacts of systems or products throughout their life cycle, that is, from raw materials and energy acquisition throughout production, use, recycling and final disposal. According to the principles and guidelines set out by the *International Organisation for Standardization* in the ISO 14044:2006 standard, an LCA has four phases: (1) *goal and scope definition*, where the aim, *system boundaries* and functional unit are defined; (2) *inventory analysis*, where a DATABASE is created to characterise all the processes within the system boundary; (3) *impact assessment*, where an evaluation of the potential environmental impacts is performed, associated with the selected inputs and outputs, to quantify the *environmental impact*, HUMAN HEALTH HAZARDS and *resources depletion*; and (4) *interpretation and evaluation* of the LCA results in accordance with the study's goals. *AJDF*

[*See also* CARBON FOOTPRINT, ENVIRONMENTAL IMPACT ASSESSMENT (EIA), ENVIRONMENTAL MANAGEMENT, INDUSTRIAL ECOLOGY, STRATEGIC ENVIRONMENTAL ASSESSMENT (SEA)]

Guinée J (ed.) (2002) *Handbook of life-cycle assessment: Operational guide to the ISO standards.* Dordrecht: Kluwer.
Hendrickson CT, Lave LB and Matthews HS (2006) *Environmental life-cycle analysis of goods and services: An input-output approach.* Washington, DC: RFF Press.

light rings A dendrochronological feature representing years of very little summer wood production, often in temperature-sensitive trees growing at high latitudes. These are equivalent to annual TREE RINGS of low maximum-latewood density and may indicate years when summer temperature was relatively low and/or *light levels* were significantly reduced. They have been shown to correspond in some cases to the dates of large explosive VOLCANIC ERUPTIONS. They may also represent years of heavy insect *defoliation*.

KRB/APP

[*See also* DENDROCHRONOLOGY, DENSITOMETRY]

Liang E and Eckstein D (2006) Light rings in Chinese pine (*Pinus tabulaeformis*) in semiarid areas of north China and their palaeo-climatological potential. *New Phytologist* 171: 783–791.
Szeicz JM (1996) White spruce light rings in northwestern Canada. *Arctic and Alpine Research* 28: 184–189.

lignin phenols *Lignin* is a major *biopolymer* in vascular plants, formed by random polymerisation of three main *monomers*: *p*-coumaryl alcohol, coniferyl alcohol and sinapyl alcohol. The *phenolic compounds* of lignin origin produced by OXIDATION of SEDIMENTS are often used to evaluate terrestrial inputs into AQUATIC ENVIRONMENTS as the ratios of syringyl, vanillyl and cinnamyl phenols can be used to differentiate between the main vascular plant types: non-woody ANGIOSPERMS, non-woody GYMNO-SPERMS, woody angiosperms and woody gymnosperms. *Vegetation changes* since the LAST GLACIAL MAXIMUM (LGM) can therefore be deduced from specific lignin phenols. The $\delta^{13}C$ variations of lignin phenols reflect the differences in their origins and enable an estimate of *C-4 plant* abundances to be calculated.   *KJF*

Hedges JI and Mann DC (1979) The characterisation of plant tissues by their cupric oxide oxidation products. *Geochimica et Cosmochimica Acta* 43: 1803–1818.

Hou J, Huang Y, Brodsky C et al. (2010) Radiocarbon dating of individual lignin phenols: A new approach for establishing chronology of Late Quaternary lake sediments. *Analytical Chemistry* 82: 7119–7126.

Huang Y, Freeman KH, Eglinton TI and Street-Perrott FA (1999) $\delta^{13}C$ analyses of individual lignin phenols in Quaternary lake sediments: A novel proxy for deciphering past terrestrial vegetation changes. *Geology* 27: 471–474.

Tareq SM, Tanaka N and Ohta K (2004) Biomarker signature in tropical peat: Lignin phenol vegetation index (LPVI) and its implication for reconstructing paleoenvironment. *Science of the Total Environment* 324: 91–103.

liman   A shallow bay with a muddy bottom.   *HJW*

limestone   A SEDIMENTARY ROCK composed entirely or dominantly of CARBONATE minerals, principally *calcite* and *dolomite*. Shells of macroscopic and microscopic marine invertebrate FOSSILS form an important constituent of many limestones, and biological and chemical processes are important in understanding their origins. This contrasts with the importance of physical processes to the formation of CLASTIC sediments and sedimentary rocks. Limestones are important indicators of PALAEOENVIRONMENT.   *TY*

[*See also* KARST]

Bathurst RGC (1975) *Carbonate sediments and their diagenesis.* Amsterdam: Elsevier.

Nichols G (2009) *Sedimentology and stratigraphy.* Oxford: Blackwell.

Swart PK, Eberli GP and McKenzie JA (eds) (2009) *Perspectives in carbonate geology.* Chichester: Wiley-Blackwell.

Tucker ME and Wright VP (1990) *Carbonate sedimentology.* Oxford: Blackwell.

Wilson JL (1975) *Carbonate facies in geologic history.* Berlin: Springer.

Wright VP and Burchette TP (1996) Shallow-water carbonate environments. In Reading HG (ed.) *Sedimentary environments: Processes, facies and stratigraphy, 3rd edition.* Oxford: Blackwell.

liming   The addition of LIMESTONE or *quicklime* (decalcified limestone in the form of calcium oxide produced in a lime kiln) to ACID SOILS in order to counteract ACIDITY.   *JAM*

Adams F (ed.) (1984) *Soil acidity and liming, 2nd edition.* Madison, WI: American Society of Agronomy and Soil Science Society of America.

Johnson DS (2010) *Liming and agriculture in the central Pennines: The use of lime in land improvement from the late thirteenth century to c. 1900.* Oxford: Archaeopress [BAR Report 525].

limiting factors   ENVIRONMENTAL FACTORS that slow down or stop the productivity, growth or reproduction of a POPULATION. Limiting factors may be physical (e.g. TEMPERATURE and moisture levels) or chemical (e.g. NUTRIENT levels). They may operate at lower and upper extremes or species' TOLERANCE. For example, a population's *tolerance range* will normally be subject to a lower temperature limit and to a higher temperature limit. Some limiting factors partly determine species' geographical RANGES; for example, the small-leafed lime tree (*Tilia cordata*) has a northern limit in England and Scandinavia corresponding with the 19°C mean July isotherm. Limiting factors also help shape *ecosystems*. Broad-leaved deciduous TEMPERATE FOREST grow in humid and mesic regions of Asia where the coldest-month mean is less than 1°C and the warmest-month mean is less than 20°C. Limiting factors also affect *agricultural systems*. Successful AGRICULTURE commonly depends on recognising limiting factors and employing techniques to accommodate or surmount them. Once a limiting factor is modified, this may act as a *trigger factor* (triggering adjustments within the ecosystem), which may lead a different environmental factor to become limiting.   *RJH*

[*See also* ABIOTIC, ECOSYSTEM CONCEPT, HOLOCOENOTIC ENVIRONMENT]

Kennedy AD (1993) Water as a limiting factor in the Antarctic terrestrial environment: A biogeographical synthesis. *Arctic and Alpine Research* 25: 308–315.

Sundareshwar PV, Morris JT, Koepfler EK and Fornwalt B (2003) Phosphorus limitation of coastal ecosystem processes. *Science* 299: 563–565.

limnic eruption   In August 1986 at least 1,700 people died of asphyxiation due to a massive release of CARBON DIOXIDE (CO_2) gas from Lake Nyos, a CRATER lake in Cameroon, West Africa (the *Lake Nyos gas disaster*). Subsequent investigations showed that the source of the gas was the sudden *decompression* of gas-rich BOTTOM WATER in the stratified lake (see LAKE STRATIFICATION AND ZONATION), leading to the catastrophic *exsolution* of CO_2 gas as a self-sustaining fountain. Although the CO_2 was of magmatic origin (see MAGMA), its release was not associated with a VOLCANIC ERUPTION, and the terms LIMNIC ERUPTION or *eruptive outgassing* have been applied to the presumed process, for which the *trigger factor* could have been a LANDSLIDE, EARTHQUAKE or STORM. The eruption, driven by the exsolution of CO_2, has similarities with water-driven HYDROVOLCANIC ERUPTIONS. Measurements have shown that CO_2 concentrations are again building up in the bottom water of Lake Nyos, and *controlled degassing* has taken place to reduce the NATURAL HAZARD represented by Lake Nyos and similar lakes.   *GO*

Kling GW, Clark MA, Wagner GN et al. (1987) The 1986 Lake Nyos gas disaster in Cameroon, West Africa. *Science* 236: 169–175.

Pérez NM, Hernández PA, Padilla G et al. (2011) Global CO_2 emission from volcanic lakes. *Geology* 39: 235–238.

Rice A (2000) Rollover in volcanic crater lakes: A possible cause for Lake Nyos type disasters. *Journal of Volcanology and Geothermal Research* 97: 233–239.

Yoshida Y (2010) An efficient method for measuring CO_2 concentration in gassy lakes: Application to Lakes Nyos and Monoun, Cameroon. *Geochemical Journal* 44: 441–448.

limnogeology   The study of modern and ancient LACUSTRINE SEDIMENTS and lake basins from an EARTH SCIENCE perspective.   *MRT*

[*See also* LAKES AS INDICATORS OF ENVIRONMENTAL CHANGE]

Ariztegui D, Anselmetti FS, Gill A and Waldmann N (2008) Late Pleistocene environmental change in eastern Patagonia and Tierra del Fuego: A limnogeological approach. *Developments in Quaternary Sciences* 11: 241–253.

limnology   The study of lakes, ponds and other standing waters, including physical and chemical characteristics of water and SEDIMENT, the *biota* and the relationship of the aquatic *ecosystem* to the *catchment*.   *HHB*

[*See also* AQUATIC ENVIRONMENT, LACUSTRINE SEDIMENTS, LAKE STRATIFICATION AND ZONATION, LAKE-LEVEL VARIATIONS, LAKES AS INDICATORS OF ENVIRONMENTAL CHANGE, PALAEOLIMNOLOGY]

Hutchinson GE (1957) *A treatise on limnology, volume 1: Geography & physics of lakes.* New York: Wiley.

Hutchinson GE (1967) *A treatise on limnology, volume 2: Introduction to lake biology and the limnoplankton.* New York: Wiley.

Hutchinson GE (1975) *A treatise on limnology, volume 3: Limnological botany.* New York: Wiley.

Hutchinson GE (1993) *A treatise on limnology, volume 4: The zoobenthos.* New York: Wiley.

O'Sullivan PE and Reynolds CS (eds) (2004) *The lakes handbook, volume 1: Limnology and limnetic ecology.* Chichester: Wiley-Blackwell.

line reduction   A *clipping technique* used in *computer graphics* which aims to determine the visible portion of lines that may be partially or totally occluded by other graphical objects. These algorithms are useful for HIDDEN SURFACE REMOVAL, for example, in the visualisation of DIGITAL ELEVATION MODELS (DEMs).   *ACF*

Foley JD, van Dam A, Feiner SK et al. (1994) *Introduction to computer graphics.* Reading, MA: Addison-Wesley.

lineage   A line of evolutionary descent, either of a single evolving species or several species descended from a common ancestor.   *KJW*

linear interpolation   A technique for determining the value of a function at a point, given the value of the function at two neighbouring points, by assuming that the function varies linearly in this neighbourhood. *PJS*

linear-mixture modelling   A technique used for spectral unmixing of optical REMOTE-SENSING data. It assumes the spectral signatures of PIXELS containing different LAND COVER types to consist of a weighted linear sum of the component land cover signatures. The weights are determined directly from the relative proportions of land cover types in the pixel. Given k land cover classes covering proportions f_i in a pixel with signature vectors m_i, the resulting expected linearly mixed signature of that pixel in the absence of sensor noise is $f_1 m_1 + ... + f_i m_i + ... + f_1 m_k$. The NOISE caused by the sensor and the NATURAL *variability* on the ground can be added as a vector of error terms. The linear mixture model assumes that each PHOTON hitting the ground is interacting with only one land cover type before being reflected. It is thus only an approximation to reality. A prerequisite for the model is that there are more significant principal components (see PRINCIPAL COMPONENTS TRANSFORM) in the satellite data than in the land cover classes.   *HB*

[*See also* OPTICAL REMOTE-SENSING INSTRUMENTS]

Ferreira ME, Ferreira LG, Sano EE and Shimabukuro YE (2007) Spectral linear mixture modelling approaches for land cover mapping of tropical savanna areas in Brazil. *International Journal of Remote Sensing* 28: 413–429.

lineation   The act of marking or outlining with *lines*, pattern of lines, configuration or arrangement of lines. For environmental applications, this involves recognition of shearing and mineral lineation in structural geology, lineation in cartographic representations (e.g. contour lines) and their digital representation in a GEOGRAPHICAL INFORMATION SYSTEM (GIS).   *VM*

[*See also* ARC, LINE REDUCTION, MAP GENERALISATION]

linguistic dating   RELATIVE-AGE DATING of languages based on *lexicostatistical analysis* of linguistic similarity. It rests on the principle that the longer the groups of people are separated, the more different their languages become. *Glottochronology* is a form of *absolute-age dating* of languages, which assumes a quantifiable rate of divergence between languages.   *JAM*

[*See also* SERIATION]

Bergsland K and Vogt H (1962) On the validity of
  glottochronology. *Current Anthropology* 3: 115–153.
Crowley T and Bowern C (2010) *An introduction to
  historical linguistics, 4th edition*. Oxford: Oxford
  University Press.

lipids Lipids are a group of naturally occurring molecules including *fats, waxes, sterols, glycerides* and
phospholipids that are insoluble in water, but soluble
in organic solvents. They are an organic chemical constituent in the BIOMASS. All organisms are composed of
lipids, *proteins* and *carbohydrates*. The main biological
functions of lipids are energy storage and as structural
components of *cell membranes*. The resistant parts of
organisms (e.g. waxes, membranes) are composed of
lipids and hence lipids tend to survive DECOMPOSITION
and are the most important constituent in the formation
of PETROLEUM. Up to a third of the chemical composition of organisms may be composed of lipids and lipid-
like components.   *KJF*

[*See also* BIOMARKER, MOLECULAR STRATIGRAPHY]

Blyth AJ, Baker A, Thomas LE and Van Calsteren P
  (2011) A 2000-year lipid biomarker record preserved
  in a stalagmite from north-west Scotland. *Journal of
  Quaternary Science* 26: 326–334.
Cranwell PA (1982) Lipids of aquatic organisms
  as potential contributors to lacustrine sediments II.
  Organic Geochemistry 11: 513–527.
Killiops S and Killops V (2005) *Introduction to organic
  geochemistry*. Malden, MA: Blackwell.
Peters KE, Walters CC and Moldowan JM (2005) *The
  biomarker guide: Biomarkers and isotopes in petroleum
  exploration and Earth history*. New York: Cambridge
  University Press.

liquefaction A temporary loss of strength that
may affect cohesionless granular materials (e.g.
SAND) related to an increase in *pore fluid pressure* to
a level that equals the *overburden pressure* so that
GRAINS are no longer supported at grain contacts, but
'float' in the pore fluid. The sediment then behaves
as a fluid, or QUICKSAND. Localised FLUIDISATION may
develop in the upper parts of BEDS that have liquefied due to the focussing of escaping pore water.
Unlike FLUIDISATION, no external source of fluid is
required for liquefaction, and grain contacts immediately begin to be re-established, restoring strength
over time spans of tens of seconds to tens of minutes.
Liquefaction is most commonly reported in loosely
packed sediments and SOILS of SILT to fine-sand grade
as a consequence of large EARTHQUAKES, giving rise
to failure of foundations and triggering *slope failures*. Liquefaction can also be triggered by waves
during STORMS, by breaking waves or by the onset of

FLOOD surges. Features such as SAND VOLCANOES (*sand
blows*) and other SEDIMENTARY STRUCTURES developed
as a consequence of liquefaction may be preserved
as SOFT-SEDIMENT DEFORMATION structures and can be
used as evidence of PALAEOSEISMICITY. It has been
suggested that the term *sand boils* be restricted to
similar features with a demonstrably aseismic origin. Such studies have suggested a potential threat
from large earthquakes along the eastern seaboard of
North America, although there is no historical record
of large earthquakes there.   *GO*

[*See also* LIQUIDISATION]

Amick D and Gelinas R (1991) The search for evidence of
  large prehistoric earthquakes along the Atlantic seaboard.
  Science 251: 655–658.
Galli P (2000) New empirical relationships between
  magnitude and distance for liquefaction. *Tectonophysics*
  324: 169–187.
Ishihara K (1993) Liquefaction and flow failure during
  earthquakes. *Géotechnique* 43: 351–415.
Li Y, Craven J, Schweig ES and Obermeier SF (1996) Sand
  boils induced by the 1993 Mississippi River flood: Could
  they one day be misinterpreted as earthquake-induced
  liquefaction? *Geology* 24: 171–174.
Obermeier SF (1996) Use of liquefaction-induced features
  for paleoseismic analysis: An overview of how seismic
  liquefaction features can be distinguished from other
  features and how their regional distribution and properties
  of source sediment can be used to infer the location and
  strength of Holocene paleo-earthquakes. *Engineering
  Geology* 44: 1–76.
Owen G and Moretti M (2011) Identifying triggers for
  liquefaction-induced soft-sediment deformation in sands.
  Sedimentary Geology 235: 141–147.

liquid limit The minimum *moisture content* at
which a soil or sediment passes from the plastic to
liquid state and can then flow under its own weight.
A sediment with a low liquid limit is likely to be susceptible to downslope movement by GELIFLUCTION.
                                                                          PW

[*See also* FAILURE, PLASTIC LIMIT, SOLIFLUCTION]

Evans DJA and Benn DI (eds) (2004) *A practical guide to
  the study of glacial sediments*. London: Arnold.

liquidisation Any mechanism that causes a sudden
reduction in strength of a SEDIMENT, including LIQUE-
FACTION and FLUIDISATION in cohesionless materials (e.g.
SAND) and THIXOTROPY in cohesive MUD (see SENSITIVE
CLAYS). While liquidised, a sediment can be deformed
by weak stresses, and may develop SOFT-SEDIMENT
DEFORMATION structures or induce *slope failure* (see
SLOPE STABILITY).   *GO*

Allen JRL (1982) *Sedimentary structures: Their character
  and physical basis*. Amsterdam: Elsevier.

lithalsa A mesoscale FROST MOUND composed entirely of minerogenic material and SEGREGATED ICE, found in areas of PERMAFROST. Unlike PALSAS, to which they are genetically related, *peat* plays no part in their origin or development. RELICT examples have been identified in regions with a TEMPERATE CLIMATE (and possibly on Mars), in the form of *ramparted depressions*. *JAM*

Harris SA (1993) Palsa-like mounds developed in a mineral substrate, Fox Lake, Yukon Territory. *Proceedings of the 6th International Conference on Permafrost, Beijing, China, volume* 1. Wushan Guangzhou: South China University Press, 238–243.
Pissart A (2000) Remnants of lithalsas of the Hautes Fagnes, Belgium: A summary of present-day knowledge. *Permafrost and Periglacial Processes* 11: 327–355.
Pissart A (2011) Pingos, palsas and lithalsas: Comparison with the Martian mounds. *Zeitschrift für Geomorphologie NF* 55: 463–473.

lithic Made of rock. In GEOLOGY, the term is used in the description of CLAST composition in SEDIMENTS and SEDIMENTARY ROCKS to distinguish rock fragments (*lithic clasts*) from MINERAL fragments. In ARCHAEOLOGY, lithics are stone materials worked by PREHISTORIC people including both the useful ARTEFACTS (especially *core tools* but also *flake tools*) and workshop debris or *debitage* (made up of unused flakes). *GO/JAM*

[*See also* CORE, MEGALITH, MESOLITHIC, MICROLITH, NEOLITHIC, PALAEOLITHIC, PROVENANCE, SANDSTONE]

Shea JJ (2006) The origins of lithic projectile point technology: Evidence from Africa, the Levant, and Europe. *Journal of Archaeological Science* 33: 823–846.

lithification The change from unconsolidated SEDIMENT to coherent SEDIMENTARY ROCK during DIAGENESIS, brought about principally by CEMENTATION and COMPACTION. INDURATION, meaning hardening, is commonly used in an equivalent sense. *TY*

lithofacies A FACIES in SEDIMENTS or SEDIMENTARY ROCKS defined primarily using characteristics of LITHOLOGY. Description may integrate biological characteristics (e.g. cross-bedded crinoidal grainstone lithofacies), or indicate depositional process, environment or PALAEOENVIRONMENT (e.g. oolitic limestone lithofacies, reef-flank lithofacies). *LC*

[*See also* BIOFACIES, ICHNOFACIES]

Anestas AS, Dalrymple RW, James NP and Nelson CS (2006) Lithofacies and dynamics of a cool-water carbonate seaway: Mid-Tertiary, Te Kuiti Group, New Zealand. In Pedley M and Carannante G (eds) *Cool-water carbonates*. Bath: Geological Society, 245–268.

lithology The general characteristics of a rock, particularly those observable in FIELD RESEARCH, *hand specimens* and CORES, including *rock type, composition* and TEXTURE (IN SEDIMENTS AND ROCKS). *TY*

[*See also* LITHOFACIES, LITHOSTRATIGRAPHY]

lithosphere The rigid outer layer of the Earth, comprising the CRUST and the MANTLE to a depth of about 100 km. The lithosphere comprises discrete slabs (see PLATE TECTONICS), at the edges of which lithosphere-bearing OCEANIC CRUST is created and destroyed (see PLATE MARGIN). *GO*

[*See also* EARTH SPHERES, EARTH STRUCTURE]

Artemieva I (2011) *The lithosphere: An interdisciplinary approach*. Cambridge: Cambridge University Press.
Kearey P, Klepeis KA and Vine FJ (2009) *Global tectonics*. Chichester: Wiley.
Stacey FD and Davis PM (2008) *Physics of the Earth*. Cambridge: Cambridge University Press.

lithostratigraphy The subdivision of the STRATIGRAPHICAL RECORD using characteristics of LITHOLOGY or LITHOFACIES and the determination of the spatial relationships of the defined units through geological mapping. This is fundamental to all other types of STRATIGRAPHY. The primary unit for geological mapping is the FORMATION (STRATIGRAPHICAL), and the hierarchy of *lithostratigraphical units* is SUPERGROUP, GROUP, *formation*, MEMBER and BED. Lithostratigraphical units are three-dimensional bodies of sediment or sedimentary, igneous or metamorphic rocks. A unit is defined and described from a *stratotype section*, or reference sections. Boundaries are drawn at horizons of lithological change, which may be abrupt or be placed at an arbitrary point in a gradational sequence. A lithostratigraphical unit characteristic of a particular DEPOSITIONAL ENVIRONMENT may migrate laterally and vertically within a rock succession (i.e. in space and time in response to changes to the PALAEOENVIRONMENT. DIACHRONISM (cutting across time lines) of lithostratigraphical units is demonstrated by BIOSTRATIGRAPHY or EVENT horizons. *LC*

Cox BM and Sumbler MG (1998) Lithostratigraphy: Principles and practice. In Doyle P and Bennett MR (eds) *Unlocking the stratigraphical record: Advances in modern stratigraphy*. Chichester: Wiley, 11–27.
Waters CN, Browne MAE, Dean MT and Powell JN (2007) *Lithostratigraphical framework of Carboniferous successions of Great Britain (onshore)* [BGS Research Report RR/07/01]. Nottingham: British Geological Survey, 1–60.

litter (1) Organic DEBRIS that accumulates, often seasonally, at the soil surface. It is usually composed mainly of leaves (*leaf litter*), which forms the uppermost SOIL HORIZON. Its original constituents are still recognisable as it has been affected by DECOMPOSITION to only a limited degree. The *litter* (*L*) horizon tends to be best developed under forests. (2) Human debris, such as food packaging, discarded in the environment where it can be an ENVIRONMENTAL NUISANCE and an animal or HUMAN HEALTH HAZARD.   *JAM*

Berg B and McClaugherty C (2010) *Plant litter: Decomposition, humus formation, carbon sequestration,* 2nd edition. Berlin: Springer.

Little Ice Age Although originally used by F. Matthes to describe the renewed glaciation associated with NEOGLACIATION earlier in the HOLOCENE, the term 'Little Ice Age' is now used for the cold interval during recent centuries when GLACIERS were considerably larger than today and many attained their Holocene maxima. Matthews and Briffa (2005) distinguished between *Little Ice Age glacierisation* over about 650 years (ca AD 1300–1950) when glaciers were larger than before or since, and *Little Ice Age climate* over about 330 years (ca AD 1570–1900) when Northern

Hemisphere summer temperatures fell significantly below the 1961–1990 mean.

The Little Ice Age appears to have been a global event, but probably began at different times in different regions. In the European Alps, where recent GLACIER VARIATIONS and CLIMATIC CHANGE are known in greatest detail from DOCUMENTARY EVIDENCE and PALAEOENVIRONMENTAL RECONSTRUCTION, the Little Ice Age began around AD 1300 as the MEDIAEVAL WARM PERIOD terminated. Subsequently, alpine glaciers attained maximum advance positions around AD 1350, 1650 and 1850 (see Figure). The same three HIGHSTANDS of the Little Ice Age did not occur elsewhere. In Scandinavia, for example, the evidence suggests one major glacier expansion episode, which began in the seventeenth century and peaked around AD 1750.

For much of the Little Ice Age, summer temperatures were likely to have been 0.5–2.0°C colder than today, at least in Europe, but this average value

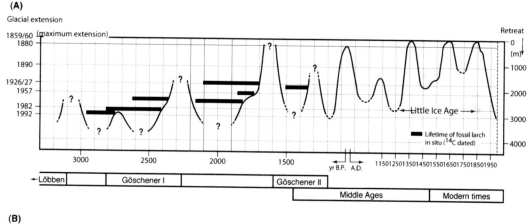

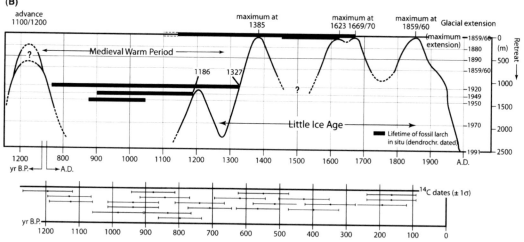

Little Ice Age *The three glacier maxima (highstands) of Little Ice Age and the preceding Mediaeval Warm Period as defined by glacier variations of (A) the Grosser Aletsch Glacier and (B) the Gorner Glacier in the Swiss Alps. A multiproxy approach has been used in the reconstruction, including documentary evidence and radiocarbon-dated fossil larch stumps (Holzhauser, 1997).*

conceals CLIMATIC VARIATIONS at various temporal and spatial scales. The largest negative temperature anomalies may have occurred in northwest-central Asia. Winter precipitation levels also varied, so that the climate of the Little Ice Age should not be viewed simply as a cold period. During the Little Ice Age, polar waters extended from Iceland, across the Norwegian Sea and south to the vicinity of the Shetland Islands, displacing warmer surface waters of the *North Atlantic drift* farther south. This indicates that the average positions of the OCEANIC POLAR FRONT (OPF) and associated atmospheric STORM tracks lay south of their present average positions. Outside Europe there is evidence of EXTREME CLIMATIC EVENTS, such as DROUGHTS in the SAHEL, being correlated to phases of the Little Ice Age, while research from equatorial east Africa has demonstrated a relatively wet climate at this time. Thus, the cause of the Little Ice Age can be linked to changes in the GENERAL CIRCULATION OF THE ATMOSPHERE. Possible ULTIMATE CAUSES, which are poorly understood, include SOLAR FORCING, VOLCANIC FORCING and AUTOVARIATION within the Earth-atmosphere-ocean system.

Little Ice Age climate is generally considered to have ended by the end of the nineteenth century. Since then, the strong worldwide trend towards glacier recession has been affected by GLOBAL WARMING. However, glacier advances at various times, such as in the 1960s–1980s in the Alps and the 1980s–1990s in Scandinavia, demonstrate the fluctuating nature of natural CLIMATIC VARIABILITY. *JAM*

[*See also* CENTURY- TO MILLENNIAL-SCALE VARIABILITY, NEOGLACIATION, NORSE GREENLAND SETTLEMENTS]

Bradley RS and Jones PD (1993) 'Little Ice Age' summer temperature variations: Their nature and relevance to recent global warming trends. *The Holocene* 3: 367–376.

Briffa KR, Osborn TJ, Schweingruber FH et al. (2001) Low frequency temperature variations from a northern tree-ring density network. *Journal of Geophysical Research* 106D: 2929–2941.

Fagen B (2001) *The Little Ice Age: How climate made history 1300-1850.* New York: Basic Books.

Grove JM (2004) *Little Ice Ages: Ancient and modern, 2 volumes.* London: Routledge.

Holzhauser H (1997) Fluctuations of the Grosser Aletsch Glacier and the Gorner Glacier during the last 3200 years: New results. In Frenzel B, Boulton GS, Gläser B and Huckriede U (eds) *Glacier fluctuations during the Holocene.* Stuttgart: Gustav Fischer, 35–58 [*Paläoklimaforschung* 24].

Luckman BH (2000) The Little Ice Age in the Canadian Rockies. *Geomorphology* 32: 357–384.

Mann ME, Zhang Z, Rutherford S et al. (2009) Global signatures and dynamic origins of the Little Ice Age and Medieval climate anomaly. *Science* 326: 1256–1260.

Matthes F (1939) Report of the Committee on Glaciers. *Transactions of the American Geophysical Union* 20: 518–523.

Matthews JA and Briffa KR (2005) The 'Little Ice Age': Re-evaluation of an evolving concept. *Geografiska Annaler* 87A: 17–36.

Mikami T (ed.) (1992) *Proceedings of the International Symposium on the Little Ice Age Climate.* Tokyo: Department of Geography, Tokyo Metropolitan University.

Nesje A and Dahl SO (2003) The 'Little Ice Age': Only temperature? *The Holocene* 13: 139–145.

Ogilvie AEJ and Jónsson T (2001) 'Little Ice Age' research: A perspective from Iceland. *Climatic Change* 48: 9–52.

Verschuren D, Laird KR and Cumming BF (2000) Rainfall and drought in equatorial east Africa during the past 1,100 years. *Nature* 403: 410–414.

littoral zone The *intertidal zone* along a SHORELINE. In Europe, the term is sometimes synonymous with *coastal zone*. It is also used in the context of lakes. The *sublittoral zone* lies between the intertidal zone and the outer edge of the CONTINENTAL SHELF. *HJW*

[*See also* SHORE ZONE]

living fossil A popular but loosely defined term, introduced by Charles Darwin, for an extant organism that has a long FOSSIL RECORD showing little evolutionary change. A classic example is the *coelacanth* (*Latimeria*), a member of a group of fish with a record as FOSSILS dating from the DEVONIAN period. They were thought to have become extinct about 80 million years ago in the Late CRETACEOUS until a modern specimen was recovered from the Indian Ocean in AD 1938. Other 'living fossils' include the *brachiopod* genus *Lingula*, with a fossil record dating to the ORDOVICIAN, the reptile *tuatara* (*Sphenodon*), the horseshoe crab *Limulus* and the *ginkgo* tree. Living fossils represent organisms that have undergone little morphological change through EVOLUTION but have escaped EXTINCTION. A variety of factors seem to be responsible, but living fossils include some organisms that are adaptable to a range of environmental conditions and others that are adapted to conditions that are hostile to most organisms. *GO*

Beerling DJ, McElwain JC and Osborne CP (1998) Stomatal responses of the 'living fossil' Ginkgo biloba L. to changes in atmospheric CO_2 concentrations. *Journal of Experimental Biology* 49: 1603–1607.

Fortey R (2011) *Survivors: The animals and plants that time has left behind.* London: Harper Press.

Lee J, Alrubaian J and Dores RM (2006) Are lungfish living fossils? Observation on the evolution of the opioid/orphanin gene family. *General and Comparative Endocrinology* 148: 306–314.

Mitchell NJ, Kearney MR, Nelson NJ and Porter WP (2008) Predicting the fate of a living fossil: How will global warming affect sex determination and hatching phenology in tuatara? *Proceedings of the Royal Society B* 275: 2185–2193.

Ward PD (1992) *On Methuselah's trail: Living fossils and the great extinctions.* New York: Freeman.

Lixisols   Soils of the seasonally dry TROPICAL and SUBTROPICAL ENIVIRONMENMTS having an accumulation of LOW-ACTIVITY CLAYS in an *argic horizon* (leached from an upper, eluvial horizon) and high BASE SATURA-TION. These strongly weathered soils, also known as *latosols* or *red-yellow podzolic soils* (SOIL TAXONOMY: oxic subgroups of *Alfisols*), may be regarded as the tropical equivalent of LUVISOLS. Lixisols are common on old landscapes in Brazil, sub-Saharan Africa and the Indian subcontinent, and are often POLYGENETIC, retaining features inherited from a wetter climate. The red-yellow colouration results from RUBIFICATION (RUBEFACTION), which is produced by dehydration of iron compounds during the long dry season. Erosion greatly reduces their already low fertility so the thin surface horizon requires careful LANDUSE PLANNING AND MANAGEMENT with use of FERTILISERS and/or LIM-ING. When dry, the eluvial horizon of Lixisols may become very hard, a condition referred to as *hard setting.*   *EMB*

[*See also* WORLD REFERENCE BASE FOR SOIL RESOURCES (WRB)]

Korodjourna O, Badion O, Ayernou A and Sedogo PM (2006) Long-term effects of ploughing and organic matter imprint on soil moisture characteristics of a ferric Lixisol in Burkina Faso. *Soil and Tillage Research* 88: 217–224.

loading   (1) The amount of a substance added to the ATMOSPHERE, HYDROSPHERE or BIOSPHERE or other parts of the EARTH SYSTEM by natural or ANTHROPOGENIC pro-cesses. (2) The weight exerted on the Earth's surface by any material, such as a mass of rock or DEBRIS, or an ICE SHEET. (3) The extent to which a particular VARI-ABLE is related to a factor or component in MULTIVARIATE ANALYSIS.   *JAM*

[*See also* ENVIRONMENTAL BURDEN, UNLOADING]

local climate   A small-scale CLIMATE, where climatic conditions are clearly different from those of nearby surrounding areas. Local climate may be the result of topography (TOPOCLIMATE, including ASPECT and *expo-sure*), vegetation cover and landuse, soil type, water availability or the presence of anthropogenic features. It includes the climate in and around mountains, hills, lakes, coasts, forests and cities. Some view it as inter-mediate between MICROCLIMATE and MESOCLIMATE (see Figure); to others it is the equivalent of mesoclimate. A precise scale is difficult to define but typically ranges from metres to kilometres.   *JBE/JAM/NP*

[*See also* MACROCLIMATE, URBAN CLIMATE]

Oke TR (1987) *Boundary layer climates, 2nd edition.* London: Methuen.

Orlanski I (1975) A subdivision of scales for atmospheric processes. *Bulletin of the American Meteorological Society* 56: 527–530.

Sturman AP, McGowan HA and Spronken-Smith RA (1999) Mesoscale and local climates in New Zealand. *Progress in Physical Geography* 23: 611–635.

Yoshino M (2005) Local climatology. In Oliver JE (ed.) *Encyclopedia of world climatology.* Dordrecht: Springer, 460–467.

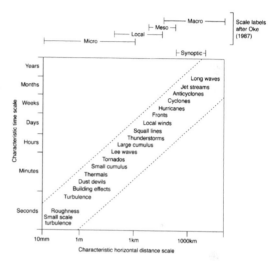

Local climate *The temporal and spatial scales of atmospheric phenomena and the definition of microclimate, local climate, mesoclimate, macroclimate and synoptic climatology (Sturman et al., 1999).*

local glaciation   Formation and existence of indi-vidual, isolated GLACIERS, mostly *cirque glaciers, ice fields, ice caps* and *plateau glaciers*, beyond the limit of larger ice masses. They are also known, sometimes inappropriately, as *mountain glaciers*. In western Scan-dinavia during the YOUNGER DRYAS STADIAL, for example, hundreds of local glaciers existed in the coastal moun-tains beyond the margin of the *continental ice sheet.* AN

[*See also* CONTINENTAL GLACIATION]

Finlayson AG (2006) Glacial geomorphology of the Creag Meagaigh Massif, western Grampian Highlands: Implications for local glaciation and palaeoclimatology during the Loch Lomond Stadial. *Scottish Geographical Magazine* 122: 293–307.

Kelly MA, Lowell TV, Hall BL et al. (2008) A [10]Be chronology of Lateglacial and Holocene mountain glaciation in the Scoresby Sund region of east Greenland: Implications for seasonality during Lateglacial time. *Quaternary Science Reviews* 27: 2273–2282.

Nesje A (2009) Latest Pleistocene and Holocene alpine glacier fluctuations in Scandinavia. *Quaternary Science Reviews* 28: 2119–2136.

local operator In IMAGE PROCESSING, any function that, in order to produce a new value for a certain co-ordinate, requires the previous one and those in a local neighbourhood (usually, those observed in the NEAREST NEIGHBOURS). *Filters* belong to this class of functions. *ACF*

Loch Lomond Stadial The Late Glacial STADIAL in Britain traditionally dated to between ca 11 and 10 ka (radiocarbon years BP) (^{14}C years). It followed the WINDERMERE INTERSTADIAL, and culminated in the termination of the DEVENSIAN. It is equivalent to the YOUNGER DRYAS STADIAL of northwest Europe and the *Greenland Stadial* 1 (GS1) dated to between 12.9 and 11.7 ka (12,896–11,703 b2k). *MHD/CJC*

[*See also* LATE GLACIAL ENVIRONMENTAL CHANGE]

Golledge NR (2009) Glaciation of Scotland during the Younger Dryas: A review. *Journal of Quaternary Science* 25: 550–566.
Macleod A, Palmer A, Lowe JJ et al. (2011) Timing of glacier response to Younger Dryas climatic cooling in Scotland. *Global and Planetary Change* 79: 264–274.
Sissons JB (1979) The Loch Lomond Stadial in the British Isles. *Nature* 280: 199–202.

lodgement till A traditional name for a SEDIMENT deposited at the base of a sliding GLACIER by plastering on to a rigid or semi-rigid bed. *MRB*

[*See also* DIAMICTON, GLACIAL SEDIMENTS, SUBGLACIAL]

Clark PU and Hansel AK (1989) Clast ploughing, lodgement and glacier sliding over a soft glacier bed. *Boreas* 18: 201–207.

loess Well-sorted, sedimentary deposits of predominantly SILT-sized particles, first correctly identified as being of AEOLIAN origin by Ferdinand van Richthofen. The material is derived from ARIDLANDS and/or outwash surfaces and has often travelled long distances from freshly deglaciated and PERIGLACIAL areas during periods of glaciation, being deposited beyond the ice margin. Loess has therefore been entrained, transported, modified and sorted by aeolian transport processes. Loess deposits cover about 10 per cent of the Earth's continental surfaces and are most extensive in the Loess Plateau region of central China where the loess has a high carbonate content (sometimes exceeding 40 per cent by weight). The Chinese loess deposits can be >300 m thick and represent deposition which began as early as 2.5 million years ago.

Successions of loess deposits and interbedded PALAEOSOLS are interpreted as representing full GLACIAL conditions and warmer INTERGLACIAL or INTERSTADIAL episodes, respectively. In the deepest Chinese deposits almost 40 distinct palaeosol units have been identified. Loess deposits are also widespread in Central Europe, central Asia and the Great Plains of North America and, in the Southern Hemisphere, form the Pampas of Argentina and Uruguay. PALAEOCLIMATIC and PALAEOENVIRONMENTAL RECONSTRUCTIONS can be inferred from lithological analysis of loess sequences, and from a range of FOSSIL and SUBFOSSIL evidence, including *pollen, mollusca* and VERTEBRATES. Loess is commonly unstratified but can show well-developed jointing and is amenable to LUMINESCENCE DATING. *DH/CJC*

[*See also* LOESS STRATIGRAPHY]

Derbyshire E (ed.) (1995) *Wind blown sediments in the Quaternary record.* Chichester: Wiley [Quaternary Proceedings 4], 119–132.
Liu T (1988) *Loess in China.* Berlin: Springer.
Markovic SB, Bokhorst MP, Vandenberghe J et al. (2008) Late Pleistocene loess-palaeosol sequences in the Vojvodina region, north Serbia. *Journal of Quaternary Science* 23: 73–84.
Mason JA (2012) Evidence of environmental change from Aeolian and hillslope sediments and other terrestrial sources. In Matthews JA, Bartlein PJ, Briffa KR et al. (eds) *The SAGE handbook of environmental change, volume 1.* London: Sage, 284–304.
Nugteren G and Vandenberghe J (2004) Spatial climatic variability on the central Loess Plateau (China) as reconstructed by grain size for the last 250 kyr. *Global and Planetary Change* 41: 185–206.

LOESS regression LOESS or LOWESS stands for *locally weighted scatterplot smoothing.* It is a method of REGRESSION ANALYSIS in which *polynomials* of degree one (linear) or two (quadratic) are used to approximate the regression function in particular 'neighbourhoods' of the space of the *predictor variables.* It uses weighted LEAST SQUARES REGRESSION with local subsets of the data so as to pay less attention to distant points. It assumes no predetermined model for the entire data set and therefore provides no explicit formula for the fitted curve. LOESS is increasingly used in environmental research as a means of 'letting the data speak for themselves' and for highlighting '*signal* from NOISE' in EXPLORATORY DATA ANALYSIS and assessing patterns in graphical displays of statistical modelling results. *HJBB*

Cleveland WS and Devlin SJ (1988) Locally-weighted regression: An approach to regression analysis by local fitting. *Journal of the American Statistical Association* 83: 596–610.
Trexler JC and Travis J (1993) Non-traditional regression analysis. *Ecology* 74: 1629–1637.

loess stratigraphy Loess sequences can be characterised and correlated using a range of sedimentary properties and analyses, including HEAVY MINERAL ANALYSIS, CLAY MINERALS, MICROMORPHOLOGICAL ANALYSIS OF SEDIMENTS and MINERAL MAGNETISM. Loess sections usually display little visual evidence of stratification, although faint bedding may be seen with careful examination. SCANNING ELECTRON MICROSCOPY (SEM) and optical MICROSCOPY of particle fabric can distinguish differences in TEXTURE, which are important for PALAEOENVIRONMENTAL RECONSTRUCTION and CLIMATIC RECONSTRUCTION. On the basis of such analyses, two types of loess can be defined: the first from areas of high ARIDITY during GLACIAL EPISODES with humid INTERGLACIALS, as in China; the second from areas with persistently high humidity, as in Western Europe.

MAGNETIC SUSCEPTIBILITY analysis of loess deposits has shown that certain sedimentary units are more strongly marked than others and can be correlated between individual loess sections across regions. Often the most distinctive features in loess sections are the darker interbedded PALAEOSOL units. Variations within palaeosol units appear to reflect CLIMATIC CHANGE, with relatively long periods of PEDOGENESIS during interglacial stages reflected by well-developed, complex palaeosols. Shorter episodes of soil formation during INTERSTADIALS produce less prominent soil units. The successions of loess and interbedded palaeosols that have been found in Eastern Europe, China and the former Soviet Central Asia provide evidence of alternating cold (loess) and warm (palaeosols) stages extending back over at least 17 GLACIAL-INTERGLACIAL CYCLES.   *DH/CJC*

Ding Z, Yu Z, Rutter NW and Liu T (1994) Towards an orbital time scale for Chinese loess deposits. *Quaternary Science Reviews* 13: 39–70.
Forman SL, Oglesby R, Markgraf V and Stafford T (1995) Paleoclimatic significance of Late Quaternary eolian deposition on the Piedmont and High Plains, central United States. *Global and Planetary Change* 11: 35–55.
Guo ZT, Ruddiman WF, Hao QZ et al. (2002) Onset of Asian desertification by 22 My ago inferred from loess deposits in China. *Nature* 416: 159–163.
Kemp RA, Zarate M, Toms P et al. (2006) Late Quaternary paleosols, stratigraphy and landscape evolution in the Western Pampa, Argentina. *Quaternary Research* 66: 119–132.
Kukla G and An Z (1989) Loess stratigraphy in central China. *Palaeogeography, Plaeoclimatology and Palaeoecology* 72: 203–225.
Verosub KL, Fine P, Singer MJ and TenPas J (1993) Pedogenesis and paleoclimate: Interpretation of the magnetic susceptibility of Chinese loess-paleosol sequences. *Geology* 21: 1011–1014.

log   In the study of SEDIMENTS and rocks, particularly SEDIMENTARY ROCKS, a descriptive record of the characteristics of a vertical succession. A log can be compiled visually from a natural or artificial exposure, from a CORE or, using instruments, from a BOREHOLE. Information is recorded in the form of a columnar diagram or a GRAPHIC LOG.   *GO*

Coe AL (ed.) (2010) *Geological field techniques*. Milton Keynes: Wiley-Blackwell and Open University.
Ellis DV and Singer JM (2007) *Well logging for Earth scientists*. Dordrecht: Springer.

logging   (1) In the context of reconstructing environmental change, the recording of a continuous record (LOG) of stratigraphic units as a function of depth, according to visible variations in the rocks, SEDIMENTS or SOILS encountered. (2) The term also refers to the felling of trees, which may involve CLEAR CUTTING or SELECTIVE CUTTING, each with distinctive environmental impacts.   *DH/JAM*

[*See also* DEFORESTATION]

Laporte NT, Stabach JA, Grosch R et al. (2007) Expansion of industrial logging in Central Africa. *Science* 316: 1451.

logical positivism   A strongly empiricist, normative approach to the philosophy of science; most closely identified with the *Vienna Circle* of philosophers in the early 1900s. Logical positivism posits that SCIENCE and *philosophy* are advanced through rational observation and measurement of entities or their properties which are inherently knowable. This is performed by positively testing HYPOTHESES or statements describing these entities or properties. As hypotheses are revised and retested, there is the logical expectation that they become more accurate reflections of reality. INDUCTION (or *inductive reasoning*) and VERIFICATION (or the *verifiability principle*) are key attributes distinguishing logical positivism from other philosophical positions, such as CRITICAL RATIONALISM and REALISM.   *MAU/CET/JAM*

[*See also* POSITIVISM]

Ayer AJ (ed.) (1959) *Logical positivism*. New York: Simon and Schuster.
Lee JA (2000) *The scientific endeavor: A primer on scientific principles and practice*. San Francisco: Addison Wesley.
Ray C (2000) Logical positivism. In Newton-Smith WH (ed.) *A companion to the philosophy of science*. Oxford: Blackwell, 243–251.

long waves   A hemispheric-scale wave pattern in the upper-air WESTERLIES, also known as *Rossby waves*. Preferred positions for troughs and ridges (wavelengths typically 3,000–6,000 km) strongly influence the formation and direction of motion of disturbances, such as mid-latitude CYCLONES. Departures from quasi-stationary long-wave conditions are described by the *index cycle* (see ZONAL INDEX) and are frequently associated with climatic ANOMALIES. Related long waves also develop in association with OCEAN CURRENTS.   *JCS*

Cipollini P, Cromwell D, Challenor P and Raffaglio S (2001) Rossby waves detected in global ocean colour data. *Geophysical Research Letters* 28: 323–326.
Dickinson RE (1978) Rossby waves: Long-period oscillations of oceans and atmospheres. *Annual Review of Fluid Mechanics* 10: 159–195.
Harman J (1991) *Synoptic climatologies of the Westerlies: Processes and patterns*. Washington, DC: Association of American Geographers.

Platzman GW (1968) The Rossby wave. *Quarterly Journal of the Royal Meteorological Society* 94: 225–248.

longitudinal dune   A long, narrow SAND DUNE, parallel to the DOMINANT WIND (or perhaps two obliquely converging PREVAILING WINDS) responsible for its formation in DESERTS with a moderate sand supply. Such dunes are also referred to as *linear dunes* and they can reach lengths of tens of kilometres, such as the *qos dune systems* of the Sudan. A *seif dune* is a longitudinal dune with a sinuous crest along which the slip face alternates from one side of the dune to the other (in response to seasonal differences in wind direction). In arid phases of the QUATERNARY, longitudinal dunes were active over much larger areas than currently.   *MAC*

[*See also* ARIDLAND: PAST ENVIRONMENTAL CHANGE, PALAEODUNES]

Bristow CS, Bailey SD and Lancaster N (2000) The sedimentary structure of linear sand dunes. *Nature* 406: 56–59.
Lancaster N (1995) *Geomorphology of desert dunes.* London: Routledge.
Mason JA, Swinehart JB and Hanson PR (2011) Late Pleistocene dune activity in the central Great Plains, USA. *Quaternary Science Reviews* 30: 3858–3870.

longshore drift   The movement of SAND and other material along the shore by *wind-* or *wave-driven currents* shoreward and within the *breaker zone.* The amount of sediment transported by longshore drift depends on WAVE ENERGY, *beach slope* and *wave approach angle.* Maximum wave-driven longshore transport is achieved when waves break at 45° to the beach. *Beach drifting* is the movement of beach material induced by the swash and backwash of waves but longshore drift also occurs within the breaker zone. It includes suspended sediment and BEDLOAD as well as the water itself.   *HJW*

[*See also* BEACH DEPOSITS, GROYNE, SHORE ZONE]

Ingle Jr JC (1966) *The movement of beach sand.* New York: Elsevier.
Komar PD (1971) The mechanism of sand transport on beaches. *Journal of Geophysical Research* 76: 713–721.

look-up table   In IMAGE PROCESSING, a memory-resident table that associates colours and numbers as used to convert DIGITAL NUMBERS (DNs) in an image into colours on a display device. It is an economic way of colouring objects in computational systems and, usually, the user has access to this object in order to specify the behaviour of the display and other graphic devices. *ACF*

loss on ignition (LOI)   A simple, widely used measure of the ORGANIC CONTENT of soils and sediments, obtained by weight difference before and after a soil sample is subjected to high-temperature ignition. It also provides a means of measuring the *minerogenic content.* Unless carried out at sufficiently low temperatures, ~550°C, the method includes weight losses through breakdown of carbonates. At higher temperatures (~800°C), it forms the basis of a method for determination of *carbonate content* of rocks and sediments.

EMB

Ball DF (1967) Loss-on-ignition as an estimate of organic matter and organic carbon in non-calcareous soils. *Journal of Soil Science* 15: 84–92.
Dean WE (1974) Determination of carbonate and organic matter in calcareous sediments and sedimentary rocks by loss on ignition: Comparison with other methods. *Journal of Sedimentary Petrology* 44: 242–248.
Heiri O, Lotter AF and Lemcke G (2001) Loss on ignition as a method for estimating organic carbonate content in sediments: Reproducibility and comparability of results. *Journal of Paleolimnology* 25: 101–110.
Smith JG (2003) Aspects of the loss-on-ignition (LOI) technique in the context of clay-rich glaciolacustrine sediments. *Geografiska Annaler* 85A: 91–97.

Low Arctic   The TUNDRA environment in the Northern Hemisphere north of the TREE LINE where there is a more-or-less predominantly complete vegetation cover, dominated by flowering plants including shrubby growth and dwarf woodland up to 2.0 m high in places. With a GROWING SEASON of 2–3 months, the Low Arctic has a richer plant and animal assemblage than the HIGH ARCTIC.   *HMF*

[*See also* ARCTIC, SUBARCTIC]

French HM (1999) Arctic environments. In Alexander DE and Fairbridge RW (eds) *Encyclopedia of environmental science.* Dordrecht: Kluwer Academic, 29–33.

low-activity clays   Clays with a low CATION EXCHANGE CAPACITY (<24 cmol$_c$/kg), associated with ACRISOLS, FERRALSOLS and LIXISOLS.   *EMB*

[*See also* HIGH-ACTIVITY CLAYS]

lowstand   The minimum of an environmental cycle, fluctuation or oscillation as reflected in, for example, GLACIER VARIATIONS, LAKE-LEVEL VARIATIONS and SEA-LEVEL CHANGE.   *JAM*

[*See also* HIGHSTAND]

Cross SL, Baker PA, Seltzer GO et al. (2000) A new estimate of the Holocene lowstand level of Lake Titicaca, central Andes, and implications for tropical palaeohydrology. *The Holocene* 10: 97–108.
Graham NE and Hughes MK (2007) Reconstructing the Mediaeval low stands of Mono Lake, Sierra Nevada, California, USA. *The Holocene* 17: 1197–1210.
Skorko K, Jewell PW and Nicoll K (2012) Fluvial response to an historic lowstand of the Great Salt Lake, Utah. *Earth Surface Processes and Landforms* 37: 143–156.

luminescence dating   A method used to determine the time elapsed since deposition of mineral grains (mainly QUARTZ and FELDSPAR) in sediments, including AEOLIAN, *fluvial* and GLACIOFLUVIAL SEDIMENTS. It is also used in ARCHAEOLOGY to date CERAMICS, baked clay and flints from fire hearths. In any geological environment, natural RADIATION induces free *electrons* in minerals that can be trapped in lattice defects (ELECTRON TRAPS).

Luminescence is the light emitted when trapped electrons are subjected to heat (*thermoluminescence* (TL)), a beam of light of visible wavelengths (*optically stimulated luminescence* (OSL)) or infrared wavelengths (INFRARED-STIMULATED LUMINESCENCE (IRSL)). The number of trapped electrons, and hence the luminescence signal, is proportional to the strength of the radioactive field (*dose rate*) and the length of time since the signal was last set to zero by exposure to light or heat (a process known as *bleaching*). The effective dating limit varies from site to site and is the time elapsed between exposure and saturation of defect sites, and depends on the radiation dose and capacity to accumulate electrons, but is generally accepted to be about 0.25 million years (Ma). Recent development of *Thermally Transferred Optically Stimulated Luminescence* (TT-OSL) and its application to fine-grained quartz in Chinese LOESS sequences offers the potential to extend the limit to around 1 Ma. The intensity of the luminescence emitted from a natural sample is compared with that from the same mineral separately irradiated in the laboratory by a dose of *beta* and/or *gamma radiation* to calculate the PALAEODOSE or *equivalent dose*. The age of deposition is determined using the equation: *age = palaeodose/dose rate*, where dose rate is derived from the concentration of radioactive components in the sediment and surrounding matrix (uranium, thorium and potassium, and to a lesser degree, rubidium) and the COSMIC-RAY FLUX.

One of the major assumptions of luminescence dating is that the signal is set to zero immediately prior to burial. However, *zeroing* will be incomplete in some environments such as fluvial or glaciofluvial sediments where exposure to sunlight may be limited. The advantage of OSL and IRSL dating is that the electron traps utilised in these techniques are more sensitive to light than traps stimulated by heating in TL dating, and only brief exposure to sunlight is likely to reduce luminescence to zero. OSL and IRSL dating are therefore applicable to a wider variety of deposits, although IRSL only occurs in feldspars. Deposits as young as a few decades in age can be dated using OSL and IRSL.   *DAR*

Aitken MJ (1998) *An introduction to optical dating: The dating of Quaternary sediments by the use of photon-stimulated luminescence.* Oxford: Oxford University Press.

Bateman MD (2008) Luminescence dating of periglacial sediments and structures. *Boreas* 37: 574–588.
Duller GAT (1996) Recent developments in luminescence dating of Quaternary sediments. *Progress in Physical Geography* 20: 127–145.
Feathers JK (1997) Application of luminescence dating in American archaeology. *Journal of Archaeological Method and Theory* 4: 1–66.
Lian OB and Huntley DJ (2001) Luminescence dating. In Last WM and Smol JP (eds) *Tracking environmental change using lake sediments, volume 1.* Dordrecht: Kluwer, 261–282.
Stokes S (1999) Luminescence dating applications in geomorphological research. *Geomorphology* 29: 153–171.
Tsukamoto S, Duller GA, Murray AS and Choi JH (eds) (2009) Application of luminescence dating in geomorphology. *Geomorphology* 109: 1–78 [Special Issue].
Wintle AG (1993) Luminescence dating of aeolian sands: An overview. In Pye K (ed.) *The dynamics and environmental context of aeolian sedimentary systems.* London: Geological Society, 49–58.
Wintle AG (2008) Fifty years of luminescence dating. *Archaeometry* 50: 276–312.
Wintle AG (2008) Luminescence dating: Where it has been and where it is going. *Boreas* 37: 471–482.

lunar cycles   Cycles due to the gravitational pull of the Moon on the Earth and its ATMOSPHERE. They range from the *semi-diurnal tides* to the 18.6-year lunar cycle. The atmospheric lunar tide is too small (0.2 hPa) to be of practical importance but, when combined with the effect of the Sun, the semi-diurnal pressure variation ranges from 4 hPa in the tropics to 1 hPa in mid-latitudes. As a result of its orbit characteristics in relation to the Earth, the Moon's declination relative to the ECLIPTIC varies with a periodicity of 18.6 years and its tidal force is greatest when its *perigee* position coincides with maximum declination. Using maximum entropy SPECTRAL ANALYSIS, the 18.6-year cycle has been found in a variety of long-term climatic records: air temperature, atmospheric pressure, Chinese DROUGHTS, European VINE HARVEST date and *fish catches*, TREE RINGS, NILE FLOODS, North Atlantic CYCLONES and *thunderstorm* occurrences. However, the physical cause-and-effect relationship is difficult to establish.   *BDG*

Currie RG (1996) M_n and S_c signals in North Atlantic tropical cyclone occurrence. *International Journal of Climatology* 16: 427–439.
Lamb HH (1972) *Climate: Present, past and future, volume 1.* London: Methuen.

Lusitanean floral element   Plant species with their main range in Portugal (latitude: Lusitania); some of which exhibit DISJUNCT DISTRIBUTIONS extending to western Ireland and/or southwest England.   *BA*

Luvisols   Soils with subsurface *argic horizon* (also known as an *argillic horizon*) into which HIGH-ACTIVITY CLAYS have migrated from higher in the soil profile. The clay is mobilised from aggregates in the surface soil, transported downwards in aqueous suspension and immobilised by reaggregation (FLOCCULATION) and filtration in the argic horizon (see LESSIVAGE). Also known as *grey-brown podzolic soils*, *pseudo-podzols*, *sols lessivés* and *parabraunerde*. Luvisols are diverse and occur widely under *cool-temperate* and *warm-temperate* (*Mediterranean-type*) climates in North America and South America, Europe and Australia where there is a dry season (SOIL TAXONOMY: *Alfisols*). They are predominantly brown in colour with some RUBIFICATION (RUBEFACTION) in the warmer regions. The presence of Luvisols can be an indication of a stable land surface and some Luvisols occur in environments that are no longer conducive to clay eluviation. Other Luvisols have formed from ALBELUVISOLS that have been truncated by erosion. With change to a wetter climate, these soils could suffer from GLEYING and ACIDIFICATION. Being characterised by an intermediate stage of WEATHERING, Luvisols tend to be inherently fertile and are widely used for crop production but are susceptible to erosion.   *EMB*

[*See also* TERRA FUSCA, TERRA ROSSA, WORLD REFERENCE BASE FOR SOIL RESOURCES (WRB)]

Růšek L, Růškova M, Voříšek K et al. (2009) Chemical and microbiological characteristics of Cambisols, Luvisols and Stagnosols. *Plant, Soil, and Environment* 55: 231–237.

Terhorst B and Ottner F (2003) Polycyclic Luvisols in northern Italy: Palaeopedological and clay mineralogical characteristics. *Quaternary International* 106: 215–231.

lynchet   A so-called *cultivation terrace* along the edge of a field boundary, which is formed by SOIL EROSION and deposition, especially by SLOPEWASH, of cultivated fields. Erosion tends to be most obvious on the downslope side of walls or banks (*negative lynchet*) whereas deposition occurs against the obstacle on the upslope side (*positive lynchet*). Lynchets are important indicators of rates of erosion and of prehistoric FIELD SYSTEMS.   *JAM*

[*See also* STRIP LYNCHET]

lysimeter   A device for measuring PERCOLATION and LEACHING losses of water and solutes from a column of soil under controlled conditions. Lysimeters can also be used to measure the EVAPOTRANSPIRATION of water from vegetated soil surfaces, net additions to the soil, and hence the water balance of the *soil-plant system*.   *SN/JAM*

[*See also* HYDROLOGICAL BALANCE/BUDGET]

Howell TA (2005) Lysimetry. In Hillel D (ed.) *Encyclopedia of soils in the environment*. Oxford: Elsevier, 379–386.

lysocline   The depth in the OCEANS at which the solution rate of calcium carbonate increases markedly. It is thought to correspond to the upper boundaries of cold, corrosive BOTTOM WATERS. The lysocline is found a few hundred metres above the CARBONATE COMPENSATION DEPTH (CCD).   *BTC*

Ma   A *mega-annum*: An abbreviation for millions of years before the present.   *GO*

maar   A broad, low-rimmed CRATER caused by a VOLCANIC ERUPTION in which MAGMA encountered GROUNDWATER, converting it to steam with explosive expansion (see HYDROVOLCANIC ERUPTION). Maars are typically around 1 km in diameter and are commonly occupied by a *maar lake*, usually 10–500 m deep. These fill with SEDIMENT of organic, local slope-wash and aeolian origin, providing excellent opportunities for *palaeoclimatic reconstruction* and PALAEOENVIRONMENTAL RECONSTRUCTION. The term was originally derived from the Eifel region of Germany. *Tuff rings* are similar features, and *tuff cones* are taller.   *JBH*

[*See also* LACUSTRINE SEDIMENTS]

Brauer A (1999) High resolution sediment and vegetation response to Younger Dryas climate change in varved lake sediments from Meerfelder Maar, Germany. *Quaternary Science Reviews* 18: 321–329.

DeBenedetti AA, Funiciello R, Giordano G et al. (2008) Volcanology, history and myths of the Lake Albano maar (Colli Albani volcano, Italy). *Journal of Volcanology and Geothermal Research* 176: 387–406.

Mingram J (1998) Laminated Eocene maar-lake sediments from Eckfeld (Eifel region, Germany) and their short-term periodicities. *Palaeogeography, Palaeoclimatology, Palaeoecology* 140: 289–305.

Negendank JFW and Zolitschka B (eds) (1993) *Palaeolimnology of European maar lakes.* Berlin: Springer.

Németh K, Goth K, Martin U et al. (2008) Reconstructing paleoenvironment, eruption mechanism and paleomorphology of the Pliocene Pula maar, (Hungary). *Journal of Volcanology and Geothermal Research* 177: 441–456.

Sohn YK and Park KH (2005) Composite tuff ring/cone complexes in Jeju Island, Korea: Possible consequences of substrate collapse and vent migration. *Journal of Volcanology and Geothermal Research* 141: 157–175.

Watts WA, Allen JRM and Huntley B (1996) Vegetation history and palaeoclimate of the Last Glacial period at Lago Grande di Monticchio, southern Italy. *Quaternary Science Reviews* 15: 133–153.

White JDL and Ross P-S (2011) Maar-diatreme volcanoes: A review. *Journal of Volcanology and Geothermal Research* 201: 1–29.

machair   Originally, calcareous COASTAL DUNE pastures of the highlands and islands of northwestern Scotland, although the term is also now applied to coastal dune PASTURES in other locations, such as Ireland and New Zealand. Though currently treeless, the Scottish machair LANDSCAPES may have had a significant woodland cover in the early Holocene.   *JAM*

Cooper A, McCann T and Ballard E (2005) The effects of livestock grazing and recreation on Irish machair grassland vegetation. *Plant Ecology* 181: 255–267.

Edwards KJ, Whittington G and Ritchie W (2005) The possible role of humans in the early stages of machair evolution: Palaeoenvironmental investigations in the Outer Hebrides, Scotland. *Journal of Archaeological Science* 32: 435–449.

Ritchie W (1976) The meaning and definition of machair. *Transactions of the Botanical Society of Edinburgh* 42: 431–440.

macroclimate   The distinguishing climatic features of large areas. Horizontal scales are typically above 100 km², at least regional and frequently subcontinental to global. Macroclimate incorporates, and is controlled by, the large-scale features of the GENERAL CIRCULATION OF THE ATMOSPHERE.   *JCS*

[*See also* LOCAL CLIMATE, MESOCLIMATE, MICROCLIMATE]

macro-ecology   An approach to ECOLOGY focusing on the emergent statistical phenomena exhibited by ASSEMBLAGES of species, for example, body size, geographical RANGE and abundance, and on mechanisms concerning how species use and divide energy, space and other resources.   *RJW*

[*See also* AUTECOLOGY, BIOGEOGRAPHY, COMMUNITY ECOLOGY, ECOLOGY, EMERGENCE AND EMERGENT PROPERTIES]

Beck J, Ballesteros-Mejia L, Buchmann CM et al. (2013) What's on the horizon for macroecology. *Ecography* 35: 673–683.

Blackburn TM and Gaston KJ (eds) (2004) *Macroecology: Concepts and consequences*. Cambridge: Cambridge University Press.

Fisher JAD, Frank KT and Leggett WC (2010) Dynamic macroecology on ecological timescales. *Global Ecology and Biogeography* 19: 1–15.

Gaston KJ and Blackburn TM (2000) *Pattern and process in macroecology*. Oxford: Blackwell.

Witman JD and Kaustov R (eds) (2009) *Marine macroecology*. Chicago: University of Chicago Press.

macro-evolution EVOLUTION of species' POPULA-TIONS with a focus on SPECIATION and higher taxonomic groups, in contrast to *micro-evolution*, which focuses on smaller-scale changes, ADAPTATION below the species level and the associated GENETICS. Macro-evolution considers, for example, the patterns in the TREE OF LIFE, ADAPTIVE RADIATION, CONVERGENCE, IN EVOLUTION and DIVERGENCE, IN EVOLUTION, which occur too slowly to be observed or subjected to EXPERIMENT but evidence of which is available from PALAEONTOLOGY and the GEO-LOGICAL RECORD, *developmental biology* and *genomics* (see GENOME).   *JAM/KJF*

[*See also* EXTINCTION, GRADUALISM, PHYLETIC GRADUALISM, PUNCTUATED EQUILIBRIA]

Barbieri M (ed.) (2008) *The codes of life: The rules of macroevolution*. Berlin: Springer.

Briggs D (2009) *Plant microevolution and conservation in human-influenced ecosystems*. Cambridge: Cambridge University Press.

Jablonski D (2007) Scale and hierarchy in macroevolution. *Palaeontology* 50: 87–109.

Levinton JS (2001) *Genetics, palaeontology and microevolution*, 2nd edition. Cambridge: Cambridge University Press.

Reiss JO (2009) How to talk about macroevolution. In Reiss JO (ed.) *Not by design: Retiring Darwin's watchmaker*. Berkeley: University of California Press, 279–311.

Resnick DN and Ricklefs RE (2009) Darwin's bridge between microevolution and macroevolution. *Nature* 457: 837–842.

macrofossil A FOSSIL or SUBFOSSIL of a size visible to the naked eye. For identification purposes, a microscope or strong binocular loop is usually needed. Examples include *fruits*, *seeds* and *leaves*. Macrofossils often have the advantage over MICROFOSSILS, particularly pollen, because identification can more readily be made to the species instead of the genus or family level. Macrofossils also experience less long-distance transport than microfossils.   *BA/CJC*

[*See also* PALAEOBOTANY, PALAEOECOLOGY, PLANT MACROFOSSIL ANALYSIS]

Mauquoy D, Hughes PD and van Geel B (2010–2011) A protocol for plant macrofossil analysis of peat deposits. *Mires and Peat* 7: 1–5.

macronutrients (1) ELEMENTS other than *carbon*, *hydrogen* and OXYGEN, which are required by organisms in substantial amounts, including CALCIUM (Ca), *chlorine*, *magnesium*, *nitrogen*, PHOSPHORUS (P), POTASSIUM (K) *and sulfur*. (2) The major classes of COMPOUNDS required by humans as food, namely, *carbohydrates*, *proteins* and *fats*.   *JAM*

[*See also* MICRONUTRIENTS, TRACE ELEMENT]

Madden-Julian oscillation (MJO) An intraseasonal fluctuation of 30–60 days within the tropical, zonal WALKER CIRCULATION, which is characterised by an eastward shift in convectional activity and cloud clusters from the Indian Ocean to the central Pacific Ocean. The Madden-Julian oscillation occurs throughout the year but its amplitude and frequency are related to the MONSOON circulation and EL NIÑO-SOUTHERN OSCILLATION (ENSO). In turn, it may influence the evolution of extra-tropical weather on timescales of months to seasons.
                              JAM

Chand SS and Walsh KJE (2010) The influence of the Madden-Julian oscillation on tropical cyclone activity in the Fiji region. *Journal of Climate* 23: 868–886.

Klotzbach PJ (2010) On the Madden-Julian oscillation-Atlantic hurricane relationship. *Journal of Climate* 23: 282–293.

Lau KM and Chan PH (1986) The 40-50 day oscillation and the El Niño/southern oscillation: A new perspective. *Bulletin of the America Meteorological Society* 67: 533–534.

Madden R and Julian PR (1971) Detection of a 40-50 day oscillation in the zonal wind in the tropical Pacific. *Journal of Atmospheric Science* 28: 702–708.

Pohl B, Fauchereau N, Reason CJC and Rouault M (2010) Relationships between the Antarctic oscillation, the Madden-Julian oscillation, and ENSO, and consequences for rainfall analysis. *Journal of Climate* 23: 238–254.

made ground A type of ARTIFICIAL GROUND, such as a road or rail embankment or LANDFILL site, where material has been placed on a pre-existing land surface.
                              JAM

magma Molten material generated in the Earth's CRUST and upper MANTLE, which solidifies to form an IGNEOUS ROCK. Magma is a mixture of silicate melt, crystals and gas. Gas is *exsolved* during *decompression* as magma approaches the Earth's surface, releasing VOLA-TILES to the ATMOSPHERE. Reservoirs of magma (*magma chambers*) underlie and feed many ACTIVE VOLCANOES. Magma temperatures in VOLCANIC ERUPTIONS lie within the range 700–1,200°C as the composition ranges from silica-rich (*acidic*) to silica-poor (*basic*).   *JBH*

Francis P and Oppenheimer C (2003) *Volcanoes*. Oxford: Oxford University Press.

Middlemost EAK (1997) *Magmas, rocks and planetary development: A survey of magma/igneous rock systems.* Harlow: Longman.

magnesium:calcium ratio (Mg:Ca)

A *proxy environmental indicator* method for determining *deep-sea temperatures* and, in combination with OXYGEN ISOTOPES, PALAEOSALINITY, based on the tests of *benthic foraminifera* and *ostracods.* Mg/Ca PALAEOTHERMOMETRY came widely into use in the 1990s and has been recently applied to *planktonic* foraminifera.   *JAM*

Elderfield H, Yu J, Arnand P et al. (2006) Calibrations for benthic foraminiferal Mg/Ca paleothermometry and the carbonate ion hypothesis. *Earth and Planetary Science Letters* 250: 633–649.

Farmer EJ, Chapman MR and Andrews JE (2011) Holocene temperature evolution of the subpolar North Atlantic recorded in the Mg/Ca ratios of surface and thermocline dwelling planktonic foraminifers. *Global and Planetary Change* 79: 234–243.

Martin PA, Lea DW, Rosenthal Y et al. (2002) Quaternary deep-sea temperature histories derived from benthic foraminiferal Mg/Ca. *Earth and Planetary Science Letters* 198: 193–209.

Rosenthal Y, Boyle EA and Slowey N (1997) Temperature control on the incorporation of magnesium, strontium, fluorine and cadmium into benthic foraminifera shells from Little Bahama Bank: Prospects for thermocline paleoceanography. *Geochimica and Cosmochimica Acta* 61: 3633–3643.

magnetic anomaly

A localised variation in the Earth's MAGNETIC FIELD superimposed on the ambient regional magnetic field which results from the deep geological character of the Earth (see GEOMAGNETISM). Magnetic anomalies can be attributed to the magnetic properties of near-surface materials, and GEOPHYSICAL SURVEYING can be used to detect features such as GEO-LOGICAL STRUCTURES, *archaeological structures* and ORE deposits. The interpretation of ocean-floor magnetic anomalies that occur as bands parallel to the MID-OCEAN RIDGE (MOR) and are symmetrical either side of the ridge as a record of GEOMAGNETIC POLARITY REVERSALS preserved by SEA-FLOOR SPREADING was critical to the development of PLATE TECTONICS theory in the early 1960s. Ocean-floor magnetic anomalies form the basis of the GEOMAGNETIC POLARITY TIMESCALE (GPTs). They are numbered from 1, the most recent, such that those with dominantly normal polarity have odd numbers and those with dominantly reversed polarity have even numbers.   *DNT*

[*See also* ARCHAEOLOGICAL PROSPECTION]

Kristjansson L and Geirfinnur J (2007) Paleomagnetism and magnetic anomalies in Iceland. *Journal of Geodynamics* 43: 30–54.

Nicolosi I, Speranza F and Chiappini M (2006) Ultrafast oceanic spreading of the Marsili Basin, southern

Tyrrhenian Sea: Evidence from magnetic anomaly analysis. *Geology* 34: 717–720.

Vine FJ (1966) Spreading of the ocean floor: New evidence. *Science* 154: 1405–1415.

magnetic excursion

A magnetic EVENT in which the geomagnetic pole (see GEOMAGNETISM) moves from one geographical pole to a position close to the other but continues moving back towards the original pole rather than stabilising in the opposite position as in a true GEOMAGNETIC POLARITY REVERSAL. Such excursions are evident in the record of PALAEOMAGNETISM, particularly in sequences of SEDIMENTARY ROCKS, and are a valuable tool in CORRELATION (STRATIGRAPHICAL).   *DNT*

[*See also* MAGNETOSTRATIGRAPHY]

Blanchet CL, Thouveny N and de Garidel-Thoron T (2006) Evidence for multiple paleomagnetic intensity lows between 30 and 50 ka BP from a western equatorial Pacific sedimentary sequence. *Quaternary Science Reviews* 25: 1039–1052.

Laj C, Kissel C and Roberts AP (2006) Geomagnetic field behavior during the Iceland Basin and Laschamp geomagnetic excursions: A simple transitional field geometry? *Geochemistry, Geophysics, Geosystems* 7: Q03004.

Valet J-P and Valladas H (2010) The Laschamp-Mono lake geomagnetic events and the extinction of Neanderthal: A causal link or a coincidence? *Quaternary Science Reviews* 29: 3887–3893.

magnetic field

It is produced by a permanent magnet or where there is electrical charge in motion and is measured in amperes per metre (A/m). It is the change in energy (H) generated in a given volume of space such that the energy gradient created produces a force which can be detected, e.g. the torque induced on a compass needle.   *AJP/WZ*

Jiles D (1991) *Introduction to magnetism and magnetic minerals.* London: Chapman and Hall.

magnetic grain size

Ferromagnetic and ferrimagnetic materials can be divided into different regions or cells of MAGNETISATION, which are known as *domains.* Larger magnetic minerals contain many domains, and are called *multidomain* (MD), because energetically it is favourable to have more than one magnetic domain. In small grains, only one domain will form, and these are then *single domain* (SD) or *stable single domain* (SSD). Grains between these two types of domain have more than one domain but exhibit the magnetic properties of SD grains; such grains are termed pseudo-single domain (PSD). Very small ferro- or ferrimagnetic grains have thermal vibrations at room temperature, which are of the same order of magnitude as their magnetic energies. Consequently, these *super-paramagnetic* (SP) materials do not exhibit stable remanent magnetisation. The size boundaries between the various domain types are a function of the mineral type and shape. For MAGNETITE, the size boundaries among SP, SD, PSD and MD are roughly 0.03, 0.1 and 10 μm, respectively. Under certain circumstances, magnetic grain size can be used as a rapid, non-destructive proxy for physical grain size.   *WZ/AJP*

[**See also** FERROMAGNETISM, MINERAL MAGNETISM]

Dekkers MJ (2007) Magnetic proxy parameters. In Gubbins D and Herrero-Bervera E (eds) *Encyclopedia of geomagnetism and palaeomagnetism*. Berlin: Springer, 525–534.

Oldfield F, Hao QZ, Bloemendal J et al. (2009) Links between bulk sediment particle size and magnetic grain-size: General observations and implications for Chinese loess studies. *Sedimentology* 56: 2091–2106.

Peters C and Dekkers MJ (2003) Selected room temperature magnetic parameters as a function of mineralogy, concentration and grain size. *Physics and Chemistry of the Earth* 28: 659–667.

magnetic hysteresis For ferromagnetic, ferrimagnetic and imperfect antiferromagnetic minerals, the relationship between MAGNETISATION (*M*) and an external MAGNETIC FIELD (*H*) is non-linear (see Figure). When a strong positive field is applied, the MAGNETISATION becomes saturated. This is known as *saturation magnetisation* (Ms). As *H* is then removed, *M* does not decline to zero but is left with a remanent magnetisation, which is called *saturation remanent magnetisation* or *saturation remanence* (Mrs). If, however, the original field was insufficient to achieve saturation, the remanent magnetisation is referred to as the *Remanence* (Mr). Then, increasing *H* in the negative direction, the field necessary to reduce *M* to zero is termed the *coercive force* or *coercivity* (Hc). By further increasing *H* in the negative direction, *M* becomes saturated in this negative direction. Repeated cycling of *H* generates a hysteresis loop. For Mrs to become zero, a negative field that is stronger than Hc is necessary. This is termed the *coercivity of remanence* (Hcr). By plotting Mrs/Ms against Hcr/Hc on a so-called *Day plot*, it is possible to estimate the *average domain state*, or MAGNETIC GRAIN SIZE, of the magnetic minerals present.   *WZ/AJP*

Day R, Fuller M and Schmidt VA (1977) Hysteresis properties of titanomagnetites: Grain size and compositional dependence. *Physics of the Earth and Planetary Interiors* 13: 260–267.

Dunlop DJ (2002) Theory and application of the Day plot (Mrs/Ms versus Hcr/Hc) 1: Theoretical curves and tests using titanomagnetite data. *Journal of Geophysical Research-Solid Earth* 107(B3). doi:10.1029/2001JB000486.

Evans ME and Heller F (2003) *Environmental magnetism: Principles and applications of enviromagnetics.* San Diego, CA: Academic Press.

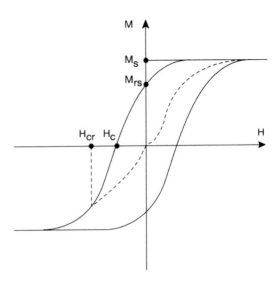

Magnetic hysteresis *Expressed in terms of MAGNETIC FIELD (H) and MAGNETISATION (M). Explanations for the terms Ms, Mrs, Hc and Hcr are given in the text (Evans and Heller, 2003).*

magnetic induction When a MAGNETIC FIELD, *H*, has been generated in a medium, the response of the medium is its magnetic induction, *B* (also known as *magnetic flux density*). Magnetic induction is measured in *tesla* (T).   *AJP/WZ*

magnetic intensity The strength (*F*) of the Earth's MAGNETIC FIELD, measured in *tesla* (T). Natural materials can acquire a MAGNETIC REMANENCE in the Earth's magnetic field and, hence, the fossil record can be used to study ancient direction and intensity.   *AJP/WZ*

[**See also** PALAEOMAGNETIC DATING]

magnetic moment The *torque* (T_{max}) produced in the presence of a MAGNETIC FIELD. It is given by the equation $m = T_{max}/B$, where *B* is the MAGNETIC INDUCTION. The unit of magnetic moment is ampere per square metre (A/m^2).   *AJP/WZ*

magnetic remanence Upon removal of an applied MAGNETIC FIELD, the MAGNETISATION of a given material may not return to zero—that is, the material is no longer unmagnetised but has a *remanent magnetisation*.
                                                      AJP/WZ

magnetic secular variation The Earth's MAGNETIC FIELD can be described according to its strength and direction. In addition to MAGNETIC INTENSITY, the direction of the field is expressed as an angle of dip below the horizontal plane, that is, the INCLINATION (I), and the angle between the horizontal component of the field and the true geographical north, that is, the DECLINATION (D). Secular variation or *palaeomagnetic secular variation* (PSV) refers to systematic changes in the Earth's magnetic field over time. The first measurement of declination at London was in ca AD 1570, when it was found that the compass needle pointed 11° E. In AD 1660, the compass pointed due north, and by AD 1820 it had swung around to 24° W. Since then, declination has decreased and is now approximately 3° W of true north, decreasing by about 0.5° every four years.

AJP/WZ

[*See also* PALAEOMAGNETIC DATING]

Evans ME and Heller F (2003) *Environmental magnetism: Principles and applications of enviromagnetics*. San Diego, CA: Academic Press.
Thompson R (1973) Palaeolimnology and palaeomagnetism. *Nature* 242: 182–184.
Thompson R and Oldfield F (1986) *Environmental magnetism*. London: Allen and Unwin.

magnetic susceptibility   The MAGNETISATION (M) acquired per unit of applied MAGNETIC FIELD (H), is the MAGNETIC SUSCEPTIBILITY (κ), where $\kappa = M/H$. Thus, it represents the extent to which any material, such as sediment or soil, can be magnetised. It is a dimensionless parameter in SI units when expressed per unit volume of M (κ is more accurately the *volume susceptibility*), but when divided by density it becomes the *mass susceptibility* (χ), measured in units of cubic metre per kilogram (m³/kg), where $\chi = \kappa/\rho$. Various measures of magnetic susceptibility (see MINERAL MAGNETISM) are widely used in CORE CORRELATION, in the characterisation of sediments and in PALAEOENVIRONMENTAL RECONSTRUCTION.   *AJP/WZ/JAM*

Blundell A, Dearing JA, Boyle JF and Hannam JA (2009) Controlling factors for the spatial variability of soil magnetic susceptibility across England and Wales. *Earth-Science Reviews* 95: 158–188.
Dearing JA (1999) *Environmental magnetic susceptibility*. Kenilworth: Chi Publishing.
Maher BA and Thompson R (1992) Palaeoclimatic significance of the mineral magnetic record of the Chinese loess and palaeosols. *Quaternary Research* 37: 155–170.

magnetisation   The MAGNETIC MOMENT per unit volume of a solid: $M = m/V$. It is measured in amperes per metre (A/m).   *AJP/WZ*

magnetite   One of the most common and perhaps the most important *iron oxide* minerals in terms of magnetism, Fe_3O_4, found in the majority of IGNEOUS ROCKS, many metamorphic and sedimentary rocks and nearly all SOILS. It is a dense, shiny black mineral with a cubic crystallographic structure, which exhibits ferrimagnetic behaviour. Low-frequency MAGNETIC SUSCEPTIBILITY values for magnetite lie in the range of $400–1,000 \times 10^{-6}$ m³/kg.   *AJP/WZ*

[*See also* FERROMAGNETISM, MINERAL MAGNETISM]

Dunlop DJ and Özdemir Ö (1997) *Rock magnetism: Fundamentals and frontiers*. Cambridge: Cambridge University Press.
Evans ME and Heller F (2003) *Environmental magnetism: Principles and applications of enviromagnetics*. San Diego, CA: Academic Press.

magnetometer   A device used for measuring the strength and direction of a MAGNETIC FIELD. Magnetometer measurements of variations in the Earth's geomagnetic field (see GEOMAGNETISM) and magnetisation of materials can be used in PALAEOMAGNETIC DATING.   *MHD*

magnetostratigraphy   The use of variations in the *geomagnetic field* (see GEOMAGNETISM) through GEOLOGICAL TIME, preserved as GEOMAGNETIC POLARITY REVERSALS, MAGNETIC EXCURSIONS and variations in MAGNETIC SUSCEPTIBILITY, to subdivide sequences of sediments and rocks in CORRELATION (STRATIGRAPHICAL) and STRATIGRAPHY. The *absolute-age dating* of specific geomagnetic polarity reversals has given rise to the GEOMAGNETIC POLARITY TIMESCALE (GPTS), or *magnetostratigraphic timescale*.   *DNT*

Hailwood EA (ed.) (1989) *Magnetostratigraphy*. Bath: Geological Society.
Pluhar CJ, Bjornstad BN, Reidel SP et al. (2006) Magnetostratigraphic evidence from the Cold Creek bar for onset of ice-age cataclysmic floods in eastern Washington during the Early Pleistocene. *Quaternary Research* 65: 123–135.
Rawson PF, Allen PM, Brenchley PJ et al. (2002) *Stratigraphical procedure*. Bath: Geological Society.
Roberts AP, Tauxe L and Heslop D (2013) Magnetic paleointensity stratigraophy and high-resolution Quaternary geochronology: Successes and future challenges. *Quaternary Science Reviews* 61: 1–16.

magnetozone   A unit of rock characterised by a specific *magnetic polarity* (i.e. normal or reversed) or other palaeomagnetic property (see PALAEOMAGNETISM). The corresponding interval of time is a POLARITY CHRON.

GO

[*See also* GEOMAGNETIC POLARITY REVERSAL, GEOMAGNETIC POLARITY TIMESCALE (GPTS)]

magnitude-frequency concepts   Used in relation to the geomorphological, hydrological and biological processes operating in LANDSCAPES, the term refers to the frequency of process events of different sizes and the relative work or impact that they achieve. The overall concept also proposes that geomorphological systems (or other features of the landscape, such as *ecosystems* or VEGETATION) may be influenced disproportionately by processes or events of a particular size and frequency (or range of sizes and frequencies), rather than by the whole range of events. Some features may be the result of relatively moderate, frequent events, whereas other features may be influenced only by high-magnitude events of rare frequency (see EXTREME EVENT).

The concept is well exemplified with respect to fluvial processes, in which relationships can be demonstrated of the relative proportions of the total geomorphic work done by different process magnitudes. Thus, analysis of the SUSPENDED LOADS of rivers has often demonstrated that most of the load is carried by a few *storm events* or flows of high magnitude but low frequency, though the *recurrence interval* of the flow varies between rivers with climatic and other factors. Other studies have focused on trying to identify which flow (or range of flows) is responsible for producing, maintaining or changing the cross sections (HYDRAULIC GEOMETRY) and CHANNEL PATTERNS of rivers. The concepts of *dominant discharge* and *effective discharge* have been developed to cover these channel-forming flows.

As the magnitude-frequency characteristics of climatic variables, such as large STORMS or FREEZE-THAW CYCLES, have important bearings in turn on the magnitude-frequency of hydrological and geomorphological processes such as FLOOD FREQUENCY and ROCKFALL activity, it is important to include them in more realistic schemes relating climate to geomorphology. It has been argued that a magnitude-frequency index of EXTREME EVENTS should be developed and included in future, more refined schemes of morphogenetic regions (see MORPHOCLIMATIC ZONES). The lack of long-term records of climatic extremes and landscape processes and, where they do exist, the difficulties often of separating anthropogenic from natural components of magnitude and frequency, prevent this objective from being realised.

There is growing evidence of changes through time in the magnitude-frequency of climatic events with climatic change even over the timescale of the past two centuries. Rainstorms capable of generating WADI flows and shallow GROUNDWATER recharge in the White Nile Province of the Sudan were significantly smaller and less frequent since AD 1965 than in the mid-twentieth century, resulting in reduced rural water supplies and MIGRATION of people to the cities. An increase in flood frequency in the rivers of South Wales since 1925 has been linked to an increase in the magnitude-frequency of large daily rainfalls. Dry periods in the otherwise perhumid equatorial environment of Borneo were longer and more frequent in the late nineteenth and early twentieth centuries and again in recent decades than in the period 1916–1967, with implications for the dynamics of the local TROPICAL RAIN FORESTS.

Human interference in the landscape can have radical impacts on the magnitude-frequency of geomorphological and hydrological processes and events.

LANDUSE CHANGES such as URBANISATION, DEFORESTATION, LAND DRAINAGE schemes and AFFORESTATION (if accompanied by *ditching*) lead to an increase in the flood magnitude-frequency, whereas DAMS and reservoirs (see RESERVOIRS: ENVIRONMENTAL EFFECTS) usually lead to a reduction in flood flows. Such changes can lead to changes in channel size, shape and pattern.

There is great current interest in changes in the magnitude-frequency of EXTREME-WEATHER EVENTS with GLOBAL WARMING. The ability of a warmer atmosphere to hold more WATER VAPOUR is considered to be the driver behind a widely predicted increase in extreme-rainfall events, and a warmer sea surface of increased spatial extent is considered to favour an increase in the magnitude-frequency of tropical cyclones. There is growing evidence that changes in the above may already be occurring. There is uncertainty, however, concerning possible future changes in EL NIÑO-SOUTHERN OSCILLATION (ENSO) events.   *RPDW*

Corominas J and Moya J (2010) Contribution of dendrochronology to the determination of magnitude-frequency relationships for landslides. *Geomorphology* 124: 137–149.

Douglas I, Bidin K, Balamurugan G et al. (1999) The role of extreme events in the impacts of selective tropical forestry on erosion during harvesting and recovery phases at Danum Valley, Sabah. *Philosophical Transactions of the Royal Society of London B* 354: 1749–1761.

Ma Y, Huang HQ, Xu JX et al. (2010) Variability of effective discharge for suspended sediment transport in a large semi-arid river basin. *Journal of Hydrology* 388: 357–369.

Rodda JC, Little MA, Rodda HJE and McSharry PE (2010) A comparative study of the magnitude, frequency and distribution of intense rainfall in the United Kingdom. *International Journal of Climatology* 30: 1776–1783.

Walsh RPD (1996) Drought frequency changes in Sabah and adjacent parts of northern Borneo since the late nineteenth century and possible implications for tropical rain forest dynamics. *Journal of Tropical Ecology* 12: 385–407.

Walsh RPD, Hulme M and Campbell M (1988) Recent rainfall changes and their impact on hydrology and water supply in the semi-arid zone of the Sudan. *Geographical Journal* 154: 181–198.

Webb BW and Walling DE (1982) The magnitude and frequency characteristics of fluvial transport in a Devon drainage basin and some geomorphological implications. *Catena* 9: 9–24.

Webster PJ, Holland GJ, Curry JA and Chang HR (2005) Changes in tropical cyclone number, duration and intensity in a warming environment. *Science* 309: 1844–1846.

Wolman MG and Miller JP (1960) Magnitude and frequency of forces in geomorphological processes. *Journal of Geology* 68: 54–74.

makatea island A South Pacific mid-plate island, with KARST-eroded, uplifted Tertiary to Late Pleistocene reef LIMESTONES (sometimes encircling an old volcanic core), suggesting regional flexure of the LITHOSPHERE. *TSp*

Stoddart DR, Woodroffe CD and Spencer T (1990) Mauke, Mitiaro and Atiu: Geomorphology of Makatea Islands in the Southern Cooks. *Atoll Research Bulletin* 341: 1–66.

malacophyllous   Soft-leaved, chiefly DECIDUOUS, low shrubs characteristic of MEDITERRANEAN-TYPE VEGETATION communities, which are subject to seasonal DROUGHT. Unlike the hard leaves of SCLEROPHYLLOUS plants, the leaves of malacophyllous plants (e.g. *Cistus* spp.) *wilt* during drought, but they are only shed under extreme drought conditions. The plants tolerate the reduced water content in the tissues rather than reduce water loss, and the high *osmotic pressure* generated assists water uptake.   *MLC/JAM*

[*See also* XEROPHYTE]

Iovi K, Kolovou C and Kyparissis A (2009) An ecophysiological approach of hydraulic performance for nine Mediterranean species. *Tree Physiology* 29: 889–900.

malaria   An infectious DISEASE endemic to most of the tropical world, caused by a protozoan PARASITE (*Plasmodium*) and spread by the female *Anopheles* mosquito, which injects the protozoan into the host during feeding. According to Stamp (1964: 38) 'it has been the greatest killer of all and when it does not kill it reduces mankind almost to incompetence'. The disease is relatively ancient, being known from Egyptian MUMMIES dating from 3000 BC. It was carried to the New World by the Spanish colonists, who imported slaves from Africa to Cuba in the sixteenth century. LAND DRAINAGE, PESTICIDE use and drug treatment have reduced its extent and impact. Although effective treatment and prevention have been available since AD 1633, it still kills more people each year than any other infectious disease except AIDS and TUBERCULOSIS (TB). Billions of people in tropical countries are at risk of infection: at least 300 million become infected each year, of whom between 1 and 3 million die. Ninety per cent of malarial deaths are in Africa. The RANGE of the insect vector and the areas in which it can persist during the cold season may be increased by GLOBAL WARMING. Where it is common, young children are most at risk because they have had insufficient time to build up immunity. Starting in Asia in the AD 1960s, the parasite developed immunity to *chloroquine*, an old and reliable drug, and now resistance to the newer drugs is developing (see Figure).   *JAM*

[*See also* HUMAN HEALTH HAZARDS, WATER-RELATED VECTOR]

Carter R and Mendis KN (2002) Evolutionary and historical aspects of the burden of malaria. *Clinical Microbiology Reviews* 15: 564–594.
Hay SI, Cox J, Rogers DJ et al. (2002) Climate change and the resurgence of malaria in the East African highlands. *Nature* 415: 905–909.
Marshall E (2000) A renewed assault on an old and deadly foe. *Science* 290: 428–430.

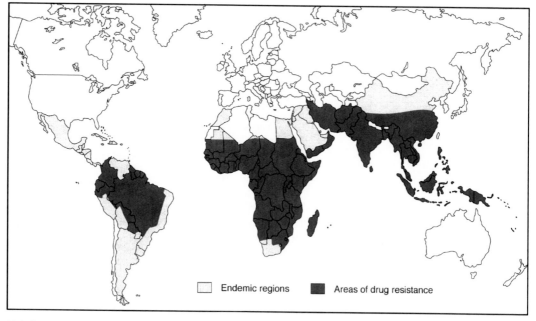

Malaria *The countries of the world where malaria is endemic and those areas where the parasite has developed drug resistance (Marshall, 2000).*

National Research Council (1991) *Malaria: Obstacles and opportunities*. Washington, DC: National Academy Press.

Sachs J and Malaney P (2002) The economic and social burden of malaria. *Nature* 415: 680–685.

Stamp LD (1964) *The geography of life and death*. London: Collins.

Webb J (2008) *Humanity's burden: A world history of malaria*. New York: Cambridge University Press.

Malthusian   The view that human population increase destroys the means of subsistence based on the principle that the population tends to increase geometrically while food production at best achieves arithmetic growth. Originally proposed by Thomas Malthus, the concept was embraced by many *Neo-Malthusian* environmentalists from the early 1970s. The Malthusian/Neo-Malthusian idea that there is a direct link between population growth and ENVIRONMENTAL DEGRADATION caused by the overuse of RESOURCES is held by its critics to be too simplistic and fails to consider social and historical factors. In particular, critics note that around one-sixth of the world's people use most of the resources and cause much of the POLLUTION. Ester Boserüp argued that rising population (if the increase is not too fast to overwhelm environment and resources) can trigger AGRICULTURAL INTENSIFICATION and technological innovation, making resource use less damaging and providing livelihood for the expanding population.

CJB

[*See also* DEMOGRAPHIC CHANGE, ENVIRONMENTALISM, OVERPOPULATION, TECHNOLOGICAL CHANGE]

Boserüp E (1965) *Population and technology*. Oxford: Basil Blackwell.

Ehrlich PR (1970) *The population bomb*. New York: Ballantine Books.

Malthus TR (1993) *An essay on the principle of population*. Oxford: Oxford University Press. [Oxford World Classics; originally published 1798]

Simon JL (1981) *The ultimate resource*. Princeton, NJ: Princeton University Press.

mammoth   An extinct *elephant* of the genus *Mammuthus*, closely related to modern elephants, although differentiated by twisted tusks and other features. The ancestral mammoth, *Mammuthus meridionalis*, evolved from a branch of the *Elephantidae* that migrated out of Africa approximately 2.5 Ma. Two separate lineages subsequently evolved, *Mammuthus trogontherii* (the *steppe mammoth*) in Eurasia and *Mammuthus colombi* (the *Colombian mammoth*) in North America. The word 'mammoth' is often used particularly to describe the descendant of *M. trogontherii*, the *woolly mammoth* (*Mammuthus primigenius*), which arose in Eurasia around 500,000 years ago and became widespread in the Northern Hemisphere. Along with other elements of the Pleistocene MEGAFAUNA, the last mammoths became extinct from the continental land masses 10,000 years ago, although RADIOCARBON DATING of mammoth remains from Wrangel Island in the Arctic Ocean has demonstrated the survival of an isolated dwarf race until about 3,700 years ago.

DCS

[*See also* MEGAFAUNAL EXTINCTION]

Haynes G (1991) *Mammoths, mastodonts, and elephants: Biology, behaviour, and the fossil record*. Cambridge: Cambridge University Press.

Lacombat F and Mol D (eds) (2012) Mammoths and their relatives I: Biotopes, evolution and human impact. *Quaternary International* 255: 1–256 [Special Issue].

Lister A and Bahn P (2007) *Mammoths: Giants of the Ice Age*. London: Frances Lincoln.

Kuzmin YV (2010) Extinction of the woolly mammoth (*Mammuthus perimagenius*) and woolly rhinoceros (*Coelodonta antiquitatis*) in Eurasia: Review of chronological and environmental issues. *Boreas* 39: 247–261.

Vartanyan SL, Garutt VE and Sher AV (1993) Holocene dwarf mammoths from Wrangel Island in the Siberian Arctic. *Nature* 362: 337–340.

managed realignment   The realignment of *sea defences* (e.g. SEA WALLS, REVETMENTS and embankments), typically to a more landward location. Hence, it was formerly referred to as *managed retreat*. It is a strategy of COASTAL ZONE MANAGEMENT for adapting to SEA-LEVEL RISE along low-lying coasts. By relocating sea defences inland or on higher ground, the tidal high water position moves inland and the *intertidal zone* is widened. The coastline is thus allowed to recede to a more landward position, often with the aim of restoring natural coastal systems such as *mudflats*, SALTMARSHES and SAND DUNES on the seaward side of COASTAL ENGINEERING STRUCTURES in recognition of the NATURAL sea-defence attributes of such landforms.

HJW/IM

Brooke JS (1992) Coastal defence: The retreat option. *Journal of the Institute of Water and Environmental Management* 6: 151–157.

French PW (1997) *Coastal and estuarine management*. London: Routledge.

Garbutt RA, Reading CJ, Wolters M et al. (2006) Monitoring the development of intertidal habitats on former agricultural land after the managed realignment of coastal defences at Tollesbury, Essex, UK. *Marine Pollution Bulletin* 53: 155–164.

Maddrell RJ (1996) Managed coastal retreat, reducing flood risks and protection costs, Dungeness nuclear power station, UK. *Coastal Engineering* 28: 1–15.

management   The human control of a system with a view to directing it for economic gain or for some other purpose such as CONSERVATION.

JAM

[*See also* ENVIRONMENTAL MANAGEMENT]

Ejarque A, Julià R, Riera S et al. (2009) Tracing the history
of highland human management in the eastern pre-
Pyrenees: An interdisciplinary palaeoenvironmental study
at the Pradell fen, Spain. *The Holocene* 19: 1241–1256.

Punnett BJ (2011) *Management: A developing country
perspective*. Abingdon: Routledge.

mangrove A diverse group of shallow-rooted tropical to subtropical trees and shrubs providing structure and HABITAT for highly productive ecosystems on
muddy coasts. Mangrove species show physiological,
morphological and life history ADAPTATION to saline,
waterlogged conditions. *Mangrove swamps* occupy
about two-thirds of the Earth's tropical coastline,
associated with ESTUARINE ENVIRONMENTS, CORAL REEFS
and SALTMARSHES. They are typically found at levels
from just below mean sea level to about 2 m above it.
They provide an important buffer to wave action during STORMS and to coastal and river FLOODS, although
their role in mitigating TSUNAMI impacts has been
contested.   *TSp*

Alongi DM (2009) Paradigm shifts in mangrove biology.
In Perillo GME, Wolanski E, Cahoon DR and Brinson
MM (eds) *Coastal wetlands: An integrated ecosystem
approach*. Amsterdam: Elsevier, 615–640.

Feller IC, Lovelock CE, Berger U et al. (2010)
Biocomplexity in mangrove ecosystems. *Annual Review
of Marine Science* 2: 395–417.

Spalding M, Kainuma M and Collins L (2010) *World atlas of
mangroves*. London: Earthscan.

Woodroffe CD and Davies G (2009) The morphology and
development of tropical coastal wetlands. In Perillo GME,
Wolanski E, Cahoon DR and Brinson MM (eds) *Coastal
wetlands: An integrated ecosystem approach*. Amsterdam:
Elsevier, 65–68.

mangrove succession Classical ECOLOGICAL SUC-
CESSION theory, the *gradient concept* and *physiographical ecology* are used to explain persistence and change
in MANGROVE communities. *Successional models*
emphasise the active role of plants in inducing both
organic and mineral accumulation. Increased surface
elevation drives community transitions until a *climax
community* is reached (see CLIMAX VEGETATION). Temporal succession at a site may be mirrored in the spatial
ZONATION of mangrove communities. The somewhat
complementary gradient concept relates surface inun-
dation and gradients in environmental factors, notably
SALINITY, to plant physiology to establish spatial limits to species groups. Physiographical ecology argues
that the evolution of mangrove vegetation patterns is
closely related to the dynamics of SHORELINE change and
that mangroves opportunistically colonise surfaces that
develop at appropriate elevations. This approach predicts mangrove community mosaics—where individual
species can recur at more than one elevation—in both

TERRIGENOUS and carbonate environments and argues
that the successional model is one case of a more general model, where sediment inputs drive shoreline PRO-
GRADATION.   *TSp*

Bunt JS (1996) Mangrove zonation: An explanation of data
from seventeen riverine estuaries in tropical Australia.
Annals of Botany 78: 333–341.

Ellison AM, Mukherjee BB and Karim, A (2000) Testing
patterns of zonation in mangroves: Scale dependence
and environmental correlates in the Sundarbans of
Bangladesh. *Journal of Ecology* 88: 813–824.

Panapitukkul N, Duarte CM, Thampanya U et al. (1998)
Mangrove colonisation: Mangrove progression over the
growing Pak Phanang (SE Thailand) mud flat. *Estuarine,
Coastal and Shelf Science* 47: 51–61.

Snedaker SC (1982) Mangrove species zonation: Why? In
Sen DN and Rajpurohit KS (eds) *Tasks for vegetation
science*. The Hague: Junk, 111–125.

Thom BG (1984) Coastal landforms and geomorphic
processes. In Snedaker SC and Snedaker JG (eds) *The
mangrove ecosystem: Research methods*. Paris: UNESCO,
3–17.

Watson JG (1928) Mangrove forests of the Malay Peninsula.
Malaya Forest Records 6: 1–275.

mangrove swamps: human impacts Traditionally, MANGROVE ecosystems have been a RENEWABLE
RESOURCE, sustaining local populations with building
materials, firewood, charcoal and medicine and, from
surrounding waters, fish and shellfish. Recently, however, large areas of mangrove have been lost to LOGGING,
agricultural impacts, land CONVERSION for industrial,
urban and tourist resort developments and replacement
by BRACKISH WATER and AQUACULTURE. Furthermore,
the removal of protective barriers, the interruption to
natural sediment supply through LAND RECLAMATION and
control of *waterways*, the acceleration of SUBSIDENCE
through oil and gas EXTRACTIVE INDUSTRY and the direct
removal of *mangrove peats* may lead to WETLAND LOSS
through EROSION and ENVIRONMENTAL DEGRADATION.

Coastal populations of nation states (including many
SMALL-ISLAND STATES (SIS)) with mangroves is predicted
to rise by 50 per cent, from 1.8 billion to 2.7 billion,
in the period between 2000 and 2025. FAO (Food and
Agriculture Organization of the United Nations) data
for 1980, 1990 and 2000 suggest that global mangrove
area declined by 26 per cent, from 198,000 km^2 to
146,500 km^2, over the period 1980–2000, a deforestation rate of 1.1 per cent per year. Forward extrapolation
suggests a global mangrove area of 111,108 km^2 by
2025, a loss of 24 per cent on the 2000 area and 44 per
cent of the coverage in 1980. However, these global
figures mask regional disparities—African percentage losses have been much less than Asian percentage
losses, for example—and time-specific losses in particular countries (e.g. 26,400 in 1980 to 11,500 in 1990,

or −8.0 per cent, in Brazil). Valuation of mangrove, and policy formulation for mangrove use, is complex and problematic.   *TSp*

Christiansen B (1983) Mangroves: What are they worth? *Unasylva* 35: 2–15.
Dodd RS and Ong JE (2008) Future of mangrove ecosystems to 2025. In Polunin NVC (ed.) *Aquatic ecosystems: Trends and global prospects.* Cambridge: Cambridge University Press, 172–187.
González C, Urrego LE, Martinez JI et al. (2010) Mangrove dynamics in the southwestern Caribbean since the 'Little Ice Age': A history of human and natural disturbances. *The Holocene* 20: 849–862.
Spalding M, Kainuma M and Collins L (2010) *World atlas of mangroves.* London: Earthscan.

mangrove swamps: impact of tropical cyclones

MANGROVE forests in the *tropical storm belts* (7–25° N and S) are vulnerable to *hurricane* damage because *storm frequencies* are often well within the lifespan of an individual tree. Periodic destruction of Caribbean mangroves by hurricanes may explain their low structural complexity as well as a lack of CLIMAX VEGETATION components in the vegetation community. Impacts, some of which may be delayed, include *defoliation* by winds and/or waves; *shearing* of branches and trunks; *uprooting*, often preferentially of large trees (see WINDTHROW); and deposition of released sediments and organic debris, the latter leading to *nutrient flushes* in mangrove-rimmed bays. In systems where organic-matter accumulation is the key process whereby surface elevation changes, the impacts of STORMS can not only cause the cessation of surface inputs but also trigger *mangrove peat* collapse, with the DECOMPOSITION of dead root material and COMPACTION in the absence of continued root growth. Whereas hurricanes with typical windspeeds of 120–150 km/h result in a patchwork of impacted and non-impacted areas, severe storms, with windspeeds in excess of 200 km/h, may overcome the structural resistance of the mangrove forest as a whole, reducing the VEGETATION CANOPY wholesale, creating environments unconducive to plant *re-establishment* and producing a hiatus to mangrove forest cover lasting for up to 50 years. REGENERATION depends upon *vegetative growth* of neighbouring plants, COLONISATION by new seedlings (requiring the delivery of new propagules and their trapping by woody debris) and/or growth of established *seedlings*. Mangrove forests are important buffers of WAVE ENERGY, and they help promote net SEDIMENTATION on CORAL REEF ISLANDS during cyclone events and aid their growth. Conversely, their removal may lead to more severe hurricane impacts on cleared coasts in terms of coastline EROSION and *coastal retreat*, and *island decline* and disappearance.   *TSp*

[See also CORAL REEFS: IMPACT OF TROPICAL CYCLONES, TROPICAL CYCLONES: IMPACT ON ECOSYSTEMS]

Cahoon DR, Hensel PR, Rybczyk J et al. (2003) Mass tree mortality leads to mangrove peat collapse at Bay Islands, Honduras after Hurricane Mitch. *Journal of Ecology* 91: 1093–1105.
Milbrandt EC, Greenawalt-Boswell JM, Sokoloff PD and Bortone SA (2006) Impact and response of southwest Florida mangroves to the 2004 hurricane season. *Estuaries and Coasts* 29: 979–984.
Smith III TJ, Anderson GH, Balentine K et al. (2009) Cumulative impacts of hurricanes on Florida mangrove ecosystems: Sediment deposition, storm surges and vegetation. *Wetlands* 29: 24–34.

mangrove swamps: impact of tsunamis

The role of MANGROVE in dissipating TSUNAMI wave impacts has come under reassessment following the Asian Tsunami of 26 December 2004. However, many statements on the protective role of mangroves have been based upon untestable anecdotal evidence, post hoc observational studies which assume causation, unvalidated mathematical modelling or flume studies where scaling issues are considerable. Furthermore, much of the debate has used REMOTE SENSING data to link changes in vegetation cover and damage levels but fails to acknowledge the role of other, covarying controls on tsunami impact, particularly elevation and distance from the sea. MODELS validated by reference to the tsunami that severely damaged the north coast of Papua New Guinea in 1998 have suggested a 90 per cent reduction in maximum tsunami flow pressure for a 100 m wide forest belt planted at a density of 3,000 trees/ha. Of the six different characteristic mangrove vegetation classes, the most effective appears to be *Rhizophora apiculata*.   *TSp*

Alongi DM (2008) Mangrove forests: Resilience, protection from tsunamis, and responses to global climate change. *Environmental Conservation* 29: 331–349.
Feagin RA, Mukherjee N, Shanker K et al. (2010) Shelter from the storm? Use and misuse of coastal vegetation bioshields for managing natural disasters. *Conservation Letters* 3: 1–11.
Kerr AM and Baird AH (2007) Natural barriers to natural disasters. *Bioscience* 57: 102–103.
Tanaka N, Sasaki Y, Mowjood MIM et al. (2007) Coastal vegetation structures and their functions in tsunami protection: Experience of the recent Indian Ocean tsunami. *Landscape Ecological Engineering* 3: 33–45.
Vermaat JE and Thampanya U (2006) Mangroves mitigate tsunami damage: A further response. *Estuarine, Coastal and Shelf Science* 69: 1–3 and 75: 564.

mantle

The part of the Earth's interior that underlies the CRUST, from which it is separated by the

Mohorovičić discontinuity (*Moho*), and overlies the CORE, from which it is separated by the *Gutenberg discontinuity*. The mantle comprises over 80 per cent of the Earth's volume. It is essentially solid and rocky, but its RHEOLOGY is such that it undergoes slow, convective movements. A partially molten layer of the outer mantle—the ASTHENOSPHERE—allows PLATE TECTONICS to operate and is the source of most of the MAGMA. *GO*

[*See also* EARTH STRUCTURE]

Kearey P, Klepeis KA and Vine FJ (2009) *Global tectonics.* Chichester: Wiley.
Stein S and Wysession M (2002) *An introduction to seismology, earthquakes, and Earth structure.* Oxford: Blackwell.

mantle plume A long-lived column of high heat flow from a deep level in the MANTLE. It causes enhanced partial melting of the ASTHENOSPHERE, giving rise to VOLCANISM at a hotspot (see HOTSPOT, IN GEOLOGY) on the Earth's surface. The link between hotspots and mantle plumes has, however, recently been challenged and is hotly debated. *GO*

Foulger GR (2010) *Plates vs plumes: A geological controversy.* Chichester: Wiley.
Foulger GR, Natland JH, Presnall DC et al. (eds) (2005) *Plates, plumes, and paradigms.* Boulder, CO: Geological Society of America.
Sheth HC (1999) Flood basalts and large igneous provinces from deep mantle plumes: Fact, fiction, and fallacy. *Tectonophysics* 311: 1–29.
Sleep NH (2006) Mantle plumes from top to bottom. *Earth-Science Reviews* 77: 231–271.

manufacturing industry The secondary sector of industry, which converts the raw materials of EXTRACTIVE INDUSTRY into fabricated products. It includes the making and assembly of parts and components, and it plays the major role in INDUSTRIALISATION. The two most important developments in the history of manufacturing from the viewpoint of environmental change were probably, first, the invention of the steam engine powered by coal, which was one of the main planks of the INDUSTRIAL REVOLUTION, and, second, the diversification of energy sources and energy end-use technologies. Particularly important in this respect were the far-reaching effects of electricity, the internal combustion engine and oil on the *transport industry*, employment and international trade. The manufacturing industry is declining relative to the SERVICE INDUSTRIES in developed countries, where former manufacturing regions are often areas of industrial blight and restructuring. Although the local, regional and global environmental impacts of manufacturing industry have been, and remain, considerable, several generic strategies for impact reduction can be recognised. These include

ECOLOGICAL MODERNISATION (adoption of ecological principles), DEMATERIALISATION (decrease in materials used per unit of output), MATERIALS SUBSTITUTION (use of different or new materials and alternative energy sources), WASTE RECYCLING sensu stricto (processing of discarded artefacts) and WASTE MINING (processing of manufacturing waste). Current world recycling rates for lead and steel are 45–50 per cent. *JAM*

[*See also* DE-INDUSTRIALISATION, INDUSTRIAL ECOLOGY, POSTINDUSTRIALISATION]

Ausubel JH (1991) Does climate still matter? *Nature* 350: 649–652.
Ayres RU and Ayres LW (1996) *Industrial ecology: Towards closing the materials cycle.* Cheltenham: Edward Elgar.
Cooke P (ed.) (1995) *The rise of the rustbelt: Revitalizing older industrial regions.* London: Routledge.
Essletzbichler J (2004) The geography of job creation and destruction in the U.S. manufacturing sector, 1967-1997. *Annals of the Association of American Geographers* 94: 602–619.
Kalpakjian S and Schmid S (2005) *Manufacturing, engineering and technology.* Upper Saddle River, NJ: Prentice Hall.

map An abstract, simplified, visual representation of geographical reality. It is normally created to scale and projected onto a flat surface (paper or screen). A major characteristic of maps is that there is a loss of detail with reduction in scale (see MAP GENERALISATION). There are many specialised types of map, such as those produced by *geomorphological mapping* and *geological mapping*. *JAM/TF*

[*See also* CARTOGRAPHY, MAP PROJECTION]

Dodge M, McDerby M and Turner M (2008) *Geographic visualization: Concepts, tools and applications.* Chichester: Wiley-Blackwell.
Hanna SP (2010) *Maps and diagrams.* In Gomez B and Jones III JP (eds) *Research methods in geography.* Chichester: Wiley, 259–278.
James LA, Walsh SJ and Bishop MP (2012) Geospatial technologies and geomorphological mapping. *Geomorphology* 137: 1–198 [Special Issue].
Smith MJ, Paron P and Griffiths JS (eds) (2011) *Geomorphological mapping: Methods and applications.* Oxford: Elsevier.

map generalisation Also known as *cartographic generalisation*, it encompasses the essential process of reducing and symbolising map information from a real-world scale to a level of detail appropriate to the map. The smaller the scale of the map, the greater the level of generalisation applied. The process of map generalisation is now largely automated and digitised. *TF*

[*See also* AREA GENERALISATION]

Longley P, Goodchild MF, Maguire DJ and Rhind DW
(2011) *Geographic information systems and science, 3rd edition.* Chichester: Wiley.
Mackaness W, Ruas A and Sarjakoski (eds) (2007)
Generalisation of geographic information: Cartographic modelling and applications. Amsterdam: Elsevier.

map projection A graphical or mathematical *transformation* used in map construction to translate the curved surface of the Earth to the plane surface of a map. There exist an infinite number of map projections, but many have little practical use. As it is impossible to retain all the characteristics of the curved global surface when translating to a plane surface, the objective of any map projection is to minimise distortion in or retain one or more of a set of desirable characteristics. Hence, projections display specific properties (see Table).

Projections can vary in their ASPECT to the globe (effectively, the way in which the projection plane impinges on the global surface): hence, projections are classed as *normal, transverse* or *oblique* (see Figure 1). Depending on the projection chosen, the image of Earth features (e.g. continents, countries, oceans or lakes) will appear slightly differently but in doing so reflect the properties of the chosen projection (see Figure 2).   *TF*

Kraak MJ and Ormeling FJ (2010) *Cartography: Visualization of spatial data, 3rd edition.* Harlow: Pearson.
Longley P, Goodchild MF, Maguire DJ and Rhind DW
(2011) *Geographic information systems and science, 3rd edition.* Chichester: Wiley.
Snyder JP (1993) *Flattening the Earth: Two thousand years of map projections.* Chicago: University of Chicago Press.

Projection type	Property	Features
Equivalent	Equal area	Relative size of features is retained (shape is lost)
Conformal or orthomorphic	Shape retention	Shape of small elements is retained (size is not)
Azimuthal or zenithal	Directions from the central point are correct	All angles and directions from the central point are retained

Map projection *Types of map projection and their characteristics.*

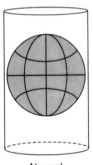 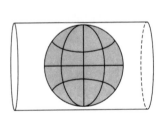 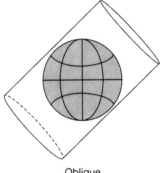

| Normal | Transverse | Oblique |

Map projection 1 *Cylindrical class projection showing aspect.*

World projections based on WGS 84 spheroid

Mercator projection -
Conformal cylindrical

Robinson projection -
Designed as effective world map,
neither conformal nor equal area

Sinusoidal projection -
Equal area

Map projection 2 *Examples of map projections: (A) Mercator projection (conformal cylindrical); (B) Robinson projection (neither conformal nor equal area, but an effective world map) and (C) sinusoidal projection (equal area). These world projections are all based on WGS 84 spheroid.*

marching desert concept   A simplistic, outdated and emotive view that LAND DEGRADATION in the form of DESERTIFICATION occurs through progressive advance of desert-like conditions across agriculturally useful land.

RAS

Forse B (1989) The myth of the marching desert. *New Scientist* 121: 31–32.
Thomas DSG and Middleton NJ (1994) *Desertification: Exploding the myth.* Chichester: Wiley.

marginal areas   Those areas where types of AGRI-CULTURE are or were practised at or near their *climatic limits*. This is *marginality* in a spatial/climatic sense, but the concept of marginality can also be considered in terms of economic marginality (when returns barely exceed costs) and *social marginality* (when population pressure is excessive). *Climatic marginal areas* may be considered in terms of both horizontal and vertical (altitudinal) distributions. The marginality is gener-ally due to insufficient water (from either PRECIPITA-TION or IRRIGATION) being available or due to either too much or too little heat for the type of agriculture being practised. The main global marginal areas are found in DEVELOPING COUNTRIES, where insufficient rainfall is combined with population pressure due either to war or to increasing birth rates (e.g. the SAHEL, southwest and southeast Asia, the Horn of Africa and the Andes). In the past, marginal lands have been the most suscep-tible to the abandonment or expansion of agriculture and settlement in the face of CLIMATIC CHANGE, and they may increase in area as a result of GLOBAL WARMING.

BDG

[*See also* CLIMATIC IMPACT ASSESSMENT, DESERTED VILLAGES, DESERTIFICATION, DRYLANDS]

Carter TR, Parry ML, Harasawa H and Nishioka S (1994) *IPCC technical guidelines for assessing climate change impacts and adaptations.* London: UCL Press.
Food and Agriculture Organization of the United Nations (FAO) (1997) *Report of the study on CGIAR research priorities for marginal lands.* Rome: FAO.
Glantz MH (ed.) (1994) *Drought follows the plow: Cultivating marginal areas.* Cambridge, UK: Cambridge University Press.
Leimgruber W (2004) *Between global and local: Marginality and marginal regions in the context of globalization and deregulation.* London: Ashgate Publishing.
Pakeman RJ, Le Duc MG and Marrs RH (2000) Bracken distribution in Great Britain: Strategies for its control and the sustainable management of marginal land. *Annals of Botany* 85, Supplement 2: 37–46.
Parry ML (1978) *Climate change, agriculture and settlement.* Folkestone: Dawson.
Parry ML (1990) *Climate change and world agriculture.* London: Earthscan.
Pollard S (1997) *Marginal Europe: The contribution of marginal land since the Middle Ages.* Oxford: Oxford University Press.

marginal basins   Small SEDIMENTARY BASINS associ-ated with DESTRUCTIVE PLATE MARGINS and ISLAND ARCS, including BACK-ARC basins, FORE-ARC basins and OCE-ANIC TRENCHES.

GO

[*See also* PLATE TECTONICS]

Allen PA and Allen JR (2004) *Basin analysis: Principles and applications.* Chichester: Wiley.
Hutchison CS (2004) Marginal basin evolution: The southern South China Sea. *Marine and Petroleum Geology* 21: 1129–1148.
Kearey P, Klepeis KA and Vine FJ (2009) *Global tectonics.* Chichester: Wiley.

mariculture   The culturing of organisms, especially fish and shellfish, in marine environments (coastal or off-shore). It has a long history, especially in Asia, but it has grown exponentially since the 1980s as the world's wild fisheries have declined. China accounts for about two-thirds of the world's mariculture by volume and 40 per cent by value. Expensive, high-quality sea-foods such as shrimp, molluscs and other shellfish are much cultured in the tropics and salmon in temperate seas and fjords. Mariculture, or *marine aquaculture*, may involve one or more of the following: feeding at various stages of the life history, construction of enclosures, provision of artificial habitats, including reefs and floating structures, and breeding and release programmes, including *ocean ranching* in the open sea. ENVIRONMENTAL ISSUES associated with maricul-ture include BIODIVERSITY LOSS, DISEASE, ENVIRONMENTAL DEGRADATION, HABITAT LOSS and MARINE POLLUTION, espe-cially in relation to wild species.

JAM

[*See also* AQUACULTURE, FISHERIES CONSERVATION AND MANAGEMENT, SUSTAINABLE FISHERIES]

Food and Agriculture Organization of the United Nations (FAO) (2010) *Integrated mariculture: A global review.* Rome: FAO.
Hargrave B (ed.) (2010) *Environmental effects of marine finfish aquaculture.* Berlin: Springer.
Naylor RI, Goldburg RJ, Primavera JH et al. (2000) Effect of aquaculture on world fish supplies. *Nature* 405: 1017–1024.
Nolan JT (ed.) (2009) *Offshore marine aquaculture.* Hauppauge, NY: Nova Science Publishers.
Tobey J, Clay J and Vergne P (1998) *Maintaining a balance: The economic, environmental and social impact of shrimp farming in Latin America.* Washington, DC: Island Press.

marine band   A thin BED rich in marine FOSSILS within a succession of sediments or sedimentary rocks dominated by deposits lacking marine fossils. The con-cept is exemplified by *goniatite*-bearing SHALE horizons within COAL-bearing CYCLOTHEMS in Upper CARBONIFER-OUS rocks. These successions represent DELTA DEPOSITS and the marine bands record periods of marine trans-gression (see TRANSGRESSION, MARINE)] or *flooding surfaces* (see SEQUENCE STRATIGRAPHY).

GO

Wells MR, Allison PA, Hampson GJ et al. (2005) Modelling ancient tides: The Upper Carboniferous epi-continental seaway of northwest Europe. *Sedimentology* 52: 715–735.

marine cloud whitening (MCW)

marine cloud whitening (MCW)   An approach to CLIMATIC ENGINEERING through SOLAR RADIATION MANAGEMENT that seeks to increase the proportion of SOLAR RADIATION reflected back to space by increasing the ALBEDO of the cloud layer. It involves producing a fine mist of sea-salt droplets, which would be released into the turbulent BOUNDARY LAYER beneath marine strato-cumulus clouds, and loft upwards, to provide a source of CONDENSATION NUCLEI for cloud droplets and further cloud growth.   *JAM*

Salter S, Sortino G and Latham J (2008) Sea-going hardware for the cloud albedo method of reversing global warming. *Philosophical Transactions of the Royal Society A* 366: 3989–4006.

marine conservation   The preservation, RESTORATION or improvement of BIODIVERSITY in *marine ecosystems*. This basic definition encompasses a broad subject that involves the integration of *marine science* into applied policy, legislation and MANAGEMENT and is holistically applied to embrace *conservation quality* and *socioeconomic value* through human exploitation of *marine resources*. Purely speaking, marine conservation is the preservation and, where possible, the improvement of biodiversity at the GENETIC, species (see SPECIES CONCEPT), community (see COMMUNITY ECOLOGY) and ecosystem levels (see ECOSYSTEM CONCEPT). Historically, conservation per se was addressed at the species level, but in recent years this has changed to an ecosystem-level approach, encompassing inter- and intraspecies interaction, as well as species interactions with their physical and chemical ENVIRONMENT. Such an approach is more practical from a management perspective, as well as being logical in conserving species through preservation and improvement of their HABITAT as a whole. Within the management aspect of marine conservation is the recognition that human exploitation of marine RESOURCES necessitates a compromise within marine conservation. SUSTAINABLE DEVELOPMENT of marine resources coupled with preservation, or improvement of conservation quality, assessed as biodiversity, is hence a viable approach to marine conservation, as opposed to a purist approach in considering conservation of marine biodiversity purely for its own intrinsic value. The clearest example of this in recent years is the aim of sustainable exploitation and management of commercial fisheries at the wider-seas and regional-seas levels, with *holistic management* (see HOLISTIC APPROACH) adopted by all stakeholders. Assessing the balance between fish stock conservation and socioeconomic value through their exploitation is controversial. However, the recognition at government level

that 'healthy seas' are a necessity for economic health is an important step towards increasing implementation of marine conservation. The adoption of *Marine Protected Areas*, linked to produce a network of areas protected to conserve *marine biodiversity*, is an essential aspect of present and future marine conservation.   *RGP*

[*See also* ECOLOGICAL RESTORATION, ECOLOGICAL VALUE, ENVIRONMENTAL MANAGEMENT, ENVIRONMENTAL PROTECTION, ENVIRONMENTAL QUALITY, FISHERIES CONSERVATION AND MANAGEMENT, PROTECTED AREAS]

Carleton RG and McCormick-Ray J (2004) *Coastal-marine conservation: Science and policy*. Oxford: Blackwell.
Edgar GJ, Russ GR and Babcock RC (2007) *Marine protected areas*. In Connell S (ed.) *Marine ecology*. Melbourne: Oxford University Press, 533–555.
Kidd S, Plater A and Frid C (eds) (2011) *The ecosystem approach to marine planning and management*. London: Earthscan.
Norse EA and Crowder CB (2005) *Marine conservation biology*. Washington, DC: Island Press.
Roff J and Zacharias M (2011) *Marine conservation ecology*. London: Earthscan.
Tittensor DP, Mora C, Jetz W et al. (2010) Global patterns and predictors of marine biodiversity across taxa. *Nature* 466: 1098–1101.

marine geology   The branch of GEOLOGY that deals with the OCEAN floor and the CONTINENTAL MARGINS, encompassing studies of submarine relief, PLATE MARGINS, the influence of physical processes such as *waves*, TIDES, STORMS and CONTOUR CURRENTS on the sea floor, the movement of SEDIMENTS on the CONTINENTAL SHELF, and to and in the deep ocean, the SEDIMENTOLOGY of SUBMARINE FANS and other deep-water CLASTIC systems, the GEOCHEMISTRY of rocks at and beneath the sea floor and the fluids moving through the CRUST. Marine geology has expanded rapidly since the advent of DEEP-SEA DRILLING in the 1960s, and through recent technological advances such as SIDE-SCAN SONAR and SEISMIC STRATIGRAPHY. Fifty years ago, the ocean floor was thought to be an inert landscape with little geological or biological activity: marine geology has shown it to be an important, dynamic, responsive and varied environment.   *BTC/JP*

[*See also* OCEANOGRAPHY]

Kennett JP (1982) *Marine geology*. Englewood Cliffs, NJ: Prentice Hall.
Pickering KT, Hiscott N, Smith R and Kenyon NH (eds) (1995) *Atlas of deep water environments: Architectural style in turbidite systems*. London: Chapman and Hall.
Steele JH, Thorpe SA and Turekian KK (eds) (2010) *Marine geology and geophysics: A derivative of the encyclopedia of ocean sciences*. London: Academic Press.

marine isotopic stage (MIS)   An interval defined by inflections in the global OXYGEN ISOTOPE stratigraphic record derived from MARINE SEDIMENTS. The oxygen

ISOTOPE RATIO of FORAMINIFERA in deep-ocean sediments is influenced primarily by global continental ice volume. Isotopic stages are assigned to GLACIAL (even numbers) and INTERGLACIAL (odd numbers) periods (see Figure). These stages may be further subdivided as higher-resolution records become available, thus the Last Interglacial, Stage 5, is divided into 5.1–5.5 or 5a–e, 'odd' letters (a, c, etc.) relating to 'warm' *substages* and 'even' letters (b, d, etc.) to cold substages. During the past 2.73 million years, 116 stages have been recognised.   *IR/CJC*

[*See also* ISOTOPES AS INDICATORS OF ENVIRONMENTAL CHANGE, MARINE SEDIMENT CORES, QUATERNARY TIMESCALE]

Lüthi D, Le Floch M, Bereiter B et al. (2008) High-resolution carbon dioxide concentration record 650,000–800,000 years before present. *Nature* 453: 379–382.

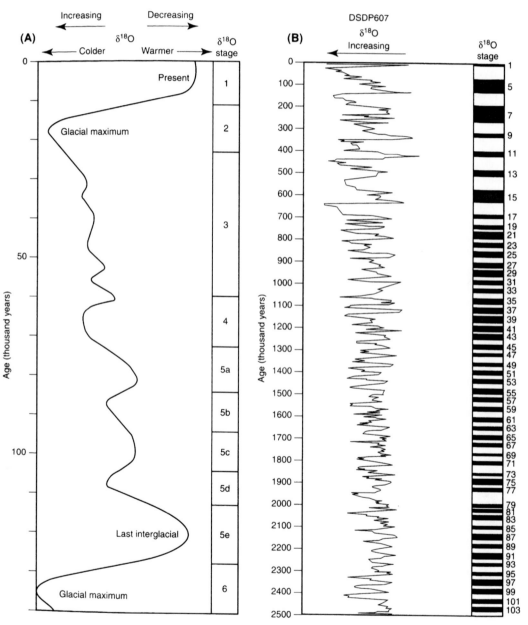

Marine isotopic stage (MIS) *Isotopic stages recognised for (A) the past 150,000 years and (B) the past 2.5 million years based on oxygen isotope variations in marine sediment cores. Odd numbers indicate warm stages (interglacials shown as black bars in B); even numbers (in A) are cold stages (Wilson et al., 2000).*

Shackleton NJ (1987) Oxygen isotopes, ice volume and sea level. *Quaternary Science Reviews* 6: 183–190.

Shackleton NJ and Opdyke ND (1973) Oxygen isotope and palaeomagnetic stratigraphy of equatorial Pacific core V28-238: Oxygen isotope temperatures and ice volumes on a 105 year and 106 year scale. *Quaternary Research* 3: 39–55.

Wilson RCL, Drury SA and Chapman JL (2000) *The Great Ice Age: Climate change and life*. London: Routledge.

marine palaeoclimatic proxies Indirect methods of estimating CLIMATE in the GEOLOGICAL RECORD using natural materials preserved in marine rocks and sediments. The highest-quality marine palaeoclimatic NATURAL ARCHIVES are derived from MARINE SEDIMENT CORES that allow scientists to reconstruct climate histories on annual to multimillion-year timescales. Information comes in the form of natural physical, biological and chemical ENVIRONMENTAL INDICATORS that are transformed to proxies using statistically and chemically defined relationships with environmental parameters. There is no clear distinction between marine palaeoclimatic proxies and *palaeoceanographic proxies*, since the latter mostly record climatically controlled variations in past OCEAN characteristics. Strictly speaking, *palaeoclimatic proxies* (see PALAEOCLIMATOLOGY) refer to atmospheric properties (e.g. concentration of CARBON DIOXIDE, wind strength and direction). However, proxies based on marine sediment composition can also indicate a wide range of climatically controlled levels of, for example, PRIMARY PRODUCTIVITY, ICE-RAFTED DEBRIS (IRD), AEOLIAN fluxes and WEATHERING processes, as well as background ocean chemistry. The FOSSILS and MICROFOSSILS in sediments provide proxy information on OCEAN PALAEOTEMPERATURE, PALAEOSALINITY, SEA ICE, pH, PRODUCTIVITY and atmospheric CARBON DIOXIDE in the form of (1) PLANKTON assemblage composition and distribution and (2) shell/sediment geochemical compositions. CORAL skeletons and the shells of FORAMINIFERA, often the most abundant components of marine BIOGENIC CARBONATE, are common sources of marine palaeoclimatic proxy data (the Table provides a full list of common proxy methods). The maximum age for climate histories produced in this way is limited by the quality of fossil and sediment preservation and the maximum age of modern OCEANIC CRUST (on which marine sediments accumulate), which is Early JURASSIC (~200 million years old). Older and/or uplifted

Climatic variable	Proxy methods
OCEAN PALAEOPRODUCTIVITY/ palaeoflux	Biogenic silica/CARBONATE/organic matter/MICROFOSSIL accumulation rate, marine barite, BIOGENIC CARBONATE/organic $\delta^{13}C$, organic BIOMARKERS, DIATOM ANALYSIS, NUTRIENT tracers (N, P, Fe, Si, Cd/Ca), elemental particle flux indicators (Al/Ti, Ba/Ti, $^{231}Pa/^{230}Th$, $^{10}Be/^{230}Th$)
SEA-SURFACE TEMPERATURE (SST)	$\delta^{18}O$, Mg/Ca and Sr/Ca in CORAL and *planktonic carbonate*, ALKENONE saturation indices, TEX86, clumped isotopes, plankton assemblages/distribution (TRANSFER FUNCTIONS)
BOTTOM WATER temperature	$\delta^{18}O$, Mg/Ca and Sr/Ca in *benthic carbonate*
WEATHERING processes	Clay minerals, sediment MAGNETIC SUSCEPTIBILITY, osmium isotopes, STRONTIUM ISOTOPES
ICE SHEETS/SEA-LEVEL CHANGES	Coupled CARBONATE $\delta^{18}O$ and Mg/Ca
ICEBERGS	Detection of ICE-RAFTED DEBRIS
SEA ICE	DIATOM ANALYSIS, DINOFLAGELLATE CYST ANALYSIS
DEEP WATER masses and OCEAN PALAEOCIRCULATION	Sediment particle size, neodymium isotopes, foraminifera $\delta^{13}C$, ^{14}C, accumulation of deep-sea *contourite drift* sediments, sediment $^{231}Pa/^{230}Th$
Surface-ocean pH/ atmospheric CO_2	ALKENONE CO_2 proxy, boron isotopes in FORAMINIFERA
Ocean-carbonate chemistry, OCEAN ACIDIFICATION	DIATOM/FORAMINIFERA ANALYSIS, FORAMINIFERA Li/Ca and B/Ca, BIOGENIC CARBONATE concentration, FORAMINIFERA shell weight, *coccolith size*
PALAEOSALINITY/PRECIPITATION	PLANKTON assemblages/distribution, marine BIOGENIC CARBONATE $\delta^{18}O$
Atmospheric circulation	Wind-blown (AEOLIAN) SEDIMENT flux

Marine palaeoclimatic proxies *Common marine palaeoclimatic proxy methods.*
$\delta^{18}O$ = the ratio of oxygen-16: oxygen-18 expressed in delta notation; $\delta^{13}C$ = the ratio of carbon-12: oxygen-13 expressed in delta notation.

oceanic sediments preserved on land can also yield valuable records (e.g. Euro-African Tethys sequences).   *HKC*

[*See also* CLIMATIC RECONSTRUCTION, OCEAN PALAEOTEMPERATURE, PALAEOCEANOGRAPHY: BIOLOGICAL PROXIES, PALAEOCEANOGRAPHY: PHYSICAL AND CHEMICAL PROXIES, PROXY CLIMATIC INDICATOR]

Cronin TM (2010) *Paleoclimates: Understanding climate change past and present.* New York: Colombia University Press.
Eiler JM (2011) Paleoclimate reconstruction using carbonate clumped isotope thermometry. *Quaternary Science Reviews* 30: 3575–3588.
Foster GL (2008) Seawater pH, pCO$_2$ and [CO32-] variations in the Caribbean Sea over the last 130 kyr: A boron isotope and B/Ca study of planktic foraminifera. *Earth and Planetary Science Letters* 271: 254–266.
Gornitz V (eds) (2009) *Encyclopedia of paleoclimatology and ancient environments.* Dordrect: Springer.
Hillaire-Marcel C and de Vernal A (eds) (2007) *Proxies in Late Cenozoic paleoceanography.* Amsterdam: Elsevier.
Kuwae M, Hayami Y, Oda H et al. (2009) Using foraminiferal Mg/Ca ratios to detect an ocean warming trend in the twentieth century from coastal shelf sediments in the Bungo Channel, southwest Japan. *The Holocene* 19: 285–294.
Lear CH (2007) Mg/Ca palaeothermometry: A new window into Cenozoic climate change. In Williams M, Hayward A, Gregory J and Schmidt DN (eds) *Deep time perspectives on climate change: Marrying the signal from computer models and biological proxies.* Bath: Geological Society, 313–322.
Pagani M (2002) The alkenone-CO$_2$ proxy and ancient atmospheric carbon dioxide. *Philosophical Transactions of the Royal Society A* 360: 609–632.

marine pollution 'The introduction by man, directly or indirectly, of substances or energy into the marine environment (including estuaries) resulting in deleterious effects such as harm to living resources, hazards to human health, hindrance to marine activities including fishing, impairment of quality in the use of seawater and reduction of amenities (GESAMP, 1989). This definition from the IMO (International Maritime Organisation), FAO (Food and Agriculture Organization of the United Nations), UNESCO (United Nations Educational, Scientific and Cultural Organization), WMO (World Meteorological Organization), WHO (World Health Organization), IAEA (International Atomic Energy Agency), UN (United Nations) and UNEP (United Nations Environment Programme) Joint Group of Experts on the Scientific Aspects of Marine Pollution summarises an environmental problem that has become worse since the 1980s, especially in coastal and ESTUARINE ENVIRONMENTS. Some 60 per cent of the world's population live within 100 km of the SHORELINE and more than half the population of DEVELOPING COUNTRIES obtain >30 per cent of their animal protein from *marine fish.* Once polluted discharge from the world's rivers reaches the sea, it is diluted along with the POLLUTANTS from other sources: however, then it is not only relatively difficult to remove but also becomes an international problem.

Types of marine pollution include oil pollution (see OIL SPILL), synthetic PERSISTENT ORGANIC COMPOUNDS (POCs) (e.g. HALOGENATED HYDROCARBONS (HALOCARBONS), METAL POLLUTION (especially HEAVY METALS), *microbial pollution* (mostly from sewage WASTE), THERMAL POLLUTION and RADIOACTIVE WASTE pollution. Artificial radioactive ISOTOPES are new to the marine environment. *Dumping* of liquid or solid radioactive substances at sea is now banned by international agreement, but discharge from the land-based nuclear industry continues as it is regulated nationally. Particular concern has been expressed over dumping of radioactive waste by the former Soviet Union in the Barents and Kara seas of the *Arctic Ocean.* Findings suggest that contamination was restricted to the immediate vicinity of the radioactive material.   *JAM*

[*See also* FISHERIES CONSERVATION AND MANAGEMENT, POLLUTION]

Abuzinada AH, Barth H-J, Krupp F et al. (eds) (2008) *Protecting the Gulf's marine ecosystems from pollution.* Basel: Birkhaüser.
Clark RB (2001) *Marine pollution, 5th edition.* Oxford: Oxford University Press.
Føyn L (1997) Marine pollution. In Brune D, Chapman DV, Gwynne MD and Pacyna JM (eds) *The global environment: Science, technology and management, volume 1.* Weinheim: VCH, 515–531.
Group of Experts on the Scientific Aspects of Marine Environmental Protection (GESAMP) (1989) *Report of Working Group 26 on the state of the marine environment.* Nairobi: United Nations Environment Programme.
Hofer TN (ed.) (2008) *Marine pollution: New research.* New York: Nova Science Publishers.
Islam MS and Taneka M (2004) Impacts of pollution on coastal and marine ecosystems including coastal and marine fisheries and approach for management: A review and synthesis. *Marine Pollution Bulletin* 48: 624–649.
O'Brine T and Thompson RC (2010) Degradation of plastic carrier bags in the marine environment. *Marine Pollution Bulletin* 60: 2279–2283.
Ofiara DD and Seneca JJ (2001) *Economic losses from marine pollution: A handbook for assessment.* Washington, DC: Island Press.

marine sediment cores Vertical sections of SEDIMENT collected from the OCEAN floor. They represent the net accumulation of sediment over time at one location, and cores can vary in length from tens of centimetres to tens of metres. Marine sediment cores are used to improve our understanding of the changes in physical, biological and chemical processes that occur within the oceans and at the sea floor. A wide range of scientific disciplines benefits from the collection of marine sediment cores, including PALAEOCEANOGRAPHY, GEOCHEMISTRY, MICROPALAEONTOLOGY, PALAEOMAGNETISM and SEDIMENTOLOGY.

Marine sediment cores are collected using different, specially designed coring equipment. Murdoch and Macknight (1991) indicate that 'CORERS are fundamental tools for obtaining sediment samples for geological and geotechnical surveys and, recently, for the investigation of historical inputs of contaminants to aquatic systems.' Typically, the collection and retrieval of marine sediment cores from the sea floor require the use of a vessel equipped with appropriate winches and cranes. This can range from small boats to large research vessels, for example, the drilling ship of the Ocean Drilling Program (ODP), the *JOIDES Resolution.* The main types of marine sediment cores that can be obtained are box, gravity, piston, vibro- and drilled cores.

Box corers, for example, a *Kastenlot corer,* collect large rectangular sediment cores (about 30 × 60 × 22 cm), into which plastic sediment core tubes can be inserted to retrieve sediment once recovered. A *gravity corer* (see Figure) has a simple, large weighted head with a tubular or box section barrel. Once deployed from a ship, the corer free-falls through the water column, and the barrel, often with an internal plastic liner, is driven into the sediment with the aid of a cutting head. The cutting head is placed on the end of the core barrel, in order to achieve better penetration into the sediment. They are normally made out of stainless steel, brass or plastic, commonly with a screw or bayonet fitting. A *catcher mechanism,* often a series of spring-loaded metal fingers, closes as the corer is pulled out of the sediment and the sediment core is retained. A tubular shaped core of up to 3 m of marine sediment can be collected with a gravity corer; however, problems of sediment compression are encountered within the core barrel. PISTON SAMPLERS are often used for studies of *deep-ocean sediment* and range from 3 to 50 m in length. The corer consists of a weighted stabilised head, a core barrel with plastic core liner, a piston, a core retainer, a cutting head and a trigger mechanism. The piston within the corer creates a partial vacuum in the core barrel and sucks the sediment into the barrel. In recent years, large-piston corers have been developed, for example, the French giant piston or *Calypso corer,* which have been utilised during the *International Marine Global Change Studies.*

Recovering loose and unconsolidated SAND and GRAVEL is far more problematic. The *vibrocorer* was developed to recover the non-cohesive sediments, which often dominate energetic, *shallow-marine environments.* Vibrocorers overcome the resistance

of sediment by a vibration action of the core barrel to retrieve marine sediment cores similar in nature to piston cores. Marine sediment cores can also be obtained from the sea floor by mechanical drilling. Improved drilling technology over the past 30 years, since the first site drilled by the *Glomar Challenger* in 1968, has enabled pioneering research to be carried out on the *JOIDES Resolution* for the ODP. This drilling ship is able to drill through soft sediment or MUD using an *advanced hydraulic piston corer (APC)* and, where more resistant sediment or rock is encountered, an *extended core barrel (XCB)* or *rotary core barrel (RCB)* method is applied. The drilling system utilised on the *JOIDES Resolution* is able to handle 9,150 m of drill pipe in water depths up to 8,235 m. The great revolution within ODP came with the advent of triple APC, allowing the construction of a continuous composite section. This allows the construction of a *metres composite depth (MCD) scale,* accommodating core expansion and drilling gaps through *interhole correlation* using continuous measurements of core physical properties (e.g. MAGNETIC SUSCEPTIBILITY).

In the case of biological and geochemical analysis, a *multicore system* is often employed for the collection of marine sediment. This consists of a series of plastic core tubes mounted on a metal frame that is lowered slowly to the seabed. On contact with the sea floor, the core tubes are pushed into the sediment, generally producing an excellent contact at the *sediment-water interface.*

Marine sediment cores have provided the fundamental data for understanding change in the global oceanographic system over the QUATERNARY period, and earlier, as examination of important MARINE PALAEOCLIMATIC PROXIES such as FORAMINIFERA, ICE-RAFTED DEBRIS (IRD) and OXYGEN ISOTOPES has provided detailed continuous evidence of changes in major parameters such as SEA-SURFACE TEMPERATURES (SSTs), *ocean volume,* OCEAN PALAEOPRODUCTIVITY and the nature of the THERMOHALINE CIRCULATION. These data can be placed within a uniform global CHRONOLOGY utilising the OXYGEN ISOTOPE stratigraphy to define MARINE ISOTOPIC STAGES, which can then be correlated to terrestrial evidence through PALAEOMAGNETIC DATING and the definition of occasional MARKER HORIZONS, as in the case of TEPHRA horizons. Correlation of the relative volumes of water held in the world's oceans and ice caps has proved possible through the reciprocal relationship seen in the oxygen isotope record of marine sediment cores and ICE CORES.   *JBE/WENA/JP*

[*See also* DEEP-SEA DRILLING, MARINE SEDIMENTS, MARINE SEDIMENT CORES: RESULTS]

Curry WB, Shackleton NJ, Richter C et al. (1995) *Proceedings of the ocean drilling programme (ODP),* initial reports 154. College Station, TX: Ocean Drilling Program.

Davis EE, Mottl MJ, Fisher AT et al. (1992) *Proceedings of the ocean drilling programme (ODP), initial reports* 139. College Station, TX: Ocean Drilling Program.

De Vernal A and Hillaire-Marcel C (2000) Sea-ice cover, sea-surface salinity and halo-/thermocline structure of the northwest North Atlantic: Modern versus full glacial conditions. *Quaternary Science Reviews* 19: 65–85.

Gersonde R and Seidenkrantz M-S (2013) Sampling marine sediment. *PAGES News* 21: 8–9.

Murdoch A and Macknight SD (1991) *CRC handbook of techniques for aquatic sediments sampling.* Boca Raton, FL: CRC Press, 29–95.

Skinner LC and McCave IN (2003) Analysis and modelling of gravity- and piston coring based on soil mechanics. *Marine Geology* 199: 181–204.

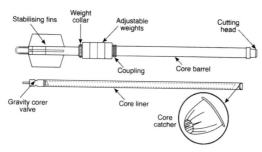

Marine sediment cores *Typical parts of a gravity corer (Murdoch and Macknight, 1991).*

marine sediment cores: results Evidence of environmental change derived from the analysis of MARINE SEDIMENT CORES. Marine sediment cores have provided fundamental PROXY DATA for global PALAEOENVIRONMENTAL RECONSTRUCTION in the GEOLOGICAL RECORD. A diverse range of MARINE PALAEOCLIMATIC PROXIES, such as FORAMINIFERA: ANALYSIS, DINOFLAGELLATE CYST ANALYSIS, DIATOM ANALYSIS, NANNOFOSSILS, GRAIN SIZE, ICE-RAFTED DEBRIS (IRD), MAGNETIC SUSCEPTIBILITY, STABLE ISOTOPES, GREY-SCALE ANALYSIS and organic BIOMARKERS, has been employed to reconstruct phenomena such as SEA-SURFACE TEMPERATURE (SST) or PALAEOTEMPERATURE PROXIES, past SALINITY, biological PRODUCTIVTY, OCEAN PALAEOCIRCULATION, oceanic BIOGEOCHEMICAL CYCLES, sediment records of continental WEATHERING, the TECTONIC drivers of deep-sea TURBIDITY CURRENTS, atmospheric circulation using AEOLIAN dust, PALAEO-ENSO, SEA-ICE extent, terrestrial ICE-SHEET GROWTH and the global ocean THERMOHALINE CIRCULATION in the past. These results can be placed within a consistent chronology by using marine MICROFOSSILS for BIOSTRATIGRAPHY, OXYGEN ISOTOPE stratigraphy and RADIOCARBON DATING, and by using iron minerals in the sediments for PALAEOMAGNETIC DATING. The marine palaeoclimate proxy records provide BOUNDARY CONDITIONS

for PALAEOCEAN MODELLING as well as evidence for the VALIDATION of model output.   *JP*

Barker S, Knorr G, Vautravers MJ et al. (2010) Extreme deepening of the Atlantic overturning circulation during deglaciation. *Nature Geoscience* 3: 567–571.

Coxall HK, Wilson PA, Pälike H et al. (2005) Rapid stepwise onset of Antarctic glaciation and deeper calcite compensation in the Pacific Ocean. *Nature* 433: 53–57.

Fralick P and Carter JE (2011) Neoarchean deep marine paleotemperature: Evidence from turbidite successions. *Precambrian Research* 191: 78–84.

Makou MC, Eglinton TI, Oppo DW and Hughen KA (2010) Postglacial changes in El Niño and La Niña behaviour. *Geology* 38: 43–46.

Pike J, Crosta X, Maddison EJ et al. (2009) Observations on the relationship between the Antarctic coastal diatoms *Thalassiosira antarctica* Comber and *Porosira glacialis* (Grunow) Jørgensen and sea ice concentrations during the Late Quaternary. *Marine Micropaleontology* 73: 14–25.

marine sediments SEDIMENTARY DEPOSITS in the SEAS and OCEANS, including OCEANIC SEDIMENTS and those that accumulate on the CONTINENTAL SHELF. Marine sediments can be CLASTIC (e.g. sand and clay), *biogenic* (e.g. calcium carbonate or siliceous OOZE) or *chemical sediments* (e.g. EVAPORITES) in nature and hold an important record of ENVIRONMENTAL CHANGE.   *JP/GO*

[*See also* BIOGENIC SEDIMENT, CORAL REEF, DEEP-SEA DRILLING, ICE-RAFTED DEBRIS (IRD), MARINE SEDIMENT CORES, PALAEOCEANOGRAPHY, PELAGIC SEDIMENT, SAPROPEL, TURBIDITE]

Burdige DJ (2002) *Geochemistry of marine sediments.* Princeton, NJ: Princeton University Press.

Gray JS and Elliott M (2009) *Ecology of marine sediments: From science to management, 2nd edition.* Oxford: Oxford University Press.

Nichols G (2009) *Sedimentology and stratigraphy, 2nd edition.* Chichester: Wiley-Blackwell.

Scourse J (ed.) (2006) The HOLSMEER Project: Late Holocene shallow marine environments of Europe. *The Holocene* 16: 931–1029 [Special Issue].

marine-based ice sheet An ICE SHEET where the majority of its bed is below sea level. Ice is anchored to its bed only because it is too thick to float and couples with the surrounding floating ICE SHELVES at the GROUNDING LINE, where the ice reaches a *critical flotation thickness*, which has significant implications for ICE-SHEET DYNAMICS. The West Antarctic Ice Sheet is the only contemporary example of a marine-based ice sheet, although they also existed in the Northern Hemisphere during the LAST GLACIAL MAXIMUM (LGM) and during earlier GLACIATIONS.

Marine-based ice sheets are thought to be less stable than land-based ones because the grounding lines tend to occur on inland-dipping slopes. Collapse, thinning or retreat of fringing ICE SHELVES would cause the rapid retreat of grounded ice downslope, particularly in West Antarctica, where the deepest part of the bed is near the centre of the ICE SHEET. This process is particularly prevalent where relatively warm OCEAN CURRENTS intrude onto the CONTINENTAL SHELF, sometimes through overdeepened troughs, and cause increased basal melting at the grounding line and overlying ice shelf. Any future thinning of the grounded ice inland in Antarctica would also change the boundary at which the ice begins to float and similarly cause retreat of the grounding line. In contrast, bedrock highs, GROUNDING-ZONE WEDGES, *ice rises* and changes in grounding-line geometry are all thought to stabilise the grounding line as well as influence the amount of warmer ocean water reaching the grounding line.   *BTIR*

[*See also* GROUNDING LINE, GROUNDING-ZONE WEDGE, ICE SHEET, ICE SHELF, ICE STREAM]

Hindmarsh RCA and Le Meur E (2001) Dynamical processes involved in the retreat of marine ice sheets. *Journal of Glaciology* 47: 271–282.
Mercer J (1978) West Antarctic Ice Sheet and CO_2 greenhouse effect: A threat of disaster. *Nature* 271: 321–325.
Schoof C (2007) Marine ice-sheet dynamics, part 1: The case of rapid sliding. *Journal of Fluid Dynamics Mechanics* 573: 27–55.
Vaughan DG (2008) West Antarctic Ice Sheet collapse: The fall and rise of a paradigm. *Climatic Change* 91: 65–79.

marker horizon   A position in a stratigraphic sequence defined by a particular horizon found over a large area and representing a SYNCHRONOUS event. Marker horizons are useful for stratigraphical CORRELATION and RELATIVE-AGE DATING. The commonest marker horizons over the QUATERNARY TIMESCALE are (1) TEPHRAS originating from single VOLCANIC ERUPTIONS, (2) PALAEOSOLS and (3) GEOMAGNETIC POLARITY REVERSALS. In the last case, this provides a horizon found on the global scale.   *CJC*

[*See also* DATING TECHNIQUES, GAUSS-MATUYAMA GEOMAGNETIC BOUNDARY, STRATIGRAPHY]

Barber K, Langdon P and Blundell A (2008) Dating the Glen Garry tephra: A widespread late-Holocene marker horizon in the peatlands of northern Britain. *The Holocene* 18: 31–43.

Markov process   A *stochastic* process in which the time parameter is discontinuous and the probability of an event is dependent only upon the state of the system and the probabilities of a change from one state to another (*transition probabilities*). In practice, this means that the state at any future time is dependent on the present state and the transition probabilities but is unaffected by any additional knowledge of the past history of the system. The usefulness of the concept,

and the corresponding *Markov chain analysis*, depends especially on the reliability of the established transition probabilities used and whether they actually remain constant through time.   *AHP/JAM*

Caskey JE (1964) Markov chain model of cold spells at London. *Meteorological Magazine* 93: 36–138.
Nicolis C (1990) Chaotic dynamics, Markov processes and climate dynamics. *Tellus* 42A: 401–412.
Waggoner PE and Stephens GR (1970) Transition probabilities for a forest. *Nature* 225: 1160–1161.

marl   A calcareous, slightly clayey LACUSTRINE SEDIMENT with >80 per cent *calcium carbonate*, deposited in the *sublittoral* and *eulittoral zone* of a lake and formed through the activity of *calcareous algae*.   *UBW*

[*See also* ALGAE, SEDIMENT TYPES, STABLE ISOTOPE ANALYSIS]

Kelts K and Hsü KJ (1978) Freshwater carbonate sedimentation. In Lerman A (ed.) *Lakes, chemistry, geology, physics.* New York: Springer Verlag, 295–323.
Walker MJC, Griffiths HI, Ringwood V and Evans JG (1993) An early-Holocene pollen, mollusk and ostracod sequence from lake marl at Llangorse Lake, South Wales, UK. *The Holocene* 3: 138–149.

marsh   A WETLAND dominated by HERBS (commonly GRAMINOIDS) rather than woody plants.   *JAM*

[*See also* CARR, FEN, MEADOW, SALTMARSH, SWAMP]

Burt W (2007) *Marshes: The disappearing Edens.* New Haven, CT: Yale University Press.

mass balance   The change in mass at any point on the surface of a GLACIER at any time. Commonly, it means the change in mass of the entire glacier in a standard unit of time (the *balance year* or measurement year). The mass balance is the result of variations in ABLATION and *accumulation*, which cause volume changes. The *net balance* is the sum of the *winter balance* (positive) and the *summer balance* (negative). If ablation exceeds accumulation, the net balance is negative; the opposite situation produces a positive net balance.   *AN*

Cogley JG (2011) Mass-balance terms revisited. *Journal of Glaciology* 56: 997–1001.
Hanna E, Navarro FJ, Pattyn F et al. (2013) Ice-sheet mass balance and climate change. *Nature* 498: 51–59.
Houghton J, Bamber JL and Payne AJ (eds) (2004) *Mass balance of the cryosphere: Observations and modelling of contemporary and future changes.* Cambridge: Cambridge University Press.
Huss M, Bauder A, Funk H and Hock R (2008) Determination of the seasonal mass balance of four alpine glaciers since 1865. *Journal of Geophysical Research* 113: F01015.
Rignot E and Thomas RH (2002) Mass balance of polar ice sheets. *Science* 297: 1502–1506.

mass extinctions   There have been a number of episodes in the history of the Earth when EXTINCTIONS have occurred at rates far greater than the BACKGROUND RATE. These episodes are sufficiently prominent that they form the basis of much of the classic subdivision of the geological column—for example, the extinction of the *dinosaurs* at the boundary between the CRETACEOUS and the TERTIARY (see K-T BOUNDARY). Some episodes are more pronounced than others, and the severity, in terms of the proportion of species that become extinct, varies widely between different groups. FAUNA, especially *marine fauna*, tend to be more severely affected than plants. Marine BIODIVERSITY tends to 'rebound' after a mass extinction, extinction rates being correlated with origination rates with a time lag of around 10 million years (see Figure). There has been, and still is, debate about the nature of these episodes, whether they are caused by CATASTROPHIC EVENTS and, in particular, whether they exhibit PERIODICITY. They tend to occur every 26–30 million years, but it is not clear whether this is a genuinely periodic frequency. It has been argued that the current wave of ANTHROPOGENIC extinctions constitutes the 'sixth mass extinction'.

It is possible, but not established, that mass extinction may have an extraterrestrial explanation, through METEORITE IMPACT, for example, and this might explain the periodicity. The Earth is certainly subjected to all manner of extraterrestrial phenomena. So, for example, the number and size of known ASTEROIDS in orbits that approach the Earth mean that on average one collision every 1.4 Ma with objects greater than 2 km in diameter can be expected, one every 330 Ma with objects greater than 8 km in diameter, and so on. Whatever the explanation, a high proportion of the Earth's fauna, and some FLORA, became extinct during these episodes, with far-reaching consequences for EVOLUTION, because the continuance of life on Earth depends on the survivors, which may be a small, randomly selected, proportion of the original fauna.

It may be that mass extinctions solve what has been termed the 'paradox of the first tier': 'our failure to find any clear vector of fitfully accumulating progress, despite expectations that processes regulating the first tier should yield such advance, represents our greatest dilemma for the study of pattern in life's history' (Gould, 1985: 4). Mass extinctions resolve the paradox because evolutionary processes at the first tier (i.e. *ecological processes*) cannot extend to the longest timescales. Mass extinctions, occurring randomly with respect to these shorter-term processes, prevent the extension of evolutionary trends. They may, thus, be of considerable importance in determining which LINEAGES comprise modern faunas and (possibly to a lesser extent) floras.   *KDB*

[*See also* EXTREME CLIMATIC EVENT, MEGAFAUNAL EXTINCTION, NEAR-EARTH OBJECT (NEO)]

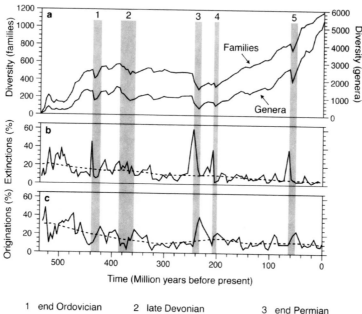

1 end Ordovician   2 late Devonian   3 end Permian
4 end Triassic   5 Cretaceous-Tertiary

Mass extinctions *The five largest mass extinctions of the geological record (shaded and numbered columns) as exemplified by the fossil record of marine animal diversity: (A) number of families and genera, and corresponding percentages of (B) extinction and (C) origination. Dotted lines are background rates estimated by curve fitting (Kirchner and Weil, 2000).*

Bambach RK, Knoll AH and Wang SC (2004) Origination, extinction, and mass depletions of marine diversity. *Paleobiology* 30: 522–542.

Barnosky AD, Matzke N, Tomiya S et al. (2011) Has the Earth's sixth mass extinction already arrived? *Nature* 471: 51–57.

Courtillot V (1999) *Evolutionary catastrophes: The science of mass extinction.* Cambridge: Cambridge University Press.

Erwin DH (1994) The Permo-Triassic extinction. *Nature* 367: 231–236.

Frankel C (1999) *The end of the dinosaurs: Chicxulub crater and mass extinction.* Cambridge: Cambridge University Press.

Gould SJ (1985) The paradox of the first tier: An agenda for paleobiology. *Paleobiology* 11: 2–12.

Kirchner JW and Weil A (2000) Delayed biological recovery from extinctions throughout the fossil record. *Nature* 404: 177–180.

Knoll AH, Bambach RK, Payne JL et al. (2007) Paleophysiology and end-Permian mass extinction. *Earth and Planetary Science Letters* 256: 295–313.

Raup DM and Sepkoski Jr JJ (1984) Periodicity of extinctions in the geologic past. *Proceedings of the National Academy of Sciences* 81: 801–805.

Traverse A (1988) Plant evolution dances to a different beat: Plant and animal evolutionary mechanisms compared. *Historical Biology* 1: 277–301.

Willis KJ and Bennett KD (1995) Mass extinction, punctuated equilibrium and the fossil plant record. *Trends in Ecology and Evolution* 10: 308–309.

mass movement processes These processes, involving the downslope displacement of rock and REGOLITH under the influence of gravity, are of global significance in the development of hillslopes. They are also of economic and social importance as they may disrupt public utilities and cause loss of life. Although not confined to a particular environment, mass movement processes probably reach their greatest intensity and efficacy in steep and mountainous PERIGLACIAL regions. A range of materials and processes is involved in mass movement and gives rise to a great variety of movement types (see LANDSLIDE). Criteria for recognition of these types include velocity and mechanism of movement, physical properties of the material, mode of DEFORMATION, geometry of both the *source area* and the displaced mass and *water content.* Common types of movement include *fall, flow, slide, rebound, sag,* SLUMP, *avalanche, topple* and *creep.* The type of movement occurring in an area depends on the local climatic factors, the lithology and structure and the local TOPOGRAPHY. *PW*

[*See also* CAMBERING, DEBRIS FLOW, FLOW SLIDE, GELIFLUCTION, MUDSLIDE, PARAGLACIAL, PLOUGHING BLOCK, ROCK CREEP, ROCK GLACIER, ROCK STREAM, ROCKFALL, SACKUNG, SNOW AVALANCHE, SOIL CREEP, SOLIFLUCTION, STURZSTRÖM, TALUS, UNLOADING, VALLEY BULGING]

Brunsden D (1993) Mass movement: The research frontier and beyond: A geomorphological approach. *Geomorphology* 7: 85–128.

Cooper RG (2007) *Mass movements in Great Britain.* Peterborough: Joint Nature Conservation Committee.

Dickau R (2004) Mass movement. In Goudie AS (ed.) *Encyclopedia of geomorphology.* London: Routledge, 644–653.

Mosher DC, Ship C, Moscadelli L et al. (eds) (2009) *Submarine mass movements and their consequences.* Berlin: Springer.

Shroder JF, Owen LA, Seong YB et al. (2011) The role of mass movements on landscape evolution in the central Karakoram: Discussion and speculation. *Quaternary International* 236: 34–47.

mass spectrometer An instrument designed to measure ISOTOPE RATIOS by separating positively charged IONS on the basis of their *mass-to-charge ratio.* *IR*

[*See also* ISOTOPES AS INDICATORS OF ENVIRONMENTAL CHANGE, STABLE ISOTOPE ANALYSIS]

Benson S, Lennard C, Maynard P and Roux C. (2006) Forensic applications of isotope ratio mass spectrometry: A review. *Forensic Science International* 157: 1–22.

mass wasting A collective term for the downslope movement of REGOLITH under the influence of gravity. Mass wasting occurs in all climatic regions and accounts for a major part of DENUDATION in many but is probably of greatest intensity in PERIGLACIAL ENVIRONMENTS. Many different types of slow and rapid MASS MOVEMENT PROCESSES are involved. *PW*

[*See also* LANDSLIDE]

Fort M (2000) Glaciers and mass wasting processes: Their influence in the shaping of the Kali Gandaki valley (higher Himalaya of Nepal). *Quaternary International* 65–66: 101–119.

Harris C (1981) *Periglacial mass wasting: A review of research.* Norwich: Geobooks.

Massenerhebung effect The effect of large mountain masses, particularly plateaus, in raising temperatures above the values found at similar heights in the free atmosphere and on isolated peaks. It arises because large upland surfaces present extensive areas for heating by SOLAR RADIATION receipts that increase with altitude, whereas isolated mountains merely protrude into the free atmosphere and temperatures fall according to the local atmospheric LAPSE RATE. The term was first used in the European Alps, where the SNOW LINE, TREE LINES and the ALTITUDINAL ZONATION of vegetation are all at higher altitude on the large central massifs, such as the Appennine and Engadine Alps, than in the more coastal alpine ranges. Similarly, on East African plateaux, tropical rain forest extends to higher altitudes in part as a result of the Massenerhebung effect. The telescoping of altitudinal limits in coastal mountains compared with farther inland in tropical and temperate areas was formerly ascribed erroneously to *Massenerhebung* inland and its absence on coastal hills, whereas the main reason for the contrast is usually linked to higher humidity and hence cloud formation at lower altitudes in coastal areas. *RPDW*

Brockmann-Jerosch H (1913) *Der Einfluss des Klimacharakters auf die Verbreitung der Pflanzen und Pflanzengesellschaften* [The influence of climate on the distribution of plants and plant communities]. *Botanische Jahrbuch (Beibl.)* 49: 19–43.
Chiou CR, Song GZM, Chien JH et al. (2010) Altitudinal distribution patterns of plant species in Taiwan are mainly determined by the northeast monsoon rather than the heat retention mechanism of Massenerhebung. *Botanical Studies* 51: 89–97.
Ellenberg H (1988) *Vegetation ecology of central Europe.* Cambridge: Cambridge University Press.
Walsh RPD (1996) Climate. In Richards PW (ed., with Walsh RPD, Baillie I and Greig-Smith P) *The tropical rain forest, 2nd edition.* Cambridge: Cambridge University Press, 159–205.

massive   BEDS or bodies of SEDIMENT lacking obvious STRATIFICATION or variations in TEXTURE (IN SEDIMENTS AND ROCKS).   *MRT*

master chronology   An average time series of TREE-RING data from many trees in a region. All of the annual values are aligned in the precise calendar year of their growth and averaged, with the resulting series expressing the pattern of common year-to-year variability contained in the trees from that region. The master chronology thus represents a 'templet' of ANNUAL VARIABILITY in growth, against which series of ring widths in other wood samples from that area may be compared, matched and so dated with absolute PRECISION. The term is also used in the context of VARVE CHRONOLOGY.   *KRB/APP*
[*See also* CHRONOLOGY, CROSSDATING, DENDROCHRONOLOGY, TREE-RING INDEX]

Baillie MGL (1995) *A slice through time: Dendrochronology and precision dating.* London: Routledge.
Speer JH (2010) *Fundamentals of tree-ring research.* Tucson: University of Arizona Press.

material culture   In the archaeological context, all the physical expressions of a culture, including not only the portable objects (ARTEFACTS) but also non-portable human-made remains (*features*) such as buildings, FIELD SYSTEMS, MIDDENS, burials and roads.   *JAM*

Hicks D and Beaudry MC (eds) (2010) *The Oxford handbook of material culture studies.* Oxford: Oxford University Press.

Thomas N (1991) *Entangled objects: Exchange, material culture, and colonialism in the Pacific.* Cambridge, MA: Harvard University Press.
Tilley C, Keane W, Kuechler-Fooden S et al. (eds) (2006) *Handbook of material culture.* London: Sage.
Woodward I (2007) *Understanding material culture.* London: Sage.

materials detoxification   (1) An aspect of MATERIALS SUBSTITUTION: the substitution of non-toxic chemicals for toxic ones (e.g. POLYCHLORINATED BIPHENYLS (PCBs), CHLOROFLUOROCARBONS (CFCs), LEAD (Pb) and CADMIUM (Cd). The search continues for inherently safer substitutes. (2) The term also describes the use of *micro-organisms* and/or high-temperature incineration for the detoxification of WASTE.   *JAM/CJB*
[*See also* INDUSTRIAL ECOLOGY, MANUFACTURING INDUSTRY]

Geiser K (2001) *Materials matter: Towards a sustainable materials policy.* Cambridge: MIT Press.

materials substitution   The substitution of one material for another in the MANUFACTURING INDUSTRY and a core phenomenon of INDUSTRIALISATION. Classic examples include the successive use of CHARCOAL, COAL, PETROLEUM and NATURAL GAS in energy production, both during and since the INDUSTRIAL REVOLUTION, and the displacement of natural materials by synthetic fibres, rubber, PLASTICS and FERTILISERS. Materials substitution can overcome resource constraints, it can result in products that are more economically efficient and/or function better, it can promote new applications, it can replace harmful substances by environmentally friendly ones and it can contribute to reduced use of materials.   *JAM*
[*See also* DEMATERIALISATION, INDUSTRIAL ECOLOGY, MATERIALS DETOXIFICATION]

Ashby MF (2009) *Materials and the environment: Eco-informed material choice.* Amsterdam: Elsevier.
Wernick IK (1996) Consuming materials: The American way. *Technological Forecasting and Social Change* 53: 111–122.

matrix   Particles in a SOIL, SEDIMENT or SEDIMENTARY ROCK that are finer grained than the nominal GRAIN SIZE: for example, MUD matrix in a SANDSTONE, SAND or mud matrix in a CONGLOMERATE or DIAMICTON. Some dispute surrounds the definition of matrix in sandstones: it is commonly defined as particles finer than 0.03 mm (i.e. medium to coarse SILT). Matrix can be formed by DEPOSITION of poorly sorted sediment, by INFILTRATION of fines after deposition, or by COMPACTION of mechanically weak LITHIC fragments during DIAGENESIS. Larger particles or objects are embedded in matrix: the matrix may support larger CLASTS or fill the interstices between them. A mud matrix in a RUDITE may indicate non-aqueous deposition (e.g. by DEBRIS FLOW or as glacial TILL), particularly where it forms a *matrix-supported* FABRIC. In an archaeological context, the matrix is the physical medium (usually a soil or sediment) that surrounds and/or supports an ARTEFACT or other archaeological find. *GO/TY*

[*See also* FLYSCH, GREYWACKE, MICROMORPHOLOGICAL ANALYSIS OF SEDIMENTS, MICROMORPHOLOGICAL ANALYSIS OF SOILS, MICROMORPHOLOGICAL ANALYSIS OF TILL, PROVENIENCE, WACKE]

Lee G-A (2012) Taphonomy and sample size estimation in paleoethnobotany. *Journal of Archaeological Science* 39: 648–655.
Pettijohn FJ (1975) *Sedimentary rocks, 3rd edition.* New York: Harper and Row.
Walker TR, Waugh B and Crone AJ (1978) Diagenesis in first-cycle desert alluvium of Cenozoic Age, southwestern United States and northwestern Mexico. *Geological Society of America Bulletin* 89: 19–32.

maturity (1) In the context of LANDSCAPES, maturity describes a stage in LANDSCAPE EVOLUTION. It is used especially as rivers develop towards the low-gradient, wide, typically highly sinuous morphology found near BASE LEVEL (see DAVISIAN CYCLE OF EROSION and the GRADE CONCEPT).

(2) In the context of SEDIMENTS and SEDIMENTARY ROCKS, maturity is a concept that combines several parameters of TEXTURE or *composition* of a SANDSTONE. The concept can be used descriptively or can be interpreted in terms of a more mature sediment having experienced more extensive WEATHERING, a longer path of SEDIMENT TRANSPORT, or a DEPOSITIONAL ENVIRONMENT of higher energy (see ENERGY (OF ENVIRONMENT)). However, the concept needs to be applied with caution. *Source inheritance* is a particular problem; for example, polycyclic or POLYGENETIC sediment derived from a quartz-rich source will necessarily be compositionally mature (see PROVENANCE). Such problems can be minimised by comparing different types of maturity for a given sediment.

Textural maturity describes the tendency towards being well sorted (see SORTING), having well-rounded GRAINS (see ROUNDNESS) and lacking MATRIX. *Compositional maturity* (also known as *mineralogical maturity*) describes the tendency towards the occurrence of only the most stable MINERAL grains (typically QUARTZ), with the loss of the more degradable (labile) mineral and LITHIC grains, which are chemically altered to CLAY MINERALS. A *maturity index* expressed as the ratio (quartz + chert): (feldspars + rock fragments) is sometimes used. *Chemical maturity* describes the evolution towards a more inert chemical composition during CHEMICAL WEATHERING (e.g. during loss of soluble components by LEACHING). The *chemical index of alteration* is calculated (on a carbonate-free basis) as the ratio of *aluminium oxide* to *total oxide* content.

(3) In the context of *organic sediments*, maturity refers to the degree to which organic material has been converted during burial to *kerogen* and to HYDROCARBONS (*maturation*). *TY/GO*

Boggs Jr S (2009) *Petrology of sedimentary rocks, 2nd edition.* Cambridge: Cambridge University Press.
Cox R and Lowe DR (1995) A conceptual review of regional-scale controls on the composition of clastic sediment and the co-evolution of continental blocks and their sedimentary cover. *Journal of Sedimentary Research* A65: 1–12.
Gluyas J and Swarbrick R (2004) *Petroleum geoscience.* Oxford: Blackwell.
Nichols G (2009) *Sedimentology and stratigraphy.* Oxford: Blackwell.
Pettijohn FJ (1975) *Sedimentary rocks, 3rd edition.* New York: Harper and Row.
Selley RC (1997) *Elements of petroleum geology, 2nd edition.* Orlando, FL: Academic Press
Summerfield MA (1991) *Global geomorphology.* Harlow: Longman.

Maunder minimum The period of very low *sunspot* activity from AD 1645 to 1715. AD 1693 was the lowest trough in a SUNSPOT CYCLE ever recorded. The coincidence of the Maunder minimum in solar activity with the coldest period of the LITTLE ICE AGE provides evidence for SOLAR FORCING of SECULAR VARIATIONS in climate. The Maunder minimum, together with other historical sunspot minima (DALTON, OORT, SPÖRER and WOLF MINIMA), and the *contemporary solar activity maximum* and *Mediaeval solar activity maximum* have been detected in various PROXY DATA sources, such as the *cosmogenic isotope* record in ICE CORES and TREE RING (see Figure). *JAM/AHP*

Damon PE, Eastoe CJ, Hughes MK et al. (1998) Secular variation of Δ14C during the Medieval Solar Maximum: A progress report. *Radiocarbon* 40: 343–350.
Diodato N and Bellocchi G (2012) Discovering the anomalously cold Mediterranean winters during the Maunder minimum. *The Holocene* 22: 589–596.
Eddy JA (1976) The Maunder minimum. *Science* 192: 1189–1202.
Lean JL (2010) Cycles and trends in solar irradiance and climate. *Wiley Environmental Reviews Climate Change* 1: 111–122.
Lockwood M (2006) What do cosmogenic isotopes tell us about past solar forcing of climate? *Space Science Reviews* 125: 95–109.

meadow A *grassland* maintained by *mowing*, as distinct from PASTURE maintained by GRAZING. Meadows are ANTHROPOGENIC ecosystems and part (sometimes an ancient part) of the CULTURAL LANDSCAPE. Distinct *flood meadows* (flooded naturally), *water meadows* (flooded artificially), *hay meadows, cornfields* and *alpine meadows* are amongst the types that may be recognised. All are characterised by a rich flora of tall HERBS. Tall herbs

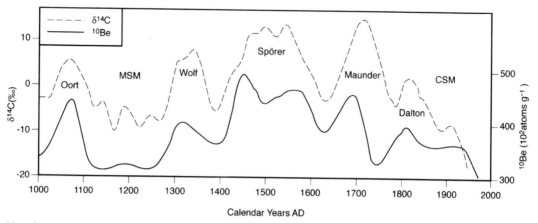

Maunder minimum *Trends in $\delta^{14}C$ and ^{10}Be over the past millennium from tree rings and ice cores compared with named minima in solar activity, including the Dalton, Maunder, Spörer, Wolf and Oort minima. The contemporary solar activity maximum (CSM) and the Mediaeval solar activity maximum (MSM) are also shown (Damon et al., 1998).*

are not killed by mowing but by grazing, and are able to flower and set seed before the meadows are mowed. Because of the decline in traditional agricultural practices, meadows are in urgent need of CONSERVATION. In Britain, for example, 97 per cent of meadows have disappeared since the 1930s. In North America, the term is often used to denote patches of grassland, regardless of the form of management.   *JAM*

[*See also* HABITAT LOSS]

Cook H and Williamson T (1999) *Water management in the English landscape: Field, marsh and meadow.* Edinburgh: Edinburgh University Press.

Feltwell J (1992) *Meadows: A history and natural history.* Stroud: Alan Sutton.

Michaud A, Plantureux S, Amiaud B et al. (2012) Identification of the environmental factors which drive the botanical and functional composition of permanent grasslands. *Journal of Agricultural Science* 150: 219–236.

Roche LM, Allen-Diaz B, Eastburn DJ and Tate KW (2012) Cattle grazing and Yosemite toad (*Bufo canorus* Camp) breeding habitat in Sierra Nevada meadows. *Rangeland Ecology and Management* 65: 56–65.

mean annual air temperature (MAAT)   Usually calculated as the average ANNUAL MEAN AIR TEMPERATURE over a standard time period of 30 years (i.e. the average of 30 annual mean temperature values). The years 1971–2000 are commonly recognised as the latest *standard period*. Data for different locations around the world can be accessed at http://www.ncdc.noaa.gov/oa/wdc/index.php.   *DAW/JAM*

[*See also* CLIMATIC NORMALS, GLOBAL MEAN SURFACE AIR TEMPERATURE]

Pearce EA and Smith CG (2005) *The Hutchinson world weather guide, 6th edition.* Oxford: Helican.

mean lethal dose (L50)   The dose of a TOXIN required to kill one-half of a target population.   *JAM*

mean monthly temperature   Usually calculated as the average MONTHLY MEAN TEMPERATURE over a standard time period of 30 years.   *AHP*

mean sea level   The average elevation of the sea surface along a SHORELINE over a given period of time.   *HJW*

[*See also* RELATIVE SEA LEVEL, SEA-LEVEL CHANGE]

Pugh DT (2004) *Changing sea levels: Effects of tides, weather and climate.* Cambridge: Cambridge University Press.

Tooley MJ (1993) Long term changes in eustatic sea level. In Warrick RA, Barrow EM and Wigley TML (eds) *Climate and sea level change: Observations, projections and implications.* Cambridge: Cambridge University Press, 81–107.

meander cut-off   A MEANDER bend abandoned by the main river channel after AVULSION or progressive EROSION of the *meander neck*. A meander cut-off appears on the floodplain as a horseshoe-shaped depression, sometimes water filled, hence the term *oxbow lake*. A sudden increase in FLOOD MAGNITUDE-FREQUENCY CHANGES with climatic or LANDUSE CHANGE may produce a large number of cut-offs over a short period. *Sediment infill*, vegetation SUCCESSION, and SOIL DEVELOPMENT follow their formation. Meander cut-offs are a likely component of CHANNEL PATTERN change resulting from predicted increases in flood magnitude-frequency changes.   *DML*

[*See also* CHANNEL CHANGE, ECOLOGICAL SUCCESSION]

Constantine JA and Dunne T (2008) Meander cutoff and the controls on the production of oxbow lakes. *Geology* 36: 23–26.

Lewis GW and Lewin J (1983) Alluvial cutoffs in Wales
and the Borderland. In Collinson JD and Lewin J (eds)
Modern and ancient fluvial systems. Oxford: Blackwell,
145–154.

meanders Sinuous winding bends in rivers. The
term originates from the Maiandros River in northwestern Turkey. Meandering is a natural physical tendency of rivers, as with all fluids and gases (e.g. OCEAN
CURRENTS and JET STREAMS) on a spinning Earth. The
development of meanders in sediment-free meltwater
streams on the surface of GLACIERS and by DISSOLUTION
in underground KARST demonstrates this. The occurrence and dimensions of meanders (wavelength, amplitude and sinuosity) are strongly influenced by *dominant
discharge, channel width, sediment supply* and *bank
strength.* Meanders occur in moderately active river
systems, and bend creation can be related to sufficient
energy for systematic bed deformation and BANK ERO-
SION. Meandering rivers can exhibit very regular or
tortuous and irregular meanders and can be active or
inactive. They tend to be best developed where rivers
have FLOODPLAINS and for this reason are particularly
associated with the middle and lower reaches of river
systems. Meander occurrence and dimensions in alluvial rivers are very sensitive to changes in climate, landuse and land management. The controls of meandering
of bedrock channels in mountainous terrain, where
larger-wavelength valley meanders are characteristically found, are in contrast not well understood. INCISED
MEANDERS have cut down into bedrock.   *RPDW/DML*

[*See also* CHANNEL CHANGE, CHANNEL PATTERN,
MEANDER CUT-OFF, MISFIT MEANDER]

Chorley RJ, Schumm SA and Sugden DA (1984)
Geomorphology. London: Methuen.
Hooke JM (2007) Complexity, self-organization
and variation in behaviour in meandering rivers.
Geomorphology 91: 236–258.
Ikeda S and Parker G (eds) (1989) *River meandering.*
Washington, DC: American Geophysical Union.
Schumm SA (1963) Sinuosity of alluvial rivers on the Great
Plains. *Bulletin of the Geological Society of America* 74:
1089–1100.
Smith CE (1998) Modeling high-sinuosity meanders in a
small flume. *Geomorphology* 25: 19–30.

measurement Quantitative *description,* normally
involving an instrument. However, this definition must
be adopted with caution, particularly in the light of the
so-called *theory of scales.* This refers to the *level of
measurement* or type of *measurement scale* on which
VARIABLES or items are described. Many physical quantities measured in SCIENCE use a *ratio scale,* on which
the ratios between numbers are meaningful and zero
is non-arbitrary (e.g. length, mass and time). *Interval-
scale* measurement uses an arbitrary zero value (e.g.

the freezing point of water is zero only when measured
in degrees Celsius) and the ratios between numbers on
the scale are not meaningful. Both ratio and interval
scales produce what are normally understood to be
quantitative data. On an *ordinal scale,* data values can
only be placed in rank order: one knows the relative
position of one value in relation to another but not by
how much they differ (e.g. MOHS' SCALE of the hardness
of minerals, or describing people as upper, middle or
lower class. *Nominal-scale* measurement produces *cat-
egorical data,* in which items are placed in categories
or classes that differ in kind rather than degree (e.g.
landuse categories and vegetation types); thus, there
are no definable CLASS INTERVALS within and between
the classes. Nominal and ordinal scales produce what
are normally understood to be *qualitative data* and are
particularly widely used in the SOCIAL SCIENCES. *Metrol-
ogy* is the science of measurement.   *JAM*

[*See also* INSTRUMENTAL RECORD, OBSERVATION, SI UNITS]

Stevens SS (1946) On the theory of scales of measurement.
Science 103: 677–680.

mechanical denudation The contribution of physical processes of EROSION and PHYSICAL WEATHERING to DEN-
UDATION. Mechanical denudation rates tend to be greatest
in regions with high RELIEF, where DRAINAGE BASINS are
highly *glacierised* and where the rocks exhibit high
ERODIBILITY. Globally, mechanical denudation accounts
for 75 per cent of the total denudation.   *JAM/RAS*

[*See also* CHEMICAL DENUDATION]

Gaillardet J (2004) Denudation. In Goudie AS (ed.)
Encyclopedia of geomorphology. London: Routledge,
240–244.

mechanical soil conservation measures Structures or *land-shaping techniques* intended to prevent
or reduce SOIL EROSION on slopes. On land under CUL-
TIVATION, they include *land formation techniques* such
as CONTOUR BUNDS and TERRACING. There are also *stabilisation structures* including GABIONS and GULLY
CONTROL DAMS. The other forms of SOIL CONSERVATION
measures on cultivated land are *agronomic measures*
(e.g. MULCHING and *crop management*) and *soil management,* which involves, for example, various types
of conservation or reduced TILLAGE.   *RAS*

Hudson NW (1995) *Soil conservation, 3rd edition.* London:
Batsford.
Mane MS, Mahadkar UV, Ayare BL and Thorat TN (2009)
Performance of mechanical soil conservation measures
on cashew plantation grown on steep slopes of Konkan.
Indian Journal of Soil Conservation 37: 181–184.

Mediaeval Warm Period Variously known as
the *Little Climatic Optimum, Mediaeval Warm Epoch*

or *Mediaeval Climatic Anomaly*, the Mediaeval Warm Period extended from approximately the late ninth to the early fourteenth centuries. It was characterised by a climate at least as warm as modern times over at least parts of the Northern and Southern Hemispheres. Ice-sheet temperature profiles (SUBSURFACE TEMPERATURES from BOREHOLES) indicate temperatures of about 1.0°C warmer than the AD 1881–1980 average during its peak. Viking settlements flourished in Greenland and Iceland at this time, correlating with the higher temperatures also indicated by OXYGEN ISOTOPES in Greenland ICE CORE records. Agriculture expanded, moreover, in MARGINAL AREAS of the uplands of northwest Europe, only to decline during the subsequent LITTLE ICE AGE. Major DROUGHTS occurred in California, Patagonia and China. However, the Mediaeval Warm Period was not uniformly warm: evidence from sources such as GLACIER VARIATIONS and DENDROCLIMATOLOGY indicates interruptions. It represents one of the short-term, SUB-MILANKOVITCH climatic variations that characterised the HOLOCENE and implies a global reorganisation of climate during part of the MIDDLE AGES. *JAM*

[*See also* CENTURY- TO MILLENNIAL-SCALE VARIABILITY, NORSE GREENLAND SETTLEMENTS, VIKING AGE]

Bradley RS, Hughes MK and Diaz HL (2003) Climate in Medieval time. *Science* 302: 404–405.

Chu GQ, Liu JQ, Sun Q et al. (2002) The 'Mediaeval Warm Period' drought recorded in Lake Huguangyan, tropical south China. *The Holocene* 12: 511–516.

Cook ER, Palmer JG and D'Arrigo RD (2002) Evidence for a 'Medieval Warm Period' in a 1,100-year tree-ring reconstruction of past austral summer temperature in New Zealand. *Geophysical Research Letters* 29: 1669. doi:10.1029/2001GL014580.

Lamb HH (1965) The early Medieval Warm Epoch and its sequel. *Palaeogeography, Palaeoclimatology, Palaeoecology* 1: 13–37.

Hughes MK and Diaz HF (eds) (1994) *The Medieval Warm Period*. Dordrecht: Kluwer.

Koch J and Clague JJ (2011) Extensive glaciers in northwest North America during Medieval time. *Climatic Change* 107: 593–613.

Stine S (1994) Extreme and persistent drought in California and Patagonia during Mediaeval time. *Nature* 369: 546–549.

Trouet V, Esper J, Graham NE et al. (2009) Persistent positive North Atlantic oscillation mode dominated the Medieval climate anomaly. *Science* 324: 78–80.

Vinther BM, Jones PD, Briffa KR et al. (2010) Climatic signals in multiple highly resolved stable isotope records from Greenland. *Quaternary Science Reviews* 29: 522–538.

Xoplaki E, Fleitman D, Diaz HF et al. (eds) (2011) Medieval climate anomaly. *PAGES News* 19: 4–31 [Special Issue].

medical science The application of scientific knowledge to the problems of human health. Ever since critical thinking came to the fore in the SCIENTIFIC REVOLUTION and the Enlightenment, medical practices have become increasingly science and evidence based. Along with education, and the development of hospitals (at first treated as charitable poorhouses), medical laboratories and public health services, medical science has led to profound DEMOGRAPHIC CHANGES and changes in ENVIRONMENT-HUMAN INTERACTION. This has not, however, removed the inequalities in health care between rich and poor or between developed and DEVELOPING COUNTRIES. In developed countries, improvements in environmental conditions and living standards today contribute more than curative medicine to longer lifespans, medical science is making only slow inroads into the diseases of aging, and the role of medicine is shifting towards satisfying lifestyle wishes. Meanwhile, in developing countries, there remains the potential of medical science to keep many more people alive and healthy.

JAM

[*See also* DISEASE, EPIDEMIOLOGY, GEOMEDICINE, HUMAN HEALTH HAZARDS, QUALITY-OF-LIFE (QOL) INDICES]

Bynum W (2008) *The history of medicine: A very short introduction.* Oxford: Oxford University Press.

Porter R (1999) *The greatest benefit to mankind: A medical history of humanity.* New York: Norton.

Porter R (2002) *Blood and guts: A short history of medicine.* London: Penguin Books.

Mediterranean environmental change and human impact

During the QUATERNARY, CLIMATIC CHANGE caused phases of SLOPE STABILITY and instability in Mediterranean lands. In general, INTERGLACIAL episodes were characterised by low SEDIMENTATION amounts, a dense vegetation cover, relatively high INFILTRATION and moderate river DISCHARGES. In contrast, GLACIAL phases tended to be cold and dry, with increased SEASONALITY of rainfall, a sparse VEGETATION cover, extensive areas of bare ground and loose SEDIMENTS, leading to extensive slope EROSION and 'flashy' RIVER REGIMES. *Coastline development* and *fluvial activity* were affected by SEA-LEVEL RISE and fall (see MEDITERRANEAN LANDSCAPE EVOLUTION). After fluctuations around 14,000 years ago, the climate became rather wetter than hitherto and forest characterised the higher-rainfall and SAVANNA the low-rainfall areas.

The Mediterranean landscape has a long history of human occupancy and modification of the landscape, people, plants and animals. The mainland has been continuously occupied since the PALAEOLITHIC, with islands (e.g. Cyprus, Crete) not inhabited until about 8,500 years ago. The construction of settlements and AGRICULTURE began at around this time in Anatolia, followed by Palestine and southern Greece.

Around 3,500 years ago, in the Late BRONZE AGE, human impact on the landscape was strongly in evidence, with TERRACING and IRRIGATION works being constructed and a range of crops grown (e.g. winter *wheat, barley, vines* and *olives*) and animals domesticated (e.g. *cattle, sheep, goats and pigs*). Much of what survived in terms of WETLANDS and forest was subject to MANAGEMENT. Large population levels were reached, with numbers in Greece peaking in the *Hellenistic period* (about 2,200 years ago), but later in Italy during the ROMAN PERIOD (about 1,800 years ago). The Romans developed *irrigation methods*, with storage DAMS and AQUEDUCTS. After the Roman Period, there was the *Muslim expansion* into the Mediterranean to include certain Mediterranean islands (Crete, Sicily and Sardinia) and most of the Iberian Peninsula. Persistence of *Arabic influence* lasted longest in southeast Spain, where new crops were introduced (e.g. *mulberry, rice, sugar cane, cotton, citrus fruits* and *aubergine*), irrigation works were elaborated and extended to include terraced hillsides and SEDIMENT TRAPS for soil eroded from BADLANDS were devised.

By Mediaeval times (see MIDDLE AGES), a typical landscape in most Mediterranean countries comprised scattered trees amongst GRASSES AND GRASSLAND, and SHRUBLAND (or savanna). The agricultural use of such terrain comprised (and still comprises in many areas) GRAZING of animals (sheep, goats and pigs) combined with *woodcutting*, SHIFTING CULTIVATION and the HARVESTING of cork. Most savannas in Italy and the Balearic Islands have become forest or scrub in the past 200 years, but retention of savannas (see DEHESA) has persisted in Spain and Portugal.

Population increases from the eighteenth and nineteenth centuries led to an increase in the area of cultivated land, with DEFORESTATION and terracing on increasingly difficult terrain. From AD 1830 to 1950, however, most European Mediterranean areas suffered periods of OVERPOPULATION and *poverty*, which led to programmes of tree planting in some countries in order to provide employment, recreate (supposedly) a former forest cover or attempt to prevent soil erosion and *flooding*. Modern agricultural practices making use of machinery, artificial fertilisers, greenhouses, piped irrigation water and pesticides have encouraged CULTIVATION on plains and on drained wetlands that had been left uncultivated because of MALARIA. These developments, together with the expansion of TOURISM in coastal locations since the 1960s, have led, for example, to the construction of dams, leading to storage of sediments that would otherwise have reached the coast to supply beach sediment, BEACH MINING for construction materials, causing increased coastal erosion, excessive extraction of GROUNDWATER in coastal areas, causing SALINISATION,

and widespread *abandonment of terracing*, leading to reversion to flammable SEMI-NATURAL VEGETATION. In some places, the vegetation has been supplemented by highly flammable *pines* and *eucalyptus*. With minimal management, these areas have been subject to an increased frequency of WILDFIRES, which have led to increased SOIL EROSION.

The long human occupancy of Mediterranean landscapes and their predisposition to erosion as a result of their subtropical location, with summer DROUGHT, *intense rainfall*, often steep terrain and often highly erodible soils and/or unconsolidated SEDIMENTARY ROCKS, have led to speculation about the extent to which indications of HOLOCENE erosion in the landscape reflect ANTHROPOGENIC or NATURAL agencies. Such evidence relates to hill slopes largely denuded of soil, badlands, FLOODPLAINS, ALLUVIAL FANS and coastal sedimentation, with ancient ports stranded far inland (e.g. Troy and Ephesus).

Spectacular GULLIES and badlands in southeast Spain would seem to indicate a recent anthropogenic origin, but archaeological evidence has shown that many of the gullies were already established some 4,000 years ago. Later vegetation clearance seems to have altered little the geomorphological patterns already in existence in the area.

An important milestone in the human versus natural debate concerning Mediterranean landscape change was the publication in 1969 of *The Mediterranean Valleys* by C. Vita-Finzi. He argued that in many Mediterranean valleys there were terraces reflecting two types of alluvial VALLEY-FILL DEPOSITS (an *Older Fill* and *Younger Fill*). The Older Fill was thought to date from the LAST GLACIAL MAXIMUM (LGM), when increased FROST WEATHERING and more intense *seasonal rainfall* led to removal of sediment from slopes and AGGRADATION in valleys. Frequently 'nested' within the Older Fill is the Younger Fill, which was interpreted as broadly synchronous throughout the region and ranging in age (based largely on archaeological evidence) from about AD 400 to 1500 or even 1800. The apparent synchroneity, the emerging evidence of CLIMATIC CHANGE in northern Europe and North America and the supposed inadequacy of an anthropogenic origin led Vita-Finzi to favour a climatic-change explanation. Subsequently, alternative views have been expressed for the origin of the previously identified Older Fill (possibly still under formation as late as the early Holocene) but there has been debate especially in respect of the Younger Fill, with suggestions of a wholly anthropogenic or a combined natural (extreme climatic events providing the necessary trigger) and anthropogenic (extensive removal of the vegetation cover) as well as more complex climatic origins (involving CENTURY- TO MILLENNIAL-SCALE VARIABILITY).

Irrespective of the cause(s), the phases of removal of soil from the steep slopes have had beneficial as well as negative effects. The extensive ALLUVIAL SOILS on the valley floors represent an important landuse resource. Nevertheless, soil erosion remains a major environmental problem in the Mediterranean and is aggravated by the gradual abandonment of terraces and a rapid increase in the incidence of wildfires since the 1960s, particularly in the Iberian Peninsula. Many of the drier regions of the Mediterranean with a plentiful supply of erodible soil or REGOLITH are prone to high estimated EROSION RATES of 10–165 t/ha, with instances of rates as high as 300 t/ha being reported. Successive annual losses of 150 t/ha for Mediterranean soils would cause serious SOIL DEGRADATION in an estimated 5 years, giving rise to fears of DESERTIFICATION. However, the long-term degradation of soils in many Mediterranean areas has left soils on hill slopes that are often thin with a high proportion of rock fragments, which tends to limit SOIL LOSS through, even when denuded of the vegetation and LITTER cover (e.g. following a fire). Thus, annual erosion rates monitored following wildfire rarely exceed 10 t/ha and most are much less. On the other hand, soil disruption through TILLAGE or the creation of BENCH TERRACES on slopes leads to high erosion rates in these soils.

GLOBAL WARMING is expected to cause increased temperatures and decreased total rainfall, though still with rainstorms of high intensity. Expected repercussions for environmental change include changes in vegetation communities, reduced availability of WATER RESOURCES and increased wildfire activity. Housing development on urban fringes will create increasing areas of *urban-wildland interfaces*, which are prone to damage by wildfire.

RAS

[*See also* MEDITERRANEAN LANDSCAPE EVOLUTION]

Benoit G and Comeau A (eds) (2005) *A sustainable future for the Mediterranean: The Blue Plan's environment and development outlook.* London: Earthscan.

Bintliff J-L (1992) Erosion in the Mediterranean lands: A reconsideration of pattern, process and methodology. In Bell M and Boardman J (eds) *Past and present soil erosion.* Oxford: Oxbow Books, 125–131.

Grove AT and Rackham O (2000) *The nature of Mediterranean Europe: An ecological history.* London: Yale University Press.

Hughes PD, Woodward JC and Gibbard PL (2006) Quaternary glacial history of the Mediterranean mountains. *Progress in Physical Geography* 30: 334–364.

Kuhlemann J, Rohling EJ, Krumrei I et al. (2008) Regional synthesis of Mediterranean atmospheric circulation during the Last Glacial Maximum. *Science* 321: 1338–1340.

Lewin J, Macklin MG and Woodward JC (eds) (1995) *Mediterranean Quaternary river environments.* Rotterdam: Balkema.

Mazzoleni S, di Pasquale G, Mulligan M et al. (eds) (2004) *Recent dynamics of the Mediterranean vegetation and landscape.* Chichester: Wiley.

Poesen J, van Wesemael B and Bunte K (1998) Soils containing rock fragments and their response to desertification. In Mairota P, Thornes JB and Geeson N (eds) *Atlas of Mediterranean environments in Europe: The desertification context.* Chichester: Wiley, 50–55.

Prentice I, Guiot J and Harrison SP (1992) Mediterranean vegetation, lake levels and palaeoclimate at the Last Glacial Maximum. *Nature* 360: 658–660.

Roberts N, Kuzucuoglu C and Karabiyikoglu M (eds) (1999) The Late Quaternary in the eastern Mediterranean. *Quaternary Science Reviews* 18: 497–716.

Roberts N, Brayshaw D, Kuzucuoğlu C et al. (2011) The mid-Holocene climatic transition in the Mediterranean: Causes and consequences. *The Holocene* 21: 3–13.

Rose J, Meng X and Watson C (1999) Palaeoclimatic and palaeoenvironmental responses in the western Mediterranean over the last 140 ka: Evidence from Mallorca, Spain. *Journal of the Geological Society of London* 156: 435–448.

Shakesby RA (2011) Post-wildfire soil erosion in the Mediterranean: Review and future research directions. *Earth-Science Reviews* 105: 71–100.

Solé Benet A (2006) Spain. In Boardman J and Poesen J (eds) *Soil erosion in Europe.* Chichester: Wiley, 311–346.

Vannière B, Colombaroli D, Chapron E et al. (2008) Climate versus human-driven fire regimes in Mediterranean landscapes: The Holocene record of Lago dell'Accesa (Tuscany, Italy). *Quaternary Science Reviews* 27: 1181–1196.

Vita-Finzi C (1969) *The Mediterranean valleys.* Cambridge: Cambridge University Press.

Vita-Finzi C (1976) Diachronism in Old World alluvial sequences. *Nature* 250: 568–570.

Wainwright J and Thornes JB (2004) *Environmental issues in the Mediterranean: Process and perspectives from the past and present.* London: Routledge.

Wise SM, Thornes JB and Gilman A (1982) How old are the badlands? A case study from south-east Spain. In Bryan R and Yair A (eds) *Badland geomorphology and piping.* Norwich: GeoBooks, 259–277.

Woodward J (ed.) (2009) *The physical geography of the Mediterranean.* Oxford: Oxford University Press.

Mediterranean landscape evolution The geological and geomorphological evolution of the modern Mediterranean began early in the Cenozoic era (65 Ma to the present) as the northerly drifting *African Plate* collided with the main *Eurasian Plate* to the north and a mosaic of small plates throughout the Mediterranean (Arabian, Adriatic and Iberian plates). Initially, the TETHYS ocean was narrow and characterised by intense deformation and volcanicity in the Apennines, Greece and Turkey. Continental impact through PLATE TECTONICS resulted in mountain-building episodes (see OROGENY) that formed the Pyrenees, Carpathians and Alps. The plate boundaries account for the pattern of SEISMICITY throughout the Mediterranean Basin. The major EARTHQUAKE zone associated with plate collision lies along the *Hellenic Arc*, stretching from the west coast of Greece to southern Turkey.

There are strong links between GEOLOGICAL STRUCTURE and TOPOGRAPHY in landscapes around the Mediterranean Sea. In essence, the Mediterranean can be viewed as a ring of mountains surrounding the sea. At a large scale, the configuration of mountain belts, island chains and mountain ranges reflects *collisional tectonics*. These areas are separated by resistant *microplates* such as Iberia, or by areas such as the western Mediterranean and the Adriatic and Dead seas. The present plan form of the Mediterranean coastline is influenced strongly by FLUVIAL PROCESSES with extensive ALLUVIAL FAN systems (e.g. Rhône, Ebro, Po and Nile deltas) and *marine processes* (e.g. deep-water channels in the straits of Gibraltar and Bosphorus).

In addition to the impacts of Pleistocene SEA-LEVEL CHANGE, CLIMATIC CHANGE and VEGETATION HISTORY (caused naturally or through human impact), drainage systems in the Mediterranean show strong links to geological structure and NEOTECTONICS. There are examples of disrupted *drainage systems* in the form of RIVER CAPTURE, reversal, diversion or ponding (e.g. the Greek islands, Turkey and Italy). Mature rivers tend to drain parallel to major mountain belts (e.g. the Po) and minor extensional basins (e.g. Rhône, Nile).

Whereas geological effects in the Mediterranean can be seen best at the large scale, the effects of Pleistocene climatic changes on Mediterranean landscapes are best displayed at the medium or small scales. Fluctuations in sea levels impacted on the coasts as well as rivers, with RAISED SHORELINES in the form of ROCK PLATFORMS common on the predominantly rocky coasts (70–75 per cent of the Mediterranean coastline is rocky), often reflecting the high eustatic sea levels (see EUSTASY) attained during INTERGLACIALS, with their altitudes affected by subsequent neotectonic action. In many places, glacio-eustatically low sea levels (see GLACIO-EUSTASY) are recorded along present-day coastlines by AEOLIANITE formation.

The shift to a MEDITERRANEAN-TYPE CLIMATE occurred during the PLIOCENE. As summer rainfall decreased, the modern natural vegetation characterised by coniferous trees and SCLEROPHYLLOUS trees and shrubs developed. Inland, away from the high mountains, Pleistocene climatic fluctuations in general led to phases of slope stability during interglacials and instability during glaciations, producing sequences of COLLUVIUM and ALLUVIUM sometimes interdigitating with PALAEOSOLS. Human action during the HOLOCENE also caused slope instability, leading to SOIL EROSION, and changed the nature of the vegetation communities.

Where the eroded sediments reached valley floors, they often formed VALLEY-FILL DEPOSITS.   *RAS*

[*See also* MEDITERRANEAN ENVIRONMENTAL CHANGE AND HUMAN IMPACT, MEDITERRANEAN-TYPE VEGETATION]

Bar-Matthews M (2012) Environmental change in the Mediterranean region. In Matthews JA, Bartlein PJ, Briffa KR et al. (eds) *The SAGE handbook of environmental change, volume 2*. London: Sage, 163–187.

Collier REL, Leeder MR, Trout M et al. (2000) High sediment yields and cool, wet winters: Test of the last glacial paleoclimates in the northern Mediterranean. *Geology* 28: 999–1002.

Combourieu Nebout N, Peyron O, Dormoy I et al. (2009) Rapid climatic variability in the west Mediterranean during the last 25 000 years from high resolution pollen data. *Climate of the Past* 5: 503–521.

Fornós JJ, Clemmensen LB, Gómez-Pujol L and Murray AS (2009) Late Pleistocene carbonate aeolianites on Mallorca, western Mediterranean: A luminescence chronology. *Quaternary Science Reviews* 28: 2697–2709.

Hughes PD and Woodward JC (2008) Timing of glaciation in the Mediterranean mountains during the last cold stage. *Journal of Quaternary Science* 23: 575–588.

Quaternary International (2008) The last 15 ka of environmental change in Mediterranean regions: Interpreting different archives. *Quaternary International* 181(1) [Special Issue].

Quaternary International (2008) Quaternary stratigraphy and evolution of the alpine region and the Mediterranean area in the European and global framework. *Quaternary International* 190 [Special Issue].

Roberts N, Meadows ME and Dodson JR (eds) (2001) The Holocene history of Mediterranean-type environments in the Eastern Hemisphere. *The Holocene* 11: 631–764 [Special Issue].

Robertson AHF (ed.) (1985) *The geological evolution of the Eastern Mediterranean*. Bath: Geological Society.

Rose J and Meng X (1999) River activity in small catchments over the last 140ka, northeast Mallorca, Spain. In Brown AG and Quine TA (eds) *Fluvial processes and environmental change*. Chichester: Wiley, 91–102.

Ruffell A (1997) Geological evolution of the Mediterranean Basin. In King R, Proudfoot L and Smith B (eds) *The Mediterranean: Environment and society*. London: Arnold, 12–29.

Woodward J (ed.) (2009) *The physical geography of the Mediterranean*. Oxford: Oxford University Press.

Mediterranean-type climate A warm-temperate CLIMATE characterised by an annual summer DROUGHT of three or more months' duration and mild, wetter winters. The former results from the annual summer extension of the nearby subtropical ANTICYCLONES. In winter, the influence of the subpolar JET STREAMS is more evident, bringing unsettled wet weather. It is often regarded as a transitional climate between the *arid* latitudes dominated throughout the year by the subtropical anticyclones and the persistently humid climates of the mid-latitudes on its polewards margins. The 'classic' Mediterranean climate is found along the coasts of the Mediterranean Sea, but smaller regions enjoying a similar regime occur on the western continental margins at similar latitudes around 35° N and S in North and South America, Australia and South Africa. Concern surrounds the European Mediterranean region: it is the largest such area, heavily populated and possessing advanced economic systems, and suggestions that the length and severity of the summer drought might intensify under different predicted CLIMATIC SCENARIOS raise questions about adequate WATER RESOURCES for domestic, agricultural and industrial needs. In addition, the already demonstrable increase in *heat wave* activity as a consequence of GLOBAL WARMING raises justifiable anxieties about HUMAN HEALTH HAZARDS and increased DEATH RATES amongst the more vulnerable members of society. *DAW*

[*See also* ARIDITY, ARIDLAND: PAST ENVIRONMENTAL CHANGE, CLIMATIC CLASSIFICATION, MEDITERRANEAN ENVIRONMENTAL CHANGE AND HUMAN IMPACT]

Bolle H-J (ed.) (2003) *Mediterranean climate:Variability and trends*. Dordrecht: Springer.
Camuffo D, Bertolin C, Diodato N et al. (2010) The western Mediterranean climate: How will it respond to global warming? *Climatic Change* 100: 137–142.
Giorgi F and Lionello P (2008) Climate change projections for the Mediterranan region. *Global and Planetary Change* 63: 90–104.
Lionello P, Malanotti-Rizzola P and Boscolo R (eds) (2006) *Mediterranean climate variability*. Amsterdam: Elsevier.
Thomson MC, Garcia-Herrera R and Beniston M (eds) (2008) *Seasonal forecasts, climatic change and human health*. Dordrecht: Springer.

Mediterranean-type vegetation

True Mediterranean-type vegetation is characterised by SCLEROPHYLLOUS woodland and SHRUBS which are well adapted to periodic fire and the MEDITERRANEAN-TYPE CLIMATE, which has hot, dry summers and mild, humid winters. Mediterranean-type vegetation is characteristic of a VEGETATION FORMATION TYPE found in five regions between 30° N and 40° S of the equator, which have a Mediterranean-type climate: the Mediterranean basin, southern Africa, southwestern and south-central Australia, central Chile and southern California.

All five Mediterranean regions have been occupied by humans for millennia but the greatest human impacts are apparent in the Mediterranean basin. The Mediterranean climatic regime is young, having developed during the PLEISTOCENE, and a certain degree of COEVOLUTION is claimed for Mediterranean-type vegetation and PALAEOLITHIC humans in the Mediterranean basin. The effects of NEOLITHIC and BRONZE AGE agriculture and PASTORALISM, in particular fire clearance and goat herding, initiated

the transformation of the Mediterranean vegetation from dense forest to low-lying SCRUB, in which highly valuable crops such as *Olea* (olive) and *Vitis* (vine) are now cultivated. Human activity has similarly altered the other Mediterranean regions, although this has occurred only within the past 500 years since European settlement and has involved invasion by weedy European plant species better adapted to grazing. Although pockets of sclerophyllous woodland remain evident, the modern vegetation consists of shrub and HEATHLAND rarely exceeding 2 m in height. The thorny *garrigue* (*phrygana* in Greece, *chapparal* in California, *mallee* in Australia and *mattoral* in Spain and Chile) consists of scattered, chew-resistant sclerophyllous shrubs and is best represented in the Mediterranean basin by *Quercus coccifera* (kermes oak), with a ground layer of herbaceous PERENNIALS. The less common maquis (*fynbos* in southern Africa) heathlands are dominated by Ericaceous scrub of *Erica arborea* (tree heath) and *Arbutus unedo* (strawberry tree) in the Mediterranean basin and *Protea-Erica* heathland in southern Africa. *ARG*

Arroyo MTZ, Zedler PH and Fox MD (1994) *Ecology and biogeography of Mediterranean ecosystems in Chile, California and Australia*. New York: Springer.
Cowling R (1992) *The ecology of fynbos: Nutrients, fire and diversity*. Oxford: Oxford University Press.
Greater W (1994) Extinctions in Mediterranean areas. *Philosophical Transactions of the Royal Society of London B* 344: 41–46.
Grove AT and Rackham O (2001) *The nature of Mediterranean Europe: An ecological history*. London: Yale University Press.
Groves RH and Di Castri F (eds) (1991) *Biogeography of Mediterranean invasions*. Cambridge: Cambridge University Press.
Kruger FJ, Mitchell DT and Jarvis JUM (eds) (1983) *Mediterranean-type ecosystems*. Berlin: Springer.
Mazzoleni S, di Pasquale G, Mulligan M et al. (eds) (2004) *Recent dynamics of the Mediterranean vegetation and landscape*. Chichester: Wiley.
Thompson JD (2005) *Plant evolution in the Mediterranean*. Oxford: Oxford University Press.

megadrought

An extreme DROUGHT, exceptional in magnitude and/or duration. Hunt and Elliott (2002) defined megadroughts as 'persistent droughts of more than a decade in duration' but such quantitative definitions are arbitrary. DENDROCLIMATOLOGY has demonstrated, for example, the occurrence of four North American megadroughts (AD 936 1034, 1150 and 1253) during the MEDIAEVAL WARM PERIOD. *JAM*

[*See also* DROUGHTS: HISTORICAL RECORDS, EXTREME EVENT, PALAEODROUGHT]

Cook ER, Anchukaitis KJ, Buckley BM et al. (2010) Asian monsoon failure and megadrought during the last millennium. *Science* 328: 486–489.

Fawcett PJ, Werner JP, Anderson RA et al. (2011) Extended megadroughts in the southwestern United States during Pleistocene interglacials. *Nature* 470: 518–521.

Hunt BG and Elliott TI (2002) Mexican megadrought. *Climate Dynamics* 20: 1–12.

Stahle DW, Cook ER, Cleaveland MK et al. (2000) Tree-ring data document 16th century megadrought over North America. *EOS, Transactions, American Geophysical Union* 81: 121–125.

megafan A very large fan-shaped LANDFORM. The term is usually used to refer to large ALLUVIAL FANS draining mountain ranges, also termed *fluvial mega-fans*.   *GO*

Chakraborty T, Kar R, Ghosh P et al. (2010) Kosi megafan historical records, geomorphology and the recent avulsion of the Kosi River. *Quaternary International* 227: 143–160.

Leier AL, DeCelles PG and Pelletier JD (2005) Mountains, monsoons, and megafans. *Geology* 33: 289–292.

megafauna The very large animal species within a FAUNA, encompassing those species attaining a body mass exceeding about 40 kg (90 lb).   *DCS*

[*See also* MASS EXTINCTIONS, MEGAFAUNAL EXTINCTION]

Farina RA, Vizcaino SF and De Iuliis G (2012) *Megafauna: Giant beasts of Pleistocene South America.* Bloomington: Indiana University Press.

megafaunal extinction The Late Quaternary extinction of many large animals (MEGAFAUNA). In South America, at least 37 genera and 85 species were lost. The timing of these extinctions in different regions, with examples of the species, is summarised in the Table. Megafaunal survived largely intact only in Africa. In the Americas, New Zealand and probably Australia, the collapse of the megafaunal populations was rapid but in Eurasia they extended over many thousands of years. Although the timing varies in the different regions, these extinctions tend to correspond with the arrival of humans, which led to the OVERKILL HYPOTHESIS of human-induced extinction. Spears of *Clovis hunters* have actually been found in association with some MAMMOTH and ground sloth kill sites in North America. Although people were undoubtedly involved in the demise of some megafauna, the question remains whether they were causal in the collapse of the populations. In New Zealand, where nine species of *Moa* became extinct within the past thousand years, at the same time as human settlement, this was clearly the case. In Australia, where 55 large *mammal species* became extinct shortly after human arrival on the continent, the evidence clearly points to the conclusion that the hunters did it by 40 ka.

The competing hypothesis involving the effects of CLIMATIC CHANGE on HABITATS is not favoured by the variable timings or by the fact that similar extinction phases did not occur at the close of previous cold periods. However, the debate over whether such extinctions were the result of overkill, climatic change or a combination of both continues unabated. It is possible, for example, that final extinction was caused in some regions by the hunting of populations that were weakened and vulnerable due to climatic change.

Beyond the question of *causation*, there remain important questions about the effects on *ecosystems* and LANDSCAPES of the elimination of large HERBIVORES. Results from Lynch's Crater in northeastern Australia suggest that human arrival rather than climate caused megafaunal extinction, which then triggered the replacement of TROPICAL RAIN FOREST by SCLEROPHYLLOUS vegetation through the direct effects of reduced GRAZING and increased FIRE FREQUENCY from the accumulation of fuel.   *JAM*

Barnosky AD (2008) Megafaunal biomass tradeoff as a driver of Quaternary and future extinctions. *Proceedings of the National Academy of Sciences* 105: 11543–11548.

Barnosky AD, Koch PL, Feranec RS et al. (2004) Assessing the causes of Late Pleistocene extinctions on the continents. *Science* 306: 70–75.

Gill JL, Williams JW, Jackson ST et al. (2009) Pleistocene megafaunal collapse, novel plant communities, and enhanced fire regimes in North America. *Science* 326: 1100–1103.

Grayson DK and Meltzer DJ (2003) A requiem for North American overkill. *Journal of Archaeological Science* 30: 585–593.

Region	Timing (ka)	Example species
Eurasia	48–23; 14–10	Giant deer, auroch, mammoth
North America	15.6–11.5	Mastodon, mammoth, camels
Southeast Australia	50–40	Giant kangaroos
South America	12–8	Ground sloth, toxodon
New Zealand	0.75	Moa, large flightless geese

Megafaunal extinction *The timing of Late Quaternary extinctions of megafauna (Barnosky, 2008; Worthy and Holdaway, 2002).*

Grayson DK and Meltzer DJ (2004) North American
    overkill continued? *Journal of Archaeological Science* 31:
    133–136.
Johnson CN (2009) Ecological consequences of Late
    Quaternary extinctions of megafauna. *Proceedings of the
    Royal Society Series B*: 2509–2519.
Martin PS and Klein RG (1984) *Quaternary extinctions:
    A prehistoric revolution*. Tucson: University of Arizona
    Press.
McGlone M (2012) The hunters did it. *Science* 335:
    1452–1453.
Rule S, Brook BW, Haberle SG et al. (2012) The aftermath
    of megafaunal extinction: Ecosystem transformation in
    Pleistocene Australia. *Science* 335: 1483–1486.
Worthy TH and Holdaway RN (2002) *The lost world of the
    Moa*. Christchurch: Canterbury University Press.

megaflood   An extreme, catastrophic *outburst flood*
from a GLACIAL LAKE. Many megafloods were associated with DEGLACIATION during the QUATERNARY and
early HOLOCENE. They produced spectacular LANDFORMS
of GLACIAL EROSION, such as the CHANNELED SCABLANDS,
and had important *climatic change impacts*, particularly on CENTURY- TO MILLENNIAL-SCALE VARIABILITY,
through their effects on the THERMOHALINE CIRCULATION
of the oceans.   *JAM*

[*See also* DANSGAARD-OESCHGER (D-O), EVENTS,
EXTREME EVENT, JÖKULHLAUP, PALAEOFLOOD]

Burr DM, Carling PA and Baker VR (eds) (2009)
    Megaflooding on Earth and Mars. Cambridge: Cambridge
    University Press.
Fisher TG, Clague JJ and Teller JT (eds) (2002) The role of
    outburst floods and glacial meltwater in subglacial and
    proglacial landform genesis. *Quaternary International* 90:
    1–115.

megafossil   A very large FOSSIL or SUBFOSSIL, such as
a tree trunk preserved in a mire beyond the TREE LINE.
Unlike MACROFOSSILS, they are too large to be manipulated by hand. Because of their size, they are normally
found in situ and hence are likely to be particularly reliable indicators of changing site conditions.   *JAM*

[*See also* MICROFOSSIL]

Kultti S, Mikkola K, Virtanen T et al. (2006) Past changes in
    the Scots pine forest line and climate in Finnish Lapland:
    A study based on megafossils, lake sediments, and GIS-
    based vegetation and climate data. *The Holocene* 16:
    381–391.

megalake   A lake of much larger proportions than
any remnant existing today within the same DRAIN-
AGE BASIN. A well-known example is the megalake
Chad in Central Africa, which lies in a basin that
covers nearly 2.5 million km². Modern Lake Chad
is a BRACKISH WATER remnant of the large lake that
existed during the early HOLOCENE (ca 10,000–5,000

years ago) and covered an area of at least 300,000–
350,000 km². Drier conditions thereafter led to its
disappearance on a number of occasions, including
AD 1450, 1550, 1750, 1850 and 1900. Evidence for
the early Holocene lake includes SHORELINES and the
bones of lake-living mammals found far from the
present-day lake. The lake has fluctuated in extent
since the 1960s, when it covered an open-water area
of 23,000 km², but overall it has undergone shrinkage. There are many other examples of so-called
megalakes worldwide, some of which were GLACIAL
LAKES.   *RAS*

Clarke G, Leverington D, Teller J and Dyke A (2003)
    Superlakes, megafloods, and abrupt climate change.
    Science 301: 922–923.
Hooke R LeB (2002) Is there any evidence of Mega-Lake
    Manly in the eastern Mojave Desert during Oxygen
    Isotope Stage 5e/6? *Quaternary Research* 57: 177–179.
Mangerud J, Astakhov V, Jacobsson M and Svendson
    JI (2001) Huge Ice-Age lakes in Russia. *Journal of
    Quaternary Science* 16: 773–777.

megalith   A large boulder used in the construction of
a monument or other structure. Megalithic structures,
such as *stone circles* or *tombs*, were mainly constructed
in the NEOLITHIC.   *JAM*

[*See also* EOLITHS, LITHIC, MICROLITH, MONOLITH
SAMPLING]

Scarre C (2007) *The megalithic monuments of Britain and
    Ireland*. London: Thames and Hudson.

megaripple   Sometimes used as a synonym for
DUNE, but best avoided because it clouds the hydrodynamic distinction between dunes and RIPPLES.   *GO*

Ashley G (1990) Classification of large scale subaqueous
    bedforms: A new look at an old problem. *Journal of
    Sedimentary Petrology* 60: 160–172.

megathrust earthquake   An EARTHQUAKE generated at a DESTRUCTIVE PLATE MARGIN by slip at the
interface between the plates (the *Benioff zone*). This
produces infrequent earthquakes of high EARTHQUAKE
MAGNITUDE, in contrast to those generated more frequently along shallow FAULTS in the overriding plate.
The largest instrumentally recorded earthquakes, of
magnitude 9 or more, have been megathrust earthquakes, including the 1960 Chile earthquake and the
December 2004 Sumatra earthquake and March 2011
Tōhoku earthquake in Japan, both of which generated
highly destructive TSUNAMIS.   *GO*

McCloskey J, Antonioli A, Piatanesi A et al. (2008) Tsunami
    threat in the Indian Ocean from a future megathrust
    earthquake west of Sumatra. *Earth and Planetary Science
    Letters* 265: 61–81.

Walter TR and Amelung F (2007) Volcanic eruptions following M ≥ 9 megathrust earthquakes: Implications for the Sumatra-Andaman volcanoes. *Geology* 35: 539–542.

megaturbidite An extremely thick TURBIDITE of regional extent. Megaturbidites, or *megabeds*, may be many tens of metres thick, implying deposition from a major EVENT such as a large EARTHQUAKE, sediment instability caused by a fall in *sea level*, a catastrophic FLOOD or a VOLCANIC ERUPTION. Megaturbidites can be used in CORRELATION (STRATIGRAPHICAL) within a SEDIMENTARY BASIN (see EVENT STRATIGRAPHY). *GO*

Brunner CA, Normark WR, Zuffa GG and Serra F (1999) Deep-sea sedimentary record of the late Wisconsin cataclysmic floods from the Columbia River. *Geology* 27: 463–466.

Mulder T, Zaragosi S, Razin P et al. (2009) A new conceptual model for the deposition process of homogenite: Application to a Cretaceous megaturbidite of the western Pyrenees (Basque region, SW France). *Sedimentary Geology* 222: 263–273.

Rothwell RG, Thomson J and Kahler G (1998) Low-sea-level emplacement of a very large Late Pleistocene 'megaturbidite' in the western Mediterranean Sea. *Nature* 392: 377–380.

meltout till A traditional term for a SEDIMENT deposited by debris melting out from GLACIER ice. *MRB*

[*See also* GLACIAL SEDIMENTS, TILL]

meltwater Water derived from melting snow or ice. *Snow meltwater* and meltwater from *river ice* and *ice-covered lakes* are most abundant in spring. *Glacial meltwater* is produced in large quantities throughout the summer season. In addition to the large annual variations in meltwater discharge from GLACIERS, strong *diurnal cycles* may be evident, resulting from variations in SOLAR RADIATION. Meltwater plays an increasingly recognised role in controlling the ICE-SHEET DYNAMICS of the Greenland and Antarctic Ice Sheets, as well as the dynamics of valley glaciers. *MJH*

[*See also* GLACIAL EROSION, GLACIOFLUVIAL LANDFORMS, GLACIOFLUVIAL SEDIMENTS, MELTWATER CHANNEL, SNOW AND SNOW COVER]

Menzies J (1995) Hydrology of glaciers. In Menzies J (ed.) *Modern glacial environments*. Oxford: Butterworth-Heinemann, 197–239.

Wingham DJ, Siegert MJ, Shephard A and Muir AS (2006) Rapid discharge connects Antarctic subglacial lakes. *Nature* 440: 1033–1036.

meltwater channel A channel eroded by glacial meltwater beneath, on the margins of, or in front of a GLACIER or ICE SHEET. In the subglacial context, meltwater channels can be cut down into the bed (*Nye* or *N-type channels*) or up into the ice (*Röthlisberger* or *R-type channels*). On DEGLACIERISATION, Nye channels may be distinguished from subaerial river or stream channels by steep gradients and undulating long profiles (the latter having been produced by meltwater flowing under HYDROSTATIC PRESSURE. *JAM/RAS*

[*See also* ICE-MARGIN INDICATORS, OVERFLOW CHANNEL, TUNNEL VALLEY, URSTROMTÄLER]

Atkins CB and Dickinson WW (2007) Landscape modification by meltwater channels at margins of cold-based glaciers, Dry Valleys, Antarctica. *Boreas* 36: 47–55.

Syverson KM and Mickelson DM (2009) Origin and significance of lateral meltwater channels formed along a temperate glacier margin, Glacier Bay, Alaska. *Boreas* 38: 132–145

member A unit in LITHOSTRATIGRAPHY which is a locally distinct subdivision of a FORMATION (STRATIGRAPHICAL). Members commonly wedge out through lateral FACIES changes. *LC*

mercury (Hg) A silver-grey HEAVY METAL (*quicksilver*), liquid at normal temperatures and highly toxic in some compounds, especially as *methyl mercury*. It was relatively rare in the environment prior to its extraction from its ore, *cinnabar* (HgS). Also known as *vermilion*, cinnabar was used as a pigment in CAVE ART dating from the PALAEOLITHIC. Environmental levels first rose significantly above background levels in the ROMAN PERIOD, or possibly even earlier (see Figure). The ELEMENT is toxic and BIO-ACCUMULATION of mercury in organisms, especially *fish*, is a potential environmental hazard. As a component of *barometers* and *thermometers*, mercury played an important role in early understanding of environmental change. *JAM*

[*See also* CHEMICAL TIME BOMB, EARLY INSTRUMENTAL METEOROLOGICAL RECORDS]

Bank MS (ed.) (2012) *Mercury in the environment: Pattern and process*. Berkeley: University of California Press.

Bargagli R (1999) Mercury in the environment. In Alexander DE and Fairbridge RW (eds), *Encyclopedia of environmental science*. Dordrecht: Kluwer, 402–405.

Harris R, Krabbenhoft DP, Mason R et al. (eds) (2007) *Ecosystem response to mercury contamination: Indicators of change*. Pensacola, FL: Society of Environmental Toxicology and Chemistry.

Martínez-Cortizas A, Pontevedra-Pombal X, García-Rodeja E et al. (1999) Mercury in a Spanish peat bog: Archive of climatic change and atmospheric metal deposition. *Science* 284: 939–942.

Selin NE (2009) Global biogeochemical cycling of mercury: A review. *Annual Review of Environmental Resources* 34: 43–63.

Zuber SL and Newman MC (2011) *Mercury pollution: A transdisciplinary treatment*. Boca Raton, FL: CRC Press.

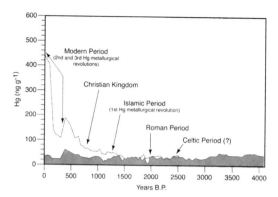

Mercury *Total mercury and its natural (darker area) and anthropogenic (lighter area) components during the past 4,000 years, as recorded in a Spanish peat bog. Major prehistoric and historic phases of mercury exploitation in Spain are indicated (Martinez-Cortizas et al., 1999).*

meridional flow A prevalence of north-south AIRFLOW TYPES often associated with weakened upper westerly winds, LONG WAVES of large AMPLITUDE and atmospheric BLOCKING. Meridional flow, parallel to the meridians (lines of longitude), is important in transporting heat and moisture from the tropics towards the poles, and hence in the maintenance of Earth's ENERGY BALANCE.   *JCM/EH*

[*See also* MERIDIONAL OVERTURNING CIRCULATION (MOC), ZONAL INDEX]

Skeie P (2000) Meridional flow variability over the Nordic Seas in the Arctic oscillation framework. *Geophysical Research Letters* 27: 2569–2572.

meridional overturning circulation (MOC) The two-dimensional meridional flow component of the THERMOHALINE CIRCULATION of the oceans. It is driven by ATMOSPHERE-OCEAN INTERACTION and is especially important for heat transfer within the oceans. Its two cells are shown in the Figure. The upper cell, essentially the *Atlantic meridional overturning* (*AMO*), is driven by the formation of *North Atlantic Deep Water* (*NADW*) in the North Atlantic, its UPWELLING in the Southern Ocean around Drake Passage (DP) and its return flow as *Antarctic Intermediate Water* (*AAIW*). The lower cell involves the formation of *Antarctic Bottom Water* (*AABW*) around Antarctica, its flow to the north, where its return flow mixes with NADW. An intensified AMO cell is associated with stronger northward oceanic heat transport. A meridional overturning circulation can also be recognised in the ATMOSPHERE as part of the GENERAL CIRCULATION OF THE ATMOSPHERE.   *JAM*

[*See also* BRINE REJECTION, DEEP CONVECTION, OCEANIC FORCING (OF CLIMATE)]

Ortega P, Hawkins E and Sutton R (2011) Processes governing the predictability of the Atlantic meridional

overturning circulation in a coupled GCM. *Climate Dynamics* 37: 1771–1782.
Sun Y, Clemens SC, Morrill C et al. (2012) Influence of Atlantic meridional overturning circulation on East Asian winter monsoon. *Nature Geoscience* 5: 46–49.
Toom MD, Dijkstra HA, Cimatoribus AA and Drijfhout SS (2012) Effect of atmospheric feedbacks on the stability of the Atlantic meridional overturning circulation. *Journal of Climate* 25: 4081–4096.
Trouet V, Scourse JD and Raible CC (2012) North Atlantic storminess and Atlantic meridional circulation during the last millennium: Synthesizing results from proxy data and climate models. *Global and Planetary Change* 84: 48–55.
Tulloch R and Marshall J (2012) Exploring mechanisms of variability and predictability of Atlantic meridional circulation in two coupled climate models. *Journal of Climate* 25: 4067–4080.
Weaver AJ and Saenko OA (2009) Thermohaline circulation. In Gornitz V (ed.) *Encyclopedia of paleoclimatology and ancient environments.* Dordrecht: Springer, 943–948.

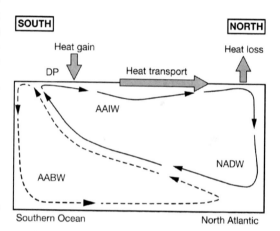

Meridional overturning circulation *Schematic representation of the meridional overturning circulation of the oceans. Thick arrows at the top of the diagram indicate interchange of heat with the atmosphere. Acronyms are explained in the text (Weaver and Saenko, 2009).*

mérokarst A KARST landscape developed on thin sequences of LIMESTONE, often interbedded with other rocks, as well as on pure carbonate formations. Also known as '*half karst*', and more or less synonymous with FLUVIOKARST, mérokarst lacks the development of many classical karst landforms and is dominated by valley forms.   *RAS/JAM*

Cvijiæ J (1925) Le mérokarst. *Comptes Rendus de l'Académie de Sciences* 180: 757–758.

meromictic lake A lake with a permanently stratified water column maintained by a density gradient caused by chemical stratification (including salinity). Meromictic lakes often preserve LAMINATIONS in their SEDIMENTS (see RHYTHMITE, VARVE).   *HHB*

[*See also* HOLOMICTIC LAKE, LAKE STRATIFICATION
AND ZONATION]

Albéric P, Viollier E, Jézéquel D et al. (2000) Interactions
between trace elements and dissolved organic matter
in the stagnant anoxic deep layer of a meromictic lake.
Limnology and Oceanography 45: 1088–1096.
Corella JP, Moreno A, Morellón M et al. (2011) Climate
and human impact on a meromictic lake during the last
6,000 years (Momtcortès Lake, Central Pyrenees, Spain).
Journal of Paleolimnology 46: 351–367.

mesa   A flat-topped upstanding LANDFORM (alternatively known as a *table mountain*) capped by a hard
rock that is subject to slower WEATHERING and ERO-
SION than in the surrounding lower ground. A *butte* is
similar but smaller and may be regarded as a RESIDUAL
LANDFORM.   *RAS*

[*See also* CAPROCK, INVERTED RELIEF]

Migoň P (2004) Mesa. In Goudie AS (ed.) *Encyclopedia of
geomorphology*. London: Routledge, 668.

mesoclimate   The climatic characteristics of a small
or local area between about 10 and 100 km^2, such as a
valley or a city. It differs from the wider MACROCLIMATE
in terms of *spatial scale*.   *JGT*

[*See also* LOCAL CLIMATE, MICROCLIMATE, URBAN
CLIMATE]

Atkinson BW (1981) *Meso-scale atmospheric circulations*.
London: Academic Press.
Blarney RC and Reason CJC (2012) Mesoscale convective
complexes over southern Africa. *Journal of Climate* 25:
753–766.
Orlanski I (1975) A rational subdivision of scales for
atmospheric processes. *Bulletin of the American
Meteorological Society* 56: 527–530.

mesocosm   An artificial model *ecosystem* constructed for the experimental study of more complex
natural ecosystems; a type of ECOLOGICAL ENGINEERING.
There is no sharp dividing line between mesocosms
and MICROCOSMS. The former tend to be more realistic
representations of the interactions and changes that
occur in nature but are more difficult to manage. MESO-
COSMS are widely used in aquatic ecology, with practical applications in relation to FISHERIES CONSERVATION
AND MANAGEMENT. *Biosphere 2* is a very large mesocosm, constructed in the Arizona Desert in conjunction
with the United States Space Program. The *Eden Project*, a major tourist attraction in southwest England,
includes several mesocosms.   *JAM*

Kangas P and Adey W (2008) Mesocosm management. In
Jorgensen S (ed.) *Encyclopedia of ecology*. Amsterdam:
Elsevier, 2308–2313.
Odum EP (1984) The mesocosm. *BioScience* 34: 558–562.

mesocratic phase   The early-temperate phase of
an INTERGLACIAL, characterised in northern Europe by
the expansion of DECIDUOUS mixed oak woodland (typically *Quercus, Ulmus, Fraxinus* and *Corylus*) on rich
forest soils.   *MJB*

[*See also* INTERGLACIAL CYCLE]

Iversen J (1958) The bearing of glacial and interglacial
epochs on the formation and extinction of plant taxa.
Uppsala Universitet Årsskrift 6: 210–215.

Mesolithic   The generally poorly defined archae-
ological period between the PALAEOLITHIC and the
NEOLITHIC. It therefore predates the development of
AGRICULTURE. In northwest Europe, it includes the
Late Glacial and the early HOLOCENE from about
10000–5000 BC.

MESOLITHIC peoples lived by HUNTING, FISHING AND
GATHERING, and their main impacts on the landscape
seem to have been the creation of small temporary
clearings and the use of fire. These have been detected
by HIGH-RESOLUTION RECONSTRUCTIONS using POLLEN
ANALYSIS combined with CHARRED-PARTICLE ANALYSIS.
Some types of vegetation seem to have been burned
systematically, possibly as a deliberate management
strategy, such as the burning of *reedswamp* at the Early
Mesolithic site of Star Carr, in northeast England, and
of hazel (*Corylus avellana*) woodland in later Mesolithic contexts in the Severn Estuary Levels, western
Britain.

In upland areas, such as the North York Moors in
England, burning of MOORLAND and open woodland in
the later Mesolithic may have lowered the height of the
TREE LINE, and even triggered BLANKET MIRE formation. It
has been suggested that burning by Mesolithic peoples
encouraged the spread of hazel and alder (*Alnus gluti-
nosa*) in the early Holocene. There is little evidence to
support this for hazel, but the population expansion of
alder is sometimes associated with evidence of burning, suggesting that the tree may have colonised areas
of Mesolithic woodland DISTURBANCE.   *SPD*

[*See also* AGRICULTURAL HISTORY, THREE-AGE
SYSTEM, VEGETATION HISTORY]

Bailey G and Spikins P (ed.) (2008) *Mesolithic Europe*.
Cambridge: Cambridge University Press.
Bell M (2007) *Prehistoric coastal communities: The
Mesolithic in western Britain*. York: Council for British
Archaeology.
Brown AG (1997) Clearances and clearings: Deforestation
in Mesolithic-Neolithic Britain. *Oxford Journal of
Archaeology* 16: 133–146.
Larsson L, Kindgren H, Knutsson K et al. (eds) (2003)
Mesolithic on the move. Oxford: Oxbow Books.
Mellars P and Dark P (1998) *Star Carr in context*.
Cambridge: McDonald Institute.

Mithen SJ (1994) *The Mesolithic Age*. In Cunliffe B (ed.) *Prehistoric Europe*. Oxford: Oxford University Press, 79–135.

Mithen SJ (1999) Mesolithic archaeology, environmental archaeology and human palaeoecology. *Quaternary Proceedings* 7: 477–483.

Simmons IG (1996) *The environmental impact of later Mesolithic cultures*. Edinburgh: Edinburgh University Press.

Simmons IG, Dimbleby GW and Grigson C (1981) The Mesolithic. In Simmons IG and Tooley M (eds) *The environment in British prehistory*. London: Duckworth, 82–124.

Smith C (1992) *Late Stone Age hunters of the British Isles*. London: Routledge.

Vermeersch PM and Van Peer P (eds) (1990) *Contributions to the Mesolithic in Europe*. Leuven: Leuven University Press.

Mesozoic   An ERA of geological time comprising three PERIODS: the TRIASSIC, JURASSIC and CRETACEOUS. Known as the *Age of Reptiles*, the Mesozoic saw the evolution and rise to dominance of the terrestrial *dinosaurs*, flying *pterosaurs* and marine *ichthyosaurs* and *plesiosaurs*, while *molluscs* became a key part of the marine *invertebrate fauna*.   *GO*

[*See also* GEOLOGICAL TIMESCALE]

Allen JRL, Hoskins BJ, Sellwood BW et al. (eds) (1994) *Palaeoclimates and their modelling: With special reference to the Mesozoic Era*. London: Chapman and Hall/Royal Society.

Ogg JG, Ogg G and Gradstein FM (2008) *The concise geologic time scale*. Cambridge: Cambridge University Press.

Schweitzer MH (2011) Soft tissue preservation in terrestrial Mesozoic vertebrates. *Annual Review of Earth and Planetary Sciences* 39: 187–216.

Messinian   The uppermost STAGE of the MIOCENE series, 7.2–5.3 million years ago (Ma). The *Messinian salinity crisis* refers to the initially controversial theory, derived from DEEP-SEA DRILLING observations, that the Straits of Gibraltar were closed by TECTONIC processes, allowing evaporation of water from the enclosed Mediterranean basin and the precipitation of thick sequences of EVAPORITES. The Atlantic waters broke through the Straits of Gibraltar again 5.33 Ma producing a catastrophic flooding of the Mediterranean basin, an event known as the *Zanclean flood*.   *JCWC*

Butler RWH, McClelland E and Jones RE (1999) Calibrating the duration and timing of the Messinian salinity crisis in the Mediterranean: Linked tectonoclimatic signals in thrust-top basins of Sicily. *Journal of the Geological Society* 156: 827–835.

Garcia-Castellanos D, Estrada F, Jiménez-Munt I et al. (2009) Catastrophic flood of the Mediterranaean after the Messinian salinity crisis. *Nature* 462: 778–782.

Hsü KJ (1983) *The Mediterranean was a desert: A voyage of the Glomar Challenger*. Princeton, NJ: Princeton University Press.

Hsü KJ, Montadert L, Bernoulli D et al. (1977) History of the Mediterranean salinity crisis. *Nature* 267: 399–403.

metadata   Data about data. Metadata provide the broader context to the nature and meaning of the data used in a scientific investigation, how the data were collected and their purpose. Metadata therefore allow the data to be understood more fully.   *JAM*

metaknowledge   Knowledge about KNOWLEDGE. Metaknowledge research involves collecting META-DATA, uncovering regularities in scientific procedures and claims, inferring the beliefs and strategies behind these regularities and investigating the influence of the scientific, institutional and publishing context on the conduct of research. Such information is relevant to understanding the nature and meaning of knowledge. According to Evans and Foster (2011), the scope and effects of metaknowledge would be apparent from differences in the information gleaned from a collection of scientific literature by a first-year undergraduate in comparison with a leading scientist in that field.   *JAM*

[*See also* CULTUROMICS, SCIENCE]

Evans JA and Foster JG (2011) Metaknowledge. *Science* 331: 721–724.

metal pollution   Contamination of soils, streams and groundwater with traces of metallic elements is widespread, but as these so-called HEAVY METALS occur naturally in the rocks that are weathered to provide the parent materials for soils, all soils contain natural BACKGROUND LEVELS. As a rule of thumb, any soil with metallic elements in excess of the background levels given in the Table may be considered as polluted.

Small quantities of *manganese*, *boron* and *molybdenum* are essential for normal plant growth, and ruminant animals require *cobalt*, the other metals are toxic particularly if the background concentrations are enhanced by ANTHROPOGENIC sources. CADMIUM (Cd) concentrations in Silurian and Lower Lias BLACK SHALE are abnormally high leading to *teart pastures*, and the use of *phosphate fertilisers* containing cadmium as a CONTAMINANT has led to enhanced concentrations in *agricultural soils*. LEAD (Pb) concentrations from *mining* and *smelting*, sewage SLUDGE and vehicle exhausts are widespread, the latter particularly affecting *roadside verges*. All urban areas and particularly sites where metal-working has taken place have enhanced metallic concentrations. The effect of *toxic metals* in soils is long-lasting as the metal ions are held by the CLAY-HUMUS COMPLEX and inhibit the beneficial activities of soil *microfauna*.   *EMB*

[*See also* ARSENIC (As), CONTAMINATED LAND, COPPER (Cu), IRON (Fe), MERCURY (Hg), MICRONUTRIENTS, TRACE ELEMENT, ZINC (Zn)]

Element	Median soil content (mg/kg) and range
As	6 (0.1–40)
B	20 (2–270)
Cd	0.35 (0.01–2)
CO	8 (0.05–65)
Cu	30 (2–250)
Hg	0.06 (0.01–0.5)
Mn	1,000 (20–10 k)
Pb	35 (2–300)
Se	0.4 (0.1–2)
Zn	90 (1–900)

Metal pollution *Natural background levels for heavy metals in soil.*

Blum WEH (1990) *Soil pollution by heavy metals.* Strasbourg: Council of Europe.
Bradi HB (2005) *Heavy metals in the environment: Origin, interaction and remediation.* Amsterdam: Elsevier.
Brown SE and Welton WC (eds) (2009) *Heavy metal pollution.* New York: Nova Science Publishers.
Hutchinson TC (ed.) (1987) *Lead, mercury, cadmium and arsenic in the environment.* New York: Wiley.
Kabata-Pendias A (2011) *Trace elements in soils and plants.* Boca Raton, FL: CRC Press.
Nriagu JO (1990) Global metal pollution: Poisoning the biosphere. *Environment* 32: 7–32.
Sánchez ML (ed.) (2009) *Causes and effects of heavy metal pollution.* New York: Nova Science Publishers.
Sigel A and Sigel H (eds) (1997) *Metal ions in biological systems, volume 34: Mercury and its effects on environment and biology.* New York: Marcel Dekker.

metals in environmental history Metals released to the atmosphere by natural processes (e.g. EROSION) or by ANTHROPOGENIC activities (e.g. ore processing) are deposited in sedimentary sequences and can be measured using geochemical analysis. Much research has focused on the postindustrial flux of metals in the atmosphere amid concerns over POLLUTION, and high concentrations of lead and mercury have been discovered in the Greenland ICE CORES for this period. The ANTHROPOGENIC flux of metals released since the COPPER AGE has shown that industrial pollution has occurred for millennia, yet new research has demonstrated that the natural flux of metals has varied over longer timescales and can be used as a PROXY CLIMATIC INDICATOR.   *ARG*

[*See also* BRONZE AGE, IRON (Fe), IRON AGE, LEAD (Pb), MERCURY (Hg), POLLUTION HISTORY]

Craddock PT (ed.) (1980) *Scientific studies in early mining and extractive metallurgy.* London: British Museum.
Maddin R (1988) *The beginning of the use of metals and alloys.* Cambridge: MIT Press.
Shepherd R (1980) *Prehistoric mining and related industries.* London: Academic Press.
Shotyk W, Weiss D, Appleby PG et al. (1998) History of atmospheric lead deposition since 12,370 ^{14}C yr BP from a peat bog, Jura Mountains, Switzerland. *Science* 281: 1635–1640.
Tylecote RF (1987) *The early history of metallurgy in Europe.* London: Longman.

metamorphic rocks Rocks that have been altered in the solid state (recrystallised) as a result of changes in temperature, pressure and/or chemical environment (e.g. HYDROTHERMAL effects). *Recrystallisation* involves changes to MINERALS and to their TEXTURE (shape or arrangement). Most metamorphic rocks have a CRYSTALLINE texture and a distinctly anisotropic FABRIC. *Contact metamorphism* converts rocks in a *metamorphic aureole* adjacent to an igneous INTRUSION into *hornfels*. *Regional metamorphism* results from heat and stress during OROGENESIS. MUDSTONE is converted into *phyllite*, *slate*, *schist* and *gneiss* with increasing degree of metamorphism. Basic IGNEOUS ROCKS are altered to *amphibolite*, SANDSTONE to *psammite* or, if rich in QUARTZ, to *metaquartzite* and LIMESTONE to *marble*. *Dynamic metamorphism* results from rock DEFORMATION and forms rocks such as *mylonite* in FAULT zones. Other categories of metamorphism are summarised in the Table.

Much of the GEOLOGICAL RECORD, particularly of the PRECAMBRIAN, is preserved in metamorphic rocks, and their interpretation yields important information about the evolution of environments, the ATMOSPHERE, past CLIMATE and *life* on Earth. *Orogenic metamorphism* in the past may have been associated with the release of CARBON DIOXIDE into the atmosphere: although the significance of this process is controversial, it suggests that metamorphism may have a long-term effect on CLIMATIC CHANGE. Although usually considered in relation to rocks, metamorphic processes also underlie the crystalline transitions between *snow*, FIRN and glacier ICE.   *JBH*

[*See also* GREENSTONE BELT, SEDIMENTARY ROCKS]

Best MG (2003) *Igneous and metamorphic petrology.* Oxford: Blackwell.
Bucher K and Grapes R (2011) *Petrogenesis of metamorphic rocks, 8th edition.* Berlin: Springer.
Fettes D and Desmons J (eds) (2007) *Metamorphic rocks: A classification and glossary of terms.* Cambridge: Cambridge University Press.
Fry N (1991) *The field description of metamorphic rocks.* Bath: Geological Society.

Location	Process	Description
Local metamorphism	Contact metamorphism	Metamorphic rocks adjacent to and clearly related to IGNEOUS ROCKS
	Dynamic metamorphism	Metamorphic rocks associated with severe deformation along FAULT or shear zones
	Impact metamorphism	Metamorphic rocks associated with high pressure-temperature regimes caused by METEORITE IMPACT
	Micro-contact metamorphism	Small-scale changes due to high-temperature lightning strikes (creating FULGURITES)
Regional metamorphism	Orogenic metamorphism	Metamorphic rocks formed in association with SUBDUCTION and collision-related zones of OROGENESIS
	Burial metamorphism	Metamorphic rocks buried in SEDIMENTARY BASINS, where higher pressures and temperatures have formed new minerals
	Oceanic metamorphism	Metamorphic rocks altered by circulating heated seawater driven by HYDROTHERMAL ACTIVITY at MID-OCEANIC RIDGES (MORs)

Metamorphic rocks *Types of metamorphism associated with the formation of metamorphic rocks.*

Kerrick DM and Caldeira K (1998) Metamorphic CO_2 degassing from orogenic belts. *Chemical Geology* 145: 213–232.

metapedogenesis   Changes in soil characteristics resulting from human activity. The outcome may be beneficial or detrimental and results in changes to the natural SOIL PROFILE. If these changes are profound, and the original soil profile is completely altered or buried, soils are classified as ANTHROSOLS.   *EMB*

[*See also* AGRICULTURAL IMPACT ON SOILS, ANTHROBLEME, HUMAN IMPACT ON SOIL]

Richter Jr D deB (2007) Humanity's transformation of Earth's soil: Pedology's new frontier. *Soil Science* 172: 957–967.
Yaalon DH and Yaron B (1966) Framework for man-made soil changes: An outline of metapedogenesis. *Soil Science* 102: 272–277.

metaphysics   The part of *philosophy* that raises questions about the nature of *reality*, and about being and knowing, and claims to deal with questions that are beyond the ability of SCIENCE to solve. Metaphysics is closely related to ONTOLOGY. It can be seen as an attempt to characterise existence or reality as a whole, instead of, as in the NATURAL ENVIRONMENTAL SCIENCES, particular parts or aspects of reality. *Materialism* and IDEALISM are examples of metaphysics in this sense.   *ART*

Conee E and Sider T (2005) *Riddles of existence: A guided tour of metaphysics.* Oxford: Oxford University Press.
Loux MJ (2006) *Metaphysics: A contemporary introduction, 3rd edition.* London: Routledge.
Mumford S (2012) *Metaphysics: A very short introduction.* Oxford: Oxford University Press.

metapopulation model   Derived from ISLAND BIOGEOGRAPHY theory of the 1960s, its fundamental assumption is that the characteristics of isolated communities and populations result from a dynamic balance between the opposing forces of *immigration* (which adds individuals and species) and EXTINCTION (which removes them). *Metapopulation theory* typically focuses on individual species. While this theory has developed into a diversity of models over the past two decades, all of these models assume that each species exists as a collection of interdependent POPULATIONS which comprise the *metapopulation*. Each population is assumed to be isolated enough to allow differentiation among populations, but dependent to some degree on immigration among populations to stave off EXTIRPATION and eventual extinction of the entire metapopulation. The concept has been expanded to *metacommunities*, whereby individual communities are subject to the same forces as metapopulations.   *MVL/JLI*

[*See also* FRAGMENTATION, ISOLATION, POPULATION DYNAMICS]

Gilpin ME and Hanski I (1991) *Metapopulation dynamics: Empirical and theoretical investigations.* San Diego, CA: Academic Press.
Hanski I and Gaggiotti OE (ed.) (2004) *Ecology, genetics and evolution of metapopulations.* Burlington, MA: Elsevier.
Holyoak M, Leibold MA and Holt RD (eds) (2005) *Metacommunities: Spatial dynamics and ecological communities.* Chicago: University of Chicago Press.

Kritzer JP and Sale PF (2006) *Marine metapopulations*. Burlington, MA: Elsevier.

Royle JA and Dorazio RM (2008) *Hierarchical modelling and inference in ecology: The analysis of data from populations, metapopulations and communities*. London: Academic Press.

meteor   A METEOROID made visible as it is heated and burns up on passing through the Earth's ATMOSPHERE. Meteors may occur in groups (*meteor showers*). Very bright meteors are termed *fireballs*.   JAM

meteorite   A small ASTEROID or METEOROID that survives passage through the ATMOSPHERE and strikes the Earth's surface. Around 180 meteoroid *impact craters* are known worldwide, most of which have been detected by SATELLITE REMOTE SENSING. It has been estimated, however, that <10 per cent of PHANEROZOIC terrestrial impact structures with diameters >10 km and <20 per cent of those with diameters >20 km have been discovered (see Figure).   JAM

[*See also* BOLIDE, METEORITE IMPACT]

McCall GJH (ed.) (1977) *Meteorite craters*. Stroudsburg, PA: Dowden, Hutchinson and Ross.

Stewart SA (2011) Estimates of yet-to-find impact crater population on Earth. *Journal of the Geological Society* 168: 1–14.

meteorite impact   The collision of debris from space (including ASTEROIDS and COMETS) with the surface of a PLANET, *satellite* or other object in the *solar system*, producing an *impact crater*. Impact CRATERS are well known from other bodies in the solar system but most were thought to have formed early in its history, and evidence of any that had formed on Earth was thought to have been removed by WEATHERING and EROSION, a notable exception being Meteor Crater (Barringer Crater) in Arizona, which is 1.2 km in diameter and formed about 49,000 years ago. The frequency of impacts is inversely proportional to the size of the impacting object (see Figure) and has declined through GEOLOGICAL TIME.

A renewed acceptance of the role of rare events in the history of the Earth (see NEOCATASTROPHISM), driven partly by the *impact hypothesis* for the K-T BOUNDARY, has led to the successful search for more meteorite impact craters on Earth (see the Figure in METEORITE entry). These can be distinguished from volcanic craters by details of their morphology, preservation of fragments of the impacting BOLIDE and evidence of shock-induced metamorphism of the surrounding rocks. They can be divided into *meteorite craters* which are relatively fresh and uneroded, and *impact structures* where the surface features have been eroded and deeper levels are now exposed. Large craters

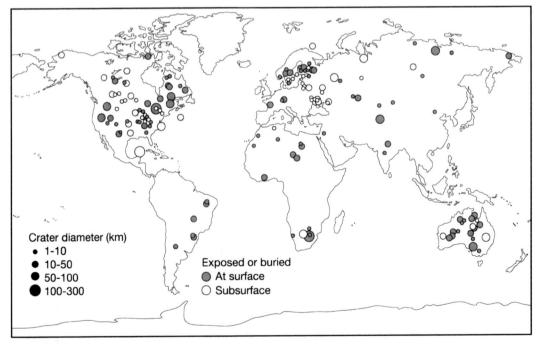

Meteorite *Known meteorite impact craters according to the Earth Impact Database (2009; n = 176). Surface and subsurface craters varying in diameter from <1 to 300 km are differentiated (Stewart, 2011).*

(> 10 km diameter) are sometimes called ASTROBLEMES, or *hydroblemes* if they are formed in water: such large meteorite impact events are capable of inducing devastating ENVIRONMENTAL CHANGE through the shock waves at impact, huge TSUNAMI waves that would be generated if the impact were in the OCEAN and the ejection of dust and gases into the ATMOSPHERE (see also NUCLEAR WINTER).

Some 180 meteorite impact events have now been identified on Earth ranging from the Sudbury structure in Canada, some 1,850 million years old and 180 km across, which is a major nickel-ore deposit, through the Chicxulub crater in Mexico identified as the 'smoking gun' for the end-Cretaceous MASS EXTINCTION, to the Sikhote Alin craters in Russia that formed in AD 1947. Additional incentive was given to the study of meteorite impacts by the observed impact of Comet Shoemaker-Levy 9 with Jupiter in 1994, and research is now being undertaken into the possibility of collisions between the Earth and NEAR-EARTH OBJECTS (NEOs).   GO

[*See also* EARTHQUAKE MAGNITUDE, EVENT, K-T BOUNDARY]

Evans KR, Wright Horton Jr J, King Jr DT and Morrow JR (eds) (2007) *The sedimentary record of meteorite impacts.* Boulder, CO: Geological Society of America.

Ferrière L, Lubala FRS, Asinski GR and Kaseli PK (2011) The newly confirmed Luizi impact structure, Democratic Republic of Congo: Insights into central uplift formation and post-impact erosion. *Geology* 39: 851–854.

Gorter JD and Glikson AY (2012) Talundilly, Western Queensland, Australia: Geophysical and petrological evidence for an 84 km-large impact structure and an Early Cretaceous impact cluster. *Australian Journal of Earth Sciences* 59: 51–73.

Grady MM, Hutchison R, McCall GJH and Rothery D (eds) (1998) *Meteorites: Flux with time and impact effects.* Bath: Geological Society.

Hodge P (1994) *Meteorite craters and impact structures of the Earth.* Cambridge: Cambridge University Press.

Jewitt D (2000) Eyes wide shut. *Nature* 403: 145–148.

Levy DH (1995) *Impact Jupiter: The crash of comet Shoemaker-Levy 9.* New York: Plenum Press.

Melosh HJ (1989) *Impact cratering: A geologic process.* New York: Oxford University Press.

Melosh HJ (2011) *Planetary surface processes.* Cambridge: Cambridge University Press.

Pierazzo E and Melosh HJ (2012) Extraterrestrial causes of environmental catastrophes. In Matthews JA, Bartlein PJ, Briffa KR et al. (eds) *The SAGE handbook of environmental change, volume 1.* London: Sage, 384–404.

Schmieder M, Buchner E, Schwarz WH et al. (2010) A Rhaetian 40Ar/39Ar age for the Rochechouart impact structure (France) and implications for the latest Triassic sedimentary record. *Meteoritics and Planetary Science* 45: 1225–1242.

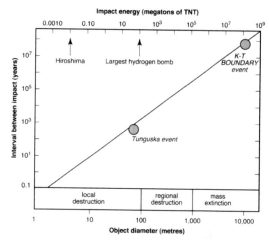

Meteorite impact *Estimated recurrence intervals for impacts with near-Earth objects of different sizes, with some key events for comparison (Jewitt, 2000).*

meteoroids Natural solid objects moving in interplanetary space that are larger than a molecule but smaller than an ASTEROID. Beech and Steel (1995) proposed that they should be defined as 100 μm to 10 m, which takes account of instrumental limitations, physical constraints on their formation and the necessity to distinguish meteoroids from both asteroids and DUST. They are thought to be fragments of ASTEROIDS. Meteoroids that impact on the Earth's surface are termed METEORITES.   JAM

Beech M and Steel DI (1995) On the definition of the term meteoroid. *Quarterly Journal of the Royal Astronomical Society* 36: 281–284.

meteorological satellites Orbital platforms, also known as *weather satellites*, for making observations of the Earth's surface, OCEAN and ATMOSPHERE for *weather forecasting* and CLIMATOLOGY. They are invaluable, for example, providing advanced warning of TROPICAL CYCLONES and other STORMS. The first purpose-built meteorological satellite, TIROS-1, was launched on 1 April 1960. Observations are now made regularly over large areas and provide 'real-time' monitoring of the *atmospheric circulation*. Weather satellites in *geostationary orbit* (*geostationary satellites*) continuously view a hemisphere of the Earth. They allow regular observations (up to every 15 minutes) of large-scale features and have a spatial *resolution* of between 1 and 5 km at the *Equator*. A network of satellites, such as *Meteosat* and *Geostationary Operational Environmental Satellites*, give global overlapping coverage. Weather satellites in *Sun-synchronous polar orbits* (*polar-orbiting satellites*) pass over the same place on the Earth's surface a number of times each day and can record images with a spatial resolution down to 1 km.

Groups of satellites, such as the National Oceanic and Atmospheric Administration (NOAA) series, collect images of the same location every 2 or 3 hours. A constellation of weather satellites in polar orbits around the Earth are operated by the *Polar Operational Environmental Satellite* project, a joint initiative between the NOAA and the *European Organisation for the Exploitation of Meteorological Satellites*. Weather satellites carry OPTICAL REMOTE-SENSING INSTRUMENTS to identify CLOUD formations, THERMAL REMOTE-SENSING instruments to measure SEA-SURFACE TEMPERATURE (SST) and PASSIVE MICROWAVE REMOTE-SENSING instruments to measure atmospheric properties. *GMS/JBE/NP*

Carleton AM (1991) *Satellite remote sensing in climatology.* London: Belhaven.

Chuvieco E (ed.) (2007) *Earth observation of global change: The role of satellite remote sensing in monitoring the global environment.* Berlin: Springer.

Kelkar RR (2007) *Satellite meteorology.* Hyderabad: BS Publications.

Kidder SQ and Vondar Haar TH (1995) *Satellite meteorology.* New York: Academic Press.

meteorology The science of phenomena and processes observed in the ATMOSPHERE, principally involving heat, motion and moisture. It focuses primarily on the study of WEATHER, including weather FORECASTING and its basis in *atmospheric chemistry* and *atmospheric physics*. Meteorology can be subdivided into *dynamic meteorology*, which covers the FLUID DYNAMICS of the atmosphere, and *physical meteorology*, which covers its *thermodynamics* (including clouds and precipitation) and optical, electrical and acoustic aspects. *JAM*

[*See also* ATMOSPHERIC SCIENCES, CLIMATOLOGY]

Ahrens CD (2006) *Meteorology today: An introduction to weather, climate and the environment, 8th edition.* Belmont, CA: Wadsworth Publishing.

Aristotle (1952) *Meteorologica* [Translated by HDP Lee]. Cambridge, MA: Harvard University Press.

Brunt D (2011) *Physical and dynamical meteorology, 2nd edition.* Cambridge: Cambridge University Press.

Holton JR (2004) *An introduction to dynamic meteorology, 4th edition.* Burlington, MA: Elsevier.

meteotsunami An ocean wave with similar characteristics to a TSUNAMI, generated by meteorological processes rather than displacement of the ocean floor. *GO*

Monserrat S, Vilibić I and Rabinovich AB (2006) Meteotsunamis: Atmospherically induced destructive ocean waves in the tsunami frequency band. *Natural Hazards and Earth System Sciences* 6: 1035–1051.

Vilibić I, Monserrat S, Rabinovich A and Mihanović H (2008) Numerical modelling of the destructive meteotsunami of 15 June, 2006 on the coast of the Balearic Islands. *Pure and Applied Geophysics* 165: 2169–2195.

methane variations Atmospheric methane (CH_4) is one of the most important atmospheric TRACE GASSES because of its role in trapping RADIATION re-emitted by the Earth (see GREENHOUSE EFFECT). Although methane only accounts for 0.2 per cent of total ATMOSPHERIC COMPOSITION, it has a very short RESIDENCE TIME (9 years compared, for example, with the 50–200 years residence time for atmospheric CARBON DIOXIDE). Small variations can therefore have a rapid effect on the global CLIMATIC SYSTEM. Major sources of atmospheric methane include emissions from FOSSIL FUELS, WETLANDS, TROPICAL PEATLANDS, domestic animals, rice PADDY SOILS, BIOMASS BURNING and TERMITES (see Figure). Evidence of past levels of atmospheric methane has been recorded in the air trapped in the Arctic and Antarctic ICE CORES. Results from these studies indicate that levels of atmospheric methane have varied considerably through GLACIAL-INTERGLACIAL CYCLES with higher concentrations occurring during interglacials. During the past 150–250 years, concentrations of atmospheric methane have, however, reached unprecedented levels due to anthropogenic activity. Present estimates suggest that atmospheric methane is now 158 per cent of its 1750 level, and after a period of relative stability between 1999 and 2006, concentrations are now again rising. Apparently anomalous rises in methane after ca 5 ka BP have been considered an indicator of early anthropogenic impacts on climate supporting the EARLY-ANTHROPOCENE HYPOTHESIS. *KJW/CJC*

Berrittella C and van Huissteden J (2011) Uncertainties in modeling CH4 emissions from northern wetlands in glacial climates: The role of vegetation parameters. *Climate of the Past* 7: 1075–1088.

Delmotte M, Chappallaz J, Brook E et al. (2004) Atmospheric methane during the last four glacial-interglacial cycles: Rapid changes and their link with Antarctic temperatures. *Journal of Geophysical Research: Atmospheres* 109: D12104.

Ferretti DF, Miller JB, White JWC et al. (2005) Unexpected changes to the global methane budget over the past 2000 years. *Science* 309: 1714–1717.

Fuller DQ, van Etten J, Manning K et al. (2011) The contribution of rice agriculture and livestock pastoralism to prehistoric methane levels: An archaeological assessment. *The Holocene* 21: 743–759.

Reay D, Smith P and van Amstel A (eds) (2010) *Methane and climate change.* London: Earthscan.

Ruddiman WF, Kutzbach JE and Vavrus SJ (2011) Can natural or anthropogenic explanations of late-Holocene CO_2 and CH_4 increases be falsified? *The Holocene* 21: 865–879.

Severinghaus JP and Brooke EJ (1999) Abrupt climate change at the end of the last glacial period inferred from trapped air in polar ice. *Science* 286: 930–934.

Singarayer JS, Valdes PJ, Friedlingstein P et al. (2011) Late Holocene methane rise caused by orbitally controlled increase in tropical sources. *Nature* 470: 82–85.

Stern DI and Kaufmann RK (1996) Estimates of global
anthropogenic methane emissions 1960-1993.
Chemosphere 33: 159–176.
Walter KM, Edwards ME, Grosse G et al. (2007)
Thermokarst lakes as a source of atmospheric CH_4 during
the last deglaciation. *Science* 318: 633–636.
Wolff E (2011) Greenhouse gases in the Earth system: A
palaeoclimate perspective. *Philosophical Transactions of
the Royal Society of London A* 369: 2133–2147.
Zhou X (2012) Asian monsoon precipitation changes and the
Holocene methane anomaly. *The Holocene* 22: 731–738.

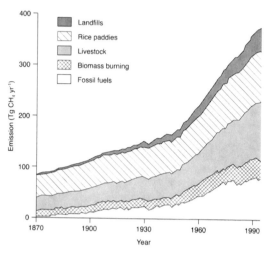

Methane variations *Increases in methane emissions from various
sources since AD 1870 (Stern and Kaufmann, 1996).*

methanogenesis   The formation of methane gas
(CH_4) by DECOMPOSITION under ANAEROBIC conditions. It
is carried out by micro-organisms (*methanogens*) and
is the final stage in decomposition, resulting in the further breakdown of the products of FERMENTATION. There
is evidence, however, of methane EMISSIONS from plants
under AEROBIC conditions by an unknown mechanism.
                                                             JAM/JP

[*See also* GAS HYDRATES, METHANE VARIATIONS,
RUMINANT]

Kepler F, Hamilton JTG, Brass M and Röckmann T (2006)
Methane emissions from terrestrial plants under aerobic
conditions. *Nature* 439: 187–191.
Khalil MAK (ed.) (2000) *Atmospheric methane: Its role in
the global environment.* Berlin: Springer.

methodology   The study and evaluation of method,
both the techniques of investigation and the approaches
to research and problem solving. It includes considering how problems are conceptualised and investigated
and the setting of standards in scientific investigation.
                                                             JAM

[*See also* EXPERIMENTAL DESIGN, RESEARCH DESIGN,
SCIENTIFIC METHOD]

Bauer BO (1999) On methodology in physical geography:
Current status, implications and future prospects. *Annals
of the Association of American Geographers* 89: 677–679.
Ford ED (2000) *Scientific method for ecological research.*
Cambridge: Cambridge University Press.
Kumar R (2010) *Research methodology, 3rd edition.*
London: Sage.

microatoll   A single CORAL colony, usually massive
and circular, with a dead, flat surface and living lateral margins. *Growth banding* and morphology records
ANNUAL VARIABILITY and DECADAL-SCALE VARIABILITY in
water depth over intertidal *reef flats*. In tectonically
active areas, microatolls have been used to reconstruct patterns of UPLIFT and SUBSIDENCE associated
with EARTHQUAKES and longer-term plate DEFORMATION
processes. Fossil microatolls have been used to reconstruct mid-late Holocene SEA-LEVEL CHANGE.   *TSp*

Briggs RW, Sieh K, Meltzner AJ et al. (2006) Deformation
and slip along the Sunda Megathrust in the great 2005
Nias-Simeulue earthquake. *Science* 311: 1897–1901.
Smithers SG, Hopley D and Parnell KE (2006) Fringing and
nearshore coral reefs of the Great Barrier Reef: Episodic
Holocene development and future prospects. *Journal of
Coastal Research* 22: 175–187.
Woodroffe CD and Gagan MK (2000) Coral microatolls
from the central Pacific record late Holocene El Nino.
Geophysical Research Letters 27: 1511–1514.
Woodroffe CD and McLean RF (1990) Microatolls and recent
sea level change on coral atolls. *Nature* 344: 531–534.

microclimate   The climate of a small space with
dimensions 2–10 m horizontally and within 10 m of
the Earth's surface. The timescale of measurements
for investigation of the microclimate is usually less
than 24 hours. These dimensions mean that it overlaps in scale with LOCAL CLIMATE. *Microclimatology* is
the study of the atmospheric BOUNDARY LAYER and the
interaction between the ATMOSPHERE and the Earth's
surface in terms of the ENERGY BALANCE. A slightly
looser definition includes the study of the climate of
the BIOSPHERE: from the crowns of trees to the bottom
of the root zones. This depends on the concept of the
active surface, which is the level where the majority
of the radiant energy is absorbed, reflected and emitted; where transformations of energy (radiant to thermal, sensible to latent) and mass (change of state of
water) take place; where PRECIPITATION is intercepted
and where most *frictional drag* on airflow occurs. *BDG*

[*See also* MACROCLIMATE, URBAN CLIMATE]

Erell E, Pearlmutter D and Williamson T (2010) *Urban
microclimates: Designing the spaces between buildings.*
London: Routledge.
Geiger R, Aron RH and Todhunter P (2003) *The climate
near the ground, 6th edition.* Lanham, MD: Rowman and
Littlefield.

Oke TR (2002) *Boundary layer climates, 2nd edition.* London: Routledge.

Rosenberg NJ, Blad BL and Verma SB (eds) (1983) *Microclimate: The biological environment, 2nd edition.* New York: Wiley.

microcosm   A *micro-ecosystem* or model ecosystem consisting of subsets of BIOTIC and ABIOTIC components, put together for experimental purposes. They are partly or wholly isolated from the external environment in laboratory or field situations and are one way of simplifying reality in order to understand it. The term has also been used in the sense of a *natural microcosm.* Examples of the latter include *phytotelmata* (aquatic ecosystems contained within single plants, such as bromeliads). GLACIER FORELANDS have also been termed microcosms, in the literal sense of 'little worlds' that are simpler than normal landscapes and are therefore conducive to investigation.   *JAM*

[*See also* MESOCOSM]

Beyers RJ and Odum HT (1993) *Ecological microcosms.* New York: Springer.

Matheson FE (2008) Microcosms. In Jorgensen S (ed.) *Encyclopedia of ecology.* Amsterdam: Elsevier, 2393–2397.

Matthews JA (1975) The gletschervorfeld: A biogeographical system and microcosm. *University of Edinburgh, Department of Geography, Research Discussion Paper* 2: 1–44.

Srivasta DS, Kolasa J, Bengtsson J et al. (2004) Are natural microcosms useful model systems for ecology. *Trends in Ecology and Evolution* 19: 379–384.

microdendroclimatology   The intra-ring study of TREE RINGS, particularly their isotopic content, for HIGH-RESOLUTION RECONSTRUCTION of palaeoclimates. The major advantage of a microdendroclimatic approach is that SEASONALITY effects can be analysed or controlled.   *JAM*

[*See also* DENDROCLIMATOLOGY, PALAEOCLIMATOLOGY]

Loader NJ, Switsur VR and Field EM (1995) High-resolution stable isotope analysis of tree rings: Implications of 'microdendroclimatology' for palaeoenvironmental research. *The Holocene* 5: 457–460.

Robertson I, Loader NJ and McCarroll D (2004) $\delta^{13}C$ of tree-ring lignin as an indirect measure of climate change. *Water, Air and Soil Pollution: Focus* 4: 531–544.

micro-erosion meter   An instrument for measuring accurately the change in the surface level of rock over time. Its three legs rest on metal studs fixed in the rock providing a stable platform from which measurements to the rock surface are made with a spring-loaded probe connected to an engineer's dial gauge. Repeat measurements enable the WEATHERING RATE and/ or amount of EROSION to be determined directly.   *RAS*

Drysdale R and Gillieson D (1997) Micro-erosion meter measurements of travertine deposition rates: A case study from Louie Creek, northwest Queensland, Australia. *Earth Surface Processes and Landforms* 22: 1037–1051.

Stephenson WJ, Kirk RM, Hemmingsen SA and Hemmingsen MA (2010) Decadal scale micro erosion rates on shore platforms. *Geomorphology* 114: 22–29.

microfossil   A FOSSIL or SUBFOSSIL of such a small size that a microscope is needed for inspection (e.g. pollen 10–100 μm).   *BA*

[*See also* MACROFOSSIL, MICROFOSSIL ANALYSIS, PALAEOBOTANY, PALAEOECOLOGY, POLLEN ANALYSIS, SPORES]

Armstrong H and Brasier MD (2004) *Microfossils, 2nd edition.* Chichester: Wiley-Blackwell.

Van Geel B (1986) Application of fungal and algal remains and other microfossils in palynological analyses. In Berglund BE (ed.) *Handbook of Holocene palaeoecology and palaeohydrology.* Chichester: Wiley, 497–505.

microfossil analysis   The investigation of a wide range of microscopic FOSSILS or SUBFOSSILS, including *pollen,* SPORES, DIATOMS, OSTRACODS, TESTATE AMOEBAE, *chironomids, cladocera* and FORAMINIFERA. Microfossil analysis is an important tool for studies in BIOSTRATIGRAPHY (where the occurrence of taxa or assemblages is used to establish the relative age of units), PALAEOECOLOGY (where microfossils are used to reconstruct past environments, plant and animal communities and interactions between them) and *evolutionary biology* (establishing the evolutionary descent of different LINEAGES of organisms).   *MJB*

[*See also* MACROFOSSIL]

Armstrong HA and Brasier MD (2004) *Microfossils, 2nd edition.* Oxford: Blackwell.

Haslett SK (2002) *Quaternary environmental micropalaeontology.* London: Arnold.

McGowran B (2008) *Biostratigraphy: Microfossils and geological time.* Cambridge: Cambridge University Press.

Smol JP, Birks HJB and Last WM (eds) (2001) *Tracking environmental change using lake sediments, volume 3: Terrestrial, algal, and siliceous indicators.* Dordrecht: Kluwer.

microhabitats   Microscopic HABITATS, especially those inhabited by *micro-organisms,* such as the pore spaces, water films and RHIZOSPHERE in SOILS. Habitats that are small-scale but not truly microscopic, such as the aquatic ecosystems contained within individual Bromeliad plants (*phytotelmata*) or the *cryoconite holes* in the surface of GLACIERS, are often included.   *JAM*

[*See also* MICROCOSM]

Forster RC (1988) Microenvironments of soil organisms. *Biology and Fertility of Soils* 6: 189–203.

Wharton RA, McKay CP, Simmons GM and Parker BC (1985) Cryoconite holes on glaciers. *BioScience* 35, 499–503.

microlith   A small stone tool (arbitrarily defined as <5 cm long) used from the PALAEOLITHIC onwards, especially to form the points of hunting weapons. *Macroliths* are larger.

JAM

[*See also* MEGALITH]

Costa LJ, Sternke F and Woodman PC (2005) Microlith to macrolith: The reasons behind the transformation of production in the Irish Mesolithic. *Antiquity* 79: 19–33.

micromorphological analysis of sediments   *Micromorphology* is the term used to describe the distinctive arrangement of MATRIX particles and *voids* making up a sediment FABRIC. This can be established by THIN-SECTION ANALYSIS under a microscope. Micromorphological analysis of sediments can reveal evidence of WEATHERING, alteration of MINERALS, orientation and packing of CLAY particles, arrangement and concentration of voids, presence and type of *calcite* crystal growth, animal excrement, clay *coatings*, rootlet PSEUDOMORPHS and other features. It can be used as both a descriptive and a diagnostic tool in *palaeopedology* and is regarded as one of the most reliable methods for detecting evidence of PEDOGENESIS, from which it is possible to distinguish sequential phases of soil formation and infer changes in environmental conditions. Widely used in analysis of PALAEOSOLS, such analyses are of value in determining characteristics of sediments on archaeological sites.

DH/CJC

[*See also* MICROMORPHOLOGICAL ANALYSIS OF SOILS, MICROMORPHOLOGICAL ANALYSIS OF TILL]

Goldberg P and Berna F (2010) Micromorphology and context. *Quaternary International* 214: 56–62.
Kemp RA (1985) *Soil micromorphology and the Quaternary.* London: Quaternary Research Association.
Kemp RA (1998) Role of micromorphology in palaeopedological research. *Quaternary International* 51: 133–141.
Kemp RA (1999) Micromorphology of loess-paleosol sequences: A record of paleoenvironmental change. *Catena* 35: 179–196.
Macphail RI, Cruise GM, Allen MJ et al. (2004) Archaeological soil and pollen analysis of experimental floor deposits: With special reference to the Butser Ancient Farm, Hampshire. *Journal of Archaeological Science* 31: 175–191.
Stoops G, Marcelino V and Mees F (eds) (2010) *Interpretation of micromorphological features of soils and regolith.* Amsterdam: Elsevier.

micromorphological analysis of soils   Examination of soils using the microscope was introduced by W.L. Kubiena in the 1930s. Techniques developed for optical mineralogy have been adapted to study the fabric of soil in THIN–SECTION ANALYSIS under plain and polarised light. Soil samples with known orientation are taken from the field, dried and set in polyester resin and subsequently cut into thin slices 0.30 μm thick, mounted on microscope slides and examined under the polarising microscope. The *soil fabric* can be seen to comprise *skeleton (mineral) grains, plasma* and *voids*. The disposition of mineral grains, voids and clay domains and the location and appearance of various segregations and concentrations of CLAY, IRON (Fe), *manganese* and *organic matter* enable the pedologist to assess the results of SOIL-FORMING PROCESSES at the microscopic level. The three-dimensional arrangement of these components provides an insight into the detailed processes that operate in soils and can supply evidence of environmental change. SCANNING ELECTRON MICROSCOPY (SEM) has carried the examination of soil fabrics into even greater detail. Micromorphology is used, for example, to confirm the presence of *clay-skins (cutans)* in, for example, LUVISOLS and the presence of *organic-iron coatings* to mineral grains in PODZOLS. Quantitative micromorphology is sometimes termed *micromorphometry*.

EMB

Bullock P, Fedoroff N, Jongerius A et al. (1985) *Handbook for the description of thin sections of soils.* Wolverhampton: Waine Research Publishing.
French C (2002) *Geoarchaeology in action: Studies in soil micromorphology and landscape evolution.* London: Routledge.
Gerasimova MI (2003) Higher levels of description: Approaches to the micromorphological characterisation of Russian soils. *Catena* 54: 319–337.
Goldberg P (1992) Micromorphology, soils and archaeological sites. In Holliday V (ed.) *Soils in archaeology: Landscape evolution and human occupation.* Washington, DC: Smithsonian Institution Press, 145–167.
Kapur S, Mermut A and Stoops G (eds) (2008) *New trends in soil micromorphology.* Berlin: Springer.
Kubiena WL (1938) *Micropedology.* Ames, IA: Collegiate Press.

micromorphological analysis of till   The qualitative and quantitative microscopic study of thin-sectioned, impregnated samples of undisturbed TILL (usually subglacially deposited glacial sediment). *Transmitted-light petrographic microscopes* are used, commonly with magnifications of up to 50 times, in both plane- and cross- (or circularly) polarised light settings. Micromorphological analysis of till normally comprises *texture, composition, microstructure, microfabric* and *plasmic fabrics* (birefringence models reflecting the organisation and orientation of clay particles), and allows interrelations between in-situ till constituents to be determined. From these, inferences can often be drawn about processes of till deposition and deformation (as a result of subglacial shearing), which may in turn assist in reconstructions of SUBGLA-CIAL environmental conditions.

The technique of micromorphology in its application to tills is particularly useful given that the subglacial DEPOSITIONAL ENVIRONMENT is dynamic and spatially variable, which tends to lead to highly diverse and complex till sequences. Furthermore, in cases where visible exposures through till are limited, such as in sediment CORES, there may be a need for specific and highly detailed information that conventional SEDIMEN-TOLOGY cannot provide.

While micromorphology of tills has traditionally been qualitative, descriptive and inventorial in nature, the past few decades have seen a gradual development into more quantitative, computer-assisted analyses. Thus far, this has led to the introduction of *semi-automated measurements* of fabrics and recognition of patterns. Ideally, such developments should eventually reduce observer BIAS, which is an inevitable element in any visual analysis.   *JFH*

[See also MICROMORPHOLOGICAL ANALYSIS OF SEDIMENTS, MICROMORPHOLOGICAL ANALYSIS OF SOILS]

Evans DJA, Phillips ER, Hiemstra JF and Auton CA (2006) Subglacial till: Formation, sedimentary characteristics and classification. *Earth-Science Reviews* 78: 115–176.
van der Meer JJM (1993) Microscopic evidence of subglacial deformation. *Quaternary Science Reviews* 12: 553–587.
van der Meer JJM and Menzies J (2011) The micromorphology of unconsolidated sediments. *Sedimentary Geology* 238: 213–232.
Phillips ER, van der Meer JJM and Ferguson A (2011) A new 'microstructural mapping' methodology for the identification, analysis and interpretation of polyphase deformation within subglacial sediments. *Quaternary Science Reviews* 30: 2570–2596.

micronutrients (1) ELEMENTS, mostly TRACE ELE-MENTS, that are required by organisms in small amounts, such as *boron*, COPPER (Cu), IRON (Fe), manganese, molybdenum and ZINC (Zn) by plants and cobalt, copper, fluorine, iodine, iron, manganese, selenium and zinc by animals. *Nutrient deficiencies* usually induce reduced growth; surpluses are often toxic. (2) The minor classes of COMPOUNDS required by humans, including the elements listed above, other elements required in larger quantities (e.g. CALCIUM, *magnesium*, PHOSPHORUS (P), *sodium* and POTASSIUM (K)) and *vitamins* and MINERALS.
JAM

[See also MACRONUTRIENTS]

Alloway BJ (ed.) (2008) *Micronutrient deficiencies in global crop production.* Berlin: Springer.
Prasad KN (2010) *Micronutrients in health and disease.* Boca Raton, FL: CRC Press.

micropalaeontology A branch of GEOLOGY that encompasses the TAXONOMY, EVOLUTION and BIODIVER-SITY of MICROFOSSILS and their geological applications, such as those in BIOSTRATIGRAPHY and PALAEOENVIRON-MENTAL RECONSTRUCTION.   *CES*

[See also DIATOM ANALYSIS, DINOFLAGELLATE CYST ANALYSIS, FORAMINIFERA: ANALYSIS, MICROFOSSIL ANALYSIS, MICROSCOPY, OSTRACOD ANALYSIS, POLLEN ANALYSIS, RADIOLARIA]

Armstrong H and Brasier M (2005) *Microfossils, 2nd edition.* Chichester: Wiley-Blackwell.
Haslett SK (ed.) (2002) *Quaternary environmental micropalaeontology, 2nd edition.* London: Arnold.
Whittaker JE and Hart MB (eds) (2010) *Micropalaeontology, sedimentary environments and stratigraphy.* Bath: Geological Society.

microresidual fraction The very fine PARTICULATE fraction remaining after PRETREATMENT of a sample for dating. For example, in the RADIOCARBON DATING of BUR-IED SOILS beneath MORAINES in New Zealand, the Himalaya, South America and Alaska, the microresidual fraction tends to be the oldest soil organic fraction and consists of the resistant remains of pioneer plants. *JAM*

[See also SOIL DATING]

Geyh MA, Röthlisberger F and Gellatly A (1985) Reliability tests and interpretation of 14C dates from palaeosols in glacier environments. *Zeitschrift für Gletscherkunde und Glazialgeologie* 21: 275–281.

microscopy The use of the *light microscope (optical microscope)* and/or the *electron microscope* to view microscopic objects and parts of larger objects that cannot be resolved by the unaided eye.   *CES*

[See also SCANNING ELECTRON MICROSCOPY (SEM)]

Mertz J (2009) *Introduction to optical microscopy.* Greenwood Village, CO: Roberts.
Wu Q, Merchant F and Castleman KR (eds) (2008) *Microscope image processing.* Burlington, MA: Academic Press.

microseism A very minor EARTHQUAKE that is detectable only by instrumentation, and is not necessarily generated by TECTONIC processes such as slip along a FAULT. Microseismic vibrations generate weak, almost continuous background SEISMIC WAVES or '*Earth noise*' detected by SEISMOMETERS. They can be generated by ocean *waves, surf, wind* and human activities such as *mining*, QUARRYING, *traffic* and *reservoir* construction. The monitoring of changes in microseismic activity (*microseismic analysis*) can be used to predict sudden events, such as mining HAZARDS and LANDSLIDES. *GO*

Alcott JM, Kaiser PK and Simser BP (1998) Use of microseismic source parameters for rockburst hazard assessment. *Pure and Applied Geophysics* 153: 41–65.

Arosio D, Longoni L, Papini M et al. (2009) Towards rockfall forecasting through observing deformations and listening to microseismic emissions. *Natural Hazards and Earth System Sciences* 9: 1119–1131.

Friedrich A, Kruger F and Klinge K (1998) Ocean-generated microseismic noise located with the Grafenberg array. *Journal of Seismology* 2: 47–64.

Roux P-F, Marsan D, Métaxian J-P et al. (2008) Microseismic activity within a serac zone in an alpine glacier (Glacier d'Argentiere, Mont Blanc, France). *Journal of Glaciology* 54: 157–168.

microvertebrate accumulations Accumulations of microvertebrate remains, composed predominantly of the remains of *small mammals* such as *voles*, *mice* and *shrews*, although small *reptiles*, *amphibians* and *bird bones* may also be present. The accumulations are the result of indigestible prey remains regurgitated as *pellets* by diurnal or nocturnal *birds of prey* (*raptors*) and are preserved most often in sediments within CAVES where the birds have eaten or roosted. The microvertebrate remains show traces of digestion which are frequently specific to particular *predators*, according to the degree of fragmentation and corrosion observed, thereby permitting the agent of accumulation to be identified. The microvertebrates themselves provide an important means of interpreting past environmental and climatic conditions and of dating the deposits, particularly in the case of small mammals.   *DCS*

[*See also* ANIMAL REMAINS, RODENT MIDDEN]

Andrews P (1990) *Owls, caves and fossils*. London: Natural History Museum.

Peterson JE, Scherer RP and Huffman KM (2011) Methods of microvertebrate sampling and their influences on taphonomic interpretations. *Palaios* 26: 81–88.

microwave radiometer A highly sensitive receiver for detecting microwave radiation emitted by material media. In the microwave region of the ELECTROMAGNETIC SPECTRUM, the spectral RADIANCE of a blackbody (see BLACKBODY RADIATION) is directly related to its physical or KINETIC TEMPERATURE (*Rayleigh-Jeans approximation* of *Planck's law*). This has led to the interchangeable use of the two terms and the radiance of the scene observed by radiometers is usually characterised by the RADIANT (or brightness) TEMPERATURE. The *brightness temperature* may vary from zero Kelvin (for a non-emitting medium) to a maximum equal to the physical temperature of the scene (for a perfect emitter or blackbody). Equivalently, the emissivity, defined as the ratio of the brightness to the physical temperature, varies between zero and unity. The emissivity characterises the medium as a signature.

Through proper choice of the radiometer parameters (wavelength, polarisation and incidence angle), it is possible to establish useful relationships between the energy received by the radiometer and specific terrestrial or atmospheric parameters of interest. Microwave radiometry is therefore used for ENVIRONMENTAL MONITORING and has found extensive use in HYDROLOGY, OCEANOGRAPHY and METEOROLOGY. Monitoring of snow and soil moisture from satellite-based radiometers is useful in hydrology. Applications of microwave radiometry in oceanography include the monitoring of SEA-SURFACE TEMPERATURE (SST), SALINITY and *wind speed*. In meteorology, *satellite*, *airborne* or *ground-based radiometers* are used for instance to measure WATER VAPOUR, OZONE and TEMPERATURE profiles. The most limiting factor in SATELLITE REMOTE SENSING with radiometers is the coarse resolution of several kilometres per PIXEL.   *TS*

Skou N and Le Vine D (2006) *Microwave radiometer systems: Design and analysis, 2nd edition*. Norwood, MA: Artech House.

microwave remote sensing (MRS) A suite of different techniques for obtaining information using instruments that measure radiated or reflected RADIATION in the microwave region of the ELECTROMAGNETIC SPECTRUM (*wavelengths* approximately from 1 mm to 1 m). At certain wavelengths (particularly lower frequencies), the microwave signal is unaffected by *atmospheric attenuation* caused by AEROSOLS, CLOUDS and ice. *Microwave sensors* fall broadly within two categories: passive and active instruments. *Passive microwave instruments*, or RADIOMETERS, measure radiated or reflected *microwave radiation* (resulting from either thermal emission or the Sun) from the scene that is being imaged. *Active microwave instruments* illuminate an area of the Earth's surface with *microwave energy* transmitted from a sensor and measure the amount scattered back to the same sensor. As active instruments are themselves the source of radiation illuminating a scene they can operate both day and night.

PJS/MEJC

[*See also* PASSIVE MICROWAVE REMOTE SENSING, RADAR, RADAR REMOTE-SENSING INSTRUMENTS]

Woodhouse IH (2006) *Introduction to microwave remote sensing*. Boca Raton, FL: Taylor and Francis.

midden An archaeological deposit, essentially a *refuse tip*, often from food preparation (*kitchen midden*). Examples include the SHELL MIDDENS at coastal sites from Scandinavia to the Mediterranean, which provide important evidence of HUNTING, FISHING AND GATHERING *economies* in the MESOLITHIC.   *JAM*

[*See also* OCCUPATION LAYER, RODENT MIDDEN]

Alvarez M, Godino IB, Balbo A and Madella M (2010) Shell middens as archives of past environments, human dispersal and specialized resource management. *Quaternary International* 239: 1–7.

Needham S and Spence T (1997) Refuse and the formation of middens. *Antiquity* 71: 77–90.

Sullivan M and O'Connor S (1993) Middens and cheniers: Implications of Australian research. *Antiquity* 67: 776–788.

Middle Ages The loosely defined archaeological and historical period from the end of the ROMAN PERIOD to the RENAISSANCE of the fourteenth to seventeenth centuries. In Britain, one narrower definition places the Middle Ages between the *Norman Conquest* (AD 1066) and the DISSOLUTION of the monasteries (AD 1540s). This period provides the subject matter of the subdiscipline of *Mediaeval archaeology*. It included the *Dark Ages*, the rise of FEUDALISM, the building of *castles* and *towns*, the *Crusades*, the *Black Death* and the introduction of *gunpowder*, which led to significant effects on LANDUSE, land ownership, POPULATION, URBANISATION and economy. In the Middle Ages, much of the European landscape consisted of agricultural land or SEMI-NATURAL VEGETATION communities that were carefully managed to maximise their provision of useful resources. Arable activity was widespread, evidence for which survives as the remains of RIDGE AND FURROW extending into upland areas now suitable only for rough grazing. The cultivation of MARGINAL AREAS in the uplands may have been favoured by the MEDIAEVAL WARM PERIOD, while LAND DRAINAGE schemes allowed agriculture to extend onto former WETLANDS. The expansion of arable land seems to have increased SOIL EROSION, resulting in ALLUVIATION in river valleys across much of northern Europe. GRASSLAND was widespread, much of which was grazed by sheep to supply the wool trade, while other areas were managed as hay meadows. WOODLAND MANAGEMENT PRACTICES were also commonplace, often involving COPPICING or WOOD PASTURE. HEATHLAND expansion was encouraged by *burning*, heavy GRAZING and the practice of *plaggen*. Hunting was a popular pastime amongst the aristocracy, who created *Royal Forests* and *deer parks*, and *warrens* in which to keep *rabbits*.   *SPD/JAM*

[*See also* AGRICULTURAL HISTORY, DESERTED VILLAGES, MIGRATION AGE, SCIENTIFIC REVOLUTION, VEGETATION HISTORY]

Andersson H, Carelli P and Ersgård L (eds) (1997) *Visions of the past: Trends and traditions in Swedish Medieval archaeology*. Stockholm: Almqvist and Wicksell.

Astill G and Grant A (eds) (1988) *The countryside of Medieval England*. Oxford: Blackwell.

Astill G and Langdon J (ed.) (1997) *Medieval farming and technology*. Leiden: Brill.

Christie N (1999) Europe in the Middle Ages. In Barker G (ed.) *Companion encyclopaedia of archaeology*. London: Routledge, 1040–1076.

Lewin J (2010) Medieval environmental impacts and feedbacks: The lowland floodplains of England and Wales. *Geoarchaeology* 25: 267–311.

Rackham O (1986) *The history of the countryside*. London: Dent.

Rackham O (2003) *Ancient woodland, 2nd edition*. Colvend: Castlepoint Press.

Stamper P (2009) Landscapes of the Middle Ages: Rural settlement and manors. In Hunter J and Ralston I (eds) *The archaeology of Britain, 2nd edition*. London: Routledge, 328–347.

Williamson T (2003) *Shaping medieval landscapes*. Macclesfield: Windgather Press.

mid-ocean ridge (MOR) A chain of submarine mountains extending some 84,000 km through the Atlantic, Arctic, Indian and South Pacific OCEANS (see Figure). The mid-ocean ridges rise 1–3 km above the ABYSSAL plains, are about 1,500 km wide and are characterised by very rugged TOPOGRAPHY. In some places (e.g. Iceland), they rise above sea level. The centre or axis of the ridge is offset laterally by TRANSFORM FAULTS and experiences shallow SEISMICITY. It is a site of active basaltic VOLCANISM, and related HYDROTHERMAL activity gives rise to submarine HYDROTHERMAL VENTS at which communities of *chemosynthetic organisms* (see CHEMOSYNTHESIS) were discovered in the 1970s. According to PLATE TECTONICS theory, the mid-ocean ridge is a constructive PLATE MARGIN and SEA-FLOOR SPREADING occurs at its centre. The ridge is elevated because its young OCEANIC CRUST is hot and buoyant. By the time it is about 110 million years old, ocean crust has spread from the ridge axis, lost excess heat and subsided to the level of the abyssal plains. SUBSIDENCE is accommodated along active FAULTS, which are responsible for the rugged topography. Where spreading is rapid, the ridge is narrow and typically has a central RIFT VALLEY, for example, the *Mid-Atlantic Ridge* (*MAR*); faster spreading leads to a wider ridge, for example, *East Pacific Rise* (*EPR*).   *BTC/JP*

[*See also* HOTSPOT, IN GEOLOGY]

Cann JR, Elderfield H and Laughton AS (eds) (1999) *Mid-ocean ridges: Dynamics of processes associated with the creation of new oceanic crust*. Cambridge: Cambridge University Press.

Decker R and Decker B (1997) *Volcanoes, 3rd edition*. Basingstoke: Freeman.

Helo C, Longpre M-A, Nobumichi S et al. (2011) Explosive eruptions at mid-ocean ridges driven by CO_2-rich magmas. *Nature Geoscience* 4: 260–263.

Pedersen RB, Rapp HT, Thorseth IH et al. (2010) Discovery of a black smoker vent field and vent fauna at the Arctic mid-ocean ridge. *Nature Communications* 1: 126.

Reston TJ and Ranero CR (2011) The 3-D geometry of detachment faulting at mid-ocean ridges. *Geochemistry Geophysics Geosystems* 12: Q0AG05.

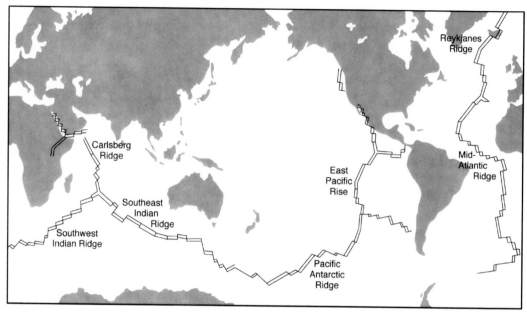

Mid-ocean ridge *The world's mid-ocean ridge system (based on Decker and Decker, 1997).*

migration The movement of an organism from one location to another. Migration may take many forms. Many species undertake *annual migration* (or *seasonal migration*), including a wide array of *insects* and *birds*, in response to seasonal climatic changes or changes in food supply. *Partial migration* is where animal POPULA-TIONS are made up of a mixture of resident and migratory individuals. While most species migrate to and from a destination, annual migration patterns may involve more than one generation, such as the one which occurs with the Monarch butterfly (*Danaus plexippus*). Migrations may also follow long-term CLIMATIC CHANGE. Following QUATERNARY ENVIRONMENTAL CHANGE, for example, spe-cies migrated in response to shifting climatic zones or suffered EXTINCTION. Such long-term *species migrations* occur individualistically because of interspecific varia-tion in migration speed and climatic TOLERANCE. Species associations and *ecosystems* did not necessarily remain intact. ANTHROPOGENIC climate change is likely to force species to migrate to remain within climatically suitable areas, but species capacity to migrate successfully may be seriously reduced by widespread habitat FRAGMENTA-TION or other ENVIRONMENTAL DEGRADATION. As a result, many species may have to be helped through *assisted migration*, but this remains controversial because of the many mistakes that have been made in the past.   *JTK*

[*See also* DISPERSAL, HUMAN MIGRATION, NOMADISM, RANGE ADJUSTMENT, TRANSHUMANCE]

Bunnefeld N, Börger L, van Moorter B et al. (2011) A model-driven approach to quantify migration patterns: Individual, regional and yearly differences. *Journal of Animal Ecology* 80: 466–476.

Chapman BB, Brönmark C, Nilsson J-Å and Hansson L-A (2011) The ecology and evolution of partial migration. *Oikos* 120: 1764–1775.

Gray LK, Gylander T, Mbogga MS et al. (2011) Assisted migration to address climate change: Recommendations for aspen reforestation in western Canada. *Ecological Applications* 21: 1591–1603.

Mila B, Smith TB and Wayne RK (2006) Postglacial population expansion drives the evolution of long-distance migration in a songbird. *Evolution* 60: 2403–2409.

Milner-Gulland EJ, Fryxell JM and Sinclair ARE (2011) *Animal migration: Synthesis.* Oxford: Oxford University Press.

Mueller T, Olson KA, Dressler G et al. (2011) How landscape dynamics link individual- to population-level movement patterns: A multispecies comparison of ungulate relocation data. *Global Ecology and Biogeography* 20: 683–694.

Newton I (2007) *The migration ecology of birds.* London: Academic Press.

Wilcove DS (2007) *No way home: The decline of the world's great animal migrations.* Washington, DC: Island Press.

Migration Age The archaeological and historical period following the ROMAN PERIOD characterised by the invasion and spread of 'barbarian' Asian and East-ern European tribes across Europe. It is also known, especially in northern Europe, as the *Germanic Iron Age*. In Britain, it lasted for about 400 years (ca AD 410–800). It was followed by the VIKING AGE.   *JAM*

Frenzel B, Reisch L and Gläser B (eds) (1994) *Evaluation of land surfaces cleared from forests by prehistoric man in Early Neolithic times and the time of migrating Germanic tribes*. Stuttgart: Gustav Fischer Verlag [Paläoklimaforschung, volume 8].

Milankovitch theory   A mathematical theory that accounts for the PERIODICITY and timing of the CLIMATIC CHANGES responsible for GLACIAL-INTERGLACIAL CYCLES in terms of ORBITAL FORCING. Milutin Milankovitch (1879–1958) was a Serbian astronomer whose theory was disputed during his lifetime. It was not until the 1960s that the theory was re-examined and later fully tested and largely accepted. The theory incorporated and expanded on the earlier ideas of J.A. Adhémar (1842) and James Croll (1875).

The Milankovitch theory, also known as the '*astronomical theory of climatic change*', suggests that glacial-interglacial cycles are linked to *quasi-periodic variations* in the Earth's orbit around the Sun, which produce changes in the quantity and/or distribution of SOLAR RADIATION received by the Earth-atmosphere system. It has been estimated that about 60 per cent of the VARIANCE in the record of global average temperature over the past million years occurs close to frequencies identified in the Milankovitch theory. This suggests that the CLIMATIC SYSTEM is not strongly chaotic but is responding in a predictable way to orbital forcing and hence that the Milankovitch theory provides a valid explanation of long-term climate changes.

The theory is based on the fact that the position and configuration of the Earth in relation to the Sun changes in a predictable way. Thus, the receipt of INSOLATION at the top of the Earth's atmosphere is equally predictable. There are three main *orbital parameters* that describe cyclical variations in the Earth's orbit with respect to the Sun: first, the 'precession of the equinoxes' (PRECESSION) with a periodicity of about 21,000 years; second, the 'obliquity of the ecliptic' (OBLIQUITY) with a periodicity of about 41,000 years; and third, the 'eccentricity of the orbit' (ECCENTRICITY) with a periodicity of about 96,000 years (see Figure 1A). These three parameters may be considered as representing the 'wobble', 'tilt' and 'stretch' of the Earth's orbit, respectively. The 'composite curve' produced by combining the variations in these three parameters (Figure 1B) is the basis of Milankovitch theory.

The distance of the Earth from the Sun varies from about 147 to 152 million miles. Precession controls the time of year when the Earth is closest to the Sun (PERIHELION). Summers receive more insolation and are relatively warm when the Earth is closer to the Sun and

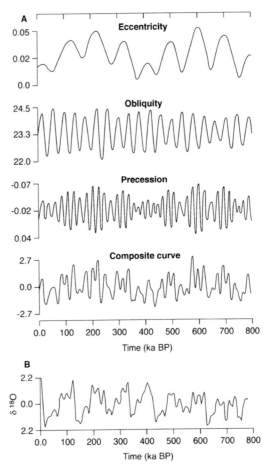

Milankovitch theory (1) *(A) Calculated variations in eccentricity, obliquity and precession over the past 800,000 years and the 'composite curve' produced by combining the three indices together after statistical normalisation; the basis of Milankovitch theory. (B) An observed (normalised and smoothed) palaeoenvironmental record of oxygen isotope variations from five marine sediment cores (Imbrie et al., 1984).*

the effects are greatest at low latitudes. We now have relatively cool summers and the conditions are conducive to ICE-SHEET growth; about 10,500 years ago, conditions were ripe for ice-sheet decay. OBLIQUITY reflects changes in the Earth's axial tilt from about 21.5° to 24.5° and controls SEASONALITY (the contrast between winter and summer conditions) with relatively low tilt leading to lower seasonality. Obliquity effects are greatest at high latitudes. ECCENTRICITY may be viewed as departures in *elipticity* of up to 6 per cent from a circular orbit (0 per cent elipticity), it accentuates the effects of precession, and it is the only one of the three parameters to affect the total heat received by the Earth.

These variations in the Earth's orbit interact and do not affect different latitudes in precisely the same way: for example, the 21,000-year precession cycle is more

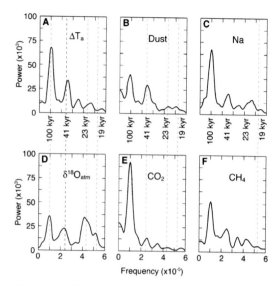

Milankovitch theory (2) *Results of spectral analysis (Blackman-Tukey technique) of time series from the Vostok ice core, Antarctica: (A) isotopic temperature of the atmosphere, (B) dust content, (C) sodium concentration, (D) $\delta^{18}O$, (E) carbon dioxide concentration and (F) methane concentration. Note correspondence of the spectral peaks with the periodicities of 100, 41, 23 and 19 ka predicted by Milankovitch theory (Petit et al. 1999).*

important at low latitudes and the 41,000-year obliquity cycle at high latitudes. GLACIAL and INTERGLACIAL conditions are nevertheless expected when appropriate combinations of the three elements coincide and lead to greatest cooling or warming of the Earth, respectively. Particularly important is the insolation received in the Northern Hemisphere at middle and high latitudes.

Ever since Milankovitch proposed the theory, the predictions of the theory have been compared with PROXY DATA from the field of PALAEOCLIMATOLOGY. The first decisive test by Hays, Imbrie and Shackleton (1976) utilised data from MARINE SEDIMENT CORES; specifically the oxygen ISOTOPE RATIO from the shells of microscopic plankton (FORAMINIFERA). The isotope ratio reflects the volume of water in the OCEANS and hence glacial-interglacial cycles. Use of SPECTRAL ANALYSIS identified dominant frequencies in the palaeoenvironmental data that were in very close agreement with the predictions of the theory. More recently, there have been further tests on data from CORAL-REEF sequences, long vegetation records reconstructed by POLLEN ANALYSIS, LOESS sequences and ICE CORES. Figure 2 shows examples from the Vostok ice core (Antarctica).

Milankovitch theory is clearly of major importance in the explanation of climatic change on timescales of 10^4–10^5 years. The timing of cold and warm stages predicted by the theory and identified in PALAEOCLIMATOLOGY provides a chronological and theoretical framework

for studies of QUATERNARY ENVIRONMENTAL CHANGE and in many ways can be viewed as a PARADIGM for the field. It must be emphasised, however, that it does not explain the onset of ICE AGES or the quantitative scale of GLOBAL COOLING during the GLACIAL EPISODES. Changes in the Earth's geography and AUTOVARIATION and other internal mechanisms involving the global CARBON CYCLE and ATMOSPHERIC DUST are likely to have played a role here. The theory does not explain why and when some periodicities are dominant. It is, nevertheless, aptly described as the 'pacemaker' or 'pulsebeat' of ICE-AGE climate.

EZ/JAM/AHP

[**See also** ENVIRONMENTAL CHANGE: HISTORY OF THE FIELD]

Adhémar JA (1842) *Revolutions de la mer* [Revolutions of the sea]. Paris: privately printed.

Berger A (1988) Milankovitch theory and climate. *Reviews of Geophysics* 26: 624–657.

Berger A, Imbrie J, Hays JD et al. (eds) (1984) *Milankovitch and climate*. Dordrecht: Reidel.

Berger A and Loutre MF (1994) Astronomical forcing through geological time. In De Boer PL and Smith DG (eds) *Long-term climatic variations, data and modelling*. Dordrecht: Reidel, 107–151.

Berger A and Yin Q (2012) Astronomical theory and orbital forcing. In Matthews JA, Bartlein PJ, Briffa KR et al. (eds) *The SAGE handbook of environmental change, volume 1*. London: Sage, 405–425.

Croll J (1875) *Climate and time in their geological relations*. New York: Appleton.

EPICA community members (2004) Eight glacial cycles from an Antarctic ice core. *Nature* 429: 623–628.

Ganopolski A, Calov C and Claussen M (2010) Simulation of the last glacial cycle with a coupled climate ice-sheet model of intermediate complexity. *Climate of the Past* 6: 229–244.

Hays JD, Imbrie J and Shackleton NJ (1976) Variations in the Earth's orbit: Pacemaker of the Ice Ages. *Science* 194: 1121–1132.

Hooghiemstra H, Melice JL, Berger A and Shackleton NJ (1993) Frequency spectra and palaeoclimatic variability of the high-precision 30-1450 ka Funza I pollen record (Eastern Cordillera, Colombia). *Quaternary Science Reviews* 12: 141–156.

Imbrie J, Berger A and Shackleton NJ (1993) Role of orbital forcing: A two-million-year perspective. In Eddy JA and Oescher H (eds) *Global changes in the perspective of the past*. New York: Wiley, 263–277.

Imbrie J, Hays JD, Martinson DG et al. (1984) The orbital theory of Pleistocene climate: Support from a revised chronology of the marine $\delta^{18}O$ record. In Berger A, Imbrie J, Hays J et al. (eds) *Milankovitch and climate*. Dordrecht: Reidel, 269–306.

Imbrie J and Imbrie KP (1979) *Ice Ages: Solving the mystery*. London: Macmillan.

Kutzbach JE and Street-Perrott FA (1985) Milankovitch forcing of fluctuations in the level of tropical lakes from 18-0 k yr BP. *Nature* 317: 1301–1304.

Milankovitch M (1941) *Canon of insolation and the Ice-Age problem*. Belgrade: Royal Serbian Academy [Jerusalem: Israel Program for Scientific Translations, 1968].

Petit JR, Jouzel J, Raynaud D et al. (1999) Climate and atmospheric history of the past 420,000 years from the Vostok ice core, Antarctica. *Nature* 399: 429–436.

Raymo M, Lisiecki LE and Nisancioglu H (2006) Plio-Pleistocene ice volume, Antarctic climate, and the global $\delta^{18}O$ record. *Science* 313: 492–495.

Wunsch C (2004) Quantitative estimate of the Milankovitch-forced contribution to observed Quaternary climate change. *Quaternary Science Reviews* 23: 1001–1012.

military impacts on environment

Military forces have the ability to exert substantial impacts on air, land and water during peacetime and conflict. It is anticipated that a modern full-scale conflict would be an environmental CATASTROPHE, as ENVIRONMENTAL PROTECTION is not a priority for warring troops. POLLUTION arising from the ignition of Kuwaiti oil wells during the Gulf War is a notable example, while the concept of a NUCLEAR WINTER was widely discussed during the Cold War. Peacetime activities, such as training exercises and the maintenance of infrastructure and equipment, also contribute to ENVIRONMENTAL DEGRADATION, DISTURBANCE and pollution. Military training areas are often very extensive and can occupy a significant proportion of a nation's land resource. The use of heavy, armoured tracked vehicles in training exercises can be particularly damaging to soil and vegetation and impacts are akin to those of severe TRAMPLING. Other damaging or disturbing consequences of training include the formation of CRATERS by explosives, discharge of WASTE from naval vessels and NOISE POLLUTION from aircraft and live firing. North Atlantic Treaty Organization (NATO) policy and the progressive loss of immunity from national conservation legislation have ensured that ENVIRONMENTAL MANAGEMENT is given greater importance in the military sector. A positive consequence of the military ownership of large training areas is that agriculture, urban development and public access have been restricted. Thus, some of these sites have become nationally or even internationally renowned as havens for rare species, habitats and archaeological features. *MLW*

[*See also* ENVIRONMENTAL SECURITY, OIL FIRE]

Al-Hassan JM (1992) *The Iraqi invasion of Kuwait: An environmental catastrophe*. Kuwait: Jassim M. Al-Hassan.

Brauer J (2009) *War and nature: The environmental consequences of war in a globalized world*. Lanham, MD: AltaMira Press.

Cuddy SM (1990) Modelling the environmental effects of training on a major Australian army base. *Mathematics and Computers in Simulation* 32: 83–88.

Machlis GE, Hanson T, Špirić Z and McKendry JE (eds) (2011) *Warfare ecology*. Dordrecht: Springer.

Mather JD and Rose EPF (eds) (2012) *Military aspects of hydrogeology*. Bath: Geological Society.

Westing AH (1992) Protected natural areas and the military. *Environmental Conservation* 19: 343–347.

Wilson SD (1986) The effects of tank traffic on prairie: A management model. *Environmental Management* 12: 397–403.

Woodward R (2004) *Military geographies*. Oxford: Blackwell.

mineral

A naturally occurring inorganic solid with a chemical composition that is fixed or varies within a limited range, and a CRYSTALLINE structure. Minerals are the building blocks of rocks. Some geological materials that do not strictly conform to this definition are commonly described as minerals or, more correctly, *mineraloids* (e.g. AMBER and AMORPHOUS silica such as *opal*). The term *mineral deposit* is commonly used to refer to materials of economic value obtained from the ground, including OIL, GAS and GROUNDWATER. *GO*

[*See also* CARBONATE, EVAPORITE, FELDSPAR, IGNEOUS ROCKS, METAMORPHIC ROCKS, MINERALISATION, MINEROGENIC SEDIMENT, QUARTZ, SEDIMENTARY ROCKS, SILICATE MINERALS]

Deer WA, Howie RA and Zussman J (1992) *An introduction to rock forming minerals, 2nd edition*. Harlow: Longman.

Gaines RV, Skinner HCW, Foord E et al. (1997) *Dana's new mineralogy, 8th edition*. Chichester: Wiley.

Nesse WD (2011) *Introduction to mineralogy*. New York: Oxford University Press.

mineral cycling

An important aspect of *ecological systems* at local to global scales. Knowledge of the nature and quantity of minerals transferred between compartments and the rate of mineral cycling within and between *ecosystems* is essential for understanding the development, functioning, maintenance and use of all types of *natural ecosystems* and *anthropogenic ecosystems*. A simple model of mineral cycling in three of the world's major forested BIOMES is summarised in the Figure. This shows that progressively smaller quantities of minerals are stored in the BIOMASS compartment of the higher-latitude forests, which store larger quantities of minerals in the LITTER compartment. The latter is largest in the BOREAL FOREST (C), where a relatively small proportion of the minerals is transferred to the SOIL compartment and made available for the trees because litter DECOMPOSITION is slowest there, especially under winter conditions. Mineral cycling is most rapid in the TROPICAL RAIN FOREST (A): this means that the rain forest soils are exhausted within a few years after DEFORESTATION (a feature that accounts for the success of traditional *slash-and-burn* in contrast to the demise of many systems of permanent CULTIVATION). *JAM*

[*See also* AGRICULTURE, BIOGEOCHEMICAL CYCLES, ENERGY FLOW, NUTRIENT]

Gersmehl PJ (1976) An alternative biogeography. *Annals of the Association of American Geographers* 66: 223–241.

Golley FB (1982) *Mineral cycling in a tropical moist forest ecosystem.* Athens: University of Georgia Press.

Marschner P and Rengel Z (eds) (2007) *Nutrient cycles in terrestrial ecosystems.* Berlin: Springer.

Matthews JA and Herbert DT (2008) *Geography: A very short introduction.* Oxford: Oxford University Press.

Sardans J and Penuelas J (2007) Drought changes phosphorus and potassium accumulation patterns in an evergreen Mediterranean forest. *Functional Ecology* 21: 191–201.

Schlesinger WH (1997) *Biogeochemistry: An analysis of global change, 2nd edition.* San Diego, CA: Academic Press.

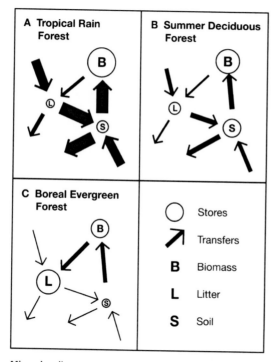

Mineral cycling *Models of mineral cycling in (A) tropical rain forest, (B) summer deciduous forest and (C) boreal evergreen forest (Matthews and Herbert, 2008; based on Gersmehl, 1976). Note that transfers (arrow thickness) represent the proportion of the source store (circle size) that is cycled each year.*

mineral magnetism In measuring the macroscopic response of a material to MAGNETIC FIELD, mineral magnetism provides evidence primarily about the iron-bearing MINERALS found in *rocks*, SEDIMENTS, SOILS and DUSTS. The data may, therefore, enable identification of the different types of minerals present, estimation of their concentrations and relative inputs,

classification of different types of materials, identification of their formation and/or processing and the determination of sediment PROVENANCE. Due to its non-destructive nature and its broad applicability in a range of environments, mineral magnetism or *environmental magnetism* is ideal as a tool in RECONNAISSANCE SURVEYS tool where a large sample set is needed to characterise an environment or to reveal changes in input, source and/or environmental conditions. Indeed, it is rare to find examples where mineral magnetism is not used in parallel with other palaeoenvironmental or analytical techniques.

Both field- and laboratory-based measurements of magnetic susceptibility and MAGNETIC REMANENCE can be used to investigate the magnetic properties of environmental samples, for example, DIAMAGNETISM, PARAMAGNETISM and FERROMAGNETISM. In addition, measurements of ISOTHERMAL REMANENT MAGNETISATION (IRM) and ANHYSTERETIC REMANENT MAGNETISATION (ARM) may be used to determine the *magnetic mineralogy*, for example, the relative contributions from magnetic minerals such as MAGNETITE or HAEMATITE, or the MAGNETIC GRAIN SIZE of a particular sample (see Table).

First-order Reversal Curve (*FORC*) diagrams (Roberts et al., 2000) obtained from measurements of MAGNETIC HYSTERESIS can be used to obtain more detailed information about the magnetic minerals in a sample than the more traditional interpretational parameters described in the Table. For example, *Single Domain* (*SD*), *Multidomain* (*MD*) and *Superparamagnetic* (*SP*) grains produce diagnostic characteristics on contour plots of hysteresis parameters. In addition, the mineralogy of magnetic minerals can be characterised further by *thermomagnetic analysis* of the dependence of MAGNETIC SUSCEPTIBILITY and/or MAGNETISATION on temperature over the range of liquid helium to about 800°C and above (Dekkers, 1997).

Mineral magnetic measurements have been used successfully in the study of lacustrine, marine and riverine sediments, with particular emphasis on temporal changes in sediment supply and provenance (e.g. Hatfield and Maher, 2009). Indeed, such studies have successfully investigated the influence of both climate and humans on *lake-sediment budgets*, catchment SEDIMENT YIELDS and WATER QUALITY (e.g. Gedye et al., 2000; Oldfield, 1991). The use of magnetic properties to differentiate the relative significance of parent material, time, climate, drainage, relief and organisms in controlling *soil formation* underpins research on obtaining long-term palaeoclimatic data from the mineral magnetic properties of LOESS STRATIGRAPHY, ICE CORES and MARINE SEDIMENT CORES, which can be interpreted in the context of ORBITAL FORCING and/or major temporal and spatial changes in world climate patterns (e.g. Bloemendal and Liu, 2005; Maher and

Magnetic Property	Interpretation
Volumetric magnetic susceptibility (κ) and mass-specific magnetic susceptibility (χ) (measured at high and low frequency).	A first-order measure of the amount of ferro(i)magnetic material (e.g. magnetite). Susceptibility is enhanced by superparamagnetic (SP) magnetite (<0.03 μm). When the concentration of ferrimagnetic material is low, susceptibility responds to antiferromagnetic (e.g. haematite), paramagnetic and diamagnetic materials.
Frequency dependent magnetic susceptibility, χfd (or κfd), is the ratio of low-frequency to high-frequency susceptibility difference expressed as a percentage of the low-frequency susceptibility.	χfd indicates the presence of ultra-fine ferro(i)magnetic (superparamagnetic) material, produced by bacteria or by chemical processes mainly in soil. SP material in high concentrations complicates the grain-size interpretations using the interparametric ratios such as $\chi ARM/\chi$.
Isothermal remanent magnetisation (IRM), commonly expressed as a saturation IRM (or SIRM) when a field of 1 T is used. A backfield IRM is that magnetisation acquired in a reversed field after SIRM acquisition.	SIRM depends primarily on the concentration of magnetic, mainly ferrimagnetic, material.
Anhysteretic remanent magnetisation (ARM), commonly expressed as susceptibility of ARM (χARM) when normalised using the biasing DC field.	χARM is a measure of the concentration of ferrimagnetic material but is also strongly grain-size dependent. χARM is highly selective of stable single-domain (SSD) ferrimagnetic grains in the range of 0.03–0.1 μm.
The 'hard' IRM (HIRM) is derived by imparting a backfield IRM, typically 100–300 mT, on a sample previously given an SIRM.	HIRM is a measure of the concentration of magnetic material with a higher coercivity than the backfield, which commonly gives the concentration of antiferromagnetic (e.g. haematite and goethite) or very fine-grained ferrimagnetic grains.
The 'S ratio' is derived by imparting a backfield of −100 mT (or −300 mT) on a sample previously given an SIRM from the ratio IRM/SIRM: S = (−IRM−100)/(SIRM1000) or (−IRM−300)/(SIRM1000)	The S ratio can be used to estimate magnetic mineralogy. Values close to −1 indicate lower coercivity and a ferrimagnetic dominant mineralogy (e.g. magnetite), whilst values closer to zero indicate a higher coercivity, possibly an antiferromagnetic mineralogy (e.g. haematite).
$\chi ARM/\chi$ (or $KARM/\kappa$)	Providing the magnetic mineralogy is dominated by magnetite and grain size above SP grains, $\chi ARM/\chi$ varies inversely with magnetic grain size.
SIRM/χ (or SIRM/κ)	Providing the magnetic mineralogy is dominated by magnetite and grain size above SP grains, SIRM/χ varies inversely with magnetic grain size.
$\chi ARM/SIRM$ (or /$\kappa ARM/SIRM$)	Providing the magnetic mineralogy is dominated by magnetite, $\chi ARM/SIRM$ increases with decreasing magnetic grain size.

Mineral magnetism *Mineral magnetic parameters and their interpretation (Dekkers, 2007; Peters and Dekkers, 2003; Stoner et al., 1996; Verosub and Roberts, 1995).*

Thompson, 1999; Stoner et al., 1996; Watkins et al., 2007). The more recent impacts of humans in the form of pollution have also been investigated in MIRES, lake, coastal and urban sites (e.g. Hansard et al., 2011; Oldfield, 1991; Shu et al., 2001). Since magnetic minerals are sensitive to early diagenesis processes, mineral magnetism is also useful in the biogeochemical study of the IRON CYCLE, including bacterial MINERALISATION of iron oxides and sulfides called *magnetosomes* (e.g. Evans and Heller, 2003). *AJP/WZ*

[*See also* NATURAL REMANENT MAGNETISM (NRM)]

Bloemendal J and Liu X (2005) Rock magnetism and geochemistry of two plio-pleistocene Chinese loess-palaeosol sequences: Implications for quantitative palaeoprecipitation reconstruction. *Palaeogeography, Palaeoclimatology, Palaeoecology* 226: 149–166.

Dekkers MJ (1997) Environmental magnetism: An introduction. *Geologie and Mijnbouw* 76: 163–182.

Dekkers MJ (2007) Magnetic proxy parameters. In Gubbins D and Herrero-Bervera E (eds) *Encyclopedia of geomagnetism and palaeomagnetism*. Berlin: Springer, 525–534.

Evans ME and Heller F (2003) *Environmental magnetism: Principles and applications of enviromagnetics*. San Diego, CA: Academic Press.

Gedye SJ, Jones RT, Tinner W et al. (2000) The use of mineral magnetism in the reconstruction of fire history: A case study from Lago di Origlio, Swiss Alps. *Palaeogeography, Palaeoclimatology, Palaeoecology* 164: 101–110.

Hansard R, Maher BA and Kinnersley R (2011) Biomagnetic monitoring of industry-derived particulate pollution. *Environmental Pollution* 159: 1673–1681.

Hatfield RG and Maher BA (2009) Fingerprinting upland sediment sources: Particle size-specific magnetic linkages between soils, lake sediments and suspended sediments. *Earth Surface Processes and Landforms* 34: 1359–1373.

Liu Q, Torrent J, Maher BA et al. (2005) Quantifying grain size distribution of pedogenic magnetic particles in Chinese loess and its significance for pedogenesis. *Journal of Geophysical Research* 110: B11102.

Maher BA and Thompson R (1999) *Quaternary climates, environments and magnetism*. Cambridge: Cambridge University Press.

Oldfield F (1991) Environmental magnetism: A personal perspective. *Quaternary Science Reviews* 10: 73–85.

Peters C and Dekkers MJ (2003) Selected room temperature magnetic parameters as a function of mineralogy, concentration and grain size. *Physics and Chemistry of the Earth* 28: 659–667.

Roberts AP, Pike CR and Verosub KL (2000) First-order reversal curve diagrams: A new tool for characterizing the magnetic properties of natural samples. *Journal of Geophysical Research* 105(B12): 28461–28475.

Shu J, Dearing JA, Morse AP et al. (2001) Determining the sources of atmospheric particles in Shanghai, China, from magnetic and geochemical properties. *Atmospheric Environment* 35: 2615–2625.

Stoner JS, Channell JET and Hillaire-Marcel C (1996) The magnetic signature of rapidly deposited detrital layers from the deep Labrador Sea: Relationship to North Atlantic Heinrich layers. *Palaeoceanography* 11: 309–325.

Thompson R and Oldfield F (1986) *Environmental magnetism*. London: Allen and Unwin.

Verosub K and Roberts AP (1995) Environmental magnetism: Past, present and future. *Journal of Geophysical Research* 100: 2175–2192.

Walden J, Oldfield F and Smith JP (eds) (1999) *Environmental magnetism: A practical guide*. London: Quaternary Research Association.

Watkins SJ, Maher BA and Bigg GR (2007) Ocean circulation at the Last Glacial Maximum: A combined modeling and magnetic proxy-based study. *Paleoceanography* 22: PA2204.

mineral resources   A loosely defined term covering MINERALS and rocks that are of value to society, either as obtained from the ground through *mining* and QUARRYING, or after processing. ENERGY RESOURCES are usually excluded. Mineral resources include metal ORE deposits, *gemstones*, *construction materials* (e.g. sand, gravel, clay and building stone) and *industrial minerals* such as GYPSUM, LIMESTONE and PHOSPHORITE.   *GO*

[*See also* NON-RENEWABLE RESOURCES]

Evans AM (1997) *An introduction to economic geology and its environmental impact*. Oxford: Blackwell.

Moon CJ, Whateley MKG and Evans AM (eds) (2006) *Introduction to mineral exploration*. Oxford: Blackwell.

mineralisation   (1) The transformation of *organic substances* into *inorganic* ones. In SOIL, it is the process by which NUTRIENT elements (e.g. *nitrogen* and PHOSPHORUS (P)) are transformed from immobile organic forms (unusable by plants) to mobile inorganic ions that are available for root uptake. In *biomineralisation*, the process is itself carried out by organisms such as DIATOMS and FORAMINIFERA. In BIOREMEDIATION, mineralisation occurs during the conversion of hazardous organic POLLUTANTS and WASTE into harmless inorganic compounds. It may also occur in this sense during the formation of FOSSILS.

(2) Mineralisation is used more generally in GEOLOGY for the PRECIPITATION of MINERALS, especially in relation to useful minerals. Examples include the formation of ORE minerals of value to society, often in *veins*, and the replacement of one mineral by another during the burial and DIAGENESIS of SEDIMENTARY DEPOSITS.   *JAM*

[*See also* BIOGEOCHEMICAL CYCLES, LITHIFICATION, NITROGEN CYCLE, TAPHONOMY]

Craig JR, Vaughan DJ and Skinner BJ (2000) *Resources of the Earth: Origin, use, and environmental impact*. New York: Prentice Hall.

Dove PM, De Yoreo JJ and Weiner S (eds) (2003) *Biomineralization*. Washington, DC: Mineralogical Society of America and Geochemical Society.

Nesse WD (2011) *Introduction to mineralogy*. New York: Oxford University Press.

minerogenic sediment   Used especially in contradistinction to ORGANIC SEDIMENT, minerogenic sediment is either (1) a sediment composed wholly or dominantly of mineral particles or (2) the mineral component of a sediment.   *JAM*

[*See also* CLASTIC]

Petterson G, Renberg I, Sjöstedt-de Luna S et al. (2010) Climatic influence on the inter-annual variability of late-holocene minerogenic sediment supply in a boreal forest catchment. *Earth Surface Processes and Landforms* 35: 390–398.

minimum critical size of ecosystems One of the most fundamental challenges for developing effective strategies to conserve BIODIVERSITY is to determine the minimum size of NATURE RESERVE necessary to maintain POPULATIONS of particular ENDANGERED SPECIES. One of the earliest and most insightful attempts to empirically estimate the 'minimum critical size of ecosystems' was initiated by T.E. Lovejoy and colleagues during the early 1980s. In the TROPICAL RAIN FOREST of Brazil, just north of Manaus, Lovejoy's team carved out fragments of rain forest that varied in size from 1 to 1,000 hectares. By following the fate of populations of NATIVE SPECIES over the past two decades, these scientists provided many important insights into the effects of *habitat reduction* and FRAGMENTATION on BIODIVERSITY. *MVL*

[*See also* HABITAT LOSS, SINGLE LARGE OR SEVERAL SMALL RESERVES (SLOSS DEBATE)]

Gurd DB, Nudds TD and Rivard DH (2001) Conservation of mammals in eastern North American wildlife reserves: How small is too small? *Conservation Biology* 15: 1355–1363.

Laurance WF and Bierregaard Jr RO (1997) *Tropical forest remnants: Ecology, management and conservation of fragmented communities.* Chicago: University of Chicago Press.

Lovejoy TE and Bierregaard Jr RO (1990) Central Amazonian forests and the Minimum Critical Size of Ecosystems Project. In Gentry AH (ed.) *Four neotropical rainforests.* New Haven, CT: Yale University Press, 60–71.

Tjorve E (2010) How to resolve the SLOSS debate: Lessons from species-diversity models. *Journal of Theoretical Biology* 264: 604–612.

minimum viable population The minimum number of individuals that will ensure the survival of a POPULATION, not just in the short term, but also in the long term, often cast in terms such as '95 per cent probability of persistence for 100 or for 1,000 years'. While some have advocated that a minimum viable population size of 5,000 individuals is required, regardless of taxonomy, there is no evidence to support such a generalisation, and the number varies by species and by situation. *RJW/JLI*

[*See also* EFFECTIVE POPULATION SIZE, POPULATION VIABILITY ANALYSIS (PVA)]

Brook BW, Traill LW and Bradshaw CJA (2006) Minimum viable population sizes and global extinction risk are unrelated. *Ecology Letters* 9: 375–382.

Flather CH, Hayward GD, Beissinger SR and Stephens PA (2011) Minimum viable populations: Is there a 'magic number' for conservation practitioners? *Trends in Ecology and Evolution* 26: 307–316.

mining and mining impacts Extraction of mineral resources from underground (*deep mining*) or near the surface (*opencast* or *strip mining*). Surface mines and *quarries* (a term sometimes reserved for the mining of building stone) are known from prehistory, but organised underground mining for salt, amber, metals and other materials was an activity developed by early civilisations. Underground mining is more hazardous than surface mining, but the latter generally presents greater land instability, LAND RECLAMATION and RESTORATION problems. These include the volume, calibre, chemical composition and instability of dumped mine WASTE and LAND SUBSIDENCE. Major potential POLLUTION problems include ACID MINE DRAINAGE and the toxicity of GROUNDWATER (particularly associated with HEAVY METALS) seeping from mine waste tips long after the mines have been abandoned. *Hydraulic mining*—the use of high-pressure jets to remove ore, transport and process—can also cause serious WATER POLLUTION in streams and rivers.

Probably the greatest environmental impact of mining has been the exploitation of FOSSIL FUELS since the mid-eighteenth century, especially COAL and then PETROLEUM, leading to elevated atmospheric CARBON DIOXIDE levels. There are also many diverse specific local impacts from, for example, ASBESTOS mining (especially *blue asbestos*), PHOSPHORITE mining (especially *guano* on some oceanic islands), SEDIMENTS from mining affecting CORALS and CORAL REEFS and OIL FIRES (e.g. those started deliberately during the invasion of Kuwait).

In the past, mining has mainly been terrestrial, although there is some dredging for gold, diamonds and tin on continental shelves (so far there has been no removal of PHOSPHORITE nodules or GAS HYDRATES from deep ocean sites). Mining by means of boreholes is common beneath the oceans to 500 and even 1,000 m depth and has resulted in spectacular OIL SPILLS and POLLUTION through borehole mud (lubricating compounds) release. *CJB/JAM*

[*See also* CADMIUM (Cd), COAL, COPPER (Cu), EXTRACTIVE INDUSTRY, LEAD (Pb), MERCURY (Hg), QUARRYING]

Bell FG and Donelly LJ (2006) *Mining and its impact on the environment.* New York: Taylor and Francis.

Carlsom CL and Swisher JH (eds) (1987) *Innovative approaches to mined land reclamation.* Carbondale: Southern Illinois University Press.

Craddock P (1996) *Early metal mining and production.* Washington, DC: Smithsonian Institution.

Graf WL (1979) Mining and channel response. *Annals of the Association of American Geographers* 69: 262–275.

Gregory CE (2001) *A concise history of mining, 2nd edition.* New York: Taylor and Francis.

Hester RE and Harrison RM (eds) (1994) *Mining and its environmental impact.* London: Royal Society of Chemistry.

Kelly M, Allison WJ, Garmon AR and Symon CJ (1988) *Mining and the freshwater environment.* Amsterdam: Elsevier.

Latifovic R (2009) *Mining and the environment: Satellite remote sensing in assessing the environmental impact of large-scale surface mining.* Saarbrücken: VDM Verlag.

Ripley EA and Redmann RE (1996) *Environmental effects of mining.* Delray Beach, FL: St. Lucie Press.

Shepherd R (1993) *Ancient mining.* Dordrecht: Kluwer.

Spitz K and Trudinger J (2008) *Mining and the environment: From ore to metal.* Boca Raton, FL: CRC Press.

Miocene   An EPOCH of the NEOGENE period, from 23.0 to 5.3 million years ago, accounting for almost 90 per cent of Neogene time (see PLIOCENE). The Miocene is associated with a warm climate (see GREENHOUSE CONDITION), which subsequently cooled, and the establishment of *grasslands* and *kelp forests.*   GO

[*See also* GEOLOGICAL TIMESCALE, TERTIARY]

Galeotti S, von der Heydt A, Huber M et al. (2010) Evidence for active El Niño Southern Oscillation variability in the Late Miocene greenhouse climate. *Geology* 38: 419–422.

Kürschner WM, Kvacek Z and Dilcher DL (2008) The impact of Miocene atmospheric carbon dioxide fluctuations on climate and the evolution of terrestrial ecosystems. *Proceedings of the National Academy of Sciences* 105: 449–453.

Mutti M, Piller WE and Betzler C (eds) (2010) *Carbonate systems during the Oligocene-Miocene climatic transition.* Chichester: Wiley-Blackwell.

mires   Used especially in Europe to denote peat-forming environments, mires include *bog,* FEN, SWAMP and CARR, and may also include *moor.* 'Mire' is all-embracing: it includes the PEAT AND PEATLAND and the peat-forming VEGETATION. In this regard, it is a useful, though not precise, term; in the United States, the term *peatland* is more often used. *Diplotelmic* (two-layered) mires have an anaerobic CATOTELM below the ACROTELM, which is AEROBIC.

CLASSIFICATION of mires can be based on one or more of a range of features. Moore (1984: 2) considered these under seven headings: (1) floristics, (2) vegetation structure and physiognomy, (3) morphology, (4) hydrology, (5) stratigraphy, (6) chemistry and (7) peat characteristics, but noted that even such a classification of taxonomic criteria is far from perfect, for many of these features are themselves closely interrelated. This wide range of criteria has resulted in a diverse range of classifications, from detailed classifications based principally on PHYTOSOCIOLOGY, such as those used in Central Europe and in Ireland, to the more generalised classification based on nutrient source, as is sometimes used in parts of northwest Europe, particularly by palaeoecologists (e.g. OMBROTROPHIC MIRE or RHEOTROPHIC MIRE). The difficulty with phytosociological classifications of mires is that they are based on the current vegetation. This can be a rather static view of what is inherently a dynamic

system. Such classifications might not always acknowledge a mire's ontogeny, or what might be growing there naturally but for the considerable human influence over recent centuries or decades (as, e.g. the underrepresentation of carr habitats, the overrepresentation of depauperate BLANKET MIRE and the relative emphasis given to *lowland wet heath* in Britain's National Vegetation Classification of mires and HEATHLANDS) or what might be able to grow there were the climate to shift perceptibly. It is clear from the analysis of PEAT STRATIGRAPHY that the surface vegetation of some mires has responded to climatic changes in the past and would be expected to do so in the future. A simpler classification, which was formerly used in Britain, is one based on *trophic status:* oligotrophic bogs and EUTROPHIC fens. However, 'poor fen' vegetation is at best *mesotrophic,* whereas some valley and basin mires are mesotrophic but support 'bog' rather than 'fen' communities.

Mires are NATURAL ARCHIVES of ENVIRONMENTAL HISTORY and can be examined using a range of palaeoecological techniques, including POLLEN ANALYSIS, PLANT MACROFOSSIL ANALYSIS, analysis of testate amoebae (RHIZOPOD ANALYSIS), STABLE ISOTOPE ANALYSIS and determination of PEAT HUMIFICATION, to give various PROXY RECORDS of vegetation or climatic history, which assist in reconstructing the environmental changes of the Late Quaternary.

The major *peat* formers in circum-boreal mires of the Northern Hemisphere are the bog mosses (*SPHAGNUM* spp.), ericaceous SHRUBS and GRAMINOIDS, including members of the *Cyperaceae* (sedge) and *Poaceae* (grass) families. Further south, and particularly in Southern Hemisphere mires, a wider range of taxa may be major peat formers, including members of the *Restionaceae.* The peat of *tropical bog forests* may largely be composed of tree remains.

In some parts of northwest Europe, the deliberate LAND DRAINAGE of mires, the cutting of peat for fuel and for use in HORTICULTURE and the AFFORESTATION of bogs have led to a rapid and catastrophic loss of mire HABITATS. The loss of RAISED MIRE habitats through drainage was particularly great in The Netherlands—losses that accumulated over recent centuries and were paralleled in Britain to a lesser degree—but in the past 50 years, mechanised *peat cutting* has led to increased loss of raised mire habitats in both Britain and Ireland, such that mire CONSERVATION and RESTORATION have become a major conservation issue, and there are growing concerns for mire habitats even in countries with abundant and extensive mires, such as Canada, Estonia and Finland. The fragility of mire vegetation is well recognised and there is increasing concern over the loss of WETLAND habitats worldwide.

Although geologically, most north-temperate bogs and fens are relatively young, having developed within the HOLOCENE, some TROPICAL PEATLANDS were initiated earlier. The present state of intact mires is the culmination of thousands of years of development, in which they may have passed through several stages of ECOLOGICAL SUCCESSION. The ontogeny of mires therefore needs to be considered carefully in plans for mire CONSERVATION and MANAGEMENT, particularly in those cases where restoration of damaged and *cut-over mires* is attempted.   FMC

Charman D (2002) *Peatlands and environmental change.* Chichester: Wiley-Blackwell.

Evans M and Warburton J (2010) *Geomorphology of upland peat: Erosion, form and landscape change.* Chichester: Wiley.

Gore AJP (ed.) (1983) *Mires: Bog, fen, swamp and moor.* Amsterdam: Elsevier.

Grootjans A, Iturraspe R, Lanting A et al. (2010) Ecohydrological features of some contrasting mires in Tierra del Fuego, Argentina. *Mires and Peat* 6: Article 1.

Heathwaite AL and Gottlich KL (eds) (1993) *Mires: Process, exploitation and conservation.* Chichester: Wiley.

Keddy PA (2010) *Wetland ecology: Principles and conservation, 2nd edition.* Cambridge: Cambridge University Press.

Moore PD (ed.) (1984) *European mires.* London: Academic Press.

Rodwell JS (ed.) (1991) *British plant communities, volume 2: Mires and heaths.* Cambridge: Cambridge University Press.

misfit meander   A meander apparently too small for its valley or drainage basin. Together with *misfit streams*, misfit meanders were studied extensively by G.H. Dury, who argued that they were globally widespread and reflected substantial postglacial shifts in *hydrological regime*, driven by CLIMATIC CHANGE. During cool, wetter, PLUVIAL episodes, Dury estimated by RETRODICTION from HYDRAULIC GEOMETRY relationships for valley dimensions that discharges up to 100 times the present-day values were generated, which carved large '*valley meanders*' (see INCISED MEANDERS). When rainfalls and temperatures assumed present-day levels, ALLUVIATION of valley meanders and 'shrinkage' of the river channels occurred as an adjustment to the much reduced discharge. These ideas proved highly controversial, especially in the 1970s, with some arguing that valley meanders were cut (1) by active estuarine processes at times of higher sea level to create *tidal palaeomorphs*, (2) during times of higher run-off caused by frozen ground in a PERIGLACIAL ENVIRONMENT, (3) under a *glaciofluvial regime* (see GLACIOFLUVIAL LANDFORMS) characterised by a much greater seasonal concentration of flow (MELTWATER) in the summer months, (4) at times of higher discharge before RIVER CAPTURE had 'beheaded' part of the contributing catchment or (5) by catastrophic glacial meltwater releases (see JÖKULHLAUP).   DML

[*See also* CHANNEL CHANGE, CHANNEL PATTERN, UNDERFIT STREAM]

Dury GH (1983) Osage-type underfitness on the river Severn near Shrewsbury, Shropshire, England. In Gregory KJ (ed.) *Background to palaeohydrology.* Chichester: Wiley, 399–412.

Dury GH (1985) Attainable standards of accuracy in the retrodiction of palaeodischarge from former channel dimensions. *Earth Surface Processes and Landforms* 10: 205–213.

Williams GP (1988) Paleofluvial estimates from dimensions of former channels and meanders. In Baker VR, Kochel RC and Patton PC (eds) *Flood geomorphology.* Chichester: Wiley, 321–334.

Mississippian   Formerly used in North America to define a PERIOD of geological time, or elsewhere a subperiod of the CARBONIFEROUS, 'Mississippian' is now agreed upon to define the earlier EPOCH of the Carboniferous period (see PENNSYLVANIAN). The Mississippian lasted from 359 to 318 million years ago and does not exactly correspond with the former use of "Early Carboniferous" in Europe. Marine *transgressions* early in the Mississippian led to flooding of CONTINENTAL MARGINS by CONTINENTAL SHELF seas, represented by extensive developments of marine LIMESTONE in northern Europe and North America (e.g. the 'Carboniferous Limestone' of Britain).   GO

[*See also* GEOLOGICAL TIMESCALE]

Falcon-Lang HJ (2004) Early Mississippian lycopsid forests in a delta-plain setting at Norton, near Sussex, New Brunswick, Canada. *Journal of the Geological Society of London* 161: 969–981.

Kammer TW and Ausich WI (2006) The "Age of Crinoids": A Mississippian biodiversity spike coincident with widespread carbonate ramps. *Palaios* 21: 238–248.

mitigation   Technological and policy measures to control or soften the IMPACTS of environmental change by focusing on ameliorating the causes. The term is commonly used in the context of GLOBAL ENVIRONMENTAL CHANGE (e.g. CLIMATIC CHANGE) and in relation to NATURAL HAZARDS. However, in many cases, mitigation is not possible because the change is not adequately perceived, is already under way or cannot be controlled. Examples of mitigation include reducing EMISSIONS of GREENHOUSE GASES, introducing CARBON TAXATION, developing sources of ALTERNATIVE ENERGY, promoting MATERIALS SUBSTITUTION and RECYCLING and slowing DEFORESTATION.

Mitigation has tended to receive greater attention than ADAPTATION from scientists and policymakers. Adaptation and mitigation can take place before, during but more often after environmental change through mechanisms such as learning (*individual learning, social learning, policy learning*). Mitigation efforts are frequently distorted by politics, procrastination, corruption and bureaucracy, so they can be piecemeal, delayed, unsustainable and unfair. Mitigation is likely to be most effective and costs more manageable if action is undertaken gradually and as soon as a challenge is perceived. That was a core message of the 2007 *Stern Review* presented to the UK Parliament: that if effective adaptation and mitigation of climate change are delayed, there will be very serious economic, environmental and social impacts later.   *CJB*

Davoudi S, Crawford J and Mehmood A (eds) (2009)
  *Planning for climate change: Strategies for mitigation
  and adaptation for spatial planners*. London: Earthscan.
Metz B, Davidson OR, Bosch PR et al. (eds) (2007) *Climate
  change 2007: Mitigation of climate change* [Contribution
  of Working Group III to the Fourth Assessment Report
  of the Intergovernmental Panel on Climate Change].
  Cambridge: Cambridge University Press.
Parry M (2009) Closing the loop between mitigation,
  impacts and adaptation. *Climatic Change* 96: 23–27.
Stern NH (2007) *The economics of climate change*.
  Cambridge: Cambridge University Press.
Stern NH and Patel IG (2009) *A blueprint for a safer planet:
  How to manage climate change and create a new era of
  progress and prosperity*. London: Bodley Head.
Vaughan NE, Lenton TM and Shepherd JG (2009) Climate
  change mitigation: Trade-offs between delay and strength
  of action required. *Climatic Change* 96: 29–43.

model   A simplified representation of a phenomenon, developed to predict a new phenomenon or to provide insights into existing phenomena. A model may be seen as a complex formulation of a HYPOTHESIS, and there are many types, including semantic, conceptual, graphic, hardware, mathematical and statistical models.   *HB*

[*See also* CLIMATIC MODELS, DETERMINISTIC MODEL, ENVIRONMENTAL MODELLING, SIMULATION MODEL, STOCHASTICITY, THEORY, VALIDATION]

Knutti R (2012) Modelling environmental change and
  developing future projections. In Matthews JA, Bartlein
  PJ, Briffa KR et al. (eds) *The SAGE handbook of
  environmental change, volume 1*. London: Sage, 116–133.
Pelletier JD (2008) *Quantitative modelling of earth surface
  processes*. Cambridge: Cambridge University Press.

moder   An acid form of HUMUS in which there is incomplete breakdown of the *organic material*. It is loose and contains many animal droppings, mineral particles and brown-stained plant remains. F and H layers (SOIL HORIZONS) are of approximately equal thickness and moder is found in moist conditions under both deciduous and coniferous forests.   *EMB*

[*See also* MOR, MULL]

Kubiena WL (1953) *The soils of Europe*. London: Thomas
  Murby.

modern   In the context of environmental change, 'modern' refers to similarity to and continuity with the present day in terms of conditions, characteristics or age. A radiocarbon date, for example, is said to be modern if the difference in age from the present (AD 1950 by convention) is not statistically significant. A fossil flora may be considered modern if the species are all extant. The *Modern Age* or the *supermodern* is studied in *modern* or *contemporary archaeology*, which focuses on the twentieth and twenty-first centuries. Some historians see the Modern Age as beginning with the RENAISSANCE, at the end of the MIDDLE AGES.   *JAM*

[*See also* HISTORICAL ARCHAEOLOGY, MODERNISM, POSTMODERNISM]

Kaser G (1999) A review of the modern fluctuations of
  tropical glaciers. *Global and Planetary Change* 22, 93–103.
Schofield J (2009) The Modern Age. In Hunter J and Ralston
  I (eds) *The archaeology of Britain: An introduction from
  the earliest times to the twenty-first century, 2nd edition*.
  Abingdon: Routledge, 390–409.

modern analogue   A modern analogue is the contemporary organism, assemblage or condition used as a benchmark for comparison with similar FOSSIL or SUBFOSSIL forms or conditions. Modern analogues are the basis for the reconstruction and interpretation of past conditions. A fundamental difficulty with geological analogues is whether strictly comparable analogues occur at present. In BIOSTRATIGRAPHY, for instance, although the same species may have survived for many thousands, or even millions, of years, they may have changed in their ecological or climatic requirements. So-called *non-analogue conditions* or *non-analogue assemblages* are frequently encountered when investigating the past. Nevertheless, careful examination of modern analogues provides an invaluable tool in PALAEOENVIRONMENTAL and CLIMATIC RECONSTRUCTION. Whilst developing TRANSFER FUNCTIONS in PALAEOCEANOGRAPHY, where no analogues could be found, the *modern analogue technique* was developed to define the degree of dissimilarity between modern and fossil faunal assemblages within defined temperature ranges.   *DH*

[*See also* ANALOGUE METHOD]

Axford Y, Briner JP, Francis DR et al. (2011) Chironomids
  record terrestrial temperature changes throughout Arctic
  interglacials of the past 200,000 years. *Bulletin of the
  Geological Society of America* 123: 125–1287.
Kullman L (1998) Non-analoguous tree flora in the
  Scandes Mountains, Sweden, during the early Holocene:
  Macrofossil evidence of rapid geographic spread and
  response to palaeoclimate. *Boreas* 27: 153–161.

Lytle DE and Wahl ER (2005) Palaeoenvironmental reconstructions using the modern analogue technique: Effects of sample size and decision rules. *The Holocene* 15: 554–566.

Pflaumann U, Duprat J, Pujol C and Labeyrie L (1996) SIMMAX: A modern analog technique to deduce Atlantic sea surface temperatures from planktonic foraminifera in deep-sea sediments. *Paleoceanography* 11: 15–35.

Soepboer W, Sugita S and Lotter AF (2010) Regional vegetation-cover changes on the Swiss Plateau during the past two millennia: A pollen-based reconstruction using the REVEALS model. *Quaternary Science Reviews* 29: 472–483.

modernism The artistic, architectural and intellectual movements of the eighteenth to twentieth centuries that challenged the conventions of REALISM and *romanticism* within the arts by exploring ideas of 'newness' and expressing them as 'new' aesthetics. Ideas of 'modern' are most commonly defined through their opposition to the old and the traditional. Hence, 'modern' is synonymous with 'newness' and 'modernity' refers to the 'post-traditional' historical epoch within which newness is produced and valued, as well as to the economic, social, political and cultural formations characteristic of that period. More generally, modernism is often associated with the twentieth-century belief in the virtues of progress, order, SCIENCE, TECHNOLOGICAL CHANGE, rationality and the planned MANAGEMENT of *social change*. Since the 1970s, there has been the rise of POSTMODERNISM, signalling a transition from modernism, representing a loss of faith in rationality and progress. *ART*

Bell M (1999) The metaphysics of modernism. In Levenson M (ed.) *The Cambridge companion to modernism.* Cambridge: Cambridge University Press, 9–32.

Butler C (2010) *Modernism: A very short introduction.* Oxford: Oxford University Press.

Childs P (2008) *Modernism, 2nd edition.* Abingdon: Routledge.

Mohs' scale A relative scale of surface hardness for MINERALS, developed by F. Mohs (1773–1839). Common or distinctive minerals represent each of 10 points on the scale: 1 (softest) = *talc*, 2 = GYPSUM, 3 = *calcite*, 4 = *fluorite*, 5 = *apatite*, 6 = *orthoclase* FELDSPAR, 7 = QUARTZ, 8 = *topaz*, 9 = *corundum*, 10 (hardest) = *diamond*. The hardness of an unknown material can be determined relative to a known mineral or to common materials like finger nail ($H \sim 2\frac{1}{2}$), copper ($H \sim 3\frac{1}{2}$) and steel nail or knife blade ($H \sim 5\frac{1}{2}$). *GO*

molasse A term originally applied in the early nineteenth century to thick sequences dominated by SANDSTONE and CONGLOMERATE of the OLIGOCENE to MIOCENE age in Switzerland. Molasse represents material eroded from the then recently formed alpine mountains and deposited in adjacent FORELAND BASINS in a range of shallow marine, freshwater and continental environments. Deposits of the Ganges-Brahmaputra SEDIMENTARY BASIN adjacent to the Himalayas represent a modern analogue. The term can be considered to represent a post-orogenic tectono-stratigraphic FACIES (see FLYSCH). It has since been applied more generally to post-orogenic deposits of other OROGENIC BELTS (OROGEN). *BTC/GO/JP*

Covault JA, Hubbard SM, Graham SA et al. (2009) Turbidite-reservoir architecture in complex foredeep-margin and wedge-top depocenters, Tertiary Molasse foreland basin system, Austria. *Marine and Petroleum Geology* 26: 379–396.

Hsü KJ (1995) *The geology of Switzerland.* Princeton, NJ: Princeton University Press.

Willett SD and Schlunegger F (2010) The last phase of deposition in the Swiss Molasse Basin: From foredeep to negative-alpha basin. *Basin Research* 22: 623–639.

molecular stratigraphy Molecular information such as BIOMARKERS, *carbon content, nitrogen content,* bulk $\delta^{13}C$ values, COMPOUND-SPECIFIC CARBON ISOTOPE ANALYSIS and *compound-specific hydrogen isotope* values can be analysed in MARINE SEDIMENT CORES, LACUSTRINE SEDIMENTS, *peat* and SOIL. Molecular stratigraphic analyses, including LIPID distributions and compound-specific $\delta^{13}C$ and δD measurements, represent changes in specific compound classes or individual compounds with depth or age. These molecular parameters form a STRATIGRAPHICAL RECORD that can be interpreted in terms of PALAEOECOLOGY and CLIMATIC CHANGE, *organic matter source* and atmospheric CARBON DIOXIDE concentration.

For the biomarker approach to be applicable, *biomarker proxies* need to be assigned for the expected major inputs of organic matter, such as higher plants in TERRESTRIAL ENVIRONMENTS, *aquatic macrophytes* and ALGAE. Compound-specific carbon isotope analysis by *gas chromatography* combined with *isotope ratio mass spectrometry* provides a tool with which to study the CARBON CYCLE at the molecular level, whereas compound-specific hydrogen isotope analysis enables ARIDITY and PRECIPITATION δD to be examined. This determination of molecule-specific $\delta^{13}C$ values of the organic compounds in sediments clarifies understanding of the *carbon fluxes* within the ecosystem. By measuring the relative abundances and the $\delta^{13}C$ values for individual compounds specific to higher plant *leaf waxes,* algae or aquatic macrophytes, changing inputs to the environments can be followed. Depth profiles of $\delta^{13}C$ values and biomarker distributions can be interpreted in terms of changes in atmospheric carbon dioxide concentration and local climate conditions. This approach is more discriminatory than conventional combustion isotope ratio mass spectrometry of *total organic carbon* or conventional biomarker analyses alone. The δD

isotopic composition of *meteoric water* influences the δD of leaf waxes (δDlw) along with other factors such as ISOTOPIC FRACTIONATION, EVAPOTRANSPIRATION from soil and leaf water, *relative humidity* (see HUMIDITY), plant LIFE FORM and physiological differences. These factors complicate the use of δDlw as a *palaeohydrological proxy*. *Algal lipids* are particularly suited to δD studies as all hydrogen is derived from water and algae do not transpire. *Lipid biomarker δD stratigraphy* is emerging as a potentially powerful *palaeoclimate proxy* for environments that do not preserve carbonates. In effect, carbon cycling and *palaeohydrological cycles* can now be studied using combined biomarker, STABLE ISOTOPE and biological data.   *KJF*

[*See also* PALAEOCLIMATOLOGY, PROXY DATA]

Brassell SC, Eglinton G, Marlowe IT et al. (1986) Molecular stratigraphy: A new tool for climatic assessment. *Nature* 320: 129–133.

Casteñeda IS and Schouten S (2011) A review of molecular organic proxies for examining modern and ancient lacustrine environments. *Quaternary Science Reviews* 30: 2851–2891.

Farrimond P and Flanagan RL (1996) Lipid stratigraphy of a Flandrian peat bog (Northumberland, UK): Comparison with the pollen record. *The Holocene* 6: 69–74.

Ficken KJ, Barber KE and Eglinton G (1998) Lipid biomarker, $\delta^{13}C$ and plant macrofossil stratigraphy of a Scottish montane peat bog over the last two millennia. *Organic Geochemistry* 28: 217–237.

Ficken KJ, Street-Perrott FA, Perrott RA et al. (1998) Glacial/interglacial variations in carbon cycling revealed by molecular and isotope stratigraphy of Lake Nkunga, Mt. Kenya, East Africa. *Organic Geochemistry* 29: 1701–1719.

Schefuβ E, Schouten S and Schneider RR (2005) Climatic controls on Central African hydrology during the past 20,000 years. *Nature* 437: 1003–1006.

Tierney JE, Russell JM, Damsté JSS et al. (2011) Late Quaternary behaviour of the East African monsoon and the importance of the Congo air boundary. *Quaternary Science Reviews* 30: 798–807.

molecule

molecule The smallest constituent part of an ELEMENT or COMPOUND that retains its characteristic chemical properties. Molecules consist of more than one ATOM held together by shared electrons (*covalent bonds*). Some involve very large numbers of atoms bonded together in complex chemical structures (e.g. HUMIC ACIDS and DNA).   *JAM*

Mollusca analysis

Mollusca analysis Mollusca are common FOSSILS and SUBFOSSILS in calcareous sediments of the PLEISTOCENE age. Terrestrial and *freshwater Mollusca* are valuable in both PALAEOENVIRONMENTAL RECONSTRUCTION and PALAEOCLIMATOLOGY, based upon analogy with their present-day ecological and climatic preferences (see

MODERN ANALOGUE). The grouping of species into ecological categories provides a useful means of observing changes in the local environment, for example, a transition from swamp to open-water conditions, or woodland to grassland. Certain species of BRACKISH WATER may also be used as indicators of former marine *transgressions*. Marine *Mollusca* are not particularly sensitive to changes in water temperature, although the appearance of genuine arctic and southern species in Pleistocene deposits may be viewed as significant. Their application stems principally from the reconstruction of past *water depths*, SALINITY and energy conditions. Mollusca are less suitable as chronostratigraphic than PALAEOENVIRONMENTAL INDICATORS, since they appear to have undergone little *evolutionary change*. Nevertheless, some diagnostic features, particularly EXTINCTIONS, may be significant.   *DCS*

[*See also* SCLEROCHRONOLOGY]

Evans JG (1972) *Land snails in archaeology.* London: Seminar Press.

Garilli V (2011) Mediterranean Quaternary interglacial molluscan assemblages: Palaeobiogeographical and palaeoceanographical responses to climate change. *Palaeogeography, Palaeoclimatology, Palaeoecology* 312: 98–114.

Martin K and Sommer M (2004) Relationships between land snail assemblage patterns and soil properties in temperate-humid forest ecosystems. *Journal of Biogeography* 31: 531–545.

Miller BB and Tevesz MJS (2001) Freshwater molluscs. In Last WM and Smol JP (eds) *Tracking environmental change using lake sediments, volume 4.* Dordrecht: Kluwer, 153–171.

Peacock JD (1989) Marine molluscs and Late Quaternary environmental studies with particular reference to the Late-Glacial period in north-west Europe: A review. *Quaternary Science Reviews* 8: 179–192.

Penkman KEH, Preece RC, Bridgland DR et al. (2011) A chronological framework for the British Quaternary based on *Bithynia opercula*. *Nature* 476: 446–449.

Rousseau DD (1992) Terrestrial molluscs as indicators of global aeolian dust fluxes during glacial stages. *Boreas* 21: 105–110.

Sparks BW (1961) The ecological interpretation of Quaternary non-marine mollusca. *Proceedings of the Linnean Society of London* 172: 71–80.

Zaarur S, Olack G and Affek HP (2011) Paleo-environmental implication of clumped isotopes in land snail shells. *Geochimica Cosmochimica et Acta* 75: 6859–6869.

monochromatic

monochromatic Any system or process using a very small range of colours or wavelengths of the ELECTROMAGNETIC SPECTRUM. The term is often used in reference to black-and-white *photographic film* (even though this system is really *panchromatic*) and REMOTE-SENSING systems which utilise only a narrow range of wavelengths.   *TS*

monoculture The practice of cultivating a single species rather than many species together (*polyculture*). Monoculture is common in modern, industrialised AGRICULTURE and may be justified in terms of short-term economic return but tends to be unjustified in the long term or ecologically as there is loss of BIODIVERSITY and ecosystem complexity, with concomitant vulnerability to PESTS and DISEASE, and to downturns in market prices. *JAM*

monogenetic An entity, such as a SOIL, SEDIMENT body, LANDFORM or LANDSCAPE, that was formed in a single time interval and ENVIRONMENT (representing a single 'generation') is said to be monogenetic. Because of ENVIRONMENTAL CHANGE, monogenetic entities are the exception rather than the rule in NATURE. *JAM*

[*See also* POLYGENETIC]

monolith sampling The sampling of a vertical column of material (often *peat*, but can be SOIL, PALAEOSOL, LOESS or even LACUSTRINE SEDIMENTS, if accessible on land) by means of a metal (or wooden) three-sided box (usually open-ended). The intact column of sediment enclosed by the monolith box then needs to be severed from the vertical section by cutting behind and beneath. Special monolith cutters have been devised for peat monolith sampling. *FMC*

De Vleeschouwer F, Chambers FM and Swindles GT (2010) Coring and sub-sampling of peatlands for palaeoenvironmental research. *Mires and Peat* 7: 1–10.
Lageard JGA, Chambers FM and Grant ME (1994) Modified versions of a traditional peat cutting tool to improve field sampling of peat monoliths *Quaternary Newsletter* 74: 10–15.

monsoon Derived from the Arabian word 'mausim', meaning season, the monsoon refers to large-scale seasonal winds in the Afro-Asian region from around 40° N to 20° S and 30° W to 180° E (see Figure). Four criteria delimit these monsoon areas: (1) the prevailing wind direction shifts by at least 120° between January and July, (2) the average frequency of prevailing directions in January and July exceeds 40 per cent, (3) the mean resultant wind speed in at least one of the months exceeds 3 m/s and (4) there is less than one cyclone-anticyclone alternation every two years in any one month in a 5° latitude-longitude rectangle.

Three factors account for the existence of monsoons: (1) differential seasonal heating of oceans and continents, which results in atmospheric pressure changes; (2) moisture processes in the tropical atmosphere and (3) the Earth's rotation and the CORIOLIS FORCE. Several regional monsoon systems are recognised: Indian (or South Asian), East Asian, Australian, African, North American and South American.

However, the American systems do not fulfil the wind reversal requirement. Over 55 per cent of humankind live in these areas and the rural economies of tropical countries are closely linked to the annual cycle, especially the seasonal occurrence of monsoon rains. EXTREME EVENTS associated with the Indian monsoon (DROUGHTS and FLOODS) have devastating effects on the Indian subcontinent.

The history of the monsoon circulation has been investigated on various timescales. On the GEOLOGICAL TIMESCALE, it is involved in the TECTONIC UPLIFT–CLIMATIC CHANGE INTERACTION. Evidence from Chinese LOESS suggests a three-step evolution of the East Asian monsoon: (1) initiation about 2.6 million years ago (Ma), (2) progressively greater variation at 1.2 Ma and (3) 0.6 Ma between the relatively dry-cold periods dominated by the northerly monsoon and the humid-warm conditions characteristic of the southerly monsoon. During the last GLACIAL-INTERGLACIAL CYCLE, there appear to have been at least six episodes of strengthened northerly monsoon, which may correspond with HEINRICH EVENTS in the North Atlantic Ocean. Part C of the Figure suggests that these changes in the strength of the monsoon have been accompanied by appreciable changes in the areas receiving monsoon rains, with greater penetration of monsoon rains into the interiors of continents during the INTERGLACIALS.

CENTURY- TO MILLENNIAL-SCALE VARIABILITY has been identified in the strength of monsoonal atmospheric circulation during the HOLOCENE, and DECADAL-SCALE VARIABILITY has also been recognised, which are influenced by EL NIÑO-SOUTHERN OSCILLATION (ENSO) phenomena, the INDIAN OCEAN DIPOLE (IOD) and SOLAR CYCLES; reduced wind strengths associated with both the winter and summer East Asian monsoons in recent decades have been attributed to GLOBAL WARMING. There has been considerable success with statistical PREDICTION of monsoons but dynamical prediction has generally been unsuccessful due to its inadequate representation in GENERAL CIRCULATION MODELS (GCMs). *JAM*

An Z, Clemens SC, Shen J et al. (2011) Glacial-interglacial Indian summer monsoon dynamics. *Science* 333: 719–723.
Chang CP (ed.) (2004) *East Asian monsoon*. Singapore: World Scientific Publishers.
Clift PD and Plumb RA (2008) *The Asian monsoon: Causes, history and effects*. Cambridge: Cambridge University Press.
Clift PD, Tada R and Zheng H (eds) (2010) *Monsoon evolution and tectonics-climate linkage in Asia*. Bath: Geological Society.
Fein JS and Stephens PL (eds) (1987) *Monsoons*. New York: Wiley.
Fu C, Freney JR and Stewart JWB (eds) (2008) *Changes in the human-monsoon system of East Asia in the context of global change*. Singapore: World Scientific Publishers.

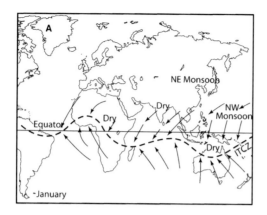

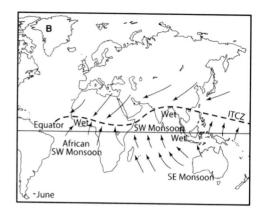

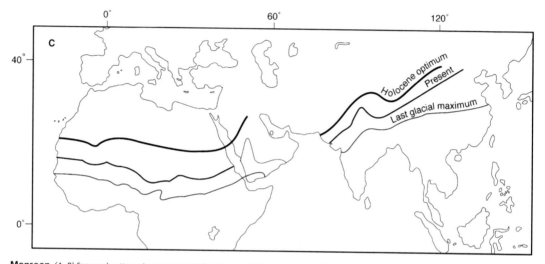

Monsoon *(A, B) Seasonal patterns in monsoon winds at present (ITCZ = Intertropical Convergence Zone) and (C) changes in the average northern limits of monsoon rains at the Last Glacial Maximum (LGM) and since the Holocene thermal maximum (Wilson et al., 2000).*

Huang C, Pang J and Zhao J (2000) Chinese loess and the evolution of the East Asian monsoon. *Progress in Physical Geography* 24: 75–96.

Kodera K, Coughlin K and Arakawa O (2007) Possible modulation of the connection between the Pacific and Indian Ocean variability by the solar cycle. *Geophysical Research Letters* 34: L03710.

Kutzbach JE and Liu Z (1997) Response of the African monsoon to orbital forcing and ocean feedbacks in the middle Holocene. *Science* 278: 440–443.

Marengo JA, Liebmann, Grimm AM et al. (2012) Recent developments on the South American monsoon system. *International Journal of Climatology* 32: 1–21.

Rajeevan M (2001) Prediction of the Indian summer monsoon: Status, problems and prospects. *Current Science* 81: 1451–1457.

Ramage CS (1971) *Monsoon meteorology*. New York: Academic Press.

Saha K (2009) *Tropical circulation systems and monsoons*. Berlin: Springer.

Shaman J and Tziperman E (2007) Summertime ENSO-North African-Asian jet teleconnections and implications for the Indian monsoon. *Geophysical Research Letters* 34: L11702.

Sirocko F, Sarnthein M, Erlenkeuser H et al. (1993) Century scale events in monsoon climate over the past 24,000 years. *Nature* 364: 322–324.

Wang B (2006) *The Asian monsoon*. Heidelberg: Springer.

Wilson, RCL, Drury SA and Chapman JL (2000) *The Great Ice Age: Climate change and life*. London: Routledge.

Xu M, Chang C, Fu C et al. (2006) Steady decline of East Asian monsoon winds, 1969-2000: Evidence from direct ground measurements of wind speed. *Journal of Geophysical Research* 111: D24111.

montane forest A forest VEGETATION FORMATION TYPE found on mountains in SUBTROPICAL and TROPICAL ENVIRONMENTS, characterised by short, gnarled, *mesophyllous trees* and abundant *bryophytes*. The formation is often split into lower and upper montane forest, depending on *tree physiognomy*. The altitude at which the formation begins is determined by the size of the mountain mass and its proximity to the sea. The term is sometimes used for any high-elevation forest growing close to the TREE LINE. *NDB*

[*See also* ALTITUDINAL ZONATION, CLOUD FOREST, ELFIN WOODLAND, LEAF PHYSIOGNOMY, MASSENERHEBUNG EFFECT, TROPICAL AND SUBTROPICAL FORESTS]

Bruijnzeel LA, Scatena FN and Hamilton LS (2011) *Tropical montane cloud forests: Science for conservation and management.* Cambridge: Cambridge University Press.
Marchant R and Taylor D (1998) Dynamics of montane forest in Central Africa during the late Holocene: A pollen-based record from western Uganda. *The Holocene* 8: 375–381.

Monte Carlo methods Named after the famous casino in the principality of Monaco, Monte Carlo methods use RANDOM numbers to study either real data sets or the behaviour of statistical methods through computer SIMULATION MODELLING. They are computer-intensive methods for finding solutions to mathematical and statistical problems by simulation or *permutation*. They are most commonly used when the analytical solution is intractable or very time-consuming. Within environmental statistics, Monte Carlo methods include *randomisation tests*, *permutation tests* and, in some instances, BOOTSTRAPPING. They can also be used to estimate CONFIDENCE INTERVALS and CONFIDENCE LIMITS for population PARAMETERS by using computer-generated data to estimate the expected amount of variation in the *sample statistics* of interest.

In many applied statistical applications, Monte Carlo methods are used to assess the *statistical significance* of an observed test statistic by comparing it with a large number of test statistics obtained by generating random data sets under some *assumed model*. If the assumed model implies that all data orderings are equally likely, the test is a randomisation test with random sampling of the randomisation distribution. The reliability of any Monte Carlo test totally depends on the generation of data sets that are equally likely under the relevant *null hypothesis*. Completely random permutations (as in a randomisation test) will yield invalid results if the observations are structured as a result of the way the data were collected, for example, *line transects*, *spatial grids*, *stratigraphical time series*, *split-plot designs*, etc. Restricted permutation tests using model-based permutations are thus required in such instances to obtain reliable results.

The basic steps in all such randomisation and permutation tests are as follows:

1. Calculate the test statistic (T_0) for the observed data set.

2. Generate K new data sets that are equally likely under the null hypothesis being tested.

3. Calculate the test statistic for each new data set, giving estimates of $T_1, T_2, \ldots , TK$.

4. Derive the exact Monte Carlo SIGNIFICANCE LEVEL by determining the proportion of values greater than or equal to T_0. The Monte Carlo level is thus the rank of T_0 among all values of T divided by $K + 1$ (1 is added because T_0 is included in the null distribution).

Monte Carlo permutation tests provide *distribution-free methods* for statistical testing and are widely used in evaluating the results of CANONICAL CORRESPONDENCE ANALYSIS (CCA) and REDUNDANCY ANALYSIS. *GMF*

[*See also* NON-PARAMETRIC STATISTICS]

ter Braak CJF and Šmilauer P (1998) *CANOCO reference manual and user's guide to canoco for windows: Software for canonical community ordination, version 4.* Ithaca, NY: Microcomputer Power.
Manly BFJ (2007) *Randomization, bootstrap and Monte Carlo methods in biology, 3rd edition.* London: CRC Press.
Roff DA (2006) *Introduction to computer-intensive methods of data analysis in biology.* Cambridge: Cambridge University Press.

monthly mean temperature The mean of the daily maximum and minimum temperatures of a particular month. The most frequent expression of general temperature and the basis, for example, of the Central England Temperature series. *JCM*

[*See also* CLIMATIC ARCHIVE, MEAN MONTHLY TEMPERATURE]

montmorillonite A 2:1 CLAY MINERAL of the SMECTITE group having a basic structure of two sheets of *silicon atoms* and one sheet of *aluminium atoms* held together by shared *oxygen atoms*. There is little attraction between the oxygen atoms in the top of one sheet and those in the bottom sheet of the next unit; consequently, the lattice can expand when wetted. ISOMORPHIC SUBSTITUTION of Mg^{2+} for Al^{3+} within the crystal lattice increases the negative charge so that it is 10–15 times that of KAOLINITE. Soils in which montmorillonite is the dominant clay mineral occur upon base-rich PARENT MATERIALS, particularly in warm temperate continental areas where VERTISOLS, CHERNOZEMS and some LUVISOLS are common. *EMB*

Kittrick JA (1971) Montmorillonite equilibria and the weathering environment. *Proceedings of the Soil Science Society of America* 35: 815–823.

moorland A VEGETATION community and/or LANDSCAPE formed on waterlogged *peat* soils, which can accumulate to considerable depths. The vegetation is typically dominated by *dwarf shrubs* (e.g. *Calluna vulgaris* and *Vaccinium myrtillus*), *grasses* and *sedges* (e.g. *Eriophorum* spp.) adapted to wet, acidic environments.   *MJB*

[*See also* HEATHLAND, MIRES, WETLAND]

Holden J, Shotbolt L, Bonn A et al. (2007) Environmental change in moorland landscapes. *Earth-Science Reviews* 82: 75–100.
Simmons IG (1990) The mid-Holocene ecological history of the moorlands of England and Wales and its relevance for conservation. *Environmental Conservation* 17: 61–69.

mor   A strongly acid HUMUS form that has three clearly recognisable layers. At the surface is an L (LIT-TER) layer of plant debris accumulated from several years as decomposition is slow. Below is an F (FERMEN-TATION) layer of increasingly fragmented plant material, matted together with *fungal hyphae* but containing few animal droppings. A thin H (HUMUS) layer of completely humified material lies on the surface of the mineral soil. Incorporation of organic matter and conversion to MULL can be achieved by LIMING and CULTIVATION.   *EMB*

[*See also* HUMIFICATION, LITTER, MODER]

Kubiena WL (1953) *The soils of Europe*. London: Thomas Murby.

moraines   Most commonly, moraines are depositional ridges formed by a range of processes (e.g. pushing, dumping, squeezing, thrusting) at the margins and termini of GLACIERS. The term was formerly used to describe GLACIAL SEDIMENTS (TILL) as well as GLACIAL LANDFORMS, but modern usage restricts its use to LANDFORMS, not SEDIMENTS. The term is also used for a range of landforms other than *moraine ridges*; these may be SUPRAGLACIAL or deposited in a *proglacial* position. They include landforms that are unstructured in planform, such as HUMMOCKY MORAINE deposited from the meltout of supraglacial debris, and *till plains* (*ground moraine*) of low relief composed largely of *subglacial till*. Supraglacial debris in transport forms a large component of *lateral moraines* along glacier margins, and *medial moraines*, which usually form along lines of glacial confluence in the ABLATION zone. Lateral moraines form along the flanks of *cirque glaciers* or *valley glaciers*, and *end moraines* (or *terminal moraines*) at their maximum extent downvalley. *Recessional moraines* form during marked stillstands or stages of renewed advance of a glacier. *Push moraines* or *glaciotectonic moraines* form during a glacier advance or re-advance by the deformation of proglacial sediment. Moraines in alpine environments are usually composed of a combination of deposits originating in debris slope processes, ENGLACIAL transport and *basal transport*. Dating of end and recessional moraines by a variety of techniques (e.g. LICHENOMET-RIC DATING, DENDROCHRONOLOGY, RADIOCARBON DATING,

SCHMIDT-HAMMER EXPOSURE-AGE DATING (SHD) and COS-MOGENIC–NUCLIDE DATING) enables reconstruction of GLACIER VARIATIONS.   *MRB*

[*See also* ANNUAL MORAINES, DE GEER MORAINES, FLUTES, ICE-CONTACT FANS/RAMPS, ICE-CORED MORAINE, ICE-MARGIN INDICATORS, PALAEOGLACIOLOGY, ROGEN MORAINES]

Bennett MR (2001) The morphology, structural evolution and significance of push moraines. *Earth-Science Reviews* 53: 197–236.
Boulton GS, van der Meer JJM, Beets DJ et al. (1999) The sedimentary and structural evolution of a recent push moraine complex: Holmstrombreen, Spitsbergen. *Quaternary Science Reviews* 18: 339–371.
Evans DJA (2007) Moraine forms and genesis. In Elias SA (ed.) *Encyclopedia of Quaternary science*. Amsterdam: Elsevier, 772–784.
Krüger J, Kjær KH and van der Meer JJM (2002) From push moraine to single-crested dump moraine during a sustained glacier advance. *Norsk Geografisk Tidsskrift* 56: 87–95.
Schlüchter Ch (ed.) (1979) *Moraines and varves: Origin, genesis, classification*. Rotterdam: Balkema.
Schomacker A (2011) Moraine. In Singh VP, Singh P and Haritashya UK (eds) *Encyclopedia of snow, ice and glaciers*. Dordrecht: Springer, 747–756.
Winkler S and Matthews JA (2010) Observations on terminal moraine-ridge formation during recent advances of southern Norwegian glaciers. *Geomorphology* 116: 87–106.

morphoclimatic zones   Regions of the Earth's surface governed by similar climatic relief-forming mechanisms. Each zone experiences dominant and subsidiary processes: the dominant processes largely determine the LANDFORM characteristics; the subsidiary processes are more limited in their action, only affecting certain rock types or only acting at certain times. According to Tricart and Cailleux (1965), delimitation of the world's morphoclimatic zones should be based primarily on present-day phenomena (see CLIMATIC GEOMORPHOLOGY). Their original classification divides the Earth's surface into 12 such zones, including glacial, periglacial, mid-latitude, desert and *tropical forest zones*. An alternative earlier approach was Peltier's (1950) system of *morphogenetic regions*, which were defined in terms of dominant *geomorphic processes* based on the *climatic parameters* of MEAN ANNUAL AIR TEMPERATURE and *mean annual precipitation*. The main drawback of morphoclimatic zone systems is that so many landforms and LANDSCAPES are the products of a past climate, a sequence of different climates and/or the transition between climates, rather than particular climates per se. Also, considerable climate-related differences in landscape variables (e.g. DRAINAGE DENSITY) and in EROSION RATES have been demonstrated within particular morphoclimatic regions (e.g. the *humid tropics* or the *humid temperate zone*). MAC/RPDW

[*See also* CLIMATOGENETIC GEOMORPHOLOGY, LANDSCAPE EVOLUTION]

Derbyshire E (ed.) (1976) *Geomorphology and climate.* London: Wiley.
Peltier LC (1950) The geographic cycle in periglacial regions as it is related to climatic geomorphology. *Annals of the Association of American Geographers* 40: 214–236.
Slaymaker O, Spencer T and Embleton-Hamann C (eds) (2009) *Geomorphology and global environmental change.* Cambridge: Cambridge University Press.
Tricart J and Cailleux A (1965) *Introduction à la géomorphologie climatique.* Paris: SEDES.
Walsh RPD and Blake WH (2009) Tropical rainforests. In Slaymaker O, Spencer T and Embleton-Hamann C (eds) (2009) *Geomorphology and global environmental change.* Cambridge: Cambridge University Press, 214–247.

morphometry   The quantitative description of form. Morphometric investigations exist in many disciplines and may be used as a basis for CLASSIFICATION and for the inference of processes and change. In GEOMORPHOMETRY (the study, via MEASUREMENT, of the Earth's form), for example, *general geomorphometry* (analysis of the form of the entire landsurface) and *specific geomorphometry* (analysis of particular LANDFORMS) have been recognised. JAM

[*See also* HYPSOMETRY]

Evans IS (2004) Geomorphometry. In Goudie AS (ed.) *Encyclopedia of geomorphology.* London: Routledge, 435–439.

morphostratigraphy   Originally developed by geologists and geomorphologists in the mapping of glacial STRATIGRAPHY, it can be defined as part of the *lithostratigraphic* record seen in the form of geomorphological features (i.e. LANDFORMS). Although complexes of such features can represent temporal relationships, as in the case of recessional MORAINES deposited by a retreating ice margin, or a RIVER TERRACE sequence, it is necessary to use other forms of evidence to provide a full record of the chronological relationships between *morphostratigraphic units.* CJC

Hughes PD (2007) Morphostratigraphy: Allostratigraphy. In Elias SA (ed.) *Encyclopedia of Quaternary science.* Amsterdam: Elsevier, 2841–2847.
Hughes PD (2010) Geomorphology and Quaternary stratigraphy: The roles of morpho-, litho, and allostratigraphy. *Geomorphology* 123: 189–199.
Lowe JJ and Walker MJC (1997) *Reconstructing Quaternary environments, 2nd edition.* Harlow: Addison Wesley Longman.
Lukas S (2006) Morphostratigraphic principles in glacier reconstructions: A perspective from the British Younger Dryas. *Progress in Physical Geography* 30: 719–736.

mosaicing   The process of combining and matching adjoining images such as those acquired by AERIAL PHOTOGRAPHY or REMOTE SENSING. It is important that elements such as *image geometry*, *edge feature matching* and *colour similarity* are addressed. TF

Capel D (2004) *Image mosaicing and super-resolution.* Berlin: Springer.

motu   An island typical of the inner reef flats of the Indian and Pacific Ocean ATOLLS, possessing a seaward SHINGLE BEACH ridge of coral and molluscan fragments, an interior depression and a sand ridge towards the LAGOON. TSp

[*See also* CAY, CORAL REEF ISLANDS]

McLean RF and Hosking PL (1991) Geomorphology of reef islands and atoll motu in Tuvalu. *South Pacific Journal of Natural Science* 11: 167–189.

mountain permafrost   Permanently frozen ground that occurs in mountainous terrain, usually but not always at elevations above the TREE LINE. It is distinguished from *latitudinal permafrost* occurring at high latitudes (e.g. in ARCTIC and SUBARCTIC regions), and may also be distinguished from *plateau permafrost* occurring at high elevations but with relatively low slope gradients (e.g. on Quinghai-Xizang [Tibet] Plateau). Mountain permafrost is also referred to as *alpine permafrost.* HMF

[*See also* PERMAFROST]

Gruber S and Haeberli W (2009) Mountain permafrost. In Margesin R (ed.) *Permafrost soils.* Berlin: Springer, 33–44.
Haeberli W, Noetzli J, Arenson L et al. (2011) Mountain permafrost: Development and challenges of a young research field. *Journal of Glaciology* 56: 1043–1058.
Isaksen K, Ødegård RS, Etzelmüller B et al. (2011) Degrading mountain permafrost in southern Norway: Spatial and temporal variability of mean ground temperatures, 1999-2009. *Permafrost and Periglacial Processes* 22: 361–377.
Lewkowicz AG and Bonnaventure PP (2011) Equivalent elevation: A new method to incorporate variable surface lapse rates into mountain permafrost modeling. *Permafrost and Periglacial Processes* 22: 153–162.

mountain regions: environmental change and human impact   Mountains are usually defined as natural elevations of the Earth's surface exceeding 600 m above sea level and rising abruptly above the surrounding terrain. They invariably possess several or all of the following characteristics: distinct summits or extensive plateaux, steep slopes, low temperatures, moderate to high precipitation, and high wind speeds. They may occur as single, isolated peaks or form substantial *mountain ranges*, and many display evidence of past or present GLACIATION and/or PERIGLACIATION.

Mountain regions cover about 20 per cent of the Earth's land surface, and about 50 per cent of the human population are dependent, to some degree, on mountain environments for *water supply, agricultural produce*, MINERAL RESOURCES or *power generation*. The past 100 years have seen accelerated use of mountain areas for AGRICULTURE, *forestry*, TOURISM, *mining* and QUARRYING and HYDROPOWER. The pace of change has been such that it poses a serious threat to the SUSTAINABILITY of mountain GEO-ECOSYSTEMS. This is particularly evident in parts of the Himalayas where population pressures have resulted in DEFORESTATION for both FUELWOOD and arable CULTIVATION. As a consequence, SOIL EROSION and LAND DEGRADATION have increased on deforested steep slopes, and increased RUNOFF has caused a greater incidence of FLOOD events and SEDIMENTATION in the lower reaches of river systems. The natural component of soil erosion and, hence, in SEDIMENT YIELD is, however, also important in mountain regions throughout the world (see Figure).

Environmental DEGRADATION in mountain regions is also a product of developments associated with tourism and RECREATION. The growth in popularity of *skiing* and other mountain-based activities in, for example, Western Europe and North America has resulted in the construction of chairlifts, cable cars and restaurants, and the bulldozing of hillsides to create *pistes*, often in ecologically and geomorphologically sensitive areas. *Footpaths* have multiplied and some are severely eroded. Associated impacts include *visual intrusion* (see AESTHETIC DEGRADATION), NOISE POLLUTION, LITTER and WASTE disposal problems, and a decline in BIODIVERSITY, as well as EROSION; but these must be set against the socioeconomic benefits that such developments can bring to small mountain communities.

In many mountain areas, *climatic warming* during the past 100–150 years has caused significant GLACIER retreat and PERMAFROST DEGRADATION, and has increased the incidence of slope FAILURE and *outburst floods* (JÖKULHLAUPS) from ICE-DAMMED LAKES and *moraine-dammed lakes*. Predictions of continued warming imply that these types of geomorphic events will increase in FREQUENCY and become a more serious threat to human activities.   *PW*

[*See also* CLIMATIC CHANGE, DEBRIS FLOW, LANDSLIDE, MASS MOVEMENT PROCESSES, OROGENY, STURZSTRÖM, TRAMPLING, UNLOADING]

Beniston M (2000) *Environmental change in mountains and uplands*. London: Arnold.
Beniston M (2003) Climatic change in mountain regions: A review of possible impacts. *Climatic Change* 59: 5–31.
Beniston M (2012) Environmental change in mountain regions. In Matthews JA, Bartlein PJ, Briffa KR et al.

(eds) *The SAGE handbook of environmental change, volume 2.* London: Sage, 262–281.
Dedkov AP and Mozzheim VI (1992) Erosion and sediment yields in mountain regions of the world. *International Association for Scientific Hydrology Publication* 209: 29–36.
Evans SG and Clague JJ (1994) Recent climatic change and catastrophic geomorphic processes in mountain environments. *Geomorphology* 10: 107–128.
Funnell D and Parish R (2001) *Mountain environments and communities*. London: Routledge.
Godde P, Price MF and Zimmermann FM (eds) (2000) *Tourism and development in mountain regions*. Wallingford: CABI Publishing.
Isaksen K, Sollid JL, Holmlund P and Harris C (2007) Recent warming of mountain permafrost in Svalbard and Scandinavia. *Journal of Geophysical Research: Earth Surface* 112: F02504.
Kalvoda J and Rosenfeld CL (eds) (1998) *Geomorphological hazards in high mountain areas*. Kluwer: Dordrecht.
Marston RA (2008) Land, life, and environmental change in mountains. *Annals of the Association of American Geographers* 98: 507–520.
Messerli B and Ives JD (eds) (1997) *Mountains of the world: A global priority*. New York: Parthenon Publishing.
Orlove B, Wiegandt E and Luckman BH (eds) (2008) *Darkening peaks: Glacier retreat, science and society*. Berkeley: University of California Press.
Owens PN and Slaymaker O (eds) (2004) *Mountain geomorphology*. London: Arnold.
Parish R (2002) *Mountain environments*. Harlow: Pearson.
Slaymaker O and Embleton-Hamann C (2009) Mountains. In Slaymaker O, Spencer T and Embleton-Mamann C (eds) *Geomorphology and global environmental change*. Cambridge: Cambridge University Press, 37–70.
Warren C (2009) *Managing Scotland's environment*. Edinburgh: Edinburgh University Press.

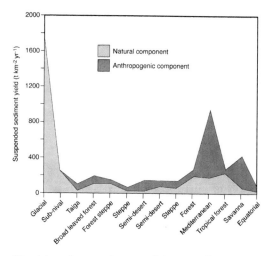

Mountain regions: environmental change and human impact *The relative importance of natural and anthropogenic contributions to suspended sediment yields of mountain river basins in various climatic regions (Dedkov and Mozzheim, 1992).*

mud Fine-grained SEDIMENT (CLASTIC or CARBON-ATE) with a GRAIN SIZE <0.063 mm, including both SILT and CLAY. Particles of fine silt and clay tend to stick together through COHESION. *TY*

[*See also* AGGREGATION, MUDROCK, MUDSTONE]

Macquaker JHS, Aplin AC and Fleet AJ (eds) (1999) *Muds and mudstones: Physical and fluid-flow properties*. Bath: Geological Society.
Potter PE, Maynard JB and Depetris PJ (2004) *Mud and mudstones: Introduction and overview*. New York: Springer.

mud cracks A polygonal network of downward-tapering cracks on a surface of cohesive sediment (MUD) that may be preserved as a SEDIMENTARY STRUC-TURE in SEDIMENTARY DEPOSITS by an infilling of SAND. Most are DESICCATION cracks (*sun cracks*), formed by shrinkage as the mud dried: they may be associated with EVAPORITE minerals or PSEUDOMORPHS and are a val-uable indicator of exposure in a PALAEOENVIRONMENT. Cracks can also form in association with the expulsion of pore water in subaqueous settings: these *synaeresis cracks* are most commonly found in lacustrine deposits and can be confused with DESICCATION cracks. *GO*

[*See also* FROST CRACK]

Allen JRL (1982) *Sedimentary structures: Their character and physical basis*. Amsterdam: Elsevier.
Parizot M, Eriksson PG, Aifa T et al. (2005) Suspected microbial mat-related crack-like sedimentary structures in the Palaeoproterozoic Magaliesberg Formation sandstones, South Africa. *Precambrian Research* 138: 274–296.
Pratt BR (1998) Syneresis cracks: Subaqueous shrinkage in argillaceous sediments caused by earthquake-induced dewatering. *Sedimentary Geology* 117: 1–10.

mud drape A thin layer of fine sediment (MUD) covering a BEDFORM, usually preserved within CROSS-BEDDING and indicative of an interruption to flow by, for example, a slack-water interval between ebb and flood TIDES. *GO*

[*See also* FLASER BEDDING, TIDAL BUNDLES]

Allen JRL (1981) Lower Cretaceous tides revealed by cross-bedding with mud drapes. *Nature* 289: 579–581.
Sato T, Taniguchi K, Takagawa T and Masuda F (2011) Generation of tidal bedding in a circular flume experiment: Formation, process and preservation potential of mud drapes. *Geo-Marine Letters* 31: 101–108.

mud volcano A conical structure formed by the extrusion of MUD onto the land surface or sea floor. Mud in the subsurface may become mobile due to elevated fluid pressures (see FLUIDISATION) as a result of movements of GROUNDWATER, sometimes exacerbated by drilling activities, or the build-up of gas under pres-sure. The resulting structure resembles a VOLCANO with a central CRATER and sloping flanks. Mud volcanoes represent a significant source of *methane* to the ATMOS-PHERE. They may cover large areas and represent a sig-nificant NATURAL HAZARD by damaging AGRICULTURE or settlements. *GO*

[*See also* SAND VOLCANO]

Camerlenghi A and Pini GA (2009) Mud volcanoes, olistostromes and Argille Scagliose in the Mediterranean region. *Sedimentology* 56: 319–365.
Davies RJ, Brumm M, Manga M et al. (2008) The East Java mud volcano (2006 to present): An earthquake or drilling trigger? *Earth and Planetary Science Letters* 272: 627–638.
Dimitrov LI (2002) Mud volcanoes: The most important pathway for degassing deeply buried sediments. *Earth-Science Reviews* 59: 49–76.

mudbelt/mudpatch A zone of fine-grained sedi-ments, for example, on the CONTINENTAL SHELF, where mudbelts are mostly of TERRIGENOUS origin and thicken in the vicinity of major river systems. *JAM*

Anthony EJ, Gardel A, Gratiot N et al. (2010) The Amazon-influenced muddy coast of South America: A review of mud-bank-shoreline interactions. *Earth-Science Reviews* 103: 99–121.
Bernárdez P, González-Álvarez R, Francés R et al. (2008) Palaeoproductivity changes and upwelling variability in the Galicia Mud Patch during the last 5000 years: Geochemical and microfloral evidence. *The Holocene* 18: 1207–1218.
Meadows ME, Dingle RV, Rogers J and Mills E (1997) Radiocarbon chronology of Namaqualand mudbelt sediments: Problems and prospects. *South African Journal of Science* 93: 321–327.

mudlump A small *diapiric structure* that forms at the mouth of a rapidly *aggrading river* where CLAYS are extruded into overlying SANDS following LOADING. They are especially common near the mouth of the Missis-sippi River, where they may interfere with navigation. *HJW*

[*See also* DIAPIR, SOFT-SEDIMENT DEFORMATION]

Coleman JM and Walker HJ (1998) Sediment instability in the Mississippi River delta. *Journal of Coastal Research* 14: 872–881.

mudrock A SEDIMENTARY ROCK (or its low-grade metamorphic derivative) entirely or dominantly com-prising material with a GRAIN SIZE <0.063 mm. The term embraces MUDSTONE, SHALE, SILTSTONE and CLAYSTONE. *TY*

[*See also* MUD]

mudslide A type of MASS MOVEMENT in which fine-grained sediment moves downslope under grav-ity, mainly by sliding on discrete shear surfaces. The movement is usually slow (1–25 m/year), although there may be episodes of more rapid movement, commonly triggered by extreme rainfall. Many of the well-documented coastal landslips or LANDSLIDES of west Dorset, UK, are of this type. There is some concern that CLIMATIC CHANGE might modify the activity of known mudslides.

The term is also used informally, for example, by the media, for any damaging slide or flow of muddy water, including DEBRIS FLOWS, LAHARS and sediment-laden river FLOODS. Flows described as mudslides were responsible for over 30,000 deaths in Venezuela in December 1999 following heavy rainfall. DEFOR-ESTATION and unsuitable *urban development* have been identified as factors contributing to the increasing frequency and severity of such NATURAL HAZARDS in recent years.   *GO*

[*See also* FLOW SLIDE, SLOPE STABILITY]

Allison RJ (ed.) (1992) *The coastal landforms of west Dorset*. London: The Geologists' Association.
Brunsden D and Ibsen M-L (1996) Mudslide. In Dikau R, Brunsden D, Schrott L and Ibsen M-L (eds) *Landslide recognition*. Chichester: Wiley, 103–119.
Dehn M, Burger G, Buma J and Gasparetto P (2000) Impact of climate change on slope stability using expanded downscaling. *Engineering Geology* 55: 193–204.
Walter M, Arnhardt C and Joswig M (2012) Seismic monitoring of rockfalls, slide quakes, and fissure development at the Super-Sauze mudslide, French Alps. *Engineering Geology* 128: 12–22.

mudstone   A SEDIMENTARY ROCK that is not FIS-SILE, entirely or dominantly comprising material with a GRAIN SIZE <0.063 mm, including both CLAYSTONE and SILTSTONE. Although the restriction to non-fissile MUDROCKS is useful in distinguishing mudstone from SHALE, the term 'mudstone' is commonly applied to any unmetamorphosed mudrock.   *TY*

Potter PE, Maynard JB and Depetris PJ (2004) *Mud and mudstones: Introduction and overview*. New York: Springer.

mulching   The spreading of a material such as straw, sawdust, leaves or plastic film onto the soil surface to prevent soil and plant roots from *freezing* or to protect the soil surface against RAINDROP IMPACT or excessive EVAPORATION. However, some effects may be detrimental, with reduced soil temperatures in cold climates or enhanced soil moisture conditions leading to GLEYING or ANAEROBIC conditions.   *SHD*

[*See also* DRY FARMING]

Chalker-Scott L (2007) Impact of mulching on landscape, plants and the environment. *Journal of Environmental Horticulture* 25: 239–249.

Hamilton C (2013) *Earthmasters: The dawn of the age of climate engineering*. New Haven, CT: Yale University Press.
White RE (2006) *Principles and practice of soil science: The soil as a natural resource, 4th edition*. Oxford: Blackwell.
Zuzel JF and Pikul JL (1993) Effects of straw mulch on runoff and erosion from small agricultural plots in northeastern Oregon. *Soil Science* 156: 111–117.

mull   A form of HUMUS that is fully incorporated into the mineral soil (see Figure). *Plant material* is completely decomposed by the action of the *soil fauna*. EARTHWORMS actively bring together CLAY and HUMUS in their alimentary canal to make a neutral, crumb-structured, humus-rich *topsoil*. Human activity may transform mull to MOR, as occurred in many areas of northwest Europe during PREHISTORY when forests were replaced by HEATHLAND.   *EMB*

[*See also* LITTER]

Andersen ST (1979) Brown earth and podzol: Soil genesis illuminated by microfossil analyses. *Boreas* 8: 59–73.

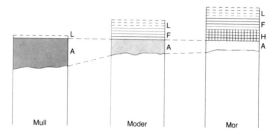

Mull *Mull, moder and mor: the thin surface forms of soil organic matter. The litter (L), fermentation (F) and humus (H) layers make up the O horizon. Fully incorporated organic matter is a component of the A horizon (the uppermost soil mineral horizon).*

multicollinearity   A situation in REGRESSION ANALY-SIS and constrained ORDINATION (CANONICAL CORRESPOND-ENCE ANALYSIS (CCA), REDUNDANCY ANALYSIS) when two or more explanatory or *predictor variables* are highly correlated, leading to *regression coefficients* with high VARIANCE. It is also known as *collinearity*.   *HJBB*

Fox J and Weisberg S (2011) *An R companion to applied regression, 2nd edition*. Thousand Oaks, CA: Sage.
Hastie TJ, Tibshirani RJ and Friedman J (2011) *The elements of statistical learning, 2nd edition*. New York: Springer.

multiconvex landscape   Low-relief LANDSCAPES formed by the presence of thick zones of partially altered SAPROLITE within weathered rock, found especially in *tropical environments* with high precipitation and rapid weathering. Piotr (2009) considers that such landscapes require that weathering operates efficiently in all *topographic settings* and that *dissection* and *downcutting* are able to proceed to maintain *relative relief*. Hence, such landscapes tend to be associated with moderate UPLIFT. He argues that there is little chance for such landscapes to survive major environmental changes towards ARIDITY or *cooling* because the accompanying *stripping* of REGOLITH would not then be compensated by saprolite renewal. This may be a reason why they are found in *humid tropical* areas of relatively low relief in South America, Southeast Asia and Africa that have experienced less wet phases, rather than semi-arid phases or arid phases during the Quaternary.   *RPDW/MAC*

[*See also* HUMID TROPICS: ECOSYSTEM RESPONSES TO ENVIRONMENTAL CHANGE, LANDSCAPE EVOLUTION]

Piotr M (2009) Are any granite landscapes distinctive of the humid tropics? Reconsidering multiconvex topographies. *Singapore Journal of Tropical Geography* 30: 327–342.

multidimensional scaling   A general term for methods that attempt to construct as accurately as possible a low-dimensional representation of a *distance, dissimilarity*, or *proximity matrix* between objects and/or variables.   *HJBB*

[*See also* CORRESPONDENCE ANALYSIS, NON-METRIC MULTIDIMENSIONAL SCALING, ORDINATION, PRINCIPAL COMPONENTS ANALYSIS (PCA), PRINCIPAL CO-ORDINATES ANALYSIS]

Everitt BS and Hothorn T (2011) *An introduction to applied multivariate analysis with R*. New York: Springer.
Everitt BS and Rabe-Hesketh S (1997) *The analysis of proximity data*. London: Arnold.

multidisciplinary   Involving a combination of several academic *disciplines*, with each discipline retaining its own METHODOLOGIES. Youngblood (2007) maintains that some disciplines, such as GEOGRAPHY and ANTHROPOLOGY, are naturally multidisciplinary *bridging disciplines*.   *FMC/JAM*

[*See also* ENVIRONMENTAL ARCHAEOLOGY, ENVIRONMENTAL SCIENCE, INTERDISCIPLINARY RESEARCH]

De Stasio G, Gilbert B, Frazer BH et al. (2001) The multidisciplinarity of spectromicroscopy: From geomicrobiology to archaeology. *Journal of Electron Spectroscopy and Related Phenomena* 114–116: 997–1003.
Youngblood D (2007) Multidisciplinarity, interdisciplinarity, and bridging disciplines. *Journal of Research Practice* 3: Article M18.

multiple working hypotheses   The method that advocates proposing a wide range of possible explanations. Such hypotheses are then tested against independent evidence, with the aim of eliminating unacceptable hypotheses and leaving the true

explanation. Proposed by T.C. Chamberlin at the end of the nineteenth century, it is still seen today as an essential part of SCIENTIFIC METHOD. The method of multiple working hypotheses is particularly appropriate in complex situations and/or where levels of UNCERTAINTY are high, such as in the early stages of a scientific investigation, in FIELD RESEARCH and in the investigation of environmental change.   *JAM*

[*See also* CRITICAL RATIONALISM, DEDUCTION, HYPOTHESIS, METHODOLOGY]

Baker VR (1999) Geosemiosis. *Bulletin of the Geological Society of America* 111: 633–645.
Chamberlin TC (1890) The method of multiple working hypotheses. *Science* 15: 92–96 [Reprinted in *Science* 148: 754–759 (1965)].
Elliott LP and Brook BW (2007) Revisiting Chamberlain: Multiple working hypotheses for the 21st century. *BioScience* 57: 608–614.
Haines-Young RH and Petch J (1983) Multiple working hypotheses: Equifinality and the study of landforms. *Transactions of the Institute of British Geographers NS* 8: 458–466.
Turner RE (1997) Wetland loss in the northern Gulf of Mexico: Multiple working hypotheses. *Estuaries and Coasts* 20: 1–13.

multiplicative model   A set of distributions used to model data obtained with coherent illumination (e.g. in REMOTE SENSING), as in the case of synthetic aperture RADAR, *sonar* and LASER images. Most of these distributions are quite different from the GAUSSIAN MODEL.   *ACF*

Frery AC, Müller H-J, Yanasse CCF and Sant'Anna SJS (1997) A model for extremely heterogeneous clutter. *IEEE Transactions on Geoscience and Remote Sensing* 35: 648–659.

multiproxy approach   The reconstruction of environmental change using many types of PROXY EVIDENCE from the same site or sediment core. Thus, the PALAEOENVIRONMENTAL RECONSTRUCTION from one source can be evaluated and extended by independent reconstructions from different proxies. PROXY CLIMATIC INDICATORS, or other ENVIRONMENTAL INDICATORS, can be physical or biological parameters. The multiproxy approach is advocated as the best approach to climate and environmental reconstruction, and also for assessing the impact of environmental changes on GEO-ECOSYSTEMS in general.

Diverse multiproxy investigations have used LACUSTRINE SEDIMENTS. Important Late-Glacial environmental physical indicators include sediment composition and chemistry, GRAIN SIZE, TEPHRA layers, VARVES, ORGANIC CONTENT and MINERAL MAGNETISM. Strong *Late-Glacial* climatic variations (ALLERØD INTERSTADIAL, YOUNGER DRYAS STADIAL, HOLOCENE) are reflected by aquatic and terrestrial biota. Comparative quantitative climate reconstructions can be made from a biostratigraphical sequence using POLLEN ANALYSIS, CHIRONOMIDS, CLADOCERA ANALYSIS and *beetles*, and vegetation analogues from PLANT MACROFOSSIL ANALYSIS data.

Multiproxy evidence for Holocene GLACIER VARIATIONS in lake sediments has been used to reconstruct summer temperature and winter precipitation. VEGETATION HISTORY and TREE-LINE VARIATIONS have been reconstructed using pollen, plant macrofossils, and SOILS. EUTROPHICATION studies use multiproxy evidence from DIATOMS, cladocera, chironomids, *aquatic macrophytes*, CYANOBACTERIA, *sediment chemistry* and *plant pigments*. SALINITY changes reflecting *precipitation:evaporation balance* have been reconstructed using diatoms, *mollusca*, OXYGEN ISOTOPES, sediment chemistry and organic content. The reconstruction of environmental change from ICE CORES is typically multiproxy. Frequently used parameters include $^{18}O{:}^{16}O$, DUST content, *electrical conductiivity*, BOREHOLE temperature (see SUBSURFACE TEMPERATURE), annual layers, measurements of CARBON DIOXIDE and METHANE content in included air bubbles. Multiproxy investigations of MARINE SEDIMENT CORES have used, for example, $\delta^{18}O$, $\delta^{13}C$, ^{14}C, sedimentology, diatoms, FORAMINIFERA, tephra and ICE-RAFTED DEBRIS (IRD).

ENVIRONMENTAL ARCHAEOLOGY also uses multiproxy evidence. Plant material gives evidence of vegetation, *building material*, *crops* and other *food*, *textiles* and *clothing*, *rope*, etc. ANIMAL REMAINS demonstrate *animal hunting* and *husbandry* and PALAEODIET, and INSECTS reveal PESTS and PARASITES. The lifestyle of a community is well reflected by the remains in waterlogged situations such as a *latrine* (e.g. Greig, 1981). Archaeology is very diverse, and past environments can also be reconstructed from human ARTEFACTS, CAVE ART and *grave contents*, and also from *human bodies* and *skeletons*.   *HHB*

[*See also* NATURAL ARCHIVES]

Birks HH, Battarbee RW and Birks HJB (2000) The development of the aquatic ecosystem at Kråkenes Lake, western Norway, during the Late Glacial and early Holocene: A synthesis. *Journal of Paleolimnology* 23: 91–114.

Ficken KJ, Wooller MJ, Swain DL et al. (2002) Reconstruction of a subalpine grass-dominated ecosystem, Lake Rutundu, Mount Kenya: A novel multiproxy approach. *Palaeogeography, Palaeoclimatology, Palaeoecology* 177: 137–149.

Gehrels WR, Roe HM and Charman DJ (2001) Foraminifera, testate amoebae and diatoms as sea-level indicators in UK saltmarshes: A quantitative multi proxy approach. *Journal of Quaternary Science* 16: 201–220.

Greig J (1981) The investigation of a medieval barrel-latrine from Worcester. *Journal of Archaeological Science* 8: 265–282.

Guiot J, Wu WB, Garreta V et al. (2009) A few prospective ideas on climate reconstruction from a statistical single proxy approach towards a multi-proxy and dynamical approach. *Climate of the Past* 5: 571–583.

Kucera M, Schneider R and Weinelt M (eds) (2005) *MARGO: Multiproxy approach for the reconstruction of the global ocean surface*. Amsterdam: Elsevier.

Li B, Nychka DW and Ammann CM (2010) The value of multiproxy reconstruction of past climate. *Journal of the American Statistical Association* 105: 883–911 [Special Issue No. 491: Approaches and Case Studies].

Lotter AF (2003) Multi-proxy climatic reconstructions. In Mackay A, Batterbee R, Birks J and Oldfield F (eds) *Global change in the Holocene*. London: Arnold, 373–383.

McCarroll D, Jalkanen R, Hicks S et al. (2003) Multiproxy dendroclimatology: A pilot study in northern Finland. *The Holocene* 13: 829–830.

McCarroll D, Tuovinen M, Campbell R et al. (2011) A critical evaluation of multi-proxy dendroclimatology in northern Finland. *Journal of Quaternary Science* 26: 7–14.

Tinner W, Ammann B and Germann P (1996) Treeline fluctuations recorded for 12,500 years by soil profiles, pollen, and plant macrofossils in the Central Swiss Alps. *Arctic and Alpine Research* 28: 131–147.

Van der Schrier G, Osborn TJ, Briffa KR and Cook ER (2007) Exploring an ensemble approach to estimating skill in multiproxy palaeoclimate reconstructions. *The Holocene* 17: 119–129.

multisol   The concept of a three-dimensional SOIL or PALAEOSOL that bifurcates into two or more layers at different levels of the SOIL PROFILE or sedimentary column. The soil has lateral continuity, whereas the layers may coalesce and bifurcate over distances of tens or hundreds of metres. The concept allows for complexity during SOIL DEVELOPMENT in dynamic LANDSCAPES.

JAM

[*See also* K-CYCLES, TRUNCATED SOIL]

Iriondo M (2009) Nultisol: A proposal. *Quaternary International* 209: 131–141.

multitemporal analysis   A technique for analysing REMOTE-SENSING data which makes use of several images acquired at different dates. It is used in relation to environmental variables for TEMPORAL CHANGE DETECTION. Multitemporal analysis is also able to yield additional thematic information about LAND COVER and VEGETATION type. By combining images from winter and summer, *coniferous* and *deciduous forests* can easily be distinguished due to loss of leaves. Multitemporal images can be used for repeat-pass *synthetic aperture radar* INTERFEROMETRY, which estimates the *coherence* and *phase difference* between two *synthetic aperture radar* images separated by a temporal baseline (usually 1–44 days). DIGITAL ELEVATION MODELS (DEMs) can be generated from the *radar phase difference*.   *HB*

Souza Jr CM, Roberts DA and Monteiro AL (2005) Multitemporal analysis of degraded forests in the southern Brazilian Amazon. *Earth Interactions* 9: 1–25.

Wegmüller U and Werner C (1997) Retrieval of vegetation parameters with SAR interferometry. *IEEE Transactions on Geoscience and Remote Sensing* 35: 18–24.

multivariate analysis A general term for the many numerical techniques now available for the analysis of large, complex data sets consisting of many observations and many variables. Each observation usually consists of values for more than one *random response variable*. In some cases, there may also be one or more predictor variables. Examples include CLUSTER ANALYSIS, PRINCIPAL COMPONENTS ANALYSIS (PCA), *multiple discriminant analysis* (*MDA*) and CANONICAL CORRESPONDENCE ANALYSIS (CCA).   *HJBB*

Borcard D, Gillet F and Legendre P (2011) *Numerical ecology with R*. New York: Springer.

Everitt BS and Hothorn T (2011) *An introduction to applied multivariate analysis with R*. New York: Springer.

Krzanowski WJ and Marriot FHC (1994–1995) *Multivariate analysis, parts 1 and 2*. London: Arnold.

Legendre P and Legendre L (1998) *Numerical ecology, 2nd edition*. Amsterdam: Elsevier.

multivariate regression trees (MRT) CLASSIFICATION AND REGRESSION TREES (CART) are *univariate* as they consider a single *response variable* (e.g. abundance of single species), whereas many environmental-change data are *multivariate* and consist of many variables. The extension of CART to the multivariate response case was made by De'Ath (2002), involving MRT. They have turned out to be surprisingly versatile and are a useful counterpart to *constrained ordination* techniques like REDUNDANCY ANALYSIS and CANONICAL CORRESPONDENCE ANALYSIS (CCA). In MRT, the aim is to find a set of simple rules from the set of *predictor variables* that best explains variation in the multivariate response-variable matrix. In conventional regression trees, the concept of sum-of-squared errors is used as the measure of node impurity. This is inherently univariate but can be extended to the multivariate case by considering sum-of-squared errors about the multivariate mean (centroid) of the observations in each tree node. Although MRT usually use *Euclidean distance*, which is not always appropriate with large, sparse ecological data, they can be adapted to work with any dissimilarity coefficient (e.g. *Bray and Curtis coefficient*) via direct decomposition of the supplied *dissimilarity matrix* to derive within-node sum-of-squared distances between node numbers. MRT are increasingly being used in PALAEOECOLOGY and PALAEOLIMNOLOGY.   *HJBB*

Davidson TA, Sayer CD, Perrow M et al. (2010) The simultaneous inference of zooplanktivorous fish and macrophyte density from sub-fossil cladoceran assemblages: A multivariate regression tree approach. *Freshwater Biology* 55: 546–564.

De'ath G (2002) Multivariate regression trees: A new technique for modeling species-environment relationships. *Ecology* 83: 1108–1117.

Herzschuh U and Birks HJB (2010) Evaluating the indicator value of Tibetan pollen taxa for modern vegetation and climate. *Review of Palaeobotany and Palynology* 160: 197–208.

Simpson GL and Birks HJB (2012) Statistical learning in palaeolimnology. In Birks HJB, Lotter AF, Juggins S and Smol JP (eds) *Tracking environmental change using lake sediments, volume 5: Data handling and numerical techniques*. Dordrecht: Springer.

mummy A corpse deliberately preserved by human agency (*mummification*), as opposed to being preserved under natural waterlogged, arid or frozen conditions (e.g. BOG PEOPLE and the Alpine ICEMAN). Mummification has been practised by many different cultures at various times but especially in extreme environments such as the Sahara, the Andes and Siberia.   *JAM*

Cockburn A and Cockburn E (eds) (1980) *Mummies, disease and ancient cultures*. Cambridge: Cambridge University Press.

Wieczorek A and Rosendahl W (eds) (2010) *Mummies of the world: The dream of eternal life*. Munich: Prestel.

museums Repositories or ARCHIVES for objects of natural and cultural HERITAGE. Museum collections provide a basis for RESEARCH and a means of EX-SITU CONSERVATION and are also important, through display and *exhibitions*, for the public understanding of SCIENCE and CULTURE.   *JAM*

Keene S (2005) *Fragments of the world: Uses of museum collections*. Oxford: Elsevier.

Knell SJ (ed.) (2007) *Museums in the material world*. Abingdon: Routledge.

Stone PG and Molyneaux BL (eds) (1994) *The presented past: Heritage, museums and education*. London: Routledge.

Sullivan LP and Childs ST (2003) *Curating archaeological collections: From the field to the repository*. Walnut Creek, CA: AltaMira Press.

muskeg A term of Cree Indian origin for the Canadian Arctic and especially subarctic *peatlands*. It is also the name of a fluvial regime which is characteristic of DRAINAGE BASINS (*catchments*) where extensive WETLANDS act to increase *water storage*, reduce rates of DISCHARGE, and extend *flood peaks*.   *DCS*

Radforth NW and Brawner CO (eds) (1977) *Muskeg and the northern environment in Canada*. Toronto: University of Toronto Press.

mutagen Any physical (e.g. ultraviolet light) or chemical (e.g. cigarette smoke) agent that alters DNA and markedly increases the frequency of mutational events above an 'average' spontaneous MUTATION rate.   *GOH*

[*See also* CARCINOGEN]

Frickel S (2004) *Chemical consequences: Environmental mutagens, scientist activism and the rise of genetic toxicology.* Piscataway, NJ: Rutgers University Press.
Russell PJ (1998) *Genetics.* Menlo Park, CA: Benjamin Cummings.

mutation An abrupt alteration of chromosomal DNA, occurring at random and inheritable. They can be caused by ENVIRONMENTAL FACTORS (e.g. RADIATION, VIRUSES and *mutagenic chemicals*) or errors in the structure of DNA produced during cell division. Mutations are often deleterious, but such changes are the basis of genetic variability, and hence the raw material for NATURAL SELECTION and EVOLUTION. Mutations favourable to survival persist due to the process of natural selection.   *KDB*

Woodruff RC and Thompson Jr JN (eds) (1998) *Mutation and evolution.* Dordrecht: Kluwer.

mutual climatic range (MCR) method A method permitting the quantitative reconstruction of past climates, using biological proxies with precise thermal requirements, such as *beetles* or *herpetofauna*. For each species in an ASSEMBLAGE, a *climatic tolerance range* is deduced from present-day climatic distributions, commonly based on the temperature of the warmest month (T_{max}) and the temperature range between the warmest and coldest months (T_{range}). When the various climatic ranges of an assemblage of species are considered, an area of overlap will be expected within the *climatic space* where all the species can coexist (see Figure). This part of the climate space is known as the mutual climatic range of that assemblage. Where a series of assemblages can be dated (e.g. by RADIOCARBON DATING), a *palaeotemperature curve* may then be constructed.   *DCS*

[*See also* BEETLE ANALYSIS, INSECT ANALYSIS, PROXY RECORDS]

Bray PJ, Blockley SPE, Coope GR et al. (2006) Refining mutual climatic range (MCR) quantitative estimates of palaeotemperature using ubiquity analysis. *Quaternary Science Reviews* 25: 1865–1876.
Horne DJ (2007) A mutual temperature range method for Quaternary palaeoclimatic analysis using European non marine Ostracoda. *Quaternary Science Reviews* 26: 1398–1415.
Lowe JJ and Walker MJC (1997) *Reconstructing Quaternary environments, 2nd edition.* Harlow: Wesley Longman.

mutualism A form of SYMBIOSIS in which all interacting organisms benefit from the association, such as the interaction between higher-order plants and MYCORRHIZA. Most plant species practice one or more forms of mutualism, especially in *reproduction* (e.g. *pollination, seed dispersal*). Environmental change may affect one species in an ASSOCIATION more than another, creating the potential for unpredictable effects within the *ecosystem* as a whole. For example, relationships between species can switch from mutualism to *pathogenism*.   *JLI*

[*See also* PATHOGENS]

Archetti M, Scheuring I, Hoffman M et al. (2011) Economic game theory for mutualism and cooperation. *Ecology Letters* 14: 1300–1312.
Burkle LA and Alarcon R (2011) The future of plant-pollinator diversity: Understanding interaction networks across time, space, and global change. *American Journal of Botany* 98: 528–538.
Crowley PH and Cox JJ (2011) Intraguild mutualism. *Trends in Ecology and Evolution* 26: 627–633.
Eaton CJ, Cox MP and Scott B (2011) What triggers grass endophytes to switch from mutualism to pathogenism? *Plant Science* 180: 190–195.
Garcia C and Grivet D (2011) Molecular insights into seed dispersal mutualisms driving plant population recruitment. *Acta Oecologia: International Journal of Ecology* 37: 632–640.
Sachs JL, Essenberg CJ and Turcotte MM (2011) New paradigms for the evolution of beneficial infections. *Trends in Ecology and Evolution* 26: 202–209.

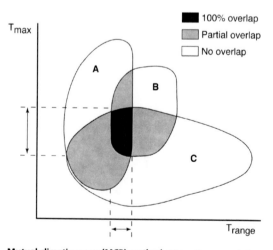

Mutual climatic range (MCR) method *Schematic representation of the mutual climatic range as the area of overlap between the climatic tolerance ranges of three species (A, B and C) based on ecological knowledge: T_{max} is the mean monthly temperature of the warmest month; T_{range} is the difference between the mean monthly temperatures of the warmest and coldest months. Arrows indicate the reconstructed climate for any site where this species assemblage is found (Lowe and Walker, 1997).*

mycorrhiza A symbiotic association between a FUNGUS and the *roots* of a higher plant that enhances plant uptake of certain NUTRIENTS (sometimes water) from the soil. Two main types are *ectomycorrhizae* (sheathing the roots) and *endomycorrhizae* (inside the roots). Mycorrhizae are part of the RHIZOSPHERE.   *NDB*

[*See also* BELOW-GROUND PRODUCTIVITY, NUTRIENT, NUTRIENT POOL, SYMBIOSIS]

Azcón-Aguilar C, Barea JM, Gianinazzi S and Gianinazzi-Pearson V (2009) *Mycorrhizas: Functional processes and ecological impacts.* Berlin: Springer.

Drigo B, Kowalchuk GA and van Veen JA (2008) Climate change goes underground: Effects of elevated atmospheric CO_2 on microbial community structure and activities in the rhizosphere. *Biology and Fertility of Soils* 44: 667–679.

Fitter AH, Heinemeyer A, Husband R et al. (2004) Global environmental change and the biology of arbuscular mycorrhizas: Gaps and challenges. *Canadian Journal of Botany* 82: 1133–1139.

Garg N and Chandel S (2010) Arbuscular mycorrhizal networks: Process and functions. A review. *Agronomy for Sustainable Development* 30: 581–599.

Van der Heijden MGA and Horton TR (2009) Socialism in soil? The importance of mycorrhizal fungal networks for facilitation in natural ecosystems. *Journal of Ecology* 97: 1139–1150.

nadir point The point on the ground directly beneath (measured as plumb line) a camera lens or digital sensor. It forms a right angle with the recording plane.   *TF*

nanism Also known as *dwarfism*, the tendency for some isolated POPULATIONS to undergo significant decreases in body size in comparison to their *conspecifics* on the mainland. Nanism and GIGANTISM form the two components of the *island rule*. However, this may simply be an artefact associated with comparisons of distantly related groups that have responded in clade-specific ways to *insularity*. Nanism is especially common in insular populations of *elephants* and MAMMOTHS, *deer* and other *ungulates* and *canids*, although the most celebrated example may be *Homo floresiensis*.   *MVL/JLI*

[*See also* FLORES MAN, GIGANTISM, ISOLATION]

Heupink TH, Huynen L and Lambert DM (2011) Ancient DNA suggests dwarf and 'giant' emu are conspecific. *PLoS ONE* 6: Article No. e18728.
Kaifu Y, Baba H, Sutikna T et al. (2011) Craniofacial morphology of *Homo floriensis*: Description, taxonomic affinities, and evolutionary implication. *Journal of Human Evolution* 61: 644–682.
Meiri S, Cooper N and Purvis A (2008) The island rule: Made to be broken? *Proceedings of the Royal Society B* 275: 141–148.

nannofossils Microscopic FOSSIL or SUBFOSSIL calcite platelets produced by unicellular marine algae, the commonest of which are *coccoliths*. They are between 2 and 15 μm in size and are PLANKTONIC (*nannoplankton*) or PELAGIC in origin, being restricted to the *photic zone* in the upper 200 m of the oceans. They are of particular value in indicating thermal conditions, although their distribution is also affected by light and salinity, and provide one of the major constituents of MARINE SEDIMENT CORES covering the past 150 million years.   *DH/CJC*

[*See also* COCCOLITHOPHORES]

Brown PR (ed.) (1998) *Calcareous nannofossil biostratigraphy*. Dordrecht: Kluwer.

Crux J (1989) *Nannofossils and their applications*. Chichester: Wiley.
Villa G, Palandri S and Wise SW (2005) Quaternary calcareous nannofossils from Periantarctic basins: Paleoecological and paleoclimatic implications. *Marine Micropaleontology* 56: 103–121.

nanoscience The study of physical, chemical and biological structures, on the scale of one-billionth of a metre. At this small scale, some materials behave differently from their larger forms and may be exploited for economic, health and environmental benefits. *Environmental nanoscience* focuses on the fate and behaviour, ECOTOXICOLOGY and ecological effects of engineering *nanoparticles* and *nanostructures* in the environment. Inevitably, nanoscience cross-cuts many scientific disciplines but also raises issues and concerns about potential environmental impacts.   *AEV*

[*See also* APPLIED SCIENCE, NANOTECHNOLOGY]

Hochella Jr MF, Lower SK, Maurice PA et al. (2008) Nanomaterials, mineral nanoparticles, and Earth systems. *Science* 319: 1631–1635.
Royal Society and Royal Academy of Engineering (2004) *Nanoscience and nanotechnology: Approaches and uncertainties*. London: Royal Society.

nanosensors Devices (*sensory points*) used to measure or collect information about the physical and chemical properties of *nanoparticles*. Nanosensors also exist in the natural world, as *biological receptors* of outside stimulation, for example, within plants (to detect sunlight) and insects (to detect sex pheromones).   *AEV*

[*See also* NANOSCIENCE, NANOTECHNOLOGY]

nanotechnology The design, manipulation and manufacture of *nanoparticles* and *nanostructures*, creating new materials and devices with a wide range of environmental applications. A good example is the use of *nanomembranes* in simple, low-cost, user-friendly water-treatment systems that do not use chemicals or energy other than SOLAR RADIATION to provide clean drinking water (POTABLE WATER). Nanotechnology is also being developed as a solution to POLLUTION, by using techniques which can bind and remove POLLUTANTS from the land, sea and air. Nanotechnology takes advantage of the fact that NANOSCIENCE produces structures with different properties that are not displayed in their bulk matter.   *AEV*

Mehta MD (2004) From biotechnology to nanotechnology: What can we learn from earlier technologies? *Bulletin of Science, Technology and Society* 24: 34–39.
Pagliaro M (2010) *Nano-Age: How nanotechnology changes our future.* Weinheim: Wiley-VCH.

National Parks Extensive areas of countryside that have been protected by national legislation from development, in order to protect and conserve the natural and cultural landscape. Yellowstone National Park (United States) became the first example in AD 1872. The first British examples were designated under the National Parks and Access to the Countryside Act (1949) to *preserve and enhance the natural beauty of the areas and promote their quiet enjoyment by the public.* The degree of protection afforded by National Park status varies across different countries and states. The first National Park in Africa—Virunga, Demographic Republic of Congo—the home of around 800 surviving mountain gorillas, is vulnerable to population growth, agricultural development, poverty and the devastation of civil war. The gorillas are being protected but at the cost of more than 130 rangers killed since the mid-1990s.   *MLW/JAM*

[*See also* CONSERVATION, ENVIRONMENTAL PROTECTION, PROTECTED AREAS, RECREATION]

Languy M and De Merode E (eds) (2009) *Virunga: The survival of Africa's first National Park.* Tielt: Lannoo.
Lowry WR (1994) *The capacity for wonder: Preserving National Parks.* Washington, DC: Smithsonian Institution Press.
Sheail J (2010) *Nature's spectacle: The world's first National Parks and protected places.* London: Earthscan.
Wright RG (ed.) (1996) *National Parks and protected areas: Their role in environmental protection.* Cambridge, MA: Blackwell Science.

native Individuals and POPULATIONS of a BIOLOGICAL SPECIES that occur within the species' natural, historic range. Native species are also known as INDIGENOUS species.   *MVL*

[*See also* ALIEN SPECIES, EXOTIC SPECIES, NATURALISED SPECIES]

Westhoff JT, Rabeni CF and Sowa SP (2011) The distribution of one invasive and two native crayfishes in relation to coarse-scale natural and anthropogenic factors. *Freshwater Biology* 56: 2415–2431.

natural Part of, or formed by NATURE; not made or caused by human agency. A narrower definition would include 'not modified' by human agency. In the context of environmental change in general and CLIMATE CHANGE in particular, the disentangling of natural from ANTHROPOGENIC variability is one of the major research problems.   *JAM*

[*See also* ARTIFICIAL, NATURAL ENVIRONMENT, UNNATURAL]

Comiti F (2012) How natural are alpine mountain rivers? Evidence from the Italian Alps. *Earth Surface Processes and Landforms* 37: 693–707.
Graf WL (1996) Geomorphology and policy for restoration of impounded American rivers: What is 'natural'? In Rhoades BL and Thorn CE (eds) *The scientific nature of geomorphology.* Chichester: Wiley, 443–473.
Science (2013) Natural systems in changing climates. *Science* 341: 473–524 [Special Section].
Willis KJ and Birks HJB (2006) What is natural? The need for a long-term perspective in biological conservation. *Science* 314: 1261–1265.

natural archives SEDIMENTS, *rocks* or naturally occurring biological remains that have accumulated continuously and chronologically as a *natural repository* and, when cored and examined with appropriate techniques, can be 'read' as an ARCHIVE of, for example, VEGETATION HISTORY, DRAINAGE BASIN HISTORY and *palaeoclimate.* Further examples are shown in the Table.   *FMC*

Bradley RS and Eddy JA (1991) Introduction. In Bradley RS (ed.) *Global changes of the past.* Boulder, CO: UCAR/Office for Interdisciplinary Earth Studies, 5–9.
Chambers FM (2012) Reconstructing and inferring past environmental change. In Matthews JA, Bartlein PJ, Briffa KR et al. (eds) *The SAGE handbook of environmental change, volume 1.* London: Sage, 67–91.
Chambers FM, Daniell JRG and ACCROTELM Members (2010) Peatland archives of Late-Holocene climate change in northern Europe. *PAGES Newsletter* 18: 4–6.

natural areas concept The concept refers to the intactness or integrity of HABITATS, *ecosystems* and LANDSCAPES that have not been changed or affected by human activity. There are few such areas remaining on Earth, but there are many areas which have been called *semi-natural*, that is relatively unmodified by human activity. A framework for evaluating naturalness of ecosystems has been described by Anderson (1991), involving three indices: (1) the degree to which the system would change if humans were removed. (2) the amount of *cultural energy* required to maintain the functioning of the ecosystem as it currently exists and (3) the complement of NATIVE or INDIGENOUS species in the area compared with that which previously existed. In the United Kingdom, it refers to

Archive	Best temporal resolution[a]	Temporal range (years)	Information derived
Historical records	Day/hour	10^3	T, H, B, V, M, L, S
Tree rings	Season/year	10^4	T, H, C_A, B, V, M, S
Lake sediments	1–20 years	10^4–10^6	T, H, C_W, B, V, M
Ice cores	1 year	10^5	T, H, C_A, B, V, M, S
Pollen	100 years	10^5	T, H, B
Loess	100 years	10^6	H, B, M
Ocean cores	1,000 years	10^7	T, C_W, B, M
Corals	1 year	10^4	C_W, L
Paleosols	100 years	10^5	T, H, C_S, V
Geomorphic features	100 years	10^7	T, H, V, L
Sedimentary rocks	1 years	10^{10}	H, C_S, V, M, L

a. Minimum sampling interval in most cases.
Abbreviations: T = temperature; H = humidity or precipitation; C = chemical composition of air (C_A), water (C_W) or soil (C_S); B = biomass and vegetation patterns; V = volcanic eruptions; M = geomagnetic field variations; L = sea levels; S = solar activity.

Natural archives *Characteristics of natural archives (Bradley and Eddy, 1991).*

an approach developed by *English Nature* based on geographical integration of floras, geological maps, landscape accounts, and so on, to derive distinct areas for CONSERVATION planning.   IFS

[*See also* CONSERVATION BIOLOGY, LANDSCAPE UNITS, NATURAL VEGETATION, PROTECTED AREAS, SEMI-NATURAL VEGETATION, WILDERNESS CONCEPT]

Anderson JA (1991) A conceptual framework for evaluating and quantifying naturalness. *Conservation Biology* 5: 347–352.
Gray JM (2001) Geomorphological conservation and public policy in England: A geomorphological critique of English Nature's 'Natural Areas' approach. *Earth Surface Processes and Landforms* 26: 1009–1023.
Spellerberg IF (1992) *Evaluation and assessment for conservation.* London: Chapman and Hall.

natural capital The stock of NATURAL RESOURCES from which ECOLOGICAL GOODS and ECOLOGICAL SERVICES can be obtained and maintained; the useful aspects of the NATURAL ENVIRONMENT. The concept involves considering the Earth's resources within the context of a capitalist mode of production, including the ways processes of capitalist production affect and are affected by the natural environment. Using natural capital requires trade-offs between its various functions, which include providing the following: sources of *food, fibres, water, medicines* and *raw materials* that are directly exploited by humans (*provisioning services*); sources of the processes that indirectly allow further exploitation of natural resources, such as PRIMARY PRODUCTIVITY and *pollination* (*supporting services*); natural mechanisms responsible for maintaining ENERGY FLOW and MINERAL CYCLING, *climate regulation*, dispersal of POLLUTANTS, DECOMPOSITION of WASTE, *water circulation*, FLOOD CONTROL MEASURES and PEST CONTROL (*regulating services*); and the benefits gained by people from the natural environment for living, working and RECREATION (*cultural services*). Another approach to natural capital recognises three main types, each of which requires a different approach to MANAGEMENT:

1. Non-renewable and exhaustible resources (e.g. fossil fuels and metals)

2. Renewable but exhaustible resources (e.g. fish stocks, water and soil)

3. Renewable and non-exhaustible resources (e.g. wind and wave energy)   JAM

[*See also* CONSERVATION, NATURAL RESOURCES MANAGEMENT, NON-RENEWABLE RESOURCES, RENEWABLE RESOURCES, SUSTAINABLE CONSUMPTION, SUSTAINABLE PRODUCTION]

Aronson J, Milton SJ and Blignaut JN (eds) (2007) *Restoring natural capital: Science, business, and practice.* Washington, DC: Island Press.
Ehrlich PR, Kareiva PM and Gretchen CD (2012) Securing natural capital and expanding equity to rescale civilization. *Nature* 486: 68–73.
European Environment Agency (EEA) (2010) *The European environment: State and outlook 2010, synthesis.* Copenhagen: EEA.

Jansson AM, Hammer M, Folke C and Costanza R (eds) (1998) *Investing in natural capital: The ecological economics approach to sustainability.* Washington, DC: Island Press.

Kareiva P, Tallis H, Ricketts TH et al. (eds) (2011) *Natural capital: Theory and practice of mapping ecosystem services.* Oxford: Oxford University Press.

Millennium Ecosystem Assessment (2005) *Ecosystems and human well-being: Synthesis report.* Washington, DC: Island Press.

Tisdell C (2011) Biodiversity conservation, loss of natural capital and interest rates. *Ecological Economics* 70: 2511–2515.

Turner WR, Brandon K, Brooks TM et al. (2012) Global biodiversity conservation and the alleviation of poverty. *BioScience* 62: 85–92.

natural capitalism A strategy for sustainable economies, which safeguards the natural resource base that ultimately underpins all economic activity. Conventional CAPITALISM is wasteful and polluting. In the United States, materials used by the 'metabolism' of industry involve over 20 times the total weight of the American population on a daily basis. Natural capitalism advocates, in particular, reduction of materials and energy used *per unit output* and the eventual elimination of WASTE by emulating the RECYCLING that characterises natural *ecosystems*. It has been suggested that to achieve sustainable economies worldwide, a 50 per cent reduction in the intensity of materials and energy use is required and that, given the inability of DEVELOPING COUNTRIES to achieve this target, developed nations should aim for a 90 per cent reduction.   *JAM*

[*See also* CAPITALISM, ENVIRONMENTAL ECONOMICS, GLOBAL CAPITALISM, INDUSTRIAL ECOLOGY, POLITICAL ECOLOGY, SUSTAINABILITY]

Hawken P, Lovins AB and Lovins LH (2005) *Natural capitalism: The next industrial revolution, 2nd edition.* London: Earthscan.

Kareiva P, Tallis H, Ricketts TH et al. (eds) (2011) *Natural capital: Theory and practice of mapping ecosystem services.* Oxford: Oxford University Press.

Porritt J (2007) *Capitalism as if the world matters, 2nd edition.* London: Earthscan.

natural change Environmental change that is not influenced by any human activity. In the context of HUMAN IMPACT ON CLIMATE, for example, it is often referred to as the natural CLIMATIC VARIABILITY or the natural CLIMATIC VARIATION. A major concern is to identify when climatic change exceeds this natural BACKGROUND LEVEL and hence when a human imprint can be discerned in the climatic record. In relation to DEMOGRAPHIC CHANGE, the natural change in a population is the difference between the BIRTH RATE and the DEATH RATE.   *AHP*

[*See also* ATTRIBUTION, DETECTION, NATURAL, NATURAL ENVIRONMENTAL CHANGE]

natural disaster The outcome of a NATURAL HAZARD where major damage is inflicted on ecosystems, landscapes or the human environment by a natural agency such as storm, flood or volcanic eruption. According to the International Strategy for Disaster Reduction (ISRD), nearly 9,000 disastrous events (excluding EPIDEMICS) accounted for 2.3 million deaths between 1975 and 2008. Although deaths were concentrated in DEVELOPING COUNTRIES, absolute economic losses were greater in developed countries. The process of recovery from a natural disaster may be characterised as a sequence of four overlapping stages: *emergency*, RESTORATION, *replacement reconstruction* and *developmental reconstruction* (see Figure).   *JAM*

[*See also* DISASTER, ENVIRONMENTAL DISASTER, RISK, VULNERABILITY]

Alexander D (1993) *Natural disasters.* London: UCL Press.

Bankoff G, Frerks G and Hilhorst T (eds) (2004) *Mapping vulnerability: Disasters, development and people.* London: Earthscan.

Etkin D, Medalye J and Higuchi K (2012) Climate warming and natural disaster management: An exploration of the issues. *Climate Change* 112: 585–599.

ISDR (2009) *Global assessment report on disaster risk reduction.* Geneva: United Nations.

Kates RW and Pijawka D (1977) From rubble to monument: The pace of reconstruction. In Haas J, Kates M and Bowden M (eds) *Disaster and reconstruction.* Cambridge: MIT Press, 1–23.

Richardson HW, Gordon P and Moore II JE (eds) (2009) *Natural disaster analysis after hurricane Katrina: Risk assessment, economic impacts and social implications.* Cheltenham: Edward Elgar.

Svensson H (2009) *The end is nigh: A history of natural disasters.* London: Reaktion Books.

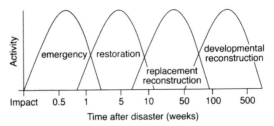

Natural disaster *A model of the stages of recovery following a natural disaster (Kates and Pijawka, 1977).*

natural environment (1) Those aspects of the ENVIRONMENT that are not made by people (but which people may influence). Arguably these include most living (*biotic environment*) and non-living (*abiotic environment*) things. (2) Environments in which HUMAN IMPACTS are minimal. A shrinking proportion of the Earth's environments qualify under this definition. *JAM*

[*See also* ANTHROPOGENIC, BUILT ENVIRONMENT, NATURAL ENVIRONMENTAL CHANGE, NATURAL ENVIRONMENTAL SCIENCES, NATURE, SEMI-NATURAL VEGETATION]

natural environmental change   (1) Those aspects of ENVIRONMENTAL CHANGE caused by natural forces rather than human agency. Natural environmental change has occurred throughout Earth's history. Before the human species evolved, all environmental changes were NATURAL, in more recent times, natural environmental changes may be considered as a background and baseline to human-induced environmental changes. (2) Any change to the NATURAL ENVIRONMENT, irrespective of cause: changes to the natural environment are nevertheless increasingly caused by human activity. *JAM*

Hulme M, Barrow EM, Arnell NW et al. (1999) Relative impacts of human-induced climate change and natural climate variability. *Nature* 397: 688–691.
Mannion AM (1999) *Natural environmental change.* London: Routledge.
Vogiatzakis IN and Pungetti G (eds) (2010) *Mediterranean island landscapes: Natural and cultural approaches.* Berlin: Springer.

natural environmental sciences   Sciences such as ECOLOGY, ENVIRONMENTAL SCIENCE, GEOLOGY and PHYSICAL GEOGRAPHY that investigate the nature, dynamics and evolution of the NATURAL ENVIRONMENT, including *human impacts.*   *JAM*

natural experiment   An 'experiment' carried out in the environment, rather than under controlled, for example, laboratory, conditions. Traditionally defined EXPERIMENTS, involving *controls* and predetermined variables, are rarely if ever possible in the 'natural' environment without large-scale intervention. Natural experiments are, necessarily, less formally defined and lack precise controls. By ingenious use of proxy-based reconstructions of past conditions as a basis for controlling the ways in which environmental processes and variables have interacted, it has sometimes been possible to design *post hoc experiments* that provide a basis for identifying common causes and eliminating locally contributing processes. In this way, it proved possible to identify the overriding cause of the ACIDIFICATION of surface waters in the United Kingdom, Europe and the United States as ACID DEPOSITION from industrial processes such as power generation from FOSSIL FUELS rather than changes in agricultural land

MANAGEMENT, commercial AFFORESTATION or long-term NATURAL soil acidification, each of which may have had a compounding effect on a local or regional scale. In the case of contemporary natural experiments, the term is often used to describe co-ordinated observational campaigns which are designed to test HYPOTHESES, to characterise and quantify major environmental processes or to improve the basis for ENVIRONMENTAL MODELLING. Such 'experiments' may involve intervention in the environmental system (e.g. OCEAN FERTILISATION experiments), though this is not always the case.   *FO*

[*See also* CHRONOSEQUENCE, EXPERIMENTAL DESIGN, FIELD RESEARCH, MESOCOSM, MICROCOSM, PALAEOENVIRONMENTAL RECONSTRUCTION, PROXY RECORDS]

Battarbee RW (ed.) (1990) *Palaeolimnology and lake acidification.* London: Royal Society.
Deevey ES (1969) Coaxing history to conduct experiments. *Bioscience* 19: 40–43.
Diamond J and Robinson JA (2010) *Natural experiments of history.* Cambridge, MA: Harvard University Press.
Strong AL, Chisholm S, Miller C and Cullen J (2009) Ocean fertilization: Time to move on. *Nature* 461: 347–348.
Strong AL, Cullen JJ and Chisholm SW (2009) Ocean fertilization, science, policy, and commerce. *Oceanography* 22: 237–261.

natural gamma radiation   The emission of *gamma rays* by naturally occurring radioactive MINERALS, notably those containing POTASSIUM (K), *uranium* and *thorium*. It is commonly measured as a continuous record from CORES and BOREHOLES, indicating the presence of CLAY MINERALS and organic-rich SHALES.   *MRT*

Hoppie BW, Blum P and Shipboard Scientific Party (1994) Natural gamma-ray measurements on ODP cores: Introduction to procedures with examples from Leg 150. *Proceedings of the Ocean Drilling Program, Initial Reports* 150: 51–59.

natural gas   The gaseous form of PETROLEUM, exploited as a FOSSIL FUEL by drilling into buried rocks, usually SEDIMENTARY ROCKS, that hold the gas in their pore spaces or other cavities, commonly overlying accumulations of CRUDE OIL. Chemically, natural gas is dominated by *methane* (CH_4). In the past, low energy prices and distance from markets led to natural gas being flared off in many oilfields, particularly in the Middle East. Rising prices since the 1970s and new developments in its liquid storage have led to increasing use. Natural gas is a cleaner ENERGY RESOURCE than crude oil. It accounted for about 24 per cent of world energy consumption at the end of 2010 (40 per cent for the UK). At present rates of consumption, global RESERVES of natural gas are sufficient to last for about 60 years, although the date for exhaustion has changed little for at least 10 years. About

72 per cent of natural gas reserves are held by the countries of the Middle East and the former Soviet Union, with 24 per cent in the Russian Federation, 16 per cent in Iran and 14 per cent in Qatar. *GO*

British Petroleum (BP) (2011) *BP statistical review of world energy, June 2011.* London: BP.

Gluyas J and Swarbrick R (2004) *Petroleum geoscience.* Oxford: Blackwell.

natural hazards Elements of the physical environment potentially harmful to humankind and caused by natural forces, which are generally extreme cases of phenomena or EXREME EVENTS. This definition excludes POLLUTION, which is normally considered to be caused only by humankind, and biological events such as EPIDEMICS infestation by PESTS (plants and animals). Consequently, natural hazards are usually considered to be either *meteorological/climatic* or *geological/geomorphological* hazards.

The most important meteorological natural hazards include FLOODS, TROPICAL CYCLONES, DROUGHTS and TORNADOS, but other extreme weather events are also natural hazards: *blizzards* (a combination of near-gale force winds with heavy snow), *fog* (defined internationally as visibility reduced to less than 1 km by suspended water droplets—in the United Kingdom it generally refers to visibility of less than 180 m), *frost* (when the air temperature is below 0°C), *hailstorms* (solid precipitation in the form of pieces of ice falling from cumulonimbus clouds), *heat waves* (extremely high temperatures accompanied by high humidity, which cause physiological stress) and *lightning strikes* (lightning is a giant spark between unlike electrical charges in clouds and on the ground). These meteorological hazards can be predicted over short time periods (usually hours or a day or two) in a particular location.

The main geological natural hazards are EARTHQUAKES, VOLCANIC ERUPTIONS, AVALANCHES, LANDSLIDES and TSUNAMIS. These natural hazards cannot be predicted in time with accuracy or location, although some places are more prone to them than others. AD 2005 was the most hazardous year in over three decades for NATURAL DISASTERS with 432 events that cost US$240 billion. AD 2010 (with 373 natural disasters) ranked the deadliest in two decades with 296,800 people killed. The relevant DATABASE of the Centre for Research on the Epidemiology of Disasters (CRED) in Louvain, Belgium, known as *EM-DAT* (*The International Disaster Database*), can be accessed at http://www.emdat.be/natural-disasters-trends. *BDG*

[*See also* FORECASTING, RAPID-ONSET HAZARDS, RISK, VULNERABILITY]

Blaikie P, Cannon T, Davis I and Wisner B (2005) *At risk: Natural hazards, people's vulnerability and disasters, 2nd edition.* London: Routledge.

Blong R (1997) A geography of natural perils. *Australian Geographer* 28: 7–28.

Bryant EA (2005) *Natural hazards, 2nd edition.* Cambridge: Cambridge University Press.

Burton I, Kates RW and White GF (1993) *The environment as hazard, 2nd edition.* New York: Guilford Press.

Dobrowsky PT (ed.) (2012) *Encyclopedia of natural hazards.* Dordrecht: Springer.

McGuire WJ, Kilburn CR and Mason IM (2002) *Natural hazards and environmental change.* London: Hodder Arnold.

Smith K and Petley DN (2009) *Environmental hazards: Assessing risk and reducing disaster, 5th edition.* Abingdon: Routledge.

Terry JP and Goff J (eds) (2012) *Natural hazards in the Asia-Pacific region: Recent advances and emerging concepts.* Bath: Geological Society.

Willetts O, Burton P, Kilburn C and McGuire WJ (2004) *World atlas of natural hazards.* London: Hodder Arnold.

natural networks A suite of high-quality sites which collectively contain the diversity and area of habitat that are needed to support species and which have ecological connections between them (Lawton et al., 2010). They are essentially functioning green areas of the LANDSCAPE, including rivers, gardens, parks, farmlands, coastlands and forests. *AEV*

Lawton JH, Brotherton PNM and Brown VK (2010) *Making space for nature: A review of England's wildlife sites and ecological network.* London: Department for Environment, Food and Rural Affairs (DEFRA).

natural remanent magnetism (NRM) Magnetism acquired by rocks and unconsolidated SEDIMENTS during formation as a result of alignment of contained ferromagnetic particles with the Earth's ambient geomagnetic field. Measurement of NRM is used in PALAEOMAGNETIC DATING. *MHD*

Thompson R (1991) Palaeomagnetic dating. In Smart PL and Frances PD (eds) *Quaternary dating methods: A users guide.* Cambridge: Quaternary Research Association, 177–198.

natural resources Aspects of NATURE that have utility (i.e. they have the capacity to meet perceived needs or wants). Utility may be determined by economic or non-economic (AESTHETICS, morals, etc.) criteria. Some natural resources are directly available (e.g. air, wind, solar energy) and others must be processed by humans first. Natural resources may be classified as follows: (1) those which are finite (NON-RENEWABLE RESOURCES). (2) those which offer sustainable use if subjected to effective MANAGEMENT (*potentially renewable resources*) or (3) those sustained easily because they are self-replacing provided they are not mismanaged (RENEWABLE RESOURCES). Some finite resources can be recycled. Natural resources are often *common property resources*—owned and controlled by no single person/body but subject to use by many and sometimes global in extent—which presents further management challenges. *CJB*

[*See also* RESERVES, WASTE MANAGEMENT]

Collier P and Venables AJ (eds) (2011) *Plundered nations? Successes and failures in natural resource extraction.* Basingstoke: Palgrave Macmillan.
Perman R, Ma Y, Common M and Maddison D (2011) *Natural resource and environmental economics, 4th edition.* Harlow: Addison-Wesley.

natural resources management   In Western societies, from roughly the eighteenth century onwards, the general view was that natural resources should be used to improve human wealth and welfare. MANAGEMENT focused on specific resources mainly for short-term gain, minimum oversight by authorities was laissez-faire, development was often wasteful and POLLUTION was rife, with little thought about consequences. The benefits went to companies, governments, special interest groups and individuals. From the 1920s, resources managers began to note broader issues, such as CONSERVATION, and some adopted bioregional (see BIOREGIONALISM) and more integrated approaches. Before the 1970s, the emphasis was on *exploitation* rather than *stewardship*; by the 1980s, there was growing awareness of the social and environmental impacts of development, the need to protect BIODIVERSITY and the desirability of achieving SUSTAINABLE DEVELOPMENT. Management could no longer be laissez-faire: in a crowded world it needed to be controlled, anticipatory and broader in outlook. Gradually a more aware form of resource stewardship began to evolve, merging into ENVIRONMENTAL MANAGEMENT. *CJB*

[*See also* ENVIRONMENTAL ECONOMICS, NON-RENEWABLE RESOURCES, RENEWABLE RESOURCES]

Anderson DA (2010) *Environmental economics and natural resource management, 3rd edition.* London: Routledge.
Hering JG and Ingold KM (2012) Water resources management: What should be integrated? *Science* 336: 1234–1235.
Rees J (1990) *Natural resources: Allocation, economics and policy, 2nd edition.* London: Routledge.
Tietenberg T and Lewis L (2008) *Environmental and natural resource economics, 8th edition.* Harlow: Pearson Education.

natural selection   An evolutionary process identified and described, independently, by Alfred Russel Wallace (1823–1913) and Charles Robert Darwin (1809–1882). Both noted that individuals differ, and that differential survival means that subsequent generations will have characteristics more suited to the ENVIRONMENT. 'This preservation of favourable variations and the rejection of injurious variations, I call Natural Selection' (Darwin, 1859: 81). *KDB*

[*See also* DARWINISM, EVOLUTION, SELECTIVE BREEDING]

Darwin C (1859) *On the origin of species by means of natural selection, or the preservation of favoured races in the struggle for life.* London: John Murray.
Godfrey-Smith P (2009) *Darwinian populations and natural selection.* Oxford: Oxford University Press.
Smith CH and Beccaloni G (eds) (2008) *Natural selection and beyond: The intellectual legacy of Alfred Russel Wallace.* Oxford: Oxford University Press.
Wallace AR (1895) *Natural selection and tropical nature.* London: Macmillan.

natural vegetation   Vegetation that has become established without human intervention and that continues to remain undisturbed by human actions. Because of the ubiquity of HUMAN IMPACT ON TERRESTRIAL VEGETATION, natural vegetation is increasingly illusive, if not impossible to find today. However, owing to natural DISTURBANCES, natural vegetation should not be viewed as characterised by an entirely closed VEGETATION CANOPY but would, even in the past, have included numerous open HABITATS, some short lived. *IFS/JAM*

[*See also* CLIMAX VEGETATION, HUMAN IMPACT ON VEGETATION HISTORY, POTENTIAL NATURAL VEGETATION, PRIMARY WOODLAND, SEMI-NATURAL VEGETATION, VEGETATION HISTORY, WILDWOOD]

Svenning JC (2002) A review of natural vegetation openness in north-western Europe. *Biological Conservation* 104: 133–148.

naturalisation   (1) The process by which an introduced species (see INTRODUCTION) adapts to local conditions and becomes integrated with the NATURAL or semi-natural ecosystem into which it has been introduced. (2) Human intervention in an environmental SYSTEM with the effect of driving the system towards greater diversity, stability and/or SUSTAINABILITY by natural processes. *JAM*

[*See also* INVASIVE SPECIES]

Lever C (2009) *The naturalized animals of Britain and Ireland.* London: New Holland Publishers.
Rhodes BL, Wilson D, Urban M and Hendricks E (1999) Interaction between scientists and non-scientists in community-based watershed management: Emergence of the concept of stream naturalization. *Environmental Management* 24: 297–308.

naturalised species   Individuals and POPULATIONS of a *species* that are descendants of those introduced to a site outside their natural historic RANGE. If these populations have persisted for a sufficient period to allow local adaptation, they may sometimes be viewed as integral components of the *natural ecosystem*. *MVL*

[*See also* EXOTIC SPECIES, INDIGENOUS PEOPLES, INTRODUCTION, NATURAL, NATURALISATION]

Coates P (2006) *American perceptions of immigrant and invasive species: Strangers on the land.* Berkeley: University of California Press.

Cox GW (2004) *Alien species and evolution: The evolutionary ecology of exotic plants, animals, microbes, and interacting native species.* Washington, DC: Island Press.

Lever C (2009) *The naturalized animals of Britain and Ireland.* London: New Holland.

nature 'Perhaps the most complex word in the [English] language' (Williams, 1976: 184), but currently most commonly used to signify the natural world, or the biophysical environment, as opposed to the ANTHROPOGENIC environment, or CULTURE. However, 'nature' has both concrete and abstract meanings and its interpretations have changed throughout history. In the Western world, five important categories of meanings can be recognised: (1) nature as a physical place, notably places 'unspoiled' by human modification; (2) nature as the Earth or the universe as a collective phenomenon, including or excluding humans; (3) nature as an essence or quality that pervades the functioning of the Earth or universe; (4) nature as an inspiration and guide to human affairs; and (5) nature as the conceptual opposite to culture. *JAM*

[*See also* BALANCE OF NATURE, CONSERVATION, ENVIRONMENTAL ETHICS, ENVIRONMENTALISM, SOCIAL CONSTRUCTION, WILDERNESS CONCEPT]

Castree N (2005) *Nature.* London: Routledge.

Coates P (1998) *Nature: Western attitudes since ancient times.* Cambridge: Polity Press.

Inglis D, Bone J and Wilkie R (eds) (2005) *Nature, 4 volumes.* London: Routledge.

Macnaghten P and Urry J (1998) *Contested natures.* London: Sage.

Williams P (1976) *Keywords: A vocabulary of culture and society.* Oxford: Oxford University Press.

nature reserve A haven for wildlife where nature CONSERVATION is prioritised. Land is usually owned by, or subject to the regulations of, international, national or non-governmental conservation organisations. MANAGEMENT plans and conservation practices are tailored towards both individual species and extensive rare habitats. *MLW*

[*See also* ENVIRONMENTAL PROTECTION, PROTECTED AREAS, WILDLIFE CONSERVATION]

Reed TM and Lawson TG (1997) *Nature reserves: Who needs them?* London: Joint Nature Conservation Committee.

NBS oxalic acid *Standard reference materials* produced by the United States National Bureau of Standards, Washington, for RADIOCARBON DATING laboratories. The first batch of material has run out and a second was prepared in 1974. The nomenclature of the first standard is variously given as NBS Oxalic acid I, NBS OxI, HOxI, SRM4990B; 0.95 times the activity is equivalent to AD 1950 with the $\delta^{13}C$ fractionation normalised to $-19.0‰$ (per mille). The new standard is described as NBS OxII, HOxII, SRM4990C; 0.7459 times the activity is equivalent to AD 1950 with the $\delta^{13}C$ fractionation normalised to $-25.0‰$. *PQD*

[*See also* REFERENCE STANDARD, STANDARD SUBSTANCE]

Stuiver M (1983) International agreements and the use of the new oxalic standard. *Radiocarbon* 25: 793–795.

Neanderthal demise Classified as *Homo neanderthalensis* (or as a subspecies of *Homo sapiens*), Neanderthal populations lived alongside ANATOMICALLY MODERN HUMANS (AMH) across Europe and the Middle East, as late as 28,000 years ago. They were the first 'humans' to colonise the Eastern European Plain and appear to have been well adapted to the cold climates of the PLEISTOCENE. According to *ancient* DNA evidence, limited interbreeding took place with AMH. However, Neanderthal demise has been primarily attributed to competition with AMH, possibly in combination with CLIMATIC CHANGE. *JAM*

[*See also* ARCHAIC HUMANS]

Chacon MG, Vaquero M and Carbonell E (eds) (2012) The Neanderthal home: Spatial and social behaviours. *Quaternary International* 247: 1–362 [Special Issue].

Finlayson C (2010) *The humans who went extinct: Why Neanderthals died out and we survived.* Oxford: Oxford University Press.

Gilligan I (2008) Neanderthal extinction and modern human behaviour: The role of climate change and clothing. *World Archaeology* 39: 499–514.

Jimenez-Espejo F, Martinez-Ruiz F, Finlayson C et al. (2007) Climate forcing and Neanderthal extinction in Southern Iberia: Insights from a multiproxy marine record. *Quaternary Science Reviews* 26: 836–852.

neap tide The TIDE of comparatively small range occurring twice each lunar month. *RAS*

[*See also* SPRING TIDE]

near-Earth object (NEO) ASTEROID, METEORITE, or cometary debris with a solar orbit approaching within 1.3 Astronomical Units of the Earth. NEOs are classed as *Earth Crossing* or *Earth Approaching*, or as *Potentially Hazardous Asteroids* (PHAs) if their minimum potential Earth distance is less than 7,480,000 km. NEOs and PHAs are potential BOLIDES. *JBH*

[*See also* COMETARY IMPACT, COMETS, METEORITE IMPACT]

Barucci MA, Yoshikawa M, Michel P et al. (2008): MARCO POLO: Near earth object sample return mission. *Experimental Astronomy* 23: 785–808.

Rabinowitz D, Helin E, Lawrence K and Pravdo S (2000) A reduced estimate of the number of kilometre-sized near-Earth asteroids. *Nature* 403: 165–166.

near-infrared reflectance spectroscopy (NIRS)

A rapid, NON-DESTRUCTIVE SAMPLING technique for the acquisition of data on the chemical constituents of organic materials (e.g. *nitrogen, phosphorus, lignin, carbon content* and ASH CONTENT) and for obtaining PROXY DATA on a potentially wide range of biological and environmental variables including, for example, the digestibility of *forage*; the BULK DENSITY, HUMIFICATION, moisture content and MACROFOSSIL content of *peat*; and lake-water pH, carbon, nitrogen and phosphorus concentration, *diatom* content and inferred mean July air temperature from LACUSTRINE SEDIMENTS. Samples are scanned using a SPECTROPHOTOMETER producing reflectance spectra, the CALIBRATION of which may involve use of a TRAINING SET of samples that have been analysed by conventional laboratory techniques and/or correlation with environmental variables.   *JAM*

Bokobza L (1998) Near infrared spectroscopy. *Journal of Near Infrared Spectroscopy* 6: 3–17.
Fuentes M, Hidalgo C, Gonzalez-Martin I et al. (2012) NIR spectroscopy: An alternative for soil analysis. *Communications in Soil Science and Plant Analysis* 43: 346–356.
Korsman T, Renberg I, Dåbakk E and Nilsson MB (2001) Varve chronology techniques. In Last WM and Smol JP (eds) *Tracking environmental change using lake sediments, volume 2*. Dordrecht: Kluwer, 299–317.
Malley DF, Lockhart L, Wilkinson P et al. (2000) Determination of carbon, carbonate, nitrogen, and phosphorus in freshwater sediments by near-infrared reflectance spectroscopy: Rapid analysis and a check on conventional analytical methods. *Journal of Paleolimnology* 24: 415–425.
Rosén P, Dåbakk E, Renberg I et al. (2000) Near-infrared spectrometry (NIRS): A new tool for inferring past climatic changes from lake sediments. *The Holocene* 10: 161–166.
Zan FY, Huo SL, Xi BD et al. (2010) Rapid determination of nutrient components in lake sediments using near infrared spectroscopy. *Spectroscopy and Spectral Analysis* 30: 2624–2627.

nearest neighbour

In IMAGE PROCESSING, local techniques such as FILTERING require the specification of the positions that will have influence in the computation of new values. The PIXELS in those positions are called the nearest neighbours, and they are usually specified using distance criteria. Nearest-neighbour techniques are also widely used in ECOLOGY for *pattern analysis*.   *ACF*

nebkha

A small sand accumulation or DUNE formed around vegetation (also known as a *phytogenetic dune*, shrub-coppice dune or *nabkha*) rather than a dune vegetated following their formation.   *SHD*

[*See also* DIKAKA]

Markley WG and Wolfe NA (1994) The morphology and origin of nabkhas, region of Mopti, Mali, West Africa. *Journal of Arid Environments* 28: 13–30.

needle ice

Long, thin, needle-like ice crystals that form perpendicular to the ground surface. Also known as *pipkrake*, it forms at night when there is excessive radiation cooling, causing *ice segregation* in the surface soil layers. It can play a role in the formation of PATTERNED GROUND and in SOIL EROSION.   *HMF*

Branson J, Lawler DM and Glen JW (1996) Sediment inclusion events during needle ice growth: A laboratory investigation of the role of soil moisture and temperature fluctuations. *Water Resources Research* 32: 459–466.
Lawler DM (1988) Environmental limits of needle ice: A global survey. *Arctic and Alpine Research* 20: 137–159.
Mackay JR and Mathews WH (1974) Needle ice striped ground. *Arctic and Alpine Research* 6: 79–84

negative feedback

An interaction within a SYSTEM, which causes a reduction or *dampening* of the response of the system to a force. Thus, a negative feedback has a stabilising effect.   *AHP*

[*See also* FEEDBACK MECHANISMS]

nekton/nektonic/nektic

Free-swimming organisms.   *GO*

[*See also* PELAGIC ORGANISM, PLANKTON/PLANKTONIC/ PLANKTIC]

nemoral forest

The TEMPERATE FORESTS dominated by frost-resistant broadleaved deciduous trees of the genus *Fraxinus, Acer, Ulmus, Tilia, Fagus* and *Quercus*. This is an alternative term for the deciduous summer forest VEGETATION FORMATION TYPE with three formations (in North America, Europe and the Far East). The *boreo-nemoral forest*, which occupies the transition zone between the nemoral forest and the BOREAL FOREST, is characterised by a mixture of deciduous and coniferous tree dominants.   *JAM*

Sjörs H (1963) Amphi-Atlantic zonation, nemoral to Arctic. In Löve Á and Löve D (eds) *North Atlantic biota and their history*. Oxford: Pergamon Press, 109–125.
Svenning JC and Skov F (2006) Potential impact of climate change on the northern nemoral forest herb flora of Europe. *Biodiversity and Conservation* 15: 3341–3356.

neocatastrophism

The principle that the GEOLOGICAL RECORD includes the products of rapid, short-lived natural processes (EVENTS), sometimes operating on scales that have not been observed in historical times. For example, most geologists accept that METE-ORITE IMPACTS have influenced Earth history, and the observed impact of *Comet Shoemaker-Levy 9* on *Jupiter* in 1994 elevated COMETARY IMPACT (albeit on another planet) to the status of an *actualistic process* (see ACTUALISM). On a regional scale, catastrophic JÖKULHLAUPS are accepted as having been a factor in the creation of many *proglacial* landforms such as the CHANNELED SCABLANDS. Neocatastrophism can be distinguished from a mere revival of CATASTROPHISM by its moderation (it allows for a background GRADUALISM and is cautious about events that are global in impact) and its rejection of direct *supernatural intervention*. Neocatastrophism can be seen as the recognition that, given the vast lengths of GEOLOGICAL TIME, the rare event is inevitable. *CDW*

Ager DV (1993) *The new catastrophism: The importance of the rare event in geological history.* Cambridge: Cambridge University Press.
Baker VR and Bunker RC (1985) Cataclysmic Late Pleistocene flooding from glacial Lake Missoula: A review. *Quaternary Science Reviews* 4: 1–41.
Hallam T (2004) *Catastrophes and lesser calamities: The causes of mass extinctions.* Oxford: Oxford University Press.
Marriner N, Morhange C and Skrimshire S (2010) Geoscience meets the four horsemen? Tracking the rise of neocatastrophism. *Global and Planetary Change* 74: 43–48.
Palmer T (2003) *Perilous planet Earth: Catastrophes and catastrophism through the ages.* Cambridge: Cambridge University Press.

neocolonialism The concept of continuing outside economic and political control over independent DEVELOPING COUNTRIES (a continuing colonial system by another name). Control may be direct or indirect and exerted, for example, through the provision of various types of aid from governments, intergovernmental organisations, or multinational companies. Neocolonial theory argues that developing countries are powerless in the face of the world economy being fixed in the interest of developed countries, especially the Western powers. *JAM*

[*See also* COLONIALISM, POSTCOLONIALISM]

Hoogvelt A (2001) *Globalization and the postcolonial world: The new political economy of development.* Baltimore: Johns Hopkins University Press.
Perkins J (2007) *The secret history of the American empire.* New York: Dutton.

neo-Darwinism The theory of EVOLUTION combining DARWINISM with modern ideas from GENETICS, that is, the so-called *modern synthesis*. *JAM*

Berry RJ (1982) *Neo-Darwinism.* London: Arnold.

neoecology The study of the interactions of extant organisms and their present-day environment. It contrasts with PALAEOECOLOGY. *SHJ*

[*See also* COMMUNITY ECOLOGY, ECOLOGY, MACRO-ECOLOGY]

Harris DR and Thomas KD (eds) (1991) *Modelling ecological change: Perspectives from neoecology, palaeoecology and environmental archaeology.* London: Institute of Archaeology.
Rull V (2010) Ecology and palaeoecology: Two approaches, one objective. *Open Ecology Journal* 3: 1–5.

neo-evolutionism A PARADIGM relating to knowledge and understanding of human behaviour and cultural change. It holds that CULTURAL CHANGE is characterised by distinct linear or multilinear sequences controlled largely by deterministic evolutionary processes involving uncontrollable factors of the human environment, e.g. DEMOGRAPHIC CHANGE, TECHNOLOGICAL CHANGE and economics. Concepts of the NEOLITHIC and *urban revolutions* are neo-evolutionist. *JAM*

Trigger BG (2006) *A history of archaeological thought, 2nd edition.* Cambridge: Cambridge University Press.

Neogene A SYSTEM of rocks; a PERIOD of GEOLOGICAL TIME from 23.0 to 2.6 million years ago, part of the CENOZOIC (*CAINOZOIC*) era, comprising the MIOCENE and PLIOCENE epochs and formerly known as the Upper or Late TERTIARY. HOMINIDS first appeared in the Neogene, and major environmental events included climatic cooling leading to the ICEHOUSE CONDITION of the QUATERNARY period, the enlargement of the Antarctic ICE SHEET and the first evidence of Arctic GLACIATION. *GO*

[*See also* GEOLOGICAL TIMESCALE]

Ogg JG, Ogg G and Gradstein FM (2008) *The concise geologic time scale.* Cambridge: Cambridge University Press.
Retallack GJ (2004) Late Miocene climate and life on land in Oregon within a context of Neogene global change. *Palaeogeography, Palaeoclimatology, Palaeoecology* 214: 97–123.
Van Dam JA (2006) Geographic and temporal patterns in the Late Neogene (12-3 Ma) aridification of Europe: The use of small mammals as paleoprecipitation proxies. *Palaeogeography, Palaeoclimatology, Palaeoecology* 238: 190–218.

neoglaciation The concept of glacier recrudescence or regrowth following partial or complete glacier melting during the HOLOCENE THERMAL OPTIMUM or *hypsithermal* of the early- to mid-HOLOCENE. Neoglaciation in this conventional sense occurred at different times in different regions in response to LATE-HOLOCENE CLIMATIC DETERIORATION. In southern Norway, the conventional onset of neoglaciation occurred at

6,500–6,000 cal BP. Before this, the largest ice-cap on mainland Europe (Jostedalsbreen) melted away. The term is also used with reference to *neoglacial events* (*Little Ice Age-type events*) whenever they occurred during the Holocene. Around 20 neoglacial events have been identified in Scandinavia and the Alps, of which over half were Europe-wide in extent. Increasing knowledge and understanding of these short-term, possibly abrupt, century- to millennial-scale GLACIER VARIATIONS suggest a hybrid model of neoglaciation in which the magnitude and frequency of neoglacial events increased in the late Holocene (see the Figure in the entry CENTURY- TO MILLENNIAL-SCALE VARIABILITY). The extent to which neoglaciation and neoglacial events were synchronous on a global scale is currently unknown. Continuous records of Holocene glacier variations are rare and few, if any, of these qualify as HIGH-RESOLUTION RECONSTRUCTIONS. Winkler and Matthews (2010) suggested that the first priority for research in this field should be to establish more firmly the spatial patterns in glacier behaviour at various scales.

JAM

[*See also* LITTLE ICE AGE, SUB-MILANKOVITCH]

Connor C, Streveler G, Post A et al. (2009) The Neoglacial landscape and human history of Glacier Bay, Glacier Bay National Park and Preserve, southeast Alaska, USA. *The Holocene* 19: 381–393.

Davis P, Menounos B and Osborn G (eds) (2009) Holocene and latest Pleistocene alpine glacier fluctuations: A global perspective. *Quaternary Science Reviews* 28(21–22): 2021–2238 [Special Issue].

Deline P and Orombelli G (2005) Glacier fluctuations in the western Alps during the Neoglacial, as indicated by the Miage moraine amphitheatre (Mont Blanc massif, Italy). *Boreas* 34: 456–467.

Holzhauser H, Magny M and Zumbühl HJ (2005) Glacier and lake-level variations in west-central Europe over the last 3,500 years. *The Holocene* 15: 789–801.

Matthews JA (2013) Neoglaciation in Europe. In Elias SA (ed.) *Encyclopedia of Quaternary science, volume 2, 2nd edition.* Amsterdam: Elsevier, 257–268.

Matthews JA and Dresser PQ (2008) Holocene glacier variation chronology of the Smørstabbtindan massif, Jotunheimen, southern Norway, and the recognition of century- to millennial-scale European Neoglacial Events. *The Holocene* 18: 181–201.

Matthews JA and Karlén W (1992) Asynchronous neoglaciation and Holocene climatic change reconstructed from Norwegian glacio-lacustrine sequences. *Geology* 20: 991–994.

Nesje A and Kvamme M (1991) Holocene glacier and climatic variations in western Norway: Evidence for early Holocene glacier demise and multiple Neoglacial events. *Geology* 19: 610–612.

Porter SC (2007) Neoglaciation in the American Cordilleras. In Elias SA (ed.) *Encyclopedia of Quaternary science.* Amsterdam: Elsevier, 1133–1142.

Porter SC and Denton GH (1967) Chronology of Neoglaciation in the North American Cordillera. *American Journal of Science* 265: 177–210.

Winkler S and Matthews JA (2010) Holocene glacier chronologies: Are 'high-resolution' global and inter-hemispheric comparisons possible? *The Holocene* 20: 1137–1147.

Neolithic   The archaeological period in which agriculture based on DOMESTICATION of animals and plants replaced HUNTING, FISHING AND GATHERING as the primary source of food. An agricultural economy required a sedentary lifestyle that created the stability for further social and technological developments, such as village societies and pottery firing. Originating in the Near East at ca 10,000 years ago, adoption of the Neolithic 'package' or *Neolithic Revolution*, spread throughout Eurasia and northern Africa, reaching the British Isles about 6,000 years ago (4000 BC). Also known as the *New Stone Age*, the Neolithic lasted until the BRONZE AGE, which began about 2200 BC in Britain.

The first landscape-scale impacts resulting from ANTHROPOGENIC activity occurred during the Neolithic as woodland was cleared to create fields for agriculture. In some *pollen diagrams*, this is indicated by a decline in ARBOREAL POLLEN and the appearance of herbaceous ANTHROPOGENIC INDICATOR species and the first appearance of CEREAL POLLEN grains, often with high concentrations of microscopic charcoal which may indicate that fire was used in FOREST CLEARANCE. Studies which employ sedimentary and chemical analyses have also shown a mineral SEDIMENT INFLUX which is interpreted as SOIL EROSION from forest clearance activity.

Early Danish studies revealed an apparent cyclicity in the pollen records of Neolithic LANDNÁM clearance phases that was related to SHIFTING CULTIVATION. In northwestern Europe, the appearance of anthropogenic indicator and cereal pollen often coincided with the ELM DECLINE. Elsewhere, particularly in southeast Europe, such clearly defined pollen responses are not evident and Neolithic impacts on the landscape are assumed to be minor, perhaps due to the use of naturally available open land.

ARG

[*See also* AGRARIAN CIVILISATION, AGRICULTURAL HISTORY, AGRICULTURAL ORIGINS, AGRICULTURAL REVOLUTION, ARCHAEOLOGICAL TIMESCALE, DEFORESTATION, FIRE IMPACTS: ECOLOGICAL, NEOLITHIC TRANSITION]

Bellwood P (2004) *First farmers: The origins of agricultural societies*: Chichester: Wiley.

Edmonds M (1999) Inhabiting Neolithic landscapes. *Quaternary Proceedings* 7: 485–492.

Milisauskas S (ed.) (2002) *European prehistory.* New York: Kluwer.

Smith AG (1981) The Neolithic. In Simmons IG and Tooley MMJ (eds) *The environment in British prehistory.* London: Duckworth, 125–209.

Van Andel TH and Runnels CN (1995) The earliest farmers in Europe. *Antiquity* 69: 481–500.

Neolithic Transition   The time interval in which a mobile MESOLITHIC lifestyle based on HUNTING, FISHING AND GATHERING was succeeded by a sedentary NEOLITHIC lifestyle based on the production of food from the DOMESTICATION of plants and animals. From origins in the Near East around 10,000 years ago, Neolithic technology radiated outwards to north Africa, southwest Asia and Europe, reaching Greece ca 9000 cal. BP and the British Isles by ca 6000 cal. BP. Recent discoveries in Southeast Asia have indicated that the Neolithic may have evolved independently at the same time in more than one location.   *ARG*

[*See also* OASIS HYPOTHESIS]

Ammerman AJ and Cavalli-Sforza LL (1984) *The Neolithic transition and the genetics of population in Europe.* Princeton: Princeton University Press.

Denham TP, Haberle SG, Lentfer C et al. (2003) Origin of agriculture at Kuk Swamp in the highlands of New Guinea. *Science* 301: 189–193.

Thomas J (2008) The Mesolithic-Neolithic transition in Britain. In Pollard J (ed.) *Prehistoric Britain.* Oxford: Blackwell, 58–89.

Willis KJ and Bennett KD (1994) The Neolithic transition: Fact or fiction? Palaeoecological evidence from the Balkans. *The Holocene* 4: 326–330.

Neoproterozoic   The latest ERA of the PROTEROZOIC eon, from 1,000 million years ago (Ma) to the beginning of the CAMBRIAN period (542 Ma). This episode in the Earth's history was characterised by several ICE AGES during which there is evidence for ice cover at low latitudes (see SNOWBALL EARTH), the existence of the SUPERCONTINENT of RODINIA and the rise of *metazoan organisms* (see EDIACARAN), setting the scene for the CAMBRIAN EXPLOSION.   *GO*

Kasemann SA, Prave AR, Fallick AE et al. (2010) Neoproterozoic ice ages, boron isotopes, and ocean acidification: Implications for a snowball Earth. *Geology* 38: 775–778.

neotectonics   Tectonic activity (DEFORMATION, UPLIFT or SUBSIDENCE) occurring during the relatively recent geological past. Some consider neotectonics to be currently active tectonics; others define a *neotectonic period* extending from as early as the mid-MIOCENE. Recent vertical crustal movements are characteristic of areas of *plate collision*, ISLAND ARCS and GLACIO-ISOSTASY. Detailed studies have enabled the reconstruction of numerous neotectonic movements taking place during recent centuries.   *AGD/JAM*

[*See also* GLACIOSEISMOTECTONICS, PLATE MARGIN, PLATE TECTONICS, TECTONICS]

Becker A (1993) An attempt to define a 'neotectonic period; for central and northern Europe. *Geologische Rindschau* 82: 67–83.

Dramis F and Tondi E (2005) Neotectonics. In Koster EA (ed.) *Physical geography of Western Europe.* Oxford: Oxford University Press, 25–38.

Olesen O, Blikra LH, Braathen A et al. (2004) Neotectonic deformation in Norway and its implications: A review. *Norwegian Journal of Geology* 84: 3–34.

Pavlides SB (1989) Looking for a definition of neotectonics. *Terra Nova* 1: 233–235.

Sieh K and Natawidjaja D (2000) Neotectonics of the Sumatran fault, Indonesia. *Journal of Geophysical Research B* 105: 28295–28326.

Slobelev SF, Hanon M, Klerx J et al. (2004) Active faults in Africa: A review. *Tectonophysics* 380: 131–137.

Terrinha P, Pinheiro LM, Henriet J-P et al. (2003) Tsunamigenic-seismogenic structures, neotectonics, sedimentary processes and slope instability on the southwest Portuguese margin. *Marine Geology* 195: 55–73.

Van Balen RT, Houtgast RF and Cloetingh SAPL (2005) Neotectonics in the Netherlands: A review. *Quaternary Science Reviews* 24: 439–454.

Vita-Finzi C (1985) *Neotectonics.* Chichester: Wiley.

nepheloid layer   A turbid, near-bottom layer in parts of the OCEANS and some large LAKES, with a relatively high concentration of very fine-grained sediment in *suspension* (average grain size 0.012 mm). The nepheloid layer is best developed in deep water with strong bottom currents, particularly along the western margins of OCEAN BASINS, where it may be several hundred metres thick.   *GO*

[*See also* CONTOUR CURRENT, PELAGIC SEDIMENT (PELAGITE), WATER MASS]

Bout-Roumazeilles V, Cortijo E, Labeyrie L and Debrabant P (1999) Clay mineral evidence of nepheloid layer contributions to the Heinrich layers in the northwest Atlantic. *Palaeogeography, Palaeoclimatology, Palaeoecology* 146: 211–228.

Open University Course Team (2005) *Marine biogeochemical cycles.* Oxford: Butterworth-Heinemann.

Ransom B, Shea KF, Burkett PJ et al. (1998) Comparison of pelagic and nepheloid layer marine snow: Implications for carbon cycling. *Marine Geology* 150: 39–50.

net primary productivity (NPP)   GROSS PRIMARY PRODUCTIVITY (GPP) less the chemical energy used in plant RESPIRATION. It is usually 80–90 per cent of GPP. Global mean NPP is 440 $g/m^2/year$. This is potentially available to *primary consumers* and/or HARVESTING.   *RJH*

Prieto-Blanco A, North PJ, Barnsley MJ and Fox N (2009) Satellite-driven modeling of net primary productivity

(NPP): Theoretical analysis. *Remote Sensing of Environment* 113: 137–147.

Vitousek PM, Ehrlich PR, Ehrlich AH and Matson PA (1986) Human appropriation of the products of photosynthesis. *BioScience* 36: 368–373.

net radiation   The difference between the radiation travelling downwards to the Earth's surface (direct and diffuse short-wave SOLAR RADIATION together with atmospheric counterradiation) and upwards from the Earth's surface (reflected short-wave radiation together with TERRESTRIAL RADIATION). Net radiation, the RADIATION BALANCE at a point on the Earth's surface, is measured using a *net radiometer* and is positive when the downward radiation exceeds the upward radiation. *JAM*

network analysis   An examination of a high-order system (or arrangement) of *nodes* interconnected by linear direction and magnitude. Network analyses are used to represent and MODEL the transport and flow of materials, for instance, water (in rivers and pipes), natural gas, sewage or vehicles from one place to another. A river network tends to form a tree (dendritic) structure, controlled by changes in terrain and water availability. Over time, direction, discharge and velocity of a river system may be calculated to determine the balance and efficiency of flow within the network. A GEOGRAPHICAL INFORMATION SYSTEM (GIS) allows these measurements to be monitored, modelled and displayed more systematically, as well as facilitate many other functions, such as *minimum cost paths*. By calculating a three-dimensional surface (see DIGITAL ELEVATION MODEL (DEM)), a GIS can determine the least resistance route for, say, a pipeline, with respect to gravity and landuse constraints. Network analysis also describes the operations of a type of DATABASE structure. It employs software pointers which explicitly link data items from one FIELD to another. More recently, network analysis has referred to distributed systems of interlinked *computers* around the world. *VM*

[*See also* NEURAL NETWORK, RELATIONAL DATABASE]

Gong JY and Xie J (2009) Extraction of drainage networks from large terrain datasets using high throughput computing. *Computers and Geosciences* 35: 337–346.

Kolaczyk ED (2009) *Statistical analysis of network data: Methods and models.* Berlin: Springer.

Newman M (2010) *Networks: An introduction.* Oxford: Oxford University Press.

Okabe A, Okunuki K and Shiode S (2006) SANET: A toolbox for spatial analysis on a network. *Geographical Analysis* 38: 57–66.

neural network   A form of *artificial intelligence* that mimics aspects of biological neural systems. It consists of a relatively large number of processing units that are highly interconnected. The most widely used

network is the *multilayer perceptron* which comprises a set of units arranged in a layered architecture with each unit connected to every unit in adjacent layers. This network learns by example to convert a set of input data (e.g. on past climate) into an output (e.g. future climate). To achieve this, a sample of cases with known inputs and outputs is used with a learning ALGORITHM to iteratively adjust the network's internal properties until it successfully predicts the output given the inputs. While each unit performs simple tasks, the entire network can solve problems that are complex, non-linear and poorly understood without making any assumptions about the data. Each type of network has its own application domain, but as a whole, neural networks are general-purpose computing tools capable of application to almost any problem. They are particularly useful in REMOTE-SENSING applications for CLASSIFICATION or as a non-parametric alternative to REGRESSION ANALYSIS. *GMF*

[*See also* NON-PARAMETRIC STATISTICS]

Bishop CM (1995) *Neural networks for pattern recognition.* Oxford: Oxford University Press.

neutron activation analysis (NAA)   A non-destructive method of analysis in which a sample bombarded with *neutrons* gives off *gamma rays* with an energy characteristic of the original ISOTOPE. NAA is an analytical method in GEOCHEMISTRY that can be used to identify major ELEMENTS and TRACE ELEMENTS. The technique has also been used to identify trace elements in ores, pottery and metal ARTEFACTS, and hence to identify their sources and affinities. A *neutron activation log* or *neutron log*, obtained using an instrument in a BOREHOLE, can be used to characterise LITHOLOGY and *porosity* and to distinguish water from oil. *GO*

[*See also* ARCHAEOLOGICAL GEOLOGY, POROUS]

Dias MI and Prudencio MI (2007) Neutron activation analysis of archaeological materials: An overview of the ITN NAA laboratory, Portugal. *Archaeometry* 49: 383–393.

Grave P, Kealhofer L, Marsh B et al. (2008) Using neutron activation analysis to identify scales of interaction at Kinet Höyük, Turkey. *Journal of Archaeological Science* 35: 1974–1992.

Hughes MJ, Cowell MR and Hook DR (eds) (1991) *Neutron activation and plasma emission spectrometric analysis in archaeology.* London: British Museum.

Wilson D (2000) Provenance of the Hillsboro Formation: Implications for the structural evolution and fluvial events in the Tualatin Basin, northwest Oregon. *Journal of Sedimentary Research* 70: 117–126.

new forestry   Forest practices that retain greater BIODIVERSITY and ecosystem complexity than traditional, intensive, monocultural practices of SILVICULTURE.
JAM

[*See also* AGRO-ECOLOGY, FOREST MANAGEMENT, SUSTAINABLE FORESTRY]

Swanson FJ and Franklin JF (1992) New forestry principles from ecosystem analysis of Pacific northwest forests. *Ecological Applications* 2: 262–274.

newly industrialising countries (NICs)

A group of states that developed substantial MANUFACTURING INDUSTRIES in the late twentieth century. Most are in Asia, where they are dominated by China. Other Asian NICs include the first generation of four 'Asian tigers'—South Korea, Taiwan, Hong Kong and Singapore—and also Thailand, Malaysia, Indonesia and India. Latin American NICs include Brazil, Mexico and Chile. Some also include South Africa and Turkey.   JAM

[*See also* DEVELOPING COUNTRIES, INDUSTRIALISATION]

Noland M (1987) Newly industrializing countries' comparative advantage in manufactured goods. *Review of World Economics* 123: 679–696.
Singer H and Hatti N (2007) *Newly industrializing countries after the Asian crisis*. New Delhi: BR Publishing.

niche Originally used to describe the role of an organism within a community as opposed to the HABITAT in which it lives, the niche of an organism may also be defined by an array of environmental factors. The *fundamental niche* is that part of the *resource availability field* in which an organism can survive and which it may utilise. The *realised niche* may be limited by *interspecific interactions*, such as COMPETITION, and is that part of the fundamental niche that is actually occupied by a species.   JLI

[*See also* LIMITING FACTORS, RESOURCE]

Chase JM and Leibold MA (2003) *Ecological niches: Linking classical and contemporary approaches*. Chicago: University of Chicago Press.
Levine JM and HilleRisLambers J (2009) The importance of niches for the maintenance of species diversity. *Nature* 461: 254–257.
Peterson AT, Soberón J, Pearson RG et al. (2011) *Ecological niches and geographic distributions*. Princeton, NJ: Princeton University Press.

Nile floods The oldest year-by-year record of flood levels for the River Nile in Egypt. The earliest known records in Cairo date from the Early Dynastic Period recording the height of every flood back to about 3090 BC. The most reliable measurements date from AD 622 onwards. These are continuous up to AD 1470 and then, with a few gaps, run up to the present day. This record is the longest continuous annual climatic series monitoring the rainfall in a large DRAINAGE BASIN. It is particularly sensitive to precipitation in the Ethiopian Highlands, which are drained by the Blue Nile. Due to the river's links to other climatic zones, the record holds key evidence of the global nature of CLIMATIC VARIABILITY and the existence of TELECONNECTIONS. Major, long-term *variations* can be identified with periods of low discharge from AD 630 to 1071 and AD 1180 to 1350. High discharge episodes occurred from AD 1070 to 1180 and AD 1350 to 1470. The annual flood is related to the summer monsoonal rains in Ethiopia, which are associated with a northward shift in the INTERTROPICAL CONVERGENCE ZONE (ITCZ).   GS

Bell B (1970) The oldest records of the Nile floods. *Geographical Journal* 136: 569–573.
Fraedrich K, Ziang J, Gerstengarbe F-W and Werner PC (1997) Multiscale detection of abrupt climatic changes: Applications to River Nile flood levels. *International Journal of Climatology* 17: 1301–1315.
Kondrashov D, Feliks Y and Ghil H (2005) Oscillatory modes of extended Nile River records (AD 622–1922). *Geophysical Research Letters* 32: L10702.
Popper W (1951) *The Cairo Nilometer*. Berkeley: University of California Press.
Said R (1993) *The River Nile: Geology, hydrology and utilization*. Oxford: Pergamon Press.

NIMBY (Not-In-My-Backyard) A widespread attitude against development adopted by citizens or NON-GOVERNMENTAL ORGANISATIONS (NGOs) fearing (real or imaginary) threats or nuisance from a particular development to be sited near them. A particularly prominent example current in the United Kingdom involves opposition to the siting of *wind farms* (see WIND ENERGY). Grounds for opposition often include reduced ENVIRONMENTAL QUALITY. Opposition to a proposal wherever it is sited is *NIABY* (*Not-In-Anyone's-Backyard*). This often applies, for example, to the siting of nuclear power plants or facilities for the disposal of HAZARDOUS WASTE. Support for a development to be sited in one's area is *YIMBY* (*Yes-In-My-Backyard*), a common example of which is demands for affordable housing. There are other, related, acronyms.   JAM/CJB

[*See also* ALTERNATIVE ENERGY, NUCLEAR ENERGY]

Devine-Wright P (2010) *Renewable energy and the public: From NIMBY to participation*. London: Earthscan.
Hunter S and Leyden KM (1995) Beyond NIMBY: Explaining opposition to hazardous waste facilities. *Policy Studies Journal* 23: 601–619.
Saint PM (2009) *NIMBY wars: The politics of land use*. Hingham, MA: Saint University Press.

Nitisols Tropical soils with a deeply extended *nitic horizon* (a clay-rich subsurface horizon having a nut-shaped, polyhedric SOIL STRUCTURE with shiny PED faces) occurring in Africa, South America and India on base-rich PARENT MATERIAL (SOIL TAXONOMY: kandic groups of *Alfisols* and *Ultisols*). Processes involved in their formation include, FERRALITISATION, *nitidisation* (production of the characteristic nut-shaped peds) and BIOTURBATION (by a rich soil fauna, especially TER-MITES). They have similarities to FERRALSOLS, but they are developed from parent materials rich in bases. As a result, these soils have a higher CATION EXCHANGE CAPACITY (CEC) and are far more productive and stable under agriculture than other tropical soils.   *EMB*

[*See also* WORLD REFERENCE BASE FOR SOIL RESOURCES (WRB)]

Kapkiyani JJ, Karanja NK, Qureshi JN et al. (1999) Soil organic matter and nutrient dynamics in a Kenyan nitisol under long-term fertilizer and organic input management. *Soil Biology and Biochemistry* 31: 1773–1782.
Sombroek WG and Siderius W (1982) *Nitosols, a quest for significant diagnostic criteria* (Annual Report). Wageningen: International Soil Reference and Information Centre, 11–31.

nitrate (NO$_3^-$)   The main source of nitrogen for plants, an important FERTILISER, the highest oxidation state for nitrogen in wastewater and a POLLUTANT in surface waters, GROUNDWATER and ESTUARINE ENVIRON-MENTS. *Nitrifying bacteria* oxidise ammonia to nitrite (NO$_2^-$) and nitrate, *nitroso-bacteria* convert ammonia to nitrite and *nitro-bacteria* convert nitrite to nitrate. A high concentration of nitrate in wastewater is generally considered to represent a relatively stable EFFLUENT. However, its high solubility in water enables and its importance as a NUTRIENT leads to excess nitrate contributing to EUTROPHICATION, ALGAL BLOOMS and potentially lethal *methaemoglobinaemia* (reduced ability of haemoglobin to carry oxygen) in infants. A *Nitrate Vulnerable Zone* (*NVZ*) is a type of PROTECTED AREA. *Saltpeter* (potassium nitrate) was an important constituent of classical *gunpowder*.   *JAM*

[*See also* NITROGEN CYCLE]

Addiscott TM (2005) *Nitrate, agriculture and the environment*. Wallingford: CABI Publishing.
Burt TP, Heathwaite AL and Trudgill ST (eds) (1993) *Nitrate: Processes, patterns and management*. Chichester: Wiley.
Laluraj CM, Thamban M, Naik SS et al. (2011) Nitrate records of a shallow ice core from East Antarctic: Atmospheric processes, preservation and climatic implications. *The Holocene* 21: 351–356.

nitrogen cycle   The cyclical progress of nitrogen through living things, air, rocks, soil and water. Relatively inactive *atmospheric nitrogen* either forms inorganic compounds in *rainwater*, or is fixed by *nitrogen-fixing bacteria* (*nitrifiers*). It is then assimilated and metabolised by animals and plants. It returns to the soil in nitrogenous *animal waste* and in *dead organisms*. Nitrogen in the soil is subject to *nitrification* (conversion to nitrates and nitrites by nitrifying micro-organisms), to MINERALISATION or *ammonification* (the release of

ammonia and ammonium ions from dead organic matter by *decomposers*) and to DENITRIFICATION (the reduction of NITRATES (NO$_3^-$) to gaseous nitrogen forms that return to the ATMOSPHERE). Practices within AGRICULTURE and INDUSTRY modify the *stores* and *fluxes* in the nitrogen cycle. Nitrogen FERTILISERS add to the *soil nitrogen pool*. Growing human and livestock populations increase nitrogenous waste volumes. Industrial activities release NITROGEN OXIDES into the atmosphere. These growing nitrogen pools create environmental problems, including EUTROPHICATION and OZONE DEPLETION.   *RJH*

[*See also* ACID RAIN, BIOGEOCHEMICAL CYCLES, CARBON CYCLE, NITROGEN FIXATION, NITROGEN ISOTOPES, SULFUR CYCLE]

Bothe H, Ferguson SJ and Newton WE (2006) *Biology of the nitrogen cycle*. Amsterdam: Elsevier.
Boyer EW and Howarth RW (eds) (2002) *The nitrogen cycle at regional and global scales*. Dordrecht: Kluwer.
Canfield DE, Glazer AN and Falkowski PG (2010) The evolution and future of Earth's nitrogen cycle. *Science* 330: 192–196.
Galloway JN, Dentener FJ, Capone DG et al. (2004) Nitrogen cycles: Past, present and future. *Biogeochemistry* 70: 153–226.
Galloway JN, Townsend AR, Erisman JW et al. (2008) Transformation of the nitrogen cycle: Recent trends, questions, and potential solutions. *Science* 320: 889–892.
Hietz P, Turner BL, Wanek W et al. (2011) Long-term change in the nitrogen cycle of tropical forests. *Science* 334: 664–666.
Huggett RJ (2007) Drivers of global change. In Douglas I, Huggett R and Perkins C (eds) *Companion encyclopedia of geography: From local to global, 2nd edition*. London and New York: Routledge, 75–91.
Jaffe DA (2000) The nitrogen cycle. In Jacobsen MC, Charlson RJ, Rodhe H and Orians GH (eds) *Earth system science: From biogeochemical cycles to global change*. London: Academic Press, 322–342.
Mosier AR, Syers JK and Freney JR (eds) (2004) *Agriculture and the nitrogen cycle: Assessing the impacts of fertilizer use on food production and the environment*. Washington, DC: Island Press.
Purvaja R, Ramesh R, Ray AK and Rixen T (2008) Nitrogen cycling: A review of the processes, transformations and fluxes in coastal ecosystems. *Current Science* 94: 1419–1438.
Sprent JI (1987) *The ecology of the nitrogen cycle*. Cambridge: Cambridge University Press.
Vitousek PM, Aber J, Howarth RW et al. (1997) Human alteration of the global nitrogen cycle: Causes and consequences. *Issues in Ecology* 1: 1–17.

nitrogen dating   A RELATIVE-AGE DATING technique based on the *post-mortem* decrease in the nitrogen content of bone and teeth as the protein, collagen, decomposes. In vivo bone and dentine contain about 5 per cent nitrogen in the collagen and the decomposition rate is influenced by soil or sediment moisture content, pH and temperature. Nitrogen dating is mainly used in parallel with the accumulating uranium and fluorine content of the bone (see FLUORINE DATING).   *JAM*

[*See also* NITROGEN-PROFILE DATING]

Oakley KB (1980) Relative dating of the fossil hominids of Europe. *Bulletin British Museum Natural History (Geology)* 34: 1–63.
Ortner DJ, von Endt DW and Robinson MS (1972) The effect of temperature on protein decay in bone: Its significance in nitrogen dating of archaeological samples. *American Antiquity* 37: 514–520.

nitrogen fixation   The natural and synthetic processes that convert atmospheric *nitrogen* (N_2) gas into a form usable by organisms. This generally involves *ammonification* (conversion to *ammonium ions*) and nitrification (conversion of ammonium ions to *nitrate ions* via *nitrite ions*). *Biological nitrogen fixation* is the natural process of nitrogen capture and conversion by organisms into organic nitrogen compounds available to plants. *Symbiotic nitrogen fixation* is carried out by heterotrophic *Rhizobium* BACTERIA that are associated with the *root nodules* of leguminous plants. *Non-symbiotic nitrogen fixation* is carried out independently by the free-living *organotrophic* (*heterotrophic*) soil bacterium *Azotobacter* and *blue-green* ALGAE which are *phototrophic* (*autotrophic*) organisms. *Chemical nitrogen fixation* is the artificial equivalent, which is used to produce FERTILISERS.   *SN*

[*See also* SYMBIOSIS]

Leigh GJ (ed.) (2002) *Nitrogen fixation at the millennium*. Amsterdam: Elsevier.
Rao NSS (ed.) (2009) *Current developments in biological nitrogen fixation*. Cambridge: Cambridge University Press.

nitrogen isotopes   Nitrogen has two STABLE ISOTOPES (natural abundances, $^{14}N = 99.64$ per cent and $^{15}N = 0.36$ per cent). The ISOTOPE RATIO is expressed as $^{15}N/^{14}N$ relative to atmospheric nitrogen. Most nitrogen is located in the ATMOSPHERE or dissolved in the ocean. The cycling of nitrogen through an *ecosystem* may be followed through the NITROGEN CYCLE. Metabolic processes fractionate nitrogen isotopes. Soil nitrogen usually exhibits ISOTOPIC ENRICHMENT in terms of ^{15}N relative to atmospheric nitrogen. Non-nitrogen-fixing plants whose primary source of nitrogen is from the SOIL have higher $\delta^{15}N$ values than plants, such as *legumes*, that have the ability to fix atmospheric nitrogen. The $\delta^{15}N$ values of NITRATE (NO_3^-) have been used to identify POLLUTION from nitrogen-based FERTILISERS. However, pollution studies are limited since the largest ISOTOPIC FRACTIONATION of nitrogen isotopes is from metabolic reactions. Nitrogen in animal tissues is isotopically enriched in ^{15}N relative to dietary inputs, owing to the preferential excretion of ^{14}N in *urea*. This enrichment

in ^{15}N is propagated through the FOOD CHAIN as $\delta^{15}N$ values increase at each TROPHIC LEVEL. In non-arid environments, marine and terrestrial *diets* may be distinguished using $\delta^{15}N$ and $\delta^{14}C$ values.

The ratio of $^{15}N:^{14}N$ found in ICE CORES has recently been used in association with other isotopes (notably $^{40}Ar:^{39}Ar$) as confirming estimates of rapid temperature change originally indicated by OXYGEN ISOTOPE results. Thus, such RAPID ENVIRONMENTAL CHANGES have been shown not to be artefacts of the incorporation of such isotopes into the ice, or of offsets in age between different proxy signals resulting from gas diffusion within the ice as it accumulates. Over recent decades, palaeolimnological studies have identified increased nitrogen deposition enhancing changes especially at high latitudes.   *IR/CJC*

[*See also* NITROGEN CYCLE, PALAEODIET]

De Pol-Holz R, Ulloa O, Lamy F et al. (2007) Late Quaternary variability of sedimentary nitrogen isotopes in the eastern South Pacific Ocean. *Paleoceanography* 22, PA2207.
Hedges REM and Reynard LM (2007) Nitrogen isotopes and the trophic level of humans in archaeology. *Journal of Archaeological Science* 34: 1240–1251.
Holmgren SU, Bigler C, Ingolfsson O and Wolfe AP (2010) The Holocene/Anthropocene transition in lakes of western Spitsbergen, Svalbard (Norwegian High Arctic). *Journal of Paleolimnology* 43: 393–412.
Huon S, Grousset FE, Burdloff D et al. (2002) Sources of fine-sized organic matter in North Atlantic Heinrich layers: $\delta^{13}C$ and $\delta^{15}N$. *Geochimica et Cosmochimica Acta* 66: 223–239.
Severinghaus JP and Brook EJ (1999) Abrupt climate change at the end of the last glacial period inferred from trapped air in polar ice. *Science* 286: 930–934.
Sponheimer M, Robinson T, Ayliffe L et al. (2003) Nitrogen isotopes in mammalian herbivores: Hair $\delta^{15}N$ values from a controlled feeding study. *International Journal of Osteoarchaeology* 13: 80–87.
Talbot MR (2001) Nitrogen isotopes in palaeolimnology. In Last WM and Smol JP (eds) *Tracking environmental change using lake sediments, volume 2*. Dordrecht: Kluwer, 401–439.

nitrogen oxides   The reactive species of nitrogen and oxygen. The most important are *nitric oxide* (NO) and *nitrogen dioxide* (NO_2), which are major contributors to urban AIR POLLUTION, and *nitrous oxide* (N_2O), which is an important GREENHOUSE GAS. Nitrogen oxides (known informally as NO_x gases) are generated when combustion occurs at high temperatures (>1000°C) so that naturally occurring nitrogen and oxygen combine to form NO. NO is relatively innocuous but is rapidly oxidised by OZONE in the air to form NO_2: it accounts for >50 per cent of the natural destruction of stratospheric ozone. NO_x gases are primary POLLUTANTS in *low air quality episodes* in winter, and in summer they promote

the formation of SECONDARY POLLUTANTS in PHOTOCHEMI-
CAL SMOG. In urban areas, acute exposure to NO_2 can
lead to coughing and sore throats, and will aggravate
emphysema and other *respiratory diseases*. Peak hourly
levels of NO_2 of 314 ppb (parts per billion) or more
have been recorded in, for example, Amsterdam, Athens, Brussels, London, Los Angeles and Munich.   GS

[*See also* NITROGEN CYCLE, OZONE DEPLETION]

Bouwman AF, Van der Hoek KW and Oliver JG (1995)
    Uncertainties in the global source distribution of nitrous
    oxide. *Journal of Geophysical Research* 100: 2785–2790.
Elsom DE (1996) *Smog alert: Managing urban air quality.*
    London: Earthscan.
Lee SD (1980) *Nitrogen oxides and their effects on health.*
    Ann Arbor, MI: Ann Arbor Science.
Richter A, Burrows JP, Nüβ et al. (2005) Increase in
    tropospheric nitrogen dioxide over China observed from
    space. *Nature* 437: 129–132.

nitrogen/nitrate analysis   A method in *soil science* to determine the nitrate/ammonium/nitrite content, in AQUATIC ENVIRONMENTS (freshwater and marine)
to measure PRODUCTIVITY changes, and in SEDIMENTS
from TERRESTRIAL ENVIRONMENTS to characterise the
ORGANIC CONTENT. Dominant forms of nitrogen in
freshwater include dissolved molecular N_2, ammonium nitrogen (NH_4^+), nitrite (NO_2^-), nitrate (NO_3^-)
and organic compounds. The organic *carbon:nitrogen*
(C:N) *ratio* indicates an approximate state of resistance
of complex mixtures of organic compounds to DECOM-
POSITION. Organic compounds from ALLOCHTHONOUS and
WETLAND sources commonly have C:N ratios from 45:1
to 50:1 and contain mainly organic matter of low nitrogen content. AUTOCHTHONOUS organic matter produced
by the decomposition of PLANKTON tends to have higher
protein content and C:N ratios of about 12:1. Changing
carbon:phosphorous (C:P) *ratios* reflect shifts in species of ALGAE. Increased loading of inorganic nitrogen
to freshwater and marine ecosystems frequently results
from *agricultural activities*, *sewage* and AIR POLLUTION.
                                                       UBW

[*See also* NITROGEN CYCLE, NITROGEN ISOTOPES,
NITROGEN OXIDES, TROPHIC LEVEL]

Bengtsson L and Enell M (1986) Chemical analysis. In
    Berglund BE (ed.) *Handbook of Holocene palaeoecology
    and palaeohydrology.* Chichester: Wiley, 423–451.

nitrogen-profile dating   Unrelated to NITROGEN
DATING, the technique of *nitrogen-profile dating* is used
in archaeological contexts on the carved surface of
non-porous stone, such as jade and flint, to estimate the
time elapsed since carving. The gradient in the nitrogen
content with increasing depth from the surface of the
stone is related to the rate of diffusion of nitrogen into
the stone.   JAM

Bowman S (1999) Nitrogen profiling. In Shaw I and
    Jameson R (eds) *A dictionary of archaeology.* Oxford:
    Blackwell, 428–429.
Wagner GA (1998) *Age determination of young rocks and
    artefacts: Physical and chemical clocks in Quaternary
    geology and archaeology.* Berlin: Springer.

nitrous oxide variations   Nitrous oxide (N_2O)
is an atmospheric TRACE GAS which is produced naturally by SOILS and OCEANS. Current estimates suggest
that N_2O is increasing at the rate of 0.3 per cent per
year and that each molecule has the potential to contribute 300 times to the GREENHOUSE EFFECT relative to
each molecule of CARBON DIOXIDE. Its impact on GLOBAL
WARMING is therefore of serious concern. Evidence from
air trapped in the ICE CORES indicate that variations in
atmospheric nitrous oxide have mirrored the GLACIAL-
INTERGLACIAL CYCLES. Lower levels during GLACIAL EPI-
SODES are thought to reflect either reduced soil activity
or less output from oceanic sources. Variation between
glacials and INTERGLACIALS is considerable: during the
last glacial-interglacial cycle, for example, nitrous
oxide levels increased by 30 per cent. Over the past 200
years, levels of atmospheric nitrous oxide have been
strongly influenced by ANTHROPOGENIC activity and in
particular EMISSIONS resulting from the use of nitrogen-
rich FERTILISERS. Current estimates suggest that this has
increased levels of atmospheric nitrous oxide to 8 per
cent higher than pre-industrial levels.   KJW

Battle M, Bender M and Sowers T (1996) Atmospheric gas
    concentrations over the past century measured in air from
    firn at the South Pole. *Nature* 383: 231–235.
Machida T, Nakazawa T, Fujii Y et al. (1995) Increase
    in atmospheric nitrous oxide concentrations during
    the last 250 years. *Geophysical Research Letters* 22:
    2921–2924.
Naqvi SWA, Jayakumar DA, Narrekar PV et al. (2000)
    Increased marine production of N_2O due to intensifying
    anoxia on the Indian Ocean shelf. *Nature* 408: 346–349.
Navishankara AR, Daniel JS and Portmann RW (2009)
    Nitrous oxide (N_2O): The dominant ozone-depleting
    substance emitted in the 21st century. *Science* 326:
    123–125.
Schlesinger WH (1997) *Biogeochemistry: An analysis of
    global change.* London: Academic Press.
Smith K (ed.) (2010) *Nitrous oxide and climate change.*
    London: Earthscan.
Wolf B, Zheng X, Brügemann N et al. (2010) Grazing-
    induced reduction of natural nitrous oxide release from
    continental steppe. *Nature* 464: 881–884.

nival zone   An *altitudinal zone* in high mountains
above the SNOW LINE or above the altitude of the lowest
semi-permanent SNOWBEDS averaged over a number of
years.   JAM

[*See also* AEOLIAN ZONE, ALPINE ZONE, ALTITUDINAL
ZONATION]

Huber E, Wanek W, Gottfied M et al. (2007) Shift in soil-plant nitrogen dynamics of an alpine-nival ecotone. *Plant and Soil* 301: 65–76.

nivation A collective term for the processes of ERO-SION and DEPOSITION associated with semi-permanent and late-lying SNOWBEDS that may give rise to such landforms as hillside *nivation hollows* and CRYOPLANA-TION TERRACES. The effectiveness of nivation processes has been much debated on the basis of often scant evidence. It may be possible to reconcile disparate views by closer examination of the range of environments in which snowbeds occur.   *PW/JAM*

[*See also* CHEMICAL WEATHERING, FROST WEATHERING, PRONIVAL RAMPART, SLOPEWASH, SNOWMELT, SOLIFLUCTION]

Berrisford MS (1991) Evidence for enhanced mechanical weathering associated with seasonally late-lying and perennial snow patches, Jotunheimen, Norway. *Permafrost and Periglacial Processes* 2: 331–340.
Christiansen HH (1998) 'Little Ice Age' nivation activity in northeast Greenland. *The Holocene* 8: 719–728.
Kariya Y (2005) Holocene landscape evolution of a nivation hollow on Gassan volcano, northern Japan. *Catena* 62: 57–76.
Thorn CE and Hall K (2002) Nivation and cryoplanation: The case for scrutiny and integration. *Progress in Physical Geography* 26: 533–550.

niveo-aeolian deposits Windblown silty sand deposited together with *snow* and subsequently reworked by snow MELTWATER. As seen in SNOWBEDS in winter and early summer, niveo-aeolian deposits consist of stratified snow and sediment layers. In summer, as the snow banks melt, the sediment is further stratified and locally reworked by SLOPEWASH processes.   *HMF*

[*See also* AEOLIAN, NIVATION, PERIGLACIAL SEDIMENTS, SNOW AND SNOW COVER]

Christiansen HH (1998) 'Little Ice Age' nivation activity in northeast Greenland. *The Holocene* 8: 719–728.
Koster EA and Dijkmans JWA (1988) Niveo-aeolian deposits and denivation forms, with special reference to the Great Kobuk Sand Dunes, Northwestern Alaska. *Earth Surface Processes and Landforms* 13: 153–170.
Seppälä M (2004) *Wind as a geomorphic agent in cold climates*. Cambridge: Cambridge University Press.

nodule A discrete mass of material, usually spherical to ellipsoidal in form, that stands out from its surroundings. Many nodules are CONCRETIONS that formed during DIAGENESIS of SEDIMENTARY DEPOSITS and differ in chemistry from their surrounding MATRIX, rendering them more resistant to WEATHERING and EROSION. Nodules composed of many different MINERALS, such as GYP-SUM, SILICA (CHERT) and *manganese* or *iron oxide*, occur in a wide range of terrestrial and marine environments.

Calcareous nodules are an important constituent of some SOILS, PALAEOSOLS and CALCRETES. The chemistry of nodules, particularly in relation to their matrix, can provide valuable *chemical indicators* of environmental change.   *GO*

Chan MA, Beitler B, Parry WT et al. (2004) A possible terrestrial analogue for haematite concretions on Mars. *Nature* 429: 731–734.
Muller KJ (1985) Exceptional preservation in calcareous nodules. *Philosophical Transactions of the Royal Society B* 311: 67–73.

noise Generally referred to as unwanted sound, noise at elevated levels can block, distort, change or interfere with the meaning of a message in human, animal and electronic communication. In *data analysis*, *computing* and *signal processing*, it can be considered random unwanted data without meaning, with the term SIGNAL-TO-NOISE commonly used to express the ratio of useful to irrelevant information in an exchange. In IMAGE PROCESSING, it can be *random noise* or *systematic noise* and correction for *image noise* is common in *pre-processing*. In contemporary environmental change, there is an increasing precedence for change in *ambient noise levels* in both the terrestrial and aquatic environments to be used as a critical first indicator of change.   *DSB*

Ilyina T, Zeebe RE and Brewer PG (2010) Future ocean increasingly transparent to low-frequency sound owing to carbon dioxide emissions. *Nature Geoscience* 3: 18–22.
Mather PM and Koch M (2011) *Computer processing of remotely-sensed images: An introduction, 4th edition*. Chichester: Wiley-Blackwell.

noise pollution Although the intensity of *sound* can be measured in decibels, and other aspects of sound (e.g. its frequency and duration) can also be measured precisely, noise (unwanted sound) is a subjective property. It is questionable whether noise should be described as a POLLUTANT or as causing POLLUTION. In certain environments, such as in some factories, in and around airports and close to urban motorways, noise may be more than an irritant or nuisance and becomes a HUMAN HEALTH HAZARD. *Noise abatement* can be achieved at source, during transmission or by receiver control.   *JAM*

Gjestland T (1999) Sound and noise. In Brune D, Chapman DV, Gwynne MD and Pacyna M (eds) *The global environment: Science, technology and management, volume 1*. Weinheim: VCH, 610–624.

nomadism A mobile community lifestyle involving more-or-less continued shifting of residence, which is often adapted to environmental constraints, particularly in marginal areas (e.g. high altitudes, ARIDLANDS). It is often associated with PASTORALISM where the population relocates to new PASTURE, and is under threat in many parts of the world.   *LD-P*

[*See also* DOMESTICATION, MIGRATION, SEDENTISM]

Barnard H and Wendrich W (eds) (2007) *The archaeology of mobility: Old World and New World nomadism.* Los Angeles: Cotsen Institute of Archaeology.
Boute P, Guillaum H and Zecchin F (1996) Nomads: Changing societies and environments. *Nature and Resources* 32: 2–10.
Cribb R (1991) *Nomads in archaeology.* Cambridge: Cambridge University Press.
Humphry C and Sneath D (1999) *The end of nomadism? Society, state and the environment in Inner Asia.* Durham, NC: Duke University Press.

nomothetic science Science that is concerned with developing general ideas or LAWS, as opposed to a detailed explanation of individual cases (IDIOGRAPHIC SCIENCE). A somewhat dated concept but an appropriate descriptor for most of the approaches adopted in the SCIENCES today. Contemporary debate over whether HUMAN GEOGRAPHY, HISTORY and many behavioural SOCIAL SCIENCES were, are, can be or should be nomothetic sciences takes many forms.   *JAM/CET*

Lyman RL and O'Brien MJ (2004) Nomothetic science and idiographic history in twentieth-century Americanist anthropology. *Journal of the History of the Behavioral Sciences* 40: 77–96.
Thomae H (1999) The nomothetic-idiographic issue: Some roots and recent trends. *International Journal of Group Tensions* 28: 187–215.

non-arboreal pollen (NAP) Pollen from dwarf SHRUBS and HERBS, but not *trees*. The NAP percentage has been used as a rough indicator of non-forested land. *SPH*

[*See also* ARBOREAL POLLEN (AP), POLLEN ANALYSIS, TREE LINE/LIMIT]

non-communicable disease (NCD) A chronic DISEASE that is non-infectious. For long the scourge of Western nations, NCDs include (in decreasing order of numbers affected) *cardiovascular disease*, CANCER, *diabetes*, chronic *respiratory* and *digestive diseases* and *dementia*. The World Health Organisation recognises four main risk factors for these diseases: tobacco smoking, physical inactivity, alcohol abuse and poor diet. They now kill more people worldwide than all other causes combined and, following the adoption of Western lifestyles, are adding to the DISEASE BURDEN of DEVELOPING COUNTRIES. Diabetes, for example, is growing at an alarming rate, especially in the developing world. Around 350 million people worldwide are afflicted with the disease, 90 million of which occur in China, 60 million in India, 24 million in the United States and 20 million in Europe. In *high-income*

countries, diabetes primarily affects those >50 years of age but in *middle-income countries*, such as China and India, the highest prevalence is in younger, more productive groups of the population. Worldwide, >80 per cent of deaths from diabetes occur in *low-income countries* and middle-income countries, where the DEATH RATE from diabetes is twice as high as for high-income countries, largely due to poorer health-care facilities. Various types of *dementia* (the most common of which is *Alzheimer's disease*) affect around 6 per cent of the world's population over 60 years of age, but the numbers are increasing due to aging populations.   *JAM*

Abbott A (2011) Dementia: A problem for our age. *Nature* 475, Supplement: S2–S4.
Diamond J (2003) The double puzzle of diabetes. *Nature* 423: 599–602.
Reardon S (2011) A world of chronic disease. *Science* 333: 558–559.
Scully T (2012) Diabetes in numbers. *Nature* 485, Supplement: S2–S3.

non-destructive sampling Scientific investigation often influences the phenomenon under investigation and sometimes results in its destruction. It is good practice, particularly in environmentally conscious disciplines, to minimise such effects, and non-destructive SAMPLING and/or measurement is a case in point. Examples include satellite REMOTE SENSING and various types of CORE SCANNING used in palaeoenvironmental investigation. Similarly, *non-destructive testing* is widely used in science, engineering and industry, particularly in searching for weaknesses in materials.   *JAM*

Caseldine C, Baker A and Barnes WL (1999) A rapid, non-destructive scanning method for detecting distal tephra layers in peats. *The Holocene* 9: 635–638.

non-governmental organisation (NGO) A non-profit-making body that is independent of state control. Around AD 1900 there were less than 200 NGOs; in 1975 there were only 1,400; now there are between 50,000 and 100,000, depending on how they are defined. They represent a huge diversity of special-interest groups, some with links to religion, others charitable organisations or champions of a cause. Some *international non-governmental organisations* (*INGOs*) command considerable power in the fields of *international aid*, SUSTAINABLE DEVELOPMENT and CONSERVATION; at the other end of the spectrum are *community-based organisations* (*CBOs*) with local 'grass-roots' interests and links. Examples of the former include *Christian Aid*, *Friends of the Earth*, *Greenpeace* and the *World Wildlife Fund*. NGOs may help negotiate agreements and covenants to control environmental problems, lobby and monitor situations and whistle-blow if problems develop. Some question their representativeness, legitimacy and accountability. *CJB*

[***See also*** CIVIL SOCIETY]

Bebbington A, Hickey S and Mitlin D (eds) (2008) *Can NGOs make a difference? The challenge of development alternatives*. New York: Zed Books.
Edwards M and Fowler A (eds) (2002) *The Earthscan reader on NGO management*. London: Earthscan.
Jordan L and van Tuijl P (eds) (2007) *NGO accountability: Politics, principles, and innovations*. London: Earthscan.
Lewis D and Kanji N (2009) *Non-governmental organizations and development*. London: Routledge.

non-linear behaviour/dynamics

A non-superposition condition where output is not directly proportional to input. Individual actions are part of a COMPLEX SYSTEM that may operate under chaotic states and unpredictable relationships. The natural environment often operates under *immanent laws* involving complex interactions that are frequently non-linear, for example, in atmospheric processes. *VM*

[***See also*** CHAOS THEORY, COMPLEXITY]

Boucharel J, Dewitte B, Penhoat Y et al. (2011) ENSO nonlinearity in a warming climate. *Climate Dynamics* 37: 2045–2065.
Wainwright J and Mulligan M (eds) (2004) *Environmental modelling: Finding simplicity in complexity*. Chichester: Wiley.

non-metric multidimensional scaling

A form of MULTIDIMENSIONAL SCALING in which the ranks of the distances or dissimilarities between objects are used to produce the required low-dimensional representation of the *distance matrix*. *HJBB*

Everitt BS and Hothorn T (2011) *An introduction to applied multivariate analysis with R*. New York: Springer.
Robbins JA and Matthews JA (2010) Regional variation in successional trajectories and rates of vegetation change on glacier forelands in south-central Norway. *Arctic, Antarctic and Alpine Research* 42: 351–361.

non-parametric statistics

Statistical procedures for testing hypotheses or estimating parameters that make no assumptions about the underlying PROBABILITY DISTRIBUTION of the variables (e.g. *normality* or *linearity*). They commonly involve the ranks of the observations rather than the observations themselves (*ordinal-scale data*; see MEASUREMENT). They are often only marginally less powerful than their parametric counterparts, even when the underlying assumptions of the latter are true. They are also known as *distribution-free methods*, although the terms are not completely synonymous. Like PARAMETRIC STATISTICS, non-parametric statistics do make particular assumptions. Non-parametric statistics are *not* no-assumption statistics. *HJBB*

Gotelli NJ and Ellison AM (2004) *A primer of ecological statistics*. Sutherland, MA: Sinauer.
Siegel S and Castellan Jr NJ (1988) *Non-parametric statistics for the behavioural sciences, 2nd edition*. New York: McGraw-Hill.

non-renewable resource

A RESOURCE that, once used, cannot be replaced, at least within a timescale to be useful. Typically, non-renewable, *depletable*, *exhaustible, finite* or *stock resources*, such as FOSSIL FUELS, MINERAL RESOURCES and biological species that, once extinct, cannot be replaced, are formed over geological timescales: rates of CONSUMPTION therefore greatly exceed regeneration rates. If the latter are too slow, other resources that are, in principle, RENEWABLE RESOURCES (e.g. tropical forests) can also be considered non-renewable. At any time, the stock or RESERVES of a non-renewable resource are finite and, with use, will eventually be exhausted within a period of time that can be estimated. Similarly, *optimal resource extraction models* attempt to define the rate at which non-renewable resources should be used. However, as a non-renewable resource is exploited and reserves are depleted, demand may change due to such factors as MATERIALS SUBSTITUTION (use of substitutes) and WASTE RECYCLING, thus delaying exhaustion. *JAM*

[***See also*** NATURAL RESOURCES, RESOURCE, SUSTAINABILITY]

Ruth M (2006) The economics of sustainability and the sustainability of economics. *Ecological Economics* 56: 332–342.
Sinding-Larsen R (ed.) (2012) *Non-renewable resource issues: Geoscientific and societal challenges*. Berlin: Springer.

noösphere

The realm or sphere of influence of the human mind or intellect in the context of the Earth and especially the GEO-ECOSPHERE. Unlike other EARTH SPHERES, the noösphere is an abstract phenomenon without physical existence. It is, however, an increasingly pervasive influence on the ECOSPHERE and may be viewed as the summation of the mental activity or human thought behind inadvertent *human impacts* as well as the conscious human use of Earth's RESOURCES, MANAGEMENT and CONSERVATION. *JAM*

[***See also*** ANTHROPOSPHERE, EARTH SYSTEM, ECUMENE, GAIA HYPOTHESIS, SUSTAINABILITY, WORLD SYSTEM]

Oldfield JD and Shaw DJB (2006) V.I. Vernadsky and the noosphere concept: Russian understanding of society-nature interactions. *Geoforum* 37: 145–154.
Samson PR and Pitt D (eds) (1999) *The biosphere and noosphere reader: Global environment, society and change*. London: Routledge.
Teilhard de Chardin P (1956) The antiquity and world expansion of human culture. In Thomas Jr WL (ed.)

Man's role in changing the face of the Earth. Chicago: University of Chicago Press, 103–114.

Teilhard de Chardin P (1964) *The future of man*. London: Collins.

Vernadsky VI (1945) The biosphere and the noösphere. *American Scientist* 33: 1–12.

normal distribution A continuous PROBABIL-ITY DISTRIBUTION used to describe continuous random variables that is assumed by many methods based on PARAMETRIC STATISTICS. It is symmetrical, unimodal and bell-shaped. Its shape and distribution are defined by the mean and standard deviation. In a normal distribution, 95 per cent of all observations lie within the mean ± 1.96 standard deviations and 99 per cent within the mean ± 2.576 standard deviations. It is also known as a *Gaussian distribution*.   *HJBB*

Sokal RR and Rohlf FJ (1995) *Biometry*. New York: Freeman.

normal polarity The present-day orientation of the dipole component of the Earth's geomagnetic field. Normal polarity and the opposite, *reverse polarity* (when the orientation changes through 180°), are fundamental components of the GEOMAGNETIC POLARITY TIMESCALE (GPTS).   *MHD*

[*See also* GEOMAGNETIC POLARITY REVERSAL]

normal science The concept, proposed by Thomas Kuhn, that SCIENCE proceeds routinely for most of the time by tackling relatively minor research problems within a particular PARADIGM or stable research framework. According to this view, periods of normal science are interrupted by relatively rare episodes of revolutionary change (*scientific revolutions*) following innovative developments in METHODOLOGY or THEORY. The term *post-normal science* is sometimes used to characterise otherwise normal science under pressure from political or other extraneous factors when facts are uncertain, values in dispute, stakes are high and decisions urgent (Funtowicz and Ravetz, 1991).   *JAM*

Funtowicz SO and Ravetz JR (1991) A new scientific methodology for global environmental issues. In Constanza R (ed.) *Ecological economics: The science and management of sustainability*. New York: Columbia University Press, 137–152.

normalisation The statistical transformation of data so they display a Gaussian or NORMAL DISTRIBU-TION, or at least approximate to a normal distribution. Data are frequently converted into *z-scores* to approximate a standard normal distribution (STANDARDISATION to zero mean and unit variance). This involves subtracting the arithmetic mean from each value of the series and dividing the resulting values by the series standard deviation. Normalisation is commonly used in order to apply PARAMETRIC STATISTICS to data that do not exhibit a normal distribution. An example is the use of a *logarithmic transformation* in relation to a distribution that exhibits positive skew.   *KRB/APP*

Patel JK and Read CB (1996) *Handbook of the normal distribution, 2nd edition*. New York: Marcel Dekker.

Read CB (1985) Normal distribution. In Kotz S and Johnson NL (eds) *Encyclopedia of statistical sciences, volume 6*. New York: Wiley, 347–359.

Norse Greenland Settlements The Norse colony in Greenland was founded by settlers from Iceland around AD 985 and lasted for about 500 years. The *Western Settlement* lasted until the mid-fourteenth century; the more southerly *Eastern Settlement* disappeared towards the end of the late fifteenth century. CLIMATIC CHANGE associated with the LITTLE ICE AGE was involved in the decline in these MARGINAL AREAS but the precise cause(s) have not been established despite intense inter-disciplinary investigation. Other factors that have been implicated include Inuit competition, increasing crop failure and soil erosion, inability to adapt, declining trade with Europe, population decline, social collapse, disease and congenital infertility. The traditional model, emphasises the effects of climatic change, which made the subsistence pastoral system submarginal to which the people were unable to adapt. Dugmore et al. (2010) have proposed an alternative, more complex model, which proposes that climatic change was met with successful adaptation but that social collapse followed declining trade with Europe for their prestige goods, particularly walrus ivory. In this model, cumulative change in climate is merely a contributory factor to the declining population and a cascading collapse of an integrated settlement system.   *JAM*

[*See also* COLLAPSE OF CIVILISATIONS, ENVIRONMENT-HUMAN INTERACTIONS]

Barlow LK, Sadler JP, Ogilvie AEJ et al. (1997) Interdisciplinary investigations of the end of the Norse Western Settlement in Greenland. *The Holocene* 7: 489–499.

Dugmore AJ, Keller C and McGovern TH (2010) Norse Greenland settlement: Reflecting on climate change, trade and the contrasting fates of human settlements in the North Atlantic islands. *Arctic Anthropology* 44: 12–36.

Hunt BG (2009) Natural climatic variability and the Norse settlements in Greenland. *Climatic Change* 97: 389–407.

North Atlantic Oscillation (NAO) A predominantly wintertime, decadal-scale, regional CLIMATIC MODE of the *atmospheric circulation*, the influence of which extends beyond the North Atlantic region. The *NAO Index* is defined as the *sea-level pressure* difference between Ponta Delgada (37°7' N, 25°7' W) in the

Azores and Stykkisholmur (65°0′ N, 22°8′ W) in Iceland. The resultant standardised monthly index values provide a useful measure of the strength and frequency of the atmospheric circulation across the eastern North Atlantic upwind of Western Europe. Positive values indicate strong WESTERLIES, with STORMS tracking farther south into Europe. Large negative values suggest BLOCKING and an easterly component to circulation. The NAO INSTRUMENTAL RECORD begins in AD 1865, the start of the Azores record, but it can be extended by PALAEO-CLIMATIC RECONSTRUCTION, for example, from TREE RINGS and other PROXY DATA. Significant statistical associations exist between the NAO and certain climatic parameters (e.g. winter PRECIPITATION and the CENTRAL ENGLAND TEMPERATURE RECORD (CET)), the THERMOHALINE CIRCULATION, *cod fisheries* of the North Atlantic and the mass balance of European glaciers.   *GS/JAM*

[*See also* ATMOSPHERE-OCEAN INTERACTION, CLIMATIC MODES, DECADAL-SCALE VARIABILITY, TELECONNECTION]

Cook ER, D'Arrigo RD and Briffa KR (1998) A reconstruction of the North Atlantic Oscillation using tree-ring chronologies from North America and Europe. *The Holocene* 8: 9–17.

Dickson RR, Osborn TJ, Hurrell JM et al. (2010) The Arctic Ocean response to the North Atlantic Oscillation. *Journal of Climate* 13: 2671–2696.

Folland CK, Knight J, Linderholm HW et al. (2009) The summer North Atlantic Oscillation: Past, present and future. *Journal of Climate* 22: 1082–1103.

Hurrell JW, Kushnir Y, Ottersen G and Visbeck M (eds) (2003) *The North Atlantic Oscillation: Climatic significance and environmental impact.* Washington, DC: American Geophysical Union.

Hurrell JW and van Loon H (1997) Decadal variations in climate associated with the North Atlantic Oscillation: Climatic change at high-elevation sites. *Climatic Change* 36: 301–326.

Manganello JV (2008) The influence of sea surface temperature anomalies on low-frequency variability of the North Atlantic Oscillation. *Climate Dynamics* 30: 621–641.

Nesje A, Lie Ø and Dahl (2000) Is the North Atlantic Oscillation reflected in Scandinavian glacier mass balance records? *Journal of Quaternary Science* 15: 587–601.

Vincente-Serrano SM and Trigo RM (2011) *Hydrological, socioeconomic and economic impacts of the North Atlantic Oscillation in the Mediterranean region.* Dordrecht: Springer.

Wanner H, Brönnimann S, Casty C et al. (2001) North Atlantic Oscillation: Concept and studies. *Surveys in Geophysics* 22: 321–382.

notch   An indentation at the base of a sea *cliff* close to *sea level*. Notches demonstrate that processes of WEATHERING and/or EROSION are more intensive in the *intertidal zone* and at *surf level* than in subtidal or other supratidal positions. Notches are best developed where

the tidal range is low and in coastal locations sheltered from intense wave action. RELICT palaeoforms, which may be emergent or submergent, are good SEA-LEVEL INDICATORS used in the reconstruction of SEA-LEVEL CHANGE. Pirazzoli (1986) recognises two main types of notches: (1) *tidal notches* are found in the intertidal zone in relatively protected sites where they may be relatively narrow slits, undercutting the cliff by up to several metres with a distinct notch roof or *visor*; (2) *surf notches* occur above high-tide level, are assymetrical in profile with generally a short roof and long floor. Both types may occur at the same site. According to Kelletat (2005), tidal notches are normally formed by BIOLOGICAL WEATHERING (or *bio-erosion*), especially in LIMESTONE rocks where DISSOLUTION (CORROSION) is ineffective due to the oversaturated nature of seawater. However, mechanical ABRASION by the sedimentary particles carried by waves (CORRASION) and/or the purely *hydraulic action* of waves (QUARRYING) must account for surf notches.   *JAM*

Furlani S, Cucchi F, Biochi S and Odorico R (2011). Notches in the North Adriatic Sea: Genesis and development. *Quaternary International* 232: 158–168.

Kelletat DH (2005) Notches. In Schwartz ML (ed.) *Encyclopedia of coastal science.* Dordrecht: Springer, 728–729.

Nunn PD (1995) Holocene tectonic histories for the south-central Lau group, South Pacific. *The Holocene* 5: 160–171.

Pirazzoli PA (1986) Marine notches. In van de Plassche O (ed.) *Sea level research: A manual for the collection and evaluation of data.* Norwich: Geobooks, 361–400.

nuclear accident   An inadvertent release of a substantial amount of radioactive material into the environment. The most serious nuclear accident, or *nuclear disaster*, to date was the explosion at the *Chernobyl* nuclear reactor in the Ukraine in 1986. RADIONUCLIDES were released over a period of 10 days and spread beyond the region to include northern and Western Europe, where soils, pasture and animals were contaminated. Over 130,000 people were evacuated from the local area but continue to be at risk from radiation-induced diseases. The nuclear accident at Fukushima, Japan, in 2012, resulted from the location of nuclear power plants in places vulnerable to a NATURAL HAZARD (TSUNAMI). *JAM*

[*See also* DISASTER, ENVIRONMENTAL DISASTER, HAZARD, NUCLEAR WAR, SOIL RADIOACTIVITY]

Anspaugh LR, Catlin RJ and Goldman M (1988) The global impact of the Chernobyl reactor accident. *Science* 242: 1513–1519.

Boronov A and Bogatov S (1997) *Consequences of Chernobyl.* New York: Plenum Press.

Kingston J (ed.) (2012) *Natural disaster and nuclear crisis in Japan.* London: Routledge.

nuclear energy   Energy released by *atomic fission* or *atomic fusion*. So far, only fission-powered gas- or water-cooled generators exploiting RADIOACTIVITY from enriched uranium are feasible. In theory, fusion promises cheap, safe, sustainable future energy with less dependence on uranium. Fission has been used since the 1940s for civil electricity production and mainly military shipping. Today, >400 commercial reactors produce 15 per cent of the world's electricity. In the United States, 20 per cent of the nation's electricity comes from nuclear energy; in France, the proportion is 80 per cent. Reactors are surrounded by containment vessels but NUCLEAR ACCIDENTS have nevertheless occurred, the latest in 2011 at Fukushima, Japan, as a result of TSUNAMI damage. Other possible ENVIRONMENTAL PROBLEMS associated with nuclear power generation include interruptions to coolant water supply, radioactive leaks into the environment, terrorist interventions (monitoring is necessary to keep track of materials that might find their way into terrorist hands), provision of safe long-term storage of RADIOACTIVE WASTE and safe DECOMMISSIONING of old reactors. Nuclear energy is already an important form of ALTERNATIVE or RENEWABLE ENERGY, but it is expensive when all costs are considered and there is much opposition to its further expansion, mainly on safety grounds. Nevertheless, safer designs (e.g. the proposed fluoride-thorium reactors) are possible, and it has been argued that current environmental change is happening so fast that other low-carbon energy options cannot cope and the nuclear option is the best *practicable* environmental option (Lovelock, 2006).   *CJB*

Bodansky D (2008) *Nuclear energy, 2nd edition*. New York: Springer.
Elliott D (2007) *Nuclear or not? Does nuclear power have a place in a sustainable energy future?* Basingstoke: Palgrave Macmillan.
Furguson CD (2011) *Nuclear energy: What everyone needs to know*. Oxford: Oxford University Press.
Lillington JN (2004) *Future of nuclear power*. Amsterdam: Elsevier.
Lovelock J (2006) *The revenge of Gaia*. London: Allen Lane.
Mahaffey J (2010) *Atomic awakening: A new look at the history and future of nuclear power*. New York: Pegasus Books.
Nuttall WJ (2005) *Nuclear renaissance: Technologies and policies for the future of nuclear power*. New York: Taylor and Francis.
Sovacool BK (2011) *Contesting the future of nuclear power: A critical global assessment of atomic energy*. Singapore: World Scientific.

nuclear war   The potential environmental consequences of a nuclear war are difficult to estimate. The Royal Swedish Academy of Sciences (1983) and others assessed the effects of a full-scale nuclear war between the United States of America and the former Soviet Union targeting urban, military and economic targets with particular emphasis on the effects of the enormous quantities of PARTICULATE matter which would cloak the Earth. Forests, agricultural land and oilfields would be ignited, adding PHOTOCHEMICAL SMOG to the reduction in light, short-term climatic cooling (NUCLEAR WINTER), direct blast effects and the hazard of RADIOACTIVITY, which would be particularly devastating on the terrestrial ecosystems and human populations of the Northern Hemisphere. They concluded that long-term and less predictable environmental effects might match or exceed these more immediate impacts.   *JAM/JET*

Glover L (2006) *Postmodern climate change*. London: Routledge.
National Research Council (1985) *The effects on the atmosphere of a major nuclear exchange*. Washington, DC: National Academy Press.
Royal Swedish Academy of Sciences (1983) *Nuclear war: The aftermath*. Oxford: Pergamon Press.

nuclear winter   The hypothesis that a NUCLEAR WAR would cause severe GLOBAL COOLING of the Earth due to the injection into the atmosphere of vast quantities of DUST and *soot*, which would intercept SOLAR RADIATION for a prolonged time interval. Global temperatures might fall by as much as 25°C. Although precise effects are difficult to predict, analogues for nuclear winter include the contrast between day and night, the *firestorms* of World War II, *forest fires* (see BIOMASS BURNING), OIL FIRES, VOLCANIC IMPACTS ON CLIMATE and the MASS EXTINCTIONS of the geological record possibly caused by METEORITE IMPACT.   *JAM*

Badash L (2009) *Nuclear winter's tale*. Cambridge: MIT Press.
Robock A (1996) Nuclear winter. In Schneider SH (ed.) *Encyclopedia of climate and weather*. New York: Oxford University Press.

nuclide   A *nuclear species* characterised by having the same *atomic number* (number of *protons*) and the same *mass number* (number of *neutrons*).   *PQD*

[*See also* COSMOGENIC-NUCLIDE DATING, ISOTOPE]

nudation   The removal of existing *ecosystems* and BIOTIC communities by major DISTURBANCE prior to the initiation of ECOLOGICAL SUCCESSION.   *LRW*

nuée ardente   A PYROCLASTIC FLOW produced by the collapse of a LAVA DOME. The term, from the French for 'glowing cloud', was coined following the AD 1902 eruption of Mont Pelée on Martinique, in which 28,000 people died. Because of some confusion in the application of the term, it is probably best avoided.   *JBH*

Druitt TH and Kokelaar BP (eds) (2002) *The eruption of Soufrière Hills volcano, Montserrat, from 1995 to 1999*. Bath: Geological Society [Memoir 21].

La Croix A (1904) *La montagne Pelée et ses éruptions* [Mount Pelée and its eruptions]. Paris: Masson.

Scarth A (2002) *La catastrophe: Mount Pelee and the destruction of Saint-Pierre, Martinique.* Harpenden: Terra Publishing.

Tanguy JC (1994) The 1902-1905 eruptions of Montagne-Pelée, Martinique: Anatomy and retrospection. *Journal of Volcanology and Geothermal Research* 60: 87–107.

number of looks In a *synthetic aperture radar* system, SPECKLE is reduced by splitting the full Doppler bandwidth of the AZIMUTH beam into a number of smaller 'looks', which effectively makes multiple measurements (images) of the same ground resolution cell. *Multilooking* (which often amounts to averaging over PIXELS in the azimuth direction) is an important component of the data *pre-processing*.   MEJC

numerical analysis Data relevant to environmental-change research are usually complex, quantitative and multivariate, consisting of many observations and many variables. Such data are often stratigraphical or geographical in character and hence the observations have a fixed order in one or two dimensions. *Biological data* (e.g. pollen or diatom stratigraphical data) usually have large numbers of variables (100–200), have many zero (absence) values, are most commonly expressed as percentages and are thus 'closed' compositional data. *Geological data* (e.g. magnetic properties) usually have relatively few variables (<50) and few zero values. *Environmental data* (e.g. lake chemistry) also have few variables (10–30), few zero values, but may have some missing unmeasured values.

Numerical analysis of such complex data sets is now a regular part of environmental research. This is a result, in part, of the many recent developments in applied and *environmental statistics* (ENVIRONMETRICS) and, in part, of the increasing availability of powerful computers with the advent of personal computers. Numerical analysis involves DATA SUMMARISATION, EXPLORATORY DATA ANALYSIS and CONFIRMATORY DATA ANALYSIS, the last usually involving specific HYPOTHESIS testing.

An essential first step in any data analysis is basic exploratory data analysis. The main purpose is to provide the data analyst with 'a feel for the data'. It involves estimation of measures of *central tendency* (e.g. *mean, median*), *dispersion* (e.g. *standard deviation, interquartile range*) and shape (e.g. SKEWNESS, KURTOSIS) of the data; simple graphical tools such as *box-and-whisker plots* and *scatter plots; outlier detection* involving influence and leverage measures and *data transformations* in an attempt to achieve a NORMAL DISTRIBUTION that is the PROBABILITY DISTRIBUTION that is assumed by many methods of PARAMETRIC STATISTICS. *Data display* is an essential step, either as two- or three-dimensional

scatter plots or as *stratigraphical diagrams*. Interactive *graphical tools* are indispensible in exploratory analysis. *Non parametric* regression techniques, such as locally weighted regression scatter plot smoothing (LOESS REGRESSION), are useful graphical tools for highlighting the '*signal*' or major trends in data in the absence of any a priori model.

Pattern detection or STRUCTURING within multivariate data can be usefully detected by means of numerical CLUSTERING (e.g. CLUSTER ANALYSIS, TWO-WAY INDICATOR SPECIES ANALYSIS (TWINSPAN)) to detect groups of observations of similar composition and/or groups of variables with similar occurrences. ORDINATION techniques (e.g. PRINCIPAL COMPONENTS ANALYSIS (PCA), CORRESPONDENCE ANALYSIS (CA), NON-METRIC MULTIDIMENSIONAL SCALING) provide useful graphical summaries in two or three dimensions of the major patterns of variation in multivariate data, can help identify latent variables and gradients in data, and can display patterns of similarity and dissimilarity between observations and variables.

When the observations come from two or more pre-defined groups (e.g. from two different bedrock types), DISCRIMINANT ANALYSIS provides a means of evaluating how distinct the groups are, of characterising the groups in terms of the variables and of providing a low-dimensional graphical representation of the within-group and between-group variations.

If there is some a priori biological or geological reason for considering some variables as responding (so-called *response* or *dependent variables*) to other variables, so-called *explanatory, predictor* or *independent variables*, statistical modelling techniques such as REGRESSION ANALYSIS and ANALYSIS OF VARIANCE (ANOVA) can be used to model the relationship between one response variable and one or more predictor variables. All such modelling techniques are a part of GENERALISED LINEAR MODELS (GLM). The non-parametric GENERALISED ADDITIVE MODELS (GAM) can provide a useful exploratory tool. If there are two or more response variables and two or more predictor variables, constrained ordination techniques such as REDUNDANCY ANALYSIS or CANONICAL CORRESPONDENCE ANALYSIS (CCA) can be used to derive a low-dimensional multivariate regression model that combines both an ordination graphical display and regression modelling. Given two sets of predictor variables (e.g. climatic variables and chemical variables), it is possible by means of a series of (partial) constrained ordinations to *partition the variation* in the response variables into four independent components—(1) variation due to climate independent of chemistry, (2) variation due to chemistry independent of climate, (3) variation due to the covariance between climate and chemistry and (4) variation not explained by climate or chemistry. *Variation decomposition* can be extended for three or more groups of predictor variables.

In palaeoenvironmental research, an important role of numerical analysis involves the estimation of modern *calibration functions* or TRANSFER FUNCTIONS. These model the relationship between modern biological assemblages (e.g. pollen, diatoms) in surface sediments and contemporary environmental variables (e.g. mean July temperature, lake-water pH). Such *calibration functions* are most commonly derived by *inverse regression* that assumes either a linear or a unimodal response model of organisms to the environment. A large number of techniques now exist for deriving calibration functions, including *weighted-averaging regression* and calibration, *weighted-averaging partial least squares regression, partial least squares regression* and GLM.

Stratigraphical data have special mathematical properties, the most important is that the observations are in a fixed stratigraphical and temporal order. Numerical analysis of such data requires taking account of this order. Stratigraphically constrained clustering methods have been developed for partitioning or zoning such data. Individual stratigraphical variables can be partitioned by sequence splitting into segments of uniform mean and variance to provide a means of detecting and testing for consistent patterns of stratigraphical change between variables. Comparison of two or more stratigraphical records (e.g. pollen, diatoms) from the same sequence can be done numerically by either summarising the different data sets independently first as the first few principal component analysis or correspondence analysis axes and then by comparing these axes by means of oscillation logs or using *constrained ordination* techniques such as REDUNDANCY ANALYSIS. Comparison and correlation of two or more stratigraphical sequences can be made by sequence slotting or by combined classification or ordination of the sequences. Rate-of-change analysis can be useful to estimate the amount of change per unit time in stratigraphical data. Such analyses require detailed chronologies in calendar years. MODERN ANALOGUE techniques in which modern analogues for fossil assemblages are sought numerically can provide a factual basis for interpretation of the stratigraphical data.

TIME-SERIES ANALYSIS can involve the *time-domain approach* based on the concept of temporal AUTOCORRELATION or the *frequency-domain approach* that focuses on bands of frequency or wavelength over which the variance of the time series is concentrated. Time-series analysis can be used to detect patterns of *temporal autocorrelation* and cross-correlation and PERIODICITY within time-ordered data. Time-series analysis makes many demanding assumptions of the data, many of which are difficult to meet in palaeoenvironmental studies.

Spatial data similarly have a fixed ordering in two dimensions. This geographical information is important in *mapping*, in *trend-surface analysis*, in *geostatistical techniques* such as KRIGING, in ORDINATION and CLASSIFICATION of spatial data and in the estimation of statistical parameters and the statistical testing of HYPOTHESES in the presence of spatial autocorrelation (see SPATIAL ANALYSIS).

The testing of hypotheses about the impacts of environmental variables on biological assemblages is an important part of confirmatory data analysis in environmental research. Given the complexity of environmental data with their spatial or temporal ordering, non-normality and many zero values, hypothesis testing requires MONTE CARLO METHODS involving *randomisation tests* or *permutation tests* and constrained ordination techniques such as redundancy analysis. Such tests also provide means of testing for spatial and/or temporal trends in data.

The numerical analysis of data associated with environmental change research is a rapidly developing field. Future developments are likely to involve CLASSIFICATION AND REGRESSION TREES (CART), NEURAL NETWORKS, and the greater incorporation of BAYESIAN STATISTICS in the analysis of biological and environmental data. *HJBB*

[*See also* CLIMATIC RECONSTRUCTION, PALAEOCLIMATIC MODELLING, STATISTICAL ANALYSIS]

Birks HJB (1998) Numerical tools in palaeolimnology: Progress, potentialities, and problems. *Journal of Paleolimnology* 20: 307–332.

Birks HJB (2007) Numerical analysis methods. In Elias SA (ed.) *Encyclopedia of Quaternary science*. Amsterdam: Elsevier, 2515–2521.

Birks HJB and Gordon AD (1995) *Numerical methods in Quaternary pollen analysis*. London: Academic Press.

Birks HJB, Heiri O, Seppä H and Bjune AE (2010) Strengths and weaknesses of quantitative climate reconstructions based on late-Quaternary biological proxies. *Open Ecology Journal* 3: 68–110.

Birks HJB, Lotter AF, Juggins S and Smol JP (eds) (2012) *Tracking environmental change using lake sediments, volume 5: Data handling and numerical techniques*. Dordrecht: Springer.

Bivand R, Pebesma EJ and Gómez-Rubio V (2008) *Applied spatial data analysis with R*. New York: Springer.

Borcard D, Gillet F and Legendre P (2011) *Numerical ecology with R*. New York: Springer.

Davis JC (1986) *Statistics and data analysis in geology*. New York: John Wiley.

Jongman RHG, ter Braak CJF and van Tongeren OFR (1987) *Data analysis in community and landscape ecology*. Wageningen: Pudoc.

Kent M (2012) *Vegetation description and data analysis, 2nd edition*. Chichester: Wiley-Blackwell.

Legendre P and Legendre L (1998) *Numerical ecology*. Amsterdam: Elsevier.

Tingley MP, Craigmile PF, Haran M et al. (2012) Piecing together the past: Statistical insights into paleoclimatic reconstruction. *Quaternary Science Reviews* 35: 1–22.

Wehrens R (2011) *Chemometrics with R: Multivariate analysis in the natural sciences and life sciences.* New York: Springer.

numerical-age dating A category of techniques used to obtain quantitative estimates of age and UNCERTAINTY based on an absolute timescale. It was formerly termed *absolute-age dating.* *DAR*

[*See also* CALIBRATED-AGE DATING, CORRELATED-AGE DATING, RELATIVE-AGE DATING]

Colman SM, Pierce KL and Birkeland PW (1987) Suggested terminology for Quaternary dating methods. *Quaternary Research* 28: 314–319.

nunatak A mountain top protruding above an ICE SHEET or GLACIER surface. During the QUATERNARY, nunataks may have provided refugia for certain flora and fauna. The term is derived from the Inuit language. *DIB*

[*See also* NUNATAK HYPOTHESIS, REFUGE THEORY IN TEMPERATE REGIONS, REFUGIUM]

Holderegger R and Thiel-Egenter C (2008) A discussion of different types of glacial refugia used in mountain biogeography and phylogeography. *Journal of Biogeography* 36: 476–480.
Paus A, Velle G and Berge J (2011) The Late glacial and Early Holocene vegetation and environment in the Dovre mountains, central Norway, as signalled in two Lateglacial nunatak lakes. *Quaternary Science Reviews* 30: 1780–1796.

nunatak hypothesis The concept that the patchy occurrence of mountain species is evidence of survival on the mountain tops above an ICE SHEET. Thus, a nunatak can be a REFUGIUM. The alternative is that the species migrated there after the ice sheet melted. *BA/RMF*

[*See also* GLACIAL-INTERGLACIAL CYCLE, REFUGE THEORY IN TEMPERATE REGIONS, TABULA RASA]

Brochman C, Gabrielson TM, Nordal I et al. (2003) Glacial survival or tabula rasa? The history of North Atlantic biota revisited. *Taxon* 52: 417–450.
Schneeweiss GM and Schonswetter P (2011) A reappraisal of nunatak survival in arctic-alpine phylogeography. *Molecular Ecology* 20: 190–192.

nurse plant A plant that facilitates the establishment and growth of another plant of a different species, often in ECOLOGICAL SUCCESSION, by providing direct *physical shelter* from, for example, excessive *wind, heat* or HERBIVORES. Nurse plants are also employed in AGRICULTURE and HORTICULTURE. *LRW*

[*See also* FACILITATION]

Callaway RM (2007) *Positive interactions and interdependence in plant communities.* New York: Springer.

Carlucci MB, Teixeira FZ, Brum FT and Duarte LDS (2011) Edge expansion of *Araucaria* forest over southern Brazilian grasslands relies on nurse plant effect. *Community Ecology* 12: 196–201.

nutrient A *raw material* required by organisms for life. Nutrients include not only the essential mineral elements (MACRONUTRIENTS and MICRONUTRIENTS) but also *water, inorganic salts* and, for animals, organic compounds (*carbohydrates, fats, proteins* and *vitamins*). *JAM*

[*See also* BIOGEOCHEMICAL CYCLES, HUMAN HEALTH HAZARDS, MINERAL CYCLING]

Marschner P and Rengel Z (eds) (2007) *Nutrient cycles in terrestrial ecosystems.* Berlin: Springer.
Miransari M (ed.) (2012) *Soil nutrients.* New York: Nova Science Publishers.
Paoletti MG, Foissner W and Coleman DC (eds) (1993) *Soil biota, nutrient cycling and farming systems.* Boca Raton, FL: Lewis.
Sharon M (2009) *Nutrients A-Z: A user's guide to foods, herbs, vitamins, minerals and supplements,* 5th edition. London: Carlton Books.

nutrient pool The store or reserve of a NUTRIENT in an *ecosystem* or its component parts (in the soil, water, air or living organisms). *RJH*

[*See also* BIOGEOCHEMICAL CYCLES, LABILE POOL, LIMITING FACTORS]

Menge DNL, Pacala SW and Hedin LO (2009) Emergence and maintenance of nutrient limitation over multiple timescales in terrestrial ecosystems. *American Naturalist* 173: 164–175.

nutrient status The concept of SOIL or other medium as a source of NUTRIENTS for plants, particularly the three main plant MACRONUTRIENTS (NITRATES (NO_3^-), PHOSPHATES (PO) and POTASSIUM (K)). The nutrient status of soils is important for *natural ecosystems,* AGRICULTURE, HORTICULTURE and FORESTRY. Soils may have nutrients in sufficient quantity for plant growth, but more often than not certain elements are deficient and need to be supplied for optimum crop yields. Plant nutrients are removed from the soil naturally by LEACHING and by crops and livestock HARVESTING and CONSUMPTION elsewhere. Unless these nutrients are replaced naturally, a long-term process, or more rapidly by chemical FERTILISERS, soil FERTILITY is reduced. The extreme situation is found in impoverished human communities who have to resort to *fertility mining* for subsistence, and the result is SOIL DEGRADATION. Excessive nutrient concentrations can cause imbalance of other essential nutrients and reduce yields. Certain MICRONUTRIENTS may also be in limited supply. The term nutrient status is also commonly used in the context of *marine* and ESTUARINE ENVIRONMENTS, LAKES, MIRES and other WETLANDS. *EMB/JAM*

[*See also* EUTROPHIC, LABILE POOL, LIMITING FACTORS, MINERAL CYCLING, OLIGOTROPHICATION, SOIL DEGRADATION]

Ahern J, Posch M, Forsius M et al. (2012) Impacts of forest biomass removal on soil nutrient status under climate change: A catchment-based modelling study for Finland. *Biogeochemistry* 107: 471–488.

BussiriRad H (ed.) (2005) *Nutrient acquisition by plants: An ecological perspective.* Berlin: Springer.

Engstrom DR, Fritz SC, Almendinger JE et al. (2000) Chemical and biological trends during lake evolution in recently deglaciated terrain. *Nature* 408: 161–166.

Gowen RJ and Stewart BM (2005) The Irish Sea: Nutrient status and phytoplankton. *Journal of Sea Research* 54: 36–50.

Havlin JL, Tisdale SL, Nelson WL and Beaton JD (2004) *Soil fertility and fertilizers: An introduction to nutrient management, 7th edition.* Upper Saddle River, NJ: Prentice Hall.

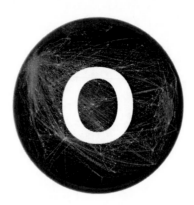

oasis hypothesis In relation to early agriculture, an obsolete hypothesis to account for DOMESTICATION whereby aggregation around reduced *Late Glacial* water resources (oases) forged new human-animal relationships. *ARG*

[*See also* AGRICULTURAL HISTORY, AGRICULTURAL ORIGINS]

Childe VG (1935) *New light on the most ancient East.* London: Kegan Paul.

object oriented An approach to programming, REMOTE SENSING and IMAGE PROCESSING based on objects. Defining spatial landscape objects in an image or in a GEOGRAPHICAL INFORMATION SYSTEM (GIS) can lead to better CLASSIFICATION results. *HB*

objective knowledge Knowledge that has been subjected to SCIENTIFIC METHOD and hence has been researched, analysed, synthesised and debated by scientists. Although it no longer represents the subjective, untested views of individual people, it still does not imply absolute or permanent knowledge. Objective knowledge evolves as investigation continues, new information is discovered and new insights are incorporated: it is simply 'the most reliable current knowledge' (Ford, 2000). However, well-established, reliable, objective knowledge (facts) may not be believed because of cultural factors. *JAM*

[*See also* KNOWLEDGE, OBJECTIVITY, SCIENCE, SOCIAL CONSTRUCTION, TRUTH]

Ford ED (2000) *Scientific method for ecological research.* Cambridge: Cambridge University Press.
Howlett P and Morgan MS (eds) (2011) *How well do facts travel? The dissemination of reliable knowledge.* Cambridge: Cambridge University Press.
Popper KR (1982) *Objective knowledge: An evolutionary approach.* Oxford: Oxford University Press.

objectivity Independence from individual BIAS and subjective personal opinion. Maintenance of objectivity is one of the distinguishing characteristics and virtues of SCIENTIFIC METHOD. Scientists believe that their conclusions about NATURE have little value unless they are objective. Objectivity does not mean the same as *consensus*: all of the people can be fooled some of the time. The proof of objectivity in scientific conclusions is often said to lie in whether results are reproducible, testable and useful. However, the difference between objectivity and *subjectivity* is not clear-cut, some philosophers even disputing its existence. *Expert judgement*, for example, is always required, both in RESEARCH DESIGN and in interpreting the results of RESEARCH. It is probably best, therefore, to say SCIENCE aims to *maximise* objectivity. *JAM*

[*See also* OBJECTIVE KNOWLEDGE, REPLICATION, TRUTH]

Archer MS and Outhwaite W (eds) (2004) *Defending objectivity.* Abingdon: Routledge.
Burge T (2010) *Origins of objectivity.* Oxford: Oxford University Press.
Daston L and Galison P (2007) *Objectivity.* New York: Zone Books.
Gay C and Estrada F (2010) Objective probabilities about future climate are a matter of opinion. *Climatic Change* 99: 27–46.

obligate Pertaining to an ENVIRONMENTAL FACTOR that is required by an organism for survival, as opposed to one that the organism may benefit from but is not necessary for survival (FACULTATIVE). Organisms may also be described as obligate in relation to particular environmental factors. Thus, an obligate HALOPHYTE is one that requires relatively high SALT concentrations, whereas a facultative halophyte can tolerate high SALINITY but does not require it. *JAM*

Wang W, Yan Z, You S et al. (2011) Mangroves: Obligate or facultative halophytes? A review. *Tree Structure and Function* 25: 953–963.

obliquity The angle between the plane of the Earth's orbit (the ECLIPTIC) and the plane of the Earth's equator. Known colloquially as *axial tilt*, the full term is *obliquity of the ecliptic*. It affects the angle of the solar beam, which depends on time of day, time of year and latitude. For example, due to the greater obliquity, INSOLATION in the polar latitudes is less concentrated and more SOLAR RADIATION is lost by atmospheric scattering, reflection and absorption. Obliquity is one of the three main orbital parameters in the MILANKOVITCH THEORY. It varies between about 22.0° and 24.5° with a PERIODICITY of about 40,000 years. *RDT/DAW*

[*See also* ECCENTRICITY, PRECESSION]

Mantsis DF, Clement AC, Broccoli AJ and Erb MP (2011) Climate feedbacks in response to changes in obliquity. *Journal of Climate* 24: 2830–2845.

obrution Smothering by rapid burial in SEDIMENT, typically resulting from STORM events, also *slumping* (see SLUMP), DEBRIS FLOWS or other types of MASS MOVEMENT PROCESSES. *LC*

[*See also* FOSSIL LAGERSTÄTTEN, TAPHONOMY]

Brett CE (1992) Obrution deposits. In Briggs DEG and Crowther PR (eds) *Palaeobiology: A synthesis, 1st edition*. Oxford: Blackwell.

observation First-hand *description*, which may be qualitative or quantitative, and may involve MEASUREMENT. It is a crucial part of SCIENTIFIC METHOD, especially in HYPOTHESIS testing. *JAM*

[*See also* EARTH OBSERVATION (EO), OBSERVATIONAL DATA, REMOTE SENSING]

observational data A record of the MEASUREMENT or *qualitative assessment* of an environmental variable at specified place(s) and time(s). Use of observational data implies a relatively direct relationship between the observer and the environment, and the maintenance of maximum levels of OBJECTIVITY during data collection so that the data are reliable, relatively independent of theoretical constructs and capable of being used to test HYPOTHESES. The credibility of such data depends both on the accuracy of the instruments used in measurement and the expertise of the individual in using the instrument or in making qualitative assessments. In CLIMATOLOGY, for example, long records of daily data are required at the same place that have been obtained using standard instruments to produce a time series. The ideal is a HOMOGENEOUS SERIES, but this is not always possible and a variety of statistical techniques have been developed in the past few decades to make allowances for site changes, breaks in the data series and to produce regional values. The same is true of other NATURAL ENVIRONMENTAL SCIENCES, each of which has its own criteria for the production and VALIDATION of its observational data. Particularly in the investigation of *biophysical environmental change*, observational data are increasingly collected by REMOTE SENSING.

However, in the investigation of some HUMAN DIMENSIONS OF ENVIRONMENTAL CHANGE, *observational studies* may be limited to qualitative assessment. *BDG/JAM*

[*See also* ENVIRONMENTAL MONITORING, INSTRUMENTAL RECORD, OBSERVATION, TIME-SERIES ANALYSIS]

Alverson K (2012) Direct observation and monitoring of climate and related environmental change. In Matthews JA, Bartlein PJ, Briffa KR et al. *The SAGE handbook of environmental change, volume 1*. London: Sage, 53–66.
Rosenbaum VR (2010) *Observational studies, 2nd edition*. New York: Springer.
Strangeways I (1997) Ground and remotely sensed measurements. In Thompson RD and Perry A (eds) *Applied climatology: Principles and practice*. London: Routledge, 13–21.
Zhang HM, Reynolds RW and Smith TM (2006) Adequacy of the in situ observing system in the satellite era for climate SST. *Journal of Atmospheric and Oceanic Technology* 23: 107–120.

obsidian hydration dating A technique using the rate and extent of chemical alteration of obsidian rock surfaces for dating purposes. Obsidian is a glassy product of VOLCANICITY, formed by the rapid cooling of silica-rich LAVAS. On exposure to water in air or surrounding SOIL, a fresh obsidian surface reacts by HYDRATION to form a WEATHERING RIND, the thickness of which is a non-linear function of time, temperature and chemical composition. CALIBRATED-AGE DATING using this technique is possible if (1) independent evidence can be found based on NUMERICAL-AGE DATING methods such as RADIOCARBON DATING or POTASSIUM-ARGON (K-Ar) DATING; or (2) hydration rate can be determined in the laboratory by heating experiments at different relative humidities. Recent, more accurate definition of the hydration profile utilising SECONDARY ION MASS SPECTROMETRY (SIMS) has offered possibilities of greater reliability of ages. The technique is most widely used in archaeological studies because obsidian was widely traded in PREHISTORIC times, although the timing of glacial events can be determined where glacial ABRASION has created fresh surfaces of obsidian in GLACIAL SEDIMENTS. *DAR/CJC*

Anovitz LM, Riciputi LR, Cole DR et al. (2006) Obsidian hydration: A new paleothermometer. *Geology* 34: 517–520.
Beck C and Jones GT (1994) Dating surface assemblages using obsidian hydration. In Beck C (ed.) *Dating in exposed and surface contexts*. Albuquerque: University of New Mexico Press, 47–76.
Friedman I, Trembour F and Hughes RE (1997) Obsidian hydration dating. In Taylor RE and Aitken MJ (eds) *Chronometric dating in archaeology*. New York: Plenum Press, 297–321.
Friedman I, Trembour F, Smith FL and Smith GI (1994) Is obsidian hydration dating affected by relative humidity? *Quaternary Research* 41: 185–190.

Liritzis I (2010) Strofilas (Andros Island, Greece): New evidence for the Cycladic final Neolithic period through novel dating methods using luminescence and obsidian hydration. *Journal of Archaeological Science* 37: 1367–1377.

Liritzis I and Laskaris N (2011) Fifty years of obsidian hydration dating in archaeology. *Journal of Non-Crystalline Solids* 357: 2011–2023.

occult deposition In the context of DEPOSITION from the ATMOSPHERE of POLLUTANTS, especially ACID PRECIPITATION, it is the material deposited by INTERCEPTION. For example, *hoar frost* and *fog droplets*, which may be many times more polluted than rain, can accumulate on vegetation. In some circumstances, occult deposition combined with DRY DEPOSITION may deliver more pollutants to the Earth's surface than WET DEPOSITION. *JAM*

[*See also* AIR POLLUTION]

Humova I, Kurfuerst P, Maznova J and Conkova M (2011) The contribution of occult precipitation to sulphur deposition in the Czech Republic. *Erdkunde* 65: 243–259.

occupation layer/level A buried horizon at or near an archaeological site with evidence of human occupation. The term is widely used, not only for levels that were occupied sensu stricto (e.g. former floors or hearths and the remains of collapsed buildings) but also for various types of dumps for *refuse* (see MIDDEN) and the soils modified by the spreading of refuse on fields as *manure* (see ANTHROPOGENIC SOIL HORIZONS). Like *ditches* and *pits*, occupation layers tend to be rich in organic material and ARTEFACTS. *JAM*

[*See also* TELL]

Davidson DA and Simpson IA (1984) The formation of deep topsoils in Orkney. *Earth Surface Processes and Landforms* 9: 75–81.

Matthews W (1995) Micromorphological characterisation and interpretation of occupation deposits and microstratigraphical sequences at Abu Salabikh, southern Iraq. In Barham AJ and Macphail RI (eds) *Archaeological sediments and soils*. London: University College London Press, 41–74.

ocean A continuous body of seawater, much of it deeper than 1,000 m, covering much of the Earth's surface, particularly the OCEAN BASINS, where the ocean is underlain by OCEANIC CRUST. The Earth's oceans are the ARCTIC, ATLANTIC, *Indian* and *Pacific Oceans*, with many people considering the *Southern* (or *Antarctic*) *Ocean* as the fifth. Ocean water is salty because of EVAPORATION. Oceans and their fringing SEAS cover about 71 per cent of the Earth's surface (total area about 360 million km^2), with an estimated volume of about 1.4 billion km^3 and an average depth of about 3,500 m (see HYPSOMETRY). The Pacific Ocean accounts for almost 40 per cent of the Earth's total sea area and is the deepest ocean. It contains the deepest parts of the oceans, in the OCEANIC TRENCHES (*Mindanao Trench*, 11,524 m; *Mariana Trench*, 11,022 m). Oceans and seas account for about 98 per cent by volume of the Earth's surface waters, the remainder being made up of ice (1.7 per cent) and lakes and rivers (<1 per cent). A growing number of ENVIRONMENTAL PROBLEMS are affecting the oceans, for example, MARINE POLLUTION, OCEAN ACIDIFICATION, OVERFISHING, HABITAT LOSS and CLIMATE CHANGE. *GO*

[*See also* CONTINENTAL MARGIN, MID-OCEAN RIDGE (MOR), OCEANOGRAPHY, SEAWATER COMPOSITION]

Couper A (ed.) (1983) *The Times atlas of the oceans*. London: Times Books.

Earle SA and Glover L (eds) (2008) *Ocean: An illustrated atlas*. Washington, DC: National Geographic Society.

Hedgpeth JW (1957) Classification of marine environments. *Geological Society of America Memoir* 67: 17–280.

Oka A, Tajika E, Abe-Ouchi A and Kubota K (2011) Role of the ocean in controlling atmospheric CO$_2$ concentration in the course of global glaciations. *Climate Dynamics* 37: 1755–1770.

Roberts C (2012) *Ocean of life: How our seas are changing*. London: Allen Lane.

Smith J, Wigginton N, Ash C et al. (eds) (2010) Changing oceans. *Science* 328: 1497–1528 [Special Section].

Sverdrup KA and Armbrust EV (2008) *An introduction to the world's oceans, 10th edition*. New York: McGraw-Hill.

ocean acidification The decrease in pH and increase in acidity of the OCEANS caused by the uptake of excess CARBON DIOXIDE (CO$_2$) from the ATMOSPHERE. Acidification occurs as CO$_2$ reacts with seawater to form *bicarbonate ions* and *protons* (i.e. free hydrogen ions) which are the agents of ACIDITY. The result of this process is consumption of CARBONATE ions and reduction in the saturation states of the biologically important calcium carbonate MINERALS, *aragonite* and *calcite*. It is estimated that in the past 200 years, the oceans have absorbed approximately half of the CO$_2$ produced by human activities. Calculations indicate that this uptake of CO$_2$ has led to a reduction of the pH of surface seawater of 0.1 pH units. If global CO$_2$ emissions continue to rise as they are currently, surface ocean pH is predicted to fall by 0.5 pH units by the year 2100, a level that has not been experienced for hundreds of thousands of years. A major concern is that this process will affect the ability of calcifying marine organisms such as CORALS, FORAMINIFERA and COCCOLITHOPHORES to produce and maintain a skeleton.

Observational data are already showing that continued ocean acidification is causing many parts of the ocean to become undersaturated with respect to aragonite and evidence of reduced calcification has been reported in some organisms. Many calcifying species exhibit reduced calcification and growth rates in laboratory experiments under high-CO_2 conditions. Culture experiments investigating the physiological response of calcifing organisms to low pH, however, have yielded contradictory results, with some species exhibiting reduced calcification and growth rates while others appear to compensate by producing larger *tests*. Improving understanding of the consequences of ocean acidification for marine communities and the potential disruption these consequences could have for the ocean-climate and marine-ecosystem service roles (e.g. sustaining FOOD CHAINS and *fisheries*) that the marine communities provide are high priorities for marine research. The studies of the history of ocean carbonate chemistry and naturally forced ocean acidification episodes in the GEOLOGICAL RECORD (e.g. the Palaeocene Eocene Thermal Maximum) are part of this effort. A range of chemical and biological *palaeoceanographic proxies* are used to accomplish this.   *HKC*

[*See also* CARBON CYCLE, ECOLOGICAL GOODS, ECOLOGICAL SERVICES, MARINE PALAEOCLIMATIC PROXIES, PALAEOCEANOGRAPHY: BIOLOGICAL PROXIES, PALAEOCEANOGRAPHY: PHYSICAL AND CHEMICAL PROXIES]

Beaufort L, Probert I, de Garidel-Thoron T et al. (2011) Sensitivity of coccolithophores to carbonate chemistry and ocean acidification. *Nature* 476: 80–83.

De Moel H, Ganssen GM, Peeters FJC et al. (2009) Planktic foraminiferal shell thinning in the Arabian Sea due to anthropogenic ocean acidification? *Biogeosciences Discussions* 6: 1811–1835.

Hoegh-Guldberg O, Mumby PJ, Hooten AJ et al. (2007) Coral reefs under rapid climate change and ocean acidification. *Science* 318: 1737–1742.

Hofmann AW and Schellnhuber H-J (2009) Oceanic acidification affects marine carbon pump and triggers extended marine oxygen holes. *Proceedings of the National Academy of Sciences* 106: 3017–3022.

Hönisch B, Ridgwell A, Schmidt DN et al. (2012) The geological record of ocean acidification. *Science* 335: 1058–1063.

Orr JC, Fabry VJ, Aumont O et al. (2005) Anthropogenic ocean acidification over the twenty-first century and its impact on calcifying organisms. *Nature* 437: 681–686.

Ries JB, Cohen AL and McCorkle DC (2009) Marine calcifiers exhibit mixed responses to CO_2-induced ocean acidification. *Geology* 37: 1131–1134.

The Royal Society (2005) *Ocean acidification due to increasing atmospheric carbon dioxide.* Cardiff: Clyvedon Press [Royal Society Policy document 12/05].

Zachos JC, Röhl U, Schellenberg SA et al. (2005) Rapid acidification of the ocean during the Paleocene-Eocene thermal maximum. *Science* 308: 1611–1615.

ocean basin   OCEAN basins originate when extension across a RIFT VALLEY on a CONTINENT allows MAGMA from the ASTHENOSPHERE to rise and form new LITHOSPHERE-bearing OCEANIC CRUST. The basin expands through the process of SEA-FLOOR SPREADING, a fundamental feature of PLATE TECTONICS. Young and mature ocean basins are exemplified by the Red Sea and the Atlantic Ocean, respectively; the oldest oceanic crust in the Red Sea is about 5 million years old, while the oldest in the Atlantic Ocean is about 160 million years old. These ocean basins are characterised by a central MID-OCEAN RIDGE (MOR) where sea-floor spreading takes place, representing a CONSTRUCTIVE PLATE MARGIN. On either side the morphology of the ocean basin is symmetrical with, in the mature ocean, an ABYSSAL plain passing into a PASSIVE CONTINENTAL MARGIN comprising CONTINENTAL RISE, CONTINENTAL SLOPE and CONTINENTAL SHELF. The true ocean-continent boundary lies close to the edge of the continental shelf (the *shelfbreak*), marking the boundary between oceanic crust and CONTINENTAL CRUST.

A DESTRUCTIVE PLATE MARGIN may develop in an ocean basin, causing oceanic crust to be destroyed at a SUBDUCTION zone. This is marked by a deep OCEANIC TRENCH and either a volcanic ISLAND ARC or an ACTIVE CONTINENTAL MARGIN, depending on whether the subduction zone is situated within the ocean basin or at the edge of a continent. If the rate of destruction of oceanic crust exceeds its rate of formation, the ocean basin will contract (e.g. Pacific Ocean), leading ultimately to a COLLISION ZONE between continents, forming an OROGENIC BELT (OROGEN). The former ocean will be represented by a *suture zone*, along which fragments of oceanic crust may be preserved as *ophiolite complexes*. The idealised cycle of opening and closure of an ocean basin is termed the WILSON CYCLE. Simple patterns of ocean basin morphology are complicated by the presence of oceanic islands, SEAMOUNTS, GUYOTS and ATOLLS, many of which represent VOLCANOES situated at HOTSPOTS, IN GEOLOGY (see Figure).

The different morphological elements of an ocean basin are characterised by distinct ranges of water depths, ranging from HADAL (the deepest, corresponding to trenches) through abyssal to BATHYAL (corresponding to the continental rise and slope). Much of the knowledge about the ocean basins has been gained since the mid-twentieth century using techniques such as DEEP-SEA DRILLING, SIDE-SCAN SONAR and SEISMIC REFLECTION SURVEYING.   *GO*

[*See also* OCEAN CURRENTS, THERMOHALINE CIRCULATION, WATER MASS]

Gleason JD, Thomas DJ, Moore Jr TC et al. (2009) Early to middle Eocene history of the Arctic Ocean from Nd-Sr isotopes in fossil fish debris, Lomonosov Ridge. *Paleoceanography* 24: PA2215.

Korenaga J (2008) Plate tectonics, flood basalts and the evolution of Earth's oceans. *Terra Nova* 20: 419–439.

Kunzig R (2000) *Mapping the deep: The extraordinary story of ocean science.* New York: Norton.

Open University Course Team (1998) *The ocean basins: Their structure and evolution, 2nd edition.* Oxford: Butterworth-Heinemann.

Rose P and Laking A (2008) *Oceans.* London: British Broadcasting Corporation.

Seton M, Muller RD, Zahirovic S et al. (2012) Global continental and ocean basin reconstructions since 200Ma. *Earth-Science Reviews* 113: 212–270.

Wilson JT (1966) Did the Atlantic close and then re-open? *Nature* 211: 676–681.

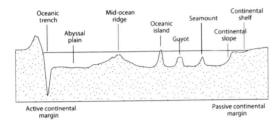

Ocean basin Cross section of an idealised ocean basin showing the principal morphological features.

ocean circulation The large-scale movement of water in the OCEAN. Ocean circulation is driven by energy from the Sun and the rotation of the Earth. The Sun drives ocean circulation through two means: (1) *direct frictional coupling* between the ocean and ATMOSPHERE at the sea surface via the atmospheric winds and (2) solar radiation heating of the surface ocean which causes variations in SEA-SURFACE TEMPERATURE (SST) and SALINITY, which in turn control seawater density and deep-ocean circulation. *Wind forcing* is responsible for driving the surface OCEAN CURRENTS and the ocean GYRE circulation. Surface currents are relatively slow moving across the Earth surface and so are significantly affected by the CORIOLIS FORCE and subsequent EKMAN MOTION. Changes in temperature are caused by fluxes of heat across the air-sea BOUNDARY LAYER and changes in salinity are caused by the addition or removal of freshwater via EVAPORATION, PRECIPITATION or SEA-ICE processes. If surface water becomes denser than the underlying water, the ocean is unstable and the denser surface water sinks. This vertical circulation, driven by changes in density (i.e. cooling and/or increased salinity) is known as the THERMOHALINE CIRCULATION. SOLAR RADIATION is not equally distributed across the Earth's surface; there is a positive RADIATION BALANCE at low latitudes and a negative radiation balance at high latitudes. Redistribution of this heat energy is carried out by the GENERAL CIRCULATION OF THE ATMOSPHERE combined with the oceanic circulation system, hence ocean circulation plays a key role in CLIMATIC CHANGE. Ocean currents contribute more to energy transfer in the tropics and *atmospheric circulation* contributes more at high latitudes. *JP*

[*See also* ATMOSPHERE-OCEAN INTERACTION, OCEAN PALAEOCIRCULATION, TROPICAL CYCLONE, WALKER CIRCULATION]

Huang RX (2009) *Ocean circulation: Wind-driven and thermohaline processes.* Cambridge: Cambridge University Press.

Miller RN (2007) *Numerical modeling of ocean circulation.* Cambridge: Cambridge University Press.

Open University Course Team (2001) *Ocean circulation.* Oxford: Butterworth-Heinemann.

Siedler G, Church J and Gould J (2001) *Ocean circulation and climate: Observing and modeling the global ocean.* San Diego, CA: Academic Press.

Timmermans M-L, Proshutinsky A, Krishfield RA et al. (2011) Surface freshening in the Arctic Ocean's Eurasian Basin: An apparent consequence of recent change in the wind-driven circulation. *Journal of Geophysical Research-Oceans* 116: C00D03.

ocean currents The surface CURRENTS of the oceans (see Figure), which move at speeds from several kilometres per day to several kilometres per hour, are driven by the surface wind systems and deflected by the distribution of continents. They form semi-closed circular patterns (GYRES), which move clockwise in the Northern Hemisphere and anticlockwise in the Southern Hemisphere. Predominantly easterly currents near the equator are driven by the TRADE WINDS, whereas the predominant westerly currents of higher latitudes are driven by the WESTERLIES. The mid-latitude gyres are centred over the subtropical high pressure zones (see warm ANTICYCLONE). Depending on the relative water temperature, the surface currents may be classified as *warm* or *cold currents*, which move away from or towards the equator, respectively, and are important in the redistribution of heat in the Earth's atmosphere-ocean system. Below the surface of ocean, low 'friction coupling' and the rotation of the Earth cause a reduction in the velocity of the currents and their deflection to the right in the Northern Hemisphere and to the left in the Southern Hemisphere (see CORIOLIS FORCE) resulting in an *Ekman spiral*, which is important in explaining UPWELLING off coasts where a current is moving parallel to a coastline and towards the equator. Changes in ocean currents also reflect and cause CLIMATIC CHANGE. *JAM/JP*

[*See also* ATMOSPHERE-OCEAN INTERACTION, EKMAN MOTION, EL NIÑO, GULF STREAM, OCEAN CIRCULATION, THERMOHALINE CIRCULATION]

Bigg GR (2003) *The oceans and climate, 2nd edition.* Cambridge: Cambridge University Press.

Binkley MS (1996) Oceans. In Schneider SH (ed.) *Encyclopedia of climate and weather.* Oxford: Oxford University Press, 547–552.

Open University Course Team (2001) *Ocean circulation, 2nd edition.* Oxford: Pergamon Press.

Thorpe SA (ed.) (2009) *Ocean currents: A derivative of the encyclopedia of ocean sciences.* London: Academic Press.

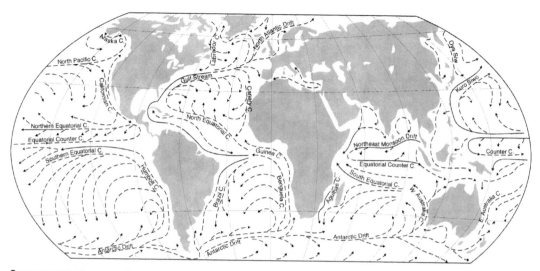

Ocean currents The major surface currents of the oceans. Note that there is seasonal variation in the positions of the currents, especially in the Indian Ocean, where the Southwest Monsoon Drift is characteristic of the Northern Hemisphere summer (Binkley, 1996).

ocean fertilisation The addition of NUTRIENTS to the OCEAN surface waters in order to promote PRIMARY PRODUCTIVITY. Ocean fertilisation occurs naturally when UPWELLING brings nutrient-rich waters into the sunlit EUPHOTIC ZONE, and PHYTOPLANKTON or BACTERIA utilise the nutrients. These regions of upwelling are among the most biologically productive in the oceans, such as the coastal upwelling regions off Peru and along the *Antarctic Polar Front* (see OCEANIC POLAR FRONT (OPF)). Natural ocean fertilisation also occurs when ATMOSPHERIC DUST is deposited in *low nutrient– high chlorophyll (LNHC) regions* of the world ocean, alleviating *iron limitation* and promoting primary productivity (see IRON CYCLE). Reproducing these natural processes of ocean fertilisation by intentionally adding dissolved iron to the surface waters (*iron fertilisation*) in high nutrient–low chlorophyll (HNLC) regions has been suggested as a way of reducing atmospheric CARBON DIOXIDE concentrations. This GEO-ENGINEERING process of intentional ocean fertilisation is controversial because the biogeochemical impacts of large-scale ecosystem PERTURBATION are not fully understood.   *JP*

[*See also* BIOGEOCHEMICAL CYCLES, ECOSYSTEM CONCEPT]

Boyd PW and Elwood MJ (2010) The biogeochemical cycle of iron in the ocean. *Nature Geoscience* 3: 675–682.
Jickells TD, An ZS, Andersen KK et al. (2005) Global iron connections between desert dust, ocean biogeochemistry, and climate. *Science* 308: 67–71.
Jin X and Gruber N (2003) Offsetting the radiative benefit of ocean iron fertilization by enhancing N₂O emissions. *Geophysical Research Letters* 30: 2249.

Keith DW (2000) Geoengineering the climate: History and perspective. *Annual Review of Energy and the Environment* 25: 245–284.

ocean freshening The recent reduced SALINITY of the North Atlantic Ocean attributed to the melting of SEA ICE in the ARCTIC and increasing freshwater DISCHARGE from rivers draining into the Arctic Ocean. Similar freshening has been observed in the Ross Sea in the ANTARCTIC. MODERN ocean freshening, and similar events in the past, has implications for the strength of the THERMOHALINE CIRCULATION.   *JAM*

Dickson R, Yashayaev I, Meincke J et al. (2002) Rapid freshening of the deep North Atlantic Ocean over the past four decades. *Nature* 416: 832–837.
Jacobs SS, Giulivi CF and Mele PA (2002) Freshening of the Ross Sea during the late twentieth century. *Science* 297: 386–389.
Morison J, Kwok R, Peralta-Ferriz et al. (2012) Changing Arctic Ocean freshwater pathways. *Nature* 481: 66–70.

ocean palaeocirculation The OCEAN CIRCULATION in the GEOLOGICAL RECORD. PLATE TECTONIC movements in the past mean that the ocean circulation has not always had the same configuration as it has today. During warm *greenhouse* intervals of the CRETACEOUS, for example, there were no subtropical or OCEANIC POLAR FRONTS (OPFs), the high- and low-latitude ocean GYRES would have been weakened, or non-existent, and the *thermal gradient* from the tropics to the poles would have been reduced. With the tectonic opening of the Southern Ocean, *ocean gateways* between South America and Antarctica, and Australia and Antarctica, in the Early CENOZOIC (*CAINOZOIC*), a more familiar pattern of surface OCEAN CURRENTS and deep water circulation (see THERMOHALINE CIRCULATION) evolved. Cenozoic and QUATERNARY reconstruction of ocean palaeocirculation is carried out using MARINE PALAEOCLIMATIC PROXIES from MARINE SEDIMENT CORES. The *protactinium/thorium ratio* ($^{231}Pa/^{230}Th$) *palaeocirculation proxy* relies on the radioactive decay (see RADIOACTIVITY, RADIONUCLIDE) of uranium in the water column and can be used to reconstruct the rate of flow of deep WATER MASSES in the past, and the strength of the *overturning circulation*. *Sortable silt mean size* (mean GRAIN SIZE of the 10–63 μm sediment fraction) is also used as a proxy for reconstructing rates of deep water flow. NUTRIENT proxies, such as CARBON ISOTOPE ratios and CADMIUM:CALCIUM RATIOS (Cd:Ca) measured in benthic FORAMINIFERA, have been used to monitor past changes in DEEP WATER and INTERMEDIATE WATER formation. Stable carbon isotopes from marine sediments are also used to reconstruct ocean VENTILATION history. Water masses formed in different source regions are imprinted with distinct *neodymium* (Nd) isotope signatures, hence neodymium isotope ratios in foraminifera, the TEETH of *fossil fish*, and *ferromanganese crusts* can all act as conservative tracers for ocean palaeocirculation in the open ocean. Neodymium isotopes have also been used to investigate the timing of opening of *ocean gateways*. *JP*

[*See also* FORAMINIFERA: ANALYSIS, MERIDIONAL OVERTURNING CIRCULATION (MOC), PALAEOCEANOGRAPHY, PALAEOCLIMATOLOGY, PROXY CLIMATIC INDICATOR, STABLE ISOTOPE, TRACERS: IN THE OCEANS]

Bostock HC, Opdyke BN, Gagan MK and Fifield LK (2004) Carbon isotope evidence for changes in Antarctic Intermediate Water circulation and ocean ventilation in the southwest Pacific during the last deglaciation. *Paleoceanography* 19: PA4013.

Came RE, Oppo DW and Curray WB (2003) Atlantic Ocean circulation during the Younger Dryas: Insights from a new Cd/Ca record from the western subtropical South Atlantic. *Paleoceanography* 18: 1086.

Hay WW (2008) Evolving ideas about the Cretaceous climate and ocean circulation. *Cretaceous Research* 29: 725–753.

Thomas AL, Henderson GL and Robinson LF (2006) Interpretation of the $^{231}Pa/^{230}Th$ paleocirculation proxy: New water-column measurements from the southwest Indian Ocean. *Earth and Planetary Science Letters* 241: 493–504.

Van de Flierdt T and Frank M (2010) Neodymium isotopes in palaeoceanography. *Quaternary Science Reviews* 29: 2439–2441.

ocean palaeoproductivity

The rate or amount of organic matter (see ORGANIC CONTENT) synthesised from inorganic carbon (e.g. dissolved CARBON DIOXIDE) in the past through marine *primary production*. This is an important focus of research in PALAEOCEANOGRAPHY and PALAEOCLIMATOLOGY because of the major role ocean PRODUCTIVITY plays in controlling the partitioning of carbon between its ATMOSPHERE, OCEAN and rock reservoirs. Organic carbon production and flux cannot be measured directly in the past but is reconstructed from MARINE SEDIMENTS using physical, chemical and biological MARINE PALAEOCLIMATIC PROXIES. Most of these methods result in qualitative estimates of ocean palaeoproductivity while others are quantitative.

(1) *Physical proxies.* The accumulation rates of *planktonic opal* (DIATOMS and RADIOLARIA) and CARBONATE (NANNOFOSSILS and FORAMINIFERA), which are symptomatic of surface ocean productivity, and *benthic organisms* (e.g. *benthic foraminifera, ostracods*), which record particulate food arriving at the sea floor, are common methods for reconstructing ocean palaeoproductivity. All accumulation-based PROXY RECORDS are highly dependent on age models, mineral preservation and the controls on this including water depth, sediment density, pore water chemistry and burial depth.

(2) *Biological proxies.* Analysis of temporal changes in the morphology and species make-up of fossilised *microfauna* and *microflora* provides qualitative insight into water column productivity and the type, amount and frequency of organic matter exported to the sea floor.

(3) GEOCHEMICAL PROXIES. Sedimentary organic carbon concentration may provide an indicator of *export productivity*, however, because of its extremely low preservation potential, geological measurements are rarely informative. Other geochemical approaches focus on measurement of sedimented NUTRIENTS (N, P, Fe and Si), often through the use of stable ISOTOPE RATIOS and or elemental ratios (e.g. Cd/Ca). Accumulation of marine *barite* is another proxy for export production while measurements of Al/Ti, Ba/Ti, $^{231}Pa/^{230}Th$ and $^{10}Be/^{230}Th$ are used as indicators of *water column particle flux*. Carbon stable isotopes ($\delta^{13}C$) in bulk sediment and foraminiferal *tests* are indicative of ocean-wide changes in organic *carbon burial*. The distribution of organic BIOMARKERS, compounds that are specific to the type of primary producer, are a growing area of research. *HKC*

[*See also* MARINE SEDIMENT CORES, PRIMARY PRODUCTIVITY]

Bolton CT, Lawrence KT, Wilson PA et al. (2010) Glacial-interglacial productivity changes recorded by alkenones and microfossils in late Pliocene eastern equatorial Pacific and Atlantic upwelling zones. *Earth and Planetary Science Letters* 295: 401–411.

Calvert SE and Pederson TF (2007) Elemental proxies for marine palaeoclimate and palaeoceanographic variability in marine sediments: Interpretation and application. In Hillaire-Marcel C and de Vernal A (eds) *Proxies in Late*

Cenozoic paleoceanography, Part 2: Biological tracers and biomarkers. Amsterdam: Elsevier, 568–639.

Jorissen FJ, Fontanier C and Thomas E (2007) Paleoceanographical proxies based on deep-sea benthic foraminiferal assemblage characteristics. In Hillaire-Marcel C and de Vernal A (eds) *Proxies in Late Cenozoic paleoceanography, Part 2: Biological tracers and biomarkers.* Amsterdam: Elsevier, 263–325.

Paytan A (2008) Ocean paleoproductivity. In Gornitz V (ed.) *Encyclopedia of paleoclimatology and ancient environments.* Dordrecht: Springer, 644–651.

ocean palaeotemperature The temperature of OCEAN water in the geological past. This is reconstructed indirectly using PALAEOTHERMOMETRY. Ocean surface *palaeotemperature*, also known as SEA-SURFACE TEMPERATURE (SST), which directly affects the behaviour of Earth's atmosphere above, is reconstructed from sedimented PLANKTONIC organisms, mainly FORAMINIFERA, and organic compounds produced by marine photosynthesisers (e.g. COCCOLITHOPHORES) living close to the ocean surface. Ocean DEEP-WATER palaeotemperatures are derived from BENTHIC organisms, mainly foraminifera, living at the sea floor. Together these palaeotemperatures provide insight into the thermal structure and mixing of the oceans in the past, which are important for understanding past variability in ocean circulation and climate (see OCEAN PALAEOCIRCULATION, MARINE PALAEOCLIMATIC PROXIES). Over GEOLOGICAL TIME, ocean temperatures have been influenced by variations in the amount of SOLAR RADIATION received due to PERIODICITIES in Earth's orbit, as well as the geographical distribution of the oceans compared with land across Earth's surface, which changes on multimillion-year timescales because of PLATE TECTONICS. Ocean palaeotemperatures serve as useful BOUNDARY CONDITIONS to initiate and evaluate CLIMATIC MODELS that predict future global warming trends.

HKC

[*See also* CLIMATIC RECONSTRUCTION, PALAEOCEANOGRAPHY, PALAEOCEANOGRAPHY: BIOLOGICAL PROXIES, PALAEOCEANOGRAPHY: PHYSICAL AND CHEMICAL PROXIES, PALAEOCLIMATOLOGY, PROXY CLIMATIC INDICATOR]

Hillaire-Marcel C and de Vernal A (eds) (2007) *Proxies in Late Cenozoic paleoceanography.* Amsterdam: Elsevier Science.

Marshall J and Plumb RA (2008) *Atmosphere, ocean and climate dynamics: An introductory text.* Amsterdam: Elsevier Academic Press.

Weinelt M (2009) Ocean paleotemperatures. In Gornitz V (ed.) *Encyclopedia of paleoclimatology and ancient environments.* Dordrecht: Springer, 651–659.

ocean warming The slow rise in SEA-SURFACE TEMPERATURES (SSTs) since AD 1955–1960, which started at a rate of about 0.1°C per decade and accelerated to 0.2–0.3°C per decade in the 1980s and 1990s. However, there was a pause in ocean warming between 2003 and 2010, which has been attributed to NATURAL variability, especially reduced RADIATION to space (45 per cent of the reduction) and a rise in ocean temperatures at depth following weakening of the MERIDIONAL OVERTURNNG CIRCULATION (35 per cent).

JAM

[*See also* ATMOSPHERE-OCEAN INTERACTION, DETECTION, GLOBAL WARMING, OCEAN FRESHENING]

Katsanan CA and van Oldenborg GT (2011) Tracing the upper ocean's "missing heat". *Geophysical Research Letters* 38: L14610.

Levitus S, Antonov JI, Boyer TP and Stephens C (2000) Warming of the world ocean. *Science* 287: 2225–2229.

oceanic anoxic event An episode of abnormally low dissolved oxygen in the OCEANS, leading to the widespread preservation of laminated, organic-rich BLACK SHALES or SAPROPELS. Factors favouring oceanic anoxic events are those associated with generally warmer ocean waters during periods of global warmth, or GREENHOUSE CONDITIONS. They include slow renewal of oxygen in BOTTOM WATERS because of the absence of ICE SHEETS (see WATER MASS), reduced levels of dissolved oxygen, increased organic PRODUCTIVITY and *marine transgression* leading to the flooding of CONTINENTAL SHELF areas. Oceanic anoxic events are known from the Mid-Cambrian, early Mid-Ordovician, Early Silurian, Late Devonian, Early Carboniferous, Early Jurassic, Late Jurassic, Mid-Cretaceous and Late Cretaceous.

GO/JP

[*See also* EUXINIA/EUXINIC, PALAEOCEANOGRAPHY]

Bottjer DJ and Savrda CE (1993) Oxygen-related mudrock biofacies. In Wright VP (ed.) *Sedimentology review 1.* Oxford: Blackwell, 92–102.

Jenkyns HC (2010) Geochemistry of oceanic anoxic events. *Geochemistry Geophysics Geosystems* 11: Q03004.

Pancost RD, Crawford N, Magness S et al. (2004) Further evidence for the development of photic-zone euxinic conditions during Mesozoic oceanic anoxic events. *Journal of the Geological Society* 161: 353–364.

Strauss H (2006) Anoxia through time. In Neretin LN (ed.) *Past and present water column anoxia.* Dordrecht: Springer, 3–19.

oceanic crust The outermost layer of the Earth that underlies the OCEANS (see Figure). Compared with CONTINENTAL CRUST, oceanic crust is relatively thin (6–8 km), dense (typically 3,000 kg/m³), basaltic (see BASALT) in composition, and young: because of the recycling of oceanic crust through its formation by VOLCANISM and igneous INTRUSION at CONSTRUCTIVE PLATE MARGINS and its destruction at DESTRUCTIVE PLATE MARGINS, there is no in-situ oceanic crust older than 200 million years. Fragments of older oceanic crust are preserved as *ophiolite* *complexes* in OROGENIC BELTS (OROGEN). The structure and composition of oceanic crust are known from studies of ophiolite complexes, DEEP-SEA DRILLING and SEISMIC exploration.　　　　　　　　　　　　　　　*GO*

[*See also* EARTH STRUCTURE]

Ding L, Kapp P and Wan XQ (2005) Paleocene-Eocene record of ophiolite obduction and initial India-Asia collision, south central Tibet. *Tectonics* 24: TC3001.

Mueller RD, Sdrolinas M, Gaina C and Roest WR (2008) Age, spreading rates, and spreading asymmetry of the world's ocean crust. *Geochemistry, Geophysics, Geosystems* 9: Q04006.

Tucholke BE (1998) Discovery of "megamullions" reveals gateways into the ocean crust and upper mantle. *Oceanus: Reports on Research from the Woods Hole Oceanographic Institution* 41: 15–19.

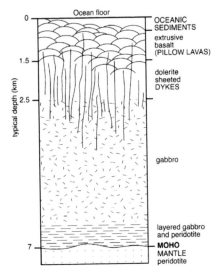

Oceanic crust Cross section through typical oceanic crust, based on information from deep-sea drilling, seismic surveying and the study of ophiolite complexes (Tucholke, 1998).

oceanic forcing (of climate)

oceanic forcing (of climate) Direct influences that the OCEAN exerts on local or global CLIMATE. The primary mechanisms of oceanic forcing involve the ocean taking up heat and releasing heat to the ATMOSPHERE, and by providing WATER VAPOUR through EVAPORATION to the atmosphere. For example, the ocean takes up a lot of heat in the summer from SOLAR RADIATION without increasing its TEMPERATURE very much because of the large THERMAL CAPACITY of water. The air above the ocean can therefore remain cooler relative to the air above land. If the oceanic air is blown towards the land it will cool the adjacent land. In the winter, the process is reversed so that the ocean warms the surrounding land. The ocean thus has a mediating effect on the surrounding land temperature (see OCEANICITY). The *heat*

reservoir of the ocean also acts to delay the warmest and coldest time of the year by a month or two after the summer and winter SOLSTICE.

OCEAN CURRENTS also affect climate by transporting heat from one location to another, including from low to high latitudes. An example is the MERIDIONAL OVERTURNING CIRCULATION (MOC), an important component of the THERMOHALINE CIRCULATION, which brings heat from the equatorial Atlantic to the North Atlantic and thereby contributes to the relatively warm temperatures of Western Europe. The amount of heat that is taken up or released by the ocean is affected by factors such as the *wind* and SEA ICE. The ocean also affects the climate in indirect ways, for instance, by acting as a reservoir and SINK for GREENHOUSE GASES such as CARBON DIOXIDE.

Enhanced oceanic forcing involved strengthening of the MOC, resulting in increased heat transport to the high northern latitudes is thought to be responsible for inhibiting significant ICE SHEET build-up and prolonging warmer conditions during MARINE ISOTOPIC STAGE (MIS) 11 approximately 415,000 years ago. It has also been suggested that reduced ocean forcing, via sudden shutdown of the thermohaline circulation, may also trigger ABRUPT CLIMATIC CHANGE and cause the Earth's CLIMATIC SYSTEM to 'flip-flop' from one relatively stable state to another (see DANSGAARD-OESCHGER (D-O) EVENTS). EL NIÑO–SOUTHERN OSCILLATION (ENSO), the most prominent mode of ANNUAL VARIABILITY of the atmosphere-ocean climate system, is driven by *tropical oceanic forcing.*　　　　　　　　　　　　*HKC*

[*See also* ATMOSPHERE-OCEAN INTERACTION, CLIMATIC RECONSTRUCTION, MARINE SEDIMENT CORES, OCEAN PALAEOTEMPERATURE, PALAEOCEANOGRAPHY, PALAEOCLIMATOLOGY, PALAEOTHERMOMETRY, PROXY CLIMATIC INDICATOR]

Broecker W (2010) *The great ocean conveyor: Discovering the trigger for abrupt climatic change.* Princeton, NJ: Princeton University Press.

Dickson AJ, Beer CJ, Dempsey C et al. (2009) Oceanic forcing of the Marine Isotope Stage 11 interglacial. *Nature Geoscience* 2: 428–433.

Joughin I, Alley RB and Holland DM (2012) Ice-sheet response to oceanic forcing. *Science* 338: 1172–1176.

Sutton RT and Hodson DLR (2005) Atlantic Ocean forcing of North American and European summer climate. *Science* 309: 115–118.

Vallis GK (2012) *Climate and the oceans.* Princeton, NJ: Princeton University Press.

Oceanic Polar Front (OPF)

Oceanic Polar Front (OPF) The OCEAN boundary separating warm water of high SALINITY flowing polewards from cold, low-salinity water flowing towards lower latitudes. In the Arctic Ocean, it is approximately defined by the winter SEA-ICE extent and is maintained by buoyant freshwater influx. It should be distinguished from the oceanic ARCTIC FRONT, which corresponds with the limit of permanent sea ice. In the Southern Ocean, it is defined by the northern limit of the *Antarctic Circumpolar Current* (*ACC*). OPFs are sharp gradients in TEMPERATURE and SALINITY, quasi-stationary and seasonally persistent, and may be accompanied by along-front CURRENTS (*oceanic jets*). Long-term changes in the location of the OPF have been reconstructed on the basis of the analysis of MARINE SEDIMENT CORES for the period from the LAST GLACIAL MAXIMUM (LGM) to the HOLOCENE (see Figure). *WENA/JP*

[*See also* OCEANOGRAPHY, POLAR FRONT]

Belkin I (2005) Ocean fronts. In Nuttall M (ed.) *Encyclopedia of the Arctic*. New York: Routledge, 1545–1547.

Kemp AES, Grigorov I, Pearce RB and Naveira Garabato AC (2010) Migration of the Antarctic Polar Front through the Mid-Pleistocene transition: Evidence and climatic implications. *Quaternary Science Reviews* 29: 1993–2009.

Koc N, Jansen E and Haflidason H (1993) Paleoceanographic reconstruction of surface ocean conditions in Greenland, Iceland and the Norwegian seas through the last 14 ka based on diatoms. *Quaternary Science Reviews* 12: 115–140.

Ruddiman WF and McIntyre A (1981) The North Atlantic Ocean during the last deglaciation. *Palaeogeography, Palaeoclimatology, Palaeoecology* 35: 145–214.

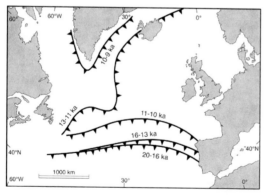

Oceanic Polar Front Schematic representation of variations in the position of the Oceanic Polar Front between the Last Glacial Maximum (20–16 ka) and the early Holocene (10–9 ka), including its retreat to Iceland during the Late Glacial Interstadial (Allerød; 13–11 ka) and its readvance to the latitude of northern Spain in the Late Glacial Stadial (Younger Dryas; 11–10 ka): dates are in radiocarbon years BP (after an early significant reconstruction by Ruddiman and McIntyre, 1981). A more recent and detailed reconstruction of palaeoceanographic conditions in the North Atlantic can be found in Koc et al. (1993).

oceanic sediments   SEDIMENTS that accumulate on the deep OCEAN floor, far from land, including PELAGIC SEDIMENT (PELAGITE) and redistributed material such as the deposits of TURBIDITY CURRENTS (TURBIDITES), DEBRIS

FLOWS and other forms of submarine MASS MOVEMENT PROCESSES. DEEP-SEA DRILLING has demonstrated that the thickness of oceanic sediments on igneous OCEANIC CRUST increases with distance from the central axes of the MID-OCEAN RIDGES (MORs), providing evidence in support of SEA-FLOOR SPREADING and PLATE TECTONICS. Oceanic sediments hold an important record of *environmental change*. *GO/JP*

[*See also* MARINE SEDIMENT CORES, MARINE SEDIMENTS]

Barker S, Diz P, Vautravers MJ et al. (2009) Interhemispheric Atlantic seesaw response during the last deglaciation. *Nature* 457: 1097–1102.

Davies A, Kemp AES and Pike J (2009) Late Cretaceous seasonal ocean variability from the Arctic. *Nature* 460: 254–258.

MacLeod KG, Isaza Londoño C, Martin EE et al. (2011) Changes in North Atlantic circulation at the end of the Cretaceous greenhouse interval. *Nature Geoscience* 4: 779–782.

oceanic trench   A narrow, deep trough in the OCEAN floor. An oceanic trench, *deep ocean trench* or *submarine trench* is the surface expression of SUBDUCTION at a DESTRUCTIVE PLATE MARGIN, and is associated with VOLCANISM and EARTHQUAKES. Oceanic trenches occur either within the ocean, with a volcanic ISLAND ARC on the overriding plate, or as part of an ACTIVE CONTINENTAL MARGIN, and may be partly or completely filled by OCEANIC SEDIMENTS such as TURBIDITES (see ACCRETIONARY COMPLEX). They are particularly well developed close to the Pacific Ocean rim (*Ring of Fire*) and include the deepest parts of the ocean such as the Mindanao Trench and the Challenger Deep of the Mariana Trench, which exceed 11,000 m depth (see Figure). *GO*

Decker R and Decker B (1997) *Volcanoes, 3rd edition.* Basingstoke: Freeman.

Kearey P, Klepeis KA and Vine FJ (2009) *Global tectonics.* Chichester: Wiley.

oceanicity   The extent to which the climate of a location is influenced by maritime influences: the equable character of climates; the converse of CONTINENTALITY. Indices of oceanicity attempt to quantify this, usually based on annual temperature range. *Oceanic climates* (*maritime climates*) exhibit high oceanicity: these include oceanic islands and the western sides of continents in mid-latitudes, which experience the WESTERLIES throughout the year. In regions of low relief, high oceanicity may extend hundreds of kilometres inland. *JAM*

Oliver JE (1996) Maritime climate. In Schneider SH (ed.) *Encyclopedia of climate and weather*. New York: Oxford University Press, 491–496.

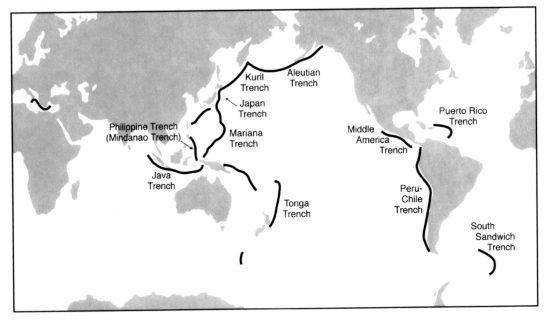

Oceanic trench The world's deep oceanic trenches (Decker and Decker, 1997).

Wolfe BB, Edwards TWD, Jiang H et al. (2003) Effect of varying oceanicity on early- to mid-Holocene palaeohydrology, Kola Peninsula, Russia: Isotopic evidence from treeline lakes. *The Holocene* 13: 153–160.

oceanography The scientific study of the world's OCEANS. In recent years, the interdisciplinary nature of oceanography, which involves MARINE GEOLOGY (*geological oceanography*), *marine physics* (PHYSICAL OCEANOGRAPHY), *marine chemistry* (*chemical oceanography*), *marine biology* (see BIOLOGICAL OCEANOGRAPHY) and METEOROLOGY of the oceans, has alerted scientists to the complex and sensitive interrelationships amongst different components of the *Earth-atmosphere-ocean system* (see EARTH SYSTEM, GEO-ECOSPHERE). Many pressing issues on the environmental agenda, including aspects of CLIMATIC CHANGE, CARBON DIOXIDE emissions, polar ICE-CAP melting, POLLUTION and RADIOACTIVE WASTE disposal involve studies of the oceans. *BTC/JP*

[***See also*** MARINE POLLUTION, PALAEOCEANOGRAPHY]

Denny M (2008) *How the ocean works: An introduction to oceanography*. Princeton, NJ: Princeton University Press.
Garrison TS (2011) *Essentials of oceanography, 6th edition*. Florence, KY: Cengage Learning.
Millero FJ (2005) *Chemical oceanography, 3rd edition*. Boca Raton, FL: CRC Press.
Open University Course Team (2005) *Marine biogeochemical cycles, 2nd edition*. Oxford: Butterworth-Heinemann.
Steele J, Turekian K and Thorpe S (eds) (2001) *Encyclopedia of ocean science, 6 volumes*. San Diego, CA: Academic Press.

Sverdrup KA, Duxbury AC and Alison B (2006) *Fundamentals of oceanography*. New York: McGraw-Hill.

off-road vehicles Vehicles designed for *cross-country travel*, including: four-wheel drive and tracked vehicles, those with low-pressure tyres, skidoos, motorcycles, mountain bikes and hovercraft. In many countries, vulnerable environments (e.g. DESERT, TUNDRA, SAND DUNES, COASTAL DUNES, salt flats and SALT-MARSHES and other areas with sensitive vegetation or soil) are seriously damaged, sometimes permanently, by the passage of off-road vehicles. The tracks from 1920s expeditions and abandoned World War II military vehicles are still visible in some desert regions. Oil exploration activities are notorious for long-lasting damage related to PERMAFROST DEGRADATION in tundra areas. Vehicles used during logging activities associated with DEFORESTATION may also cause lasting damage. In affluent nations, recreational use of these vehicles can be a source of serious EROSION problems and they also emit exhaust POLLUTANTS. Farm vehicles also cause damage, especially soil COMPACTION, in the long term but are designed to minimise this, and are usually driven with more caution than military, transport or recreational vehicles. *CJB*

Abele G, Brown J and Brewer MC (1984) Long-term effects of off-road vehicle traffic on tundra terrain. *Journal of Terramechanics* 21: 283–294.
Ayres PD (2003) Environmental damage from tracked vehicle operation. *Journal of Terramechanics* 31: 173–183.

Ouren DS, Haus C, Melcher CP et al. (2009) *Environmental effects of off-highway vehicles.* Hauppauge, NY: Nova Science Publishers.

oil fire   A source of AIR POLLUTION and a potential ENVIRONMENTAL DISASTER. Oil fires in Kuwait during the Gulf War, 1990, which resulted from the ignition of over 700 oil wells, probably introduced more anthropogenic pollutants into the atmosphere than any other single event.   *JAM*

Husain T (1995) *Kuwaiti oil fires: Regional environmental perspectives.* Oxford: Elsevier.

oil shale   A BLACK SHALE or fossil SAPROPEL that is rich in carbon. Oil shales are the source rocks for conventional PETROLEUM accumulations, releasing HYDROCARBONS upon burial and heating. Oil shales at the Earth's surface that have not released their hydrocarbons can be mined as an ENERGY RESOURCE or FOSSIL FUEL (SYNFUEL), but they must be processed to yield petroleum. They are presently exploited in Brazil, China and Estonia, and there are RESERVES in Australia, Jordan and the United States, the largest being the Green River Formation covering parts of Colorado, Wyoming and Utah. Their exploitation involves large-scale *opencast mining* leading to ENVIRONMENTAL DEGRADATION. Oil shale was worked in central Scotland from AD 1850 to 1964 and many of the spoil tips are prominent features in the landscape today. It has been estimated that the total energy that could be obtained by the exploitation of oil shales is equal to, or even greater than, the total obtained from CRUDE OIL.   *GO*

[*See also* SHALE GAS, TAR SANDS]

Jaber JO and Probert SD (1997) Exploitation of Jordanian oil-shales. *Applied Energy* 58: 161–175.
Kahru A and Pollumaa L (2006) Environmental hazard of the waste streams of Estonian oil shale industry: An ecotoxicological review. *Oil Shale* 23: 53–93.
Vandenbroucke M and Largeau C (2007) Kerogen origin, evolution and structure. *Organic Geochemistry* 38: 719–833.

oil spill   A leakage of PETROLEUM (oil) into the marine environment as a result of an ENVIRONMENTAL ACCIDENT. Although more oil reaches the marine environment from land-based discharges, and the largest single oil spill yet recorded (>550,000 tonnes) followed the 2011 explosion at the Deepwater Horizon oil rig in the Gulf of Mexico, the term is more frequently associated with tanker accidents at sea, such as the wrecks of the *Amoco Cadiz* (220,000 tonnes) in the English Channel, the *Braer* (80,000 tonnes) off the Shetland Islands, Scotland, and the *Exxon Valdez* (40,000 tonnes) in Prince William Sound, Alaska. The environmental impact depends on many factors including the properties of the oil, the characteristics of the wind, waves and currents, the sea temperature and salinity, the air temperature, and the SENSITIVITY and VULNERABILITY of affected ecosystems. Damage tends to be intensive in the short term but not long-lasting because of natural DEGRADATION (*weathering of oil*) and the capacity of wildlife to recolonise (RESILIENCE). RESTORATION methods include chemical-assisted dispersion, containment booms, mechanical recovery, burning and BIOTECHNOLOGY (BIOREMEDIATION using oil-degrading bacteria).   *JAM*

[*See also* ENVIRONMENTAL DISASTER, MARINE POLLUTION]

Adler E and Inbar M (2007) Shoreline sensitivity to oil spills, the Mediterranean coast of Israel: Assessment and analysis. *Ocean and Coastal Management* 50: 24–34.
Parker JG (1998) Oil spill response. In Brune D, Chapman DV, Gwynne MD and Pacyna JM (eds) *The global environment: Science, technology and management.* Weinheim: VCH, 955–972.
Peterson CH, Rice SD, Short JW et al. (2003) Long-term ecosystem response to the Exxon Valdez oil spill. *Science* 302: 2082–2086.
Shroder T and Konrad J (2011) *Fire on the horizon: The untold story of the Gulf disaster.* New York: HarperCollins.
Teal JM and Howarth RW (1984) Oil spill studies: A review of ecological effects. *Environmental Management* 8: 27–44.

Older Dryas Stadial   A brief period of relatively cold conditions in Europe during the Late Glacial, following the BØLLING INTERSTADIAL and preceding the ALLERØD INTERSTADIAL. Evidence for this can be found from Norway to Spain. It is dated to between ca 12 and 11.8 ka (radiocarbon years BP), equivalent to GI-1b.

                                                                  MHD/CJC

[*See also* LATE GLACIAL ENVIRONMENTAL CHANGE]

Lowe JJ, Rasmussen SO, Bjorck S et al. (2008) Synchronisation of palaeoenvironmental events in the North Atlantic region during the Last Termination: A revised protocol recommended by the INTIMATE group. *Quaternary Science Reviews* 27: 6–17.
Mangerud J, Andersen ST, Berglund BE and Donner JJ (1974) Quaternary stratigraphy of Norden: A proposal for terminology and classification. *Boreas* 10: 109–127.

old-growth forest   An imprecise term emphasising the large, old trees in forests that have reached a mature stage of development and are particularly worthy of CONSERVATION for their economic and/or cultural value. There is no generally accepted age at which old growth is identified: it varies significantly between forest types.   *JAM*

[*See also* ANCIENT WOODLAND, ECOLOGICAL VALUE, PRIMARY FOREST]

Kaufmann MR, Binkley D, Fule PZ et al. (2007) Defining old growth for fire-adapted forests of the western United States. *Ecology and Society* 12: Article No. 15.

Luyssaert S, Schulze E-D, Börner A et al. (2008) Old-growth
forests as global carbon sinks. *Nature* 455: 213–215.
Spies TA and Duncan SL (eds) (2009) *Old growth in a
new world: A Pacific Northwest icon re-examined.*
Washington, DC: Island Press.
Wirth C, Gleixner G and Heimann M (eds) (2009)
Old-growth forests: Function, fate and value. Berlin:
Springer.

Olduvai event   A short interval of NORMAL POLAR-
ITY when there was a return to normal polarity in the
Earth's magnetic field within the longer *Matuyama
chron* characterised by reverse polarity. It has been
dated by POTASSIUM-ARGON (K-Ar) DATING to between
1.87 and 1.67 million years ago, or 1.95 and 1.78 Ma
BP based upon ORBITAL TUNING of the OXYGEN ISOTOPE
record of MARINE SEDIMENTS. As it lies close to the traditional PLIOCENE-PLEISTOCENE TRANSITION, it has been an
important chronological MARKER HORIZON in the study
of QUATERNARY ENVIRONMENTAL CHANGE.   *DH*

[*See also* GEOMAGNETIC POLARITY TIMESCALE (GPTS)]

Hilgen FJ (1991) Astronomical calibration of Gauss
to Matuyama sapropels of the Mediterranean and
implications for the geomagnetic polarity time scale.
Earth and Planetary Science Letters 104: 226–244.

Oligocene   An EPOCH of the PALAEOGENE period, from
33.9 to 23.0 million years ago (see TERTIARY). The Oligocene is associated with a cooling trend, including
the beginning of glaciation in the ANTARCTIC. *Grasses*
expanded and forests contracted.   *GO*

[*See also* GEOLOGICAL TIMESCALE]

Davies BJ, Hanbrey MJ, Smellie JL et al. (2012) Antarctic
Peninsula Ice Sheet evolution during the Cenozoic Era.
Quaternary Science Reviews 31: 30–66.
Scher HD, Bohaty SM, Zachos JC and Delaney ML (2011)
Two-stepping into the icehouse: East Antarctic weathering
during progressive ice-sheet expansion at the Eocene-
Oligocene transition. *Geology* 39: 383–386.
Sun XJ and Wang PX (2005) How old is the Asian
monsoon system? Palaeobotanical records from China.
Palaeogeography, Palaeoclimatology, Palaeoecology
222: 181–222.

oligotrophication   *Nutrient depletion* over time.
The term signifies the opposite of EUTROPHICATION and
is used especially in the context of the NUTRIENT STATUS
of waterbodies such as LAKES.   *JAM*

Burkholder JM (2001) Eutrophication and oligotrophication.
In Levin SA (ed.) *Encyclopedia of biodiversity, volume 2.*
San Diego, CA: Academic Press, 649–670.
Evans MA, Fahnenstiel G and Scavia D (2011) Incidental
oligotrophication of North American Great Lakes.
Environmental Science and Technology 45: 3297–3303.
Mozetic P, Solidaro C, Cossarini G et al. (2010) Recent
trends towards oligotrophication of the Northern Adriatic:
Evidence of chlorophyll *a* time series. *Estuaries and
Coasts* 33: 362–375.
Stich HB and Brinker A (201) Oligotrophication outweighs
effects of global warming in a large, deep, stratified lake
ecosystem. *Global Change Biology* 11: 877–888.

ombrotrophic mire   A MIRE, the only water source
which is from PRECIPITATION and as a result is mineral
poor or *oligotrophic*.   *MJB*

[*See also* OLIGOTROPHICATION, RAISED MIRE/BOG,
WETLAND CLASSIFICATION]

Amesbury MJ, Barber KE and Hughes PDM (2011) The
methodological basis of fine resolution, multi proxy
reconstructions of ombrotrophic peat bog surface wetness.
Boreas 40: 161–174.
Van Bellen S, Garneau M and Booth RK (2011) Holocene
carbon accumulation rates from three ombrotrophic
peatlands in boreal Quebec, Canada: Impact of climate-
driven ecohydrological change. *The Holocene* 21:
1217–1231.

ontogeny/ontogenesis   The development of an
individual organism from its earliest fertilised form
(zygote) through to maturity.   *KJW*

ontology   The branch of METAPHYSICS concerned
with the study of the nature of *reality* and its relationship to *human consciousness*. Ontology is concerned
with how the world works in terms of what exists, how
it does so and what can be known. It is usually contrasted with EPISTEMOLOGY, which asks how it is possible to know the world.   *ART*

Oort minimum   The minimum in solar *sunspot*
activity between AD 1010 and 1050.   *JAM*

[*See also* MAUNDER MINIMUM, SUNSPOT CYCLES]

ooze   Fine-grained PELAGIC SEDIMENT (PELAGITE)
dominantly of biogenic origin. *Carbonate ooze* or
calcareous ooze includes *pteropod ooze*, found in
water depths ranging from 1,000 to 2,500 m, and
nannofossil ooze and *globigerinid ooze* in water
depths from 2,000 to 4,000 m. One-third of the ocean
floor is covered with globigerinid ooze. *Siliceous
ooze*, composed of organisms with tests of *opaline
silica*, includes *diatom ooze*, found between 1,100
and 4,000 m in Antarctic regions and in low-latitude
parts of the Pacific, and *radiolarian ooze* in water
depths exceeding 5,000 m in isolated parts of the
equatorial Pacific and Indian Oceans. The distribution of calcareous and siliceous oozes is controlled
by organic PRODUCTIVITY and water depth. Calcareous
ooze is today found only in shallow, warm oceans, as
most ocean basins have average depths of 4–5 km,
which is below the CARBONATE COMPENSATION DEPTH
(CCD).   *BTC/JP*

Davies A, Kemp AES and Pike J (2009) Late Cretaceous seasonal ocean variability from the Arctic. *Nature* 460: 254–258.

Denis D, Crosta X, Zaragosi S et al. (2006) Seasonal and subseasonal climate changes recorded in laminated diatom ooze sediments, Adélie Land, East Antarctica. *The Holocene* 16: 1137–1147.

Sheldon E, Ineson J and Bown P (2010) Late Maastrichtian warming in the Boreal Realm: Calcareous nannofossil evidence from Denmark. *Palaeogeography, Palaeoclimatology, Palaeoecology* 295: 55–75.

Smith AJ and Gallagher SJ (2003) The Recent foraminifera and facies of the Bass Canyon: A temperate submarine canyon in Gippsland, Australia. *Journal of Micropalaeontology* 22: 63–83.

open system   A SYSTEM that may exchange both energy and matter with its surroundings (its ENVIRONMENT); consequently, an open system is the most realistic simplification of the real world.   *CET*

Huggett RJ (1985) *Earth surface systems*. Berlin: Springer-Verlag.

optical emission spectroscopy (OES)   A technique, also known as *atomic emission spectroscopy* (*AES*), used for elemental analysis in GEOCHEMISTRY and related fields. It involves analysis of the optical *emission spectrum* (170–780 nm wavelength) of a vaporised sample. The vapour may be held in various sources including the relatively low temperature of a flame (in *flame emission spectroscopy*) or, in more recent approaches, at much higher temperatures in a *plasma* (see INDUCTIVELY COUPLED PLASMA ATOMIC EMISSION SPECTROMETRY (ICP-AES)).   *TY*

[*See also* ATOMIC ABSORPTION SPECTROPHOTOMETRY (AAS)]

optical remote-sensing instruments   Instruments which record reflected RADIATION at wavelengths between 300 and 5,000 nm. This region of the ELECTROMAGNETIC SPECTRUM is characterised by ABSORPTIONS due to electron transitions and changes in molecular rotation and vibration within MOLECULES. These instruments cover the *wavelengths* visible to the human eye and the peak of the Sun's *emission spectrum* as well as overlapping at longer wavelengths with emitted radiation from high-temperature surfaces recorded by THERMAL REMOTE SENSING. The instruments available for optical remote sensing range from simple *cameras* to complex *hyper spectral digital scanners*.

The cameras used for AERIAL PHOTOGRAPHY were the first and simplest REMOTE SENSING instruments. They have a similar design to conventional cameras with a lens system focusing radiation from the surface on to piece of film in the focal plane. The radiation causes a chemical response in the emulsion of the film and thus records the image. *Black-and-white films* have a single layer of emulsion while *colour films* have three layers of emulsion separated by filters so that each layer records a different wavelength range.

The first electronic remote sensing instruments were the *return-beam vidicon* (RBV) cameras, used on early METEOROLOGICAL SATELLITES (*Nimbus series*) and *Earth resources satellites* (LANDSAT 1-3). *Electro-optical instruments*, or *multispectral scanners*, use solid-state detectors to convert radiation from the surface to electrical energy in a manner similar to RADIOMETERS. They build up RASTER images from lines of discrete measurements or PIXELS values. Radiation from the surface is focussed onto a mirror via a fore optic and then split up into narrow waveband by a diffraction grating or a set of filters. The brightness of the radiation in each waveband is measured by a detector. *Whiskbroom scanners*, such as the AVIRIS (*Airborne Visible-Infrared Imaging Spectrometer*), *Thematic Mapper*, AVHRR (*Advanced Very High Resolution Radiometer*) and MSS (*Multi-Spectral Scanner*), measure one pixel at a time with linear photo-diode arrays. *Pushbroom scanners*, such as the CASI (*Compact Airborne Spectrographic Imager*), AIS (*Airborne Imaging Spectrometer*) and HRVIR (*High Resolution Visible and Infrared*) measure a line of pixels at a time with a two-dimensional array of detectors or *charge-coupled device* (CCD). As *digital cameras* are now in use where the film at the focal plane has been replaced by a large CCD, AERIAL PHOTOGRAPHY can be acquired directly in digital form.   *GMS*

Elachi C and van Zyl JJ (2006) *Introduction to the physics and techniques of remote sensing, 2nd edition*. New York: Wiley.

Kerekes J (2009) Optical sensor technology. In Warner TA, Nellis MD and Foody GM (eds) *The SAGE handbook of remote sensing*. London: Sage, 95–107.

Ryerson RA (1998) *Manual of remote sensing*. New York: Wiley.

Schowengerdt RA (2006) *Remote sensing: Models and methods for image processing, 3rd edition*. New York: Academic Press.

optically stimulated luminescence (OSL) dating   A form of LUMINESCENCE DATING utilising the emission of luminescence from visible-light sensitive ELECTRON TRAPS to assess time elapsed since deposition of mineral grains. It is particularly valuable because it utilises QUARTZ and allows improved measurement precision over THERMOLUMINESCENCE (TL) DATING.   *DAR/CJC*

[*See also* INFRARED-STIMULATED LUMINESCENCE (IRSL) DATING]

Haskell E, Difley R, Kenner G et al. (1999) A comparison of optically stimulated luminescence dating methods applied to eolian sands from the Mojave Desert in southern Nevada. *Quaternary Science Reviews* 18: 235–242.

Huntley DJ, Godfrey-Smith DI and Thewalt MLW (1995) Optical dating of sediments. *Nature* 313: 105–107.

Madsen AT and Murray AS (2009) Optically stimulated luminescence dating of young sediments: A review. *Geomorphology* 109: 3–16.

Rhodes EJ (2011) Optically stimulated luminescence dating of sediments over the last 200,000 years. *Annual Review of Earth and Planetary Sciences* 39: 461–486.

Stone AEC and Thomas DSG (2008) Linear dune accumulation chronologies from the southeast Kalahari, Namibia: Challenges of reconstructing Late Quaternary palaeoenvironments from aeolian landforms. *Quaternary Science Reviews* 27: 1667–1681.

optimisation   A process of searching for the 'best' solution to a problem. The search requires the *maximisation* or *minimisation* of a mathematical function (known as the *objective function*) which itself encodes what is meant by 'best' and is often subject to constraints that an acceptable solution must satisfy.   *PJS*

[*See also* EARTH-SYSTEM ANALYSIS (ESA)]

oral history   The individual memories of people that are collected by interview techniques. Oral histories can contribute data on recent ENVIRONMENTAL HISTORY and HISTORICAL ARCHAEOLOGY. They often have the advantage of first-hand experience but limitations include small SAMPLE sizes, distinguishing fact from *myth* or *legend* and minimising *subjectivity* (of both the interviewee and the interviewer). There are a growing number of ARCHIVES for storage of oral histories.   *JAM*

Perks R and Thomson A (eds) (2006) *The oral history reader, 2nd edition.* London: Routledge.

Thompson P (2000) *Voice of the past: Oral history, 3rd edition.* Oxford: Oxford University Press.

Wilkie LA (2002) Oral history. In Orser Jr CE (ed.) *Encyclopedia of historical archaeology.* London: Routledge, 403–404.

orbital forcing   The effect of the Earth's orbital parameters (see ECCENTRICITY, OBLIQUITY, PRECESSION) on the RADIATION BALANCE of the Earth. In particular, orbital forcing is the main cause of GLACIAL-INTERGLACIAL CYCLES.   *JAM/EZ*

[*See also* FORCING FACTOR, MILANKOVITCH THEORY]

Bush MB, Miller MC, De Oliveira et al. (2002) Orbital forcing signal in sediments of two Amazonian lakes. *Journal of Paleolimnology* 27: 341–352.

Ganopolski A and Calov R (2011) The role of orbital forcing, carbon dioxide and regolith in 100 kyr glacial cycles. *Climate of the Past* 7: 1415–1426.

Khodri M, Leclainche Y, Ramsten G et al. (2001) Simulating the amplification of orbital forcing by ocean feedbacks in the last glaciation. *Nature* 410: 570–574.

orbital tuning   A method which allows global *correlation* by applying MILANKOVITCH THEORY to age-model development and palaeoclimatic research. Originally developed around the observed synchrony of changes in the OXYGEN ISOTOPE composition of FORAMINIFERA from MARINE SEDIMENT CORES throughout the world OCEANS, it is based upon the assumption that variations in the isotopic composition of the oceans result from changes in global ice that were driven by changes in solar INSOLATION arising from the Milankovitch PERIODICITIES. These are recorded in MARINE SEDIMENTS and can be examined using TIME-SERIES ANALYSIS, such as SPECTRAL ANALYSIS. The SPECTRAL MAPPING PROJECT TIMESCALE (SPECMAP) provides a global reference composite curve with which other oxygen isotope records can be compared. The method makes several assumptions—most importantly, that the degree of coherency between the Milankovitch rhythms and the CLIMATIC SYSTEM depends on the degree of linearity of the ice-volume response to external forcing.   *WENA/JP*

[*See also* MARINE ISOTOPIC STAGE (MIS), MARINE SEDIMENT CORES: RESULTS, MILANKOVITCH THEORY, NON-LINEAR BEHAVIOUR/DYNAMICS, ORBITAL FORCING]

Hays JD, Imbrie J and Shackleton NJ (1976) Variations in Earth's orbit: Pacemaker of the ice ages. *Science* 194: 1121–1132.

Huybers P and Aharonson O (2010) Orbital tuning, eccentricity, and the frequency modulation of climatic precession. *Paleoceanography* 25: PA4228.

Marinson DG, Pisias NG, Hays JD et al. (1987) Age dating and the orbital theory of the Ice Ages: Development of a high-resolution 0 to 30,000-year chronostratigraphy. *Quaternary Research* 27: 1–29.

Prokopenko AA, Hinnov LA, Williams DF and Kuzmin MI (2006) Orbital forcing of continental climate during the Pleistocene: A complete astronomically tuned climatic record from Lake Baikal, SE Siberia. *Quaternary Science Reviews* 25: 3431–3457.

ordinate   The scale value measured along the *y*-axis of a system of CARTESIAN CO-ORDINATES, from the origin of the system to the desired point location.   *TF*

[*See also* ABSCISSA, CO-ORDINATE SYSTEM]

ordination   The arrangement of objects and/or variables along *gradients* on the basis of composition, occurrence or environmental preferences. The term comes from the German *ordnung* and involves putting things into order. Ordination and CLASSIFICATION (CLUSTERING) are the main approaches for the analysis of large, complex multivariate data sets (STRUCTURING), particularly ecological and palaeoecological data, and, if available, associated environmental data. *Indirect ordination* or *indirect gradient analysis* (see GRADIENT

ANALYSIS) involves the biological data only and uses explicitly or implicitly a measure of the differences in composition between the objects or variables. Commonly used methods are PRINCIPAL COMPONENTS ANALYSIS (PCA) and CORRESPONDENCE ANALYSIS (CA) to derive a small number of *composite variables* or *ordination axes* that represent as much of the information in the original data as possible according to stated mathematical criteria (*dimensionality* is reduced). A low (two- or three-) dimensional representation of the differences between objects and/or variables is produced as a means of summarising patterns in the data, investigating possible data structure and generating HYPOTHESES about the environmental characteristics of the objects and possible underlying causal factors that may influence the observed patterns. Objects that are close to one another in an ordination diagram are inferred to resemble one another in terms of their composition; objects that are far apart are inferred to be dissimilar in composition and variables that are close together are inferred to have similar environmental preferences. It is assumed that objects with similar composition reflect similar environments.

Direct ordination or *direct gradient analysis* requires known environmental or other predictor variables as well as biological data. It involves modelling and summarising the responses of variables to *environmental gradients* by REGRESSION ANALYSIS (one response variable or species) or constrained or *canonical ordination* (CANONICAL CORRESPONDENCE ANALYSIS (CCA), REDUNDANCY ANALYSIS) (two or more response variables). The latter simultaneously ordinates objects, variables and predictor environmental variables, thereby often simplifying the interpretation of the ordination results. Being primarily a correlative approach, ordination can aid in hypothesis generation but it can rarely demonstrate CAUSAL RELATIONSHIPS. *HJBB*

[*See also* DISCRIMINANT ANALYSIS, FACTOR ANALYSIS, MULTIDIMENSIONAL SCALING, NON-METRIC MULTIDIMENSIONAL SCALING, PRINCIPAL CO-ORDINATES ANALYSIS (PCA)]

Borcard D, Gillet F and Legendre P (2011) *Numerical ecology with R*. New York: Springer.
Legendre P and Birks HJB (2012) From classical to canonical ordination. In Birks HJB, Lotter AF, Juggins S and Smol JP (eds) *Tracking environmental change using lake sediments, volume 5: Data handling and numerical techniques*. Dordrecht: Springer, 201–248.
Legendre P and Legendre L (1998) *Numerical ecology*. Amsterdam: Elsevier.
Lepš J and Šmilauer P (2003) *Multivariate analysis of ecological data using CANOCO*. Cambridge: Cambridge University Press.
Ter Braak CJF (1987) Ordination. In Jongman RHG, ter Braak CJF and van Tongeren OFR (eds) *Data analysis in community and landscape ecology*. Wageningen: Pudoc, 78–173.
Ter Braak CJF and Prentice IC (1988) A theory of gradient analysis. *Advances in Ecological Research* 18: 271–317.
Whittaker RH (ed.) (1973) *Ordination and classification of communities*. The Hague: Junk.

Ordovician   A SYSTEM of rocks, and a PERIOD of geological time from 488 to 444 million years ago and part of the Early PALAEOZOIC Era. Land plants first evolved in the Mid-Ordovician, and the Late Ordovician (*Hirnantian*) was a time of GLACIATION.   *GO*

[*See also* GEOLOGICAL TIMESCALE]

Finnegan S, Bergmann K, Eller JM et al. (2011) The magnitude and duration of late Ordovician-early Silurian glaciation. *Nature* 331: 903–906.
Finney SC and Berry WBN (eds) (2010) *The Ordovician Earth system*. Boulder, CO: Geological Society of America.
Lenton TM, Crouch M, Johnson M et al. (2012) First plants cooled the Ordovician. *Nature Geoscience* 5: 86–89.
Marshall JD, Brenchley PJ, Mason P et al. (1997) Global carbon isotopic events associated with mass extinction and glaciation in the Late Ordovician. *Palaeogeography, Palaeoclimatology, Palaeoecology* 132: 195–210.
Servais T, Owen A, Harper DAT et al. (2010) The Great Ordovician Biodiversification Event (GOBE): The palaeoecological dimension. *Palaeogeography, Palaeoclimatology, Palaeoecology* 294: 99–119.
Turner BR, Makhlouf IM and Armstrong HA (2005) Late Ordovician (Ashgillian) glacial deposits in southern Jordan. *Sedimentary Geology* 181: 73–91.

ore   An aggregate of MINERALS from which one or more useful constituents, principally *metals*, can be extracted. The main ore of ALUMINIUM (Al), for example, is BAUXITE.   *GO*

[*See also* RESOURCE]

Moon CJ, Whateley MKG and Evans AM (eds) (2006) *Introduction to mineral exploration*. Oxford: Blackwell.

organic content   *Organic matter* is one of the major constituents of the Earth's terrestrial and marine systems. In SOILS, it is of vital importance for SOIL BIOTA, and SOIL QUALITY. It is derived from the residues of the *shoots* and *roots* of plants that are gradually altered into HUMUS through, for example, *physical fragmentation*, *faunal interaction*, DECOMPOSITION and MINERALISATION. In LACUSTRINE SEDIMENTS and MARINE SEDIMENTS, the organic content reflects not only in-situ production from plants and PHYTOPLANKTON but also inputs of eroded material (organic and mineral) from DRAINAGE BASINS and other TERRESTRIAL ENVIRONMENTS. Measurement of the organic content of soils and sediments is either by LOSS ON IGNITION (LOI) or determination of *organic carbon* content by a *wet oxidation* technique. *Total organic content* is commonly measured in the analysis of lake and MARINE SEDIMENT CORES. The terrestrial organic content of vegetation, soils and sediments is a major *carbon pool* that may be influenced by human activities to increase or decrease the atmospheric content of CARBON DIOXIDE, and hence the rate of GLOBAL WARMING.

EMB

[*See also* CARBON BALANCE/BUDGET, CARBON CONTENT OF SOILS, CARBON CYCLE]

Buttman D and Raymond PA (2011) Significant efflux of carbon dioxide from streams and rivers in the United States. *Nature Geoscience* 4: 839–842.

Eswaran H, van den Bwerg E and Reich P (1993) Organic carbon in soils of the world. *Soil Science Society of America Journal* 57: 192–194.

Meyers PA and Teranes JL (2001) Sediment organic matter. In Last WM and Smol JP (eds) *Tracking environmental change using lake sediments, volume 2.* Dordrecht: Kluwer, 239–269.

Stubbins A, Hood E, Raymond PA et al. (2012) Anthropogenic aerosols as a source of ancient dissolved organic matter in glaciers. *Nature Geoscience* 5: 198–210.

organic farming An expanding form of farming, also known as *ecological farming*, which relies on alternatives to synthetic chemical FERTILISERS and PESTICIDES, TRANSGENIC ORGANISMS, ANTIBIOTICS or other AGRICHEMICALS. Non-polluting modern methods are often acceptable. Organic farmers build and maintain soil FERTILITY by a number of practices, such as the use of CROP ROTATION with *legumes, mixed crop-animal systems, manuring, cover crops* and reduced TILLAGE, which tend to mimic natural *ecosystems*. Similarly, PEST CONTROL may involve the use of mechanical and BIOLOGICAL CONTROL, intercropping and STRIP CROPPING, crop rotation and pest-resistant varieties. Some of these practices are used in related types of farming (e.g. *biodynamic farming, biological farming* and *integrated farming*).

What is now termed 'organic farming' had its origins in traditional, highly sustainable agricultural systems in Asia. The organic farming movement as such was founded in the AD 1930s. Lady Eve Balfour raised support for it in the United Kingdom in the 1940s and today the Soil Association (UK) supports and promotes it. Issues associated with organic farming include its costs of conversion, the maintenance of high yields, the need for greater areas to be brought under AGRICULTURE to compensate for lower yields, the questionable nutritional benefits, the problems of marketing organic foods at generally higher cost than non-organic foods and the problems associated with INDUSTRIALISATION. Various studies have shown that organic farming yields are consistently lower (on average ~20 per cent lower) than those from *conventional farming* but the yield differences vary considerably from 3 to 5 per cent lower for organic fruits and rain-fed legumes

(e.g. soya beans), to 30–40 per cent lower when only the most comparable conventional and organic systems are compared. Best organic management practices produced yields that are closer (~13 per cent) to conventional yields. Thus, organic systems nearly rival conventional yields in some cases but often they do not. The term *permaculture* is used for organic agriculture that seeks SUSTAINABILITY.

CJB/JAM

[*See also* CONSUMPTION, CULTIVATION, SUSTAINABLE AGRICULTURE, YIELD GAP]

Dabbert S, Haring AM and Zanoli R (2004) *Organic farming: Policies and prospects.* London: Zed Books.

De Ponti T, Rijk B and van Ittersum MK (2012) The crop yield gap between organic and conventional agriculture. *Agricultural Systems* 108: 1–9.

Francis CA (2005) Organic farming. In Hillel D (ed.) *Encyclopedia of soils in the environment.* Amsterdam: Elsevier, 77–84.

Lotter D (2002) Organic farming. *Journal of Sustainable Agriculture* 21: 59–128.

Seufert V, Ramankutty N and Foley JA (2012) Comparing the yields of organic and conventional agriculture. *Nature* 485: 229–232.

organic sediment A sediment or deposit containing plant and/or animal DETRITUS.   *UBW*

[*See also* BIOGENIC SEDIMENT, LACUSTRINE SEDIMENTS, MINEROGENIC SEDIMENT, SEDIMENT TYPES]

organochlorides Also known as *chlorinated hydrocarbons*, they are organic MOLECULES containing chlorine ATOMS. Organochlorides are persistent, mobile, synthetic, organic PESTICIDES (e.g. *DDT, dieldrin*). They are highly soluble in fatty tissues, hence easily absorbed by a wide range of organisms, in addition to the main target species.   *GOH*

[*See also* BIOLOGICAL MAGNIFICATION, CHLOROFLUOROCARBONS (CFCs), HALOGENATED HYDROCARBONS (HALOCARBONS), PERSISTENT ORGANIC COMPOUNDS (POCs)]

Carson R (1962) *Silent spring.* Boston: Houghton Mifflin.

organophosphates Esters of phosphoric acid, in which organic COMPOUNDS are linked to PHOSPHORUS (P) atoms via oxygen ATOMS. Organophosphates include some biodegradable, non-persistent, synthetic, organic PESTICIDES (e.g. *parathion, malathion*), some of which have been banned. Although they are not subject to BIOLOGICAL MAGNIFICATION, they are quick acting and extremely toxic (10–100 times more poisonous than ORGANOCHLORIDES) to *mammals, birds* and *fish*.   *GOH*

Costa LG (2006) Current issues in organophosphate toxicology. *Clinica Chimica Acta* 366: 1–13.

oribatid mites   Minute (200–800 μm) arthropods (*Acari*), which dominate the arthropod fauna in *subtropical* and TEMPERATE FORESTS. They are important for DECOMPOSITION and MINERAL CYCLING, and reconstructing local habitat conditions: unlike beetles they do not fly and unlike *pollen grains* they are not blown long distances. Development of their potential as proxies for temperature and moisture in LACUSTRINE SEDIMENTS and MIRES is at an early stage.   *JAM*

Erickson JM and Platt Jr RB (2007) Oribatid mites. In Elias SA (ed.) *Encyclopedia of Quaternary science*. Amsterdam: Elsevier, 1547–1566.
Solhøy T (2001) Oribatid mites. In Smol JP, Birks HJB and Last WM (eds) *Tracking environmental change using lake sediments, volume 4*. Dordrecht: Kluwer, 81–104.

oriented thaw lake   A lake elongate in form usually associated with PERMAFROST areas, although oriented lakes are also found elsewhere. In permafrost areas, the long axes of such lakes, which range in size from small ponds to large lakes, are aligned typically at right angles to the wind direction. The precise reason for the alignment is a matter of debate but it appears to relate to differential THERMAL EROSION.   *RAS*

[*See also* THERMOKARST]

Pelletier JD (2005) Formation of oriented thaw lakes by thaw slumping. *Journal of Geophysical Research: Earth Surface* 110: F02018.

originality   Novel ideas in the search for KNOWLEDGE and understanding of phenomena. As such, originality is a necessary part of SCIENCE.   *JAM*

Medaware PB (1967) *Art of the soluble: Creativity and originality in science*. London: Heinemann.

orogenesis   The process of mountain building on the Earth, associated with the DEFORMATION and UPLIFT of rocks (see TECTONICS) in an OROGENIC BELT (OROGEN) during a period of OROGENY. Orogenesis occurs through convergent processes at DESTRUCTIVE PLATE MARGINS and continent-continent COLLISION ZONES (see PLATE TECTONICS), in contrast to EPEIROGENY.   *GO*

Cawood PA, Leitch EC, Merle RE and Nemchin AA (2011) Orogenesis without collision: Stabilizing the Terra Australis accretionary orogen, eastern Australia. *Geological Society of America Bulletin* 123: 2240–2255.
Johnson MRW and Harley SL (2012) *Orogenesis*. Cambridge: Cambridge University Press.
Potter PE and Szatmari P (2009) Global Miocene tectonics and the modern world. *Earth-Science Reviews* 96: 279–295.

orogenic belt (orogen)   A linear belt of rocks with a common history of SEDIMENTATION, DEFORMATION, VOLCANISM, igneous INTRUSION and *metamorphism*.

Young orogenic belts comprise high *mountains*; older orogenic belts have undergone DENUDATION, exposing deformed and metamorphosed rocks (see GEOLOGICAL STRUCTURES, METAMORPHIC ROCKS). The classic PLATE TECTONICS model for the formation of an orogenic belt is at a continent-continent COLLISION ZONE (*collision orogen*, e.g. Himalayas, Alps). Orogenic belts can also be formed by compression at an ocean-continent DESTRUCTIVE PLATE MARGIN, representing an ACTIVE CONTINENTAL MARGIN or *continental margin orogen* (e.g. Andes), or through TERRANE assembly (*accretionary orogen*, e.g. Rockies). These models relating OROGENY to plate tectonics have replaced GEOSYNCLINE theory.   *GO*

Cawood PA (2005) Terra Australis Orogen: Rodinia breakup and development of the Pacific and Iapetus margins of Gondwana during the Neoproterozoic and Paleozoic. *Earth-Science Reviews* 69: 249–279.
Collins AS and Pisarevsky SA (2005) Amalgamating eastern Gondwana: The evolution of the Circum-Indian Orogens. *Earth-Science Reviews* 71: 229–270.
Murphy JB, Keppie JD and Hynes AJ (eds) (2009) *Ancient orogens and modern analogues*. Bath: Geological Society.
Schmid SM, Fügenschuh B, Kissling E and Schuster R (2004) Tectonic map and overall architecture of the alpine orogen. *Eclogae Geologicae Helvetiae* 97: 93–117.
Sears JW, Harms TA and Evenchick CA (eds) (2007) *Whence the mountains? Inquiries into the evolution of orogenic systems: A volume in honor of Raymond A. Price*. Boulder, CO: Geological Society of America.

orogeny   An episode of OROGENESIS (*mountain building*), giving rise to an OROGENIC BELT (OROGEN), and involving UPLIFT associated with DEFORMATION (folding and faulting) of the Earth's CRUST and metamorphism. Relatively recent episodes of orogeny are represented by high *mountain chains*, such as the Himalayas (*Himalayan orogeny*, EOCENE collision) and the Alps (*alpine orogeny*, EOCENE-OLIGOCENE deformation). The products of older episodes of orogeny have experienced DENUDATION and are represented by belts of deformed and metamorphosed rocks. Examples include the *Hercynian* or *Variscan orogeny* (Late PALAEOZOIC events in Central Europe) and the *Caledonian orogeny* (Early Palaeozoic events affecting northwest Europe and northeastern North America).   *GO*

[*See also* GEOLOGICAL STRUCTURES, METAMORPHIC ROCKS, THRUST FAULT]

Burg J-P and Ford M (eds) (1997) *Orogeny through time*. Bath: Geological Society.
Prave AR, Kessler II LG, Malo M et al. (2000) Ordovician arc collision and foredeep evolution in the Gaspé Peninsula, Québec: The Taconic Orogeny in Canada and its bearing on the Grampian Orogeny in Scotland. *Journal of the Geological Society of London* 157: 393–400.

Reiners PW and Brandon MT (2006) Using thermochronology to understand orogenic erosion. *Annual Review of Earth and Planetary Sciences* 34: 419–466.

orographic　Pertaining to the influence of topography, especially large-scale relief features such as mountains and hills, acting as barriers to air flow and affecting WEATHER and CLIMATE. Orographic effects are evident from local to global in scales in CYCLOGENESIS, LONG WAVES (*Rossby waves*), RAIN SHADOWS and strong, downslope winds such as the föhn, bora and mistral (in Europe) and the chinook and Santa Ana (in North America). *Orographic precipitation*—a precipitation type generated by the forced uplift of air flow over a topographic barrier—tends to be concentrated on the windward side of mountains and uplands.　　*JAM*

[*See also* TOPOCLIMATE]

Barry RG (1992) *Mountain weather and climate, 2nd edition.* London: Routledge.
Broccoli AJ and Manabe S (1992) The effect of orography on midlatitude Northern Hemisphere dry climates. *Journal of Climate* 5: 1181–1201.
Hide R and White PW (eds) (1981) *Orographic effects in planetary flows.* Geneva: World Meteorological Organization.

orography　A little used term for the form, especially the RELIEF, of mountains.　　*JAM*

osmosis　The DIFFUSION of a solvent through a semi-permeable membrane, such as a cell wall, from a solution of relatively low concentration to one of relatively high concentration. The SOLUTES do not pass through the membrane. The process is of fundamental importance to the uptake of water by plants, where the concentration of salts (and hence *osmotic pressure*) inside plant root cells is normally greater than in the surrounding soil water, and in *osmoregulation* of the water and salt content of organisms generally. *Reverse osmosis* is used in some DESALINATION plants and in the treatment of some wastewaters.　　*JAM*

osteology　The study of BONES and TEETH.　　*JAM*

[*See also* ANIMAL REMAINS, PALAEODIET]

Lambert JB and Grupe G (eds) (1993) *Prehistoric human bone: Archaeology at the molecular level.* Berlin: Springer.
O'Connor T (2000) *The archaeology of animal bones.* Stroud: Sutton Publishing.

ostracod analysis　The study of a group of organisms, or MICROFOSSILS, comprising part of the *zoobenthos* (see BENTHOS) and ZOOPLANKTON in fully marine (see SEAWATER COMPOSITION) through BRACKISH WATER to terrestrial freshwater AQUATIC ENVIRONMENTS for BIOSTRATIGRAPHY, PALAEOENVIRONMENTAL RECONSTRUCTION and

GEO-ARCHAEOLOGY. *Ostracods* (or *ostracodes*) are a group of bivalved *crustaceans* (*arthropods*) with an external chitinous or more commonly calcitic *carapace* (a shell of two valves) that is typically 0.5–2 mm in length (range: 0.2–30 mm), and a well-documented fossil record from the Early CAMBRIAN to the HOLOCENE (*Recent*). Parameters such as ASSEMBLAGE composition, abundance, distribution, preservation, ornamentation variation, shell shape (morphometrics), molecular approaches and shell chemistry are used in their analysis.

Ostracod analysis in marginal marine (PARALIC) to fully MARINE SEDIMENTS is important for reconstructing past SALINITY, OCEAN CIRCULATION and WATER-MASS movement, SEA-SURFACE TEMPERATURE (SST), SEA-LEVEL CHANGE, *palaeobathymetry, palaeobiogeography* (see BIOGEOGRAPHY) and CLIMATIC CHANGE, often at orbital and suborbital resolution. *Non-marine ostracods* in freshwater sediments, *peat* and SOIL are used for localised QUATERNARY and HOLOCENE reconstructions of ALKALINITY, EUTROPHICATION, SALINITY, *lake-water circulation patterns* and OVERTURNING, STABLE-ISOTOPE ANALYSIS (e.g. $\delta^{18}O$, $\delta^{13}C$) and the isotope ratio $^{87/86}Sr$ of valves are frequently used in the study of terrestrial and marginal marine sediments to reconstruct local PALAEO-SALINITY and runoff (WEATHERING). TRACE ELEMENT ratios (Mg, Sr and Na/Ca) of BENTHIC species are widely used to reconstruct both *bottom-water marine temperatures* and *lake-water temperatures*. Ostracods have also been employed in the use of AMINO ACID DATING of both marine and terrestrial sediments.　　*AWM*

[*See also* MAGNESIUM:CALCIUM RATIO (Mg:Ca), MICROFOSSIL ANALYSIS, STRONTIUM ISOTOPES].

Frenzel P and Boomer I (2005) The use of ostracods from marginal marine, brackish waters as bioindicators of modern and Quaternary environmental change. *Palaeogeography, Palaeoclimatology, Palaeoecology* 225: 68–92.
Griffiths HI and Holmes JA (2000) *Non-marine ostracods and Quaternary palaeoenvironments.* London: Quaternary Research Association [Technical Guide 8].
Holmes JA (2001) Ostracoda. In Smol JP, Birks HJB and Last WM (eds) *Tracking environmental change using lake sediments, volume 4.* Dordrecht: Kluwer, 125–151.
Holmes JA and Chivas AR (eds) (2002) *The Ostracoda: Applications in Quaternary research.* Washington, DC: American Geophysical Union.
Holmes JA and Engstrom DR (2003) Non-marine ostracod records of Holocene environmental change. In Mackay A, Battarbee R, Birks J and Oldfield F (eds) *Global change in the Holocene.* London: Arnold, 310–327.
Kauffman DS (2000) Amino acid racemization in ostracodes. In Goodfriend GA, Collins MJ, Fogel ML et al. (eds) *Perspectives in amino acid and protein geochemistry.* Oxford: Oxford University Press, 145–160.
Park LE and Smith AJ (eds) (2003) *Bridging the gap: Trends in ostracode biological and geological sciences.* New Haven, CT: Palaeontological Society.

otolith A calcareous concretion from the ear of a vertebrate. Fish otoliths, also known as *ear stones*, can be indicative of the species, fish size, season caught and SEA-SURFACE TEMPERATURE (SST). *JAM*

[*See also* SEASONALITY, SEASONALITY INDICATOR]

Andrus CFT, Crowe DE, Sandweiss DH et al. (2002) Otolith δ^{18}O record of Mid-Holocene sea surface temperatures in Peru. *Science* 295: 1508–1511.

Hufthammer AK, Hoie H, Folkvord A et al. (2010) Seasonality of human site occupation based on stable oxygen isotope ratios of cod otoliths. *Journal of Archaeological Science* 37: 78–83.

outbreak The occurrence of a large number of cases (in excess of BACKGROUND LEVELS) of a communicable, *infectious disease* in a community in a short period of time. Large-scale outbreaks, when many people in a wide geographic area are simultaneously affected, are termed EPIDEMICS. *MLW*

Holmgren A and Borg G (eds) (2010) *Handbook of disease outbreaks: Prevention, detection and control.* Hauppauge, NY: Nova Science Publishers.

outgassing (1) The release of VOLATILES (DEGASSING) from molten MAGMA due to *decompression*. Outgassing is believed to have been the main process in the formation of secondary *planetary atmospheres*. The ATMOSPHERE of the early Earth was affected by three phases of outgassing initiated by (1) bombardment by METEORITES, (2) internal differentiation of rocks within the Earth and (3) VOLCANISM.

(2) The term is also used to describe the release or *evasion* of CARBON DIOXIDE from *rivers* and WETLANDS, an important process of *carbon loss* in the *humid tropical environments*. The carbon probably originates from *organic matter* transported from *uplands* and *flooded forests*. *JAM/GO*

[*See also* EARTH REVOLUTIONS, HADEAN (*PRISCOAN*)]

Mayorga E, Aufdenkampe AK, Masiello CA et al. (2005) Young organic matter as a source of carbon dioxide outgassing from Amazonian rivers. *Nature* 436: 538–541.

Tian F, Toon OB, Pavlov AA and De Sterk H (2005) A hydrogen-rich early Earth atmosphere. *Science* 308: 1014–1017.

Tobie G, Lunine JI and Sotin C (2006) Episodic outgassing as the origin of atmospheric methane on Titan. *Nature* 440: 61–64.

outlet glacier A stream of fast-flowing ice discharging from an ICE SHEET or *ice cap* via a valley, or between areas of slow-moving ice as an ICE STREAM. In terms of *frontal variations* (*glacier advance* or *retreat*), outlet glaciers are more sensitive to climate than other parts of the *ice margin* of ice caps or ice sheets. The GLACIER FORELANDS of outlet glaciers have therefore been widely

used in the reconstruction of GLACIER VARIATIONS from ICE-MARGIN INDICATORS such as MORAINES. Within their margins, ice streams leave a strong streamlined imprint on the LANDSCAPE and on CONTINENTAL SHELVES. *JAM/MJH*

[*See also* GLACIER]

Bickerton RJ and Matthews JA (1993) 'Little Ice Age' variations of outlet glaciers from the Jostedalsbreen ice-cap, southern Norway: A regional lichenometric-dating study of ice-marginal moraine sequences and their climatic implications. *Journal of Quaternary Science* 8: 45–66.

Clague JJ, Koch J and Geertsema M (2010) Expansion of outlet glaciers of the Juneau Icefield in northwest British Columbia during the past two millennia. *The Holocene* 20: 447–461.

Howat IM, Joughin I and Scambos TA (2007) Rapid changes in ice discharge from Greenland outlet glaciers. *Science* 315: 1559–1561.

outlier (1) In NUMERICAL ANALYSIS, an anomalous data point, which appears not to belong to the same population as the remainder of the distribution. (2) In GEOLOGY, an area of younger rock completely surrounded by older: it contrasts with the normal situation where a rock outcrop is bounded by older rocks on one side and younger on the other. Outliers can form as remnants left by EROSION, or through DEFORMATION. *GO/JAM*

[*See also* INLIER]

outwash deposit Where MELTWATER emerges at a GLACIER margin, high DEPOSITION rates lead to the deposition of an *outwash plain* built of outwash deposits (see SANDUR). A reduction of SEDIMENT load may cause *downcutting* and the original outwash plain may be abandoned to leave an *outwash terrace*. Sequences of terraces and associated MORAINES in lowland valleys of rivers issuing from the Alps were the basis for the widely applied model of four Pleistocene glaciations put forward by A. Penck and E. Brückner in the early twentieth century. *MRB/JS*

[*See also* GLACIAL THEORY, GLACIOFLUVIAL LANDFORMS, GLACIOFLUVIAL SEDIMENTS, ICE-CONTACT FAN/RAMP]

Hein AS, Dunai TJ, Hulton NRJ et al. (2011) Exposure dating outwash gravels to determine the age of the greatest Patagonian glaciations. *Geology* 39: 103–106.

Salamon T (2009) Origin of Pleistocene outwash plains in various topographic settings, southern Poland. *Boreas* 38: 362–378.

outwelling The outflow of NUTRIENTS and organic carbon from WETLAND or ESTUARINE ENVIRONMENTS into SEAS of the CONTINENTAL SHELF. The term has been used in relation to potential losses from highly productive ecosystems, especially after heavy PRECIPITATION events and coastal flooding. *JAM*

[*See also* DOWNWELLING, UPWELLING]

Odum EP (2002) Tidal marshes as outwelling/pulsing
systems. In Weinstein MP and Kreeger DA (eds) *Concepts
and controversies in tidal marsh ecology*. New York:
Kluwer, 3–7.
Ridd P, Sandstrom M and Wolanski E (1988) Outwelling
from tropical salt flats. *Estuarine, Coastal and Shelf
Science* 26: 243–253.
Wolanski E (2007) *Estuarine ecohydrology*. Amsterdam:
Elsevier.

overbank deposit *Flood sediments*, usually silts
and clays but occasionally sands and gravels, deposited
on river FLOODPLAINS during OVERBANK FLOW events.
They tend to decrease in thickness and grain size away
from a channel. Rates and patterns of overbank deposition along rivers are greatly affected by CHANNELISATION
schemes.   *DML*

Walling DE and He Q (1999) Changing rates of overbank
sedimentation on the floodplains of British rivers during
the past 100 years. In Brown AG and Quine TA (eds)
Fluvial processes and environmental change. Chichester:
Wiley, 207–222.

overbank flow A river-flow event that overtops
the bank, thereby exceeding BANKFULL DISCHARGE and
CHANNEL CAPACITY and spreading water across the
FLOODPLAIN. It can lead to OVERBANK DEPOSITS.   *DML*

[*See also* FLOOD, FLOODOUT]

Beven K and Carling PA (eds) (1989) *Floods: Hydrological,
sedimentological and geomorphological implications*.
Chichester: Wiley.

overcultivation Changes in CULTIVATION practices
causing or likely to cause LAND DEGRADATION. In DRY-
LANDS, it is viewed as a classic cause of DESERTIFICATION.
Changes may include the introduction of inappropriate
cultivation techniques unsuited to the local conditions
(e.g. inappropriate PLOUGHING or TILLAGE, and reduction
of FALLOW periods).   *RAS*

[*See also* OVERGRAZING]

Zeng D and Jiang F (2006) Deterring "three excesses"
(over-cultivation, overgrazing and deforestation) is the
only way from the source to control the desertification in
ecologically frangible regions in China: Taking Keerqin
sandy land as an example. *Chinese Journal of Ecology* 12:
Article No. 18.

overdeepening The process whereby an eroding
GLACIER or ICE SHEET excavates a preglacial river valley
well below river BASE LEVEL. The long profiles of overdeepened valleys have steep gradients near their heads
and gentler, sometimes reverse slopes near their mouths.
For example, in its overdeepened section, the Sognefjord
glacial valley descends to 1,308 m below sea level.   *JS*

[*See also* GLACIAL EROSION, U-SHAPED VALLEY]

Dehnert A, Lowick SE, Preusser F et al. (2012) Evolution of
an overdeepened trough in the northern alpine foreland
at Niederweningen, Switzerland. *Quaternary Science
Reviews* 34: 127–145.
Nesje A, Dahl SO, Valen V and Ovstedal J (1992)
Quaternary erosion in the Sognefjord drainage basin,
western Norway. *Geomorphology* 5: 511–520.

overfishing The reduction of fish stocks to numbers
below the maximum SUSTAINABLE YIELD, that is the maximum number of fish available each year for HARVEST-
ING without diminishing the longer-term stocks. Three
types of overfishing can be distinguished: (1) *recruitment overfishing*, where harvesting impairs the ability
of a species to reproduce back to preharvest conditions;
(2) *growth overfishing*, where harvesting involves fish
of suboptimal maturity in terms of size or value (the
latter impinges on the former if juvenile fish are caught
before spawning); and (3) *ecosystem overfishing*, where
harvesting damages the ecosystem by creation of an
imbalance between the species populations. Numerous cases in recent history demonstrate an inability to
prevent overfishing, with disastrous consequences for
fisheries and their dependent economies. However,
the Maine lobster fishery off the northeastern USA has
proved an exception.   *JAM*

[*See also* FISHERIES CONSERVATION AND MANAGEMENT,
SUSTAINABLE FISHERIES, TRAGEDY OF THE COMMONS]

Acheson J (2003) *Capturing the commons: Devising
institutions to manage the Maine lobster industry*.
Lebanon, NH: University Press of New England.
Allen JD, Abell R, Hogan Z et al. (2005) Overfishing of
inland waters. *BioScience* 55: 1041–1051.
Clover C (2005) *The end of the line: How overfishing is
changing the world and what we eat*. Berkeley: University
of California Press.
Kurlansky M (1997) *Cod: A biography of the fish that
changed the world*. New York: Walker.

overflow channel In glacial geomorphology, a
channel (or notch) draining water from an ICE-DAMMED
LAKE at its lowest point. A former *glacial overflow* (or
overspill) *channel* may be indicated by evidence of
washed bedrock surfaces along the channel route and
ALLUVIAL FANS where the channel gradient declines
sufficiently to promote BEDLOAD deposition, together
with SHORELINES, LACUSTRINE SEDIMENTS and/or DELTAS
relating to the associated former lake. Following a
two-part review by Sissons (1960, 1961), many former
ice-dammed lakes in Britain reconstructed primarily or
solely on the basis of supposed glacial overflow channels were subsequently rejected through lack of supporting evidence.   *RAS*

[*See also* MELTWATER CHANNEL]

Sissons JB (1960) Some aspects of glacial drainage channels in Britain. Part I: *Scottish Geographical Magazine* 76: 131–146.

Sissons JB (1961) Some aspects of glacial drainage channels in Britain. Part II: *Scottish Geographical Magazine* 77: 15–36.

overgrazing   The rate of removal of vegetation by GRAZING and *browsing* animals that exceeds the capacity of the vegetation to recover. Consequently, there are changes in the vegetation community and typically there is soil exposure leading to SOIL EROSION. The combined impact of deteriorating PASTURE and TRAMPLING may reduce soil moisture and increase runoff. Overgrazing in some regions has contributed to DESERTIFICATION.   *IFS*

[*See also* CARRYING CAPACITY, SUSTAINABILITY, SUSTAINABLE YIELD]

Gordon IJ and Prins HHT (eds) (2010) *The ecology of browsing and grazing*. Berlin: Springer.

Mace R (1991) Overgrazing overstated. *Nature* 349: 280–281.

Mysterud A (2006) The concept of overgrazing and its role in management of large herbivores. *Wildlife Biology* 12: 129–141.

Urbanska KM, Webb NR and Edwards PJ (1997) *Restoration ecology and sustainable development*. Cambridge: Cambridge University Press.

overhunting   The *overexploitation* of species for subsistence, economic or recreational purposes. Historically, overhunting was the leading cause of EXTINCTION, although now it has probably been overtaken by HABITAT LOSS. Most oceanic fisheries, exploited for all three purposes, are now in steep decline and several have already collapsed, raising serious ecological, economic and FOOD SECURITY issues. OVERFISHING rarely causes extinction of the *target species* but secondary ecological consequences of massive reductions in fish abundance may be severe. While hunting by INDIGENOUS PEOPLES has popularly been viewed as far less detrimental than *commercial hunting*, fossil evidence suggests that traditional hunters have caused the extinction of thousands of species, including perhaps 20 per cent of all *birds*, in Madagascar, North America, the Mediterranean, and on many *oceanic islands*. Overhunting, coupled with other forms of human ENVIRONMENTAL DEGRADATION, continues to threaten many species in terrestrial, marine and aquatic ecosystems and has led to the phenomenon of *empty forests*. Overhunting and the subsequent loss of species can have serious knock-on effects, especially if *zoochorous* (animal-dispersed) trees suffer from reduced *seed dispersal*.   *JTK*

[*See also* BUSH MEAT, ECOSYSTEM COLLAPSE, FISHERIES CONSERVATION AND MANAGEMENT, KEYSTONE SPECIES, RECREATION, SUSTAINABILITY, SUSTAINABLE YIELD, TRAGEDY OF THE COMMONS, WHALING AND SEALING]

Brodie JF, Helmy OE, Brockelman WY and Maron JL (2009) Bushmeat poaching reduces the seed dispersal and population growth rate of a mammal-dispersed tree. *Ecological Applications* 19: 854–863.

Harrison RD (2011) Emptying the forest: Hunting and the extirpation of wildlife from tropical nature reserves. *Bioscience* 61: 919–924.

Jones TL, Porcasi JF, Erlandson JM et al. (2008) The protracted Holocene extinction of California's flightless sea duck (*Chendytes lawi*) and its implications for the Pleistocene overkill hypothesis. *Proceedings of the National Academy of Sciences* 105: 4105–4108.

Pauly D (2009) Beyond duplicity and ignorance in global fisheries. *Scientia Marina* 73: 215–224.

Scott E (2010) Extinctions, scenarios, and assumptions: Changes in latest Pleistocene large herbivore abundance and distribution in western North America. *Quaternary International* 217: 225–239.

overkill hypothesis   As human populations expanded their geographic range across the globe, they brought with them an unparalleled ability to transform *native landscapes* and directly or indirectly devastate populations of NATIVE or INDIGENOUS PEOPLES species. According to Paul Martin and other palaeoecologists, this 'overkill' may have been responsible for the waves of EXTINCTION of *native faunas* (especially the MEGAFAUNA) that coincided with the arrival of environmentally significant humans in new lands. While strongly disputed by some, especially those that favour a climatic explanation for the extinctions, more and more dating is supporting the overkill hypothesis.   *MVL/JLI*

[*See also* MAMMOTH, MASS EXTINCTIONS, MEGAFAUNAL EXTINCTION]

Gillespie R (2008) Updating Martin's global extinction model. *Quaternary Science Reviews* 27: 2522–2529.

Leakey LSB (1966) Africa and Pleistocene overkill? *Nature* 212: 1615–1616.

Martin PS (1966) Africa and Pleistocene overkill. *Nature* 212: 339–342.

Martin PS (1984) Prehistoric overkill. In Martin PS and Klein RG (eds) *Quaternary extinctions: A prehistoric revolution*. Tucson: University of Arizona Press, 345–403.

Martin PS (2005) *Twilight of the mammoths*. Berkeley: University of California Press.

overland flow   Water flow over the land surface but not in stream channels. It is also known as *sheet flow, sheet wash* and *interrill flow*.   *JAM*

[*See also* HORTONIAN OVERLAND FLOW, SATURATION OVERLAND FLOW]

Parsons AJ and Abrahams AD (eds) (1992) *Overland flow: Hydraulics and erosion mechanics*. London: UCL Press.

overlay analysis   The operation of stacking MAPS on top of each other so that information is compared at common locations. It is normally associated with a

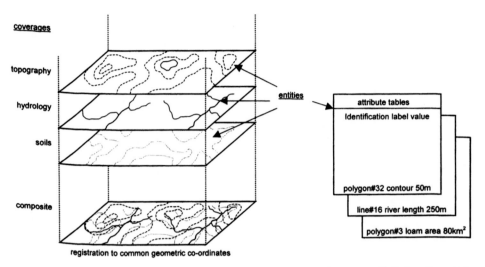

Overlay analysis The overlay principle and the fundamental units of data representation used in a geographical information system.

GEOGRAPHICAL INFORMATION SYSTEM (GIS) where multiple digital COVERAGES are superimposed and ENTITIES are compared at common *geometric co-ordinates* (see Figure). For example, an overlay analysis examining possible environmental hazards arising from the location of a new power station would involve coverages such as relief, hydrology, vegetation, soils and climatic indicators. By stacking these coverages on top of each other, and ensuring common *geometric registration*, entities such as low-lying land, nearby water supplies, conservation areas of woodland, nearby agricultural land and prevailing winds could all be used within a DECISION-MAKING process to determine the least cost to environmental stability. Through time, overlays would be able to reveal changes to the environment, such as whether soils have become acidic, water supplies contaminated or habitats damaged. Overlay analysis is now part of SPATIAL ANALYSIS and modelling in GIS. *VM*

[*See also* POLYGON ANALYSIS]

Longley PA, Goodchild MF, Maguire DJ and Rhind DW (eds) (2010) *Geographic information systems and science, 3rd edition.* Chichester: Wiley.

overpopulation   Usually applied to animal populations, including humans, overpopulation is the condition that results where the number or density of individuals exceeds the CARRYING CAPACITY of the environment. It may be followed rapidly by a *population decline* or *population crash*.

At the end of the eighteenth century, Thomas Malthus observed the poverty and malnourishment around him and concluded that the world could only support a limited number of human beings as population when unchecked increases in a geometric ratio.

Subsistence increases in an arithmetical ratio (Malthus, 1798). Many *neomalthusian* variants of this idea have appeared in scientific literature and in policymaking ever since. Up to the present day, Malthus's prediction has not been proven correct as scientific development of the means of food production has kept ahead of the world's increasing population, although considerable numbers remain undernourished (see HUNGER). At the same time, it must be appreciated that the world's capacity is finite and that the soils most suited for agriculture are already being used.

Overpopulation implies that there are too many people present for the land to support by the provision of food and water. As a result, these people exert an undue pressure upon the land, extending cultivation into areas unsuited to cropping, removing natural vegetation, causing SOIL EROSION and initiating LAND DEGRADATION. However, the carrying capacity of the land is not a fixed value as countries have exploited their resources to enable greater or fewer people to live and at different standards of living. In Kenya, a study of Machakos District demonstrated that a sixfold increase in the rural population over about 50 years was actually accompanied by a reduction in land degradation. This was attributed to good land MANAGEMENT combined with financial investment in traditional SOIL CONSERVATION practices, such as TERRACING.

It has been argued that overpopulation and its effects have tended to be kept in check throughout PREHISTORIC time and much of HISTORY by FAMINE, DISEASE and warfare. Today, lower BIRTH RATES that accompany improved *quality of life* play a similar role. Nevertheless, in the face of continuing rapid growth in the world population (see DEMOGRAPHIC CHANGE) and its effects, it cannot be certain that the carrying capacity of the environment will not be exceeded in the near future. Meanwhile, it is obvious that a balance must be struck between wise use of the environment and its overexploitation for reasons of ECOSYSTEM HEALTH, ECOLOGICAL INTEGRITY and ENVIRONMENTAL SECURITY.   *EMB/JAM*

[*See also* LAND DEGRADATION, MALTHUSIAN, SUSTAINABILITY, SUSTAINABLE DEVELOPMENT, TECHNOLOGICAL CHANGE]

Greenland DJ, Gregory PJ and Nye PH (eds) (1997) *Land resources: On the edge of the Malthusian precipice?* Wallingford: CABI Publishing.
Malthus TR (1798) *An essay on the principle of population as it affects the future improvement of society.* London: J. Johnson [Oxford World's Classics edition, 2008].
Meadows D, Randers J and Meadows D (2004) *Limits to growth: The 30-year update.* London: Chelsea.
Tiffen M and Mortimore M (2008) Environment, population growth and productivity in Kenya: A case study of Machakos District. *Policy Development Review* 10: 359–387.
United Nations Population Fund (UNFPA) (1991) *Population, resources and the environment: The critical challenges.* New York: UNFPA.

overshoot   The tendency of a SYSTEM to fail to maintain a dynamic equilibrium (see EQUILIBRIUM CONCEPTS) due to delays in FEEDBACK MECHANISMS.   *JAM*

[*See also* VICIOUS CIRCLE PRINCIPLE (VCP)]

overtopping/overwashing   Overtopping is the process by which coastal BARRIER BEACHES and BARRIER ISLANDS are built up by deposition during STORMS from swash flows of insufficient magnitude to extend fully across the crest. It contrasts with *overwashing*, during which higher-magnitude events may erode the barrier crest leading to *washover deposition* on the *backslope* and shoreward movement of the barrier. The balance between overtopping and overwashing controls the stability and migration of *gravel barriers* whereas, in the case of *sand barriers*, AEOLIAN deposition may be the more effective control.   *JAM*

Laudier NA, Thornton EB and MacMahon H (2011) Measured and modelled water overtopping on a natural beach. *Coastal Engineering* 58: 815–825.
Orford JD and Carter RWG (1982) The structure and origins of recent sandy gravel overtopping and overwashing features at Carnsore Point, southeast Ireland. *Journal of Sedimentary Petrology* 52: 265–278.
Roelvink D, Reniers A, van Dongeren A et al. (2009) Modelling storm impacts on beaches, dunes and barrier islands. *Coastal Engineering* 56: 1133–1152.
Switzer AD and Jones BG (2008) Large-scale overwash sedimentation in a freshwater lagoon from the southeast Australian coast: Sea-level change, tsunami or exceptionally large storm? *The Holocene* 18: 787–803.

overturning   Complete water circulation that destroys LAKE STRATIFICATION AND ZONATION, and is caused by the water density differential breaking down during autumn cooling or ice melting in spring.   *HHB*

[*See also* HOLOMICTIC LAKE, LIMNOLOGY, MERIDIONAL OVERTURNING CIRCULATION (MOC), MEROMICTIC LAKE]

Balistrieri LS, Tempel RW, Stillings LC and Shevenell LA (2006) Modeling spatial and temporal variations in temperature and salinity during stratification and overturn in Dexter Pit Lake, Tuscarora, Nevada, USA. *Applied Geochemistry* 21: 1184–1203.

oxidation   The chemical definition is the increase in the positive valence or the decrease in the negative valence of an element, or the loss of an *electron* from an element: ferrous iron (Fe^{2+}), for example, is oxidised to ferric (Fe^{3+}) iron. In CHEMICAL WEATHERING, this commonly means the combination of a mineral with oxygen to form *oxides* or *hydroxides* and usually involves atmospheric or soil oxygen dissolved in water. Well-drained, AEROBIC conditions, the DECOMPOSITION of organic matter and relatively high temperatures all favour this important process of chemical weathering with IRON (Fe), *manganese, sulfur* and *titanium* being among the major elements involved.   *RPDW*

[*See also* REDUCTION]

Brunsden D (1979) Weathering. In Embleton C and Thornes JB (eds) *Process in geomorphology.* London: Arnold, 73–129.

oxygen   A colourless, odourless, two-ATOM *gas* (O_2), the third most abundant ELEMENT in the *universe* (after *hydrogen* and *helium*), the most abundant in the Earth's CRUST (comprising almost 50 per cent by weight), and the second most abundant in the ATMOSPHERE (after *nitrogen*). The element is formed as part of the *fuel cycle* that provides the energy of *stars*. The Earth's crust is mainly SILICATE MINERALS (MINERALS of SILICON (Si) and oxygen). Earth is the only PLANET with an oxygen-rich atmosphere, which was produced by plants as a by-product of PHOTOSYNTHESIS. Oxygen is essential for all forms of life as it is a constituent of DNA, CARBON DIOXIDE and *water*, and it is required by all *plants* and *animals* for RESPIRATION. Oxygen is a highly reactive gas, permitting COMBUSTION and forming *oxides* with almost all other elements. It is also used widely in industry where most commercially produced oxygen is used in the manufacture of steel (>50 per cent) and chemicals (~25 per cent).   *JAM*

[*See also* ATMOSPHERIC COMPOSITION, BIOCHEMICAL OXYGEN DEMAND (BOD), EARTH REVOLUTIONS, OXYGEN CYCLE, OXYGEN ISOTOPES, OXYGEN VARIATIONS, OZONE, SILICA]

Emsley J (2001) *Nature's building blocks: An A-Z guide to the elements.* Oxford: Oxford University Press.

Lane N (2002) *Oxygen: The molecule that made the world.* Oxford: Oxford University Press.

oxygen cycle The BIOGEOCHEMICAL CYCLE that describes the movement of oxygen within and between its three main reservoirs: (1) the ATMOSPHERE, (2) the BIOSPHERE and (3) the LITHOSPHERE. The interaction of biological, chemical and physical processes on and beneath the Earth's surface determines the concentration of oxygen and OXYGEN VARIATIONS through time. OXYGEN gas is one of the major COMPOUNDS found in the atmosphere and is a component of *coupled biogeochemical cycles* of many other elements (see CARBON CYCLE, IRON CYCLE, NITROGEN CYCLE, PHOSPHORUS CYCLE, SILICON CYCLE, SULFUR CYCLE).

Almost all living things need oxygen as it is used in the process of creating energy in living cells. PHOTOSYNTHESIS by *land plants* and PHYTOPLANKTON in LAKES and OCEANS drives the oxygen cycle by producing oxygen and *sugars* from CARBON DIOXIDE and *water*. Photosynthesis accounts for most of the atmospheric oxygen but a small amount of it is produced from the breakdown of atmospheric water and *nitrous oxide* by *ultraviolet radiation* in a process called *phytolysis*.

Oxygen is removed from the atmosphere by aerobic RESPIRATION and decay as organisms consume oxygen and release CO_2. Oxidation of *sulfide minerals*, the formation of *iron oxides* and the OXIDATION of reduced VOLCANIC GASES also consume atmospheric oxygen. The lithosphere is the main reservoir or SINK of oxygen, with the oxygen held in a form that is unavailable for use by organisms within the *silicate minerals* and *oxide minerals*. LIMESTONE, for example, is rich in oxygen as it is composed of the *calcium carbonate* shells of *marine organisms*. WEATHERING processes release the oxygen from the lithosphere.

Waterbodies such as oceans, lakes and rivers contain large amounts of *dissolved oxygen* as oxygen from the atmosphere can transfer into water very quickly due to the movement of the water. Gas solubility is temperature dependent and so oxygen concentrations are greater in colder, high-latitude surface waters than in waters near the Equator. *Aquatic organisms* breathe the dissolved oxygen by filtering it out of solution and the dissolved oxygen is replenished by the process of RE-AERATION. As organisms are constantly using up the oxygen in the water and oxygen is constantly re-entering the water from the air, the amount of dissolved oxygen remains relatively constant. However, if consumption exceeds supply, such as during an ALGAL BLOOM, ANOXIA can develop and the oxygen cycle fails.

Small changes in oxygen PRODUCTION and CONSUMPTION, resulting in large fluxes of oxygen to and from the atmosphere, have the potential to generate large shifts in atmospheric oxygen concentration over geologically short periods of time. Although oxygen levels have varied over geological time due to an imbalance of oxygen production and consumption, evidence suggests that oxygen levels were stable over wide spans of geological time; the reasons for this are still poorly understood. OXYGEN ISOTOPES derived from palaeo records (ICE CORES, LACUSTRINE SEDIMENTS, MARINE SEDIMENTS, etc.) can be used to infer changes in the oxygen cycle and to reconstruct past climates. *KJF*

Berner RA, Petsch ST, Lake JA et al. (2000) Isotope fractionation and atmospheric oxygen: Implications for Phanerozoic O_2 evolution. *Science* 287: 1630–1633.

Bianucci L and Denman KL (2012) Carbon and oxygen cycles: Sensitivity to changes in environmental forcing in a coastal upwelling system. *Journal of Geophysical Research-Biosciences* 117: G01020.

Heimann M (2001) The cycle of atmospheric molecular oxygen and its isotopes. In Schulze E-D, Heimann M, Harrison S et al. (eds) *Global biogeochemical cycles in the climate system.* San Diego, CA: Academic Press, 235–244.

Lasaga AC and Ohmoto H (2002) The oxygen geochemical cycles: Dynamics and stability. *Geochimica et Cosmochimica Acta* 66: 361–381.

Najjar RG and Keeling RF (2000) Mean annual cycle of the air-sea oxygen flux: A global view. *Global Biogeochemical Cycles* 14: 573–584.

Palastanga V, Slomp CP and Heinze C (2011) Long-term controls on ocean phosphorus and oxygen in a global biogeochemical model. *Global Biogeochemical Cycles* 25: GB3024.

Petsch ST (2005) The global oxygen cycle. In Schlesinger WH (ed.) *Biogeochemistry.* Amsterdam: Elsevier, 515–555.

oxygen isotopes Oxygen has three STABLE ISOTOPES (natural abundances, ^{16}O = 99.762 per cent, ^{17}O = 0.038 per cent and ^{18}O = 0.200 per cent). The ISOTOPE RATIO is expressed as $^{18}O/^{16}O$ relative to the *Vienna Pee Dee Belemnite (VPDB) standard* for CARBONATES and the *Vienna Standard Mean Ocean Water (VSMOW) standard* for all other samples. The $\delta^{18}O$ value of FORAMINIFERA in deep-ocean sediments is influenced primarily by global continental ice volume and thus reflects global climate (see Figure). The age of these sediments cannot be dated directly beyond the range of RADIOCARBON DATING: however, ages have been determined indirectly by matching properties of the sediments with changes in ORBITAL FORCING indicated by the MILANKOVITCH THEORY. A composite 780 ka record obtained from five MARINE SEDIMENT CORES forms the basis of the SPECTRAL MAPPING PROJECT TIMESCALE (SPECMAP) (the standard against which the generally fragmentary terrestrial records have been compared). Thus, MARINE ISOTOPIC STAGES (MIS) have been assigned to GLACIAL and INTERGLACIAL episodes.

The terrestrial record of past climate includes the physico-chemical properties of ICE CORES taken from Antarctica, the Greenland ice-sheet and high-altitude ice caps. Oxygen isotopes from the EPICA Droning Maud Land (EDML) ice core in Antarctica are directly coupled with the *North Greenland Ice Core Project* (NGRIP) ice core through the Atlantic MERIDIONAL OVERTURNING CIRCULATION (MOC). The DECADAL-SCALE VARIABILITY and CENTURY- TO MILLENNIAL-SCALE VARIABILITY in the EDML record can be unambiguously linked to corresponding DANSGAARD-OESCHGER (D-O) EVENTS in NGRIP after methane SYNCHRONISATION of age scales. In terrestrial plants, the ISOTOPE composition of *xylem water* is identical to the water absorbed by the roots. Mixing occurs as water ascends the plant to the site of PHOTOSYNTHESIS through the *apoplastic pathway* or the slower *symplastic pathway*. However, there is no ISOTOPIC FRACTIONATION until the water reaches tissues undergoing water loss, where EVAPORATION causes ISOTOPIC ENRICHMENT of ^{18}O in the remaining water. The ISOTOPE RATIO of *leaf water* is imparted to the intercellular CARBON DIOXIDE utilised in carbohydrate synthesis. *Biochemical fractionation* during photosynthesis causes an enrichment of ^{18}O of the resulting *carbohydrate*. Post-photosynthetic isotopic exchange occurs between the oxygen ATOMS of intermediate carbohydrates and *metabolic water*, with the extent of the exchange determined by *carbohydrate recycling*. Despite this complexity, the $\delta^{18}O$ values of organic matter have been used to reconstruct past climates using a TRANSFER FUNCTION or mechanistic approach.   *IR*

[*See also* CORAL, LACUSTRINE SEDIMENTS, OSTRACOD ANALYSIS, PEAT AND PEATLANDS, SPELEOTHEMS]

Craig H (1961) Isotopic variations in meteoric waters. *Science* 133: 1702–1703.

Daley TJ, Barber KE, Street-Perrott FA et al. (2010) Holocene climate variability revealed by Sphagnum cellulose oxygen isotope analyses from Walton Moss, northern England. *Quaternary Science Reviews* 29: 1590–1601.

EPICA Community Members (2006) One-to-one coupling of glacial climate variability in Greenland and Antarctica. *Nature* 444: 195–198.

Farquhar GD, Barbour MM and Henry BK (1998) Interpretation of oxygen isotope composition of leaf material. In Griffiths H (ed.) *Stable isotopes: Integration of biological, ecological and geochemical processes*. Oxford: Bios Scientific Publishers, 27–62.

Imbrie J, Hays JD, Martinson DG et al. (1984) The orbital theory of Pleistocene climate: Support from a revised chronology of the marine $\delta^{18}O$ record. In Berger A, Imbrie J, Hays J et al. (eds) *Milankovitch and climate: Understanding the response to astronomical forcing. Part I*. Dordrecht: Reidel, 269–305.

Shackleton NJ and Opdyke ND (1973) Oxygen isotope and palaeomagnetic stratigraphy of equatorial Pacific core

V28-238: Oxygen isotope temperatures and ice volumes on a 10^5 year and 10^6 year scale. *Quaternary Research* 3: 39–55.

Treydte K, Schleser GH, Helle G et al. (2006) Millennium-long precipitation record from treering oxygen isotopes in northern Pakistan. *Nature* 440: 1179–1182.

Wilson RCL, Drury SA and Chapman JL (2000) *The great ice age: Climate change and life*. London: Routledge.

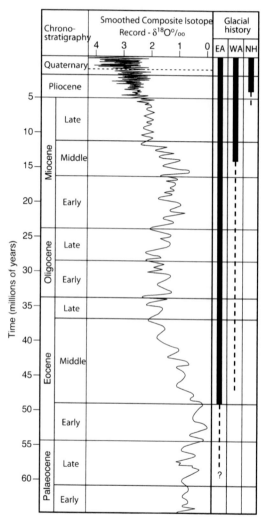

Oxygen isotopes Composite oxygen isotope record for the Cenozoic derived from marine sediment cores: the columns on the right indicate the presence of the East Antarctic Ice Sheet (EA) from ca 50 Ma, the West Antarctic Ice Sheet (WA) from ca 15 Ma and ice sheets in the Northern Hemisphere (NH) from ca 5 Ma, based on the occurrence of ice-rafted debris (Wilson et al., 2000).

oxygen variations   The current ambient concentration of atmospheric oxygen (O_2) is ~21 per cent. The stability of this O_2 level relies on a critical balance between the CONSUMPTION of O_2 by *aerobic respiration*,

COMBUSTION and other processes and its PRODUCTION from oxygenic PHOTOSYNTHESIS and *photolytic dissociation* (break down of water and CARBON DIOXIDE molecules in the Earth's atmosphere by high-energy *ultraviolet radiation*). These essential processes have not been balanced over the whole of geological time and, as a result, O_2 concentrations have varied markedly. The earliest *prebiotic atmosphere* was devoid of O_2 as the small quantities produced by photolytic dissociation were rapidly consumed by *ferrous iron* and reduced VOLCANIC GASES. As the main source of atmospheric O_2 is from plant and algal photosynthesis, the evolution of photosynthesising life on land in conjunction with increased *carbon burial* and a reduction in the supply of reduced material to the ATMOSPHERE resulted in the *oxygenation* of the atmosphere at the end of the ARCHAEAN and into the PROTEROZOIC. Models of oxygen variations indicate that atmospheric O_2 levels have remained between a minimum of 15 per cent and a maximum of 35 per cent throughout the Phanerozoic, with a large spike of 35 per cent occurring between about 350 and 250 Ma, attributed to excessive carbon burial. *JCMcE*

[*See also* ATMOSPHERIC COMPOSITION, EARTH REVOLUTIONS, OXIDATION, OXYGEN ISOTOPES, REDUCTION]

Belcher CM and McElwain JC (2008) Limits for combustion in low O_2 redefine paleoatmospheric predictions for the Paleozoic. *Science* 321: 1197–1200.

Bergman NM, Lenton TM and Watson AJ (2004) COPSE: A new model of biogeochemical cycling over Phanerozoic time. *American Journal of Science* 304: 397–437.

Berner RA (2006) GEOCARBSULF: A combined model of Phanerozoic almospheric O_2 and CO_2. *Geochimica et Cosmochimica Acta* 70: 5653–5664.

Berner RA, Beerling DJ, Dudley R et al. (2003) Phanerozoic atmospheric oxygen. *Annual Review of Earth and Planetary Sciences* 31: 105–134.

Berner RA and Canfield DE (1989) A new model for atmospheric oxygen over Phanerozoic time. *American Journal of Science* 289: 333–361.

Chaloner WG (1989) Fossil charcoal as an indicator of palaeoatmospheric oxygen level. *Journal of the Geological Society* 146: 171–174.

Lovelock JE and Whitfield M (1982) Life span of the biosphere. *Nature* 296: 561–563.

ozone   Ozone (O_3) is a form of oxygen (O_2) which contains three atoms rather than the two atoms of ordinary oxygen. It is constantly being formed and destroyed in the STRATOSPHERE by a range of natural processes including the absorption of *ultraviolet radiation* by oxygen and ozone, respectively, and by subsequent collision processes. The relatively high concentration of ozone in the stratosphere, with maximum concentrations between 20 and 25 km, is referred to as the *ozone layer*. Since the late 1970s, significant stratospheric OZONE DEPLETION over the ANTARCTIC each

Southern Hemispheric spring has created what is commonly referred to as the *ozone hole*. It is less prominent and more variable over the ARCTIC.

In the 1970s, widespread international concern for OZONE DEPLETION centred around CHLOROFLUOROCARBONS (CFCs). These inert, non-toxic and inexpensive chemicals became widely used as a cooling agent for refrigerators and air conditioners, a propellant in aerosol sprays, a solvent in the electronics industry and a blowing agent to create packing and insulating foams. When a long-lived CFC molecule reaches the stratosphere, it is decomposed by ultraviolet radiation to produce an unattached chlorine atom. The chlorine atom initiates an ozone-destroying reaction sequence before emerging unchanged and capable of repeating the reaction many thousands of times before eventually being removed by other substances. Other ozone-depleting substances that act as CATALYSTS in this way include *halons* (containing variously bromine, chlorine and fluorine), *carbon tetrachloride*, *methyl chloroform*, *methyl bromide* and even *hydrochlorofluorocarbons* (*HCFCs*), which were introduced as substitutes for CFCs as their ozone-depletion potential was much less.

Stratospheric ozone depletion by CFCs and other compounds was first hypothesised in the early 1970s but the proof did not come until 1985 when measurements above Halley Bay, on the Antarctic coast, revealed that large depletions had occurred there each spring since the late 1970s. By October 1987, ozone concentrations were around half the levels in the 1970s and they have continued to fall since then. Concern over ozone depletion resulted in an international CONVENTION, the 1987 *Montreal Protocol on Substances Which Deplete the Ozone Layer*, to begin phasing out the production and consumption of ozone-depleting substances. Subsequent amendments to the Protocol have forwarded this process.

Loss of stratospheric ozone results in more *ultraviolet radiation* reaching the surface and may lead to an increase in sunburn and *skin cancer*, reduction in some *crop yields*, damage to oceanic PLANKTON and DEGRADATION of many materials such as paints and fabrics. Given the long *atmospheric residence time* of CFCs, the depletion of ozone is expected to last for several decades. *DME/PU/CW*

Bodeker GE, Shiona H and Eskes H (2005) Indicators of Antarctic ozone depletion. *Atmospheric Chemistry and Physics* 5: 2603–2615.

Gégo E, Porter PS, Gilliland A and Rao ST (2007) Observation-based assessment of the impact of nitrogen oxides emissions reductions on ozone air quality over the Eastern United States. *Journal of Applied Meteorology and Climatology* 46: 994–1008.

Hart M, De Dear R and Hyde R (2006) A synoptic climatology of tropospheric ozone episodes in Sydney, Australia. *International Journal of Climatology* 26: 1635–1649.

Martens P (2009) *Health and climate change: Modelling the impacts of global warming and ozone depletion.* London: Earthscan.

Metz B, Kuijpers L, Solomon S et al. (eds) (2005) *Safeguarding the ozone layer and the global climate system: Issues related to hydrofluorocarbons and perfluorocarbons.* Cambridge: Cambridge University Press.

Rowland FS (2009) Stratospheric ozone depletion. In Zerefos C, Contopoulos G and Skalkeas G (eds) *Twenty years of ozone decline: Proceedings of the Symposium for the 20th Anniversary of the Montreal Protocol.* Berlin: Springer Verlag, 23–66.

Urbach F (1989) Potential effects of altered solar ultraviolet radiation on human skin cancer. *Photochemistry and Photobiology* 50: 507–514.

Weatherhead B, Tanskanen A and Stevermer A (2005) Ozone and ultraviolet radiation. In Arctic Climate Impact Assessment (eds) *ACIA scientific report.* Cambridge: Cambridge University Press, 151–182.

World Meteorological Organization (WMO) (2011) S*cientific assessment of ozone depletion: 2010.* Geneva: WMO.

ozone depletion   The natural and pollution-forced decrease in the OZONE (O_3) layer, which is found in the STRATOSPHERE at heights between 20 and 40 km (see Figure). Ozone is maintained by a series of chemical and photochemical reactions involving atomic (O) and molecular (O_2) oxygen, other gases present and short-wave *ultraviolet* (UV) SOLAR RADIATION. These reactions are usually shown in the following equations:

$$O_2 + UV \rightarrow 2O$$
$$O_2 + O + M \rightarrow O_3 + M$$
$$O_3 + UV \rightarrow O_2 + O$$
$$O + O_3 \rightarrow 2O_2$$

where M is any other gas molecule that acts as a catalyst, causing ozone to be produced and destroyed. Ozone is environmentally important because of the harmful effects of UV on most life forms. The ozone layer shields the Earth's surface from this radiation. If M is a POLLUTANT (e.g. a CHLOROFLUOROCARBON, represented by $CFCL_3$), three other reactions occur continuously and ozone destruction is accelerated:

$$CFCL_3 + UV \rightarrow CL + CFCL_2$$
$$CL + O_3 \rightarrow CLO + O_2$$
$$CLO + O \rightarrow CL + O_2$$

There is a seasonal variation in ozone, particularly a diminution in late winter and early spring due to the lack of ultraviolet radiation in that season. The loss over Antarctica is more prominent and due partly to the strength of the circumpolar WESTERLIES, which prevent ozone from being imported from lower latitudes. In the late 1970s and early 1980s, it was noticed that the spring depletions were increasing in amount (by up to 70 per cent); this was, and is, the *ozone hole.*

In 1993, the decrease was 99 per cent. Loss of ozone over the Arctic is generally less and more variable but in 2011 exceeded 80 per cent for the first time. It is now accepted that CFCs are mainly responsible for the depletion. The production of CFCs was sharply limited by the Montreal Protocol (see CONVENTION) which, by 2010, had been ratified by 196 states. Recent assessments of ozone depletion suggest that ozone will reach pre-1980 levels outside the polar regions by about 2050 but in polar regions it may be later.   *BDG*

[*See also* HALOGENATED HYDROCARBONS (HALOCARBONS)]

Anderson SO and Madhava Sarma K (2002) *Protecting the ozone layer: The United Nations history.* London: Earthscan.

Christie M (2001) *The ozone layer: A philosophy of science perspective.* Cambridge: Cambridge University Press.

Crutzen PJ (1974) Estimates of possible variations in total ozone due to natural causes and human activities. *Ambio* 3: 201–210.

Dessler A (2000) *Chemistry and physics of stratospheric ozone.* London: Academic Press.

Farman JC, Gardiner BG and Shanklin JD (1985) Large losses of total ozone in Antarctica reveal ClO_x/NO_x interaction. *Nature* 315: 207–210.

Gribbin J (1993) *The hole in the sky, 2nd edition.* New York: Bantam Books.

Manney GL, Santee ML, Rex M et al. (2011) Unprecedented Arctic ozone loss in 2011. *Nature* 478: 469–475.

Molina MJ and Rowland FS (1974) Stratospheric sink for chlorofluoromethanes: Chlorine atom catalysed destruction of ozone. *Nature* 249: 810–812.

Noone KJ (2012) Human impacts on the atmosphere. In Matthews JA, Bartlein PJ, Briffa KR et al. (eds) *The SAGE handbook of environmental change, volume 2: Human impacts and responses.* London: Sage, 95–110.

United Nations Environment Programme (UNEP), Ozone Secretariat (2010) *The 2010 assessment of the Scientific Assessment Panel.* Nairobi: UNEP. [Available at http://ozone.unep.org/Assessment_Panels/SAP/Scientific_Assessment_2010/index.shtml]

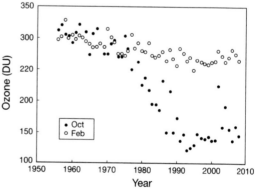

DU = Dobson unit.

Ozone depletion Mean ozone levels at Halley Research Station, Antarctica, for the months of February and October since AD 1950 (Noone, 2012).

Pacific Decadal Oscillation (PDO) A CLIMATIC OSCILLATION in SEA-LEVEL TEMPERATURE and sea-level pressure in the northern Pacific Ocean. The PDO exhibits strong multidecadal persistence with two full cycles in the twentieth century: relatively cool phases in AD 1890–1924 and AD 1947–1976, and warm phases in 1925–1946 and from 1977 onwards. PDO amplitudes are weaker than those for EL NIÑO-SOUTHERN OSCILLATION (ENSO) in the tropics but stronger in extratropical latitudes, where there are significant effects on the climate of Alaska and northwestern Canada.   *JAM*

[*See also* ATMOSPHERE-OCEAN INTERACTION, CLIMATIC MODES]

MacDonald GM and Case RA (2005) Variations in the Pacific Decadal Oscillation over the past millennium. *Geophysical Research Letters* 32(8): L08703.
Mantua NJ, Hare SR, Zhang Y et al. (1997) A Pacific interdecadal climate oscillation with impacts on salmon production. *Bulletin of the American Meteorological Society* 78: 1069–1079.

Pacific-North American (PNA) teleconnection A large-scale TELECONNECTION between the climates of the North Pacific Ocean and the continent of North America, which is apparent in the 500 mb geopotential height field. The *PNA Index* is a weighted average of 500-mb normalised height anomaly differences between the centres of four distinct cells over North Pacific, Hawaii, Alberta and the southeastern USA. This is correlated with the EL NIÑO-SOUTHERN OSCILLATION (ENSO) and accounts for some of the DECADAL-SCALE VARIABILITY in the northwestern Pacific coastal salmon fisheries. The PNA and its index cover a larger region than the *North Pacific Oscillation* and the corresponding *North Pacific Index*, which is based on areally averaged sea-level pressure over a large area of the North Pacific Ocean. The PACIFIC DECADAL OSCILLATION (PDO) and the *West Pacific Oscillation* are other relatively small North Pacific CLIMATIC MODES.
                          JAM

Leathers DJ, Yarnal B and Palecki MA (1991) The Pacific/ North American teleconnection pattern and United States climate. Part I: Regional temperature and precipitation associations. *Journal of Climate* 4: 517–528.
Mantua NJ, Hare SR, Zhang Y et al. (1997) A Pacific interdecadal climate oscillation with impacts on salmon production. *Bulletin of the American Meteorological Society* 78: 1069–1079.
National Research Council (1999) *Global environmental change: Research pathways for the next decade.* Washington, DC: National Academies Press.

pack ice SEA ICE floating free under the influence of currents and wind. It comprises *ice floes* centred on relatively resistant ice; *leads* of open water, which may be subject to rapid freezing in winter; and chaotically fractured and crumpled ice-forming features (*pressure ridges* or *keels*).   *RAS*

Hanna E (1996) The role of Antarctic sea ice in global climate change. *Progress in Physical Geography* 20: 371–401.

paddy soils The soils used in WETLAND *rice cultivation*. The management of these soils requires PUDDLING, which destroys all surface attributes in the upper 25–40 cm of the soil. These soils are artificially seasonally flooded, and the anthropogenic influences significantly alter the innate SOIL-FORMING PROCESSES. The combination of a puddle layer at the surface with a compacted plough pan reduces the rate of SOIL DRAINAGE and conserves water. The ANAEROBIC conditions produced may result in losses of *nitrogen* and EMISSIONS of *methane*. These soils are difficult to classify because of the overriding influence of the soil management, but most should be considered ANTHROSOLS.   *SN*

[*See also* EARLY-ANTHROPOCENE HYPOTHESIS]

Kyuma K (2004) *Paddy soil science.* Melbourne: Trans Pacific Press.
Wassman R, Lantin RS and Neue H-U (eds) (2000) *Methane emissions from major rice ecosystems in Asia.* Dordrecht: Kluwer.

Witt C and Haefele SM (2005) Paddy soils. In Hillel D (ed.) *Encyclopedia of soils in the environment.* Amsterdam: Elsevier, 141–150.

pahoehoe
A Hawaiian term for *ropy lava*, used to describe a LAVA FLOW with a smooth, undulating surface, caused by the deformation of a surface crust formed over a fast-moving, low-viscosity BASALT flow.   *GO*

[*See also* AA]

Guest JE, Duncan AM, Stofan ER and Anderson SW (2012) Effects of slope on development of pahoehoe flow fields: Evidence from Mount Etna. *Journal of Volcanology and Geothermal Research* 219: 52–62.

palaeobiology
The study of FOSSILS as organisms from a biological perspective, including studies of TAXONOMY and EVOLUTION. This contrasts with the use of fossils in PALAEOECOLOGY and STRATIGRAPHY although there is much overlap.   *GO*

[*See also* PALAEONTOLOGY]

Benton MJ and Harper DAT (2009) *Introduction to paleobiology and the fossil record.* Chichester: Wiley-Blackwell.
Briggs DEG and Crowther PR (eds) (2001) *Palaeobiology II.* Malden, MA: Blackwell.
Brusatte SL (2012) *Dinosaur paleobiology.* Chichester: Wiley-Blackwell.
Hodkinson TR, Jones MB, Waldren S and Parnell JAN (eds) (2011) *Climate change, ecology and systematics.* Cambridge: Cambridge University Press.
Sepkoski D and Ruse M (eds) (2009) *The paleobiological revolution: Essays on the growth of modern paleontology.* Chicago: University of Chicago Press.

palaeobotany
The study of FOSSIL plants. Palaeobotany is a wide field of science involving many disciplines including PALAEOCLIMATOLOGY, PALAEOECOLOGY, POLLEN ANALYSIS, *palaeofloristics*, plant EVOLUTION, *phylogeny* and ONTOGENY, BIOSTRATIGRAPHY and BIOGEOGRAPHY to name a few. Traditionally, palaeobotany research was concerned with the evolution and phylogeny of different plant groups from studies of the plant fossil record. However, over the past century, the use of fossil plants as palaeoenvironmental, palaeoclimatic and palaeoatmospheric indicators has been realised and widely utilised. Examples include the use of LEAF PHYSIOGNOMY, *fossil wood* characteristics and stable CARBON ISOTOPE composition as indicators of palaeoclimate and the use of fossil leaf STOMATA and fossil CHARCOAL as indicators of palaeoatmospheric CARBON DIOXIDE VARIATIONS and OXYGEN VARIATIONS, respectively. *Palaeofloristics*, the study of assemblages of fossil plants in space and time, is another important facet of palaeobotany which has provided evidence of PLATE TECTONICS, CONTINENTAL DRIFT and CLIMATIC CHANGE.   *JCMcE*

[*See also* CHARRED-PARTICLE ANALYSIS, PALAEOBIOLOGY, PALAEONTOLOGY, STOMATAL ANALYSIS, WOOD ANATOMY]

Taylor TN, Taylor EL and Krings M (2009) *Paleobotany: The biology and evolution of fossil plants.* Cambridge: Cambridge University Press.
Willis KJ and McElwain JC (2002) *The evolution of plants.* Oxford: Oxford University Press.

palaeocean modelling
The use of computerised numerical MODELS to reconstruct past OCEAN CIRCULATION (OCEAN PALAEOCIRCULATION) and ocean BIOGEOCHEMICAL CYCLES. Three-dimensional *ocean general circulation models* (OGCMs) are used to predict the horizontal and vertical velocity of OCEAN CURRENTS and the TEMPERATURE and SALINITY gradients for the full depth of the oceans. They consist of realistic *physical oceanographic processes*, a representation of continental boundaries, layers in the ocean-depth domain and a surface spatial resolution range of between kilometres to a few degrees, dependent upon the objectives of the experiment and the available computational power. In palaeocean modelling, these models are run for particular time slices (e.g. the LAST GLACIAL MAXIMUM (LGM)) in the GEOLOGICAL RECORD and are forced with realistic boundary conditions using observations and PROXY DATA. Model outputs are then tested against further modern observations and proxy data. OGCMs can have tracers (see TRACERS: IN THE OCEANS) added, such as stable oxygen isotope ratios, to trace the evolution of water masses, for example, fresh water fluxes into the oceans. Fully coupled climate models (*Earth system models*) have three-dimensional ocean models coupled with atmospheric models, sea ice and land and vegetation models for investigating ATMOSPHERE-OCEAN INTERACTIONS. These models can be forced, for example, with variations in GREENHOUSE GASES or global ice volumes. *Ocean biogeochemical models* are used to investigate interactions between biology and ocean chemistry in the fossil record, such as the impact of variations in atmospheric CARBON DIOXIDE concentrations on the surface water chemistry and biology. Biogeochemical ocean models consist of biological *functional types*, such as the different sizes of PHYTOPLANKTON groups, and NUTRIENTS in realistic interactions, such as PHOTOSYNTHESIS, and are often coupled with OGCMs to investigate PALAEOCEANOGRAPHY and PALAEOCLIMATOLOGY.   *JP*

[*See also* CLIMATIC MODELS, COUPLED OCEAN-ATMOSPHERE MODELS, GENERAL CIRCULATION MODELS (GCMs)]

Knorr G and Lohmann G (2003) Southern Ocean origin for the resumption of Atlantic thermohaline circulation during deglaciation. *Nature* 424: 532–536.

Renssen H, Goosse H, Fichefeti T et al. (2005) Holocene climate evolution in the high-latitude Southern Hemisphere simulated by a coupled atmosphere-sea ice-ocean-vegetation model. *The Holocene* 15: 951–964.

Ridgwell A and Schmidt DN (2010) Past constraints on the vulnerability of marine calcifiers to massive carbon dioxide release. *Nature Geoscience* 3: 196–200.

Tagliabue A, Bopp L, Roche DM et al. (2009) Quantifying the roles of ocean circulation and biogeochemistry in governing ocean carbon-13 and atmospheric carbon dioxide at the Last Glacial maximum. *Climate of the Past* 5: 695–706.

palaeoceanography The study of the physical, chemical and biological characteristics of the OCEANS in the past (also see OCEANOGRAPHY). The aims are to determine former ocean characteristics, including OCEAN BASIN size and shape, past OCEAN CURRENT patterns and WATER MASS behaviour and changes in *sea level*. Palaeoceangraphy also investigates past events in the oceans, such as OCEANIC ANOXIC EVENTS, UPWELLING, DESICCATION and ICEBERG discharge and melting events (HEINRICH EVENTS) and reconstructs past *seawater* parameters such as SALINITY, SEA-SURFACE TEMPERATURE, pH and the extent and duration of SEA ICE. Much information comes from the study of marine MICROFOSSILS, such as FORAMINIFERA, NANNOFOSSILS, DIATOMS, *dinoflagellate cysts* and RADIOLARIA, from OCEANIC SEDIMENTS obtained by DEEP-SEA DRILLING. CORAL skeletons are also commonly used for palaeoceangraphy in the low-latitude oceans.

As well as providing direct evidence of past environments using PALAEOECOLOGY, the microfossils and coral skeletons provide PROXY DATA via isotopic and geochemical analysis of their skeletons. For example, the measurement of stable oxygen ISOTOPE RATIOS in the shells of foraminifera are used to reconstruct past seawater temperatures and the volume of ice stored in the terrestrial ice sheets. In CENOZOIC (*CAINOZOIC*) ocean sediments, particularly those of the QUATERNARY, these stable isotope measurements are used to define MARINE ISOTOPIC STAGES (MIS) and *isotope stratigraphies* that allow the dating of the sediments and identification of past glacial and interglacial climates. The past abundance of marine diatom skeletons and dinoflagellate cysts in high-latitude ocean sediments has been used to reconstruct past sea-ice extent and duration. Measurements of the organic *carbon content*, calcium CARBONATE content, *biogenic silica* (*biogenic opal*) content and various other chemical ELEMENT abundances in ocean sediments also provide information about past ocean PRIMARY PRODUCTIVITY and sea-floor dissolution events. Rapid changes in the percentage of calcium carbonate in deep ocean sediments have been used to identify variation in the position of the calcium CARBONATE COMPENSATION DEPTH (CCD) in relation to deep OCEAN ACIDIFICATION.

The reconstruction of past ocean characteristics allows the investigation of long-term relationships between ocean circulation and climate, such as the past record of processes such as EL NIÑO-SOUTHERN OSCILLATION (ENSO) and GLACIAL-INTERGLACIAL CYCLES relating to MILANKOVITCH THEORY. For example, the ratios of magnesium to calcium and strontium to calcium (Mg/Ca and Sr/Ca) in annual growth rings of coral skeletons from the Mekong Delta, South China Sea, have been used to reconstruct ENSO and the East Asian MONSOON variability in the past. The timing of changes in sea level during past INTERGLACIALS, determined using URANIUM-SERIES DATING of coral skeletons from the South Pacific, has been used to test the validity of the Milankovitch theory in forcing climate changes. Other data sources for palaeoceanography, including the study of magnetic stripes from SEA-FLOOR SPREADING, and the known age-depth relationship for cooling OCEANIC CRUST (see MID-OCEAN RIDGE (MOR)), are used to reconstruct ocean-basin morphology. Most palaeoceanography is carried out on MESOZOIC and CENOZOIC (*CAINOZOIC*) sediments that are still found on the ocean floor, but many of the methods can be applied to older SEDIMENTARY ROCKS preserved on land. *JP/GO*

[*See also* ATMOSPHERE-OCEAN INTERACTION, CORAL AND CORAL REEFS: ENVIRONMENTAL RECONSTRUCTION, ICE-RAFTED DEBRIS (IRD), ISOTOPES AS INDICATORS OF ENVIRONMENTAL CHANGE, MAGNESIUM:CALCIUM RATIO (Mg:Ca), ORGANIC CONTENT, QUATERNARY TIMESCALE, SILICA]

Allen CS, Pike J and Pudsey CJ (2011) Last Glacial-Interglacial sea-ice cover in the SW Atlantic and its potential role in global deglaciation. *Quaternary Science Reviews* 30: 2446–2458.

Andersen MB, Stirling CH, Potter E-K et al. (2010) The timing of sea-level high-stands during Marine Isotope Stages 7.5 and 9: Constraints from the uranium-series dating of fossil corals from Henderson Island. *Geochimica et Cosmochimica Acta* 74: 3598–3620.

Barker S, Cacho I, Benway H and Tachikawa K (2005) Planktonic foraminiferal Mg/Ca as a proxy for past oceanic temperatures: A methodological overview and data compilation for the Last Glacial Maximum. *Quaternary Science Reviews* 24: 821–834.

Cartapanis O, Tachikawa K and Bard E (2011) Northeastern Pacific oxygen minimum zone variability over the past 70 kyr: Impact of biological production and oceanic ventilation. *Paleoceanography* 26: PA4208.

Christiansen JL and Stouge S (1999) Oceanic circulation as an element in palaeogeographical reconstructions: The Arenig (Early Ordovician) as an example. *Terra Nova* 11: 73–78.

De Vernal A, Eynaud F, Henry M et al. (2005) Reconstruction of sea-surface conditions at middle to high latitudes of the Northern Hemisphere during the Last Glacial Maximum (LGM) based on dinoflagellate cyst assemblages. *Quaternary Science Reviews* 24: 897–924.

Hsü KJ, Garrison RE, Montadert L et al. (1977) History of the Mediterranean salinity crisis. *Nature* 267: 399–403.

Mitsuguchi T, Dang PX, Kitagawa H et al. (2008) Coral Sr/Ca and Mg/Ca records in Con Dao Island off the Mekong Delta: Assessment of their potential for monitoring ENSO and East Asian monsoon. *Global and Planetary Change* 63: 341–352.

Perner K, Moros M, Jennings A et al. (2013) Holocene palaeoceanographic evolution off West Greenland. *The Holocene* 23: 374–387.

Pike J, Crosta X, Maddison EJ et al. (2009) Observations on the relationship between the Antarctic coastal diatoms *Thalassiosira Antarctica* Comber and *Porosira glacialis* (Grunow) Jørgensen and sea ice concentrations during the Late Quaternary. *Marine Micropaleontology* 73: 14–25.

Shackleton NJ (2006) Formal Quaternary stratigraphy: What do we expect and need? *Quaternary Science Reviews* 25: 3458–3461.

palaeoceanography: biological proxies

Methods for reconstructing OCEAN history, for example, PALAEOTEMPERATURE, PALAEOPRODUCTIVITY and *palaeochemistry*, and the GEOLOGICAL RECORD using biological remains in the form of FOSSIL CARBONATE, SILICA or *refractory organic matter* (see ORGANIC CONTENT) and organic BIOMARKERS. The principal MICROFOSSIL groups used are FORAMINIFERA, NANNOFOSSILS, DIATOMS and *dinocysts*, but other materials, such as RADIOLARIA, *ostracods* and *icthyoliths* (fossil fish teeth and scales) also contribute valuable information. PLANKTON assemblages in the surface ocean respond to SEA-SURFACE TEMPERATURE (SST) and NUTRIENT concentrations, with distinctive combinations of species appearing and disappearing depending on the OCEAN BASIN, latitude, season, SEA-ICE cover, PRODUCTIVITY and/or UPWELLING conditions. Similarly, BENTHIC assemblages reflect the supply of organic matter to the sea floor. Changes in species composition, abundance, foraminifera *test* size and thickness provide insight into organic supply rate and ocean chemistry (see OCEAN ACIDIFICATION). Other environmental factors such as OXYGEN concentration, SALINITY and temperature also play a role in assemblage composition and cannot always be easily unravelled. Methods that include CALIBRATION of modern distributions against environmental data (e.g. TRANSFER FUNCTIONS) can provide quantitative estimates of past climate parameters, whereas deeper time biological proxies, where there is no living analogue, will be qualitative.

Organic biomarkers are fossilised biosynthates that can be traced to specific algal groups. When occurring in MARINE SEDIMENTS, biomarkers can provide insight into SEA-SURFACE TEMPERATURE (SST) and surface ocean CARBON DIOXIDE as well as OCEAN PALAEOPRODUCTIVITY. Preservation is a major issue as most organic compounds degrade rapidly. The presence of specific biomarkers is useful for distinguishing marine from terrestrial organic matter and, where preservation permits, allows reconstruction of PHYTOPLANKTON community structure which is, in turn, linked to productivity conditions. *HKC*

[*See also* ANALOGUE METHOD, CLIMATIC RECONSTRUCTION, DIATOMS/DIATOM ANALYSIS, FORAMINIFERA: PALAEOENVIRONMENTAL RECONSTRUCTIONS, MARINE PALAEOCLIMATIC PROXIES, MARINE SEDIMENT CORES, OCEAN ACIDIFICATION, OCEAN PALAEOTEMPERATURE, PALAEOCEANOGRAPHY, PALAEOCEANOGRAPHY: PHYSICAL AND CHEMICAL PROXIES, PALAEOCLIMATOLOGY]

Hillaire-Marcel C and de Vernal A (2007) Methods in Late Cenozoic paleoceanography: Introduction. In Hillaire-Marcel C and de Vernal A (eds) *Proxies in Late Cenozoic paleoceanography.* Amsterdam: Elsevier, 2–15.

Paytan A (2008) Ocean paleoproductivity. In Gornitz V (ed.) *Encyclopedia of paleoclimatology and ancient environments.* Dordrecht: Springer, 644–651.

Rosell-Melé A and McClymont EL (2007) Biomarkers as paleoceanographic proxies. In Hillaire-Marcel C and de Vernal A (eds) *Proxies in Late Cenozoic paleoceanography.* Amsterdam: Elsevier, 441–490.

palaeoceanography: physical and chemical proxies

Methods for reconstructing OCEAN history in the GEOLOGICAL RECORD using physical and chemical tracers preserved in MARINE SEDIMENTS (see the Table in the entry MARINE PALAEOCLIMATIC PROXIES). *Physical tracers* include properties of the sediment such as GRAIN SIZE, MINERAL content, DENSITY and magnetic properties (e.g. MAGNETIC SUSCEPTIBILITY) that reflect climatically controlled aspects of WEATHERING, DEEP WATER flow speed, *ice rafting* (see ICE-RAFTED DEBRIS (IRD)), AEOLIAN *dust flux* and TURBIDITE deposition as well as processes of DIAGENESIS in the PLEISTOCENE and Early CENOZOIC (*CAINOZOIC*). Various physical properties of sediments, such as magnetic susceptibility, *gamma density*, NATURAL GAMMA RADIATION, *p-wave velocity*, *electrical resistivity* and *colour reflectance* can be measured continuously at centimetre-scale using a *multisensor core logger* system. This approach provides fast, non-destructive time series (see TIME-SERIES ANALYSIS) that are widely used as chronological and palaeoceanographic tracers, including proxies for CARBONATE content/dissolution and burial history. X-RADIOGRAPHY methods can provide complementary data for assessing BIOTURBATION and depositional characteristics.

Chemical tracers include geochemical and isotopic properties of DETRITAL and BIOGENIC SEDIMENTS. These include *trace metal ratios* in foraminiferal *calcite* such as Mg/Ca, Sr/Ca (OCEAN PALAEOTEMPERATURE), Li/Ca, B/Ca (pH, ALKALINITY), Cd/Ca (PRODUCTIVITY), stable CARBON ISOTOPES in foramininiferal *calcite* ($\delta^{18}O$ and $\delta^{13}C$; temperature, ice volume and productivity), *boron isotopes* in foraminifera (ocean pH and surface ocean carbon dioxide), detrital *osmium*, STRONTIUM and *neodymium* isotopes in sediment (weathering and palaeocirculation) and ^{230}Th and ^{231}Pa (water column particle flux).   *HKC*

[*See also* CLIMATIC RECONSTRUCTION, MARINE PALAEOCLIMATIC PROXIES, MARINE SEDIMENT CORES, OCEAN ACIDIFICATION, PALAEOCEANOGRAPHY, PALAEOCEANOGRAPHY: BIOLOGICAL PROXIES, PALAEOCLIMATOLOGY, PROXY CLIMATIC INDICATOR]

Bayon G, Dennielou B, Etoubleau J et al. (2012) Internsifying weathering and land use in Iron Age Central Africa. *Science* 335: 1219–1222.

Hillaire-Marcel C and de Vernal A (2007) Methods in Late Cenozoic paleoceanography: Introduction. In Hillaire-Marcel C and de Vernal A (eds) *Proxies in Late Cenozoic paleoceanography*. Amsterdam: Elsevier, 2–15.

Kleiven HKF, Hall IR, McCave IN et al. (2011) Coupled deep-water flow and climate variability in the middle Pleistocene North Atlantic. *Geology* 39: 343–346.

McCave IN (2007) Deep-sea sediment deposits and properties controlled by currents. In Hillaire-Marcel C and de Vernal A (eds) *Proxies in Late Cenozoic paleoceanography*. Amsterdam: Elsevier, 19–62.

St-Onge G, Mulder T, Francus P and Long B (2007) Continuous physical properties of cored marine sediments. In Hillaire-Marcel C and de Vernal A (eds) *Proxies in Late Cenozoic paleoceanograph*. Amsterdam: Elsevier, 63–98.

Palaeocene   The earliest EPOCH of the PALAEOGENE period (SERIES of the PALAEOGENE SYSTEM), from 65.5 to 55.8 million years ago. At the start of the Palaeocene, many groups of organisms, notably the *mammals*, evolved to fill the vacant ecological NICHES resulting from the MASS EXTINCTION event at the end of the CRETACEOUS period. The North Atlantic began to open during the Palaeocene, and the end of the epoch is notable for the *hothouse conditions* of the *Palaeocene-Eocene Thermal Maximum*.   *GO*

[*See also* GEOLOGICAL TIMESCALE, TERTIARY]

Anell I, Thybo H and Artemieva IM (2009) Cenozoic uplift and subsidence in the North Atlantic region: Geological evidence revisited. *Tectonophysics* 474: 78–105.

Giusberti L (2009) Perturbation at the sea floor during the Paleocene-Eocene Thermal Maximum: Evidence from benthic foraminifera at Contessa Road, Italy. *Marine Micropaleontology* 70: 102–119.

Sluijs A, Brinkhuis H, Crouch EM et al. (2008) Eustatic variations during the Paleocene-Eocene greenhouse world. *Paleoceanography* 23: PA4216.

palaeochannel   A *river channel* that was active in the past. Palaeochannels may represent former channel positions in systems that are still active, such as channel segments on a river FLOODPLAIN that have been abandoned through channel migration or AVULSION,

or whole systems of channels that no longer exist as routeways for flows, such as palaeochannels in the STRATIGRAPHICAL RECORD. Palaeochannels can be recognized by means of surface GEOMORPHOLOGY, by the geometry of BEDDING in SEDIMENTARY ROCKS or by BEDS of distinctive LITHOLOGY. The analysis of palaeochannels can yield important information about former environments and environmental change, particularly when integrated with techniques such as PALAEOHYDROLOGY, FACIES ANALYSIS and ALLUVIAL ARCHITECTURE.
                                                        GO

[*See also* PALAEOFLOOD]

Cairncross B, Stanistreet IG, McCarthy TS et al. (1988) Paleochannels (stone-rolls) in coal seams: Modern analogs from fluvial deposits of the Okavango Delta, Botswana, southern Africa. *Sedimentary Geology* 57: 107–118.

De Smedt P, Van Meirvenne M, Meerschman E et al. (2011) Reconstructing palaeochannel morphology with a mobile multicoil electromagnetic induction sensor. *Geomorphology* 130: 136–141.

Kemp J and Rhodes EJ (2010) Episodic fluvial activity of inland rivers in southeastern Australia: Palaeochannel systems and terraces of the Lachlan River. *Quaternary Science Reviews* 29: 732–752.

Page K, Frazier P, Pietsch T and Dehaan R (2007) Channel change following European settlement: Gilmore Creek, southeastern Australia. *Earth Surface Processes and Landforms* 32: 1398–1411.

Rajani MB and Rajawat AS (2011) Potential of satellite based sensors for studying distribution of archaeological sites along palaeo channels: Harappan sites, a case study. *Journal of Archaeological Science* 38: 2010–2016.

palaeoclimatic modelling   The simulation of past climates with numerical CLIMATIC MODELS. These models may be the same as the GENERAL CIRCULATION MODELS (GCMs) used for FUTURE CLIMATE projections, but due to the long computing time required to simulate long periods, they more often are simplified model versions, for example, containing only some model components, such as the ATMOSPHERE or the OCEAN. Alternatively, they may be complete but simplified *Earth system models of intermediate complexity* (see INTERMEDIATE-COMPLEXITY MODELLING). These models have a coarser spatial and temporal resolution than GCMs and represent small-scale processes in a more simplified form, but can include other processes, such as ICE-SHEET DYNAMICS, that may be considered more relevant for the long timescale simulations.

Palaeoclimatic modelling may fulfil two different purposes. One is to serve as numerical laboratories to test HYPOTHESES about the physical mechanisms behind past CLIMATIC VARIABILITY and CLIMATIC CHANGE. Notable examples include the intriguing very weak meridional temperature gradient that PROXY DATA suggest for the Early EOCENE about 55 million years ago, in which

ARCTIC annual mean surface temperatures may have reached 10–15°C. Simulations with climate models with elevated concentrations of GREENHOUSE GASES, suggested by *marine proxies*, may indicate whether or not those elevated concentrations can produce such high Arctic temperatures. Another important and still debated example is the origin of the PLEISTOCENE glacial-interglacial difference in atmospheric CARBON DIOXIDE VARIATIONS (see GLACIAL-INTERGLACIAL CYCLE). Many processes, ranging from cardon dioxide solubility in ocean waters, CARBON SEQUESTRATION in corals and BIOSPHERE and OCEAN PALAEOPRODUCTIVITY may contribute to offset this difference. Series of simulations in which one or some of these processes are artificially switched off in the model can be very helpful to identify the main responsible mechanisms. The physical mechanism leading to DEGLACIATION from a glacial state are just beginning to be unravelled by the comparison of multiproxy data sets (see MULTIPROXY APPROACH) and very long palaeosimulations with GCMs.

A second purpose for palaeoclimatic modelling is to test the capability of state-of-the art climate models to realistically simulate climate changes (e.g. the *Paleoclimate Model Intercomparison Project*). In this case, it is important that the models are similar to those used for future climate projections. Since the computing time required by these models is large, the simulations are restricted to a few hundred years length. In this category, the more adequate periods are also close to the present climate: the past millennium or the HOLOCENE THERMAL OPTIMUM. Models are then benchmarked against existing proxy-based CLIMATIC RECONSTRUCTIONS, taking into account the UNCERTAINTIES present on both sides, models and reconstructions.

To conduct a palaeoclimatic simulation, the climate model has to be driven by estimates of past CLIMATIC FORCING. These may include concentrations of greenhouse TRACE GASES, configurations of the Earth's orbit around the Sun (see ORBITAL FORCING), *solar irradiance* (see SOLAR RADIATION), vegetation and LAND COVER, distribution of the CONTINENTS on the globe, ICE SHEETS, SEA-SURFACE TEMPERATURE (SST), etc. The forcings that need to be prescribed depend on the nature of the model used and on the purpose of the numerical experiment. For instance, a simulation to investigate the effect of reconstructed sea-surface temperatures on the GENERAL CIRCULATION OF THE ATMOSPHERE would use an atmosphere-only model, whereas for the investigation of the effect of carbon dioxide FEEDBACK MECHANISMS on the onset of GLACIATION, a coupled atmosphere-ocean-ice sheet-vegetation model with an interactive CARBON CYCLE would be required.

There are two main types of palaeoclimate simulations: *time-slice simulations* and *transient simulations*. In time-slice simulations, the external forcings are kept constant throughout the simulation, fixed to their values at a particular moment in the past. This type of simulation does not aim to represent as realistically as possible the past climate evolution but to analyse the statistical properties of past climates. One example would be the study of the different characteristics in duration, intensity and frequency of EL NIÑO-SOUTHERN OSCILLATION (ENSO) events in the mid-Holocene. To generate a statistically large sample size of such events and better understand the mechanisms by which the orbital forcing may affect the ENSO phenomenon, a climate model may be run for a few thousand years with the (fixed) orbital configuration corresponding to year 6000 BP (although in reality the orbital configuration would be continuously changing over time).

In transient simulations, the external forcing is continuously changed to mimic their assumed evolution in the past. Here, the aim is to analyse the dynamical interactions between the external forcing and all components of the climate systems, detect LAG TIMES and identify mechanisms that may trigger ABRUPT CLIMATIC CHANGES. One of the most important open questions in PALAEOCLIMATOLOGY is the ultimate origin of ICE AGES, and understanding the carbon cycle and sea-level evolution through glacial-interglacial cycles. This type of study requires transient simulations due to the complex feedbacks and time lags between the climatic subsystems.   *EZ*

[*See also* COUPLED OCEAN-ATMOSPHERE MODELS, EARTH-SYSTEM ANALYSIS (ESA), FORCING FACTOR, PALAEOCLIMATOLOGY, PROXY CLIMATIC INDICATOR, SIMULATION MODEL]

Bartlein PJ and Hostetler SW (2004) Modeling paleoclimates. In Gillespie AR, Porter SC and Atwater BF (eds) *The Quaternary period in the United States.* Amsterdam: Elsevier, 563–582.

Crucifix M (2008) Modeling the climate of the Holocene. In Battarbee RW and Binney HA (eds) *Natural climate variability and global warming: A Holocene perspective.* Chichester: Wiley-Blackwell, 98–122.

Crucifix M (2012) Traditional and novel approaches to palaeoclimatic modeling. *Quaternary Science Reviews* 57: 1–16.

Ganopolsk A, Calov R and Claussen M (2010) Simulation of the Last Glacial cycle with a coupled climate ice-sheet model of intermediate complexity. *Climate of the Past* 6: 229–244.

Harrison SP, Bracannot P, Joussaume S et al. (2002) Comparison of palaeoclimate simulations enhances confidence in models. *Eos Transactions of the American Geophysical Union* 83: 447.

Huber M and Caballero R (2011) The Early Eocene equable climate problem revisited. *Climate of the Past* 7: 603–633.

Knutti R (2012) Modelling environmental change and developing future projections. In Matthews JA, Bartlein PJ, Briffa KR et al. (eds) *The SAGE handbook of environmental change, volume 1.* London: Sage, 116–133.

Shakun JD, Clark PU, Shaun FH et al. (2012) Global
    warming preceded by increasing carbon dioxide
    concentrations during the last deglaciation. *Nature* 484:
    49–55.
Washington WM and Parkinson CL (2005) *An introduction
    to three-dimensional climate models, 2nd edition.*
    Herndon, VA: University Science Books.
Zorita E, Moberg A, Leijonhufvud L et al. (2010) European
    temperature records of the past five centuries based
    on documentary information compared to climate
    simulations. *Climatic Change* 101: 143–168.

palaeoclimatology The study of past CLIMATE. It includes CLIMATE CHANGE throughout geological time, but most palaeoclimate research is carried out in the context of QUATERNARY ENVIRONMENTAL CHANGE, including HOLOCENE ENVIRONMENTAL CHANGE and HISTORICAL CLIMATOLOGY. However, some confine palaeoclimatology to the study of climate prior to the period of the INSTRUMENTAL RECORD.

Model-based predictions concerning the scale and magnitude of FUTURE CLIMATE CHANGE due to *anthropogenic impact* constitutes, as yet, an inexact science. It is clear that human-induced CLIMATIC FORCING will be superimposed on a climate system which has as one of its inherent characteristics a natural propensity for change on all timescales. Palaeoclimatology is concerned with uncovering the range of past CLIMATIC VARIATION in order that the context for present and future climate variation may be better understood and the MANAGEMENT of climate as a RESOURCE in its own right may be optimised.

Palaeoclimatic reconstructions are most satisfactory when based on standard instrumental observations. When allowance is made for past problems such as non-standard exposure, the quality of manufacture (especially glassware), a multiplicity of measurement scales and several other difficulties, sufficient coverage can be provided from the instrumental record to permit the construction of daily weather maps—in the case of Europe, from the early eighteenth century. Elsewhere, CLIMATIC RECONSTRUCTIONS based on direct observations are more difficult. The longest continuous record of temperature is Manley's CENTRAL ENGLAND TEMPERATURE RECORD (CET) from AD 1659, but for the Southern Hemisphere, the longest record dates only from AD 1832 (Rio de Janeiro). Instrumental records therefore frequently provide information on relatively poor spatial and temporal *resolution* and encompass a period which may be quite exceptional in the longer climatic context. Careful treatment of historical documents, such as SHIP LOG BOOK RECORDS, Mediaeval *manorial records*, *grain prices*, CHRONICLES and other early manuscripts may provide patchy information on palaeoclimates back into the first millennium AD. Before this, PROXY DATA sources from the NATURAL world are necessary.

NATURAL ARCHIVES of the effects of CLIMATIC VARIABILITY, such as contained in TREE RINGS, LACUSTRINE SEDIMENTS, MARINE SEDIMENT CORES, CORAL AND CORAL REEFS and ICE CORES have, with the use of advanced technologies and statistical techniques, enabled palaeoclimate reconstruction with annual, and sometimes seasonal resolution (see Figure). Changes in ocean SEAWATER COMPOSITION and in ATMOSPHERIC COMPOSITION, such as that due to VOLCANIC ERUPTIONS, may also be recovered. Plant and animal remains, such as pollen grains (see POLLEN ANALYSIS), beetle assemblages (BEETLE ANALYSIS) and planktonic foraminifera (FORAMINIFERA: ANALYSIS), also enable TRANSFER FUNCTIONS to be derived from which palaeoclimate information can be inferred.

On long timescales, palaeoclimate reconstruction is complicated by the changing location of places due to PLATE TECTONICS. Before about 2.4 million years ago, conditions were generally warmer than at present. Since the beginning of the QUATERNARY, however, the climate system has been characterised by two recurring modes: GLACIAL and INTERGLACIAL, during which GLOBAL MEAN SURFACE AIR TEMPERATURE may have fluctuated by about 5–7°C. Considerable climate instability has characterised glacial episodes with abrupt periods of relatively short-lived warming. Since the LAST GLACIAL MAXIMUM (LGM), approximately 20,000 years ago, climate has warmed sporadically to the HOLOCENE THERMAL OPTIMUM when temperatures in the North Atlantic and Europe were about 2°C above those of today. During more recent times, significant warming occurred around AD 900–1200 (the MEDIAEVAL WARM PERIOD), though this may have been largely confined to the North Atlantic region. The LITTLE ICE AGE, which reached its nadir in Europe in the late seventeenth century brought, on average, global cooling of about 1°C that persisted into the nineteenth century. Since then, GLOBAL WARMING of about 0.8°C has occurred. Almost all of the world's governments have signed acceptance of the overwhelming consensus among atmospheric scientists that most of the increase in global average temperatures since the mid-twentieth century is very likely attributable to increased concentrations of GREENHOUSE GASES in the atmosphere. Modulation of past climates as a result of changes in BOUNDARY CONDITIONS is believed to result from intrinsic variations in SOLAR RADIATION and to orbitally induced changes in insolation receipt. Firstly, though the SOLAR CONSTANT has varied less than 0.2 per cent during recent decades, greater variability in the near ultraviolet wavelengths may produce changes in the absorption of SOLAR RADIATION by stratospheric OZONE. These in turn may propagate downwards to affect the LONG-WAVE pattern in the WESTERLIES and the strength of the JET STREAM. Secondly, orbitally induced changes in the seasonal receipt of insolation at various latitudes are now widely accepted as the driver of the GLACIAL-INTERGLACIAL CYCLES of the Quaternary. These changes occur as a result of PERIODICITIES of 21,000 years (PRECESSION of the equinoxes), 41,000 years (OBLIQUITY of the ecliptic) and 100,000 years (ECCENTRICITY), which act in conjunction to alter the receipt of insolation at key latitudes and critical seasons of the year.

On shorter timescales, SECULAR VARIATION in climate and CLIMATIC FLUCTUATIONS of various types result from various FORCING FACTORS, some internal to the present CLIMATIC SYSTEM. Detecting these fluctuations and relating them to specific factors entails discriminating any 'signal' from the background 'noise' or RANDOM variability, which can be of a similar or even greater magnitude. Among the key factors which may operate at these timescales are changes in ATMOSPHERIC COMPOSITION (especially TRACE GASES such as CARBON DIOXIDE, *methane* and *nitrous oxide*), VOLCANIC ERUPTIONS, fluctuations in the behaviour of the OCEANS and CRYOSPHERE and the dynamics of the BIOSPHERE, particularly involving biological responses to physical climatic change which may show complex feedback relations. High-frequency variability, such as is provided by the EL NIÑO-SOUTHERN OSCILLATION (ENSO) phenomenon, is increasingly recognised as a major factor in past and present climate variability in the 2- to 10-year range. ENSO is influential both within the tropics and in extratropical areas where strong TELECONNECTIONS have been established in several locations. Palaeoclimatic variations may also show associations between annual, decadal and century scales as a result of ATMOSPHERE-OCEAN INTERACTION and FEEDBACK MECHANISMS.

Palaeoclimatic research, particularly that based on ICE CORES from the polar regions, has revealed the existence of several instances of ABRUPT CLIMATIC CHANGES during the Last Glacial-Interglacial cycle. Rapid TERMINATIONS of glacial stages involving temperature rises of several degrees within a few decades have been identified. Equally dramatic GLOBAL COOLING phases during early interglacial times in the northern oceans may relate to large-scale freshwater influx from greatly increased iceberg outflow (see HEINRICH EVENTS) which may have induced an adjustment in the THERMOHALINE CIRCULATION. The response of the ocean to changes in the composition of the atmosphere and to insolation changes may be more rapid than hitherto believed. The lesson of palaeoclimatology is that the climatic system is probably a *quasi-transitive system* which adjusts to FORCING FACTORS by first resisting change and then moving rapidly to a new *equilibrium*. This renders the predictability of possible human impacts more difficult and their outcomes potentially more hazardous.   JCS

[*See also* CENTURY- TO MILLENNIAL-SCALE VARIABILITY, CLIMATE OF THE LAST MILLENNIUM, DECADAL-SCALE VARIABILITY, EARTH SYSTEM, HIGH-RESOLUTION RECONSTRUCTIONS, PALAEOENVIRONMENTAL RECONSTRUCTION]

Alverson KD, Bradley RS and Pedersen TF (eds) (2003) *Paleoclimate, global change and the future.* Berlin: Springer.
Alverson KD, Oldfield F and Bradley RS (eds) (2000) Past global changes and their significance for the future. *Quaternary Science Reviews* 19(1–5): 1–479 [Special Issue].

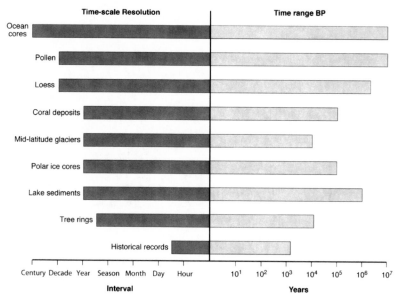

Palaeoclimatology *The temporal resolution and range of some palaeoclimatic data.*

Beltrami H (2002) Paleoclimate: Earth's long-term memory. *Science* 297: 206–207.

Bradley RS (1999) *Paleoclimatology: Reconstructing climates of the Quaternary, 2nd edition.* San Diego, CA: Academic Press.

Cronin TM (2010) *Paleoclimates: Understanding climate change past and present.* New York: Columbia University Press.

D'Arrigo R, Wilson R and Jacoby G (2006) On the long-term context for late twentieth century warming. *Journal of Geophysical Research* 111: D03103.

Francis J, Haywood AM, Hill D et al. (2012) Environmental change in the geological record. In Matthews JA, Bartlein PJ, Briffa KR et al. (eds) *The SAGE handbook of environmental change, volume 1.* London: Sage, 165–180.

Jansen E, Overpeck J, Briffa KR et al. (2007) Palaeoclimate. In Solomon S, Qin D and Manning M (eds) *Climate change 2007: The physical science basis* [Contribution of Working Group I to the Fourth Assessment Report of the Intergovernmental Panel on Climate Change]. Cambridge: Cambridge University Press, 433–497.

Jones PD, Briffa KR, Osborn TJ et al. (2009) High-resolution palaeoclimatology of the last millennium: A review of current status and future prospects. *The Holocene* 19: 3–49.

Mann ME and Jones PD (2003) Global surface temperatures over the past two millennia. *Geophysical Research Letters* 30: 1820.

Ruddiman WF (2009) *Earth's climate: Past and future, 2nd edition.* New York: Freeman.

Stocker TF (2000) Past and future reorganizations of the climate system. *Quaternary Science Reviews* 19: 301–319.

Tingley MP, Craigmile PF, Haran M et al. (2012) Piecing together the past: Statistical insights into paleoclimatic reconstruction. *Quaternary Science Reviews* 35: 1–22.

palaeocommunity   A past *community* of organisms that may be reconstructed from FOSSIL or SUBFOSSIL evidence.   *GO*

[*See also* DEATH ASSEMBLAGE, LIFE ASSEMBLAGE, PALAEOECOLOGY]

palaeocurrent   The direction of a past wind or water current, commonly inferred from the orientation of SEDIMENTARY STRUCTURES such as CROSS-STRATIFICATION.   *MRT*

Collinson JD, Mountney N and Thompson DB (2006) *Sedimentary structures, 3rd edition.* Harpenden: Terra Publishing.

Potter PE and Pettijohn FJ (1977) *Paleocurrents and basin analysis, 2nd edition.* Berlin: Springer-Verlag.

palaeodata   Data relating to past environmental events and/or past environmental conditions, prior to the availability of INSTRUMENTAL RECORDS and DOCUMENTARY EVIDENCE. Palaeodata are derived from NATURAL ARCHIVES and relate primarily to the *effects* of former events or conditions from which the events or conditions themselves are reconstructed rather than observed or measured.   *JAM*

[*See also* ABDUCTION, PROXY DATA/PROXY EVIDENCE]

palaeodiet   The nature of the food used by humans or other animals in the past. A wide variety of sources are used to reconstruct palaeodiets. Archaeological dietary reconstruction for humans and HOMINIDS includes the following: the compilation of an inventory of foods known to be available, including on-site evidence such as STOMACH CONTENT or *intestine content* (rarely available), BONES and TEETH, COPROLITE, ARTEFACTS and RESIDUES; the application of a potentially wide range of analytical techniques (see SCIENTIFIC ARCHAEOLOGY); and the interpretation of likely environmental constraints and dietary choices. Data on palaeodiets are relevant for understanding FOOD WEBS, *nutrition* and DISEASE, CULTURAL CHANGE and *palaeoenvironmental change*.   *JAM*

[*See also,* ENVIRONMENTAL ARCHAEOLOGY, PALAEO-ETHNOBOTANY, RODENT MIDDEN, ZOOARCHAEOLOGY]

Ambrose SH (1993) Isotopic analysis of paleodiets: Methodological and interpretive considerations. In Sandford MK (ed.) *Investigations of ancient human tissue: Chemical analyses in anthropology* (pp. 59–130). Langhorne, PA: Gordon and Breach.

Casey MM and Post DM (2011) The problem of isotopic baseline: Reconstructing diet and trophic position of fossil animals. *Earth-Science Reviews* 106: 131–148.

Emslie SD and Woehler EJ (2005) A 9000-year record of Adelie penguin occupation and diet in the Windmill Islands, East Antarctica. *Antarctic Science* 17: 57–66.

Knudson KJ, Williams HM, Buikstra JE et al. (2010) Introducing delta(88/86)Sr analysis in archaeology: A demonstration of the utility of strontium isotope fractionation in paleodietary studies. *Journal of Archaeological Science* 37: 2352–2364.

Mays S and Beavan N (2012) An investigation of diet in early Anglo-Saxon England using carbon and nitrogen stable isotope analysis of human bone collagen. *Journal of Archaeological Science* 39: 867–874.

Reynard LM and Hedges REM (2008) Stable hydrogen isotopes of bone collagen in palaeodietary and palaeoenvironmental reconstruction. *Journal of Archaeological Science* 35: 1934–1942.

Sobolik KD (ed.) (1994) *Paleonutrition: The diet and health of prehistoric Americans.* Carbondale: Southern Illinois University Press.

Ungar PS and Sponheimer M (2011) The diet of early hominins. *Science* 334: 190–193.

palaeodose   In LUMINESCENCE DATING, the laboratory dose of *beta* (β) and/or *gamma* (γ) *radiation* needed to induce a luminescence equal to that accumulated in a SEDIMENT sample since the most recent exposure to light or significant heating ('*bleaching*' event). The unit of dose is the *gray* ($1 \text{ Gy} = 1 \text{ J kg}^{-1}$). The palaeodose is also termed the *equivalent dose* or *accumulated dose*.   DAR

palaeodrought   A DROUGHT in the past. Various PROXY RECORDS of drought are available from NATURAL ARCHIVES, such as TREE RINGS, LAKE-LEVEL VARIATIONS, LACUSTRINE SEDIMENTS, SPELEOTHEMS and AEOLIAN SEDIMENTS. DENDROCLIMATOLOGY, and POLLEN, DIATOM and STABLE ISOTOPE ANALYSES are amongst the methods used for CLIMATIC RECONSTRUCTION involving droughts. Benson et al. (2002) provide an example of the HIGH-RESOLUTION RECONSTRUCTION of droughts based on stable isotope analysis of the sediments in Pyramid Lake, Nevada. This record, which spans the whole of the HOLOCENE, shows that droughts occurred at intervals of 80–230 years, had a duration of 20–100 years and were most frequent between 6,500 and 3,800 years ago. Palaeodroughts have had profound impacts on early human societies and have been implicated in the COLLAPSE OF CIVILISATIONS, such as the Akkadian collapse in the Near East about 4.2 ka and the Maya collapse around AD 800–1000 in Central America. There are numerous other examples, which are relevant in considerations of the future availability of WATER RESOURCES. On longer timescales, increasing ARIDITY and drought frequency during GLACIATIONS is believed to have been one of the main environmental factors in HOMINID evolution.   JAM

[*See also* DROUGHTS: HISTORICAL RECORDS, HUMAN EVOLUTION: CLIMATIC INFLUENCES, MEGADROUGHT, PALAEOPRECIPITATION PROXIES]

Benson L, Kashgarian M, Rye R et al. (2002) Holocene multidecadal and multicentennial droughts affecting northern California and Nevada. *Quaternary Science Reviews* 21: 659–682.

Cook ER, Seager R, Cane MA and Stahle DH (2007) North American drought: Reconstructions, causes and consequences. *Earth-Science Reviews* 81: 93–134.

Cullen MM, deMenocal PB, Hemming S et al. (2000) Climate change and the collapse of the Akkadian Empire: Evidence from the deep sea. *Geology* 28: 379–382.

Laird KR, Cumming BE, Wunsam S et al. (2003) Lake sediments record large-scale shifts in moisture regimes across the northern praries of North America during the past two millennia. *Proceedings of the National Academy of Sciences* 100: 2483–2488.

Medina-Elizade M and Rohling EJ (2012) Collapse of classic Maya Civilisation related to modest reduction in precipitation. *Science* 335: 956–959.

Muhs DR and Zarate M (2001) Late Quaternary eolian records of the Americas and their paleoclimatic significance. In Markgraf V (ed.) *Interhemispheric climate linkages*. San Diego, CA: Academic Press, 183–216.

palaeodunes   Ancient RELICT dune systems or FOSSIL dunes. Preservation of palaeodunes takes place through three processes, all of which may be associated with environmental change: (1) a reduction in wind velocity, (2) a change in wind direction and (3) a change in rainfall regime. These process variations allow SOIL DEVELOPMENT and surface stabilisation, which can then resist surface reactivation in any later return to drier or windier conditions.

Active dunes covered large areas of present-day temperate lands during the LAST GLACIATION. Over 10^5 km² of Europe and America are now covered in stabilised dune areas including much of eastern England, Sweden and Finland in Europe, and the western Great Plains and eastern seaboard states of North America. The formation of SAND SEAS and *dunefields* in these now temperate zones was stimulated by three features of the GLACIAL EPISODES in high latitudes: ARIDITY, plentiful supplies of sand from glacial OUTWASH DEPOSITS and strong winds blowing off the glaciers (*glacier winds*). In low latitudes, the expansion of the *arid zone* at the height of the glaciations brought dune-forming conditions into areas that were previously, and have since become, less arid. Around 18,000 years ago, stabilised sand seas covered as much as 50 per cent of the land surface between 30° N and 30° S; today, the proportion stands at only 10 per cent. Extensive systems of relict dunes, many of which are covered by SAVANNA or *grassland* vegetation, have been recognised from the margins of modern, active sand seas in the SAHEL, southern Africa, India, Australia and throughout the western USA. Relict dunes systems have been identified from AERIAL PHOTOGRAPHY and SATELLITE REMOTE SENSING in areas where current annual precipitation is as high as 1,000 mm. Lithified, fossil dunes can be recognised in the GEOLOGICAL RECORD.   MAC

[*See also* AEOLIANITE, BIOSTASY, RHEXISTASY]

Clemmensen LB, Fornós JJ and Rodriquez-Perea A (1997) Morphology and architecture of a Late Pleistocene cliff-front dune, Mallorca, western Mediterranean. *Terra Nova* 9: 251–254.

Goring-Moris AN and Goldberg P (1990) Late Quaternary dune incursions in the southern Levant: Archaeology, chronology and paleoenvironments. *Quaternary International* 5: 115–137.

Karpeta WP (1990) The morphology of Permian palaeodunes: A reinterpretation of the Bridgnorth sandstone around Bridgnorth, England, in the light of modern dune studies. *Sedimentary Geology* 69: 59–75.

Lancaster N (2009) Deserts. In Slaymaker O, Spencer T and Embleton-Hamann C (eds) *Geomorphology and global environmental change*. Cambridge: Cambridge University Press, 276–296.

Mason JA, Swinehart JB and Hanson PR (2011) Late Pleistocene dune activity in the central Great Plains, USA. *Quaternary Science Reviews* 30: 3858–3870.

palaeoecology   The ECOLOGY of the past. This discipline involves the study of past *ecosystems* and

ENVIRONMENTS and, where possible, the interaction between organisms and their environment in the past. Both biological and geological evidence is used to reconstruct long-term environmental conditions, and to determine the distribution and abundance of organisms in the past. Palaeoecology can be studied in any period of Earth history in which life was present. However, the main relevance of this subject to the study of contemporary ENVIRONMENTAL CHANGE lies in the QUATERNARY. This geological time interval covers the past 2 million years or so and is characterised by regular GLACIAL-INTERGLACIAL CYCLES and the evolution of ANATOMICALLY MODERN HUMANS (AMH), both of which had profound palaeoecological effects.

The FLORA and/or FAUNA that occurred at a particular time and place in the past can be reconstructed by analysing assemblages of FOSSIL and SUBFOSSIL organisms. Preservation of BIOTIC remains usually requires deposition in ANAEROBIC environments (e.g. LAKES, WETLANDS and the OCEAN floor), DESICCATION or *freezing*. Plant MICROFOSSILS, such as *pollen*, and MACROFOSSILS, such as *seeds*, *leaves* and *wood*, provide information on the composition of past vegetation. Similarly, analysis of animal remains from organisms such as OSTRACODS, CHIRONOMIDS, BEETLES (*Coleoptera*), CLADOCERA and MOLLUSCA, as well as larger organisms such as *mammals*, allow a reconstruction of past *faunal assemblages* (see ANIMAL REMAINS). In addition, palaeoenvironmental conditions (e.g. climatic conditions, water chemistry and nutrient status, watershed processes, lake levels and water temperature) can be inferred from both sediments and the contained plant and animal assemblages, assuming that the ecology and physiology of the MODERN species is the same as it was in the past. In order to understand the interaction between organisms and their environment, independent environmental data are essential. Additional palaeolimnological data can be inferred from the composition of DIATOMS, *fungal hyphae*, PHYTOLITHS, TESTATE AMOEBAE and *chrysphyte cysts* that are found within the sedimentary matrix.

Fossil biotic data can be qualitative and/or quantitative depending on the sampling techniques, fossil production, dispersal and preservation. Interpretation of past community composition, structure and function based on fossil material is more difficult farther back in time as *ecological tolerances* may change and taxa recovered may be extinct. Taphonomic processes (see TAPHONOMY) can alter the composition of fossil assemblages and must always be considered when interpreting fossil data.

Data on past environments and environmental history can also be obtained from physical and CHEMICAL ANALYSES OF SEDIMENTS AND SOILS and, in cases where the sediments are fossiliferous, inferences based on lithological changes can be supported by those based on fossil evidence. A wide variety of materials and methods are employed depending on the site and nature of the stratigraphic material. These include describing the sediment STRATIGRAPHY (for classification and correlation purposes), sediment GRAIN-SIZE distribution (reflecting changes in the DEPOSITIONAL ENVIRONMENT), organic content (a measure of biological PRODUCTIVITY), properties of MINERAL MAGNETISM (characterising sedimentary sequences and useful for core correlation), HEAVY MINERAL ANALYSIS (reflecting the PROVENANCE of inorganic deposits), *elemental analyses* (indication of changing erosional history of lake catchments) and STABLE ISOTOPE ANALYSIS of the sediments or fauna contained with the sediments (indirect evidence of water temperature and hydrology). In addition, LANDFORMS that developed under a previous environmental regime, and have survived, can provide information on the nature of the environmental conditions in which they evolved, for example, climatic, glacial and fluvial activity.

A sound CHRONOLOGY is essential in palaeoecological investigations to determine the timing of past events, rates of change and for correlation among sites. Various DATING TECHNIQUES are used to provide a chronology for palaeoecological studies, such as RADIOMETRIC DATING, PALAEOMAGNETIC DATING, TEPHROCHRONOLOGY, DENDROCHRONOLOGY and VARVE CHRONOLOGY.

In addition to documenting past ecosystems and environments, the *palaeoecological record* provides a long-term context for current phenomena, landscape patterns and processes. Many ecosystem processes (relating to both BIOTIC and ABIOTIC ecosystem components) occur over relatively long timescales of decades to millennia (e.g. ECOLOGICAL SUCCESSION, species MIGRATION and SOIL DEVELOPMENT). A long temporal perspective is therefore essential. Long-term environmental data are critical to place current GLOBAL ENVIRONMENTAL CHANGE, including CLIMATIC CHANGE, into context and to determine whether current trends are part of NATURAL ENVIRONMENTAL CHANGE. In addition, an understanding of past environmental and climatic change, and the response of organisms to those changes, can be used in the PREDICTION of future *ecosystem dynamics*.   JLF

[*See also* CHRYSOPHYTE CYST ANALYSIS, DIATOM ANALYSIS, PALAEOCOMMUNITY, PALAEOENVIRONMENT, PALAEOGEOGRAPHY, PALAEOHYDROLOGY, PALAEOLIMNOLOGY, POLLEN ANALYSIS]

Battarbee RW, Bennion H, Gell P and Rose N (2012) Human impacts on lacustrine ecosystems. In Matthews JA, Bartlein PJ, Briffa KR et al. (eds) *The SAGE handbook of environmental change, volume 2.* London: Sage, 47–70.

Behrensmeyer AK, DAmuth JD, DiMichele WA et al. (eds) (1992) *Terrestrial ecosystems through time: An evolutionary palaeoecology of terrestrial plants and animals.* Chicago: University of Chicago Press.

Birks HJB and Birks HH (1980) *Quaternary palaeoecology.* London: Arnold.

Brenchley PJ, Brenchley P and Harper D (1998) *Palaeoecology: Ecosystems, environments and evolution.* London: Chapman and Hall.

Delcourt HR and Delcourt PA (1991) *Quaternary ecology: A paleoecological perspective.* London: Chapman and Hall.

Rull V (2004) Biogeography of the 'Lost World': A palaeoecological perspective. *Earth-Science Reviews* 67: 125–137.

Seppa H and Bennett KD (2003) Quaternary pollen analysis: Recent progress in palaeoecology and palaeoclimatology. *Progress in Physical Geography* 27: 548–579.

Smith AJ (2012) Evidence of environmental change from terrestrial and freshwater palaeoecology. In Matthews JA, Bartlein PJ, Briffa KR et al. (eds) *The SAGE handbook of environmental change, volume 1.* London: Sage, 254–283.

Vegas-Vilarrubia T, Rull V, Montoya E and Safont E (2011) Quaternary palaeoecology and nature conservation: A general review with examples from the neotropics. *Quaternary Science Reviews* 30: 2361–2388.

palaeo-ENSO   The EL NIÑO-SOUTHERN OSCILLATION (ENSO) circulation pattern in the GEOLOGICAL RECORD. Palaeo-ENSO can be investigated with FOSSIL corals, LAKE and MARINE SEDIMENT CORES using a variety of MARINE PALAEOCLIMATIC PROXIES. GREY-SCALE ANALYSIS has been used on lake cores from Ecuador to reconstruct ENSO in the past and revealed that ENSO variability increased markedly approximately 5,000 years ago. Organic BIOMARKERS from marine sediments off the coast of Chile have also shown a marked increase in ENSO activity since approximately 5,000 years ago. Here, two different organic molecules, *dinosterol* and *cholesterol*, were used to reconstruct EL NIÑO events and LA NIÑA events, respectively. GRAIN SIZE and sediment GEOCHEMISTRY records from a Galápagos lake have been used to show a similar increase in ENSO activity since the middle Holocene, related to increased PRECIPITATION and migration of the INTERTROPICAL CONVERGENCE ZONE (ITCZ).   *JP*

Conroy JL, Overpeck JT, Cole JE et al. (2008) Holocene changes in eastern tropical Pacific climate inferred from a Galápagos lake sediment record. *Quaternary Science Reviews* 27: 1166–1180.

Makou MC, Eglinton TI, Oppo DW and Hughen KA (2010) Postglacial changes in El Niño and La Niña behavior. *Geology* 38: 43–46.

Moy CM, Seltzer GO, Rodbell DT and Anderson DM (2002) Variability of El Niño/Southern Oscillation activity at millennial timescales during the Holocene epoch. *Nature* 420: 162–165.

palaeo-entomology   The study of FOSSIL and SUB-FOSSIL insects—an important source of natural and anthropogenic palaeoenvironmental information.   *JAM*

[*See also* BEETLE ANALYSIS, CHIRONOMID ANALYSIS, ELM DECLINE, INSECT ANALYSIS]

Dinnin MH and Sadler JP (1999) 10,000 years of change: The Holocene entomofauna of the British Isles. *Quaternary Proceedings* 7: 545–562.

Elias SA (2010) *Advances in Quaternary entomology.* Amsterdam: Elsevier.

palaeoenvironment   A past ENVIRONMENT.   *MRT*

[*See also* PALAEOENVIRONMENTAL RECONSTRUCTION]

palaeoenvironmental indicator   Any biological or physical indicator of environmental conditions in the past. Interpretation of palaeoenvironmental indicators generally relies heavily on MODERN ANALOGUES, which are used to establish the sensitivity of the indicator to present environmental conditions prior to their use in reconstructing past conditions.   *JAM*

[*See also* ANALOGUE METHOD, ENVIRONMENTAL INDICATOR, SEASONALITY INDICATOR]

Booth RK, Sullivan ME and Sousa VA (2008) Ecology of testate amoebae in a North Carolina pocosin and their potential as environmental and paleoenvironmental indicators. *Ecoscience* 15: 277–289.

Guthrie RD (1982) Mammals of the mammoth steppe as paleoenvironmental indicators. In Hopkins DM, Matthews Jr JV, Schweger CE and Young SB (eds) *Paleoecology of Beringia.* New York: Academic Press, 307–326.

Mitusov AV, Mitusova OE, Pustovoytov K et al. (2009) Palaeoclimatic indicators in soils buried under archaeological monuments in the Eurasian steppe: A review. *The Holocene* 19: 1153–1160.

palaeoenvironmental reconstruction   The deduction of the characteristics of past environments from PROXY EVIDENCE such as FOSSILS, SEDIMENTOLOGY, GEOMORPHOLOGY and GEOCHEMISTRY. It is the key step in identifying ENVIRONMENTAL CHANGE in the GEOLOGICAL RECORD. NATURAL ARCHIVES such as MARINE SEDIMENT CORES and MARINE PALAEOCLIMATIC PROXIES are used in PALAEOCEANOGRAPHY and in reconstructing OCEAN PALAEOCIRCULATION. PROXY CLIMATIC INDICATORS, such as MICROFOSSIL assemblages, STABLE ISOTOPES and CHARCOAL, from LAKES, CAVES and BLANKET MIRES are used to reconstruct CLIMATIC CHANGE in TERRESTRIAL ENVIRONMENTS, as well as changes in the HYDROLOGICAL CYCLE. *Chemical proxies* from ICE CORES, such as STABLE ISOTOPES, VOLCANIC AEROSOLS and ATMOSPHERIC DUST are used to reconstruct past ATMOSPHERIC COMPOSITION, such as concentrations of GREENHOUSE GASES, and changes in the GENERAL CIRCULATION OF THE ATMOSPHERE, over GLACIAL-INTERGLACIAL CYCLES. LANDFORMS, SEDIMENTS and LANDSYSTEMS from a wide range of environments (e.g. GLACIAL LANDFORMS, GLACIAL SEDIMENTS and *glacial landsystems*) are used to reconstruct past terrestrial environments. DINOFLAGELLATE CYST

ANALYSIS, DIATOM ANALYSIS and proxies from ICE CORES are used to reconstruct past SEA-ICE concentrations and extent.   *JP/MRT*

[*See also* CHARRED-PARTICLE ANALYSIS, CHIRONOMID ANALYSIS, FACIES ANALYSIS, FORAMINIFERA: ANALYSIS, GEOMORPHOLOGY AND ENVIRONMENTAL CHANGE, GLACIER VARIATIONS, GLACIERS AND ENVIRONMENTAL CHANGE, HOLOCENE ENVIRONMENTAL CHANGE, NATURAL ARCHIVES, PALAEOCEANOGRAPHY, PALAEOCLIMATOLOGY, PALAEOECOLOGY, PALAEOLIMNOLOGY, PALAEO-ENTOMOLOGY, PALAEOSCIENCES, POLLEN ANALYSIS, QUATERNARY ENVIRONMENTAL CHANGE, SEDIMENTOLOGICAL EVIDENCE OF ENVIRONMENTAL CHANGE]

Elias SA (ed.) (2007) *Encyclopedia of Quaternary science, 4 volumes.* Amsterdam: Elsevier.
Lowe JJ and Walker MJC (1997) *Reconstructing Quaternary environments, 2nd edition.* Harlow: Longmans.
Mackay A, Battarbee R, Birks J and Oldfield F (eds) (2003) *Global change in the Holocene.* London: Arnold.
Matthews JA, Bartlein PJ, Briffa KR et al. (eds) (2012) *The SAGE handbook of environmental change, 2 volumes.* London: Sage.
Reading HG (ed.) (1996) *Sedimentary environments: Processes, facies and stratigraphy, 3rd edition.* Oxford: Blackwell.

palaeo-ethnobotany   The identification, analysis and interpretation of plant remains associated with archaeological sites. Also known as *archaeobotany* and *phytoarchaeology*, it is important in PALAEOENVIRONMENTAL RECONSTRUCTION.   *JAM*

[*See also* CEREAL POLLEN, CHARCOAL, CHARRED-PARTICLE ANALYSIS, ENVIRONMENTAL ARCHAEOLOGY, ETHNOBIOLOGY, PHYTOLITHS, PLANT MACROFOSSIL ANALYSIS]

Brooks RR and Johannes D (1990) *Phytoarchaeology.* Leicester: Leicester University Press.
Gale R and Cutler D (2000) *Plants in archaeology.* Otley: Westbury.
Gremillion KJ (ed.) (1997) *People, plants, and landscapes: Studies in paleoethnobotany.* Tuscaloosa: University of Alabama Press.
Hather JG (ed.) (1994) *Tropical archaeobotany: Applications and new developments.* London: Routledge.
Renfrew C (ed.) (1991) *New light on early farming: Recent developments in palaeoethnobotany.* Edinburgh: Edinburgh University Press.
Zeist W van, Wasylikowa K and Behre KE (eds) (1991) *Progress in Old World palaeoethnobotany.* Rotterdam: Balkema.

palaeoflood   A FLOOD, usually of a river, that occurred in the past, prior to historical records. The term is commonly applied to severe EVENTS, sometimes also described as MEGAFLOODS, that may have exceeded in magnitude any that have been experienced at the same locality in historical times. Floods from the pre-Quaternary GEOLOGICAL RECORD, that are inferred from GEOLOGICAL EVIDENCE OF ENVIRONMENTAL CHANGE, tend not to be referred to as palaeofloods, since by definition they predate human experience. Palaeofloods can be reconstructed using evidence from SEDIMENTOLOGY and GEOMORPHOLOGY. Of particular value is the preservation of fine-grained deposits (*slackwater deposits*) in areas of reduced flow velocity. Techniques such as PALAEOHYDROLOGY can be used to reconstruct flood HYDROLOGY, DISCHARGE and magnitude, and the principles of GEOCHRONOLOGY and STRATIGRAPHY can be used to reconstruct *flood chronologies*.

The recognition and reconstruction of palaeofloods can provide important evidence of ENVIRONMENTAL CHANGE and CLIMATIC CHANGE. The clustering in time of large-magnitude palaeofloods, for example, may provide evidence for the HOLOCENE record of phenomena such as EL NIÑO or the MONSOON systems. Palaeoflood chronologies can be used to assess potential HAZARDS from severe floods in the future, and estimates of the magnitudes of palaeofloods are critical in evaluating the significance of severe floods that have occurred in recent times (see NATURAL HAZARDS).

Probably the most impressive examples of palaeofloods are those associated with the catastrophic drainage of ICE-DAMMED LAKE Missoula at the end of the Last GLACIATION (see also JÖKULHLAUP). Giant PALAEOCHANNELS form the CHANNELED SCABLANDS of the Columbia River Basin, Washington, USA, and are associated with giant BEDFORMS, dry WATERFALLS and transported blocks. It has been estimated that the peak DISCHARGE may have been up to 20 times the mean discharge of all the world's rivers today. Holocene and QUATERNARY palaeofloods are also well documented from other areas, many of which are today ARIDLANDS, including southwestern USA, Australia, South Africa, the Indian peninsula and the Middle East. Surface features on other planets have also been interpreted in terms of palaeofloods (see PLANETARY ENVIRONMENTAL CHANGE).   *GO*

Baker VR (2008) Paleoflood hydrology: Origin, progress, prospects. *Geomorphology* 101: 1–13.
Baker VR (2009) The Channeled Scabland: A retrospective. *Annual Review of Earth and Planetary Sciences* 37: 393–411.
Benito G, Thorndycraft VR, Rico M et al. (2008) Palaeoflood and floodplain records from Spain: Evidence for long-term climate variability and environmental changes. *Geomorphology* 101: 68–77.
Bøe A-G, Dahl SO, Lie Ø and Nesje A (2006) Holocene river floods in the upper Glomma catchment, southern Norway: A high-resolution multiproxy record from lacustrine sediments. *The Holocene* 16: 445–455.
Burr DM (2010) Palaeoflood-generating mechanisms on Earth, Mars, and Titan. *Global and Planetary Change* 70: 5–13.

Ely LL (1997) Response of extreme floods in the southwestern United States to climatic variations in the late Holocene. *Geomorphology* 19: 175–201.

Greenbaum N, Schick AP and Baker VR (2000) The palaeoflood record of a hyperarid catchment, Nahal Zin, Negev Desert, Israel. *Earth Surface Processes and Landforms* 25: 951–971.

Hanson MA, Lian OB and Clague JJ (2012) The sequence and timing of large Late Pleistocene floods from glacial Lake Missoula. *Quaternary Science Reviews* 31: 67–81.

Huang CC, Pang J, Zha X et al. (2010) Extraordinary floods of 4100-4000 a BP recorded at the Late Neolithic ruins in the Jinghe River gorges, middle reach of the Yellow River, China. *Palaeogeography, Palaeoclimatology, Palaeoecology* 289: 1–9.

Thorndycraft VR, Benito G, Sánches-Moya Y and Sopeña A (2012) Bayesian age modelling applied to palaeoflood geochronologies and the investigation of Holocene flood magnitude and frequency. *The Holocene* 22: 13–22.

Palaeogene A system of rocks, and a period of geological time from 65.5 to 23.0 million years ago, part of the cenozoic (*cainozoic*) era, comprising the palaeocene, eocene and oligocene epochs, and formerly known as the Lower or Early tertiary. Notable events were the evolution and diversification of *mammals* following the mass extinction at the end of the cretaceous period; the first appearances of *primates*, *grasses*, and *horses*; the opening of the *North Atlantic Ocean*; the growth of the *Himalayas*; and the initial phases of orogeny that formed the Alps (see plate tectonics).   GO

[*See also* geological timescale]

Bijl PK, Schoutens S, Sluijs A et al. (2009) Early Palaeogene temperature evolution of the southwest Pacific Ocean. *Nature* 461: 776–779.

Cooper MR, Walsh JJ, Van Damm CL et al. (2012) Palaeogene alpine tectonics and Icelandic plume-related magmatism and deformation in Northern Ireland. *Journal of the Geological Society of London* 169: 29–36.

Davis SJ, Mulch A, Carroll AR et al. (2009) Palcogene landscape evolution of the central North American Cordillera: Developing topography and hydrology in the Laramide foreland. *Geological Society of America Bulletin* 121: 100–116.

Wing SL, Gingerich PD, Schmitz B and Thomas E (eds) (2003) *Causes and consequences of globally warm climates in the Early Paleogene.* Boulder, CO: Geological Society of America.

palaeogeography The geography of the geological past (see Figure). The reconstruction of palaeogeography involves many aspects of geology and must

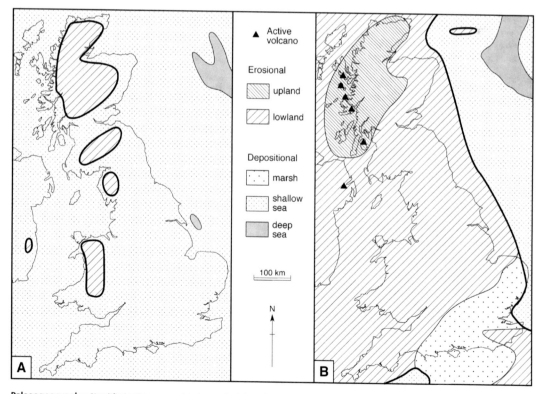

Palaeogeography *Simplified palaeogeographical maps for Britain: (A) Late Cretaceous (ca 75 million years ago) and (B) Late Palaeocene (ca 60 million years ago) (based on Cope et al., 1999).*

consider the distribution of land and sea, the TOPOG-
RAPHY of the land surface and the seabed, the position
of major *river systems*, the form of any COASTLINE and
the relative configuration of the CONTINENTS and their
PALAEOLATITUDE. Many SEDIMENTARY DEPOSITS and FACIES
are sensitive to PALAEOCLIMATE and hence, assuming a
similar climatic ZONATION to the present day, can help
reconstruct several of these aspects of palaeogeography. These include COAL-bearing deposits, DESERT
deposits, EVAPORITES, fossil REEFS, RED BEDS, PALAEOSOLS
and TILLITES. Other aspects of palaeogeography can be
reconstructed using PALAEOCURRENT indicators, FOSSIL
distributions and FAUNAL PROVINCES, ISOPACHS, PALAE-
OCEANOGRAPHY and PALAEOMAGNETISM and by matching
present-day CONTINENTAL MARGINS and the lineaments of
TECTONICS. To be strictly comparable with present-day
geography, palaeogeographical maps must remove the
effects of any *tectonic shortening* that has taken place
through *orogenic* movement (see PALINSPASTIC MAP).
The term is also applicable to the most recent geological past, on the ARCHAEOLOGICAL TIMESCALE.   *JCWC/GO*

[*See also* CONTINENTAL DRIFT]

Christiansen JL and Stouge S (1999) Oceanic circulation as
    an element in palaeogeographical reconstructions: The
    Arenig (Early Ordovician) as an example. *Terra Nova* 11:
    73–78.
Cocks LRM (2000) The Early Palaeozoic geography of
    Europe. *Journal of the Geological Society* 157: 1–10.
Cope JCW, Ingham JK and Rawson PF (eds) (1999) *Atlas of
    palaeogeography and lithofacies, 2nd edition.* London:
    Geological Society.
Gifford JA, Rapp Jr G and Vitali V (1992) Palaeogeography
    of Carthage (Tunisia): Coastal change during the first
    millenium BC. *Journal of Archaeological Science* 19:
    575–596.
Harper DAT and Servais T (eds) (2012) *Lower Palaeozoic
    biogeography.* London: Geological Society.
Scotese CR (2011) *Paleomap project* [Available at http://
    www.scotese.com/earth.htm].
Smith AG, Smith DG and Funnell BM (1994) *Atlas
    of Mesozoic and Cenozoic coastlines.* Cambridge:
    Cambridge University Press.

palaeoglaciology The study of ancient GLACIERS
and ICE SHEETS from the LANDFORMS and SEDIMENTS left
behind. The process is referred to as *inversion modelling*, which involves reconstructing the horizontal
and vertical extent of a former ice body and its internal flow patterns from the landforms and sediments.
Palaeoglaciology involves the application of the principles of GLACIOLOGY, and it is a key research avenue
in understanding the response of past ice sheets to
CLIMATE CHANGE.   *MRB*

Evans DJA (2009) Controlled moraines: Origins,
    characteristics and palaeoglaciological implications.
    Quaternary Science Reviews 28: 183–208.

Kleman J, Hättestrand C, Stroeven AP et al. (2007)
    Reconstruction of palaeo-ice sheets: Inversion of their
    glacial geomorphological record. In Knight PG (ed.)
    Glacier science and environmental change. Malden, MA:
    Blackwell, 192–198.

palaeohydrology The study of aspects of the
HYDROLOGICAL CYCLE that operated in the past, in contrast
to studies of present-day HYDROLOGY. The aims of palaeohydrology are to reconstruct hydrological parameters such as annual PRECIPITATION, EVAPORATION, RUNOFF
and DISCHARGE; variations in water storage capacity in
lakes and WETLAND areas; EXTREME EVENTS such as PAL-
AEOFLOODS; changes in BASE LEVEL such as LAKE-LEVEL
VARIATIONS and SEA-LEVEL CHANGE; and changes in
GROUNDWATER conditions. Where hydrological parameters in the past can be shown to have been different
from those that exist today, palaeohydrology provides
direct evidence of ENVIRONMENTAL CHANGE. Palaeohydrological relationships can also be used in GENERAL
CIRCULATION MODELS to forecast the effects of future
CLIMATIC CHANGE, such as changes in the frequency of
FLOODS or DROUGHTS. The discipline of palaeohydrology
is sometimes referred to as *palaeohydraulics*.

The concept of palaeohydrology has most significance when applied to the reconstruction of quantitative parameters. This can be achieved through
empirical and theoretical relationships between hydrological parameters, such as discharge, and preserved
features, such as the characteristics and thickness
of river channel deposits. Variations in hydrological
parameters can in turn be related to climatic and other
environmental factors. The term palaeohydrology is
commonly used, however, in a more general, qualitative sense, in which case there is considerable blurring
of the boundaries between palaeohydrology, *fluvial
sedimentology*, *fluvial geomorphology*, PALAEOLIMNOL-
OGY and PALAEOECOLOGY.

Methods used in the reconstruction of palaeohydrological parameters include GEOMORPHOMETRY; SEDIMEN-
TOLOGY, particularly FACIES ANALYSIS, GRAIN-SIZE analysis
and the interpretation of SEDIMENTARY STRUCTURES; and
analytical techniques such as STABLE ISOTOPE ANALY-
SIS. Interpretations are placed within a chronological
framework using the techniques of GEOCHRONOLOGY and
STRATIGRAPHY.

The important field of *continental palaeohydrology* is
concerned primarily with reconstructions of *river environments*, such as the depth, width and morphology of
PALAEOCHANNELS, flow rates and discharge regimes. The
former behaviour of rivers that still exist can be reconstructed using aspects of fluvial GEOMORPHOLOGY, such as
RIVER TERRACES or abandoned MEANDERS. The characteristics of ALLUVIAL DEPOSITS, particularly their grain size,
can be used to reconstruct hydrological parameters from
deposits in present-day river basins or in the GEOLOGICAL

RECORD. These approaches to reconstructions of continental palaeohydrology, using geomorphology, sediment characteristics or the geometry of sedimentary deposits, can be combined to provide detailed reconstructions of former hydrological conditions.

Palaeohydrological interpretations have been applied to landforms and deposits of all ages in the STRATIGRAPHICAL RECORD. Studies of HOLOCENE rivers have been used to identify PALAEOFLOODS and determine *flood chronologies* and the effects of climatic change and anthropological factors on FLOOD MAGNITUDE-FREQUENCY CHANGES. In the QUATERNARY, the relative effects of TECTONICS and climatic change, particularly GLACIAL-INTERGLACIAL CYCLES, on CHANNEL PATTERNS, CHANNEL CHANGE and *drainage evolution* have been investigated. Further back in the GEOLOGICAL RECORD, palaeohydrological methods have been applied to SEDIMENTARY ROCKS to infer environmental characteristics and changes over GEOLOGICAL TIME, such as channel dimensions and geometry, WETLAND ecology and the behaviour of rivers prior to the covering of land surfaces by terrestrial plants, with associated consequences for SOIL DEVELOPMENT.   *GO*

Benito G, Baker VR and Gregory KJ (eds) (1998) *Palaeohydrology and environmental change.* Chichester: Wiley.

Daniels JM and Knox JC (2005) Alluvial stratigraphic evidence for channel incision during the Mediaeval Warm Period on the central Great Plains, USA. *The Holocene* 15: 736–747.

Davies NS and Gibling MR (2010) Paleozoic vegetation and the Siluro-Devonian rise of fluvial lateral accretion sets. *Geology* 38: 51–54.

Eriksson P, Long D, Bumby A et al. (2008) Palaeohydrological data from the c. 2.0 to 1.8 Ga Waterberg Group, South Africa: Discussion of a possibly unique Palaeoproterozoic fluvial style. *South African Journal of Geology* 111: 281–304.

Foix N, Allard JO, Paredes JM and Giacosa RE (2012) Fluvial styles, palaeohydrology and modern analogues of an exhumed, Cretaceous fluvial system: Cerro Barcino Formation, Canadon Asfalto Basin, Argentina. *Cretaceous Research* 34: 298–307.

Gregory KJ and Benito G (eds) (2003) *Palaeohydrology: Understanding global change.* Oxford: Blackwell.

Hoek WZ (2012) Evidence of environmental change from terrestrial palaeohydrology. In Matthews JA, Bartlein PJ, Briffa KR et al. *The SAGE handbook of environmental change, volume 1.* London: Sage, 239–253.

Knox JC (2000) Sensitivity of modern and Holocene floods to climate change. *Quaternary Science Reviews* 19: 439–457.

Stone AEC, Thomas DSG and Viles HA (2010) Late Quaternary palaeohydrological changes in the northern Namib sand sea: New chronologies using OSL dating of interdigitated aeolian and water-lain interdune deposits. *Palaeogeography, Palaeoclimatology, Palaeoecology* 288: 35–53.

Thorndycraft VR, Benito G and Gregory KJ (2008) Fluvial geomorphology: A perspective on current status and methods. *Geomorphology* 98: 2–12.

Van Bocxlaer B, Verschuren D, Schettler G and Kröpelin S (2011) Modern and early Holocene mollusc fauna of the Ouniangа lakes (northern Chad): Implications for the palaeohydrology of the central Sahara. *Journal of Quaternary Science* 26: 433–447.

Wolfe BB, Edwards WD, Arvena R et al. (2000) Holocene paleohydrology and paleoclimate at treeline, north-central Russia, inferred from oxygen isotope records in lake sediment cellulose. *Quaternary Research* 53: 319–329.

palaeokarst   Karst features that are not undergoing further modification by DISSOLUTION as a result, for example, of burial or isolation. The presence of palaeokarst features (e.g. CAVE PASSAGES and SPELEOTHEMS) can be a useful indicator of palaeoenvironmental conditions conducive to dissolutional weathering. The term *relict karst* has been used for karst systems that are not completely decoupled from the *hydrogeochemical environment* of formation. Palaeokarst is common in the GEOLOGICAL RECORD; relict karst is common in the contemporary LANDSCAPE and may result from CLIMATIC CHANGE, especially an increase in ARIDITY.   *SHD/JAM*

Bosák P, Ford DC, Glazek J and Horacek I (eds) (1989) *Paleokarst: A systematic and regional review.* Amsterdam: Elsevier.

Williams PW (2004) Palaeokarst and relict karst. In Goudie AS (ed.) *Encyclopedia of geomorphology.* London: Routledge, 754–755.

palaeolatitude   The latitude of a CONTINENT at a given time in geological history. Palaeolatitude can be established using evidence from PALAEOMAGNETISM or PALAEOCLIMATOLOGY, and changes in palaeolatitude through GEOLOGICAL TIME contribute to the reconstruction of CONTINENTAL DRIFT.   *DNT*

[*See also* APPARENT POLAR WANDER (APW), COAL, EVAPORITE, PLATE TECTONICS]

palaeolimnology   The study of physical and biological parameters of LACUSTRINE SEDIMENTS to reconstruct environmental and *ecosystem* changes in the past—usually a MULTIDISCIPLINARY study. Lake sediments generally accumulate at a rate of around 1 mm/yr, forming a NATURAL ARCHIVE—a window on the past, a key to our future (Smol, 2008). Sediments can be sampled by a variety of CORERS from marginal fens, open water or ice. A range of biological (FOSSIL or SUBFOSSIL), chemical (inorganic, organic) and physical (e.g. GRAIN SIZE, PALAEOMAGNETISM) parameters can be analysed to reconstruct past changes in the lake ecosystem including the *catchment*, which can be interpreted in terms of environmental change, NATURAL and ANTHROPOGENIC.

CHRONOLOGY can be controlled by RADIOCARBON DATING, LEAD-210 DATING in young sediments and VARVES, if present, allowing the timing and RATES OF ENVIRONMENTAL CHANGE to be estimated. CORE CORRELATION within one or from several lakes can be carried out using PALAEOMAGNETISM, LOSS ON IGNITION (LOI), sedimentary features, fossil stratigraphy or other techniques.

Sediments have an AUTOCHTHONOUS component (produced within the lake) and an ALLOCHTHONOUS component, derived from the catchment or the ATMOSPHERE (see Figure). Changes in the catchment are controlled largely by temperature. In combination with precipitation, it affects the vegetation, soil types and stability; the HYDROLOGICAL BALANCE (see LAKE-LEVEL VARIATIONS); the degree of snow cover and frost activity; and sometimes glacier formation. Changes in the catchment affect the allochthonous input. The proportions of CLASTIC (including carbonate) and organic material (measured by LOI) reflect catchment stability. The MINEROGENIC SEDIMENT content increases with DISTURBANCE that can result from EROSION, FROST HEAVE and CRYOTURBATION, glacier activity, changes in lake level and water inflow and human activity (e.g. DEFORESTATION). Base *cations* and the chemistry of insoluble components reflect catchment geology. Nitrogen (N), PHOSPHOROUS (P) and *dissolved organic carbon (DOC)* originate in soils. N and P increase with EUTROPHICATION, often caused by *human impact* (e.g. FOREST CLEARANCE, FERTILISER application and INDUSTRIAL WASTE). DOC increases with forest development and natural soil ACIDIFICATION (see PODZOLISATION, PALUDIFICATION) but decreases with acidification below pH 5 resulting from, for example, ACID RAIN. Other organic materials derive from vegetation or soils and comprises plant and animal MACROFOSSILS and MICROFOSSILS as well as decayed material. Atmospheric inputs include POLLEN RAIN (reflecting regional vegetation), DUST, LOESS, VOLCANIC ASH (a chronological marker), CHARRED PARTICLES (from BIOMASS BURNING), SPHEROIDAL CARBONACEOUS PARTICLES (SCP) (yielding an industrial chronology), *radioisotopes* (Pb, Cs, etc.), strong acids in rain, HEAVY-METAL residues (Pb, Zn, Cu, Cr, Mn, Hg) that are industrial POLLUTION indicators and PESTICIDE derivatives in AEROSOLS.

Autochthonous sediments are composed of decayed remains of aquatic organisms and lake precipitates (e.g. DY, MARL, *saline deposits*). A lake ecosystem is affected by internal processes such as sediment accumulation and SEDIMENT TYPE; the organisms present; their *immigration*, ECOLOGICAL SUCCESSION and EXTINCTION; and COMPETITION between them for light (plants), food (phytophagous or zoophagous), NUTRIENTS, HABITAT, dissolved oxygen, seasonal OVERTURNING and degree of ANOXIA in the *hypolimnion* and winter *ice-cover* (see Figure). Thus, organisms are valuable INDICATOR SPECIES of past lake conditions. In a pioneer situation,

aquatic organisms react very fast to climate changes, by changes in abundance if they are already present and by immigration. DIATOMS are the most abundant plant fossils, but other algae, mosses and higher plant remains can be identified (plant MACROFOSSILS and *pollen*). Common animal fossils include *chironomidae, cladocera, coleoptera, Trichoptera*, other *insects* and ORIBATID MITES. *Fish* remains are less common. Diatoms are sensitive pH INDICATORS, and lake-water pH can be reconstructed through time by application of modern TRANSFER FUNCTIONS to the fossil assemblage. Other diatom and *chrysophyte cyst* transfer functions have been constructed for PHOSPHATE (PO_4^{3-}) and SALINITY. Similarly, Chironomidae and Cladocera have been used to quantitatively reconstruct summer and winter temperatures. Qualitative reconstruction of lake-water quality and chemistry and lake and catchment HABITATS can be made from all plant and animal groups depending upon their known ecology. *Plant pigments* derived from ALGAE and CYANOBACTERIA can be preserved in ANAEROBIC conditions and used to reconstruct EUTROPHICATION and POLLUTION. STABLE ISOTOPE fractionation (of e.g. CARBON, NITROGEN, OXYGEN and HYDROGEN ISOTOPES) in different sediment and organism components also yields environmental information (see Figure).

Many lake-sediment sequences span thousands of years. Some include the *Late Glacial*, the majority cover the HOLOCENE and other sequences preserve INTERGLACIAL records. Long-term palaeolimnological studies can assess climate change, the reaction of lake ecosystems to natural, external driving climate changes (e.g. Younger Dryas-Holocene transition) and natural lake ecosystem and catchment development in relation to environmental changes and internal lake processes. *Human impact* has gradually increased from mid-Holocene time, registering effects in lake sediments. Over the past century, AIR POLLUTION, WATER POLLUTION, EUTROPHICATION, DAMS and water ABSTRACTION have affected WATER QUALITY increasingly severely. The continued existence of many lakes in ARIDLANDS is under severe threat. Palaeolimnology can be used to provide a baseline of natural BIODIVERSITY and *ecosystem structure* against which impacts of modern TECHNOLOGICAL CHANGE can be registered and evaluated. ENVIRONMENTAL MONITORING programs take several years to register trends, but HIGH-RESOLUTION RECONSTRUCTIONS from recent sediments can readily provide evidence of trends over longer periods of time and can indicate appropriate action for ecosystem quality improvement. Palaeolimnology thus interfaces with LIMNOLOGY and ECOSYSTEM MANAGEMENT and CONSERVATION).   *IIIIB*

[*See also* LAKE STRATIFICATION AND ZONATION, LAKES AS INDICATORS OF ENVIRONMENTAL CHANGE, SEDIMENTOLOGICAL EVIDENCE OF ENVIRONMENTAL CHANGE]

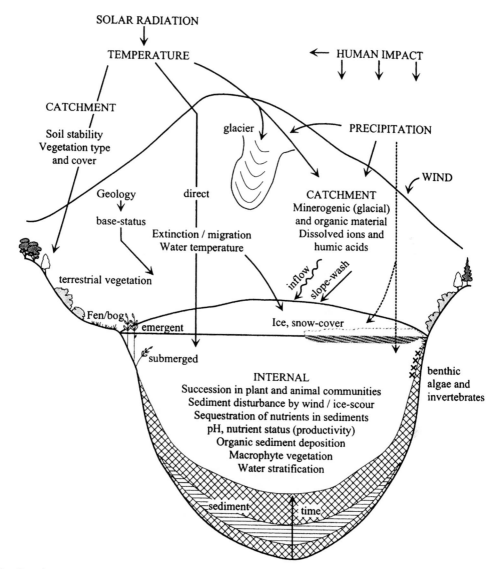

Palaeolimnology *Features of a lake ecosystem, which can cause or be influenced by environmental changes that can be subsequently registered in the sediments (Birks et al., 2000).*

Battarbee RW (2000) Palaeolimnological approaches to climate change, with special regard to the biological record. *Quaternary Science Reviews* 19: 107–124.

Battarbee RW, Bennion H, Gell P and Rose N (2012) Human impact on lacustrine ecosystems. In Matthews JA, Bartlein PJ, Briffa KR et al. (eds) *The SAGE handbook of environmental change, volume 2.* London: Sage, 47–70.

Binford MW, Deevey ES and Crisman TL (1983) Paleolimnology: An historical perspective on lacustrine ecosystems. *Annual Review of Ecology and Systematics* 14: 255–286.

Birks HH, Battarbee RW and Birks HJB (2000) The development of the aquatic ecosystem at Kråkenes

Lake, western Norway, during the Late Glacial and early Holocene. *Journal of Paleolimnology* 23: 91–114.

Charles DF, Smol JP and Engstrom DR (1994) Paleolimnological approaches to biological monitoring. In Loeb SL and Spacie A (eds) *Biological monitoring of aquatic systems.* Boca Raton, FL: CRC Press, 233–293.

Cohen AS (2003) *Paleolimnology: The history and evolution of lake systems.* Oxford: Oxford University Press.

Pienitz R, Douglas MSV and Smol JP (2004. *Long-term environmental change in Arctic and Antarctic lakes. Developments in Paleoenvironmental Research, volume 8.* Dordrecht: Springer.

Smith AJ (2012) Evidence of environmental change from terrestrial and freshwater palaeoecology. In Matthews JA,

Bartlein PJ, Briffa KR et al. (eds) *The SAGE handbook of environmental change, volume 1*. London: Sage, 254–283.

Smol JP (2008) *Pollution of lakes and rivers: A paleoenvironmental perspective, 2nd edition*. Oxford: Blackwell.

Smol JP, Cumming BF, Douglas MSV and Pienitz R (1996) Inferring past climatic changes in Canada using palaeolimnological techniques. *Geoscience Canada* 21: 113–118.

Palaeolithic The *Old Stone Age*, traditionally defined as the archaeological period during which stone implements were used. The opening of the period equates roughly with the opening of the PLEISTOCENE, although early use of stone tools may now go back as far as 3.4 million years ago (Ma). The close is usually taken for simplicity as the opening of the HOLOCENE. It is separated into the *Lower, Middle* and *Upper Palaeolithic*, with the Lower/Middle boundary around 250–200 ka BP and the Middle/Upper at ca 40 ka BP. During the Palaeolithic, HUNTING, FISHING AND GATHERING societies frequently moved location (NOMADISM) to obtain RESOURCES. Although stone was the principal material used for making tools, antler, bone and wood were commonly utilised. *LD-P/CJC*

[*See also* ARCHAEOLOGICAL TIMESCALE, CAVE ART, STONE AGE, THREE-AGE SYSTEM]

Camps M and Chauhan PR (eds) (2012) *Sourcebook of Palaeolithic transitions: Methods, theories and interpretations, 2nd edition*. New York: Springer.

Dennell R (2009) *The Palaeolithic settlement of Asia*. Cambridge: Cambridge University Press.

Gamble C (1999) *The Palaeolithic societies of Europe*. Cambridge: Cambridge University Press.

Palaeolithic: human–environment relations

PALAEOLITHIC societies were based on a HUNTING, FISHING AND GATHERING economy and modes of food and resource procurement which depended on the environment for their success. ENVIRONMENTAL CHANGE during the Palaeolithic had significant impacts on society, which must be viewed against a background of HUMAN EVOLUTION and increasing development of *technology*. For example, CLIMATIC CHANGE and the retreat of ice sheets from many lowland areas of Europe at the end of the PLEISTOCENE influenced the MIGRATION of Palaeolithic groups. The environment would also have determined the availability of food (game, plant) and non-food (stone, wood) RESOURCES. Although the environment influenced where and how Palaeolithic groups lived, people also had the ability to modify and adapt to the environment. For example, changes in the tool types demonstrate behavioural or ETHOLOGICAL ADAPTATION to the environment, and changing methods of food procurement and the use of *fire* demonstrate the ability to modify wooded landscapes. *LD-P/RMF*

[*See also* HUMAN EVOLUTION: CLIMATIC INFLUENCES, HUMAN MIGRATION: 'OUT OF AFRICA', NEANDERTHAL DEMISE]

Barton N (1991) Technological innovation and continuity at the end of the Pleistocene in Britain. In Barton N (ed.) *The Late Glacial in North West Europe*. London: Council for British Archaeology [Research Report 77], 234–245.

Dolukhanov PM (1997) The Pleistocene-Holocene transition in northern Eurasia: Environmental changes and human adaptations. *Quaternary International* 41/42: 181–191.

Van Andel TH and Tzedakis PC (1996) Palaeolithic landscapes of Europe and environs, 150,000–25,000 years ago: An overview. *Quaternary Science Reviews* 15: 481–500.

palaeomagnetic dating A broad term that includes ARCHAEOMAGNETIC DATING and describes *age-equivalence* dating techniques based on the NATURAL REMANENT MAGNETISM (NRM) of archaeological materials, SEDIMENTS or rocks. The DECLINATION (D), INCLINATION (I) and MAGNETIC INTENSITY of the Earth's *geomagnetic field* (see GEOMAGNETISM) vary continuously over time. Temporal variations are preserved in stratigraphic sequences. As molten volcanic rocks and baked CERAMICS cool, ferromagnetic particles within them align with the ambient geomagnetic field, acquiring THERMOREMANENT MAGNETISATION (TRM). Similarly, ferromagnetic sediment particles settling out of air, water or saturated sediments can acquire DETRITAL REMNANT MAGNETISATION (DRM), providing a proxy record of the ambient geomagnetic field at the time of formation.

On geological timescales, variations in the Earth's geomagnetic field include 180° shifts in direction of the Earth's geomagnetic pole, that is, between NORMAL POLARITY and *reverse polarity*. Such GEOMAGNETIC POLARITY REVERSALS vary in duration. POLARITY CHRONS or epochs lasting 10^5–10^7 years (e.g. the *Bruhnes chron* present normal polarity) are punctuated by relatively short-lived *polarity events* (or *subchrons*) of 10^4–10^5 years duration (e.g. six normal polarity events occur within the *Matuyama chron*). The NRM signal of geomagnetic reversal boundaries in sediments such as slowly accumulating MARINE and LACUSTRINE SEDIMENTS, LOESS and LAVA sequences provide important age-equivalent MARKER HORIZONS. The global and broadly SYNCHRONOUS nature of geomagnetic polarity reversals means that reversal stratigraphy can be used to correlate between stratigraphic sequences on a global scale. Furthermore, POTASSIUM-ARGON (K-Ar) DATING of field reversal boundaries identified in lava sequences enables their use as global chronostratigraphic marker horizons. Recently, an alternative palaeomagnetic timescale has been provided by ORBITAL TUNING of *marine sediment records*. Palaeomagnetic dating of field reversals has aided *correlation* and dating of the GEOLOGICAL RECORD, including records of QUATERNARY ENVIRONMENTAL CHANGE preserved in different MARINE SEDIMENT CORES. It has also facilitated correlation of marine records with terrestrial loess and lake sequences. For example, the GAUSS-MATUYAMA GEOMAGNETIC BOUNDARY occurs close to the PLIOCENE-PLEISTOCENE TRANSITION.

NRM records of MAGNETIC SECULAR VARIATIONS (i.e. short term, 10^3 years or less) may also be preserved in lake sediments, in LOESS sequences or in baked archaeological materials (e.g. pottery, kiln walls and tiles). Matching of palaeomagnetic signals from different stratigraphic sequences or samples provides a basis for establishing age-equivalence. Age estimates can be derived by correlation with detailed master geomagnetic curves derived from lake sequences that have been independently dated or calibrated by, for example, RADIOCARBON DATING. Regional differences in secular geomagnetic variations means that this palaeomagnetic dating method is limited to intraregional correlation.

A third type of palaeomagnetic dating is based on the MINERAL MAGNETISM of sediment sequences arising from variations in the magnetic mineral content. These magnetic properties are independent of the geomagnetic field and may reflect environmental changes. MAGNETIC SUSCEPTIBILITY, ISOTHERMAL REMANENT MAGNETISM (IRM) and *coercivity* of IRM profiles have been used to correlate marine and terrestrial stratigraphic sequences. Mineral magnetic stratigraphy has been used, for example, to correlate and establish age equivalence of Chinese loess sequences and MARINE ISOTOPIC STAGES in marine records.   *MHD*

Creer KM and Kopper JS (1974) Paleomagnetic dating of cave paintings in Tito Bustillo Cave, Asturias, Spain. *Science* 186: 348–350.

Hagstrum JT and Champion DE (1994) Paleomagnetic correlation of Late Quaternary lava flows in the lower east rift zone of Kilawea Volcano, Hawaii. *Journal of Geophysical Research* 99: 21679–21690.

King J and Peck J (2001) Use of palaeomagnetism in studies of lake sediments. In Last WM and Smol JP (eds) *Tracking environmental change using lake sediments, volume 1*. Dordrecht: Kluwer, 371–389.

Kukla GJ, Heller F, Liu XM et al. (1988) Pleistocene climates in China dated by magnetic susceptibility. *Geology* 16: 811–814.

Mankinen KE and Dalrymple GB (1979) Revised geomagnetic polarity time scale for the interval 0-5 My B.P. *Journal of Geophysical Research* 84: 615–626.

Tarling DH (1983) *Palaeomagnetism: Principles and applications in geology, geophysics and archaeology*. London: Chapman and Hall.

Tarling DH (1990) Archaeomagnetism and palaeomagnetism. In Göksu HY, Oberhofer M and Regulla D (eds) *Scientific dating methods*. Dordrecht: Kluwer, 217–250.

Thompson R (1991) Palaeomagnetic dating. In Smart PL and Frances PD (eds) *Quaternary dating methods: A users guide*. Cambridge: Quaternary Research Association, 177–198.

Thompson R and Oldfield F (1986) *Environmental magnetism*. London: Allen and Unwin.

Valet J-P and Meynadier L (1993) Geomagnetic field intensity and reversals during the past four million years. *Nature* 366: 234–238.

Verosub KL (1988) Geomagnetic secular variations and the dating of Quaternary sediments. *Geological Society of America* 227: 123–138 [Special Paper].

palaeomagnetism   The study of the characteristics of the Earth's magnetic field (see GEOMAGNETISM) in the past, and its application to geological, archaeological and environmental issues. Rocks, sediment and soil retain a permanent record of the Earth's magnetic field at the time of their formation in magnetic MINERALS such as *iron oxides*. This MAGNETIC REMANENCE can be isolated by laboratory techniques. Of particular importance are variations in the INCLINATION (I) of the preserved field and GEOMAGNETIC POLARITY REVERSALS. The science of palaeomagnetism expanded in the 1950s and is now a technologically sophisticated research field in EARTH SCIENCE.

Two major applications of palaeomagnetism are in reconstructing PALAEOGEOGRAPHY and in GEOCHRONOLOGY. The inclination of the preserved magnetic field is related to PALAEOLATITUDE, and changes through time can be used to construct a trajectory of APPARENT POLAR WANDER. The recognition that MAGNETIC ANOMALIES in IGNEOUS ROCKS of the OCEANIC CRUST had been produced by SEA-FLOOR SPREADING was vital in the acceptance of the theory of PLATE TECTONICS. Both apparent polar wander paths and ocean-floor magnetic anomalies are invaluable in reconstructing PALAEOGEOGRAPHY and CONTINENTAL DRIFT. The dating of geomagnetic polarity reversals has led to the development of the GEOMAGNETIC POLARITY TIMESCALE (GPTS). Other applications include the refinement of stratigraphical CORRELATION (MAGNETOSTRATIGRAPHY), PALAEOMAGNETIC DATING, determining the PROVENANCE of sediments and sedimentary rocks (see MAGNETIC SUSCEPTIBILITY), reconstructions of PALAEOCLIMATE and PALAEOENVIRONMENT and the reorientation of drill CORES by comparison with present field DECLINATIONS (D).   *DNT*

Courtillot V, Gallet Y, Le Mouël J-L et al. (2007) Are there connections between the Earth's magnetic field and climate? *Earth and Planetary Science Letters* 253: 328–339.

Dinares-Turell J, Orti F, Playa E and Rosell L (1999) Palaeomagnetic chronology of the evaporitic sedimentation in the Neogene Fortuna Basin (SE Spain): Early restriction preceding the 'Messinian Salinity Crisis'. *Palaeogeography, Palaeoclimatology, Palaeoecology* 154: 161–178.

Gibbard PL, Boreham S, Andrews JE and Maher BA (2010) Sedimentation, geochemistry and palaeomagnetism of the West Runton Freshwater Bed, Norfolk, England. *Quaternary International* 228: 8–20.

McElhinny MW and McFadden PL (1999) *Paleomagnetism: Continents and oceans.* London: Academic Press.

Théveniaut H, Quesnel F, Wyns R and Hugues G (2007) Palaeomagnetic dating of the "Borne de Fer" ferricrete (NE France): Lower Cretaceous continental weathering. *Palaeogeography, Palaeoclimatology, Palaeoecology* 253: 271–279.

palaeontology The study of FOSSILS as a field of GEOLOGY. It overlaps with many other fields, particularly PALAEOBIOLOGY and ARCHAEOLOGY. Major subdivisions include *vertebrate palaeontology*, *invertebrate palaeontology* and MICROPALAEONTOLOGY.   *GO/JP*

[*See also* BIOSTRATIGRAPHY, PALAEOBOTANY, PALAEOCLIMATOLOGY, PALAEO-ECOLOGY, PALAEOENTOMOLOGY]

Wynn Jones R (2006) *Applied palaeontology.* Cambridge: Cambridge University Press.

Wyse Jackson PN (2010) *Introducing palaeontology: A guide to ancient life.* Edinburgh: Dunedin Academic Press.

palaeoprecipitation proxies Data, evidence or records of past PRECIPITATION obtained indirectly (rather than measured or observed). There is a growing range of such proxies based, for example, on ICE CORES, TREE RINGS, SPELEOTHEMS, LAKE-LEVEL VARIATIONS, GLACIER VARIATIONS and PEAT HUMIFICATION. Many palaeoprecipitation records relate to summer DROUGHT; glacier variations provide one of the few approaches to the reconstruction of winter precipitation. However, reconstructing past precipitation and moisture regimes independently of palaeotemperature is often difficult because of covariation at large spatial scales and long timescales. At local to regional spatial scales and SUB-MILANKOVITCH timescales, the breakdown of this covariation is often a key to understanding the mechanisms of CLIMATIC CHANGE.   *JAM*

[*See also* CLIMATE OF THE LAST MILLENNIUM, CLIMATIC RECONSTRUCTION, PALAEOCLIMATOLOGY, PROXY CLIMATIC INDICATOR, PROXY DATA]

Nesje A, Matthews JA, Dahl SO et al. (2001) Holocene glacier fluctuations of Flatebreen and winter-precipitation changes in the Jostedalsbreen region, western Norway, based on glaciolacustrine sediment records, *The Holocene* 11: 267–280.

Verschuren D and Charman DJ (2008) Latitudinal linkages in late-Holocene moisture-balance variation. In Battarbee RW and Binney HA (eds) *Natural climate variability and global warming: A Holocene perspective.* Chichester: Wiley-Blackwell, 189–231.

palaeosalinity The SALINITY (the concentration of dissolved salts) of any past AQUATIC ENVIRONMENT (LAKE, OCEAN, LAGOON, GROUNDWATER, etc.) or past SOIL environment. Palaeosalinity is recorded and 'measured' by PROXY EVIDENCE (biological, lithological and chemical) in NATURAL ARCHIVES. Past fluctuations in salinity reflect a variety of changing environmental conditions, and this forms the basis of palaeosalinity reconstructions. Salinity increased during times when net EVAPORATION exceeded PRECIPITATION or other inputs such as via rivers or melting ice, whereas salinity decreases occurred during times when the opposite conditions prevailed. DIATOM ANALYSIS, MOLLUSCA ANALYSIS and OSTRACOD ANALYSIS (including TRACE ELEMENT ratio analysis of *carapaces*) are widely employed in palaeosalinity reconstructions. Recently, calibration data sets generated from diatom (and ostracod) salinity TRANSFER FUNCTIONS have been used for quantitative reconstruction of LACUSTRINE SEDIMENTS. In SALINE LAKES, stable OXYGEN ISOTOPE RATIOS ($\delta^{18}O$) are used as a proxy for EVAPORATION intensity, and therefore palaeosalinity. Chemical analysis of lake sediment cores can indicate changing salinity, for example, a sediment sequence of freshwater *carbonates* through *sulfates* to *chlorides* implies increasing salinity through time.

Quantitative *oceanic palaeosalinity* can be reconstructed using the stable OXYGEN ISOTOPE ratio ($\delta^{18}O$), stable HYDROGEN ISOTOPE ratios (δ^2H or δD) (see STABLE ISOTOPE ANALYSIS), paired $\delta^{18}O$ and MAGNESIUM:CALCIUM RATIO (Mg/Ca), analysis of carbonate, molecular organic proxies and SEA-SURFACE TEMPERATURES (SSTs). In turn, palaeosalinity can provide a history of the HYDROLOGICAL CYCLE, *basin hydrography*, *moisture transport*, THERMOHALINE CIRCULATION, SEA-LEVEL CHANGE, water-column STRATIFICATION, SALINISATION of soil and *lake evolution* (e.g. inferred freshwater to saline conditions indicates a change from an open to a closed lake setting).

SALINITY is critical to the pattern of ocean thermohaline circulation and past changes in salinity have resulted in significant oscillations in the behaviour of OCEAN CURRENTS and surface climate. The GEOLOGICAL RECORD OF ENVIRONMENTAL CHANGE contains evidence of PALAEOSALINITY in the form of EVAPORITE deposits that are indicative of past periods of ARIDITY. Thick evaporate deposits of Late MIOCENE age in the western Mediterranean are evidence for partial or nearly complete cyclical desiccation of the Mediterranean Sea at this time ('The MESSINIAN salinity crisis'). The effects of lower sea levels during the LAST GLACIAL MAXIMUM (LGM), coupled with environmental change to drier conditions in mid-latitudes, are believed to have caused *hypersaline conditions* in both the Red Sea and the Mediterranean Sea. Terrestrial evidence of palaeosalinity is also found in the geological record relating to past DESERT conditions.   *CES*

[*See also* MARINE PALAEOCLIMATIC PROXIES, PALAEOCEANOGRAPHY: BIOLOGICAL PROXIES,

PALAEOCEANOGRAPHY: PHYSICAL AND CHEMICAL PROXIES]

Castañeda IS and Schouten S (2011) A review of molecular organic proxies for examining modern and ancient lacustrine environments. *Quaternary Science Reviews* 30: 2851–2891.

Fritz SC, Cumming BF, Gasse F and Laird KR (2010) Diatoms as indicators of hydrologic and climatic change in saline lakes. In Smol JP and Stoermer EF (eds) (2010) *The diatoms: Applications for the environmental and Earth sciences, 2nd edition.* Cambridge: Cambridge University Press, 186–208.

LeGrande A and Schmidt G (2011) Water isotopologues as a quantitative paleosalinity proxy. *Paleoceanography* 26: PA3225.

Rohling EJ (2007) Progress in paleosalinity: Overview and presentation of a new approach. *Paleoceanography* 22: PA3215.

Turich C and Freeman KH (2011) Archaeal lipids record paleosalinity in hypersaline systems. *Organic Geochemistry* 42: 1147–1157.

palaeosciences The *historical sciences* or, more appropriately, those branches of the NATURAL ENVIRONMENTAL SCIENCES (e.g. PALAEOBIOLOGY, PALAEOCLIMATOLOGY, PALAEOECOLOGY, PALAEOENTOMOLOGY, PALAEOHYDROLOGY, PALAEOLIMNOLOGY and PALAEOCEANOGRAPHY) that focus on the reconstruction and modelling of past events rather than direct observation and experiment. However, they use MODERN ANALOGUES in the interpretation of past events and provide a long-term perspective for knowledge and understanding of present and future environmental change.   *JAM*

[*See also* PALAEODATA]

Jakobsson M, Ingólfsson Ó, Kjær KH et al. (eds) (2010) APEX: Arctic palaeoclimate and its extremes. *Quaternary Science Reviews* 29(25–26): 3349–3675 [Special Issue].

palaeoseismicity The study of past EARTHQUAKES, with the aim of determining their dates, EPICENTRES and magnitudes, in order to understand the TECTONIC evolution of *fault systems* and to contribute to EARTHQUAKE PREDICTION through reconstructing the RETURN PERIODS of potentially damaging events by extending the evidence beyond the INSTRUMENTAL RECORD and the duration of HISTORICAL EVIDENCE. This is particularly important in the New World where the detailed historical record may be shorter than the return periods of major events, particularly intraplate earthquakes. For example, the New Madrid earthquakes of AD 1811–1812 had the largest EARTHQUAKE MAGNITUDE of any historic event in the United States outside Alaska, although the area does not lie close to a PLATE MARGIN and no other large earthquakes have occurred in the area since Western colonisation of the Americas.

Methods that can be used include the study of SEDIMENTARY STRUCTURES indicative of seismically induced LIQUEFACTION or FLUIDISATION, such as *sand blows* and SOFT-SEDIMENT DEFORMATION structures; the recognition of seismically generated EVENT DEPOSITS within successions of sediments or sedimentary rocks; the analysis of GEOMORPHOLOGY; and cross-cutting features such as upward terminations of FAULTS in cross sections. A common technique is *trenching*, which involves excavating a trench across known FAULTS.   *GO/DNT*

Akciz SO, Ludwig LG, Arrowsmith JR and Zielke O (2010) Century-long average time intervals between earthquake ruptures of the San Andreas fault in the Carrizo Plain, California. *Geology* 38: 787–790.

Amick D and Gelinas R (1991) The search for evidence of large prehistoric earthquakes along the Atlantic seaboard. *Science* 251: 655–658.

Audemard FA, Michetti AM and McCalpin JP (eds) (2011) *Geological criteria for evaluating seismicity revisited: Forty years of paleoseismic investigations and the natural record of past earthquakes.* Boulder, CO: Geological Society of America.

Galli P, Galadini F and Pantosti D (2008) Twenty years of paleoseismology in Italy. *Earth-Science Reviews* 88: 89–117.

Galli P and Scionti V (2006) Two unknown M >6 historical earthquakes revealed by palaeoseismological and archival researches in eastern Calabria (southern Italy): Seismotectonic implications. *Terra Nova* 18: 44–49.

Manighetti I, Boucher E, Chauvel C et al. (2010) Rare earth elements record past earthquakes on exhumed limestone fault planes. *Terra Nova* 22: 477–482.

Obermeier SF (1996) Use of liquefaction-induced features for paleoseismic analysis: An overview of how seismic liquefaction features can be distinguished from other features and how their regional distribution and properties of source sediment can be used to infer the location and strength of Holocene paleo-earthquakes. *Engineering Geology* 44: 1–76.

Reicherter K, Michetti AM and Silva PG (eds) (2009) *Palaeoseismology: Historical and prehistorical records of earthquake ground effects for seismic hazard assessment.* Bath: Geological Society.

Sintubin M, Stewart IS, Niemi TM and Altunel E (eds) (2010) *Ancient earthquakes.* Boulder, CO: Geological Society of America.

Tuttle MP, Shweig III ES, Campbell J et al. (2005) Evidence for New Madrid earthquakes in AD 300 and 2350 BC. *Seismological Research Letters* 76: 489–501.

palaeoslope A slope that no longer exists. Its presence can be inferred by geologists using PALAEOCURRENT indicators and other SEDIMENTARY STRUCTURES such as SLUMP folds.   *GO*

Strachan LJ and Alsop GI (2006) Slump folds as estimators of palaeoslope: A case study from the Fisherstreet Slump of County Clare, Ireland. *Basin Research* 18: 451–470.

palaeosol A soil formed in a past environment not representative of the present phase of soil formation. It may be a FOSSIL SOIL, completely buried by later sedimentary deposition and possibly lithified (see LITHIFICATION), or it may be a RELICT soil at the land surface containing features indicating it formed in a different climate or beneath other vegetation than those influencing the current conditions of soil formation. Palaeosols are clear evidence of environmental change as they usually possess characteristics that are different from present soils. The full VERIFICATION of palaeosol status in a single section may not be possible. Ideally, soil morphological variation should be examined laterally as well as vertically, to verify the presence of a *palaeocatena* (see CATENA).

In southern Britain, soils in pre-DEVENSIAN PARENT MATERIALS have been described as palaeo-argillic *brown earths* (LUVISOLS), characterised by *decalcification, clay movement* (see ELUVIATION) and reorganisation of clayey soil matrices. Relict features of CRYOTURBATION from the Devensian are commonly seen (see INVOLUTIONS) in the subsoils of CAMBISOLS in northern Europe. CHERNOZEMS have been changed to LUVISOLS or PODZOLS by a change to more humid and forested conditions. Thus, the pedology of palaeosols has great potential for elucidating and reconstructing past climatic change, and possibly for predicting future pedological changes as environmental change occurs.

The most useful palaeosols are those forming time-transgressive CHRONOSEQUENCES such as in LOESS deposits containing several episodes of soil formation. Palaeosol sequences have also been recognised in volcanic, alluvial and slope deposits. Sequences of RIVER TERRACES may have soils of successively decreasing age. Although palaeosols are less likely to be well preserved in glacial deposits, buried INTERGLACIAL soils of Central Europe have been correlated with loess sequences for the Middle and Late Pleistocene. It is almost impossible to prove that a palaeosol has not been altered after its burial, and the RADIOCARBON DATING of palaeosols is often problematic because the turnover of carbon in soils occurs at varying rates (see APPARENT MEAN RESIDENCE TIME (AMRT)) according to climate and other factors, and the possibility of *contamination* by recent root material is great.   EMB

[*See also* EXHUMED SOIL, GEOSOL, SOIL STRATIGRAPHY]

Catt JA and Bronger A (eds) (1998) Reconstruction and climatic implications of palaeosols. *Catena* 34: 1–207 [Special Issue].

Chen Q, Li C, Li P et al. (2008) Late Quaternary palaeosols in the Yangtze Delta, China, and their palaeoenvironmental implications. *Geomorphology* 100: 465–483.

Ellis S and Matthews JA (1984) Pedogenetic implications of a ¹⁴C-dated palaeopodzolic soils at Haugabreen, southern Norway. *Arctic and Alpine Research* 16: 77–91.

Firman JB (2006) Ancient weathering zones, pedocretes and palaeosols on the Australian Precambrian shield and in adjoining sedimentary basins. *Journal of the Royal Society of Western Australia* 89: 57–82.

Gall Q (1999) Precambrian palaeosols: A view from the Canadian shield. In Thiry M and Simon-Coinçon R (eds) *Palaeoweathering, palaeosurfaces and related continental deposits.* Oxford: Blackwell, 207–224.

Jacobs P and Sedov S (eds) (2012) Paleosols in soilscapes of the past and present. *Quaternary International* 265: 1–204 [Special Issue].

Kraus MJ (1999) Paleosols in clastic sedimentary rocks: Their geologic applications. *Earth-Science Reviews* 47: 41–70.

Mauz B and Felix-Henningsen P (2005) Palaeosols in Saharan and Sahelian dunes of Chad: Archives of Holocene North African climatic changes. *The Holocene* 15: 453–458.

Retallack GJ (2001) *Soils of the past: An introduction, 2nd edition.* Oxford: Blackwell.

Wright VP (ed.) (1986) *Palaeosols: Their recognition and interpretation.* Oxford: Blackwell.

Yaalon DH (ed.) (1971) *Palaeopedology: Origin, nature and dating of palaeosols.* Jerusalem: ISSS and Jerusalem University Press.

Zhao H, Chen F-H, Li S-H et al. (2007) A record of Holocene climate change in the Guanzhong Basin, China, based on optical dating of a loess-palaeosol sequence. *The Holocene* 17: 1015–1022.

palaeotemperature proxies Data, evidence or records of past temperatures obtained indirectly; especially referring to AIR TEMPERATURE at the surface of the Earth reconstructed from NATURAL ARCHIVES. There is a vast array of such proxies, ranging from ICE CORES, TREE RINGS and SPELEOTHEMS to GLACIER VARIATIONS, MARINE SEDIMENT CORES and LACUSTRINE SEDIMENTS. In high-latitude land areas, for example, dense networks of tree-ring records enable HIGH-RESOLUTION RECONSTRUCTION, especially of summer temperatures. As temperature variations tend to exhibit greater SPATIAL COHERENCE than precipitation variations, regional- to global-scale patterns in palaeotemperature can be reconstructed from fewer sites.   JAM

[*See also* CLIMATE OF THE LAST MILLENNIUM, CLIMATIC RECONSTRUCTION, DENDROCLIMATOLOGY, MARINE PALAEOCLIMATIC PROXIES, OCEAN PALAEOTEMPERATURE, PALAEOCLIMATOLOGY, PROXY CLIMATIC INDICATOR, PROXY DATA]

Briffa KR, Osborn TJ, Schweingruber FH et al. (2002) Tree-ring width and density data around the Northern Hemisphere. Part 1: Local and regional climatic signals. *The Holocene* 12: 737–757.

Cronin TM (2009) Paleotemperatures and proxy reconstructions. In Gornitz V (ed.) *Encyclopedia of paleoclimatology and ancient environments.* Dordrecht: Springer, 757–764.

Jouzel J (1999) Calibrating the isotopic paleothermometer. *Science* 286: 910–911.

National Research Council (2006) *Surface temperature reconstructions for the last 2,000 years*. Washington, DC: National Academies Press.

palaeotempestology   The study of past TROPICAL CYCLONE activity before the INSTRUMENTAL RECORD. It focuses on DOCUMENTARY EVIDENCE, coastal SEDIMENTO-LOGICAL EVIDENCE OF ENVIRONMENTAL CHANGE (e.g. sand layers in lagoons, lakes and marshes) and evidence from DENDROCHRONOLOGY, and CORAL (see SCLERO-CHRONOLOGY). Use of the term could be appropriately extended to PALAEOENVIRONMENTAL RECONSTRUCTION associated with all types of past 'tempest' including STORMS, STORM SURGES, TSUNAMI, etc.   *JAM*

Liu K-B (2004) Paleotempestology: Principles, methods and examples from the Gulf Coast lake sediments. In Murname RJ and Liu K-B (eds) *Hurricanes and typhoons: Past present and future*. New York: Columbia University Press, 13–57.

palaeothermometry   The estimation of temperature in the GEOLOGICAL RECORD. The methodology involves using a PROXY found in natural records such as SEDI-MENT, ICE CORES and FOSSIL assemblages. Common and emerging geochemical *palaeothermometers* are oxygen STABLE ISOTOPES, Mg/Ca (see MAGNESIUM:CALCIUM RATIO (Mg:Ca)) and Sr/Ca (see STRONTIUM/CALCIUM PALAEO-THERMOMETRY), the *alkenone unsaturation index* (see ALKENONES), TEX_{86} and CLUMPED ISOTOPE PALAEOTHER-MOMETRY. Fossil assemblage and morphology-based approaches include quantitative assemblage analyses (involving TRANSFER FUNCTIONS, e.g. the CLIMAP study), nearest living relative analogy/coexistence analysis and leaf-margin analysis. Each method has its associated assumptions and uncertainties.   *HKC*

[*See also* PALAEOCEANOGRAPHY: PHYSICAL AND CHEMICAL PROXIES]

Eiler JM (2011) Paleoclimate reconstruction using carbonate clumped isotope thermometry. *Quaternary Science Reviews* 30: 3575–3588.
Hillaire-Marcel C and de Vernal A (eds) (2007) *Proxies in Late Cenozoic paleoceanography*. Amsterdam: Elsevier Science.
Jordan GJ (2011) A critical framework for the assessment of biological palaeoproxies: Predicting past climate and levels of atmospheric CO_2 from fossil leaves. *New Phytologist* 192: 29–44.
Lear CH (2007) Mg/Ca palaeothermometry: A new window into Cenozoic climate change. In Williams M, Hayward A, Gregory J and Schmidt DN (eds) *Deep time perspectives on climate change: Marrying the signal from computer models and biological proxies*. Bath: Geological Society, 313–322.

palaeovalley   A valley that no longer exists, or a valley that was formed in the past under different conditions than those pertaining at present (and there-fore is RELICT or buried). Examples include valleys that have been infilled by SEDIMENTARY DEPOSITS or LAVA FLOWS, or valleys whose existence in the past can be inferred from GEOLOGICAL EVIDENCE OF ENVIRONMEN-TAL CHANGE. The term is commonly used in relation to *incised palaeovalleys* in the context of SEQUENCE STRATIGRAPHY. These are river valleys that experienced major incision during a rapid fall in BASE LEVEL and were infilled by AGGRADATION during a subsequent base-level rise such as a *marine transgression*.   *GO*

Clark-Lowes DD (2005) Arabian glacial deposits: Recognition of palaeovalleys within the Upper Ordovician Sarah Formation, Al Qasim district, Saudi Arabia. *Proceedings of the Geologists' Association* 116: 331–347.
Delgado J, Boski T, Nieto JM et al. (2012) Sea-level rise and anthropogenic activities recorded in the Late Pleistocene/ Holocene sedimentary infill of the Guadiana Estuary (SW Iberia). *Quaternary Science Reviews* 33: 121–141.
Hou B, Frakes LA, Sandiford M et al. (2008) Cenozoic Eucla Basin and associated palaeovalleys, southern Australia: Climatic and tectonic influences on landscape evolution, sedimentation and heavy mineral accumulation. *Sedimentary Geology* 203: 112–130.

palaeovelocity   The velocity of a past FLOOD flow (or other type of flow) estimated from geomorphologi-cal evidence, such as the depth of the flow, and the slope and roughness of the former channel.   *JAM*

Church M, Wolcott J and Maizels J (1990) Palaeovelocity: A parsimonious proposal. *Earth Surface Processes and Landforms* 15: 475–480.

palaeowind   A prevailing wind that occurred in the past. The direction and strength of palaeowinds may be reconstructed from RELICT landforms and SEDIMENTARY STRUCTURES. There is also the possibility of reconstruct-ing *atmospheric circulation* patterns for specific peri-ods of the GEOLOGICAL RECORD. Since AEOLIAN processes tend to leave their mark on the landscape under *arid* or SEMI-ARID conditions (see ARIDITY), evidence of ancient airflows may provide a clear indication of CLIMATIC CHANGE. The erosional effectiveness of wind depends on weak binding agencies in the soil, a SOIL MOISTURE DEFICIT (SMD) and a bare soil surface, which are all associated with ARIDITY.   *RC/RDT*

[*See also* PALAEODUNES, VENTIFACT]

Muhs DR and Bettis EA (2000) Geochemical variations in Peoria loess of western Iowa indicate paleowinds of midcontinental North America during Last Glaciation. *Quaternary Research* 53: 9–61.
Thomas DSG and Shaw PA (1991) 'Relict' desert dune systems: Interpretations and problems. *Journal of Arid Environments* 20: 1–14.

Palaeozoic An ERA of geological time incorporating the CAMBRIAN, ORDOVICIAN, SILURIAN (*Early/Lower Palaeozoic*), DEVONIAN, CARBONIFEROUS and PERMIAN (*Late/Upper Palaeozoic*) PERIODS. Prior to the start of the Palaeozoic, the Earth's continents were assembled in the SUPERCONTINENT known as RODINIA. During the Palaeozoic Era, the Earth's continents drifted apart from their previous SUPERCONTINENT configuration known as RODINIA, and their assembly into the subsequent supercontinent PANGAEA (PANGEA) in the Late Palaeozoic was associated with a series of episodes of OROGENY. The Late Ordovician and Late Carboniferous to Permian were episodes of ICEHOUSE CONDITIONS when polar and temperate regions experienced ICE AGES and associated SEA-LEVEL CHANGES. The end of the Permian period is marked by the most severe MASS EXTINCTION event in Earth's history.   *GO*

[*See also* GEOLOGICAL TIMESCALE]

Linnemann U, Nance RD, Kraft P and Zulauf G (eds) (2007) *The evolution of the Rheic Ocean: From Avalonian-Cadomian active margin to Alleghenian-Variscan collision.* Boulder, CO: Geological Society of America.
López-Gamundi OR and Buatois LA (eds) (2010) *Late Paleozoic glacial events and postglacial transgressions in Gondwana.* Boulder, CO: Geological Society of America.
Scheffler K, Buehmann D and Schwark L (2006) Analysis of Late Palaeozoic glacial to postglacial sedimentary successions in South Africa by geochemical proxies: Response to climate evolution and sedimentary environment. *Palaeogeography, Palaeoclimatology, Palaeoecology* 240: 184–203.
Tarver JE, Braddy SJ and Benton MJ (2007) The effects of sampling bias on Palaeozoic faunas and implications for macroevolutionary studies. *Palaeontology* 50: 177–184.

palimpsest A 'multilayered', RELICT feature (e.g. LANDSCAPE, CONTINENTAL MARGIN, SEDIMENT distribution and rock TEXTURE) in which the effects of successive generations of formative processes can still be deciphered. The environmental use of the term derives, by analogy, from the reuse of parchment after earlier writing has been rubbed out.   *JAM/GO*

Marrs R (2008) Landscape as a palimpsest: Grassland sustainability in Sweden. *Biological Conservation* 141: 1445–1446.
Rivers JM, James NP, Kyser TK and Bone Y (2007) Genesis of palimpsest cool-water carbonate sediment on the continental margin of southern Australia. *Journal of Sedimentary Research* 77: 480–494.

palinspastic map A reconstruction of PALAEOGEOGRAPHY that removes the effects of DEFORMATION due to TECTONICS, and represents the original spatial disposition of features on the Earth's surface. The term is from the Greek, 'again pulling'.   *JCWC*

Cope JCW, Ingham JK and Rawson PF (eds) (1999) *Atlas of palaeogeography and lithofacies, 2nd edition.* London: Geological Society.

palsa A dome-shaped or oval peat-covered perennial cryogenic mound (FROST MOUND) commonly 1–10 m high, 10–30 m in width and 15–150 m in length. Palsas occur in MIRES in areas of *subpolar climates* with discontinuous PERMAFROST. Ice layers in palsas have been interpreted as lenses of SEGREGATION ICE but recently have been reinterpreted as ice growth at the base of a frozen core that is effectively floating. Palsas require large quantities of water, are often found in groups and are related to more extensive *palsa plateaux*.   *RAS/JAM*

[*See also* LITHALSA]

Gurney SD (2001) Aspects of the genesis, geomorphology and terminology of palsas: Perennial cryogenic mounds. *Progress in Physical Geography* 25: 249–260.
Matthews JA, Dahl SO, Berrisford MS and Nesje A (1997) Cyclic development and thermokarstic degradation in palsas in the mid-alpine zone at Leirpullen, Dovrefjell, southern Norway. *Permafrost and Periglacial Processes* 8: 107–122.
Seppälä M (2011) Synthesis of studies of palsa formation underlining the importance of local environmental and physical characteristics. *Quaternary Research* 75: 366–370.

paludification The process whereby BLANKET MIRES or MIRES expand into formerly dry, often forested upland as a result of changes in hydrological conditions of the surrounding area. This can result from natural ENVIRONMENTAL CHANGE or ANTHROPOGENIC influence. The rising water level which causes bog expansion may result from increased PRECIPITATION or a decrease in EVAPO-TRANSPIRATION as a result of a cooler and wetter or more humid environment. Paludification can also be caused by a change in drainage characteristics of the area and DEFORESTATION and *forest fires* have been implicated. SPHAGNUM moss species often act as pioneer communities within areas undergoing paludification, stimulating rapid colonisation by other bog or mire communities. This results in organic matter accumulation and an increase in thickness of PEAT. CLIMATIC CHANGE has often been seen as responsible for paludification during the HOLOCENE.   *MLC*

Hulme PD (1994) A paleobotanical study of a paludifying pine forest on the island of Hailuolo, Finland. *New Phytologist* 126: 153–162.
Morris PJ, Belyea LR and Baird AJ (2011) Ecohydrological feedbacks in peatland developments: A theoretical modelling study. *Journal of Ecology* 99: 1190–1201.
Simard M, Bernier P, Bergeron Y et al. (2009) Paludification dynamics in the boreal forest of the James Bay Lowlands:

Effects of time since fire and topography. *Canadian Journal of Forest Research* 39: 546–552.

Ugolini FC and Mann DH (1979) Biopedological origin of peatland in south east Alaska. *Nature* 281: 366–368.

palustrine   Pertaining to the marginal areas of LAKES and SWAMPS. Many lakes are sensitive to variations in HYDROLOGICAL BUDGET, so ancient palustrine deposits are potentially valuable indicators of LAKE-LEVEL VARIATIONS.   *MRT*

[*See also* PARALIC]

Djarnali M, Sonlie-Marsche I, Esu D et al. (2006) Palaeoenvironment of a Late-Quaternary lacustrine-palustrine carbonate complex: Zarand Basin, Saveh, central Iran. *Palaeogeography, Palaeoclimatology, Palaeoecology* 237: 315–334.

Seton M, Muller RD, Zahirovic S et al. (2012) Global continental and ocean basin reconstructions since 200Ma. *Earth-Science Reviews* 113: 212–270.

Verrecchia EP (2007) Lacustrine and palustrine geochemical sediments. In Nash DJ and McLaren SJ (eds) *Geochemical sediments and landscapes.* Malden, MA: Blackwell, 298–329.

Wright VP and Platt NH (1995) Seasonal wetland carbonate sequences and dynamic catenas: A re-appraisal of palustrine limestones. *Sedimentary Geology* 99: 65–71.

palynomorph   Pollen-sized organic microfossils (*pollen grains* and *non-pollen palynomorphs*), which are often encountered during POLLEN ANALYSIS and may be of interest in their own right for PALAEOENVIRON-MENTAL RECONSTRUCTION. Examples include SPORES, the *guard cells* of STOMATA, ALGAE, BACTERIA and the eggs of *rotifers*.   *JAM*

Haas JN (ed.) (2010) Fresh insights into the palaeoecological and palaeoclimatological value of Quaternarty non-pollen palynomorphs. *Vegetation History and Archaeobotany* 19: 1–558 [Special Issue].

Van Geel B (2001) Non-pollen palynomorphs. In Smol JP, Birks HJB and Last WM (eds) *Tracking environmental change using lake sediments, volume 3.* Dordrecht: Kluwer, 99–119.

pan   A closed ARIDLAND depression up to about 1,000 km^2 in area, which may contain an EPHEMERAL lake or have contained a permanent lake in the past. Pans have been attributed to the combination of a variety of processes, including DEFLATION by wind, GROUNDWA-TER variations and SALINISATION. In some regions, they are termed *nor* (Mongolia), *playa* (North America) and *sebkha* (North Africa to the Middle East). The term 'pan' has also been applied to SALINE LAKES and/or their basins. The term *panfan* has been used to describe an extensive plain produced by the amalgamation of ALLUVIAL FANS.   *JAM*

[*See also* CHOTT, HARDPAN, SABKHA]

Goudie AS and Wells GL (1995) The nature, distribution and formation of pans in the arid zone. *Earth-Science Reviews* 38: 1–69.

Neal JT (ed.) (1975) *Playas and dried lakes.* Stroudsburg, PA: Dowden, Hutchinson and Ross.

Shaw PA and Thomas DSG (1997) Pans, playas and salt lakes. In Thomas DSG (ed.) *Arid zone geomorphology: Process, form and change in drylands, 2nd edition.* Chichester: Wiley, 293–317.

panarchy   The concept of a hierarchical dynamical system (*panarchical organisation*) with multiple interrelationships between elements used in relation to biological, ecological and social SYSTEMS and ENVI-RONMENT-HUMAN INTERACTION. *Stability* and SUSTAINABIL-ITY is maintained by multiple interactions amongst a nested set of ADAPTIVE CYCLES, while both incremental and abrupt changes can occur.   *JAM*

[*See also* DYNAMICAL SYSTEMS, EQUILIBRIUM CONCEPTS, HIERARCHY THEORY, STABILITY CONCEPTS: ECOLOGICAL CONTEXT]

Gunderson LH and Holling CS (eds) (2002) *Panarchy: Understanding transformations in human and natural systems.* Washington, DC: Island Press.

pandemic   An outbreak of DISEASE that extends beyond its country of origin and may become global in extent. Major pandemics include those of the historical past, associated with PLAGUE and SMALLPOX, the INFLU-ENZA pandemic of AD 1918–1919 and the current AIDS pandemic.   *JAM*

[*See also* EPIDEMIC]

Barry JM (2009) Pandemics: Avoiding the mistakes of 1918. *Nature* 459: 324–325.

Hays JN (2005) *Epidemics and pandemics: Their impacts on human history.* Santa Barbara, CA: ABC-CLIO.

Pangaea (Pangea)   A single CONTINENT (SUPERCON-TINENT) assembled from all the present-day continents that existed for up to 70 million years from the Late-CARBONIFEROUS or Early-PERMIAN to the Late-TRIASSIC periods. Pangaea formed when the continents of GOND-WANA (Western Europe, Africa, Antarctica, Australia, peninsular India and South America) joined with those of *Laurasia* (Asia, Europe and North America). Its PALAEOGEOGRAPHY consisted of a continent stretching from northern to southern polar regions with a narrow equatorial part, to the east of which the TETHYS Ocean formed an embayment. The concept of Pangaea ('all lands') was first proposed by Alfred Wegener as part of his theory of CONTINENTAL DRIFT. It is unclear whether such amalgamations of continents are a random or cyclical feature of Earth history, but it seems likely that at least one previous supercontinental assemblage has occurred, during the Late PROTEROZOIC.

The break-up of Pangaea and its consequences dominates the MESOZOIC and CENOZOIC (*CAINOZOIC*) history of the Earth. SEA-FLOOR SPREADING first developed between North America and Africa/South America, initiating the southern part of the North Atlantic Ocean. East of Africa another rift separated Africa/South America from other Gondwanan continents and a further rift parted India from Antarctica. Progressive extension of these spreading centres through the JURASSIC and CRETACEOUS periods led to opening of the southern Atlantic Ocean and widening of the northern Atlantic, which remained closed to the north. As Africa and India separately moved northwards, the Tethys OCEAN was squeezed out of existence and during the Late PALAEOGENE and Early NEOGENE this compressional movement culminated in the formation of the great mountain chains that cross Europe and Asia, including the Alps and the Himalayas (see OROGENIC BELT (OROGEN)). The Atlantic finally connected with the Arctic Ocean and continues to grow at the expense of the Pacific. In 200 or 300 million years' time, the Pacific may have disappeared, forming a *neopangaean supercontinent* with Europe and Africa on its western margin.   JCWC

[*See also* PLATE TECTONICS]

Embry AF, Beauchamp B and Glass DJ (eds) (1994) *Pangea: Global environments and resources.* Canadian Society of Petroleum Geologists, Memoir 17.

Parrish JT (1993) Climate of the supercontinent Pangea. *Journal of Geology* 101: 215–233.

Piper JDA and Zhang QR (1999) Palaeomagnetic study of Neoproterozoic glacial rocks of the Yangzi Block: palaeolatitude and configuration of South China in the Late Proterozoic supercontinent. *Precambrian Research* 94: 7–10.

Scotese CR (2011) *Paleomap project* [Available at http://www.scotese.com/earth.htm].

Shi GR and Grunt TA (2000) Permian Gondwana-Boreal antitropicality with special reference to brachiopod faunas. *Palaeogeography, Palaeoclimatology, Palaeoecology* 155: 239–263.

Windley BF (1995) *The evolving continents, 3rd edition.* Chichester: Wiley.

parabiosphere   That part of the BIOSPHERE where only dormant life exists, including areas at too high an altitude, too dry, too cold or too hot to support metabolising organisms (except technically equipped human explorers).   JAM

Hutchinson GE (1970) The biosphere. *Scientific American* 223: 45–53.

parabolic dune   An elongate SAND DUNE ranging up to kilometres in length and in plan view approximating in shape to a parabola in which the horns point upwind (to windwards). Alternative names are *blowout dune* and *hairpin dune*. The horns are anchored by vegetation (grasses, shrubs or trees). Such dunes are associated particularly with SEMI-ARID regions, but they are also found as COASTAL DUNES, and they are the commonest type of dune in PERIGLACIAL ENVIRONMENTS.   RAS/JAM

Forman SL, Sagintayev Z, Sultan M et al. (2008) The twentieth-century migration of parabolic dunes and wetland formation at Cape Cod National Sea Shore, Massachusetts, USA: Landscape response to a legacy of environmental disturbance. *The Holocene* 18: 765–774.

Reitz MD, Jerolmack DJ, Ewing RC and Martin RL (2010) Barchan-parabolic dune pattern transition from vegetation stability threshold. *Geophysical Research Letters* 37: L19402.

Seppälä M (2004) *Wind as a geomorphic agent in cold climates.* Cambridge: Cambridge University Press.

Wolfe SA and David PP (1997) Parabolic dunes: Examples from the Great Sand Hills, southwestern Saskatchewan. *Canadian Geographer* 41: 207–213.

paradigm   A term promoted by Thomas Kuhn to embrace a combination of THEORY and METHODOLOGY that permits scientists to select, criticise and evaluate KNOWLEDGE. For a while, scientists may operate within one paradigm which may eventually be overthrown during a *paradigm shift* by another in a 'SCIENTIFIC REVOLUTION'. The term was so loosely used by Kuhn that he was forced into redefinition of the term and subdivision of it into *exemplars* and *disciplinary matrixes*. It is still widely used in a very loose fashion with a meaning akin to disciplinary matrix or *Weltanschauung* (*world view* or *world outlook*). Thus, a paradigm may be defined as a framework not only for existing knowledge and understanding but also for the conduct of RESEARCH, which pervades the problems posed by scientists and the approaches used in problem solving.

Matthews et al. (2012) identified 49 paradigms or milestones in ENVIRONMENTAL CHANGE: HISTORY OF THE FIELD (milestones being not sufficiently revolutionary across the whole field to be termed 'paradigms'). Birks (2008) recognised 14 paradigms in the development of Holocene palaeoclimatic research alone. Chambers and Brain (2002) considered whether the increased emphasis on CLIMATE CHANGE and GLOBAL WARMING that was evident in citation statistics during the 1990s constituted a paradigm shift. Although they identified the importance not only of new scientific findings from the 1970s and 1980s in triggering the change of emphasis in research but also much earlier precursors, the change was arguably sufficiently rapid and pervasive to constitute a paradigm shift.   JAM/CET

Birks HJB (2008) Holocene climate research: Progress, paradigms and problems. In Battarbee RW and Binney HA (eds) *Natural climate variability and global warming: A Holocene perspective.* Chichester: Wiley-Blackwell, 7–57.

Chambers FM and Brain SA (2002) Paradigm shifts in late-Holocene climatology? *The Holocene* 12: 239–249.

Kuhn T (1996) *The structure of scientific revolutions, 3rd edition.* Chicago: Chicago University Press.

Matthews JA, Bartlein PJ, Briffa KR et al. (2012) Background to the science of environmental change. In Matthews JA, Bartlein PJ, Briffa KR et al. (eds) *The SAGE handbook of environmental change, volume 1.* London: Sage, 1–33.

paraglacial   Church and Ryder (1972: 3059) applied this term to both 'nonglacial processes that are directly conditioned by GLACIATION' and the period 'during which paraglacial processes occur'. The *paraglacial concept* was developed with respect to the formation of valley-floor ALLUVIAL FANS following DEGLACIATION in the Late PLEISTOCENE. Although originally used to describe the gravitational (*mass movement*) and *fluvial* reworking of GLACIAL SEDIMENTS in North America, the term has since been employed in other formerly glaciated areas and has been extended to encompass sediment reworking by AEOLIAN and coastal-zone processes; recent and present-day analogues are also encompassed by the term.

The processes recognised as being responsible for the reworking of *glacigenic sediments* are not unique to recently deglaciated terrain but usually occur there with greater frequency and intensity than elsewhere because of the relative abundance of readily entrainable or unstable *glacial deposits*. The *paraglacial readjustment* of slopes of *glacigenic sediments* leads to *gullying*, and COLLUVIAL FANS or DEBRIS CONES develop at and beyond the downslope limit of gullies. DEBRIS FLOW is a particularly common paraglacial slope process, and bedrock slopes are prone to FAILURE as a result of glacial-steepening and UNLOADING. *Paraglacial activity* in FLUVIAL systems may create ALLUVIAL FANS and VALLEY-FILL DEPOSITS, and AEOLIAN reworking of GLACIAL SEDIMENTS often leads to thick and extensive accumulations of LOESS and COVERSAND. MARINE TRANSGRESSION of glaciated terrain in the coastal zone produces distinctive features reflecting the abundance and type of glacially derived SEDIMENT and the presence of glacial LANDFORMS.

The *paraglacial period*, characterised by high rates of sediment delivery, begins with the onset of DEGLACIATION and ends when SEDIMENT YIELDS fall to rates typical of unglaciated terrain. Gravitational processes may rework unstable valley-side debris within a relatively short time (decades to centuries) following ice removal (see REDEPOSITION), whereas FLUVIAL PROCESSES and FAILURE of rock slopes may operate over substantially longer periods (millennia). It is therefore difficult to define the end of a paraglacial period because some of these LANDSCAPES may experience delayed slope responses.

Paraglacial activity should be distinguished from PERIGLACIAL processes, which are associated with cold, nonglacial environments, although some processes and forms may be common to both zones.   *PW*

[*See also* DENUDATION RATE, LANDSLIDE, SEA-LEVEL CHANGE, STURZSTRÖM]

Ballantyne CK (2002) A general model of paraglacial landscape response. *The Holocene* 12: 371–376.

Ballantyne CK (2002) Paraglacial geomorphology. *Quaternary Science Reviews* 21: 1935–2017.

Church M and Ryder JM (1972) Paraglacial sedimentation: A consideration of fluvial processes conditioned by glaciation. *Geological Society of America Bulletin* 83: 3059–3072.

Curry AM, Cleasby V and Zukowskyj P (2006) Paraglacial response of steep, sediment-mantled slopes to post-'Little Ice Age' glacier recession in the central Swiss Alps. *Journal of Quaternary Science* 21: 211–225.

Knight J and Harrison S (eds) (2009) *Periglacial and paraglacial processes and environments.* Bath: Geological Society.

Meigs A, Krugh WC, Davis K and Bank G (2006) Ultra-rapid landscape response and sediment yield following glacier retreat, Icy Bay, southern Alaska. *Geomorphology* 78: 207–221.

Mercier D, Étienne S, Sellier D and André MF (2009) Paraglacial gullying of sediment-mantled slopes: A case study of Colletthøgda, Kongsfjorden area, West Spitsbergen (Svalbard). *Earth Surface Processes and Landforms* 34: 1772–1789.

Wilson P (2005) Paraglacial rock-slope failures in Wasdale, western Lake District, England: Morphology, styles and significance. *Proceedings of the Geologists' Association* 116: 349–361.

paraglaciation   The modification of a LANDSCAPE by PARAGLACIAL processes and the extent of that modification. Paraglaciation is most intense following DEGLACIATION when paraglacial processes are most active. The term *neoparaglaciation* refers to similar modification of recent landscapes following DEGLACIERISATION of GLACIER FORELANDS since the LITTLE ICE AGE, or following earlier NEOGLACIATION.   *JAM*

[*See also* GLACIATION, PERIGLACIATION]

Embleton-Hamann C and Slaymaker O (2012) The Austrian Alps and paraglaciation. *Geografiska Annaler* 94A: 7–16.

Matthews JA and Shakesby RA (2004) A twentieth-century neoparaglacial rock topple on a glacier foreland, Ötztal Alps, Austria. *The Holocene* 14: 454–458.

paralic   Pertaining to marginal marine ENVIRONMENTS, such as *intertidal zones, supratidal zones,* LAGOONS and DELTAS, and the organisms therein. *Paralic environments* are non-marine environments landwards of the SHORELINE, often characterised by *intertonguing* of marine and TERRESTRIAL deposits. Seawards of the *paralic zone* is the shallow-marine environment of the

neritic zone. ENVIRONMENTAL STRESS for *paralic organisms* comes from variations in SALINITY, OXYGEN, DESICCATION, TEMPERATURE and DISTURBANCE of substrates.
LC/JAM

[*See also* PALUSTRINE]

Debenay J-P and Guillou J-J (2002) Ecological transitions indicated by foraminiferal assemblages in paralic environments. *Estuaries and Coasts* 25: 1107–1120.
Guelorget O and Perthuisot J-P (1992) Paralic ecosystems: Biological organisation and functioning. *Vie et Milieu* 42: 215–251.

paralithic contact The zone of weathered BEDROCK at the base of a SOIL, which differs from the unmodified (LITHIC) rock beneath. It may be altered chemically, contain fissures and allow root penetration but is not itself part of the SOIL.
JAM

[*See also* REGOLITH, SAPROLITE, WEATHERING PROFILE]

parallax The change in the apparent relative positions of two points when viewed from different positions.
TS

parallel lamination A SEDIMENTARY STRUCTURE comprising thin sheets (usually <1 mm) approximately parallel to the depositional horizontal, as represented by BEDDING (cf. CROSS-LAMINATION). Parallel lamination implies deposition on a flat sediment surface. When developed in fine- to medium-grained SAND, the structure indicates upper-stage PLANE BED conditions, indicative of a strong current or, in SEDIMENTARY ROCKS, PALAEOCURRENT. *Parting lineation* is commonly preserved on bedding surfaces, giving an indication of palaeocurrent direction, and the structure is also known as *planar lamination* or *horizontal lamination*. Thin sheets of alternating fine sand and MUD imply sediment settling from a SUSPENDED LOAD under conditions of repeatedly alternating sediment movement and still water; such deposits accumulate over longer time intervals and are better described as *horizontal bedding*. They may represent *cyclic sedimentation* associated with seasons or TIDES.
GO

[*See also* TIDAL RHYTHMITE]

Allen JRL (1982) *Sedimentary structures: Their character and physical basis*. Amsterdam: Elsevier.
Bridge JS and Demicco RV (2008) *Earth surface processes, landforms and sediment deposits*. New York: Cambridge University Press.

parallel slope retreat The view that DENUDATION of a LANDSCAPE occurs by backwearing of *escarpment* slopes and hills while slope angles are maintained. Developed by L.C. King to explain the African escarpments and INSELBERGS, it is one of the three classic SLOPE EVOLUTION MODELS.
RAS

King LC (1962) *The morphology of the Earth*. Edinburgh: Oliver and Boyd.

paramagnetism A weak positive MAGNETISATION, that occurs when individual ATOMS, IONS or MOLECULES possessing a permanent *magnetic dipole moment*, align themselves parallel with the direction of an applied MAGNETIC FIELD. Paramagnetism is lost once the field is removed.
AJP/WZ

parameteorology The study of weather-dependent natural phenomena (e.g. GLACIER VARIATIONS, DROUGHTS, FLOODS, freeze-up and break-up of rivers and lakes and the extent of SEA ICE) but usually excluding the timing of the seasonal biological phenomena that constitutes the separate field of PHENOLOGY.
JAM

Bradley RS (1999) *Paleoclimatology*, 2nd edition. San Diego, CA: Academic Press.

parameter (1) A characteristic of a statistical POPULATION, such as the population mean (μ) or the population standard deviation (σ), as opposed to *sample estimates* of the same, such as the SAMPLE mean ($\bar{x}$) or the sample standard deviation (s). (2) A quantity related to one or more VARIABLES in such a way that it remains constant even though the values of the variable(s) may change.
JAM

Aster RC, Borchers B and Thurber CH (2012) *Parameter estimation and inverse problems*, 2nd edition. Waltham, MA: Elsevier.

parameterisation The technique used in model CALIBRATION, also known as *initialisation* or *tuning*, whereby PARAMETERS used in the model are estimated from VARIABLES that can be measured and used to represent a complex process as a simplified function. Thus, the aim of parameterisation or *parameter estimation* is to make the model fit the data. Once the MODEL has been parameterised, CONFIRMATION or VALIDATION procedures should be applied to test it before it is used to predict the workings of the real world. Parameter estimation can be performed graphically or mathematically.
JAM

parametric statistics Statistical procedures for testing HYPOTHESES or estimating PARAMETERS (e.g. CONFIDENCE INTERVALS) that make assumptions about the underlying PROBABILITY DISTRIBUTION of the variables (e.g. NORMAL DISTRIBUTION) and their variability. Examples include ANALYSIS OF VARIANCE (ANOVA) and REGRESSION ANALYSIS.
HJBB

[*See also* NON-PARAMETRIC STATISTICS]

parasite An organism that obtains a material benefit from another organism to the detriment of the second organism. *Parasitism* differs from PREDATION in that there is a long-lasting relationship with the host organism. *Endoparasites* live inside the host and include the *microparasites*, which are causative agents for many human DISEASES, such as *measles*, *leprosy* and MALARIA. The term *parasitoid* is used for an organism where the adult is free living but one or more immature stages of the life cycle live on or in the host. *JLI*

[*See also* ANIMAL DISEASES, EPIDEMIC, EPIPHYTE, HUMAN HEALTH HAZARDS, PATHOGENS, PREDATOR-PREY RELATIONSHIPS, SYMBIOSIS, ZOONOSES]

Hatcher MJ and Dunn AM (2011) *Parasites in ecological communities: From interactions to ecosystems.* Cambridge: Cambridge University Press.
Morand S and Krasnov BR (eds) (2010) *The biogeography of host-parasite interactions.* Oxford: Oxford University Press.
Poulin R (2006) *Evolutionary ecology of parasites, 2nd edition.* Princeton, NJ: Princeton University Press.
Thomas F, Renaud F and Guégan J-F (2005) *Parasitism and ecosystems.* Oxford: Oxford University Press.
Woo PTK and Buchmann K (eds) (2012) *Fish parasites: Pathobiology and protection.* Wallingford: CABI Publishing.

parent material   The mineral material from which a soil has been derived. It is one of the SOIL-FORMING FACTORS and can be characterised by the subsoil or the C horizon. In terms of SOIL DEVELOPMENT, it represents the initial state of the soil system. Parent material may be the BEDROCK below the SOIL PROFILE, but often thin veneers of REGOLITH, including AEOLIAN, glacial, colluvial or other *sedimentary deposits*, are the major contributors of mineral material to soils. Young soils may have many features of the parent material, but older soils retain little influence of parent material except where it is of special composition such as QUARTZ sand or extremely CLAY-rich materials. Organic deposits are the parent materials for HISTOSOLS. *EMB*

[*See also* ALLUVIUM, COLLUVIUM, GLACIAL SEDIMENTS, PEAT AND PEATLAND]

Chesworth W (1973) The parent rock effect in the genesis of soil. *Geoderma* 10: 215–225.
Gellatly AF (1987) Establishment of soil covers on tills of variable texture and implications for interpreting palaeosols: A discussion. In Gardiner V (ed.) *International geomorphology 1986, Part II.* London: Wiley, 775–784.
Mason JA, Milfred CJ and Nater EA (1994) Distinguishing soil age from parent material effects on an Ultisol of north-central Wisconsin, USA. *Geoderma* 61: 165–189.
Paton TR (1978) *The formation of soil material.* London: Allen and Unwin.

parietal art   Paintings or engravings on rock walls as opposed to smaller, movable art objects such as sculpture (*mobiliary art*). *JAM*

[*See also* CAVE ART, PETROGLYPH]

parna   A deposit of DUST, of AEOLIAN origin, distinguishable from LOESS by its high CLAY content (30–70 per cent). The term is of Australian aboriginal origin. In southeastern Australia, parna layers were deposited during arid climatic phases of the QUATERNARY. *JAM*

Butler BE (1956) Parna: An aeolian clay. *Australian Journal of Science* 18: 145–151.
Hesse PP and McTainsh (2003) Australian dust deposits: Modern processes and the Quaternary record. *Quaternary Science Reviews* 22: 2007–2035.

parsimony   The *principle of parsimony, simplicity* or economy states that preference should be given to explanations invoking the fewest assumptions or explanatory contingencies. It concerns ontological simplicity and the theory of choice and is also known as *Ockam's (Occam's) Razor*. *MAU/CET*

[*See also* MULTIPLE WORKING HYPOTHESES, ONTOLOGY]

Baker A (2003) Quantitative parsimony and explanation. *British Journal for the Philosophy of Science* 54: 245–259.

partial correlation coefficient   A measure of the *correlation* between *dependent* (response) and *independent* (explanatory) *variables* while the statistical effects of one or more other independent variables are 'held constant'. Partial correlation and *partial regression* are of use in differentiating the relative importance of the effects of interacting independent variables. The *order* of a partial correlation coefficient refers to the number of independent variables held constant. *JAM*

Matthews JA (1987) Regional variation in the composition of Neoglacial end moraines, Jotunheimen, Norway: An altitudinal gradient in clast roundness and its possible palaeoclimatic significance. *Boreas* 16: 173–188.

particle-induced gamma-ray emission (PIGE or PIGME)   A sensitive, non-destructive technique for elemental analysis that analyses *gamma rays* emitted when a *proton beam* is applied to an object. Like PROTON-INDUCED X-RAY EMISSION (PIXE), it is a method of *ion-beam analysis* (*IBA*). Unlike PIXE, however, it relies on excitation of the *nucleus* and is better for detecting light elements (lighter than *sodium*). It is widely used in combination with PIXE to obtain a 'total analysis' of ARTEFACTS, especially those made of flint, obsidian and pottery, with minimal sample preparation. *JAM/GO*

[*See also* GEOCHEMISTRY]

Chadefaux C, Vignaud C, Menu M and Reiche I (2008) Multianalytical study of palaeolithic reindeer antler. Discovery of antler traces in Lascaux pigments by TEM. *Archaeometry* 50: 516–534.
Mueller K and Reiche I (2011) Differentiation of archaeological ivory and bone materials by micro-PIXE/

PIGE with emphasis on two Upper Palaeolithic key sites: Abri Pataud and Isturitz, France. *Journal of Archaeological Science* 38: 3234–3243.

particulates   Small solid particles of material; particulate matter. The larger particles suspended in the ATMOSPHERE are mainly of natural origin (e.g. soil particles, sea salts) whereas the finer particles are mostly of COMBUSTION origin.   *DME/PU/CW*

[*See also* AEROSOLS, DRY DEPOSITION, WET DEPOSITION]

Antuña JC, Robock A, Stenchikov G et al. (2003) Spatial and temporal variability of the stratospheric aerosol cloud produced by the 1991 Mount Pinatubo eruption. *Journal of Geophysical Research* 108: 4624.

Durant AJ, Bonadonna C and Horwell CJ (2010) Atmospheric and environmental impacts of volcanic particulates. *Elements* 6:235–240.

Engling G and Gelencsér A (2010) Atmospheric brown clouds: From local air pollution to climate change. *Elements* 6: 223–228.

Gieré R and Querol X (2010) Solid particulate matter in the atmosphere. *Elements* 6: 215–222.

Yang F and Schlesinger M (2002) On the surface and atmospheric temperature changes following the 1991 Pinatubo volcanic eruption: A GCM study. *Journal of Geophysical Research* 107: D8, 4073.

passive continental margin   A CONTINENTAL MARGIN characterised by a wide CONTINENTAL SHELF, separated by a CONTINENTAL SLOPE and CONTINENTAL RISE from the deep-ocean ABYSSAL plain, and lacking active SEISMICITY or VOLCANISM. A passive continental margin does not correspond to a PLATE MARGIN (see Figure). The continental margins of northwest Europe and eastern North America are of this type, and they are also known as *Atlantic-type continental margins*

or *aseismic continental margins*. The true edge of the CONTINENT—the junction between CONTINENTAL CRUST and OCEANIC CRUST—corresponds approximately to the outer margin of the continental shelf (the *shelf break*).

A passive continental margin originates as a continental RIFT VALLEY. Spreading as a CONSTRUCTIVE PLATE MARGIN leads to the injection of oceanic crust at a MID-OCEAN RIDGE (MOR), allowing two mirror-image continental margins separated by an OCEAN to migrate to a tectonically inactive mid-plate position (see WILSON CYCLE). After the phase of *rifting*, the thinned continental crust of passive continental margins cools, allowing considerable SUBSIDENCE to make space for thick sequences of SEDIMENTARY DEPOSITS to accumulate. Modern passive continental margins have accumulated sediment piles 10–15 km thick over the past 200 million years. These deposits contain important SEDIMENTOLOGICAL EVIDENCE OF ENVIRONMENTAL CHANGE associated with CONTINENTAL DRIFT, as well as important PETROLEUM reserves. It is increasingly being realised that the term 'passive' may be misleading as these margins may continue to show various types of activity associated with TECTONICS. *CDW*

Bradley DC (2008) Passive margins through Earth history. *Earth-Science Reviews* 91: 1–26.

Burbank DW and Anderson RS (2011) *Tectonic geomorphology, 2nd edition*. Chichester: Wiley-Blackwell.

Busby C and Azor A (2011) *Tectonics of sedimentary basins: Recent advances*. Chichester: Wiley-Blackwell.

Kearey P, Klepeis KA and Vine FJ (2009) *Global tectonics, 3rd edition*. Chichester: Wiley.

Leeder MR (2011) *Sedimentology and sedimentary basins: From turbulence to tectonics, 2nd edition*. Chichester: Wiley-Blackwell.

Roberts DG and Bally AW (eds) (2012) *Passive margins, cratonic basins and global tectonic maps*. Amsterdam: Elsevier.

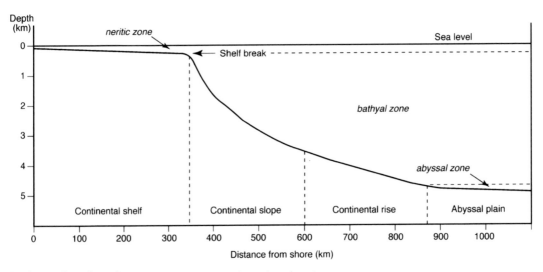

Passive continental margin *Schematic section across a passive continental margin.*

passive microwave remote sensing   The observation of the Earth's surface with a microwave RADIOM-
ETER that receives energy only (and does not transmit).
Observations in some parts of the microwave region
(<40 GHz) are almost completely independent of
solar illumination and largely unaffected by the presence of CLOUDS and atmospheric AEROSOLS. *Passive
microwave instruments*, such as NASA's *Advanced
Microwave Scanning Radiometer—Earth Observing
System* AMSR-E instrument (launched in 2002) generally operate over several frequencies (12 in the case of
AMSR-E) and can also receive different POLARISATIONS
(horizontal and vertically polarised microwave radiation). The spatial resolution of microwave radiometers
typically ranges between 5 and 60 km depending upon
frequency. These instruments measure microwave
emission from objects (*brightness temperature*), which
depend upon the temperature, as well as the object's
geometry and dielectric properties. This information is
useful for applications including study of the OCEANS
(SALINITY, SEA-SURFACE TEMPERATURE (SST) and *windspeed*), SEA-ICE and SNOW AND SNOW COVER, SOIL MOIS-
TURE and the WATER VAPOUR and liquid water content of
the ATMOSPHERE.   *MEJC*

[*See also* MICROWAVE REMOTE SENSING (MRS)]

Belchansky GI (2004) *Arctic ecological research from
    microwave satellite observations*. Boca Raton, FL: CRC
    Press.
Cherny V and Raizer VY (1998) *Passive microwave remote
    sensing of oceans*. Chichester: Wiley: Blackwell.
Sharkov EA (2003) *Passive microwave remote sensing of the
    Earth: Physical foundations*. Berlin: Springer-Praxis.

Pastoral Neolithic   Pre-Iron Age East African cultures based on PASTORALISM with advanced technologi-
cal and economic features.   *JAM*

Collett DP and Robertshaw PT (1983) Dating the Pastoral
    Neolithic of East Africa. *African Archaeological Review*
    1: 57–74.

pastoralism   An agricultural system whereby flocks
and herds of animals are grazed in an open landscape
or enclosed fields. Such systems focuses on the use of
domesticated or semi-domesticated animals (e.g. *goats,
sheep, cattle, camels, llamas* and *yaks*) as a source of
food and other materials such as hide, but vary widely
in their degrees of movement of people and herds from
nomadic to *semi-sedentary*. There are an estimated 100
million pastoral people worldwide, around 20 million
of whom are in Africa, including, for example, the
Maasai in East Africa, *Bedouin* in the Middle East and
Mongols in north-central Asia. Pastoralism tends to
occur in MARGINAL AREAS with little potential for CUL-
TIVATION, including DRYLANDS, RANGELANDS, SAVANNAS,
STEPPES, TUNDRA and *mountain regions*. As a system

of NATURAL RESOURCES MANAGEMENT, pastoralism is an
effective strategy in these environments, but its VUL-
NERABILITY has increased in recent years for environmental, economic, demographic and political reasons.
CLIMATIC VARIABILITY and the frequency of DROUGHTS in
drylands are important issues, and there is increasing
evidence of LAND DEGRADATION at local scales. *JAM/LD-P*

[*See also* DOMESTICATION, GRAZING, NOMADISM,
OVERGRAZING, RANGE MANAGEMENT, SEDENTISM,
TRANSHUMANCE]

Foerch W (2007) Pastoralism. In Robbins P (ed.)
    Encyclopedia of environment and society. Thousand Oaks,
    CA: Sage, 1345–1348.
Food and Agriculture Organization of the United Nations (FAO)
    (2001) *Pastoralism in the new millennium*. Rome: FAO.
Fratkin EM (2003) *Ariaal pastoralists of Kenya: Studying
    pastoralism, drought and development in Africa's arid
    lands, 2nd edition*. London: Pearson.
Hanotte O, Bradley DG, Ochieng JW et al. (2002) African
    pastoralism: Genetic imprints of origins and migrations.
    Science 296: 336–339.
Harris DR (ed.) (1996) *The origins and spread of agriculture
    and pastoralism in Eurasia*. London: University College
    London Press.
Porter A (2012) *Mobile pastoralism and the formation
    of Near Eastern civilizations*. Cambridge: Cambridge
    University Press.
Western D and Finch V (1996) Cattle and pastoralism: Survival
    and production in arid lands. *Human Ecology* 14: 77–94.

pasture   A *grassland* maintained by GRAZING of
domestic livestock. *Grasses* are the DOMINANT SPECIES
but low-growing HERBS, such as *rosette*, tufted and
mat-like GROWTH FORMS, which tolerate grazing, are
also characteristic. When not subject to OVERGRAZING,
grasses are stimulated by grazing. Pastures tend to be
grazed throughout the GROWING SEASON of the grasses.
A pasture differs from a MEADOW, which is maintained
by *mowing* and typically only used for grazing at the
end of the summer. Pasture, which may be classified as
improved pasture or *unimproved pasture*, cover about
26 per cent of the Earth's ice-free surface. Improved
pasture involves human inputs in the form of NUTRI-
ENTS, the INTRODUCTION of new species and/or other
MANAGEMENT practices.   *JAM*

[*See also* GRAZING HISTORY, GRAZING: IMPACTS
ON ECOSYSTEM, PUGGING, RANGE MANAGEMENT,
RANGELAND, TRAMPLING]

Hilbrunner D, Schulze S, Hagedorn F et al. (2012) Cattle
    trampling alters soil properties and changes soil microbial
    communities in a Swiss sub-alpine pasture. *Geoderma*
    170: 367–377.
Lovett JV and Scott JM (eds) (1997) *Pasture production and
    management*. Melbourne: Butterworth-Heinemann.
Prochazka NT (ed.) (2011) *Pastures: Dynamics, economics
    and management*. New York: Nova Science Publishers.

patch dynamics The nature and causes of change in landscape patches within the LANDSCAPE MOSAIC. Some patches form mosaics of vegetation in different successional stages that undergo cyclical changes about a steady state. English HEATHLANDS, for instance, contain heather (*Calluna*) patches of different age. Over time, through CYCLIC REGENERATION, the phases change places, but the PEATLAND ecosystem as a whole remains the same. Some patches follow directional changes. Depending on the force driving the change, an abandoned field on a hillside may become a wood (through natural ECOLOGICAL SUCCESSION), a new field (through PLOUGHING), a quarry (through digging and blasting) or a GULLY (through SOIL EROSION). DISTURBANCE patches result from the action of disturbing agents. Some disturbing agents, such as *fire* and *wind*, are likely to become more frequent and more severe in certain ecosystems if GLOBAL WARMING occurs during the next century.   *RJH*

[*See also* FRAGMENTATION]

Busing RT (2007) *A spatial landscape model of forest patch dynamics and climate change.* Reston, VA: US Department of the Interior [United States Geological Survey Scientific Investigations Report 2007–5040].
Pickett STA, Cadenasso ML and Grove JM (2005) Biocomplexity in coupled natural-human systems: A multidimensional framework. *Ecosystems* 8: 1–8.
Pickett STA and White PS (eds) (1985) *The ecology of natural disturbance and patch dynamics.* New York: Academic Press. Shorrocks B and Swingland I R (1990) *Living in a patchy environment.* Oxford: Oxford University Press.

pathogens Living organisms, usually BACTERIA, VIRUSES, FUNGI, *protozoa* or *parasitic worms*, that cause DISEASE. Exposure is usually through inhalation, ingestion, direct contact or transmission via an insect vector. In human populations, the proliferation of pathogenic microbes is often linked to standards of hygiene as well as changes in temperature, humidity or precipitation. The term is also in the context of ANIMAL DISEASES and PLANT DISEASES.   *MLW*

[*See also* HUMAN HEALTH HAZARDS, VECTOR, WATERBORNE DISEASES]

Guerrant RL, Walker DH and Weller PF (eds) (2011) *Tropical infectious diseases: principles, pathogens and practice, 3rd edition.* Amsterdam: Elsevier.
Mottet NK (ed.) (1985) *Environmental pathology.* New York: Oxford University Press.
Waller M (2013) Drought, disease, defoliation and death: Forest pathogens as agents of past vegetation change. *Journal of Quaternary Science* 28: 336–342.

patterned ground A general term used to describe the variety of small-scale, more or less symmetrical forms, such as circles, polygons, nets and stripes, that characterise the ground surface in PERIGLACIAL ENVIRONMENTS and are particularly widespread in the ARCTIC. RELICT forms occur in former periglacial environments. Relatively small active forms are also common in ALPINE ZONES at mid-latitudes and some of these forms are active on *tropical-alpine mountains*. Most periglacial patterned ground is thought to be produced by FREEZE-THAW CYCLES but some forms of patterned ground also occur in hot, SEMI-ARID regions.

Traditionally, patterned ground is classified on the basis of (a) its geometric form and (b) the presence or absence of sorting. The main geometric forms are circles, polygons, stripes, nets and steps, all of which may be sorted and unsorted with respect to particle sizes. Stripes and steps are usually confined to sloping terrain but can occur on very low gradients. If large-scale thermal contraction cracking is excluded, most investigators who have studied the small-scale forms in cold regions conclude that (a) much patterned ground is polygenetic and (b) similar forms can be created by different processes. DESICCATION CRACKING rather than THERMAL CONTRACTION CRACKING, for example, is regarded as important in development of polygonal forms and relatively unimportant in development of circular forms. The latter are most likely the result of CRYOTURBATION.

The wide range of cold-climate patterned ground phenomena means that their use as a diagnostic feature of such environments is limited. Patterned ground is not limited to PERMAFROST regions, or to those experiencing only SEASONALLY FROZEN GROUND. Moreover, lithology, grain size, moisture availability, vegetation and a host of other site-specific factors are relevant. On the other hand, patterned ground phenomena are best developed and most widespread in the cold regions of the world and there is scope for further investigation of the nature and climatic relations of the different types.   *HMF*

[*See also* PERIGLACIAL LANDSCAPE EVOLUTION]

Feuillet T (2011) Statistical analysis of active patterned ground occurrences in the Taillon Massif (Pyrénées, France/Spain). *Permafrost and Periglacial Processes* 22: 228–238.
Gleason KJ, Krantz WB, Caine N et al. (1986) Geometrical aspects of sorted patterned ground in recurrently frozen soil. *Science* 232: 216–220.
Grab SW (1997) Annually re-forming miniature sorted patterned ground in the high Drakensberg, southern Africa. *Earth Surface Processes and Landforms* 22: 733–745.
Hallet B, Anderson SP, Stubbs CW and Gregory EC (1988) Surface soil displacements in sorted circles, Western Spitzbergen. In *Proceedings, 5th International Conference on Permafrost, volume 1.* Trondheim: Tapir Publishers, 770–775.
Mackay JR (1980) The origin of hummocks, western Arctic coast. *Canadian Journal of Earth Sciences* 17: 996–1006.

Matthews JA, Shakesby RA, Berrisford MS and McEwen LJ (1998) Periglacial patterned ground on the Styggedalsbreen glacier foreland, Jotunheimen, southern Norway: Micro-topographic, paraglacial and geoecological controls. *Permafrost and Periglacial Processes* 9: 147–166.

Van Vliet-Lanoë B (1991) Differential heave, load casting and convection: Converging mechanisms. A discussion of the origin of cryoturbations. *Permafrost and Periglacial Processes* 2: 123–139.

Van Vliet-Lanoë B and Seppälä M (1991) Stratigraphy, age and formation of peaty earth hummocks (pounus), Finnish Lappland. *The Holocene* 12: 187–199.

Warburton J and Caine N (1999) Sorted patterned ground in the English Lake District. *Permafrost and Periglacial Processes* 10: 193–197.

Washburn AL (1956) Classification of patterned ground and review of suggested origins. *Geological Society of America Bulletin* 67: 823–865.

Washburn AL (1997) Plugs and plug circles: A basic form of patterned ground, Cornwallis Island, Arctic Canada: Origin and implications. *Geological Society of America, Memoir* 190: 1–94.

Wilson P (1995) Forms of unusual patterned ground: Examples from the Falkland Islands, South Atlantic. *Geografiska Annaler* 77A: 159–165.

payload   That which a spacecraft (e.g. a satellite or aircraft) carries over what is necessary for the operation of the vehicle in flight. It includes any REMOTE-SENSING instruments that are accommodated on board for ENVIRONMENTAL MONITORING.   *TS*

peat and peatlands   The undecayed remains of vegetation that accumulates in MIRES and HEATHLAND and the LANDSCAPES characterised by the accumulation of peat. Peat can range from unhumified, yellow- or orange-brown *fibrous peat* of RAISED MIRES, with visible plant remains, to highly humified black *amorphous peat* of HEATHLANDS. Peat is widespread in the *peatlands* of north-temperate and circum-boreal regions, especially in Canada, Estonia, Finland, Ireland and Russia; in southerly latitudes in Tierra del Fuego, Falkland Islands and Tasmania; also at lower latitudes in TROPICAL PEATLANDS. In the British Isles, peat was traditionally cut (as 'turf'), dried and burnt for fuel; it is used in HORTICULTURE as a *soil conditioner* or growing medium. PEAT STRATIGRAPHY of OMBROTROPHIC MIRES can be interpreted as a PROXY CLIMATIC INDICATOR.   *FMC*

[*See also* HUMIFICATION, PEAT HUMIFICATION, PEAT STRATIGRAPHY]

Charman D (2002) *Peatlands and environmental change.* Chichester: Wiley-Blackwell.

Charman D and Chambers F (eds) (2004) Peatlands and Holocene environmental change. *The Holocene* 14: 1–143 [Special Issue].

De Vleeschouwer F, Hughes PDM, Nichols JE and Chambers FM (2010) A review of protocols in peat palaeoenvironmental studies. *Mires and Peat* 7: Article 00, 1.

Jackson ST, Charman D, Newman L and Kiefer T (eds) (2010) Peatlands: Paleoenvironments and carbon dynamics. *PAGES Newsletter* 18 [Special Issue].

peat erosion   Peat deposits erode if the vegetation cover is damaged allowing attack by WIND EROSION and WATER EROSION. Both lowland and upland MIRES are often very vulnerable to DISTURBANCE by ANTHROPOGENIC or NATURAL environmental change. Damage to upland BLANKET MIRES may start as localised GULLY erosion or *blowouts* that concentrate RUNOFF or are vulnerable to the wind and become more widespread. Peat deposits may degrade following reduced PRECIPITATION or increased EVAPOTRANSPIRATION during CLIMATIC CHANGE, or following LAND DRAINAGE operations. DEGRADATION may involve erosion damage or shrinkage and oxidation or, especially in some TROPICAL PEATLANDS, *below-ground burning*. Peat and its vegetation cover can also be damaged by ACID DEPOSITION, GRAZING, TRAMPLING, OFF-ROAD VEHICLES and cutting for fuel (peat is a FOSSIL FUEL) or use in HORTICULTURE. Considerable amounts of CARBON DIOXIDE and *methane* may be released as peat degrades (a potential positive FEEDBACK MECHANISM process in GLOBAL WARMING).   *CJB*

Evans M and Lindsay J (2010) Impact of gully erosion on carbon sequestration in blanket peatlands. *Climate Research* 45: 31–41.

Evans M and Warburton J (2007) *Geomorphology of upland peat: Erosion, form and landscape change.* Oxford: Wiley-Blackwell.

Wilson L, Wilson J, Holden J et al. (2011) Ditch blocking, water chemistry and organic carbon flux: Evidence that blanket bog restorsation reduces erosion and fluvial carbon loss. *Science of the Total Environment* 409: 2010–2018.

peat humification   The extent to which PEAT has degraded (or decomposed, to produce humic acid and humin) in situ. It can be determined variously, and with varying degrees of precision, by physical and chemical methods. Early field methods used a visual scale, based on the colour of water extruded from a hand sample of peat, to record PEAT STRATIGRAPHY. Aaby and Tauber (1975) later used the laboratory technique of COLORIMETRY (on an alkali extract) to determine the degree of peat humification from a Danish raised mire, and considered it to give a reliable indication of the *hydrological environment* at the time of peat deposition. As a consequence, peat humification is now used, particularly in northwest Europe, as a PROXY CLIMATIC INDICATOR, and colorimetric and other methods (e.g. FLUORESCENCE SPECTROSCOPY) can produce HIGH-RESOLUTION RECONSTRUCTIONS with decadal to centennial resolution, provided the peat can be dated reliably.   *FMC*

Aaby B and Tauber H (1975) Rates of peat formation in relation to degree of humification and local environment, as shown by studies of a raised bog in Denmark. *Boreas* 4: 1–17.

Borgmark A (2005) Holocene climate variability and periodicities in south-central Sweden, as interpreted from peat humification analysis. *The Holocene* 15: 387–395.

Chambers FM, Beilman DW and Yu Z (2010–2011) Methods for determining peat humification and for quantifying peat bulk density, organic matter and carbon content for palaeostudies of climate and peatland carbon dynamics. *Mires and Peat* 7 1–10.

peat stratigraphy The visible horizons, or layers, of a MIRE, such as can be seen in a peat CORE or in a vertical column of PEAT taken by MONOLITH SAMPLING from an EXCAVATION. Peat stratigraphy is most evident in a fast-growing mire in which the peat is relatively unhumified, but in which past climatic or hydrological changes caused variations in peat growth, giving rise to changes in VEGETATION (which can be demonstrated through PLANT MACROFOSSIL ANALYSIS) and/or to changes in PEAT HUMIFICATION. Peat laid down in periods of dry or warmer climate will be more highly humified and so appear darker in colour than peat laid down under cooler or wetter climatic conditions.

Gross variations in the peat stratigraphy of Scandinavian bogs prompted Blytt and Sernander to postulate past climatic change, and gave rise to the BLYTT-SERNANDER TIMESCALE of postglacial climatic periods, which was then adopted over much of northwest Europe as a CLIMOSTRATIGRAPHY for the HOLOCENE. This is now viewed as too simplistic, but the names used in the Blytt-Sernander timescale (PREBOREAL, BOREAL, ATLANTIC, SUBBOREAL and SUBATLANTIC), which derived from peat stratigraphy, live on as a CHRONOSTRATIGRAPHY, particularly in Scandinavia. In northwest Europe, changes from dark peat to lighter coloured peat in RAISED MIRES gave rise to the notion of recurrences of bog growth in times of wetter climate, producing supposed synchronous RECURRENCE SURFACES. The major postglacial peat stratigraphic change in Europe is termed the GRENTZHORIZONT, and its age in many mires may well equate with a claimed major climate shift noted at about 2,650 radiocarbon years BP. Peat stratigraphy, particularly when analysed in detail and dated precisely, can yield valuable PROXY CLIMATIC INDICATORS.

FMC

Barber KE (1981) *Peat stratigraphy and climatic change: A palaeoecological test of the theory of cyclic bog regeneration*. Rotterdam: Balkema.

Barber KE, Chambers FM and Maddy D (2003) Holocene palaeoclimates from peat stratigraphy: Macrofossil proxy climate records from three oceanic raised bogs in England and Ireland. *Quaternary Science Reviews* 22: 521–539.

Blytt A (1876) *Essays on the immigration of Norwegian flora during alternating rainy and dry periods*. Christiana: Cammermayer.

De Vleeschouwer F, Chambers FM and Swindles GT (2010) Coring and sub-sampling of peatlands for palaeoenvironmental research. *Mires and Peat* 7: 1–10.

Sernander R (1908) On the evidences of postglacial changes of climate furnished by the peat mosses of Northern Europe. *Geologiska Föreningens i Stockholm Förhandlingar* 30: 467–478.

Speranza A, van Geel B and van der Plicht J (2002) Evidence for solar forcing of climate change at ca. 850 cal BC from a Czech peat sequence. *Global and Planetary Change* 35: 51–65.

pebble A sedimentary particle of GRAIN SIZE 4–64 mm.

TY

[*See also* GRAVEL, SEDIMENT]

ped A natural aggregate of soil particles: An important feature of SOIL STRUCTURE.

JAM

pedalfer A major category of ZONAL SOIL in which downward movements of soil water dominate the SOIL PROFILE leading to LEACHING and ACIDIFICATION of upper SOIL HORIZONS and enrichment in iron and aluminium sesquioxides relative to silica. Pedalfers occur especially in humid climates, where PRECIPITATION exceeds EVAPORATION.

JAM

pedestal rock Relatively large caps or table-shaped rock masses supported by essentially narrow *stems* of rock. Many pedestal rocks are structurally homogeneous. They are thought to form in a two-stage process: the pillar or stem being formed through WEATHERING under moist conditions just beneath the surface. The CAPROCK remains intact sometimes because of its massive structure, but its exposure, relatively dry state and protection by *lichens, mosses* and *chemical crusts* may also enhance its resistance to weathering and EROSION. Exhumation of the stem occurs through removal of the weathered bedrock, by *wash* and *stream action* and in places by *wind* and *waves*. Alternative names are *hoodoo rocks, mushroom rocks, stone* or *rock babies* and various terms in other languages, such as *Pilzfelsen, Tischfelsen, roches champignons* and *rocas fungiformas*. They differ from EARTH PILLARS, in that the latter are formed in *unconsolidated sediments* where coarse CLASTS provide the caprock and protect the underlying sediment from erosion. They are related but differ from YARDANGS and ZEUGEN.

RAS

[*See also* BORNHARDT, TOR]

Twidale CR and Campbell EM (1992) On the origin of pedestal rocks. *Zeitschrift für Geomorphologie NF* 36: 1–13.

pediment A gently sloping surface (0.5–7°, but usually 2–4°) connecting eroding steeper slopes or scarps with areas of *sediment deposition* at lower levels. It is developed across bedrock that may or may not be thinly veneered with ALLUVIUM and/or in-situ weathered rock. The pediment is separated from the steeper upper slope (often associated with an INSELBERG) by a relatively rapid change of slope angle in a transitional zone, in which the change of gradient is termed the *piedmont angle*. It is an element of a *piedmont zone* (see PIEDMONT), which may include depositional elements such as ALLUVIAL FANS and PLAYAS. Coalescing pediments create PEDIPLAINS (EROSION SURFACES produced by *pediplanation*).

Pediments are found in a wide range of climatic environments and there are numerous theories regarding the origin of pediments, only some of which invoke environmental change (normally in the form of CLIMATIC CHANGE, tectonic UPLIFT or *tilting*) as being necessary in their formation. Some consider that pediment formation and maintenance are favoured by a highly seasonal (wet-dry) or SEMI-ARID climate, whereas others consider that they can form in a wide range of climates.

Theories proposing a semi-arid or arid climatic origin argue that the rate of erosion on the pediment must be greater than or equal to the rate of SOIL DEVELOPMENT in order to create or maintain an *erosional surface* that has at most a thin SOIL or REGOLITH; this, they argue, is favoured by high ARIDITY, because of its low rate of soil production. The alternate *mantling* and *stripping* of pediments is a key element of most theories invoking environmental change. In the Mojave Desert, Oberlander (1974) demonstrated the importance of more humid climates in pediment formation, and Busche (1976) identified cycles of deep WEATHERING and stripping on pediment surfaces in Chad. In such theories, pediments are often regarded as indicators of past more humid conditions when found in arid zones, and as indicators of past drier conditions when they occur in the humid tropics. MAC/RPDW

[*See also* STRIPPING PHASE]

Busche D (1976) Pediments and climate. *Palaeoecology of Africa, the surrounding islands and Antarctica* 11: 20–24.
Cooke RU, Warren A and Goudie AS (1993) *Desert geomorphology*. London: UCL Press.
Dohrenwend JC and Parsons AJ (2009) Pediments in arid environments. In Parsons AJ and Abrahams AD (eds) *Geomorphology of desert environments, 2nd edition*. Berlin: Springer, 377–412.
Oberlander TM (1974) Landscape inheritance and the pediment problem in the Mojave Desert of southern California. *American Journal of Science* 274: 849–875.
Pelletier JD (2010) How do pediments form? A numerical modelling investigation with comparison to pediments in southern Arizona, USA. *Geological Society of America Bulletin* 122: 1815–1829.
Whitaker CE (1979) The use of the term 'pediment' and related terminology. *Zeitschrift für Geomorphologie* 23: 427–439.

pediplain/pediplanation A pediplain is an extensive, thinly alluviated EROSION SURFACE found generally in DESERT, SEMI-ARID and SAVANNA regions. It is considered to be formed by the coalescence of two or more adjacent PEDIMENTS and to represent the end result of the mature stage of the *arid cycle of erosion*. *Pediplanation* is the action or process of formation and development of a pediplain by *scarp retreat* and *pedimentation*. It was proposed originally by L.C. King as an alternative to the DAVISIAN CYCLE OF EROSION and invoked on a continental scale to explain widespread *multiconcave erosion surfaces* with steep-sided hills in the non-humid tropical environments of Africa. The cycle is initiated by the UPLIFT of a pediplain, and existing streams rapidly cut downwards to the new BASE LEVEL. Eventually, as *downcutting* becomes less active, small PEDIMENTS appear in valley bottoms that become more extended as interfluve and upland areas are eroded by scarp retreat. Interfluve areas are converted to INSELBERGS and BORNHARDTS and, ultimately, to a landscape of low relief. Although there are similar limitations to other concepts linked to cycles of erosion, particularly in relation to the relative rates of EROSION and UPLIFT, pediplains and pediplanation continue to be recognised, albeit in a more dynamic framework than they were originally proposed. MAC/JAM

[*See also* ETCHPLAIN, PENEPLAIN]

King LC (1962) *The morphology of the Earth*. Edinburgh: Oliver and Boyd.
Leroux JS (1991) Is the pediplanation cycle a useful model: Evaluation in the Orange-Free-State (and elsewhere) in South Africa. *Zeitschrift für Geomorphologie* 35: 175–185.
Peulvast JP and Claudino-Sales V (2004) Stepped surfaces and palaeolandforms in the northern Brazilian "Nordeste": Constraints on models of morphotectonic evolution. *Geomorphology* 62: 89–122.
Ufimstsev GF (2010) Pediments of Asia. *Russian Journal of Pacific Geology* 4: 250–259.
Veldkamp A and Oosterom AP (1994) The role of episodic plain formation and continuous etching and stripping processes in the end-Tertiary landform development of SE Kenya. *Zeitschrift für Geomorphologie* 38: 75–90.

pedocal A major category of ZONAL SOIL in which upward movement of soil water (by CAPILLARY ACTION) dominates the SOIL PROFILE. Pedocals occur especially in ARID and SEMI-ARID regions where evaporation exceeds precipitation and calcium carbonate accumulates in upper SOIL HORIZONS.   *JAM*

pedogenesis The formation and dynamic evolution of SOILS. Soils are formed by the operation of SOIL-FORMING PROCESSES upon the PARENT MATERIAL over a period of time within the framework of the SOIL-FORMING FACTORS. The processes, for example, LEACHING, PODZOLISATION, GLEYING, FERALLITISATION and SALINISATION, are names given to 'bundles' of chemical, physical and biological actions. However, pedogenesis may be simplified as additions to, and subtractions from, a parent material within which various transformations and translocations take place (see Figure). Fresh *organic matter* is incorporated into the surface, and raw mineral material is incorporated by WEATHERING at the base of the soil. Soluble compounds are leached out of the soil in humid climates and drawn towards the surface by CAPILLARY ACTION in DRYLANDS. Transformation of minerals, especially iron and manganese, occur and translocations of organic matter, CLAY, ALUMINIUM (Al) and IRON (Fe) take place. The processes eventually result in a SOIL PROFILE with recognisable SOIL HORIZONS. Most soils, even when mature, are dynamic and continue to evolve, possibly never attaining an *equilibrium state*.   *EMB/JAM*

[*See also* EQUILIBRIUM CONCEPTS IN SOILS, SOIL DEVELOPMENT]

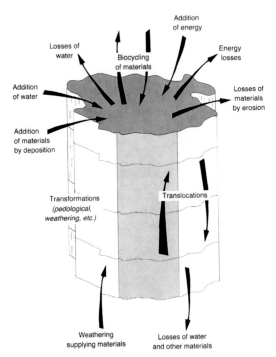

Pedogenesis *A soil pedon showing additions, translocations, transformations and losses during pedogenesis (Simonson, 1959).*

Johnson DL and Watson-Stegner D (1987) Evolution model of pedogenesis. *Soil Science* 143: 349–366.
Kemp RA (2001) Pedogenic modification of loess: Significance for palaeoclimatic reconstructions. *Earth-Science Reviews* 54: 145–156.
Minasny B, McBratney AB and Salvador-Blanes S (2008) Quantitative models for pedogenesis: A review. *Geoderma* 144: 140–157.
Samouelian A and Cornu S (2008) Modelling the formation and evolution of soils, towards an initial synthesis *Geoderma* 145: 401–409.
Schaetzl RJ and Anderson S (2005) *Soils: Genesis and geomorphology*. Cambridge: Cambridge University Press.
Simonson RW (1959) Outline of a generalized theory of soil genesis. *Soil Science Society of America Proceedings* 23: 152–156.
Simonson RW (1978) A multi-process model of soil genesis. In Mahaney WC (ed.) *Quaternary soils*. Norwich: Geoabstracts, 1–25.
Stockmann U, Minasny B and McBratney AB (2011) Quantifying processes of pedogenesis. *Advances in Agronomy* 113: 1–71.

pedology The scientific study of SOILS; *soil science* is a synonym. Pedology integrates studies of soil chemistry, physics and biology to create the study of soils as a natural phenomenon in their own right. It focuses upon the appearance, mode of formation, distribution, classification and use of soils. *Palaeopedology* investigates the soils of past environments (PALAEOSOLS).   *EMB*

Arnold RW (2008) Pedology and pedogenesis. In Chesworth W (ed.) *Encyclopedia of soil science*. Dordrecht: Springer, 512–516.
Retallick GJ (2002) *Soils of the past: An introduction to paleopedology*. Oxford: Blackwell.
White RE (2005) *Principles and practice of soil science: The soil as a natural resource, 4th edition*. Oxford: Blackwell.

pedometrics The application of NUMERICAL ANALYSIS to the study of the pattern and distribution of soils and their development. Pedometrics is an interdisciplinary science integrating soil science, applied statistics/mathematics and GEOGRAPHICAL INFORMATION SYSTEMS (GIS). It also includes numerical SOIL CLASSIFICATION and *automated soil mapping*. Pedometrics is a rapidly growing field of soil science incorporating the results of new technological developments such as REMOTE SENSING and *close-range sensing* techniques. These, coupled with the rapid advances in computer technology, have enabled major advances in the prediction of the distribution and behaviour of soils. *Digital soil mapping* is the part of pedometrics dealing with computer-assisted production of digital maps of soils and soil properties. Information from field and laboratory OBSERVATION of soils is combined with environmental data, both spatial and non-spatial, from a wide range of sources using mathematical and statistical procedures.   *SN*

Burrough PB, Bouma J and Yates SR (1994) State of the art in pedometrics. *Geoderma* 62: 311–326.
Odeh IOA and McBratney AB (2005) Pedometrics. In Hillel D (ed.) *Encyclopedia of soils in the environment*. Oxford: Elsevier, 166–175.

pedon   The smallest volume of soil that includes all significant features necessary for SOIL CLASSIFICATION. It varies in area from 1 to 10 m² and is large enough to permit the study of the SOIL PROFILE and its SOIL HORI-ZONS. The depth of a pedon is not strictly defined: it is governed primarily by the depth of the genetic soil horizons (see SOLUM). Other terms that have been pro-posed are *soil area* and *pedo-unit* (see Figure in the SOIL PROFILE entry).   *EMB/JAM*

Johnson DW (1963) The pedon and the polypedon. *Proceedings of the Soil Science Society of America* 27: 212–215.
Soil Survey Staff (1999) *Soil Taxonomy: A basic system of soil classification for making and interpreting soil surveys, 2nd edition*. Washington, DC: US Department of Agriculture.

pedosphere   The relatively thin but essential soil cover of the terrestrial parts of the Earth.   *EMB*

[*See also* EARTH SPHERES]

Bronstert A, Carrera J, Kabat P and Lütkemeier S (eds) (2005) *Coupled models for the hydrological cycle: Integrating the hydrosphere, biosphere and pedosphere*. Berlin: Springer.

pedotransfer function (PTF)   A numerical MODEL for predicting soil physical and chemical properties from data readily available from soil survey. Data on SOIL TEXTURE or CLAY content, for example, may be used to predict properties related to SOIL MOISTURE content, ERODIBILITY, INFILTRATION and WATER REPELLENCY but should not be extrapolated beyond the geographical area or particular soil type(s) from which they were derived. They are increasingly used in LAND EVALUATION and are an important tool to overcome the dearth of soil data in the tropics.   *JAM*

[*See also* TRANSFER FUNCTION]

McBratney AB, Minasny B, Cattle SR and Vervoort RW (2002) From pedotransfer functions to soil inference systems. *Geoderma* 109: 41–73.
Minasny B and Hartemink AE (2011) Predicting soil properties in the tropics. *Earth-Science Reviews* 106: 52–62.

pedoturbation   A process of *soil formation* in which soil materials undergo mixing by biological and/or physical (freezing and thawing, wetting and drying) churning and cycling of soil materials, thereby homog-enising the SOLUM to varying degrees. Where 2:1 CLAY MINERALS are present and there is a distinct wet and dry season, expandable clays will swell on wetting and shrink on drying. This is a characteristic SOIL-FORMING PROCESS in VERTISOLS and may also produce an undu-lating GILGAI landscape. Where the physical processes involve FREEZE-THAW CYCLES, the process is described as CRYOTURBATION and PATTERNED GROUND may result.   *SN*

[*See also* BIOTURBATION, CRYOTURBATION, EARTHWORMS, FROST HEAVE, FROST SORTING, TERMITES]

Schaetzl RJ (2008) Pedoturbation. In Chesworth W (ed.) *Encyclopedia of soil science*. Dordrecht: Springer, 516–522.
Whitford WG and Kay FR (1999) Biopedoturbation by mammals in deserts: A review. *Journal of Arid Environments* 41: 203–230.

pelagic organism   An organism living or originat-ing in the water column. Pelagic organisms live in the open-water areas of OCEANS and LAKES, in contrast to the bottom-dwelling *benthic organisms* and those living close to the SHORELINE. Free-floating pelagic organisms are *planktonic* (see PLANKTON/PLANKTONIC/PLANKTIC); swimming organisms are *nektonic* (see NEKTON).   *LC*

[*See also* BENTHIC, BENTHOS, PELAGIC SEDIMENT (PELAGITE)]

pelagic sediment (pelagite)   Fine-grained SEDI-MENT (MUD) that has its origin in the water column, usually in the OCEANS, and settles to the deep sea floor, usually as composite particles or aggregates formed as faecal material or through FLOCCULATION. It is some-times described as *marine snow* or ABYSSAL deposits. Pelagic sediment includes material of biological ori-gin (OOZE) dominated by tests of PLANKTONIC organ-isms, and material of non-biological origin, which includes VOLCANIC ASH and AEOLIAN (windblown) ATMOSPHERIC DUST. The finest fraction of land-derived (TERRIGENOUS) sediment is termed *hemipelagic sedi-ment (hemipelagite)*.

The present-day distribution of pelagic sediments is controlled by organic PRODUCTIVITY, OCEAN CUR-RENTS and the dissolution of carbonate material below the CARBONATE COMPENSATION DEPTH (CCD). *Carbon-ate* or *calcareous ooze* forms the dominant sediment on shallow parts of the MID-OCEAN RIDGES (MORs) and is one of the largest SINKS for calcium carbonate on the Earth. *Siliceous ooze* is found in areas of deeper, colder water and high productivity. *Red clays* accumu-late very slowly (0.0001–0.001 mm/year) below the carbonate compensation depth in areas of low surface productivity for siliceous plankton. Organic-rich sedi-ment (SAPROPEL) accumulates in areas of high organic input or reduced BOTTOM-WATER oxygenation. ICE-RAFTED DEBRIS is deposited in and around polar regions.

Pelagic sediments are often associated with *manganese nodules*. The distribution of older pelagic sediments, derived from the analysis of MARINE SEDIMENT CORES obtained through DEEP-SEA DRILLING, is an important tool in PALAEOCEANOGRAPHY. *GO/JP*

Bernoulli D and Jenkyns HC (2009) Ancient oceans and continental margins of the Alpine-Mediterranean Tethys: Deciphering clues from Mesozoic pelagic sediments and ophiolites. *Sedimentology* 56: 149–190.
Hsu KJ and Jenkyns HC (eds) (1974) *Pelagic sediments: On land and under the sea.* Oxford: Blackwell [International Association of Sedimentologists, Special Publication 1].
Ziegler CL, Murray RW, Hovan SA and Rea DK (2007) Resolving eolian, volcanogenic, and authigenic components in pelagic sediment from the Pacific Ocean. *Earth and Planetary Science Letters* 254: 416–432.

peneplain   A *planation surface* of gentle RELIEF formed through long-term DENUDATION. The term is associated in particular with the DAVISIAN CYCLE OF EROSION concept in which it was viewed as the final stage of *subaerial denudation* in the absence of a change of BASE LEVEL. *Monadnocks* are residual hills that rise above a peneplain and hence have survived *peneplanation*. Their existence is still disputed, particularly on the grounds that tectonic UPLIFT, CLIMATIC CHANGE and landscape *dissection* interrupts any long-term tendency to *planation*. *AJDF/JAM*

[*See also* DENUDATION CHRONOLOGY, EROSION SURFACE, GRADE CONCEPT, SLOPE DECLINE, SLOPE EVOLUTION MODELS]

Johansson M (1999) Analysis of digital elevation data for palaeosurfaces in south-western Sweden. *Geomorphology* 26: 279–295.
Ollier CD (1991) *Ancient landforms.* London: Belhaven.
Phillips JD (2002) Erosion, isostatic response, and the missing peneplains. *Geomorphology* 45: 225–241.
Twidale CR (1983) Pediments, peneplains and ultiplains. *Revue de Géomorphologie Dynamique* 32: 1–35.

penetrability   The extent to which a soil, unconsolidated sediment or other material can be penetrated. It can be measured using a *penetrometer*, which measures the depth to which a rod penetrates under a given force. *JAM*

Pennsylvanian   Formerly used in North America as a PERIOD of geological time, or elsewhere as a subperiod of the CARBONIFEROUS, Pennsylvanian is now agreed to define the later EPOCH of the Carboniferous period (see MISSISSIPPIAN). The Pennsylvanian lasted from 318 to 299 million years ago, and does not exactly correspond with the former use of "Late Carboniferous" in Europe. The Pennsylvanian saw the occurrence of COAL-forming environments across much of what is now northwest Europe and northeast North America, associated with marine *transgressions* and *regressions* driven by the eustatic effects of an ICE AGE (see ICEHOUSE CONDITION). *GO*

[*See also* GEOLOGICAL TIMESCALE]

Bethoux O (2009) The earliest beetle identified. *Journal of Paleontology* 83: 931–937.
Falcon-Lang HJ, Benton MJ, Braddy SJ and Davies SJ (2006) The Pennsylvanian tropical biome reconstructed from the Joggins Formation of Nova Scotia, Canada. *Journal of the Geological Society of London* 163: 561–576.
Falcon-Lang HJ, Benton MJ and Stimson M (2007) Ecology of the earliest reptiles inferred from basal Pennsylvanian trackways. *Journal of the Geological Society of London* 164: 1113–1118.

percolation   The downward vertical flow of water through the VADOSE ZONE or *aeration zone* of soil and rock towards the WATER TABLE. *RPDW*

[*See also* INFILTRATION]

perennial   A plant that lives for several to many years, or a stream that flows throughout the year. *JLI*

periglacial   Referring to a wide range of cold, non-glacial environmental conditions. The term was coined by W. Łoziński to refer to a zone peripheral to former GLACIERS that experienced an ARCTIC-type climate. He used the term in relation to the palaeoenvironmental context of a *periglacial facies* attributed to mechanical WEATHERING. However, modern usage of 'periglacial' implies no necessary temporal or spatial proximity to a GLACIER or ICE SHEET. *JAM/RAS*

[*See also* FREEZE-THAW CYCLES, FROST WEATHERING, PERIGLACIAL ENVIRONMENTS, PERIGLACIAL LANDFORMS AND PROCESSES, PERIGLACIAL SEDIMENTS, PERIGLACIATION, PERMAFROST, SOLIFLUCTION]

French HM (2000) Does Lozinski's periglacial realm exist today? A discussion relevant to modern usage of the term 'periglacial'. *Permafrost and Periglacial Processes* 11: 35–42.
Karte J and Liedtke H (1981) The theoretical and practical definition of the term 'periglacial' in its geographical and geological meaning. *Biuletyn Peryglacjalny* 28: 123–135.
Slaymaker O (2009) Proglacial, periglacial or paraglacial? In Knight J and Harrison S (eds) *Periglacial and paraglacial processes and environments.* Bath: Geological Society, 71–84.
Worsley P (2004) Periglacial geomorphology. In Goudie AS (ed.) *Encyclopedia of geomorphology.* London: Routledge, 772–776.

periglacial environments   Environments characterised by cold nonglacial conditions. There are two broad categories: (1) those that experience frequent freeze-thaw oscillations and/or deep seasonal frost

(SEASONALLY FROZEN GROUND) but lack PERMAFROST and (2) those that experience both frost action and permafrost. The *permafrost zone* of the Earth is sometimes termed the *cryolithozone*. These environments range from POLAR DESERTS and TUNDRA to BOREAL FOREST and ALPINE ZONES, all of which are most extensive in the Northern Hemisphere. Around one-quarter of the Earth's land surface experiences periglacial conditions and up to another quarter experienced similar conditions during the PLEISTOCENE. In the Northern Hemisphere, about 24 per cent of the land area experiences permafrost, almost half of which is continuous permafrost (defined as covering >90 per cent of the landscape), while about 58 per cent experiences seasonal frozen ground.   *HMF*

[*See also* PERIGLACIAL LANDFORMS AND PROCESSES, PERIGLACIAL LANDSCAPE EVOLUTION, PERIGLACIAL SEDIMENTS]

French HM (2007) *The periglacial environment, 3rd edition.* Chichester: Wiley.

French HM (2011) *Periglacial environments.* In Gregory KJ and Goudie AS (eds) *The SAGE handbook of geomorphology.* London: Sage, 393–411.

Marshall SJ (2012) *The cryosphere.* Princeton, NJ: Princeton University Press.

Zhang T, Barry RG, Knowles K et al. (2008) Statistics and characteristics of permafrost and ground ice distribution in the Northern Hemisphere. *Polar Geography* 31: 47–68.

periglacial landforms and processes The field of *periglacial geomorphology*: the landforms and processes found in cold but essentially nonglacial environments that range from humid and relatively mild (e.g. Svalbard) to extremely cold and arid (e.g. ANTARCTIC). The distinctiveness of many LANDFORMS owes much to the processes involved. These include the formation of PERMAFROST and PERMAFROST DEGRADATION, including the development of THERMAL CONTRACTION CRACKING, the thawing of permafrost (THERMOKARST), the *creep* of ice-rich permafrost and the formation of various forms of GROUND ICE (e.g. *wedge ice* and *intrusive ice.* Other processes, not necessarily restricted to permafrost environments, are important because of their high magnitude or frequency. These relate to *freeze-thaw* processes that affect soil and bedrock, including the disintegration of exposed rock by FROST WEATHERING or other poorly understood cryogenic weathering processes and, in soils, FROST HEAVE and *ice segregation* (see SEGREGATION ICE).

The most distinct periglacial landforms are associated with permafrost. Most widespread are ICE-WEDGE polygons, 15–30 m in diameter, formed by the thermal contraction cracking of the frozen ground in winter. The cracks may fill with water in early summer and after a number of years ice wedges may form. Less commonly, the cracks may fill with sand. Far less widespread, but equally distinctive, are various perennial ice-cored FROST MOUNDS. The largest of these are PINGOS, which sometimes exceed 40 m in height. Other aggradational permafrost landforms are associated with the preferential growth of segregated ice lenses: these include PALSAS (formed in *peat*), LITHALSAS (palsa-like features formed in mineral soil), and *peat plateaux*, 1–8 m in height. *Ground-ice slumps, thaw lakes and depressions* (e.g. ALASES) and *thermokarst mounds (baydjarackii)* result from the degradation of ice-rich permafrost.

In relatively unconsolidated sedimentary rock, the in-situ creep of permafrost may cause non-diastrophic structures. These include up-arching beneath valley bottoms (VALLEY BULGING) and the bending, deformation and sliding of strata on slopes, leading to CAMBERING and joint widening, or GULL formation. Within the ACTIVE LAYER, local conditions of SOIL MOISTURE saturation and high PORE WATER PRESSURES can induce rapid shallow movements confined to the active layer, with the top of permafrost acting as a lubricated slip plane for movement, thereby controlling the depth of the FAILURE plane. These failures, generally referred to as *active-layer detachments*, are frequent in the summer months on terrain underlain by ice-rich and unconsolidated shales and siltstones. In most cases, failure is initiated when the LIQUID LIMIT is exceeded; this happens during years of rapid spring thaw and/or following periods of unusually heavy summer precipitation.

There are also periglacial landforms related to the cryogenic weathering of exposed bedrock. Coarse, angular rock-rubble, commonly termed BLOCKFIELDS in North America and Europe, and *kurums* in Siberia, occurs widely over large areas of the HIGH ARCTIC polar deserts and semi-deserts, and in the *high alpine belts* of some alpine zones. They surround outcrops of more resistant rock that form isolated hills or TORS. In regions of extreme aridity, rectilinear, debris-veneered bedrock-controlled slopes (*Richter denudation slopes*) may evolve.

In upland areas, below bedrock cliffs, frost-shattered TALUS may accumulate as sheets and cones. Modification of talus may occur through SNOW-AVALANCHE activity, DEBRIS FLOWS, ROCK GLACIER development and PRONIVAL RAMPART formation.

At a smaller scale, many periglacial landscapes possess a diversity of PATTERNED GROUND phenomena resulting from differential FROST HEAVE and ice segregation in the active layer. These features include sorted and non-sorted circles, stripes and nets. On debris-mantled slopes, MASS-WASTING processes may result in sorted and non-sorted stripes. Depending upon moisture supply, SOLIFLUCTION lobes and SOLIFLUCTION TERRACES may form by FROST CREEP and GELIFLUCTION, especially at locations below late-lying or perennial SNOWBEDS. *HMF*

[*See also* FROST MOUNDS, ICE WEDGE, PERIGLACIAL LANDSCAPE EVOLUTION, PERIGLACIAL SEDIMENTS]

André M-F (2009) From climatic to global change geomorphology: Contemporary shifts in periglacial geomorphology. In Knight J and Harrison S (eds) *Periglacial and paraglacial processes and environments.* Bath: Geological Society, 5–28.

Ballantyne CK and Harris C (1994) *The periglaciation of Great Britain.* Cambridge: Cambridge University Press.

Barsch D (1993) Periglacial geomorphology in the 21st century. In Vitek JD and Giardino JR (eds) *Geomorphology: The research frontier and beyond.* Amsterdam: Elsevier, 141–163.

Boardman J (ed.) (1987) *Periglacial processes and landforms in Britain and Ireland.* Cambridge: Cambridge University Press.

Clarke MJ (ed.) (1988) *Advances in periglacial geomorphology.* Chichester: Wiley.

French HM (2007) *The periglacial environment, 3rd edition.* Chichester: Wiley.

French HM and Thorn CE (2006) The changing nature of periglacial geomorphology. *Géomorphologie* 3: 165–174.

Harris C (1981) *Periglacial mass-wasting: A review of research.* Norwich: Geo Books.

Washburn AL (1979) *Geocryology: A survey of periglacial processes and environments.* London: Edward Arnold.

periglacial landscape evolution   The evolution of landscape under the control of cold-climate rock WEATHERING and associated MASS WASTING of the REGOLITH. Typical slope evolution is thought to involve a progressive and sequential reduction of relief with the passage of time. This progression, termed CRYOPLANATION, consists of SLOPE REPLACEMENT from below, with the formation of *Richter denudation slopes* that are ultimately replaced by low-angle PEDIMENTS (*cryopediments*). *Richter slopes* are rectilinear, debris-veneered, bedrock surfaces on which the rate of production of weathered debris is equal to, or less than, the ability of gravity-controlled transport processes to remove it.

There are several reasons why it is difficult to generalise about the evolution of PERIGLACIAL LANDFORMS AND PROCESSES. Firstly, certain lithologies are more prone to FROST WEATHERING than others; equally, some are more capable of preserving distinct periglacial slope morphology, once formed. Secondly, the variety of *periglacial climates* (see PERIGLACIAL ENVIRONMENTS) existing today means that periglacial landform assemblages may also vary. For example, depending upon the degree of ARIDITY, running water may, or may not, be an important landscape-modifying process. Thirdly, many areas experiencing periglacial conditions today have only recently emerged from beneath continental ICE SHEETS. Therefore, these PARAGLACIAL landscapes

cannot be regarded as being in true geomorphological equilibrium. Finally, there are relatively few studies that detail, in quantitative terms, the manner and speed of periglacial *slope evolution*.

The typical slope forms found in periglacial regions today can be summarised as follows: (1) rectilinear debris-mantled (Richter) slopes, (2) free-face (i.e. exposed bedrock) and associated debris (i.e. TALUS) slope profiles, (3) smooth convexo-concave debris-mantled slopes, (4) stepped profiles and (5) pediment-like forms. Whilst (2) is frequently associated with glacially oversteepened valleys, as in Svalbard and northern Scandinavia, the other slope forms fit the cryoplanation concept reasonably well.

Richter denudation slopes are best developed in very arid periglacial regions, such as Antarctica, central Siberia, north-central Alaska/interior northern Yukon Territory and in the Canadian HIGH ARCTIC. They indicate a LANDSCAPE EVOLUTION model that involves slow frost weathering of bedrock combined with gravity-controlled free-face retreat and slope replacement from below. Cryopediments are the end result. In more humid environments, especially those recently deglaciated, not only is the rate of debris production greater due to the increased efficacy of frost weathering on exposed and oversteepened rock walls but SOLIFLUCTION, SLOPEWASH, SNOW AVALANCHES and DEBRIS FLOWS are more common. As a result, slope forms are more varied.

It is sometimes assumed, though not proven, that slopes evolve more rapidly under periglacial conditions than under non-periglacial conditions. The few available data suggest that this applies only to relatively humid and recently deglaciated regions. In the equally extensive arid environments, there is no evidence that landscape evolution is faster than elsewhere. Instead, evidence points to a similarity of forms with hot DESERTS and the SEMI-ARID regions of the world. It is probable that many slope forms in the temperate mid-latitudes of North America and Europe are essentially RELICT periglacial forms dating from the PLEISTOCENE.   *HMF*

[*See also* EQUILIBRIUM CONCEPTS IN GEOMORPHOLOGICAL AND LANDSCAPE CONTEXTS, SLOPE EVOLUTION MODELS]

André M-F (203) Do periglacial landscapes evolve under periglacial conditions? *Geomorphology* 52: 149–164.

Ballantyne CK and Harris C (1994) *The periglaciation of Great Britain.* Cambridge: Cambridge University Press.

Boelhouwers JC and Meklejohn KI (2002) Quaternary periglacial and glacial geomorphology of southern Africa: review and synthesis. *South African Journal of Science* 98: 47–55.

French HM and Harry DG (1992) Pediments and cold-climate conditions, Barn Mountain, unglaciated northern Yukon, Canada. *Geografiska Annaler* 74: 145–157.

Halina P (2011) Periglacial evolution of slopes: Rock control versus climate factors (Cracow Uplands, S. Poland). *Geomorphology* 132: 139–152.

Peltier LC (1950) The geographical cycle in periglacial regions as it related to climatic geomorphology. *Annals of the Association of American Geographers* 40: 214–236.

Rapp A (1960) Recent development of mountain slopes in Karkevagge and surroundings, northern Sweden. *Geografiska Annaler* 42: 71–200.

Schirrmeister L, Grosse G, Kunitsky V et al. (2008) Periglacial landscape evolution and environmental changes of Arctic Lowland areas for the last 60,000 years (western Laptev Sea coast, Cape Momontov Klyk). *Polar Research* 27: 249–272.

Selby MJ (1971) Slopes and their development in an ice-free, arid area of Antarctica. *Geografiska Annaler* 53: 235–245.

Vandenberghe J and Czudek T (2008) Pleistocene cryopediments on variable terrain. *Permafrost and Periglacial Processes* 19: 71–83.

periglacial sediments   Characteristics diagnostic of these sediments are difficult to identify because they are often confused with the processes, landforms and environments with which these sediments are linked. In theory, periglacial sediments sensu stricto are the result of (1) the FROST WEATHERING (*cryogenic weathering*) of bedrock in situ, (2) the deposition of sediment under climatic conditions that favour freezing and the subsequent aggradation of PERMAFROST or (3) the seasonal thawing of this perennially frozen ground or SEASONALLY FROZEN GROUND and its subsequent reworking and redeposition (see THERMOKARST).

So-called *periglacial facies* were first identified by the Polish geologist Walery von Lozinski in 1909 when he described the rock-rubble deposits that mantle the slopes of the Carpathian Mountains of Central Europe. He assumed they were similar to the 'stone runs' and blockfields described earlier from the Falkland Islands by J.G. Andersson in 1906. It quickly became conventional wisdom that these widespread blocky deposits were the result of mechanical weathering and, specifically, the disintegration of bedrock due to FREEZE-THAW CYCLES. It is now understood that the growth of SEGREGATION ICE is the main mechanism of rock disintegration due to freeze-thaw processes. However, other mechanisms of bedrock disintegration are also possible, such as *hydration shattering* and *thermal stress*. Moreover, some field investigations suggest that these openwork boulder accumulations are not due primarily to cold-climate processes but may have formed during Tertiary times and were merely refashioned by cold-climate processes during the Quaternary. Therefore, the recognition of rock-rubble slope mantles (e.g. BLOCKFIELDS) as being periglacial sediments, or periglacial FACIES, is open to debate.

The *Yedoma Suite* constitutes unambiguous periglacial sediments. First described in the Russian literature as occurring in central and northern Siberia, *Yedoma* is a frozen slope deposit. It is characterised by ice-rich and organic-rich silty sediment containing large SYNGENETIC ice wedges. These ICE WEDGES may exceed 30 m in vertical height and 1–10 m in horizontal extent. Yedoma is clearly sediment that has been deposited under cold nonglacial conditions. *Nival lithogenesis* is thought to be the cause: sediment accumulation is the result of wind-blown snow together with plant and mineral detritus being subject to repeated thawing and the transport of clastic and organic detritus by snow MELTWATER run-off (see NIVEO-AEOLIAN DEPOSITS). Fine-grained debris is subsequently reworked by alluvial, fluvial and aeolian transport to PIEDMONT plains, CRYOPLANATION TERRACES or ALLUVIAL FANS. Concurrent with sediment accumulation is GROUND-ICE segregation, syngenetic ice-wedge growth, *sediment reworking*, *peat* aggradation, CRYOSOL formation and CRYOTURBATION. The formation of large polygonal ice-wedge systems together with thick sequences of frozen sediment is closely related to the persistence of stable, poorly drained, low-gradient accumulation plains. In the northwestern Arctic of North America, the term *muck* is frequently used to describe localised yedoma-like sediments.

Sediments that are the result of THERMOKARST are also clearly periglacial in nature. However, it is difficult to generalise about thermokarst sediments because, being formed when icy permafrost degrades and ultimately thaws, these sediments represent a range of redeposited and heterogeneous materials or DIAMICTONS that reflect the nature of the previously frozen rock or soil. Typically, thermokarst sediments incorporate clumps of organic matter and portray SOFT-SEDIMENT DEFORMATION structures indicative of differential LOADING and density readjustments. Where thaw LAKES and basins form, LACUSTRINE SEDIMENTS are deposited, and where the permafrost is exceptionally icy, COLLUVIAL FANS and small DELTA structures may form.

LOESS is a homogeneous unstratified silt of wind-blown origin that is commonly associated with periglacial environments. It has a uniform grain size (0.01–0.05 mm in diameter) and is particularly widespread in the marginal zones of the cold SEMI-ARID deserts of China, Siberia and interior Yukon-Alaska. Loess deposits are often tens of metres thick, and nearly always calcareous. Because (1) loess shrinks when water is applied (2) large syngenetic sand-filled thermal contraction cracks are sometimes present, and (3) as it often contains cold-climate land snails, loess must have been deposited in arid, STEPPE-like environments, sufficiently cold for permafrost to occur. The majority of loess is PLEISTOCENE in age. However, it is not a periglacial sediment sensu stricto. This is because it occurs not only around the margins of hot and cold deserts but also in the temperate regions of North America and Europe. Moreover, because it has been transported, it need not indicate a periglacial environment as regards either the area of deposition or, equally, the source area.

SOLIFLUCTION and SLOPEWASH deposits are common in many present-day periglacial environments of moderate to high humidity. Solifluction deposits, involving FROST CREEP and GELIFLUCTION, are heterogeneous in nature and include coarse TALUS, ALLUVIUM (silt and clay) and DIAMICTONS of both glacial and periglacial origin. Their FABRIC is often organised such that elongate particles tend to be aligned with their long axes pointing in the direction of flow. Solifluction may result in the formation of lobes, tongues and sheets. *Slopewash deposits* are layers of sand and silt that are usually concentrated locally beneath and downslope of SNOWBEDS. In areas of especially heavy snowfall and/or frequent freeze-thaw cycles, and where the bedrock is suited to cryogenic disintegration, *stratified slope deposits* may form (see GRÈZES LITÉES).   *HMF*

[*See also* PERIGLACIAL LANDFORMS AND PROCESSES]

Andre M-F, Hall K, Bertran P and Arocena J (2008) Stone runs in the Falkland Islands: Periglacial or tropical? *Geomorphology* 95: 524–543.

French HM (2007) *The periglacial environment, 3rd edition.* Chichester: Wiley.

Murton JM, Peterson R and Ozouf J-C (2006) Bedrock fracture by ice segregation in cold regions. *Science* 314: 1127–1129.

Schirrmeister L, Meyer H, Wetterich S et al. (2008) The Yedoma Suite of the northeastern Siberian Shelf region: Characteristics and concept of formation. In Kane DI and Hinkel KM (eds) *Proceedings of the 9th International Conference on Permafrost, Fairbanks, Alaska, volume 2.* Fairbanks: Institute of Northern Engineering, University of Alaska, 1595–1600.

Vandenberghe J (2011) Periglacial sediments: Do they exist? In Martini AP, French HM and Perez-Alberti A (eds) *Ice-marginal and periglacial processes and sediments.* Bath: Geological Society, 205–212.

periglacial trimline   The upper level to which GLACIAL EROSION has removed, or "trimmed" a pre-existing REGOLITH cover in glaciated mountain environments (Ballantyne et al., 2011). They can be interpreted in two ways: firstly, they represent the upper limit of a warm-based GLACIER or ICE SHEET at its maximum thickness; or, secondly, they represent an ENGLACIAL transition within a former ice sheet from erosive *warm-based ice* occupying the valley bottoms and *cold-based ice* frozen to the upper slopes. Arguably, the term is a misnomer and the phenomenon would be better termed a *glacial trimline* (because the glacier does the trimming).   *JAM*

[*See also* GLACIER THERMAL REGIME]

Ballantyne CK, McCarroll D and Stone JO (2011) Periglacial trimlines and the extent of the Kerry-Cork Ice Cap, SW Ireland. *Quaternary Science Reviews* 30: 3834–3845.

periglaciation   The degree to which processes and landforms in a LANDSCAPE are formed, modified or maintained by a PERIGLACIAL ENVIRONMENT (i.e. non-glacial cold climate). It is the periglacial equivalent of the concept of GLACIATION inasmuch as the latter is used to refer to the periodic glacial modification of the landscape.   *HMF*

[*See also* PERIGLACIAL LANDFORMS AND PROCESSES, PERIGLACIAL LANDSCAPE EVOLUTION]

Ballantyne CK and Harris C (1994) *The periglaciation of Great Britain.* Cambridge: Cambridge University Press.

Worsley P (1977) Periglaciation. In Shotton FW (ed.) *British Quaternary studies: Recent advances.* Oxford: Clarendon Press, 203–219.

periglaciofluvial system   The concept of a process-sediment-landform association associated with PERIGLACIAL ENVIRONMENTS that is different from both *glaciofluvial systems* and the *fluvial systems* characteristic of temperate landscapes. McEwen and Matthews (1998) showed that the *alpine periglaciofluvial system* is particularly distinctive in terms of channel form, bed material characteristics, sediment sources (including FROST WEATHERING in the river channel and SNOW AVALANCHE material from the valley sides) and strong linkages between the river channel and the valley slopes.   *JAM*

McEwen LJ and Matthews JA (1998) Channel form, bed material and sediment sources of the Sprongdøla, southern Norway: Evidence for a distinct periglacio-fluvial system. *Geografiska Annaler* 80(A): 17–36.

perihelion   The point on the orbit of the Earth (or that of another planet) at which it is closest to the Sun.   *JAM*

perimarine   The HABITATS, *sedimentary environments* and FACIES of low-lying coastal areas protected by BARRIER ISLANDS, comprising *fluvial, lagoonal* and WETLAND (e.g. peat marsh) environments.   *RAS*

Plater A and Kirby J (2006) The potential for perimarine wetlands as an ecohydrological and phytotechnological management tool in the Guadiana estuary, Portugal. *Estuarine Coastal and Shelf Science* 70: 98–108.

period   An interval of time in GEOCHRONOLOGY that is equivalent to a SYSTEM in CHRONOSTRATIGRAPHY. Period and system names are the same (e.g. SILURIAN period and system).   *LC*

periodicities   Systematic recurrence of events after fixed periods of time or at a fixed FREQUENCY. Thus, the 24-hour day and 365.24 days in a year repeat themselves year after year. In a similar way, TIDES can be calculated many years in advance. In the field of environmental change, the term is also used to mean recurrence at nearly but not exact intervals and with different intensities, although these are better termed QUASI-PERIODIC PHENOMENA. The terms *cycle* and *cyclicities* are often used in the same way.

There have been many attempts to find periodicities, partly in order to predict future events, but often they have been shown to be a product of the technique used (e.g. FILTERING and SPECTRAL ANALYSIS). Quasi-periodicities that have been found in a variety of climatic data include the MADDEN-JULIAN OSCILLATION (MJO, frequency of 40–50 days), the NORTH ATLANTIC OSCILLATION (NAO, weeks to months), the ARCTIC OSCILLATION (AO, weeks to months), the ANTARCTIC OSCILLATION (AAO, weeks to months), the Indian Ocean Dipole (IOD, 2 or more years), the QUASI-BIENNIAL OSCILLATION (QBO, 2.5 years), EL NIÑO-SOUTHERN OSCILLATION (ENSO, 2–7 years), SUNSPOT CYCLES (11 years), LUNAR CYCLES (18.6 years), the HALE CYCLE (22 years), the *Interdecadal Pacific Oscillation* (IPO, 40–60 years) and the ATLANTIC MULTIDECADAL OSCILLATION (AMO, 50–80 years). The strongest periodicities in longer-term PALAEOCLIMATOLOGY are those associated with the MILANKOVITCH THEORY. Amongst the weaker periodicities reported on intermediate or SUB-MILANKOVICH timescales of centuries to millennia are those at ca 550, 1,000 and 1,600 years.   BDG

[*See also* CLIMATIC FLUCTUATION, CLIMATIC MODES, CYCLICITY, EARTH CYCLES, EPISODIC EVENTS, TIME-SERIES ANALYSIS]

Baldwin WP, Gray LJ and Dunkerton TJ (2001) The quasi-biennial oscillation. *Reviews of Geophysics* 32: 179–229.
Burroughs TJ (2005) Cycles and periodicities. In Oliver JE (ed.) *Encyclopedia of world climatology*. Dordrecht: Springer, 309–314.
Chapman MR and Shackleton NJ (2000) Evidence of 550 year and 1000 year cyclicities in North Atlantic circulation patterns during the Holocene. *The Holocene* 10: 287–291.
Loehle C and Singer SF (2010) Holocene temperature records show millennial-scale periodicity. *Canadian Journal of Earth Sciences* 47: 1327–1336.
Salinger MJ, Renwick JA and Mullan AB (2001) Interdecadal Pacific Oscillation and South Pacific climate. *International Journal of Climatology* 21: 1705–1721.
Stuiver M, Grootes PM and Braziunas TF (1995) The GISP2 $\delta^{18}O$ record of the past 16,500 years and the role of the sun, oceans and volcanoes. *Quaternary Research* 44: 341–354.
Swindles GT, Patterson HM, Roe HM and Galloway JM (2012) Evaluating periodicities in peat-based climate proxy records. *Quaternary Science Reviews* 41: 94–103.

peri-urban   Areas fringing towns and cities which are subject to influences like POLLUTION and human DISTURBANCE. While there may also be problems of vandalism and theft, these areas have the advantage of being close to the market for *peri-urban agriculture*. In poor countries, rural-urban MIGRATION has led to heavy settlement in peri-urban areas by people with little chance of employment. Agriculture in these areas can employ and feed the poor and provide produce for the city. Productivity can be high and, in the future, peri-urban agriculture is likely to be of increasing importance. There are risks from irrigation or contamination with sewage and proximity of livestock (especially pigs and poultry) to people, which can lead to DISEASE and EPIDEMICS. In richer nations, peri-urban areas can support recreational uses (e.g. allotments and green belts). Until recently, authorities seldom supported peri-urban activities but that is changing and there is interest in integrating urban WASTE MANAGEMENT (refuse and sewage) with agriculture.   CJB

[*See also* ANIMAL DISEASES, URBAN AGRICULTURE, URBAN AND RURAL PLANNING, URBANISATION]

McGregor D, Simon D and Thompson D (eds) (2005) *The peri-urban interface: Approaches to sustainable natural and human resource use*. London: Earthscan.
Mougeot LJA (ed.) (2005) *Agropolis: The social, political and environmental dimensions of urban agriculture*. London and Ottawa: Earthscan and International Development Research Centre.

permafrost   According to the International Permafrost Association, permafrost is ground (soil, sediment or rock) that remains at or below 0°C for at least two consecutive years. Permafrost that forms after deposition of the host sediment or rock is termed EPIGENETIC. The time lag between accumulation and perennial freezing of epigenetic permafrost reaches thousands or millions of years. By contrast, permafrost that forms at the same time as continued cold-climate sedimentation is termed SYNGENETIC. By definition, syngenetic permafrost is of the same age (approximately) as the sediment in which it is formed.

To differentiate between the thermal (i.e. temperature) and state (i.e. frozen or unfrozen) conditions of permafrost, the terms *cryotic* and *non-cryotic* have been proposed. These terms refer solely to the temperature of the material independent of its water or ice content. Therefore, *perennially frozen ground* is not always synonymous with permafrost because it may be 'unfrozen', 'partially frozen' or 'frozen' depending upon the state of the ICE and water content.

The PERMAFROST TABLE is the upper surface of the permafrost, and the ground above the permafrost table is called the *suprapermafrost layer*. The ACTIVE LAYER is that part of the suprapermafrost layer that freezes in the winter and thaws during the summer; that is, it is

SEASONALLY FROZEN GROUND. The base of the active layer (i.e. the depth of annual thaw) may vary on annual decadal and millennia timescales. Accordingly, the *transient layer* includes the typically ice-rich layer marking the long-term contact between the active layer, as defined above, and the upper part of permafrost. Although *seasonal frost* usually penetrates to the permafrost table in most areas, in some areas it does not; instead, an unfrozen zone exists between the bottom of seasonal frost and the permafrost table (a TALIK). Unfrozen zones within and below the permafrost are also termed 'taliks'.

Permafrost thickness depends on the balance between the internal heat gain with depth and heat loss from the surface. *Heat flow* from the Earth's interior normally results in a temperature increase of approximately 1°C per 30–60 m increase in depth (the *geothermal gradient*). If climatic conditions at the ground surface alter, permafrost thickness will change appropriately. For example, an increase in mean surface temperature will result in a decrease in permafrost thickness, while a decrease in surface temperature will give the reverse. Typically, near-surface permafrost temperatures range between −10°C and −15°C in the high latitudes ('cold' permafrost) to near 0°C in areas of marginal ('warm') permafrost.

Nearly 25 per cent of the Earth's land surface is underlain by permafrost (see Figure), the vast majority of which occurs in the Northern Hemisphere. Russia possesses the largest area of permafrost (11.0 million km^2), followed by Canada (5.7 million km^2) and China (2.1 million km^2). In parts of Siberia and interior Alaska, permafrost has existed for several hundred thousand years; in other areas, such as the modern Mackenzie Delta, permafrost is young and currently forming under the existing cold climate.

Permafrost occurs in two contrasting and, in places, overlapping localities, namely, high latitudes and high altitudes. Accordingly, permafrost can be classified into one of the following categories: (1) *polar* (or *latitudinal*) *permafrost* (i.e. permafrost in the ARCTIC and ANTARCTIC), (2) *alpine permafrost* (i.e. MOUNTAIN PERMAFROST), (3) *plateau permafrost* (i.e. extensive permafrost at high elevations, such as on the Tibet (Quinghai-Xizang) Plateau of China) and (4) *submarine permafrost* (i.e. on the CONTINENTAL SHELVES of the Laptev, Siberian and Beaufort seas).

Permafrost is also classified according to whether it is continuous or discontinuous in nature (see Figure). In areas of *continuous permafrost*, frozen ground is present at all localities except for localised taliks existing beneath lakes and river channels. In *discontinuous permafrost*, bodies of frozen ground are separated by areas of unfrozen ground. At the southern limit of this zone, permafrost becomes restricted to isolated 'islands', typically occurring beneath peaty organic sediments (sometimes termed *sporadic permafrost*).

At the local level, variations in permafrost conditions are determined by a variety of terrain and other factors. Of widespread importance are the effects of relief and aspect and the nature of the physical properties of soil and rock. More complex is the control exerted by vegetation, snow cover, waterbodies, drainage and fire. GLOBAL WARMING is causing permafrost to warm in areas of 'cold' permafrost and leading to its thaw and disappearance in areas of marginal or 'warm' permafrost.

Deep permafrost probably represents the oldest ice on Earth. At the surface, it shapes the landscape, controls the HYDROLOGY and poses major challenges to society, especially for infrastructure engineering. Examples include the SUBSIDENCE problems associated with thawing permafrost beneath buildings and oil pipelines. Some of the world's highest rates of *coastal erosion* are found along the Arctic coast, which is affected by SEA-LEVEL RISE and reduced SEA ICE. Permafrost also interacts with the CLIMATIC SYSTEM through its role as a long-term carbon store in the global CARBON CYCLE.   *HMF*

[*See also* GEOMORPHOLOGY AND ENVIRONMENTAL CHANGE, HUMAN IMPACT ON LANDFORMS AND GEOMORPHIC PROCESSES, PERIGLACIAL LANDSCAPE EVOLUTION, PERIGLACIAL SEDIMENTS, SUBSURFACE TEMPERATURE]

Brown J, Ferrians OJ, Heginbottom JA and Melnikov ES (1998) *Circum-Arctic map of permafrost and ground ice*. Boulder, CO: National Snow and Ice Data Center.

Burn CR (2012) Permafrost distribution and stability. In French HM and Slaymaker O (eds) *Changing cold environments: A Canadian perspective*. Chichester: Wiley, 126–146.

Dobinski W (2011) Permafrost. *Earth-Science Reviews* 108: 158–169.

French HM (2007) *The periglacial environment, 3rd edition*. Chichester: Wiley.

Harris C, Arenson LU, Christiansen HH et al. (2009) Permafrost and climate in Europe: Monitoring and modeling thermal, geomorphological and geotechnical responses. *Earth-Science Reviews* 92: 117–171.

Harris C, Volder Mühll D, Isaksen K et al. (2003) Warming permafrost in European mountains. *Global and Planetary Change* 39: 215–225.

Harris SA (1986) *The permafrost environment*. London: Croom Helm.

Jorgenson MT, and Osterkamp TE (2005) Response of boreal ecosystems to varying modes of permafrost degradation. *Canadian Journal of Forestry Research* 35: 2100–2111.

Kneisel C (2010) The nature and dynamics of frozen ground in alpine and subarctic periglacial environments. *The Holocene* 20: 423–445.

Marshall SJ (2012) *The cryosphere*. Princeton, NJ: Princeton University Press.

Shur YL, Hinkel KL and Nelson FE (2005) The transient layer: Implications for geocryology and climate-change science. *Permafrost and Periglacial Processes* 16: 5–18.

Zhang T, Barry RG, Knowles K et al. (2008) Statistics and characteristics of permafrost and ground ice distribution in the Northern Hemisphere. *Polar Geography* 31: 47–68.

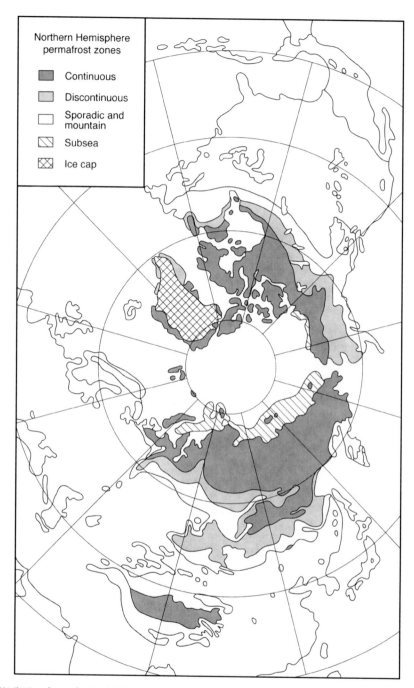

Permafrost *Distribution of permafrost in the Northern Hemisphere (French, 2007).*

permafrost degradation   A series of events which occur when the *thermal equilibrium* of PERMAFROST is perturbed either by climatic changes or changes in ground surface condition. In the latter case, DISTURBANCE to overlying insulating VEGETATION and/or SOIL can lead to a rise in GROUND TEMPERATURE and a lowering of the PERMAFROST TABLE. LAND SUBSIDENCE occurs as the GROUND ICE thaws, the depth of subsidence depending on both the depth of thaw and the ice content of the upper layers of permafrost. Degradation can also be caused by the development of small ponds that last until refreezing in autumn when LATENT HEAT is released locally. Human disturbance of the ground surface by *fire, construction* or *vehicles* can cause permafrost degradation. Ground temperature rise caused by GLOBAL WARMING is expected to reduce areas of permafrost worldwide. For example, in Siberia, a reduction of 10 per cent in permafrost area over a 50-year period has been predicted. *RAS*

[*See also* THERMOKARST]

Avis CA, Weaver AJ and Meissner KJ (2011) Reduction in areal extent of high-latitude wetlands in response to permafrost thaw. *Nature Geoscience* 4: 444–448.
French HM (2007) *The periglacial environment, 3rd edition.* Harlow: Longman.
Harris C, Davies MCR and Etzelmüller (2001) The assessment of potential geotechnical hazards associated with mountain permafrost in a warming global climate. *Permafrost and Periglacial Processes* 12: 1345–156.
Isaksen K, Sollid JL, Holmlund P and Harris C (2007) Recent warming of mountain permafrost in Svalbard and Scandinavia. *Journal of Geophysical Research* 112: F02S04.
Nelson FE, Anisimov OE and Shiklomonov OI (2001) Subsistence risk from thawing permafrost. *Nature* 410: 889–890.

permafrost table   In PERIGLACIAL environments, the boundary between the ACTIVE LAYER and the underlying PERMAFROST. Because of winter freezing penetrating downwards at unequal rates, TALIKS may occur below the permafrost table. *RAS*

[*See also* CRYOSOLS]

Swanson DK, Ping CL and Michaelson GJ (1999) Diapirism in soils due to thaw of ice-rich material near the permafrost table. *Permafrost and Periglacial Processes* 10: 349–367.

permanent drought   A type of DROUGHT characteristic of ARIDLANDS, where there is normally insufficient moisture for AGRICULTURE without IRRIGATION, and only XEROPHYTES can grow unaided. *JAM/JET*

[*See also* CONTINGENT DROUGHT, INVISIBLE DROUGHT, SEASONAL DROUGHT]

Bonan GB (2008) *Ecological climatology.* Cambridge: Cambridge University Press.

permeable   Having a structure or texture that permits the transmission of liquids or gases. In relation to rocks, sediments and soils, the distinction should be made between *primary permeability*, which allows water to pass through the pores or matrix by DIFFUSION and *secondary permeability*, which allows the flow of water through fissures, joints or cracks. Whereas some make no distinction between permeable and PERVIOUS, others restrict usage of the term 'permeable' to what has been defined above as primary permeability, and reserve 'pervious' for secondary permeability. *JAM/ADT*

[*See also* IMPERMEABLE, IMPERVIOUS, POROUS]

Stamp LD (ed.) (1961) *A glossary of geographical terms.* London: Longmans.

Permian   A SYSTEM of rocks, and a PERIOD of geological time from 299 to 251 million years ago. The supercontinent PANGAEA (PANGEA) had assembled, and a Southern Hemisphere ICE AGE that began during the CARBONIFEROUS period continued into the Permian. The end of the Permian is marked by a major MASS EXTINCTION. *GO*

[*See also* GEOLOGICAL TIMESCALE, PERMO-TRIASSIC]

Angiolini L, Jadoul F, Leng MJ et al. (2009) How cold were the Early Permian glacial tropics? Testing sea-surface temperature using the oxygen isotope composition of rigorously screened brachiopod shells. *Journal of the Geological Society of London* 166: 933–945.
Heydari E, Arzani N and Hassanzadeh J (2008) Mantle plume: The invisible serial killer: Application to the Permian-Triassic boundary mass extinction. *Palaeogeography, Palaeoclimatology, Palaeoecology* 264: 147–162.
Heydari E, Wynn TC and Chen ZQ (eds) (2010) Late Permian-Early Triassic Earth. *Global and Planetary Change* 73 [Special Issue].
Şengör AMC and Atayman S (2009) *The Permian extinction and the Tethys: An exercise in global geology.* Boulder, CO: Geological Society of America.
Shen S-Z, Crowley Jl, Wang Y et al. (2011) Calibrating the end-Permian mass extinction. *Science* 334: 1367–1372.

Permo-Triassic   The PERMIAN and TRIASSIC geological SYSTEMS which, in northern Europe, share many characteristics resulting from a time of DESERT conditions. The term is commonly shortened to *Permo-Trias*. *GO*

Kidder DL and Worsley TR (2004) Causes and consequences of extreme Permo-Triassic warming to globally equable climate and relation to the Permo-Triassic extinction and recovery. *Palaeogeography, Palaeoclimatology, Palaeoecology* 203: 207–237.
Mader D (1992) *Evolution of palaeoecology and palaeoenvironment of Permian and Triassic fluvial basins in Europe.* Stuttgart: Gustav Fischer.
Xie S, Pancost RD, Wang Y et al. (2010) Cyanobacterial blooms tied to volcanism during the 5 m.y. Permo-Triassic biotic crisis. *Geology* 38: 447–450.

peroxyacetyl nitrate (PAN)   An oxidant in urban PHOTOCHEMICAL SMOG produced from the interaction of HYDROCARBONS and NITROGEN OXIDES in vehicle exhaust. It is an eye irritant, a contributor to FOREST DECLINE and other plant damage and a reservoir or SINK for nitrogen oxides in the upper TROPOSPHERE, where it is relatively stable. *JAM/JET*

[*See also* OZONE, PHOTOCHEMICAL OXIDANT, PHOTOCHEMISTRY]

Finlayson-Pitts BJ and Pitts Jr JN (2000) *Chemistry of the upper and lower atmosphere: Theory, experiment and applications.* San Diego, CA: Academic Press.

persistence   (1) In the context of organisms, the developmental stage at which the organism will maintain itself indefinitely. (2) For chemicals in the environment, the length of time taken for the chemical to be broken down to a point whereby it is no longer measurable. (3) In meteorology, the duration of a particular synoptic situation. (4) More generally, the term refers to the STABILITY or RESISTANCE to a SYSTEM (e.g. an individual, population, ecosystem or the CLIMATIC SYSTEM) in relation to a DISTURBANCE or PERTURBATION.   *JLI*

[*See also* AUTOREGRESSIVE (AR) MODELLING, EQUILIBRIUM CONCEPTS, PERSISTENT ORGANIC COMPOUNDS (POCs), STABILITY CONCEPTS: ECOLOGICAL CONTEXTS]

Gilarranz LJ and Bascompte J (2012) Spatial network structure and metapopulation persistence. *Journal of Theoretical Biology* 297 11–16.
Megharaj M, Ramakrishnan B, Venkateswarlu K et al. (2011) Bioremediation approaches for organic pollutants: A critical perspective. *Environment International* 37: 1362–1375.
Purschke O, Sykes MT, Reitalu T et al. (2011) Linking landscape history and dispersal traits in grassland plant communities. *Oecologia* 168: 773–783.
Uezu A and Metzger JP (2011) Vanishing bird species in the Atlantic Forest: Relative importance of landscape configuration, forest structure and species characteristics. *Biodiversity and Conservation* 20: 3627–3643.
Weber J, Halsall CJ, Muir D et al. (2010) Endosulfan, a global pesticide: A review of its fate in the environment and occurrence in the Arctic. *Science of the Total Environment* 408: 2966–2984.

persistent organic compounds (POCs)   Complex organic compounds that are resistant to destruction and therefore persist in the environment without change for long periods. Persistence in the environment can be taken to mean a HALF-LIFE of >6 months, which represents a RISK of long-time exposure, but many POCs are much more recalcitrant. Many are POLLUTANTS (*persistent organic pollutants*). Most are either manufactured by humans or enter the environment inadvertently as a result of human activities. The number of synthetic organic compounds currently present in the global environment is probably 60,000–100,000 and about 1,000 new ones are produced per year. It is the chlorinated hydrocarbons (ORGANOCHLORIDES) manufactured for use as PESTICIDES or released during industrial processing that have caused most environmental concern.

The main structural classes of POCs in organochloride pesticides are *polychlorohydrocarbons* (e.g. DDT), *polychlorinated cyclodienes* (e.g. aldrin,

dieldrin), *hexachlorocyclohexanes* (e.g. lindane) and *polychlorinated monoterpenes* (e.g. toxaphene). Industrial processing produces POCs in several other structural classes including POLYCHLORINATED BIPHENYLS (PCBs), *polychlorinated dibenzo-p-dioxins* (PCDDs), *polychlorinated dibenzofurans* (PCDFs) and *polycyclic aromatic hydrocarbons* (PAHs). Most of the organochloride pesticides have been banned in industrialised countries, but they continue to be used in tropical DEVELOPING COUNTRIES for MALARIA control, and much of past production is still involved in BIOGEOCHEMICAL CYCLES. Some PCBs are manufactured for use as liquid electrical insulators and in plastics and, although they are not highly toxic for most species, they have been implicated in the decline of seal populations through affecting reproduction. Some PCDDs and PCDFs are not manufactured for use but are produced as by-products or trace constituents and are highly toxic to fish, particularly in their early stages of development.

Many PAHs are CARCINOGENS or MUTAGENS and are produced in very small quantities as a consequence of incomplete combustion of organic matter in, for example, burning FOSSIL FUELS and cooking foodstuffs. Persistent organic compounds tend to concentrate in body fat tissues as well as in blood and milk, exhibit BIO-ACCUMULATION in animals at the top of FOOD CHAINS and elicit a broad spectrum of biochemical and toxic responses. The levels of POCs found in the larger marine mammals, including whales, dolphins and porpoises, tend to be an order of magnitude greater than the levels in terrestrial birds and mammals including humans. Environmental concentrations are generally decreasing in air, water, soil, biota and food.   *JAM*

[*See also* CHLOROFLUOROCARBONS (CFCs), ECOTOXICOLOGY, HALOGENATED HYDROCARBONS (HALOCARBONS), VOLATILES, XENOBIOTICS]

Fishbein L (1998) Organochlorine and polycyclic aromatic hydrocarbon contaminants. In Brune D, Chapman DV, Gwynne MD and Pacyna JM (eds) *The global environment: Science, technology and management, volume 1.* Weinheim: VCH, 481–497.
Kamrin MA and Ringer RK (1994) PCB residues in mammals: A review. *Toxicology and Environmental Chemistry* 41: 63–84.
Korte FW and Coulston F (1994) Some consideration of the impact of energy and chemicals on the environment. *Ecotoxicology and Environmental Safety* 29: 243–250.
Loganthan BG and Lam PKS (eds) (2011) *Global contaminant trends of persistent organic chemicals.* Boca Raton, FL: CRC Press.
Muir DCG and Howard PH (2006) Are there other persistent organic pollutants? A challenge for environmental chemists. *Environmental Science and Technology* 40: 7157–7166.
Oehme M (1991) Further evidence for long-range air transport of polychlorinated aromatics and pesticides:

North America and Eurasia to the Arctic. *Ambio* 20: 293–297.

Swedish Environmental Protection Agency (SEPA) (1997) *Persistent organic pollutants: A Swedish view of an international problem*. Stockhom: SEPA [Monitor 16].

Tanabe S, Iwata H and Tatsukawa R (1994) Global contamination by persistent organochlorides and their toxicological impact on marine mammals. *Science of the Total Environment* 15: 163–177.

United Nations Environment Programme (UNEP) (2003) *Regionally based assessment of persistent organic substances*. Geneva: UNEP.

perturbation   Any small-scale departure of a SYS-TEM from a *steady-state equilibrium* due to an external DISTURBANCE. In the general context of environmental systems and ENVIRONMENTAL CHANGE, perturbations may be considered NATURAL or ANTHROPOGENIC. The behaviour of such systems in the face of such perturbations and the extent to which environmental systems are capable of returning to the same or a different equilibrium (with or without human assistance) are major concepts in the field of environmental change. A major current concern, for example, is whether the perturbation of the CLIMATIC SYSTEM being caused by the burning of FOSSIL FUELS and the enhanced GREENHOUSE EFFECT is leading the system away from the present steady-state equilibrium and towards a TIPPING POINT.

Weather disturbances, such as the *easterly waves* that are the precursors to TROPICAL CYCLONES, provide another specific example. They may be described mathematically as the difference between the motion of an undisturbed wave or current in a fluid and a small superimposed motion, which is variable in space and time. The perturbation may then be viewed as the correction equating the actual motion in the fluid with a simple theoretical motion. It is an important concept in *dynamical meteorology*, especially in numerical weather FORECASTING. *BDG/JAM*

[*See also* CLIMATIC FLUCTUATION, EQUILIBRIUM CONCEPTS, TIPPING POINT]

Brierley CM, Collins M and Thorpe AJ (2010) The impact of perturbations to ocean-model parameters on climate and climate change in a coupled model. *Climate Dynamics* 34: 325–343.

Godske CL, Bergeron T, Bjerknes J and Bundgaard RC (1957) *Dynamic meteorology and weather forecasting*. Boston: American Meteorological Society.

Holton JR (2004) *An introduction to dynamical meteorology*. 4th edition. Burlington, MA: Elsevier.

pervection   The mechanical movement or down-washing of solid particles (especially silt) through interconnected pores in soil. It occurs, for example, in the earliest stages of SOIL DEVELOPMENT on GLACIER FORELANDS, where surface soil layers rapidly lose their

silt MATRIX. It may also be responsible for movement of microfossils, such as DIATOMS and PHYTOLITHS, down the SOIL PROFILE. It should be distinguished from LEACHING of chemicals in solution and ELUVIATION of clays and organic material. *JAM*

[*See also* LESSIVAGE]

Frenot Y, Van Vliet Lanoë B and Gloaguen J-C (1995) Particle translocation and initial soil development on a glacier foreland, Kerguelen Islands, subantarctic. *Arctic and Alpine Research* 27: 107–115.

Paton TR (1978) *The formation of soil material*. London: Allen and Unwin.

pervious   (1) Synonymous with PERMEABLE. (2) It is also used in a more restricted sense of materials (e.g. rocks, sediments or soils) that permit the flow of fluids through fissures, joints or cracks (*secondary permeability*) despite being non-porous. *ADT/JAM*

[*See also* IMPERVIOUS, POROUS]

pest   Any organism detrimental to human health, welfare and comfort. Strictly, any organism (e.g. *insect*, PARASITE, *plant* or *animal* that reduces yield in *managed ecosystems* by competing with or destroying the harvestable crop. *GOH*

[*See also* BIOLOGICAL CONTROL, COMPETITION, DISEASE, HARVESTING, PARASITE, PATHOGENS, PEST MANAGEMENT, PESTICIDES, WEED]

Alford DV (2011) *Plant pests*. London: Collins.

Kocmankova E, Trnka M, Eitzinger J et al. (2010) Estimating the impact of climate change on the occurrence of selected pests in the Central European region. *Climate Research* 44: 95–105.

pest control   Reducing the abundance of a PEST and/or its EXTIRPATION. The most commonly used approach is through PESTICIDES but other methods include (1) modification or removal of breeding sites; (2) release of irradiated males (to prevent breeding) or hormones (to prevent maturation); (3) INTRODUCTION of predators, PARASITES or DISEASE; and (4) selective breeding of pest-resistant varieties. *JAM*

[*See also* BIOLOGICAL CONTROL, PEST MANAGEMENT]

Van Emden HF (1992) *Pest control, 2nd edition*. Cambridge: Cambridge University Press.

pest management   The control and regulation of pest populations using, especially, chemical PESTI-CIDES and/or BIOLOGICAL CONTROL methods. *Integrated pest management* may employ both methods together with additional techniques, such as the use of resistant CULTIVARS, intercropping and the rationalisation of CULTIVATION practices. It is an integral part of modern AGRICULTURE. *GOH*

Koul O and Cuperus GW (2007) *Ecologically-based integrated pest management*. Wallingford: CABI Publishing.

Radcliffe EB, Hutchison WD and Cancelado RE (eds) (2009) *Integrated pest management: Concepts, targets, strategies and case studies*. Cambridge: Cambridge University Press.

pesticides   Chemicals that kill PESTS. Some prefer the term *biocide*, which avoids the necessity to define 'pest'. Pesticides may be 'general' (*broad-spectrum pesticides*) or 'specific' (*narrow-spectrum pesticides*). Examples of the latter include the following:

- *acaricide* (mites)

- *algacide* (algae)

- ANTIBIOTIC (bacteria)

- *avicide* (birds)

- *fungicide* (fungi)

- *herbicide* (herbs, weeds)

- *insecticide* (insects)

- *molluscicide* (slugs and snails)

- *rodenticide* (rodents)

Use of pesticides, such as sulfur and ARSENIC (As), has a long history, especially in HORTICULTURE, but modern approaches to PEST CONTROL originated in the mid-nineteenth century with the discovery of natural insecticides, such as *derris dust* and *pyrethrum*. From the 1930s, however, attention focused on *synthetic organic pesticides*, which were widely used in continuing intensification of horticulture, AGRICULTURE and SILVICULTURE. There are advantages and disadvantages of pesticide use. The main advantage is undoubtedly the increase in food production that results from the elimination of competitors for human food resources.

The non-specificity and PERSISTENCE of some pesticides remains a global problem both in the environment and in terms of HUMAN HEALTH HAZARDS, especially when accompanied by BIO-ACCUMULATION and BIOLOGICAL MAGNIFICATION. Lessons have been learned, however: use of persistent ORGANOCHLORIDES has been banned in many developed countries and increasing attention is being paid to alternatives, such as greater use of naturally occurring *pyrethrins*, synthetic analogues of these (*pyrethroids*), BIOLOGICAL CONTROL by natural enemies, biological control by purpose-built organisms (products of GENETIC ENGINEERING and *integrated pest management*.

Thus, since the mid-twentieth century, there have been changes in the types of pesticides used and the types of problems identified and the solutions adopted, at least in highly industrialised countries (see Figure). A major dilemma for DEVELOPING COUNTRIES, where

rapid AGRICULTURAL INTENSIFICATION is currently taking place, is whether large-scale use of available, cheap but potentially environmentally damaging pesticides can be justified.   *JAM*

[*See also* AGROCHEMICALS, DEFOLIANT, PEST MANAGEMENT, SUSTAINABLE AGRICULTURE]

Coates JR and Yamamoto H (eds) (2003) *Environmental fate and effects of pesticides*. New York: Oxford University Press and American Chemical Society.

Galt RE (2008) Beyond the circle of poison: Significant shifts in the global pesticide complex. *Global Environmental Change* 18: 786–799.

Horn DJ (1988) *Ecological approach to pest management*. New York: Guilford Press.

Pettersson O (1994) Swedish pesticide policy in a changing environment. In Pimentel D and Lehman H (eds) *The pesticide question: Environment, economics and ethics*. New York: Chapman and Hall, 182–205.

Stensersen J and Strauss S (2004) *Chemical pesticides: Modes of action and toxicology*. Boca Raton, FL: CRC Press.

Vorley W and Keeney D (eds) (1998) *Bugs in the system: Redesigning the pesticide industry for sustainable agriculture*. London: Earthscan.

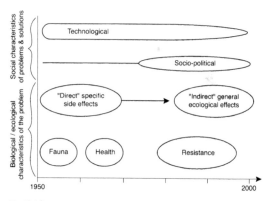

Pesticides *Schematic representation of problems and solutions in pesticide use since the mid-twentieth century (Pettersson, 1994).*

petroglyph   A type of *rock art*; a carving or engraving in a CAVE or on an open rock surface. The carving process removes any WEATHERING RIND (*patina*) or ROCK VARNISH to reveal a fresh rock surface, on which patina forms anew. Methods such as CATION-RATIO DATING can be used to date prehistoric and historic petroglyphs. Animals or other features depicted may indicate different environmental conditions from today, such as greater moisture availability. A recent survey in Egypt of the Eastern Desert and Nile Valley, for example, found 324 giraffe images at 66 sites, indicating that these animals were widely distributed before the AFRICAN HUMID PERIOD came to an end in the mid-HOLOCENE.   *JAM*

[*See also* CAVE ART, GEOGLYPH, PARIETAL ART]

Judd A (2006) Presumed giraffe petroglyphs in the Eastern Desert of Egypt: Style, location and Nubian comparisons. *Rock Art Research* 23: 59–70.

Kröpelin S (1993) *Zur Rekonstruktion der spätquartären Umwelt amk Unteren Wadi Howar (Südöstliche Sahara/ NW-Sudan)* [The reconstruction of the Late Quaternary environmental Lower Wadi Howar (Southeastern Sahara/ NW Sudan] *(Berliner Geographische Abhandlungen, volume 54)*. Berlin: Geomorphologisches Laboratorium der Freien Universität.

Lanteigne MP (1991) Cation-ratio dating of rock engravings: a critical appraisal. *Antiquity* 65: 292–295.

petrography The description of rocks in hand specimens and, particularly, thin sections.   *GO*

[*See also* PETROLOGY, THIN-SECTION ANALYSIS]

Harwood G (1988) Microscopic techniques: II. Principles of sedimentary petrography. In Tucker ME (ed.) *Techniques in sedimentology*. Oxford: Blackwell, 108–173.

MacKenzie WS and Adams AE (1994) *A colour atlas of rocks and minerals in thin section*. London: Manson.

petroleum Naturally occurring HYDROCARBON compounds in liquid (CRUDE OIL), gas (NATURAL GAS) or semi-solid (*asphalt*, *bitumen*) form that can be exploited as an ENERGY RESOURCE. Petroleum forms over geological time spans through the burial and partial decay of marine ALGAE in organic-rich mudrocks (see BLACK SHALE, OIL SHALE): it is therefore a NON-RENEWABLE RESOURCE, or FOSSIL FUEL. Crude oil and natural gas collect in the pore spaces of SEDIMENTARY ROCKS and are extracted by drilling into rocks buried as much as several kilometres beneath the land surface or sea floor. The first oil well was drilled in Pennsylvania, USA, in AD 1859. Note that *petrol* (*gasoline*) is a processed product of petroleum; others include *diesel, paraffin* (*kerosene*) and *lubricating oils*.

At the end of 2010, petroleum accounted for 57 per cent of world energy consumption (down from 64 per cent in 1998, with much of the difference accounted for by an increase in global use of COAL). RESERVES-to-production ratios indicate how long known supplies should last and stand at 46 years for crude oil and 59 years for natural gas, but these estimates have changed little, or even increased, since 1989 as new discoveries and revisions to reserves estimates have matched PRODUCTION and CONSUMPTION, and it is estimated that supplies of *conventional petroleum* should last through much of the twenty-first century. Sources of UNCONVENTIONAL PETROLEUM include HEAVY OIL, OIL SHALES, TAR SANDS, SHALE GAS, COAL-BED METHANE and GAS HYDRATES: together these represent a far greater energy resource than the total available from conventional petroleum sources.   *GO*

British Petroleum (BP) (2011) *BP statistical review of world energy, June 2011*. London: BP.

Evans AM (1997) *An introduction to economic geology and its environmental impact*. Oxford: Blackwell.

Gluyas J and Swarbrick R (2004) *Petroleum geoscience*. Oxford: Blackwell.

petrology The study of rocks, principally their composition, *texture* and origin.   *GO*

[*See also* LITHOLOGY, PETROGRAPHY]

Best MG (2003) *Igneous and metamorphic petrology, 2nd edition*. Oxford: Blackwell.

Tucker ME (2001) *Sedimentary petrology, 3rd edition*. Oxford: Blackwell.

P-forms Smooth, small-scale depressions in the form of potholes, bowls, assemblages of sickle-shaped depressions (*Sichelwannen*), grooves and channels eroded into bedrock (for depths of centimetres to metres) by glacial or glaciofluvial erosion. Precise mechanisms of formation are unclear but probably involve SUBGLACIAL meltwater under pressure (possibly heavily laden with SEDIMENT) and/or SOFT-SEDIMENT DEFORMATION.   *RAS/JAM*

[*See also* GLACIAL LANDFORMS, GLACIOFLUVIAL LANDFORMS]

Dahl R (1965) Plastically sculptured detail forms on rock surfaces in northern Nordland, Norway. *Geografiska Annaler* 47: 83–140.

Glasser NF and Bennett MR (2004) Glacial erosional landforms: Origins and significance for palaeoglaciology. *Progress in Physical Geography* 28: 43–75.

pH The negative logarithm of the *hydrogen-ion concentration* in an aqueous solution: $pH = -\log[H^+]$. Solutions are termed *neutral* when pH = 7.0; those below 7.0 are *acidic*, and those above are *alkaline*.   *AWM*

[*See also* ACIDITY, ALKALINITY]

Phaeozems Soils with a dark-coloured surface *mollic horizon* of moderate thickness, rich in organic matter, but with evidence of LEACHING of carbonates from the profile and, in some cases, accumulation of clay in the lower horizons. These soils, also termed *Brunizems, Greyzems* or degraded CHERNOZEMS, are characteristic of the forest-steppe ECOTONE in North America, South America and Eurasia under relatively humid *grasslands*. Phaeozems are inherently fertile and are widely used for arable agriculture and ranching despite possible problems associated with DROUGHT, WIND EROSION and WATER EROSION. RELICT Phaeozems occur on the drier areas of the High Plains in Texas, formed originally under moister climatic conditions (SOIL TAXONOMY: *Udolls* or *Aquolls*).   *EMB*

[*See also* WORLD REFERENCE BASE FOR SOIL RESOURCES (WRB)]

Fischer-Zujkov U, Schmidt R and Brande A (1999)
Phaeozems of northeastern Germany and their position
in Holocene landscape development. *Journal of Plant
Nutrition and Soil Science* 162: 443–449.

Miedema R, Koulechova IN and Gerasimova MI (1999)
Soil formation in Greyzems in Moscow district:
Micromorphology, chemistry, clay mineralogy and
particle size distribution. *Catena* 34: 315–347.

Phanerozoic   An EON of geological time from the
beginning of the CAMBRIAN period to the present day,
characterised by abundant FOSSILS of organisms with
hard parts (shells, bones). The term is derived from the
Greek for 'visible life'. More is known about the Phanerozoic eon than about the PRECAMBRIAN, although the
Precambrian accounts for almost 90 per cent of GEO-
LOGICAL TIME.   *GO*

[*See also* GEOLOGICAL TIMESCALE]

Glasspool IJ and Scott AC (2010) Phanerozoic
concentrations of atmospheric oxygen reconstructed from
sedimentary charcoal. *Nature Geoscience* 3: 627–630.

Hay WW, Migdisov A, Balukovski AN et al. (2006)
Evaporites and the salinity of the ocean during the
Phanerozoic: Implications for climate, ocean circulation
and life. *Palaeogeography, Palaeoclimatology,
Palaeoecology* 240: 3–46.

McGhee GR, Sheehan PM, Bottjer DJ and Droser ML
(2012) Ecological ranking of Phanerozoic biodiversity
crises: The Serpukhovian (Early Carboniferous) crisis
had a greater ecological impact than the end-Ordovician.
Geology 40: 147–150.

Ogg JG, Ogg G and Gradstein FM (2008) *The concise
geologic time scale.* Cambridge: Cambridge University
Press.

phenology   The scientific study of the life-cycle
response of plants and animals to SEASONALITY, which
involves OBSERVATION, EXPERIMENT and/or ENVIRONMEN-
TAL MODELLING of the timing of naturally occurring phenomena such as plant growth, flowering and fruiting;
seed times and harvesting; the migration of animals
and birds; and the first appearance of insects. Comparing similar observations from year to year allows an
analysis of temporal change. Of particular interest are
the dates of the first and the last events such as the first
leaf, the first flower and the first and last appearances
of birds and animals. For parts of Europe, long-term
phenological data are available for a variety of deciduous tree species and for *viticulture*, in the form of VINE
HARVEST records. Such observations have been used as
PROXY CLIMATIC INDICATORS, thereby converting historical phenological material into climatic data. However,
predicting species' phenological responses to climatic
change is a major challenge. A recent review of the
response of >1,600 plant species in so-called warming
experiments (whereby plants are artificially warmed in

the field by, for example, use of cloches) found that
observed plant flowering and leafing dates occurred
significantly earlier than the PREDICTIONS calculated
from the warming experiments. This *underprediction*
of the advances in timing of flowering and leafing
compared with long-term observation introduces considerable UNCERTAINTY into *ecosystem models* informed
solely by experiments.   *JBE/JAM/NP*

[*See also* SEASONALITY]

Fretwell SD (1972) *Populations in a seasonal environment.*
Princeton, NJ: Princeton University Press.

Hudson IL (2010) Interdisciplinary approaches: Towards
new statistical methods for phonological studies. *Climatic
Change* 100: 143–171.

Menzel A, Sparks TH, Estrella N et al. (2006) European
phenological response to climate change matches the
warming pattern. *Global Change Biology* 12: 1969–1976.

Noormets A (ed.) (2009) *Phenology of ecosystem processes:
Applications to global change research.* Dordrecht: Springer.

Schwartz MD (2003) *Phenology: An integrative
environmental science.* Berlin: Springer.

Wolkovich EM, Cook BI, Allen JM et al. (2012) Warming
experiments underpredict plant phenological responses to
climate change. *Nature* 485: 494–497.

phenomenology   Philosophies concerned with particular phenomena with an emphasis on direct intuition
and a focus on meanings and experiences of the self.
In emphasising human experience, it contrasts with the
approaches developed by the SCIENCES based on POSI-
TIVISM. In the environmental context, phenomenologi-
cal statements may thus be considered non-empirical
descriptions of environmental phenomena. This philosophical approach evolved within HUMANISM.   *ART*

Embree L (ed.) (1997) *Encyclopedia of phenomenology.*
Dordrecht: Kluwer.

Seamon D (2000) Phenomenology in environment-behavior
research. In Wapner S (ed.) *Theoretical perspectives in
environment-behavior research.* New York: Plenum Press,
157–178.

phenotype   The characteristics of an individual
organism, resulting from the interaction of its GENOTYPE
and ENVIRONMENT. Whereas the genotype is genetically
determined, the phenotype is the sum total of an organism's characteristics, some of which were acquired
during its lifetime as a result of its environmental
relations and cannot be inherited by offspring. Nevertheless, EVOLUTION by means of NATURAL SELECTION
requires *phenotypic variation* between individuals. The
term *phenome* is sometimes used for the material basis
of the phenotype (analogous to the GENOTYPE as the
material basis of the GENOTYPE). *Phenotypic plasticity*
refers to the organism's ability to respond to changes in
its environment during its lifetime.   *JAM/KDB*

[*See also* LAMARCKISM]

Mahner M and Kay M (1997) What exactly are genomes, genotypes and phenotypes? And what about phenomes? *Journal of Theoretical Biology* 186: 55–63.

phi (φ)   A dimensionless expression of GRAIN SIZE in SEDIMENTS and SEDIMENTARY ROCKS calculated as $\varphi = -\log_2(d/d_0)$, where d is the grain diameter and d_0 is the diameter of a 1-mm grain. High positive values on the phi scale denote small grain sizes.   *GO*

Krumbein WC (1964) Some remarks on the phi notation. *Journal of Sedimentary Petrology* 34: 195–196.

-phile, -philous   Suffix for an organism that tolerates, and hence is an indicator of the *presence* of, a particular ENVIRONMENTAL FACTOR: thus, a *thermophile* is an organism tolerant of relatively high temperatures; it is THERMOPHILOUS.   *JAM*

[*See also* EXTREMOPHILE, -PHOBE, -PHOBOUS]

Kullman L (1998) The occurrence of thermophilous trees in the Scandes Mountains during the early Holocene: Evidence for a diverse tree flora from macroscopic remains. *Journal of Ecology* 86: 421–428.

-phobe, -phobous   Suffix for an organism that is intolerant of, and hence is an indicator of the *absence* of, a particular ENVIRONMENTAL FACTOR: thus, a *chionophobe* is a plant intolerant of snow cover; it is *chionophobous*.   *JAM*

[*See also* -PHILE, -PHILOUS]

phosphate (PO$_4^{3-}$)   The main source of phosphorus for plants, an important FERTILISER (especially in the form of *superphosphates*) and a POLLUTANT. It is often a LIMITING FACTOR, especially in terrestrial freshwater environments. Supplies of phosphates for AGRICULTURE are currently quite restricted and could become problematic. As a former constituent of detergents (in the form of *polyphosphates*), phosphates were a major cause of EUTROPHICATION in rivers and lakes.   *JAM*

[*See also* PHOSPHORUS CYCLE]

Griffiths EJ (ed.) (1973) *Environmental phosphorus handbook*. New York: Wiley.
Jahnke RA (1992) The phosphorus cycle. In Butcher SS, Carlson RJ, Orians GH and Wolfe GV (eds) *Global biogeochemical cycles*. London: Academic Press, 301–316.
Toy ADF and Walsh EN (1987) *Phosphorus chemistry in everyday living*. Washington, DC: American Chemical Society.

phosphate analysis   Phosphorus levels in soils can be increased markedly by human activity. This enrichment results from three sources: *organic refuse* (e.g. bones and plants in MIDDENS or *burials*), *faeces* and *urine* from people and their *livestock* and deliberate use of *manure* or FERTILISER application. Samples are collected by soil *augering* and *total soil phosphate* determined, usually in the laboratory. Spatial mapping of phosphate levels may help locate *refuse dumps*, *cess pits*, burials, areas where livestock were penned or areas where soils have been improved by manuring or fertilising. The method is, however, most effective if used in tandem with other techniques of ARCHAEOLOGICAL PROSPECTION, such as MAGNETIC SUSCEPTIBILITY. Phosphate analysis may also be used to estimate *phosphate retention*, which has been suggested as a basis for SOIL DATING.   *MJB*

[*See also* CHEMICAL ANALYSIS OF SOILS AND SEDIMENTS]

Craddock PT, Gurney D, Proyor F and Hughes MJ (1985) The application of phosphate analysis to the location and interpretation of archaeological sites. *Archaeological Journal* 142: 361–376.
Dockrill SJ and Simpson IA (1994) The identification of prehistoric anthropogenic soils in the northern Isles using an integrated sampling methodology. *Archaeological Prospection* 1: 75–92.
Holliday VT and Gartner WG (2007) Methods of soil P analysis in archaeology. *Journal of Archaeological Science* 34: 301–333.
Hutson SR, Magnoni A, Beach T et al. (2009) Phosphate fractionation and spatial patterning in ancient ruins: A case study from Yucatan. *Catena* 78: 260–269.
Luzzader-Beach S, Beach T, Terry RE and Doctor KZ (2011) Elemental prospecting and geoarchaeology in Turkey and Mexico. *Catena* 85: 119–129.
Sjöberg A (1976) Phosphate analysis of anthropic soils. *Journal of Field Archaeology* 3: 447–454.

phosphorite   A SEDIMENTARY DEPOSIT rich in PHOSPHATE (PO$_4^{3-}$) minerals. Phosphorites are an important RESOURCE, used mainly as a raw material in FERTILISERS, and are significant in PALAEOENVIRONMENTAL RECONSTRUCTION, representing former CONTINENTAL SHELF sites of UPWELLING, episodes of sediment *reworking* (see REDEPOSITION) or sites of *guano* accumulation.   *GO*

Abed AM and Amireh BS (1999) Sedimentology, geochemistry, economic potential and palaeogeography of an Upper Cretaceous phosphorite belt in the southeastern desert of Jordan. *Cretaceous Research* 20: 119–133.
Burnett WC and Riggs SR (eds) (1990) *Phosphate deposits of the world, volume 3: Neogene to Modern phosphorites*. Cambridge: Cambridge University Press.
Dornbos SQ, Botjer DJ, Chen JY et al. (2006) Environmental controls on the taphonomy of phosphatized animals and animal embryos from the Neoproterozoic Doushantuo Formation, Southwest China. *Palaios* 21: 4–37.
Gnandi K and Tobschall HJ (1999) The pollution of marine sediments by trace elements in the coastal region of Togo

caused by dumping of cadmium-rich phosphorite tailing into the sea. *Environmental Geology* 38: 13–24.

Nelson GJ, Pufahl PK and Hiatt EE (2010) Paleoceanographic constraints on Precambrian phosphorite accumulation, Baraga Group, Michigan, USA. *Sedimentary Geology* 226: 9–21.

phosphorus (P)   The tenth most abundant ELEMENT in the Earth's CRUST but only rarely found as concentrated PHOSPHATE (PO_4^{3-}) mineral deposits, which are mined for use in FERTILISERS. It is a MACRONUTRIENT in both terrestrial and marine environments but is nevertheless an important source of POLLUTION and a major cause of EUTROPHICATION. A lack of available phosphorus is a major limitation to plant growth and is widespread, particularly in *tropical environments*. This is largely because the phosphate present in soil is in an insoluble form and there is a low mobility between this *insoluble pool* and the LABILE POOL which is available to plants.   *SN/JAM*

[*See also* BIOGEOCHEMICAL CYCLES, PHOSPHATE ANALYSIS, PHOSPHORITE]

Emsley J (2001) *The shocking history of phosphorus: A biography of the Devil's element.* London: Pan Books.

Montagna P, McCulloch M, Taviani M et al. (2006) Phosphorus in cold-water corals as a proxy for seawater nutrient chemistry. *Science* 312: 1788–1791.

Tiessen H, Ballester MV and Salcedo I (2011) Phosphorus and global change. In Bünemann EK, Oberson A and Frossard E (eds) *Phosphorus in action: Biological processes in soil phosphorus cycling.* Berlin: Springer, 459–472.

Turner BL, Frossard E and Baldwin DS (eds) (2005) *Organic phosphorus in the environment.* Wallingford: CABI Publishing.

phosphorus cycle   Circulation of PHOSPHORUS (P) in various forms through the LITHOSPHERE, HYDROSPHERE and BIOSPHERE. Phosphorus is the scarcest of all elements cycled through the biosphere and therefore one of the main LIMITING FACTORS in *ecological systems*. Phosphorus is an essential NUTRIENT for all plants and animals and is fundamental in biochemical reactions and structural support (*membranes* and BONES). PHOSPHATES (PO_4^{3-}) move quickly through plants and animals, driven by *microbial activity*. However, the major transfers in the global cycle of phosphorus through the soil and oceans are slow, driven by TECTONICS over GEOLOGICAL TIME, thus making the phosphorus cycle one of the slowest biogeochemical cycles.

Unlike in other BIOGEOCHEMICAL CYCLES, the ATMOSPHERE does not play a significant role in the phosphorus cycle. This is because phosphorus and phosphorus-based COMPOUNDS are usually solids at the temperatures and pressures found on Earth. Much of the phosphorus is locked up in BEDROCK, SOILS and SEDIMENTS and is therefore not directly available to organisms unless released by WEATHERING, LEACHING or *mining*. Phosphorus is weathered from bedrock by the dissolution of phosphorus minerals such as *apatite*. As phosphorus is highly reactive, it exists in combined form with many other ELEMENTS. *Soluble phosphate*, produced by *micro-organisms* from insoluble forms, is available for uptake by *terrestrial plants* and returned to the soil by DECOMPOSITION of leaf LITTER. In the soil, however, *solution phosphate* is low as phosphorus is incorporated into soil particles by sorption with *ferric iron* and *aluminium hydroxides*. SORPTION is considered the most important process controlling *bio-available phosphorus*. Plants have developed different physiological strategies for obtaining phosphorus despite low soil solution concentrations and can minimise loss by resorbing much of their phosphorus prior to leaf fall.

Phosphorus is delivered to the OCEANS and LAKES by means of SOIL EROSION and *river transport*. Within rivers there are two main forms of phosphorus: *particulate phosphorus* and *dissolved phosphorus*, although the majority is in the particulate form. The particulate form is not active in the *phosphorus biogenic cycle*, and this is the same when it is delivered to the oceans and so much of the eroded phosphorus is deposited unchanged onto the CONTINENTAL MARGINS until tectonic processes returns the phosphorus to the land. Phosphorus can also be brought back to the land through the HARVESTING of *fish* and the collection of *guano*.

Phosphorus that is absorbed onto soil surfaces can provide an additional source of phosphorus to the oceans as it becomes displaced by the high ionic strength of seawater. ANOXIA is favourable for oxide dissolution and the release of incorporated phosphorus. Phosphorus is the critical limiting nutrient for *aquatic organisms* and concentrations are near zero in most surface waters as it is taken up by PHYTOPLANKTON, but is more enriched in deeper, older waters. Phosphorus input and output are driven to steady-state mass balance in the ocean by biological PRODUCTIVITY.

ANTHROPOGENIC activities have nearly doubled the amount of phosphorus in the oceans and as such it is difficult to determine the pre-anthropogenic RESIDENCE TIME of phosphorus in the ocean. Due to the steady diversion of phosphorus into the oceans, it is mined and added to soils as a fertiliser to maintain the productivity of AGRICULTURE, but in addition to this, DEFORESTATION and the disposal of DOMESTIC and INDUSTRIAL WASTE have enhanced phosphorus transport from the TERRESTRIAL to AQUATIC ENVIRONMENTS. Overenrichment of phosphorus can lead to ALGAL BLOOMS, which in turn lead to anoxic conditions.   *KJF*

Bünemann EK, Oberson A and Frossard E (eds) (2011) *Phosphorus in action: Biological processes in soil phosphorus cycling*. Berlin: Springer.

De Schrijver A, Vesterdal L, Hansen K et al. (2012) Four decades of post-agricultural forest development have caused major redistributions of soil phosphorus fractions. *Oecologia* 169: 221–234.

Filippelli GM (2002) The global phosphorus cycle. *Reviews in Mineralogy and Geochemistry* 48: 391–425.

Fillippelli GM (2008) The global phosphorus cycle: Past, present and future. *Elements*: 4: 89–95.

Grizzetti B, Bouraoui F and Aloe A (2012) Changes of nitrogen and phosphorus loads to European seas. *Global Change Biology* 18: 769–782.

Radcliffe DE and Cabrera ML (2006) *Modeling phosphorus in the environment*. Boca Raton, FL: CRC Press.

Ruttenberg KC (2005) The global phosphorus cycle. In Schlesinger WH (ed.) *Biogeochemistry*. Amsterdam: Elsevier, 585–643.

Stevenson FJ and Cole MA (1999) *Cycles of soils: Carbon, nitrogen, phosphorus, sulfur, micronutrients*. Oxford: Blackwell.

photochemical oxidant A secondary POLLUTANT produced in the atmosphere by a complex series of chemical reactions involving volatile organic pollutants, NITROGEN OXIDES, oxygen and sunlight. Photochemical oxidants include aldehydes, nitrogen dioxide, OZONE and PEROXYACETYL NITRATE (PAN). *JAM/JET*

[*See also* PHOTOCHEMICAL SMOG, PHOTOCHEMISTRY]

Wardle B (2009) *Principles and applications of photochemistry*. Chichester: Wiley-Blackwell.

photochemical smog SMOG produced by photochemical processes. It is characterised by the presence of relatively high concentrations of various atmospheric POLLUTANTS, notably OZONE, formed by the action of sunlight on NITROGEN OXIDES and volatile hydrocarbons. The phenomenon was first recognised in Los Angeles but is now a widespread aspect of URBAN CLIMATE. The main irritants are *formaldehyde, acrolein* and PEROXYACETYL NITRATE (PAN). *DME/PU/CW*

[*See also* AIR POLLUTION, PHOTOCHEMISTRY, URBAN CLIMATE]

Fenger J (2009) Air pollution in the last 50 years: From local to global. *Atmospheric Environment* 43: 13–22.

Harrison RM and Hester RE (eds) (2009) *Air quality in urban environments*. Cambridge: Royal Society of Chemistry.

Hess GD, Tory KJ, Cope ME et al. (2004) The Australian air quality forecasting system. Part II: Case study of a Sydney 7-day photochemical smog event. *Journal of Applied Meteorology* 43: 663–679.

photochemistry The study of chemical reactions in the ATMOSPHERE initiated, assisted or accelerated by ABSORPTION of SOLAR RADIATION, specifically photons of light in the visible or ultraviolet wavelengths. *DME/PU/CW*

[*See also* OZONE, OZONE DEPLETION, PHOTOCHEMICAL SMOG]

Brimblecombe P (1996) *Air composition and chemistry*. Cambridge: Cambridge University Press.

Moortgat GK (2001) Important photochemical processes in the atmosphere. *Pure and Applied Chemistry* 73: 487–490.

photo-electronic erosion pin (PEEP) An instrument based on photo-sensitive cells capable of monitoring SOIL EROSION or river BANK EROSION and DEPOSITION automatically and continuously within individual STORM HYDROGRAPHS. The PEEP system thus gives clearer pictures of the magnitude, frequency and timing of erosion and deposition events. *DML*

[*See also* CHANNEL CHANGE, EROSION PIN]

Lawler DM (1991) A new technique for the automatic monitoring of erosion and deposition rates. *Water Resources Research* 27: 2125–2128.

Lawler DM (2008) Advances in the continuous monitoring of erosion and deposition dynamics: Developments and applications of the new PEEP-3T system. *Geomorphology* 93: 17–39.

photogrammetry Techniques for making reliable and accurate spatio-geographic environmental measurements from *photographs*. Although generally associated with the analysis of AERIAL PHOTOGRAPHY for mapping applications, the same techniques are also used in a wide range of other applications, for instance, to record building facades. Photogrammetry for *mapping* applications requires aerial photographs to be acquired as close to vertical as possible using cameras calibrated for high geometric accuracy and containing *fiducial marks*, which identify the centre of the photograph. Measurements made from a single photograph require knowledge of the elevation of the points being measured to correct for terrain-induced changes in scale. If multiple photographs acquired in stereo format with a large amount of overlap are available, the PARALLAX effect caused by viewing the same point from different locations can be used to calculate elevation and therefore correct the scale. *Stereo aerial photographs* are therefore used to produce topographic maps and DIGITAL ELEVATION MODELS (DEMs). DEMs are now used to produce *orthophotographs* with terrain effects removed. *GMS*

Avery TE and Berlin GL (1992) *Fundamentals of remote sensing and airphoto interpretation, 5th edition*. New York: Prentice Hall.

Linder W (2009) *Digital photogrammetry, 3rd edition*. Berlin: Springer.

Wolf PR and Dewitt BA (2000) *Elements of photogrammetry with applications in GIS*. New York: McGraw-Hill.

photon   An *elementary particle* of all electromagnetic RADIATION.   *HB*

[*See also* ELECTROMAGNETIC SPECTRUM]

photoperiod   The duration of darkness and light, normally described in terms of *day length*. Many plants and animals respond and are adapted to the photoperiod. For example, many *long-day plants* in the Northern Hemisphere flower as day length becomes longer in the spring and early summer. However, both long-day plants and *short-day plants* are responding to *night length* rather than day length. Some exhibit OBLIGATE photoperiodism; others are FACULTATIVE.   *RJH/JAM*

[*See also* LIMITING FACTORS, PHENOLOGY, PHOTOSYNTHESIS, SEASONALITY]

Nelson RJ, Denlinger DL and Somers DE (eds) (2010) *Photoperiodism: The biological calendar*. New York: Oxford University Press.
Thomas B and Vince-Prue D (1997) *Photoperiodism in plants, 2nd edition*. San Diego, CA: Academic Press.

photosynthesis   The process by which AUTOTROPHIC ORGANISMS synthesise carbohydrates from light energy, CARBON DIOXIDE and water in the presence of CHLOROPHYLL. OXYGEN is released during photosynthesis.   *JAM*

Lawlor DW (2000) *Photosynthesis, 3rd edition*. Oxford: Bios.
Raghavendra AS (ed.) (2000) *Photosynthesis: A comprehensive treatise*. Cambridge: Cambridge University Press.

photosynthetically active radiation (PAR)   The spectral range (waveband) of incident SOLAR RADIATION (400–700 nm) that photosynthetic organisms are able to use in the process of PHOTOSYNTHESIS. It is defined as the number of *photons* within the PAR spectrum incident per unit time on a unit surface area (normally quantified as μmol photons/m^2/second) and changes seasonally and varies depending on the latitude and time of the day. There is a simple relationship between the number of plant MOLECULES changed photochemically and the number of photons absorbed within the waveband and this reveals the efficiency of a system. PAR can be measured over space and time with regional to global measurements provided by REMOTE SENSING.   *DSB/GMS*

Frouin R and Pinker RT (1995) Estimating photosynhetically active radiation (PAR) at the Earth's surface from satellite observations. *Remote Sensing of Environment* 51: 98–107.
Gates DM (1980) *Biophysical ecology*. New York: Springer.

phreatic zone   The saturated GROUNDWATER zone below the permanent WATER TABLE. *Phreatic water*, which completely fills the rock interstices and exists under HYDROSTATIC PRESSURE in the phreatic zone, is the water that feeds WELLS.   *JAM*

[*See also* PHREATOPHYTE, VADOSE ZONE]

Dingman SL (2008) *Physical hydrology*. Long Grove, IL: Waveland Press.

phreatophyte   A plant that derives its water supply from GROUNDWATER. The term, which is from the Greek for *well plant*, implies that the plant possesses deep *roots* that can reach the PHREATIC ZONE.   *JAM*

[*See also* HYDROPHYTE, HYGROPHYTE]

Sommer B and Froend R (2011) Resilience of phreatophytic vegetation to groundwater drawdown: Is recovery possible under a drying climate? *Ecohydrology* 4: 67–82.

phyletic gradualism   EVOLUTION seen as gradual transformation of entire species POPULATIONS by means of NATURAL SELECTION. The rate of change is even and slow in explicit contrast to the concept of PUNCTUATED EQUILIBRIA. *Punctuated gradualism* is an intermediate theory that states that SPECIATION is not needed for phases of rapid evolution between long phases of slow evolution.   *KDB/JAM*

[*See also* GRADUALISM]

Eldredge N and Gould SJ (1972) Punctuated equilibria: An alternative to phyletic gradualism. In Schopf TJM (ed.) *Models in paleobiology*. San Francisco: Freeman, Cooper, 82–115.
Malmgren BA, Berggren WA and Lohmann GP (1983) Evidence for punctuated gradualism in the Late Neogene *Globoratalia tumida* lineage of planktonic foraminifera. *Paleobiology* 9: 377–389.
Sheldon PF (1987) Parallel gradualistic evolution of Ordovician trilobites. *Nature* 330: 561–563.

phylogeography   Study of the principles and processes governing the geographical distribution of genealogical LINEAGES, especially those of animals at the intraspecific level.   *JAM*

[*See also* DNA: ANCIENT]

Avise JC (2000) *Phylogeography: The history and formation of species*. Cambridge, MA: Harvard University Press.
Riddle BR and Hafner DJ (2004) The past and future roles of phylogeography in historical biogeography. In Lomolino MV and Heaney LR (eds) *Frontiers of biogeography: New directions in the geography of nature*. Sunderland, MA: Sinauer, 93–110.
Rutgers DS (2011) *Phylogeography: Concepts, intraspecific patterns and speciation processes*. New York: Nova Science Publishers.

physical analysis of soils and sediments   Analytical methods used to describe and measure physical properties. For example, common physical properties investigated in SOILS are BULK DENSITY, WATER CONTENT, *water-holding capacity*, HYDRAULIC CONDUCTIVITY, *porosity, pore-size distribution* and *particle-size* *distribution*. In the context of SEDIMENTS, additional properties include, for example, aspects of the *ground thermal regime* and the *subsurface properties* measured by various types of GEOPHYSICAL SURVEYING, which are used to detect subsurface structures. *EMB*

[*See also* CHEMICAL ANALYSIS OF SOILS AND SEDIMENTS, PALAEOCEANOGRAPHY: PHYSICAL AND CHEMICAL PROXIES]

Dane JH and Topp GK (eds) (2002) *Methods of soil analysis. Part 4: Physical methods*. Madison, WI: American Society of Agronomy.

Dirksen C (1999) *Soil physics measurements*. Reisenkirchen: Catena Verlag.

Last WM and Smol JP (eds) (2001) *Tracking environmental change using lake sediments, volume 2: Physical and geochemical methods*. Dordrecht: Kluwer.

Menounos B (1997) The water content of lake sediments and its relationship to the physical parameters: An alpine case study. *The Holocene* 7: 202–212.

Smith KA and Mullins CE (1991) *Soil analysis: Physical methods*. New York: Marcel Dekker.

Thorn CE, Darmody RG and Allen CE (2008) Ground temperature variability on a glacier foreland, Storbreen, Jotunheimen, Norway. *Norsk Geografisk Tisdsskrift* 62: 290–302.

physical degradation

The natural or anthropogenic DEGRADATION of materials, land or soil by physical processes, such as PHYSICAL WEATHERING, WATER and WIND EROSION. SOIL DEGRADATION by physical processes, for example, may result in detrimental changes to SOIL STRUCTURE, COMPACTION, CRUSTING OF SOIL and SOIL LOSS. *JAM*

Barros E, Grimaldi M, Sarrazin M et al. (2004) Soil physical degradation and changes in macrofaunal communities in central Amazon. *Applied Soil Ecology* 26: 157–168.

physical geography

A subdiscipline of GEOGRAPHY and a NATURAL ENVIRONMENTAL SCIENCE that focuses on understanding the biophysical environments and natural LANDSCAPES of the Earth's surface at local to global scales. This encompasses spatial patterns in the GEOECOSPHERE, their controlling processes and dynamics, the evolution of the natural landscape and its interactions with human activity. Physical geographers employ a wide range of methodological strategies, including FIELD RESEARCH, REMOTE SENSING, LABORATORY SCIENCE, NUMERICAL ANALYSIS, *modelling* (see MODEL) and GEOGRAPHICAL INFORMATION SYSTEMS (GIS). Many physical geographers specialise in the pure or applied aspects of BIOGEOGRAPHY, CLIMATOLOGY or GEOMORPHOLOGY, but there are important integrative themes, one of which is ENVIRONMENTAL CHANGE. In this context, physical geography is most concerned with environmental change on timescales relevant to understanding present landscapes as the domain of human societies, past, present and future: particularly important aspects include HOLOCENE ENVIRONMENTAL CHANGE and current and potential future HUMAN IMPACT ON ENVIRONMENT. *JAM*

[*See also* ANTHROPOCENE, ANTHROPOSPHERE, HUMAN GEOGRAPHY]

Gregory KJ (2000) *The changing nature of physical geography*. London: Arnold.

Gregory KJ (ed.) (2005) *Physical Geography, 4 volumes*. London: Sage.

Gregory KJ, Gurnell AM and Petts GE (2002) Restructuring physical geography. *Transactions of the Institute of British Geographers* 27: 136–154.

Haines-Young RH and Petch JR (1986) *Physical geography: Its nature and method*. London: Harper and Rowe.

Huggett R (2010) *Physical geography: The key concepts*. London: Routledge.

Inkpen R (2005) *Science, philosophy and physical geography*. London: Routledge.

Slaymaker O and Spencer T (1998) *Physical geography and global environmental change*. London: Addison Wesley Longman.

Trudgill S and Roy A (2003) *Contemporary meanings in physical geography: From what to why?* London: Arnold.

physical oceanography

The branch of OCEANOGRAPHY dealing with observations and computer modelling of physical aspects of the OCEANS, including their temperature, density, OCEAN CURRENTS, UPWELLING, TIDES, waves and acoustic properties. *BTC/JP*

Knauss JA (2005) *Introduction to physical oceanography, 2nd edition*. Long Grove, IL: Waveland Press.

physical weathering

The mechanical disintegration of rocks and minerals in situ, without chemical alteration. It is also known as *mechanical weathering* and is often closely related to CHEMICAL WEATHERING, which may reduce the mechanical resistance of rocks to physical weathering. Thus, determining the relative contribution of both types of processes in overall rock breakdown is often difficult. Physical processes include water-based mechanisms such as FROST WEATHERING, HYDRATION and *hydraulic pressure weathering*. The last is associated especially with *waves*, which compress air in cracks and rock fissures. Also relevant are SALT WEATHERING, which involves the expansion of salt crystals, and INSOLATION WEATHERING, which reflects the thermal expansion and shrinkage of minerals often at different, mineral-specific rates. Fire can be particularly effective in leading to rapid thermal expansion of rock. The resulting spalls have been called *ignifracts*. UNLOADING is caused by the reduction of compressive stress owing to the removal of overlying rock. In general, physical weathering is most effective in climates of high diurnal and/or annual temperature amplitudes and areas where the temperature oscillates around 0°C. *SHD*

[*See also* EXFOLIATION, FREEZE-THAW CYCLES,
FREEZING INDEX, NIVATION]

Bland W and Rolls D (1998) *Weathering: An introduction to
the scientific principles*. London: Arnold.
Hall K, Thorn CE, Matsuoka N and Prick A (2002)
Weathering in cold regions: Some thoughts and
perspectives. *Progress in Physical Geography* 26:
577–603.
Lerman A and Meybeck M (eds) (1988) *Physical and
chemical weathering in geochemical cycles*. Berlin:
Springer.
Shakesby RA and Doerr SH (2006) Wildfire as a
hydrological and geomorphological agent. *Earth-Science
Reviews* 74: 269–307.

physiography   A largely obsolete term that has
been used in a variety of ways ranging from synonyms
for GEOMORPHOLOGY and LANDSCAPE to the holistic study
of PHYSICAL GEOGRAPHY conceived in the broadest possible way.   *JAM*

Huxley TH (1877) *Physiography: An introduction to the
study of nature*. London: Macmillan.
Stoddart DR (1975) 'That Victorian science': Huxley's
physiography and its impact on geography. *Transactions
of the Institute of British Geographers* 66: 17–40.

phytogeography   Also known as PLANT GEOGRAPHY,
phytogeography focuses on the study of the distribution of different types of plants or *taxonomic* groups of
plants (e.g. families and genera). The objective is often
to analyse the geographical range of particular taxa or
FLORAS and to explain them in terms of origin, DISPERSAL
and EVOLUTION of the type or group (see Figure).   *MJB*

[*See also* BIOGEOGRAPHY, FLORAL PROVINCES/
REALMS/REGIONS, ISLAND BIOGEOGRAPHY,
WALLACE'S LINE, ZOOGEOGRAPHY]

Dahl E (2007) *The phytogeography of northern Europe:
British Isles, Fennoscandia and adjacent areas*.
Cambridge: Cambridge University Press.
Stott PA (1981) *Historical plant geography*. London: Allen
and Unwin.
Tivy J (1993) *Biogeography, 3rd edition*. London: Longman.

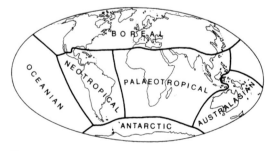

Phytogeography *Phytogeographical regions (floral provinces) of
the world (Tivy, 1993).*

phytogeomorphology   Study of the interrelationships between plants, landforms and geomorphological
processes.   *JAM*

Howard JA and Mitchell CW (1985) *Phytogeomorphology*.
New York: Wiley.
Korkalainen T (2005) Using phytogeomorphology,
cartography and GIS to explain forest site productivity
expressed as tree height in southern and central Finland.
Geomorphology 74: 271–284.

phyto-indication   Use of plants as indicators of
environmental change, for ENVIRONMENTAL MONITOR-
ING or as a DATING TECHNIQUE. Russian researchers recognise three main subdivisions of phyto-indication
methods: (1) *phytocoenotic methods* (based on plant
community properties), (2) *dendro-indication* (including DENDROCHRONOLOGY and DENDROCLIMATOLOGY) and
(3) *lichenometry* (including LICHENOMETRIC DATING).
Phytocoenotic methods include the use of changes in
species composition and community structure through
time for dating surfaces (see PHYTOMETRIC DATING) and
for identifying the extent and timing of debris flows,
snow avalanches and landslides. Also included is the
use of epiphytic lichens as monitors of AIR POLLUTION in
cities. Such EPIPHYTES are sensitive to SULFUR DIOXIDE air
pollution in urban areas, where zones of distinct lichen
communities provide a semi-quantitative scale of pollution levels.   *JAM*

[*See also* BIO-INDICATORS]

Hawkesworth DL and Rose F (1976) *Lichens as pollution
monitors*. London: Arnold.
Lekhatinov AM (1988) Phytoindication methods for
studying landslide and mudflow regimes. In Kozlovski
EA (ed.) *Landslides and mudflows, volume 1*. Moscow:
UNESCO, 94–97.

phytoliths   The opalaetic bodies, also known as
*silica bodies, biogenic opals, biogenic silica, plant
opals, opal phytoliths, silica cells* or *micrometric
hydrated opal-A particles* ($SiO_2 \cdot nH_2O$), which constitute the 'stony' parts of a plant and range in size
from 20 to 200 μm. The term *opal* is often used in
describing phytoliths in order to distinguish them
from other mineral substances secreted by plants
(e.g. calcium carbonate). Soluble silica is carried
passively through the plant vascular system until the
evaporation of water leads to chemical PRECIPITATION
of silica in or around the plant cells. Phytoliths are
most abundant in the leaves where plants lose most
water as a result of EVAPOTRANSPIRATION. Although the
purpose of phytoliths remains unclear, it has been suggested that they function as structural skeletons, anti-
wilting devices, or that they provide plant resistance
to *herbivory*. Phytoliths are valuable BIO-INDICATORS
since they are relatively inert, are extractable from

SEDIMENTS and PALAEOSOLS and permit identification of the source plants to the family or subfamily level. Phytolith analysis has found particular application in the reconstruction of past GRASSLANDS, where phytoliths provide complementary micromorphological evidence to GRASS CUTICLE ANALYSIS.   *MJW*

[*See also* PALAEOBOTANY, PALAEOECOLOGY, SILICA]

Albert RM and Madella M (eds) (2008) Perspectives on phytolith research. *Quaternary International* 193: 1–191 [Special Issue].

Meunier JD and Colin F (eds) (2001) *Phytoliths: Applications in Earth science and human history.* Lisse: Balkema.

Morris LR, West NE and Ryel RJ (2010) Testing soil phytolith analysis as a tool to understand vegetation change in the sagebrush steppe and pinyon-juniper woodlands of the Great Basin Desert, USA. *The Holocene* 20: 697–710.

Pearsall DM (1989) *Paleoethnobotany: A handbook of procedures.* San Diego, CA: Academic Press.

Piperno DR (2001) Phytoliths. In Smol JP, Birks HJB and Last WM (eds) *Tracking environmental change using lake sediments, volume 3.* Dordrecht: Kluwer, 235–251.

Piperno DR (2006) *Phytoliths: A comprehensive guide for archaeologists and palaeoecologists.* Lanham, MD: AltaMira Press.

phytometric dating   Use of plant community properties (e.g. species composition and community structure) for dating purposes. It relies on these properties reflecting ECOLOGICAL SUCCESSION.   *JAM*

[*See also* PHYTO-INDICATION]

Matthews JA (1978) Plant colonisation patterns on a gletschervorfeld, southern Norway: A meso-scale geographical approach to vegetation change and phytometric dating. *Boreas* 7: 155–178.

phytoplankton   Small, free-floating, photosynthesizing organisms (largely ALGAE) that inhabit the EUPHOTIC ZONE of most oceans and bodies of freshwater and are agents of *primary production*. Phytoplankton draw NUTRIENTS from the water and form the base of FOOD CHAINS in AQUATIC ENVIRONMENTS. Common groups include CYANOBACTERIA, *green algae*, COCCOLITHOPHORES, DIATOMS and *dinoflagellates*. FOSSIL phytoplankton are a source of PROXY DATA in PALAEOCLIMATOLOGY, PALAEOLIMNOLOGY and PALAEOCEANOGRAPHY studies. *HKC*

[*See also* DINOFLAGELLATE CYST ANALYSIS, MARINE PALAEOCLIMATIC PROXIES, OCEAN PALAEOPRODUCTIVITY, PALAEOCEANOGRAPHY: BIOLOGICAL PROXIES, PLANKTON/PLANKTONIC/ PLANKTIC, PRODUCTIVITY]

Boyce DG, Lewis MR and Worm B (2010) Global phytoplankton decline over the past century. *Nature* 466: 591–596.

Reynolds CS (2006) *The ecology of phytoplankton.* Cambridge: Cambridge University Press.

Sournia A (1978) *Phytoplankton manual.* Paris: UNESCO.

Tomas CR (ed.) *Identifying marine phytoplankton.* San Diego, CA: Academic Press.

phytosociology   An approach to the study of VEGETATION, which focuses on the CLASSIFICATION of *plant communities* based on their species composition, as typified by the *Zürich-Montpellier school*.   *JLI*

[*See also* COMMUNITY CONCEPTS, INDIVIDUALISTIC CONCEPT]

Braun-Blanquet J (1964) *Pflanzensoziologie: Grundzüge der Vegetationskunde, 3rd edition* [Plant sociology: Broad vegetation science]. Berlin: Springer.

Keller W, Wohlgemuth T, Kuhn N et al. (1998) Waldgesellschaften der Schweiz auf floristischer Grundlage [Forest communities in Switzerland on the basis of floristic]. *Mitteilungen der Eidgenössischen Forschungsanstalt für Wald, Schnee und Landschaft* 73: 93–357.

Pott R (2011) Phytosociology: A modern geobotanical method. *Plant Biosystems* 145: 9–18.

Sribille T and Chytry M (2002) Vegetation surveys in the circumboreal forests: A review. *Folia Geobotanica* 37: 365–382.

piedmont   An area of relatively low relief at the base of an upland or mountain range. The piedmont often consists of a PEDIMENT, separated from the mountain by a sharp break of slope, the *piedmont angle*, which is particularly prominent in SEMI-ARID and arid INSELBERG landscapes. Piedmonts, or *piedmont zones*, often reveal erosional and/or depositional evidence of LANDSCAPE EVOLUTION.   *MAC*

[*See also* ARIDLAND, BAJADA, GLACIS]

Baker SE, Gosse JC, McDonald EV et al. (2009) Quaternary history of the piedmont reach of Rio Diamante, Argentina. *Journal of South American Earth Sciences* 28: 54–73.

Bourne JA and Twidale CR (1998) Pediments and alluvial fans: Genesis and relationships in the western piedmont of the Flinders Range, South Australia. *Australian Journal of Earth Science* 45: 123–135.

Gosvani PK, Pant CC and Pandey S (2009) Tectonic controls on the geomorphic evolution of alluvial fans in the piedmont zone of Ganga Plain, Uttarakhand, India. *Journal of Earth System Science* 118: 245–259.

pigments: fossil   In sediments, pigments such as CHLOROPHYLLS, *caronenoids* and their derivatives are used in palaeolimnological studies. ALGAE, *microorganisms* and *higher plants* synthesise a variety of pigments which are used during PHOTOSYNTHESIS. PALAEOLIMNOLOGY utilises fossil pigments preserved in sediments to indicate changes in the community composition of algae and BACTERIA, changes in *food-web* interaction, *trophic status*, lake ACIDIFICATION, past levels of *ultraviolet radiation* and ANTHROPOGENIC impacts on AQUATIC ENVIRONMENTS, such as EUTROPHICATION, FISHERIES CONSERVATION AND MANAGEMENT, LANDUSE practices and CLIMATIC CHANGE.

While some pigments are widely distributed (e.g. chlorophyll), others provide a high degree of taxonomic specificity, for example, carotenoids, which, in turn, can be used to infer changes in specific algal classes. Unique pigments, such as *scytonemin*, are produced by some *benthic algae* to protect against cellular damage when exposed to UV radiation and have the potential to act as proxies for past *solar irradiance*. Other pigments, produced by green and purple *sulfur bacteria*, for example, *okenone* and *isorenieratene*, when found in lake sediments can be used to infer water-column ANOXIA and stratified conditions. Recently, past phosphorous concentrations in lakes have been inferred from *sedimentary pigments* due to the positive relationship between total carotenoid concentration and total phosphorous concentrations in lake water. *KJF*

[*See also* PLANT PIGMENT ANALYSIS]

Castañeda IS and Schouten S (2011) A review of molecular organic proxies for examining modern and ancient lacustrine sediments. *Quaternary Science Reviews* 30: 2851–2891.
Leavitt PR and Hodgson DA (2001) Sedimentary pigments. In Smol JP, Birks HJB and Last WM (eds) *Tracking environmental change using lake sediments, volume 3. Terrestrial, algal and siliceous indicators.* Dordrecht: Kluwer, 295–325.
Leavitt PR, Vinebrooke RD, Donald DB et al. (1997) Past ultraviolet radiation environments in lakes derived from fossil pigments. *Nature* 388: 457–459.
Sinninghe Damsté JS and Schouten S (2006) Biological markers for anoxia in the photic zone of the water column. In Volkman JK (ed.) *The handbook of environmental chemistry, volume 2.* Berlin: Springer, 128–163.
Soma Y, Tani Y, Soma M et al. (2007) Sedimentary steryl chlorin esters (SCEs) and other photosynthetic pigments as indicators of paleolimnological change over the last 28,000 years from the Buguldeika Saddle of Lake Baikal. *Journal of Paleolimnology* 37: 163–175.

pillow lava   A body of LAVA made up of bulbous, rounded masses up to a few metres across. The base of each pillow is moulded to the shapes of those beneath. Pillow lavas are characteristic of lava erupted into water and are useful in PALAEOENVIRONMENTAL RECONSTRUCTION. Pillow lavas make up the surface layer of OCEANIC CRUST, which forms through SEA-FLOOR SPREADING at MID-OCEAN RIDGES (MORs). *GO*

Walker GPL (1992) Morphometric study of pillow-size spectrum among pillow lavas. *Bulletin of Volcanology* 54: 459–474.

pilot survey   A preliminary survey or trial run used as a basis either for an informed decision on whether a full study would be worthwhile or for deciding on a suitable SAMPLING design for a later, more definitive survey. *JAM*

[*See also* FIELD RESEARCH, RECONNAISSANCE SURVEY]

Smith DB, Woodruff LG, O'Leary RM et al. (2009) Pilot studies for the North American Soil Geochemical Landscape Project: Site selection, sampling protocols, analytical methods and quality control protocols. *Applied Geochemistry* 24: 1357–1368.

pine decline   A mid-HOLOCENE decline in the abundance of the coniferous tree species *Pinus sylvestris* L. (Scots pine). Pollen data from sites in the north and west of the British Isles record a marked decline in the abundance of pine, where it still persisted, around 5,000 years ago. Pine stumps can be found in abundance in blanket PEAT throughout the northwestern British Isles, but pine appears not to have persisted within this habitat after about 4,500 years ago. However, pine did not become extinct at this time, surviving at low levels in Ireland until the MIDDLE AGES, and in Scotland until the present day. The cause of this decline is still not clear. Hypotheses that have been proposed include CLIMATIC CHANGE, VOLCANIC ERUPTION, HUMAN IMPACT ON ENVIRONMENT and changes in FIRE FREQUENCY. *JLF*

[*See also* ELM DECLINE, HUMAN IMPACT ON VEGETATION HISTORY, POLLEN ANALYSIS]

Tipping R (2008) Storminess as an explanation for the decline of pine woodland ca 7400 years ago at Loch Tulla, western Scotland. *Vegetation History and Archaeobotany* 17: 345–350.
Willis KJ, Bennett KD and Birks HJB (1998) The Late-Quaternary dynamics of pines in Europe. In Richardson DM (ed.) *Ecology and biogeography of Pinus.* Cambridge: Cambridge University Press, 107–121.

pingo   A typically dome-shaped, ice-cored hill up to 70 m high and up to 600 m in diameter in PERMAFROST areas formed by the freezing of water moving under a pressure gradient towards the pingo. Based on the origin of the GROUNDWATER necessary for their growth, they can be classified into *hydraulic* (or *open-system*) *pingos* and *hydrostatic* (or *closed-system*) *pingos*: the former being located on valley floors or lower valley-side slopes, and fed by UPWELLING water generated by a *hydraulic head*, and the latter being located on the floors of drained LAKES, and fed by *pore water* that is expelled from the sediments ahead of *aggrading permafrost*. A pingo is a *periglacial landform* and a type of FROST MOUND. *RAS/JAM*

Burr DM, Tanaka KL and Yoshikawa K (2009) Pingos on
Earth and Mars. *Planetary and Space Science* 57: 541–555.

Gurney SD (1998) Aspects of the genesis and
geomorphology of pingos: Perennial permafrost mounds.
Progress in Physical Geography 22: 307–324.

Mackay JR (1998) Pingo growth and collapse, Tuktoyaktuk
Peninsula, western Arctic coast, Canada: A long-term field
study. *Géographie Physique et Quaternaire* 52: 271–323.

Worsley P and Gurney SD (1996) Geomorphology and
hydrological significance of the Holocene pingos in the
Karup Valley, Trail Island, northern East Greenland.
Journal of Quaternary Science 11: 249–262.

pinning point (1) For MARINE-BASED ICE SHEETS, a
high point on the sea floor consisting of bedrock hills or
upstanding masses of sediment that exert an influence
on ICE SHELF dynamics by increasing locally the basal
drag and reducing losses by CALVING. In applying the
GLACIOMARINE HYPOTHESIS to DEGLACIATION of the Irish
Sea basin, for example, the Pembrokeshire Peninsula
(Wales) and Wexford (Ireland) have been viewed as
pinning points preventing GLACIER SURGE and collapse
of the marine-based ICE DOMES. This deglaciation model
for the Irish Sea basin is hotly disputed. (2) Evidence
in marine and/or subaerial SEDIMENTS thought to be useful in reconstructing former RELATIVE SEA LEVEL. In this
sense, a *pinning point curve* can provide an illustration
of relative sea-level history. *RAS*

Eyles N and McCabe AM (1989) The Late Devensian
(<22,000 BP) Irish Sea basin: The sedimentary record of
a collapsed ice sheet margin. *Quaternary Science Reviews*
8: 307–351.

Goldstein RH and Franseen EK (1995) Pinning-points:
A method providing quantitative constraints on relative
sea-level history. *Sedimentary Geology* 95: 1–10.

pioneers Plants and animals that colonise disturbed
areas, thereby initiating ECOLOGICAL SUCCESSION. *LRW*

[**See also** COLONISATION, DISTURBANCE, PRIMARY
SUCCESSION, R-SELECTION, RUDERAL]

Robbins JE and Matthews JA (2009) Pioneer vegetation
on glacier forelands in southern Norway: Emerging
communities? *Journal of Vegetation Science* 20: 889–902.

piosphere An approximately circular area of land
centred on a watering point (WELL or borehole) for people and/or animals in a DRYLAND area. As cattle need
to drink every second day, the radius of a piosphere in
theory is equivalent to one day's walk from the water
source, although piospheres of up to 50 km in radius
have been reported. Continual GRAZING pressure on the
piospheres is seen by most researchers as leading to
LAND DEGRADATION, and coalescence of piospheres as a
cause of DESERTIFICATION over large areas (see Figure). It
has been argued, however, that it is difficult to separate
vegetation loss by grazing from that caused naturally

from low rainfall, that piosphere coalescence is theoretical only and that the high input of nutrients around
the watering point may actually lead to improved SOIL
QUALITY. *RAS/JAM*

Andrew MH (1988) Grazing impacts in relation to livestock
watering points. *Trends in Ecology and Evolution* 3:
336–339.

Heshmatti GA, Facelli JM and Conran JG (2002) The
piosphere revisited: Plant species patterns close to
waterpoints in small, fenced paddocks in chenopod
shrublands of South Australia. *Journal of Arid
Environments* 51: 547–560.

Thomas DSG and Middleton NJ (1994) *Desertification:
Exploding the myth.* Chichester: Wiley.

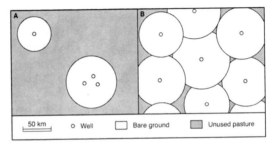

Piosphere *(A) Piospheres centred around wells. (B) Theoretical
coalescence of piospheres, regarded by some as a process of
desertification (Thomas and Middleton, 1994).*

pipeflow The downslope movement of water through
natural tunnels or passages (*soil pipes*) in SOIL or REGO-
LITH. Their role in generating stream RUNOFF varies.
Shallow pipes with surface connections tend to respond
quickly to rainfall and can be major QUICKFLOW providers,
but deep pipes tend to respond slowly and may be important in providing BASEFLOW. *Piping* may have a biotic
origin, such as *animal burrows* or voids left after roots
decay, or may be produced by other pedological/hydro-
logical processes, notably with cracking associated with
seasonal *wetting and drying cycles*. Pipe development in
PEAT in the British Isles may have been enhanced by the
drying effects of AFFORESTATION. Pipeflow can also play
a major role in *solute dynamics*, in physical erosional
processes (*pipe erosion* or *tunnelling*) and in DRAINAGE
NETWORK initiation and development. *RPDW/ADT*

[**See also** RUNOFF PROCESSES, THROUGHFLOW]

Jones JAA (2012) Soil piping and catchment response.
Hydrological Processes 24: 1548–1566.

Jones JAA and Cottrell CI (2007) Long-term changes in stream
bank soil pipes and the effects of afforestation. *Journal of
Geophysical Research: Earth Surface* 112: F01010.

Sayer AM, Walsh RPD and Bidin K (2006) Pipeflow
suspended sediment dynamics and their contribution to
stream sediment budgets in small rainforest catchments,
Sabah. *Forest Ecology and Management* 224: 119–130.

piston sampler   These samplers or CORERS are more effective than CHAMBER SAMPLERS in soft sediments or under water. They consist of a hollow tube fitted with a moveable piston. The piston creates a negative pressure within the tube, thus preventing compression of sediments and helping to keep the material within the tube when the corer is retrieved. The commonest example is probably the *Livingstone corer*.   *MJB*

Mingram J, Negendank JFW, Brauer A et al. (2007) Long cores from small lakes: Recovering up to 100m long lake sediment sequences with a high-precision rod-operated piston corer (Usinger-corer). *Journal of Paleolimnology* 37: 517–528.

Nesje A (1992) A piston corer for lacustrine and marine sediments. *Arctic and Alpine Research* 24: 257–259.

Wright HE (1980) Cores of soft lake sediments. *Boreas* 9: 107–114.

pixel   The smallest element of information in an image (e.g. that formed by REMOTE SENSING), consisting of a position and a value. Allowed values are real numbers (*grey-level images*), vectors of real numbers (*multispectral images*) and classes (CLASSIFICATION), among others.   *ACF*

placer deposit   A type of ORE deposit in which dense minerals of economic value have been eroded from a source material and concentrated by mechanical sorting during SEDIMENTATION. Most placer deposits occur in SEDIMENTARY DEPOSITS in *river*, *beach* and COASTAL DUNE environments. *Palaeoplacers* are lithified placer deposits in SEDIMENTARY ROCKS and include the Witwatersrand *gold* deposits in South Africa of ARCHAEAN age. Much of the world's gold, TIN (from tin oxide, *cassiterite*), *titanium*, *diamonds* and other *gemstones* are obtained from placer deposits. Valuable minerals can be separated from placer deposits by means of *panning*, as in the gold rushes of the mid-nineteenth century in North America, or, on a larger, commercial scale, by DREDGING. The term is derived from the Spanish *plaza*, meaning a place.   *GO*

[*See also* MINERAL RESOURCES]

Carling PA and Breakspear RMD (2006) Placer formation in gravel-bedded rivers: A review. *Ore Geology Reviews* 28, 377–401.

Craw D, Youngson JH and Koons PO (1999) Gold dispersal and placer formation in an active oblique collisional mountain belt, southern Alps, New Zealand. *Economic Geology and the Bulletin of the Society of Economic Geologists* 94: 605–614.

Frimmel HE (2005) Archaean atmospheric evolution: Evidence from the Witwatersrand gold fields, South Africa. *Earth-Science Reviews* 70: 1–6.

Jewett SC, Feder HM and Blanchard A (1999) Assessment of the benthic environment following offshore placer gold mining in the northeastern Bering Sea. *Marine Environmental Research* 48: 91–122.

Le Roux JP (2005) Grains in motion: A review. *Sedimentary Geology* 178: 285–313.

Wendland C, Fralick P and Hollings P (2012) Diamondiferous Neoarchean fan-delta deposits, western Superior Province, Canada: Sedimentology and provenance. *Precambrian Research* 196: 46–60.

plagioclimax   The vegetation that wholly or partly replaces or modifies a supposed CLIMAX VEGETATION after an environmental DISTURBANCE. It may be maintained by disturbances of various types, including, for example, GRAZING, HARVESTING, LOGGING, *planting* or *fire*. In this context, the disturbances are *arresting factors*, which prevent directional change by ECOLOGICAL SUCCESSION. The alternative term *disclimax* (short for *disturbed climax*) is often used in the USA. Most of the so-called *natural grasslands* in the British Isles lying below 500 m are plagioclimax vegetation maintained by *cattle grazing* and *sheep grazing*. If grazing were to cease, the grasslands would mostly revert to deciduous forest. In the arid and semi-arid United States, ecological succession around some ghost towns is permanently altered by an annual grass—cheatgrass (*Bromus tectorum*)—introduced from the Mediterranean. The cheatgrass will probably stay dominant at the expense of other, native species: SECONDARY SUCCESSION has produced plagioclimax vegetation. Logging in the Peace River Lowlands, Wood Buffalo National Park, Canada, is creating a plagioclimax vegetation in BOREAL FOREST: convergence towards the original white spruce and mixed-wood forests is not apparent, and a long-term deciduous disclimax (with balsam poplar and lesser amounts of Alaska birch and aspen) is predicted.   *RJH*

[*See also* CONVERGENCE AND DIVERGENCE, IN SUCCESSION, TEMPERATE GRASSLANDS, TROPICAL GRASSLAND]

Gammage B (2008) Plain facts: Tasmania under aboriginal management. *Landscape Research* 33: 241–254.

Knapp PA (1992) Secondary plant succession and vegetation recovery in two western Great Basin desert ghost towns. *Biological Conservation* 6: 81–89.

Miller GR, Geddes C and Mardon DK (2010) Effects of excluding sheep from an alpine dwarf-herb community. *Plant Ecology and Diversity* 3: 87–93.

Neba NE (2009) Cropping systems and post-cultivation vegetation successions: Agro-ecosystems in Ndop, Cameroon. *Journal of Human Ecology* 27: 27–33.

Timoney KP, Peterson G and Wein R (1997) Vegetation development of boreal riparian plant communities after flooding, fire, and logging, Peace River, Canada. *Forest Ecology and Management* 93: 101–120.

plague   An infectious DISEASE caused by BACTERIA (*Yersinia pestis*). *Bubonic plague* is the commonest form, characterised by swollen lymph nodes (bubo) and typically spread by fleas carried by rodents, such as the black rat (*Rattus rattus*). *Pneumonic plague* is contagious, spreading from an infected person by coughing or sneezing. Although the earliest recorded plague EPIDEMIC occurred in China in 224 BC, there have been three main PANDEMICS, each of which experienced several waves of infection with high fatality rates. The first, the *Justinian plague*, which killed several million people, started in Egypt and affected mainly the Mediterranean world between AD 541 and 750. The second, the *Black Death*, which caused an estimated 25 million deaths in Europe and the Middle East, may have begun in Central Asia in the AD 1330s, included the *Great Plague of London* in 1665 and lasted at least into the eighteenth century. The third, which mainly affected Asia, started in China in the mid-nineteenth century and led to some 12 million deaths, mainly in India. Plague is still a HUMAN HEALTH HAZARD, with a few thousand cases per year, mainly in Africa, Asia and South America, of which around 10 per cent are fatal. It should be noted that the term 'plague' is often used, incorrectly, to refer to any disease. Thus, the *Plague of Athens* in 430–426 BC and the *Antonine Plague* of AD 164–180 probably referred to SMALLPOX.   *JAM*

[*See also* ZOONOSES]

Aberth J (2011) *Plagues in world history*. Lanham, MD: Rowman and Littlefield.

Benedictow OJ (2004) *The Black Death, 1346-1353: The complete history*. Woodbridge: Boydell Press.

Gage KL and Kosoy MY (2005) Natural history of plague: Perspectives from more than a century of research. *Annual Review of Entomology* 50: 505–528.

Little LK (ed.) (2007) *Plague and the end of antiquity: The pandemic of 541-750*. Cambridge: Cambridge University Press and the American Academy in Rome.

Nutton V (ed.) (2008) *Pestilential complexities: Understanding Medieval plague*. London: Welcome Trust.

plane bed   A BEDFORM comprising a flat, or nearly flat, surface across which sand or fine gravel is transported. A *lower-stage plane bed* forms in response to weak currents, replacing CURRENT RIPPLES in coarse sand. An *upper-stage plane bed* forms in strong currents, especially where flows are shallow. Delicate flow-parallel ridges on the surface are termed *parting lineation* or *primary current lineation*. Deposition under upper-stage plane bed conditions gives rise to PARALLEL LAMINATION with parting lineation. Plane bed conditions also exist for strong oscillating currents associated with *waves*.   *GO*

[*See also* BEDFORM STABILITY DIAGRAM]

Allen JRL (1982) *Sedimentary structures: Their character and physical basis*. Amsterdam: Elsevier.

Bridge JS and Demicco RV (2008) *Earth surface processes, landforms and sediment deposits*. New York: Cambridge University Press.

Fielding CR (2006) Upper flow regime sheets, lenses and scour fills: Extending the range of architectural elements for fluvial sediment bodies. *Sedimentary Geology* 190: 227–240.

Yagishita K, Ashi J, Ninomiya S and Taira A (2004) Two types of plane beds under upper-flow-regime in flume experiments: Evidence from grain fabric. *Sedimentary Geology* 163: 229–236.

planet   A large celestial body orbiting a star: in the case of our *solar system*, the Sun. The planets of our solar system include *rocky planets* (*Mercury*, *Venus*, *Earth* and *Mars*), of which Earth and Venus are surrounded by a gaseous ATMOSPHERE, and *gaseous planets* (*Jupiter*, *Saturn*, *Uranus* and *Neptune*). Other objects in the solar system include PLUTO (formerly considered a planet) and other ASTEROIDS, which orbit the Sun but are too small to become spherical through their own gravity, and *satellites* such as the *Moon*, which orbit planets. Planets have now been identified as surrounding other stars in our galaxy (*exoplanets*).   *GO*

Cox B and Cohen A (2011) *Wonders of the universe*. London: HarperCollins.

Howard AW (2013) Observed properties of extrasolar planets. *Science* 340: 572–576.

Mayor M and Frei P-Y (2003) *New worlds in the cosmos: The discovery of exoplanets*. Cambridge: Cambridge University Press.

planetary boundaries   The concept that there is a SAFE OPERATING SPACE for humanity with respect to the effects of *human impact* on the global environment, which can be defined in terms of key EARTH-SYSTEM processes and their control variables. The boundaries are set at a 'safe' distance from intrinsic THRESHOLDS beyond which the Earth system may become unstable (see TIPPING POINT). Safety in this sense requires value judgements in relation to acceptable RISK and UNCERTAINTY. Six proposed planetary boundaries are shown in the Table, two of which (those for CLIMATIC CHANGE and the NITROGEN CYCLE) have already been crossed. Others are yet to be defined. An important aim of the planetary boundaries approach is to return the Earth system to the range of values characteristic of the HOLOCENE DOMAIN.   *JAM*

[*See also* ANTHROPOCENE, (THE) GREAT ACCELERATION]

Rockström J, Steffen W, Noone K et al. (2009) A safe operating space for humanity. *Nature* 461: 472–475.

Steffen W, Grinevald J, Crutzen P and McNeill J (2011) The Anthropocene: Conceptual and historical perspectives. *Philosophical Transactions of the Royal Society* 369: 842–867.

Earth-system process	Control variable	Proposed boundary	Current status	Pre-industrial value
Climatic change	Atmospheric CO_2 concentration (ppmv)	350	387	280
Nitrogen cycle	N_2 removed for use from the atmosphere (million tonnes per year)	35	121	0
Ozone depletion	Stratospheric ozone concentration (Dobson units)	276	283	290
Ocean acidification	Global mean saturation state of aragonite in surface seawater	2.75	2.90	3.44
Global freshwater use	Human consumption (km^3/year)	4,000	2,600	415
Landuse change	Percentage of global land cover converted to cropland	15	11.7	Low

Planetary boundaries *Six proposed planetary boundaries are shown in relation to their Earth-system processes, control variables, current values and pre-industrial values (Rockström et al., 2009; Steffen et al., 2011).*

planetary environmental change Discussion beyond the highly speculative on the geological/geo-morphological history and CLIMATIC CHANGE (where there is an ATMOSPHERE) of planetary bodies in the *solar system* has depended on REMOTE-SENSING imagery and other data sent back from orbiter and lander *spacecraft* launched since AD 1959. The most detailed GEOLOGICAL RECORD OF ENVIRONMENTAL CHANGE has been assembled for the *Moon*, as a result of its proximity to Earth, but particularly because of repeated visits by uncrewed and crewed spacecraft and rock samples returned to Earth. According to RADIOMETRIC DATING, the heavily cratered highlands date from ca 4.4 to 4.0 billion years ago, but the smooth-surfaced, relatively CRATER-free *maria* (infilled by LAVA FLOWS issuing from partially molten *lunar mantle*) formed ca 3.9–3.2 billion years ago. The maria give an idea of the *cratering rate* from impacting objects (BOLIDES) over the past 3 million years or so, providing a means of assessing the lengths of exposure of the surfaces of other planets.

Imagery of *Mercury* resembles that obtained from the Moon with ubiquitous impact craters, though apparently no VOLCANISM. A further difference lies in *compressional cliffs* or FAULTS postdating most craters, suggesting that the planet's interior may have shrunk slightly (perhaps by only 1 km or less in diameter) many hundreds of million years after crustal solidification.

Venus and *Mars* are Earth-like in two important respects: both have dynamic atmospheres and rocky surfaces. With little EROSION and SEDIMENTATION on Venus, its cratered surface seems to have been periodically obliterated by new lava flows every few hundreds of million years. This is suggested by the comparatively

small number of identified craters (~900). There is also clear evidence of TECTONICS in a broad belt with comparatively small-scale *blob tectonics* (cf. PLATE TECTONICS). The atmosphere may well have resembled that of Earth about 4 billion years ago, but it is now composed largely of CARBON DIOXIDE (96 per cent) with little OXYGEN or WATER VAPOUR. It has been argued that it is at or near a state of *unstable equilibrium* and that only moderate PERTURBATIONS of radiatively active VOLATILES might cause climatic change. The massive carbon dioxide atmosphere means that Venus experiences an extreme GREENHOUSE EFFECT, leading to surface temperatures of around 450°C.

A series of orbiting and robotic landing spacecraft sent to Mars for more than 30 years has provided a considerable amount of information (remote and *ground-surface imagery*, and geochemical and mineralogical measurements) on the environmental changes undergone by the planet. Relatively recent availability of *high-resolution imagery* and other evidence has led to an improved appreciation of the surface processes and LANDFORMS compared with what was possible with information available during the 1970s and 1980s. A simple view of an uninterrupted dry-surface environment replacing early wetness has been contested in recent years, with researchers arguing for a hydrologically more dynamic history, with episodic hydrological activity. The atmospheric composition of Mars is close to that of Venus but the low *atmospheric pressure* (<1 per cent that of Earth) combined with lower solar input mean that it is a cold (down to below −125°C at the poles), desiccated planet probably with GROUNDWATER trapped beneath a PERMAFROST layer. Approximately half the surface is moderately cratered and, from the evidence of extensive, mainly WIND EROSION, is probably at least 4 billion years old. Younger surfaces are volcanic in character, as in the case of the *Tharsis bulge* (a huge uplifted area the size of North America), which is thought to be between 1 and 3 billion years old. Very different climates and surface processes seem to have operated during the past. Early in its history, Mars was apparently much warmer than now, with rainfall, flowing rivers and the development of *valley networks*. Later-formed valleys on volcanoes may be explained by shallow surface ice melted by warm ejecta from the impact craters. The intricate dendritic pattern of many channels in the older (cratered) uplands implies *rain-fed* rather than *spring-fed flow*, and the existence of widespread liquid water indicates much higher atmospheric pressures in the past. There is apparent morphological evidence of large waterbodies (e.g. possible SHORELINES, extensive, flat smooth surfaces) covering the northern plains, but there is less controversial evidence of LAKES occupying highland craters with associated fluvial DELTAS formed during the heavy-bombardment epoch. Water did not remain on the surface for geologically long periods. There would have been some atmospheric loss, but much water remained within or beneath semi-permanent ice-rich permafrost. Numerous small gullies on crater rims and valley walls suggest an origin by water-rich DEBRIS FLOWS involving melting of this near-surface GROUND ICE. A series of PERIGLACIAL (e.g. PINGOS, PATTERNED GROUND) and glacial (e.g. erosional grooves, DRUMLINS, *horns*, CIRQUES, MORAINES) LANDFORMS were identified from high-resolution imagery. Different origins have been suggested for huge *outflow channels* resembling those of the CHANNELED SCABLANDS. They might have resulted from catastrophic FLOODS caused by the sudden release of ARTESIAN water from beneath the frozen upper permafrost layer triggered by internal geological activity. Alternative mechanisms are debris flows, liquefied crustal material or low-viscosity lava flows. Early Martian warmth and wetness have led to speculation about whether some simple forms of life may have developed. Episodic climatic change over the past few million years may be recorded in layers apparently present throughout the polar regions, which probably represent AEOLIAN deposition of sediment transported from lower latitudes.

The so-called *giant outer planets* (Jupiter, Saturn, Uranus and Neptune) are much larger than the *inner terrestrial planets* and have thick atmospheres. *Jupiter* has no crustal surface, but has a spectacular kaleidoscopic system of CLOUDS with notable disturbances. The largest of these, the *Great Red Spot* (diameter 30,000 km), has existed at least since detailed telescopic observations began some 300 years ago, but the three *White Ovals* formed only as recently as AD 1940. No similar continuity of observations exists for the other giant outer planets.

The *moons* of the giant planets have proved unexpectedly varied. Images of the *Jovian moons* were obtained during Voyager spacecraft flybys in 1979. *Ganymede* is composed partly of ice. Its surface has craters but at some time in the past was warm enough for eruptions of water to flood the surface and obliterate the old craters. *Callisto* has a cratered surface modified by deformation of slightly plastic ice. The smaller inner moons, *Europa* and *Io*, are very different. Europa is geologically unique, with a smooth-surface *ice crust* possibly floating on a global OCEAN. Io has a close orbit with Jupiter, resulting in a heat-generating *tidal bulge* causing very active volcanism, which renews the surface comparatively frequently. This explains the lack of impact craters. One of the satellites of *Saturn, Titan*, is unique amongst moons in the solar system in having an atmosphere. Instruments on the Cassini spacecraft have shown that the landscape has evidence of *river channels*, impact craters and *dry basins*, possibly representing *lake beds*. The largest of the moons of *Neptune, Triton*, is the coldest place in the solar system, such that most potential atmospheric gas is frozen. There are some impact craters, but many regions have undergone resurfacing, possibly through flowage of *methane* and *nitrogen* 'ice' precipitated out of Triton's thin atmosphere. A *polar cap* possibly comprising frozen nitrogen covers much of the *Southern Hemisphere*, apparently evaporating through solar warming along its northern edge via *geysers* or volcano-like plumes of nitrogen ejected up to 10 km from the surface.   RAS

[*See also* ATMOSPHERIC COMPOSITION, GLOBAL WARMING, METEORITE IMPACT, NEAR-EARTH OBJECT]

Baker VR (2006) Geomorphological evidence for water on Mars. *Elements* 2: 139–143.

Balme MR, Bargery AS, Gallagher CJ and Gupta S (eds) (2011) *Martian geomorphology*. Bath: Geological Society.

Balme MR and Gallagher C (2009) An equatorial periglacial landscape on Mars. *Earth and Planetary Science Letters* 285: 1–15.

Bullock MA and Grinspoon DH (1996) The stability of climate on Venus. *Journal of Geophysical Research: Planets* 101: 7521–7529.

Carr MH (1995) The Martian drainage system and the origin of valley networks and fretted channels. *Journal of Geophysical Research: Planets* 100: 7479–7507.

Dressler O and Sharpton VL (1999) Large meteorite impacts and planetary evolution II. *Geological Society of America Special Paper* No. 339.

Greeley R (1995) Geology of terrestrial planets with dynamic atmospheres. *Earth, Moon and Planets* 67: 13–29.

Greeley R (2013) *Introduction to planetary geomorphology*. Cambridge: Cambridge University Press.

Kleinhans MG (2010) A tale of two planets: Geomorphology applied to Mars surface, fluvio-deltaic processes and landforms. *Earth Surface Processes and Landforms* 35: 102–117.

Levy JS, Marchant DR and Head JW (2010) Thermal contraction crack polygons on Mars: A synthesis from HiRISE, Phoenix, and terrestrial analog studies. *Icarus* 45: 264–303.

Matsubara Y, Howard AD and Drummond SA (2011) Hydrology of early Mars: Lake basins. *Journal of Geophysical Research: Planets* 116: E04001.

Morgan GA, Head JW, Forget F et al. (2010) Gully formation on Mars: Two recent phases of formation suggested by links between morphology, slope orientation and insolation history. *Icarus* 208: 658–666.

Pepin RO (1994) Evolution of the Martian atmosphere. *Icarus* 111: 289–304.

Radebaugh J, Lorenz RD, Wall SD et al. (2011) Regional geomorphology and history of Titan's Xanadu province. *Icarus* 211: 672–685.

Schmitt B, de Bergh C and Festou M (eds) (1998) *Solar system ices.* Dordrecht: Kluwer.

Tanaka KL (1997) Sedimentary history and mass flow structures of Chryse and Acidalia Planitiae, Mars. *Journal of Geophysical Research: Planets* 102: 4131–4149.

Treiman AH, Fuks KH and Murchie S (1995) Diagenetic layers in the upper walls of Valles Marineris, Mars: Evidence for drastic climate change since the mid-Hesperian. *Journal of Geophysical Research: Planets* 100: 26339–26344.

Wells GL and Zimbelman JR (1997) Extraterrestrial arid surface processes. In Thomas DSG (ed.) *Arid zone geomorphology: Process, form and change in drylands, 2nd edition.* Chichester: Wiley, 659–690.

planetary geology   The study of the *rocky celestial bodies* (*terrestrial planets, moons* and ASTEROIDS). This large and recent discipline has considerable potential for studies of environmental change. Firstly, these bodies can provide independent control on whether Earth's environmental changes are driven by internal forces (e.g. PLATE TECTONICS) or external forces (e.g. SOLAR FORCING). Secondly, they provide further MODERN ANALOGUES for testing MODELS for features such as the GENERAL CIRCULATION OF THE ATMOSPHERE and processes of TECTONICS. The history of climate change on Mars, for example, is a rapidly developing field of strong relevance to terrestrial studies of CLIMATIC CHANGE.   *CDW*

[*See also* ASTROGEOLOGY, GEOLOGY, PLANETARY ENVIRONMENTAL CHANGE]

Carroll, M (2011) *Drifting on alien winds: Exploring the skies and weather of other worlds.* New York: Springer.

Chan MA, Beitler B, Parry WT et al. (2004) A possible terrestrial analogue for haematite concretions on Mars. *Nature* 429: 731–734.

De Pater I and Lissauer J (2010) *Planetary sciences, 2nd edition.* Cambridge: Cambridge University Press.

Greeley R (1994) *Planetary landscapes.* London: Chapman and Hall.

Lewis SR and Read PL (2006) *Evidence for climate change on Mars.* Berlin: Springer.

Rossi A and Van Gasselt S (2010) Geology of Mars after the first 40 years of exploration. *Research in Astronomy and Astrophysics* 10: 621–652.

Shirley JH and Fairbridge RW (eds) (1997) *Encyclopedia of planetary sciences.* London: Chapman and Hall.

Vita-Finzi C (2005) *Planetary geology: An introduction.* Harpenden: Terra Publishing.

planetary interest   The concept that it is necessary to consider the world as a single system for DECISION-MAKING in relation to major ENVIRONMENTAL PROBLEMS. Planetary interest is a concept of ENVIRONMENTAL GOVERNANCE that embraces GLOBAL ENVIRONMENTAL CHANGE and ENVIRONMENTAL SECURITY, and recognises the inadequacy of both *national sovereignty* as a principle and the *nation-state* as the decision-making unit. It may be seen as adding a political dimension to the idea of a *common interest* beyond the *national interest*, which, if not recognised, will put planet Earth in jeopardy in terms of both its *biophysical integrity* and its ability to satisfy human needs. Graham (1999: 9) identifies these two aspects as the *vital planetary interest* (necessary for survival and viability) and the *normative planetary interest* (necessary for human well-being and *quality of life*).   *JAM*

[*See also* HUMAN DIMENSIONS OF ENVIRONMENTAL CHANGE, HUMAN ENVIRONMENT, STEWARDSHIP CONCEPT, SUSTAINABILITY, SUSTAINABLE DEVELOPMENT, WORLD SYSTEM]

Graham K (ed.) (1999) *The planetary interest.* London: UCL Press.

Graham K (2008) 'Survival research' and the 'planetary interest': Carrying forward the thoughts of John Herz. *International Relations* 22: 417–472.

plankton/planktonic/planktic   Small or microscopic free-floating PELAGIC ORGANISMS and larvae in marine or freshwater, including ALGAE and *protozoans*. PHYTOPLANKTON (AUTOTROPHS, algae) in SURFACE WATERS are the primary producers in the FOOD CHAIN; ZOOPLANKTON are consumers. *Planktonic* or *planktic* organisms such as FORAMINIFERA, RADIOLARIA, DIATOMS, COCCOLITHOPHORES and the extinct *graptolites* have good potential as INDEX FOSSILS in BIOSTRATIGRAPHY. Smaller plankton are classified by size as *nannoplankton* (5–60 μm), *ultraplankton* (≤5 μm) and *picoplankton* (≤1 μm).   *LC*

[*See also* DINOFLAGELLATE CYST ANALYSIS, FORAMINIFERA: ANALYSIS, MARINE SEDIMENT CORES]

Huisman J and Weissing FJ (1999) Biodiversity of plankton by species oscillations and chaos. *Nature* 402: 407–410.

Kiørboe T (2008) *A mechanistic approach to plankton ecology.* Princeton, NJ: Princeton University Press.

Suthers IM and Rissik D (eds) (2008) *Plankton: A guide to their ecology and monitoring water quality.* Collingwood: CSIRO Publishing.

Planosols Soils with a silty or loamy grey surface or shallow subsurface horizon, which are signs of periodic wetness, abruptly overlying a dense subsoil, upon which downward percolating water stagnates. Breakdown of clay by FERROLYSIS leads to the abrupt change of SOIL TEXTURE in the soil profile. These are poor soils that occur under light forest on low plateau surfaces in tropical and warm-temperate Latin America, eastern United States, eastern Africa, Southeast Asia and Australia, where there is a marked alternation between wet and dry seasons (SOIL TAXONOMY: *Albaqualfs, Albaquults* and *Argialbolls*). These soils may have been derived from VERTISOLS following a change to a wetter climate. *EMB*

[*See also* WORLD REFERENCE BASE FOR SOIL RESOURCES (WRB)]

Feutel TC, Jongmans AG, Van Breemen N and Miedema R (1988) Genesis of two Planosols in the Massif-Central, France. *Geoderma* 43: 249–269.

Parahyba RDV, dos Santos MC, Neto FCR and Jacomine PKT (2010) Pedogenesis of Planosols in a toposequence of the Agreste region of Pernambuco, Brazil. *Revista Brasileira de Ciecia do Solo* 34: 1991–2000.

plant diseases Damaging effects brought about when plants are infected by PATHOGENS, including VIRUSES, BACTERIA, FUNGI and higher-plant PARASITES. Plant diseases include various types of *rusts, mildews, galls, leaf curls, scabs, cankers, blights, rots, spots, flecks,* CHLOROSIS, *necrosis* and DIEBACK. Although they should be viewed as having positive as well as negative roles (see ECOLOGICAL GOODS and ECOLOGICAL SERVICES), plant diseases tend to be regarded as wholly negative in relation to ANTHROPOGENIC ecosystems and have contributed to, if not caused, some major detrimental events in history. The Irish potato FAMINE of the AD 1840s, for example, was caused by the INTRODUCTION of a parasitic fungus (*Phytophthora infestans*) native to Peru, which devastated the potato crop. Further effects included an estimated 1.5 million human deaths and the migration of 1 million people to form the Irish DIASPORA in North America, Australia and other parts of the British Empire. Hence, the economic, social and political effects of the 'potato blight' changed the course of history. Plant diseases continue to cause massive losses in relation to cultivated plants, accounting, for example, for 20 per cent of the annual world cereal crop. *JAM*

[*See also* ANIMAL DISEASES, DISEASE, HUMAN HEALTH HAZARDS]

Agros GN (2005) *Plant pathology. 5th edition.* San Diego, CA: Academic Press.

Brown JKM and Hovmøller MS (2002) Aerial dispersal of pathogens on the global and continental scales and its impact on plant disease. *Science* 297: 537–541.

Gisi U, Chet I and Gullino ML (eds) (2010) *Recent developments in management of plant diseases.* Dordrecht: Springer.

Gray DR (2008) The relationship between climate and outbreak characteristics of the spruce budworm in eastern Canada. *Climatic Change* 89: 447–449.

Ingram D and Robertson N (1999) *Plant disease: A natural history.* London: HarperCollins.

plant geography The study of the spatial relationships and distribution patterns of plants, which is also known as PHYTOGEOGRAPHY. It includes the geography of VEGETATION as well as the geography of *taxa* (e.g. *species, genera* and *families*). Pioneering studies were undertaken in the early nineteenth century by the German geographer Alexander von Humboldt, who, in describing the correspondence between global vegetation patterns and climate, laid the foundations for *vegetation geography.* These studies were seminal in the development of understanding of the global relationships between plant form (*vegetation physiognomy*) and CLIMATE. The study of *species distributions,* in historical or current ecological contexts, is now a fundamental part of modern plant geography, often encompassing multidisciplinary approaches involving systematics, molecular genetics and PALAEONTOLOGY.

Plant geography is a subdiscipline of BIOGEOGRAPHY, which includes historical and ecological schools. *Historical plant geography* seeks explanations for distribution patterns, such as those which are non-continuous (see DISJUNCT DISTRIBUTION) or isolated (see ENDEMIC), in terms of DISPERSAL and/or VICARIANCE events. Classic studies include the one by Pole (1994) on the origin of the current flora of New Zealand. Historical plant geography is more analytical than experimental. *Ecological plant geography* studies the BIOTIC and ABIOTIC factors that influence current species' distribution patterns. It is often experimental in approach and, among other things, is a fundamental tool in studies of CONSERVATION, BIODIVERSITY and INVASIVE SPECIES. *CRH*

Kilic DE and Ikiel C (2012) Vegetation geography of western part of Elmacik mountain, Turkey. *Journal of Environmental Biology* 33: 293–305.

Maestre FT, Quero JL, Gotelli NJ et al. (2012) Plant species richness and ecosystem multifunctionality in global drylands. *Science* 335: 214–218.

Moreira Munoz A (2011) *Plant geography of Chile.* Berlin: Springer.

Pole M (1994) The New Zealand flora: Entirely long distance dispersal? *Journal of Biogeography* 21: 625–635.

plant macrofossil analysis Plant MACROFOSSILS are plant parts preserved in Quaternary sediments that are large enough to be manipulated by hand, thus contrasting with MICROFOSSILS. They usually comprise FRUITS AND SEEDS but can be *vegetative* parts (e.g. *leaves*, CUTICLES, *bud-scales*, *rhizomes*, *twigs* and *wood*). Macrofossils can be identified using a stereomicroscope, a high-power, *light* or *optical microscope* or SCANNING ELECTRON MICROSCOPY (SEM). Wood identification requires thin-section *wood analysis*.

Macrofossils supplement POLLEN ANALYSIS for past vegetation and PALAEOENVIRONMENTAL RECONSTRUCTION. They can often be identified to the *species* level. They often occur in small quantities, and hence require larger sediment samples than pollen. Percentages are not useful, and stratigraphic diagrams are best presented as concentrations (numbers/unit volume sediment). Macrofossils usually have limited dispersal from their source and hence represent local vegetation development. They may be frequent from species that produce little or poorly preserved pollen, or cryptogams (e.g. mosses, including *SPHAGNUM* and *Charophytes*). Thus, they illuminate the 'blind spots' in a pollen assemblage. Because there is no macrofossil rain equivalent to POLLEN RAIN, they are limited for regional correlation. However, they can indicate the local presence of a large pollen producer, for example, many tree species, and thus they are valuable in reconstructing TREE-LINE VARIATIONS.

Macrofossil analyses from LACUSTRINE SEDIMENTS provide information on aquatic flora and vegetation, and can thus reconstruct changes in lake environments, for example, pH, water conductivity and chemistry, water depth and EUTROPHICATION. Interpretation is aided by the few available surface-sample representation studies. Macrofossils can demonstrate PEAT development, as plant remains form the peat. They register the local vegetation, and can reconstruct changes in *surface wetness*, and hence PRECIPITATION VARIATIONS. In the context of ENVIRONMENTAL ARCHAEOLOGY, macrofossils can reconstruct the living environment of a settlement from nearby deposits in lakes or MIRES, or directly from the settlement, where they are preserved in relation to ARTEFACTS (e.g. CERAMICS, pits, latrines, fireplaces etc.). Remains of plant foods are found in latrines and MIDDENS, and in the stomachs and faeces (see COPROLITE) of both humans and animals. A particularly elegant study using plant remains has been made of the lifestyle of the Italian-Austrian 5,200-year-old ICEMAN.   *HHB*

Birks HH (2000) Aquatic macrophyte vegetation development in Kråkenes Lake, western Norway, during the Late-glacial and early-Holocene. *Journal of Paleolimnology* 23: 7–19.

Birks HH (2001) Plant macrofossils. In Smol JP, Birks HJB and Last WM (eds) *Tracking environmental changes in lake sediments, volume 3*. Dordrecht: Kluwer, 49–74.

Birks HH and Birks HJB (2000) Future uses of pollen analysis must include plant macrofossils. *Journal of Biogeography* 27: 31–35.

Birks HH and Bjune AE (2010) Can we detect a west-Norwegian tree-line from modern samples of plant remains and pollen? Results from the DOORMAT project. *Vegetation History and Archaeobotany* 19: 325–340.

Fowler B (2002) *Iceman: Uncovering the life and times of a prehistoric man found in an alpine glacier*. Oxford: PanMacmillan.

Jacomet S (2007) Use in environmental archaeology. In Elias SA (ed.) *Encyclopedia of Quaternary science*. Amsterdam: Elsevier, 2384–2412.

Mortensen MF, Birks HH, Christensen C et al. (2011) Lateglacial vegetation development in Denmark: New evidence based on macrofossils and pollen from Slotseng, a small-scale site in southern Jutland. *Quaternary Science Reviews* 30: 2534–2550.

Tinner W (2007) Treeline studies. In Elias SA (ed.) *Encyclopedia of Quaternary science*. Amsterdam: Elsevier, 2374–2384.

Van Geel B, Aptroot A, Baittinger C et al. (2008) The ecological implications of a Yakutian mammoth's last meal. *Quaternary Research* 69: 361–376.

plant pigment analysis Carotenoids, CHLOROPHYLLS and their derivatives extracted from ALGAE, BACTERIA and *higher plants* are measured using *high-performance liquid chromatography* (HPLC). Plants synthesise a variety of pigments for use in PHOTOSYNTHESIS. *Sedimentary plant pigment analysis* can be used to address changes in algal and bacterial community composition, TROPHIC LEVEL, *food-web interactions*, lake ACIDIFICATION and anthropogenic impacts on *aquatic ecosystems*. Fossil pigment compositions can discriminate between most algal groups and are, therefore, useful ENVIRONMENTAL INDICATORS in PALAEOECOLOGY. Changes in PHYTOPLANKTON composition, algal SEDIMENTATION and PRIMARY PRODUCTIVITY can be determined by pigment analysis. Chlorophyll concentrations are frequently used to estimate phytoplankton BIOMASS and PRODUCTIVITY, while the products of DECOMPOSITION are diagnostic indicators of *physiological status*, *detrital content* and GRAZING processes in natural populations of phytoplankton. When used in conjunction with MICROFOSSIL ANALYSIS of ZOOPLANKTON, plant pigments can help reconstruct PREDATOR-PREY RELATIONSHIPS in whole-lake FOOD WEBS. Some BENTHIC algae produce unique pigments when exposed to *ultraviolet (UV) radiation* and these pigments can be used to examine past changes in UV radiation. Sedimentary pigments have been proposed as a means of inferring past *phosphorous* concentrations in lakes. The preservation of fossil pigments varies with the water environment and is reduced under conditions of high light, oxygen, temperature and turbulence. Fossil pigments from *siliceous algae* are poorly preserved and, therefore, fossil pigment interpretations should only be based on historical changes in pigment concentration.

KJF

[*See also* PIGMENTS: FOSSIL]

Guilizoni P, Marchetto A, Lami A et al. (2011) Use of sedimentary pigments to infer past phosphorous concentration in lakes. *Journal of Paleolimnology* 45: 433–445.

Hodgson DA, Vyverman W, Verleyen E et al. (2005) Late Pleistocene record of elevated UV radiation in an Antarctic lake. *Earth and Planetary Science Letters* 236: 765–772.

Leavitt PR and Hodgson DA (2001) Sedimentary pigments. In *tracking environmental change using lake sediments, volume 3*. Dordrecht: Kluwer, 1–32.

Mantoura RFC and Llewellyn CA (1983) The rapid determination of algal chlorophyll and carotenoid pigments and their breakdown products in natural waters by reverse-phase high-performance liquid chromatography. *Analytica Chimica Acta* 151: 297–314.

Von Gunten L, Grosjean M, Rein B et al. (2009) A quantitative high-resolution summer temperature reconstruction based on sedimentary pigments from Laguna Aculeo, central Chile, back to AD 850. *The Holocene* 19: 873–881.

plantation An extensive area of planted forest or agricultural land that is usually an intensively cropped MONOCULTURE and part of an industrial-scale production system. Historically, plantations were widely associated with COLONIALISM and slavery, and grew in parallel with the growth of international trade. Today, they remain highly dependent on GLOBALISATION. In the past decade, the emergence of BIO-ENERGY based, for example, on sugar cane, corn, oil palm and cassava has led to a new plantation boom.

JAM

[*See also* CROPLAND]

Brockerhoff EG, Jactel H, Parrotta JA, Quine CP et al. (eds) (2010) *Plantation forests and biodiversity*. Berlin: Springer.

Menard RR (2006) *Sweet negotiations: Sugar, slaves, and plantation agriculture in early Barbados*. Charlottesville: University of Virginia Press.

plastic limit The *moisture content* at which a SEDIMENT transforms from a brittle to a plastic or ductile solid. The *plasticity index* is the difference between the LIQUID LIMIT and plastic limit, denoting the range of moisture contents over which the soil behaves as a plastic solid.

PW

[*See also* FAILURE, LIQUID LIMIT]

Rea BR (2004) Engineering properties. In Evans DJA and Benn DI (eds) *A practical guide to the study of glacial sediments*. London: Arnold, 182–208.

plastics Substances capable of plastic flow or plastic deformation under certain conditions or at some stage of manufacture. They include *natural resins*, but the two main classes of plastics are manufactured from *synthetic resins*, which are long-chain polymers: (1) *thermoplastics*, such as *polyethylene, polypropylene* and *polyvinyl chloride*, which retain potential plasticity after manufacture and can be remoulded by heating, and (2) *thermosetting plastics*, such as *epoxy resins* and *bakelite*, which undergo non-reversible chemical changes on heating and many of which are difficult, if not impossible, to recycle. Plastics are cheap to produce (relative to more traditional materials, i.e. iron and steel and other metals) and they are energy efficient in terms of transport costs and durability. Lightness combined with durability has made plastics the second most important *packaging* material (after cardboard and paper). However, durability becomes a disadvantage when plastics are discarded. Although they present well-known environmental problems on land, associated, for example, with LANDFILL and LITTER, they present even greater problems in the sea: plastics account for ~80 per cent of *ocean debris*. A recently invented *biodegradable* plastic made from *poly-3-hydroxybutrate*, a polymer that occurs naturally in the cells of certain BACTERIA, may be the first of the environmentally friendly plastics of the future.

JAM

[*See also* BIOPLASTICS]

Andrady AL (ed.) (2003) *Plastics and the environment*. Hoboken, NJ: Wiley.

Porteous A (2000) *Dictionary of environmental science and technology, 4th edition*. Chichester: Wiley.

Stevens ES (2002) *Green plastics: An introduction to a new science of biodegradable plastics*. Princeton, NJ: Princeton University Press.

Thompson RC, Moore CJ, von Saal FS and Swan SH (eds) (2009) *Plastics, the environment and human health*. London: Royal Society Publishing.

plate margin The edge of one of the Earth's plates of LITHOSPHERE. The three types of plate margin are CONSTRUCTIVE (divergent), DESTRUCTIVE (convergent) and CONSERVATIVE PLATE MARGINS (see Figure). *Plate boundary* is a broadly equivalent term, although it is sometimes restricted to the junction between the plates, whereas plate margin also includes the part of the plate bordering the junction.

GO

[*See also* PLATE TECTONICS]

plate tectonics The theory, or PARADIGM, that the Earth's LITHOSPHERE consists of discrete rigid slabs or *plates* that move laterally over the weak ASTHENOSPHERE. At PLATE MARGINS, plates can move apart, creating new lithosphere through SEA-FLOOR SPREADING

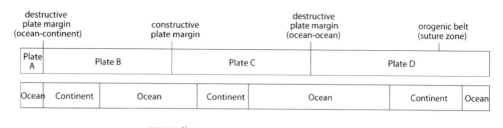

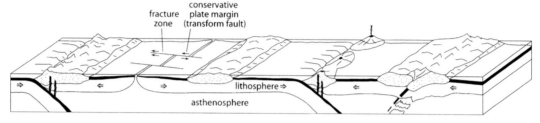

Plate margin *Schematic diagram showing the different kinds of plate margins in relation to continents and oceans.*

(*divergent* or CONSTRUCTIVE PLATE MARGIN); move together, consuming old lithosphere through SUB-DUCTION (*convergent* or DESTRUCTIVE PLATE MARGIN); or slide past each other, conserving the lithosphere (CONSERVATIVE PLATE MARGIN). Most of the Earth's TECTONIC activity, VOLCANOES and EARTHQUAKES are confined to plate margins, and plate interiors are essentially stable. Plate outlines can be defined using the distribution of earthquakes, volcanoes and certain major features of global GEOMORPHOLOGY: MID-OCEAN RIDGES (MORs) coincide with constructive plate margins, and deep OCEANIC TRENCHES, volcanic ISLAND ARCS and ACTIVE CONTINENTAL MARGINS with destructive plate margins. There are seven major plates—North American, South American, Eurasian, African, Indo-Australian, Pacific and Antarctic—and numerous minor plates (see Figure). Most include areas of both CONTINENT and OCEAN, so the edges of continents (CONTINENTAL MARGINS) do not necessarily correspond to plate margins: PASSIVE CONTINENTAL MARGINS occur within plates. Plates move at rates of centimetres to tens of centimetres per year, giving rise to CONTINENTAL DRIFT.

Only lithosphere-bearing, relatively thin and dense OCEANIC CRUST is created and destroyed in plate tectonics. Where oceanic crust and CONTINENTAL CRUST converge, the oceanic crust is subducted and destroyed, creating an active continental margin or *continental margin orogen*. If two fragments of continental crust are brought together, a continent-continent COLLISION ZONE develops: SEDIMENTARY DEPOSITS at the continental margins are subjected to compression, DEFORMATION and UPLIFT, forming high *fold mountains* in an OROGENIC BELT (OROGEN). This plate-tectonic model for OROGEN-ESIS—the WILSON CYCLE—has completely replaced *geosyncline* theory.

The driving force behind plate tectonics is not fully understood. Igneous processes at plate margins contribute to a convective release of GEOTHERMAL heat, but this is unlikely to drive plate movements. It seems more likely that the sinking of old, dense oceanic lithosphere drags the remainder of the plate behind it. It is now widely accepted that some form of plate tectonics operated in the ARCHAEAN, but because the Earth's interior was hotter then, the rates, scales and products are likely to have been significantly different (see GREENSTONE BELT, KOMATIITE). During PERMIAN and TRIASSIC times, the continental masses were assembled as a single SUPERCONTINENT (PANGAEA (PANGEA)), which has fragmented through MESOZOIC and CENOZOIC (*CAINOZOIC*) times. A previous supercontinent configuration, RODINIA, existed in the Late PROTEROZOIC. It is not known if plate tectonics operates on any other bodies in the *solar system*.

Although a theory of continental drift was formulated by Alfred Wegener in the early twentieth century, based largely on geological evidence, the development of the plate-tectonics paradigm in the early 1960s came about through the coincidence of several major discoveries. After the Second World War, accurate *echo sounders* were used to map the ocean floor, leading to the detailed charting of features such as mid-ocean ridges and oceanic trenches. RADIOMETRIC DATING of rock samples from the ocean floor gave surprisingly young ages, nowhere greater than 200 million years, and thus younger than most continental rocks. The theory of SEA-FLOOR SPREADING, developed in the late 1950s, was substantiated by the discovery of ocean-floor MAGNETIC ANOMALIES that were symmetrical either side of the mid-ocean ridges: these were interpreted in terms of ocean-floor forming at mid-ocean ridges and spreading laterally, while the polarity of the Earth's magnetic field repeatedly switched between normal and reverse states (see GEOMAGNETIC POLARITY REVERSAL). DEEP-SEA DRILLING showed that the thickness of OCEANIC SEDIMENTS increased with distance from the mid-ocean ridge and hence with the age of the oceanic crust. EARTHQUAKE studies showed an inclined zone of earthquake generation in the Earth's interior on the same side of oceanic trenches as active volcanoes occur: this was termed a *Benioff zone*. The locations of earthquake EPICENTRES were found to lie along narrow zones that coincided with the sites of active volcanoes. Studies of SEISMIC WAVES showed that the sense of displacement on FAULTS that cut the mid-ocean ridges was consistent with CRUST spreading on each side of the mid-ocean ridge: these were called TRANSFORM FAULTS. The rapid development and acceptance of plate tectonics led to a genuine revolution in the EARTH SCIENCES.

Amendments to the plate-tectonics theory since the 1960s have been relatively minor. Volcanic activity in plate interiors (*intraplate volcanism*) is attributed to HOTSPOTS that overlie MANTLE PLUMES. Some orogenic belts are now understood to have assembled over long periods of time through the accretion of small pieces of continental crust (TERRANES). The discovery of HYDROTHERMAL VENTS (see BLACK SMOKER) at mid-ocean ridges has led to some revision of ideas affecting EVOLUTION and ECOLOGY.

Plate tectonics has many implications for ENVIRONMENTAL CHANGE. As a consequence of continental drift, parts of the Earth's surface have experienced changing PALAEOLATITUDE and PALAEOCLIMATE through GEOLOGICAL TIME, offering an explanation for climate-sensitive sediment FACIES such as COAL or EVAPORITES in temperate latitudes. Changes in PALAEOGEOGRAPHY due to the joining and splitting of land masses have important implications for OCEAN CURRENTS and the evolution of terrestrial organisms. Long-term changes in the rate of plate activity influence global sea levels over geological timescales. Newly formed oceanic crust is elevated as a mid-ocean ridge because it has excess heat which is lost at a steady rate, leading to a predictable *age-depth relationship for ocean floor*. Rapid spreading produces wide ridges that occupy a greater volume of the OCEAN BASINS, leading to a rise in *sea level* and flooding of the continental margins. Increased volcanic activity with more rapid spreading influences the supply of VOLATILES to the ATMOSPHERE. Regional uplift during episodes of orogeny can disrupt ATMOSPHERIC CIRCULATION and change MONSOON climates. Orogenic activity affects the volume of rocks exposed to WEATHERING: since many weathering reactions involve CARBON DIOXIDE, this can disrupt the CARBON CYCLE and influence the amount of carbon dioxide in the atmosphere.

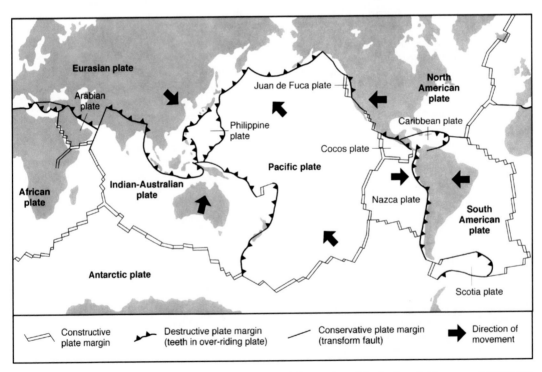

Plate tectonics *Outlines of the Earth's tectonic plates, showing the types of plate margins and the directions of plate movements (Decker and Decker, 1997).*

In summary, the central role of plate tectonics in our understanding of the entire EARTH SYSTEM, including ENVIRONMENTAL CHANGE, cannot be overemphasised. *GO*

Ali JR and Aitchison JC (2008) Gondwana to Asia: Plate tectonics, paleogeography and the biological connectivity of the Indian sub-continent from the Middle Jurassic through latest Eocene (166–5 Ma). *Earth-Science Reviews* 88: 145–166.

Bye JAT (1998) Sea level change due to oscillations in seafloor spreading rate. *Physics of the Earth and Planetary Interiors* 109: 151–159.

Condie KC and Pease V (eds) (2008) *When did plate tectonics begin on Planet Earth?* Boulder, CO: Geological Society of America.

Decker R and Decker B (1997) *Volcanoes, 3rd edition.* Basingstoke: Freeman.

Gurnis M (1988) Large-scale mantle convection and the aggregation and dispersal of supercontinents. *Nature* 332: 695–699.

Hastie AR, Kerr AC, McDonald I et al. (2010) Do Cenozoic analogues support a plate tectonic origin for Earth's earliest continental crust? *Geology* 38: 495–498.

Iaffaldano G, Bunge H-P and Dixon TH (2006) Feedback between mountain belt growth and plate convergence. *Geology* 34: 893–896.

Kearey P, Klepeis KA and Vine FJ (2009) *Global tectonics.* Chicester: Wiley.

McKenzie D (1999) Planetary science: Plate tectonics on Mars? *Nature* 399: 307–308.

Rasmussen CMO and Harper DAT (2011) Did the amalgamation of continents drive the end Ordovician mass extinctions? *Palaeogeography, Palaeoclimatology, Palaeoecology* 311: 48–62.

Redfern R (2000) *Origins: The evolution of continents, oceans and life.* London: Weidenfeld and Nocolson.

Stacey FD and Davis PM (2008) *Physics of the Earth.* Cambridge: Cambridge University Press.

Stadler G, Gurnis M, Burstedde C et al. (2010) The dynamics of plate tectonics and mantle flow: From local to global scales. *Science* 329: 1033–1038.

Stanley SM (2009) *Earth system history.* New York: Freeman.

Pleistocene   The Pleistocene is the EPOCH which follows the PLIOCENE, covers the greater part of the QUATERNARY and ended ca 11,700 years ago with the opening of the HOLOCENE. The beginning of the Pleistocene has been recently formally defined as 2.58 million years ago. West (1977) proposed that the term should include the whole Quaternary up to and including the Holocene, regarding the current temperate episode as merely the latest INTERGLACIAL in the Pleistocene epoch. However, this has not been generally accepted because the Holocene has other characteristics that make it distinctive. *DH/JAM*

[*See also* GEOLOGICAL TIMESCALE, PLIOCENE-PLEISTOCENE TRANSITION, QUATERNARY TIMESCALE]

Gibbard P and van Kelfschotten T (2004) The Pleistocene and Holocene epochs. In Gradstein FM, Ogg JG and Smith AG (eds) *A geological time scale 2004.* Cambridge: Cambridge University Press, 441–452.

Nilsson T (1993) *The Pleistocene: Geology and life in the Quaternary Ice Age.* Dordrecht: Kluwer.

West RG (1977) *Pleistocene geology and biology: With especial reference to the British Isles, 2nd edition.* London: Longman.

Pleniglacial   Used by European continental stratigraphers for much of the LAST GLACIATION (*Weichselian*), in which some distinguish Lower, Middle and Upper Pleniglacial. *FMC*

Behre K-E (1989) Biostratigraphy of the last glacial period in Europe. *Quaternary Science Reviews* 8: 25–44.

Moine O, Rousseau D-D and Antoine P (2008) Terrestrial molluscan records of Weichselian Lower to Middle Pleniglacial climatic changes from the Nussloch loess series (Rhine Valley, Germany): The impact of local factors. *Boreas* 34: 363–380.

Wasylikowa K (2005) Palaeoecology of Lake Zeribar, Iran, in the Pleniglacial, Lateglacial and Holocene, reconstructed from plant macrofossils. *The Holocene* 15: 720–735.

plinthic horizon   A clayey, iron-rich, humus-poor, mottled, subsurface DIAGNOSTIC SOIL HORIZON, the material of which will irreversibly harden on exposure to the atmosphere. This grey clay with varying amounts of reddish iron concretions, known in modern SOIL CLASSIFICATION as *plinthite*, was first described as LATERITE in India by Buchanan (1807). In the moist soil, it can be cut with a spade but on exposure to repeated wetting and drying it hardens as *petroplinthite*, *lateritic ironstone*, *murram* or FERRICRETE, in which state it is no longer considered to be plinthite. Once hardened, it is impossible to cultivate and is resistant to erosion and may form an abrupt edge to plateau features. It gradually weathers and releases *ironstone nodules* (or *pisoliths*) into contemporary soils. *EMB*

[*See also* DURICRUST]

Aleva GJJ and Creutzberg D (1994) *Laterites: Concepts, geology, morphology and chemistry.* Wageningen: International Soil Reference and Information Centre.

Buchanan F (1807) *A journey from Madras through the countries of Mysore, Canara and Malabar, etc.,* 3 volumes. London: East India Company.

Mohr ECJ, Van Baren FA and Schuylenborgh J (1972) *Tropical soils, 3rd edition.* The Hague: Mouton-Bary.

Plinthosols   Soils affected by groundwater, containing a mixture of iron, clay and quartz, in which the iron has segregated to form a mottled horizon, the PLINTHIC HORIZON (or *plinthite*). SOIL-FORMING PROCESSESS include both the accumulation of iron and aluminium by removal of silica (FERRALITISATION) and the segregation of iron by alternation of REDUCTION and OXIDATION. These soils occur mainly under TROPI-CAL RAIN FOREST and more widely in the intertropical regions of Africa, South America and Southeast Asia (SOIL TAXONOMY: *Plinthaquox*). There are severe physical and chemical limitations to agricultural use associated with waterlogging, the possibility of drought, stoniness and low fertility. Drainage, change to a drier climate or loss of vegetation cover can result in hardening of the plinthite, producing *petroplinthite* by crystallisation of amorphous iron and aluminium and subsequent dehydration. Plinthite is a useful material for making building bricks, and in some places, it is exploited as a source of iron and other metals. Plinthosols have also been termed LATERITES, *groundwater laterites* or *latosols*.   *EMB*

[*See also* WORLD REFERENCE BASE FOR SOIL RESOURCES (WRB)]

Anifowose AYB (2000) Stabilization of lateritic soils as a raw material for building blocks. *Bulletin of Engineering Geology and Environment* 58: 151–157.
Dos Anjos LHD, Franzmeier DP and Schulze DG (1995) Formation of soils with plinthite on a toposequence in Maranhão State, Brazil. *Geoderma* 64: 257–279.
Fritsch E, Herbillon AJ, Do Nascimento NR et al. (2007) From Plinthic Acrisols to Plinthisols and Gleysols: Iron and groundwater dynamics in the tertiary sediments of the upper Amazon basin. *European Journal of Soil Science* 58: 989–1006.

Pliocene An EPOCH of the NEOGENE (formerly Late TERTIARY) period, from 5.3 to 2.6 million years ago), the agreed date for the start of the QUATERNARY period. The Pliocene experienced TECTONIC changes including the joining of the North and South American plates, the rise of OROGENIC BELTS (OROGEN), including the Alps and the Himalayas, and a gradual global cooling towards the Quaternary ICE AGE.   *GO*

[*See also* GEOLOGICAL TIMESCALE]

Gibbard PL, Head MJ, Walker MJC and the Subcommission on Quaternary Stratigraphy (2010) Formal ratification of the Quaternary System/Period and the Pleistocene Series/Epoch with a base at 2.58 Ma. *Journal of Quaternary Science* 25: 96–102.
Haywood AM, Dowsett HJ, Valdes PJ et al. (2009) Pliocene climate, processes and problems. *Philosophical Transactions of the Royal Society A* 367: 3–17.
Lunt DJ, Valdes PJ, Haywood A and Rutt IC (2008) Closure of the Panama Seaway during the Pliocene: Implications for climate and Northern Hemisphere glaciation. *Climate Dynamics* 30: 1–18.
Zhang YG, Ji J, Balsam W et al. (2009) Mid-Pliocene Asian monsoon intensification and the onset of Northern Hemisphere glaciation. *Geology* 37: 599–602.

Pliocene-Pleistocene transition The Pliocene-Pleistocene boundary represents the beginning of the QUATERNARY, but the exact date of this transition has been a subject of debate. The boundary is usually located at the point in the STRATIGRAPHICAL RECORD where there are indications of climatic cooling, reflected by FOSSIL evidence or another climatic proxy. The reference site for the Pliocene-Pleistocene boundary originally accepted by the International Commission on Stratigraphy in 1985 and the International Union of Geological Sciences (IUGS) is located in Vrica in southern Italy, and is based upon the first appearance of the cold-water marine *ostracod—Cytheropteron testudo*. This was dated to ca 1.64 million years ago (Ma BP) on the basis of palaeomagnetic evidence, just below the top of the OLDUVAI EVENT. The identification of *C. testudo* at this site has been questioned, and hence its value as a major stratigraphic marker; furthermore, the dating of the top of the Olduvai event was revised to 1.81 Ma BP. Following much debate, the boundary has now been formally ratified at 2.58 Ma, the base of the *Gelasian Stage*, within MARINE ISOTOPIC STAGE (MIS) 103. Gradual cooling had already been taking place through the TERTIARY; so it is clear that any position is likely to be arbitrary, but association with a significant SYNCHRONOUS feature, such as a palaeomagnetic reversal, and a better reflection of the change in climate between TERTIARY and QUATERNARY conditions has proved a definition that is supported by Quaternary scientists.   *DH/CJC*

[*See also* GEOLOGICAL TIMESCALE, QUATERNARY TIMESCALE]

Aguirre E and Passini G (1985) The Pliocene-Pleistocene boundary. *Episodes* 8: 116–120.
Gibbard PL, Head MJ, Walker MJC and the Subcommission on Quaternary Stratigraphy (2010) Formal ratification of the Quaternary System/Period and the Pleistocene Series/Epoch with a base at 2.58 Ma. *Journal of Quaternary Science* 25: 96–102.
Jenkins DG (1987) Was the Pliocene-Pleistocene boundary placed at the wrong stratigraphic level? *Quaternary Science Reviews* 6: 41–42.
Suc JP, Bertini A, Leroy SAG and Suballyova D (1997) Towards lowering of the Pliocene-Pleistocene boundary to the Gauss-Matuyama reversal. *Quaternary International* 40: 37–42.

plough layer The upper part of a SOIL, which is disturbed and mixed by being turned over during PLOUGHING.   *JAM*

plough pan/plow pan A compacted layer of soil, caused by repeated tilling of the soil to a constant depth. It is produced by TILLAGE implements such as *tines*, *discs* and *moulboard ploughs*, which tend to smear and compact the soil immediately below their operating depth.   *SHD*

[*See also* PLOUGHING/PLOWING]

Bertolino AVFA, Fernandes NF, Miranda JPL et al. (2010) Effects of plough pan development on surface hydrology and on soil physical properties in southeastern Brazilian plateau. *Journal of Hydrology* 393: 94–104.

ploughing block   A BOULDER that moves downslope pushing soil and vegetation into a mound on its downslope side and leaving a depression or groove upslope. It is a type of SOLIFLUCTION phenomenon.   *PW*

[*See also* GELIFLUCTION]

Ballantyne CK (2001) Measurement and theory of ploughing boulder movement. *Permafrost and Periglacial Processes* 12: 267–288.
Berthling I, Eiken T, Madsen H and Sollid JL (2001) Downslope displacement rates of ploughing boulders in a mid-alpine environment: Finse, southern Norway. *Geografiska Annaler* 83A: 103–116.

ploughing/plowing   In AGRICULTURE, the practice of cutting, turning or breaking the SOIL to prepare it for planting or other agricultural purposes. It can have detrimental effects on the soil, such as reducing its AGGREGATE STABILITY and increasing its ERODIBILITY, leading to accelerated soil or organic matter losses, or the development of PLOUGH PANS. The PLOUGH LAYER is the total depth of soil disturbed and mixed by ploughing. *Deep ploughing* is often used in AFFORESTATION, prior to planting, to break up HARDPANS and/or improve drainage.   *SHD*

[*See also* CONTOUR PLOUGHING, HARROWING, ICEBERG PLOUGH MARKS, SOIL EROSION, TILLAGE]

Martinez-Casasnovas JA and Ramos MC (2009) Soil alteration due to erosion, ploughing and levelling of vineyards in north east Spain. *Soil Use and Management* 25: 183–192.

plume   A natural, elongated structural form, usually resulting from the interaction of gases or fluids, that has the shape of a feather. An *atmospheric plume* can form in the ATMOSPHERE when, for example, smoke or another substance is released as POLLUTION from a point source. A plume can also form when a fluid of one density flows into a fluid of a lower density or along a *density surface* in the OCEANS. *Sediment plumes* form, for example, when glacial MELTWATER containing ROCK FLOUR flows into a freshwater lake. Sediment plumes are also produced when a DENSITY FLOW or particle-laden fluid is emitted from a BLACK SMOKER. A column of high heat flow from the MANTLE, causing partial melting of the ASTHENOSPHERE, is known as a MANTLE PLUME.   *JP*

Vanderploeg HA, Johengen TH, Lavrentyev PJ et al. (2007) Anatomy of the recurrent sediment plume in Lake Michigan and its impact on light climate, nutrients and plankton. *Journal of Geophysical Research:-Oceans* 112: C03590.
Walker ND, Wiseman WJ, Rouse LJ and Babin A (2005) Effects of discharge, wind stress and slope eddies on circulation and the satellite-observed structure of the Missisipi River plume. *Journal of Coastal Research* 21: 1228–1244.

pluridiscipliniarity   Studies using ideas from a number of disciplines but not retaining the methodologies of those disciplines. ENVIRONMENTAL IMPACT ASSESSMENTS (EIAs), for example, are carried out in this way and generally involve the combined efforts of experts and analysts from a wide range of backgrounds working as an integral team to address a specific ENVIRONMENTAL PROBLEM.   *JGS*

[*See also* INTERDISCIPLINARY RESEARCH, MULTIDISCIPLINARY]

O'Riordan T (2000) Environmental science on the move. In O'Riordan T (ed.) *Environmental science in environmental management, 2nd edition.* Harlow: Prentice Hall, 1–28.

pluvial   At low latitudes, a QUATERNARY time interval characterised by relatively high moisture availability. Some arid and semi-arid regions (ARIDLANDS), such as parts of northern Africa, the Mediterranean and south-western USA, experienced pluvials that were synchronous with GLACIAL episodes at high latitudes. However, on account of regional patterns in climatic change related to the GENERAL CIRCULATION OF THE ATMOSPHERE, there was no general time equivalence between glacials and pluvials. Indeed, at the time of the LAST GLACIAL MAXIMUM (LGM), tropical Africa and Australasia experienced 'glacial aridity'.   *JAM*

[*See also* LAKE-LEVEL VARIATIONS, SAPROPEL]

Deuser WG, Ross EH and Waterman LS (1976) Glacial and pluvial periods: Their relationship revealed by Pleistocene sediments of the Red Sea and Gulf of Aden. *Science* 191: 1168–1170.
Fleitmann D, Burn SJ, Pekala M et al. (2011) Holocene and Pleistocene pluvial periods in Yemen, southern Arabia. *Quaternary Science Reviews* 30: 783–787.
Kallel N, Duplessssy JC, Labeyrie L et al. (2000) Mediterranean pluvial periods and sapropel formation over the last 200,000 years. *Palaeogeography, Palaeoclimatology, Palaeoecology* 157: 45–58.

poaching   The TRAMPLING of soil by animals, leading to DISTURBANCE of the vegetation cover and modification (e.g. COMPACTION) of the SOIL STRUCTURE and SOIL DRAINAGE.   *RAS*

Drewry JJ, Cameron KC and Buchan GD (2008) Pasture yield and soil physical property responses to soil comaction from treading and grazing: A review. *Australian Journal of Soil Research* 46: 237–256.

podzolisation   A SOIL-FORMING PROCESS taking place in sandy soils under a moist climate associated with BOREAL FOREST and HEATHLAND. An extremely acid humus form, MOR, is developed on the soil surface. Soluble organic breakdown products from plants percolating down the profile form complex links with IRON (Fe) compounds and disrupt CLAY MINERALS to leave a sandy, bleached, *albic horizon*. Further down the soil profile, accumulation occurs in the *spodic horizon* that becomes enriched in iron, organic matter and aluminium.   *EMB*

[*See also* CHELATION, COMPLEXING, PODZOLS]

Anderson HA, Berrow ML, Farmer VC et al. (1982) A reassessment of podzol formation processes. *Journal of Soil Science* 33: 125–136.
Buurman P and Jongmans AG (2005) Podzolisation and organic matter dynamics. *Geoderma* 125: 71–83.
Davidson DA (1987) Podzols: Changing ideas on their formation. *Geography* 72: 122–128.
Lundstrom US, van Beerman N and Bain D (2000) The podzolosation process: A review. *Geoderma* 94: 91–107.
Stutzer A (1999) Podzolisation as a soil forming process in the alpine belt of Rondane, Norway. *Geoderma* 91: 237–248.
Wilson P (2001) Rate and nature of podzolisation in Aeolian sands in the Falkland Islands, South Atlantic. *Geoderma* 101: 77–86.

Podzols   ACID SOILS with a distinct brownish or blackish subsurface *spodic horizon* (containing illuvial iron-aluminium-organic compounds) lying beneath a bleached *albic horizon* from which iron and aluminium have been leached (SOIL TAXONOMY: *Spodosols*). PODZOLISATION combines at least two specific SOIL-FORMING PROCESSES: (1) CHELATION, the formation of water-soluble metal-humus complexes, which move downwards through the soil, and their subsequent (2) *chilluviation* (accumulation of these chelates in the spodic horizon). Podzols occur typically under BOREAL FOREST and HEATHLAND in northern areas of North America and Eurasia. They may also form elsewhere on suitable PARENT MATERIALS, for example, on deep quartz sands in the tropics, where podzols occur with an albic horizon, many metres deep (so-called *giant podzols*). Visual characteristics of a podzol may develop within a few 100 years and a typical podzol within about 1,000 years under the most favourable conditions but radiocarbon dating of mature Arctic-alpine podzols indicates formation over several thousands of years. Prehistoric FOREST CLEARANCE by humans resulted in the expansion of heathland and the development of podzols during the HOLOCENE. Use for agriculture is limited by acidity, low fertility, ALUMINIUM (Al) toxicity and the low availability of PHOSPHORUS (P), which is fixed by aluminium and iron in the spodic horizon.   *EMB*

[*See also* LEACHING, PODZOLISATION, WORLD REFERENCE BASE FOR SOIL RESOURCES (WRB)]

Buurman P (ed.) (1984) *Podzols*. New York: Van Nostrand Reinhold.
Buurman P and Jongmans AG (2005) Podzolisation and soil organic matter dynamics. *Geoderma* 125: 71–83.
Dimbleby GW (1962) *The development of British heathlands and their soils*. Oxford: Department of Forestry [Forestry Memoir 23].
Ellis S and Matthews JA (1984) Pedogenic implications of a ^{14}C-dated palaeopodzolic soil at Haugabreen, southern Norway. *Arctic and Alpine Research* 16: 77–91.
Lundström US, van Breemen N and Bain N (2000) The podzolization process: A review. *Geoderma* 94: 91–107.
Mokma DL, Yli-Halla M and Linqvist K (2004) Podzol formation in sandy soils of Finland. *Geoderma* 120: 259–272.
Thompson CH (1992) Genesis of podzols on coastal dunes in southern Queensland 1. Field relationships and profile morphology. *Australian Journal of Soil Research* 30: 593–613.

point   An ENTITY in a GEOGRAPHICAL INFORMATION SYSTEM (GIS) of geometric zero dimension. A single pair of CO-ORDINATES and an ATTRIBUTE can spatially represent the location of a point (e.g. a building or a spring). Together with lines and areas, points make up the fundamental units of spatial representation in a GIS.   *VM*

[*See also* ARC]

pointer years   In DENDROCHRONOLOGY, noticeably different TREE RINGS, identifiable in the context of those that precede and follow them. The 'difference' may be defined with respect to a variety of different parameters, for example, the total width of the ring, the presence of some distinct anatomical feature such as resin ducts or a marked change in latewood density. This variety of parameters implies that a wide range of ecological interpretational possibilities is offered by the recording and dating of such phenomena.   *KRB/APP*

[*See also* CROSSDATING, DENDROECOLOGY]

Neuwirth B, Schweingruber FH and Winiger M (2007) Spatial patterns of central European pointer years from 1901 to 1971. *Dendrochronologia* 24: 79–89.
Schweingruber FH (1990) Dendroecological information in pointer years and abrupt growth changes. In Cook ER and Kairiukstis LA (eds) *Methods of dendrochronology: Applications in environmental sciences*. Dordrecht: Kluwer, 277–283.

polar amplification   The concept, which tends to be supported by CLIMATIC MODELS and PALAEODATA (*palaeoclimatic data*), that temperature changes (warming or cooling TRENDS) tend to be exaggerated at relatively high latitudes (see Figure). Polar amplification is stronger in wintertime and over the continents. The main mechanism is the *ice-albedo feedback* (with warmer temperatures, retreat of ice-covered areas reduces the reflectivity and further increases temperatures). Furthermore, with warmer temperatures, CLIMATIC MODELS additionally display a stronger atmospheric and oceanic poleward *heat transport* and a changed *cloud cover* that contributes to enhanced polar warming. However, models are unable to simulate this accurately.   *EZ/JAM*

[***See also*** ALBEDO, ANTARCTIC COOLING AND WARMING, ARCTIC WARMING]

Anderson P, Bermike O, Bigelow N et al. (2006) Last interglacial warmth confirms polar amplification of climate change. *Quaternary Science Reviews* 25: 1383–1400.
Arctic Climate Impact Assessment (ACIA) (2004) *ACIA Scientific Report*. Cambridge: Cambridge University Press.
Mason-Delmotte V, Kageyama M, Braconnot P et al. (2006) Past and future polar amplification of climate change: Climate model intercomparisons and ice-core constraints. *Climate Dynamics* 27: 437–440.
Spielhagen RF, Werner K, Sørensen SA et al. (2011) Enhanced modern heat transfer to the Arctic by warm Atlantic water. *Science* 331: 450–453.

polar desert   Areas where the cold, dry *polar climate* generally prohibits the growth of plants. Some very low-growing herbaceous or dwarf-shrub taxa may occur along sheltered brooksides or where water seeps from snowbeds.   *SPH*

[***See also*** ANTARCTIC, ARCTIC, BIOME, CRYOSOLS, LANDSCAPE, TUNDRA]

Alexandrova VD (1988) *Vegetation of the Soviet polar deserts*. Cambridge: Cambridge University Press.
Priscu JC (ed.) (1998) *Ecosystem dynamics in a polar desert: The McMurdo Dry Valley, Antarctica*. Washington, DC: American Geophysical Union.

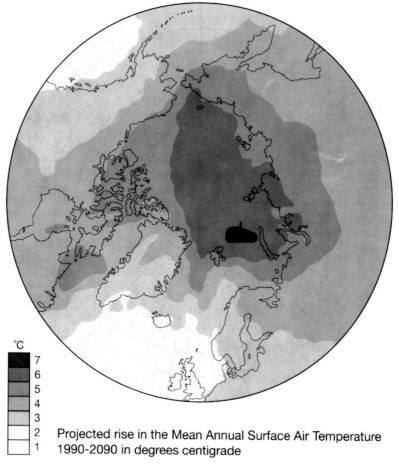

°C

7
6
5
4
3
2
1

Projected rise in the Mean Annual Surface Air Temperature 1990-2090 in degrees centigrade

Polar amplification *Mean annual air temperature projected for northern latitudes between AD 1990 and 2090 according to the Arctic Climate Impact Assessment, based on the average of five climatic models (ACIA, 2004).*

polar front   The narrow zone that separates polar and tropical AIR MASSES. The term originated in 1922 with the publication of the classic Norwegian model of a mid-latitude DEPRESSION or CYCLONE. The *atmospheric polar front* is a narrow zone of strong temperature, potential temperature, HUMIDITY and wind differences, where the warm, light tropical air is forced to rise over the dense, cold polar air. It is clearly seen on surface *synoptic charts*, stretching for thousands of kilometres across the world's oceans between 45° and 60° latitudes, but is usually disrupted over the main continental areas. SATELLITE REMOTE SENSING since the 1960s has shown that polar front can be identified in satellite cloud photographs. It is now closely associated with the polar-front JET STREAM. The OCEANIC POLAR FRONT (OPF) separates cold and warm WATER MASSES in an analogous way.   *BDG*

[*See also* ANTARCTIC FRONT, ARCTIC FRONT, FRONT]

Carlson TN (1991) *Mid-latitude weather systems*. London: Routledge.
Giles BD (1972) A three-dimensional model of a front. *Weather* 27: 352–363.
Machalett B, Oches EA and Frechen M (2008) Aeolian dust dynamics in Central Asia during the Pleistocene: Driven by the long-term magnitude, seasonality, and permanence of the Asiatic polar front. *Geochemistry, Geophysica, Geosystems* 9: Q08Q09.

polar shore erosion   The processes responsible for EROSION along SHORELINES (coastal and lacustrine) in cold-climate environments. In polar regions, these locations may be characterised by extensive coastal ROCK PLATFORMS and backing *cliffs*. Elsewhere along the shores of lakes in PERIGLACIAL environments, rock platforms and BOULDER PAVEMENTS composed of angular boulders are indicative of the efficacy of FROST WEATHERING processes in the absence of appreciable *wave action*. In coastal areas, the occurrence of shore platforms reflects the removal of large volumes of rock from the coastal zone and the transport of debris offshore by waves, currents and SEA ICE. Polar shore platforms are often well developed in sheltered FJORDS and they may also be protected from erosional processes in winter due to the presence of an ICE FOOT that forms at the cliff-platform junction. Polar shore processes may therefore be dominated by the effects of frost and largely unrelated to the effects of waves. In areas affected by GLACIO-ISOSTATIC REBOUND, shore platforms produced by polar shore erosion occur above sea level.   *AGD/JAM*

[*See also* STRANDFLAT]

Aarseth I and Fossen H (2004) Holocene lacustrine rock platform around Storavatnet, Osterøy, western Norway. *The Holocene* 14: 589–596.
Dawson AG (1980) Shore erosion by frost: An example from the Scottish Lateglacial. In Lowe JJ, Gray JM and Robinson JE (eds) *Studies in the Lateglacial of North-West Europe*. Oxford: Pergamon Press, 45–53.
Hinzman LD, Bettez ND, Bolton WR et al. (2005) Evidence and implications of recent climate change in northern Alaska and other Arctic regions. *Climatic Change* 72: 251–298.
Matthews JA, Dawson AG and Shakesby RA (1986) Lake shoreline development, frost weathering and rock platform erosion in an alpine periglacial environment. *Boreas* 15: 33–50.

polarimetry   The measurement of the POLARISATION of electromagnetic RADIATION. It is carried out using a *polarimeter*. Some RADAR REMOTE-SENSING INSTRUMENTS can transmit and receive *microwave radiation* with different polarisations and so are referred to as *polarimetric radars*.   *MEJC*

polarisation   The property of *electromagnetic fields* relating to the plane in which the electric field oscillates. A propagating electromagnetic wave has an electric field vector that oscillates in a direction at right angles to the direction of propagation. The direction of the oscillation plane is called the *polarisation plane* (or polarisation) and is determined by the orientation of the transmitting antenna. In general, polarisation can be decomposed into two orthogonal components that are often chosen to be 'horizontal' and 'vertical' with respect to the direction of propagation. The polarisation plane can also rotate continuously around the propagation direction (circular or elliptical polarisation). The propagation speed of an electromagnetic wave in a medium and the way in which it scatters from objects can depend on the polarisation of the field.   *PJS*

polarity chron   In relation to the GEOMAGNETIC POLARITY TIMESCALE (GPTS), a time interval of dominantly normal or reversed polarity of the Earth's magnetic field, typically lasting approximately 10^5–10^6 years, also known as a *magnetic chron* and formerly as a *magnetic epoch*. The rock unit corresponding to a polarity chron is a MAGNETOZONE. A shorter interval of opposing polarity within a chron is a *subchron*. The four most recent polarity chrons, named after pioneering geomagnetists, are the *Brunhes normal* (to 730,000 years ago), *Matuyama reversed* (to 2.48 million years ago), *Gauss normal* (to 3.40 million years ago) and *Gilbert reversed polarity chron*.   *DNT*

[*See also* GAUSS-MATUYAMA GEOMAGNETIC BOUNDARY, GEOMAGNETISM]

Ogg JG, Ogg G and Gradstein FM (2008) *The concise geologic time scale*. Cambridge: Cambridge University Press.

poliomyelitis   An infectious DISEASE of probable ancient lineage caused by a VIRUS. Until the late nineteenth century it was called *infantile paralysis*. At that time, EPIDEMICS that affected adults began to occur. This change in the EPIDEMIOLOGY of polio was a 'side-effect' which resulted from the generally beneficial public health measures that were taken to prevent diseases such as CHOLERA and *typhoid* fever. The near eradication of polio from a disease of worldwide extent in the AD 1960s has been a triumph of modern medicine, which owes its success to the widespread use of the oral poliovirus vaccine. In 1988, the year of the launch of a global initiative to eliminate the disease, there were >350,000 cases: by the beginning of the twenty-first century, there were <1,000 cases per year. The disease remains ENDEMIC in a number of countries in Africa and South Asia.   *JAM*

Nathanson N and Fine P (2002) Poliomyelitis eradication: A dangerous endgame. *Science* 296: 269–270.
Paul JR (1971) *A history of poliomyelitis*. New Haven, CT: Yale University Press.

political ecology   The interdisciplinary study of the interrelationships between power relationships in society and the NATURAL ENVIRONMENT. All ENVIRONMENTAL ISSUES are inevitably political as power relations in society control access to and use of NATURAL RESOURCES, ENVIRONMENTAL DEGRADATION and CONSERVATION at various levels. There are, for example, political ecologies of GLOBAL WARMING, HUMAN HEALTH HAZARDS, PASTORALISM and WATER SECURITY. Political ecology has its roots in a number of subdisciplines, most notably *political economy* (political and economic DECISION-MAKING), *cultural ecology* (ecology of the cultural LANDSCAPE) and HUMAN ECOLOGY, and is highly critical of *apolitical ecology* and apolitical approaches to environmental research generally. Until the 1990s, political ecology tended to focus on rural and agrarian societies, and DEVELOPING COUNTRIES, where ecological concepts and dynamics were explicitly applied. Since then, it has diversified and followed a more sociopolitical agenda but with a particular focus on the connections between politics and environmental change.   *JAM*

[*See also* CAPITALISM, ENVIRONMENTAL ECONOMICS, ENVIRONMENTAL ETHICS, ENVIRONMENTAL GOVERNANCE, ENVIRONMENTAL POLICY, ENVIRONMENTALISM, ENVIRONMENT-HUMAN INTERACTIONS, GREEN POLITICS]

Baer H and Singer M (2009) *Global warming and the political ecology of health*. Walnut Creek, CA: Left Coast Press.
Kane S (2012) *Where rivers meet the sea: The political ecology of water*. Philadelphia: Temple University Press.
Latour B (2004) *Politics of nature: How to bring the sciences into democracy*. Cambridge, MA: Harvard University Press.
Peet R, Robbins P and Watts M (eds) *Global political ecology*. Abingdon: Routledge.
Robbins P (2011) *Political ecology: A critical introduction, 2nd edition*. Chichester: Wiley.
Salzman PC (2003) *The political ecology of pastoralism*. Boulder, CO: Westview Press.
Walker PA (2005) Political ecology: Where is the ecology? *Progress in Human Geography* 29: 73–82.
Zimmerer K and Bassett TJ (2003) *Political ecology: An integrative approach to geography and environment-development studies*. New York: Guilford Press.

polje   The largest type of *enclosed depression* in KARST landscapes (0.5–500 km^2 in area). Poljes are flat floored, partially structurally controlled and partially formed by DISSOLUTION and they exhibit *subsurface drainage*. They are common in mature karst landscapes: in the Dinaric karst region of Bosnia, Herzegovina, Montenegro and Slovenia, for example, there are over 130 poljes. They are rarely completely hydrologically closed (*closed poljes*), in the sense that surface streams may enter or leave a polje, or both (*open poljes*). The flat floor is often covered in ALLUVIUM, which is often flooded annually, and the sharp angle with the surrounding hillslopes may be accentuated by SPRING SAPPING.   *JAM/SHD*

Gams I (1978) The polje: The problem of definition. *Zeitschrift für Geomorphologie NF* 22: 170–181.
Mijatović N (1984) *Hydrogeology of the Dinaric karst*. Hannover: Heise.
Vott A, Bruckner H, Zander AM et al. (2009) Late Quaternary evolution of Mediterranean poljes: The Vatos case study (Akarnania, NW Greece) based on geoscientific core analyses and IRSL dating. *Zeitschrift für Geomorphologie NF* 53: 145–170.

pollen accumulation rate   Also known as *pollen influx* or *absolute pollen frequency*, pollen accumulation rate is the estimated number of *pollen grains* (and SPORES) accumulated in a sedimentary environment per unit area per unit time. FOSSIL or SUBFOSSIL pollen influx is often estimated as part of palaeoecological studies of past vegetation and environments. In order to estimate pollen accumulation rate, POLLEN CONCENTRATION (the number of grains per unit volume) and ACCUMULATION RATE of the fossil-bearing SEDIMENT (depth of sediment accumulated per unit time) must first be determined:

$$\text{Pollen accumulation rate} = \text{Pollen concentration} \times \text{Sediment accumulation rate}$$

The main technique adopted by palynologists to estimate pollen concentrations is to add an *exotic marker* of known concentration and volume to a known volume of fossil-bearing material. Markers may be an exotic pollen taxon, spores, or pollen-sized polystyrene spheres. Following sample preparation, the markers are counted with the fossil pollen grains and spores.

Estimation of the age-depth relationship (based on dating of the sediments) is required to determine sediment accumulation rates. *JLF*

[*See also* ABSOLUTE COUNTING, PALAEOECOLOGY, PALAEOENVIRONMENTAL RECONSTRUCTION, POLLEN ANALYSIS, POLLEN RAIN]

Faegri K, Kaland PE and Kryzwinski K (2000) *Textbook of pollen analysis, 4th edition.* Caldwell, NJ: Blackburn Press.

Giesecke T and Fontana SL (2008) Revisiting pollen accumulation rates from Swedish lake sediments. *The Holocene* 18: 293–306.

pollen analysis The main research technique used in investigating VEGETATION HISTORY, involving the chemical extraction of pollen (and SPORES) in stratigraphical order from a SEDIMENT or PEAT sequence, and their identification back to the plants and *plant communities* which produced them. The technique is based on the fact that many plants produce large quantities of pollen (grain size: 0.01–0.1 mm), which is widely dispersed, frequently identifiable to genus and sometimes to species, and well preserved under ANAEROBIC conditions (e.g. in peat and LACUSTRINE SEDIMENTS). The technique was first introduced by Lennart von Post in 1916 for investigating climatic history from Swedish MIRES. Pollen analysis is one aspect of the wider research field of palynology (the study of pollen grains), which also includes taxonomic and genetic studies, honey studies, forensic work and AEROBIOLOGY.

Pollen analysis, as a geologically based tool, allows the tracing through time of changes in vegetation, caused by CLIMATE, ANTHROPOGENIC impact and other factors such as SEA-LEVEL CHANGE. However, the exact timing of these changes requires independent dating (e.g. RADIOCARBON DATING or TEPHROCHRONOLOGY) and the *temporal resolution* of the reconstruction depends on the rate at which the deposit accumulated and the sampling interval. The *spatial resolution* depends on the size of the pollen catchment area, which in turn is related to the size of the SEDIMENTARY BASIN. Increased spatial resolution may be achieved by analysis of more than one sequence from the same basin. The pollen assemblage in any single sample has elements which have originated from both local and regional vegetation communities, and sometimes even from quite alien communities through long-distance transportation.

Pollen analysis results are illustrated in *pollen diagrams,* which show the variations of each identified pollen taxon through the sediment profile. The taxa are usually grouped ecologically and variations expressed as percentages of the POLLEN SUM or, where ABSOLUTE COUNTING is undertaken, as pollen concentrations or POLLEN INFLUX. For ease of interpretation, the diagram is divided into POLLEN ASSEMBLAGE ZONES (PAZs), usually on the basis of statistical analysis of the data.

Increasingly, stratigraphically based pollen analysis data sets, which may stretch back to cover several INTERGLACIALS (see Figure), are stored in a pollen DATABASE, where they are available for further analyses, such as making MAPS of vegetation communities at the regional or global scale for specific points in time. TRAINING SETS based on SURFACE POLLEN are also being developed, both to provide MODERN ANALOGUES for interpreting past vegetation more exactly and for the CALIBRATION of pollen assemblages in terms of climate parameters. *SPH/SPD*

[*See also* CHAMBER SAMPLER, CORER, SURFACE SEDIMENT SAMPLER]

Bennett KD and Willis KJ (2001) Pollen. In Smol JP, Birks HJBB and Last WM (eds) *Tracking environmental change using lake sediments, volume 3.* Dordrecht: Kluwer, 5–32.

Berglund BE and Ralska-Jasiewiczowa M (1986) Pollen analysis and pollen diagrams. In Berglund BE (ed.) *Handbook of Holocene palaeoecology and palaeohydrology.* Chichester: Wiley, 455–484.

Broström A, Nielsen AB, Gaillard M-J et al. (2008) Pollen productivity estimates of key European plant taxa for quantitative reconstruction of past vegetation: A review. *Vegetation History and Archaeobotany* 17: 461–478.

Colinvaux PA, Oliveira PE and Morenõ JE (1999) *Amazon pollen manual and atlas.* New York: Harwood Academic Press.

Faegri K and Iversen J (1989) *Textbook of pollen analysis, 5th edition.* Chichester: Wiley.

Gaillard M-J, Sugita S, Bunting J et al. (2008) Human impact on terrestrial ecosystems, pollen calibration and quantitative reconstruction of past land-cover. *Vegetation History and Archaeobotany* 17: 415–418.

Gaillard M-J, Sugita S, Bunting MJ et al. (2008) The use of modelling and simulation approaches in reconstructing past landscapes from fossil pollen data: A review and results from the POLLANDCAL network. *Vegetation History and Archaeobotany* 17: 419–443.

Hooghiemstra H and Van't Veer R (1999) A 0.6 million year pollen record from the Colombian Andes. *PAGES Newsletter* 99: 4–5.

Moore PD, Webb JA and Collinson ME (1991) *Pollen analysis, 2nd edition.* Oxford: Blackwell.

Prentice IC (1985) Pollen representation, source area, and basin size; towards a unified theory of pollen analysis. *Quaternary Research* 23: 76–86.

Traverse A (1988) *Palaeopalynology.* Boston: Allen and Unwin.

Xu Q, Li Y, Bunting MJ et al. (2010) The effects of training set selection on the relationship between pollen assemblages and climate parameters: Implications for reconstructing past climate. *Palaeogeography, Palaeoclimatology, Palaeoecology* 289: 123–133.

pollen assemblage zone (PAZ) A sediment package of more or less homogeneous composition in

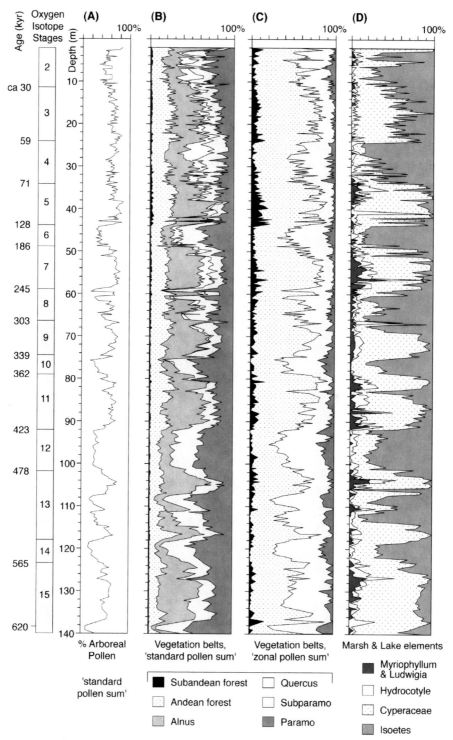

Pollen analysis *Summary pollen diagram from the upper 140 m of the 357–m-long core Funza 1, penetrating lacustrine sediments in the Columbian Andes. The history of the vegetation through many glacial-interglacial cycles is depicted in several different ways to the right of the timescale shown in terms of both thousands of years BP and isotopic stages: (A) tree-pollen percentage, (B) altitudinal vegetation belts based on the standard pollen sum, (C) vegetation belts based on the pollen sum excluding certain insensitive species and (D) marsh and lake species grouped according to characteristic water depth (increasing across the groups from left to right), which in general increases during glacial episodes (Hooghiemstra and Van't Veer, 1999).*

terms of pollen types considered to be characteristic of either the local or the regional sequence of vegetation. It is used in the basic description of sections in a pollen stratigraphy and shown in pollen diagrams.   *BA/CJC*

[*See also* ASSEMBLAGE ZONE, POLLEN ANALYSIS]

Bennett KD (1996) Determination of the number of zones in a biostratigraphical sequence. *New Phytologist* 132: 155–170.
Giesecke T, Bennett KD, Birks HJB et al. (2011) The pace of Holocene vegetation change: Tests for synchronous developments. *Quaternary Science Reviews* 30: 2805–2814.

pollen concentration   A measure of pollen abundance, usually presented as pollen grains per volume (e.g. number of grains in 1 cm³ of sediment), but sometimes per unit of weight. *Pollen concentration diagrams* have the advantage of not creating internal dependence of percentages (closed data), in which a single strong pollen producer or a local pollen source may diminish the apparent significance of the other taxa present. *Pollen concentration values* are very sensitive to sediment changes. If good time control is available, POLLEN ACCUMULATION RATE can be derived.   *BA*

[*See also* POLLEN ANALYSIS, POLLEN ASSEMBLAGE ZONE (PAZ)]

Prentice IC and Webb III T (1986) Pollen percentages, tree abundances and the Fagerlind effect. *Journal of Quaternary Science* 1: 35–44.

pollen rain   Pollen settling from the atmosphere onto a surface, such as a SEDIMENT surface, where it becomes SURFACE POLLEN. Since 'rain' implies vertical movement, which is not always the case, the term *pollen deposition* is preferred.   *SPH*

[*See also* POLLEN ANALYSIS]

Weng C, Bush MB and Silman MR (2004) An analysis of modern pollen rain on an elevation gradient in southern Peru. *Journal of Tropical Ecology* 20: 113–124.

pollen sum   The pollen count from which percentages are calculated when constructing a *pollen diagram*. Taxa included in the sum vary with the question of the investigation.   *SPH*

[*See also* ARBOREAL POLLEN (AP), NON-ARBOREAL POLLEN (NAP), POLLEN ANALYSIS]

pollutant   A substance introduced into a natural system by human agency *and* which impairs the system or harms organisms. A CONTAMINANT becomes a pollutant when there is damage or adverse effects. The adverse effects of a pollutant depend on its chemical composition, its concentration and its persistence in the environment.   *JAM*

[*See also* PERSISTENT ORGANIC COMPOUNDS (POCs), SECONDARY POLLUTANT, TOXIN]

Mansfield TA (ed.) (1976) *Effects of air pollutants on plants.* Cambridge: Cambridge University Press.
Walker CH (2001) *Organic pollutants: An ecotoxicological perspective.* London: Taylor and Francis.

polluter-pays principle (PPP)   A principle of ENVIRONMENTAL PROTECTION, ENVIRONMENTAL POLICY, ENVIRONMENTAL ECONOMICS and ENVIRONMENTAL LAW, that those who pollute the environment should be liable to pay the cost of the damage. It has come increasingly into play since it was recommended by the Organization for Economic Cooperation and Development in the 1970s to encourage polluters to act more responsibly, avoid POLLUTION (see the complementary PRECAUTIONARY PRINCIPLE) or be prepared to pay the full costs involved. The agricultural sector has lagged behind others in adopting the PPP.   *CJB*

O'Connor M (1997) The internationalisation of environmental costs: Implementing the polluter pays principle in the European Union. *International Journal of Environment and Pollution* 7: 450–482.
Organisation for Economic Co-operation and Development (OECD) (1975) *The polluter pays principle: Definition, analysis and implementation.* Paris: OECD.
Tobey JA and Smets H (1996) The polluter pays principle in the context of agriculture and the environment. *World Economy* 19: 63–87.

pollution   The contamination of an ENVIRONMENT to the extent that organisms are harmed or the functioning of the ecosystem is impaired. Some researchers regard natural agencies (e.g. volcanic eruptions) as potential causes of pollution, but normal usage is to restrict the term to the effects of substances resulting from human activities. Here, the distinction is made between ANTHROPOGENIC pollution and NATURAL DISTURBANCE of ecosystems.

Most human activities generate environmental CONTAMINANTS (potential pollutants), either deliberately in WASTE disposal or as ENVIRONMENTAL ACCIDENTS. POLLUTION HISTORY is a long one, perhaps first becoming a global environmental problem during the INDUSTRIAL REVOLUTION. Although the scale, complexity and seriousness of pollution were hardly recognised before the twentieth century, they are now a major issue of GLOBAL ENVIRONMENTAL CHANGE. Pollutants enter the environment from point sources (e.g. sewage outfalls) or diffuse sources (e.g. automobile exhaust) and as particular pollution events (e.g. oil spills in the oceans) or continuously (e.g. agricultural chemicals into rivers and groundwater). Pollution may involve naturally occurring substances increased to harmful or damaging levels (e.g. HEAVY METALS from mining activities and GREENHOUSE GASES from FOSSIL FUELS); a wide range of new, synthetic substances in gaseous, liquid or solid form (e.g. PERSISTENT ORGANIC COMPOUNDS (POCs), including some PESTICIDES); and RADIOACTIVE FALLOUT. Some organisms have been termed *biological pollutants* (e.g. the CHOLERA bacterium and other disease-causing micro-organisms), as have some forms of energy (as in NOISE POLLUTION and THERMAL POLLUTION).

Today, ENVIRONMENTAL PROTECTION agencies enforce or advise on various interpretations of what are acceptable levels of pollution, such as the World Health Organization's guidelines for drinking water. Measures may be introduced in order to encourage pollution reduction, such as the use of the POLLUTER-PAYS PRINCIPLE (PPP), although there are problems with enforcement, especially in international incidents like the Chernobyl NUCLEAR ACCIDENT. This principle, along with pollution QUOTAS—rights to release certain quantities of contaminants—acknowledges the need to minimise pollution, even though it cannot be eliminated because of the ubiquity, complexity and mobility of pollutants, together with economic and political constraints on clean-up.

Integrated pollution control (IPC) recognises the need to control particular pollutants in the broader environmental context, and especially to take account of the movement of pollutants from one *environmental compartment* (e.g. atmosphere, soil and sediment, water or biota) to another. An integrated approach attempts not only *pollution prevention* at source but also optimal *pollution control* of unavoidable pollution, using, for example, the *best available technique not entailing excessive costs* (BATNEEC). This unfortunately does not often coincide with the *best environmental option* (BEO).   *JAM*

[*See also* AIR POLLUTION, DISTURBED ECOSYSTEMS, ECOTOXICOLOGY, MARINE POLLUTION, SOIL POLLUTION, WATER POLLUTION]

Fellenberg G (1999) *The chemistry of pollution.* Chichester: Wiley.

Freedman B (1995) *Environmental ecology: The ecological effects of pollution, disturbance and other stresses.* San Diego, CA: Academic Press.

Gerdes LI (2011) *Pollution: Opposing viewpoints.* San Diego, CA: Greenhaven Press.

Hall MK (2010) *Understanding environmental pollution, 3rd edition.* Cambridge: Cambridge University Press.

Harrison RM (ed.) (2001) *Pollution: Causes, effects and control, 4th edition.* Cambridge: Royal Society of Chemistry.

Markham A (1994) *A brief history of pollution.* London: Earthscan.

Markowitz G and Rosner D (2004) *Deceit and denial: The deadly politics of industrial pollution.* Berkeley: University of California Press.

Newson M (1992) The geography of pollution. In Newson M (ed.) *Managing the human impact on the natural environment: Patterns and processes.* London: Belhaven Press, 14–35.

Visser MJ (2007) *Cold, clear and deadly: Unraveling a toxic legacy.* East Lansing: Michigan State University Press.

pollution adaptation   Genetic change in organisms as a result of environmental POLLUTION. It occurs by changes in GENE POOL composition (*genotypic adaptation*) or in outward physical appearance (*phenotypic adaptation*). Adaptation involving the GENOTYPE takes two forms: (1) the development of internal physiological or structural mechanisms, which enables species to detoxify or resist pollution, for example, pollution-resistant grasslands on the hills around Manchester, UK, and (2) modifications to outward physical appearance. Genetically based adaptation of the PHENOTYPE is displayed by many insects showing *industrial melanism*, such as the darkened melanic variety of the peppered moth (*Biston betularia*), which lives in blackened industrial areas in the UK. The camouflaged darker colouration affords protection against PREDATION by *birds*. Many species are unable to adapt to pollution stress except by avoiding it and growing elsewhere. Thus, highly polluted areas have low BIODIVERSITY and *genetic diversity*. Differential response/adaptation of species to pollution allows the use of INDICATOR SPECIES for ENVIRONMENTAL MONITORING of pollution.   *GOH*

[*See also* BIO-INDICATORS, ECOLOGICAL INDICATOR, ENVIRONMENTAL INDICATOR]

Cook LM and Turner JRG (2008) Decline of melanism in two British moths: Spatial, temporal and inter-specific variation. *Heredity* 101: 483–489.

Durrant CJ, Stevens JR, Hogstrand C and Bury NR (2011) The effect of metal pollution on the population genetic structure of brown trout (*Salmo trutta* L.) residing in the river Hayle, Cornwall, UK. *Environmental Pollution* 159: 3595–3603.

Fritsch C, Coeurdassier M, Gimbert F et al. (2011) Investigations of responses to metal pollution in land snail populations (*Cantareus aspersus* and *Cepaea nemoralis*) from a smelter-impacted area. *Ecotoxicology* 20: 739–759.

Whitehead A, Galvez F, Zhang SJ et al. (2011) Functional genomics of physiological plasticity and local adaptation in killfish. *Journal of Heredity* 102: 499–511.

pollution history   Trends in past POLLUTANT deposition reconstructed from NATURAL ARCHIVES or, for more recent years, measured. Industrial activities such as non-ferrous metal smelting, FOSSIL FUEL combustion and iron and steel manufacture emit a wide range of substances to the environment, and particularly the ATMOSPHERE. These include the HEAVY METALS and industrial PARTICULATES deposited and incorporated into accumulating lacustrine, peat or ice sediments. HEAVY

MINERAL ANALYSIS, MAGNETIC SUSCEPTIBILITY and SPHE-ROIDAL CARBONACEOUS PARTICLE (SCP) analysis of intact sedimentary sequences may then provide evidence of past industrial activity and its EMISSIONS. DOCUMENTARY EVIDENCE may be used to ascribe causes to and dates for the changes in pollutant deposition.

Pollution histories reconstructed from Great Britain and northwest Europe usually exhibit a number of common features. Pre-INDUSTRIAL REVOLUTION episodes of elevated heavy metal that presumably reflect prehistoric, Roman or Mediaeval metalworking are sometimes apparent. The majority of pollutants, however, show marked increases in concentration from the mid- to late nineteenth century, corresponding to the onset of major INDUSTRIALISATION. Concentrations rise rapidly from the turn of the twentieth century until the mid-twentieth century with widespread industrial development. Recent falls in pollutant concentration have been ascribed to declines in heavy industry and EMISSION CONTROLS. Deviations from these general trends have nonetheless been observed and are thought to reflect variations in the nature and timing of regional industrialisation. As well as elucidating trends in past industrial activity, pollution histories provide important evidence of the causes of past *ecological change*. A multiproxy approach to reconstructing pollution histories, in conjunction with DIATOM and POLLEN ANALYSIS, has been particularly useful in assessing the causes of recent lake ACIDIFICATION.

The variable retention and unpredictable behaviour of the pollutants within the sedimentary system limits the confidence that can be placed in the reliability and accuracy of pollution histories reconstructed from natural archives. Chemical, biological and physical transformations may occur after deposition, whilst the atmospheric signal may be obscured by catchment or terrestrial inputs. An appraisal of the nature of sediment accumulation processes, and of the sediments themselves, is therefore necessary for the valid interpretation of pollution histories. *DZR*

[*See also* AIR POLLUTION, CADMIUM (Cd), LEAD (Pb), MERCURY (Hg), WATER POLLUTION]

Battarbee RW, Bennion H, Gell P and Rose N (2012) Human impacts on lacustrine ecosystems. In Matthews JA, Bartlein PJ, Briffa KR et al. (eds) *The SAGE handbook of environmental change, volume 2*. London: Sage, 47–70.
Battarbee RW, Mason J, Renberg I and Talling JF (eds) (1990) *Palaeolimnology and lake acidification*. London: Royal Society.
Livett EA (1988) Geochemical monitoring of atmospheric heavy metal pollution: Theory and applications. *Advances in Ecological Research* 18: 65–177.
Mil-Homens M, Stevens RL, Boer W et al. (2006) Pollution history of heavy metals on the Portuguese shelf: A

^{210}Pb-gochronology. *Science of the Total Environment* 367: 466–480.
Mosley S (2008) *The chimney of the world: A history of smoke pollution in Victorian and Edwardian Manchester*. London: Routledge.
Nriagu JO (1996) A history of global metal pollution. *Science* 272: 223–224.
Renberg I, Brännval ML, Bindle R and Emterid O (2002) Stable lead isotopes and lake sediments: A useful combination for the study of atmospheric lead pollution history. *Science of the Total Environment* 292: 45–54.

polychlorinated biphenyls (PCBs) Synthetic hydrocarbons (ORGANOCHLORIDES) composed of two phenyl rings with substituted chlorine atoms manufactured for use in the electricity and plastics industries. First produced in the 1930s, production and use of these PERSISTENT ORGANIC COMPOUNDS (POCs) have been banned in most industrial countries since the 1970s because of adverse health effects attributed at least in part to impurities such as *polychlorinated dibenzofurans*. The outbreak of the so-called *Yusho disease* in Japan in 1968, which resulted from rice oil used for cooking being contaminated by a PCB that leaked from a heat exchanger, was important in identifying such effects. *JAM*

Dracos TM (2012) *Biocidal: Confronting the poisonous legacy of PCBs*. Boston: Beacon Press.
Hutzinger O, Safe S and Zitko V (1974) *The chemistry of PCBs*. Cleveland, OH: CRC Press.

polygenetic An entity (e.g. as a soil, sediment, sediment body, landform or landscape) is said to be polygenetic if it was formed by a sequence of different environmental conditions and processes, rather than by a single environmental condition or climate. Most entities investigated in the field of ENVIRONMENTAL CHANGE are polygenetic to some extent, having, by definition, experienced varying intensities of the same processes and RATES OF ENVIRONMENTAL CHANGE (if not different processes or complete changes in environmental régime) during the time span of their existence. It has been suggested, for example, that all SOILS are polygenetic and that the older they are, the more polygenetic they become. Hence, it may be more appropriate to refer to the *evolution* rather than the *development* of such polygenetic entities. In terms of LANDFORMS and LANDSCAPES, the larger the landform and the more stable the landscape, the more likely it is to be polygenetic in character: for example, within-channel forms, such as CHANNEL BARS, tend to respond quickly to an environmental change, whereas the landscape of Sierra Leone (see Figure) has developed through numerous climates over geological time. *JAM/RPDW*

[*See also* CLIMATIC GEOMORPHOLOGY, MONOGENETIC, PALIMPSEST]

Johnson DL, Keller EA and Rockwell TK (1990) Dynamic pedogenesis: New views on some key concepts, and a model for interpreting Quaternary soils. *Quaternary Research* 33: 306–319.

Thomas MF (1994) *Geomorphology in the tropics: A study of weathering and denudation in low latitudes.* Chichester: Wiley.

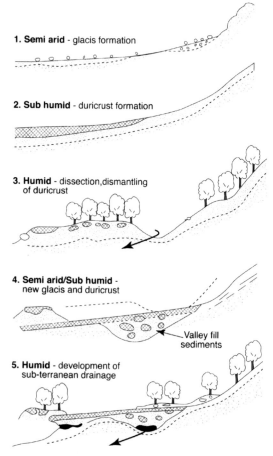

1. Semi arid - glacis formation

2. Sub humid - duricrust formation

3. Humid - dissection,dismantling of duricrust

4. Semi arid/Sub humid - new glacis and duricrust

Valley fill sediments

5. Humid - development of sub-terranean drainage

Polygenetic *Schematic reconstruction of five stages in the Quaternary evolution of the present polygenetic landscape (Stage 5) in Sierra Leone. Two semi-arid phases of GLACIS formation, two subhumid phases of DURICRUST (ferricrete) formation and two humid episodes of dissection are depicted. Note that present-day glacis formation linked to the subterranean drainage channels is not shown (Thomas, 1994).*

polygon analysis   The study and manipulation of areal ENTITIES within a GEOGRAPHICAL INFORMATION SYSTEM (GIS). A polygon is composed of at least three POINTS completely connected by at least three lines (see ARC), along with a LABEL or IDENTIFICATION (ID) tag, and ATTRIBUTE, such as an areal measurement or LANDUSE (agricultural field, forest, urban extent, etc.). Polygons are usually part of a COVERAGE in a GIS and can be interrogated by OVERLAY ANALYSIS or used to generate BUFFER ZONES. Polygon analysis examines the relationships between two or more polygons, and includes a number of operations, including finding whether one polygon is *contained* within (completely surrounded by) another or whether one polygon is *adjacent* (immediately next) to another by determining shared boundaries. Consider multiple polygons representing forest stands: over time the shape, size and location of each polygon will change, and by using polygon analysis, the changes can be summarised, evaluated and displayed using a GIS.   *VM*

[*See also* SPATIAL ANALYSIS]

Dong PL (2008) Generating and updating multiplicatively weighted Voroni diagrams for point, line and polygon features in GIS. *Computers and Geosciences* 4: 411–421.

Longley PA, Goodchild MF, Maguire DJ and Rhind DW (eds) (2010) *Geographic information systems and science, 3rd edition.* Chichester: Wiley.

polymorphism   A condition in which a POPULATION possesses more than one ALLELE at a locus (the location in the DNA occupied by a particular GENE). One common measure of *genetic diversity* is the proportion of loci in a population that are polymorphic.   *MVL*

[*See also* GENOTYPE, HETEROZYGOSITY]

Polynesian colonisation   Descendants of the early peoples who colonised Indonesia, New Guinea and Australia between 40,000 and 100,000 years ago became highly skilled sailors who set out on sophisticated vessels to colonise much of *Micronesia* and *Melanesia* in the western Pacific. Colonisation of the more isolated islands of Polynesia (east of Fiji) was much more recent, and was probably limited to the past 2,000–3,000 years. Yet archaeological evidence indicates that Polynesians eventually colonised the vast majority of these islands, including some of the most isolated ones on Earth. Data from ENVIRONMENTAL ARCHAEOLOGY and PALAEOECOLOGY indicate that these colonisations were far from benign, and often devastated the *native flora* and *native fauna*.   *MVL*

[*See also* ECOLOGICAL COLLAPSE, MIGRATION]

Athens JS (2009) *Rattus exulans* and the catastrophic disappearance of Hawai'i's native lowland forest. *Biological Invasions* 11: 1489–1501.

Kirch PV (2010) Peopling of the Pacific: A holistic anthropological perspective. *Annual Review of Anthropology* 39: 131–148.

Mieth A and Bork HR (2010) Humans, climate or introduced rats: Which is to blame for the woodland destruction on prehistoric Rapa Nui (Easter Island)? *Journal of Archaeological Science* 37: 417–426.

Rieth TM, Hunt TL, Lipo C and Wilmshurst JM (2011) The 13th century Polynesian colonization of Hawai'i Island. *Journal of Archaeological Science* 38: 2740–2749.

polyploidy The state of having three or more sets of homologous *chromosomes*. The norm for *eukaryotic organisms* is two sets (diploid). *Polyploids* with odd numbers of chromosome sets (3, 5, etc.) are usually sterile, but multiplication to even numbers of sets may restore fertility, and may be the basis of some rapid SPECIATION events, especially in plants. Polyploidy is particularly common in Arctic-alpine TUNDRA environments, accounting for ~75 per cent of the FLORA on some islands in the ARCTIC. *KDB*

Soltis DE, Buggs RJA, Doyle JJ and Soltis PS (2010) What we still don't know about polyploids. *Taxon* 59: 1387–1403.

Tate JA, Soltis DE and Soltis PS (2005) Polyploidy in plants. In Gregory TR (ed.) *The evolution of the genome*. San Diego, CA: Elsevier, 371–426.

population (1) In the statistical context, the population is all the individuals of a particular sort: in practice, however, the *sampled population* (the population that is actually sampled) may differ from the *target population* (the population that is aimed at or should be sampled for valid conclusions to be reached in relation to the research problem). (2) In the ecological and human context, it is a group of individual organisms belonging to the same species: *local*, *regional* and *global populations* of species may be recognised. *JAM*

[See also DEMOGRAPHIC CHANGE, METAPOPULATION MODEL]

Livi-Bacci M (2001) *A concise history of world population, 3rd edition*. Oxford: Blackwell.

population cycles Fluctuations in animal POPULATION size. All populations fluctuate due to the relative rates of births, deaths and, in most animals, immigration and emigration. Some fluctuations show PERIODICITIES that exhibit true CYCLICITY, due either to *extrinsic factors* (largely density-independent effects of other species and the environment) or *intrinsic factors* (largely reflecting DENSITY DEPENDENCE on factors that relate to the *life-cycle* and *life-history* strategies of the species). Without environmental constraints, a population would grow at a rate governed by its *intrinsic rate of natural increase* (*r*). However, the CARRYING CAPACITY of the environment limits the population size, and may be exceeded only temporarily. *Time lags* and *overcompensation* in the *density-dependent response* mean that the size of many populations fluctuates in a specific manner (termed *stable limit cycle*, *damped oscillation* and *monotonic return*), with a specific periodicity (typically, cycles of 3–4 and 9–10 years for small and large New World HERBIVORES, respectively). Because even minor and random environmental fluctuations can induce *oscillations* in population size, the impact of widespread and rapid ENVIRONMENTAL CHANGE and/or of climatic DISTURBANCES may disrupt the established pattern of population cycles, adversely affecting *crops*, PESTS, PREDATOR-PREY RELATIONSHIPS and PRIMARY PRODUCTIVITY. *SHJ*

[See also ALGAL BLOOM, FEEDBACK MECHANISMS, FOOD CHAIN/WEB, HOMEOSTASIS, NEGATIVE FEEDBACK, OVERFISHING, OVERHUNTING, POPULATION DYNAMICS, POPULATION EXPLOSION]

Berryman A (ed.) (2002) *Population cycles: The case for trophic interactions*. Oxford: Oxford University Press.

Kausrud KL, Mysterud A, Steen H et al. (2008) Linking climate change to lemming cycles. *Nature* 456: 93–97.

Kendall BE, Eliner SP, McCauley E et al. (2005) Population cycles in the pine looper moth: Dynamical tests of mechanistic hypotheses. *Ecological Monographs* 75: 259–276.

Klemola T, Tanhuanpaa M, Korpimaki E and Ruohomäki K (2002) Specialist and generalist natural enemies as an explanation for geographical gradients in population cycles of northern herbivores. *Oikos* 99: 83–94.

Krebs CJ (2011) Of lemmings and snowshoe hars: The ecology of northern Canada. *Proceedings of the Royal Society B* 278: 481–489.

Rockwood LL (2006) *Introduction to population ecology*. Malden, MA: Blackwell.

population dynamics Variations through time in the POPULATIONS of organisms, particularly in the sizes of the populations. Interplay between temporal and spatial variation is important in understanding population size, which may exhibit DENSITY DEPENDENCE and be affected by numerous ENVIRONMENTAL FACTORS. *JLI*

[See also BIRTH RATE, DEATH RATE, DEMOGRAPHIC CHANGE, POPULATION ECOLOGY]

Kaiser E, Burger J and Schier W (eds) (2012) *Population dynamics in prehistory and early history: New approaches using stable isotopes and genetics*. Berlin: Walter de Gruyter.

Leather SR, Watt AD, Mills NJ and Walters KFA (eds) (1994) *Individuals, populations and patterns in ecology*. Andover: Intercept.

Rhodes Jr OE, Chesser RK and Smith MH (eds) (1999) *Population dynamics in ecological space and time*. Chicago: University of Chicago Press.

Turchin P (2003) *Complex population dynamics: A theoretical/empirical synthesis*. Princeton, NJ: Princeton University Press.

United Nations (2009) *Population dynamics and climate change*. New York: UN Population Fund.

population ecology The study of groups of organisms of the same BIOLOGICAL SPECIES, the interactions within and between POPULATIONS and the relationships between populations and ENVIRONMENTAL FACTORS. *JAM*

[See also AUTECOLOGY, BIOLOGICAL SPECIES]

Gibson DJ (2002) *Methods in comparative plant population ecology*. Oxford: Oxford University Press.

Loreau M (2010) *From populations to ecosystems: Theoretical foundations for a new ecological synthesis.* Princeton, NJ: Princeton University Press.

Ranta E, Lundberg P, and Kaitala V (2006) *Ecology of populations.* Cambridge: Cambridge University Press.

Rockwood LL (2006) *Introduction to population ecology.* Malden, MA: Blackwell.

population explosion The massive and rapid increase in the POPULATION of an organism when the various natural checks on its growth and reproduction, such as limited food supply, COMPETITION or PREDATION, are absent. Population explosions can be localised and short-lived, taking place within the HABITAT of a NATIVE species (e.g. ALGAL BLOOMS in the OCEAN and *lemmings* in the Arctic TUNDRA) or spread temporarily beyond the original habitat (e.g. *locusts* in semi-arid Africa). Delayed DENSITY-DEPENDENCE relationships allow such population explosions to OVERSHOOT the CARRYING CAPACITY of their HABITAT. They are eventually controlled by NEGATIVE FEEDBACK mechanisms or HOMEOSTASIS. Algal blooms, for example, subside because of NUTRIENT shortages, and many animal populations decline once the available food is exhausted. Such explosions are also reversed by heavy predation (e.g. lemming numbers may be reduced by *weevil* attack as well as by poor-quality forage). Ecological explosions can also be widespread and have long-lasting effects following BIOLOGICAL INVASION.   *GOH/JLI*

Greenslade P (2008) Climate variability, biological control and an insect pest outbreak on Australia's Coral Sea islets: Lessons for invertebrate conservation. *Journal of Insect Conservation* 12: 333–342.

Lin YC and Augspurger CK (2008) Long-term spatial dynamics of *Acer saccharum* during a population explosion in an old-growth remnant forest in Illinois. *Forest Ecology and Management* 256: 922–928.

Meshaka WE, Smith HT, Golden E et al. (2007) Green iguanas (*Iguana iguana*): The unintended consequence of sound wildlife management practices in a South Florida park. *Herpetological Conservation and Biology* 2: 149–156.

Wetterer JK and Keularts JLW (2008) Population explosion of the hairy crazy ant, *Paratrechina pubens* (Hymenoptera: Formicidae), on St. Croix, US Virgin Islands. *Florida Entomologist* 91: 423–427.

population growth An increase in the number of individuals of any species over time. When not limited by available RESOURCES, species POPULATIONS would be expected to grow exponentially (see EXPONENTIAL GROWTH). As resources become limiting, this intrinsic rate of increase declines. In the human context, population growth is an aspect of DEMOGRAPHIC CHANGE, and an important driver of environmental change. Unprecedented growth in the world's human population resulted in a doubling between AD 1960 and 2000. The current world population of 7 billion is likely to reach at least 9 billion by AD 2050.   *JAM*

Bloom DE (2011) 7 billion and counting. *Science* 333: 562–569.

Gornall JL, Wiltshire AJ and Betts RA (2012) Anthropogenic drivers of environmental change. In Matthews JA, Bartlein PJ, Briffa KR et al. (eds) *Handbook of environmental change, volume 1.* London: Sage, 517–535.

Lee R (2011) The outlook for population growth. *Science* 333: 569–573.

Sibley RM, Hone J and Clutton-Brock TH (eds) (2003) *Wildlife population growth rates.* Cambridge: Cambridge University Press.

population pyramid A graphical representation of the proportion of a POPULATION in each age class. Population pyramids are widely used in the study of DEMOGRAPHIC CHANGE and differ greatly between developed and DEVELOPING COUNTRIES (see Figure).   *JAM*

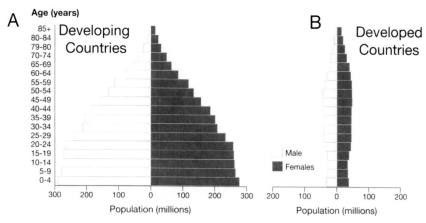

Population pyramid *Contrasting population pyramids for (A) developing and (B) developed countries in AD 2010: In developing countries, the large proportion of young people ensures rapid population growth, whereas the aging population with fewer prospective parents predicts low future growth in developed countries (based on data from the United Nations Population Division).*

population viability analysis (PVA) Techniques to determine the SUSTAINABILITY of a species POPULATION. Given some basic information on key demographic and genetic parameters, population biologists can use a variety of PVA models to estimate the viability of a population, in terms of either the expected period of PERSISTENCE or the PROBABILITY that a population will persist for a given time period. PVA has become a common part of many recovery plans for ENDANGERED SPECIES. *MVL*

[*See also* CONSERVATION BIOLOGY, EFFECTIVE POPULATION SIZE, MINIMUM VIABLE POPULATION]

Bode M and Brennan KEC (2011) Using population viability analysis to guide research and conservation actions for Australia's threatened malleefowl *Leipoa ocellata*. *Oryx* 45: 513–521.

Garcia-Ripolles C and Lopez-Lopez P (2011) Integrating effects of supplementary feeding, poisoning, pollutant ingestion and wind farms of two vulture species in Spain using a population viability analysis. *Journal of Ornithology* 152: 879–888.

Kindall JL, Muller LI, Clark JD et al. (2011) Population viability analysis to identify management priorities for reintroduced elk in the Cumberland Mountains, Tennessee. *Journal of Wildlife Management* 75: 1745–1752.

Tian Y, Wu JG, Smith AT et al. (2011) Population viability of the Siberian tiger in a changing landscape: Going, going and gone? *Ecological Modelling* 222: 3166–3180.

Virillo CB, Martins FR, Tamashiro JY and dos Santos FAM (2011) Is size structure a good measure of future trends of plant populations? An empirical approach using five woody species from the Cerrado (Brazilian savanna). *Acta Botanica Brasilica* 25: 593–600.

pore ice Ice occurring in the pores of soil, sediment and rocks. Pore ice does not include SEGREGATION ICE and, on melting, pore ice does not yield water in excess of the pore volume of the same soil when unfrozen. *HMF*

[*See also* PERMAFROST]

Harris SA, French HM, Heginbottom JA (eds) (1988) *Glossary of permafrost and related ground-ice terms.* Ottawa: National Research Council of Canada.

pore water pressure The pressure of water held within SOIL, SEDIMENT or rock in the interstices or pore spaces between particles or grains. When a soil or rock is fully saturated with water, pore water pressure is positive. The pressure is zero when the soil or rock pores are filled with air. Below the permanent WATER TABLE (see PHREATIC ZONE), the void spaces are saturated or nearly saturated: there the pore water pressure is a function of the density of the water and the hydraulic BOUNDARY CONDITIONS, and is measured using a *piezometer*. In contrast, pressure is negative when the soil or rock pores are part filled with water in the unsaturated or VADOSE ZONE, where pore pressure is determined by *capillarity* (see CAPILLARY ACTION) and pore water pressures are measured with a TENSIOMETER. The build-up of pore water pressure can cause LIQUEFACTION of soils and sediments, and *slope failure*, during EARTHQUAKES. *LJMcE*

[*See also* GROUNDWATER, SLOPE STABILITY]

Harris C, Kern-Luetschg M, Smith F and Isaksen K (2008) Solifluction processes in an area of seasonal ground freezing, Dovrefjell, Norway. *Permafrost and Periglacial Processes* 19: 31–47.

Zhang LL, Zhang J, Zhang LM and Tang WH (2011) Stability analysis of rainfall-induced slope failure: A review. *Proceedings of the Institution of Civil Engineers: Geotechnical Engineering* 164: 299–316.

porous Containing interconnected voids or interstices (*pores*), which may or may not allow the passage of fluids or gases by DIFFUSION. The extent to which a *rock*, SEDIMENT or SOIL is porous (*porosity*) is measured by the ratio of the volume of the void space to the total volume of material plus voids. Porosity should be distinguished from *permeability* (the extent to which liquids or gases can pass through the material): many porous materials (e.g. *chalk* and *sandstone* in relation to water) are PERMEABLE; some may be porous but IMPERMEABLE (e.g. clay and ARGILLACEOUS rocks, in which the pore spaces are too small and water is held firmly by SURFACE TENSION); others are non-porous but may be permeable or PERVIOUS (e.g. fissured *granite*). Hence, useful aquifers are often permeable but only moderately porous. *JAM/RPDW*

[*See also* IMPERVIOUS]

Davis SN (1969) Porosity and permeability of natural materials. In De Wiest RJM (ed.) *Flow through porous media*. New York: Academic Press, 54–89.

Stamp LD (1961) *A glossary of geographical terms*. London: Longman.

positivism A philosophical movement associated particularly with the early-nineteenth-century writings of Auguste Comte, the goal of which was the reorganisation of society and KNOWLEDGE founded on the 'certainties' provided by SCIENCE. Positive science is concerned with observable phenomena and the establishment of causal relations between them through OBSERVATION, EXPERIMENT and comparison (see COMPARATIVE METHOD). Thus, authentic knowledge is that which allows positive VERIFICATION. It acknowledges that our understanding of such phenomena is provisional and imperfect and, while science may come ever closer to TRUTH, it will not lead to absolutist forms of truth. It was explicitly applied by Comte to the study of human affairs as well as the *natural sciences*, and it was further developed as LOGICAL

POSITIVISM and *neo-positivism* in the twentieth century. Many in the HUMANITIES and SOCIAL SCIENCES equate positivism with a scientific approach, especially RESEARCH that employs NUMERICAL ANALYSIS, but this is an oversimplification of both SCIENTIFIC METHOD and its relation to positivism.   *MAU/JAM*

[*See also* CRITICAL RATIONALISM, EMPIRICISM, REALISM]

Gane M (2006) *Auguste Comte*. London: Routledge.
Singer M (2005) *The legacy of positivism*. Basingstoke: Palgrave Macmillan.

possibilism   Developed as a counterpoint to environmental DETERMINISM, possibilism argued that environments offer a range of possibilities or opportunities and that the people who occupied them had choices. Early debate centred on the limits to that choice and with identifying the constraints. At its extreme, possibilism maintained that people had the potential to control environments and that the will of 'man' was a basic factor. This latter form with its proposition that people could completely override the constraints of environments became as untenable as cruder forms of determinism.   *DTH*

[*See also* ENVIRONMENTAL IMPACT ON PEOPLE, ENVIRONMENTALISM, ENVIRONMENT-HUMAN INTERACTIONS]

Mayhew S and Penny A (2009) Possibilism. In *Oxford dictionary of geography*. Oxford: Oxford University Press.
Spate OHK (1957) How determined is possibilism? *Transactions of the Institute of British Geographers* 17: 1–12.
Tatham G (1953) Environmentalism and possibilism. In Taylor G (ed.) *Geography in the twentieth century, 2nd edition*. London: Methuen, 128–162.

postcolonialism   Not merely the investigation of the postcolonial period, postcolonialism is a critical perspective that highlights the misrepresentation of the nature of former colonial peoples and cultures by the colonisers both during the period of COLONIALISM and since. Postcolonialism considers the ways in which knowledge was 'manufactured' and used by European peoples during colonisation; the continuing legacy of the widespread acceptance of 'Western' views about the 'East' and 'South'; the need to investigate and celebrate alternative views and the possibility of corrective action. It has implications for other social groups that are viewed as different and excluded, and may be considered as a dimension of POSTMODERNISM.   *JAM*

[*See also* IMPERIALISM, NEOCOLONIALISM]

Ashcroft W, Griffiths G and Tiffin H (2007) *Post-colonial studies: The key concepts, 2nd edition*. Oxford: Routledge.

Escobar A (1995) *Encountering development: The making and unmaking of the Third World*. Princeton, NJ: Princeton University Press.
Said E (1978) *Orientalism: Western conceptions of the Orient*. New York: Pantheon Books.
Young RJC (2001) *Postcolonialism: An historical introduction*. Oxford: Blackwell.

postenvironmentalism   A concept of ENVIRONMENTAL ETHICS which holds that the ENVIRONMENTAL CRISIS facing humanity at the start of the twenty-first century is qualitatively different from that of the late twentieth century. According to postenvironmentalists, the environmental problems of CLIMATE CHANGE and HABITAT LOSS in particular require greater development of regional and global political structures.   *JAM*

Young J (1992) *Postenvironmentalism*. London: Belhaven.

posthole   A small pit (usually infilled), or marks in the ground, possibly accompanied by the remains of a wooden post (the *post pipe*) and/or packing material. Postholes indicate the former sites of archaeological structures. They are also known as *postmolds*. The term *stakehole* is reserved for smaller features produced by stakes of smaller diameter, which have simply been pushed into the ground without a prepared hole.   *JAM*

Shaw I (1999) Posthole. In Shaw I and Jameson R (eds) *A dictionary of archaeology*. Oxford: Blackwell.

postindustrialisation   The general decline in MANUFACTURING INDUSTRY characteristic of developed countries in the late twentieth century, accompanied by a continuing rise in the proportion of the population employed in the SERVICE INDUSTRIES and an emphasis on information and innovation.   *JAM*

[*See also* DE-INDUSTRIALISATION, INDUSTRIALISATION]

Bell D (1973) *The coming of post-industrial society*. New York: Basic Books.
Florida R (2004) *The rise of the creative class*. New York: Basic Books.
Savitch H (1988) *Post-industrial cities: Politics and planning in New York, Paris and London*. Princeton, NJ: Princeton University Press.

postmediaeval archaeology   The study of the MATERIAL CULTURE of the period between the late MIDDLE AGES and the onset of INDUSTRIALISATION. It was a period of cultural and technological transition.   *JAM*

[*See also* INDUSTRIAL REVOLUTION, SCIENTIFIC REVOLUTION]

Crossley D (1990) *Post-Medieval archaeology in Britain*. Leicester: Leicester University Press.
Gaimster DRM (2002) Post-Medieval archaeology. In Orser Jr CE (ed.) *Encyclopedia of historical archaeology*. London: Routledge, 441–444.

postmodernism An ill-defined concept developed by a wide range of artists, intellectuals and scientists. It is seen as a postindustrial, postcapitalist and post-Fordian cultural logic or world view that has countered MODERNISM since roughly the 1960s. Applied to environmental studies, it implies an adoption of a MULTIDISCIPLINARY, INTERDISCIPLINARY and *holistic* rather than a disciplinary and *reductionist* approach.   *CJB*

Blaikie P (1996) Post-modernism and global environmental change. *Global Environmental Change* 6: 81–85.
Cosgrove D (1990) Environmental thought and action: Pre-modern and post-modern. *Transactions of the Institute of British Geographers,* New Series 15: 344–358.
Gare AE (1995) *Postmodernism and the environmental crisis.* London: Routledge.
Harvey D (1989) *The condition of postmodernity: An enquiry into the origins of cultural change.* Oxford: Basil Blackwell.

postnormal science Recent SCIENCE that recognises the limitations of REDUCTIONISM and the procedures of 'normal' science in understanding COMPLEXITY. Postnormal science confronts limited information and high levels of UNCERTAINTY and is arguably applicable to many of the systems involved in ENVIRONMENTAL CHANGE, particularly those involving ENVIRONMENT-HUMAN INTERACTION.   *JAM*

Funtowicz SO and Ravetz JR (1990) *Uncertainty and quality in science for policy.* Amsterdam: Kluwer.
Ravetz JR (2006) Postnormal science and the complexity of transitions towards sustainability. *Ecological Complexity* 3: 275–284.

post-structuralism A philosophy that emerged during the 1960s as a reaction against and further development of STRUCTURALISM: it criticised the 'scientific' approach of structuralism, and is closely related to POSTMODERNISM. A leading post-structuralist was the French philosopher Jacques Derrida, who stressed that the meaning of a word arises from past use in other texts and contexts and cannot be understood simply by analysing its formal position within a system of structures. The French philosopher Michel Foucault strongly influenced post-structuralist thought in the 1970s, stressing that communicated knowledge is structured into 'discursive formations', and therefore helps shape society and social institutions. Thus, post-structuralism stresses the significance of linguistic and cultural constructions, and its principal methodology is *deconstruction.*   *ART*

Belsey C (2002) *Poststructuralism: A very short introduction.* Oxford: Oxford University Press.
Doel MA (1999) *Poststructuralist geographies: The diabolical art of spatial science.* Edinburgh: Edinburgh University Press.

Sarup M (1993) *An introductory guide to post-structuralism and post-modernism, 2nd edition.* Harlow: Pearson Education.

potable water Water intended for drinking, cooking or other high-quality uses. Although only 2 L of water is required per day to sustain human life, it has been estimated that each individual requires a minimum of 50 L/day to remain healthy (including water for bathing and washing clothes, etc.). Daily per capita usage of piped water in the United Kingdom has reached about 180 L/day, while in the United States it is about 280 L/day. With 1,500 million of the world's poorest people without access to a source of safe drinking water, this is a major contributory factor to high mortality rates in DEVELOPING COUNTRIES, where, for example, around 5 million children under the age of five years die from WATERBORNE DISEASES each year. Variation in the availability of potable water is shown in the Figure.   *JAM*

[*See also* DISEASE, ENVIRONMENTAL QUALITY, HUMAN HEALTH HAZARDS]

Horan N (1997) Collection, treatment and distribution of potable water. In Brune D, Chapman DV, Gwynne MD and Pacyna JM (eds) *The global environment: Science, technology and management.* Weinheim: VCH, 758–773.
Keller AZ and Wilson HC (1992) *Hazards to drinking water supplies.* New York: Springer.
Millennium Ecosystem Assessment (2005) *Ecosystems and human well-being: Synthesis.* Washington, DC: Island Press.

potassium (K) An abundant alkaline metal CATION with a concentration of around 26 mg/kg in the Earth's CRUST. It is a MACRONUTRIENT for all living organisms. K is the major cation in plants (5–50 g/kg) and is involved in a large number of physiological processes. In SOILS, K is a major nutrient and is the most abundant nutrient, with concentrations ranging from 0.1 to 40 g/kg, with a mean of the order of 14 g/kg. A substantial part of the K in soils occurs as structural K in FELDSPARS and interlayer K in *micaceous minerals,* and is considered to be *non-exchangeable.* A small proportion (less than 1 per cent) of the K in soils is held on *exchange sites* (see CATION EXCHANGE CAPACITY (CEC)) on CLAYS and *organic matter.* A small amount of the K in soils is in the *soil solution.* The release of the fixed K from feldspars is enhanced under acidic conditions and from *interlayer sites* by exchange. The supply of K from these non-exchangeable sites contributes significantly (80–100 per cent of the total) to the K supply provided for plant uptake; consequently, measures of exchangeable K often indicate very poorly the soil's ability to supply K to plants. LEACHING of K in some soils may constitute a significant loss from the system.   *SN*

Kabata-Pendias A (2011) *Trace elements in soils and plants.* Boca Raton, FL: CRC Press.

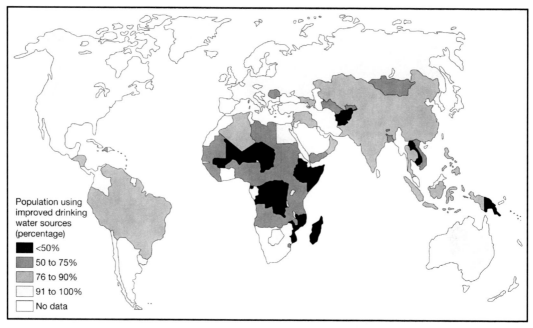

Potable water *Proportion of the human population with access to improved drinking water in 2002, via household connection, public standpipe, borehole, protected dug well and protected spring water or rainwater collection, based on World Health Organization data (Millennium Ecosystem Assessment, 2005).*

potassium-argon (K-Ar) dating A RADIOMETRIC DATING technique based on the decay of ^{40}K to the DAUGHTER ISOTOPE ^{40}Ar, by *electron capture*. Together with ARGON-ARGON (Ar-40/Ar-39) DATING, K-Ar dating is principally used to obtain the ages of potassium-rich minerals in volcanic LAVAS and TUFFS. K-Ar and Ar-40/Ar-39 techniques are mainly applied to geological materials older than 1 million years, because of the long HALF-LIFE of K-40 (1.31×10^9 years), although there have been attempts to look at very young, past-millennium material. However, there have been numerous applications in the field of QUATERNARY ENVIRONMENTAL CHANGE, notably in the determination of the GEOMAGNETIC POLARITY TIMESCALE (GPTS), in ascertaining the age of the early HOMINIDS, in dating TEPHRA layers that are found in MARINE SEDIMENT CORES, ICE CORES and LACUSTRINE SEDIMENTS and in provenancing ICE-RAFTED DEBRIS. The fundamental assumptions of K-Ar and Ar-40/Ar-39 techniques are that no argon was present in the mineral immediately after crystallisation and that the system remained closed since this event.   DAR/CJC

Deino AL, Renne PR and Swasher CC (1998) Ar-40/Ar-39 dating in paleoanthropology and archaeology. *Evolutionary Anthropology* 6: 63–75.

Deino AL, Scott GR, Saylor B et al. (2010) ^{40}Ar/^{39}Ar dating, paleomagnetism and tephrochemistry of Pliocene strata of the hominid-bearing Woranso-Mille area, west-central Afar Rift, Ethiopia. *Journal of Human Evolution* 58: 111–126.

Hemming SR, Vorren TO and Klemen J (2002) Provinciality of ice rafting in the North Atlantic: Application of Ar-40. Ar-39 dating of individual ice rafted hornblende grains. *Quaternary International* 95–96: 75–85.

Leakey LSB, Evernden JA and Curtis GH (1961) Age of Bed 1, Olduvai Gorge, Tanganyika. *Nature* 191: 478–479.

McDougall I (1995) Potassium-argon dating in the Pleistocene. In Rutter NW and Catto NR (eds) *Dating methods for Quaternary deposits.* St John's: Geological Association of Canada, 1–14.

Quidelleur X, Gilot PY, Soler V and Lefevre JC (2001) K/Ar dating extended to the last millennium: Application to the youngest effusive episode of the Teide volcano (Spain). *Geophysical Research Letters* 28: 3067–3070.

Shaeffer OA and Zäringer J (1966) *Potassium argon dating.* New York: Springer.

Walter RC (1997) Potassium-argon/argon-argon dating methods. In Taylor RE and Aitken MJ (eds) *Chronometric dating in archaeology.* New York: Plenum Press, 97–126.

potential evapotranspiration (PET) The maximum combined amount of EVAPORATION and *transpiration* (*water loss* by plants) that can take place from a surface where there is no limitation on water supply. It is a theoretical value for a particular vegetation type, which normally differs from the *actual evapotranspiration* (*AET*), which is the observed amount of EVAPOTRANSPIRATION. Some definitions define PET in terms of an extensive surface of a short green crop (to facilitate comparison), in which case AET can exceed PET for vegetation types with large leaf areas, especially forests. PET may also exceed the *potential evaporation* (PE), which is the amount of evaporation from an open water surface. *JAM/JET*

[*See also* EVAPOTRANSPIRATION]

Bonan GB (2008) *Ecological climatology.* Cambridge: Cambridge University Press.
Morton FI. (1983) Operational estimates of areal evapotranspiration and their significance to the science and practice of hydrology. *Journal of Hydrology* 66: 1–76.

potential natural vegetation The vegetation community that would develop if human activities were removed from an area and were the resulting ECOLOGICAL SUCCESSION completed in an instant. NATURAL VEGETATION occurs where the intactness or integrity of HABITATS and *ecosystems* is free of human influence. In effect, there are few such areas remaining on Earth, but there are many areas which have been called *semi-natural*, that is, somewhat modified by human activity. Restoring or managing vegetation to achieve 'natural vegetation' has been a common objective in CONSERVATION. Plant communities are affected by human activities but also undergo natural change. An issue therefore arises as to how natural vegetation can be defined with reference to different states throughout time, which has led to recent questioning of the validity and usefulness of the potential natural-vegetation concept. *IFS*

[*See also* CLIMAX VEGETATION, COMMUNITY CONCEPTS, NATURAL AREAS CONCEPT, RESTORATION, SEMI-NATURAL VEGETATION, VEGETATION HISTORY, WILDWOOD]

Chiarucci A, Araujo MB, Decocq G et al. (2010) The concept of potential natural vegetation: An epitaph? *Journal of Vegetation Science* 21: 1172–1178.
Hemsing LØ and Bryn A (2012) Three methods for modelling potential natural vegetation (PNV) compared: A methodological case study from south-central Norway. *Norsk Geografisk Tidsskrift* 66: 11–29.
Hickler T, Vohland K, Feehan J et al. (2012) Projecting the future distribution of European potential natural vegetation zones with a generalized, tree species-based dynamic vegetation model. *Global Ecology and Biogeography* 21: 50–63.
Hobbs RJ and Norton DA (1996) Towards a conceptual framework for restoration ecology. *Restoration Ecology* 4: 93–111.
Loidi J, del Arco M, Perez de Paz PL et al. (2010) Understanding properly the 'potential natural vegetation' concept. *Journal of Biogeography* 37: 2209–2211.
Sprugel DG (1991). Disturbance, equilibrium, and environmental variability: What is 'natural' vegetation in a changing environment? *Biological Conservation* 58: 1–18.

potentiation The process by which a substance is made more toxic by the presence of another substance. *JAM*

[*See also* ANTAGONISM, SYNERGISM]

potentiometric surface The level to which water would rise under HYDROSTATIC PRESSURE in a well tapping a confined AQUIFER: it can be viewed as a theoretical water table. The term is sometimes confused with *piezometric surface* (see WATER TABLE). *JAM*

[*See also* ARTESIAN]

Freeze RA and Cherry JA (1979) *Groundwater.* Englewood Cliffs, NJ: Prentice Hall.

potholes Cylindrical, smooth-sided pits in bedrock river channels that vary in size from a few centimetres to several metres in diameter and depth. Potholes are formed by EVORSION, when pebbles and other clasts transported by the river are rotated on the river bed, so eroding the bedrock. Over time, potholes may widen and join with others to form larger potholes. In this way, the whole river bed may be deepened. They are also known from SHORELINE and SUBGLACIAL environments. *LJMcE*

[*See also* CORRASION, P-FORMS]

Springer GS, Tooth S and Wohl EE (2005) Dynamics of pothole growth as defined by field data and geometrical description. *Journal of Geophysical Research: Earth Surface* 110: F04010.
Wang W, Liang M and Huang S (2009) Formation and development of stream potholes in a gorge in Guangdong. *Journal of Geographical Sciences* 19: 118–128.

Preboreal A CHRONOZONE of the early HOLOCENE (ca 11,500–10,000 years ago) characterised by rising but fluctuating temperatures. In former glaciated areas in northern Europe, DEGLACIATION led to the rapid establishment of vegetation and soil development, the expansion of juniper (*Juniperus*) and, in favourable locations, colonisation by birch (*Betula*) and pine (*Pinus*) trees. The PREBOREAL OSCILLATION (PBO) and the *Erdalen Event* are two relatively cold events near the beginning and end of the Preboreal, respectively. *JAM*

[*See also* BLYTT-SERNANDER TIMESCALE, HOLOCENE TIMESCALE]

Bjork S, Rundgren M, Ingolfson O and Funder S (1997) The Preboreal oscillation around the Nordic Seas: Terrestrial and lacustrine responses. *Journal of Quaternary Science* 12: 415–465.
Dahl SO, Nesje A, Lie Ø et al. (2002) Timing, equilibrium-line altitudes and implications of two early-Holocene readvances during the Erdalen Event at Jostedalsbreen, western Norway. *The Holocene* 12: 17–25.

Preboreal Oscillation (PBO) One or more cold events around 11,000 years ago, early in the Preboreal CHRONOZONE, thought to reflect the catastrophic drainage of GLACIAL LAKES into the North Atlantic Ocean

and affecting the CLIMATIC SYSTEM via the THERMOHA-
LINE CIRCULATION. *JAM*

[*See also* CLIMATIC OSCILLATION]

Bos JAA, van Geel B, van der Plight J and Bohncke SJP
(2007) Preboreal climatic oscillations in Europe: Wiggle
match dating of Dutch high-resolution multi-proxy
records. *Quaternary Science Reviews* 26: 1927–1950.
Fisher TG, Smith DG and Andrews JT (2002) Preboreal
oscillation caused by a glacial Lake Agassiz flood.
Quaternary Science Reviews 21: 873–878.

Precambrian An informal name for that part of the
GEOLOGICAL RECORD older than the earliest CAMBRIAN,
characterised by the absence of metazoan FOSSILS with
hard parts. It is divided into two EONS: the ARCHAEAN,
from the formation of the first rocks (~3,900 million
years ago) to 2,500 million years ago, and the PROTERO-
ZOIC, from 2,500 million years ago to the beginning of
the Cambrian period (542 million years ago). The time
before the formation of the first stable crust is known
as the HADEAN (*PRISCOAN*). The Hadean and the Precambrian together account for almost 90 per cent of the
Earth's history. Important events occurred during the
Precambrian relating to the formation and EVOLUTION
of life and the ATMOSPHERE. *GO*

[*See also* GEOLOGICAL TIMESCALE]

Catling DC and Claire MW (2005) How Earth's atmosphere
evolved to an oxic state: A status report. *Earth and
Planetary Science Letters* 237: 1–20.
Eriksson PG, Catancanu O, Sarkar S and Tirsgaard H (2005)
Patterns of sedimentation in the Precambrian. *Sedimentary
Geology* 176: 17–42.
Kah LC and Riding R (2007) Mesoproterozoic carbon
dioxide levels inferred from calcified cyanobacteria.
Geology 35: 799–802.
Kasting JF (2005) Methane and climate during the
Precambrian era. *Precambrian Research* 137: 119–129.
Knauth LP and Kennedy MJ (2009) The Late Precambrian
greening of the Earth. *Nature* 460: 728–732.
Sheldon ND (2006) Precambrian paleosols and atmospheric
CO_2 levels. *Precambrian Research* 147: 148–155.

precautionary principle Where there are threats
of *environmental damage*, the lack of scientific data
should not be used to postpone measures to prevent
ENVIRONMENTAL DEGRADATION. It is essentially a 'better
safe than sorry' approach that has evolved with changing social, economic and political ideas. *IFS*

[*See also* ECOSYSTEM COLLAPSE, GLOBAL WARMING]

Dabelko GD (2009) Planning for climate change: The
security community's precautionary principle. *Climatic
Change* 96: 13–21.
Dolan M and Rowley J (2009) The precautionary principle in
the context of mobile phone and base station radiofrequency
exposure. *Environmental Health Perspectives* 117:
1329–1332.

O'Riordan T and Cameron J (eds) (1994) *Interpreting the
precautionary principle.* London: Earthscan.
O'Riordan R, Cameron J and Jordan A (eds) (2000)
Reinterpreting the precautionary principle. London:
Cameron May.
Whiteside KH (2006) *Precautionary politics: Principle and
practice in confronting environmental risk.* Cambridge:
MIT Press.

precession One of the three main orbital parameters
of the MILANKOVITCH THEORY of GLACIAL-INTERGLACIAL
CYCLES, the 'precession of the equinoxes' relates to the
combined effect of 'wobble' of the Earth about its axis
and 'rotation' of the orbit around the Sun. It results in
the coincidence of PERIHELION (the position on the orbit
when the Earth is closest to the Sun) with the summer
EQUINOX about every 21,000 years; a phenomenon that
last occurred about 11,000 years ago. *JAM*

[*See also* ECCENTRICITY, MILANKOVITCH THEORY,
OBLIQUITY, ORBITAL FORCING]

Khodri M, Cane MA, Kukla G et al. (2005) The impact
of precession changes on the Arctic climate during the
last interglacial-glacial transition. *Earth and Planetary
Science Letters* 236: 285–304.

precipitation (1) A chemical reaction whereby an
insoluble substance (*precipitate*) is formed from a solution. (2) The deposition of water from the ATMOSPHERE in
liquid (*rain*) or solid (*snow, hail*) form. Normally, precipitation is a term reserved in METEOROLOGY for the drops
or solid particles of water that fall to the ground. The
more general term HYDROMETEOR includes other forms of
removal of WATER VAPOUR from the atmosphere (e.g. *dew,
frost* and *rime*). Three types of precipitation are recognised based on the mechanism of cooling inducing CON-
DENSATION: (1) *convectional precipitation* results where
moist air rises following local heating of the Earth's surface (see CONVECTION); (2) *frontal* or *cyclonic precipitation* results from the uplift of moist air at a FRONT and
(3) *orographic precipitation* occurs where moist air is
forced to rise over a *topographic barrier*. *JAM*

[*See also* ACID RAIN, CONDENSATION, OROGRAPHIC]

Elbert J, Grosjean M, von Gunten L et al. (2012) Quantitative
high-resolution winter (JJA) precipitation reconstruction
from varved sediments at Lago Plomo 47° S, Patagonian
Andes, AD 1530-2002. *The Holocene* 22: 465–474.
Min S-K, Zhang X, Zwiers FW and Hegerl GC (2011)
Human contribution to more-intense precipitation
extremes. *Nature* 470: 378–381.
Rasmussen LA, Andreassen LM, Baumann S and Conway
H (2010) 'Little Ice Age' precipitation in Jotunheimen,
southern Norway. *The Holocene* 20: 1039–1045.
Strangeways I (2011) *Precipitation: Theory, measurement
and distribution.* Cambridge: Cambridge University Press.
Sumner G (1988) *Precipitation: Process and analysis.*
Chichester: Wiley.

precipitation variations Precipitation exhibits greater spatial and temporal variability than temperature. There was a small overall increase in precipitation in the high latitudes of the Northern Hemisphere during the twentieth century and a decrease over low latitudes (notably the SAHEL since the 1960s). Whilst winter records may have been underestimates due to inefficient snow-catch, widespread evidence of wetter winters extends from the high to the mid-latitudes. Late-twentieth-century increases in *winter precipitation* have been unprecedented over recent centuries in locations such as Scotland and Norway (see Figure). Conversely, *summer precipitation* has declined over much of Europe and some other mid-latitude regions. These seasonal changes are consistent with CLIMATIC SCENARIOS for the twenty-first century that incorporate a slight poleward movement of *atmospheric pressure* zones. The larger number of high daily rainfalls evident in some recent records (notably in North America) is also replicated in scenarios of mid-twenty-first-century climate. Longer-term variations in precipitation form an important aspect of longer-term CLIMATIC CHANGE and PALAEOCLIMATOLOGY.   *JCM/EH*

[*See also* DROUGHT, LAKE-LEVEL VARIATIONS, PLUVIAL]

Førland EJ, van Engelen A, Hansen-Bauer et al. (1996) *Changes in 'normal' precipitation in the North Atlantic region.* Oslo: Det Norske Meteorologiske Institutt.

Hennessy KJ, Gregory JM and Mitchell JFB (1997) Changes in daily precipitation under enhanced greenhouse conditions. *Climate Dynamics* 13: 667–680.

Jones P, Conway D and Briffa K (1997) Precipitation variability and drought. In Hulme M and Barrow E (eds) *Climates of the British Isles: Present, past and future.* London: Routledge, 197–219.

Karl TR and Knight RW (1998) Secular trends of precipitation amount, frequency and intensity in the United States. *Bulletin of the American Meteorological Society* 79: 231–241.

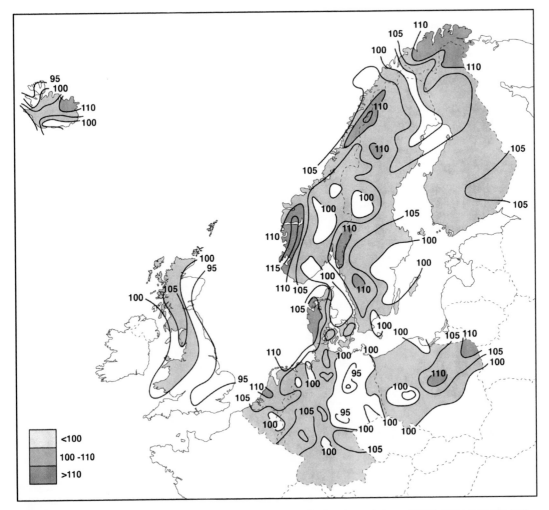

Precipitation variations *Percentage change in annual precipitation over northwestern Europe, between AD 1931–1960 and AD 1961–1990 (Førland et al., 1996).*

Nesje A, Matthews JA, Dahl S-O et al. (2001) Holocene glacier variations of Flatebreen and winter-precipitation changes in the Jostedalsbreen region, western Norway, based on glaciolacustrine sediment records. *The Holocene* 11: 267–280.

Trenbirth KE, Jones PD, Ambenje P et al. (2007) Observations: Surface and atmospheric climate change. In Solomon S, Qin D, Manning M et al. (eds) *Climate change 2007: The physical science basis* [Contribution of Working Group 1 to the Fourth Assessment Report of the Intergovernmental Panel on Climate Change]. Cambridge: Cambridge University Press, 235–336.

Verschuren D, Laird KR and Cumming BF (2000) Rainfall and drought in equatorial east Africa during the past 1,100 years. *Nature* 403: 410–414.

precision   The degree of mutual agreement among individual measurements as a result of repeated determination under the same conditions. Precision defines potential error regardless of ACCURACY (see the Figure in the Accuracy entry) and may be defined in terms of statistical UNCERTAINTY associated with MEASUREMENT and expressed in *standard deviations* away from the measured value. A technique may produce very precise results but they need not be an accurate representation of the real phenomenon under study, as in the case, for instance, of DATING TECHNIQUES. The closer values are to one another and the finer the resolution of measurement, the more precise measures or estimates become.   *DAR/MAU*

predation   The killing of one species (*predator*) by another (*prey* or victim), often but not always for food.
                                                                JAM

[***See also*** FOOD CHAIN/WEB, PREDATOR-PREY RELATIONSHIPS]

Barbosa P and Castellanos I (eds) (2005) *Ecology of predator-prey interactions*. Oxford: Oxford University Press.
Jedrzejewska B and Jedrzejewski W (2010) *Predation in vertebrate communities: The Bialowieza primeval forest as a case study*. Berlin: Springer.

predator-prey relationships   The interactions between prey and predators, particularly in relation to POPULATION DYNAMICS and COEVOLUTION. Such relationships illustrate *mutual dependence* and *cyclical phenomena*. *JLI*

[***See also*** BIOLOGICAL CONTROL, CYCLICITY, MUTUALISM, POPULATION CYCLES, PREDATION]

Barbosa P and Castellanos I (eds) (2005) *Ecology of predator-prey interactions*. Oxford: Oxford University Press.
Conover MR (2007) *Predator-prey dynamics: The role of olfaction*. Boca Raton, FL: CRC Press.
Terborgh J and Estes JA (eds) (2010) *Trophic cascades: Predators, prey, and the changing dynamics of nature*. Washington, DC: Island Press.

prediction   The casting of results into the future. In SCIENCE, predictions are based on THEORY about what is likely to happen under specified conditions, usually at a specified level of UNCERTAINTY. Theory that is valid produces accurate predictions. Although the ability to predict with ACCURACY is the hallmark of a good MODEL, the goodness of a prediction scheme depends also on the user's needs: for instance, a *conservative model* that generally is less accurate may be still preferable if it better warns of the future occurrence of damaging or catastrophic events.   *JAM/EZ*

[***See also*** ABDUCTION, FORECASTING, PROJECTION, RETRODICTION, VALIDATION]

Beven K (2009) *Environmental modeling: An uncertain future? An introduction to techniques for uncertainty estimation in environmental prediction*. Abingdon: Routledge.

prehistoric   Relating to the period before written records.   *SPD*

[***See also*** THREE-AGE SYSTEM]

prescribed fire   Low-intensity fire used intentionally to control or destroy VEGETATION, reduce a build-up of potential fuel, add NUTRIENTS to soil, or control PESTS or PLANT DISEASES. Globally, the area being burnt in this way is estimated to be only one fifth of what it was in the late fifteenth century. In fire-prone areas, it is generally carried out in winter months, under damp, cool conditions. Temperatures at the soil surface are considerably lower than those of WILDFIRES. Amounts of CARBON DIOXIDE released into the ATMOSPHERE by prescribed fire are thought to be negligible. This *land management technique* is usually performed by burning successive parallel stripes of vegetation along the contours, from the top to the bottom of slopes. Prescribed fire is also used for smoke generation in the management of MICROCLIMATE (especially in fruit growing).   *AJDF*

[***See also*** CONTROLLED FIRE, FIRE FREQUENCY, FIRE HISTORY]

Elliott KJ, Hendrick RL, Major AE et al. (1999) Vegetation dynamics after a prescribed fire in the southern Appalachians. *Forest Ecology and Management* 114: 199–213.
Garski CJ and Farnsworth A (2000) Fire weather and smoke management. In Whiteman CD (ed.) *Mountain meteorology: Fundamentals and applications*. New York: Oxford University Press, 237–272.
Smith HG, Sheridan GJ, Lane PNJ and Sherwin CB (2010) Paired Eucalyptus forest catchment study of prescribed fire effects on suspended sediment and nutrient exports in south-eastern Australia. *International Journal of Wildland Fire* 19: 624–636.

presentism   The undesirable practice of distorting the interpretation of historical events as a result of the inappropriate application of present knowledge and understanding. Presentism or *Whiggism* is a complex historiographic issue, especially important in the history of science. It includes, for example, criticising earlier scientists for not knowing what scientists know today, assuming earlier interpretations were invalid merely because they differ from present interpretations, omitting past scientists or their interpretation because these do not fit preconceptions, and assuming current scientific views will always be valid. *JAM*

Good GA (1998) Presentism. In Good GA (ed.) *Sciences of the Earth: An encyclopedia of events, people and phenomena.* New York: Garland Publishing, 708–709.
Hull DL (1979) In defence of presentism. *History and Theory* 18: 1–15.

pressure group An official or unofficial organisation that aims to stimulate public and political concern over any isssue. A wide variety of pressure groups focus on ENVIRONMENTAL ISSUES. Their status ranges from highly influential NON-GOVERNMENTAL ORGANISATIONS (NGOs), such as *Greenpeace* and *Friends of the Earth*, to less organised, radical movements that will pursue their objectives by any means necessary, including illegal activities. *JGS*

[*See also* ENVIRONMENTAL DISOBEDIENCE, ENVIRONMENTAL JUSTICE, ENVIRONMENTAL MOVEMENT, GREEN POLITICS, GREENING OF SOCIETY]

Watts D (2007) *Pressure groups.* Edinburgh: Edinburgh University Press

pressure melting point The *melting point* of a solid at a given pressure. In glaciology, it is the temperature at which GLACIER ice melts under pressure. This can occur at temperatures appreciably below 0°C. It has important implications for glacier movement, GLACIAL EROSION and GLACIAL DEPOSITION. *RAS*

[*See also* COLD-BASED GLACIERS, GLACIER THERMAL REGIME, REGELATION]

Wohlleben T, Sharp M and Bush A (2009) Factors influencing the basal temperatures of a High Arctic polythermal glacier. *Annals of Glaciology* 50: 9–16.

pressure release An alternative term for DILATION/DILATATION in the context of the expansion of BEDROCK. *Pressure-release joints* or fractures tend to form parallel to the ground surface in response to UNLOADING. *RAS*

pretreatment Cleaning and removal of extraneous material or CONTAMINANTS from a sample to be analysed, which might otherwise affect the result. For instance, samples for RADIOCARBON DATING should contain carbon only from the organism that is contemporary with the event being dated. Extraneous matter such as *rootlets*, HUMIC SUBSTANCES, etc. must usually be excluded as these will have a younger carbon age. The ideal pretreatment chemically purifies and isolates

the substance being analysed without affecting it, for example, *cellulose extraction* from *wood* removes *lignins* and *resins* as well as material foreign to the wood. For radiocarbon dating of *bone*, the protein *collagen* is extracted, but when the yield is low due to organic decay, the relative amount of contamination by soil *amino acids* and *proteins* is greater, so further purification using techniques such as ULTRAFILTRATION is required. *PQD*

Higham TFG, Jacobi RM and Bronk Ramsey C (2006) AMS dating of ancient bone using ultrafiltration. *Radiocarbon* 48: 179–195.
Hoper ST, McCormac FG, Hogg AG et al. (1998) Evaluation of wood pretreatments on oak and cedar. *Radiocarbon* 40: 45–50.

prevailing wind The wind direction (from where the wind is coming) with the greatest frequency, as opposed to the direction of the strongest (DOMINANT WIND). *JAM*

primary forest A forest that has never been extensively felled for timber or cleared for another LANDUSE (e.g. AGRICULTURE and estate PLANTATIONS). The term is sometimes used to denote a forest perceived as undisturbed by humans. *Secondary forest* results from REGENERATION following FOREST CLEARANCE. *NDB*

[*See also* PRIMARY WOODLAND, TROPICAL RAIN FOREST]

Gibson L, Lee TM, Koh LP et al. (2011) Primary forests are irreplaceable for sustaining tropical biodiversity. *Nature* 478: 378–381.

primary mineral (1) In geology, a mineral formed at the same time as the surrounding rock by crystallisation of MAGMA (see IGNEOUS ROCKS). (2) In soils, a mineral not formed by WEATHERING or PEDOGENESIS. *JAM*

primary pollutant A POLLUTANT discharged directly into the environment from an identifiable source. *JAM*

[*See also* POLLUTION, SECONDARY POLLUTANT]

primary productivity The BIOMASS production rate of AUTOTROPHIC ORGANISMS (*primary producers*). It is expressed as units of energy or as units of dry organic matter per unit area per unit time. *RJH*

[*See also* AUTOTROPHIC ORGANISM, GROSS PRIMARY PRODUCTIVITY (GPP), NET PRIMARY PRODUCTIVITY (NPP), PRODUCTIVITY, SECONDARY PRODUCTIVITY]

Cias Ph, Reichstein M, Viovy N et al. (2005) Europe-wide reduction in primary productivity caused by heat and drought in 2003. *Nature* 437: 529–533.
Roy J, Saugier B and Mooney HA (eds) (2001) *Terrestrial primary productivity.* San Diego, CA: Academic Press.

Scurlock JMO, Johnson K and Olson RJ (2002) Estimating
net primary productivity from grassland biomass dynamics
measurements. *Global Change Biology* 8: 736–753.

primary succession   A sequential change in species composition (or other ecosystem characteristics)
on surfaces with no residual biological legacy. This
definition generally implies succession on a newly
exposed, often dry and nearly *sterile substrate* with little or no organic matter present, low NUTRIENT STATUS,
and no *seed banks* or other *propagules* initially present.
Primary succession can also occur on freshly exposed
surfaces under water (e.g. overturned rocks). Primary
succession can be considered the biological response to
NUDATION, an extreme ALLOGENIC CHANGE. It differs from
SECONDARY SUCCESSION, where the initial disturbance
leaves some SOIL intact. Typical primary SERES follow
disturbances such as VOLCANIC ERUPTIONS, EARTHQUAKES,
LANDSLIDES, GLACIATION, FLOODS and abandoned ARTI-
FICIAL GROUND (e.g. *mine spoil*). However, the degree
of biological legacy forms a continuum between truly
primary seres and truly *secondary seres*. The spatial
variability of primary seres is large, from volcanoes,
glaciers, floods, or roads that can cover thousands of
hectares to single rock outcrops or small landslides.
The rate of recovery in primary succession is generally slower than in secondary succession, but pre-
disturbance plant communities can form in <100 year
in some cases (e.g. on *tropical landslides* where nutrients, light and water are all abundant).

Initial COLONISATION of denuded sites is often by
RUDERALS that disperse long distances. However, adjacent plant communities have the strongest impact on
what establishes, as most DISPERSAL by wind, water or
animals is local. Some sites are only slowly colonised
by *clonal expansion* from the surrounding vegetation.
Ecesis, or establishment on nearly sterile substrates, is
often a gradual and stepwise, although not generally
predictable, process. The particular set of early colonists that establishes the PIONEERS may be determined
from a suite of potential colonisers by chance events.
Site modification and *successional trajectories* are then
strongly influenced by the characteristics of these first
colonists. A second group of species may colonise,
or grow vigorously, only after soils and vegetation
develop. This process is called FACILITATION and may
be most important early in ECOLOGICAL SUCCESSION in
severe environments. In less harsh environments or
later in succession, COMPETITION may be a more critical process than facilitation in driving successional
change. A third possible scenario is minimal interaction among co-occuring plant species in succession,
where the growth rates and life spans of each species
determine the basic pattern of species replacements
(sometimes termed the *initial floristic composition*

model). Primary succession is certainly driven by a
combination of these three factors (competition, facilitation and *life-history attributes*). Tests of the relative
importance of each factor in many primary seres are
under way and will aid in the MANAGEMENT and RESTO-
RATION of disturbed lands. Much less is known about
the primary succession of animals and microbes.   *LRW*

[*See also* CLIMAX VEGETATION, LAND RESTORATION,
SOIL RECLAMATION]

Bråten AT, Flo D, Hågvar S et al. (2012) Primary succession
of surface active beetles and spiders in an alpine glacier
foreland, central south Norway. *Arctic, Antarctic and
Alpine Research* 44: 2–15.
Chapin III FS, Walker LR, Fastie CL and Sharman LC
(1994) Mechanisms of primary succession following
deglaciation at Glacier Bay, Alaska. *Ecological
Monographs* 64: 149–175.
Del Moral R (2009) Increasing deterministic control of
primary succession on Mount St. Helens, Washington.
Journal of Vegetation Science 20: 1145–1154.
Hågvar S (2012) Primary succession in glacier forelands:
How small animals conquer new land around melting
glaciers. In Young SS and Silvern SE (eds) *International
perspectives on global environmental change*. Intech
Open Access Publisher, 151–172 [Available at www.
intechopen.com].
Marleau JN, Jin Y, Bishop JG et al. (2011) A stoichiometric
model of early plant primary succession. *The American
Naturalist* 177: 233–245.
Matthews JA (1992) *The ecology of recently-deglaciated
terrain: A geoecological approach to glacier forelands
and primary succession*. Cambridge: Cambridge
University Press.
Miles J and Walton DH (eds) (1993) *Primary succession on
land*. Oxford: Blackwell.
Robbins JE and Matthews JA (2010) Regional variation in
successional trajectories and rates of vegetation change
in south-central Norway. *Arctic, Antarctic and Alpine
Research* 42: 351–361.
Thornton I (1996) *Krakatau: The destruction and reassembly
of an island ecosystem*. Cambridge, MA: Harvard
University Press.
Vater AE (2012) Insect and arachnid colonization on the
Storbreen glacier foreland, Jotunheimen, Norway:
Persistence of taxa suggests an alternative model of
succession. *The Holocene* 22: 1123–1133.
Walker LR and del Moral R (2003) *Primary succession
and ecosystems rehabilitation*. Cambridge: Cambridge
University Press.
Walker LR, Shiels AB (2013) *Landslide ecology*. Cambridge:
Cambridge University Press.

primary woodland   Woodlands of any type that
originated from *primaeval woods* (woods that established naturally prior to the onset of human interference)
and which occupy sites that have been continuously
wooded. They tend to be targets for CONSERVATION, being
highly valued particularly for their high BIODIVERSITY

and associated *ecosystem* characteristics, such as complex FOOD CHAINS/WEBS. True primary woodland is becoming increasingly scarce.   *IFS*

[*See also* ANCIENT WOODLAND, CLIMAX VEGETATION, NATURAL VEGETATION, PRIMARY FOREST, SECONDARY WOODLANDS]

Kitahara M, Yumoto M and Kobayashi T (2008) Relationship of butterfly diversity with nectar plant species richness in and around Aokigahara primary woodland, Mount Fuji, central Japan. *Biodiversity and Conservation* 17: 2713–2734.

principal components analysis (PCA)   An indirect ORDINATION technique that transforms the original variables in a multivariate data set into new, *composite variables* or axes that are uncorrelated and account for the maximum possible proportion of the VARIANCE in the data. Axes are selected in decreasing order of importance subject to the constraint that they are uncorrelated to all previous axes. It assumes that one or more underlying *latent variables* or *gradients* exist and that there is a linear response of the individual variables to these gradients. If these assumptions are valid, it can provide a useful low-dimensional representation of the original data.   *HJBB*

[*See also* CORRESPONDENCE ANALYSIS (CA), GRADIENT ANALYSIS, PRINCIPAL CO-ORDINATES ANALYSIS, REDUNDANCY ANALYSIS, TEMPORAL CHANGE DETECTION]

Borcard D, Gillet F and Legendre P (2011) *Numerical ecology with R.* New York: Springer.
Jolliffe IT (2002) *Principal components analysis, 2nd edition.* New York: Springer-Verlag.

principal components transform   A mathematical procedure to convert a set of measurements of possibly correlated variables into a set of values of uncorrelated variables, called principal components, using an *orthogonal transformation*. This technique is often used in IMAGE PROCESSING and REMOTE SENSING to remove the frequently encountered problem of *interband correlation* of multi- and hyperspectral data, which leads to *data redundancy*. Principal components transform generates a "new" set of bands (components) in which the information content is concentrated and that are uncorrelated. The principal components are simply linear combinations of the original data values. The first principal component (PC1) contains the greatest variation of the information in a scene. Succeeding components (PC2, PC3, etc.) contain less scene variation. Once derived, these "new" bands are then used in further analyses such as image CLASSIFICATION.   *DBS/GMS*

Jia X and Richards JA (1999) Segmented principal components transformation for efficient hyperspectral remote-sensing image display and classification. *IEEE Transactions on Geoscience and Remote Sensing* 37: 538–542.

Mather PM and Koch M (2011) *Computer processing of remotely-sensed images: An introduction, 4th edition.* Chichester: Wiley-Blackwell.

principal co-ordinates analysis   An ORDINATION technique in which the distances between objects in the ordination plot are maximally correlated with the original dissimilarities or distances between objects. Almost any *dissimilarity measure* can be used. If *Euclidean distance* is used, the technique is PRINCIPAL COMPONENTS ANALYSIS (PCA). It is also known as classical or *metric scaling*.   *HJBB*

Legendre P and Birks HJB (2012) From classical to canonical ordination. In Birks HJB, Lotter AF, Juggins S and Smol JP (eds) *Tracking environmental change using lake sediments, volume 5: Data handling and numerical techniques.* Dordrecht: Springer, 201–248.
Lepš J and Šmilauer P (2003) *Multivariate analysis of ecological data using CANOCO.* Cambridge: Cambridge University Press.

prion   A protein of unusual and misfolded form, which is responsible for the transmission of certain DISEASES, such as *bovine spongiform encephalopathy* or '*mad cow disease*' in cattle and *Creutzfeldt-Jakob disease* in humans. All known prion diseases are currently incurable.   *JAM*

[*See also* ANIMAL DISEASES, EMERGING DISEASE]

Dobson CM (2001) The structural basis of protein folding and its links with human disease. *Philosophical Transactions of the Royal Society of London B* 356: 133–145.

probability   A quantitative description of the chance or likelihood of occurrence of a particular event expressed between 0 and 1. The higher the probability value, the more likely it is that the event will occur. If the probability is 0, the event cannot happen; if the probability is 1, the event will certainly happen. Probability is a particularly important concept in SCIENCE because it is the basis of measures of UNCERTAINTY. *HJBB*

probability distribution   For a continuous *random variable*, it is the curve described by a mathematical formula that specifies, by the areas under the curve, the PROBABILITY that the variable falls within a particular interval (e.g. NORMAL DISTRIBUTION). For a discrete random variable (e.g. counts), it is a mathematical formula that gives the probability of each value of the variable (e.g. *binomial distribution*). It is also known as the *probability density*.   *HJBB*

Gotelli NJ and Ellison AM (2004) *A primer of ecological statistics.* Sutherland, MA: Sinauer.

prod mark   A type of TOOL MARK comprising an asymmetrical gouge caused by an object transported in a strong current becoming temporarily embedded in MUD.   *GO*

production   The quantity of *carbohydrates* (*sugars*) produced by PHOTOSYNTHESIS in green plants and other AUTOTROPHIC ORGANISMS, some of which are used to run metabolic processes and some of which are incorporated in BIOMASS.   *RJH*

[*See also* CARBON ASSIMILATION, NET PRIMARY PRODUCTIVITY (NPP), PRODUCTIVITY]

Clark DA, Brown S, Kicklighter DW et al. (2001) Measuring net primary production in forests: Concepts and field methods. *Ecological Applications* 11: 356–370.
Haberl H, Erb KH, Krausmann F et al. (2007) Quantifying and mapping human appropriation of net primary production in the Earth's terrestrial ecosystems. *Proceeding of the National Academy of Sciences* 281: 237–240.

productivity   The rate at which organisms manufacture new BIOMASS. It is expressed as units of energy or as units of dry organic matter per unit area per unit time (e.g. kJ/ha/year or kg/ha/year). It consists of PRIMARY PRODUCTIVITY (the productivity of *producers*) and SECONDARY PRODUCTIVITY (the productivity of *consumers*). Primary productivity is influenced by ENVIRONMENTAL FACTORS, some of which may be LIMITING FACTORS. Light commonly limits productivity in green plants. Less than 2 per cent of the light falling on the VEGETATION CANOPY of a TROPICAL RAIN FOREST reaches the *forest floor*, and few plants thrive in the deep shade. Water shortage may limit primary productivity where light is not limiting. GROSS PRIMARY PRODUCTIVITY increases with increasing temperature, but so does the RESPIRATION rate. Consequently, increasing temperatures do not necessarily boost NET PRIMARY PRODUCTIVITY (NPP). In the oceans, PHYTOPLANKTON are the powerhouse of primary production. Light, nutrients and the grazing rate of ZOOPLANKTON influence their productivity.   *RJH*

[*See also* ABOVE-GROUND PRODUCTIVITY, BELOW-GROUND PRODUCTIVITY, CARRYING CAPACITY, LIMITING FACTORS]

Connor DJ, Loomis RS and Cassman KG (2011) *Crop ecology: Productivity and management in agricultural systems, 2nd edition.* Cambridge: Cambridge University Press.
Field CB, Behrenfeld MJ, Randerson JT and Falkowski P (1998) Primary production of the biosphere: Integrating terrestrial and oceanic components. *Science* 281: 237–240.
Kanniah KD, Beringer J and Hutley LB (2010) The comparative role of key environmental factors in determining savanna productivity and carbon fluxes: A review, with special reference to northern Australia. *Progress in Physical Geography* 34: 459–490.
Roy J, Saugier B and Mooney HA (eds) (2001) *Terrestrial primary productivity.* San Diego, CA: Academic Press.
Williams PJ leB, Thomas DN and Reynolds CS (eds) (2002) *Phytoplankton productivity: Carbon assimilation in marine and freshwater ecosystems.* Oxford: Blackwell.

proglacial lake   A lake that abuts against the front of a GLACIER. Some are ICE-DAMMED LAKES; some are *moraine-dammed lakes*; some are neither. GLACIAL LAKES that are located close to glaciers and receive MELTWATER downstream of glaciers (*proximal lakes*) are only proglacial if they are in contact with the glacier.   *JAM*

Thomas E and Briner J (2009) Climate of the past millennium inferred from varved proglacial lake sediments on northeast Baffin Island, Arctic Canada. *Journal of Paleolimnology* 41: 209–224.

progradation   The lateral migration of environments as SEDIMENTARY DEPOSITS accumulate, which contrasts with the vertical build-up of deposits known as AGGRADATION. Progradation is particularly characteristic of shallow *subaqueous environments*, where the water level sets a limit on the level to which sediment can accumulate. A classic example is the seaward migration of a DELTA as sediment is deposited, causing a local *regression*. In successions of sediments or SEDIMENTARY ROCKS, the progradation of laterally adjacent environments gives rise to vertically stacked, gradationally bounded FACIES, a concept commonly expressed as WALTHER'S LAW.   *GO*

[*See also* AUTOCYCLIC CHANGE, FACIES SEQUENCE]

Bristow CS and Pucillo K (2006) Quantifying rates of coastal progradation from sediment volume using GPR and OSL: The Holocene fill of Guichen Bay, south-east South Australia. *Sedimentology* 53: 769–788.
Eilertsen RS, Corner GD, Aasheim O and Hansen L (2011) Facies characteristics and architecture related to palaeodepth of Holocene fjord-delta sediments. *Sedimentology* 58: 1784-1809.
Skene KI, Piper DJW, Aksu AE and Syvitski PM (1998) Evaluation of the global oxygen isotope curve as a proxy for Quaternary sea level by modeling of delta progradation. *Journal of Sedimentary Research* 68: 1077–1092.
Thampanya U, Vermaat JE, Sinsakul S and Pinapitukki N (2006) Coastal erosion and mangrove progradation of southern Thailand. *Estuarine, Coastal and Shelf Science* 68: 75–85.

progressive desiccation   The concept, first developed in the early AD 1900s in relation to Central Asia and later applied to southern and West Africa, that there had been a postglacial drying up of the environment. It was assumed that, at a global scale, GLACIATIONS were characterised by wet conditions and ARIDITY had increased since the waning of the last PLEISTOCENE ice sheets. It is now known that glaciations, at a global scale, were characterised generally by aridity rather than by increased wetness, although certain locations did indeed experience higher rainfall.   *RAS*

[*See also* DESERTIFICATION, PLUVIAL]

Goudie AS (1972) *The concept of post-glacial progressive desiccation.* Oxford: University of Oxford, School of Geography [Research Paper No. 4].

Goudie AS (1990) Desert degradation. In Goudie AS (ed.) *Techniques for desert reclamation*. Chichester: Wiley, 1–34.

projection   A PREDICTION of what may happen in the future given certain assumptions. Often a range of projections is present, each representing a different scenario. MODELS are used to produce projections, which enable systems, such as the CLIMATIC SYSTEM, to be understood. ENVIRONMENTAL POLICY may be guided by projections of what is likely to happen if certain actions are taken, such as the adoption of different EMISSIONS scenarios. The Figure shows projections of future atmospheric CARBON DIOXIDE levels, global temperature change and SEA-LEVEL RISE following particular levels of carbon emissions until AD 2010, with zero emissions thereafter.   *JAM*

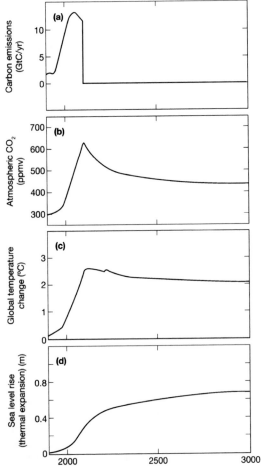

Projection *Various environmental changes (b–d) that would be expected given certain levels of carbon emissions (a). These projections were produced by an Earth-system model of intermediate complexity (EMIC; see INTERMEDIATE-COMPLEXITY MODELLING) forced with an SRES A1B emission scenario (Knutti, 2011).*

[*See also* MAP PROJECTION]

Knutti R (2011) Modelling environmental change and developing future projections. In Matthews JA, Bartlein PJ, Briffa KR et al. (eds) *The SAGE handbook of environmental change*. London: Sage, 116–133.

projective approach   The inferential framework of environmental change that research has changed in recent decades. Frank Oldfield (1993) suggested that these changes involved a sequence of three stages: from inductive, through deductive, to projective modes of *problem definition* and research development. The projective approach involves responding to a research agenda set, not by what has come to light as a result of changes in the past but by what may happen in the future. The way the theme of enhanced GLOBAL WARMING permeates research on environmental change is a prime example. The projective approach has advantages and limitations: the former including greater relevance to human society and the latter including the difficulties, if not dangers, attendant on gearing one's research to uncertain predictions.   *JAM*

[*See also* DEDUCTION, INDUCTION, INTERNATIONALISATION OF RESEARCH]

Oldfield F (1993) Forward to the past: Changing approaches to Quaternary palaeoecology. In Chambers FM (ed.) *Climatic change and human impact on the landscape*. London: Chapman and Hall, 13–21.

pronival rampart   An accumulation of debris forming a ramp or ridge(s) along the downslope margin of a perennial or semi-permanent snowbed located typically near the base of a steep bedrock slope in PERIGLACIAL ENVIRONMENTS. The formerly more common descriptor *protalus rampart* has been shown to be inappropriate. Other terms used for features thought to be formed in the same way as pronival ramparts include *nival ridge, nivation moraine, winter talus ridge, moraine de névé, bourrelet de névé* and *Hangblockwulst*. An assumed formation only by ROCKFALL debris moving over a snowbed has been recognised as inadequate, with both other *supranival processes* (e.g. SNOW AVALANCHES and DEBRIS FLOWS) and *subnival processes* (e.g. SOLIFLUCTION and *snow-push*) also contributing to their formation. RELICT ramparts can be mistaken for other landforms such as MORAINES, ROCK GLACIERS, LANDSLIDES and SNOW AVALANCHE IMPACT LANDFORMS. If correctly identified in RELICT form, they can have a limited though useful role in PALAEOENVIRONMENTAL RECONSTRUCTION, not only by indicating the former presence of long-lived snowbeds and PALAEOWIND direction but also, where there are sufficient in an area, in reflecting any gradient in the SNOW LINE.   *RAS*

[*See also* NIVATION, TALUS]

Ballantyne CK and Stone JO (2009) Rock-slope failure at Baosbheinn, Wester Ross, NW Scotland: Age and interpretation. *Scottish Journal of Geology* 45: 177–181.

Matthews JA, Shakesby RA, Owen G and Vater AE (2011) Pronival rampart formation in relation to snow-avalanche activity and Schmidt-hammer exposure-age dating (SHD): Three case studies from southern Norway. *Geomorphology* 130: 280–288.

Shakesby RA (1997) Pronival (protalus) ramparts: A review of forms, processes, diagnostic criteria and palaeoenvironmental implications. *Progress in Physical Geography* 21: 394–418.

Shakesby RA, Matthews JA, McEwen LJ and Berrisford MS (1999) Snow-push processes in pronival (protalus) rampart formation: Geomorphological evidence from southern Norway. *Geografiska Annaler* 81A: 31–45.

Wilson P (2009) Rockfall talus slopes and associated talus-foot features in the glaciated uplands of Great Britain and Ireland: Periglacial, paraglacial or composite landforms? In Knight J and Harrison S (ed.) *Periglacial and paraglacial processes and environments*. Bath: Geological Society, 133–144.

protected area approach   Conservation of BIODIVERSITY or other aspects of *ecosystems* and LANDSCAPES undertaken by the establishment and MANAGEMENT of individual protected areas (e.g. NATURE RESERVES, BIOSPHERE RESERVES and NATIONAL PARKS) or via networks of protected areas. This *nature-protection approach* to CONSERVATION is increasingly under pressure from the *social-conservation approach*, which is development oriented, with an emphasis on *sustainable use* and the alleviation of POVERTY.   *JFS/JAM*

[*See also* BIOREGIONAL MANAGEMENT, CONSERVATION BIOLOGY, NATURAL AREAS CONCEPT, PROTECTED AREAS, SINGLE LARGE OR SEVERAL SMALL RESERVES (SLOSS DEBATE), SUSTAINABLE DEVELOPMENT]

Miller TR, Minteer B and Milan LC (2011) The new conservation debate: The view from practical ethics. *Biological Conservation* 144: 948–957.

protected areas   Areas of LANDSCAPES that differ greatly in the degree of protection afforded. They range from strictly protected NATIONAL PARKS (American model) and nature RESERVES, which seek to maximise BIODIVERSITY, to *multiple-use management areas* that accommodate the economic and social requirements of society in various ways while providing for SUSTAINABLE DEVELOPMENT. The modern history of protected areas began in the mid-nineteenth century with the development of major *urban parks* in Western cities and the first calls for the preservation of *natural wonders* such as Niagra Falls and Yellowstone. Wildlife and migratory bird *sanctuaries* followed in the late nineteenth and early twentieth centuries. By the mid-twentieth century, protected areas became a global phenomenon and, since the 1960s a wide range of types has developed (see below). Protected area status implies a legislative basis and the existence of a management strategy. There are over 3,000 such protected areas in over 120 different countries. Ten categories of protected areas are recognised by the International Union for Conservation of Nature and Natural Resources (IUCN):

- *Biosphere Reserve:* An area containing unique communities with unusual natural features, harmonious landscapes resulting from traditional landuse and modified or degraded ecosystems that are capable of restoration.

- *Multiple-Use Management Area/Managed Resource Area:* An area designed for sustained production of or access to resources (e.g. water, timber, game, pasture, marine products or outdoor recreation).

- *National Park:* A relatively large natural area containing representative samples of landscapes and where species, habitats and scenery are of special scientific, educational or recreational interest.

- *Natural Biotic Area/Anthropological Reserve:* A natural area where modern humans have not significantly influenced the traditional ways of life of the inhabitants.

- *Natural Monument/Natural Landmark:* One or more relatively small-scale natural features rated as of outstanding national significance because of uniqueness or rarity.

- *Nature Conservation Reserve/Managed Nature Reserve/Wildlife Sanctuary:* Special sites or habitats protected for the continued well-being of resident or migratory fauna of national or global significance.

- *Protected Landscape or Seascape:* A natural or scenic area along coastlines, lakeshores or riverbanks, sometimes adjacent to visitor-use areas or population centres, with potential for development for recreational use.

- *Resource Reserve/Interim Conservation Unit:* Normally a relatively isolated and inaccessible natural area under pressure for colonisation or resource development.

- *Scientific Reserve/Strict Nature Reserve:* A natural area possessing outstanding and/or fragile ecosystems, features or species of at least national scientific importance. *World Heritage Site:* An area of true international significance.

The United Kingdom provides an example of the many types of protected areas that may exist at the national level. Local, national and international legislation provides the basis, mostly under the broad heading of '*nature conservation*', for a wide variety of statutory and non-statutory designations (32 types are listed below). Several government departments (e.g. Department of the Environment, Ministry of Agriculture, Fisheries and Food and Department of National Heritage), conservation agencies (e.g. Countryside Council for Wales: Welsh Historic Monuments, English Heritage, English Nature, Joint Nature Conservation Committee, National Trust and Scottish Natural Heritage) and local authorities hold responsibility for the system:

- *Area of Outstanding Natural Beauty* (*AONB*): In England and Wales, an area for the conservation of natural beauty, where agriculture, forestry and other rural industries are safeguarded and where the economic and social needs of local communities are also taken into account.

- *Area of Special Protection* (*AOSP*): An area for the protection of birds (previously known as Bird Sanctuaries).

- *Area of Special Scientific Interest* (*ASSI*): In Northern Ireland, the equivalent of a SITE OF SPECIAL SCIENTIFIC INTEREST (SSSIs).

- *Biogenic Reserve:* An area for the protection of genetic diversity (part of the Council of Europe programme for a European network).

- *Biosphere Reserve:* See above.

- *Country Park:* An area managed by local authorities for recreation and leisure close to population centres. Country Parks do not necessarily have any nature conservation interest.

- *Environmentally Sensitive Area* (*ESA*): An area of high landscape and conservation value in which farmers may enter into voluntary agreements to pursue special management procedures.

- *Geological Conservation Review Site* (*GCR*): A non-statutory designated area of national or international importance for Earth science conservation (see GEOLOGICAL CONSERVATION (GEOCONSERVATION)). GCRs are also designated as *Sites of Special Scientific Interest* (SSSIs).

- *Heritage Coast:* In England and Wales, a non-statutory designated section of undeveloped coast exceeding 1 mile in length that is of exceptionally fine scenic quality.

- *Limestone Pavement Order* (*LPO*) *land:* Areas of limestone pavement considered of special interest are protected by an LPO.

- *Local Authority Nature Reserve* (*LANR*): In Northern Ireland, the equivalent of a Local Nature Reserve.

- *Local Nature Reserve* (*LNR*): An area of habitat of local importance because of its natural or semi-natural ecosystem.

- *Marine Nature Reserve* (*MNR*): An area of special importance for marine flora and fauna and geological or geomorphological features and their study.

- *Marine Consultation Area* (*MCA*): In Scotland, a non-statutory marine area where the quality and sensitivity of the environment is sufficient to encourage awareness of the conservation issues.

- *National Nature Reserve* (*NNR*): An area nationally or internationally important for its natural or semi-natural ecosystems and habitats.

- *National Park:* In England and Wales, an extensive area of countryside of the highest landscape quality that is preserved and enhanced while being promoted for the enjoyment of the public. There are 10 National Parks.

- *National Scenic Area* (*NSA*): In Scotland, an area of landscape of outstanding natural beauty that is protected in the national interest. NSAs replaced earlier designations equivalent to the AONBs and National Parks of England and Wales.

- *National Trust Property:* Land of scenic value, buildings of historical importance, etc., held for the nation by charities, the National Trust and the National Trust for Scotland.

- *Natural Heritage Area* (*NHA*): In Scotland, areas of countryside of outstanding national heritage value and of wide nature conservation and landscape interest where integrated management will be encouraged (none yet designated).

- *Nature Conservation Review* (*NCR*) *Site*: A non-statutory designated area (although all are or will be designated as SSSIs) of national or international importance for biological conservation

- *Nitrate Sensitive Area* (*NSA*): In England only, an area where there are voluntary incentives to regulate landuse and/or reduce nitrate leaching.

- *Nitrate Vulnerable Zone* (*NVZ*): A compulsory regulated area where agricultural production contributes to degradation of drinking-water quality or eutrophication of aquifers.

- *Preferred Conservation Zone* (*PCZ*): In Scotland, non-statutory coastal areas of national scenic, environmental or ecological importance where new oil- or gas-related developments would be inappropriate and where tourism and recreation take priority.

- *Ramsar Site:* An area designated in response to the Ramsar CONVENTION on Wetlands, to conserve wetlands.

- *Regional Landscape Designation (RLD)*: In Scotland, a non-statutory designation to identify areas where there should be a strong presumption against development because of unexploited potential for tourism and local community benefit.

- *Regionally Important Geological and Geomorphological Sites (RIGS)*: A non-statutory designated site of regional or local importance to Earth science for scientific, educational, historical or aesthetic reasons.

- *Sensitive Marine Area (SMA)*: In England, a non-statutory area important nationally for its marine animal and plant communities and/or which provides ecological support to adjacent statutory sites.

- *Site of Importance for Nature Conservation (SINC)*: An area of local conservation value described in various ways in county and local plans but not notified as an SSSI.

- *Site of Special Scientific Interest (SSSI)*: In England, Scotland and Wales, an exceptional or representative area of nature conservation interest (defined broadly to include biological and geological/geomorphological aspects) according to a published set of quality and rarity criteria.

- *Special Area of Conservation (SAC)*: A site established according to the European Directive on the Conservation of Natural Habitats and of Wild Fauna and Flora (the 'Habitats Directive') as representative of important natural habitat types and as habitats for important species (see Special Protection Area, below).

- *Special Protection Area (SPA)*: A site established under the European Directive on the Conservation of Wild Birds (the 'Birds Directive') to ensure the survival of important bird species. Together with SACs, SPAs form a network of sites termed 'Natura 2000'.

- *World Heritage Site:* An area nominated by the government as a signatory to the UNESCO Convention on the protection of the world's most important areas of cultural and natural heritage. *JAM*

[*See also* CONSERVATION, ECOSYSTEM MANAGEMENT, ENVIRONMENTAL MANAGEMENT, ENVIRONMENTAL PROTECTION, HERITAGE, LANDSCAPE MANAGEMENT, NATIONAL PARKS, REGIONALLY IMPORTANT GEOLOGICAL AND GEOMORPHOLOGICAL SITES (RIGS)]

Aubertin C and Rodary E (2011) *Protected areas, sustainable land.* Aldershot: Ashgate Publishing.
Bushell R and Eagles PFJ (2006) *Tourism and protected areas: Benefits beyond boundaries.* Wallingford: CABI Publishing.
Claudet J (ed.) (2011) *Marine protected areas: A multidisciplinary approach.* Cambridge: Cambridge University Press.
Hanna KS, Clark DA and Slocombe DS (eds) (2008) *Transforming parks and protected areas: Policy and governance in a changing world.* New York: Routledge.
Joint Nature Conservation Committee (1998) *Nature conservation in the UK.* London: Her Majesty's Stationery Office.
Lockwood M, Worboys G and Kothari A (2006) *Managing protected areas: A global guide.* London: Earthscan.
McNeely JA and Miller KR (eds) (1984) *National parks conservation and development.* Washington, DC: Smithsonian Institution Press.
Mose I (ed.) (2007) *Protected areas and regional development in Europe: Towards a new model for the 21st century.* Aldershot: Ashgate Publishing.
Noss RF (1987) Protecting natural areas in fragmented landscapes. *Natural Areas Journal* 7: 2–13.
Sheail J (2010) *Nature's spectacle: The world's first National Parks and protected places.* London: Earthscan.
Wright RG and Mattson DJ (1996) The origin and purpose of National Parks and protected areas. In Wright RG (ed.) *National parks and protected areas: Their role in environmental protection.* Cambridge, MA: Blackwell Science, 3–14.

Proterozoic   An EON of geological time from 2,500 million years ago (Ma) to the beginning of the CAMBRIAN period, divided into three ERAS: (1) *Palaeoproterozoic* (2,500–1,600 Ma), (2) *Mesoproterozoic* (1,600–1,000 Ma) and (3) NEOPROTEROZOIC (1,000–542 Ma). These eras are divided into PERIODS although, in contrast to those in the subsequent PHANEROZOIC eon, the periods are not subdivided using BIOSTRATIGRAPHY, due to the inadequate FOSSIL RECORD. Organisms were simple, unicellular BACTERIA and ALGAE until the Late Neoproterozoic. The ATMOSPHERE became AEROBIC during the Proterozoic, and the first evidence of *metazoan invertebrate* organisms is from the Late Proterozoic. There may have been several ICE AGES, especially in the Neoproterozoic, when all PALAEOLATITUDES appear to have been affected (see SNOWBALL EARTH). The Proterozoic and ARCHAEAN eons make up PRECAMBRIAN time.   *GO*

[*See also* BANDED IRON FORMATION (BIF), GEOLOGICAL TIMESCALE]

Fralick P and Zaniewski K (2012) Sedimentology of a wet, pre-vegetation floodplain assemblage. *Sedimentology* 59: 1030–1049.
Halverson GP, Hoffman PF, Schrag DP et al. (2005) Toward a Neoproterozoic composite carbon-isotope record. *Geological Society of America Bulletin* 117: 1181–1207.
Hawkesworth CJ and Kemp AIS (2006) Evolution of the continental crust. *Nature* 443: 811–817.
Hoffman PF, Kaufman AJ, Halverson GP and Schrag DP (1998) A Neoproterozoic snowball Earth. *Science* 281: 1342–1346.

Kennedy MJ, Runnegar B, Prave AR et al. (1998) Two or four Neoproterozoic glaciations? *Geology* 26: 1059–1063.

Li ZX, Bogdanova SV, Collins AS et al. (2008) Assembly, configuration, and break-up history of Rodinia: A synthesis. *Precambrian Research* 160: 179–210.

Ogg JG, Ogg G and Gradstein FM (2008) *The concise geologic time scale.* Cambridge: Cambridge University Press.

Protista   Single-celled organisms with both plant and animal characteristics, which range in size from VIRUSES (30–300 nm) to *protozoa* (500–50,000 nm). They are sometimes classified as *prokaryotes* (no separation of genetic material from the rest of the cell) and *eukaryotes* (with a nucleus), though some include only eukaryotes and modern classifications, which are still developing, recognise many subdivisions.   *JAM*

Scamardella JM (1999) Not plants or animals: A brief history of the origin of Kingdoms Protozoa, Protista, Protoctista. *International Microbiology* 2: 207–221.

Simonite T (2005) Protists push animals aside in rule revamp. *Nature* 438: 8–9.

protohistory   Relating to the transitional period between *prehistory* and HISTORY, when a CULTURE may have been literate but not yet providing DOCUMENTARY EVIDENCE. Protohistory relates, for example, to the IRON AGE peoples, Celts and Germanic tribes.   *JAM*

protohominid   A primate in a stage of evolution between apes and *Australopithecus*. Protohominids lived ca 8–4 million years ago.   *JAM*

[*See also* HOMINID]

proton-induced X-ray emission (PIXE)   A highly sensitive, non-destructive technique for elemental analysis, also known as *particle-induced X-ray emission*, that analyses X-rays emitted when a proton beam is applied to an object. The method is appropriate for 'mapping' element concentrations in samples at the submicron scale and is widely used in GEO-ARCHAEOLOGY, often in combination with PARTICLE-INDUCED GAMMA-RAY EMISSION (PIGE or PIGME).   *GO*

[*See also* ARCHAEOLOGICAL GEOLOGY, GEOCHEMISTRY]

Agha-Aligol D, Oliaiy P, Mohsenian M et al. (2009) Provenance study of ancient Iranian luster pottery using PIXE multivariate statistical analysis. *Journal of Cultural Heritage* 10: 487–492.

Montero ME, Aspiazu J, Pajon J et al. (2000) PIXE study of Cuban Quaternary paleoclimate geological samples and speleothems. *Applied Radiation and Isotopes* 52: 289–297.

provenance   The nature of the source material or area from which a DETRITAL SEDIMENT, SEDIMENTARY ROCK or archaeological ARTEFACT has been derived.

Provenance can be determined by analysis of the rock types represented in LITHIC fragments in a SANDSTONE or RUDITE; by analysing the relative proportions of QUARTZ, FELDSPAR and lithic fragments in a sandstone or *rudite*; or by tracing the distribution of geochemical or mineral components (e.g. by HEAVY MINERAL ANALYSIS). Provenance studies can yield, for example, important information about the DENUDATION history of continental areas, the TECTONIC setting of a DEPOSITIONAL ENVIRONMENT or CULTURAL CHANGE at archaeological sites.   *GO*

[*See also* PROVENIENCE]

Arribas J, Johnsson MJ and Critelli S (eds) (2007) *Sedimentary provenance and petrogenesis: Perspectives from petrography and geochemistry.* Boulder, CO: Geological Society of America.

Degryse P and Schneider J (2008) Pliny the Elder and Sr-Nd isotopes: Tracing the provenance of raw materials for Roman glass production. *Journal of Archaeological Science* 35: 1993–2000.

Dunajko AC and Bateman MD (2010) Quaternary landscape response to climate change: Sediment provenance of the Wilderness Barrier dunes, southern Cape Coast, South Africa. *Terra Nova* 22: 417–423.

Ridgway KD, Trop JM and Jones DE (1999) Petrology and provenance of the Neogene Usibelli Group and Nenana Gravel: Implications for the denudation history of the central Alaska Range. *Journal of Sedimentary Research* 69: 1262–1275.

Sircombe KN and Freeman MJ (1999) Provenance of detrital zircons on the Western Australia coastline: Implications for the geologic history of the Perth Basin and denudation of the Yilgarn craton. *Geology* 27: 879–882.

Tite MS (2008) Ceramic production, provenance and use: A review. *Archaeometry* 50: 216–231.

Vital H, Stattegger K and Garbe-Schonberg CD (1999) Composition and trace-element geochemistry of detrital clay and heavy-mineral suites of the lowermost Amazon River: A provenance study. *Journal of Sedimentary Research* 69: 563–575.

provenience   The three-dimensional position of an archaeological find within the MATRIX at the time of discovery. *Horizontal provenience* is usually recorded in relation to a geographical grid system; *vertical provenience* is usually related to sea level or, failing this, a local datum.   *JAM*

[*See also* PROVENANCE]

Ashmore W and Sharer RJ (2009) *Discovering our past: A brief introduction to archaeology, 5th edition.* New York: McGraw-Hill.

proximate cause   In a chain or sequence of phenomena, the immediate action or event precipitating some subsequent response. PROXIMATE CAUSES may appear to link events, while the ULTIMATE CAUSE is attributed to more distant factors that are not as conspicuous.

MAU/CET

Geist HJ and Lambin EF (2002) Proximate causes and underlying driving forces of tropical deforestation. *BioScience* 52: 143–150.

proximity principle In the context of WASTE MAN-AGEMENT, the principle that waste should be treated close to where it is generated. Waste disposal at distant LANDFILL sites should therefore be avoided. It has wider applications in relation to SUSTAINABLE CONSUMPTION.

JAM

proxy climatic indicator A non-climatic variable or event that, because it is strongly influenced by a climatic variable, can be used to infer values of that climatic variable. In the absence of climatic records, therefore, a long series of a proxy climatic indicator can potentially be used to reconstruct a parallel series for a climatic variable. Such PROXY-DATA series allow the extension backwards of INSTRUMENTAL-DATA series (which tend to extend back only 100–200 years) and provide not only knowledge of *climatic history* but also the long-term *climatic series* with which both to develop and test more rigorously theories on recent climatic change and the validity of GENERAL CIRCULATION MODELS used in the PREDICTION of FUTURE CLIMATE. Proxy climatic indicators may be divided into *biological indicators* (grain, tree rings, phenological data) and *physical indicators* (isotopes, sediment geochemistry, ice formation, floods) types. Proxy data include both written records of phenomena, such as HARVEST RECORDS, *phenological series* (see PHENOLOGY), *flood-level chronologies* (see FLOOD HISTORY), LAKE-LEVEL VARIATIONS, and the positions, dates of formation and disappearance of SEA ICE, and sedimentological and morphological indicators in the landscape, such as pollen records, geochemical sediment properties and glacial MORAINES.

Thus, proxy indicators provide information not only for the HOLOCENE but over the whole GEOLOGICAL TIMESCALE, and are one of the most valuable investigative tools that climatologists have at their disposal. They have done much in the past three decades to further the understanding of CLIMATIC CHANGE beyond the range of INSTRUMENTAL DATA. Although some proxy records allow HIGH-RESOLUTION RECONSTRUCTIONS on an annual or even subannual basis, others provide a far less detailed picture and can be used only to interpret the broader features of CLIMATIC VARIATION. Studies in HISTORICAL CLIMATOLOGY have benefited from these sources where they have corroborated and extended DOCUMENTARY EVIDENCE of environmental change.

Proxy data can be quantitative or qualitative in nature. Many indicators (particularly those forming ANNUALLY RESOLVED RECORDS) are amenable to computerised statistical procedures which may aid their objective use in generating reliable climatic series. An important step in all studies involving proxy data is to establish the relationship between the proxy parameter and a climatic variable or variables. This usually involves using current data and assuming that current relationships also applied in the past, that is, the principle of ACTUALISM. The most useful indicators are those which correlate unambiguously with a single climatic parameter. Very often, however, indicators respond to more than one climatic parameter. Thus, a lake-level rise may reflect either increased wetness (via increased streamflow) or decreased temperatures (and reduced evapotranspiration), a combination of the two, or even a preponderance of one over a contrary change in the other. Interplay with human activities and human response may make inferences even more difficult, particularly if LANDUSE CHANGES are involved. *Climatic indicators* such as FAMINE and HARVEST RECORDS can be influenced more by non-climatic factors (e.g. *war, economic factors* and improvements in *agricultural technology*) than by DROUGHT and may therefore be very unreliable indicators of climate unless accompanied by more detailed historical information.

One of the most well-known proxy indicators are TREE-RING data from the sciences of DENDROCHRONOLOGY and DENDROCLIMATOLOGY. By this means, some aspects of the annual climatic record (rainfall in some areas, temperature in others) have been extended back to 3000 BC in the southwest USA, where *Bristlecone pines* (*Pinus aristata*) have life spans measured in millennia. LACUSTRINE SEDIMENTS, including VARVE deposits, are another well-established means of determining the general features of rainfall and RUNOFF with annual and sometimes seasonal resolution. In recent years, changes in geochemical, particle size and other properties of floodplain, lake-bottom and deltaic deposits (and their dating) have also yielded valuable data on climatic changes and their impacts on DENUDATION at a wide range of timescales. POLLEN ANALYSIS also provides information on past plant communities as environmental indicators. These techniques are used most commonly to provide information for the period since the most recent retreat of permanent ice cover. For longer time periods, those embracing, for example, the LAST GLACIATION and the preceding INTERGLACIAL, analysis of ICE CORES based on OXYGEN ISOTOPES provide a detailed proxy temperature record over at least the past 160,000 years. These records, from Arctic and Antarctic sites, have helped confirm the MILANKOVITCH THEORY of SOLAR FORCING. MARINE SEDIMENT CORES provide similar information, but the timescale detail is blurred by sediment mixing and only the longer-term climatic changes can be discerned (rarely at higher temporal *resolution* than CENTURY- TO MILLENNIAL-SCALE VARIABILITY). This disadvantage is, however, offset by the length of the record embraced by these slowly accumulating sea-floor deposits.

RG/DAW

[*See also* CLIMATE OF THE LAST MILLENNIUM,
HOLOCENE ENVIRONMENTAL CHANGE, MARINE
PALAEOCLIMATIC PROXIES, MARINE SEDIMENT CORES:
RESULTS, NATURAL ARCHIVES, PALAEOCLIMATOLOGY,
QUATERNARY ENVIRONMENTAL CHANGE]

Adams JB, Mann ME and Ammann CM (2003) Proxy
evidence for an El Niño-like response to volcanic forcing.
Nature 426: 274–278.
Birks HJB (2005) Fifty years of Quaternary pollen analysis
in Fennoscandia, 1954-2004. *Granna* 44: 1–22.
Bradley RS (ed.) (1991) *Global changes of the past.* Boulder,
CO: UCAR/Office for Interdisciplinary Earth Studies.
Chambers FM (2012) Reconstructing and inferring past
environmental change. In Matthews JA, Bartlein
PJ, Briffa KR et al. (eds) *The SAGE handbook of
environmental change, volume 1.* London: Sage, 67–91.
Glaser R (1996) Data and methods of climatological
evaluation in historical climatology. *Historical Social
Research* 21: 56–88.
Lindholm M, Ogurtsov M, Aalto T et al. (2009) A summer
temperature proxy from height increment of Scots pine
since 1561 at the northern timberline in Fennoscandia.
The Holocene 19: 1131–1138.
Schellekens J, Buurman P, Fraga I and Martinez-Cortizas
A (2011) Holocene vegetation and hydrologic changes
inferred from molecular vegetation markers in peat,
Penido Vello (Galicia, Spain). *Palaeogeography,
Palaeoclimatology, Palaeoecology* 299: 56–99.
Van Geel B, Buurman J and Waterbolk HT (1996)
Archaeological and palaeoecological indications of an
abrupt climate change in the Netherlands, and evidence
for climatological teleconnections around 2650 BP.
Journal of Quaternary Research 11: 451–460.
Wahl ER and Frank D (2012) Evidence of environmental
change from annually resolved proxies with particular
reference to dendrochronology and the last millennium.
In Matthews JA, Bartlein PJ, Briffa KR et al. (eds) *The
SAGE handbook of environmental change, volume 1.*
London: Sage, 320–344.
Williams M, Cook E, van der Kaars S et al. (2009)
Glacial and deglacial climatic patterns in Australia and
surrounding regions from 35,000 to 10,000 years ago
reconstructed from terrestrial and near-shore proxy data.
Quaternary Science Reviews 28: 2398–2419.
Woodbridge J and Roberts N (2011) Late Holocene of the
Eastern Mediterranean inferred from diatom analysis of
annual-laminated lake sediments. *Quaternary Science
Reviews* 30: 3381–3392.

**proxy data/proxy evidence/proxy indicators/
proxy records**   Data, evidence, indicators or records
relating to an environmental variable that were derived
indirectly or reconstructed from NATURAL ARCHIVES
(proxy sources) rather than directly by OBSERVATION or
MEASUREMENT.   *JAM*

[*See also* MULTIPROXY APPROACH, ORBITAL TUNING,
PALAEOCLIMATOLOGY, PALAEOLIMNOLOGY,
PALAEOPRECIPITATION PROXIES, PALAEOTEMPERATURE

PROXIES, PROXY CLIMATIC INDICATOR, PSEUDO-PROXY
APPROACH]

Blaauw M (2012) Out of tune: The dangers of aligning proxy
archives. *Quaternary Science Reviews* 36: 38–49.
Blaauw M, Christen JA, Mauquoy D et al. (2007) Testing
the timing of radiocarbon-dated events between proxy
archives. *The Holocene* 17: 283–288.
Chambers FM (2012) Reconstructing and inferring past
environmental change. In Matthews JA, Bartlein
PJ, Briffa KR et al. (eds) *The SAGE handbook of
environmental change, volume 1.* London: Sage, 67–91.
Jones PD, Briffa KR, Osborn TJ et al. (2009) High-
resolution paleoclimatology of the last millennium:
A review of present status and future prospects. *The
Holocene* 19: 3–49.
Oldfield F and Thompson R (2004) Archives and proxies
along the PEP III transect. In Battarbee RW, Gasse F and
Stickley CE (eds) *Past climate variability through Europe
and Africa.* Dordrecht: Springer, 7–29.

pseudokarst   Karst-like landforms produced by
processes other than *solutional weathering* or solution-
induced subsidence and collapse. Examples include
CAVES in GLACIERS, which are caused by a change in
phase, not DISSOLUTION, and *vulcanokarst* which comprises caves (LAVA TUBES) within LAVA FLOWS. Halliday
(2007) recognises eight types: (1) *rheogenic pseudokarst* (in lava), (2) *glacier pseudokarst* (in ice),
(3) *badlands and piping pseudokarst* (including pseudokarst on LOESS), (4) *permafrost psudokarst* (in areas of
PERMAFROST, (5) *talus pseudokarst* ((including BOULDER-
FIELDS and *roofed stream courses*), (6) *crevice pseudokarst* (including *littoral pseudokarst*), (7) *compaction
pseudokarst* (in LANDSLIDE and avalanche deposits)
and (8) *consequent pseudokarst* (associated with LAND
SUBSIDENCE in shallow mines and quarries). Karst-like
features in rocks of low solubility (e.g. QUARTZITES)
have in the past also been termed 'pseudokarst.' It is,
however, now commonly accepted that these features
should be termed KARST, provided they are of mainly
solutional origin.   *SHD/JAM*

[*See also* BIOKARST, CHEMICAL WEATHERING]

Doerr SH (1999) Karst-like landforms and hydrology in
quartzites of the Venezuelan Guyana shield: Pseudokarst
or 'real' karst? *Zeitschrift für Geomorphologie NF* 43:
1–17.
Halliday WR (2007) Pseudokarst in the 21st century. *Journal
of Cave and Karst Studies* 69: 103–113.
Wray RAL (1997) A global review of solutional weathering
forms on quartz sandstones. *Earth-Science Reviews* 42:
137–160.

pseudomorph   The replacement of one MINERAL by
another, or a CAST formed by SEDIMENT replacement of
a mineral, in both cases retaining the morphology of
the original mineral. EVAPORITE minerals are readily

dissolved and replaced during DIAGENESIS, and such *halite pseudomorphs* (*halite, NaCl*) and *gypsum pseudomorphs* (GYPSUM, $CaSO_4 \cdot 2H_2O$) are important indicators of ARIDITY in the GEOLOGICAL RECORD.   *GO*

Ghysels G and Heyse I (2006) Composite-wedge pseudomorphs in Flanders, Belgium. *Permafrost and Periglacial Processes* 17: 145–161.

Selleck BW, Carr PF and Jones BG (2007) A review and synthesis of glendonites (pseudomorphs after ikaite) with new data: Assessing applicability as recorders of ancient coldwater conditions. *Journal of Sedimentary Research* 77: 980–991.

pseudo-proxy approach   Synthetic climatic PROXY RECORDS simulated by CLIMATIC MODELS (or, in some cases, derived from INSTRUMENTAL RECORDS) used to investigate the SENSITIVITY of *palaeoclimatic reconstructions*. Pseudo-proxies are used to examine the potential errors inherent in individual proxies and proxy networks, and the performance of various reconstruction algorithms. The pseudo-proxy approach samples a limited number of GRID CELLS from the global fields of GENERAL CIRCULATION MODELS (GCMs), adds noise iteratively to represent the error in real proxy data and compares the results of such experiments with the results from the GCM models. Pseudo-proxies provide insights into the relative performance of reconstruction techniques but are limited by both the quality of the pseudo-proxies and the unknown structure of noise and bias in real-world proxies.   *JAM*

Jones PD, Briffa KR, Osborn TJ et al. (2009) High-resolution palaeoclimatology of the last millennium: A review of current status and future prospects. *The Holocene* 19: 3–49.

Mann ME and Rutherford S (2002) Climate reconstruction using 'pseudoproxies'. *Geophysical Research Letters* 29: 1501. doi:10.1029/2001GL014554.

puddling   A process resulting from TILLAGE or *trafficking* of wet soil by which soil aggregates are broken down by mechanical manipulation. This produces a *puddled soil*, chartacterised by a homogeneous, massive, dense structure and reduced pore space, which is plastic when wet but hardens when dry. In PADDY SOILS, it is the result of MANAGEMENT practices to improve moisture retention, enabling farmers to work the soil and discourage WEEDS, but in most agricultural systems, it is an unintentional process of SOIL DEGRADATION.   *JAM*

Sanchez PA (1976) Soil management in rice cultivation systems. In Sanchez PA (ed.) *Properties and management of soils in the tropics*. New York: Wiley, 413–477.

pugging   The process of COMPACTION and waterlogging of SOIL caused by the TRAMPLING of grazing animals.   *JAM*

pull-apart basin   A SEDIMENTARY BASIN that develops along a FAULT with strike-slip (*shearing*) movement, commonly a TRANSFORM FAULT or CONSERVATIVE PLATE MARGIN. Bends in such faults give rise to localised areas of *compression* (*transpression*) and *stretching* and SUBSIDENCE (*transtension*) of the CRUST. Pull-apart basins develop at sites of transtension, and may be supplied by rapid UPLIFT and EROSION in closely adjacent areas of transpression. They are characterised by rectangular shapes, small dimensions (kilometres to tens of kilometres) and rapid subsidence and infilling (several millimetres per year). Present-day examples include the Dead Sea basin and onshore and offshore basins along the San Andreas Fault in California.   *GO*

Allen PA and Allen JR (2004) *Basin analysis: Principles and applications*. Chichester: Wiley.

Armijo R, Pondard N, Meyer B et al. (2005) Submarine fault scarps in the Sea of Marmara pull-apart (North Anatolian Fault): Implications for seismic hazard in Istanbul. *Geochemistry, Geophysics, Geosystems* 6: Q06009.

Holdsworth RE, Strachan RA and Dewey JF (eds) (1998) *Continental transpressional tectonics and transtensional tectonics*. Bath: Geological Society.

Lazar M, Ben-Avraham Z and Schattner U (2006) Formation of sequential basins along a strike-slip fault: Geophysical observations from the Dead Sea Basin. *Tectonophysics* 421: 53–69.

Moreno DG, Hubert-Ferrari A, Moernaut J et al. (2011) Structure and recent evolution of the Hazar Basin: A strike-slip basin on the East Anatolian Fault, Eastern Turkey. *Basin Research* 23: 191–207.

pumice   VESICLE-rich VOLCANIC GLASS of high-silica composition, with up to 90 per cent *porosity* and DENSITY <1.0 g/cm^3. Pumice fragments float on water and hence may be transported very long distances from their source, occasionally to be utilised as tools that can later be recovered from archaeological remains. Pumice fragments may break to produce angular glass *shards*.   *JBH*

[*See also* POROUS, SCORIA, WELDED TUFF]

Bryan SE, Cook JP, Evans PW et al. (2004) Pumice rafting and faunal dispersion during 2001-2002 in the southwest Pacific: A record of dacitic submarine explosive eruption from Tonga. *Earth and Planetary Science Letters* 227: 135–154.

punctuated equilibria   EVOLUTION seen as patterns of SPECIATION in time: periods of STASIS are 'punctuated by episodic events of allopatric speciation' (Eldredge and Gould, 1972: 96).   *KDB*

[*See also* ISOLATION, PHYLETIC GRADUALISM]

Eldredge N and Gould SJ (1972) Punctuated equilibria: An alternative to phyletic gradualism. In Schopf TJM (ed.)

Models in paleobiology. San Francisco: Freeman, Cooper, 82–115.

Gould SJ (2007) *Punctuated equilibrium.* Cambridge, MA: Harvard University Press.

pyroclastic fall   PYROCLASTIC MATERIAL that settles under gravity from a VOLCANIC PLUME or from fine particles rising out of a PYROCLASTIC FLOW after an explosive VOLCANIC ERUPTION, forming a BED of TEPHRA or TUFF. The geometry, thickness and extent of pyroclastic fall deposits depend on GRAIN SIZE, plume height, and wind velocity and direction.   *GO*

[*See also* AIRFALL, ASH FALL]

Abrams LJ and Sigurdsson H (2007) Characterization of pyroclastic fall and flow deposits from the 1815 eruption of Tambora Volcano, Indonesia, using ground-penetrating radar. *Journal of Volcanology and Geothermal Research* 161: 352–361.

Macias JL, Capra L, Arce JL et al. (2008) Hazard map of El Chichon Volcano, Chiapas, Mexico: Constraints posed by eruptive history and computer simulations. *Journal of Volcanology and Geothermal Research* 175: 444–458.

Sharma K, Self S, Blake S et al. (2008) The AD 1362 Oraefajokull eruption, SE Iceland: Physical volcanology and volatile release. *Journal of Volcanology and Geothermal Research* 178: 719–739.

pyroclastic flow   A high-concentration dispersion of PYROCLASTIC MATERIAL and hot gases (VOLATILES) that flows as a low-turbulence, gravity-controlled DENSITY FLOW. Flows may be generated by the collapse of a VOLCANIC PLUME or a LAVA DOME (see NUÉE ARDENTE). Temperatures in pyroclastic flows are in the range 300–800°C, velocities have been estimated at up to 100 m/s, and studies of the deposits of some QUATERNARY flows have shown that they travelled distances of more than 100 km. Pyroclastic flows are potentially extremely destructive, and have been associated with most major explosive VOLCANIC ERUPTIONS, including Vesuvius in AD 79 and Mount St. Helens in AD 1980. Pyroclastic flows are controlled by TOPOGRAPHY and their deposits tend to infill topographic depressions. They are poorly sorted, with few SEDIMENTARY STRUCTURES, and may be preserved as WELDED TUFFS. An extensive, PUMICE-rich pyroclastic flow deposit is known as an IGNIMBRITE.   *JBH*

[*See also* PYROCLASTIC FALL, PYROCLASTIC SURGE]

Belousov A, Voight B and Belousova M (2007) Directed blasts and blast-generated pyroclastic density currents: A comparison of the Bezymianny 1956, Mount St Helens 1980, and Soufriere Hills, Montserrat 1997 eruptions and deposits. *Bulletin of Volcanology* 69: 701–740.

Cas RAF, Wright HMN, Folkes CB et al. (2011) The flow dynamics of an extremely large volume pyroclastic flow, the 2.08-Ma Cerro Galan Ignimbrite, NW Argentina, and

comparison with other flow types. *Bulletin of Volcanology* 73: 1583–1609.

Druitt TH and Kokelaar BP (eds) (2002) *The eruption of Soufriere Hills volcano, Montserrat, from 1995 to 1999.* Bath: Geological Society [Memoir 21].

Lipman PW and Mullineaux DR (eds) (1981) *The 1980 eruptions of Mount St. Helens, Washington.* Boulder, CO: Geological Society of America [*United States Geological Survey Professional Paper* 1250].

Saucedo R, Macías JL, Sheridan MF et al. (2005) Modeling of pyroclastic flows of Colima Volcano, Mexico: Implications for hazard assessment. *Journal of Volcanology and Geothermal Research* 139: 103–115.

pyroclastic material   Fragmentary material produced by VOLCANISM (cf. LAVA, VOLATILES), mostly in explosive VOLCANIC ERUPTIONS, but also by processes such as the collapse of a LAVA DOME. Pyroclastic products are divided by GRAIN SIZE into VOLCANIC ASH, LAPILLI and VOLCANIC BLOCKS (see Table). Pyroclastic deposits can be emplaced by PYROCLASTIC FALL, or by *pyroclastic density flows*, which include PYROCLASTIC FLOW and PYROCLASTIC SURGE processes. They share many of the characteristics of SEDIMENTARY ROCKS. The term is derived from the Greek *pur* meaning 'fire' and *klastos* meaning 'rock'.   *JBH*

[*See also* ASH FALL, SCORIA, TEPHRA, TUFF, VOLCANIC BLOCK, VOLCANIC BOMB, VOLCANICLASTIC, WELDED TUFF]

Allen SR and Cas RAF (1998) Rhyolitic fallout and pyroclastic density current deposits from a phreatoplinian eruption in the eastern Aegean Sea, Greece. *Journal of Volcanology and Geothermal Research* 86: 219–251.

Davies SM, Larsen G, Wastegård S et al. (2010) Widespread dispersal of Icelandic tephra: How does the Eyjafjöll eruption of 2010 compare to past Icelandic events? *Journal of Quaternary Science* 25: 605–611.

Fisher RV and Schmincke HU (1984) *Pyroclastic rocks.* Berlin: Springer.

Halvorson JJ and Smith JL (2009) Carbon and nitrogen accumulation and microbial activity in Mount St. Helens pyroclastic substrates after 25 years. *Plant and Soil* 315: 211–228.

Levrit H and Montenat C (2000) *Volcaniclastic rocks: From magmas to sediments.* Amsterdam: OPA.

Sulpizio R, De Rosa R and Donato P (2008) The influence of variable topography on the depositional behaviour of pyroclastic density currents: The examples of the Upper Pollara eruption (Salina Island, southern Italy). *Journal of Volcanology and Geothermal Research* 175: 367–385.

Wright JV, Smith AL and Self S (1980) A working terminology of pyroclastic deposits. *Journal of Volcanology and Geothermal Research* 8: 315–336.

Grain size (mm)	Pyroclastic fragments	Pyroclastic deposits	
		Unconsolidated tephra	*Consolidated pyroclastic rock*
>64	Bomb (*fluidally rounded shape*) *or* block (*angular*)	Block *or* bomb tephra; agglomerate (*bombs present*)	Pyroclastic breccias *or* agglomerate (*bombs present*)
2–64	Lapilli	Lapilli tephra	Lapillistone (*or lapilli tuff or tuff-breccia*)
0.063–2	Coarse ash	Coarse ash	Coarse tuff
<0.063	Fine ash	Fine ash	Fine tuff

Pyroclastic material *Classification of pyroclastic fragments and deposits by grain size.*

pyroclastic surge A low-concentration dispersion of *pyroclastic* fragments and hot gases that flows as a turbulent DENSITY FLOW. They are hot, gas-rich, fast-moving and do not always follow topographic depressions: hence, they represent a significant NATURAL HAZARD. *Base surges* are particularly associated with HYDROVOLCANIC ERUPTIONS. *Ground surges* and *ash-cloud surges* are associated with VOLCANIC PLUME collapse and with PYROCLASTIC FLOWS. Pyroclastic surge deposits tend to mantle topography, although they are thickest in topographic depressions. They may contain well-developed SEDIMENTARY STRUCTURES, such as CROSS-STRATIFICATION.

GO

[*See also* PYROCLASTIC FALL, WELDED TUFF]

Baxter PJ (2005) The impacts of pyroclastic surges on buildings at the eruption of the Soufrière Hills volcano, Montserrat. *Bulletin of Volcanology* 67: 292–313.

Fujii T and Nakada S (1999) The 15 September 1991 pyroclastic flows at Unzen Volcano (Japan), a flow model for associated ash-cloud surges. *Journal of Volcanology and Geothermal Research* 89: 159–172.

Herd RA, Edmonds M and Bass VA (2005) Catastrophic lava dome failure at Soufrière Hills volcano, Montserrat, 12-13 July 2003. *Journal of Volcanology and Geothermal Research* 148: 234–252.

pyrolysis The partial DECOMPOSITION of organic molecules at high temperature in the absence of oxygen. It is used in WASTE MANAGEMENT and in the production of BIOCHAR and BIO-ENERGY.

JAM

Moldoveanu SC (2009) *Pyrolysis of organic molecules: Applications to health and environmental issues.* Amsterdam: Elsevier.

pyrotechnology The intentional, controlled use of fire by humans, ranging from its early use in hunting and FOREST CLEARANCE to the advanced technologies of the INDUSTRIAL REVOLUTION and contemporary societies.

JAM

[*See also* CONTROLLED FIRE]

Goren Y and Goring-Morris AN (2008) Early pyrotechnology in the Near East: Experimental lime-plaster production at the Pre-Pottery Neolithic B site, Kfar HaHoresh, Israel. *Geoarchaeology* 23: 779–798.

Head L (1996) Rethinking the prehistory of hunter-gatherers, fire and vegetation change in northern Australia. *The Holocene* 6: 481–487.

Perlès C (1977) *Préhistoire du feu* [Prehistory of fire]. Paris: Masson.

qanat Especially in Iran and neighbouring regions of the Middle East, an ancient means of extracting GROUNDWATER. A qanat, *karez* or *fogarra* comprises a gently sloping tunnel (a subterranean AQUEDUCT) that conducts water from the PHREATIC ZONE (beneath the WATER TABLE) to the ground surface by *gravity flow*. Qanats are commonly located on large ALLUVIAL FANS in foothill regions where annual rainfall is low (100–300 mm). Most qanats are 1–5 km long but exceptionally are over 50 km in length.   *RAS*

Beaumont P, Blake GH and Wagstaff JM (1988) *The Middle East: A geographical study, 2nd edition.* London: David Fulton.
Lightfoot DR (2000) The origin and diffusion of qanats in Arabia: New evidence from the northern and southern peninsula. *Geographical Journal* 166: 215–226.

qoz An Arabic term for an area of SAND DUNES and SAND SHEETS. It is commonly applied to the LONGITUDINAL DUNE system west of the Nile in the Sudan. The qoz areas are thought to have formed mainly during drier phases of the Pleistocene.   *MAC*

[*See also* ERG]

Mitchell CW (1990) Physiography, geology, and soils. In Craig GM (ed.) *The agriculture of the Sudan.* Oxford: Oxford University Press, 1–18.

quad tree A hierarchical data structure whereby geographical space is recursively subdivided into quarters until all quadrants indicate ATTRIBUTE homogeneity, or until a predetermined cut-off depth is reached. Quad-tree representation in a GEOGRAPHICAL INFORMATION SYSTEM (GIS) is not as widespread as RASTER (fixed grid-based) and VECTOR representations (interconnected lines or ARCS).   *VM*

quality-of-life (QOL) indices Quality of life has been defined by Constanza et al. (2008) as 'the extent to which objective human needs are fulfilled in relation to personal or group perceptions of subjective well-being'.

Efforts to develop a truly comprehensive measure of *human well-being* have been stimulated, at least in part, by the fact that Bhutan has decided to use improvement in *Gross National Happiness* as its explicit policy aim. Currently, at least five imperfect QOL indices are in common use at regional, national and international levels.

First, the *gross domestic product* (*GDP*) and, second, the related *gross national product* (*GNP*) are imperfect measures of *wealth* related to economic activity derived from the resident population (GDP) or including those living abroad (GNP). GDP *per capita*, which indicates the average wealth of individuals, varied in 2007 between $40,000 per year in Scandinavia to $300 per year in Mozambique, Congo and Niger. However, this measure excludes the informal economy, subsistence agriculture, unpaid work (especially that carried out by women) and the cost of living. In Mozambique, for example, in 2006, where 3.2 million farm households and over 80 per cent of the farms used no source of power than hand tools and human labour and >90 per cent used neither chemical FERTILISERS nor PESTICIDES, 75–95 per cent of the grain crops produced was used for subsistence. Disparities between different countries are also reflected in the annual GDP growth rate per capita (see Figure). In some countries, such as China and India, this is growing rapidly; in most developed countries, it is growing slowly; while in some African counties and Russia, there is a negative growth rate.

Third, the *human poverty index* (*HPI*), devised by the United Nations Development Programme (UNDP) aggregates measures of low life expectancy, illiteracy and access to health services, clean water and adequate nutrition. Fourth, the UNDP also uses the *human development index* (*HDI*), which is a composite measure of life expectancy at birth (health), adult literacy rate and school enrolment rate (education) and GDP (standard of living). High HDI indices are characteristic of countries in North America, northern and western Europe, temperate South America and Oceania, and low indices are mostly

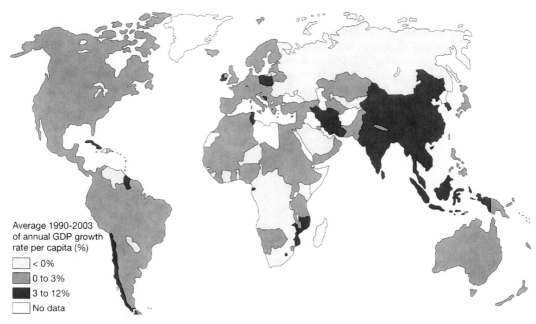

Average 1990-2003
of annual GDP growth
rate per capita (%)

☐ < 0%
▨ 0 to 3%
■ 3 to 12%
☐ No data

Quality-of-life (QOL) indices Average annual GDP growth rate per capita (AD 1990–2003) based on World Bank data (Millennium Ecosystem Assessment, 2005).

African nations on either side of the Equator. Intermediate-index states include tropical South America, northern and southern Africa and most of Eurasia. Since 1970, most regions have increased their HDI scores.

The fifth index, the *environmental performance index (EPI)* was developed at Yale University and unveiled at the World Economic Forum in Davos in 2006. Like the HDI, it supplements the GDP, in this case by measuring environmental practices and outcomes. A very wide range of data are included, relating to greenhouse gas EMISSIONS, agricultural subsidies, fisheries, forestry, BIODIVERSITY, ENVIRONMENTAL PROTECTION, WATER QUALITY, AIR POLLUTION and the environmental DISEASE BURDEN. Iceland, Switzerland, Costa Rica, Sweden, Norway and Mauritius had the highest values of the EPI in 2010.   *JAM*

[*See also* DEVELOPING COUNTRIES, DISEASE, MEDICAL SCIENCE]

Constanza R, Fisher B, Ali S et al. (2008) An integrative approach to quality of life measurement, research and policy. *Surveys and Perspectives Integrating Environment and Society (SAPIENS)* 1: 11–15.
Cuff DJ (2009) Economic levels. In Cuff DJ and Goudie AS (eds) *The Oxford companion to global change.* Oxford: Oxford University Press, 198–200.
Glaeser E (2011) Cities, productivity and quality of life. *Science* 333: 592–594.
Millennium Ecosystem Assessment (2005) *Ecosystems and human well-being: Synthesis report.* Washington, DC: Island Press.

Warf B (2010) Gross domestic product/gross national product. In Warf B (ed.) *Encyclopedia of geography.* Thousand Oaks, CA: Sage, 1382–1384.

quarrying (1) The open surface excavation of rocks or sediments by humans (e.g. slate, granite, sand and gravel). (2) A process of GLACIAL EROSION by which rock fragments greater than about 1 cm are removed from bedrock, involving (a) FAILURE of the rock leading to the fragments being loosened from the bedrock and (b) *evacuation* whereby fragments are moved from their original position. *Plucking* is an alternative term for the process. (3) In the context of FLUVIAL PROCESSES and *wave action* on SHORELINES, the term quarrying has been used for erosion by the *hydraulic action* of water alone. This includes the process of *cavitation*, which involves the effects of shock waves generated through the collapse of the 'vapour pockets' or 'airless bubbles' that form in a fast, TURBULENT FLOW with marked pressure changes. It excludes erosion by ABRASION and CORRASION.   *RAS/JAM*

Barnes HL (1956) Cavitation as a geological agent. *American Journal of Science* 254: 493–505.
Benn DI and Evans DJA (2010) *Glaciers and glaciation, 2nd edition.* London: Arnold.
Knighton AD (1998) *Fluvial forms and processes: A new perspective, 2nd edition.* London: Arnold.

quartz A SILICATE MINERAL of composition SiO_2 (SILICA). Because of its resistance to WEATHERING, quartz is common in SEDIMENTARY ROCKS, particularly SANDSTONE.   *GO*

[*See also* PHYTOLITHS, SILICON CYCLE]

quartz grain surface textures   High-resolution magnification made possible with SCANNING ELECTRON MICROSCOPY (SEM) has shown that sand grains (especially QUARTZ) can retain surface markings or *textures* diagnostic of depositional history. GLACIAL, AEOLIAN and subaqueous (*fluvial* and *littoral*) processes can give rise to distinctive suites of mechanically formed textures. The evidence of more than one depositional environment may be identifiable in a sample of grains or even on a single sand grain. Analysis typically involves noting for each of a sample of grains the presence/absence of recognised features from a standard list. However, CHEMICAL WEATHERING may render any mechanically produced textures unidentifiable.   *RAS*

Culver SJ, Bull PA, Campbell S et al. (1983) A statistical investigation of operator variance in quartz grain surface texture studies. *Sedimentology* 30: 129–136.
Madhavaraju J, Barragan JCGY, Hussain SM and Mohan SP (2009) Microtextures on quartz grains in the beach sediments of Puerto Penasco and Bahia Kino, Gulf of California, Sonora, Mexico. *Revista Mexicana de Ciencias Geologicas* 26: 367–379.
Tengberg A (1995) Sediment sources of nebkhas in the Sahel zone, Burkino-Faso. *Physical Geography* 16: 259–275.

quartzite   A SANDSTONE (*orthoquartzite*) or its derivative METAMORPHIC ROCK (*metaquartzite*) composed almost entirely of QUARTZ. A sandstone with at least 95 per cent quartz CLASTS and little or no MATRIX, but without a quartz CEMENT, is a *quartz arenite*: thus, an orthoquartzite is a quartz arenite with a quartz cement.   *TY*

Quasi-Biennial Oscillation (QBO)   A well-defined *oscillation* of the zonal wind component of the equatorial STRATOSPHERE. The period of the oscillation is 20–36 months with a mean of around 28 months and the reversal begins at high levels, taking approximately 12 months to descend from 30 to 18 km (10–60 hPa). Observations of the QBO indicate that the Easterlies are stronger than the Westerlies, that the time between the easterly and westerly maxima is much shorter than the reverse and that there is considerable variability in the PERIODICITY of the oscillation. There is evidence that the QBO may modulate other elements of TROPICAL climate: for instance, it is widely used as a prognostic in Atlantic *hurricane* forecasts. Increased hurricane activity occurs for westerly (or positive) zonal wind anomalies and reduced hurricane activity for easterly (or negative) zonal wind anomalies. Furthermore, the QBO is thought to affect rainfall pattern in the African SAHEL, and is used in forecasts for the region.   *GS*

[*See also* CLIMATIC MODES, TELECONNECTION, TEN- TO TWELVE-YEAR OSCILLATION]

Baldwin MP, Gray LJ, Dunkerton TJ et al. (2001) The Quasi-Biennial Oscillation. *Reviews of Geophysics* 39: 179–229.
Huang B, Hu ZZ, Kinter JL et al. (2012) Connection of stratospheric QBO with global atmospheric general circulation and tropical SST. Part I: Methodology and composite life cycle. *Climate Dynamics* 38: 1–23.

quasi-periodic phenomena   Environmental variables that display approximate cyclical patterns or PERIODICITIES in time series. For instance, a CLIMATIC OSCILLATION with a period of 4–12 years has been observed in the EL NIÑO-SOUTHERN OSCILLATION (ENSO). As the Earth spins around its axis and orbits around the Sun, several quasi-periodic variations have been observed to occur due to gravitational interactions. MILANKOVITCH THEORY identifies periodicities in the Earth's *orbital parameters*, which are important in the CLIMATIC FORCING of GLACIAL-INTERGLACIAL CYCLES.   *GS*

[*See also* CLIMATIC MODES, CYCLICITY, QUASI-BIENNIAL OSCILLATION (QBO)]

Burroughs WS (2003) *Weather cycles: Real or imaginary? 2nd edition.* Cambridge: Cambridge University Press.
Meyers SR and Pagani M (2006) Quasi-periodic climate teleconnections between northern and southern Europe during the 17th-20th centuries. *Global and Planetary Change* 54: 291–301.

Quaternary   The most recent PERIOD of the GEOLOGICAL TIMESCALE, continuing to the present day and comprising the PLEISTOCENE and HOLOCENE epochs. There has been much debate concerning the date for the start of the Quaternary period, which is now agreed at 2.6 million years ago. This date reflects an intensification of Northern Hemisphere GLACIATION in Eurasia and North America.   *GO*

[*See also* GEOLOGICAL TIMESCALE, PLIOCENE-PLEISTOCENE TRANSITION, QUATERNARY TIMESCALE]

Gibbard PL, Head MJ, Walker MJC et al. (2010) Formal ratification of the Quaternary System/Period and the Pleistocene Series/Epoch with a base at 2.58 Ma. *Journal of Quaternary Science* 25: 96–102.
Ogg JG, Ogg G and Gradstein FM (2008) *The concise geologic time scale.* Cambridge: Cambridge University Press.

Quaternary environmental change   The Quaternary Era, which covers about the past 2 million years, has been a time of dramatic environmental change, with Earth's climate lurching between GLACIALS and INTERGLACIALS. During glacials, ICE SHEETS expanded to cover large parts of the Northern Hemisphere, global sea levels fell and tropical latitudes became colder and drier. During interglacials, the ice shrank back to high latitudes and altitudes, sea levels rose and the tropics became warmer and wetter. This CYCLICITY of CLIMATIC CHANGE, recognised first in the

OXYGEN ISOTOPE signal recovered from MARINE SEDI-MENT CORES, and later confirmed using other proxies including LOESS deposits and ICE CORES, suggests that the driving force for the GLACIAL-INTERGLACIAL CYCLES is small fluctuations in the Earth's orbit. The resulting small fluctuations in the amount of energy received by each hemisphere, with cycles of about 100, 41 and 23 ka, have pertained throughout Earth's history, so some critical THRESHOLD must have been crossed for ORBITAL FORCING of climatic changes with the amplitude experienced during the Quaternary. A possible cause is the closure of the *Isthmus of Panama* once the continents had reached their present positions (see CONTINENTAL DRIFT). This triggered the diversion of warm waters into the North Atlantic and so large transfers of heat into high latitudes. The warm water supplied the precipitation needed to initiate the build-up of ice sheets. A small decline in energy received could have been amplified by the high ALBEDO of the growing ice sheets, so that larger proportions of SOLAR RADIATION were reflected away to space, leading to further cooling. Warming and DEGLACIATION are not so easy to explain, since it is more difficult to amplify the small increase in energy received when large areas are already covered in ice. One possibility is that increased insolation delivers large amounts of MELTWATER and ICEBERGS into the North Atlantic, severing the supply of warm water from the south and leading to reduced evaporation and starvation of the ice sheets.

The most complete records of Quaternary environmental change come from studies of deep ocean cores, and particularly the oxygen ISOTOPE RATIOS of foraminifera, which are related to ocean volume and temperature. These have been used to identify more than 100 MARINE ISOTOPIC STAGES (MIS), with even numbers representing cold intervals, odd numbers warm intervals and letters indicating smaller variations. The present interglacial is MIS 1, the LAST GLACIAL MAXIMUM (LGM) is MIS 2 and the last full interglacial is MIS 5e. These stages have been correlated with similar fluctuations in climate reflected in loess sequences and ice cores, allowing the marine record for at least part of the Quaternary to be correlated with events on the continents.

Although these long PROXY RECORDS provide a clear framework of climatic changes during the Quaternary, and give an indication of changes in the volume of ice on land, the location of the ice, the patterns of ice sheet growth and decay and the influence of climatic changes on *ecosystems* is more difficult to establish. A variety of approaches have been used to reconstruct ice sheets during the Last Glacial Maximum, including field mapping, analysis of remotely sensed images and computer modelling. Long lake cores have provided pollen and other proxy records of changes in vegetation over long timescales but

with much higher temporal *resolution* than can be obtained from slowly accumulating ocean sediments. Sediments that accumulated on the continents during the last glacial-interglacial cycle are widespread and a very wide range of proxies have been applied to study climatic and environmental changes based on HIGH-RESOLUTION RECONSTRUCTIONS.

Quaternary environmental change may have been one of the driving forces in the evolution of HOMINIDS and eventually modern Humans. HUMAN MIGRATION: 'OUT OF AFRICA' is thought to have occurred twice. *Homo erectus* (or possibly a predecessor) migrated early in the Pleistocene, evolving in Europe into the *Neanderthals*. It is generally accepted that ANATOMI-CALLY MODERN HUMANS (AMH) evolved in Africa and migrated in the Late Pleistocene, reaching Europe possibly as early as 45–50 ka, though there are still some proponents of the multiregional hypothesis, suggesting genetic continuity with earlier hominids within regions. North America was probably first occupied by MAMMOTH hunters crossing the Bering Strait, perhaps as early as 30 ka. Australia may have been occupied earlier than Europe, possibly as early as 65 ka.

The Late Pleistocene was also a time of MASS EXTINC-TION, particularly of large animals such as mammoth, *woolly rhinoceros*, *sabre-tooth tiger*, *giant sloth* and large *flightless birds*. There is continuing debate over the role played by hunting, natural climatic change and human-induced vegetation changes in causing the extinctions. The fact that the MEGAFAUNA survived the many large swings in climate of the Pleistocene, only to succumb to the last when humans were actively hunting them, does suggest some degree of culpability. Human-induced extinctions and large-scale alterations of natural environments are not new but the pace of change has accelerated enormously over the past century, coupled with changes in the chemistry of the atmosphere that may be driving changes in global climate. Thus, the study of Quaternary environmental change may help inform us of the likely consequences of our actions.   *DMcC*

[*See also* DATING TECHNIQUES, HOLOCENE ENVIRONMENTAL CHANGE, HUMAN EVOLUTION: CLIMATIC INFLUENCES, MEGAFAUNAL EXTINCTION, MILANKOVITCH THEORY, NEANDERTHAL DEMISE, OVERKILL HYPOTHESIS, QUATERNARY TIMESCALE]

Anderson DE, Goudie AS and Parker AG (2007) *Global environments through the Quaternary: Exploring environmental change.* Oxford: Oxford University Press.

Elias SA (ed.) (2007) *Encyclopedia of Quaternary science, 4 volumes.* Amsterdam: Elsevier.

Fagan B (ed.) (2009) *The complete Ice Age.* London: Thames and Hudson.

Imbrie J and Imbrie KP (1979) *Ice ages: Solving the mystery.* London: Macmillan.

Lowe JJ and Walker MJC (1997) *Reconstructing Quaternary environments, 2nd edition.* Harlow: Longman.

Metcalfe SE and Nash DJ (eds) (2012) *Quaternary environmental change in the tropics.* Chichester: Wiley-Blackwell.

Ruddiman WF (2008) *Earth's climate: Past present and future, 2nd edition.* New York: Freeman.

Shackelton NJ and Opdyke ND (1973) Oxygen isotope and palaeomagnetic evidence stratigraphy of equatorial Pacific core V28-238: Oxygen isotope temperatures and ice volumes on a 10^5 and a 10^6 scale. *Quaternary Research* 3: 39–55.

Stringer C (2011) *The origin of our species.* London: Penguin Books.

Williams M, Dunkerley D, De Decker P et al. (1998) *Quaternary environments.* London: Arnold.

Quaternary science   The interdisciplinary study of the QUATERNARY, the most recent geological PERIOD. Its main focus is on the reconstruction and dating of QUA-TERNARY ENVIRONMENTAL CHANGE with PALAEOCLIMATOL-OGY as the main unifying theme. *Human impacts* and their interaction with natural ENVIRONMENTAL CHANGE is another important theme (see HOLOCENE ENVIRONMEN-TAL CHANGE). Quaternary scientists are increasingly involved with the modelling community to understand the mechanisms of environmental change and to test CLIMATIC MODELS.   *JAM*

[*See also* DATING TECHNIQUES, ENVIRONMENTAL CHANGE: HISTORY OF THE FIELD, PALAEOENVIRONMENTAL RECONSTRUCTION]

Elias SA (ed.) (2007) *Encyclopaedia of Quaternary science, 4 volumes.* Amsterdam: Elsevier.

Quaternary timescale   The Quaternary is the most recent major subdivision, or PERIOD, of the geo-logical record and includes the present day. It is part of the CENOZOIC (*CAINOZOIC*), the fourth geological era, which also includes the TERTIARY. There has been some controversy about the exact timing of the beginning of the Quaternary at the PLIOCENE-PLEISTOCENE TRANSITION, with some workers favouring a 'long' Quaternary, and others a much shorter period. Advocates of the shorter timescale use the Pliocene-Pleistocene boundary accepted by the International Commission on Stratigra-phy in 1985 and the International Union of Geological Sciences (IUGS). It is based upon the first appearance of the cold-water marine *ostracod Cytheropteron tes-tudo* in a reference site at Vrica in Italy. This has been dated to ca 1.64 million years ago (Ma) on the basis of palaeomagnetic evidence, just below the end of the OLDUVAI EVENT (K-Ar dated to 1.87–1.67 Ma), although revisions in dating this event have led to earlier sug-gested dates of between 1.88 and 1.81 Ma. Workers who favour a longer timescale (as in the Table) argue for the boundary to be placed where there are the first global indications of cooling, such as the build-up of CONTINENTAL GLACIATION as defined by ICE-RAFTED DEBRIS (IRD) in MARINE SEDIMENTS, and LOESS deposition. Thus, some place the boundary at around 2.4–2.6 Ma close to the GAUSS-MATUYAMA GEOMAGNETIC BOUNDARY (ca 2.58 Ma). The division has now been formally ratified with a base at 2.58 Ma, the base of the *Gelasian Stage*, within MARINE ISOTOPIC STAGE (MIS) 103.

The Quaternary is divided into two EPOCHS: the PLEIS-TOCENE, which ended at 11,700 years b2k the opening of the HOLOCENE, which is the present warm stage. The Pleistocene may be further subdivided into the *Lower Pleistocene* (2.58–0.75 Ma), the *Middle Pleistocene* (750–125 ka) and the *Upper Pleistocene* (125–10 ka). These boundaries are arbitrary to a certain extent and are not sacrosanct. The Quaternary has revealed a high frequency of climatic oscillations and intensely cold phases. The exact number of GLACIAL-INTERGLACIAL CYCLES occurring during the Quaternary remains to be established, but the Earth may have experienced 30–50 GLACIAL EPISODES and the same number of INTERGLA-CIAL phases depending upon the age assigned to the Pliocene-Pleistocene boundary.

Division of the Quaternary timescale is predomi-nantly founded on CLIMOSTRATIGRAPHY (different *cli-matostratigraphic units* distinguishing between cold glacials or STADIALS and warm interglacials or INTERSTA-DIALS) in areas away from glaciation as in China where there are extensive LOESS records. Climatostratigraphic units are defined there in terms of PALAEOSOLS and LOESS STRATIGRAPHY. With increasing age, it has proved extremely difficult to differentiate individual STADIALS and INTERSTADIALS because of their short duration. By comparing ICE-CORE records from Greenland with high-resolution records from MARINE SEDIMENTS, it has been possible to subdivide the past 100,000 years into 24 different stages, mostly unnamed but given individual numbers (see MARINE ISOTOPIC STAGE (MIS)).

Although the Quaternary timescale can be applied on a global basis, different countries and regions tend to use different terms or names for particular *chron-ostratigraphic units*. Thus the LAST INTERGLACIAL is known in the United Kingdom as the *Ipswichian*, in Ireland as the *Fenitan*, in Europe as the *Eemian* and in North America as the *Sangamon*. Because of the prob-lems this causes for CORRELATION (STRATIGRAPHICAL), workers now tend to relate stratigraphies to the oxygen isotope stages derived from the record found in MARINE SEDIMENT CORES, as it is globally applicable. Hence the Last Interglacial is known as MIS 5e.

Assigning ages to the Quaternary timescale has been achieved using PALAEOMAGNETIC DATING whereby dates for distinctive horizons such as GEOMAGNETIC POLARITY REVERSALS derived from K-Ar dating of associated vol-canic rocks enabled a broad temporal framework to be obtained. Ages between such MARKER HORIZONS were then estimated by extrapolation. Marine sediments, also having a record of the palaeomagnetic timescale then provided more detail with the development of the stratigraphy based on changes in the oxygen isotope record in FORAMINIFERA. This record has since been further refined by ORBITAL TUNING to provide a universal timescale accepted by all workers (see SPECTRAL MAPPING PROJECT TIMESCALE (SPECMAP)). Although ages still have to be extrapolated between stage boundaries or marker horizons, and further refining is still being undertaken, the level of PRECISION now available is acceptable considering the length of the Quaternary as a whole.

The Table sets out the Quaternary timescale according to a number of sources set against the oxygen

Timescale Ma. BP	Marine oxygen isotope stages	Palaeo-magnetic chrons	Northern Europe	The Netherlands	British Isles	European Russia	Northern Alps	North America	Cold or temperate
0.01	1	Brunhes	Holocene	Holocene	Flandrian	Holocene	Holocene	Holocene	T
0.08	2.4d		Weichselian	Weichselian	Devensian	Devensian	Würm	Wisconsian	C
0.13	5e		Eemian	Eemian	Ipswichian	Mikulino	Riss-Würm	Sangamon	T
0.19	6		Warthe		"Wolstonian"	Moscow	Penultimate Glacial Late Riss?	Illinoian (Late)	C
0.25	7		Saale / Drenthe			Odintsovo (Dneipr Glaciation)		Illinoian	T
0.30	8		Drenthe			Dneipr	Antepenultimate Glac. Early Riss / Mindel?	(Early)	C
0.34	9		Domnitz (Wacken)			Romny			T
0.35	10		Fuhne (Mehleck)		Hoxnian	Pronya	Pre-Riss?	Pre-Illinoian A	C
0.43	11		Holsteinian (Muldsberg) [Holsteinian Interglacial]			Lichvin			T
0.48	12		Elster 1		Anglian		Late Mindel?/Donau	B	C
0.51	13		Elster 1/2	Elster		Oka			T
0.56	14		Elster 1		Cromerian		Early Mindel?/Donau	C	C
0.63	15		Cromerian IV	Cromerian IV (Noordbergum)					T
0.69	16		Glacial C	Glacial C	～?～			D	C
0.72	17		Interglacial III	Interglacial III (Rosmalen)					T
0.78	18		Glacial B	Glacial B				E	C
0.79	19		Interglacial II	Interglacial II (Westerhoven)					T
	20	Matuyama	Helme (Glacial A)	Glacial A			Early Gunz?	F	C
	21		Astern Interglacial I	Interglacial I (Waardenburg)					T
0.90	22			Dorst (Bavelian)				G	C
				Leerdam (Bavelian)					T
0.97				Linge (Bavelian)					C
				Bavel (Bavelian)					T
				Menapian					T/C
				Waalian					T
									C
				Eburonian				H	T/C
									T
1.65	6.0			Beestonian ～?～				I	C
				Tiglian — C5-6	Pastonian				T
				C-4c	Pre-Pastonian/ Baventian				C
				Cl-4b	Bramertonian/ Antian				T
				B	Thurnian			J	C
	103			A	Ludhamian				T
	104			Praetiglian	Pre-Ludhamian				C
2.60		Gauss		Pliocene	Pliocene				

Quaternary timescale The Quaternary timescale (left) in millions of years before present, based on orbital tuning of the marine oxygen isotope stages (even-numbered stages are relatively cold; odd numbers denote warm stages). The palaeomagnetic timescale is represented by the three chrons that occur in the Quaternary. The remaining columns show selected regional subdivisions based on terrestrial stratigraphic records (Lowe and Walker, 1997; Šibrava, 1986).

isotope stratigraphy and the palaeomagnetic time-scale. It should be emphasised that although the general sequence and terminology shown in the Table are widely accepted, there will still be workers who would argue for variations: the debate between the 'long' and 'short' Quaternary timescales for instance.　　*CJC*

Barker S, Knorr G, Edwards RL et al. (2011) 800,000 years of abrupt climate variability. *Science* 334: 347–351.

Bowen DQ (ed.) (1999) *A revised correlation of Quaternary deposits in the British Isles.* Bath: Geological Society.

Gibbard PL, Head MJ, Walker MJC and the Subcommission on Quaternary Stratigraphy (2010) Formal ratification of the Quaternary System/Period and the Pleistocene Series/Epoch with a base at 2.58Ma. *Journal of Quaternary Science* 25: 96–102.

Lowe JJ and Walker MJC (1997) *Reconstructing Quaternary environments,* 2nd edition. Harlow: Longman.

Prentice AJ and Kroon D (1991) Oxygen isotope chronostratigraphy. In Smart PL and Frances PD (eds) *Quaternary dating methods: A user's guide.* Cambridge: Quaternary Research Association, 199–228.

Shackleton NJ and Opdyke ND (1973) Oxygen isotope and palaeomagnetic stratigraphy of equatorial Pacific core V28-238: Oxygen isotope temperatures and ice volumes on a 10^5 year and 10^6 year scale. *Quaternary Research* 3: 39–55.

Šibrava V (1986) Correlation of European glaciations and their relation to the deep-sea record. *Quaternary Science Reviews* 5: 433–442.

Valet J-P and Meynadier L (1993) Geomagnetic field intensity and reversals during the past four million years. *Nature* 366: 234–238.

Walker MJC (2005) *Quaternary dating methods.* Chichester: Wiley.

Williams MAJ, Dunkerley DL, De Deckker P et al. (1998) *Quaternary environments,* 2nd edition. London: Arnold.

Wilson RCL, Drury SA and Chapman JL (2000) *The great ice age: Climate change and life.* London: Routledge.

query language　A software interface for defining, accessing and interrogating data in a DATABASE. *SQL* (*structured query language*) is a standard interface for a RELATIONAL DATABASE.　　*VM*

Lutz M and Klien E (2006) Ontology-based retrieval of geographic information. *International Journal of Geographical Information Systems* 20: 223–260.

quick clay　CLAY that, when disturbed, can dramatically lose its strength and flow like a liquid. Strength is not subsequently restored. Many quick clays represent marine deposits within which SALT allowed a very loose packing of clay particles to develop. Terrestrial WEATHERING has since removed the salt, leaving an unstable deposit. Quick clays have been the cause of catastrophic LANDSLIDES (see NATURAL HAZARDS) in Norway, Canada and the Amazon Basin.　　*GO*

[*See also* QUICKSAND]

Hansen L, Eilertsen RS, Solberg I-L et al. (2007) Facies characteristics, morphology and depositional models of clay-slide deposits in terraced fjord valleys, Norway. *Sedimentary Geology* 202: 710–729.

Ter-Stepanian G (2000) Quick clay landslides: Their enigmatic features and mechanism. *Bulletin of Engineering Geology and the Environment* 59: 325–348.

quickflow　That part of PRECIPITATION that takes a rapid route to a stream channel and constitutes the *flood peak* of a stream. It may be provided by a number of RUNOFF PROCESSES, notably HORTONIAN OVERLAND FLOW, SATURATION OVERLAND FLOW, PIPEFLOW and THROUGHFLOW.　　*ADT/RPDW*

[*See also* BASEFLOW, STORM HYDROGRAPH]

Jones JAA (1997) *Global hydrology: Processes, resources and environmental management.* Harlow: Addison Wesley Longman.

Jones JAA (2012) Soil piping and catchment response. *Hydrological Processes* 24: 1548–1566.

Ward RC and Robinson M (2000) *Principles of hydrology,* 4th edition. London: McGraw-Hill.

quicksand　An informal description of sand that has lost its strength and behaves like a liquid as a result of LIQUEFACTION or FLUIDISATION, resulting in flow or the sinking of massive objects. EARTHQUAKE-triggered quicksand conditions can lead to foundation failure and collapse of buildings and other infrastructure (see NATURAL HAZARDS); this was a major problem following the AD 2011 earthquake in Christchurch, New Zealand. Quicksand conditions can also be generated by movements of GROUNDWATER. Evidence of former quicksand conditions can be preserved in SEDIMENTARY DEPOSITS in the form of SOFT-SEDIMENT DEFORMATION structures, which provide important evidence of PALAEOSEISMICITY.　　*GO*

[*See also* QUICK CLAY]

Kaiser A, Holden C, Beavan J et al. (2012) The Mw 6.2 Christchurch earthquake of February 2011: Preliminary report. *New Zealand Journal of Geology and Geophysics* 55: 67–90.

quota　A limit on the quantity of a commodity that can be exported or imported. The term is increasingly used for environmentally sensitive 'commodities', such as those associated with trade in animal species and in the 'export' of POLLUTANTS.　　*JAM*

[*See also* CARBON EMISSIONS TRADING (CET), TRADEABLE EMISSIONS PERMITS]

Wolf A (1997) *Quotas in international environmental agreements.* London: Earthscan.